W9-DFG-330

FLAVINS AND FLAVOPROTEINS

FLAVINS AND FLAVOPROTEINS

Proceedings of the Fifth International Symposium
on Flavins and Flavoproteins
San Francisco, March 31–April 3, 1975

EDITED BY

T.P. SINGER

Molecular Biology Division
Veterans Administration Hospital
and Department of Biochemistry and
Biophysics University of California
San Francisco, Calif. 94121
U.S.A.

ELSEVIER SCIENTIFIC PUBLISHING COMPANY
Amsterdam — Oxford — New York 1976

ELSEVIER SCIENTIFIC PUBLISHING COMPANY
335 Jan van Galenstraat
P.O. Box 211, Amsterdam, The Netherlands

Distributors for the United States and Canada:

ELSEVIER/NORTH-HOLLAND INC.
52, Vanderbilt Avenue
New York, N.Y. 10017

Library of Congress Cataloging in Publication Data

International Symposium on Flavins and Flavoproteins,
5th, San Francisco, 1975.
Flavins and flavoproteins.

Includes index.
1. Flavins--Congresses. 2. Flavoproteins--Congresses. I. Singer, Thomas Peter, 1920- II. Title.
[DNLM: 1. Flavins--Congresses. 2. Flavoproteins--Congresses. W3 IN981C 1975f / QQU135 I5895 1975f]
QP601.I515 1975 574.1'9258 76-18849
ISBN 0-444-41458-4

Printed in The Netherlands

Organizing Committee: Thomas P. Singer (San Francisco) *Chairman;* R. Abeles (Waltham); T. Bruice (Santa Barbara); A. Ehrenberg (Stockholm); P. Hemmerich (Konstanz); H. Kamin (Durham); V. Massey (Ann Arbor); T. Nakamura (Osaka)

Honorary Chairman: Hugo Theorell

SPONSORS

International Union of Biochemistry
National Institutes of Health
National Science Foundation
University of California
Veterans Administration

Burroughs Wellcome Company
Calbiochem
Fisher Scientific Company
Hoffman-La Roche, Inc.
National Dairy Council

LIST OF PARTICIPANTS

Dr. Robert Abeles
Dr. Brian Ackrell

Dr. Adelbert Bacher
Dr. Thomas Baldwin
Dr. David Ballou
Dr. Robert Bartsch
Dr. James Becvar
Dr. Helmut Beinert
Dr. Gunter Blankenhorn
Dr. Karl Brandt
Dr. Robert Bray
Dr. Harold Bright
Dr. Margarete Brühmüller
Dr. Paula Bruice
Dr. Thomas C. Bruice
Mr. Thomas W. Bruice

Mr. John Cashman
Dr. Paolo Cerletti
Dr. T.C. Chan
Dr. Andrew Clifford
Dr. Michael Coughlan
Dr. Bruno Curti

Dr. Karl Decker
Dr. Daniel DerVartanian
Dr. Richard DeSa
Dr. William Drell
Dr. George Drysdale
Dr. Edward Duke
Dr. Abraham Dury

Dr. Dale Edmondson
Dr. Barrie Entsch

Dr. George Faini
Dr. Vincent Favaudon
Dr. Helmut Fenner
Mr. Jed Fisher
Dr. J.L. Fox
Dr. Katsuhiko Fujii

Dr. Sandro Ghisla
Dr. Shlomo Grossman
Dr. Menachem Gutman

Dr. Carole Hall
Dr. J.W. Hastings
Dr. Peter Hemmerich
Dr. Louis Hersh
Dr. Thomas Höpner
Dr. Richard Huston
Dr. J.S. Hyde

Dr. Lloyd Ingraham

Dr. Susan C. Jackels
Dr. Marilyn Schuman Jorns

Dr. Henry Kamin
Dr. Zbigniew Kaniuga
Mr. Michael Kazarinoff
Dr. Edna Kearney
Mr. Cemal Kemal
Dr. William Kenney
Dr. Tokuji Kimura
Dr. Masahiko Koike
Dr. Alvin Kwiram

Dr. Françoise Labeyrie

Mr. David Lambeth
Dr. P.W. Lancaster
Dr. Neal Langerman
Dr. Florence Lederer
Dr. John Lee
Dr. Martha Ludwig

Dr. Martha MacKnight
Dr. Donald McCormick
Dr. H.I.X. Mager
Dr. Bengt Mannervik
Dr. Vincent Massey
Dr. Bettie Sue Masters
Dr. Kunio Matsui
Dr. Rowena Matthews
Dr. Alan Maycock
Dr. Stephen Mayhew
Dr. Thomas Moore
Dr. Patrick Mowery
Dr. Franz Müller

Dr. Satoshi Nakamura
Dr. Takao Nakamura
Dr. Halina Neujahr
Dr. Miriam Nicoli

Dr. Guillermo Oestreicher
Dr. Eugene Oliver
Dr. Raul Ondarza
Dr. Shohei Otani

Dr. Graham Palmer
Dr. Israel Pecht
Dr. Harry Peck
Dr. Geoffrey Penzer
Dr. Richard Perham
Dr. Meir Perl
Dr. Gerhard Plaut
Dr. David Porter
Mr. Robert Presswood
Dr. Russell Prough

Dr. K.V. Rajagopalan
Dr. Douglas Ribbons
Dr. Richard Rivlin
Dr. Hellmuth Rössler

Dr. James Salach
Dr. Agnes Schonbrunn
Mr. Larry Schopfer
Dr. Siegfried Schreiner
Dr. Hirofumi Shoun
Dr. Lewis Siegel
Dr. Thomas Singer
Dr. Pill-Soon Song

Dr. Shigeki Takemori
Mr. J. Taulane
Dr. Hugo Theorell
Dr. Colin Thorpe
Dr. Gordon Tollin
Dr. Shiao-chun Tu

Dr. Cees Veeger
Ms. Janice Vermilion

Dr. Christopher Walsh
Dr. Keith Watenpaugh
Mr. William Waud
Dr. Gregorio Weber
Mr. Keith Wilkinson
Dr. Charles Williams, Jr.
Dr. Robert Williams
Dr. Lowell Wilson

Dr. Kunio Yagi
Dr. Shozo Yamamoto
Dr. Tateo Yamanaka
Dr. Kerry Yasunobu
Dr. Duane Yoch
Dr. Kwan-sa You
Dr. Moussa Youdim

TABLE OF CONTENTS

PART VII. DEAZAFLAVINS AND OTHER MODEL FLAVINS

PART VIII. FLAVODOXIN

PART IX. LIPOYL DEHYDROGENASE AND GLUTATHIONE REDUCTASE

PART X. COMPLEX FLAVOPROTEINS

PART XI. ADRENODOXIN, FERREDOXIN AND CYTOCHROME REDUCTASES AND ELECTRON TRANSFERRING FLAVOPROTEIN

PART XII. STRUCTURE AND FUNCTION OF FLAVOPROTEIN OXIDASES

PART XIII. BIOSYNTHESIS, METABOLISM AND TRANSPORT OF RIBOFLAVIN

PART XIV. SPECIAL TOPICS

Part I

Introduction

Flavins and Flavoproteins, edited by T.P. Singer

INTRODUCTION

It has become customary for editors of the published versions of this series of symposia to state in the preface why they think that flavins and flavoproteins are special. These have ranged from factual statements of the obvious, i.e. that the chromophoric and fluorescent properties of flavins and flavoproteins are representative of their state of protonation or oxidation and thus provide excellent handles for the study of these states and of transitions between them, to sort of personal credos, such as the somewhat anthropomorphic revelations in the preface to a recent volume, in which a flavin was described as a "coy, but not prudish" yellow lady of "labile blushes". Far from wishing to ridicule the paragraph quoted, I should like to justify it for the reader with a credo of my own for those who are newcomers or strangers to the field, since those in it need no explanation.

Is there anything special about flavins and flavorproteins or is it a field of expertise which practitioners of it exercise, as was the case with free electrophoresis or ultracentrifugation some years ago, because they are adapt at it or because "it is a living"? I believe there is something special. Let met try to explain it from both extremes, that of the personal credo and that of the objective reporter.

Speaking as a prodigal who returned some years ago to the fold, after a few years probably ill-advised total preoccupation with the treacherous field of electron transport and energy conservation, I can say that flavins are special to me because I am fascinated by them and feel comfortable with them. I am fascinated because I know of no other molecule as versatile, as capable of undergoing a variety of intriguing, often unpredictable reactions, or one offering as many intellectual challenges. What other class of compounds do you know which, after nearly half a century of intensive research, offers opportunities and surprizes, such as the discovery of the natural occurence of roseoflavin, 8-hydroxyflavin, and of four clauses of covalently bound flavins, each with a chemistry of its own, yet to be discovered, yet to be systematized? What other class of compounds forms the fascinating array of adducts, many of which mimic the enzyme-substrate complexes of living cells? What other compounds do you know which have been as thoroughly studied as flavins and which, nevertheless, are practically certain to yield as yet undiscovered secrets to any competent investigator?

It is more difficult to state why I am comfortable with flavins and flavoproteins. Perhaps it is because you can *see* what is happening. There is an undeniable emotional element in research. Atavistic feeling or not, we all feel more confident investigating a phenomenon we have observed with our own eyes, than one the existence of which can only be deduced from a recorder trace. Perhaps it is because it is a seemingly inexhaustable storehouse of new information, so that one can feel confident in steering a student or young postdoctoral to exploring a new observation in the field, because past experience has shown that there is much left to be discovered, and much of it is unexpected, some even surprising. Perhaps it is, in seeming contradiction to the foregoing, becuase one has the subjective, often unvoiced feeling that here is a chance of understanding a subject within one's lifetime, a feeling one probably cannot have in dealing with genetic information or structure-function relations in proteins. But most probably I feel comfortable in the field because there are so many truly capable workers in it and, probably as a direct result, reputations in the field are usually easy to trace to solid, lasting achievements. This is in vivid contrast to other fields I have worked in where, all too often, great reputations are more readily correlated with skill in promotion or politics than with important discoveries.

On the objective side, it is relatively easy to show that flavins and flavoproteins have been for many years and remain at the crossroads of several fields of biochemistry just as biochemistry is at the crossroads of many fields of biology and has become the tool par excellence, even the lingua franca of biology. Flavoproteins play a prominent role in just about every aspect of energy metabolism. They remain optimal proving grounds to test the applicability of ideas derived from the study of model reactions to enzyme mechanisms. They have been widely used in studies of protein structure and protein evolution. Because of their special physical features, ranging from readily quenched fluorescence to the prominent EPR signal of the radical, phyicists and physical biochemists are apt to use them in their initial excursions into the realm of biology and often remain afficionados for years to come. And, as will be evident from this volume, geneticists, cell physiologists, endocrinologists and experts in the emerging field of metabolite transport have joined the biochemist and biophysicist in their predilection for the subject.

As to the scientific contents of this Symposium, the Organizing Committee decided to arrange papers around certain topics which appeared most timely or in which major breakthroughs had occurred. The division of this volume into sections reflects, largely, these prearranged topics. As is usually the case, some investigations could not be readily classified into such convenient pigeonholes: these are collected in the final section.

A major function of the Symposium was to generate new ideas and provide meaningful interaction among workers in diverse fields of science. Only time will tell to what extent we have been successful in this. Judging by the success of past Symposia in this series, which have generated and fostered most of the significant research in this field during the past decade, and judging by the probing, often excellent discussions which took place during the meeting, there is little doubt that the future research of participants and readers of this volume will greatly benefit from this meeting.

As mentioned, the San Francisco Symposium was considerably larger than any of its predecessors, with 132 formal participants and some 15 local observers from the Berkeley and San Francisco campuses of the University of California present. Although the Organizing Committee was anxious to keep the meeting small, so as to encourage informal discussion and interchange, and tried to restrict invitations to participants at least as much interested in the subject of the meeting as in the obvious attractions of the Bay Area, it seems clear that the number of active workers in the field has more than doubled during the past decade and that much of the increase in the number of participants is due to this. While this circumstance has created problems in providing adequate support and suitable accommodations for all participants, it is impressive evidence of the viability of the field and of the attraction it retains for young investigators, at a time of general attrition in science in general, and biochemistry in particular, because of economic factors. Of the 92 papers presented, 89 are included in this volume. Regretfully, three have been omitted in order not to delay publication further. The few months delay in publication, as compared with the original optimistic projections, is, as usual, the result of a few — but important — tardy manuscripts.

Until one is charged with organizing a major meeting, acknowledgements of help and support in prefaces of the published proceedings may seem like a formal courtesy. Having gone through the experience, I realize how difficult it is to express one's gratitude in a convincing manner. All I can say is that, in retrospect, I feel that the help and support of all of the individuals and organizations acknowledged below have been essential in contributing to its success.

Thanks are first due to the Organizing Committee, whose names appear in front of this volume. They contributed in a decisive way to the difficult decisions in apportioning available time and financial resources among participants, called attention to new concepts and new names emerging in the field, helped materially in providing adequate financial support, and were the source of the hard core of expertise needed to arrange a cohesive and representative program.

Prominent thanks are due to Dr. D.E. Edmondson of this Department for his invaluable help in many phases of arranging the program and editing the manuscripts. Special thanks are due to our Administrative Assistant, Ms. George O'Keefe, or, as known to flavinologists the world over, "George". Her sheer competence was solely responsible for the smooth running of the meeting and for the timely distribution of travel support in the face of seemingly insurmountable bureaucratic obstacles. Her undeniable sex-appeal and legendary tolerance for alcohol assured that, after endless conferences with hotel managers and catering directors, the amenities provided by San Francisco's Westbury Hotel were memorable and far beyond what was justified by the price charged. The fact that shortly after the meeting this excellent hotel changed hands because of financial difficulties is purely a coincidence. She and her three highly decorative assistants made the cocktail hours, banquet, and unscheduled social activities highly enjoyable and many a male participant wish that more meetings were held in San Francisco.

In the last analysis no scientific meeting can succeed without adequate financial support. Sincere thanks are due to the international and national, public and private organizations who have contributed funds for this meeting. With its limited resources but great prestige, the Symposium Committee of the International Union of Biochemistry was the first organization to sponsor the meeting, thus paving the way for support by others. The co-operation of Dr. P.N. Campbell and Dr. E.C. Slater in assuring the timely distribution of funds has been invaluable. Generous grants from the National Institute of General Medical Sciences, NIH, the National Science Foundation, and the Veterans Adminstration have assured that young as well as established participants could attend from many corners of the world. The fact that this was one of the extremely few scientific meetings supported by these organizations in 1975, at a time of profound economic uncertainties and fiscal retrenchment, bespeaks the faith of the advisory panels and administrators in the importance and the viability of the field of flavins and their wisdom in recognizing that, in contrast to other branches of biochemistry in which meetings occur frequently under many titles, this series of symposia is the only means of direct communication among workers in the field of flavins and flavoproteins. Equally important has been the support from the many industrial sources acknowledged on the front pages.

Lastly, special thanks are due to Dr. Julius R. Krevans, Dean of the School of Medicine, University of California, San Francisco and to Mr. Hugh Vickerstaff, Director, Veterans Administration Hospital, San Francisco, for their helpfulness and hospitality and for providing the extensive clerical and administrative help needed in organizing this meeting.

Part II

Oxygen activation by reduced flavins and flavoproteins

Flavin and Flavoproteins, edited by T.P. Singer

CHAPTER 1

THE STRUCTURAL CHEMISTRY OF FLAVIN-DEPENDENT OXYGEN ACTIVATION

P. HEMMERICH and A. WESSIAK

Fachbereich Biologie der Universität Konstanz, D-775 Konstanz, Postfach 7733 (G.F.R.)

Introduction

The existence of short-lived "complexes" made up from reduced flavin and "molecular" oxygen, having a dihydroflavin hydroperoxide structure HFlOOH has become obvious in flavin-dependent oxidase and oxygenase rapid kinetics [1] as well as in flavin chemistry, when Mager and Berends [2] discovered that, under unpolar conditions, no H_2O_2 was formed in dihydroflavin autoxidation. In an early oversimplification of the problem, these intermediates were thought to have the structure 1-H-Fl-10-OOH [2, 3], a proposal sharply contradicted by the present authors, who postulated 5-H-Fl-4a-OOH for chemical reasons. Later on, Massey et al. [4] admitted "10a- or 4a-isomers". The first and obviously relevant chemical model was 5-alkyl-Fl-4a-OH obtained by Walker et al. [5] and absorbing at 356 nm. The second model, which obviously was also biologically relevant, has been presented by Mueller [3], absorbing at 420 nm. Its structure had to be 1/2α-alkyl-Fl-?-OCH_3, where the question mark stands for the choice between the formally possible addition sites 6, 8, 9a and 10a. Irrespective of this much disputed question mark, it was obvious that there must at least exist two classes of flavin-oxygen "complexes", namely a "5-blocked" and a "1/2α-blocked" type, where in the chemical models the blocking is achieved by introduction of alkyl substituents, while in flavoproteins, such blocking is clearly originating from the build-up of specific hydrogen bridges between the apoprotein and the coenzyme. By conformational change, a 5-blocked flavin might be transformed into a 1/2α-blocked flavin within a given flavoprotein, while such "class interchanges" are, of course, not feasible with chemical models.

In the present paper, we want to point out that these two main types of HFlOOH isomers coincide with a systematisation of flavoproteins propagated by Massey et al. [4], which depends largely on the nature of the radicals formed by a given flavoprotein. Table I gives a survey of these amazing coincidences. We also want to introduce a change in nomenclature: The antagonism between "dehydrogenase" and "oxidase" is misleading, as is the

TABLE I

FLAVOPROTEIN SUBCLASSES

Flavoprotein data largely taken from Massey et al. [4]. Massey and Hemmerich [6], Entsch et al. [7], Hastings [8–11] and coworkers. For chemical data cf. Hemmerich [12] and **his references.** As mentioned in the text, the system might have to be completed by adding a class of transhydrogenases = $2e^-$ input/$2e^-$ output, "nonradical" flavoproteins. Metalloflavoproteins are regarded as dehydrogenase/e^--transferases with internal $1e^-$-acceptor. Flavodoxins must be considered as zero input/$1e^-$ output or "output only" e^--transferases.

Block position	Redox equivalent		Class name	Radical		O_2-Reduction		$Fl\text{-}O_2$ Complex	
	Input	Output		Type	Yield	Rate	Product	Structure	Absorption (nm)
5		1	e^--transferase	blue	high	slow	O_2^-	5-HFL-4a-OOH	≈ 370
		2	oxidase	red		fast	HOO^-	$^-$Fl-9a-OOH	≈ 420
1 or 2α	2	2	without substrate	blue	low	slow	H_2O_2	5-HFl-4a-OOH	≈ 370
			oxygenase						
		4	with substrate	red		fast	$[O] + H_2O$	2α-HFl-9a-OOH	≈ 420
1?		2	resting (cold)	none		slow	H_2O_2	5-HFl-4a-OOH	≈ 370
			luciferase					?	
		4	active (20°C)			fast	$[O] + H_2O$	1-FL-10a-OOAc	≈ 610 ?
1 or 2α	2	2 (preferred)		red	low	slow	polar H_2O_2 unpolar $O + H_2O$	1,2α-RFl-9-OOH	420
5	1	1	free flavin	blue	high	slow	pH > 8 : O_2^- pH < 8 : H_2O_2	5-R-FL-4a-OOH	365
Unblocked	1 or	2		blue (pH 5–7) red (pH 9–11)	low	fast	O_2^- H_2O_2	isomer equilibrium pH and polarity dependent	

identification of "dehydrogenases" with "blue radical flavoproteins" versus oxidases as "red radical flavoproteins". For example, flavodoxins [13] are the most prominent blue radical flavoproteins and still they do not have dehydrogenase properties. We, therefore, want to discriminate between "input" and "output" sites of flavoproteins (Table I). Counting the number of electron equivalents transferred by input and output, we may differentiate the following classes: Pure e^--transferases (flavodoxins and, perhaps, electron-transferring flavoproteins) [14]; dehydrogenase/e^--transferase [15] flavoproteins; dehydrogenase/oxidase and dehydrogenase/oxygenase flavoproteins [6]; and, finally, pure dehydrogenases (transhydrogenases) [15], but which still undergo artificial e^--transfer (diaphorase). Since some dehydrogenase activity is common to all flavoproteins with very few exceptions, it seems more advisable, therefore, to classify according to "output" activity, using "input" and "output" in the "respiratory sense" and consciously neglecting the reversal of both sites in e.g. photosynthetic flavoproteins.

Among the thus-established classes of e^--transferase, oxidase and oxygenase flavoproteins, the oxidase type remains the most common one. This is due to the fact that oxygenases turn into oxidases in the presence of inhibitors [6], while e^--transferases may turn into oxidases by protein modification or partial denaturation accompanying separation of individual flavoproteins from multi-enzyme systems. This, in general, leaves the "input" site intact while modifying the "output" site. Free flavins (Table I), on the other hand, combine oxidase and e^--transferase action, and they may assume oxygenase properties under suitable conditions [16], while their ability to effect dehydrogenation is very weak and can only be efficiently simulated photochemically [12].

The chemical models, namely 5- and 1/2α-alkylated flavins, mimic radical color and stability of flavoprotein radicals amazingly well. The rate of O_2 reduction, however, is slowed down in the alkylated models throughout, owing to the enhanced degree of bending encountered in alkyldihydroflavins as compared to free flavins [17]. The activation of O_2 is brought about by the flat conformation of reduced flavin. In all other respects the agreement between model and flavoprotein behaviour is satisfactory, as far as "output functions" are concerned. In considering the chemical blocking of the flavin nucleus by alkylation versus the biological blocking by specifically directed hydrogen bridges of the apoprotein, one should take N(1) and O(2α) as one site to a first approximation, since proton shifts between these neighbouring atoms can be assumed to require no activation and even only minimal conformation change.

The formation of HFlOOH

Gibson and Hastings [18] were first to detect the autocatalytic course of flavohydroquinone autoxidation. They explained it by the assumption of obligatory 1e^--reduction of O_2 with the formation of two autocatalytically active species, superoxide anion and flavin radical:

$$Fl_{red}H_2 + O_2 \rightarrow H\dot{F}l + \dot{O}_2^- + H^+$$

Massey et al. [19] have partially verified this assumption, when they analyzed for superoxide by ESR and, furthermore, were able to suppress autocatalysis partially upon removal of O_2^- from the system by means of superoxide dismutase. But they had to admit that apparently this radical course of the autoxidation reaction was never fully stoichiometric and would cease with decreasing pH. In fact, since $\dot{F}l^-$ has been demonstrated to be a potent autocatalyst [20, 21], the overall radical autoxidation must read at pH > 8:

$$Fl_{red}H_2 + 2O_2 \rightarrow Fl_{ox} + 2\dot{O}_2^- + 2H^+$$

if occurring stoichiometrically. While Massey et al. [19] attributed this pH dependence solely to the well-known decrease of superoxide stability at neutral to acid pH, owing to self-dismutation

$$\dot{O}_2^- + H^+ \xrightleftharpoons{pK \approx 4} H\dot{O}_2, \quad H\dot{O}_2 + \dot{O}_2^- \rightarrow O_2 + OOH^-,$$

Hemmerich et al. [22] were pointing out, at the same time, that a non-radical course of the autoxidation

$$O_2 + Fl_{red}H_2 \rightarrow Fl_{ox} + H_2O_2$$

yeilding H_2O_2 directly, was competing with superoxide formation and even prevailing at lower pH and in such model flavin systems, which would suppress radical intermediates. In spite of this radical suppression, there was still autocatalysis to be observed.

Furthermore, it should be pointed out that flavosemiquinones are not powerful O_2 reductants by themselves, just because they are radicals and would, therefore, not involve a spin-conservation problem when reacting with triplet oxygen. On the contrary, the blue neutral radical 5-H$\dot{F}$l has been shown, first by Vaish and Tollin [21] in flavin flash photolysis and later on by Faraggi et al. [20] in flavin pulse radiolysis, to require either deprotonation ($pK \approx 8$) [23, 24] or disproportionation as initial step of autoxidation. If, therefore, both these primary reactions are blocked, either by chemical 5-alkylation or biological 5-hydrogen bridging with a concomitant pK-shift towards the alkaline range, the blue H$\dot{F}$l will become exceedingly air stable, as shown in the chemical 5-R$\dot{F}$l species [25] as well as in e.g. flavodoxin blue radicals [13].

It appears rather a mystery why the red anion $\dot{F}l^-$ is such a potent O_2 activator. Owing to this capacity, failure reactions of reduced oxidase flavoproteins with $1e^-$-acceptors are easily healed by O_2 in the presence of the dismutase as a repair enzyme.

The autoxidation kinetics of Massey et al. [19] as well as the autocatalysis found in the above non-radical autoxidation by Hemmerich et al. [22] and, last but not least, the stoichiometry of dihydroflavin autoxidation in unpolar media,

$$2Fl_{red}H_2 + O_2 \rightarrow 2Fl_{ox} + 2H_2O,$$

with exclusion even of intermediate H_2O_2 formation as observed by Mager and Berends [2]: all these data required the occurrence of intermediate "flavin-oxygen complexes", whose structure could be guessed to be H-Fl-OOH, if compared with well-known intermediates in chemical autoxidation of enamines [26], for indeed flavohydroquinone exhibits an enamine (even enediamine) structure [27].

The occurrence of HFlOOH once established, two problems remain to be solved:

(1) How can a fast reaction of triplet oxygen with singlet dihydroflavin to yield singlet HFlOOH be visualized, taking spin conservation into account?

(2) What is the position of H and OOH fixation in HFlOOH, taking the various modes of decay into account:

$$\mathrm{HFlOOH} \rightleftarrows \begin{cases} \dot{\mathrm{F}}\mathrm{l}^- + \dot{\mathrm{O}}_2^- + 2\mathrm{H}^+ & \text{at pH} > 8 \text{ prevailing } (1e^-) \\ \mathrm{Fl}_{ox} + \mathrm{H}_2\mathrm{O}_2 & \text{at pH} < 8 \text{ prevailing } (2e^-) \\ \mathrm{Fl}_{ox} + [\mathrm{O}] + \mathrm{H}_2\mathrm{O} & \text{prevailing in unpolar media } (4e^-) \end{cases}$$

The spin conservation dilemma

A most simple solution of this problem is the postulation of obligatory primary radical formation and secondary build up of HFlOOH via reversal of the (at pH < 8 surely reversed) HFl-OOH homolysis as proposed by Hamilton [28]. However, this would imply that the obligatory intermediate radical pair ($H\dot{F}l$, $\dot{O}_2^-$) would stay "complexed" to allow for spin correlation in the absence of charge correlation without either of both relatively stable partner radicals diffusing away and being recognised by efficient trapping agents such as e.g. superoxide dismutase. At least under polar conditions this is rather improbable, while, if indeed it happens in the required quantitative way, the formulation of the intermediate "biradical" complex turns into semantics. Hence, the suspicion is obvious that HFlOOH formation might, spin conserved or not, be the primary reaction and superoxide formation, under its conditions of occurrence, the secondary one.

Hemmerich and Schuman-Jones [29] have tried to explain this apparent dilemma by proposing a low-lying triplet state for the flat conformation of flavohydroquinone. Indeed, as already mentioned, the rate of $Fl_{red}H_2$ autoxidation seems to parallel the degree of planarity found in the reduced flavin molecule. Since all bigger substituents increase the bending, the unsubstituted anion $Fl_{red}H^-$ is fastest to react with O_2 [19]. The center pyrazine subnucleus of this species exhibits 8 π-electrons and is, therefore, antiaromatic in the flat state. Hence, the assumption of a low-lying triplet state, which might be thermally populated in contact with a paramagnetic species, seems to be plausible. A dioxetane structure for the primary contact between O_2 and $Fl_{red}H_2$ followed by proton shift to yield HFlOOH, might thus give a correct picture [30].

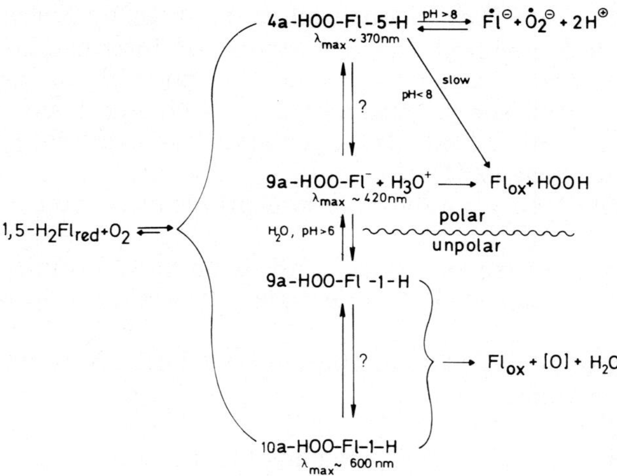

Scheme 1. Formation and decay of Fl-O_2-"complexes" HFlOOH.

H-Fl-OOH isomers and their decay modes (see Scheme 1)

As mentioned briefly in the Introduction, two main types of HFlOOH isomers have to be differentiated, viz. 5-H-Fl-4a-OOH and 1/2α-H-Fl-?-OOH. To verify these structures and to solve the problem of the question mark, we revert to chemical models R-Fl-X, where R is a small alkyl group and X equals a nucleophile which is as electronegative, but less reactive than OOH, e.g. OH or OCH_3. Thus we can trust that the chromophores of H-Fl-OOH and R-Fl-X will be to a first approximation identical.

The most stable and only possible isomer in the 5-blocked system, i.e. 5-H-Fl-4a-OOH, is well simulated by the 4a-pseudobases of 5-alkylflavoquinonium salts ($\lambda_{max} \approx 370$ nm) first described by Walker et al. [5]. These models tend towards slow homolytic C(4a)-X cleavage, if $\dot{X}$ is a suitably stabilized radical residue, e.g. mercaptyl [31]. This tendency is due to the high spin density encountered at C(4a) in R$\dot{F}$l [25]. If $\dot{X}$ is less stable, as e.g. $\dot{O}$H, acid-catalyzed splitting of XH is preferred. Hence, if X = OOH, homolysis is preferred under alkaline pH conditions suitable to ensure formation of the stable superoxide radical, while formation of the less stable neutral $\dot{O}$OH will be overcome at lower pH by slow heterolytic splitting of HOOH (Table I). Hence, 5-H-Fl-4a-OOH may serve as "failure HFlOOH" in flavoprotein e^--transferases or as "storage HFlOOH" in oxygenases devoid of substrate. In luciferase [8–11], its decay becomes exceedingly slow upon cooling.

The HFlOOH isomers of the 1/2α-blocked system are a domain of Müller [32]. The present authors' contribution to this is largely confined to a claim against position 10a being involved in OOH fixation, for the following reasons:

(1) 10a is an "amidine" position, whereas the vinylogous (and therefore

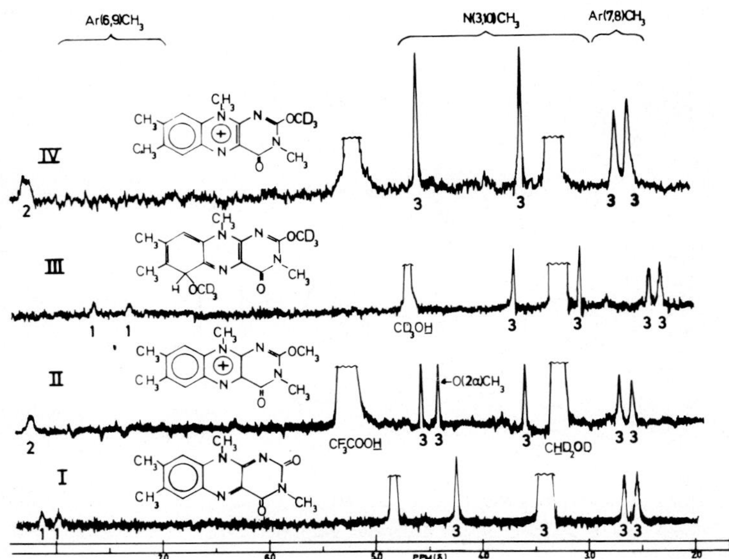

Fig. 1. ^{1}H NMR of 2α-methyllumiflavoquinonium perchlorate in $C^2H_3O^2H$ (II), its C^2H_3O "Adduct" (2α-CH_3-Fl-9a-OCH_3 cf. Müller [32]), (III) and the product of reacidification with CF_3COO^2H (IV), which is identical with II, apart from 2α-OCH_3 exchanged for 2α-OC^2H_3. The absorption maximum of III is 440 nm, ≈ 8000 $M^{-1} \cdot cm^{-1}$. For comparison, 3-methyl-lumiflavin is shown (I) under the same conditions. The high field shift and the large separation of the C(6,9)H peaks has given rise [27] to misinterpretation as a 6-adduct instead of a 9a-adduct. (cf. formula indicated). In the figure, ^{2}H is represented as "D".

equally eligible) addition sites are "aldimine" (position 6) and "ketimine" (positions 8 and 9a). If nucleophilic addition at carbon centers is discussed, thermodynamic stability of adducts will increase in the order amidine ≪ ketimine < aldimine. On the other hand, 1,2-additions (as in 10,9a and 10,10a) will be favoured over 1,4-additions (as in 10,6 and 10,8), or in other words, the nucleophiles should preferably add to a carbon having a nitrogen next to it. If one applies these two rules at once, the choice is left between the aldimine-1,4-addition in position 6 and the ketimine-1,2-addition in position 9a. One of the present authors preferred earlier position 6 over 9a in an erroneous interpretation of NMR data (Fig. 1) [27]. We express our thanks to Dr. G. Binsch, University of Munich, for privately pointing out that the peak at 7.3 ppm was definitely too far on the low field side to represent a 6-methine proton in a 6-adduct (Fig. 1).

Dr. Müller, while earlier favouring 10a-addition [3] will in the present volume present final NMR evidence for 1/2α-R-Fl-9a-OCH_3.

(2) Apart from being an amidine position, C(10a) would have to be ruled out for three more reasons:

(a) The 1/2α-R-Fl-10a-X chromophore could not absorb around 420 nm and be light red, as experimentally observed for 1/2α-R-Fl-?-X. It should instead be similar to an alkyl-amino-alloxane-anil [27], absorbing around 600 nm (deep blue), see also below (Scheme 5).

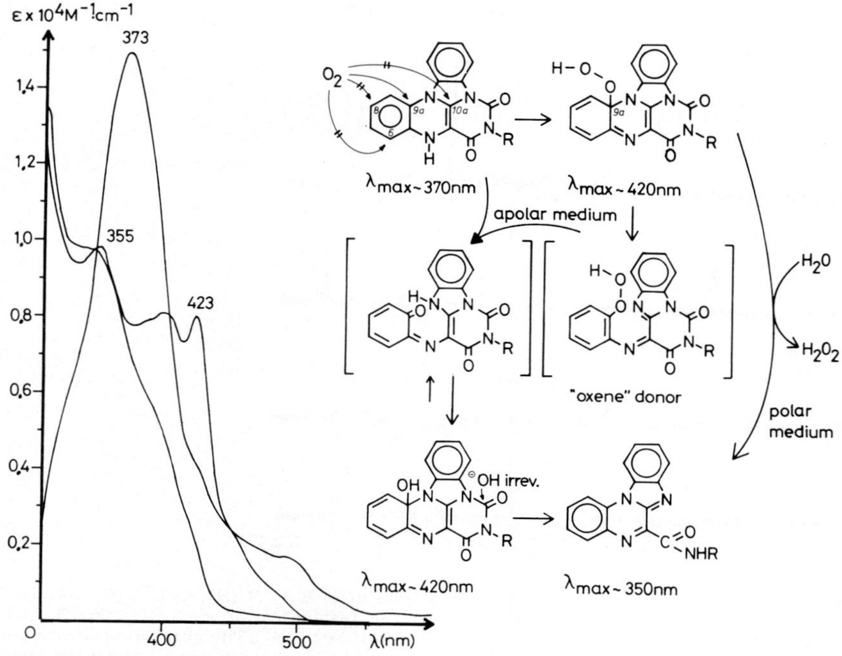

Fig. 2. 1,10-Benzo-1,5-dihydroflavin autoxidation in dry xylene, yielding red 9a-intermediate 1-R-Fl-9a-X, X being OOH or OH (according to the reaction course given). The chromophore should be, to a first approximation, independent of the nature of X [33]. The overall reaction is a $4e^-$-reduction of O_2 by 1,10-benzo-1,5-dihydroflavin via 1-R-Fl-9a-OOH as autocatalyst and decay of 1-R-Fl-9a-OH product by 2,3-cleavage. In the enzyme case, the latter reaction is prevented by deblocking N(1).

(b) Any R-Fl-10a-X chromophore must not depend to the observed extent of 25 nm upon shifting R from N(1) towards O(2α) [27], since a 10a-adduct separates the 1/2α-amide subgroup from the main chromophore.

(c) Knappe [33] has synthesized a 1,10-bridged flavin, which exhibits additional benzimidazole resonance through position 10a. For this derivative, addition at C(10a) in favour of C(9a) appears entirely improbable, since the amidine resonance characteristic for normal C(10a) is even strongly enhanced, whereas C(9a) is unaltered. This 1,10-benzo-1,5-dihydroflavin yields the most stable red intermediates upon oxidation (Fig. 2) as well as the "Mager effect" of $4e^-$-reduction of O_2 without intermediate H_2O_2 formation. 1-R-Fl-9a-OOH must, therefore, act as an oxene generator. Bruice and Smith [34] disclaim 9a as well as 10a-additions with reference to a flavin model bulkily substituted at N(10), where the autoxidation appears to occur unhindered. Since this model is, however, unsubstituted at 1/2α, it is obvious that its autoxidation proceeds via the unhindered position 4a. Thus, this argument against 9a-additions is unvalid. Summarizing, the question mark in 1/2α-R-Fl-?-X seems to be solved in favour of X = 9a.

These HFlOOH-models are, as already mentioned, light red and exhibit absorption peaks at 420 nm for 1-R-Fl-9a-X and 440 nm for 2α-R-Fl-9a-X with extinctions ranging from 5000 to 9000 $M^{-1} \cdot cm^{-1}$. If R = H, they must be assumed after all we know from analogous pteridines (lumazines) [35] to have a pK around 6, since any 1,9a-dihydroflavin system is in fact electronically identical with a 7,8-dihydro-6-butadienyllumazine. Hence, at neutral to alkaline pH, the anion Fl^--9a-X must prevail. Such an anion would obviously prefer a decay mode which splits X as anionic X^-. Hence, for X = OOH, cleavage with fast and exclusive formation of peroxide (and not superoxide) is the consequence.

If, on the other hand, the proton remains fixed at N(1) in 1-H-Fl-9a-OOH, we may again learn from pteridine chemistry that 9a,10-ring opening is facile (7,8 in the pteridine numbering). For example, riboflavin biosynthesis as described by Plaut [36], also implies such pteridine precursor ring opening.

If such ring opening occurs with 1-H-Fl-9a-OOH, we obtain, as earlier postulated in principle by McCapra [37] for the (wrong) 10a-isomer, the sp^2-carbon in position 9a required for activation of the O-O bond! This would lead to immediate "oxene" transfer towards any available C=C bond in the neighbourhood, followed by reclosure of the flavin ring with splitting of water (cf. Fig. 2). This scheme must, according to the most recent state of knowledge, replace the 10a-analogous scheme [6].

This scheme is analogous to the one proposed by Hamilton [28], but for the (again wrong) position 4a! If a 4a,5-ring opening would ever occur, we would be left with a strongly reducing phenylenediamine half and a strongly oxidizing alloxane half of what has been the "storage isomer" 5-H-Fl-4a-OOH, the two halves being linked by the trigonal N(10), which cannot bring the two halves in "through bond" resonance. Instead, electron transfer would occur through space and irreversibly, even in the absence of a 4a-carbonyl-oxide. Such electron transfer between strongly oxidizing alloxane and strongly reducing phenylenediamine is the well-known side reaction, which causes the poor yields in the good old flavin synthesis of Richard Kuhn [38].

The 9a,10-ring opening (Fig. 2), however, creates a vinylogous per-acid amide, where the two halves are in full resonance with each other. Hence, internal $1e^-$-oxidoreduction cannot take place; no oxidizing center being left except for the O-O bond. The latter, however, appears to be a most powerful "oxene gun", and if 1-H-Fl-9a-OOH was ever formed in the absence of any ternary substrate on which the gun might aim, there is no doubt that occasionally a mischief would occur by insertion of oxene into either of CH-positions 6 or 8 of the flavin itself. This is, in the present authors' opinion, the way "hydroxyflavin" admixtures in flavoprotein preparations might arise [39, 40] (cf. Scheme 2).

Hence, it appears very well understandable that flavoprotein oxygenases, in the absence of substrate to be oxygenated, do indeed form "storage HFlOOH", $\lambda_{max} \approx 370$ nm, as described in the present volume by Entsch et al. [7], which rearranges upon addition of ArH-substrate to (1-H-Fl-9a-OOH, ArH) λ_{max} 420 nm, which in turn undergoes ring opening and oxene insertion to yield ArOH, while the residual OH group at the flavin is returning

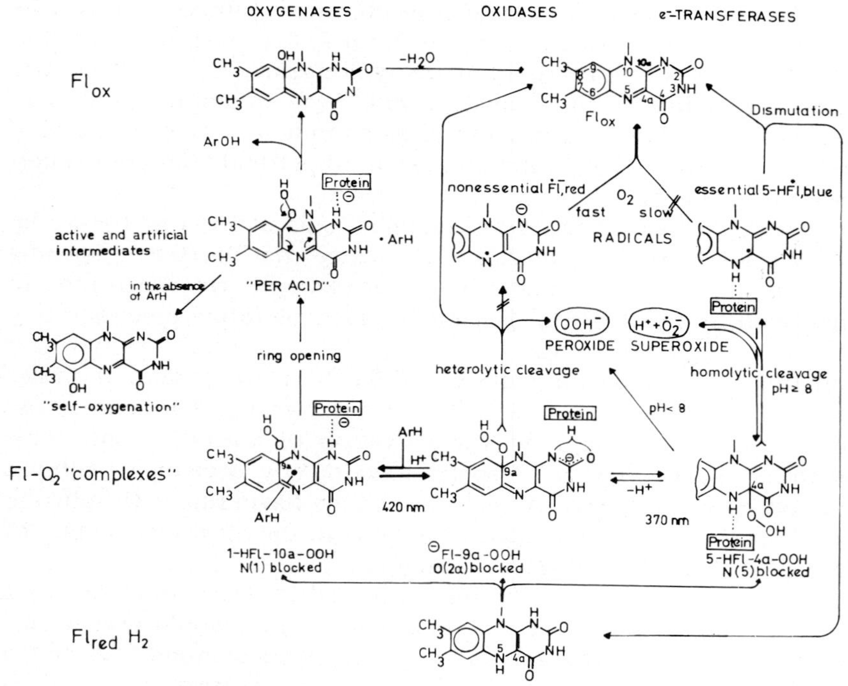

Scheme 2. Schematic summary of flavoprotein subclass action mechanisms. Attention should be paid to the fact that the HFlOOH isomers (second line from bottom) exist, in principle, in a chemical equilibrium, from which the apoprotein will "select" the "suitable" isomer depending on the polarity and hydrogen activity of the active site.

into the "storage position" 5-H-Fl-4a-OH, ($\lambda \approx 370$ nm) awaiting dehydration to return Fl_{ox}, (cf. Scheme 2).

Chemically, this reaction sequence will be hard to prove, since any 1-R-Fl-9a-OOH formed upon autoxidation of a 1-blocked dihydroflavin will most rapidly react further with either the starting material or an extraneous substrate, as is outlined, in this volume, by Mager [41]. But in the present authors' opinion, the main path of these reactions is "non-radical", though of course, and particularly in a polar environment, radical side paths might play a role. If the "oxene gun" succeeds, however, in catching a reduced flavin during its lifetime, the oxene will be transferred to the 4a,10a-double bond with subsequent splitting of water from this highly reactive epoxide.

1-CH_3-Fl-O-O-Fl-1-CH_3: A blue, inactive 10a-adduct

It has been mentioned above that 1,10a-dihydroflavin should be expected to exhibit blue colour. Indeed, Hemmerich et al. [42] found as early as 1960 a labile blue derivative arising from autoxidation of 1-CH_3-Fl_{red}H in benzene in small yield, which at this time was taken to be a radical. With the help of Professor Ehrenberg, University of Stockholm (Ehrenberg, A. and Hemmerich, P., unpublished results), it was demonstrated

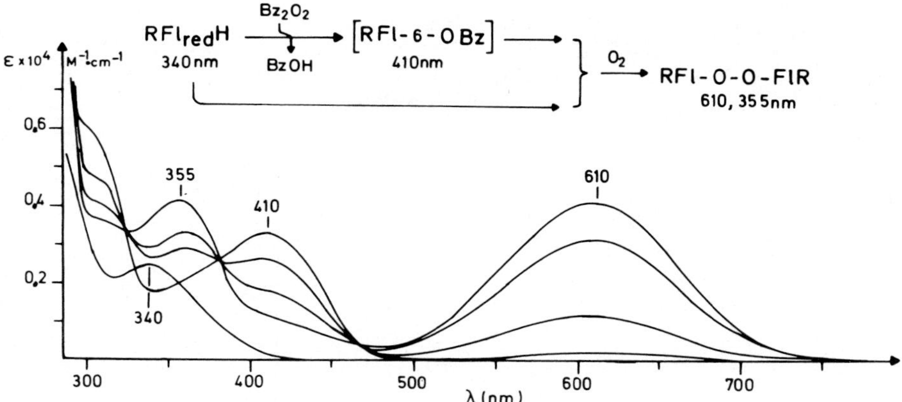

Fig. 3. Dibenzoylperoxide titration of 1,3-dimethyl-1,5-dihydrolumiflavin [43] in dry aerobic benzene. Formation of the light red 1-CH_3-Fl-9a-OBz intermediate is followed by reaction with 1-CH_3-$Fl_{red}H$ starting material and one molecule O_2 to form blue 1-CH_3-10a-O-O-10a-Fl-1-CH_3, within minutes. (cf. Scheme 5). The data of this figure are a generous gift of Dr. F. Müller [32].

in the early sixties that this compound behaved in a diamagnetic manner with regard to ESR and magnetic microbalance. Recently, Hemmerich and Hartmann [43] resumed the study of this chromophore when it had become probable that a blue colour might mean 10a-addition. The following new data were collected:

(1) Formation of the blue compound is accelerated in aerobic benzene solution, if 1-CH_3-$Fl_{red}H$ is titrated with dibenzoylperoxide (Fig. 3). While maximum formation of normal oxidized product, 1-$CH_3Fl^+_{ox}$ or (in the absence of protons) its rearranged pseudobase CH_3-Fl_{ox}-OH (a "spiro-hydantoin" known from the work of Mager and Berends [2] and Dudley [44]), required the expected 1 mol of Bz_2O_2, maximum formation of the blue compound required only one half of it (Fig. 4).

(2) Formation of the blue compound is strictly dependent on the presence of molecular oxygen in addition to the titrating agent Bz_2O_2 (Fig. 5). Its decay is accelerated by flushing the solution with argon. The spectral course of the autoxidation, yielding at first the light red 1-CH_3-Fl-9a-OBz, which turns over to the blue compound by reaction with one molecule of starting 1-CH_3-$Fl_{red}H$ and one molecule of O_2 (cf. Scheme 3).

(3) Addition of acetic acid to the once formed blue compound yields up to one molecule of oxidized 1-CH_3-Fl^+_{ox} (λ_{max} 410 nm) and one molecule of starting 1-$CH_3Fl_{red}H$, which in turn can be titrated by Bz_2O_2 to yield more oxidized flavin. The losses encountered can be accounted for quantitatively by formation of 1-$CH_3Fl_{ox}OH$, (λ_{max} 305 nm).

(4) Prolonged standing of the blue solution for several weeks results in the formation of a blue crystalline material, which cannot be split by addition of acetic acid. This material has the composition 1-CH_3-Fl-Fl-1-CH_3 and shows by NMR all flavin protons being retained. We assume, therefore, that a direct sigma-bond is established between the carbon atoms 10a of two

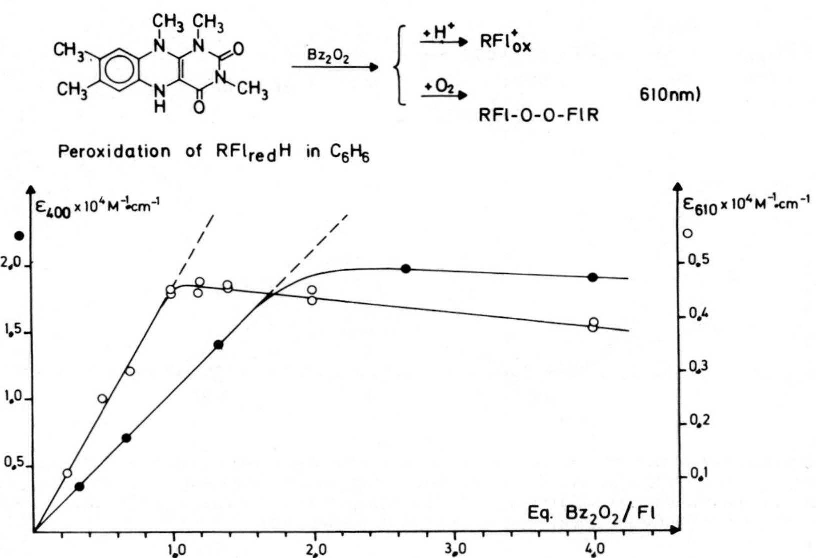

Fig. 4. Stoichiometry of blue product formation versus formation of the red intermediate, abscissa in electron equivalents (not mol) of Bz_2O_2, open circles applying to 610 nm, full circles to 410 nm absorption, cf. Fig. 3, in dry benzene.

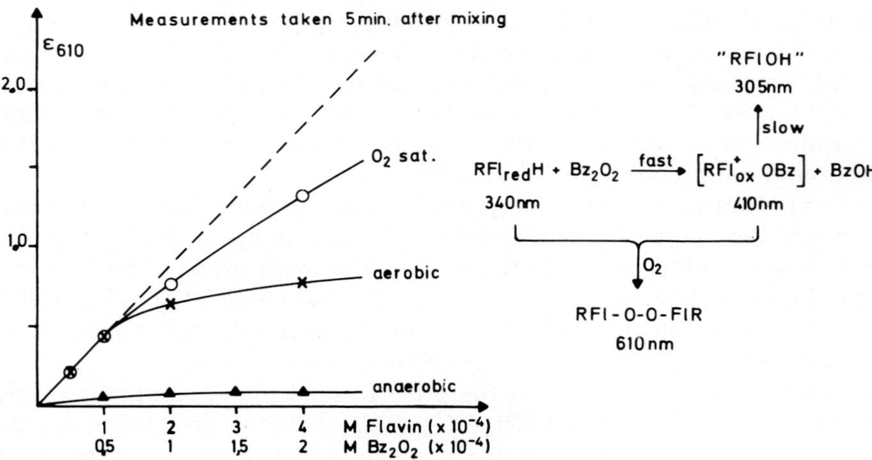

Fig. 5. Dependence of formation of blue product upon O_2 concentration in dry benzene.

flavin molecules by loss of O_2 from the preformed peroxide, since both flavin halves are chemically equivalent.

The blue chromophore seems to be biologically inessential except for, perhaps, bacterial luciferase, which implies a 1-$RFl_{ox}{}^{+*}$-type emitter [45] and would require the oxene donor next to the substrate residue R.

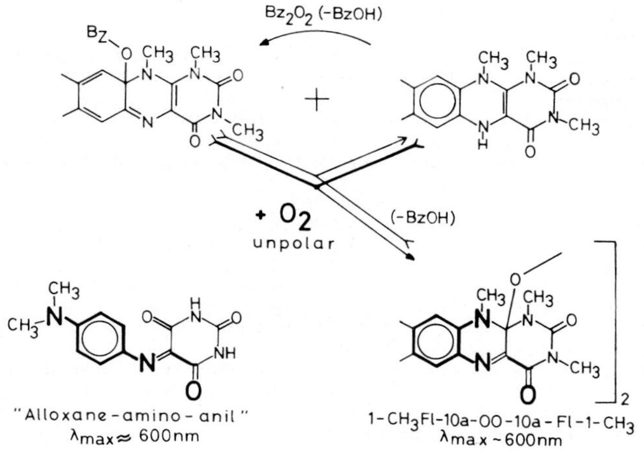

Scheme 3. Chemistry of "blue intermediate" formation and decay. Note that the 9a-benzoate (above left) splits benzoate reversibly upon acidification, yielding the stable cation 1-$CH_3Fl^+_{ox}$. With traces of OH^-, 9a-OBz is replaced by 10a-OH, which yields irreversible (but biologically irrelevant) ring contraction (cf. Mager, this volume) [41].

References

1 Massey, V., Ghisla, S., Ballou, D.P., Walsh, C.T., Cheung, Y.T. and Abeles, R.H. (1976) Flavins and Flavoproteins, Proceedings of the 5th International Symposium (Singer, T.P., ed.), Chapt. 20, pp. 199–212, Elsevier, Amsterdam

2 Mager, H.I.X. and Berends, W. (1966) Biochim. Biophys. Acta 118, 440–441

3 Müller, F. (1971) in Flavins and Flavoproteins (Kamin, H., ed.), pp. 363–373, University Park Press, Baltimore

4 Massey, V., Müller, F., Feldberg, R., Schuman, M., Sullivan, P.A., Howell, L.G., Mayhew, S.G., Matthews, R.G. and Foust, G.P. (1969) J. Biol. Chem. 244, 3999–4006

5 Walker, W.H., Hemmerich, P. and Massey, V. (1967) Helv. Chim. Acta 50, 2269–2279

6 Massey, V. and Hemmerich, P. (1975) in The Enzymes (Boyer, P.D., ed.), 3rd edn., Vol. XII, Part B, pp. 191–252, Academic Press, New York

7 Entsch, B., Ballou, D.P. and Massey, V. (1976) Flavins and Flavoproteins, Proceedings of the 5th International Symposium (Singer, T.P., ed.), Chapt. 11, pp. 111–123, Elsevier, Amsterdam

8 Hastings, J.W., Balny, C. and Douzou, P. (1976) Flavins and Flavoproteins, Proceedings of the 5th International Symposium (Singer, T.P., ed.), Chapt. 4, pp. 53–61, Elsevier, Amsterdam

9 Nicoli, M.Z., Baldwin, T.O., Becvar, J.E. and Hastings, J.W. (1976) Flavins and Flavoproteins, Proceedings of the 5th International Symposium (Singer, T.P., ed.), Chapt. 8, pp. 87–93, Elsevier, Amsterdam

10 Becvar, J.E., Baldwin, T.O., Nicoli, M.Z. and Hastings, J.W. (1976) Flavins and Flavoproteins, Proceedings of the 5th International Symposium (Singer, T.P., ed.), Chapt. 9, pp. 94–100, Elsevier, Amsterdam

11 Tu, S.-C. and Hastings, J.W. (1976) Flavins and Flavoproteins, Proceedings of the 5th International Symposium (Singer, T.P., ed.), Chapt. 10, pp. 101–107, Elsevier, Amsterdam

12 Hemmerich, P. (1975) in Progress in the Chemistry of Organic Natural Products (Herz, W., Grisebach, H. and Kirby, G.W., eds.), in press, Springer Verlag, Berlin

13 Ludwig, M.L., Burnett, R.M., Darling, G.D., Jordan, S.R., Kendall, D.S. and Smith, W.W. (1976) Flavins and Flavoproteins, Proceedings of the 5th International Symposium (Singer, T.P., ed.), Chapt. 41, pp. 393–404, Elsevier, Amsterdam

14 Whitfield, C.D. and Mayhew, S.G. (1974) J. Biol. Chem. 249, 2801–2810

15 Hemmerich, P., Nagelschneider, G. and Veeger, C. (1970) FEBS Lett. 8, 69–83

16 Mager, H.I.X. and Berends, W. (1974) Tetrahedron 30, 917–927

17 Kierkegaard, P., Norrestam, R., Werner, P., Csöregh, I., Glehn, M., Karlsson, R., Leijonmarck, M., Rönnquist, O., Stensland, B., Tillberg, O. and Torbjörnsson, L. (1971) in Flavins and Flavoproteins (Kamin, H., ed.), pp. 1–21, University Park Press, Baltimore

18 Gibson, Q.H. and Hastings, J.W. (1962) Biochem. J. 83, 368–377

19 Massey, V., Palmer, G. and Ballou, D. (1973) in Oxidases and Related Redox Systems (King, T.E., Mason, H.S. and Morrison, M., eds.), Vol. 1, pp. 25–43, University Park Press, Baltimore

20 Faraggi, M., Hemmerich, P. and Pecht, I. (1975) FEBS Lett. 51, 47–51
21 Vaish, S.P. and Tollin, G. (1970) Bioenergetics 1, 181–192
22 Hemmerich, P., Bhaduri, A.P., Blankenhorn, G., Brüstlein, M., Haas, W. and Knappe, W.R. (1973) in Oxidases and Related Redox Systems (King, T.E., Mason, H.S. and Morrison, M., eds), Vol. 1, pp. 3–24, University Park Press, Baltimore
23 Draper, R.D. and Ingraham, L.L. (1968) Arch. Biochem. Biophys. 125, 802–808
24 Ehrenberg, A., Müller, F. and Hemmerich, P. (1967) Eur. J. Biochem. 2, 286–293
25 Müller, F., Brüstlein, M., Hemmerich, P., Massey, V. and Walker, W.H. (1972) Eur. J. Biochem. 25, 573–580
26 Zigeuner, F. and Gübitz, G. (1970) Monatsh. Chem. 101, 1547–1558
27 Hemmerich, P. and Müller, F. (1973) Ann. N. Y. Acad. Sci. 212, 13–26
28 Hamilton, G.A. (1971) in Progress in Bioorganic Chemistry (Kaiser, E.T. and Kezdy, F.J., eds.), Vol. 1, pp. 83–157, Wiley — Interscience, New York
29 Hemmerich, P. and Schuman-Jorns, M. (1973) in Enzymes: Structure and Function (Drenth, J., Oosterbaan, R.A. and Veeger, C., eds.), pp. 95–118, North Holland, Amsterdam
30 Orf, H.W. and Dolphin, D. (1974) Proc. Natl. Acad. Sci. U.S. 71, 2646–2650
31 Yokoe, I. and Bruice, T.C. (1975) J. Am. Chem. Soc. 97, 450–451
32 Müller, F., Grande, H.J. and Jarbandhan, T. (1976) Flavins and Flavoproteins, Proceedings of the 5th International Symposium (Singer, T.P., ed.), Chapt. 3, pp. 38–50, Elsevier, Amsterdam
33 Knappe, W.-R. (1974) Chem. Ber. 107, 1614–1636
34 Smith, S.B. and Bruice, T.C. (1974) J. Am. Chem. Soc. 96, 3696
35 Eisele, R. (1974) Ph. D. Thesis, University of Konstanz
36 Plaut, G.W. and Beach, R.L. (1976) Flavins and Flavoproteins, Proceedings of the 5th International Symposium (Singer, T.P., ed.), Chapt. 82, pp. 737–746, Elsevier, Amsterdam
37 McCapra, F. and Hysert, D.W. (1973) Biochem. Biophys. Res. Commun. 52, 298–305
38 Kuhn, R. and Weygand, F. (1934) Berichte 67, 1409–1413
39 Mayhew, S.G., Whitfield, C.D., Ghisla, S. and Schuman-Jorns, M. (1974) Eur. J. Biochem. 44, 579–591
40 Ghisla, S. and Mayhew, S.G. (1973) J. Biol. Chem. 248, 6568–6570
41 Mager, H.I.X. (1976) Flavins and Flavoproteins, Proceedings of the 5th International Symposium (Singer, T.P., ed.), Chapt. 2, pp. 23–37, Elsevier, Amsterdam
42 Hemmerich, P., Prijs, B. and Erlenmeyer, H. (1960) Helv. Chim. Acta 43, 372–394
43 Hemmerich, P. and Hartmann, U. unpublished data
44 Dudley, K.H. and Hemmerich, P. (1967) J. Org. Chem. 32, 3049–3045
45 Eley, M., Lee, J., Lhoste, J.-M., Lee, C.Y., Cormier, M.J. and Hemmerich, P. (1970) Biochemistry 9, 2902–2908

Flavins and Flavoproteins, edited by T.P. Singer

CHAPTER 2

NONENZYMIC ACTIVATION AND TRANSFER OF OXYGEN BY REDUCED ALLOXAZINES

H.I.X. MAGER

Biochemical and Biophysical Laboratory of the University of Technology, 67 Julianalaan, Delft (The Netherlands)

Summary

In monooxygenation one atom of the O_2 molecule is incorporated into a substrate and the other reduced to H_2O.

Several monooxygenases contain a tetrahydropteridine or a dihydroflavin as cofactor and it is postulated that these cofactors activate O_2 by a covalent linking to a bridgehead carbon. We try to verify this theory by:

(a) imitating the monooxygenation in rather simple systems, in the absence of enzymes and without addition of metal ions and

(b) studying autoxidative rearrangements of blocked dihydroalloxazines and tetrahydropteridines which may provide information on the structure of the transients.

It was of further interest to find out whether these substituted derivatives could also give rise to nonenzymic hydroxylations.

On acting as oxidizing agent in "coupled" processes or as the source of H_2O_2 in "decoupled" processes, the peroxy transient (A^R–OOH) is converted into a hydroxy-dihydroalloxazine (A^R–OH) or a cationic molecule (A^+–R). The important consequence of a simultaneous occurrence of the above processes is a varying O_2 uptake and H_2O_2 production in a quantitative oxidation of the dihydroalloxazine (AH_2).

Other compounds added to the alloxazine system can also be oxidized in a "competitive" coupled oxidation, influencing also the final O_2 uptake and the net H_2O_2 production. Aromatic hydroxylation is regarded as a competitive coupled oxidation.

In a systematic research, conditions were found favouring the competitive coupled oxidation of phenylalanine to a mixture of *o*-, *m*- and *p*-hydroxyphenylalanine. In acid media yields were obtained consistent with radical mechanisms.

Homolysis of the O–O bond of the hydroperoxide gives rise to two oxygenating species: HO• and A^R–O• radicals.

If the dihydroalloxazine/O_2 system does not contain another reducing agent, we must distinguish two functions for dihydroalloxazine:

(a) activation of O_2 by hydroperoxide formation and

(b) supporting the homolysis of the O–O bond by reactions termed "assisted homolysis" and "free-radical displacement on O–O".

On adding an excess of H_2O_2 all the available A^R–H and $A^R\cdot$ (semiquinone) can be used to accomplish homolytic reactions. In others words, the peroxide-forming role of A^R–H can be completely suppressed.

A summary of the various theoretical yields of hydroxylation is discussed.

The system alloxazinium cation/H_2O_2 has been established as a hydroxylating system.

A^R–O· radicals play a greater role than we have so far assumed, considering the high yields of hydroxylation obtained and the recent finding of labile intermediates, probably aryloxyalloxazines (Sub-O–A^R).

It is concluded that a radical mechanism could also lead to an "enzymic" stoichiometry.

Introduction: incorporation of molecular oxygen

The essential feature of Wieland's theory on biological oxidation was the rejection of the direct addition of molecular oxygen to the substrate [1]. This view was generally accepted until 1955 when mono- and dioxygenation were established by Mason et al. [2] and Hayaishi et al. [3], respectively.

One of the most interesting aspects of oxygenation concerns the mechanism of activation and transfer of oxygen. Since several monooxygenases have a dihydroflavin or tetrahydropterin as cofactor, it soon became tempting to postulate their role in the activation of O_2 [4].

In a chemical approach of this problem we try to imitate monooxygenation in nonenzymic systems containing a dihydroalloxazine or tetrahydropteridine without addition of metal ions.

Concept of the activation of O_2

The fact that in enzymic monooxygenation one atom of the O_2 molecule is incorporated into the substrate and the other reduced to H_2O ($\text{Sub-H} + DH_2 + {}^*O_2 \xrightarrow{\text{Enzyme}} \text{Sub-}^*OH + H_2^*O + D$) already indicates that O_2 has been converted into a more reactive species, formally by the uptake of 1, 2 or 3 electrons. The ultimate cleavage of the O–O bond suggests the intermediacy of peroxides.

The proposal of dihydroalloxazines as O_2-activating intermediates [4] was based upon comparative studies of partially reduced pyrazines, quinoxalines, pteridines and alloxazines, which have been carried out in our laboratory since 1954. The similar behaviour suggested that all these compounds can reduce O_2 in a similar way, by a covalent linking at a carbon

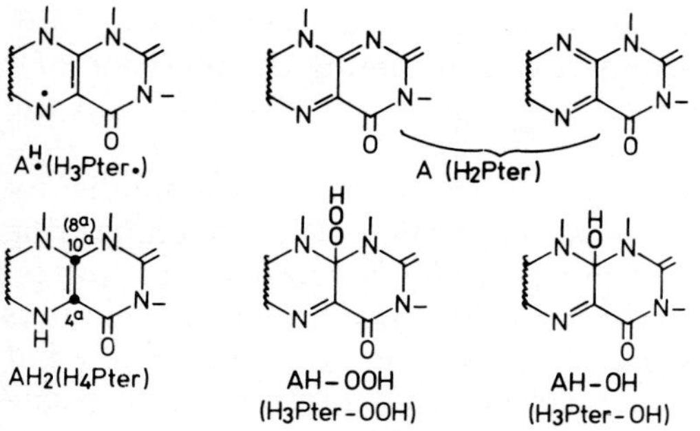

Scheme 1. First proposal on the structures of transient hydroperoxides and carbinolamines. Abbreviations: A, (iso)alloxazine (flavin); H_2Pter, *o*- or *p*-quinonoid dihydropteridine; AH• (H_3Pter•), semiquinone; AH_2, dihydroalloxazine; H_4Pter, tetrahydropteridine; AHOOH (H_3Pter-OOH), hydroperoxide; AHOH (H_3Pter-OH), hydroxy-dihydroalloxazine (hydroxy-tetrahydropteridine).

atom. The $10^a(8^a)$- and 4^a-bridgehead carbon atoms (Scheme 1) were considered to be the most probable positions for dihydroalloxazines and tetrahydropteridines [4, 5, 6]. It is expected that such hydroperoxides may give rise to carbinolamines carrying the -OH group at a bridge position also.

The presence of substituents at the nitrogen atoms may retard reactions of the perioxide or the carbinolamine or may induce irreversible conversions providing valuable information on the structure of the transients.

A summary of some important reactions shown by partially reduced pyrazines is given in Scheme 2. It represents the starting point for our studies on dihydroalloxazines and tetrahydropteridines.

Tetrahydropyrazines are very sensitive to oxygen. Spontaneous oxidation results in labile peroxides. In mono N-substituted series some of the perioxides could be isolated in crystalline form. The first peroxide was

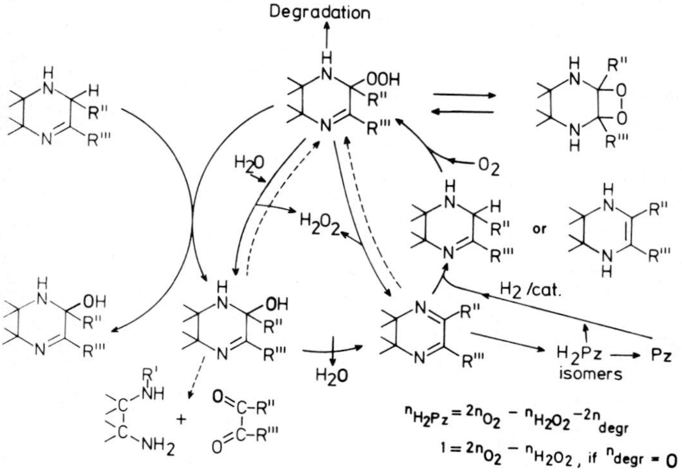

$$n_{H_2Pz} = 2n_{O_2} - n_{H_2O_2} - 2n_{degr}$$

$$1 = 2n_{O_2} - n_{H_2O_2}, \text{ if } n_{degr} = 0$$

Scheme 2. Coupled and decoupled processes in the autoxidation of tetrahydropyrazines.

obtained by Lunsford et al. [7] who described it as a dioxetane. We think that the hydroperoxide structure predominates considering the presence of a C=N double bond absorption and the absence of a N–H absorption in the infrared spectra.

Peroxides are also characterized by intramolecular nucleophilic rearrangements [5]. The finding of a typical degradation (ring opening, ring concentration) is of diagnostic value when a peroxide is too labile to isolate. For instance, characteristic rearrangements were established in reactions between H_2O_2 and a dihydropyrazine from which we concluded that addition of H_2O_2 to a C=N double bond has first given a tetrahydropyrazine hydroperoxide [8].

On acting as the oxidizing agent in a "coupled" oxidation or as the source of H_2O_2 in a "decoupled" process the tetrahydropyrazine hydroperoxide is converted into a hydroxytetrahydropyrazine ($H_3Pz–OH$). (The oxygen in $H_3Pz–OH$, arisen from hydrolytic H_2O_2 formation is derived from the solvent). The important consequence of the simultaneous occurrence of the above reactions is a varying O_2 uptake and H_2O_2 production in a quantitative autoxidation of H_4Pz to $H_3Pz–OH$. The O_2 uptake per mol of H_4Pz may vary between 0.5 and 1, while the nett production of H_2O_2 will vary between 0 and 1. Other compounds added to the reaction mixture can be oxidized in "competetive" coupled oxidations, influencing also the final O_2 uptake and nett H_2O_2 production. Suitable conditions can be found to favour each of the reactions.

Spontaneous dehydration is the most important feature of $H_3Pz–OH$ giving H_2Pz and with that the possibility to effect cyclic processes.

N-substitution may break off the process at carbinolamine level, for instance, leading to the isolation of hydrolytic ring cleavage products.

The autoxidative behaviour of dihydroalloxazines and tetrahydropteridines suggests the occurrence of the same reactions (Scheme 3). Their catalytic effect in the nonenzymic oxidation of other compounds like

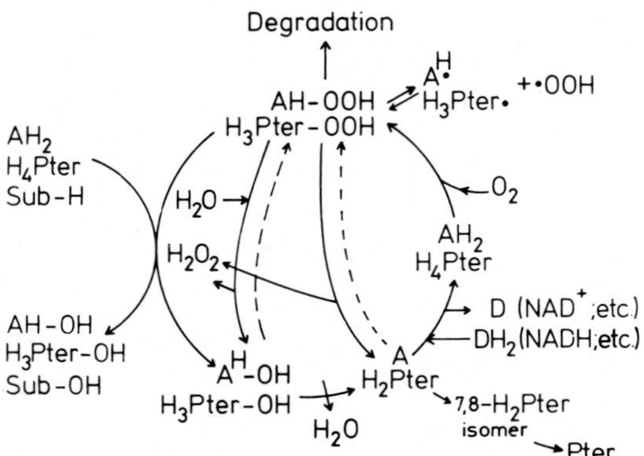

Scheme 3. Coupled and decoupled processes in the autoxidation of dihydroalloxazines and tetrahydropteridines.

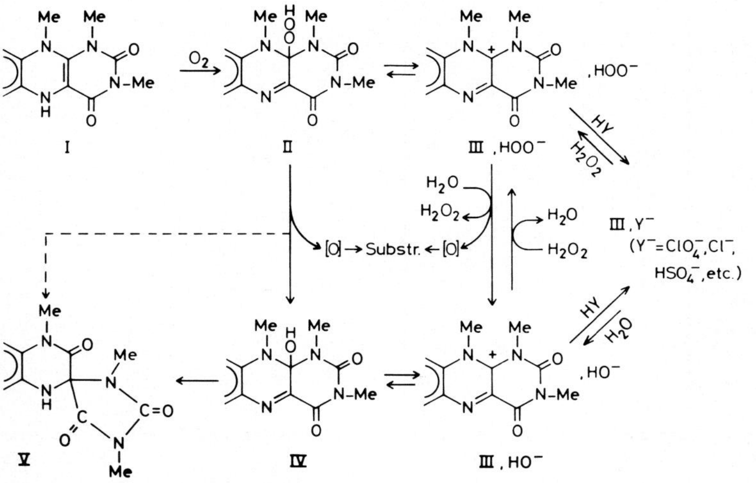

Scheme 4. Formation of spirohydantoin (V) and alloxazinium cation (III) in neutral and acid solution, respectively.

NADH, 2-mercaptoethanol, cysteine, etc. results from a continuous recycling [4]. Blocking the N_1- and N_{10}-positions prevents a dehydration in neutral media breaking off the process at the level of the carbinolamine A^R–OH [9, 10, 11]. The products were identified as spirohydantoins V (Scheme 4), being ring contracted isomers of the carbinolamines IV arising in the oxygen transfer or the H_2O_2 formation. (The oxygen introduced into a spirohydantoin may be derived from either the gasphase or the solvent depending on the way of formation and the nature of the medium [12]).

The spirohydantoin formation is an unbiological rearrangement, but deliberately looked for!! With the purpose of finding additional information further studies have been extended to the $N_{1,3,5}$-methylated series [13]. It

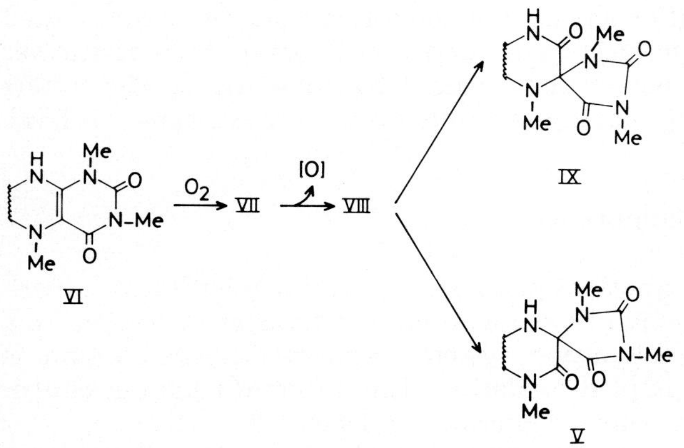

Scheme 5. Autoxidative ring contractions of $N_{1,3,5}$-methylated dihydroalloxazines and tetrahydropteridines.

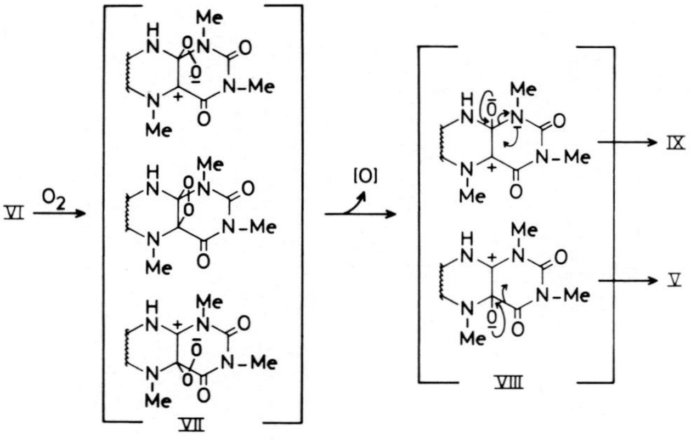

Scheme 6. Participation of both bridgehead carbons in peroxide formation and in subsequent conversions into monooxy transients (epoxide, carbinolamines).

was then established that not only the original C_{4a} but also the $C_{10^a\,(8^a)}$ may become a spiro center (Scheme 5).

These results suggest that in $N_{1,3,5}$-methylated derivatives both bridgehead carbons may be involved in the activation of O_2, giving either a$10^a\,(8^a)$- or 4^a-hydroperoxide, or perhaps a dioxetane (Scheme 6). This idea is further supported by recent findings of ring rearrangements on peroxide level [5]. Hemmerich, who postulated in succession 4^a-, 8- and 6-hydroperoxides [6, 14], now favours position 9^a as the biologically relevant addition site [15].

The presence of alkyl groups at the $N_{1,3,10(5)}$ positions does not block the occurrence of a cyclic process in acid solution. Autoxidation then produces the alloxazinium cation A^+–R (cf. III in Scheme 4). Apparently, the methylated carbinolamine A^R–OH (cf. IV) and the peroxide A^R–OOH (cf. II) can act as a pseudobase (III, HO^-) and a pseudosalt (III, HOO^-), respectively [16].

In view of the information already obtained from the heavily substituted dihydroalloxazines it was interesting to find out whether these derivatives could also give rise to nonenzymic aromatic hydroxylations. The results given by $N_{1,3,10}$- and $N_{1,3,5}$-methylated dihydroalloxazines agree on basic points [17, 13].

Search for oxygen transfer conditions

In view of the various reactions possible, systematic quantitative experiments are required to find conditions favouring the transfer of oxygen to a compound added to the alloxazine system. Aromatic hydroxylation is regarded as a competitive coupled oxidation. The different reactions contributing to the material balance are summarized in Scheme 7.

The alloxazinium molecules are formed in three reactions (Eqns. 1, 2 and 3) leading to a final production as represented by the sum Eqn. 5. If

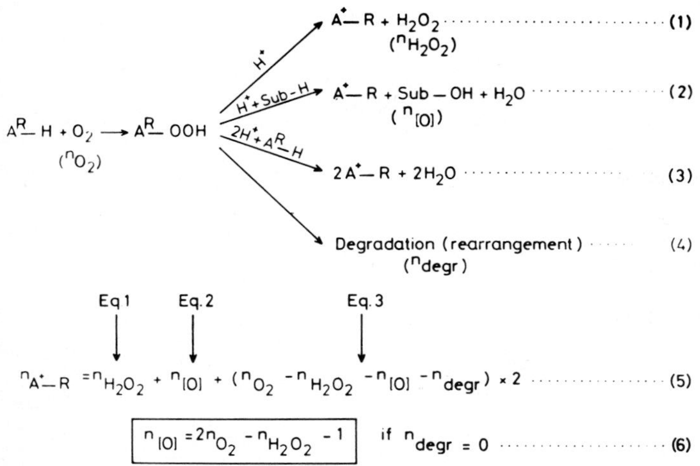

Scheme 7. Competitive oxygen transfer in relation to O_2 uptake and nett production of H_2O_2. The various quantities per mol of $A\overset{R}{—}H$ are: n_{O_2}, O_2-uptake; $n_{H_2O_2}$ = nett H_2O_2 production; $n_{[O]}$ = competitive oxygen transfer; n_{A^+-R} = production of A^+–R; n_{degr} = degradation on peroxide level. (General autoxidative termination reactions like dimerization etc. are not taken up in the material balance scheme, because they are not of practical importance under the experimental conditions used).

$n_{A^+-R} < 1$ and if this appears to be connected mainly with degradation on peroxide level, we may substitute $n_{degr} = (1 - n_{A^+-R})$ in Eqn. 5 transforming it into $n_{[O]} = 2n_{O_2} - n_{H_2O_2} + n_{A^+-R} - 2$. However, degradation on peroxide level can often be neglected ($n_{degr} = 0 \rightarrow n_{A^+-R} = 1$) resulting in an almost general applicability of the simple Eqn. 6 as a criterion for oxygen transfer, independent of the nature of the substrate!

Simple manometric determinations of n_{O_2} and $n_{H_2O_2}$, combined with a spectrophotometric check of n_{A^+-R} rapidly provide information on the degree of oxygen transfer. Such experiments have given information on inhibiting or stimulating effects of some media, the O_2 pressure, etc. [16, 17, 18]. When we have succeeded in finding conditions which make the result of Eqn. 6 greater than zero, product analysis becomes the following object. In this way we have recently worked out the nonenzymic hydroxylation of phenylalanine to *o*-, *m*-, and *p*-hydroxyphenylalanine [17]. The yields were near quantitative in acid media, but less optimal in media of pH 2–7, decreasing with increasing pH. (The theoretical values for aromatic hydroxylation are 1, for a polar mechanism, but 1/3, 1/2, 2/3, 1 and 2 for radical mechanisms, as will be discussed below).

Reaction mechanisms for nonenzymic hydroxylations

The reactivity of a peroxide can be associated with the fact that the O–O bond can be broken in reactions of several kinds. There are two types of unimolecular O–O bond cleavages: a heterolytic and a homolytic one (Scheme 8). The first is expected to be much less facile than the latter. In a review by Edwards et al. [19] this has been illustrated by the thermodynamic

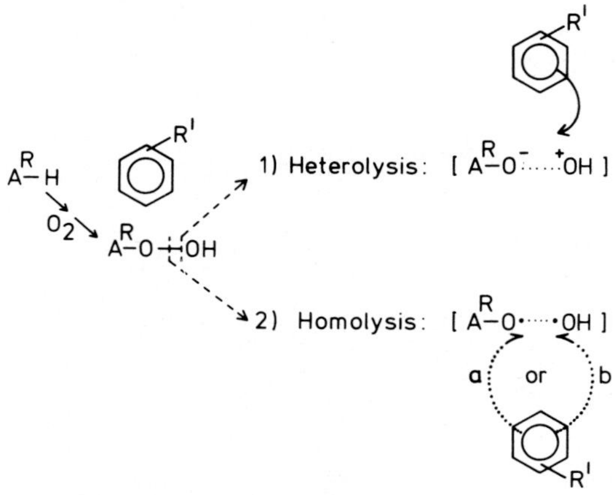

Scheme 8. Unimolecular O—O bond cleavage.

data for H_2O_2 in the gasphase at 298°K; the energy requirement for the heterolytic cleavage is about 6 times higher than for the homolytic cleavage.

The structural influence of the organic part of an organic peroxide and the ionizing power of a solvent could make the heterolytic pathway compete with the homolytic one. For a process in solution, the role of a solvent can cover a wide range of interactions going from the simple dielectric constant effect to highly specific bond formation with the peroxide fragments. However, the homolytic pathway can also be favoured, e.g. by the presence of various other compounds. In his review Hiatt [20] distinguished four general types of homolytic reaction (Scheme 9).

Hiatt pointed out that: (a) compared with dialkylperoxides and acylperoxides, hydroperoxides undergo "assisted homolysis" much more readily than unimolecular homolysis; (b) the "free-radical abstraction" though unique for hydroperoxides is much less favoured than the "free-radical displacement on oxygen-oxygen".

1. Unimolecular homolysis:

$$RO\text{-}OH \xrightarrow[\text{or } h\nu]{\Delta} RO\cdot + \cdot OH$$

2. Assisted homolysis:

$$RO\text{-}OH + X \longrightarrow RO\cdot + (XOH)\cdot$$

3. Free-radical abstraction:

$$RO\text{-}OH + Y\cdot \longrightarrow ROO\cdot + Y\text{-}H$$

4. Free-radical displacement on O-O:

$$RO\text{-}OH + Y\cdot \longrightarrow RO\text{-}Y + \cdot OH$$

$$RO\text{-}OH + Y\cdot \longrightarrow RO\cdot + Y\text{-}OH$$

Scheme 9. Four general types of homolytic reaction.

Assisted homolysis:

$$A^R\text{-}O\text{-}OH + A^R\text{-}H \xrightarrow{H^+} \begin{cases} A^R\text{-}O\cdot + A^R\cdot + H_2O \\ H\text{-}O\cdot + A^R\cdot + A^+\text{-}R + H_2O \end{cases}$$

Free - radical displacement on O - O:

$$A^R\text{-}O\text{-}OH + A^R\cdot \begin{cases} \xrightarrow{H^+} A^R\text{-}O\cdot + A^+\text{-}R + H_2O \\ \xrightarrow{2H^+} H\text{-}O\cdot + 2A^+\text{-}R + H_2O \end{cases}$$

Overall Result:

$$3A^R\text{-}H \begin{cases} \rightarrow 2A^R\text{-}O\cdot \\ \rightarrow 2H\text{-}O\cdot \end{cases}$$

Scheme 10. Homolytic reactions in dihydroalloxazine autoxidation.

Our results on dihydroalloxazine autooxidation are consistent with the above conclusions. "Assisted homolysis", accomplished by a molecule of A^R–H and "free-radical displacement on O–O", effected by $A^R\cdot$ are the two reactions providing the oxygenating species HO· or A^R–O· (Scheme 10)! (The extent of "free-radical displacement on O–O" will be dependent on the

TABLE I

THEORETICAL QUANTITIES FOR AUTOXIDATIVE HYDROXYLATIONS

The maximal production of HO· radicals on autoxidation is given by Eqn. I. The same overall hydroxylation would be obtained if A^R–O· radicals were the oxygenating species, generated in accordance with:

$$3\,A^R\text{–}H + 2\,O_2 + H^+ \rightarrow A^+\text{–}R + 2\,H_2O + 2\,A^R\text{–}O\cdot$$

Quantitiative formation of the substrate-radical adduct followed by one of the various final oxidations should give the theoretical values mentioned in: Column 1, the O_2 uptake. It deviates from the value 2/3 when O_2 is involved in the further oxidation of the adduct (Ib–Ie). Column 2, the nett production of H_2O_2. Column 3, the maximal production of oxygenating radicals. It deviates from the value 2/3 if A^R–H + O_2 are also involved in the final oxidation of the adduct (Ib) or, if H_2O_2 produced in the oxidation of the adduct is further consumed (Ie). Column 4, the maximal yield of hydroxylation. It deviates from the value 2/3 in the case of: (Ia) disproportionation; (Ib) the oxidation of the adduct by A^R–H + O_2; (Ie) a further consumption of H_2O_2 produced in the oxidation of the adduct. Column 5, characteristic ratio's ("R_0-values") to be derived from the experimental data. The R_0-values enable the rapid drawing of a conclusion concerning the type of adduct oxidation.

I. On Autoxidation.	n_{O_2}	$n_{H_2O_2}$	$n_{HO\cdot}$	$n_{[O]max}$	$\frac{n_{[O]max}}{n_{O_2}-n_{H_2O_2}}$
$3A^R\text{-}H + 2O_2 + 3H^+ \rightarrow 3A^+\text{-}R + 2H_2O + 2HO\cdot$ ···I)···	2/3	0	2/3		0
2 (R'-substituted OH/H adduct, a_1) → R'-C₆H₄-OH ····a)···	2/3	0	2/3	1/3	1/2
+ A^R-OOH + H^+ ··· → 2 R'-C₆H₄-OH ····b)···	3/4	0	1/2	1/2	2/3
+ 1/2 O_2 → 2 R'-C₆H₄-OH ····c)···	5/6	0	2/3	2/3	4/5
+ O_2 → 2 R'-C₆H₄-OH ····d)···	1	1/3	2/3	2/3	1
H_2O_2 + II, II d ····e)···	1	→0	→1	→1	1

disproportionation of $A^R\cdot$ into A^+-R and A^R-H). No discrimination between the two oxygenating species can be made on performing stoichiometric studies, but the overall result: "two oxygenating species per three molecules of A^R-H" has been established beyond doubt. In other words, in simple nonenzymic systems which do not contain other reducing agents there are two functions for the reduced alloxazine: (1) in peroxide formation and (2) in peroxide homolysis. The fact that peroxide formation is stimulated, but O-transfer is inhibited on increasing the O_2-pressure is a consequence of suppressing "assisted homolysis" and "free-radical displacement on O–O".

It is now easily understood that the theoretical yield of hydroxylation can vary. For instance, two HO· radicals can give two cyclohexadienyl radical adducts, which may either disproportionate to give one phenol or be oxidized to two phenols by H_2O_2, A^R-OOH or O_2. These figures are divided by the number of A^R-H molecules involved in the overall process leading to the values of columns 4, in Tables I and II. In acid media, several of these theoretical quantities have been experimentally approached within a few per cent [17].

On adding an excess of H_2O_2 all the available A^R-H and $A^R\cdot$ molecules can be used in the reaction types: "Assisted homolysis" and in "free-radical displacement on O–O". In other words, the peroxide-formation role of A^R-H can be completely suppressed! This has been confirmed by hydroxylations taking place under anaerobic conditions. It is beyond doubt that in

TABLE II

THEORETICAL QUANTITIES FOR HYDROXYLATIONS UNDER CONDITIONS SUPPRESSING THE DIRECT PEROXIDE-FORMATION FROM $A^R-H + O_2$

Quantitative formation of the substrate-radical adduct followed by one of the various final oxidations should give the theoretical values mentioned in: Column 1, the overall O_2 uptake = 0, unless O_2 is involved in the further oxidation of the adduct (IIc and IId). Column 2, the nett consumption of H_2O_2 = 2 unless H_2O_2 is involved in the further oxidation of the adduct (IIb), or unless the adduct is oxidized by O_2 with formation of H_2O_2 (IId). Column 3, the maximal production of oxygenating radicals = 2. Column 4, the maximal yield of hydroxylation = 2, except in case of disproportionation of the adduct (IIa). Column 5, it is striking that the same characteristic R_0-values are obtained as mentioned in Table I.

II. On Adding an Excess of H_2O_2.	n_{O_2}	$n_{H_2O_2}$	$n_{HO\cdot}$	$n_{[O]\,max}$	$\frac{n_{[O]max}}{n_{O_2}-n_{H_2O_2}}$
$A^R-H + 2H_2O_2 + H^+ \rightarrow A^+-R + 2H_2O + 2HO\cdot$ ··· II)	0	-2	2		0
2 (R'-cyclohexadienyl-OH,H) a_1 → R'-phenyl-OH ··· a)	0	-2	2	1	1/2
→ (H_2O_2) → 2 R'-phenyl-OH ··· b)	0	-3	2	2	2/3
→ (1/2 O_2) → 2 R'-phenyl-OH ··· c)	1/2	-2	2	2	4/5
→ (O_2 → H_2O_2) → 2 R'-phenyl-OH ··· d)	1	-1	2	2	1

the presence of an excess of H_2O_2 two oxygenating radicals can be produced per mol of A^R-H, e.g. a maximal production of HO· radicals in agreement with Eqn. II (Table II).

In connection with Eqn. IIb (Table II) it is emphasized that in the hydroxylation of phenylalanine the impressive yield of 2 mol of tyrosine was realized and sometimes exceeded by 5–30% [17]!

Doubts have arisen recently with regard to the question which of the two oxygenating radicals (A^R-O· and HO·) is the most important. We now think that the alloxazine-oxy radical A^R-O· plays a greater role than we have so far assumed, considering the high yields of hydroxylation and the recent finding of rather labile, but interesting products which probably are aryloxy-alloxazines:

$$R'\text{-}C_6H_4\text{-}O\text{-}{}^{10a}A\text{-}R \xrightarrow{H^+} R'\text{-}C_6H_4\text{-}OH + \overset{+}{A}\text{-}R$$

The spectral changes, confirming the formation of the alloxazinium cation, are now being subjected to further studies in connection with Müller's "10ª-alcohol adducts" [21], which are now favoured as 9ª-adducts [22].

It is concluded that alloxazine-oxy radicals can also arise under strictly anaerobic conditions in the system $A^R-H + H_2O_2$. A small amount of A^+-R is required to start the process (if not added, a starting amount must be formed by peroxidation of some A^R-H). The equilibrium: $A^+-R + H_2O_2 \rightleftarrows A^R-OOH + H^+$ can then be shifted to the right by the subsequent reactions: "assisted homolysis" and "free-radical displacement on O–O" in which practically all the reduced alloxazine is consumed. The overall reaction below should be considered instead of Eqn. II (Table II):

$$(A^+\text{-}R + H_2O_2 \rightleftarrows A^R\text{-}OOH + H^+) \times 2$$

$$A^R\text{-}H + A^R\text{-}OOH \longrightarrow A^R\cdot + H_2O + A^R\text{-}O\cdot$$

$$A^R\cdot + A^R\text{-}OOH + H^+ \longrightarrow A^+\text{-}R + H_2O + A^R\text{-}O\cdot$$

$$2A^R\text{-}O\cdot + 2\,R'\text{-}C_6H_5 + 2H^+ \longrightarrow 2A^+\text{-}R + 2\,R'\text{-}(\dot{C}_6H_5)(H)(OH)$$

$$\overline{A^R\text{-}H + 2H_2O_2 + 2\,R'\text{-}C_6H_5 + H^+ \longrightarrow A^+\text{-}R + 2H_2O + 2\,R'\text{-}(\dot{C}_6H_5)(H)(OH)}$$

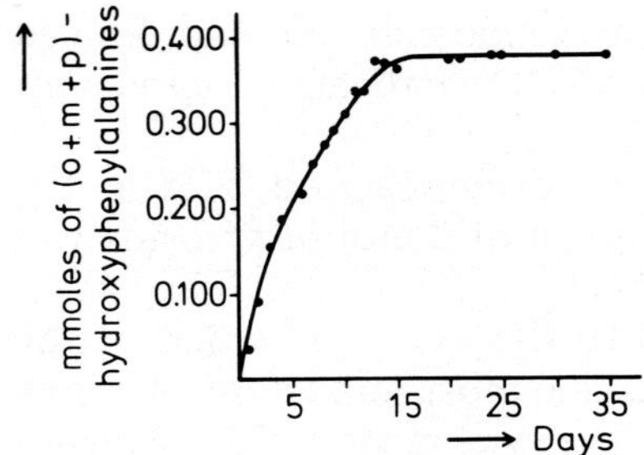

Fig. 1. Slow hydroxylation of phenylalanine effected by A^+–R (1.0 mmol) + H_2O_2 (15 mmol) in 0.25 M H_2SO_4 (50.0 ml) at 23°C.

The system $A^+{-}R + H_2O_2 \rightleftarrows A^R{-}OOH + H^+$ has been established as a hydroxylating system in the absence of reducing species also. The hydroxylation of a substrate then occurs at a relatively slow speed (cf. Fig. 1) which, however, is consistent with the absence of "assisted homolysis" and "free-radical displacement on O–O". The hydroxylation presented in the figure comes to an end, because of a slow decomposition of A^+–R taking place in dilute acid both in the presence and absence of H_2O_2.

The oxidations of the substrate-radical adduct a_1 by O_2, H_2O_2 or A^R–OOH are questionable from a mechanistic point of view. Some points should be discussed:

(1) Since the pulse radiolytic studies of Dorfman et al. [23] it has been generally accepted that an adduct like a_1 is oxidized by O_2 to a peroxy radical a_2, which decomposes into a phenol, H_2O_2 and O_2.

$$a_1 \text{ (H, OH; R')} + O_2 \longrightarrow a_2 \text{ (H, OH; R'; } O_2\cdot) \longrightarrow \text{OH-phenol (R')} + \tfrac{1}{2}H_2O_2 + \tfrac{1}{2}O_2$$

In our studies the above pathway could result only in Eqns. Id, Ie or IId. Consequently, there must be different pathways at least to account for the other relationships found.

(2) In the peroxidation of a_1 the possibility of one-electron processes has been considered but concluded to be unimportant in view of the realizations of Eqns. Ib and IIb.

$$a_1 \text{ (H, OH; R')} + H_2O_2\,(A^R\text{-}OOH) \longrightarrow \text{OH-phenol (R')} + H_2O + HO\cdot\,(A^R\text{-}O\cdot)$$

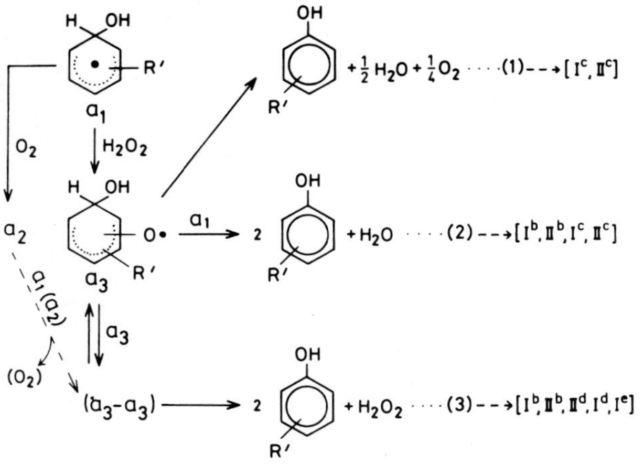

Scheme 11. Postulated intermediacy of the oxy-radical a_3

Such one-electron processes (free-radical displacements on O–O, induced by a_1) producing further oxygenating radicals should perpetuate aromatic hydroxylation as long as peroxides are present!

Sometimes, the theoretical yield of hydroxylation (IIb) was exceeded by 5–30% which, however, was not much regarding the excess of H_2O_2 used [17].

(3) We now propose the formation of a transient oxy-radical a_3 (Scheme 11) both in the oxidation of a_1 by H_2O_2 (A^R–OOH) and O_2. (The occurrence of Dorfman's peroxy radical a_2 remains a possibility either in a self-reaction: $a_2 + a_2 \rightarrow 2\,a_3 + O_2$ or in a reaction with another a_1: $a_2 + a_1 \rightarrow 2\,a_3$). Subsequent conversions of a_3 according to the overall Eqns. 1, 2 and 3 (Scheme 11) would lead to the several stoichiometric relationships. Reactions 2 and 3 are most likely on account of the various experiments which have well established the stoichiometric relationships (Ib), (IIb) and (Ic) and more or less (IIc) and (IId).

Conclusion

Biochemists seem to prefer a non-radical mechanism for enzymic mono-oxygenation. A good reason for this could be the NIH-shift which is not found to be promoted by radicals [24].

We have now established radical mechanisms for nonenzymic aromatic hydroxylations accomplished by some alloxazine systems. The biological relevancy of these results is questionable. However, the most important feature is, that heavily substituted dihydroalloxazines, which have initially been studied in connection with the reactivity of the bridge positions, are also able to effect very efficient hydroxylations. Apparently, the influences of the organic moiety (A^R–) of the peroxide and those of the solvents are not sufficient to favour heterolysis (cf. the upper part of Scheme 12).

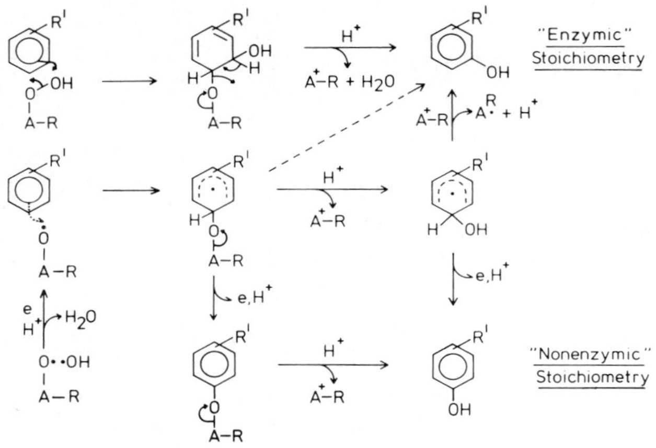

Scheme 12. Radical mechanisms leading to "enzymic" and "nonenzymic" stoichiometries.

We would like to point out that a radical mechanism (starting from the left lowest part of Scheme 12) could also lead to an "enzymic" stoichiometry. In the absence of another reducing agent, at least some quantity of the reduced alloxazine will be consumed in "assisted homolysis" and in "free-radical displacement on O–O". Nevertheless, an "enzymic" stoichiometry would still be obtained, if further oxidation of the substrate-radical adduct were coupled with the reduction of the alloxazinium cation to the

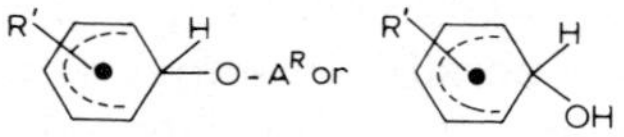

semiquinone ($A^+–R \xrightarrow{e} A^R\cdot$). In that case no reduced alloxazine would be lost for peroxide formation! The alternative disproportionation, peroxidation and oxidation by O_2 apparently are much more facile in the nonenzymic systems described in the present paper.

The presence of a suitable reducing agent could relieve the dihydroalloxazine from its supporting function in accomplishing O–O homolysis and could make all the dihydroalloxazine available for peroxide formation. It should be kept in mind that various organic compounds, which are usually not defined as reducing agents, may bring about an "assisted homolysis", while the arising radicals may cause a "free-radical displacement on O–O". In this light we explain the remarkable stimulation of nonenzymic aromatic hydroxylation by some aliphatic compounds e.g. EDTA, in the absence of metal ions (it was shown that this stimulation was not caused by traces of metal ions, acting with EDTA and H_2O_2 as a "masked" Fenton's reagent [17]).

References

1 (1974) Molecular Mechanisms of Oxygen Activation (Hayaishi, O., ed) Academic Press, New York
2 Mason, H.S., Fowlks, W.L. and Peterson, E. (1955) J. Am. Chem. Soc. 77, 2914–2915
3 Hayaishi, O., Katagiri, M. and Rothberg, S. (1955) J. Am. Chem. Soc. 77, 5450–5451
4 Mager, H.I.X. and Berends, W. (1965) Rec. Trav. Chim. 84, 1329–1343
5 Mager, H.I.X. (1975) Autoxidative Conversions of Tetrahydropteridines and some Related Ring Systems; paper presented at the 5th International Symposium on Chemistry and Biology of Pteridines and Folic Acid, to be published
6 Hemmerich, P. (1968) in Biochemie des Sauerstoffs (Hess, B. and Staudinger, Hj., eds.), pp. 249–261, Springer-Verlag, Berlin
7 Lunsford, C.D., Lutz, R.E. and Bowden, E. (1955) J. Org. Chem. 20, 1513–1530
8 Mager, H.I.X. and Berends, W. (1965) Rec. Trav. Chim. 84, 314–333
9 Mager, H.I.X. and Berends, W. (1966) Biochim. Biophys. Acta 118, 440–441
10 Mager, H.I.X., Addink, R. and Berends, W. (1967) Rec. Trav. Chim. 86, 833–851
11 Mager, H.I.X. and Berends, W. (1972) Rec. Trav. Chim. 91, 1137–1150
12 Mager, H.I.X. and Berends, W. (1973) Tetrahedron Lett. 4051–4052
13 Mager, H.I.X. and Berends, W., to be published
14 Hemmerich, P. and Müller, F. (1973) Ann. N.Y. Acad. Sci. 212, 13–26
15 Hemmerich, P and Wessiak, A. (1976) Flavins and Flavoproteins, Proceedings of the 5th International Symposium (Singer, T.P., ed.), Chapt. 1, pp. 9–22, Elsevier, Amsterdam
16 Mager, H.I.X. and Berends, W. (1972) Rec. Trav. Chim. 91, 611–629
17 Mager, H.I.X. and Berends, W. (1974) Tetrahedron 30, 917–927
18 Mager, H.I.X. and Berends, W. (1972) Rec. Trav. Chim. 91, 630–642
19 Curci, R. and Edwards, J.O. (1970) in Organic Peroxides (Swern, D., ed.), Vol. 1, pp. 199–264, Wiley-Interscience
20 Hiatt, R. (1971) in Organic Peroxides (Swern, D., ed.), Vol. II, pp. 1–151, Wiley-Interscience
21 Müller, F. (1971) in Flavins and Flavoproteins (Kamin, H., ed.), pp. 363–373, University Park Press, Baltimore
22 Müller, F., Grande, H.J. and Jarbandhan, T. (1976) Flavins and Flavoproteins, Proceedings of the 5th International Symposium (Singer, T.P., ed.), Chapt. 3, pp. 38–50, Elsevier, Amsterdam
23 Dorfman, L.M., Taub, I.A. and Buhler, R.E. (1962) J. Chem. Phys. 36, 3051–3061
24 Jerina, D.M. and Daly, J.W. (1973) in Oxidases and Related Redox Systems (King, T.E., Mason, H.S. and Morrison, M., eds.), Vol. 1, pp. 143–161, University Park Press, Baltimore

Flavins and Flavoproteins, edited by T.P. Singer

CHAPTER 3

ON THE INTERACTION OF FLAVINS WITH OXYGEN IONS AND MOLECULAR OXYGEN

FRANZ MÜLLER, HANS J. GRANDE and TJAN JARBANDHAN

Department of Biochemistry, Agricultural University, Wageningen (The Netherlands)

Introduction

From kinetic studies of the reoxidation reaction of 1,3,7,8,10-pentamethyl-1,5-dihydroisoalloxazine with molecular oxygen, Mager and Berends [1] proposed that the oxidation reaction occurs through the formation of an intermediate flavin hydroperoxyde. These authors proposed that molecular oxygen interacts with the flavin at the 10a position. Hemmerich [2], on the other hand, suggested that oxygen adds to the 4a position of the flavin nucleus. In the context of the reactivity of various flavoproteins towards sulfite we offered a possible explanation for the differences in oxygen metabolism of flavoprotein oxidases, dehydrogenases, and hydroxylases, assuming a common intermediate [3]. More recently we have been able to obtain physical evidence for the formation of a transient flavin-oxygen adduct during the reoxidation reaction of a reduced flavin [4]. At the same time we found that methoxide forms a complex with oxidized flavin and that the light absorption spectrum of the methoxide adduct was identical with that of the transient formed during the reoxidation reaction. This fact indicates that both products possess the same structure. Since the methoxide complex is rather stable as compared to the oxygen adduct we have investigated the former complex in detail and have been able to elucidate its structure by NMR technique. This paper describes the physical properties and the structure of the methoxide adduct and gives some further information on the various products formed during the reoxidation reaction of reduced flavins in the presence of various oxidants. Some results on the latter problem have been published elsewhere [5].

Results and Discussion

The physical properties and the structure of the flavin-methoxide adduct

Fig. 1 shows the light absorption spectrum of 2α,3,7,8,10-pentamethyl-isoalloxazinium perchlorate in methanolic solution (curve 1) and the spectral changes observed upon addition of various amounts of sodium methoxide to

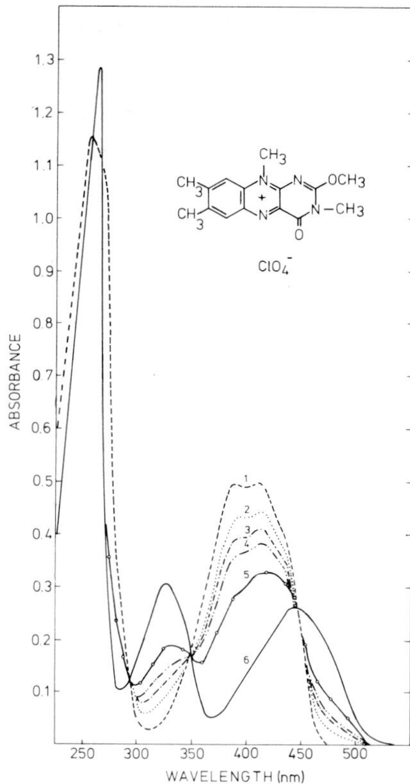

Fig. 1. Light absorption spectrum of 2α,3,7,8,10-pentamethyl-isoalloxazinium perchlorate (32 μM) in methanolic solution in the absence and presence of sodium methoxide. Curve 1: in the absence of sodium methoxide; curves 2–6: in the presence of 24, 40, 56, 128 and 600 μM sodium methoxide, respectively.

the solution of curve 1. The starting material exhibits maxima at 410, 390 and 263 nm. The methoxide adduct shows maxima at 445, 325 and 256 nm. The isosbestic points observed at 448, 350 and 288 nm indicate the presence of two species in the solution. From these results it was also ascertained that the adduct consists of one mol methoxide per mol flavin. The absorption spectrum of the methoxide adduct of 1,3,7,8,10-pentamethyl-isoalloxazinium perchlorate is shifted hypsochromically by about 20 nm [4], as compared to that shown in Fig. 1. Based on the absorption spectrum of the methoxide adduct of 1,3,7,8,10-pentamethyl-isoalloxazinium perchlorate we proposed that the addition had occurred at position 10a of the flavin nucleus [4]. It is known from crystallographic work [6] and light absorption measurements [7] conducted with 1,3,7,8,10-pentamethyl-isoalloxazinium perchlorate that the two methyl groups in positions 1 and 10 perturb each other, i.e. they exert a steric strain in the flavin molecule. The abovementioned difference in the absorption spectra of the two methoxide adducts could be caused by steric perturbation. To ascertain if such steric effects are the reason for the difference observed we have investigated models carrying various more or less bulky substituents at position 10 of the flavin nucleus. The results are summarized in Table I. It is interesting to note that the absorption spectra of the methoxide adducts of compound II and III are almost identical, indicating that both adducts are planar and that little, if any, steric strain is present. That this conclusion is correct becomes obvious when bulky groups are introduced into position 10, affecting the absorption spectra drastically, as demonstrated with compounds IV and V.

TABLE I

The light absorption maxima and the corresponding molar extinction coefficients of the methoxide adducts of various isoalloxazinium perchlorates in methanolic solution.

No.	Compound	λ_{max}(nm), ϵ($mM^{-1}\,cm^{-1}$)
I	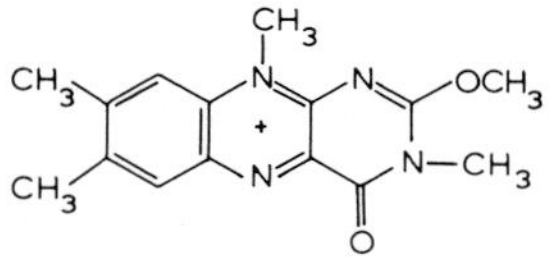 ClO_4^-	445, 8.3; 326, 9.7; 256, 36.1
II	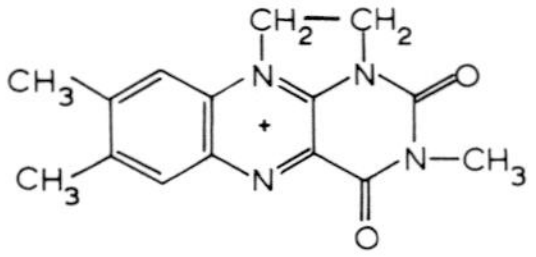 ClO_4^-	425, 4.5; 315, 5.6; 250, 19.7
III	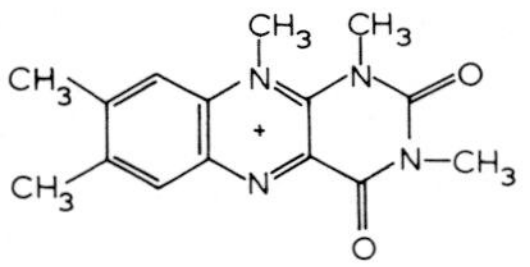 ClO_4^-	423, 4.7; 310, 8.0; 260, 13.4
IV	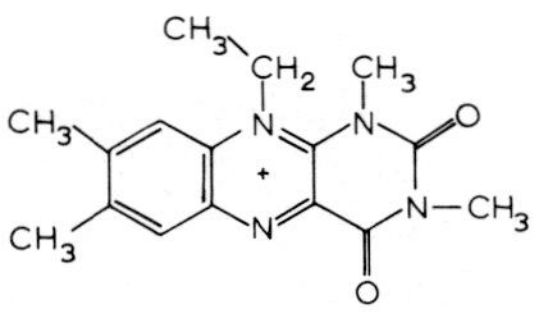 ClO_4^-	425 (shoulder), 3.9; 365, 6.6; 307, 5.6; 235, 15.5
V	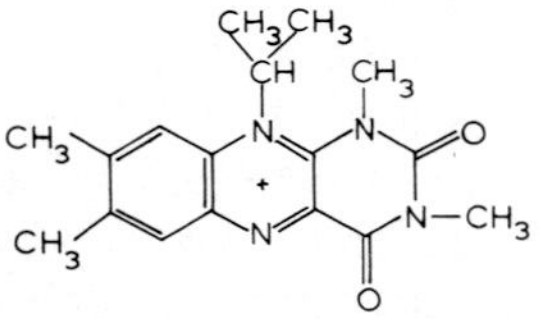 ClO_4^-	372, 7.8; 250, 23.2

Thus in going from III to V the visible spectrum changes from a two-band spectrum to a one-band spectrum. From these results it can be concluded that the difference in the absorption spectra between compound I and II, III is not due to steric effects. On the other hand, the NMR spectra show that there is no structural difference between the three methoxide adducts, i.e. in all three cases methoxide has added to the same position of the flavin nucleus. Therefore the only possibility left to explain this spectral difference is to assume a more extended chromophoric system for adduct I than for the adducts II and III. This is in agreement with the fact that the first absorption band of the adducts II and III is shifted to a wavelength about 20 nm shorter than that of the adduct I. This difference roughly corresponds to one double bond. From these results we proposed that methoxide adds to position 8 or 6 of the flavin nucleus [8, 9]. On the other hand employing compounds VI ($R' = H$, $R'' = CH_3$) and VII we found that these compounds form very

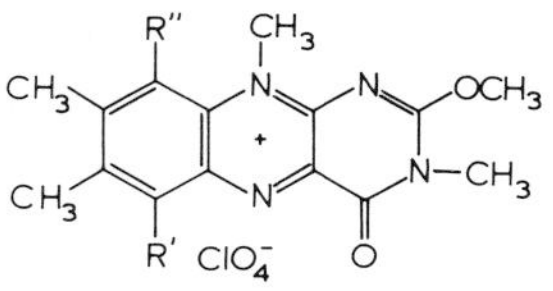

VI

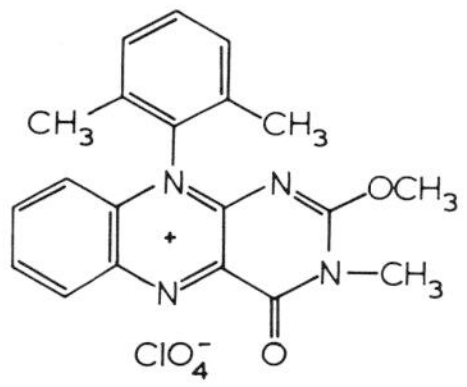

VII

labile adducts with methoxide. The final product formed in this reaction employing compound VI has been identified as the corresponding lumichrome derivative, i.e. demethylation at position 10 had occurred. Employing the isomer VI ($R'' = H$, $R' = CH_3$) a stable methoxide adduct was obtained. These results suggest that methoxide adds in the vicinity of the N(10) of the flavin nucleus. The structure of the methoxide adduct has been elucidated by NMR technique (see below). In employing 3,7-dimethyl-1,10-ethylene-isoalloxazinium perchlorate, which is rather stable towards solvolysis and dimerization reactions, we were able to add other oxygen ions such as hydroxide and hydroperoxide to the flavin nucleus. The absorption spectra of the various adducts are compared in Fig. 2. The spectrum of the starting material in aqueous solution (curve 1) exhibits maxima at 425, 369 and 262 nm. The spectrum of the methoxide addition product (curve 2) in methanolic solution shows maxima at 425, 300 and 243 nm. The spectrum of the hydroperoxide adduct (curve 3) was obtained by dissolving the compound in anhydrous acetonitrile and adding peroxide and triethylamine to the solution. If the compound was dissolved in 0.01 M NaOH an absorption spectrum showing maxima at 456, 308 and 250 nm (curve 4) was obtained. It should be mentioned that the latter two products are less stable than the methoxide adduct and undergo slow secondary reactions. However, all mentioned reactions are fully reversed upon acidification provided that acidification occurs within a few minutes of preparation of the solutions. In

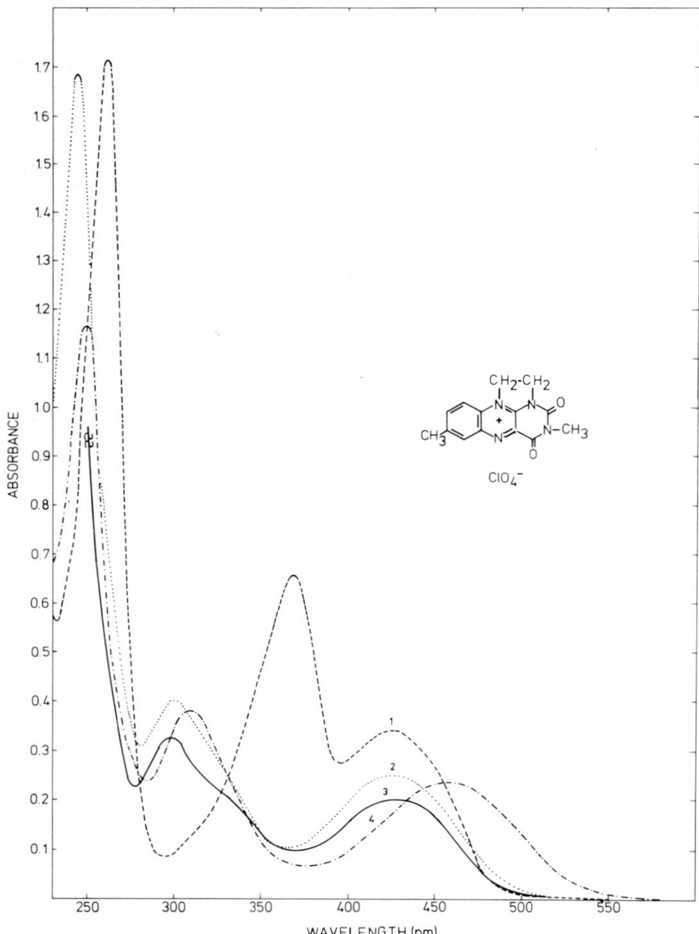

Fig. 2. The light absorption spectra of 3,7-dimethyl-1,10-ethylene-isoalloxazinium perchlorate (50 μM) and its methoxide, hydroxide and peroxide adducts. Curve 1: in distilled water; curve 2: in methanolic solution in the presence of 500 μM sodium methoxide; curve 3: in acetonitrile in the presence of 5 mM *t*-butylperoxide and 5 mM triethylamine; curve 4: in 0.01 M NaOH.

addition all adducts are weakly fluorescent as reported earlier [5]. As already mentioned, the methoxide adducts are quite stable compared to the other adducts. For this reason we have so far elucidated only the structure of the methoxide adduct using NMR techniques. However, the light absorption studies indicate that all oxygen ions mentioned above interact with the flavin at the same position. The assignment of the various proton resonances to the various flavin protons has been ascertained by comparison with the resonances of various models, by selective deuteration or by using double resonance techniques. Fig. 3 shows the NMR spectra of 2α,3,7,10-tetramethyl-isoalloxazinium perchlorate and derivatives thereof. These compounds have been selected for the simple reason that the aromatic ring contains by approximation an ABX system. Such a system has the advantage

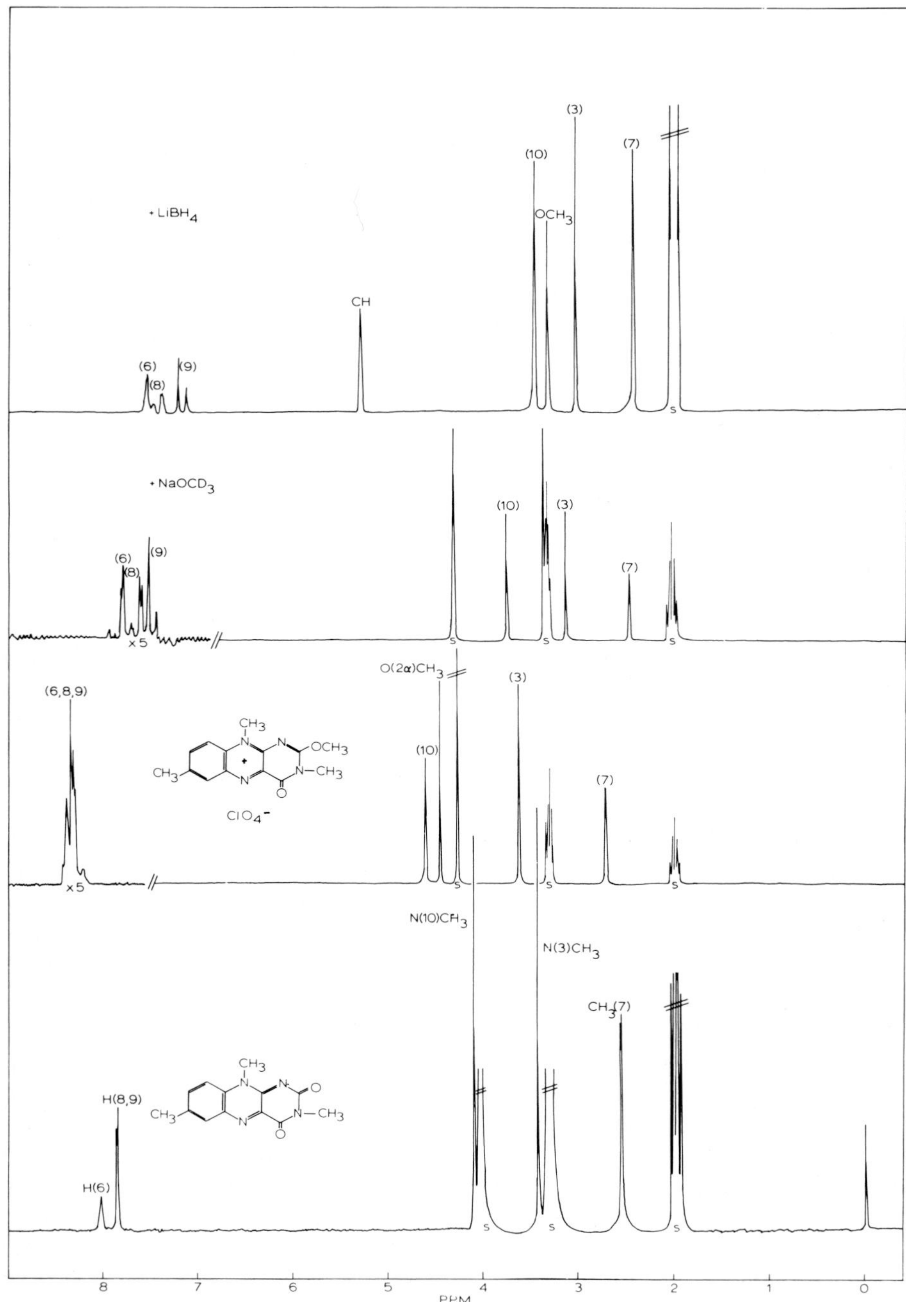

Fig. 3. ^{1}H NMR spectra of 3,7,10-trimethyl-isoalloxazine (bottom curve), 2α,3,7,10-tetramethyl-isoalloxazinium perchlorate (2nd curve from bottom) and its methoxide addition product (3rd curve from bottom). The solvent mixture used was $C^2H_3CN/C^2H_3O^2H$ (1:3, v:v). The top curve represents the spectrum of the product obtained by $LiBH_4$ reduction of the isoalloxazinium salt in C^2H_3CN solution. The numbers in parentheses refer to the flavin ring system. S stands for solvent resonances. In all cases Me_3Si has been used as an internal standard. The spectra have been recorded on a Varian XL-100 Mc instrument. ($D = {}^2H$).

that any changes, e.g. an addition reaction, occurring at the atoms involved in the AB system are easily recognized. The NMR spectrum of 2α,3,7,10-tetramethyl-isoalloxazinium perchlorate (Fig. 3) exhibits resonances at $\delta = 2.74$, 3.65, 4.48 and 4.68 ppm, respectively. These resonances are due to the methyl groups. The aromatic protons show a complex pattern. The AB pattern, due to the protons at position 8 and 9 of the flavin nucleus, consists of a quartet which is not resolved and is centered at about $\delta =$ 8.36 ppm. With the aid of double resonance techniques the absorption at $\delta = 8.39$ could be assigned unambiguously to the C(6)H. For comparison the spectrum of 3,7,10-trimethyl-isoalloxazine is also given in Fig. 3. It is similar to that of its cationic analog except that all resonances are shifted to higher fields as expected (Fig. 3). The NMR spectrum of the methoxide adduct of 2α,3,7,10-tetramethyl-isoalloxazinium salt (Fig. 3) shows resonances at $\delta = 2.47$, 3.14 and 3.75 ppm and originate, in the order of increasing δ values, from the methyl groups in position 7, 3 and 10 respectively. The resonance of the 2α-CH_3 group is hidden under the solvent peak located at $\delta = 3.35$ ppm. The aromatic region of this spectrum is much better resolved than that of the spectrum of the parent compound. The AB system consists of the doublets at $\delta = 7.41$, 7.51 ppm (C(9)H) and at $\delta = 7.58$, 7.60, 7.68 ppm (C(8)H). The doublet at $\delta = 7.77$, 7.79 ppm represents the C(6)H. The latter doublet and the doublet at $\delta = 7.58$, 7.60 ppm and at $\delta = 7.60$ ppm, which is resolved to a lesser degree, are due to a small meta coupling between the protons at positions 6 and 8. This meta coupling can only be observed if the CH_3(7) resonance is decoupled. The aromatic resonances presented in Fig. 3 for the methoxide adduct has been obtained under such resonance decoupling conditions. Since the AB system in this spectrum is fully retained it can be concluded that methoxide has not been added to position 8 of the flavin nucleus. In a similar way it has also been proven that addition does not occur at position 6 using the isomeric compound 2α,3,8,10-tetramethyl-isoalloxazinium perchlorate. Thus these results did not yield the desired information concerning the structure of the methoxide adduct. However, to our surprise we found that reduction of flavinium salts by $LiBH_4$ does not yield the expected corresponding 1,5-dihydro analog but a product which absorption spectrum is almost identical to that of the methoxide adduct. The NMR spectrum of the reduction product of 2α,3,7,10-tetramethyl-isoalloxazinium perchlorate is given in Fig. 3. The absorptions observed at $\delta = 2.37$, 2.88, 3.27 and 3.41 ppm, respectively, correspond to the methyl resonances. The integration of the additional resonance at $\delta = 5.22$ ppm, compared to the spectrum of the methoxide adduct, corresponds to one proton and thus originates from $LiBH_4$. The aromatic protons absorb at $\delta = 7.05$, 7.14 (C(9)H), 7.31, 7.40 (C(8)H) and 7.46 ppm (C(6)H), respectively. The resonances at low fields due to the CH(6, 8) are broader than those observed with the methoxide adduct. The reason for this is that the spectrum of the reduction product was obtained under normal resonance conditions and that the small coupling between the 7-CH_3 group and the two protons broadens the resonance lines of the latter ones. For the same reason the meta couplings cannot be observed. To our disappointment the additional proton observed in the spectrum of the reduction product does not show any coupling to other groups in the molecule. From this fact we were not able either to confirm nor to exclude the 10α position as a possible candidate for the addition reaction. At first glance there exists only a small difference between the spectrum of the methoxide and that of the reduction product in relation to the spectrum of the original compound, i.e. the resonances of the methoxide product are somewhat less displaced towards higher field values. However, a close inspection of the two spectra of the different products reveals that in the case of the methoxide adduct the resonances due to $CH_3(10)$ and H(9) are shifted considerably less to higher field values than the other resonances of the spectrum. This is a strong indication that methoxide adds at the position between the C(9) and N(10) atoms, i.e. 9a. In order to determine whether or not the 10a position is involved in the formation of the adduct we synthesized a flavinium salt enriched with ^{13}C at position 4 and 10a. The ^{13}C NMR spectrum of this compound and those of the corresponding addition products were almost identical indicating that the hybridization of the two ^{13}C atoms remained unchanged. If addition had occurred at the 10a position a large shift of the $^{13}C(10a)$ absorption was expected, compared to that of the starting material, because upon addition to 10a the hybridization of the carbon atom should change from sp^2 to sp^3. These data definitely exclude the 10a position as a candidate for involvement in the addition reaction. The final support for the structure of the addition product came from ^1H NMR spectra obtained with the ^{13}C-enriched flavin. The NMR spectra of this compound and its methoxide and $LiBH_4$ reduction products are shown in Fig. 4. The absorptions of the methyl groups in position 3 and 10 are split into doublets due to coupling with the ^{13}C atoms in the neighbourhood. In comparing the spectrum of the methoxide adduct with that of the $LiBH_4$ product it is seen even more clearly (cf. Fig. 3) that the H(9) and $CH_3(10)$ absorptions of the methoxide adduct are much less displaced towards higher field values than the corresponding resonances of the $LiBH_4$ product. In addition the CH absorption at about $\delta = 5.28$ ppm shows some fine structure and is less intense than that shown in Fig. 3 for the natural compound. The reduced intensity is explained easily by a weak coupling of this proton with the ^{13}C atom at position 10a. Thus the NMR data prove that addition of methoxide or hydride to flavinium salts occurs at position 9a:

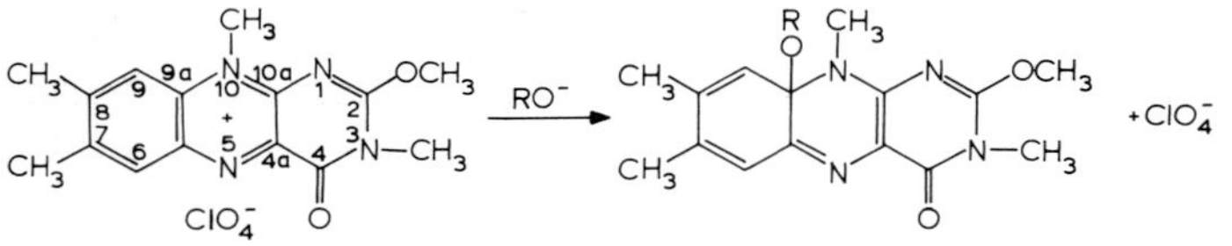

I $\xrightarrow{RO^-}$ Ia $+ ClO_4^-$

The structure Ia is also in full agreement with the spectral data.

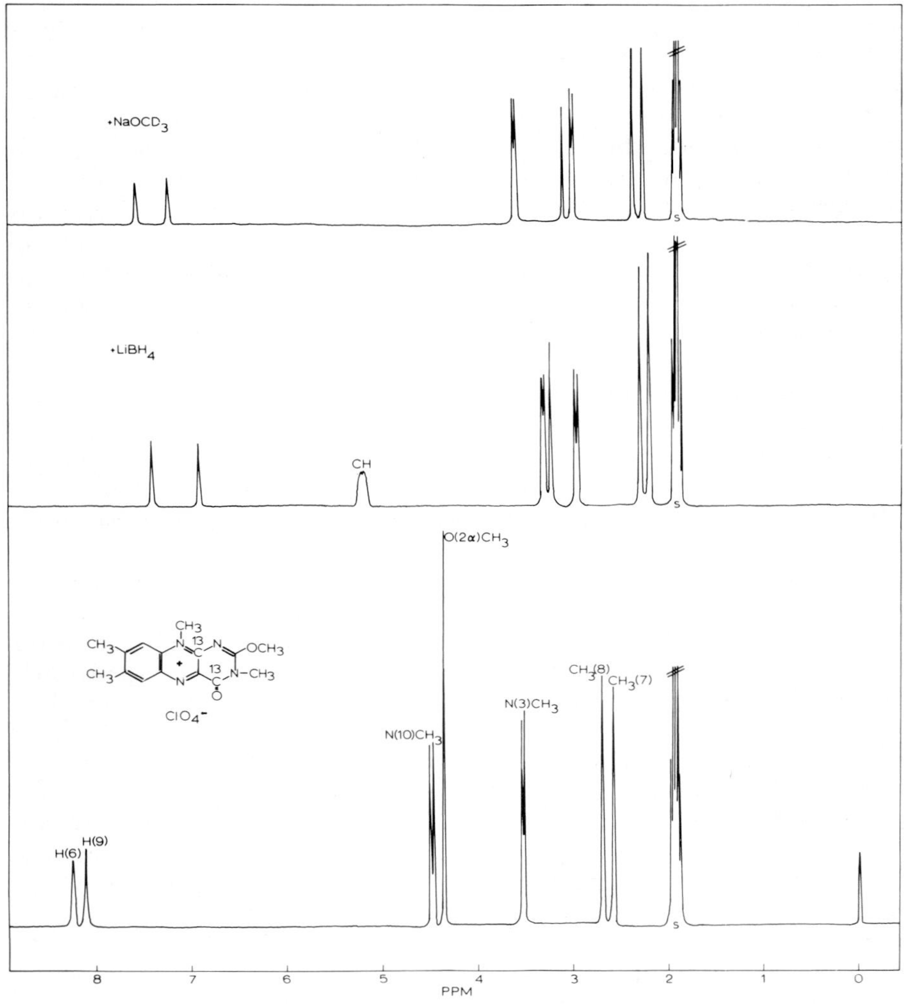

Fig. 4. ^{1}H NMR spectra of 2α,3,7,8,10-pentamethyl-isoalloxazinium perchlorate, enriched with ^{13}C (90 atom%) at position 4 and 10a, in C^2H_3CN solution (bottom curve), its methoxide adduct (top curve) and its $LiBH_4$ reduction product (middle curve). All other conditions as in Fig. 3.

Reoxidation studies with reduced flavin

In a previous paper [4] we have shown that a transient species can be observed upon careful oxidation of a methanolic solution of 1,3,7,8,10-pentamethyl-1,5-dihydroisoalloxazine with molecular oxygen. It has been shown that the light absorption spectrum of this intermediate is identical to that obtained from the corresponding oxidized compound and methoxide [4]. There remained some uncertainty if the spectrum of the intermediate observed during the oxidation reaction is due to an oxygenated flavin or results from the interaction of oxidized flavin generated during the oxidation reaction and methoxide. To solve this question we have conducted these

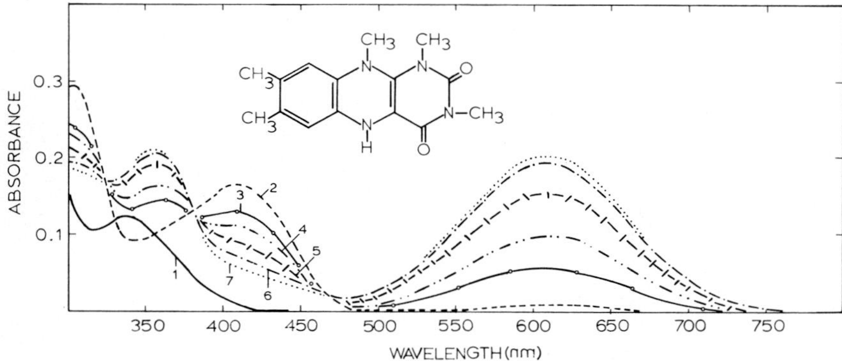

Fig. 5. Spectral changes observed on reoxidation of 40 μM 1,3,7,8,10-pentamethyl-1,5-dihydroisoalloxazine in benzene solution in the presence of 40 μM dibenzoylperoxide. Curve 1: in the absence of dibenzoylperoxide; curve 2: immediately after addition of dibenzoylperoxide, curves 3–7 were recorded 11, 22, 47, 83 and 123 min later. Before the reoxidation reaction was started the solution was flushed with N_2 for 3 min.

oxidation experiments using acetonitrile instead of methanol as a solvent. We found that under these conditions also an intermediate is formed and that its spectrum is practically identical to that obtained in methanolic solutions. 1,3,7,8,10-pentamethyl-1,5-dihydroisoalloxazine is rather reactive towards molecular oxygen in polar solvent. In apolar solvents the reactivity of this compound towards oxygen is drastically decreased. Thus a benzene solution of this model is oxidized very slowly and takes hours to complete. The final product is the spirohydantoin derivative [10]. Mager et al. [10] concluded from this fact that oxygen interacts with the 10a position of the flavin nucleus. The similarity of the absorption spectra of the methoxide adduct and the oxygenated flavin intermediate mentioned above indicate that both products exhibit the same structure. Thus it seems likely that molecular oxygen interacts with the flavin nucleus at position 9a. Nevertheless 1,3,7,8,10-pentamethyl-1,5-dihydroisoalloxazine can be easily oxidized in apolar solutions in the presence of dibenzoylperoxide (Bz_2O_2). This oxidation reaction yields two intermediates as shown in Fig. 5. Curve 1 is the absorption spectrum of a partially anaerobic benzene solution in the absence of Bz_2O_2 showing a maximum at 340 nm. Upon addition of Bz_2O_2 to this solution a yellow colored species is generated rapidly. The absorption spectrum of this species (Fig. 5, curve 2) exhibits maxima at 410 and 305 nm. This spectrum resembles those obtained from the methoxide adduct and the intermediate observed during the reoxidation reaction conducted in polar solvents. This yellow species is stable under anaerobic conditions but undergoes a secondary reaction in the presence of molecular oxygen yielding a blue colored species with absorption maxima at 610 and 355 nm (Fig. 5, curve 7). The transformation of the yellow species into the blue one (Fig. 5, curves 2–7) is isosbestic at 470, 382 and 325 nm. Furthermore, the blue species itself also undergoes a very slow reaction yielding the spirohydantoin derivative. We have published some preliminary results

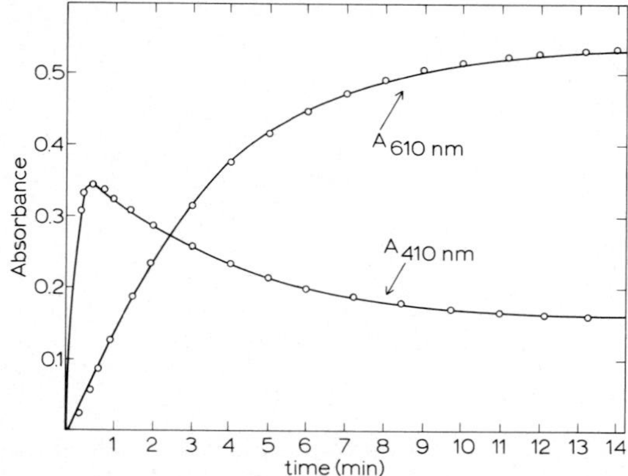

Fig. 6. Experimental (——) and calculated (ooo) time dependent absorbance changes at 410 and 610 nm occurring during the reoxidation reaction of 1,3,7,8,10-pentamethyl-1,5-dihydro-isoalloxazine in benzene solution in the presence of dibenzoylperoxide. The solvent was air-equilibrated before use.

[5] of the kinetic investigation of these reactions. It was found that the formation of the yellow species is first-order in both the flavin and the Bz_2O_2. This is in agreement with stoichiometric investigation of the reaction where it was found that one mol of reduced flavin reacts with one mol of Bz_2O_2. This and the kinetic results are not influenced by the presence or absence of molecular oxygen in the reaction solution. Furthermore, the amount of blue species formed under aerobic conditions is only dependent on the concentration of the yellow species and practically independent on the Bz_2O_2 concentration. This is demonstrated in Table II where several reoxidation experiments in the presence of various concentrations of Bz_2O_2 are summarized. It is seen that the ratio of the developed colors of the two species is quite constant whereas the ratio of flavin to Bz_2O_2 varies about 40-fold. These results are in contradiction to the

TABLE II

Different kinetic experiments of the oxidation reaction of 1,3,7,8,10-pentamethyl-1,5-dihydro-alloxazine (Fl_{red}) in the presence of dibenzoylperoxide (Bz_2O_2) and the corresponding absorbance changes at 410 and 610 nm. The reactions were carried out in benzene solutions at 25°C.

Exp.	$[Fl_{red}]$ (μM)	$[Bz_2O_2]$ (μM)	Ratio	A_{410nm}	A_{610nm}	Ratio
1	100	25	4.0	0.140	0.200	0.70
2	100	50	2.0	0.205	0.360	0.57
3	100	75	1.33	0.240	0.465	0.52
4	100	100	1.0	0.300	0.545	0.55
5	100	150	0.67	0.335	0.545	0.62
6	100	200	0.50	0.335	0.535	0.63
7	100	250	0.40	0.340	0.530	0.64
8	100	300	0.33	0.340	0.530	0.64
9	100	400	0.25	0.340	0.520	0.65
10	100	500	0.20	0.340	0.520	0.65
11	100	700	0.15	0.340	0.510	0.67
12	100	900	0.11	0.340	0.510	0.67

mechanism of the formation of the blue species as proposed by Hemmerich (cf. this volume p. 9). Working under pseudo first-order reaction conditions it can be shown that the formation of the yellow species and its transformation into the blue one follows simple parallel first-order reaction kinetics. This is shown in Fig. 6 where the experimental and calculated absorbance changes vs. time are given. The experimental curve was obtained by oxidation of a 0.1 mM aerobic benzene solution of 1,3,7,8,10-pentamethyl-1,5-dihydro-isoalloxazine in the presence of 0.9 mM Bz_2O_2. The theoretical curve was calculated with the following rate constants: 14.04 min^{-1} for the formation of the yellow species and 0.29 min^{-1} for the second reaction. Mass spectral analysis of the yellow intermediate obtained by Bz_2O_2 oxidation of a solution in acetone of 1,3,7,8,10-pentamethylisoalloxazine, under which conditions the blue species is not formed, indicates a molecular weight of 302. This would indicate the addition of one atom of oxygen to the flavin molecule. From the spectral similarity of this species with that of the methoxide adduct it might be suggested that the oxygen atom has been added to the 9a position of the flavin. About the structure of the blue species we are not ready to make a proposal.

Concluding remarks

We have established that methoxide and hydride ions add to the 9a position of the flavin nucleus. It is likely that also oxygen activation by

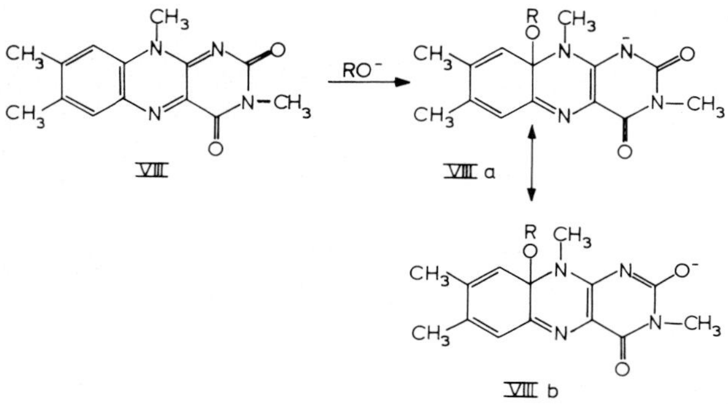

VIII $\xrightarrow{RO^-}$ VIII a $\leftrightarrow$ VIII b

reduced flavins occurs through the same position. This would mean that the oxygen intermediates recently observed during the catalysis of several flavoproteins (cf. for example Massey et al. and Hastings et al., this volume pp. 199, 53) would have structures like VIII a or b, where RO = OOH. On the basis of our spectral data it can be expected that the absorption spectra of the oxygen intermediates of the flavoproteins could vary considerably depending on which structure, VIII a or b, was preferently stabilized by the apoprotein and whether or not some steric strain is exerted on the bound flavocoenzyme by the apoprotein or by specific amino acid residues.

Acknowledgement

We are indebted to Drs. S. Ghisla and P. Hemmerich for valuable discussions, to Mr. W.J.H. van Berkel for skilful technical assistance and to Mr. B.J. Sachteleben for preparing the Figures. These studies were initiated in the laboratory of Dr. V. Massey during the tenure of a Career Development Award from the National Institutes of Health, U.S. Public Health Service (K4-GM 42599) (to F.M.) and have been supported by the Netherlands Foundation for Chemical Research (S.O.N.) with financial aid from the Netherlands Organization for the Advancement of Pure Research (Z.W.O.).

References

1 Mager, H.I.X. and Berends, W. (1966) Biochim. Biophys. Acta 118, 440
2 Hemmerich, P. (1968) in 19th Mosbacher Colloquium, p. 249, Springer Verlag, Heidelberg
3 Massey, V., Müller, F., Feldberg, R., Schuman, M., Sullivan, P.A., Howell, L.G., Mayhew, S.G., Matthews, R.G. and Foust, G.P., (1969) J. Biol. Chem. 244, 3999
4 Müller, F. (1971) in Flavins and Flavoproteins (Kamin, H., ed.), p. 363, University Park Press, Baltimore
5 Müller, F., Jarbandhan, T., Gast, R. and Grande, H.J. (1975) in Reactivity of Flavins (Yagi, K., ed.), p. 51, University of Tokyo Press, Tokyo
6 Norrestam, R., Torbjörnsson, L. and Müller, F. (1972) Acta Chem. Scand. 26, 2441
7 Dudley, K.H., Ehrenberg, A., Hemmerich, P. and Müller, F. (1964) Helv. Chim. Acta 47, 1354
8 Müller, F., Hartmann, U. and Hemmerich, P. (1973) Hoppe Seyler's Z. Physiol. Chem. 354, 215
9 Hemmerich, P. and Müller, F. (1973) Ann. N.Y. Acad. Sci. 212, 13
10 Mager, H.I.X., Addink, R. and Berends, W., (1967) Rec. Trav. Chim. 86, 833

Part III

Luciferase

Flavins and Flavoproteins, edited by T.P. Singer

CHAPTER 4

THE OXYGENATED LUCIFERASE-FLAVIN INTERMEDIATE

J. WOODLAND HASTINGS*, CLAUDE BALNY and PIERRE DOUZOU

The Biological Laboratories, Harvard University, Cambridge, Mass. 02138 (U.S.A.), INSERM, U 128, BP 5051, Montpellier and Ecole des Hautes Etudes, Institut de Biologie Physico-Chimique, 13 rue Pierre et Marie Curie, Paris 5 (France)

Summary

Bacterial luciferase ($\alpha\beta$ dimer, mol. wt. 79 000) is a flavoenzyme with unusual properties. First, one of the reaction products is light ($\lambda_{max} \approx$ 490 nm). Secondly, the turnover is slow, involving an enzyme intermediate having a lifetime of the order of 10 s at 20°C. This intermediate, identified with the product of oxygen addition to the luciferase-bound reduced flavin, has been purified by Sephadex chromatography and characterized by absorption and fluorescence spectroscopy, all at − 20 to − 30°C. It has absorption in the visible range, peaking at 372 nm and a fluorescence emission centered at about 490 nm. The fluorescence emission spectrum of the intermediate is the same as the bioluminescence emission spectrum. Although the intermediate cannot itself be the emitter, its structure may provide a clue to that of the emitter. The isolated intermediate forms products upon warming to 20°C, either with or without aldehyde. In the absence of aldehyde very little bioluminescence occurs and 1 mol each of FMN and H_2O_2 is produced per mol of luciferase intermediate. In the presence of aldehyde the formation of an aldehyde-enzyme-flavin intermediate complex can be detected by absorption at − 30°C; upon warming, bioluminescence with high quantum yield occurs with the formation of 1 mol of FMN but no H_2O_2.

Introduction

It has become increasingly evident that the reaction of bacterial luciferase is closely analogous to certain flavoprotein hydroxylases and oxidases, including *p*-hydroxybenzoate hydroxylase [1], melilotate hydroxylase [2, 3] lactate oxidase [4, 5] and salicylate hydroxylase [6, 7]. In both luciferase [8, 9] and the other enzymes cited, a reduced flavin-enzyme intermediate can react with oxygen, even in the absence of the second

* To whom correspondence should be addressed at Harvard University.

substrate, to give an oxygenated form which may be designated as the active oxidizing species.

Under certain conditions, e.g., with "pseudosubstrate" or in the absence of the second substrate, this active intermediate may break down to give H_2O_2 and oxidized flavin. Because the reducing equivalents of the flavin come from NADH + H^+, this represents in effect an enzymatic pathway for flavin-mediated, reduced pyridine nucleotide oxidation. Since the peroxide-producing reaction is not linked to the oxidation of the second substrate, it may be referred to as the "uncoupled" pathway. If the second substrate is available these enzymes function as monooxygenases, with one oxygen atom incorporated in the oxidized substrate and the second in water.

Luciferase, of course, is distinguished from other monooxygenases by the fact that it emits light with a reasonably high quantum yield; about 0.1. The energy produced by the oxidative reaction is thus specifically channeled so that an intermediate or product in the reaction is formed in an electronically excited state. Given that other monooxygenases have analogous reaction mechanisms and presumably similar free energies, it is quite conceivable that they also emit light, a possibility which has not been investigated rigorously.

The details of the alternative reaction pathways for bacterial luciferase are given in Fig. 1 [8, 9]. The second substrate is a long chain aldehyde which is oxidized to the corresponding acid, as reported by several laboratories [10–13]. We have utilized recently low temperature (− 20 to − 40°C) technology to isolate, purify and characterize the enzyme intermediate designated as II in Fig. 1 [14–16]. Its spectral and chemical properties have suggested that it should be identified with an oxygen adduct or oxygenated, reduced-flavin intermediate. Knowledge concerning it may contribute significantly to our understanding of the reactions involving flavin oxygenases more generally.

Materials and Methods

The luciferase was isolated and purified from the luminous bacterium *Beneckea harveyi* [17] as described previously [18]. Intermediate II was prepared by mixing luciferase (80–95% pure, 0.8 ml, 10 mg • ml^{-1}) and $FMNH_2$ (0.2 ml, 2 mM) at 5°C in 50% ethylene glycol/phosphate buffer, pH 7.0 [19, 20]. After 5 s the temperature was lowered quickly to below − 20°C by mixing with 1 ml of the same buffer at − 45°C and chromatographed on a Sephadex LH-20 column at −25°C [21]. The fractions (2.0–3.0 ml each) were stored between − 30 and − 45°C and assayed for bioluminescence activity by injecting 100 μl into phosphate buffer (pH 7.0) with aldehyde (50 μM octanal or dodecanal) at 23°C.

Bioluminescence intensity was measured with a calibrated photometer [22], absorbance in an Aminco Chance DW-2 spectrophotometer and fluorescence in an Aminco–Bowman spectrophotofluorometer, the latter two modified to permit temperature control of the sample down to as low as − 65°C [23].

Results

The luciferase system is favorable for the isolation of intermediates. Compared to the turnover rate (about 4 min^{-1} at 20°C), the initial steps in

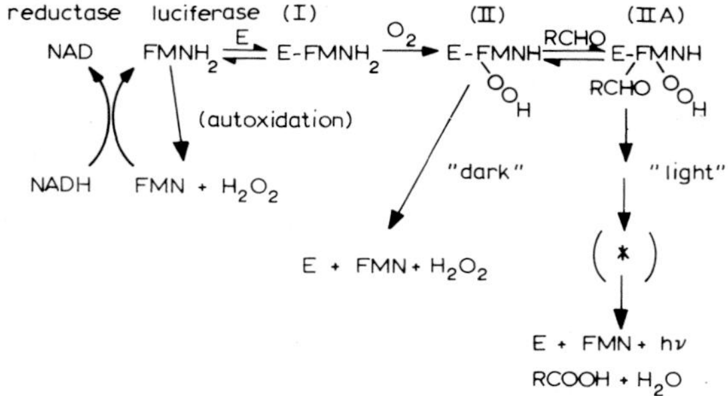

Fig. 1. The reaction sequence and intermediates postulated for the luciferase-mediated bioluminescent oxidation of $FMNH_2$ as well as nonenzymatic and dark pathways. The luciferase intermediate designated II is the one which has been isolated. With NADH, turnover occurs in a crude system or with added reductase.

the reaction (Fig. 1) are rapid, so that a long-lived intermediate (II) is formed in a small fraction of the time required for a catalytic cycle. The light emission of the reaction then ensues as this intermediate decays exponentially, in the same manner as a radioactive isotope, with concomitant emission.*

Light emission with a high quantum yield requires a long chain aldehyde, but formation of intermediate II does not. Therefore, we prepared and purified the intermediate without aldehyde. The reaction was initiated by mixing $FMNH_2$ with luciferase and O_2 at 4°C; the intermediate formed during the first few seconds was then trapped by rapidly lowering the temperature to −20°C or below, using a 50% ethylene glycol/phosphate buffer. Column chromatography (Fig. 2) at − 20°C or below allowed the separation of the enzyme intermediate (mol. wt. ≈ 79 000) from low molecular weight products, notably FMN and H_2O_2) which were eluted in the later fractions (not shown). The bioluminescence activity coincides with the protein peak and also with a 373 nm absorbance peak attributed to the oxygenated intermediate (Fig. 2, inset).

As postulated (Fig. 1), the reaction goes to completion at 20°C in the absence of aldehyde. After the intermediate is warmed, one observes the spectrum of free FMN (Fig. 2, inset), 1 mol of flavin per mol of luciferase (Table I).

The flavin produced in the reaction was identified as FMN by thin-layer chromatography, with development in 1-butanol/acetic acid/water (2:1:1; v/v). A less ambiguous identification of the products involved testing their ability to rereact (Table I), since luciferase is quite specific for $FMNH_2$. The product of the reaction obtained in the absence (II) and in the presence (IIA) of added aldehyde was tested for its ability to emit light in the luciferase reaction, compared with that of a reaction mixture composed of unreacted luciferase and stock FMN in amounts equal to those in the product mixtures. The product is fully capable of turning over a second time; in fact, the reacted material may be more active than the original material.

* Turnover does not occur in the in vitro system because the $FMNH_2$ is injected only once, and if not bound to the enzyme, it is oxidized non-enzymatically in less than 1 s [39].

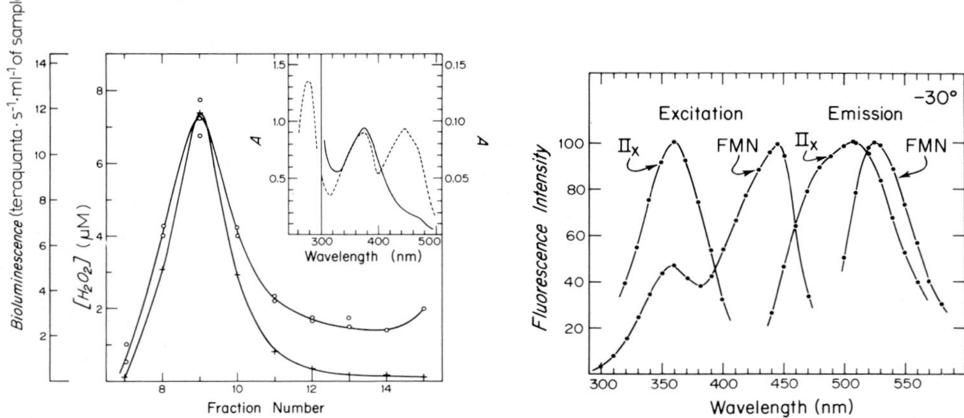

Fig. 2. Chromatography on Sephadex LH-20 at $-20°C$ of the luciferase-flavin intermediate showing correlation of bioluminescence activity (+) (measured with added aldehyde) and H_2O_2 production (○) (determined without aldehyde) of luciferase intermediate in each tube. The spectrum of tube No. 9 is shown in the inset both before (——) and after (----) warming, both measurements at $-20°C$. The H_2O_2 content of each tube after warming in the absence of aldehyde was measured and is plotted in μM concentration units.

Fig. 3. Excitation and emission spectra of intermediate(s), IIx, measured at $-30°C$ in 50% ethylene glycol, 0.1 M phosphate buffer, pH 7.0, just after elution from Sephadex LH-20 at $-25°C$. For comparison the excitation and emission spectra for FMN are given under the same conditions. The excitation spectra for IIx and FMN emission were monitored at 485 nm and at 525 nm, respectively. The emission spectra for IIx and FMN were obtained via excitation at 370 nm and 450 nm respectively.

The reaction without aldehyde results in the formation of H_2O_2. An experiment showing that 1 mol of H_2O_2 is produced per mol of intermediate II upon its breakdown is illustrated in Fig. 2. The intermediate, which eluted from the LH-20 column principally in fractions 8–10, was assayed for its bioluminescence activity by injection of an aliquot into aldehyde-containing buffer at 22°C. To measure H_2O_2 production, aliquots of the individual tubes were rapidly warmed to 20°C, permitting the reaction

TABLE I

To 2 ml of 10 mM phosphate buffer, pH 7, at 23°C, without (II) or with (IIA) 50 μM dodecanal were added 0.1 ml aliquots of intermediate to allow the reaction to go to completion. A mixture with the same absorbance was prepared by mixing luciferase and FMN. All three were assayed by adding 50 μl of 0.2 M sodium dithionite to reduce the FMN, followed by the injection of 1 ml of aerated buffer (10 mM phosphate, pH 7, with 100 μM dodecanal) to initiate the reaction. Difficulty with reproducibility was experienced in these experiments; the activity differences between the different mixtures may not be significant.

Reaction mixture derived from	Absorbance at 23°C before dilution		Bioluminescence activity
	280 nm	450 nm	(teraquanta · s^{-1})
II products	0.93	0.12	0.34 0.60
IIA products	0.93	0.12	0.34 0.28 0.24
Luciferase + FMN	0.98	0.12	0.18 0.22

to go to completion without light emission, and assayed for H_2O_2 by the chemiluminescence of luminol [15]. The concentration of the intermediate in the peak fraction (No. 9) was measured by the flavin absorbance at 450 nm after warming (Fig. 2, inset) and found to be 7.5 μM, which is very close to the amount of H_2O_2 (7.2 μM) produced in the absence of aldehyde. The H_2O_2 was stable for at least 30 min in the reaction mixture at 22°C, but was destroyed by catalase. Although the initial chemiluminescence of luminol was abolished by catalase, there was an unexplained secondary rise which occurred after about 2 min.

In the reaction via the pathway with aldehyde it is postulated that no H_2O_2 is produced. An experiment was therefore carried out in which H_2O_2 was measured following injection of aliquots (0.6 or 0.8 ml) of the intermediate kept at −25°C into 10 mM phosphate buffer, pH 7.0, with 50 μM aldehyde (octanal or decanal) at 20°C in a final volume of 2.0 ml. The results (Table II) show that H_2O_2 is not produced with aldehyde, but without added aldehyde, the amount of H_2O_2 produced is again equal to the amount of flavin in the intermediate complex. The amount of FMN formed is also the same in the reaction with aldehyde.

The scheme of Fig. 1 postulates the reversible binding of aldehyde to the intermediate II, and thus the existence of at least two species at this stage, II and IIA. With 1 ml of intermediate II in each of two matched cuvettes, 1 μl of a 1% ethanolic solution of octanal was added to one and the difference spectrum was measured. IIA exhibits a decrease in absorbance peaking at 362 nm and an increase which is maximal at 421 nm. In effect, the spectrum is slightly red shifted. The direct isolation of the intermediate with aldehyde bound was also attempted. However, intermediate IIA breaks down at a lower temperature than does II, so even with the column temperature lowered to −30°C, the protein fractions obtained contained mixtures of intermediate and product.

The fluorescence of the long-lived intermediate is also of interest. A

TABLE II

Intermediate II was prepared and chromatographed at −20°C on Sephadex LH-20. Aliquots from each fraction were assayed for bioluminescence both in the presence and absence of aldehyde (50 μM) by mixing with 0.01 M phosphate buffer, pH 7 at 22°C. Upon completion of the reaction, each of these samples was assayed for H_2O_2 using luminol. The flavin concentration of tube No. 6 after warming was 13 μM.

Tube No.	Bioluminescence activity (teraquanta · s^{-1} · ml^{-1})			H_2O_2 after reaction (μM)		
	No aldehyde	Octanal	Dodecanal	No aldehyde	Octanal	Dodecanal
4	0.003	0.11	0.10	6.0	1.2	2.5
5	0.22	6.7	4.8	2.0	1.2	2.0
6	1.0	41.0	30.0	12.0	1.0	1.2
7	0.41	10.0	7.9	2.5	1.0	0.7
8	0.12	3.5	2.3	1.0	0.7	0.8
9	0.05	2.0	1.1	0.6	1.0	0.1

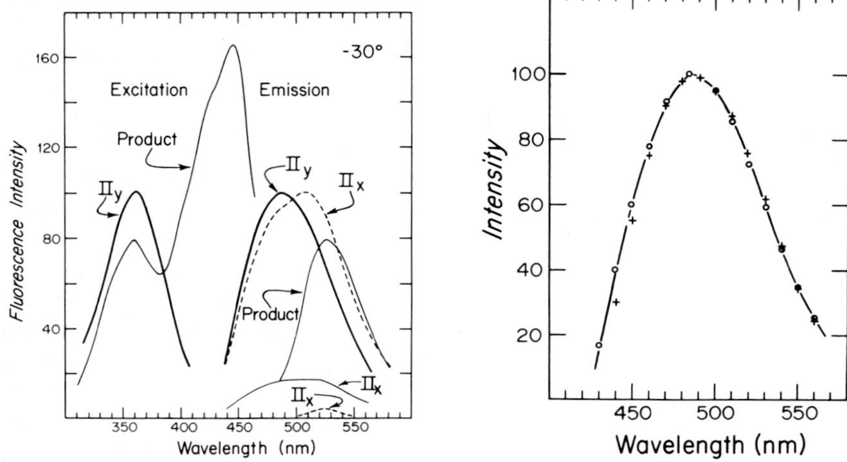

Fig. 4. Excitation and emission spectra for the intermediate(s) IIy (——), formed after irradiation of IIx in the cuvette for 15 min ($\lambda_{excitation}$, 370 nm; $\lambda_{emission}$, 485 nm). The spectrum of IIx from Fig. 3 is shown again at its correct relative maximum intensity (about 18%) and also normalized (----) to facilitate comparison with IIy. The excitation spectrum of IIx is the same as that of IIy. The fluorescence emission spectrum of the product (actually measured by excitation at 450 nm) is plotted normalized to excitation at 375 nm to permit a comparison of relative fluorescence yields of IIx, IIy and product (FMN). The excitation spectrum of the product is also shown. The amount of apparent FMN fluorescence (excited at 450 nm but based on 375 nm) which was measured in the initial IIx preparation is shown as a dashed line whose maximum is about 4% at 525 nm. This increased to about 6% following irradiation.

Fig. 5. Emission spectrum of bioluminescence (○), measured directly from the cuvette during warming of IIy in the presence of aldehyde, plotted together with the fluorescence emission spectrum of IIy (+) taken from Fig. 4. Ordinate, intensity of bioluminescence and fluorescence, normalized at the peak.

sample of the material taken from a tube with maximum bioluminescence activity (here designated IIx) exhibits fluorescence excitation and emission spectra at −30°C as shown in Fig. 3. The excitation spectrum closely matches the absorbance of II (Fig. 2). A shoulder in the emission spectrum, however, suggests the presence of at least two components, one with its maximum emission in the vicinity of 485 nm and the second with a maximum about 505 nm, but not identical with that of oxidized FMN (maximum, ≈ 525 nm). Excitation and emission spectra for a sample of FMN measured under the same conditions are presented in the same figure to facilitate comparison.

As the measurements are being carried out, with irradiation at 370 nm at −30°C, the fluorescence emission at 485 nm increases dramatically. The increase is dependent upon continued irradiation and the product formed does not decay in the dark, at least within a few minutes. An increase of more than five-fold may occur within 10 min, while the emission spectrum shifts to the blue, with a final maximum in the vicinity of 485 nm (Fig. 4). For purposes of discussion we refer to the sample prior to irradiation as IIx, and that afterwards as IIy, but it should be understood that the symbols IIx and IIy are not intended to refer to molecular species, but only to the samples.

Warming to 23°C results in the breakdown of the intermediate(s), yielding FMN. The fluorescence excitation and emission spectra of this product* (Fig. 4) are indeed the same as those of authentic FMN (Fig. 3). The amount of FMN formed was the same in the presence or absence of aldehyde, whether starting with IIx or IIy. The bioluminescence potentials of IIx and IIy were also found to be about the same, assayed by injection into buffer at 23°C with aldehyde [16]. Although there was a considerable variability in the determinations, it was certainly clear that the bioluminescence capacity of IIy was not five-fold greater than that of IIx, as was its fluorescence. The addition of aldehyde at − 30°C prior to irradiation had no major effect upon either the increase in fluorescence or the bioluminescence potential.

When the intermediate, irradiated or not, was warmed up in the presence of aldehyde in the cuvette of the spectrophotofluorometer, bioluminescence occurred. During warming, the bioluminescence emission began to occur at about − 20°C and rose to a peak around 4°C, decaying thereafter. During this period the fluorescence emission of the intermediate declined, while that of the product FMN at 525 nm appeared. The intensity and duration of the bioluminescence emission were adequate to permit determinations of the emission spectrum of bioluminescence, which was the same as the fluorescence of IIy (Fig. 5).

Discussion

Although an energy-transfer step may be involved in some cases**, the specific emitter in a chemi- or bioluminescent reaction is generally an intermediate or product whose excited state is populated during the reaction [24]. While oxidized FMN is a product of the bacterial bioluminescence reaction, and is also a highly fluorescent compound (quantum yield ≈ 0.3), it seems that it cannot be the emitter. Free flavin in aqueous solution has a fluorescence emission peaking at 525 nm, while bacterial luminescence both in vivo and in vitro emits well to the blue with λ_{max} typically in the vicinity of 495 nm [25, 26]***, a shift not readily attributable to enzyme and/or polarity effects [27]. More significantly, it has been found recently that oxidized FMN when bound to luciferase is actually nonfluorescent [18]. Unless the site where free FMN binds the apoenzyme is different from the one involved in the reaction, this finding essentially eliminates FMN itself as the emitter.

---

* In our first report on the isolation of the oxygenated reduced flavin luciferase intermediate [14], we reported that its fluorescence yield was about one third that of free FMN. It is evident that this estimate could have been based on material which had been subjected to irradiation for only a short while. After irradiation the fluorescence quantum yield of II is about the same as FMN.

** In the chemiluminescence of small synthetic dioxetanes, such as tetramethyl dioxetane, the excited product has a very low quantum yield of emission, but can efficiently transfer its energy to brightly fluorescent acceptors [36, 37]. Similar processes occur in certain coelenterates, where the in vivo light emission involves intermolecular energy transfer [38].

*** When appropriate corrections are applied to the spectral data presented in this publication (Figs. 3, 4 and 5) the emissions peaking at about 485 nm shift about 10 nm to the red and correspond very closely to the published corrected spectra for the emission of bioluminescence.

It has also been proposed that the N(1)-protonated form of oxidized FMN is the emitting species, since its fluorescence spectrum is reasonably close to that of the bioluminescence [28]. However, the fact that an actual intermediate in the reaction has a fluorescence which is virtually identical with the bioluminescence suggests that this should be viewed as an even better clue to the structure of the emitter. Thus, although the intermediate II cannot itself be the emitter, its structure, postulated to involve a luciferase-bound, reduced-flavin oxygen adduct, may be related to that of the emitter.

With regard to the possible structure of such an adduct, both the 1a and 4a positions have been suggested as reactive positions [29, 30] and several authors have suggested that oxygen may react at one or another of these two positions to form a flavin peroxide (FMNH—OOH). In fact, the spectral properties (absorption and fluorescence) of 4a-substituted flavins do resemble those of the luciferase intermediate [31–33]. Orf and Dolphin [34] made the interesting proposal that the formation of the 4a-flavin peroxide involves the intermediacy of a 1a-4a dioxetane. It is possible that the light-induced fluorescence increase might be due to a photochemical conversion from some such intermediate form to another. Or, IIx and IIy might be conformational isomers. In any event, Orf and Dolphin postulate further that it is neither the hydroperoxide, nor the carbonyl oxide of Hamilton [35] which acts as the hydroxylating agent, but the flavin oxaziridine formed between the 4a and N-5 positions. Studies with luciferase might allow us to test some of these proposals.

Acknowledgements

J.W.H. was the recipient of an NIH Special Research Fellowship (1F03 GM53274-01). The research was supported in part from NSF (GB 31977X), the CNRS (équipe associée 262), the INSERM, and the DGRST (71-7-3091).

References

1 Spector, T. and Massey, V. (1972) J. Biol. Chem. 247, 4679–4687
2 Strickland, S. and Massey, V. (1973) J. Biol. Chem. 248, 2944–2952
3 Strickland, S. and Massey, V. (1973) J. Biol. Chem. 248, 2953–2962
4 Takemori, S., Nakai, Y., Katagiri, M. and Nakamura, T. (1969) FEBS Lett. 3, 214–216
5 Lockridge, O., Massey, V. and Sullivan, P. (1972) J. Biol. Chem. 247, 8097–8106
6 White-Stevens, R.H. and Kamin, H. (1972) J. Biol. Chem. 247, 2358–2370
7 White-Stevens, R.H., Kamin, H. and Gibson, Q.H. (1972) J. Biol. Chem. 247, 2371–2381
8 Hastings, J.W. and Gibson, Q.H. (1963) J. Biol. Chem. 238, 2537–2554
9 Gibson, Q.H., Hastings, J.W., Weber, G., Duane, W. and Massa, J. (1966) in Flavins and Flavoproteins (Slater, E.C., ed.), pp. 341–359, Elsevier, Amsterdam
10 Shimomura, O., Johnson, F.H. and Kohama, Y. (1972) Proc. Natl. Acad. Sci. U.S. 69, 2086–2089
11 McCapra, F. and Hysert, D.W. (1973) Biochem. Biophys. Res. Commun. 52, 298–304
12 Dunn, D.K., Michaliszyn, G.A., Bogacki, I.G. and Meighen, E.A. (1973) Biochemistry 12, 4911–4918
13 Vigny, A. and Michelson, A.M. (1974) Biochimie 56, 171–176
14 Hastings, J.W., Balny, C., Le Peuch, C. and Douzou, P. (1973) Proc. Natl. Acad. Sci. U.S. 70, 3468–3472
15 Hastings, J.W. and Balny, C. (1975) J. Biol. Chem. 250, 7288–7293
16 Balny, C. and Hastings, J.W. (1976) Biochemistry 14, 4719–4723

17 Reichelt, J.L. and Baumann, P. (1973) Arch. Mikrobiol. 94, 283–330
18 Baldwin, T.O., Nicoli, M.Z., Becvar, J.E. and Hastings, J.W. (1975) J. Biol. Chem. 250, 2763–2768
19 Douzou, P. (1973) Mol. Cell. Biochem. 1, 15–27
20 Douzou, P. (1974) in Methods of Biochemical Analysis, (Glick, D., ed.), Vol. 22, pp. 401–512, John Wiley and Sons, Inc., Chichester, New York
21 Balny, C., Le Peuch, C. and Debey, P. (1975) Anal. Biochem. 63, 321–330
22 Mitchell, G. and Hastings, J.W. (1971) Anal. Biochem. 39, 243–250
23 Maurel, P., Travers, F. and Douzou, P. (1974) Anal. Biochem. 57, 555–563
24 Hastings, J.W. (1968) Annu. Rev. Biochem. 37, 597–630
25 Hastings, J.W., Riley, W.H. and Massa, J. (1965) J. Biol. Chem. 240, 1473–1481
26 Seliger, H.H. and Morton, R.A. (1968) in Photophysiology (Giese, A.C., ed.), Vol. 4, pp. 253–314, Academic Press, New York
27 Mitchell, G.W. and Hastings, J.W. (1969) J. Biol. Chem. 244, 2572–2576
28 Eley, M., Lee, J., Lhoste, J.-M., Lee, C.Y., Cormier, M.J. and Hemmerich, P. (1970) Biochemistry 9, 2902–2908
29 Hemmerich, P., Massey, V. and Weber, G. (1967) Nature 213, 728–730
30 Müller, F. (1971) in Flavins and Flavoproteins (Kamin, H., ed.), pp. 349–361, University Park Press, Baltimore
31 Ghisla, S., Massey, V., Lhoste, J.-M. and Mayhew, S.G. (1974) Biochemistry 13, 589–597
32 Jefcoate, C.R., Ghisla, S. and Hemmerich, P. (1971) J. Chem. Soc. C 1689–1694
33 Hemmerich, P., Ghisla, S., Hartman, U. and Müller, F. (1971) in Flavins and Flavoproteins, Third International Symposium (Kamin, H., ed.), pp. 83–105, University Park Press, Baltimore
34 Orf, H.W. and Dolphin, D. (1974) Proc. Natl. Acad. Sci. U.S. 71, 2646–2650
35 Hamilton, G.A. (1971) in Progress in Bioorganic Chemistry (Kaiser, E.T. and Kezdy, T.J., eds.), pp. 83–157, Wiley-Interscience, New York
36 Turro, N.J., Lechtken, P., Shore, N.E., Schuster, G., Steinmetzer, H.C. and Yekta, A. (1973) Acc. Chem. Res. 7, 97–105
37 Wilson, T. (1975) in Chemical Kinetics, Series 2 (Herschbach, D., ed.), MTP International Review of Science, London, in press
38 Morin, J.G. and Hastings, J.W. (1971) J. Cell. Physiol. 77, 313–318
39 Gibson, Q.H. and Hastings, J.W. (1962) Biochem. J. 83, 368–377

Flavins and Flavoproteins, edited by T.P. Singer

CHAPTER 5

REACTION MECHANISM OF BACTERIAL LUCIFERASE FROM *PHOTOBACTERIUM PHOSPHOREUM*

TAKAHIDE WATANABE, KENZABURO YOSHIDA, MASAMI TAKAHASHI, GIITI TOMITA[a] and TAKAO NAKAMURA

Department of Biology, Faculty of Science, Osaka University, Toyonaka, Osaka 560 and [a] *Institute of Biophysics, Faculty of Agriculture, Kyushu University, Fukuoka 812 (Japan)*

Summary

Mechanism of reaction of bacterial luciferase from *Photobacterium phosphoreum* was studied on each elementary step of the luminescent reaction. (1) $FMNH_2$ and O_2 reacted with luciferase with a stoichiometry of 1:1:1. (2) On binding of $FMNH_2$ to luciferase a red shift of the ultraviolet absorption bands of $FMNH_2$, and a quenching of protein fluorescence were observed. The quenching occurred in a two step reaction. The affinity of $FMNH_2$ for luciferase ($1-5 \cdot 10^{-7}$ M) was 10^5 times more than that of FMN. (3) It was shown by a rapid spectrophotometric method that an intermediate was formed on reaction of the luciferase · $FMNH_2$ complex with O_2. In this intermediate FMN was in a state close to that of the oxidized form. Results on aldehyde-initiated luminescence indicated that the intermediate was obligatory in the luminescent reaction.

Bioluminescent bacteria possess a reaction system for light emission composed of luciferase and NAD(P)H-FMN reuctase (EC 1.6.99.1, NADPH: (acceptor) oxidoreductase) [1–5]. In the luminescent reaction FMN is reduced by the reductase and the $FMNH_2$ thus formed reacts with luciferase [6] (Scheme 1). The luciferase reaction also requires molecular oxygen and a saturated, normal, long chain aliphatic aldehyde (RCHO) as substrates [8–11]:

$$FMNH_2 + O_2 + RCHO \xrightarrow{\text{Luciferase}} h\nu\ (500\text{–}510\ \text{nm}) + FMN + RCOOH$$

Hastings and Gibson [12] showed that a "long-lived intermediate" was formed on reaction of luciferase with $FMNH_2$ and O_2, and that the intermediate emitted light when allowed to react with aldehyde. Hastings et al. [13] determined the absorbance spectrum of the intermediate, which stabilized at −20°C. FMN in the intermediate is most likely to be the

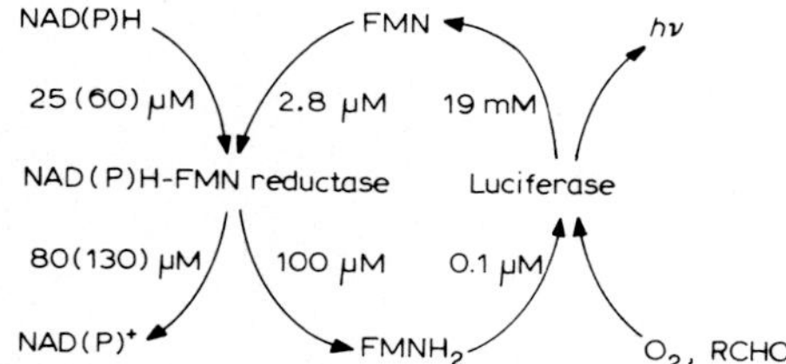

Scheme 1. Luciferase-NAD(P)H-FMN reductase luminescent reaction system of *Photobacterium phosphoreum* [7]. The K_m values for each reaction step are shown. Values in parentheses represent those for $NADP^+$ and NADPH.

emitter group of the reaction, and is thought to be in a specific state favorable for efficient conversion of chemical to light energy. However, little is known about the structure of the intermediate.

This paper is on the mechanism of the reaction of luciferase from *Photobacterium phosphoreum*. We studied the nature of the luciferase · $FMNH_2$ complex which is formed in the first step of the luminescent reaction, and the transformation of the complex on reaction with O_2 to the obligatory intermediate of the reaction by titrimetric and rapid spectrophotometric measurements. All the reactions and measurements were carried out at 20°C in 0.1 M phosphate buffer, pH 7.0.

Results

Nature of the luciferase · $FMNH_2$ complex

Absorbance spectrum of luciferase. Fig. 1 shows the absorbance spectra of luciferase (stripped luciferase [14]) and native luciferase. The latter was obtained [15] by extraction of the cells of *Photobacterium phosphoreum*

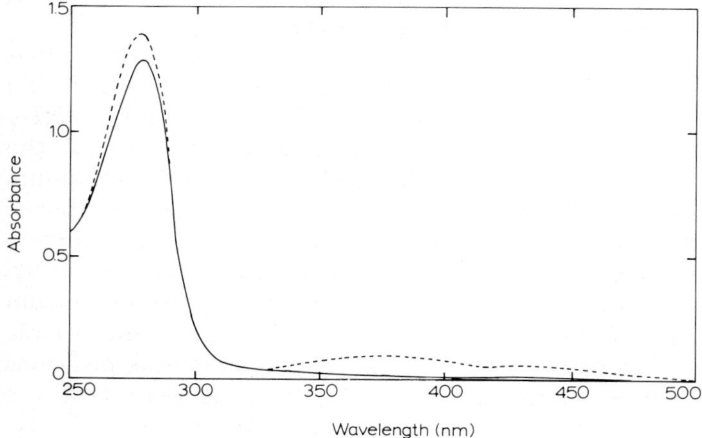

Fig. 1. Absorbance spectra of luciferase obtained from *Photobacterium phosphoreum*. – – –, native luciferase, ——, stripped luciferase. Concentration of the enzyme, 1.18 mg/ml.

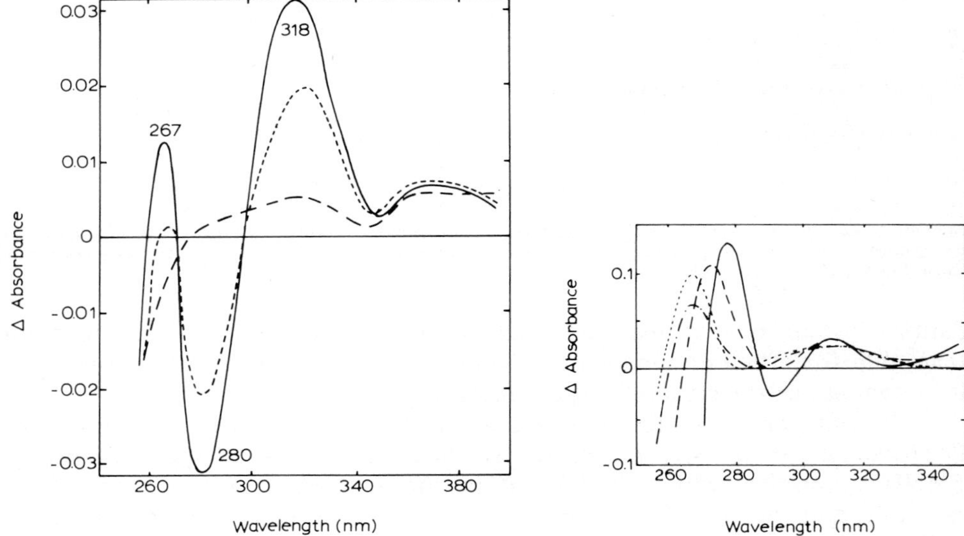

Fig. 2. Difference absorbance spectra of the luciferase · $FMNH_2$ complex [17]. $FMNH_2$ concentration: ——, 0 μM; -------, 10 μM; ——, 32.5 μM. Luciferase concentration, 13.5 μM. The wavelengths of the peak positions are shown in the figure.

Fig. 3. Difference spectra of $FMNH_2$ in various solvents (20% in buffer) against that in buffer [17]. The solvents used were: ——, dioxane; — —, dimethylsulfoxide; —·—, ethanol; -----, ethyleneglycol. Concentration of $FMNH_2$, 40 μM.

and purification by column chromatography on Sephadex A-50. This sample contained an average of 0.19 molecules of p-flavin, a flavin derivative, per one molecule of luciferase [16]. The pigment was removed from the enzyme by reversible denaturation with 8 M urea or 5 M guanidine-HCl without modifying the enzyme activity [14]. The luciferase thus obtained (stripped luciferase, molecular weight 82 000) which is colorless and non-fluorescent in the visible region, was used in the following experiments.

Difference spectrum of the luciferase · $FMNH_2$ complex [17]. Luciferase was titrated with $FMNH_2$ under anaerobic conditions and the difference absorbance spectrum of a mixture of luciferase and $FMNH_2$ against free enzyme and $FMNH_2$ was determined. As shown in Fig. 2, the difference spectrum exhibited maxima at 267 and 318 nm and a minimum at 280 nm. $FADH_2$ and reduced riboflavin also produced similar absorbance changes, indicating that luciferase and a reduced flavin form a complex. Similar spectroscopic changes of $FMNH_2$ were observed in solvent perturbation experiments, as shown in Fig. 3. The observed difference spectrum (Fig. 2) differs from that of FMN against $FMNH_2$ since it lacks an appreciable absorbance difference at around 380 nm and also has different peak positions in the 280–340 nm region. These facts exclude the possibility that the observed spectral change is due to partial oxidation of the bound $FMNH_2$. The ΔA between the 318 nm peak and 280 nm trough in Fig. 2 is plotted as a function of the total $FMNH_2$ concentration in Fig. 4. This titration curve

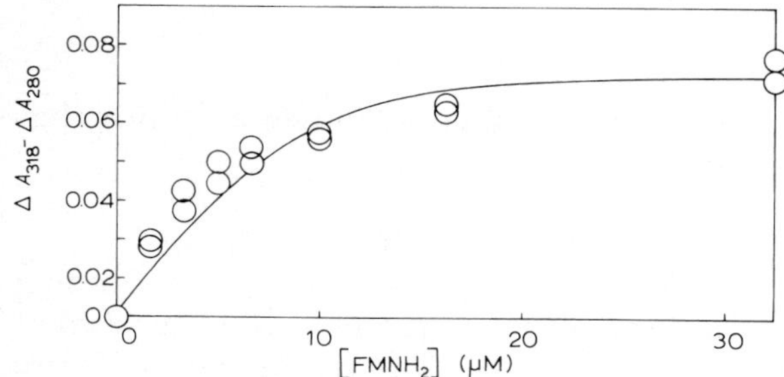

Fig. 4. Plot of the absorbance difference $\Delta A_{318} - \Delta A_{280}$ in the difference spectra of the luciferase · $FMNH_2$ complex against the total $FMNH_2$ concentration [17]. Luciferase concentration, 9.5 μM. The solid curve was drawn assuming that $K_1 = 0.5$ μM.

indicated 1:1 binding of $FMNH_2$ to luciferase, and the solid curve in the figure, fitted to the experimental points, is the theoretical one calculated assuming that the dissociation constant (K_1) was 0.5 μM.

$$\text{Luciferase} + \text{FMNH}_2 \xrightleftharpoons{K_1} \text{Luciferase} \cdot \text{FMNH}_2$$

Kinetic measurements of the stoichiometry of the reaction of luciferase with $FMNH_2$ [17]

The luciferase reaction was assayed by the stopped flow method using various concentrations of $FMNH_2$, and the time course of the light emission was recorded. The relative values of I_0, the peak intensity of the emitted light, estimated from the records in a series of the experiments are plotted against the ratio of total [$FMNH_2$] to [luciferase] in Fig. 5. As can be seen, the I_0 values increased with increase in the $FMNH_2$ concentration, and attained a saturation value at [$FMNH_2$]/[luciferase] =1.0. Thus the luciferase molecule reacts with one molecule of $FMNH_2$ to form a complex, which then undergoes the light emitting reaction:

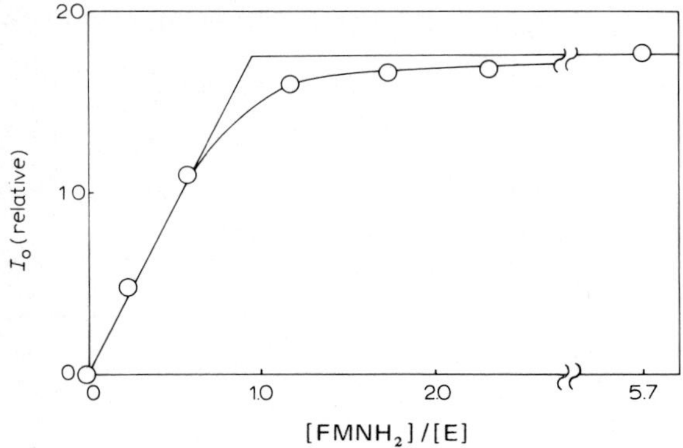

Fig. 5. Kinetic determination of the stoichiometry of the reaction of $FMNH_2$ with luciferase [17]. Concentration of luciferase, 4.7 μM.

$$\text{Luciferase} + \text{FMNH}_2 \underset{}{\overset{K_1}{\rightleftharpoons}} \text{Luciferase} \cdot \text{FMNH}_2 \dashrightarrow h\nu.$$

The solid curve in Fig. 5 which is a theoretical curve drawn assuming that $K_1 = 0.1\ \mu M$, gives the best fit for the experimental points.

Inhibition of the luciferase reaction by moieties of $FMNH_2$ [17]

The luminescent reaction of luciferase is inhibited by molecules forming parts of $FMNH_2$. In their reduced form, the compounds listed in Table I were all found to be competitive inhibitors of $FMNH_2$ in the luciferase reaction. The I_0 values were determined in the assay system supplemented with various concentrations of the inhibitor, and the values of K_i (inhibitor constant) estimated from plots of $1/I_0$ against the inhibitor concentrations (Dixon plot), are also summarized in Table I.

Fluorimetric titrations of luciferase with $FMNH_2$ and FMN [17]

Upon addition of $FMNH_2$ to a solution of luciferase under anaerobic conditions the protein fluorescence at 335 nm (excited at 288 nm which is the isosbestic point of the absorbance spectra of FMN and $FMNH_2$) was quenched without shift of the emission peak (Fig. 6). The data are replotted as a titration curve in Fig. 7, and the inflection point in the titration curve was observed at the point of equivalent binding of $FMNH_2$ to luciferase, indicating again that 1 mol of $FMNH_2$ was bound to 1 mol of the enzyme. The dissociation constant of the luciferase · $FMNH_2$ complex (K_1), estimated as the best fit for the experimental points, was 0.1 μM. No indication of FMN binding to luciferase was obtained in the fluorimetric titration.

The time course of the fluorescence quenching was followed by the stopped flow method. When luciferase in the flow apparatus was mixed with excess $FMNH_2$ under anaerobic conditions, the time course of the quenching was found to follow apparent first-order kinetics. The experiment was repeated at different $FMNH_2$ concentrations, and the apparent first-order rate constants (k_{app}) estimated from the experimental traces are plotted against the $FMNH_2$ concentrations (Fig. 8). In these experiments it is

TABLE I

K_i VALUES OF $FMNH_2$ DERIVATIVES IN THE LUCIFERASE REACTION [17]

Flavin derivative	K_i (μM)
FAD*	15
Riboflavin*	2.0
Lumiflavin*	40
Lumichrome*	100
Uracil	16 000

* In reduced form.

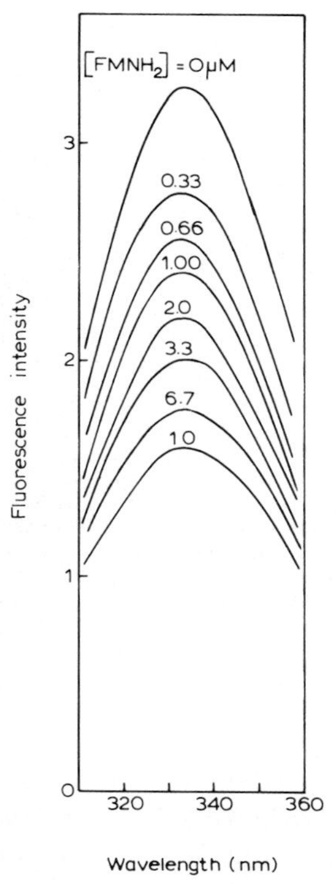

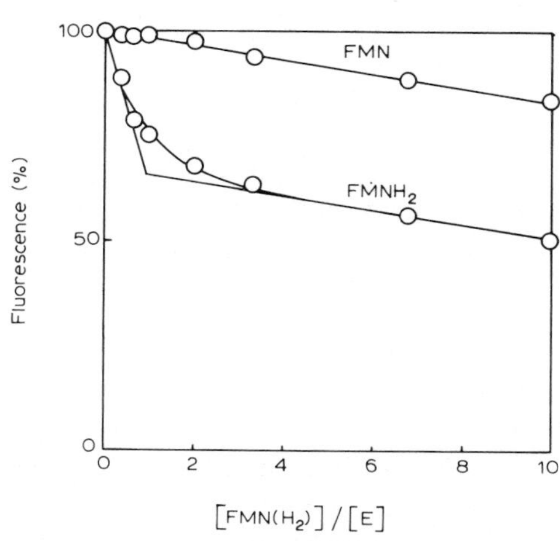

Fig. 6. Fluorescence emission spectra of luciferase in the absence or presence of added $FMNH_2$. Excitation wavelength, 288 nm. Concentration of luciferase, 1.2 μM. The $FMNH_2$ concentration added is shown in the figure.

Fig. 7. Fluorimetric titrations of luciferase by $FMNH_2$ and FMN [17]. Excited at 288 nm. Emission wavelength, 335 nm. Concentration of luciferase, 1.2 μM. With FMN, only the apparent decrease in protein fluorescence due to screening by added FMN was observed. The solid curve for $FMNH_2$ was drawn assuming that $K_1 = 0.1\ \mu$M.

reasonable to assume that the $FMNH_2$ concentrations were constant during the course of the reaction. As seen in Fig. 8, the plot of k_{app} against $FMNH_2$ concentration did not give a straight line but a saturation value was attained at higher $FMNH_2$ concentrations. This is not consistent with a simple one step mechanism and may most simply be ascribed to a two step, reversible reaction in which the first step is a rapid equilibrium between the enzyme and $FMNH_2$, and the second step is a slower isomerization process [18, 19]:

$$\text{Luciferase} + \text{FMNH}_2 \overset{K_1'}{\rightleftharpoons} \text{Luciferase} \cdots \text{FMNH}_2 \underset{k''_{-1}}{\overset{k''_{+1}}{\rightleftharpoons}} \text{Luciferase} \cdot \text{FMNH}_2$$

where luciferase and luciferase $\cdots$ $FMNH_2$ exhibit the same fluorescence spectrum. Assuming a two step reaction, the values of K_1', k''_{+1} and k''_{-1} were estimated by a reported method [19] to be 16 μM, 6 s^{-1} and 0.1 s^{-1}, respectively, as the best fit for the experimental points (Fig. 8).

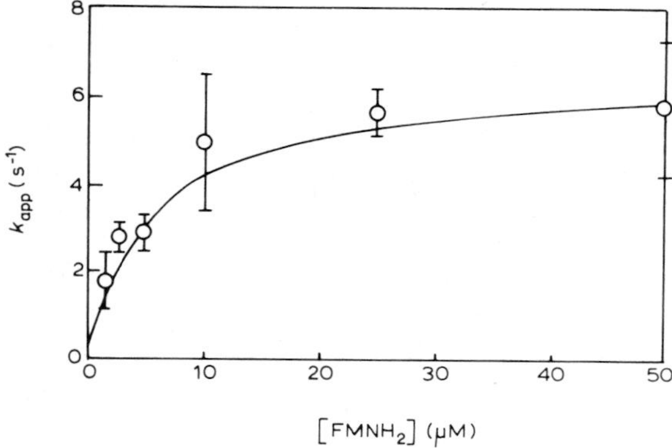

Fig. 8. Kinetics of binding of $FMNH_2$ to luciferase obtained by stopped-flow fluorimetric measurements. k_{app}, the apparent rate constant of the binding of $FMNH_2$ to luciferase was obtained from the time course of the quenching of fluorescence of the protein, and plotted against the $FMNH_2$ concentration under the condition that the $FMNH_2$ concentration was much greater than that of luciferase. The solid curve is the theoretical one calculated using the values of the equilibrium and rate constants given in the text. Luciferase concentration, 0.12 μM.

Discussion

Results on the stoichiometry of $FMNH_2$ binding to luciferase determined by the kinetic method, and by spectrophotometric and fluorimetric titrations are all in agreement, indicating the binding of 1 mol of $FMNH_2$ per mol of luciferase. Values for the dissociation constant of the luciferase · $FMNH_2$ complex (K_1) obtained by these methods were in fairly good agreement (0.1–0.5 μM). The K_m value of $FMNH_2$ in the luminescent reaction has also been determined to be 0.6 μM [17]. The difference absorbance spectrum of the luciferase · $FMNH_2$ complex against free enzyme and $FMNH_2$ is ascribable to a red-shift of the absorbance bands of $FMNH_2$ at 250 and 295 nm [20] but no particular difference attributable to a perturbation of the protein absorbance was detected. The spectroscopic features of the luciferase · $FMNH_2$ difference spectrum are similar to those of the difference spectra of $FMNH_2$ caused by solvent perturbation, indicating that the isoalloxazine nucleus of $FMNH_2$ is brought into a hydrophobic environment on formation of the luciferase · $FMNH_2$ complex. The binding of $FMNH_2$ to luciferase caused simultaneous quenching of a fluorescence which is probably due to tryptophan residues in the protein. The reaction of the fluorescence quenching appeared to follow a two step mechanism, suggesting that a conformation change of the luciferase molecule occurred on binding $FMNH_2$. Other contributions to luciferase · $FMNH_2$ binding are bondings between the anionic phosphate group and the ribitol moiety of $FMNH_2$ and the protein. The former is suggested by the stronger binding of $FMNH_2$ ($K_1 = 0.1\,\mu$M) than of reduced riboflavin ($K_i = 2\,\mu$M) and the latter is suggested by the competitive inhibition of the luminescent reaction by reduced riboflavin and the gradual decrease of the inhibition observed on stepwise fragmentation of the molecule of riboflavin (lumiflavin and lumichrome). The affinity of FMN for luciferase (dissociation constant) has been estimated to be in the order of 10 mM by the method of gel filtration [17]. The reason for the striking difference (more than 10^5 fold) between the affinities of FMN and $FMNH_2$ for luciferase is unknown. However, the isoalloxazine nucleus in a flavin derivative in the reduced form is known to be bent at an angle of 140°–160° along a line through N(5) and N(10) while that in the oxidized form has a planar configuration [21]. It seems possible that the binding site for $FMNH_2$ in luciferase is specific for the bent molecule, and that stereochemical as well as electronic differences between FMN and $FMNH_2$ are responsible for the difference in their affinities for luciferase.

Reaction intermediate of luciferase

Stoichiometry of the reaction of O_2 with the luciferase · $FMNH_2$ complex [17]. The stoichiometry of the reaction of oxygen with the luciferase · $FMNH_2$ complex in the luminescent reaction was examined by essentially the same kinetic method as that for the stoichiometry of $FMNH_2$, except that the oxygen concentration was varied from 0 to 18.6 μM while the $FMNH_2$ concentration was kept constant. To obtain all the $FMNH_2$ (calculated to be >95%) in the enzyme-bound state, the enzyme concentration was raised to twice that of $FMNH_2$. Fig. 9 shows plots of I_0 against the ratio of the initial [oxygen] to [luciferase · $FMNH_2$]. It was again estimated that in the presence of aldehyde one molecule of the luciferase · $FMNH_2$ complex reacts with one molecule of O_2 in the luminescent reaction.

Intermediate formed on reaction of the luciferase · $FMNH_2$ complex with O_2 [22]. The luciferase · $FMNH_2$ complex formed in the presence of excess luciferase was mixed with an equal volume of O_2 solution in the absence of aldehyde, and the time course of changes in absorbance of the

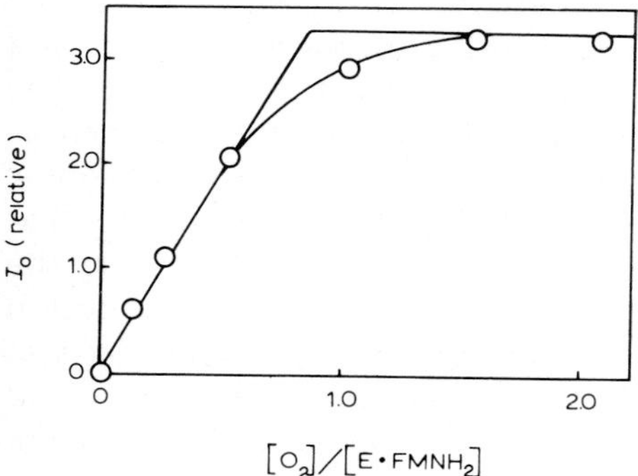

Fig. 9. Kinetic determination of the stoichiometry of reaction of oxygen with the luciferase · $FMNH_2$ complex in the luminescent reaction [17]. Concentration of luciferase · $FMNH_2$ complex, 9 μM.

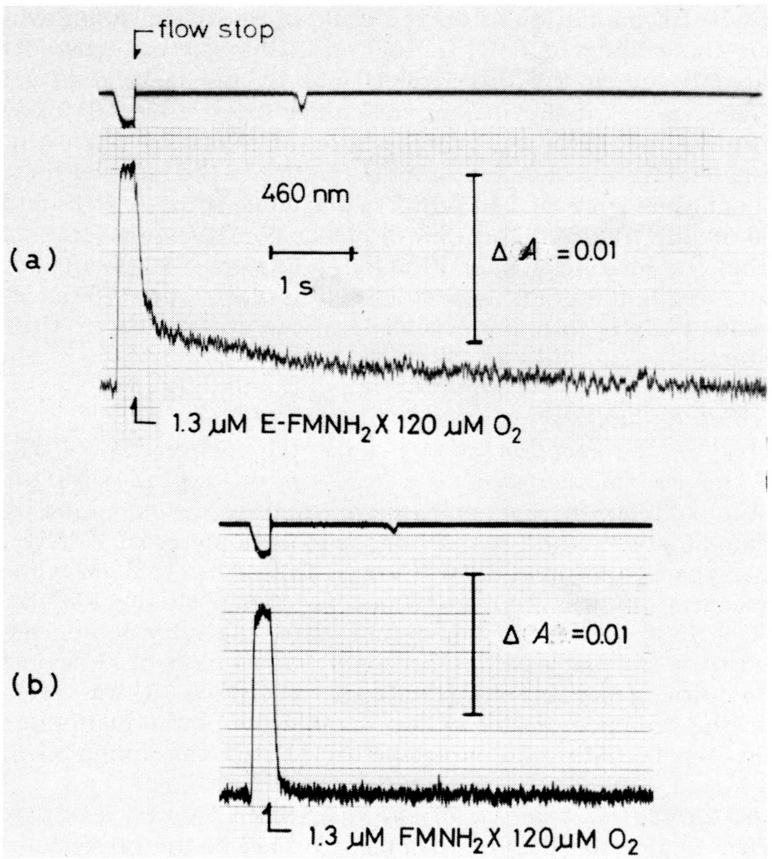

Fig. 10. Stopped-flow record of the time course of absorbance change upon reaction of the luciferase · $FMNH_2$ complex (a) or free $FMNH_2$ (b) with oxygen in the absence of aldehyde [22]. Downward deflection in the trace corresponds to increase of absorbance.

mixture was followed by the stopped-flow method. As shown in Fig 10(a), a biphasic change of the absorbance, a rapid initial increase followed by a much slower increase, was observed at 460 nm after stopping the flow. The same experiment was repeated at different wavelengths, and the absorbances of the reaction mixture at different wavelengths at intervals after stopping the flow are plotted against the wavelengths in Fig. 11. It can be seen that an intermediary spectrum rapidly appeared within 0.2 s after mixing the luciferase · $FMNH_2$ complex with O_2. The $t_{1/2}$ value of this rapid reaction was 60 ms at 20°C at each wavelength and was apparently independent of the O_2 concentration within the range tested (24–120 μM). The compound giving this intermediary spectrum had an absorbance band at 430–470 nm and was designated as X_1. After completion of the slower reaction, which was found to be a first order process with a rate constant $k = 0.25\ s^{-1}$, the absorbance spectrum of the mixture was the same as that of free FMN

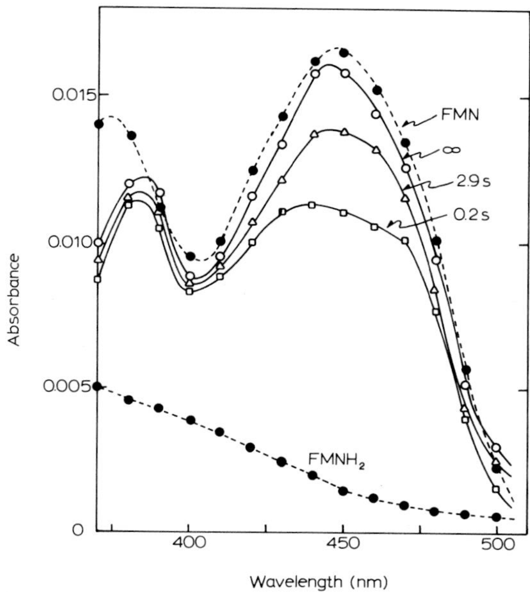

Fig. 11. Absorbance spectra of the reaction mixture as exemplified in Fig. 10(a) at intervals after stopping flow as indicated in the figure. The dotted curves represent the absorbance spectra of 1.35 μM free FMN(H_2).

(Fig. 11). Since the affinity of FMN for luciferase is more than 10^5 times less than that of $FMNH_2$, the slower process may correspond to release of FMN from X_1. In contrast to the biphasic nature of the time course of the reaction of the luciferase · $FMNH_2$ complex with O_2, aerobic oxidation of free $FMNH_2$ to FMN followed a single monophasic reaction (Fig. 10(b)). Fluorimetric experiments were conducted on the same reaction system in the flow apparatus, and the time course of change in fluorescence intensity of the mixture was recorded. The biphasic nature of the time course was again observed at each emission wavelength, but the fluorescence emission spectrum of the intermediate, X_1, thus obtained at 0.2 s after stopping the flow was indistinguishable from that of free FMN [22]. This is in contrast to the marked difference between the absorbance spectrum of X_1 and that of free FMN.

Life of the intermediate measured by aldehyde-initiated luminescence [22]. A solution of the luciferase · $FMNH_2$ complex was mixed with O_2 solution in the flow apparatus, and a 0.8 ml aliquot of the resulting X_1 was directly led into a small test tube fixed in front of a photomultiplier. At a known time after mixing, 0.2 ml of aldehyde solution from an injection syringe was rapidly mixed with the solution of X_1, and the change in luminescence intensity of the mixture with time was recorded. The same experiment was repeated with different time intervals, and results are summarized in Fig. 12. It can be seen that on addition of aldehyde there was a rapid increase of light intensity followed by a first-order decay process, the

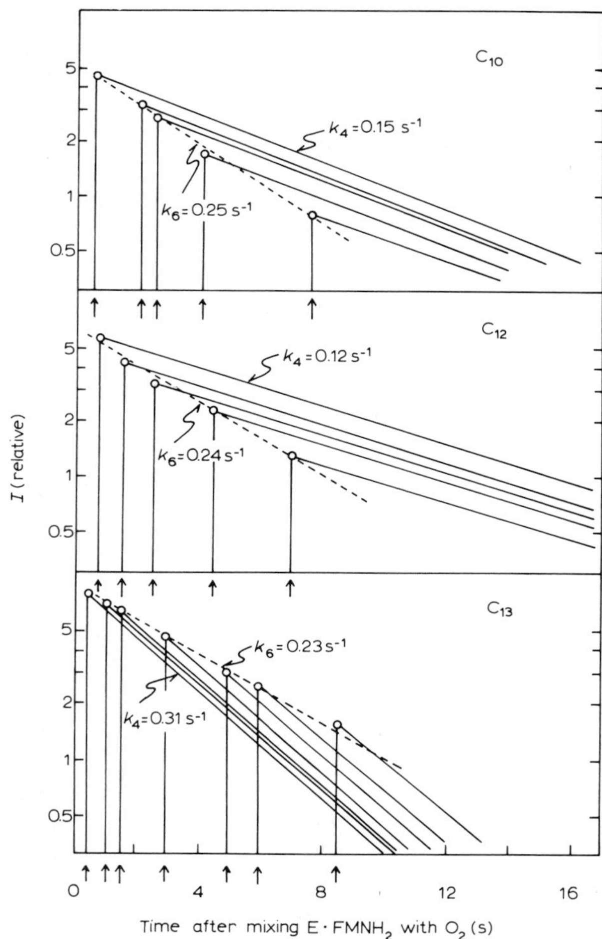

Fig. 12. Semi-log plot of the time course of the aldehyde-initiated luminescent reaction [22]. Luciferase and O_2 were mixed with $FMNH_2$ in the flow apparatus at zero time, and C_{10}-, C_{12}- or C_{13}-aldehyde was added to the reaction mixtures at the times indicated by arrows. ○, peak light intensity observed on addition of aldehyde; — — —, decay of the peak light intensity. The first-order rate constant estimated for each decay process is shown in the figure.

rate of which depended on the chain length of the aldehyde used ($k = 0.15$, 0.12 and 0.31 s^{-1} for C_{10}-, C_{12}- and C_{13}-aldehydes, respectively). The same decay rate constants were obtained when the reaction was initiated by mixing $FMNH_2$ with samples of enzyme which had previously been mixed with the various aldehydes.

Luciferase undergoes a single turnover under the present experimental conditions since free $FMNH_2$, one of the substrates, is rapidly exhausted (oxidized) upon reaction with excess O_2. With each sample, the peak height of the light intensity, I_0, is proportional to the amount of the enzyme intermediate capable of giving luminescence on reaction with aldehyde.

Decrease of I_0 with time after mixing luciferase · $FMNH_2$ with O_2 was again found to be a first order process (Fig. 12) and the values of the rate constant (k_6) determined using aldehydes of various chain lengths were in good agreement ($k_6 = 0.25$, 0.24 and 0.23 s^{-1} determined using C_{10}-, C_{12}- and C_{13}-aldehyde, respectively). The values of k_6 obtained were again in agreement with the rate constant of the reaction $X_1 \rightarrow$ Luciferase + FMN observed in the preceding rapid spectrophotometric experiments.

Effects of dithionite and ferricyanide on the time course of the luminescent reaction [22, 23]

In the following experiments the luciferase reaction was initiated in the flow apparatus by mixing the enzyme plus aldehyde and O_2 with $FMNH_2$. A 0.8 ml aliquot of the resultant mixture was led into a small test tube in front of a photomultiplier, as in the preceding experiments. While the luminescence was decaying, 0.2 ml of dithionite solution from an injection syringe was rapidly added to the reaction mixture. As can be seen in Fig. 13, an abrupt acceleration of the decay process was observed upon addition of dithionite when C_{10}-aldehyde was used, while up to 2.0 mM dithionite had a much smaller effect on the time course in reaction mixtures containing C_{13}-aldehyde. Dithionite had an intermediate effect when the reaction mixture contained C_{12}-aldehyde. In contrast to these results, when ferricyanide solution (final concentration, 1 mM) was added to the reaction mixture with an injection syringe, no appreciable change in the time course of the reaction was observed [23] demonstrating the stability of the intermediate to ferricyanide even in assay mixture containing C_{10}-aldehyde. Among the complexes of X_1 with C_{10-13}-aldehydes, the complex with C_{10}-aldehyde is the most sensitive to dithionite. These findings indicate that the flavin mononucleotide in X_1 is in a state similar to that of the oxidized form of FMN.

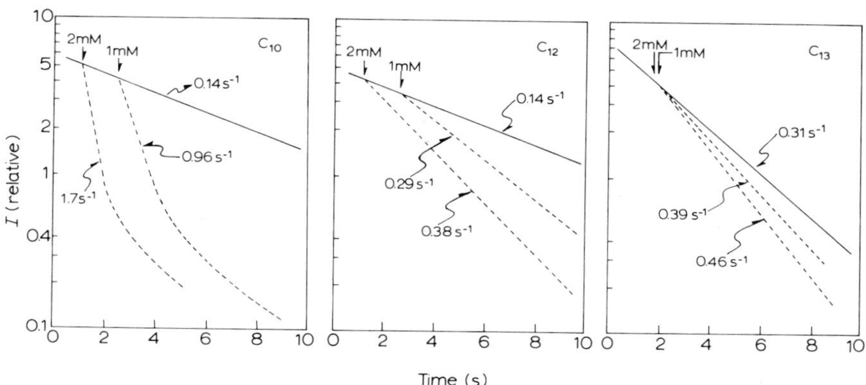

Fig. 13. Effect of dithionite on the time course of the luciferase reaction [22]. Dithionite was added at the times indicated by arrows. The concentrations of dithionite after mixing are shown in the figure. C_{10}-, C_{12}- or C_{13}-aldehyde was used. The first-order rate constant for each decay process is shown in the figure.

Conclusion

The absorbance spectrum of the luciferase intermediate (here denoted as X_1) determined in the present study by rapid mixing of the luciferase · $FMNH_2$ complex with O_2 represents an obligatory intermediate of the luciferase reaction since it fulfills the following requirements. (1) It is rapidly formed and the $t_{1/2}$ of its formation (60 ms) is shorter than that of the onset of the luminescent reaction. (2) In the absence of aldehyde the absorbance spectrum of X_1 changed to that of free FMN following a first order process, and the rate constant of this process is in agreement with the decay rate of an obligatory intermediate determined by aldehyde-initiated luminescence. A constant value was obtained for the latter irrespective of the chain length of the aldehyde used. The rate of formation of X_1 from luciferase · $FMNH_2$ and O_2 ($12\ s^{-1}$) was apparently independent of the O_2 concentration and this rate may represent that of the rate determining step in the reaction of luciferase · $FMNH_2 + O_2 \rightarrow X_1$. It is suggested that this reaction proceeds in at least two steps including rapid formation of an unstable luciferase · $FMNH_2$ · O_2 complex and then change of the complex to X_1 at a rate of $12\ s^{-1}$.

The molecular structure of X_1 cannot be deduced from available information on its nature. However, since the fluorescence spectrum of X_1 was almost identical with that of FMN [22], it is likely that the electronic state of FMN in X_1 is not very different from that of free FMN. This is supported by the fact that the time course of the luminescent reaction was insensitive to an oxidant such as ferricyanide. If so, FMN in X_1 is in a stereochemically unfavorable (unstable) state, since as discussed previously in this paper, the structure of the binding site of $FMNH_2$ in luciferase may only fit the bent isoalloxazine nucleus of FMN. The absorbance spectrum of the intermediate, which was isolated by Hastings et al. by the method of low temperature gel filtration in 50% ethyleneglycol only showed a peak at 372 nm [13]. This spectral feature is not in agreement with our results on X_1 obtained in aqueous medium at 20°C. This disagreement can not yet be explained, even if we consider differences in solvent composition and temperature and also that our transient spectrum may include that of a small amount of contaminating free FMN.

The emission maximum of the luciferase reaction is at a wavelength about 30 nm shorter than that of the fluorescence of X_1 or free FMN in aqueous medium. A shift of the emission maximum of the bound FMN in X_1 by 30 nm to a shorter wavelength must occur when a molecule in the excited state is formed as a consequence of specific binding of an aldehyde molecule to X_1 in close proximity to FMN. This possibility was also supported by results on the effect of dithionite on the luciferase reaction, as will be discussed below.

The decay rate of the luminescent reaction was dependent on the chain length of the aldehyde used. The emission maximum of the luminescent reaction was also slightly dependent on the chain length [14]. These facts indicate that there are differences between the structures of $X_1 \cdot RCHO$ complexes of different chain lengths. In fact, dithionite had different effects on the luminescent reaction in the presence of aldehydes of different chain lengths. In the presence of C_{10}-aldehyde the luminescence was rapidly extinguished when excess dithionite was added to the reaction mixture. On the other hand, dithionite caused only slight decrease in luminescence intensity when C_{13}-aldehyde was present in the reaction mixture. These results indicate that X_1-RCHO could be reduced back to luciferase$\cdot FMNH_2$ (or luciferase$\cdot FMNH_2 \cdot RCHO$) by dithionite in competition with the formation of the excited state molecule, and the larger hydrophobic pocket formed by binding aldehyde of longer chain length in the luciferase molecule protected X_1 more effectively against the reductant. These facts indicate a rather direct interaction between FMN and aldehyde in the luciferase molecule.

All these results may be summarized in the following reaction mechanism in which $FMNH_2 \rightarrow O_2 \rightarrow RCHO$ is the obligatory order of the reaction. This mechanism is compatible with that previously proposed by Hastings and Gibson [12] and X_1 in this scheme corresponds to their "long-lived intermediate".

$$\text{Luciferase} + FMNH_2 \overset{K_1}{\rightleftharpoons} \text{Luciferase} \cdot FMNH_2\,, K_1 = 0.1\,\mu M,$$

$$\text{Luciferase} \cdot FMNH_2 + O_2 \xrightarrow{k_2} X_1\,, \qquad k_2 = 12\,s^{-1}.$$

In the presence of aldehyde,

$$X_1 + RCHO \rightleftharpoons X_1 \cdot RCHO,$$

$$X_1 \cdot RCHO \xrightarrow{k_4} X_2^* \text{ (excited state)},$$

$$X_2^* \xrightarrow{k_5} \kappa h\nu + \text{products}.$$

In the absence of aldehyde,

$$X_1 \xrightarrow{k_6} \text{Luciferase} + FMN, \quad k_6 = 0.25\,s^{-1}.$$

The values of K_m for aldehyde and k_4 are 0.28 mM and 0.15 s^{-1} for C_{10}-aldehyde, 0.18 mM and 0.12 s^{-1} for C_{12}-aldehyde, and 0.12 mM and 0.31 s^{-1} for C_{13}-aldehyde, respectively. The factor $\kappa (0 < \kappa < 1)$ was tentatively introduced into the reaction scheme since the maximum quantum yield observed for the bioluminescent reaction was 0.2 per luciferase molecule. However, it is still premature to discuss the mechanism of energy dissipation in a form other than light energy.

References

1 Strehler, B.L. (1953) J. Am. Chem. Soc. 75, 1264

2 McElroy, W.D., Hastings, J.W., Sonnenfeld, V. and Coulombre, J. (1953) Science 118, 385–386

3 Totter, J.R. and Cormier, M.J. (1955) J. Biol. Chem. 216, 801–811
4 Cormier, M.J. and Kuwabara, S. (1965) Photochem. Photobiol. 4, 1217–1225
5 Hastings, J.W., Riley, W.H. and Massa, J. (1965) J. Biol. Chem. 240, 1473–1481
6 Strehler, B.L., Harvey, E.N., Chang, J.J. and Cormier, M.J. (1954) Proc. Natl. Acad. Sci. U.S. 40, 10–12
7 Watanabe, H., Mimura, N., Takimoto, A. and Nakamura, T. (1975) J. Biochem. 77, 1147–1155
8 Cormier, M.J. and Strehler, B.L. (1953) J. Am. Chem. Soc. 75, 4864–4865
9 Cormier, M.J. and Totter, J.R. (1957) Biochim. Biophys. Acta 25, 229–237
10 McCapra, F. and Hysert, D.W. (1973) Biochem. Biophys. Res. Commun. 52, 298–304
11 Dunn, D.K., Michaliszyn, G.A., Bogacki, I.G. and Meighen, E.G. (1973) Biochemistry 12, 4911–4918
12 Hastings, J.W. and Gibson, Q.H. (1963) J. Biol. Chem. 238, 2537–2554
13 Hastings, J.W., Balny, C., LePeuch, C. and Douzou, P. (1973) Proc. Natl. Acad. Sci. U.S. 70, 3468–3472
14 Yoshida, K. and Nakamura, T. (1973) J. Biochem. 74, 915–922
15 Nakamura, T. and Matsuda, K. (1971) J. Biochem. 70, 35–44
16 Yoshida, K. and Nakamura, T. (1972) J. Biochem. 72, 951–955
17 Watanabe, T., Tomita, G. and Nakamura, T. (1974) J. Biochem. 75, 1249–1255
18 Gutfreund, H. (1972) Enzymes: Physical Principles, p. 205, John Wiley & Sons, New York
19 Umemiya, Y. and Nakamura, T. (1974) FEBS Lett. 48, 192–195
20 Beinert, H. (1956) J. Am. Chem. Soc. 78, 5323–5328
21 Kierkegaad, P., Norrestam, R., Werner, P.E., Csöregh, I., Glehn, M.V., Karlsson, R., Leijonmarck, M., Rönnguist, O., Stensland, B., Tillberg, O. and Törbjornsson, L. (1970) in Flavins and Flavoproteins (Kamin, H., ed.), p. 1, Univ. Park Press, Baltimore
22 Yoshida, K., Takahashi, M. and Nakamura, T. (1974) J. Biochem. 75, 583–589
23 Yoshida, K. and Nakamura, T. (1974) J. Biochem. 76, 985–990

Flavins and Flavoproteins, edited by T.P. Singer

CHAPTER 6

DIRECT CALORIMETRIC MEASUREMENT OF THE BINDING OF FLAVIN MONONUCLEOTIDE TO *BENECKEA HARVEYII* LUCIFERASE

NEAL LANGERMAN

Utah State University, Department of Chemistry, Logan, Utah 84322 (U.S.A.)

Luciferase, isolated from a wide variety of photobacteria, catalyzes the simultaneous oxidation of reduced flavin mononucleotide ($FMNH_2$) and a long chain aldehyde (RCHO) as indicted by the in vitro scheme proposed by Hastings, et al. [1]:

$$\begin{array}{lllll}
FMNH_2 + E \rightleftharpoons E - FH_2 & \xrightarrow{O_2} & E - \underset{\substack{| \\ O \\ \quad \diagdown O^-}}{FH_2} & \overset{RCHO}{\rightleftharpoons} & E - FH_2 \;\; (RCHO,\ O{\diagdown}O^-) \\
\downarrow & & \swarrow & & \downarrow \\
FMN + H_2O_2 & & E + FMN + H_2O_2 \ (\text{dark}) & & [*] \\
 & & & & \downarrow \\
 & & & & E + FMN + RCOOH + H_2O + h\nu_{490}
\end{array} \qquad (1)$$

Recent studied in this laboratory [2] have demonstrated that this oxidation occurs with an enthalpy change of at least 22 $kcal \cdot mol^{-1}$ in excess of the energy of the emitted photons (490 nm radiation corresponds to the 59 $kcal \cdot einstein^{-1}$). As part of an overall study of the physical chemistry of chemiluminescence and the enzymes of bioluminescence, I am reporting data used to deduce the free energy change, enthalpy change and entropy change for the binding of the product molecule, FMN, to the enzyme, i.e.

$$FMN + E \rightleftharpoons [FMN - E] \qquad (2)$$

This binding was demonstrated in Hastings' laboratory [3, 4, 13]. The further characterization of the thermochemistry of reaction [2] is a prelude to the technically more difficult study of the binding of the reduced substrate, $FMNH_2$, to the enzyme.

Materials and Methods

Luciferase was isolated from *Beneckea harveyii* (formerly strain MAV) by the method of Gunsalus-Miguel et al. [5], using the light assay described therein. Protein concentration was determined from $A_{280} = 0.94$ $ml\,(mg \cdot cm)^{-1}$ and flavin (Sigma chemicals) concentration using $\epsilon = 12.2\ mM^{-1} \cdot cm^{-1}$. Calorimetric measurements were made using a flow microcalorimeter [6] calibrated by the reaction of Tris with HCl [7]. The syringe reagent delivery system was replaced by two LKB peristaltic pumps operated at a flow rate of 0.50 ml$\cdot$ min^{-1}. The flavin was diluted exponentially in time, as described by Mountcassel and Biltonen [8]. This method is very convenient as it allows one to obtain a complete binding isotherm on a single sample of protein. All experiments were performed in 0.05 M Hepes, pH 7.00, at 25.45°C.

Results and Discussion

The experimental data for the dilution of FMN, the reaction of 0.135 mM luciferase with FMN and the dilution of the protein are shown in Fig. 1. The initial flavin concentration, $[FMN]_0$, was 7.54 mM. The concentration of flavin in the reaction tubing of the calorimeter at any time was calculated from the relation

$$[FMN]_t = [FMN]_0 \exp\{-f/v(t-t_0)\} \tag{3}$$

where the dilution factor is determined by the flow rate of the flavin, f, ml$\cdot$ s^{-1}, and the volume, v, of the exponential dilution system ($\approx$ 1.33 ml). The effective time of the start of dilution is t_0.

These data reveal several interesting points. The most obvious is the shape of the FMN dilution isotherm. Each point on the curve represents the heat for isothermal process

$$FMN(aq, c) \rightarrow FMN(aq, c/2) \tag{4}$$

Analogous, though not as extensive, data for the same process, for reduced flavin, have previously been described by Beaudette [9] in his study of $\Delta H_{\text{oxidation}}$ of $FMNH_2$ in phosphate buffer, pH 7.4. Undoubtedly, this

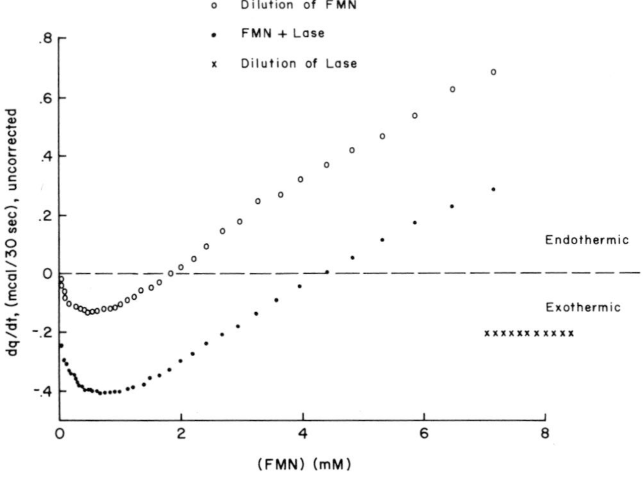

Fig. 1. Calorimetrically determined heat flux for the dilution of FMN (○), dilution of luciferaze (×), and overall reaction of luciferase with FMN(●). Reactions were performed in 0.05 M Hepes buffer, pH 7.00 at 25.45°C. The luciferase concentration was 0.135 mM and the initial FMN concentration was 7.54 mM.

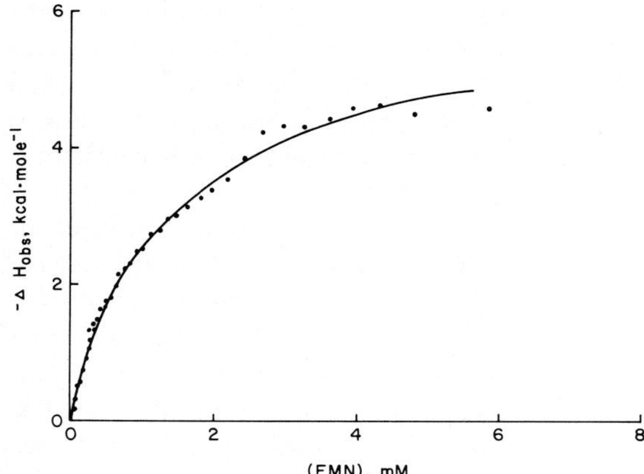

Fig. 2. Binding isotherm of FMN to luciferase in 0.05 M Hepes buffer, pH 7.00, at 25.45°C. Data were derived from those in Fig. 1 as explained in the text. The solid curve was calculated using Eqn. 7 and the parameters $\Delta H' = -6.1$ kcal·mol^{-1} (-25.0 kJ·mol^{-1}) and $K = 0.745$ mM^{-1}.

bimodal behaviour represent different contributions of the rather hydrophobic dimethyl isoalloxazine portion of the molecule and the charged phosphate to the observed heat of dilution.

Data for the dilution of the enzyme,

$$\text{E(aq, 0.135 mM)} \rightarrow \text{E(aq, 0.0675 mM)} \tag{5}$$

are also presented. Using a molecular weight $74 \cdot 10^3$ daltons (Langerman, unpublished data) for the enzyme, this observed heat flux converts to an enthalpy change for reaction 5 of -6 kcal·mol^{-1}. The source of this heat is currently not established but is most certainly not due to the dissociation of the dimeric enzyme. This latter statement is supported by sedimentation equilibrium data obtained in this laboratory.

The binding isotherm $\dot{q}$(overall) is obviously more exothermic than the FMN dilution isotherm $\dot{q}$(FMN, dil) by an amount in excess of the heat of dilution of the protein, $\dot{q}$(E, dil). The heat flux due to the binding reaction, $\dot{q}$(rx), is extracted from these data from the relation

$$\dot{q}(\text{rx}) = \dot{q}(\text{overall}) - \dot{q}(\text{FMN, dil}) - \dot{q}(\text{E, dil}) \tag{6}$$

Quite obviously, the derived heat flux is very sensitive to the stability and reproducibility of the instrument which have both been repeatedly demonstrated to be very good [10, 11].

Data derived in this manner are readily converted into a binding isotherm by dividing each point by the protein flux and plotting as in Fig. 2. These data were used to derive the thermodynamic parameters $\Delta G'$, $\Delta H'$, $\Delta S'$ by fitting them to the relation

$$\frac{\Delta H_{\text{obs}}^2}{\Delta H'}[\text{Lase}]_0 - \frac{\Delta H_{\text{obs}}}{\Delta H}([\text{Lase}]_0 + K^{-1} + [\text{FMN}]_t) + [\text{FMN}]_t = 0 \tag{7}$$

where $[\text{Lase}]_0$ is the luciferase concentration and ΔH_{obs} is the enthalpy change at any flavin concentration, $[FMN]_t$. The derived parameters were used in Eqn. (7) to generate the solid curve in Fig. 2. The values obtained in this manner are $K = 0.745\ mM^{-1}$, $\Delta G' = -3.9\ kcal \cdot mol^{-1}$ ($-16.3\ kJ \cdot mol^{-1}$), $\Delta H'^{*} = -6.1\ kcal \cdot mol^{-1}$ ($-25.0\ kJ \cdot mol^{-1}$) and $\Delta S' = -8$ e.u. ($33.5\ J \cdot mol^{-1} \cdot deg^{-1}$. These values suggest that no major conformational change is occurring to the protein, when FMN is bound [12].

It is useful to compare these data to those of Baldwin [3], Nicoli et al. [4] and Baldwin et al. [13], i.e. $K = 4.5\ mM^{-1}$, $\Delta H_H = -10\ kcal \cdot mol^{-1}$. The latter were obtained in Bis-Tris buffer, pH 7.0 over the temperature range 3–30°C. The most obvious interpretation is reduced binding affinity as a result of a positive heat capacity change for the reaction. A more complex interpretation is that the Van't Hoff enthalpy (Nicoli et al. [13]) is indeed different from the calorimetric value. This would reflect processes more involved than expressed by reaction 2. This situation is currently being investigated by direct calorimetric studies of reaction 2 at 1.65°C.

In conclusion, the data reported here have established several useful thermodynamic parameters, while raising exciting questions. Specifically, what is the source of the bimodal character of ΔH_{dil} (FMN) and of the exothermic heat of dilution of the protein? Furthermore, are the temperature dependent changes observed for the thermodynamic parameters of the binding process due to reduced binding affinity (i.e. thermodynamic) or are they actual changes (conformational, chemical) in the protein? These questions are actively being studied in this laboratory and will be subject of subsequent communications.

Acknowledgement

This work was in part supported by NIH Grant GM 18637.

References

1 Hastings, J.W., Eberhard, A., Baldwin, T.O., Nicoli, M.Z., Cline, T.W. and Nealson, K.H., in Chemiluminescence and Bioluminescence (Cormier, M.J., Hercules, D.M. and Lee, J., eds.), pp.369–380, Plenum Publishing Corportation, New York
2 Langerman, N. and Mangold, A. (1974) Fed. Proc. 33, 1229 (Abstr. 30)
3 Baldwin, T.O. (1974) Biochem. Biophys. Res. Commun. 57, 1000–1005
4 Nicoli, M.Z., Meighen, E.A. and Hastings, J.W. (1974) J. Biol. Chem. 249, 2385–2392
5 Gunsalus-Miguel, A., Meighen, E.A., Nicoli, M.Z., Nealson, K.H. and Hastings, J.W. (1972) J. Biol. Chem. 247, 398–404
6 Lyons, P.A. and Sturtevant, J.M. (1969) J. Chem. Thermodyn. 1, 201–208
7 Öjelund, G. and Wädso, I. (1968) Acta Chem. Scand. 22, 2691–2700
8 Mountcassel, D.M. and Biltonen, R.L. (1976) Biophys. J., in the press
9 Beaudett, N.V. (1973) Master thesis, Tufts University, Medford, Mass.

* This value has not been corrected for changes in the ionization state of the system. Thus, the actual process to which this value applies is best represented as

$$FMN + Lase + HEPES \rightleftharpoons [FMN - Lase] + n_H\ HEPES{-}H^+$$

where n_H, the number of protons, may be either positive or negative.

10 Rothman, S.W., Kisliuk, R.L. and Langerman, N.R. (1973) J. Biol. Chem. 248 7845–7851
11 Beaudette, N.V. and Langerman, N.R. (1974) Arch. Biochem. Biophys. 161, 125–133
12 Allewell, N.H., Friedland, J. and Niekamp, K (1975) Biochemistry 14, 224–230
13 Baldwin, T.O., Nicoli, M.Z., Becvar, J.E. and Hastings, J.W. (1975) J. Biol. Chem., in the press

Flavins and Flavoproteins, edited by T.P. Singer

CHAPTER 7

RAPID-SCANNING STOPPED-FLOW STUDY OF THE OXIDATION OF $FMNH_2$ BY O_2 CATALYZED BY BACTERIAL LUCIFERASE

GEORGE J. FAINI, RICHARD J. DESA and JOHN LEE

Department of Biochemistry, University of Georgia, Athens, Ga. 30602 (U.S.A.)

Summary

1. A computer-controlled, rapid-scanning, stopped-flow spectrophotomer has been built. It was used to study the bacterial luciferase catalyzed oxidation and autooxidation of $FMNH_2$.

2. A scanning stopped-flow study of autooxidation of $FMNH_2$ revealed no intermediates (time resolution was 6 ms).

3. In the oxidation of $FMNH_2$ catalyzed by bacterial luciferase, several spectral intermediates corresponding to luciferase-flavin complexes were observed.

Introduction

In the oxidation of $FMNH_2$ by molecular oxygen catalyzed by bacterial luciferase, an enzymatic intermediate is formed with flavin. When this intermediate is reacted with a long-chain aliphatic aldehyde, light is produced [1]. In the absence of aldehyde, the intermediate formed with luciferase from the cell type *Beneckea harveyi* has a lifetime of tens of seconds at room temperature, minutes at 0°C, and hours at − 20°C. Utilizing the long lifetimes at low temperatures, spectral properties of the *B. harveyi* luciferase-flavin intermediate have been studied upon separation of free flavin from the intermediate by column chromatography [2, 3]. However, intermediates formed with luciferases from cell types *Photobacterium fischeri* and *Achromabacter fischeri* have lifetimes on the order of 1 min at 0°C. Due to these shorter lifetimes, column chromatography at 0°C yields only small amounts of separated intermediate so that standard spectrophotometers cannot be used to obtain absorbance spectra of the intermediate. Thus, a rapid-scanning, stopped-flow spectrophotometer was built to study the kinetics and absorption spectra of these intermediates. A stopped-flow instrument interfaced to a general-purpose mini-computer [4] has been modified to scan the excitation light over a 400 nm range in 6 ms. Spectra can be acquired, stored, and

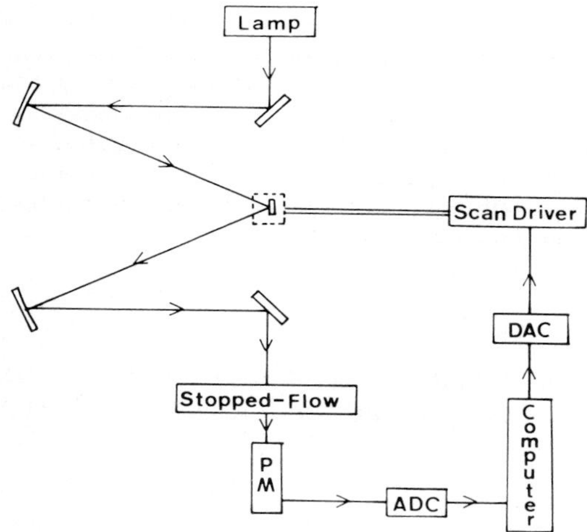

Fig. 1. Schematic diagram of a rapid-scanning, stopped-flow spectrophotometer. The vibrating grating (dashed box) is controlled by the computer program and synchronized with collection of data from the photomultiplier tube (PM).

manipulated by a variety of computer programs. Absorbance vs. wavelength and/or time can be displayed in printed or graphic forms. Results are reported for the autooxidation of $FMNH_2$ by molecular oxygen, and the luciferase-catalyzed oxidation at pH 7.0 at 2°C.

Materials and Instrumentation

FMN, obtained from Fluka, A.G., Buchs, Switzerland, was purified on a DEAE column [5]. $FMNH_2$ was prepared by photoreduction of FMN in 0.5 M phosphate buffer pH 7, containing 0.02 M EDTA. Prior to photoreduction at room temperature, the FMN buffer solution was deoxygenated by bubbling with Argon for 30 min in the dark. The solution was held in a vessel attached to the holding syringe of the stopped-flow instrument which had been previously flushed with anaerobic buffer. Luciferase was obtained from the two types of bacteria, *P. fischeri* and *A. fischeri*, and was purified as previously described [6]. In these experiments, the specific activity was usually $15 \cdot 10^{13}\, h\nu\, s^{-1} \cdot mg^{-1}$ for both types of luciferase.

A computer-controlled, single-wavelength stopped-flow spectrometer designed by DeSa [4] has been modified (Fig. 1) so that selected regions of the electromagnetic spectrum can be rapidly scanned in time. All single wavelength capabilities were retained. The fixed diffraction grating in the excitation monochromator was replaced by a small grating attached to a vibration galvanometer (dashed box, Fig. 1) which scans the spectrum across the exit slit. The angular velocity of the grating is controlled by the frequency of a waveform processed by the scan driver to provide the drive current. This waveform is provided by the digital-to-analog converter (DAC) in the computer system and consists of a "staircase" approximation of a triangular wave generated by the software. As the spectrum is scanned, the resulting signals from the photomultiplier tube (PM) enter the analog-to-digital converter (ADC). They are read under software control providing a direct correlation of the PM signal to the wavelength position of the grating. This system is capable of collecting a 128 point, 400 nm absorbance spectrum of 6 ms with a repetition of about 50 Hz. Thus, a stopped-flow experiment can be followed by measuring absorption spectra in time, yielding both kinetic and spectral data from one reaction.

The time dependent spectra are acquired in real time, stored, and processed either immediately or at any later time in a variety of ways. In particular, the assembly-language program written to date can display the absorption spectra on an oscilloscope in a split-screen mode. The upper half displays absorbance vs. wavelength, while absorbance vs. time at any selected wavelength is seen on the lower half of the screen. The difference between any two spectra can be obtained. Data can be graphed on an XY recorder either as a linear or log plot. Obviously, since the system contains a general-purpose computer, many different functions can be performed on the data once acquired.

Results and Discussion

The first step in the reaction of an air-saturated luciferase solution with $FMNH_2$ involves the competition between O_2 and luciferase for $FMNH_2$ (Eqn. 1).

$$FMNH_2 \begin{cases} \xrightarrow{E} E\text{—}FMNH_2 \rightarrow \text{light} \\ \xrightarrow{O_2} FMN + H_2O_2 \end{cases} \quad (1)$$

Although the autooxidation is rapid [7], Lee and Murphy [8] have shown that the quantum yield of bioluminescence with respect to $FMNH_2$ reaches a constant value for luciferase concentrations of 30 μM and greater. This implies that all the $FMNH_2$ is oxidized by the luciferase. Stopped-flow experiments, where the change in absorbance at a wavelength = 450 nm was measured (Fig. 2), clearly show that luciferase successfully competes with the autooxidation pathway. Thus, luciferase concentrations of at least 30 μM were used in the scanning stopped-flow experiments.

In order to compare the nature of the interactions in autooxidation with that of luciferase-catalyzed oxidation of $FMNH_2$, scanning stopped-flow studies of autooxidation at pH 7, 2°C, were performed. Over an $FMNH_2$ concentration range of 8 to 80 μM, a single isosbestic point in time was observed (Fig. 3) indicating that only two reacting species were present.

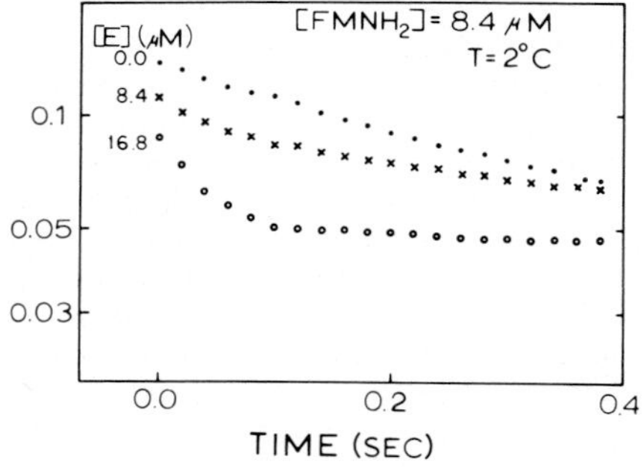

Fig. 2. *P. fischeri* luciferase catalyzed oxidation of $FMNH_2$. The luciferase solution (0.05 M P_i, 0.005 M EDTA, pH 7) was air-saturated at 2°C. The FMN solution (0.05 M P_i, 0.02 M EDTA, pH 7) was made anaerobic in the dark at 20°C, photoreduced, and then cooled to 2°C. Measurements taken at 450 nm.

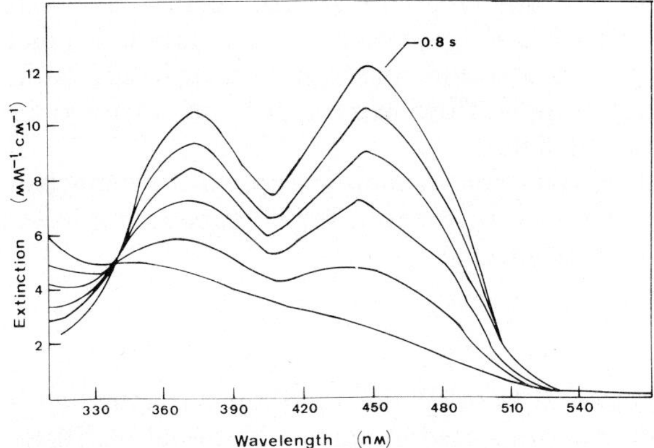

Fig. 3. Autooxidation of $FMNH_2$. $FMNH_2$ was prepared as described in Fig. 2 and reacted with air-saturated buffer (0.05 M P_i, 0.005 M EDTA, pH 7). Spectrum collection time was 6 ms. 20 spectra were collected at 50 ms intervals; not all spectral are shown. First spectrum is that of $FMNH_2$.

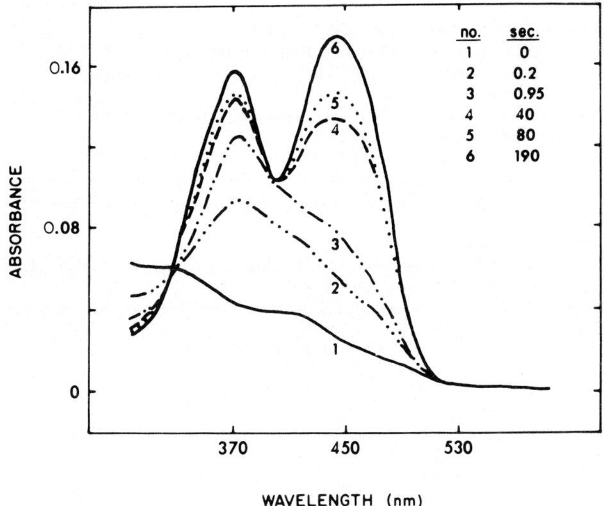

Fig. 4. Scanning stopped-flow reaction of *P. fischeri* luciferase with $FMNH_2$. Solutions were prepared as described in Fig. 2. Time resolution is 6 ms.

Massey et al. [9] observed an intermediate species at 3 ms in this reaction; however, in our experiments the first spectrum required 6 ms to be collected and no intermediate was observed.

Scanning stopped-flow experiments using both *P. fischeri* and *A. fischeri* luciferase concentrations of 30 μM to 60 μM in the enzyme catalyzed oxidation of $FMNH_2$ consistently indicated the presence of a luciferase-flavin intermediate (Fig. 4). Absorbance spectrum 3 (Fig. 4) is similar to the spectrum observed by Hastings et al. [3] for an intermediate in the *B. harveyi* luciferase catalyzed reaction. This intermediate has been suggested to be the final species to react with aldehyde and produce light [3]. However, Lee and Murphy [10] reported that the active intermediate is instead an oxidized flavoprotein having an absorption spectrum similar to spectrum 5 (Fig. 4). Additional scanning stopped-flow experiments are planned to elucidate the nature of these intermediates.

In these preliminary studies, this rapid-scanning stopped-flow spectrophotometer has been shown to be a powerful tool for characterizing intermediates in the luciferase catalyzed reactions.

Acknowldegements

We thank Dr. Onofre Paguio and Carol White for their assistance with enzyme purification. This work was supported in part by National Institutes of Health grants GM 02619-01 to G.J. Faini, GM 18634-06 to R.J. DeSa, and GM 19163-01 to J. Lee.

References

1 Hastings, J.W. and Gibson, Q.H. (1963) J. Biol. Chem. 238, 2537–2554
2 Murphy, C.L., Faini, G.J., and Lee, J. (1974) Biochem. Biophys. Res. Commun. 58, 119–125
3 Hastings, J.W., Balny, C., LePeuch, C. and Douzou, P. (1973) Proc. Natl. Acad. Sci. U.S. 70, 3468–3472
4 DeSa, R.J. (1972) in Computers in Chemical and Biochemical Research (Klepfenstein, C.E. and Wilkins, C.L., eds.), pp. 83–106, Academic Press, New York, N.Y.
5 Massey, V. and Swoboda, B.E.P. (1963) Biochem. Z. 338, 474–484
6 Lee, J., Murphy, C.L., Faini, G.J., and Baucom, T.L. (1974) in Liquid Scintillation Counting: Recent Developments (Stanley, P.E. and Scoggins, B.A., eds.), pp. 403–420, Academic Press, New York, N.Y.
7 Gibson, Q.H. and Hastings, J.W. (1962) Biochem, J. 83, 136–377
8 Lee, J. and Murphy, C.L. (1973) in Chemiluminescence and Bioluminescence (Cormier, M.J., Hercules, D.M. and Lee, J., eds.), pp. 381–386, Plenum Press, New York
9 Massey, V., Palmer, G. and Ballou, D. (1971) in Flavins and Flavoproteins (Kamin, H., ed.), pp. 349–361, University Park Press, Baltimore, Md.
10 Lee, J. and Murphy, C.L. (1973) Biochem. Biophys. Res. Commun. 53, 157–163

Flavins and Flavoproteins, edited by T.P. Singer

CHAPTER 8

THE INTERACTION OF OXIDIZED FLAVIN MONONUCLEOTIDE WITH BACTERIAL LUCIFERASE

MIRIAM Z. NICOLI, THOMAS O. BALDWIN, JAMES E. BECVAR and J. WOODLAND HASTINGS

The Biological Laboratories, Harvard University, Cambridge, Mass. 02138 (U.S.A.)

Summary

Bacterial luciferase is a heterodimeric enzyme ($\alpha\beta$) which catalyzes the bioluminescent oxidation of reduced flavin mononucleotide; the products include oxidized FMN and blue-green light. Luciferase binds the product FMN in a 1:1 molar ratio with an apparent dissociation constant of 0.12 mM at 4.5°C in 0.05 M Bis-Tris, pH 7.0. Luciferase-bound FMN is nonfluorescent, indicating that this is not the emitting species in the bioluminescence reaction. The optical absorbance and circular dichroism spectra of luciferase-bound FMN show considerable structure, with one major peak at 366 nm and resolution of the 445 nm band into two distinct peaks at 434 and 458 nm. Two mutant luciferases with active center lesions resulting in decreased $FMNH_2$ binding affinities also have altered absorbance spectra for bound FMN; furthermore, one of these mutants exhibits a 12 nm red shift in the bioluminescence emission spectrum, both in vivo and in vitro. Luciferase with the reactive "essential" sulfhydryl modified by 1 molar equivalent of *N*-octylmaleimide still binds FMN. Binding of either FMN or inorganic phosphate makes the α subunit less susceptible to proteolytic attack, apparently by causing a conformational change.

Bacterial luciferase is a 79 000 dalton dimer composed of two nonidentical subunits, α and β, which have been shown to be functionally distinct by both chemical and mutational modification [1–5]. All substrate binding ($FMNH_2$ and aldehyde) and catalytic properties appear to reside on the α subunit; the function of the β subunit, which is absolutely required for luminescence activity [6], remains undefined. However, the recent indication from amino- and carboxy-terminal sequence data that α and β may be homologous [7] suggested the possibility that vestigial substrate or product binding sites might remain on β despite the apparent lack of any catalytic function for that subunit.

Until recently, no interaction of luciferase with the product FMN had been demonstrated. It had been reported that if FMN were bound, the dissociation constant must be greater than 10^{-4} M, 1000-fold greater than the K_d for $FMNH_2$ [8, 9]. However, when a highly reactive sulfhydryl residue on the α subunit, whose modification has been shown to decrease $FMNH_2$ binding affinity by at least 10-fold, was found to be protected by FMN from alkylation [5], a more rigorous investigation of the interaction of the enzyme with FMN was undertaken [10, 11].

The properties of the luciferase-oxidized flavin complex are of interest for several reasons. First, although the identity of the emitting species has not been established, it has been shown to involve an enzyme-bound flavin species in some state [12], possibly the product FMN itself in an environment sufficiently nonpolar to blue-shift its fluorescence emission from 525 nm to the 495 nm maximum characteristic of the bioluminescence reaction. Secondly, the amino acid sequence similarity between the α and β subunits [7], together with the suggestion that oxidation of two $FMNH_2$ molecules might be required for each enzyme turnover in the light-producing pathway [13–15] made the stoichiometry of binding of the product FMN particularly interesting. Finally, since the product might be expected to bind in the vicinity of the active center of the enzyme, bound FMN should provide an excellent probe of the active center environment.

Experimental procedure

Luciferase was purified from *Beneckea harveyi* [16] and from two mutants of this strain (AK-6 and AK-2H) [3] and enzyme concentrations were determined as previously described [11, 17]. Inactivations of luciferase by *N*-octylmaleimide (Pfaltz and Bauer) and by TPCK-trypsin (Worthington) were carried out in 0.02 M phosphate, unless otherwise noted, by previously published methods [5, 18]. All binding studies with FMN (Sigma) were done in 0.05 M Bis-Tris • HCl, pH 7.0. Equilibrium dialysis was performed as previously described [10] with 0.5 mM dithiothreitol added to the buffer.

Results and Discussion

We have previously demonstrated by two independent equilibrium methods, Scatchard analysis of equilibrium dialysis data [10] and circular dichroism data analyzed by the Job method [11], that there is a single binding site on bacterial luciferase for the product of the bioluminescence reaction, oxidized FMN. A single site has also been shown for the substrate, $FMNH_2$ [19].

Spectral characteristics of enzyme-bound FMN

The most striking feature of the optical absorption spectrum of FMN bound to bacterial luciferase is its highly resolved structure (Fig. 1a), with well defined maxima at 458, 434 and 366 nm, distinct shoulders at about

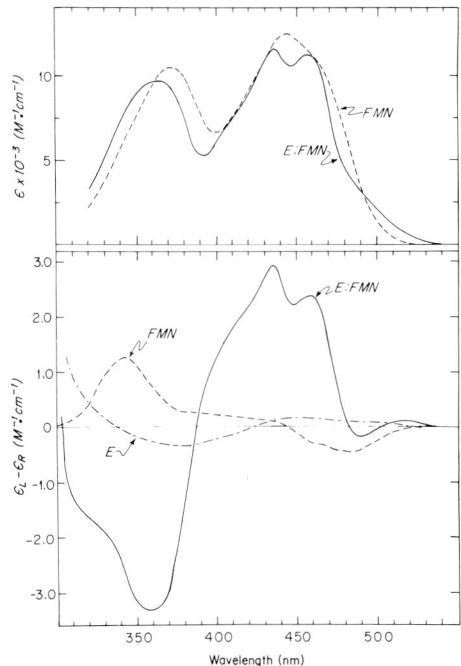

Fig. 1. (a) Optical absorption spectra of free (– – –) and luciferase-bound (——) FMN at 2 ± 2°C. (b) Circular dichroism spectra of free (– – –) and luciferase-bound (——) FMN at 6 ± 1°C. The luciferase-bound FMN spectrum was obtained by correcting for contributions of protein (—·—).

490, 410, and 355 nm, and minima at 446 and 392 nm. The spectrum of free FMN shows maxima at 444 and 373 nm with a shoulder at about 466 nm, and a minimum at 399 nm. Furthermore, binding of FMN to luciferase generates substantial optical activity not exhibited by the unbound flavin (Fig. 1b). The circular dichroism spectrum shows structure which correlates well with that observed in the optical absorbance spectrum.

This spectral detail reflects resolution of vibrational structure which is blurred in polar solvents but may be observed for free FMN in nonpolar solvents and at low temperatures [20–22]. Such spectral structure, which is consistent with the proposed hydrophobic nature of the active center of the enzyme [23], is not common among flavoproteins, and presumably reflects the special protein environment which permits generation of an excited flavin species in the light-producing reaction. Furthermore, the proposal that interaction of FMN with luciferase results in the generation of a new chromophore with the spectral characteristics of FMN protonated at the N-1 position [24, 25] is not supported by our spectral results; the flavin cation has no significant absorbance at 445 nm [26], whereas the enzyme · FMN complex absorbs maximally in this region.

Numerous luciferase mutants of *B. harveyi* having active center lesions (altered turnover rate, altered binding affinities for $FMNH_2$ and/or aldehyde) have been functionally characterized by Cline and Hastings [3]. Optical

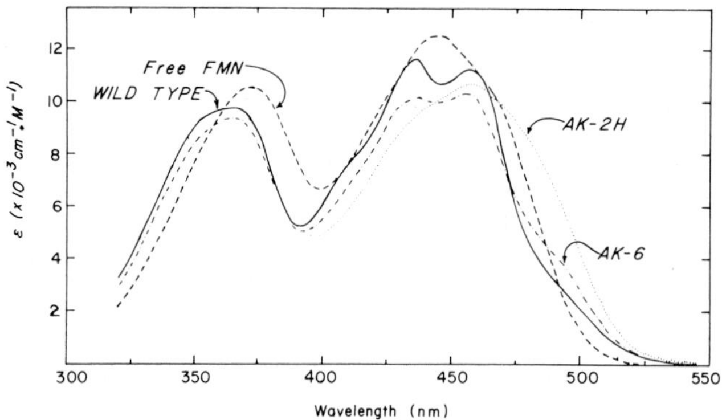

Fig. 2. Optical absorption spectra of wild type and mutant luciferase-bound FMN at 2 ± 2°C.

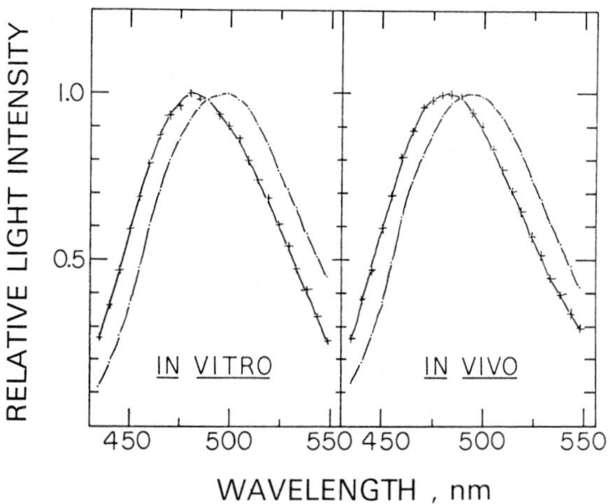

Fig. 3. Bioluminescence emission spectra comparing the wild type (— + —) with the mutant, AK-6 (— · —), uncorrected for the grating efficiency and phototube sensitivity. Upon applying these corrections, the wavelengths of peak emission were shifted 10 nm to the red for all curves to the following positions. In vitro: wild type, 492 nm; AK-6, 507 nm. In vivo: wild type, 492 nm; AK-6, 504 nm. (From Cline and Hastings, 1974) [27].

absorption spectra for FMN bound to two such mutant enzymes, AK-6 and AK-2H, both of which were originally selected for their altered turnover rates, are shown in Fig. 2. The spectrum of the AK-6 enzyme · FMN complex qualitatively resembles that of FMN bound to the wild type enzyme, while the spectrum of the AK-2H enzyme · FMN complex is quite different, being more like that of other flavoproteins. The lesion in AK-2H decreases the binding affinity for $FMNH_2$ about 4-fold and that for aldehyde about 60 fold [3]. The lesion in the AK-6 enzyme has resulted in a 500-fold decrease in $FMNH_2$ binding affinity, but unaltered aldehyde binding, and a 12 nm red shift in the bioluminescence emission spectrum, both in vivo and in vitro [27] (Fig. 3). In fact, the observation [27] that some mutant enzymes with altered affinities for reduced flavin have altered bioluminescence emission spectra provided the first direct evidence for the involvement of the enzyme with the excited state at the time of light emission.

Binding of FMN to N-octylmaleimide-modified luciferase

It has been shown that the binding of FMN to luciferase causes dramatic quenching of the flavin fluorescence emission [11], permitting a sensitive measurement of the extent of flavin binding during titration of a low concentration of FMN by increasing enzyme concentrations. The observed fluorescence at 525 nm, F_{obs}, is related to the dissociation constant for the enzyme-flavin complex (K_d), and f_{FMN} and $f_{E:FMN}$, the intrinsic fluorescence of free and bound FMN, respectively, by Eqn. 1

$$\frac{F_{FMN}}{F_{FMN} - F_{obs}} = \frac{f_{FMN}}{f_{FMN} - f_{E:FMN}} \frac{K_d}{[E]} + \frac{f_{FMN}}{f_{FMN} - f_{E:FMN}} \tag{1}$$

where F_{FMN} is the fluorescence of the flavin solution in the absence of enzyme. The total concentration of enzyme $[E_t]$ is much greater than that of FMN, so $[E] \simeq [E_t]$. Plots of $F_{FMN}/(F_{FMN} - F_{obs})$ vs. $1/[E_t]$ for native luciferase have been shown to be linear, with an ordinate intercept of 1.0, indicating that the intrinsic fluorescence of the enzyme-flavin complex, $f_{E:FMN}$, is zero ($<5\%$ that of free FMN) [11]. The observation that luciferase-bound FMN is nonfluorescent strongly implies that excited FMN is not the emitting species in bacterial bioluminescence.

Luciferase which has been 99.7% inactivated by reaction with 1.0 mol *N*-octylmaleimide per mol enzyme also quenches FMN fluorescence, with a K_d for the flavin · modified enzyme complex of 0.18 mM at 4.5°C (Fig. 4), only slightly greater than that of the native enzyme-flavin complex at this temperature ($K_d = 0.12$ mM) [11]. Alkylation of this reactive sulfhydryl group, which resides on the α subunit, results in at least 10-fold decrease in the $FMNH_2$ binding affinity; furthermore, binding of oxidized flavin protects the sulfhydryl from modification [5], although inorganic phosphate alone has no effect (Baldwin, T.O., unpublished). Since the modified enzyme still binds FMN nearly as tightly as does the native enzyme, it appears that while binding of FMN may block access to the sulfhydryl, there is no direct interaction between FMN and the thiol group.

Conformational changes upon binding of FMN or phosphate

Meighen and MacKenzie suggested that interaction of the phosphate group of the flavin with luciferase is very important [28]. They showed that a terminal negative charge (carboxylate or phosphate) which is at least 4 carbon atoms removed from the N-10 position of the isoalloxazine ring is required for tight binding of reduced flavins, and that the quantum yield of the bioluminescence reaction initiated by the injection of reduced riboflavin is greatly enhanced by inorganic phosphate. Furthermore, they found that high concentrations of inorganic phosphate appear to compete with $FMNH_2$ for binding to the enzyme, indicating the existence of a phosphate binding site on luciferase.

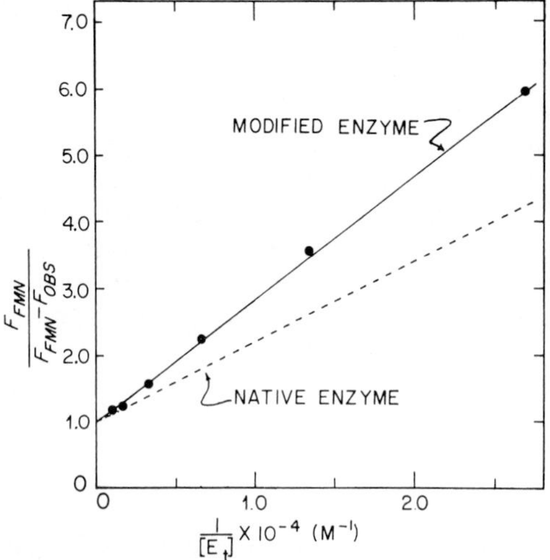

Fig. 4. Quenching of the fluorescence of 8.2 μM FMN as a function of luciferase concentration. Samples in 0.05 M Bis-Tris, pH 7.0, at 4.5°C, were excited at 450 nm and fluorescence (F_{obs}) was measured at 525 nm. K_d for the *N*-octylmaleimide-modified luciferase-FMN complex was calculated from the slope to be 0.18 mM. K_d for the native enzyme was similarly determined to be 0.12 mM at this temperature [11].

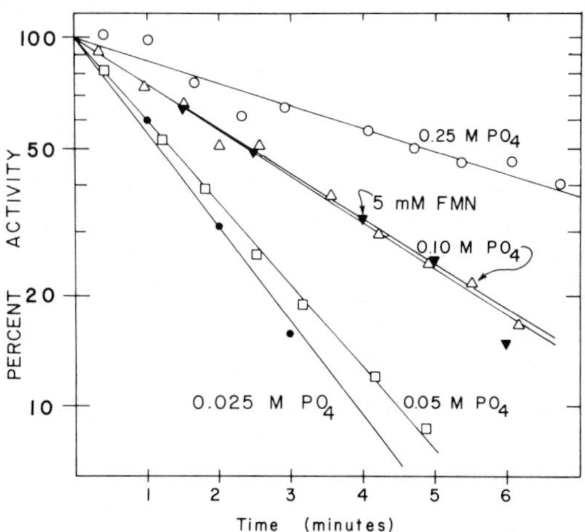

Fig. 5. FMN and phosphate protection of luciferase against proteolytic inactivation by trypsin. Trypsin (50 μg) was added at time zero to 1.0 ml of 5.6 μM luciferase containing 0.025 M phosphate + 5.0 mM FMN (▾), 0.025 M phosphate alone (•), 0.05 M phosphate (□), 0.10 M phosphate (△), or 0.25 M phosphate (○), all at 25°C, pH 7.0.

A relatively small region of the α subunit of luciferase is exceedingly labile to proteolytic digestion, as demonstrated by sodium dodecyl sulfate gel electrophoresis of the products of proteolysis with seven different proteases. In all experiments the apparent first-order rate of activity loss is paralleled by the rate of cleavage within a relatively confined region of the α subunit; β appears to be unaltered by the proteolytic treatment [4]. The binding of inorganic phosphate or FMN results in protection from proteolytic inactivation by trypsin (Fig. 5). Similar results have been obtained with six other proteases; no protection by FMN or phosphate was observed for hydrolysis of heat-denatured casein. In view of the diverse specificities of the different proteases, the protection of luciferase is unlikely to result from direct steric interference with the approach of the proteases to the susceptible peptide bonds; rather, it appears that binding of FMN or inorganic phosphate results in a change in the α subunit to a conformation which is less susceptible to proteolytic attack. The observations of Meighen and MacKenzie [28] imply that this conformational change is in the reduced flavin binding site and that the altered interactions of the amino acid side chains with the flavin result in a greater quantum yield for enzymatic oxidation of flavins with uncharged side chains, such as reduced riboflavin.

Acknowledgments

This work was supported in part by National Science Foundation Research Grant BMS74-23651 (to J.W.H.), by United States Public Health Service Postdoctoral Fellowships (to T.O.B. and J.E.B.), and by a Milton Fund Grant (to T.O.B.).

References

1 Meighen, E.A., Nicoli, M.Z. and Hastings, J.W. (1971) Biochemistry 10, 4062–4068
2 Meighen, E.A., Nicoli, M.Z. and Hastings, J.W. (1971) Biochemistry 10, 4069–4073
3 Cline, T.W. and Hastings, J.W. (1972) Biochemistry 11, 3359–3370
4 Baldwin, T.O. (1974) Fed. Proc. 33, 1441 Abstr.
5 Nicoli, M.Z., Meighen, E.A. and Hastings, J.W. (1974) J. Biol. Chem. 249, 2385–2392
6 Friedland, J.M. and Hastings, J.W. (1967) Proc. Natl. Acad. Sci. U.S. 58, 2336–2342
7 Baldwin, T.O., Nicoli, M.Z., Powers, D.A. and Hastings, J.W. (1975) Biophys. J. 15, 55a
8 Gibson, Q.H., Hastings, J.W., Weber, G., Duane, W. and Massa, J. (1966) in Flavins and Flavoproteins (Slater, E.C., ed.), pp. 341–359, Elsevier, Amsterdam
9 Meighen, E.A. and Hastings, J.W. (1971) J. Biol. Chem. 246, 7666–7674
10 Baldwin, T.O. (1974) Biochem. Biophys. Res. Commun. 57, 1000–1005
11 Baldwin, T.O., Nicoli, M.Z., Becvar, J.E. and Hastings, J.W. (1975) J. Biol. Chem. 250, 2763–2768
12 Mitchell, G. and Hastings, J.W. (1969) J. Biol. Chem. 244, 2572–2576
13 McElroy, W.D. and Green, A.A. (1955) Arch. Biochem. Biophys. 56, 240–255
14 Strehler, B.L. (1961) in Light and Life (McElroy, W.D. and Glass, B., eds.), p. 306, Johns Hopkins University Press, Baltimore
15 Lee, J. (1972) Biochemistry 11, 3350–3359
16 Reichelt, J.L. and Baumann, P. (1973) Arch. Mikrobiol. 94, 283–330
17 Gunsalus-Miguel, A., Meighen, E.A., Nicoli, M.Z., Nealson, K.H. and Hastings, J.W. (1972) J. Biol. Chem. 247, 398–404
18 Njus, D., Baldwin, T.O. and Hastings, J.W. (1974) Anal. Biochem. 61, 280–287
19 Becvar, J.E., Baldwin, T.O., Nicoli, M.Z. and Hastings, J.W. (1976), Flavins and Flavoproteins, Proceedings of the 5th International Symposium (Singer, T.P., ed.), Chapt. 9, pp. 94–100, Elsevier, Amsterdam
20 Palmer, G. and Massey, V. (1968) in Biological Oxidations (Singer, T.P., ed.), pp. 263–300, Interscience, New York
21 Koziol, J. (1969) Photochem. Photobiol. 9, 45–53
22 Sun, M., Moore, T.A. and Song, P.-S. (1972) J. Am. Chem. Soc. 94, 1730–1740
23 Nicoli, M.Z. and Hastings, J.W. (1974) J. Biol. Chem. 249, 2393–2396
24 Eley, M., Lee, J., Lhoste, J.-M., Lee, C.Y., Cormier, M.J. and Hemmerich, P. (1970) Biochemistry 9, 2902–2908
25 Lee, J. and Murphy, C.L. (1973) Biophys. Soc. Annu. Meet. Abstr. 13, 274a
26 Dudley, K.H., Ehrenberg, A., Hemmerich, P. and Müller, F. (1964) Helv. Chim. Acta 47, 1354–1383
27 Cline, T.W. and Hastings, J.W. (1974) J. Biol. Chem. 249, 4668–4669
28 Meighen, E.A. and MacKenzie, R.E. (1973) Biochemistry 12, 1482–1491

Flavins and Flavoproteins, edited by T.P. Singer

CHAPTER 9

THE FLAVIN STOICHIOMETRY OF THE BACTERIAL BIOLUMINESCENCE REACTION

JAMES E. BECVAR, THOMAS O. BALDWIN, MIRIAM Z. NICOLI and J. WOODLAND HASTINGS

The Biological Laboratories, Harvard University, Cambridge, Mass. 02138 (U.S.A.)

Summary

Recent reports have revived an hypothesis that the bacterial bioluminescence reaction involves two reduced flavins per enzyme turnover. A two-flavin mechanism requires that the two flavins bind simultaneously or sequentially to the same or different sites on luciferase during a catalytic cycle. In agreement with the kinetic analysis of Meighen and Hastings [4], we find by equilibrium techniques that the luciferase dimer has only a single $FMNH_2$ binding site. Quantum yield results demonstrate that one reduced flavin per luciferase is required for optimum bioluminescence and negate mechanisms involving either two reduced flavins or one reduced flavin plus one oxidized flavin per catalytic cycle. The fluorescence yield of $FMNH_2$ bound to enzyme indicates that this species is not the emitter in bacterial bioluminescence.

Introduction

Bacterial luciferase catalyzes the overall reaction:

$$FMNH_2 + RCHO + O_2 \xrightarrow{\text{luciferase}} FMN + RCOOH + H_2O + \approx 0.2\,h\nu_{495\text{ nm}}$$

Although this formulation now appears correct, it was earlier uncertain. For some years it was suggested that the aldehyde might have only a catalytic function and not serve as substrate in an energy-yielding step in the reaction [1]. If so, then aldehyde oxidation would not be available as an energy source, so the energy from the oxidation of a single reduced flavin (≈ 40 kcal) would be inadequate to populate an excited state at 495 nm (≈ 55 kcal), and the possible oxidation of two flavins was discussed [2]. When the $\alpha\beta$-heterodimeric structure of luciferase was elucidated [3], the participation of two reduced flavins per dimer seemed an interesting possibility. However,

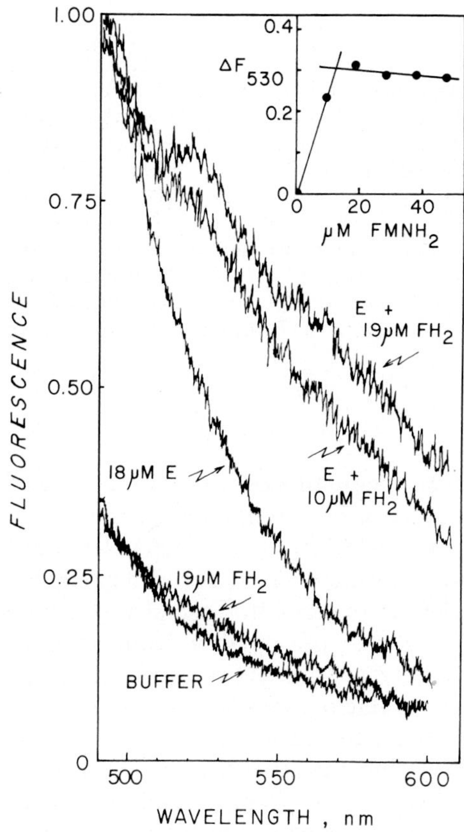

Fig. 1. Fluorescence emission spectra of the luciferase-$FMNH_2$ complex (E + FH_2), free enzyme (E), and free $FMNH_2$ (FH_2), in the presence of excess sodium dithionite, 0.1 M phosphate buffer and 0.2 mM dithiothreitol at 22°C, pH 7.0. Excitation was at 400 nm. Luciferase was from the aldehyde-requiring mutant M16. Inset. The fluorescence emission (F) at 530 nm of 18 mM luciferase upon titration with $FMNH_2$.

binding-site determinations by a kinetic method showed that luciferase possesses only a single $FMNH_2$ binding site per dimer [4]. Furthermore, chemical modification studies [5, 6] and mutant analyses [7] indicated that only the α subunit participates in the catalytic steps. A reaction mechanism involving the oxidation of a single $FMNH_2$ and the concomitant oxidation of aldehyde was thus proposed [4, 8] and has been supported by a variety of experimental results since then [9, 10, 11], including the demonstration that aldehyde is indeed oxidized to acid [12–15]. However, based on the observation that the quantum yield with respect to $FMNH_2$ is approximately half that for aldehyde, it has again been proposed that two reduced flavins are required per catalytic cycle [16, 17]. The experiments reported here confirm by means of equilibrium techniques the earlier observation by kinetic measurements [4] that luciferase has a single $FMNH_2$ binding site per $\alpha\beta$ dimer and furthermore show that only one $FMNH_2$ molecule is consumed per enzyme turnover in the bioluminescence reaction.

Materials and Methods

Bacterial luciferase was purified and its concentration determined as previously described [18] from *Beneckea harveyi* [19] and from dark mutants of this strain, designated as M16 and M18, which require exogenous aldehyde for bioluminescence. Luciferase activity was measured in a calibrated photometer [20], using either the standard assay or the dithionite assay [4]. The specific activities (decanal, 22°C) of the wild type, M16 and M18 preparations were 2.0, 2.3 and $3.1 \cdot 10^{14}$ quanta/s/mg respectively. FMN (Sigma) was used directly, or where stated, purified by chromatography on DEAE-cellulose. Comparable results were obtained with both preparations. Absorbance values were read with a Cary 15 spectrophotometer, CD spectra with a Jasco J-20 spectropolarimeter and fluorescence spectra with a Perkin-Elmer MPF-4 with corrected spectra accessory.

Results

Fig. 1 demonstrates the weak fluorescence emission of $FMNH_2$ bound to luciferase. The emission from the complex has an excitation maximum near 380 nm and a fluorescence quantum yield, $\phi_{E:FMNH_2}$, equal to about 0.01 ϕ_{FMN}, in reasonable agreement with published values for other reduced flavoenzymes [22]. The high gain needed to observe this signal reveals the tailing emission spectrum from enzyme alone, presumably from a minor impurity. Titration of enzyme with $FMNH_2$ (Fig. 1 inset) produces increasing fluorescence up to a plateau reached when flavin and enzyme are approximately equimolar. This, and the fact that subsequent addition of excess dithionite produces no change in intensity, indicate that the observed signal results from an $FMNH_2$-luciferase complex and not from a small amount of free FMN. Although these data show that one site exists on luciferase which enhances the fluorescence yield of $FMNH_2$ when bound, they do not exclude the possibility of additional sites with lower fluorescence yield.

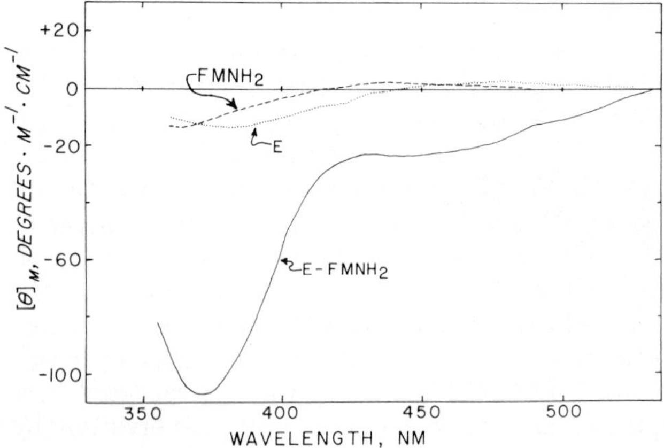

Fig. 2. Circular dichroism spectra of $FMNH_2$ (— — —), wild type luciferase (· · · · ·) and the luciferase-$FMNH_2$ complex (——) at 6 ± 1°C, pH 7.0, in the presence of excess sodium dithionite, 0.1 M bis-Tris, 0.1 M NaCl and 0.2 mM dithiothreitol, pH 7.0.

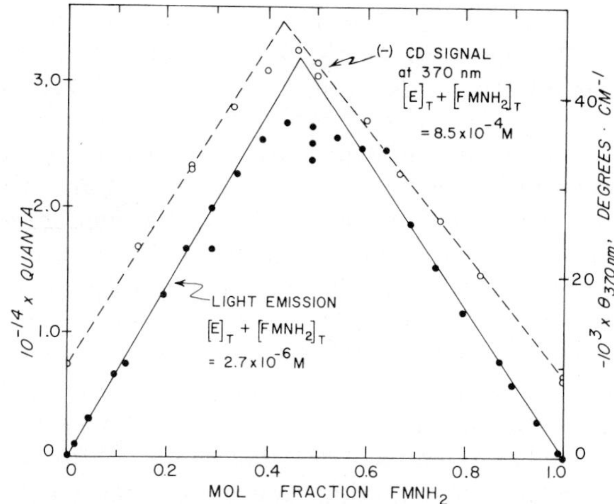

Fig. 3. Analysis of the $FMNH_2$: luciferase stoichiometry by a Job plot of CD and light-emission measurements. Wild type luciferase and FMN solutions, each at 0.85 mM, were mixed in various proportions in the presence of excess dithionite. The negative 370 nm CD signal (○, — — —, right ordinate) which results from the mixtures at 6 ± 1°C is plotted as a function of the mol fraction of $FMNH_2$ in the mixture. The light-emission data (●, ———, left ordinate) was obtained by mixing mutant (M18) luciferase and purified FMN solutions in various proportions in 50 mM phosphate buffer, pH 7.0, 0.2% bovine serum albumin and a slight excess of dithionite in a volume of 1.2 ml. The luminescent reaction was then initiated by injecting 1.0 ml of an aerated, sonicated suspension of decanal (0.5 mM) giving a final total concentration of flavin plus luciferase of 2.7 μM. Total quanta were calculated as I_0/k, where I_0 is the peak intensity and k the decay constant of light emission.

The unconditional demonstration that the luciferase molecule has only one binding site for $FMNH_2$ with an affinity relevant to activity is provided by CD studies of bound $FMNH_2$. Fig. 2 gives molar ellipticity spectra for free $FMNH_2$, free luciferase in the presence of dithionite and an equimolar mixture of $FMNH_2$ and luciferase. The relatively large signal at 370 nm for the mixture is indicative of a complex between reduced flavin and luciferase. The stoichiometry of this complex was examined by the method of continuous variation of mol fraction according to Job [23]. When equimolar solutions of $FMNH_2$ and luciferase are mixed in different proportions, the CD signal at 370 nm varies as a function of the relative mol fraction of the reactants (Fig. 3). Maximum signal develops for a mol fraction of flavin (and enzyme) near 0.5, indicating that luciferase has one binding site for $FMNH_2$. Although these data are not ideal for the calculation of a K_d for the E:$FMNH_2$ complex, they are consistent with the published value of 0.8 μM [4].

If luciferase had two equal and independent binding sites for $FMNH_2$, the maximum CD signal would develop for an $FMNH_2$ mol fraction of 0.67. Even if a second binding site were different in character, for example producing no ellipticity at 370 nm for bound $FMNH_2$ or ellipticity of a different sign or magnitude, the shape or apex position of the observed Job plot would be altered. The data thus show that there is no second binding site for $FMNH_2$ with a K_d less than about 1 mM.

Equilibrium considerations do not rule out the possibility that sequential binding of two or more $FMNH_2$ molecules to luciferase occurs during a catalytic cycle. This was examined by experiments in which the bioluminescence from enzyme-$FMNH_2$ mixtures upon addition of excess O_2 and decanal was measured as a function of mol fraction of $FMNH_2$ (Fig. 3). The data extrapolate to a maximum in light emission at mol fraction near 0.5, where $FMNH_2$ concentration equals enzyme concentration. We have obtained similar results using a variety of bacterial luciferase preparations, different preparations of FMN, and under different buffer and temperature conditions. Thus, only one reduced flavin molecule per enzyme molecule is required for light emission. If two reduced flavins per enzyme were required for bioluminescence, or if one of the two flavins consumed on a given enzyme molecule during light emission were oxidized via a totally "dark" pathway, then maximal light emission would occur at a flavin mol fraction of 0.67. Moreover, maximal light emission at a flavin mol fraction of 0.67 would also result from a mechanism requiring luciferase to react with one $FMNH_2$ molecule plus one FMN molecule during turnover. At this mol fraction, one of every two reduced flavins will bind instantaneously with luciferase (with a rate constant of $>10^8$ $M^{-1} \cdot s^{-1}$ [24]). After addition of excess oxygen, the second $FMNH_2$, which remains free, will autoxidize, in the presence of dithionite, with a halftime less than 50 ms [25] and thus be available as FMN before the development of bioluminescence (peak emission at about 1 s). Maximal bioluminescence occurs at a flavin mol fraction of 0.5, and thus both these two flavin mechanisms can be excluded.

The light-emission data in Fig. 3 can also be considered in terms of the bioluminescence quantum yield, calculated with respect to the limiting reagent, either $FMNH_2$ or luciferase (Fig. 4). At the concentrations used, and based on the known values of K_d, nearly all (>90%) luciferase molecules were complexed with $FMNH_2$ in the region of excess reduced flavin (right side of figure), while nearly all $FMNH_2$ molecules were bound

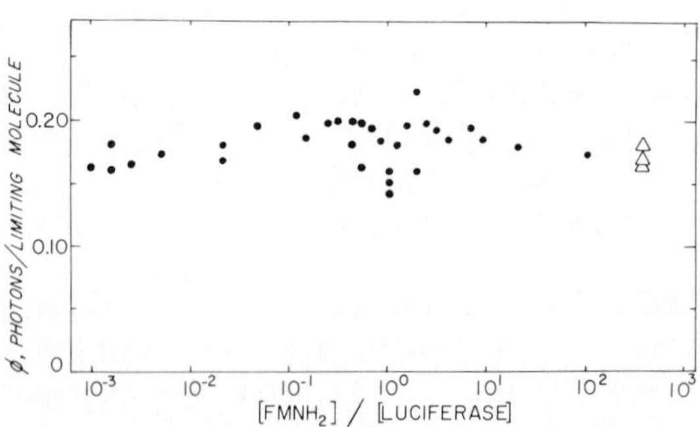

Fig. 4. Bioluminescence quantum yield of the bacterial luciferase reaction calculated per $FMNH_2$ molecule (with excess luciferase; left half) and per luciferase molecule (with excess $FMNH_2$; right half). The sum of the concentrations of luciferase and purified $FMNH_2$ is constant (2.7 μM, ●). This plot includes data points (△) obtained by the standard assay (rapid injection of 50 μM catalytically reduced $FMNH_2$). In agreement with an earlier study [4], these results demonstrate that neither the presence of dithionite nor the dithionite assay method affects adversely the quantum-yield measurements.

to enzyme in the region of luciferase excess (left side of figure). At these very low $FMNH_2$ and high enzyme concentrations, the probability that a single luciferase molecule could capture two reduced flavins becomes very small. If two reduced flavins were required for light emission, many unproductive complexes should be formed, and the bioluminescence quantum yield per $FMNH_2$ should decrease dramatically at lower $FMNH_2$ concentration. Yet the quantum yield with respect to the limiting reagent ($FMNH_2$ for $[F]/[E] < 1$, luciferase for $[F]/[E] > 1$) is relatively invariant over many orders of magnitude in $[F]/[E]$ and is the same whether enzyme or flavin is in excess. Thus, only one $FMNH_2$ is required for bioluminescence.

Discussion

Previous evidence has indicated that the emitter in bacterial bioluminescence involves a flavin species [26]. Actually, the emitting species in bioluminescence need not be at the oxidation level of FMN since the oxidation of aldehyde to acid is a sufficiently exothermic reaction to yield a photon of 495 nm without the prior oxidation of $FMNH_2$. One might thus envision the population of an excited enzyme-bound, substituted, reduced flavin species, which would fall to its ground state during emission and subsequently be oxidized to FMN and released from luciferase. The weak fluorescence of the E:$FMNH_2$ complex indicates strongly that bound, reduced flavin itself is not the emitting species in bacterial bioluminescence.

Although conjectures concerning the functions of the two, possibly homologous [27], subunits and two potential flavin binding sites are interesting, we measure only one $FMNH_2$ binding site per $\alpha\beta$ dimer, in agreement with the work of Meighen and Hastings [4]. Actually, their results were based on kinetic analyses which required making certain assumptions regarding the relative rates of forward and backward reactions. Moreover, their method would have been unable to detect additional binding sites for reduced flavin, with affinity one tenth or less the affinity for the first flavin, unless the second site was absolutely required for bioluminescence. Our conclusions about the stoichiometry of binding sites, however, are based on equilibrium studies and are independent of assumptions regarding rate constants. The concentrations used in these studies were sufficiently high to detect an interaction having a $K_d \leqslant 1$ mM, four orders of magnitude weaker than the affinity for the $FMNH_2$ involved in the bioluminescence reaction. The finding of a single $FMNH_2$ binding site is consistent with the demonstration that luciferase has one FMN binding site [18, 28].

The possibility of a reaction mechanism involving the sequential addition of two $FMNH_2$ molecules or one $FMNH_2$ molecule plus one FMN molecule during a single enzyme turnover has also been eliminated by our quantum-yield measurements. The maximum light yield results when reduced flavin concentration is equal to luciferase concentration. Furthermore, the quantum yield with respect to the limiting reagent does not vary over almost six orders of magnitude, showing that luciferase requires and consumes only one $FMNH_2$ per enzyme turnover.

Acknowledgements

We are grateful to Dr. Shiao-chun Tu for helpful discussions. This work was supported in part by National Science Foundation Research Grant BMS74-23651 (to J.W.H.) and by United States Public Health Service Postdoctoral Fellowships (to T.O.B. and J.E.B.).

References

1 Strehler, B.L. (1955) in The Luminescence of Biological Systems (Johnson, F.H., ed.), pp. 209–244, A.A.A.S. Press, Washington, D.C.
2 Strehler, B.L. (1961) in Light and Life (McElroy, W.D. and Glass, B., eds.), pp. 306–308, Johns Hopkins University Press, Baltimore
3 Friedland, J.M. and Hastings, J.W. (1967) Biochemistry 6, 2893–2900
4 Meighen, E.A. and Hastings, J.W. (1971) J. Biol. Chem. 246, 7666–7674
5 Meighen, E.A., Nicoli, M.Z. and Hastings, J.W. (1971) Biochemistry 10, 4062–4068
6 Meighen, E.A., Nicoli, M.Z. and Hastings, J.W. (1971) Biochemistry 10, 4069–4073
7 Cline, T.W. and Hastings, J.W. (1972) Biochemistry 11, 3359–3370
8 Eberhard, A. and Hastings, J.W. (1972) Biochem. Biophys. Res. Commun. 47, 348–353
9 Hastings, J.W., Balny, C., LePeuch, C. and Douzou, P. (1973) Proc Natl. Acad. Sci. U.S. 70, 3468–3472
10 Becvar, J.E., Baldwin, T.O. and Hastings, J.W. (1974) Fed. Proc. 33, 1589 Abstr.
11 Watanabe, T., Tomita, G. and Nakamura, T. (1974) J. Biochem. Tokyo 75, 1249–1255
12 Shimomura, O., Johnson, F.H. and Kohama, Y. (1972) Proc. Natl. Acad. Sci. U.S. 69, 2086–2089
13 McCapra, F. and Hysert, D.W. (1973) Biochem. Biophys. Res. Commun. 52, 298–304
14 Dunn, D.K., Michaliszyn, G.A., Bogacki, I.G. and Meighen, E.A. (1973) Biochemistry 12, 4911–4918
15 Vigny, A. and Michelson, A.M. (1974) Biochimie 56, 171–176
16 Lee, J. (1972) Biochemistry 11, 3350–3359
17 Lee, J. and Murphy, C.L. (1973) in Chemiluminescence and Bioluminescence (Cormier, M.J., Hercules, D.M. and Lee, J., eds.), pp. 381–386, Plenum Press, New York
18 Baldwin, T.O., Nicoli, M.Z., Becvar, J.E. and Hastings, J.W. (1975) J. Biol. Chem. 250, 2763–2768
19 Reichelt, J.L. and Baumann, P. (1973) Arch. Mikrobiol. 94, 283–330
20 Mitchell, G. and Hastings, J.W. (1971) Anal. Biochem. 39, 243–250
21 Hastings, J.W., Spudich, J.A. and Malnic, G. (1963) J. Biol. Chem. 238, 3100–3105
22 Ghisla, S., Massey, V., Lhoste, J.-M. and Mayhew, S.G. (1974) Biochemistry 13, 589–597
23 Job, P. (1928) Ann. Chimie Paris 10e, 113–134
24 Hastings, J.W. and Gibson, Q.H. (1963) J. Biol. Chem. 238, 2537–2554
25 Gibson, Q.H. and Hastings, J.W. (1962) Biochem. J. 83, 368–377
26 Mitchell, G. and Hastings, J.W. (1969) J. Biol. Chem. 244, 2572–2576
27 Baldwin, T.O., Nicoli, M.Z., Powers, D.A. and Hastings, J.W. (1975) Biophys. J. 15, 55a
28 Baldwin, T.O. (1974) Biochem. Biophys. Res. Commun. 57, 1000–1005

Flavins and Flavoproteins, edited by T.P. Singer

CHAPTER 10

THE FLAVIN CHROMOPHORE OF PHOTOEXCITABLE BACTERIAL LUCIFERASE

SHIAO-CHUN TU and J. WOODLAND HASTINGS

The Biological Laboratories, Harvard University, Cambridge, Mass. 02138 (U.S.A.)

Summary

A prosthetic group, designated B, isolated from bacterial photoexcitable luciferase, possesses spectral and photochemical properties characteristic of substituted reduced flavins. Luciferase-bound B has a fluorescence excitation maximum at 375 nm, correlating well with the absorption spectrum, and fluorescence emission peaking at 495 nm. However, free B is nonfluorescent in aqueous solution at ambient temperature. The addition of free B to pure luciferase generates a reconstituted photoexcitable luciferase whose properties are the same as those of native photoexcitable luciferase. In the luciferase reaction itself, initiated by $FMNH_2$, B is an inhibitor competitive with $FMNH_2$. Irradiation of photoexcitable luciferase converts B to FMN. The latter has been identified spectrally, enzymatically, and chromatograpically. These findings lead to the suggestion that B is a 4a-substituted reduced FMN. The luciferase-bound B resembles but is not identical with the normal flavin intermediate obtainable by reaction of luciferase with $FMNH_2$ and oxygen. It is hypothesized that B is a false intermediate of the bacterial bioluminescence reaction, and a mechanism for the photoexcited bioluminescence reaction is suggested.

Introduction

An interesting feacture of bacterial bioluminescence is the existence of two types of in vitro luminescence activity attributable to two distinct protein species. One is luciferase whose activity is initiated by adding $FMNH_2$ in the presence of aldehyde and oxygen. The second is photoexcitable luciferase; its light emission can be induced by flash irradiation with no dependence upon the presence of either FMN or $FMNH_2$ [1, 2]. Both reactions are oxygen- and aldehyde-dependent and utilize these substrates in the same sequence. Both have the same rates for light decay when tested with aldehydes of different chain lengths and emit at the same wavelengths.

Photoexcitable luciferase has been shown to be a luciferase molecule modified by a non-covalently bound flavin chromophore designated B. Reconstituted photoexcitable luciferase can be prepared by combining chromophore-free luciferase with the isolated B. In this work the properties of photoexcitable luciferase and reconstituted photoexcitable luciferase are shown to be similar, and the isolation and spectral characterization of B are described. The chemical nature of B closely resembles substituted reduced flavins, and a reaction mechanism for photoexcited bioluminescence is proposed.

Experimental Procedures

The bacterium *Beneckea harveyi* [3] was used for the preparation of both luciferase and photoexcitable luciferase [4, 5, 6]. The purified luciferase was at least 95% pure while the photoexcitable luciferase obtained was 30 to 50% pure, containing 10 to 15% of luciferase. Since activities of photoexcitable luciferase and luciferase are initiated by different means, the presence of one does not interfere with the assay of the other. Luciferase was also purified from an aldehyde-requiring mutant, which is dark in vivo due to the lack of endogenous aldehyde factor or factors [7]. This mutant does not contain any photoexcitable luciferase, and the luciferase isolated from the mutant was used to prepare the reconstituted photoexcitable luciferase samples. All photoexcitable luciferase and reconstituted photoexcitable luciferase samples were dialyzed to remove oxidized flavin contamination [8].

The initial maximum intensity, expressed in quanta $\cdot$ s^{-1}, of in vitro luminescence was measured at 23°C with a photometer [9] calibrated with a liquid light standard [10]. The luciferase activity was measured either by the standard assay [4] or by the dithionite assay [11], both using decanal as a cosubstrate. The purified luciferase had a specific activity of $1.4 \cdot 10^{14} q \cdot s^{-1} \cdot mg^{-1}$ as determined by the standard assay with decanal at 23°C. The photoexcitable luciferase activity was measured by the previously described flash assay [2], also using decanal as a cosubstrate. In this flash assay, the partially purified photoexcitable luciferase sample had a specific activity of $1.4 \cdot 10^{11} q \cdot s^{-1} \cdot mg^{-1}$ at 23°C.

Absorption spectra were measured with a Cary 15 and circular dichroism with a Jasco (J-20). Fluorescence was measured with an Aminco-Bowman (1200 lines/mm gratings; slit arrangement 2, plotted uncorrected).

Results

B was isolated by denaturing photoexcitable luciferase in urea followed by gel filtration to separate it from protein, and chromatography on florisil to resolve the flavins (Fig. 1). Peak I was shown to be FMN by its fluorescence excitation and emission spectra, by thin-layer chromatography, and by its activity with bacterial luciferase. This flavin was partially resolved from II and well separated from the major B activity (III). Fractions in III were pooled and used as B stock solution. In ascending thin-layer chromatography in 1-butanol: acetic acid: water (2:1:1), B moves as a single nonfluorescent but iodine-stainable spot above a weakly fluorescent spot identified as a trace amount of FMN.

Upon titration of luciferase with B, increasing amounts of photoexcited bioluminescence activity are generated, with a concomitant loss of luciferase

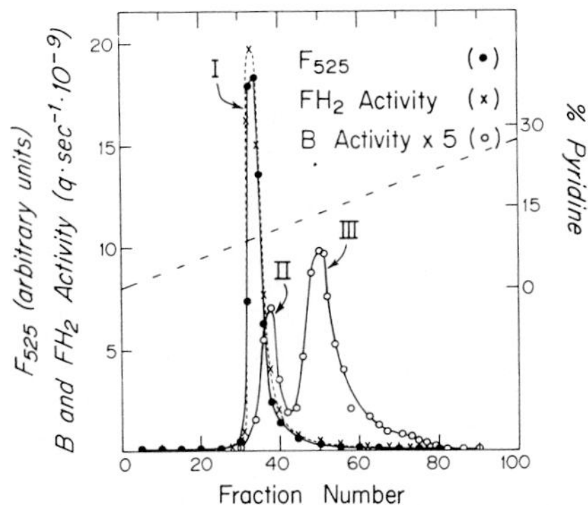

Fig. 1. Isolation of B by florisil chromatography at 23° C. Approximately 50 ml of a solution 5 M in urea, containing B activity, obtained by gel filtration in urea of 500 mg of photoexcitable luciferase were applied to a florisil column (1 × 17 cm) preequilibrated with H_2O. After washing with 50 ml H_2O a gradient of pyridine in H_2O served to elute the B. Fluorescence (•) at 525 nm was excited at 380 nm. B (○) and reduced flavin (FH_2, ×) activities were measured at 23° C by flash assay and dithionite assay, respectively, after adding 20-μl aliquots to 30 μg of luciferase in 2 ml of 0.02 M phosphate (pH 7) and 1 ml of the same buffer containing 25 mM 2-mercaptoethanol, respectively.

activity. B is competitive with $FMNH_2$ in the luciferase reaction, but decanal does not reverse or otherwise affect the inhibition by B. For the reconstituted photoexcitable luciferase samples, the degree of saturation of luciferase with B is defined as the percent inhibition of the initial luciferase activity.

The properties of photoexcitable luciferase were compared with those of luciferase and reconstituted photoexcitable luciferase. Both photoexcitable luciferase and luciferase were shown by disc gel electrophoresis [12, 13] to have the same molecular weight and similar charges at pH 8. Both can be inactivated, with comparable efficiencies, by antibodies against either pure luciferase or individual subunits thereof. Luciferase has an optimal activity near pH 6 in 0.05 M phosphate buffer, while photoexcitable luciferase shows an optimal activity plateau over the range from pH 6 to 7.5. At pH 4.5, luciferase was more labile than photoexcitable luciferase, while at pH 8.5 photoexcitable luciferase was inactivated much faster than luciferase. In the presence of 3 to 5 M urea, luciferase was inactivated more rapidly than was photoexcitable luciferase at 23°C. Thermal inactivation experiments show that photoexcitable luciferase is more stable than luciferase over the range of 42.5–55°C [6]. Bacterial luciferase is rapidly inactivated by proteases, such as trypsin [14, 15]. Trypsin also inactivates photoexcitable luciferase but less rapidly than it inactivates luciferase. The properties characteristic of photoexcitable luciferase are restored in reconstituted photoexcitable luciferase. Moreover, photoexcitable luciferase from mutants selected to have heat labile luciferase are also thermally unstable [6, 7]. It is thus concluded that photoexcitable luciferase actually consists of a luciferase:B complex which can be conformationally distinct from luciferase.

It was necessary to exclude the highly fluorescent FMN in order to obtain accurate spectral measurements of B. Spectral measurements of

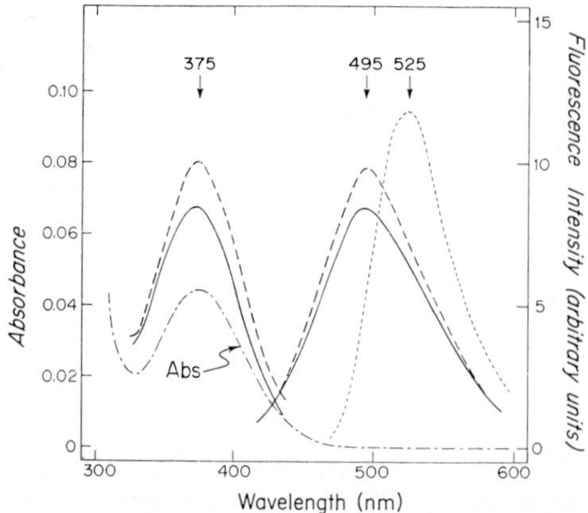

Fig. 2. Absorption spectrum of luciferase-bound B (—·—) and fluorescence spectra of photoexcitable and reconstituted photoexcitable luciferase. Absorption spectrum of the luciferase-bound B was taken with the reconstituted photoexcitable luciferase sample (26% saturated, 1.1 mg protein/ml) using the same concentration of luciferase in the reference cell. Excitation and emission spectra of reconstituted photoexcitable luciferase (——) and photoexcitable luciferase (1 mg protein/ml) (- - -) were measured with fluorescence monitored at 495 nm for the former measurements and using 375-nm excitation light for the latter. Fluorescence emission spectrum of FMN (· · ·) in the same buffer is included for comparison.

luciferase-bound B were obtained with dialyzed reconstituted photoexcitable luciferase or photoexcitable luciferase while those of free B were determined immediately after the addition of 5 M urea to exhaustively dialyzed reconstituted photoexcitable luciferase samples. Photoexcitable luciferase and reconstituted photoexcitable luciferase have the same fluorescence excitation and emission spectra, with λ_{max} at 375 nm and 495 nm respectively (Fig. 2). Free B (III in Fig. 1) is nonfluorescent at room temperature; this has also been found with B obtained from reconstituted photoexcitable luciferase treated with 5 M urea. By contrast, FMN is highly fluorescent in aqueous solution (λ_{max} at 525 nm) but is nonfluorescent when bound to luciferase [5]. The CD spectrum of the reconstituted photoexcitable luciferase sample was found to resemble closely that of the luciferase-bound $FMNH_2$ with a negative peak near 371 nm, quite different from that of luciferase-bound FMN [5, 16].

To examine the consequences of the photoexcitation of luciferase-bound B, a fresh dialyzed reconstituted photoexcitable luciferase sample was subjected to repeated flash irradiation in the absence of aldehyde. With flash irradiation, there was a displacement of the fluorescence emission maximum from 495 to 525 nm and an increase in intensity (Fig. 3). These changes are attributed to a photochemical modification of B. Successive flashes result in decreases in the remaining B activity in the reconstituted photoexcitable luciferase sample accompanied by parallel increases in fluorescence at 525 nm (Fig. 4).

FMN was found to be present in a flash-treated sample (Fig. 3), but not in a control, determined by thin-layer chromatography in two solvent systems, and by assays with three FMN-dependent enzymes: flavin reductase

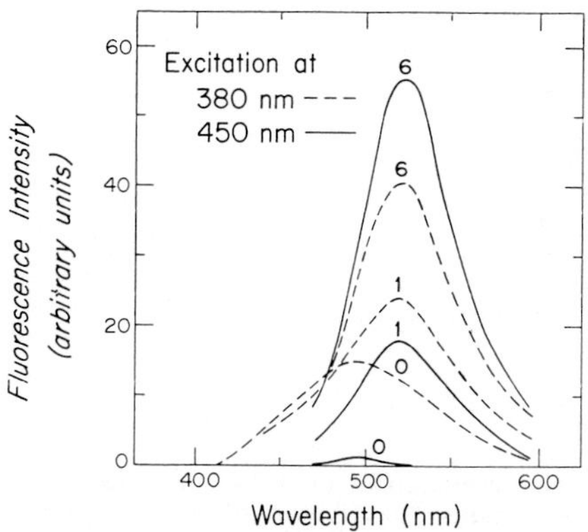

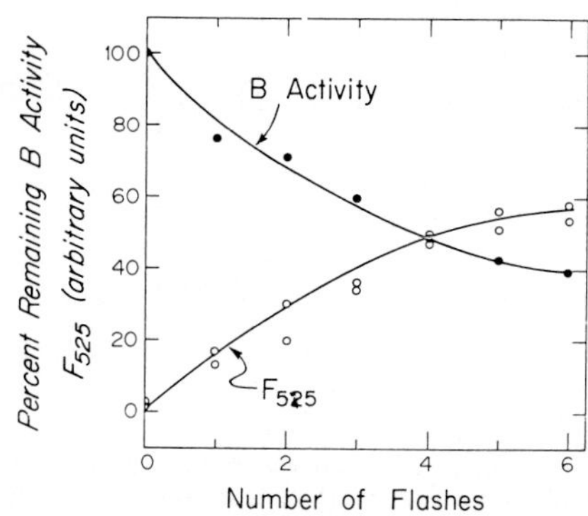

Fig. 3. Fluorescence emission spectra of reconstituted photoexcitable luciferase before and after flash irradiations. 2 ml of 0.02 M phosphate solution (pH 7) containing 0.55 mg reconstituted photoexcitable luciferase (99% saturated) was flash irradiated repeatedly. Emission spectra were determined after each flash using excitation light at either 380 nm (---) or 450 nm (——). For simplicity, only the spectra of samples irradiated once and 6 times are shown and compared with those of the starting reconstituted photoexcitable luciferase sample. The number of flashes is indicated above each spectrum.

Fig. 4. Correlation of the decrease in B activity and increase in fluorescence at 525 nm of a reconstituted photoexcitable luciferase sample subjected to repeated flashing. The reconstituted photoexcitable luciferase sample was the same as that described in Fig. 3. Fluorescence intensities at 525 nm were measured using 450 nm excitation light. For the determination of the remaining quantity of B after each consecutive flash, 10 μl of sample solution were added to 1.2 ml 0.02 M phosphate, pH 7, containing 0.15 mg of photoexcitable luciferase-free luciferase, and incubated at 23°C for 3 min. The photoexcited bioluminescence activity of each sample, which reflects the remaining quantity of B, was subsequently assayed.

[17], pyridoxine 5′-phosphate apooxidase [18], and bacterial luciferase [11, 19]. After correcting for enzyme purity and percent conversion, a ratio of 1 FMN per 1.3 luciferases was estimated, not too far, perhaps, from the expected 1 to 1 ratio.

Since luciferase with B bound is inactive in the $FMNH_2$ initiated bioluminescence reaction [2], it might be expected that the photochemical reaction in which B is converted to FMN would involve the concomitant release of free luciferase. However, no increase in luciferase activity was observed upon flash irradiation. In fact, this flash irradiation was found to inactivate free luciferase, and thus might similarly destroy the luciferase component of reconstituted photoexcitable luciferase.

Discussion

In the proposed mechanism for photoexcitable bioluminescence (Fig. 5), B is postulated to be a substituted reduced FMN (4a–R–FMN) bound to luciferase in the $FMNH_2$ site, and incapable of reacting further via the normal pathway. Photochemically, however, chromophore B can be converted to $FMNH_2$ so that the reaction can proceed via the conventional pathway and result in bioluminescence. This is consistent with the fact that light-induced bioluminescence is similar to $FMNH_2$-initiated bioluminescence in all respects except for the requirement for flavin and the mode of initiation [1], and the fact that oxygen is not required during photoexcitation. This

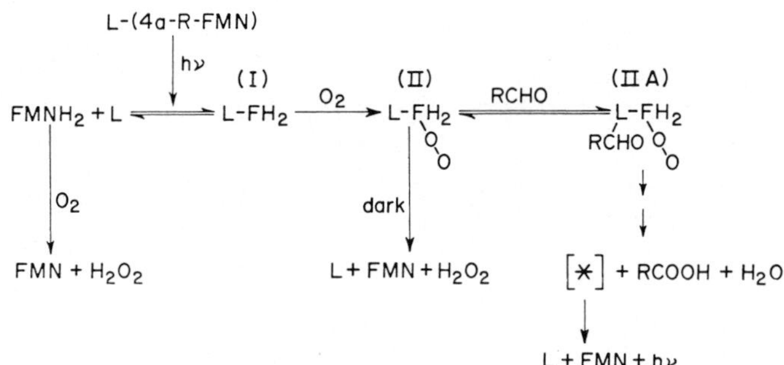

Fig. 5. Hypothetical scheme depicting the pathways and intermediates in the luciferase-catalyzed ($FMNH_2$-initiated) and the photoexcited bioluminescence reactions. The structure of the chromophore B associated with the photoexcitable luciferase is proposed to be as 4a—R—FMN.

must be added immediately afterwards, within a half-time of less than one second [20]. Since the concentration of the photochemically generated intermediate (I) in the experiments of Hastings and Gibson [20] was well below the K_D for the binding of $FMNH_2$ to luciferase (approx. $2 \cdot 10^{-7}$M), the half-time of about one second may be a measure of the off constant for the dissociation of $FMNH_2$ from luciferase. Based on a second-order rate constant of $10^7 M^{-1} \cdot sec^{-1}$ [21] and a K_D of $2 \cdot 10^{-7}$M [22], the expected first-order rate constant for the dissociation of the $FMNH_2 \cdot$ luciferase complex would be 2 sec^{-1}, in reasonable agreement with the value of 0.9 sec^{-1} in the experiment of Hastings and Gibson [20].

The spectral and photochemical properties of the chromophore B resemble those of reduced flavins, and those of 4a-substituted reduced FMN in particular, as follows: (1) Resemblance between the absorption of luciferase-bound B and 4a-substituted reduced flavins [23, 24]. (2) As with 4a- and/or 5-substituted reduced flavins [24], free B does not fluoresce in aqueous solution at room temperature but is fluorescent in enzyme-bound form. (3) Oxidized flavins can be formed from 5-substituted reduced flavins by autooxidation under acidic conditions, and from 4a-substituted reduced flavins by illumination in the presence of oxygen [25]. With luciferase-bound B we have demonstrated that FMN is formed by aerobic flash irradiation.

Although the properties of B appear to resemble those of 4a-substituted reduced FMN, its exact molecular structure cannot yet be specified. The luciferase intermediate II (Fig. 5) recently isolated by low temperature gel filtration (−20 to −30°C) has been postulated to be the 4a-peroxy FMN anion bound to luciferase [26]. But II cannot be identical with photoexcitable luciferase or reconstituted photoexcitable luciferase, since II rapidly breaks down in the dark at 20°C to give FMN, H_2O_2 and free luciferase, or, with aldehyde present, to give bioluminescence, FMN, free luciferase, but no H_2O_2 [27].

Aldehyde-requiring mutants contain normal amounts of luciferase but no photoexcitable luciferase. They are nonluminescent in vivo for want of the "natural" aldehyde [28]. However, photoexcitable luciferase activity was found in the mutant if the in vivo bioluminescence was stimulated by the additon of aldehyde to the culture during growth, indicating that B is not explicable as an artifact of the isolation procedures or of other flavoprotein systems. Instead, B can be considered as a false intermediate of the bacterial bioluminescence reaction, arising in vivo either as a by-product of the luciferase-catalyzed luminescence reaction or as a flavin adduct generated by a unique photochemical addition reaction using the bacterial bioluminescence as the light source.

Acknowledgments

This research was supported in part by National Science Foundation Research Grant GB 31977X. We are grateful to Dr. C.A. Waters for the preparation of antiluciferases. We also thank Drs. T.O. Baldwin, J.E. Becvar, S. Ghisla, V.D. Gooch, and Patricia Dobson for their help in numerous ways.

References

1 Gibson, Q.H., Hastings, J.W. and Greenwood, C. (1965) Proc. Natl. Acad. Sci. U.S. 53, 187–195
2 Mitchell, G.W. and Hastings, J.W. (1970) Biochemistry 9, 2699–2707
3 Reichelt, J.L. and Baumann, P. (1973) Arch. Mikrobiol. 94, 283–330
4 Gunsalus-Miguel, A., Meighen, E.A., Nicoli, M.Z., Nealson, K.H. and Hastings, J.W. (1972) J. Biol. Chem. 247, 398–404
5 Baldwin, T.O., Nicoli, M.Z., Becvar, J.E. and Hastings, J.W. (1975) J. Biol. Chem., 250, 2763–2768
6 Tu, S.-C., Waters, C.A. and Hastings, J.W. (1975) Biochemistry 14, 1970–1974
7 Cline, T.W. (1973) Ph.D. Thesis, Harvard University, Cambridge, Mass.
8 Tu, S.-C. and Hastings, J.W. (1975) Biochemistry 14, 1975–1980
9 Mitchell, G.W. and Hastings, J.W. (1971) Anal. Biochem. 39, 243–250
10 Hastings, J.W. and Weber, G. (1963) J. Opt. Soc. Am. 53, 1410–1415
11 Meighen, E.A. and MacKenzie, R.E. (1973) Biochemistry 12, 1482–1491
12 Gabriel, O. (1971) Methods Enzymol. 22, 565–578
13 Hedrick, J.L. and Smith, A.J. (1968) Arch. Biochem. Biophys. 126, 155–164
14 Njus, D., Baldwin, T.O. and Hastings, J.W. (1974) Anal. Biochem. 61, 280–287
15 Baldwin, T.O. (1974) Fed. Proc. 33, 1441
16 Becvar, J.E., Baldwin, T.O., Nicoli, M.Z. and Hastings, J.W. (1976) in Flavins and Flavoproteins, Proceedings of the 5th International Symposium (Singer, T.P., ed.), Chapt. 9, pp. 94–100, Elsevier, Amsterdam
17 Duane, W. and Hastings, J.W. (1975) Mol. Cell. Biochem. 6, 53–64
18 Kazarinoff, M.N., and McCormick, D.B. (1974) Biochim. Biophys. Acta 359, 282–287
19 Mitchell, G. and Hastings, J.W. (1969) J. Biol. Chem. 244, 2572–2576
20 Hastings, J.W. and Gibson, Q.H. (1967) J. Biol. Chem. 242, 720–726
21 Gibson, Q.H., Hastings, J.W., Weber, G., Duane, W. and Massa, J. (1966) in Flavins and Flavoproteins (Slater, E.C., ed.), pp. 341–359, Elsevier, Amsterdam
22 Nicoli, M.Z., Meighen, E.A. and Hastings, J.W. (1974) J. Biol. Chem. 249, 2385–2392
23 Hemmerich, P., Ghisla, S., Hartmann, U. and Müller, F. (1971) in Flavins and Flavoproteins, Proceedings of the Third International Symposium (Kamin, H., ed.), pp. 83–105, University Park Press, Baltimore, Md.
24 Ghisla, S., Massey, V., Lhoste, J.-M. and Mayhew, S.G. (1974) Biochemistry 13, 589–597
25 Hemmerich, P., Massey, V. and Weber, G. (1967) Nature 213, 728–730
26 Hastings, J.W., Balny, C., LePeuch, C. and Douzou, P. (1973) Proc. Natl. Acad. Sci. U.S. 70, 3468–3472
27 Hastings, J.W. and Balny, C. (1975) J. Biol. Chem. 250, 7288–7293
28 Cline, T. and Hastings, J.W. (1971) Proc. Natl. Acad. Sci. U.S. 68, 500–504

Part IV

Oxygenases

Flavins and Flavoproteins, edited by T.P. Singer

CHAPTER 11

ROLE OF OXYGENATED FLAVINS IN THE CATALYTIC REACTION OF *p*-HYDROXYBENZOATE HYDROXYLASE

BARRIE ENTSCH, DAVID P. BALLOU and VINCENT MASSEY

Department of Biological Chemistry, The University of Michigan, Ann Arbor, Mich. 48104 (U.S.A.)

Introduction

Since it was first discovered, about 20 years ago, that organisms can utilize molecular oxygen directly as an oxidant by inserting it into the molecular structure of metabolites [1], there has been much interest in the mechanisms by which oxygen is utilized so specifically. The enzymes responsible, which are called oxygenases, are usually classified on the basis of the redox-active prosthetic group(s) associated with the enzyme.

One class of oxygenases contains only FMN or FAD as the prosthetic group. The enzyme studied in this work is *p*-hydroxybenzoate hydroxylase from *Pseudomonas fluorescens* [2]. This enzyme has been purified from a number of organisms and has been studied fairly extensively [3]. For mechanistic studies, the enzyme from *P. fluorescens* has some advantages. It is readily available in fair quantities, it is stable enough to be handled and stored over a period of months, it contains only one FAD per molecule, and a variety of substrate analogues are available for study.

p-hydroxybenzoate hydroxylase is involved in the β-ketoadipate pathway for the degradation of certain aromatic metabolites as a source of carbon and energy [4]. It catalyzes in vivo the reaction

$$\text{NADPH} + H^+ + \text{(}p\text{-hydroxybenzoate: ring with OH, COO}^-\text{)} + O_2 \longrightarrow \text{NADP}^+ + H_2O + \text{(3,4-dihydroxybenzoate: ring with OH, OH, COO}^-\text{)}$$

As is probably the case in most flavoprotein hydroxylases (i.e. oxygenases that insert one atom of molecular oxygen into an aromatic ring), the reaction follows an ordered and effectively irreversible sequence of interactions of enzyme with substrates and products [5].

$$E_{ox} \xrightarrow{S\downarrow} E_{ox}S \xrightarrow{\text{NADPH}\downarrow} \underset{\text{NADPH}}{E_{ox}S} - \underset{\text{NADP}^+}{E_RS} \xrightarrow{\text{NADP}^+\uparrow} E_RS \xrightarrow{O_2\downarrow\;\; H_2O,P\uparrow} E_{ox}$$

Work in this laboratory [6, 7] has shown that interaction with oxygen resulting in hydroxylation of the substrate occurs after the reduction of the complex between enzyme and substrate by pyridine nucleotide. To study the reactions of the enzyme with oxygen as reported in this work, the portion of the catalytic cycle involving oxygen was isolated. The complex of reduced enzyme and substrate was reacted with oxygen in a stopped-flow spectrophotometer. Previous reports from this laboratory have described the detection of intermediates with distinct absorption spectra in the reaction of *p*-hydroxybenzoate [7] and melilotate hydroxylases [8] with oxygen. An intermediate has also been detected with phenol hydroxylase (Massey and Neujahr, unpublished results).

This paper describes oxygenated flavins that have been detected in reactions of reduced enzyme with oxygen and their relationship to the hydroxylation process.

Methods

Analysis of kinetic traces for intermediates

To determine the absorption spectra of intermediates in a reaction, a number of simple techniques was employed. (1) Conditions were selected so that intermediate reactions were well separated in rates. The most useful techniques have been low temperature and N_3^- in the reaction. Temperatures near $0^\circ C$ slowed all reactions without much relative change in rates. With azide in the reaction, some reaction rates were selectively slower without any disruption of the hydroxylation process. Ideally, only two reactions required consideration for each intermediate; its rate of formation and decay. (2) Systematic examination of the absorbance changes due to reaction progress was made as a function of wavelength. When intermediates differed in absorbance, systematic variations occurred in reaction traces with wavelength, and these variations always involved the same sequence of reaction rates. Hence, the number of detectable spectral intermediates present in a particular reaction could be determined. (3) When the rate of a particular step in a reaction differed by a factor of 40 or more from its following reaction, this reaction rate could be determined from absorption changes at all wavelengths where spectral changes occurred. This condition also permitted the direct measurement of more than 90% of the absorbance of the enzyme in a single intermediate species. (4) When two consecutive reaction rates differed by too small a factor to permit direct measurement (as in 3 above), then it was assumed that the reaction was following a two-step, irreversible, consecutive reaction. Since application of the integrated rate equations for this system required knowledge of the rates involved, these rates were usually determined by searching out appropriate isosbestic points between intermediates, where clean first-order reactions could be observed. The percentage of the enzyme in a particular intermediate stage of the reaction at any one instant after the start of the reaction could then be calculated from the rate equations. By selecting a specific time near the optimum level of an intermediate of interest, and correction for the amounts of other intermediates present, a spectrum could be obtained from the wavelength dependence of the reaction.

Results

Initial interaction with oxygen — formation of a flavin-oxygen derivative

In contrast to the multiple reactions of reduced free flavins with oxygen in aqueous solution, resulting in complex kinetics [9], no kinetic

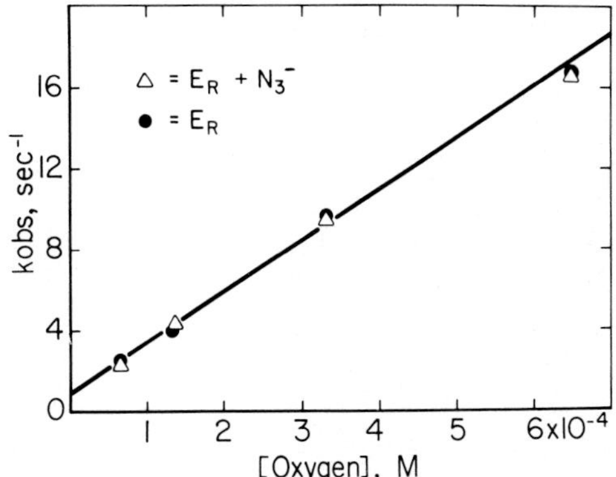

Fig. 1. Effect of oxygen concentration on the rate of reaction of reduced enzyme with oxygen, without substrate or effector present. Reduced enzyme was reacted with oxygenated buffer solution in the stopped-flow spectrophotometer. The reaction solution contained 17 μM enzyme, 15 mM EDTA, and 50 mM potassium phosphate, pH 6.55 at 4.0°C. The pseudo-first order rate of appearance of oxidized enzyme was followed at a number of wavelengths. The second-order rate constant for the reaction ($2.55 \cdot 10^4\ M^{-1} \cdot s^{-1}$) was obtained from the slope of the plot. When 0.1 M sodium azide was present in the reaction solution, oxidized enzyme was formed in a biphasic process. The oxygen dependence of the first reaction was followed at 395 nm.

complexity was observed when free reduced enzyme reacts with oxygen. At any one oxygen concentration, the same pseudo first-order reaction was observed at all wavelengths, and the reaction proceeded from reduced to oxidized enzyme without any intermediate which could be spectrally observed. This behavior has been observed also with other hydroxylases [8, 10]. However, the oxygen-dependence of these reactions indicated formation of a reduced enzyme-oxygen complex in the reaction since a plot of k_{obs} vs. oxygen concentration (Fig. 1) showed a clear intercept when extrapolated to zero oxygen concentration. This is consistent with a second-order reversible reaction with oxygen to form a complex, followed by formation of oxidized enzyme:

$$E_R + O_2 \underset{k_{-1}}{\overset{k_1}{\rightleftharpoons}} E_R - O_2 \xrightarrow{k_2} E_{ox} + H_2O_2$$

The lack of a directly observable intermediate would imply that $k_2 \gg k_1$, k_{-1}.

When azide (0.10 M) was added to the reduced enzyme before reaction with oxygen, or simultaneously with it, the reaction was biphasic, and the time course of absorbance changes varied with wavelength. The faster reaction rate was the same as the reaction with oxygen in the absence of azide (Fig. 1). The slower reaction was independent of oxygen concentration and yielded enzyme in the oxidized form. This behavior demonstrated the presence of an intermediate between reduced and oxidized enzyme. The techniques described in Methods were applied, and the apparent spectrum of the intermediate was calculated (Fig. 2). A control spectrophotometric study of the binding of azide to the reduced enzyme indicated that azide modified the

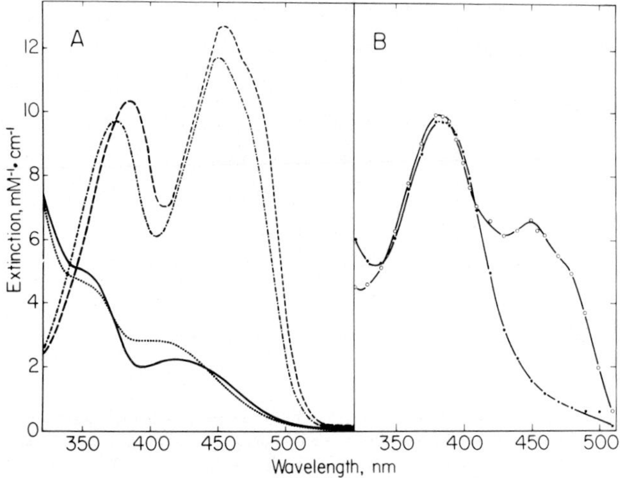

Fig. 2. The stabilization of an oxygenated flavin intermediate by azide during the reaction of reduced *p*-hydroxybenzoate hydroxylase with oxygen. Upon reaction of reduced enzyme with O_2-saturated buffer in the presence of 0.1 M azide, a transient intermediate was observed. Applying the techniques described in Methods, a spectrum of this intermediate was calculated at measured wavelengths and plotted (○ — ○). The spectrum of the intermediate was corrected (by assuming that 45% of the total enzyme was in the oxidized form) to give the spectrum of the oxygenated flavin (● — ●), with a single peak of absorbance (λ_{max} 383–385 nm). Other reaction conditions as in Fig. 1. Part A shows the shifts in the absorption spectra of reduced and oxidized enzyme caused by interaction with azide at pH 6.5 and 4–5°C; —— and · — ·, no azide; · · · and — — —, with 0.49 M sodium azide.

environment of the flavin, but that 0.1 M azide was not a saturating concentration for this interaction. This observation, combined with the indication of oxidized enzyme in the observed spectrum (Fig. 2, 440 to 450 nm peak) showed that the reaction of only a portion of the enzyme molecules was influenced by bound azide. By correcting for the presence of 45% oxidized enzyme in the observed intermediate, a characteristic new spectrum of flavin was obtained (Fig. 2). This transient intermediate is (for reasons unknown) stabilized kinetically by azide.

When the enzyme is complexed with a substrate or effector (i.e. a compound that stimulates NADPH oxidation without itself being hydroxylated), basically the same initial reaction occurs with oxygen as with the free enzyme, but with differences in detail. A second-order reversible reaction occurs, based on plots of k_{obs} vs. O_2 concentration similar to those shown in Fig. 1; however, the rate is some ten-fold greater (cf. Table I). This reaction results in the production of a spectroscopically distinct intermediate, which we shall refer to as intermediate I.

The similarity between the spectra obtained (Fig. 3) is clear. The spectrum of the intermediate formed by the free enzyme in the presence of azide (Fig. 2) is clearly indicative of the same species. With the effector 6-hydroxynicotinate at pH 6.55, I was readily detected (Fig. 3), and it decayed in a first-order reaction (10.7 s^{-1} at 5°C) to form oxidized enzyme and H_2O_2, while the effector molecule was unchanged.

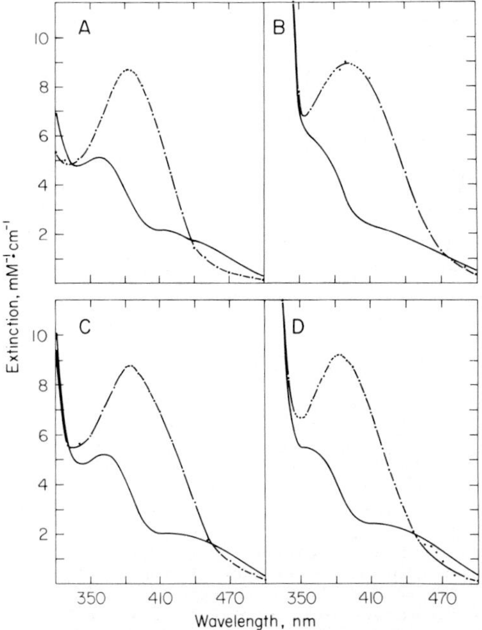

Fig. 3. Absorption spectra of intermediate I. The spectra represent the first transient intermediate observed upon reaction of reduced *p*-hydroxybenzoate hydroxylase in complex with substrates or effectors with buffer saturated with oxygen at atmospheric pressure. In each case the intermediate was formed in a second order reaction between reduced enzyme and oxygen. The lower spectrum in each section is that of the complex of reduced enzyme and substrate (or effector). The spectra of I were obtained as detailed in Methods. Reaction conditions were similar to those of Fig. 2 except that 0.1 M azide was present only in the reaction with the reduced enzyme complexed with 2,4-dihydroxybenzoate. A, with *p*-hydroxybenzoate; B, 2,4-dihydroxybenzoate; C, *p*-aminobenzoate; D, 6-hydroxynicotinate.

Hydroxylation pathway

When any of the three substrates studied with this enzyme is complexed with the reduced enzyme, intermediate I no longer decays directly to oxidized enzyme, but proceeds through further intermediates to eventually liberate hydroxylated product and H_2O. It has been reported [7] that in the hydroxylation of 2,4-dihydroxybenzoate, I was converted to an intermediate with a totally different spectrum (called intermediate II) and that oxygen transfer to the substrate probably occurred in this reaction because product could be isolated with a rate of formation coincidental with the formation of II. These results were obtained by quenching the reaction with acid using the rapid quench technique of Ballou and Palmer [11]. It has subsequently been found that the spectrum of II formed with 2,4-dihydroxybenzoate changes with pH (Fig. 4). Because this intermediate has not been observed with *p*-hydroxybenzoate as substrate, the possibility existed that the unusual spectrum of II reflected the presence of some unusual form of product absorption. With the discovery of a comparable intermediate with *p*-aminobenzoate (Fig. 4), it would appear that II represents some transient structure in the active site which is essential for hydroxylation, but is kinetically invisible with *p*-hydroxybenzoate, i.e. that with

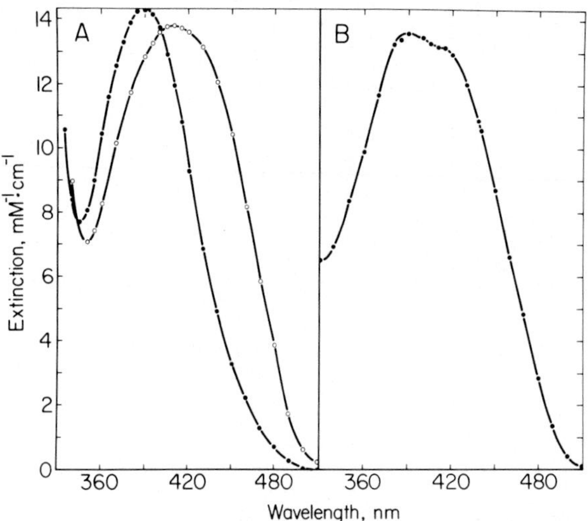

Fig. 4. Absorption spectra of intermediate II. The spectra were obtained from reactions of reduced *p*-hydroxybenzoate hydroxylase-substrate complexes with buffer containing oxygen. This transient intermediate in the reaction is formed in a first order reaction from intermediate I. (A) Spectra of intermediates formed in the hydroxylation of 2,4-dihydroxybenzoate. One spectrum (● — ●, λ_{max} = 390 nm) was derived from data collected with a reaction mixture containing 17.6 μM reduced enzyme, 10 mM dihydroxybenzoate, 0.65 mM O_2, 25 mM EDTA, 0.10 M sodium azide, and 50 mM potassium phosphate, pH 6.55 at 3.5°C. The other spectrum (○ — ○, λ_{max} 410 nm) came from data collected with a reaction mixture containing 17.5 μM reduced enzyme, 10 mM dihydroxybenzoate, 25 mM EDTA, 0.13 mM O_2, and 34 mM Tris · Cl^-, pH 8.6, at 2.0°C. (B) Spectrum formed during the hydroxylation of *p*-aminobenzoate. It has a maximum absorbance between 385 and 390 nm and a distinct shoulder between 410 and 420 nm. The spectrum appears to contain components of both spectra recorded for 2,4-dihydroxybenzoate in (A). The reaction mixture contained 17.5 μM reduced enzyme, 2.5 mM *p*-aminobenzoate, 0.13 mM O_2, 25 mM EDTA, and 50 mM potassium phosphate, pH 6.5 at 25°C.

p-hydroxybenzoate as substrate, II is formed much more slowly than it is converted to the next intermediate.

Further evidence in support of the conclusion that reaction I to II involves transfer of oxygen to the substrate came from the observation that a substrate could also be an effector under special conditions. It was found at pH 6.5 that with concentrations of 2,4-dihydroxybenzoate well above those necessary to saturate the substrate binding site, the extent of hydroxylation in catalytic assays decreased as the substrate concentration was increased. For example, at 10 mM substrate, 29% of the oxygen consumed was converted to H_2O_2, and 70% was utilized in the formation of product. In companion stopped-flow experiments, both the rate and extent of formation of I was found to be independent of 2,4-dihydroxybenzoate concentration above that necessary to saturate the substrate site. The rate of conversion of I to II was unaffected; however, the extent of formation of II was dependent upon concentration. With 10 mM substrate, only 68% of the total enzyme was converted to II. Taken in conjunction with the extent of hydroxylation in catalytic turnover, this result indicates strongly that oxygen transfer to substrate is strictly dependent upon conversion of I to II.

Intermediate II decays in a first-order reaction with a rate independent

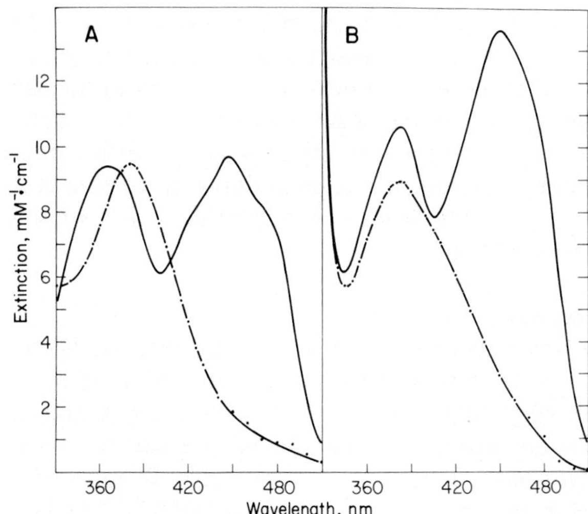

Fig. 5. Absorption spectra of intermediate III. The spectra were obtained from absorption changes in reactions of reduced *p*-hydroxybenzoate hydroxylase-substrate complexes with oxygen. This transient intermediate decays in a first order process to form the oxidized enzyme in complex with the substrate used (spectra recorded in the figure). (A) Spectrum (λ_{max} = 380 nm) of species formed by the enzyme during hydroxylation of *p*-hydroxybenzoate. The reaction solution contained 17.6 μM reduced enzyme, 1.0 mM *p*-hydroxybenzoate, 0.65 mM O_2, 0.1 M sodium azide, 25 mM EDTA, and 50 mM potassium phosphate, pH 6.55 at 3.5°C. (B) Spectrum (λ_{max} = 382 nm) of species formed by enzyme during hydroxylation of 2,4-dihydroxybenzoate. The reaction solution contained 17.6 μM reduced enzyme, 10 mM dihydroxybenzoate, 0.65 mM O_2, 0.1 M sodium azide, 25 mM EDTA, and 50 mM potassium phosphate, pH 6.55 at 3.5°C.

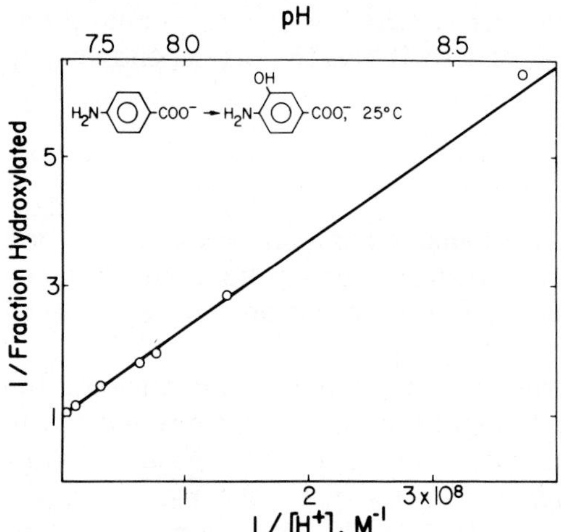

Fig. 6. The pH dependence of the degree of hydroxylation of *p*-aminobenzoate in catalytic turnover. The fraction hydroxylated was obtained by measuring the amount of product formed in the reaction after a measured amount of NADPH had been oxidized by the enzyme. Reaction mixtures contained buffer, 0.5 mM EDTA, 2.6 mM *p*-aminobenzoate, 0.25 mM NADPH, and 2.0 μM enzyme in a total of 2.0 ml at 25°C. Betweeen pH 6.5 and 7.5 the buffer was 50 mM potassium phosphate, and between pH 7.5 and 9.0, the buffer was 50 mM Tris with phosphate and sulphate.

of the substrate concentration to form another transient intermediate (called intermediate III) with a distinctive absorption spectrum (Fig. 5). This intermediate has been observed with all three substrates examined and the spectra of III are very similar to the spectra of I. It was with the decay of III that the enzyme returned to the oxidized form in a first-order reaction with a rate independent of the substrate concentration. It has been determined that this reaction involves the release of product from the enzyme as well as the observed rearrangement of flavin structure.

pH-dependence of hydroxylation pathway

The stoichiometry of hydroxylation in relation to NADPH oxidation has been examined from pH 6.5 to 8.6. It should be noted that for the phenolic substrates, except at very high concentrations, hydroxylation is very nearly quantitative, i.e. one product molecule is formed for each NADPH molecule oxidized. However, with *p*-aminobenzoate as substrate, the fraction of NADPH oxidation that forms product varies with H^+ concentration. A reciprocal plot of this dependence is shown in Fig. 6. This result suggests that some step in hydroxylation of this compound requires a proton from the medium.

In a companion study of the intermediate reactions in hydroxylation of *p*-aminobenzoate it was found that intermediate I was formed quantitatively at both pH 6.5 and 8.6. At pH 6.5, the subsequent conversion of I to II appeared quantitative. In contrast, at pH 8.6, the decay of I gave rise to a mixture of oxidized enzyme and another component, which subsequently formed oxidized enzyme. When a correction was made for 70% oxidized enzyme in the mixture, it was clear that the other component was intermediate II, and that reaction I to II had partly failed to occur. It could therefore be concluded that the conversion of I to II was the pH-dependent step in hydroxylation.

Discussion

The data in Table I are presented to enable comparison of the reaction rates observed when complexes between various compounds and reduced enzyme are reacted with oxygen under identical conditions and give rise to detectable transient intermediates.

Amongst known structural forms of the isoalloxazine nucleus, the absorption spectrum of intermediate I (Fig. 3) resembles in form, but not in molar extinction or wavelength of maximum absorption, the spectra of some C(4a)–N(5) dihydroflavins and oxidized flavins that have the electronic structure of C(4a)–N(5) dihydroflavins [12, 13]. It can now be concluded that I probably has the electronic structure of C(4a)–N(5) dihydroflavins with oxygen substituted in the C(4a) position. Dr. Ghisla in this laboratory synthesized N(5)-ethyl-C(4a)-hydroxy-FAD and attached it to apo-D-aminoacid oxidase to form a stable complex. In free solution, the spectral properties of this form of FAD (with absorption maxima at 355 and

TABLE I

A SELECTION OF KINETIC CONSTANTS MEASURED FOR REACTIONS OF REDUCED *p*-HYDROXYBENZOATE HYDROXYLASE WITH OXYGEN

Compound complexed with reduced enzyme[a]	Minimum scheme with rate constants for each reaction from observed kinetics[b]								
		$M^{-1}s^{-1}$		s^{-1}		s^{-1}		s^{-1}	
Azide	$E_R + O_2$	$\rightleftarrows$ $2.55 \cdot 10^4$	I	$\rightarrow$ 3.45	$E_{ox} + H_2O_2$				
				fast[c] $\searrow$	$E_{ox} + H_2O_2$				
6-Hydroxynicotinate	$E_R - A + O_2$	$\rightleftarrows$ $1.82 \cdot 10^5$	I	$\rightarrow$ 10.7	$E_{ox} - A + H_2O_2$				
p-Hydroxybenzoate	$E_R - S + O_2$	$\rightleftarrows$ $2.6 \cdot 10^5$	I	$\rightarrow$ 48	(II)[d]	$\rightarrow$ fast	III	S $\searrow$ $\rightarrow$ 14.5	$E_{ox} - S + P + H_2O$
2,4-Dihydroxybenzoate	$E_R - S + O_2$	$\rightleftarrows$ $4.8 \cdot 10^5$	I	$\rightarrow$ 30	II	$\rightarrow$ 1.03	III	S $\searrow$ $\rightarrow$ 0.075	$E_{ox} - S + P + H_2O$
p-Aminobenzoate	$E_R - S + O_2$	$\rightleftarrows$ $3.3 \cdot 10^5$	I	$\rightarrow$ 4.9	II	$\rightarrow$ 0.031	III	S $\searrow$ $\rightarrow$ 0.0068	$E_{ox} - S + P + H_2O$
						slow[e] $\searrow$	(III)	S $\searrow$ $\rightarrow$ fast	$E_{ox} - S + P + H_2O$

[a] The reaction solution contained the complex with enzyme in 20 mM EDTA and 50 mM potassium phosphate, pH 6.55, and between 3.5 and 4.5°C.
[b] The following abbreviations are used: E_R, reduced enzyme; E_{ox}, oxidized enzyme; S, substrate; A, effector, and P, product.
[c] "Fast" indicates a reaction that cannot be detected because it is preceded by a much slower reaction.
[d] Parentheses indicate that an intermediate is kinetically invisible.
[e] "Slow" indicates that this reaction rate has to be similar to its competing reaction, II → III, but no successful measurement has been made.

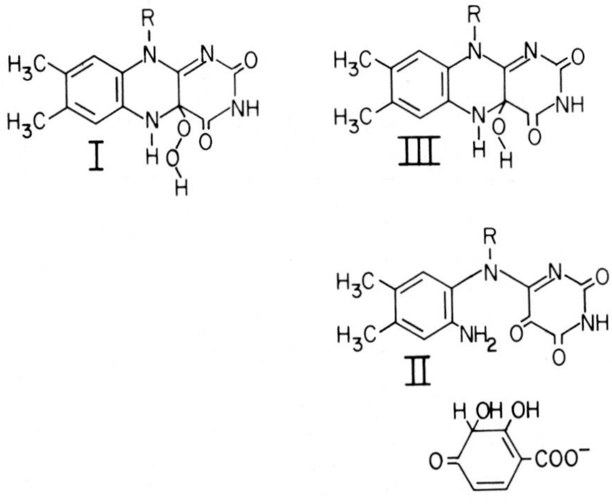

Fig. 7. Proposed structures for the isoalloxazine nucleus responsible for the observed spectra of intermediates in reactions of reduced *p*-hydroxybenzoate hydroxylase with oxygen.

310 nm) did not appear to be related to intermediate I. When it was attached to the protein, the constraints imposed by the protein changed the absorption spectrum considerably to a single peak beyond the region of protein absorption, with the maximum at 374 nm and extinction of approximately $8500\,M^{-1} \cdot cm^{-1}$. Apart from a small difference in absorption maximum (I has its maximum mostly at 382–383 nm), the spectra of the protein-bound model and I are almost identical.

The fact that intermediate I in all reactions examined is formed directly in a second order reaction with oxygen, and that it can decay to H_2O_2 and oxidized flavin, coupled with its characteristic spectrum, provides convincing evidence that I must be a peroxide of FAD which is at a higher energy level than oxidized flavin plus H_2O_2. In view of the spectroscopic argument advanced above, it would seem most likely that I is a C(4a) peroxyflavin (Fig. 7). It has been demonstrated that when the enzyme luciferase utilizes O_2 in its catalytic reaction [14], it also forms an oxygenated flavin bound to the protein as the first product of reaction with oxygen. Since this intermediate has a spectrum so similar to I for *p*-hydroxybenzoate hydroxylase and the synthetic model described, it probably has the same structure.

The fact that I is a peroxyflavin indicates that the essential first step in the activation of oxygen for insertion into substrate is reduction to the peroxide oxidation state. The process of hydroxylation then becomes a problem of the mechanism for the activation of a peroxide.

The striking feature about the spectra of III (Fig. 5) is their similarity to intermediate I. All evidence agrees that one oxygen atom has been transferred to the substrate at this stage of the reaction, implying that the 4 electron reduction of oxygen is complete. Therefore, III is almost certainly the same structural form of flavin as I, but would be C(4a) hydroxy-FAD (Fig. 7). The heterolytic fission of water from this unstable intermediate would account for the second oxygen atom and would return the enzyme flavin to the oxidized structure, ready for the next turnover.

It has been proposed previously [7] that intermediate II may be a mixture of a flavin derivative like I and III and some precursor of product which together might account for the observed spectrum. The product precursor would then require an absorption band with a peak in the region of 400 to 430 nm and an extinction of approximately $6000\,M^{-1} \cdot cm^{-1}$. Possible non-aromatic transients have been considered but the spectra of suitable models do not account for the spectral properties of II (e.g. cyclohexadienone [15] and benzene oxides and their oxepin tautomers [16]). It would therefore seem likely that the observed spectra of II are due to some as yet unprecedented structure of the flavin. Perhaps this could be an isomer of III (Fig. 7) formed in the process of oxygen transfer to substrate where the ring is opened between C(4a) and N(5). This adds a carbonyl group to the chromophore of III. Hamilton [17] has previously proposed such a structure as an intermediate in hydroxylation but via a quite different mechanism.

There appear to be two basic interpretations of the dependence of hydroxylation of *p*-aminobenzoate on pH (Fig. 6). It could represent a simple competition between the rate of breakdown of I to oxidized enzyme and H_2O_2 and the rate of uptake of a proton by I, which then leads to oxygen insertion in the substrate. The observed kinetics at pH 6.5 and 8.6 do not fit this interpretation. Alternatively, the shift in pH could displace the rapid equilibrium of a group on the enzyme vital to the process of activating the flavin peroxide. The apparent p*K* of such a group depends on the relationship of the ionization to the overall reaction. More data are required to establish the relationship with certainty, but the following scheme would fit the available data.

$$E_{ox} + H_2O_2 \xleftarrow{\text{fast}} I + H^+ \underset{}{\overset{\text{very fast}}{\rightleftharpoons}} I^+H \xrightarrow{\text{slow}} II$$

With this scheme, the true p*K* would be higher than the observed p*K* of 7.9, obtained from the slope in Fig. 6.

A comparison of the structures of *p*-aminobenzoate and *p*-hydroxybenzoate immediately suggests a reason for the different behavior of the enzyme with each. It has always appeared likely that hydroxylation by flavoprotein hydroxylases involves an oxenoid rather than a radical mechanism because hydroxylation occurs specifically at the ortho and para positions relative to phenolic groups on the substrate. When I is formed in the presence of a phenolic substrate, the group requiring a proton in the presence of *p*-aminobenzoate could extract the proton from the phenol, forming a stabilized salt bridge. The protonated group may then be involved in polarization of the O—O bond of the peroxide, thus making the terminal O more susceptible to reaction with the activated phenoxide ring.

Fig. 8 presents a hypothetical representation of the simplest reaction scheme for hydroxylation that fits the experimental data collected for *p*-hydroxybenzoate hydroxylase. It takes into account three observed

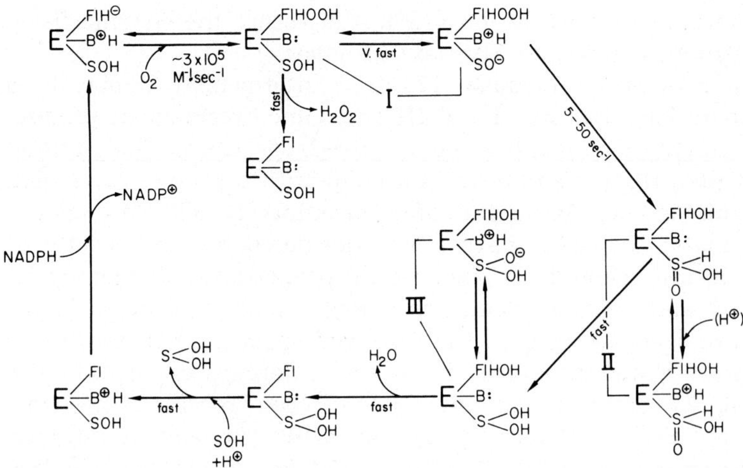

Fig. 8. Hypothetical scheme to explain the relationship between intermediates observed and hydroxylation in reactions of reduced *p*-hydroxybenzoate hydroxylase with oxygen. A description of the scheme is given in Discussion.

intermediates and the dependence of the kinetics of the reaction on the substrates and effectors studied and the pH of the medium. The scheme is constructed of interactions between flavin, oxygen, substrate, and a basic group on the enzyme which is envisioned as an essential component of catalysis.

Oxygen interacts with the flavin ring in the complex between reduced enzyme and substrate and accepts a proton from the protonated base to form the C(4a) peroxyflavin (intermediate I). The base immediately abstracts a proton from the substrate or solvent. The form of I with protonated base (which predominates about neutral pH) then transfers the terminal OH of the peroxide to the substrate (in an electrophilic substitution) to form a non-aromatic intermediate of the product and leaves the other atom of oxygen on the flavin (possibly in the ring-opened form of Fig. 7). In a competing reaction the unprotonated base catalyzes the heterolytic fission of H_2O_2 from the flavin to give oxidized flavin and unchanged substrate. When a compound does not have a favourable site for electrophilic attack (an effector), then the reaction proceeds no further than intermediate I and the enzyme liberates H_2O_2.

With *p*-hydroxybenzoate as substrate, the base after oxygen transfer is left unprotonated, and catalyzes in a fast reaction the rearrangement of product to the aromatic form (and may also be involved in a rearrangement of the flavin). However, when the substrate is left with a rapidly exchangeable proton ([H^+] in Fig. 8) upon oxygen transfer (the products from *p*-aminobenzoate and 2,4-dihydroxybenzoate would have such a proton), the base is protonated again, by (H^+) from the product; this form of the enzyme does not readily rearrange product or flavin (resulting in stabilization of intermediate II). The decay of II through the unprotonated base (like for *p*-hydroxybenzoate) forms C(4a) hydroxyflavin and aromatic product bound to the enzyme (intermediate III). The unprotonated base catalyzes the heterolytic fission of H_2O from the flavin in a moderately fast reaction to give oxidized flavin. This process would normally involve a rearrangement of the enzyme so that it is then capable of the rapid exchange of substrate for product. However, III is in rapid equilibrium with the form containing protonated base. In the protonated form, the rearrangement of flavin does not readily occur. The decay of III is most frequently the rate-limiting step in the hydroxylation process.

Acknowledgement

This work was supported in part by a Grant from the U.S.P.H.S., No. GM-11106.

References

1 Hayaishi, O., Katagiri, M. and Rothberg, S. (1955) J. Am. Chem. Soc. 77, 5450–5451
2 Howell, L.G., Spector, T. and Massey, V. (1972) J. Biol. Chem. 247, 4340–4350
3 Flashner, M.S. and Massey, V. (1974) in Molecular Mechanisms of Oxygen Activation (Hayaishi, O., ed.), pp. 245–283, Academic Press, New York
4 Stanier, R.Y. and Ornston, L.N. (1973) in Advances in Microbial Physiology (Rose, A.H. and Tempest, D.W., eds.), pp. 89–151, Academic Press, New York
5 Massey, V. and Hemmerich, P. (1975) in The Enzymes, (Boyer, P.D., ed.), vol. on Oxidation-Reduction, Vol. XII, pp. 191–252, Academic Press, New York
6 Spector, T. and Massey, V. (1972) J. Biol. Chem. 247, 5632–5636
7 Entsch, B., Massey, V. and Ballou, D.P. (1974) Biochem. Biophys. Res. Commun. 57, 1018–1025
8 Strickland, S. and Massey, V. (1973) J. Biol. Chem. 248, 2953–2962
9 Massey, V., Palmer, G. and Ballou, D.P. (1973) in Oxidases and Related Redox Systems (King, T.E., Mason, H.S. and Morrison, M., eds.), pp. 25–49, University Park Press, Baltimore
10 White-Stevens, R.H. and Kamin, H. (1972) J. Biol. Chem. 247, 2371–2381
11 Ballou, D.P. and Palmer, G.A. (1974) Anal. Chem. 46, 1248–1253
12 Ghisla, S., Massey, V., Lhoste, J.-M. and Mayhew, S.G. (1974) Biochemistry 13, 589–597
13 Walker, W.H., Hemmerich, P. and Massey, V. (1970) Eur. J. Biochem. 13, 258–266
14 Hastings, J.W., Balny, C., LePeuch, C. and Douzou, P. (1973) Proc. Natl. Acad. Sci. U.S. 70, 3468–3472
15 Derkosch, J. and Kaltenegger, W. (1957) Monat. Chem. 88, 779–792
16 Vogel, E. and Günther, H. (1967) Angew. Chem. Int. Ed. 6, 385–401
17 Hamilton, G.A. (1974) in Molecular Mechamisms of Oxygen Activation (Hayaishi, O., ed.), pp. 405–451, Academic Press, New York

Flavins and Flavoproteins, edited by T.P. Singer

CHAPTER 12

STUDIES ON THE REACTION MECHANISM OF *p*-HYDROXYBENZOATE HYDROXYLASE: FUNCTIONAL GROUPS ESSENTIAL FOR THE ENZYMATIC REACTION

HIROFUMI SHOUN, NAOKI HIGASHI, TERUHIKO BEPPU, KEI ARIMA and SATOSHI NAKAMURA[a]

Department of Agricultural Chemistry, Faculty of Agriculture, The University of Tokyo, Bunkyo-ku, Tokyo and [a] Department of Biochemistry, Kitasato University School of Medicine, Sagamihara, Kanagawa (Japan)

Summary

1. Studies on the pH dependence of *p*-hydroxybenzoate hydroxylase-catalyzed reactions indicate that two dissociating groups, with p*K* values of 6.5 and 8.2 respectively, participate in the interaction of the enzyme and NADPH. One is considered to be the 2′-phosphate of NADPH and the other to be the ϵ-amino group of lysine or the sulfhydryl group of cysteine, judging from both their p*K* values and dissociation enthalpy.

2. Studies on the effects of pH also suggested that some change occurs in *p*-hydroxybenzoate hydroxylase near pH 7, when the pH is varied.

3. The results of chemical modifications of *p*-hydroxybenzoate hydroxylase with 2,4,6-trinitrobenzene sulfonate and phenylglyoxal suggested that lysine is involved in the NADPH-binding site and arginine in the *p*-hydroxybenzoate-binding site.

Introduction

Although significant advances have been made in recent years in the investigation of flavoenzymes, the interactions between the coenzyme and apoenzyme which result in enzymatic activity are still not clear. Studies on the role of the protein moiety, such as essential functional groups in the active center, have lagged behind the recent advances in the reaction mechanisms involving the flavin moiety. So far, the involvement of histidine,

Abbreviations: K_s^{pOHB} and K_s^{BA}, the dissociation constants of the enzyme-*p*-hydroxybenzoate and the enzyme-benzoate complexes; K_m^{NADPH} and $K_m^{O_2}$, the Michaelis constants for NADPH and molecular oxygen; and k_{red}^{max} and k_{ox}^{max}, the maximum reaction rates of the reduction step and the oxidation step, respectively, expressed as first-order rate constants; N_3Bzs; 2, 4, 6-trinitrobenzene sulfonate a sulfhydryl group or a persulfide in catalysis has been shown only for a few oxidases, for arginine monooxygenase and for thioredoxin reductase [1–9]. However, for "external flavoprotein monooxygenases", such as *p*-hydroxybenzoate hydroxylase or salicylate hydroxylase, we have little information on the role of the protein moiety.

In this context, the pH dependence of some of the *p*-hydroxybenzoate hydroxylase-catalyzed reactions has been investigated. The roles of lysine and arginine are discussed on the basis of the results of chemical modifications of *p*-hydroxybenzoate hydroxylase with 2, 4, 6-trinitrobenzene sulfonate and phenylglyoxal.

Materials and Methods

p-Hydroxybenzoate hydroxylase from *Pseudomonas desmolytica* IAM 1123 was purified according to Yano et al. [10]. NADPH was obtained from Kyowa Hakko Kogyo Co., Tokyo, and all other compounds were of the purest grade commercially available. Overall enzymatic reactions were assayed by oxygen uptake and spectrophotometrically [11, 12]. The degree of hydroxylation of the substrate and substrate analogues was estimated by the method of White-Stevens and Kamin [13]. The dissociation constants were determined as reported previously [12]. The stopped-flow measurements of the reduction and reoxidation reactions of the enzyme have been described earlier [11]. The amount of trinitrophenylated residues was estimated spectrophotometrically, assuming a molecular extinction coefficient of $1.25 \cdot 10^4\ M^{-1} \cdot cm^{-1}$ at 345 nm [14]. The reaction with phenylglyoxal was followed by both amino acid analyses and radioactivity measurements of the modified enzyme. The modified enzyme was obtained by passing it through a Sephadex G-25 column, which had been equilibrated with 0.05 M phosphate buffer pH 8.0, at 5°C, after being reacted with the reagent for adequate periods [15]. ^{14}C-phenylglyoxal was kindly provided by Dr. Takahashi of Kyoto University. Amino acid analyses were made by the standard technique using a JEOL amino acid analyzer, model JLC-6Ah. Radioactivity measurements were made with a dioxane-POPOP system as the scintillator, using an Aloka liquid scintillation spectrometer, model LSC-651. All the experiments were carried out at 25°C unless otherwise stated.

Results and Discussion

pH profile of the overall oxygen uptake

pH profiles of the overall NADPH oxidation measured by oxygen uptake in the presence of the substrate and substrate analogues are shown in Fig. 1. The reactions in the presence of *p*-hydroxybenzoate, 2, 4- and 3, 4-dihydroxybenzoate had the optimum pH values in the alkaline region, but those in the presence of 6-hydroxynicotinate and benzoate in the acidic region. Thus, the pH optimum does not correlate with the presence of absence of the hydroxyl group in the substance tested, or with whether the substance is hydroxylated or not [12]. However, it is related to the structure of the aromatic compounds; i.e. all the compounds with the optimum pH in the alkaline region have the *p*-hydroxybenzoate structure in common.

pH profiles of the partial reactions

As previously reported, the overall hydroxylation is composed of three partial reactions: (1) the formation of the enzyme-substrate complex

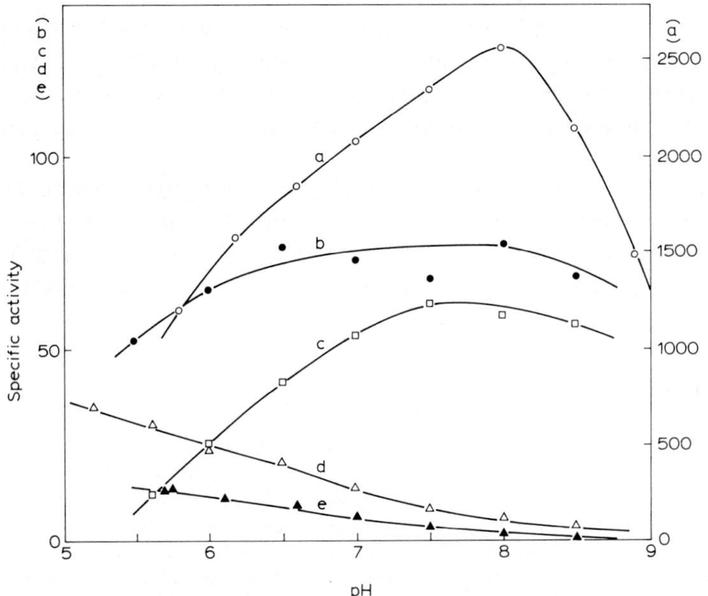

Fig. 1. pH dependence of the overall NADPH oxidation of *p*-hydroxybenzoate hydroxylase in the presence of the substrate and substrate analogues. The reaction mixture contained 0.15 mM NADPH, 5.7 mM substrate analogue with the exception of *p*-hydroxybenzoate (0.15 mM) in a 0.05 M phosphate buffer with a final volume of 3.5 ml. The reaction was initiated by adding the enzyme and the initial rate of the oxygen uptake was measured. Substrate or substrate analogues tested: (a) *p*-hydroxybenzoate, (b) 3, 4-dihydroxybenzoate, (c) 2, 4-dihydroxybenzoate, (d) 6-hydroxynicotinate, (e) benzoate. Specific activities were expressed in terms of mol of O_2 uptake per min per mol of the enzyme.

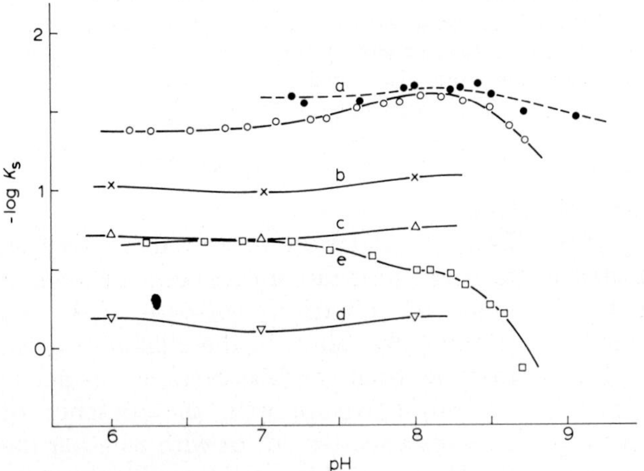

Fig. 2. pH dependence of the dissociation constants of the enzyme-effector complexes measured by the fluorescence method. 2.2 μM of the enzyme in 0.05 M phosphate buffer was excited at 450 nm and the change in the emission at 520 nm by the addition of the following substrates or substrate analogues was recorded, (a) *p*-hydroxybenzoate, closed circles were obtained by the titrations in 0.05 M Tris-phosphate buffer, (b) 3,4-dihydroxybenzoate, (c) 2,4-dihydroxybenzoate, (d) 6-hydroxynicotinate, (e) benzoate.

(E_{ox} + S ⇄ E_{ox} · S); (2) the reduction of the flavin moiety with NADPH (E_{ox} · S + NADPH → E_{red} · S + NADP); and (3) the reoxidation of the reduced enzyme-substrate complex with molecular oxygen to produce the hydroxylated product and water (E_{red} · S + O_2 → E_{ox} + P + H_2O) [11, 16]. The pH dependence of each of these reactions was investigated and the kinetic parameters are plotted according to Dixon and Webb [17].

pH profiles of the complex formation fo the five effectors with the enzyme are shown in Fig. 2. With *p*-hydroxybenzoate the optimum pH was in the alkaline region, while with benzoate it was in the acidic region. These phenomena are in parallel with the pH profiles observed with the overall oxygen uptake in the presence of these compounds (Fig. 1). Three other effectors showed almost constant values in the pH range studied. In a previous paper [12], we reported the remarkable increase of K_s^{pOHB} and K_s^{BA} at above pH 8.5 in 0.05 M phosphate buffer and ascribed it to the deprotonation of the ϵ-amino group of lysine. In this paper, we have studied the effect of pH on K_s^{pOHB} in the alkaline region in 0.05 M Tris phosphate buffer. As is shown in Fig. 2, K_s^{pOHB} die not increase as remarkably as in phosphate buffer at above pH 8.5. The increase in K_s^{pOHB} at the higher pH in phosphate buffer might be due to some effects of pH on the enzyme conformation. In these fluorometric titration experiments, only 6-hydroxy-nicotinate showed increased intensity of the fluorescence emission spectrum of the enzyme-bound FAD by forming the complex with the enzyme at pH 8.0, as observed with the *p*-hydroxybenzoate hydroxylas from *Pseudomonas fluorescence* [18]. However, at pH 7.0 and 6.0, 6-hydroxynicotinate decreased the intensity, as other effectors did (Table I). The results of Fig. 1, Fig. 2 and Table I imply some change occurring in the enzyme near pH 7.

The effect of pH on the kinetic parameters of the reduction step of the enzyme in the presence of *p*-hydroxybenzoate are shown in Fig. 3. The k_{red}^{max} showed no pH dependence throughout the region studied, while K_m^{NADPH} was found to have two inflection points, at pH 6.5 and 8.2, suggesting that two dissociating groups are participating in the interaction of the enzyme with NADPH.

Kinetic parameters obtained in the reoxidation step are plotted against pH in Fig. 4. The pH profiles of $K_m^{O_2}$, k_{ox}^{max} and $k_{ox}^{max}/K_m^{O_2}$ are all quite

TABLE I

PH DEPENDENCE OF THE % QUENCHING OR ENHANCEMENT OF FLAVIN FLUORESCENCE OF THE ENZYME FOUND ON COMPLEX FORMATION WITH EFFECTORS

Conditions were the same as those in Fig. 2. The values were obtained by the addition of saturating amounts of the compounds. Emission; 520 nm, excitation; 450 nm.

	pH 6	pH 7	pH 8
p-Hydroxybenzoate	22.2	27.5	60.0
2, 4-Dihydroxybenzoate	76.9	71.3	67.5
3, 4-Dihydroxybenzoate	85.0	80.0	73.8
6-Hydroxynicotinate	6.0	6.3	17.6*
Benzoate	55.0	55.0	42.5

* Enhancement.

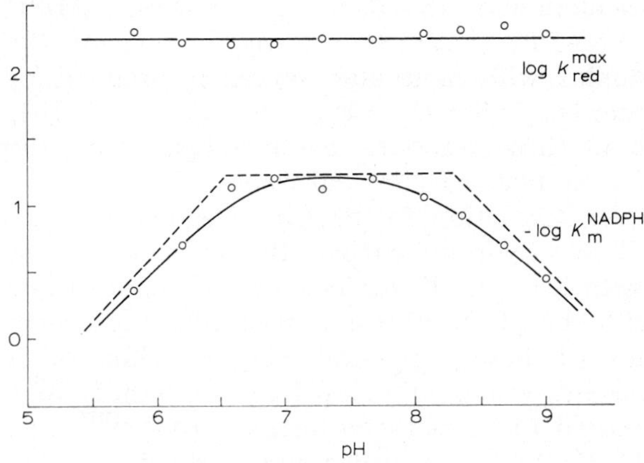

Fig. 3. pH dependence of the kinetic parameters of the reduction step. This experiment was made in 0.05 M Tris/maleate buffer.

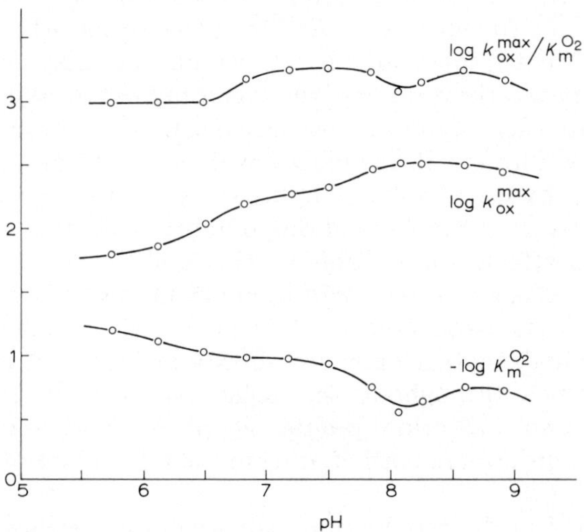

Fig. 4. pH dependence of the kinetic parameters of the reoxidation step. The buffer used was 0.05 M Tris/maleate buffer.

complex, so no conclusion can be drawn from these results as to the functional groups involved in the oxidation step.

Determination of the dissociation enthalpy of the two groups implicated in the reduction step

For the purpose of identifying the two dissociating groups suspected from the pH dependence of K_m^{NADPH} as shown in Fig. 3, the enthalpy of dissociation was also determined from the temperature dependence of the pK values (Fig. 5). The values of the enthalpy were −3200 cal/mol for the group on the acidic side and + 8800 cal/mol for that on the basic side.

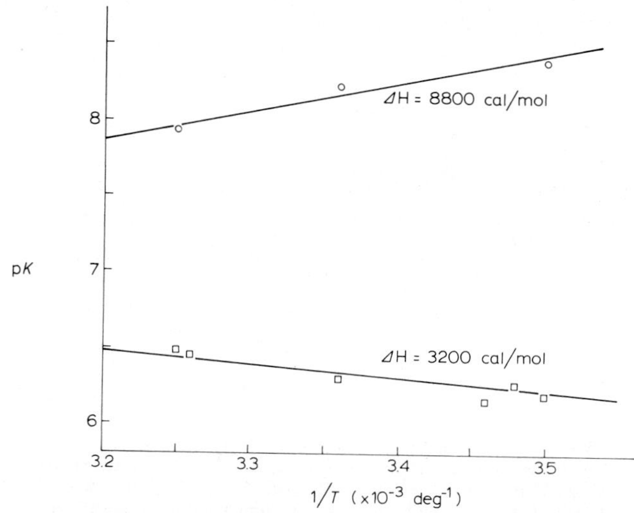

Fig. 5. Temperature dependence of pK values of the dissociating groups obtained in Fig. 3. Curves a and b are for the dissociating groups on the acidic side and on the basic side, respectively.

On the basis of both pK and enthalpy values, the dissociating group on the basic side is considered to be the ϵ-amino group of lysine or the SH group of cysteine. However, no group corresponding to the one on the acidic side (pK = 6.5, ΔH = $-$3200 cal/mol) was found in any of the amino acid residues. On the other hand, many phosphate compounds are known to possess a pK of around 7 with a negative value of the dissociation enthalpy [19, 20]. It is thus likely that the 2′-phosphate group of NADPH is responsible for the pK on the acidic side as suggested by Steennis et al. [21].

Kinetics of the inhibition of the overall enzymatic reaction by Cl^-

The inhibition of Cl^- was competitive with respect to NADPH and noncompetitive or mixed type with respect to *p*-hydroxybenzoate. K_i with NADPH was $2.0 \cdot 10^{-2}$ M, which is rather large compared with that obtained by Steennis et al. [21].

Chemical modification of p-hydroxybenzoate hydroxylase with N_3Bzs and phenylglyoxal

p-Hydroxybenzoate hydroxylase was inactivated by modification with N_3Bzs or phenylglyoxal and the presence of excess Cl^- or *p*-hydroxybenzoate showed a protective effect on the inactivation (Figs. 6 and 7). As seen in the figures, Cl^- is more protective than *p*-hydroxybenzoate against the N_3Bzs modification, while *p*-hydroxybenzoate is more protective than Cl^- against the phenylglyoxal modification.

N_3Bzs is used for the modification of free amino groups in proteins and has been reported not to react with other basic amino acid residues, such as histidine and arginine [14], while the most probable site for Cl^- binding is considered to be a basic amino acid residue which is in the NADPH-binding

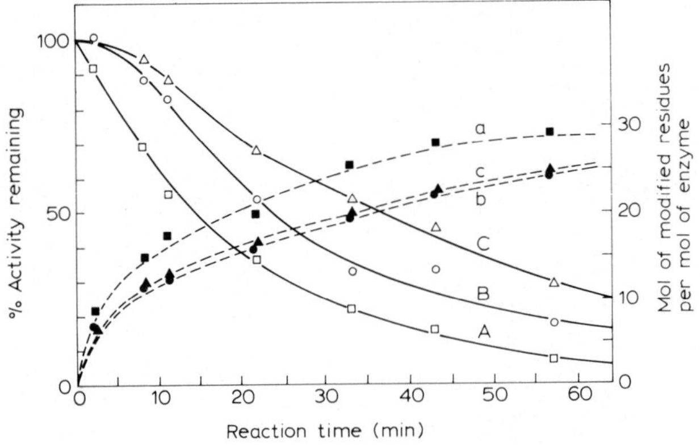

Fig. 6. Chemical modification of *p*-hydroxybenzoate hydroxylase with N_3Bzs 9.8 μM. enzyme was incubated with 4.0 mM N_3Bzs in 0.05 M phosphate buffer (pH 8.0), at 5°C, in the presence of: (A) and (a) no additive, (B) and (b) 270 μM *p*-hydroxybenzoate, (C) and (c) 0.1 M NaCl. Solid lines represent % activities remaining and dotted lines, numbers of the modified residues, respectively. All inactivation curves were obtained by comparing remaining activities with each of the controls which were exposed to the same treatments without N_2Bzs.

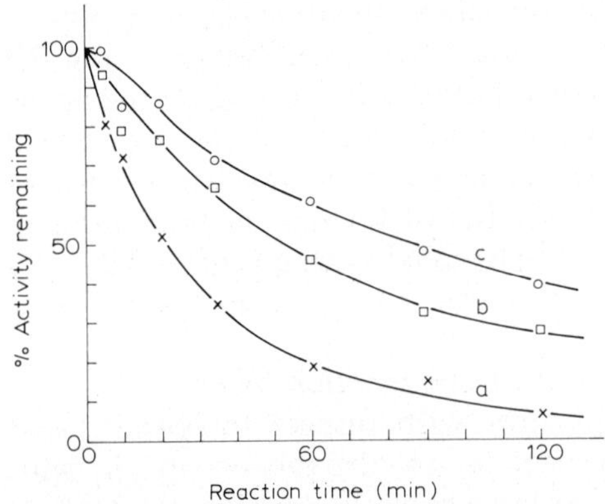

Fig. 7. Chemical modification of *p*-hydroxybenzoate hydroxylase with phenylglyoxal. 9.9 μM of the enzyme was reacted with 1.5 mM phenylglyoxal in 0.05 M phosphate buffer (pH 8.0), at 25°C, in the presence of (a) no additive, (b) 0.1 M NaCl, (c) 270 μM p-hydroxybenzoate, respectively.

site. Thus, the strong protective effect of Cl^- in the N_3Bzs modification suggests that Cl^- binds to the lysine residue which normally binds NADPH. As previously described, the carboxylate anion of *p*-hydroxybenzoate is considered to play an important role in the interaction of the substrate with the enzyme, having an ionic interaction with a basic amino acid residue in protein [12]. The results of Figs. 6 and 7 suggest that arginine rather than lysine, might be the *p*-hydroxybenzoate-binding site.

Effect of p-hydroxybenzoate on the incorporation of [^{14}C]phenylglyoxal into p-hydroxybenzoate hydroxylase

It has been reported by Takahashi that phenylglyoxal also reacts with lysine in the alkaline region [15]. *p*-Hydroxybenzoate hydroxylase contains more than 40 mol of arginine and more than 30 mol of lysine per mol (unpublished data), so it is impossible to follow the loss of a small number of these residues, in the early stage of the modification, only by amino acid analysis. However, we can distinguish to some extent the modification of arginine and lysine in the early stage, because the stoichiometry of the reaction differs between arginine and lysine; i.e. arginine reacts with 2 mol of phenylglyoxal but lysine reacts with 1 mol of the reagent [15]. To identify the *p*-hydroxybenzoate-binding site, the effect of the substrate on the incorporation of [^{14}C]phenylglyoxal into the enzyme was investigated. As shown in Table II, there is a rapid incorporation of about 2 mol of the reagent when *p*-hydroxybenzoate is absent. This corresponds to the modification of 1 mol of arginine, accompanied by the loss of about 70% of the initial enzymatic acitivity. After an incubation of 60 min, the remaining activity approaches zero when *p*-hydroxybenzoate is absent, while it is about half of the initial one in the presence of *p*-hydroxybenzoate. At the reaction time, about 2 mol of the label are masked by the addition of *p*-hydroxybenzoate. The result also suggests that an arginine rather than a lysine residue might be masked by the addition of *p*-hydroxybenzoate. Phenylglyoxal is an analogue of benzoate and so it is considered to be incorporated selectively in the active center, especially the *p*-hydroxybenzoate binding site. Numbers of the lost arginine residues obtained from amino acid analyses show good agreement with those from the radioisotope analyses in the later stages of the modification. As is also shown in the next experiment, there seems to be some sites other than the *p*-hydroxybenzoate-binding site, modification of which lead to inactivation, and this is why *p*-hydroxybenzoate does not show perfect protection.

Fluorometric titration of some inactivated enzyme-species with p-hydroxybenzoate

p-Hydroxybenzoate hydroxylase was inactivated by phenylglyoxal to the extent that residual enzymatic activities were 0–2.5% of the initial one,

TABLE II

CHEMICAL MODIFICATION OF *p*-HYDROXYBENZOATE HYDROXYLASE WITH [^{14}C] PHENYLGLYOXAL

Experimental conditions were fundementally the same as in Fig. 7. Incorporation of the label into the enzyme was measured in the presence and absence of *p*-hydroxbenzoate and the number of modified residues per mol of the enzyme was calculated as arginine (*n* in table). *N*, number of lost arginine residues per mol of the enzyme obtained from the amino acid analyses.

Reaction time (min)	*p*-Hydroxybenzoate	% Activity remaining	Enzyme (cpm/µmol)	*n*	*N*
20	+	79.9	5 500	0.48	—
	—	30.6	13 286	1.15	—
60	+	48.6	14 356	1.25	1.0
	—	6.3	26 937	2.30	3.2
120	+	26.1	21 667	1.85	2.1
	—	1.3	40 672	3.50	3.3

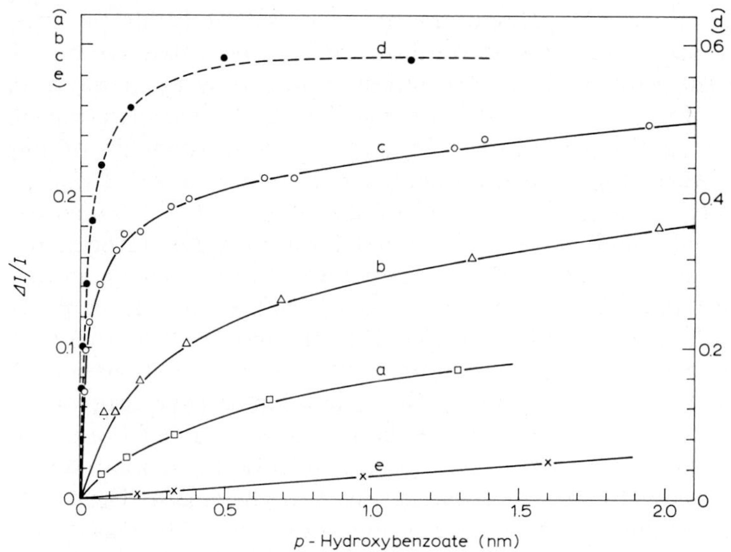

Fig. 8. Fluorometric titration of phenylglyoxal-treated enzymes with *p*-hydroxybenzoate. 30 μM *p*-hydroxybenzoate hydroxylase was treated with 4.0 mM of phenylglyoxal in the absence (a) or presence of 0.1 M NaCl (b) or 270 μM *p*-hydroxybenzoate (c) in 0.05 M phosphate buffer (pH 8.0), at 25°C, and when the enzymatic activities were almost lost, the inactivated enzymes were obtained by the gel filtration-method as described in Materials and Method. % remaining activities of them were: (a) 0%, (b) 2.5% and (c) 2.3%, respectively. The times required for the inactivations were: (a) 105, (b) 170 and (c) 210 min, respectively. The three inactivated enzyme species were titrated with *p*-hydroxybenzoate, respectively (Curves a–c). Conditions of the titration were as same as in Fig. 2. Curves d and e are the titration curves of the native enzyme and free FAD, respectively.

in the absence or presence of *p*-hydroxybenzoate or Cl^-, and the abilities of the inactivated enzymes to bind *p*-hydroxybenzoate were measured by fluorometric titration (Fig. 8). The inactivated enzyme modified in the presence of *p*-hydroxybenzoate showed a far stronger ability to bind the substrate as compared with that modified in the absence of *p*-hydroxybenzoate. The enzyme species modified in the presence of Cl^- was intermediate between the two. This result strongly suggests that the *p*-hydroxybenzoate-binding site is modified by phenyglyoxal.

Conclusions

Two dissociating groups have been shown to play important roles in the interaction of *p*-hydroxybenzoate hydroxylase with NADPH. One is considered to be the 2′-phosphate of NADPH and the other to be the ε-amino group of lysine or the sulfhydryl group of cysteine, from the results of Figs. 3 and 5. It is just reasonable to consider that the third phosphate of NADPH is important generally for enzymes to distinguish between the pyridine nucleotides, NADH and NADPH. Thus, we have good reason to believe that enzymes cannot interact with NADPH when the charge of the 2′-phosphate or that of the basic amino acid residue in the binding site interacting with the minus charge of the phosphate, is lost. The change of K_m^{NADPH} in Fig. 3 might represent such a phenomenon. This assumption is also supported by the conclusion that Cl^- binds to lysine. It is thus likely that the dissociating group in the alkaline region of Fig. 3 is the ϵ-amino group of lysine rather than the sulfhydril group of cysteine.

Chemical modification studies have suggested that the guanido group of arginine rather than the amino group might be involved in the *p*-hydroxybenzoate binding site.

The pH-effect studies also implied that some change occurs in the enzyme near pH 7 (Figs. 1 and 2 and Table I). We are not sure whether this change is due to some dissociating group or changes in protein conformation. We should investigate the possibility that some functional group with a pK value of near 7, such as histidine or cysteine, is participating in the enzyme catalysis of *p*-hydroxybenzoate hydroxylase. Further studies are required.

Acknowledgements

The authors gratefully acknowledge the kind gift of the labeled reagent and helpful discussions by Professor Kenji Takahashi of Kyoto University. We also thank Professor Ketaro Hiromi of Kyoto University, Professor Kunio Yagi of Nagoya University, Professor Kazutomo Imahori, Assistant Professor Keiji Yano and Dr. Koichi Suzuki of Tokyo University for their kind discussions. Thanks are due also to Kaken Kagaku Co., for culturing the microbes.

References

1 Kearney, E.B. and Singer, T.P. (1951) Arch. Biochem. Biophys. 33, 414
2 Zanetti, G. and Williams, Jr., C.H. (1967) J. Biol. Chem. 242, 5232
3 Kok, A.De. and Veeger, C. (1968) Biochim. Biophys. Acta 117, 35
4 Page, D.S. and van Etten, R.L. (1969) Biochim. Biophys. Acta 191, 38
5 Massey, V. and Edmonson, D. (1970) J. Biol. Chem. 245, 6595
6 Tatyanenko, L.V., Gvozdev, R.I., Lebedeva, O.I., Vorobyov, L.V., Gorkin, V.Z. and Yakovlev, V.A. (1971) Biochim. Biophys. Acta 242, 23
7 Beau, F.T., Olomucki, A. and van Thoai, N. (1971) Eur. J. Biochem. 19, 270
8 Branzoli, U. and Massey, V. (1974) J. Biol. Chem. 249, 4346
9 Steenkamp, D.J., Schabort, J.C., Holzapfel, C. and Ferreira, N.P. (1974) Biochim. Biophys. Acta 358, 126
10 Yano, K., Higashi, N. and Arima, K. (1969) Biochem. Biophys. Res. Commun. 34, 1
11 Nakamura, S., Ogura, Y. Yano, K., Higashi, N. and Arima, K. (1970) Biochemistry 9, 3235
12 Shoun, H., Higashi, N., Beppu, T., Arima, K. and Nakamura, S. (1974) FEBS lett. 49, 13
13 White-Stevens, R.H. and Kamin, H. (1970) Biochem. Biophys. Res. Commun. 38, 882
14 Okuyama, T. and Satake, K. (1960) J. Biochem. Tokyo 47, 454
15 Takahashi, K. (1968) J. Biol. Chem. 243, 6171
16 Yano, K., Higashi, N., Nakamura, S. and Arima, K. (1969) Biochem. Biophys. Res. Commun. 34, 277
17 Dixon, M. and Webb, E.C. (1964) Enzymes, 2nd edn., p. 116, Longmans Green, London
18 Howell, L.G., Spector, T. and Massey, V. (1972) J. Biol. Chem. 247, 4340
19 Edsall, J.T. and Wyman, J. (1958) Biophysical Chemistry, Vol. 1, p. 450, Academiz Press, New York
20 Sober, H.A. (1968) Handbook of Biochemistry Part J, p. 49, Chemical Rubber Co., Cleveland
21 Steenis, P.J., Cordes, M.M., Hilkens, J.G.H. and Muller, F. (1973) FEBS Lett. 36, 177

Flavin and Flavoproteins, edited by T.P. Singer

CHAPTER 13

ORCINOL AND RESORCINOL HYDROXYLASES FROM *PSEUDOMONAS PUTIDA* ORC

YOSHIYUKI OHTA* and DOUGLAS W. RIBBONS

Department of Biochemistry, University of Miami School of Medicine Miami, Fla. 33152 (U.S.A.)

Summary

In *Pseudomonas putida* ORC different flavoprotein hydroxylases are induced for growth on orcinol and resorcinol, unlike *Ps. putida* 01, which is able to grow on resorcinol by expression of mutants constitutive for orcinol hydroxylase and subsequent enzymes of the orcinol pathway. A comparison of orcinol and resorcinol hydroxylases from *Ps. putida* ORC has now been made. The enzymes are indistinguishable in several properties, namely (a) molecular size by gel filtration, sodium dodecyl sulphate electrophoresis, and sedimentation ($M_r \approx 70\,000$, native and denatured); (b) electrophoretic mobility; (c) absorption spectra; (d) FAD content ($\approx$ 1 FAD per 70 000); (e) circular dichroism spectra. They are, however, easily distinguished by several of their chemical, catalytic and immunological properties, as well as the control of their synthesis. Chemical differences include: (a) amino acid composition, (b) stability to oxidation (orcinol hydroxylase readily loses about 50% activity in dilute solutions); (c) reactivity with mercaptoethanol (resorcinol hydroxylase is readily reduced). Catalytic differences were observed in (a) reactivity with and preference for substrates and effectors (resorcinol hydroxylase is far more versatile than orcinol hydroxylase); (b) NADH oxidase activity without effectors; (c) K_m values for NADH with different effectors. Immunologically these flavoproteins are differentiated by the anti-sera obtained. Identity of the two hydroxylases was not observed with anti-sera prepared for each protein.

Introduction

Orcinol hydroxylase (EC 1.14.13.6) is the first inducible enzyme of the pathway that enables *Pseudomonas putida* 01 and ORC to grow on orcinol

---

* Present address: Faculty of Fisheries and Animal Husbandry, University of Hiroshima, Fukuyama-shi, Hiroshima-ken, Japan

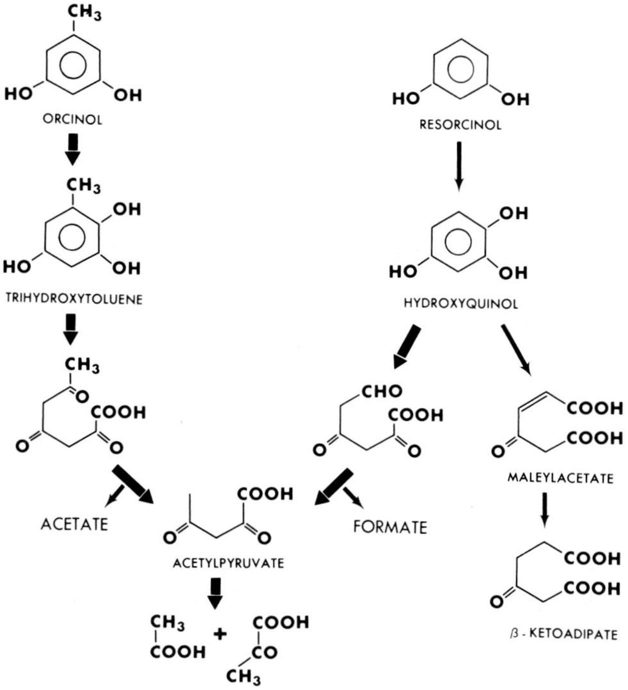

Scheme 1. Catabolism of orcinol and resorcinol by *Ps. putida* ORC. Broad arrows indicate the enzymic activities found in cells after growth on orcinol. Enzymes for the conversion of resorcinol to β-ketoadipate only occur in resorcinol grown cells.

[1, 2]. It catalyzes the introduction of a third hydroxyl group into the benzene nucleus to give 2,3,5-trihydroxytoluene. Growth of these strains on the lower homologue, resorcinol, is also possible but whereas *Ps. putida* 01 does so only by selection of mutants (010C) that constitutively express the enzymes of the orcinol pathway, *Ps putida* ORC expresses also enzymes of another catabolic route that is initiated by a reaction homologous to both pathways, but which diverges when an intradiol cleavage of hydroxyquinol occurs (Scheme 1).

Orcinol hydroxylase from *Ps. putida* 01 has been purified to homogeneity and several of its properties examined [3, 4, 5]. It shows a wide substrate and effector specificity, and resorcinol is hydroxylated to hydroxyquinol though with low efficiency since H_2O_2 is also formed in substantial quantities [5]. Resorcinol was shown to be a substrate for hydroxylation by extracts of orcinol grown and resorcinol grown *Ps. putida* ORC. However, the ratios of the activities towards orcinol and resorcinol were dependent on the substrate used for growth [2] indicating that distinct hydroxylases or catabolic sequences were specifically induced during growth on each substrate. Flavoprotein hydroxylases have now been obtained in homogeneous form from orcinol and resorcinol grown *Ps. putida* ORC and a comparison of some of the properties of the hydroxylases induced during growth of

Ps. putida ORC on orcinol and resorcinol are made in this report. Details of the purification procedures are reported elsewhere [18].

Results

Physical properties of orcinol and resorcinol hydroxylases

Orcinol and resorcinol hydroxylases were purified to apparent homogeneity from extracts of orcinol gorwn and resorcinol grown *Ps. putida* ORC. Each protein preparation gave single bands after disc-gel electrophoresis in both native and denaturing conditions. The electrophoretic mobility of each protein was indistinguishable, and electrophoresis of a mixture of the proteins in polyacrylamide gels did not resolve them. The protein bands in gels observed coincided with the enzymic activity zone for each protein. Solutions of both proteins were yellow and exhibited absorption spectra that did not significantly differ (Fig. 1). The chromophore was dissociated from each enzyme by acidic ammonium sulphate treatment and shown to be FAD by absorption spectroscopy and chromatography [5]. The sedimentation characteristics of the native proteins and the molecular weight determinations in sodium dodecyl sulphate gels indicated that the proteins were monomeric ($M_r \approx 70\,000$) and contained 1 mol of FAD per 70 000 daltons. These properties are summarized in Table I.

Amino acid analyses

Results of the amino acid analyses of orcinol and resorcinol hydroxylases are given in Table II. Except for the large differences shown by alanine and valine, the overall composition of the two proteins is very similar. Summations of the basic, acidic (+ amide) and lipophylic residues for the two proteins reinforce this conclusion (Table II).

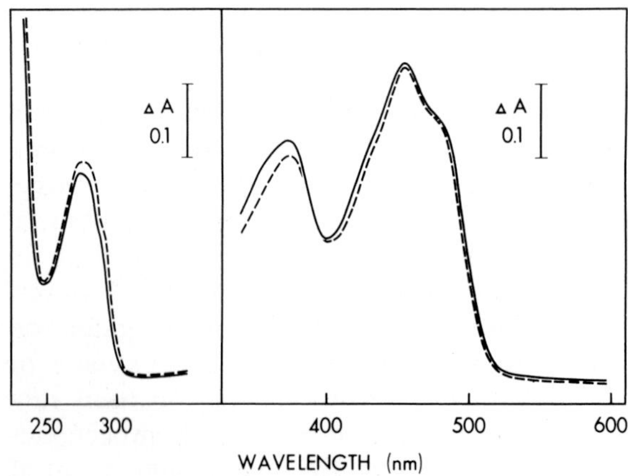

Fig. 1. Absorption spectra of orcinol (——) and resorcinol (---) hydroxylases.

Substrate hydroxylation

Confirmation that resorcinol was hydroxylated by orcinol hydroxylase and vice-versa was obtained by following the ultraviolet absorption changes that occurred during catalysis. Reaction mixtures turned brick-red indicating that a quinone had been formed. Chromatography of the ether extracts of acidified reaction mixtures showed that the substrates had been converted to more polar products with R_F values similar to hydroxyquinol and 2,3,5-trihydroxytoluene. Further characterization of the products was not attempted. Fig. 2 shows an example of substrate analogue hydroxylation. In the left panel 4-ethylresorcinol is provided as a substrate for orcinol hydroxylase and

TABLE I

PROPERTIES OF ORCINOL AND RESORCINOL HYDROXYLASES

Property	Orcinol hydroxylase	Resorcinol hydroxylase
Molecular Weight		
(a) Sephadex filatration	68 000	67 000
(b) Dodecyl sulphate gel electrophoresis	70 000	70 000
Sedimentation $S_{20,W}$ value	4.0–4.2 S	3.9–4.2 S
Absorption coefficient	10.4	10.0
λ_{max} 453 nm		
Flavin content	FAD	FAD
Substrate specificity	Restricted	Wide
Effector specificity	Restricted	Wide
Sensitivity to air	Sensitive	Insensitive
Reduction by 2-mercaptoethanol	—	+

TABLE II

AMINO ACID COMPOSITION OF ORCINOL AND RESORCINOL HYDROXYLASES

Amino acid	Residues	
	Orcinol hydroxylase	Resorcinol hydroxylase
Lysine	17	10
Histidine	17	17
Arginine	51	54
Cyst(e)ine	10	8
Aspartyl*	55	56
Threonine	19	21
Serine	27	27
Glutamyl*	68	69
Proline**	38	36
Glycine	61	63
Alanine**	65	30
Valine**	50	85
Methionine**	13	10
Isoleucine**	27	30
Leucine**	66	70
Tyrosine	16	16
Phenylalanine**	19	17
Asx + Glx	123	125
Basics	85	81
Lipophylics**	278	278

* Asp + Asn or Glu + Gln.

** Lipophylics summed.

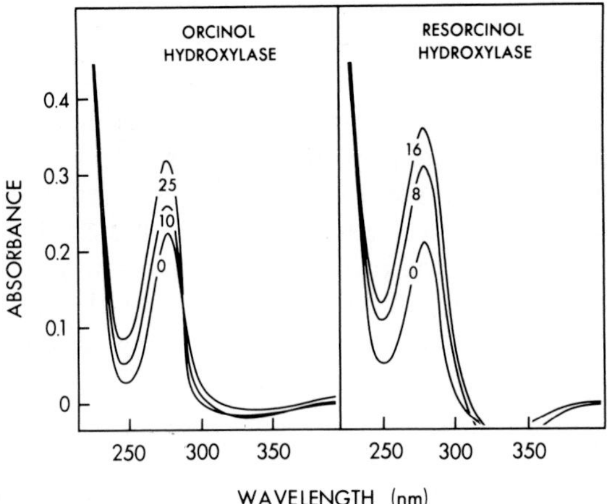

Fig. 2. Progress of hydroxylation shown by ultraviolet spectral changes. Left: hydroxylation of 4-ethylresorcinol by orcinol hydroxylase, and right: hydroxylation of 4-chlororesorcinol by resorcinol hydroxylase. Numbers refer to the time (in min) when the spectra were recorded.

in the right panel 4-chlororescorcinol is hydroxylated by resorcinol hydroxylase. While 4-ethylresorcinol is a substrate for both enzymes, 4-chlororesorcinol is an effector only for orcinol hydroxylase.

The stoichiometric relationship of the consumption of the three substrates (aromatic substrate—nucleotide—O_2) was not investigated due to the technical difficulties previously encountered with orcinol hydroxylase from *Ps. putida* [4, 5], as well as the generation of hydrogen peroxide, particularly when resorcinol was used as the substrate for orcinol hydroxylase.

Substrate (effector) specificities and kinetic constants

The non-identity of the two hydroxylases is most clearly revealed by their catalytic properties. Thus resorcinol hydroxylase is much more versatile in the range of substrates it can hydroxylate, as well as for the analogues that function only as effectors. Tables III and IV summarize some of the data obtained. Although both enzymes hydroxylate orcinol and resorcinol the ratio of the oxidation rates observed for the substrates is markedly different, orcinol being almost as good a substrate for resorcinol hydroxylase, and resorcinol being a poor substrate for orcinol hydroxylase. NADH is the preferred electron donor over NADPH under these conditions (Fig. 3).

Steady state kinetic analysis of the enzymes reveals quantitative differences in their ability to accommodate various substrates. A comparison of the apparent K_m values obtained for some of the reactants is shown in Table V.

Reduction and oxidation of orcinol and resorcinol hydroxylase

It was observed during the purification procedures that resorcinol hydroxylase was readily reduced by 2-mercaptoethanol, a component of the buffers used, but that orcinol hydroxylase remained oxidized. Fractions of column effluents possessing resorcinol hydroxylase were often almost colourless, but on shaking in air, the solutions regained their yellow colour. Orcinol hydroxylase however eluted in bright yellow fractions. Accordingly, a comparison of the rate of reduction of the flavin of both proteins by NADH was undertaken. Whereas 75 per cent reduction of resorcinol hydroxylase occurred in 1 min, orcinol hydroxylase was only 12 per cent reduced in the same time. These experiments were conducted under N_2 at 25°C with a molar excess of NADH: protein of 3:1. In the presence of effectors the bleaching of the spectrum was complete within this time for both enzymes.

Restoration of the oxidized flavin spectrum was readily achieved with aeration, as well as with ferricyanide and 2,6-dichlorophenol indophenol blue. Fig. 4 shows the reoxidation of resorcinol hydroxylase after reduction with dithionite, by 2,6-dichlorophenolindophenol blue.

Effector-induced perturbation of absorption spectra

Difference spectra of enzyme-effector minus enzyme reveal changes in the absorption spectra in the visible region similar to those observed for other flavoprotein hydroxylases (Fig. 5). A strong positive band is given at

TABLE III

SUBSTRATE-EFFECTOR SPECIFICITIES OF ORCINOL AND RESORCINOL HYDROXYLASES

S, substrate; E, effector; N, neither; ( ), minor role.

Aromatic compound	Orcinol hydroxylase	Resorcinol hydroxylase
Orcinol	S	S
Resorcinol	S, E	S
m-Cresol	E	E
Phloroglucinol	E	S, E
4-Methylresorcinol	S	S
4-Ethylresorcinol	S	S, (E)
4-Chlororesorcinol	E	S, (E)
4-Bromoresorcinol	E	S, E
4-Formylresorcinol	N	E
3,4-Dimethylphenol	N	E

TABLE IV

RELATIVE OXIDATION RATES OBSERVED FOR ORCINOL AND RESORCINOL HYDROXYLASES WITH SUBSTRATE ANALOGUES

Aromatic compound	Orcinol hydroxylase ($nmol\ O_2 \cdot min^{-1}$)	Resorcinol hydroxylase ($nmol\ O_2 \cdot min^{-1}$)
Orcinol	29	39
Resorcinol	13.5	35
m-Cresol	8.0	24.5
Phloroglucinol	8.0	28.5
4-Bromoresorcinol	4.0	39
4-Ethylresorcinol	5.0	29.5
3-Trifluoromethylphenol	41.0	29

385 nm and a strong negative band at 495 nm. The details and intensity of the shifts differ for the two enzymes with the same effectors as might be expected from the kinetic analysis. K_D values obtained by measurement of these perturbations of the enzymes' spectra in the presence of varying effector concentrations correlate reasonably well with the K_m values obtained.

Effect of sulphydryl group reagents on enzymic activity

Four sulphydryl group reagents, *p*-chloromercuribenzoate, dithionitrobenzoic acid, *N*-ethylmaleimide and iodoacetic acid, inhibited the catalytic activity of the two enzymes to varying degrees. *p*-Chloromercuribenzoate was the most effective reagent, followed by *N*-ethylmaleimide. Whereas dithionitrobenzoic acid completely inactivated both enzymes after 12–24 h incubation, iodoacetic acid only partially inhibited enzymic activity (Siegmann, D. and Ribbons, D.W. unpublished results).

Sulphydryl groups of the enzymes that reacted most readily with dithio-

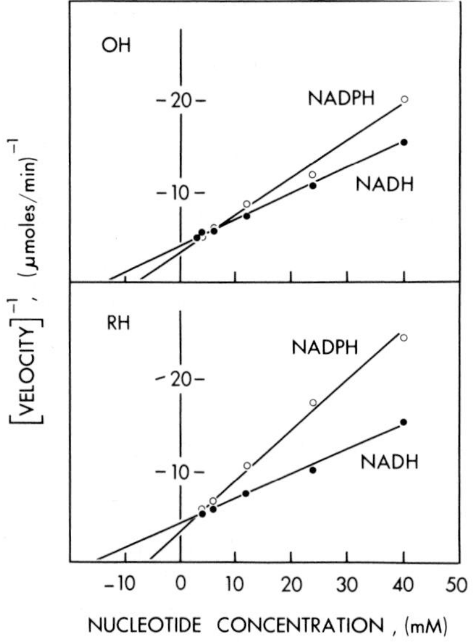

Fig. 3. Lineweaver-Burk plots of NADH and NADPH concentration and reaction velocity. Top: orcinol hydroxylase; bottom, resorcinol hydroxylase.

TABLE V

COMPARISON OF APPARENT K_m VALUES FOR THE SUBSTRATES (EFFECTORS) OF ORCINOL AND RESORCINOL HYDROXYLASES

Aromatic compound	Orcinol hydroxylase K_m (mM)	Resorcinol hydroxylase K_m (mM)
Orcinol	0.01	0.01
Resorcinol	0.33	0.007
m-Cresol	0.04	0.2
3-Trifluoromethylphenol	0.025	0.13

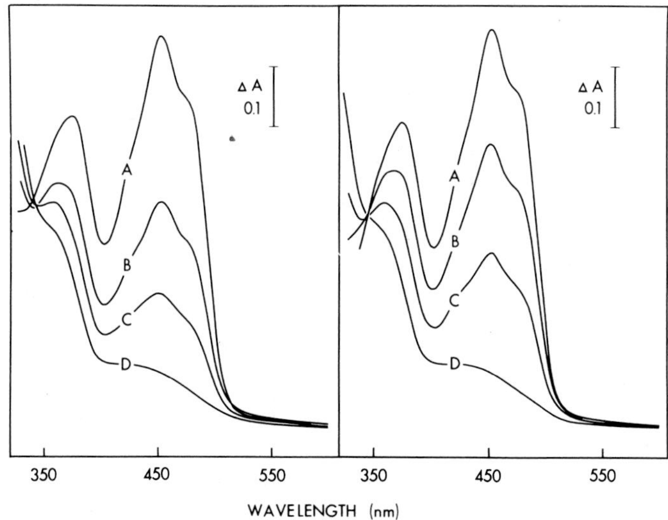

Fig. 4. Reduction of resorcinol hydroxylase by dithionite (right) and its reoxidation by 2,6-dichlorophenolindophenol blue (left).

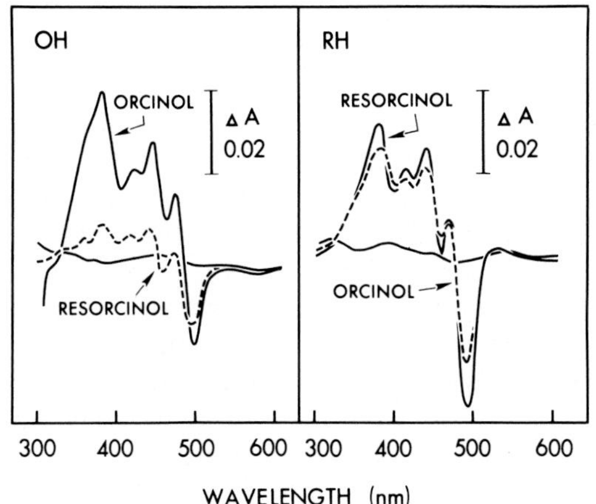

Fig. 5. Effector dependent spectral perturbation of the visbile absorption spectra of orcinol (left) and resorcinol (right) hydroxylases.

nitrobenzoic acid did not appreciably affect catalytic activity (Siegmann, D. and Ribbons, D.W., unpublished results). This did not appear to be the case when *p*-chloromercuribenzoate was used as the reagent for the titration curves of sulphydryl group modification. Comparison of the inactivation of the enzymes by dithionitrobenzoic acid is shown in Fig. 6. The control incubations also show the rapid loss of enzymic activity of orcinol hydroxylase on dilution, a feature that readily distinguishes it from resorcinol hydroxylase. Futhermore the presence of dithionitrobenzoic acid prevents this initial loss of orcinol hydroxylase activity.

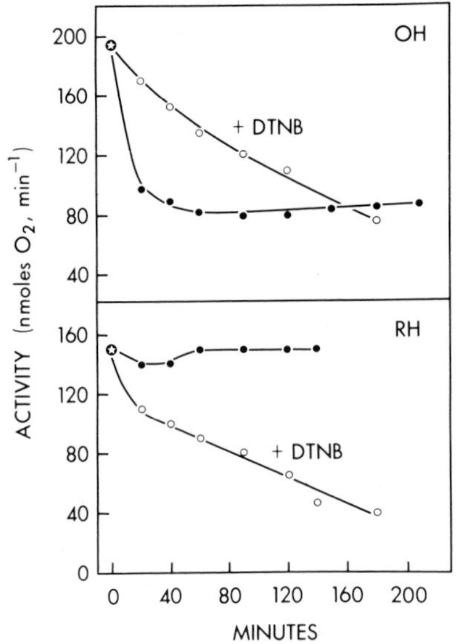

Fig. 6. Inactivation of orcinol hydroxylase (top) and resorcinol hydroxylase (bottom) by dithionitrobenzoic acid (DTNB).

Discussion

The observation that *Ps. putida* ORC possessed mechanistically dissimilar enzyme sequences after growth on orcinol and resorcinol for the oxidation of these compounds to common cellular metabolites suggested that genetically distinct hydroxylases were induced for the initial homologous reaction of these peripheral catabolic pathways (Scheme 1). It had been shown previously that the flavoprotein oricnol hydroxylase from *Ps. putida* 01 was recruited for growth on resorcinol by mutation of a regulatory gene so that it was expressed constitutively. However to explain growth of *Ps. putida* ORC on resorcinol by the use of its orcinol hydroxylase seemed unlikely, since similar viable counts were obtained on resorcinol media, unlike strain 01. Thus the present study was undertaken to demonstrate that homologously functional hydroxylases were induced in *Ps. putida* ORC, and if so, to compare their properties. Isofunctional flavoprotein hydroxylases have been compared from different species of Pseudomonas and other bacteria, but we are unaware of a study of such enzymes elaborated in the same strain.

Conventional procedures were used for the purification of orcinol and resorcinol hydroxylase activities from *Ps. putida* ORC to apparently homogenous protein preparations. It was not unambiguously ascertained that these preparations were devoid of their isofunctional counterpart, i.e. that orcinol hydroxylase was free of resorcinol hydroxylase since growth on orcinol may also allow synthesis of resorcinol hydroxylase, and vice-versa. The only evidence available to discount this possibility is (a) the failure to obtain an N-terminal analysis of resorcinol hydroxylase, while the N-terminus of orcinol hydroxylase was shown to be threonine; and (b) that anti-sera prepared to each preparation did not show precipitin lines of identity to both enzymes. However the purpose of this work was to establish whether or not different flavoprotein hydroxylases are induced for use in the orcinol and resorcinol pathways. In spite of the gross similarities shown in physicochemical properties of these enzymes, they are easily distinguished by amino acid compositions and by their reactivity to some oxidizing (air) and reducing (2-mercaptoethanol) reagents. Additionally, the spectra of substrates and effectors employed by the two enzymes is radically different, resorcinol hydroxylase being much more versatile in both respects. Significant differences in nucleotide preference were however not observed, although resorcinol hydroxylase is much more rapidly reduced by NADH in the absence of effectors than orcinol hydroxylase.

Previously six flavoprotein hydroxylases have been obtained as homogeneous preparations from several Gram-negative aerobic bacteria, usually of the genus Pseudomonas [3, 6–15], and one from the yeast *Trichosporon cutaneum* [16]. Furthermore isofunctional enzymes from different species of Pseudomonas have been studied, e.g. 4-hydroxybenzoate—3-hydroxylase [9–11, 17]. The comparison of similar flavoprotein hydroxylases from the same strain has to our knowledge not been made. From the limited studies made here, orcinol and resorcinol hydroxylases from *Ps. putida* ORC share properties similar to other flavoprotein hydroxylases including flavin content per polypeptide chain, substrate (effector) control of flavin reduction and substrate (effector) induced perturbations of the flavin absorption spectra. The existence of both orcinol and resorcinol hydroxylases in *Ps. putida* ORC, and the similarities of their properties and amino acid composition suggests that they have only recently diverged for their respective functions after gene duplication, and that negative immunological evidence for the relatedness of proteins is unsatisfactory.

Acknowldegements

The authors are grateful to Edye E. Groseclose, John L. Michalover and Keiko Ohta for the assitance they provided. Helpful discussions and the culture of *Ps. putida* ORC were provided by Dr. P.J. Chapman. This work was mainly supported by an NIH grant No. R01 GM 20172-01; the initial observations were made when Y.O. (Visiting) and D.W.R. were Howard Hughes Medical Institute Investigators. Ms. Debbie Siegmann conducted the sulphydryl group inhibition experiments during tenure of a predoctoral traineeship from NIH Training Grant No. NIH 1 T01 GM02011-05.

References

1 Chapman, P.J. and Ribbons, D.W. (1967) Biochem, J. 106, 44P
2 Chapman, P.J. and Ribbons, D.W. (1976) J. Bacteriol. 125, No. 3
3 Ohta, Y. and Ribbons, D.W. (1970) FEBS. Lett. 11, 189

4 Ribbons, D.W., Ohta, Y. and Higgins, J. (1971) Miami Winter Symposia Vol. 4, p. 251, Academic Press, Inc., New York
5 Ohta, Y., Higgins, I.J. and Ribbons, D.W. (1975) J. Biol. Chem. 250, 3814
6 Yamamoto, S., Katagiri, M., Maeno, S. and Hayaishi, O. (1965) J. Biol. Chem. 240, 3408
7 White-Stevens, R.H. and Kamin, H. (1972) J. Biol. Chem. 247, 2358
8 Takemori, S., Nakamura, M., Katagiri, M. and Nakamura, T. (1971) Proc. 3rd Int. Symp. Flavins and Flavoproteins (Kamin, H., ed.), p. 463, University Park Press, Baltimore
9 Hosokawa, K. and Stanier, R.Y. (1969) J. Biol. Chem. 241, 2453
10 Nakamura, S., Ogura, Y., Yano, K., Higashi, N. and Arima, K. (1970) Biochemistry 9, 3235
11 Howell, L.G., Spector, T. and Massey, V (1972) J. Biol. Chem. 247, 4340
12 Levy, C.C. (1967) J. Biol. Chem. 242, 747
13 Strickland, S. and Massey, V. (1973) J. Biol. Chem. 248, 2944
14 Groseclose, E.E. and Ribbons, D.W. (1973) Biochem. Biophys. Res. Commun. 55, 807
15 Michalover, J.L. and Ribbons, D.W. (1973) Biochem. Biophys. Res. Commun. 55, 1102
16 Neujahr, H.Y. and Gaal, A. (1973) Ero. J. Biochem. 35, 386
17 Hesp, B., Calvin, M. and Hosokawa, K. (1969) J. Biol. Chem. 244, 5644
18 Ohta, Y. and Ribbons, D.W. (1976) Eur. J. Biochem. 61, 259

Flavins and Flavoproteins, edited by T.P. Singer

CHAPTER 14

REACTIONS OF THE REDUCED FORM OF SALICYLATE HYDROXYLASE

ROBERT P. PRESSWOOD and HENRY KAMIN

Department of Biochemistry, Duke University Medical Center, Durham, N.C. 27710 (U.S.A.)

Salicylate hydroxylase is a mixed-function oxidase, which converts salicylate into catechol, using O_2 and NADH. The enzyme described previously [1–5] by our laboratory was induced in a soil bacterium, and contains 2 mol FAD and two apparently identical subunits per mol protein. One of the major conclusions of mechanistic studies [3, 4] was that no kinetically significant adduct of oxygen and reduced flavin could be identified, and that the rate of oxidation of reduced enzyme was independent

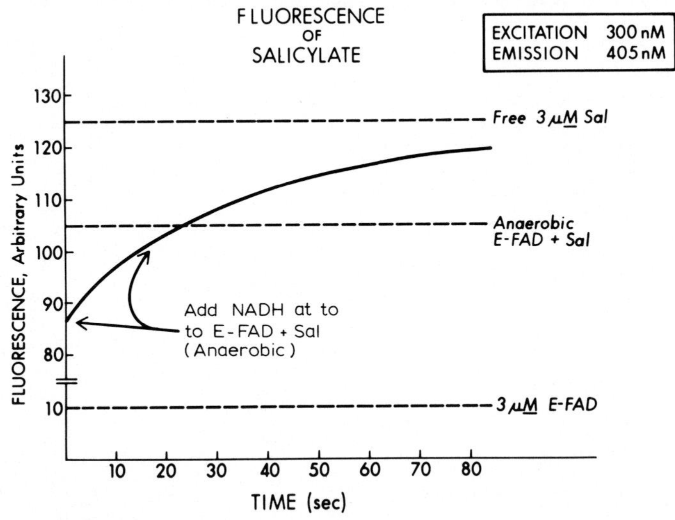

Fig. 1. Changes observed in silicylate fluorescence upon anaerobic reduction of salicylate-oxidized hydroxylase complex. Oxidized enzyme was complexed with salicylate (both 3 μM after mixing) in an anaerobic fluorescence cell in a volume of 3 ml and one equivalent of NADH per enzyme-FAD in a volume of 0.05 ml or less was placed in the side arm. Anaerobiosis was achieved by evacuation of the system (to less than 0.050 mm Hg), fluorescence was determined, and the hydroxylase was reduced by mixing in the contents of the side arm. Changes in fluorescence were then observed and recorded with a Farrand Mark I spectrofluorometer. Excitation was at 300 nm, with 5 nm slit widths; emission was at 405 nm, with 10 nm slit widths. Experiments were performed at room temperature, in 20 mM potassium phosphate, pH 7.6.

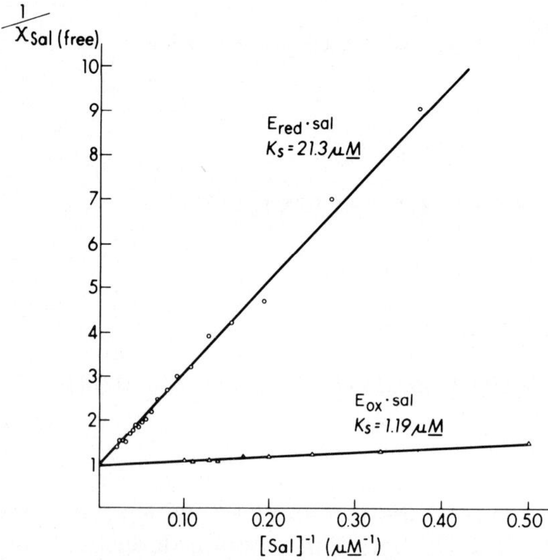

Fig. 2. Binding of salicylate to oxidized and reduced forms of salicylate hydroxylase. Polarization of salicylate fluorescence was observed vs. salicylate concentration in a Farrand Mark I spectrofluorometer and the data treated as described by Rawitch and Weber [8]. Excitation was at 300 nm, with 5 nm slit widths; emission was at 405 nm, with 10 nm slit widths. Oxidized hydroxylase was 3 μM in enzyme-FAD; reduced hydroxylase was 50 μM in enzyme-FAD. In the experiments with the anaerobic apparatus, anaerobiosis was achieved by alternate evacuation and flushing with argon bubbled through Fieser's solution. Enzyme was reduced by a slight excess of NADH from the side arm, and salicylate introduced as titrant. All solutions contained 10 mM glucose and 11 units per ml glucose oxidase to ensure anaerobiosis. Experiments were performed at room temperature in 20 mM potassium phosphate, pH 7.6.

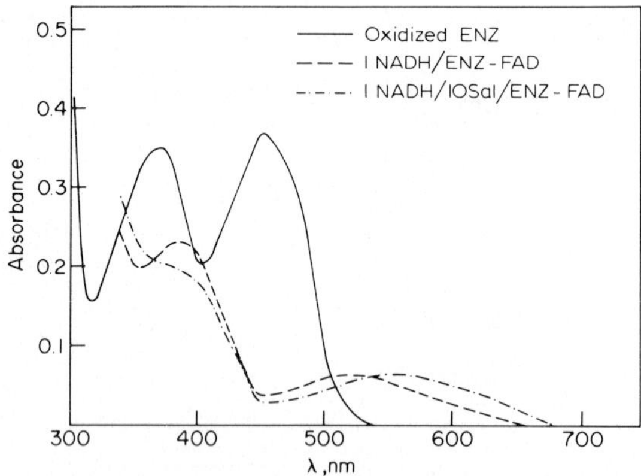

Fig. 3. Spectra obtained for recent preparations of salicylate hydroxylase in the uncomplexed oxidized and reduced states and the salicylate-complexed reduced state. The reduced spectra were obtained in an anaerobic cuvette following careful anaerobiosis by evacuation, and tipping in the indicated NADH from a side arm. The hydroxylase was 32.5 μM in enzyme-FAD in 3 ml volume and the spectra are corrected for volume changes resulting from addition of NADH-containing buffer. Spectra were recorded on a Cary 14 spectrophotometer at room temperature. All solutions were in 20 mM potassium phosphate buffer, pH 7.6.

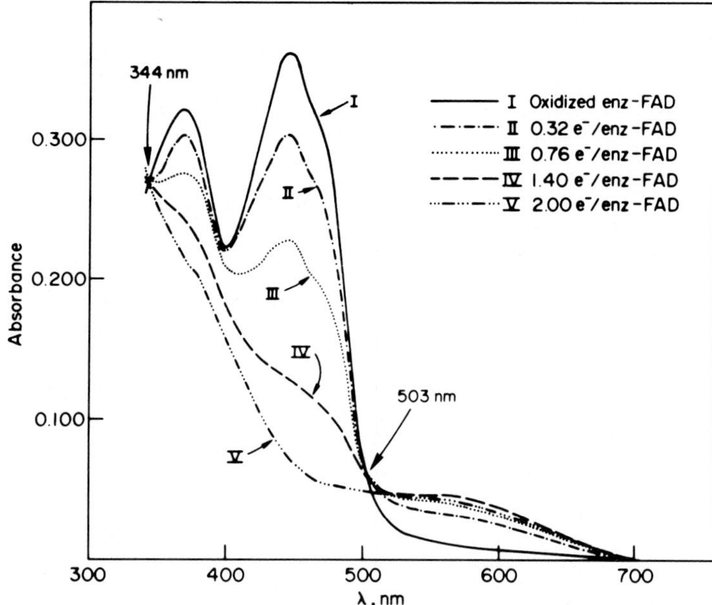

Fig. 4. Titration of salicylate-oxidized hydroxylase complex with NADH. Hydroxylase was 32.0 μM in enzyme-FAD, in the presence of 300 μM salicylate. Anaerobiosis was achieved as described in the legend to Fig. 2, with the exception that the enzyme-substrate complex was not prepared in the presence of the glucose–glucose oxidase system, although the NADH solution did contain this oxygen trap. Consequently, the first few additions of NADH serve to consume residual oxygen in the enzyme solution. Calculation of the number of electrons added per enzyme-FAD therefore is corrected for this consumption of residual oxygen. Spectra were recorded on a Cary 14 spectrophotometer at room temperature and are corrected for dilution. All solutions were in 20 mM potassium phosphate buffer, pH 7.6.

of the nature or presence of substrate or pseudosubstrate. We report a re-assessment of these results, based on newly observed properties of the reduced enzyme-salicylate complex, and also the observation of a long-wavelength absorbing species in the reduced form of recent salicylate hydroxylase preparations. A preliminary report of some of these observations has been presented [5].

When salicylate is complexed with oxidized salicylate hydroxylase, its fluorescence is partially quenched, as illustrated in Fig. 1. After anaerobic reduction of the oxidized enzyme-salicylate complex with one equivalent of NADH per FAD, salicylate fluorescence is rapidly quenched even further. This quenching then undergoes a slow-first-order relaxation, ($t_{1/2}$ of about 35 s), to yield fluorescence almost equal to that of free salicylate. If oxygen is readmitted to the system at various times during this relaxation, less hydroxylation of substrate is found to occur as time progresses; after 360 s, only about 20–40% of the expected hydroxylation is observed, as measured by salicylate disappearance (fluorescence) or catechol production (pyrocatechase-catalyzed production of *cis-cis*-muconic acid, 260 nm) [6].

Polarization of salicylate fluorescence was monitored as a function of salicylate concentration to ascertain the dissociation constants of complexes

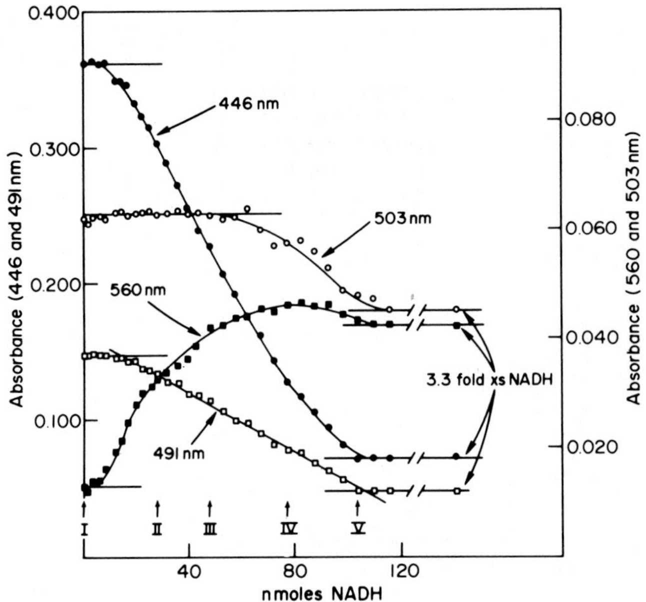

Fig. 5. Titration of salicylate-oxidized hydroxylase complex with NADH. Data represent the same experiment as described in Fig. 4, with the course of the tiration followed at selected wavelengths. Data are corrected for dilution.

of salicylate with the oxidized and the reduced forms of the hydroxylase, as shown in Fig. 2. The apparatus of Foust et al. [7], was utilized for the anaerobic experiments. These data were corrected by the method of Rawitch and Weber [8] for differing relative quantum yields of salicylate fluorescence in the free and the bound states and a non-zero polarization of free salicylate and are shown as a double-reciprocal plot of mol fraction of free salicylate vs. salicylate concentration. The dissociation constant of the salicylate-reduced enzyme complex is significantly greater than that for the salicylate-oxidized enzyme complex. Comparable results were obtained, using fluorescence quenching, by Takemori et al. [9], with a different salicylate hydroxylase.

Oxidation of reduced hydroxylase in the presence of 300 μM (as contrasted to 30 μM) salicylate yields the theoretically expected hydroxylation, as determined by catechol formation. Corresponding stopped-flow experiments with horseradish peroxidase (Complex I) as an indicator for H_2O_2 production, showed peroxide formation at 30 μM, but not at 300 μM salicylate concentration. We conclude, from both the stoichiometry and fluorescence data, that reduction of the oxidized enzyme-salicylate complex at low salicylate concentration leads to slow dissociation of substrate. Normal turnover is much more rapid than this dissociation, so that the same concentration of substrate in turnover yields total hydroxylation. Thus, we have an interesting example of a thermodynamically unstable intermediate form of an ES complex, which, nevertheless, has sufficient kinetic stability to permit

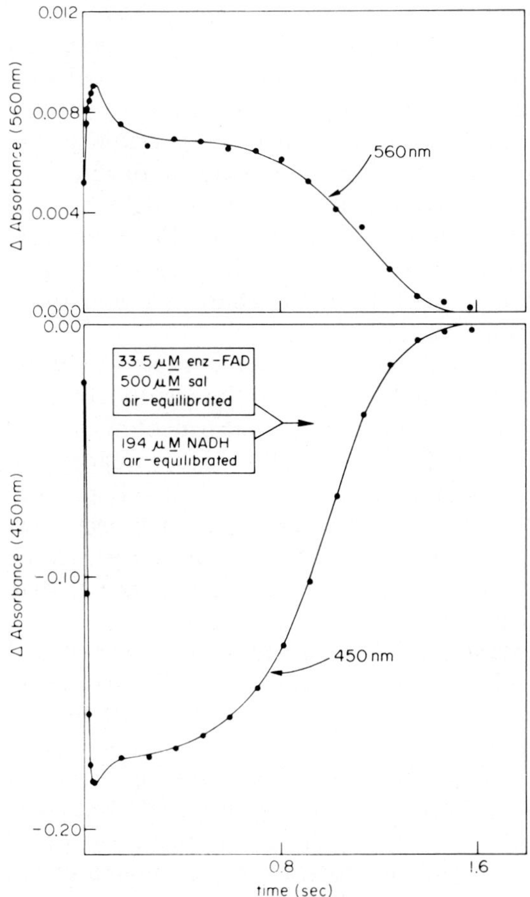

Fig. 6. Stopped-flow turnover experiment, mixing salicylate-complexed oxidized hydroxylase with NADH. Concentrations and wavelengths monitored are as illustrated. Points represent the actual data. Experiments were at 20°C in 20 mM potassium phosphate, pH 7.6. (The experiments illustrated in Figs. 6–9 were performed in collaboration with Dr. Quentin H. Gibson of Cornell University. Procedures are described in detail elsewhere [3]).

catalysis. Similar considerations have arisen recently in the case of decay of the central complex of "internal" monooxygenases [10, 11].

Recent preparations of salicylate hydroxylase have shown, upon anaerobic reduction by NADH, an unusual long-wavelength species, with an absorbance maximum at 520 nm in the absence of salicylate and at 560 nm in the presence of saturating salicylate (Fig. 3). Addition of salicylate to uncomplexed reduced enzyme results in a shift of the 520 nm band to 560 nm. Maximal absorbance changes are reached at 1 NADH/FAD, and a 100-fold excess of NADH does not diminish this absorbance. The same spectra are seen upon addition of dithionite to either complexed or uncomplexed enzyme. Enzyme-salicylate complex reduced with 1 NADH/enzyme-FAD gave no EPR signal.

Anaerobic titration of ES with NADH gives an isosbestic point at

344 nm (Fig. 4). At 503 nm, isosbesticity is observed for approximately the first half of the titration. This is better illustrated in Fig. 5, which shows the same data at selected wavelengths for the course of titration. The absorbance changes at 560 nm suggest that the contributing species is formed essentially during the first half of the titration. These results open the possibility that the reduced hydroxylase may contain two forms of reduced flavin, both absorbing at 450 nm, but only one absorbing at 560 nm. It might be conjectured that the 560 nm-absorbing species has fully-reduced flavin engaged in a charge-transfer interaction. The shift of absorption maximum upon addition of salicylate indicates that the nature of the charge-transfer interaction is sensitive to substrate binding [12, 13].

The possibility that the species is a charge-transfer complex between flavin and pyridine nucleotide was rule out not only by the results obtained in dithionite reduction, but also by experiments utilizing the alcohol–alcohol dehydrogenase system with catalytic NAD^+ instead of a reagent quantity of NADH. The long-wavelength absorbance obtained is identical to that observed using stoichiometric NADH. We have conducted a search* for adsorbed small molecules, and have found none. We therefore entertain the hypothesis that the species observed is a charge-transfer complex between the flavin and a group from the protein.

Treatment of the hydroxylase (30 μM enzyme-FAD) with 0.1 M H_2O_2 for 10–30 min or with 1 μM Cu^{2+} (aerobically) for 24 h results in enzyme preparations which are competent catalytically at about 40–50% of the original rate, but which yield no long-wavelength absorbance upon anaerobic reduction. Both H_2O_2- and Cu^{2+}-treated enzyme preparations contain about 10% of the flavin in a form unreducible by NADH but reducible by dithionite. The H_2O_2-treated enzyme shows modification of all methionines and of two cysteines per mol of protein. The enzyme does not appear to have disulfide crosslinks. We tentatively suggest that this new reduced species may be a charge-transfer complex of fully reduced flavin and a protein sulfur moiety. A charge-transfer complex between oxidized flavin and a sulfhydryl anion has been suggested for dihydrolipoyl dehydrogenase [14].

The reduced spectrum of this "new" hydroxylase is different than that observed for the enzyme preparations reported previously [2–4], which yielded no long-wavelength absorbance upon anaerobic reduction by NADH or dithionite. All recent cell cultures have been found to yield the "new" hydroxylase, and we have been unable to induce the "old" enzyme from such cultures. Nevertheless, we do still have older cultures from which we can purify the "old" hydroxylase. Both types of enzyme are purified by identical procedures and have the same specific activities at each stage of

* These studies included (with the native hydroxylase) exhaustive dialysis, repeated salt precipitations, and chomatography with a number of systems, including passage over a bed of activated charcoal and celite. The supernatants resulting from denaturation of the hydroxylase by a variety of techniques were also examined, with negative results.

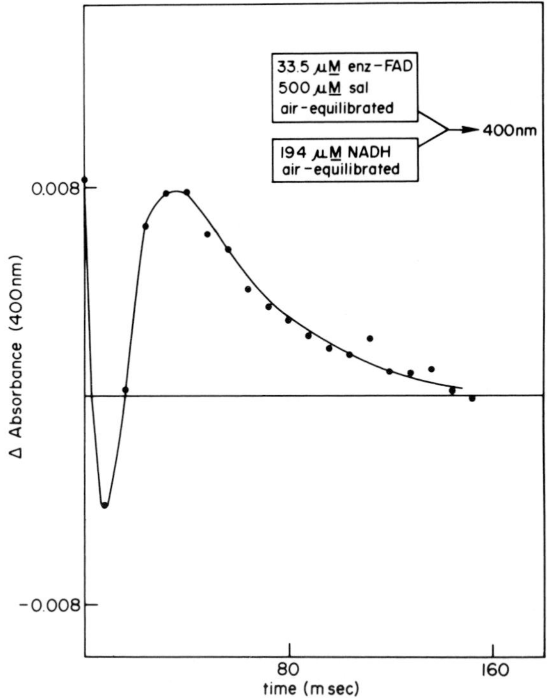

Fig. 7. Stopped-flow turnover experiment, mixing salicylate-complexed oxidized hydroxylase with NADH. Concentrations and wavelength monitored are as illustrated. Points represent the actual data. Note the expanded time scale relative to Fig. 6. Other conditions were as described in the legend to Fig. 6.

purification. They also have native and subunit molecular weights which are indistinguishable, as well as amino acid compositions and tryptic peptide maps which are strikingly similar. Both contain two apparently identical subunits and two mol of FAD per mol of protein. The "new" hydroxylase exhibits similar behavior towards the various substrates and pseudosubstrates as described for the "old" enzyme [1–4].

Stopped-flow turnover experiments, reacting air-equilibrated enzyme-salicylate complex with air-equilibrated NADH, as illustrated in Fig. 6, show that long-wavelength absorbance appears, stabilizes and disappears on the same time scale as the changes observed at 450 nm. The spectrum of the reaction mixture after entry of the enzyme into the steady-state indicates that the long-wavelength species in turnover is spectrally similar to that described in the static experiments. More recent stopped-flow studies with the "old-type" enzyme show that a small quantity of long-wavelength intermediate (less than half the concentration seen with "new-type" enzyme) can be detected during turnover in the stopped flow, but this form still is absent in static titration experiments. Thus, unexplained differences persist in the quantity and temporal stability of the long wavelength intermediate forms of the two types of enzyme preparations.

At shorter wavelengths (400 or 410 nm) two events, opposite in

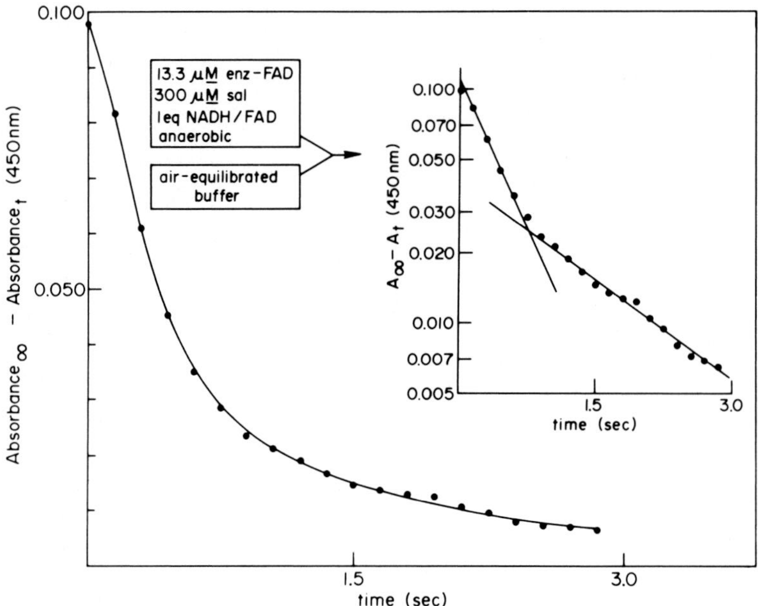

Fig. 8. Oxidation of anaerobically reduced salicylate-complexed hydroxylase in the stopped-flow apparatus. Concentrations were as illustrated. This experiment was monitored at 450 nm. Points represent the actual data. The insert shows the same data plotted semi-logarithmically. Experimental conditions were as described in the legend to Fig. 6.

direction of absorbance change, precede entry into the steady state (Fig. 7). The first event is NADH-dependent, the second is oxygen-dependent. This would be the behavior expected if the enzyme-substrate complex is first reduced, and then reduced flavin forms the type of adduct with oxygen described by Massey's laboratory [15–18].

Fig. 8 illustrates an experiment reaction reduced enzyme-salicylate complex with oxygen-containing buffer, with a concentration of salicylate sufficient to prevent the slow dissociation of reduced enzyme-salicylate complex. No peroxide was produced. A slight lag was observed initially; this lag was shorter at higher oxygen tensions. The reoxidation is biphasic at this wavelength, but if the experiment is monitored at 560 nm, only the slow phase is evident. This may reflect two reduced flavin species, whose reactivities with O_2 are different. A double-reciprocal plot of the rate of the more rapid absorbance change observed at 450 nm, corrected for contribution of the slower event, vs. oxygen tension is shown in Fig. 9. A limiting rate at "infinite" oxygen is extrapolated, with a K_m comparable to that observed in normal initial-velocity kinetic studies. The slower phase also extrapolates to approximately the same rate constant, but the apparent K_m for O_2 is relatively high (1.5–2.0 mM). With an observed turnover number of 15–19 s^{-1} for the hydroxylase, reoxidation appears to be rate-limiting for the overall reaction.

When the same stopped-flow reoxidation experiment is performed at a

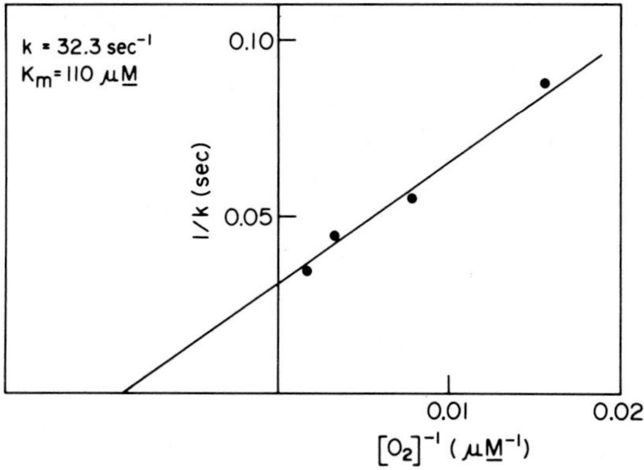

Fig. 9. Double-reciprocal plot of the apparent rates observed at 450 nm for the rapid phase of oxidation vs. oxygen tension, for the oxidation in the stopped-flow apparatus of anaerobically reduced salicylate-complexed hydroxylase. Concentrations were as illustrated in Fig. 8, except that oxygen tension was varied. The observed absorbance changes were corrected for contribution by the slower phase in derivation of the rates as plotted. Conditions of temperature, pH, and buffer concentration were as described in the legend to Fig. 6.

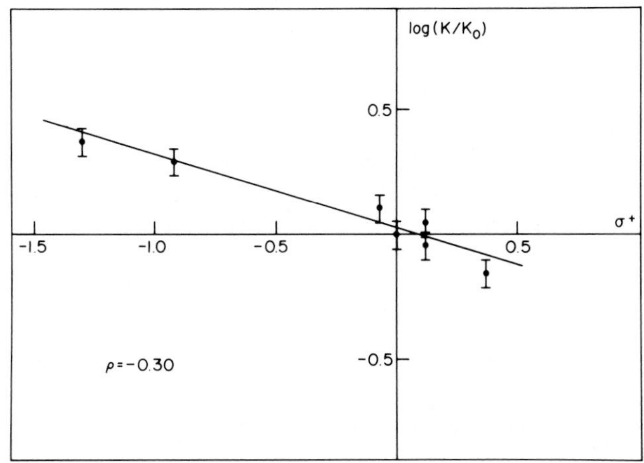

Fig. 10. Hammett plot of the derived first-order oxidative rate constants for a number of substituted salicylates based on the appropriate substituent constants for electrophilic aromatic substitution [20]. First-order rate constants were derived as illustrated in Fig. 9. Compounds utilized, with their respective substituent constants, were 4-aminosalicylate (— 1.3), 4-hydroxysalicylate (— 0.92), 3-methylsalicylate (— 0.07), salicylate (0.0), 3-hydroxysalicylate (0.12), 5-hydroxysalicylate (0.12), and 5-chlorosalicylate (0.40). Bars represent the estimated error in determination of the constants. The experiments from which these results were derived were performed at 20° C, in 20 mM potassium phosphate, pH 7.6, in a Durham—Gibson stopped-flow apparatus.

substrate concentration insufficient to prevent dissociation of reduced enzyme-salicylate complex, the observed reoxidation rates become proportional to oxygen tension [3, 4]. This is because the process observed is the reoxidation of uncomplexed reduced flavin, rather than of the salicylate-complexed reduced form.

Oxidation of reduced enzyme-salicylate complex was also examined at a number of wavelengths under varying oxygen tension, and these data were used to construct an approximate spectrum of the intermediate reduced flavin-oxygen adduct. The intermediate has an absorbance maximum at approximately 390 nm, with an extinction coefficient of about $8600\ M^{-1} \cdot cm^{-1}$. This is in good agreement with the intermediates described by Massey's laboratory [15–18]. It is also similar to an intermediate described in a bacterial luciferase system [19].

Fig. 10 illustrates the results obtained for the oxidation of reduced hydroxylase complexed with a number of substituted salicylates. These compounds were the 4-amino-, the 4-hydroxyl-, the 3-methyl-, the 3-hydroxyl-, the 5-hydroxyl-, and the 5-chloro-analogues of salicylate, as well as salicylate itself. In all cases, the rates of oxidation were based on the observed rapid phase of oxidation at 450 nm, after correction for the contribution of the slow phase. Data are plotted using Hammett substituent constants for electrophilic aromatic substitution, although a good fit is also obtained using constants for reaction at the carboxyl function [20]. This plot suggests that the observed reaction passes through a transition state in which the aromatic molecule plays a functional role, and that the species which attacks the substrate is electron-deficient in character.

Acknowledgment

We wish to thank Dr. Quentin H. Gibson for his collaboration in many of the stopped-flow studies, and Dr. K.V. Rajagopalan for collaboration in EPR studies. We also acknowledge the expert technical assistance of Mrs. Marie Hall. These studies were supported by N.I.H. Grant GM-21226.

References

1 White-Stevens, R.H. and Kamin, H. (1970) Biochem. Biophys. Res. Commun. 38, 882
2 White-Stevens, R.H. and Kamin, H. (1972) J. Biol. Chem. 247, 2358
3 White-Stevens, R.H. and Kamin, H. (1972) J. Biol. Chem. 247, 2371
4 White-Stevens, R.H. and Kamin, H. (1972) Oxidation Reduction Enzymes (Akeson, A. and Ehrenberg, A., eds.), p. 453, Pergamon Press, New York
5 Presswood, R.P. (1973) Fed. Proc., Abstr. 1444
6 Kojima, Y., Fugisawa, H., Nakazawa, J., Kanetsuna, F., Taniuchi, H., Nazaka, M. and Hayaishi, O. (1967) J. Biol. Chem. 242, 3270
7 Foust, G.P., Burleigh, Jr., B.D., Mayhew, S.G., Williams, Jr., C.H. and Massey, V. (1969) Anal. Biochem. 27, 530
8 Rawitch, A.B. and Weber, G. (1972) J. Biol. Chem. 247, 680
9 Takemor, S., Nakamura, M., Suzuki, K., Katagiri, M. and Nakamura, J. (1972) Biochim. Biophys. Acta 284, 382
10 Takemori, S., Nakai, Y., Katagiri, M. and Nakamura, J. (1969) FEBS Lett. 3, 214
11 Lockridge, O., Massey, V. and Sullivan, P. (1972) J. Biol. Chem. 247, 8097
12 Kosower, E.M. (1962) Molecular Biochemistry, p. 187, McGraw Hill, New York
13 Kosower, E.M. (1966) Flavins and Flavoproteins (Slater, E.C., ed.), p. 1, Elsevier, Amsterdam
14 Searls, R.L., Peters, J.M. and Sanadi, D.R. (1961) J. Biol. Chem. 236, 2317
15 Spector, T. and Massey, V. (1972) J. Biol. Chem. 247, 5632
16 Spector, T. and Massey, V. (1972) J. Biol. Chem. 247, 7123
17 Strickland, S. and Massey, V. (1973) J. Biol. Chem. 248, 2953
18 Entsch, B., Massey, V. and Ballou, D.P. (1974) Biochem. Biophys. Res. Commun. 57, 1018
19 Hastings, J.W., Balny, C., LePeuch, C. and Douzou, P. (1973) Proc. Natl. Acad. Sci. U.S. 70, 3468
20 Brown, H.C. and Okamoto, Y. (1958) J. Am. Chem. Soc. 80, 4979

Flavins and Flavoproteins, edited by T.P. Singer

CHAPTER 15

INHIBITION, PRELIMINARY SEQUENCE AND X-RAY STUDIES ON *p*-HYDROXYBENZOATE HYDROXYLASE FROM *PSEUDOMONAS FLUORESCENS*

JILLERT S. SANTEMA, FRANZ MULLER, PIETER J. STEENNIS, TJAN JARBANDHAN, JAN DRENTH[a] and RIK K. WIERENGA[a]

Department of Biochemistry, Agricultural University, Wageningen, and [a]Department of Structural Chemistry, University of Groningen, Zernikelaan, Groningen (The Netherlands)

Summary

p-Hydroxybenzoate hydroxylase is competitively inhibited with respect to NADPH by halogen ions and adenine nucleotides. The enzyme contains 5 sulfhydryl groups. Their chemical reactivity and catalytic importance have been investigated. Some preliminary sequence studies are also reported. Single crystals of the enzyme have been obtained. The unit cell parameters have been determined by X-ray techniques and three different heavy atom derivatives of the enzyme are described.

Introduction

The reaction mechanism of *p*-hydroxybenzoate hydroxylase (EC 1.14.13.2, 4-hydroxybenzoate, NADPH: oxygen oxidoreductase) has been studied in detail by Massey and coworkers employing the stopped flow technique [1–4]. The catalytic activity of the enzyme is strictly dependent on NADPH. The substrate *p*-hydroxybenzoate acts also as an effector facilitating the binding of NADPH. Thus the affinity of the enzyme for NADPH is increased by about a factor of ten in the presence of the substrate [1]. The substrate effects in the same manner the binding of molecular oxygen to the reduced form of the enzyme [2].

Spector and Massey [2] have tested a variety of aromatic compounds with respect to their ability to inhibit the enzyme at the *p*-hydroxybenzoate binding site. From these studies combined with circular dichroism measurements it follows that the enzyme undergoes a conformational change upon binding of the substrate or inhibitor. Recently we have shown that halogen ions inhibit the enzyme reaction competitively with respect to NADPH [5]. The halogen ion inhibition is of mixed type with respect to the substrate *p*-hydroxybenzoate. From fluorimetric data we concluded that a

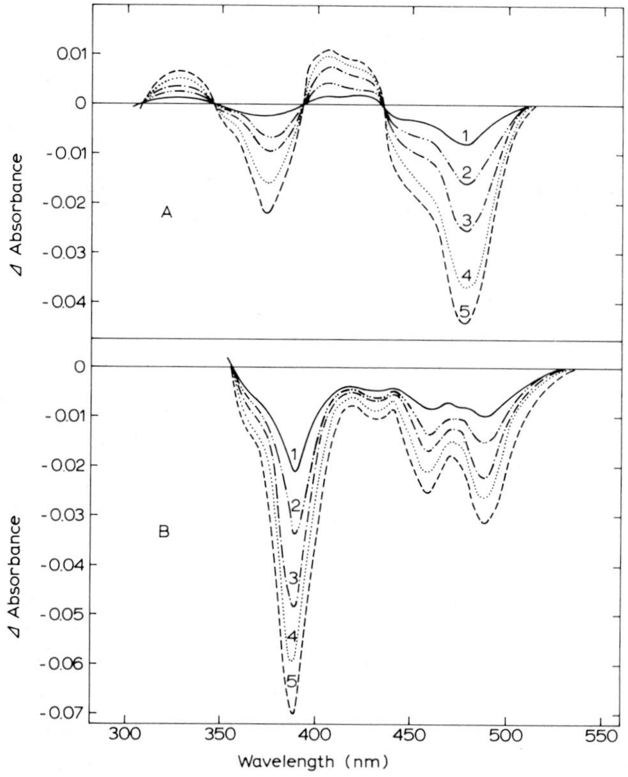

Fig. 1. Difference light absorbance spectra obtained by the titration of *p*-hydroxybenzoate hydroxylase with chloride. (A) free enzyme, 57 μM, in 30 mM potassium phosphate, pH 6.95; curves 1–5 were obtained in the presence of 2.5 mM, 5 mM, 9.9 mM, 19.6 mM and 29.1 mM chloride, respectively. (B) *p*-hydroxybenzoate-complexed enzyme, 43 μM, as in (A), the chloride concentration was 1 mM, 1.9 mM, 3.8 mM, 5.5 mM and 8.9 mM (curves 1–5), respectively.

conformational change occurs upon binding of halogen ions to the enzyme. We have extended these studies and report in this paper on other competitive inhibitors, on some chemical modification and preliminary sequence and X-ray studies.

Results and Discussion

In a previous paper [5] we have shown that halogen ions inhibit *p*-hydroxybenzoate hydroxylase competitively with respect to NADPH. Among the halogen ions, chloride is the most effective inhibitor. The competitive character of the chloride inhibition suggests that chloride should also bind to the free enzyme in a similar manner as NADPH. A chloride titration of free enzyme employing light absorbance difference spectrometry revealved that chloride indeed binds also to the free enzyme. The difference spectra (Fig. 1A) exhibit two rather broad negative absorbances at 477 and 373 nm. The first band is more intense than the second one. In addition both bands are associated with shoulders located at about 440 and 350 nm. On the other hand, when *p*-hydroxybenzoate-complexed enzyme was titrated with chloride, difference spectra as shown in Fig. 1B were obtained. These spectra are much better resolved than those of Fig. 1A and exhibit maxima at 487, 457, 432 and 388 nm. The dissociation constants calculated from these spectra according to Benesi and Hildebrand [6] are 20 mM and 5 mM for the free enzyme and the *p*-hydroxybenzoate-enzyme complex, respectively.

Previously we have shown [5] that K_I and V are independent of pH in the studied region of 6–8.3. However K_m increased below pH 7. This finding was explained by the protonation of a phosphate group of NADPH and led us to investigate whether or not adenine nucleotides also interact with the enzyme. Indeed *p*-hydroxybenzoate hydroxylase is inhibited by various adenine nucleotides and the inhibition is competitive with respect to NADPH. A comparison of the inhibitory effect of various adenine nucleotides on *p*-hydroxybenzoate hydroxylase is given in Table I. Among the nucleotides tested adenosine 2′, 5′-diphosphate and adenosine 2′-monophosphate are the strongest inhibitors. This fact indicates that the 2′-phosphate group plays an important role in the binding of NADPH to the enzyme and might explain why NADH cannot act as an electron donor for the catalytic reaction. Shoun et al. [7] have recently shown that the pH vs log K_m profile of *p*-hydroxybenzoate hydroxylase from *Pseudomonas putida* exhibits two ionization constants at 6.5 and 8.2. They suggested that these pK values are due to a SH group and an ϵ-amino group, respectively. Our inhibition studies indicate that the pK value of 6.5 is due to the (de)protonation of one phosphate group of NADPH. This phosphate group is probably the 2′-group.

In context with started X-ray and sequence studies the amino acid composition of *p*-hydroxybenzoate hydroxylase has been determined. The results will be published elsewhere. The enzyme consists of only one chain with a molecular weight of 54 000 as determined with the aid of sodium dodecyl sulphate gel-electrophoresis. The molecular weight of 55 000 of the holoenzyme is in disagreement with the value of 65 000 published by Howell

TABLE I

PERCENTAGE OF INHIBITION OF *p*-HYDROXYBENZOATE HYDROXYLASE BY VARIOUS ADENOSINE NUCLEOTIDES

The percentage of inhibition was determined by comparing the velocities of the NADPH oxidation in the presence and absence of the nucleotides. The assay mixture consisted of 0.1 mM NADPH, 0.3 mM *p*-hydroxybenzoate, 3 mM nucleotide, enzyme and 30 mM potassium phosphate buffer to 3 ml. The temperature was 25°C.

Compound	Percentage of inhibition
AMP	22
Adenosine 3′-phosphate	16
Adenosine 2′-phosphate	44
Cyclic adenosine 5′-phosphate	22
Adenosine 2′, 5′-diphosphate	75

et al. [1]. From our best enzyme preparations a molecular weight of 56 000 was also calculated from the light absorbance spectrum, assuming an extinction coefficient of 11 300 $M^{-1} \cdot cm^{-1}$ at 450 nm of the bound FAD, and the protein concentration. The enzyme does not contain any disulphides since independent SH group determinations using Ellmans reagent and denaturing the enzyme with sodium dodecyl sulphate revealed 4.6–4.8 SH groups/FAD which is in excellent agreement with the amino acid composition. The native enzyme possesses one rather reactive SH group. Employing Ellmans reagent a second order rate constant of 7000 $M^{-1} \cdot s^{-1}$ was calculated. Substitution of this SH group by maleimide does not influence much the catalytic activity of the enzyme. Labeling the enzyme with nitroxide maleimide yields an ESR spectrum which is of the immobilized type indicating that the spinlabel and thus the SH group is located in a crevice. However, we noticed that upon substitution of this SH group the dissociation constant of *p*-hydroxybenzoate was slightly increased (90 μM) as compared to the untreated enzyme (30 μM). Furthermore, inhibition studies revealed that the enzyme is slowly inactivated by *p*-chloromercuribenzoate indicating that SH groups are involved in the catalytic reaction.

Preliminary sequence studies (Beintema, J.J., unpublished results) showed that N-terminal amino acid is methionine. Incubation of the enzyme with carboxypeptidase A or carboxypeptidase B did not yield any information concerning the C-terminal amino acid. *p*-Hydroxybenzoate hydroxylase contains 7 methionines. This suggests that cyanogen bromide cleavage of the peptide chain could result in 8 different peptide chains. CNBr-treatment of the enzyme yields 7 fractions with different molecular weights.

Single crystals of *p*-hydroxybenzoate hydroxylase were grown and

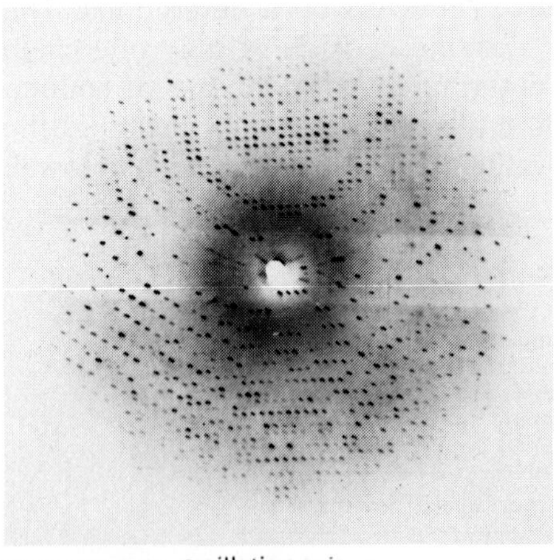

Fig. 2. Oscillation photograph of a single crystal of *p*-hydroxybenzoate hydroxylase. The oscillation axis is parallel to the b-axis. Oscillation angle: 2.5°. Oscillation speed: 4 h/degree and crystal-to-film distance 7.5 cm.

TABLE II

RESULTS OF HART REFINEMENTS IN (100) ZONE AT 4.0 Å RESOLUTION OF VARIOUS DERIVATIVES OF *p*-HYDROXYBENZOATE HYDROXYLASE

Derivative	Soaking time[a] (days)	Site	y[b]	z[b]	Occupancy[c] Z	Temperature Factor B (Å^2)	R[d] (%)	Number of reflexions
$NaAu(CN)_2$ (3 mM)	11	1	0.12	0.59	22.8	27.4	48.1	152
*p*Cl—$HgBzO^-$ (2 mM)	5	1	0.17	0.73	28.0	17.2	41.6	167
		2	0.10	0.37	21.0	7.7		
$HgCl_2$ (0.5 mM)[e]	5	1	0.17	0.73	21.5	52.0	45.0	152
		2	0.02	0.69	9.2	21.9		
		3	0.09	0.73	6.3	0.0		

[a] Soaking of single crystals was carried out in a mother liquor equal to a solution of the same composition as the crystallization medium.

[b] y and z are relative coordinates.

[c] The occupancy is in arbitrary units.

[d] $R = 100 \times \frac{\Sigma / k / F_H / \pm / F_P / - F_{calc} /}{\Sigma / k / F_H / \pm / F_P //}$.

[e] In this case EDTA, glutathione and FAD were omitted from the crystallization medium.

investigated. The space group is $C222_1$ with unit cell parameters a = 71.9 Å, b = 146.1 Å and c = 88.3 Å. The crystals, grown in glass capillaries, usually had their (100) face normal to the smallest dimension which is about 0.1 mm. The other dimensions are approximately 0.3 × 0.3 mm. Fig. 2 shows an oscillation photograph with an oscillation angle of 2.5° [8]. The diffraction pattern extends to at least 2.8 Å. There is a remarkable change in the diffraction pattern when crystals are soaked in a p-hydroxybenzoate free solution. The unit cell dimensions change slightly: a = 72.7 Å, b = 143.7 Å, c = 91.1 Å. This finding is in agreement with a conformational change due to substrate binding [2, 5]. Until now three heavy atom derivatives have been obtained which are suitable for three dimensional data collection. These crystal derivatives have been evaluated in projection studies. The intensities of reflexions in the b^*c^*-plane of the reciprocal lattice were collected on precession photographs ($\mu = 11°$) and were determined with an Optronics microdensitometer. Heavy atom positions were refined according to Hart [9]. The results are summarized in Table II. It is seen that $NaAu(CN)_2$ labels one site of the enzyme whereas the mercury salts occupy different positions on the enzyme.

It has been pointed out [10] that a structural similarity exists between the nucleotide binding sites of some protein. It is interesting to find out if this similarity extends to the FAD and NADPH binding sites of p-hydroxybenzoate hydroxylase.

Acknowledgement

We are grateful to Drs. W. Gaastra and J.J. Beintema, University of Groningen, for the communication of the preliminary sequence data. We are also indebted to Mr. B.J. Sachteleben for preparing the figures, to Mr. W.J.H. van Berkel and Mrs. J. Koopman-Horlings for skillful technical assistance, to Dr. W. de Jong, Biochemical Department, Catholic University, Nijmegen, for the amino acid analysis. The oscillation photograph was taken with an Arndt-Wonacott oscillation camera kindly made available to us by Enraf-Nonius Delft.

This study has been carried out under the auspices of the Netherlands Foundation for Chemical Research (S.O.N.) with financial aid from the Netherlands Organization for the Advancement of Pure Research (Z.W.O.).

References

1 Howell, L.G., Spector, T. and Massey, V. (1972) J. Biol. Chem. 247, 4340
2 Spector, T. and Massey, V. (1972) J. Biol. Chem. 247, 4679
3 Spector, T. and Massey, V. (1972) J. Biol. Chem. 247, 7123
4 Spector, T. and Massey, V. (1972) J. Biol. Chem. 247, 5632
5 Steennis, P.J., Cordes, M.M., Hilkens, J.G.H. and Müller, F. (1973) FEBS Lett. 36, 177
6 Benesi, H.A. and Hildebrand, J.H. (1949) J. Am. Chem. Soc. 71, 2703
7 Shoun, H., Higashi, N., Beppu, T., Arima, K; and Nakamura, S. (1974) FEBS Lett. 49, 13
8 Arndt, U.W., Champness, J.N., Phizackerley, R.P. and Wonacott, A.J. (1973) J. Appl. Crystallogr. 6, 417
9 Hart, R.G. (1961) Acta Crystalogr. 14, 1194
10 Rossmann, M.G., Moras, D. and Olsen, K.W. (1974) Nature 250, 19

Flavins and Flavoproteins, edited by T.P. Singer

CHAPTER 16

PHENOL HYDROXYLASE FROM YEAST

HALINA Y. NEUJAHR

Department of Pure and Applied Biochemistry, The Royal Institute of Technology, S 100 44 Stockholm 70 (Sweden)

Summary

Phenol hydroxylase from yeast is a monoxygenase with an absolute requirement for NADPH. It contains two FAD moieties per 148 000 daltons. The enzyme hydroxylates phenol, diphenols, monohalogenated phenols and cresols, but it does not act on hydroxybenzoates.

Phenol hydroxylase contains 16 SH groups of variable reactivity. The two most reactive are not essential for activity. The next four SH groups seem to be involved in FAD attachment and hydroxylase activity.

The enzyme exhibits a dual nature acting as a hydroxylase in the presence of phenol and as a mercaptoethanol-dependent NADPH oxidase in it absence.

TABLE I

PROPERTIES OF PHENOL HYDROXYLASE FROM *TRICHOSPORON CUTANEUM*

Molecular properties	
Mol. wt.	148 000
Subunits	2
FAD content per 148 000 daltons	1.7–1.9
Isoelectric point	5.2
Catalytic properties	
Substrate specificity	Broad
Cofactor requirement	NADPH
No effect of:	Iron chelators, copper chelators, KCN, CO
Inhibition by:	Heavy metals (reversed by SH), dithionite, $NaBH_4$ (bleaching), formaldehyde, glutaraldehyde, acetylation

Abbreviation: Nbs_2 (Ellman reagent), 5,5′-dithiobis(2-nitrobenzoic acid).

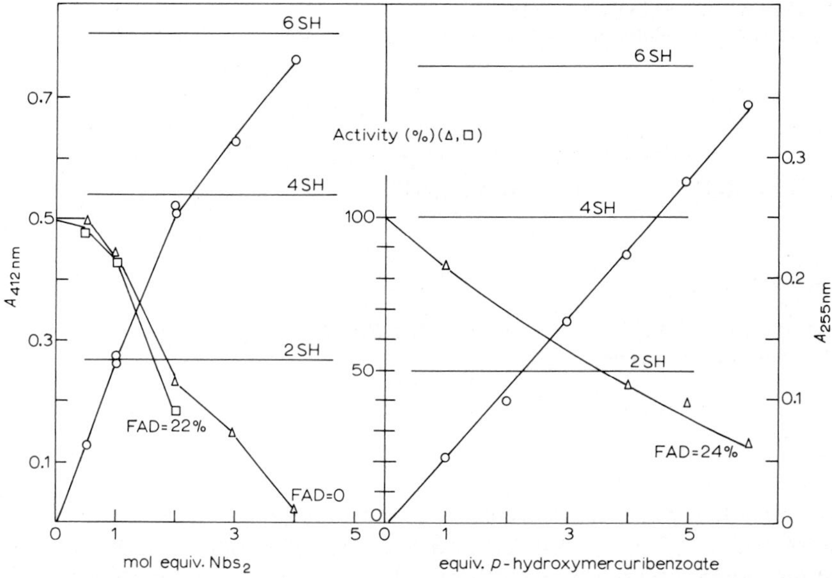

Fig. 1. Stepwise blocking of SH groups in phenol hydroxylase using Nbs_2 or *p*-hydroxymercuribenzoate and the concomitant changes in enzyme activity and FAD content. Protein concentration approximately 1 mg/ml. Enzyme assay using 20 μl aliquots (19 μg protein) in 3 ml of 0.033 M Tris sulphate pH 7.6 containing $1.7 \cdot 10^{-4}$ M NADPH and $1.7 \cdot 10^{-4}$ M phenol. Preincubation for 10 min before addition of phenol.

Phenol hydroxylase catalyses the conversion of simple phenols to their *o*-diol derivatives. The enzyme was first isolated from *Trichosporon cutaneum* [1, 2]. Later, it was also found in *Candida tropicalis* [3]. Phenol hydroxylase from yeast is a monooxygenase, with two FAD moieties per 148 000 daltons. It has an absolute requirement for NADPH. Table I shows some of its molecular and catalytic properties. Inhibitor studies show no participation of copper or iron. The activity disappears when the enzyme reacts with aldehydes or acetic anhydride. This means the ϵ-NH_2 of lysine may be involved in the active site.

Heavy metals also inhibit the enzyme, but their effects are counteracted by an excess of a thiol compound. Reduced sulfhydryl compounds are always needed to protect the activity of the enzyme and to promote reactivation of the isolated apoenzyme by FAD [2, 4].

Since the native holoenzyme contains as many as 16 SH groups per mol, we studied their effect on the attachment of FAD and the activity of the enzyme [4]. Fig. 1 shows the stepwise blocking of SH groups, using Nbs_2 (the Ellman reagent), or *p*-hydroxymercuribenzoate. The concomitant changes in enzyme activity and FAD content are also indicated.

The two SH reagents differ in their stoichiometry and their effect on enzyme activity, but they both dislocate FAD. With Nbs_2, each of the two first-added gram molecules reacts with nearly two SH groups. Thus, both nitrobenzoate moieties of Nbs_2 take part, and so the two SH groups probably lie very close together. As titration proceeds, the stoichiometric ratio

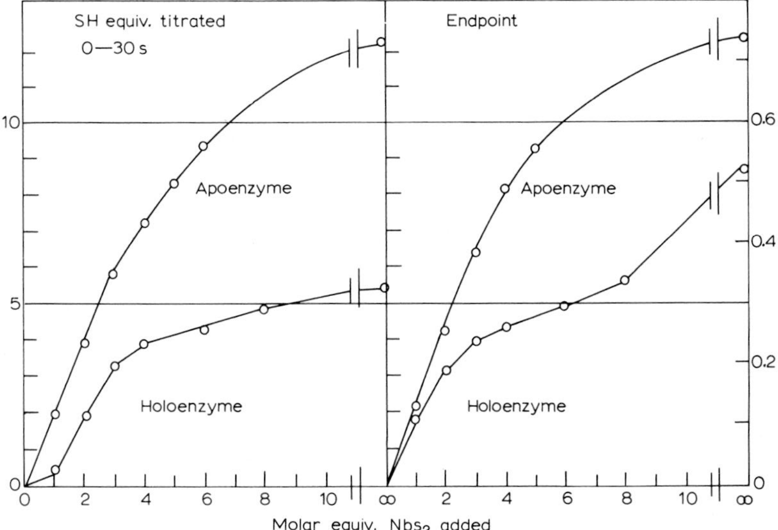

Fig. 2. The reactivity of SH groups in holo- and apoenzyme of phenol hydroxylase towards Nbs_2. Protein concentration 0.65 mg/ml; ∞ denotes excess Nbs_2.

of Nbs_2 to SH falls off. With *p*-hydroxymercuribenzoate there is only one equivalent to react with each SH group. The straight line shows that *p*-hydroxymercuribenzoate readily penetrates to far more SH groups than does Nbs_2. Using Nbs_2 to block the first two SH groups scarcely affects the activity. In contrast, using *p*-hydroxymercuribenzoate significantly reduces activity. Using either reagent to block 6–8 SH groups cuts the activity almost to zero. This agrees with the dislocation of FAD when determined by the proportion of residual FAD.

This property was used to prepare the apoenzyme via its mercaptide derivatives. These were then reduced and separated from the reagents on a Sephadex G-25 column. Fig. 2 shows the reactivity towards Nbs_2 of the SH groups in the apoenzyme and the holoenzyme.

The apoenzyme has more very reactive SH groups than does the holoenzyme. In the apoenzyme, at least 8 SH groups seem to react with both nitrobenzoate moieties of Nbs_2. The holoenzyme may have only 3 or 4 such groups. So, at least two pairs of SH groups seem involved in the attachment of FAD to the enzyme.

Phenol cuts the reactivity of the holoenzyme towards Nbs_2. Table II shows this. In the absence of phenol, the first SH group is highly reactive. This reactivity is greatly delayed by just two molecules of phenol per molecule of enzyme. A 1000-fold molar excess of phenol gives an enzyme activity not far from *V*. However, the titration of SH groups then proceeds extremely slowly.

Phenol hydroxylase acts on many other substrates; resorcinol, catechol, quinol, cresols and monohalogenated phenols. It does not act on *p*-hydroxybenzoate or salicylate. The substrates for phenol hydroxylase function as

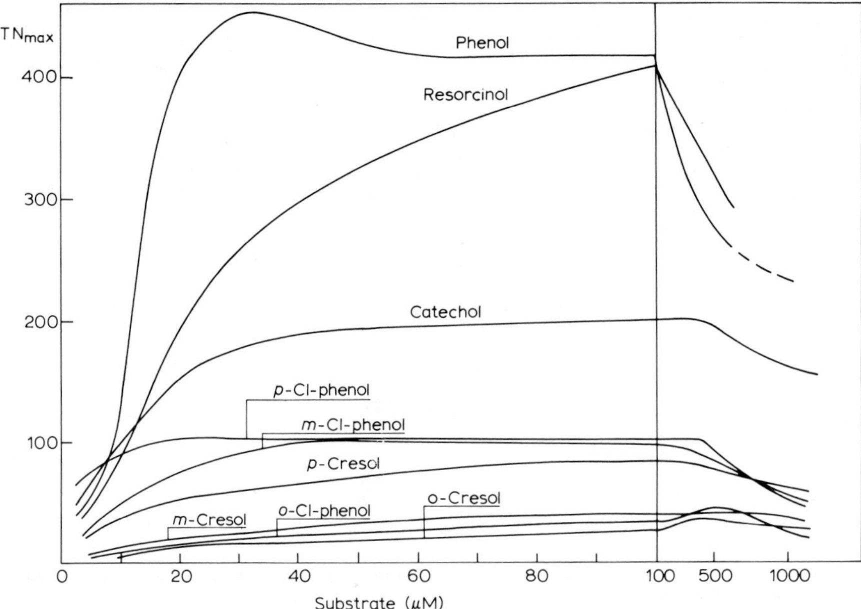

Fig. 3. Affinity of phenol hydroxylase towards various substrates. Note the compressed scale on the substrate axis. TN_{max} denotes turnover number at V.

effectors; they enhance the anaerobic reduction of FAD by NADPH. All of them inhibit at high concentrations. Their K_m values are in the range 3–45 μM. Fig. 3 shows the oxidation of NADPH in the presence of these substrates (Neujahr and Kjeller, forthcoming work). *O*-Diols are formed with all these substrates, as shown earlier by thin-layer chromatography [2]. Recently we have also studied the extent of hydroxylation or, rather, the extent of oxygen uptake not coupled to hydroxylation. The latter was determined by measuring oxygen return upon adding catalase, according to the methods of either White-Stvens and Kamin[5] or Ribbons [6]. So far, the results show that uncoupling of NADPH oxidation from hydroxylation can occur with any substrate for phenol hydroxylase, including phenol itself.

TABLE II

THE EFFECT OF SUBSTRATE ON THE REACTIVITY OF SH GROUPS IN PHENOL HYDROXYLASE

Phenol (M)	Phenol/enzyme (molar ratio)	SH/Nbs_2 (molar ratio*)	Time required (min)
0	0	1.9	14
$8 \cdot 10^{-6}$	2	1.6	20
$4 \cdot 10^{-5}$	10	1.5	35
$4 \cdot 10^{-3}$	1000	1.0	60

* Number of SH equivalents reacting with first-added molar equivalent of Nbs_2.

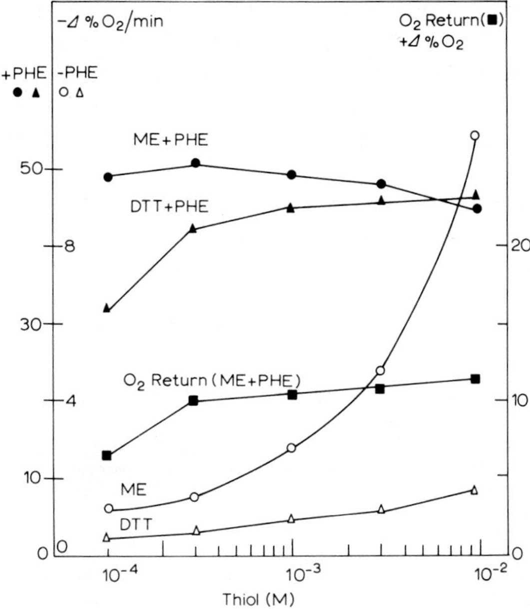

Fig. 4. Oxidation of NADPH by phenol hydroxylase in the presence and absence of phenol (PHE), at varying concentrations of mercaptoethanol (ME) or dithiothreitol (DTT) and O_2 return after the addition of catalase. Note the different scales for various parameters. The reaction mixture contained in 2 ml of 0.05 M potassium phosphate pH 7.6: 0.5 μmol NADPH, 100 μg phenol hydroxylase, 0.5 μmol phenol and a thiol compound as indicated. 450 units of crystalline catalase were added at the end of the reaction.

TABLE III

ACTIVITY OF PHENOL HYDROXYLASE TOWARDS RESORCINOL AND CATECHOL AND O_2 RETURN UPON ADDITION OF CATALASE

The reaction mixture contained in 2 ml of 0.05 M potassium phosphate pH 7.6: $5 \cdot 10^{-6}$ M FAD, $5 \cdot 10^{-4}$ M mercaptoethanol, $5 \cdot 10^{-5}$ M EDTA, 100 μg phenol hydroxylase, substrate and NADPH as indicated. 450 units of crystalline catalase (5 μl) were added at the end of the reaction.

Substrate (mM)	NADPH (mM)	$-\Delta\%O_2$/min	$+\Delta\%O_2$
Resorcinol			
0.25	0.25	48	13
5.0	0.25	25	41
5.0	5.0	31	11
Catechol			
0.25	0.25	29	24
5.0	0.25	17	34
5.0	5.0	18	17

The uncoupling seems to be affected by several different factors. Table III shows the case of resorcinol and catechol, which increase oxygen return when the enzyme is inhibited by excess substrates. However, this effect is counteracted by increasing the NADPH level. Thus, the uncoupling in this case may depend on a weak effector function rather than on excess substrate inhibition.

Table IV shows *p*-cresol. Here, the oxygen returned by catalase increases with the level of mercaptoethanol in the assay mixture. The same

TABLE IV

ACTIVITY OF PHENOL HYDROXYLASE TOWARDS *p*-CRESOL AND O_2 RETURN UPON ADDITION OF CATALASE

The reaction mixture contained in 2 ml of 0.05 M potassium phosphate pH 7.6: $5 \cdot 10^{-6}$ M FAD, $5 \cdot 10^{-4}$ M mercaptoethanol, $5 \cdot 10^{-5}$ M EDTA, 100 μg phenol hydroxylase, substrate and NADPH as indicated. 450 units of crystalline catalase (5 μl) were added at the end of the reaction.

NADPH/*p*-cresol	*p*-Cresol (mM)			
	0.25	5.0	0.25	5.0
(molar ratio)	$-\Delta\%O_2$/min		$+\Delta\%O_2$	
At 0.5 mM mercaptoethanol:				
1:1	10	38	8	4
10:1	50	—	7	—
1:20	—	4	—	6
At 10 mM mercaptoethanol:				
1:20	—	1	—	11

TABLE V

EXCESS SUBSTRATE INHIBITION OF PHENOL HYDROXYLASE AND O_2 RETURN UPON ADDITION OF CATALASE

The reaction mixture contained in 2 ml of 0.05 M potassium phosphate pH 7.6: $5 \cdot 10^{-6}$ M FAD, $5 \cdot 10^{-4}$ M mercaptoethanol, $5 \cdot 10^{-5}$ M EDTA, 100 μg phenol hydroxylase, substrate and NADPH as indicated. 450 units of crystalline catalase (5 μl) were added at the end of the reaction.

NADPH/phenol	Phenol (mM)			
	0.25	5.0	0.25	5.0
(molar ratio)	$-\Delta\%O_2$/min		$+\Delta\%O_2$	
At 0.5 mM mercaptoethanol:				
1:1	46	21	7	6
10:1	50	—	5	—
1:20	—	20	—	6
At 10 mM mercaptoethanol:				
1:20	—	22	—	12

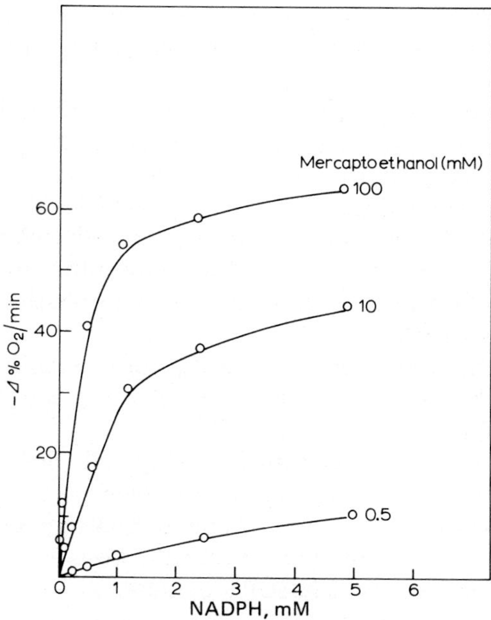

Fig. 5. NADPH oxidase activity of phenol hydroxylase. The reaction mixture contained 100 μg phenol hydroxylase in 2 ml of 0.05 M potassium phosphate pH 7.6. Mercaptoethanol and NADPH as indicated.

thing was observed with phenol (Table V). Again, oxygen return increases from 6 to 12% as mercaptoethanol in the assay mixture is increased from 0.5 to 10 mM. With neither phenol nor *p*-cresol the oxygen return seems linked to the excess substrate inhibition or to the ratio NADPH : substrate.

The effect of mercaptoethanol on NADPH oxidation was studied in the presence and in the absence of phenol and compared to the effect of other thiols. Fig. 4 compares mercaptoethanol and dithiothreitol. Both stimulate the overall reaction with phenol. However, optimum concentrations seem different. Without phenol, increasing mercaptoethanol increases drastically the oxidation of NADPH. The effect of dithiothreitol is very small, though similar in essence. Cysteine and reduced glutathione gave effects similar to that of dithiothreitol.

TABLE VI

NADPH AFFINITY OF PHENOL HYDROXYLASE IN THE ABSENCE OF PHENOL

The reaction mixture contained in 2 ml of 0.05 M potassium phosphate pH 7.6: $5 \cdot 10^{-6}$ M FAD, $5 \cdot 10^{-4}$ M mercaptoethanol, $5 \cdot 10^{-5}$ M EDTA, 100 μg phenol hydroxylase, substrate and NADPH as indicated. 450 units of crystalline catalase (5 μl) were added at the end of the reaction.

Mercaptoethanol (mM)	K_m NADPH
0.5	4.2
10	2.5
100	1.0

The oxygen return upon addition of catalase, is, of course, complete in the absence of phenol. In the presence of phenol it increases slowly with increasing mercaptoethanol level.

Fig. 5 shows how mercaptoethanol affects the affinity of phenol hydroxylase towards NADPH, in the absence of phenol. There is a dramatic stimulation of this activity. Table VI gives the corresponding K_m values. Increasing the level of mercaptoethanol reduces the K_m for NADPH. This points to an allosteric transition in the enzyme. However, the molecular weight of the enzyme is not affected by incubation with high concentrations of mercaptoethanol. Neither is there a disulfide bridge in the molecule that may be essential for the hydroxylase activity [4].

Mercaptoethanol shows also saturation kinetics. The K_m for mercaptoethanol is 15 mM, i.e. several times higher than for NADPH. The K_m is not affected by increasing levels of NADPH.

Mercaptoethanol itself is not likely to occur in the living cell. Some structural analogue of it may be the physiologically active substance.

Summing up, these results indicate a dual nature of phenol hydroxylase from yeast. It acts as a hydroxylase in the presence of phenol, while in its absence it works as a mercaptoethanol-dependent NADPH oxidase.

Experimental procedure

Trichosporon cutaneum was grown and disrupted as described earlier [1, 2]. The enzyme was purified and assayed according to procedures described elsewhere [2, 4]. Thiol determinations were carried out with Nbs_2, essentially as described by Habeeb [7, 8] or with *p*-hydroxymercuribenzoate according to Boyer [9]. Details of the apoenzyme preparation and the various thiol titrations appear elsewhere [4]. Enzyme activity in experiments relating the uncoupling of hydroxylation to other factors was measured by means of a Clark oxygen electrode (YSI), adding 450–1500 units of catalase at the end of the respective reactions. The effect of mercaptoethanol on NADPH oxidation was measured by both spectrophotometry and the oxygen uptake method.

References

1 Neujahr, H.Y. and Varga, J.M. (1970) Eur. J. Biochem. 13, 37
2 Neujahr, H.Y. and Gaal, A. (1973) Eur. J. Biochem. 35, 386
3 Neujahr, H.Y., Lindsjö, S. and Varga, J.M. (1974) Antonie van Leeuwenhoek 40, 209
4 Neujahr, H.Y. and Gaal, A. (1975) Eur. J. Biochem. 58, 351
5 White-Stevens, R.H. and Kamin, H. (1970) Biochem. Biophys. Res. Commun. 38, 882
6 Ribbons, D.W., Ohta, Y. and Higgins, J. (1971) J. Bacteriol. 106, 702
7 Habeeb, A.F.S.A. (1966) Biochim. Biophys. Acta 115, 440
8 Habeeb, A.F.S.A. (1972) Methods in Enzymology (Colowick, S.P. and Kaplan, N.O., eds.), vol. 25, p. 457, Academic Press, London
9 Boyer, P.D. (1954) J. Am. Chem. Soc. 76, 4331

Flavins and Flavoproteins, edited by T.P. Singer

CHAPTER 17

OXYGENATION AND OXIDATION CATALYZED BY LYSINE MONOOXYGENASE, A FLAVOPROTEIN

SHOZO YAMAMOTO, TAKAKO OHNISHI, KIYOFUMI MARUYAMA*, TAKASHI YAMAUCHI** and OSAMU HAYAISHI

Department of Medical Chemistry, Kyoto University Faculty of Medicine, Kyoto, (Japan)

It is now known that oxygenases constitute one of the major groups of flavoprotein enzymes. Lysine monooxygenase is one of these flavooxygenases. The enzyme was isolated from a pseudomonad which was grown with lysine as its sole carbon source. About 100-fold purification yielded a crystalline enzyme preparation. Based on a molecular weight of 191 000, it was found that 2 mol of FAD were contained per mol of enzyme. The absorbance spectrum of the enzyme is characteristic of a flavoprotein with maxima at 385 and 460 nm [1, 2].

As shown in Eqn. 1, the enzyme catalyzes the monooxygenation of lysine. Only one atom of molecular oxygen as marked with an asterisk is incorporated onto the α-carbon atom of lysine.

$$\begin{array}{l} \underset{\displaystyle NH_2}{CH_2}-CH_2-CH_2-CH_2-\underset{\displaystyle NH_2}{CH}-COOH + O_2^* \\ \qquad\qquad\qquad\qquad\qquad\downarrow \\ \underset{\displaystyle NH_2}{CH_2}-CH_2-CH_2-CH_2-\underset{\displaystyle NH_2}{C}=O^* + CO_2 + H_2O^* \end{array} \qquad (1)$$

Concomitantly, the carboxyl group is lost as carbon dioxide, and δ-aminovaleramide is produced. It is believed that the other atom of oxygen is reduced to water and that the hydrogen atoms necessary for this reduction are supplied by the dehydrogenation of lysine. Thus, lysine undergoes three reactions: oxygenation, dehydrogenation and decarboxylation. The oxygen molecule serves both as an oxidant for oxygenation and as a hydrogen acceptor for dehydrogenation.

Investigations in our laboratory over the past several years have revealed a unique property of the enzyme. During the course of studies on the substrate specificity, it was found that the enzyme functions either as an oxygenase or as an oxidase depending on the kind of substrate. Lysine and

---

* On leave from Department of Chemistry, Kyoto University Faculty of Science, Kyoto, Japan.
** Present address, Department of Biochemistry, Asahikawa Medical College, Asahikawa, Hokkaido, Japan.

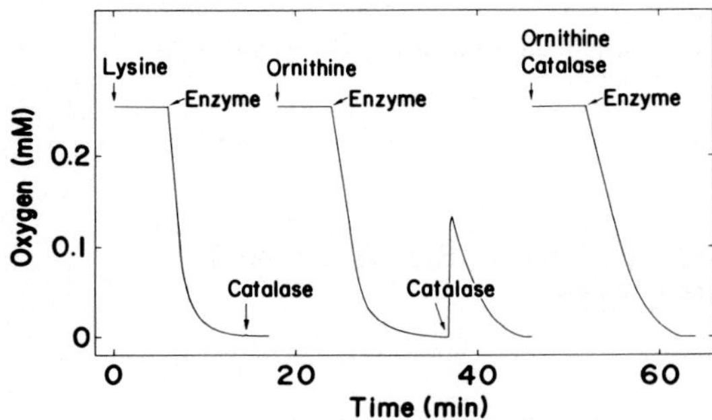

Fig. 1. Oxygenation of lysine and oxidation of ornithine by lysine monooxygenase.

certain analogues undergo oxygenative decarboxylation while other analogues undergo oxidative deamination.

Since oxygen was consumed during both the oxygenative and oxidative reactions, the activity of the enzyme was determined by an oxygen electrode. To distinguish whether the oxygen consumption was due to the oxygenation or the oxidation of the substrate, catalase was used. The catalase would decompose any hydrogen peroxide which was produced as a result of the oxidative reaction, and the oxygen evolved would be equivalent to 50% of the hydrogen peroxide.

As shown in Fig. 1, the oxygen evolution upon the addition of catalase was hardly detected with lysine, as was expected. Ornithine, which was also a diamino acid but one carbon atom shorter than lysine, was tested as a substrate analogue. The enzyme showed an oxygen consumption at a rate approx. 7% of that of lysine. When the oxygen in the reaction mixture was exhausted, the addition of catalase caused an oxygen evolution to the extent approx. 50% of the oxygen consumption (Fig. 1). The presence of catalase in the mixture during the reaction decreased the rate of oxygen consumption by approx. 50% (Fig. 1). These results indicate that hydrogen peroxide was produced concomitant with the oxygen consumption. Further studies to identify the reaction products indicated that ornithine underwent an oxidative deamination rather than an oxygenative decarboxylation. As shown in Eqn. 2, ornithine was converted to α-keto-δ-aminovalerate with the release of ammonia.

$$\begin{array}{l} \underset{\displaystyle NH_2}{CH_2}-CH_2-CH_2-\underset{\displaystyle NH_2}{CH}-COOH + O_2^* + H_2O \\ \quad\quad\quad\quad\quad\quad\downarrow \\ \underset{\displaystyle NH_2}{CH_2}-CH_2-CH_2-\underset{\displaystyle \| \atop O}{C}-COOH + NH_3 + H_2O_2^* \end{array} \qquad (2)$$

Molecular oxygen was not incorporated into the substrate but was merely reduced to hydrogen peroxide [3]. Thus, with ornithine as a substrate, lysine monooxygenase behaved like D- or L-amino acid oxidase, which is also a flavoprotein.

A series of α, ω-diamino acids of varying carbon chain lengths was examined (Table IA). Neither lysine nor a homologue with 7 carbon atoms produced hydrogen peroxide in a significant quantity. On the other hand, ornithine as well as a homologue with 8 carbon atoms produced hydrogen peroxide in an amount nearly stoichiometric with the oxygen consumption [3]. Thus, varying the carbon chain length of the substrate affected the type of reaction catalyzed by lysine monooxygenase.

These findings made earlier by Dr. Nakazawa in our laboratory, prompted us to examine other compounds with various modifications. Several lysine analogues with a methyl group attached at various positions, were synthesized and provided by Professor Shiba and his associate of Osaka University (Table IB). Analogues with methyl groups at positions distant from the α-carbon, such as the N^{ϵ}-methyl and δ-methyl derivatives, were predominantly oxygenated, as demonstrated by a small quantity of hydrogen peroxide formed. The corresponding acid amides were identified as reaction products by paper electrophoresis and thin-layer chromatography. In contrast, the enzyme catalyzed predominantly an oxidative reaction with γ-methyllysine. Its conversion to an α-keto acid was shown by paper electrophoresis and thin-layer chromatography. Lysine analogues with a methyl group closer to the α-carbon, such as β-methyllysine and N^{α}-methyllysine, were inactive as substrate (Ohnishi, T., unpublished).

It was known for some time that only basic amino acids bearing an amino or a guanidino group at the end of the carbon chain were active substrates [2, 3]. Since the terminal basic group seemed to be necessary for the enzyme activity, several analogues of lysine or arginine were tested whose terminal basicity was abolished by modification of the amino or guanidino group (Fig. 2). The amino group of lysine was acetylated, and the guanidino group of arginine was nitrated. The guanidino group of arginine or homoarginine was replaced by a ureido group, producing citrulline or homocitrulline, respectively. None of these compounds by themselves was active as

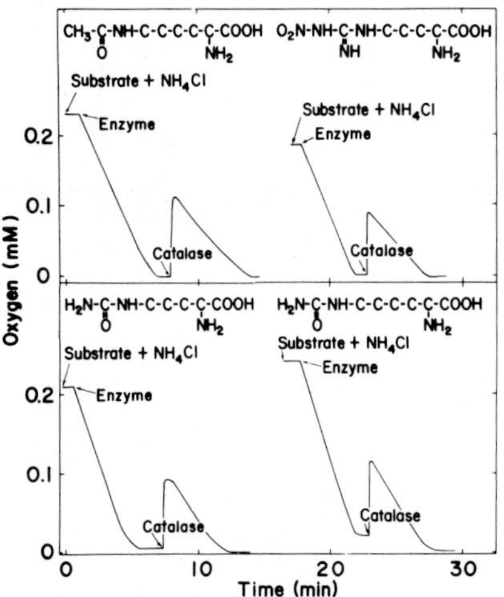

Fig. 2. Ammonium-dependent oxidation of analogues of lysine and arginine by lysine monooxygenase.

TABLE I

REACTIONS OF LYSINE MONOOXYGENASE WITH VARIOUS SUBSTRATE ANALOGUES

Specific activity was determined at a saturating concentration of each compound in the presence of 0.25 mM oxygen.

	Analogues	Spec. act. (μmol/min/mg)	H_2O_2/O_2 (%)	$A_{550}/A_{460\,nm}$ (%)
(A)	$CH_2(NH_2)$—CH_2—CH_2—CH_2—CH_2—CH_2—CH_2—$CH(NH_2)$—COOH	< 0.01	—	—
	$CH_2(NH_2)$—CH_2—CH_2—CH_2—CH_2—CH_2—$CH(NH_2)$—COOH	0.61	75	4.8
	$CH_2(NH_2)$—CH_2—CH_2—CH_2—CH_2—$CH(NH_2)$—COOH	3.27	12	—
Lysine	$CH_2(NH_2)$—CH_2—CH_2—CH_2—$CH(NH_2)$—COOH	11.9	4.3	13.6
Ornithine	$CH_2(NH_2)$—CH_2—CH_2—$CH(NH_2)$—COOH	0.87	106	0.9
	$CH_2(NH_2)$—CH_2—$CH(NH_2)$—COOH	< 0.01	—	—
(B)	$CH_2(NH$—$CH_3)$—CH_2—CH_2—CH_2—$CH(NH_2)$—COOH	6.3	6.8	—
	$CH_2(NH_2)$—$CH(CH_3)$—CH_2—CH_2—$CH(NH_2)$—COOH	0.76	7.6	5.7
	$CH_2(NH_2)$—CH_2—$CH(CH_3)$—CH_2—$CH(NH_2)$—COOH	11.0	90	9.1
	$CH_2(NH_2)$—CH_2—CH_2—$CH(CH_3)$—$CH(NH_2)$—COOH	0	—	—
	$CH_2(NH_2)$—CH_2—CH_2—CH_2—$CH(NH$—$CH_3)$—COOH	0	—	—

(C)	$CH_2(NH_2)-CH(OH)-CH_2-CH_2-CH(NH_2)-COOH$	2.8	8.0	2.4
	$CH_2(NH_2)-CH_2-CH(OH)-CH_2-CH(NH_2)-COOH$	> 0.13	38	—
	$CH_2(NH_2)-CH_2-CH(Cl)-CH_2-CH(NH_2)-COOH$	1.5	33	—
	$CH_2(NH_2)-CH_2-S-CH_2-CH(NH_2)-COOH$	0.10	12	2.9
Arginine	$(HN{=})(H_2N)CH-NH-CH_2-CH_2-CH_2-CH(NH_2)-COOH$	1.2	22	3.0
	$(HN{=})(H_2N)CH-NH-CH_2-CH_2-CH_2-CH_2-CH(NH_2)-COOH$	3.1	78	—

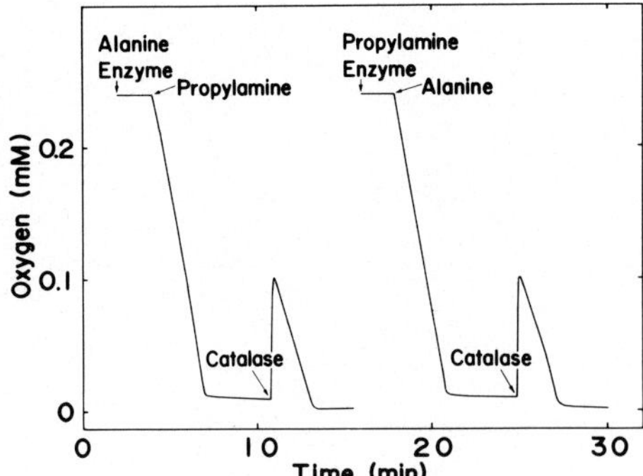

Fig. 3. Propylamine-dependent oxidation of alanine by lysine monooxygenase.

substrates, unless ammonium ion, methylamine or ethylamine was also supplied. As judged from the formation of hydrogen peroxide, these compounds underwent an oxidative deamination in the presence of ammonium ion (Ohnishi, T., unpublished).

The requirement for a terminal basic group was further demonstrated with the fragmented substrates of lysine monooxygenase. The enzyme activity was examined in the simultaneous presence of alanine and proylamine, which represent the two fragments of the lysine molecule. As shown in Fig. 3, either alanine or propylamine alone was almost inactive, but the addition of the counterpart caused a consumption of oxygen. By the use of catalase it was found that hydrogen peroxide was produced during the oxygen consumption. Formation of pyruvate and ammonia was also demonstrated. Thus, as in Eqn. 3, in the presence of propylamine the enzyme catalyzed an oxidative deamination of alanine. Propylamine remained unchanged [4].

$$\begin{array}{l} \underset{\displaystyle NH_2}{CH_2}-CH_2-CH_3 + CH_3-\underset{\displaystyle NH_2}{CH}-COOH + O_2^* + H_2O \\ \qquad\qquad\qquad\qquad\downarrow \\ \underset{\displaystyle NH_2}{CH_2}-CH_2-CH_3 + CH_3-\underset{\displaystyle \| \atop O}{C}-COOH + NH_3 + H_2O_2^* \end{array} \tag{3}$$

We extended the investigations on the fragmented substrates to include various combinations of α-monoamino acids and alkylamines [4, 5]. The results are shown in Fig. 4. Several rules seemed to govern the enzyme activity. First, with each amino acid the enzyme was most active when an appropriate alkylamine was supplied to make the total carbon number of the two fragments nearly equal to that of lysine. Secondly, when a series of α-monoamino acids was examined, maintaining the total carbon number at six, the enzyme was more active with longer amino acids. Thirdly, in each combination hydrogen peroxide was produced and the corresponding α-keto acid was identified. However, the ratio of hydrogen peroxide formation to oxygen consumption varied to some extent, and in some cases, such as alanine plus ethylamine, α-aminobutyrate plus methylamine and α-aminobutyrate plus ethylamine, the H_2O_2/O_2 values were lower than the other cases. This

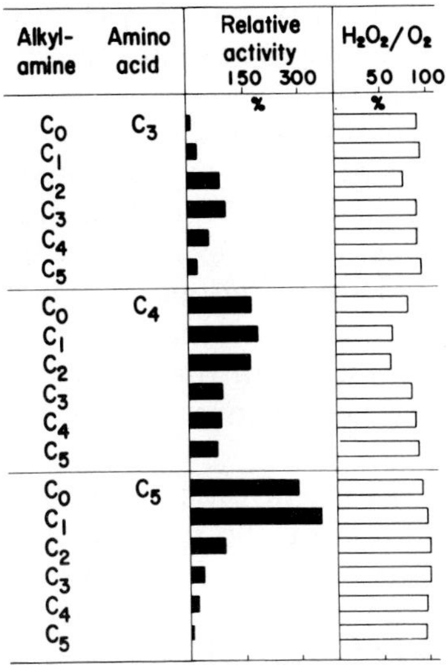

Fig. 4. Reactions of lysine monooxygenase with various fragmented substrates.

result suggests that in these combinations a part of the oxygen consumption was attributable to the oxygenation of substrate. Thus, both the oxygenative and oxidative reactions occurred concomitant with the same substrate. This was confirmed by the identification of both α-ketobutyrate and propionamide produced from α-aminobutyrate as separated by thin-layer chromatography [8].

The concomitant oxygenation and oxidation of the same substrate were also observed with other analogues. As shown in Table IC, γ-hydroxylysine, γ-chlorolysine, arginine and homoarginine showed H_2O_2/O_2 values falling between 0 and 100 (Ohnishi, T., unpublished). Actually, paper electrophoresis of reaction products from arginine showed the formation of both the acid amide and the α-keto acid (Yamauchi, T., unpublished).

The experimental results described above showed that lysine monooxygenase had a broad substrate specificity as demonstrated with a variety of substrate analogues. The carbon chain length of the substrate was varied. Several modifications were introduced at various positions on the lysine molecule. The basicity of the terminal amino or guanidino group was abolished. The lysine molecule was fragmented and the fragments were used as substrates. With a fresh preparation of the enzyme, lysine underwent almost specifically an oxygenative reaction. There was little, if any, production of hydrogen peroxide. With 2,7-diaminoheptanoate, N^{ϵ}-methyllysine, δ-methyllysine, δ-hydroxylysine and thialysine, a little oxidase activity was observed but the oxygenase activity was predominant. Lysine and these compounds are termed as "oxygenase-type substrates". On the other hand, ornithine and γ-methyllysine were predominantly oxidized. This was also the case of various combinations of α-monoamino acids and alkylamines. These analogues are "oxidase-type substrates". Furthermore, the enzyme showed both the oxygenase and oxidase activities with γ-hydroxylysine, γ-chlorolysine, arginine and homoarginine. These could be referred to as "mixed-type substrates".

A question as to whether both the oxygenation and the oxidation reactions were catalyzed by the same enzyme, was answered by the following experiments. During the enzyme purification, the ratio of the oxygenase activity (as examined with lysine) to the oxidase activity (as examined with ornithine or alanine) remained essentially constant [3, 5]. Various analogues including ornithine [3] and propylamine [5] were competitive inhibitors to the oxygenation of lysine. These results indicated that both the oxygenase and oxidase activities were associated with the same active site of the enzyme.

One approach that could now be thought of was to investigate whether any difference might be detectable in the behavior of the enzyme as it reacted with various types of substrate. This approach would be substantiated by the spectrophotometric studies on the enzyme-bound flavin.

It was earlier reported that a long-wavelength absorbance species appeared during the anaerobic reduction of the enzyme-bound flavin with lysine [6, 7]. As shown in Table I, immediately after the interaction of enzyme with lysine, a broad absorbance band in the long-wavelength region above 520 nm developed rapidly and reached a maximum. Then, this absorbance species gradually disappeared and changed to the so-called fully reduced form of the enzyme-flavin complex. It still remains to be elucidated whether this species is actually involved in the aerobic over-all reaction. However, it was intriguing to note that this species was hardly detectable when lysine was replaced by ornithine (Yamauchi, T. and Hirata, F., unpublished). Since lysine is an oxygenase-type substrate and ornithine is an oxidase-type substrate, this finding prompted us to investigate whether this long-wavelength absorbing band is specific for the oxygenase-type substrates. The absorbance change at 550 nm was followed by the stopped-flow technique. The ratio of the maximum height of the absorbance at 550 nm to the absorbance at 460 nm was determined as shown in the last column of Table I. The type of the catalyzed reaction can be judged from the H_2O_2/O_2 values. In contrast to ornithine, γ-methyllysine and 2,8-diaminooctanoate, both of which are oxidase-type substrates, showed appreciable absorbance at 550 nm. Thus, there was no correlation between the long wavelength absorbance band and the type of reaction (Maruyama, K., unpublished).

We have earlier proposed a "misfit-substrate hypothesis" to explain the bifunctional activity of lysine monooxygenase [5]. Presumably, lysine or its analogue is either oxygenated or oxidized depending on whether or not it precisely fits the active site. As described above, the extent of oxygenation and oxidation is not a matter of an "all-or-none" situation, but varies with each substrate analogue. Therefore, it seems that the interaction of the flavin with the substrate or the oxygen molecule at the catalytic site is not fundamentally different in the oxygenative and oxidative reactions.

Acknowldegements

This work has been supported in part by the Scientific Research Fund of the Ministry of Education of Japan, and by a grant from the Tanabe Amino Acid Research Foundation. The authors wish to express their thanks to Dr. C.M. Edson and Dr. S.C. Hubbard for their critical readings of this manuscript.

References

1 Takeda, H. and Hayaishi, O. (1966) J. Biol. Chem. 241, 2733
2 Takeda, H., Yamamoto, Y., Kojima, Y. and Hayaishi, O. (1969) J. Biol. Chem. 244, 2935
3 Nakazawa, T., Hori, K. and Hayaishi, O. (1972) J. Biol. Chem. 247, 3439
4 Yamamoto, S., Yamauchi, T. and Hayaishi, O. (1972) Proc. Natl. Acad. Sci. U.S. 69, 3723
5 Yamamoto, S. and Hayaishi, O. (1975) in Reactivity of Reduced Flavins (Yagi, K., ed.), University of Tokyo Press, Tokyo, p. 81
6 Yamamoto, S., Hirata, F., Yamauchi, T., Nozaki, M., Hiromi, K. and Hayaishi, O. (1971) J. Biol. Chem. 246, 5540
7 Yamamoto, S., Hirata, F., Yamauchi, T., Nozaki, M., Hayaishi, O. and Hiromi, K. (1972) Z. Naturforsch. 27b, 1056
8 Yamamoto, S., Yamauchi, T., Ohnishi, T., Maruyama, K. and Hayashi, O. (1975) Arch. Biochem. Biophys. 171, 316

Flavins and Flavoproteins, edited by T.P. Singer

CHAPTER 18

VARIABLE CATALYTIC FUNCTIONS OF FLAVOPROTEIN OXYGENASES

SHIGEKI TAKEMORI, KENZI SUZUKI and MASAYUKI KATAGIRI

Department of Chemistry, Faculty of Science, Kanazawa University, Ishikawa 920 (Japan)

We have analyzed the action mechanism of two representative types of flavoprotein oxygenases, salicylate hydroxylase from *Pseudomonas putida* [1–6] and lactate oxygenase from *Mycobacterium phlei* [7–9]. Salicylate hydroxylase catalyzes the decarboxylation-hydroxylation of salicylate to catechol with the stoichiometric consumption of molecular oxygen and NADH [10]. The enzyme (mol. wt. 54 000) is a single-chain protein containing one FAD. Lactate oxygenase, which contains FMN as the prosthetic group, catalyzes the oxygenative decarboxylation of L-lactate to acetate [11]. The enzyme (mol. wt. 350 000) exists in an oligomer structure composed of six subunits. In this paper, we would like to present our findings on some additional properties of these oxygenases.

1-Substituted analogs of salicylate as substrate of salicylate hydroxylase

In addition to salicylate, compounds such as 2, 3-, 2, 4-, 2, 5- and 2, 6-dihydroxybenzoate and *p*-aminosalicylate served as the substrate of salicylate hydroxylase. The stoichiometry was estimated to be an NADH: O_2 ratio of 1.0. On the other hand, a stoichiometric ratio of 2.0 was observed with *o*-nitro-, *o*-iodo- and *o*-aminophenol. These findings suggest that salicylate hydroxylase can catalyze a new type of reaction when the latter compounds are the substrates.

When *o*-nitrophenol was used as substrate, the identity of catechol as the product of the reaction was established using metapyrocatechase [2] and thin-layer chromatography. Nitrite was identified by the Griess-Ilosvay reaction as the another product, while nitrate [12] could not be detected in the reaction mixture, indicating that the overall NADH: O_2 ratio should not be 1. Analyses of the stoichiometry of this reaction were carried out in the two experimental systems shown in Table I. A limited amount of NADH was incubated aerobically with *o*-nitrophenol in the presence of the enzyme. Since *o*-nitrophenol has an absorbance around 405 nm, its disappearance could be detected by a decrease in absorbance at this wavelength ($\epsilon = 1.6 \cdot 10^3\ M^{-1} \cdot cm^{-1}$). After the complete oxidation of NADH had occurred, the products, catechol and nitrite, were analyzed. The

TABLE I

STOICHIOMETRY OF THE HYDROXYLATION REACTION WHEN *o*-NITROPHENOL WAS USED AS SUBSTRATE (μM)

(A) Experiments were carried out in the cell of a spectrophotometer at 20°C. The reaction mixture (2 ml) contained 1 mM *o*-nitrophenol, 470 nM salicylate hydroxylase, 30 mM potassium phosphate buffer, pH 7.0 and indicated amounts of NADH. (B) Experiments were carried out in a Clark polarograph electrode at 25°C. The reaction mixture (3.5 ml) contained 1 mM *o*-nitrophenol, 470 nM salicylate hydroxylase, 30 mM potassium phosphate buffer, pH 7.0 and indicated amounts of NADH.

System	NADH added	O_2 consumed	*o*-N:trophenol used	Catechol formed	HNO_2 formed
(A)					
Complete assay mixture	181		93	73	82
Complete assay mixture	106		57	47	45
Complete, enzyme boiled at 100°C	181		0	0	2.3
Without NADH	0		0	0	0.8
Without *o*-nitrophenol	181		0	0	0.5
(B)					
Complete assay mixture	206	108		76	92
Complete assay mixture	88	43		34	40
Complete, enzyme boiled at 100°C	181	14		0	4.5
Without NADH	0	9.4		0	5.5
Without *o*-nitrophenol	181	120		0.5	4.5

stoichiometry between NADH, o-nitrophenol, catechol and nitrite was estimated to be 2:1:1:1. A parallel experiment was carried out in a Clark electrode apparatus. The stoichiometry between NADH, O_2, catechol and nitrite in this system was found to be 2:1:1:1. The reaction systems without either NADH or o-nitrophenol and with the boiled enzyme were found to produce neither appreciable quantities of catechol nor nitrite. The fact that the enzyme did not catalyze either the hydroxylation of phenol to catechol in the presence of NADH and O_2 or reduction of nitrate to nitrite in the presence of NADH excludes the production of a potential intermediate of either phenol plus nitrite or catechol plus nitrate, respectively.

When *o*-iodo- and *o*-aminophenol were used as substrate, similar stoichiometry was estimated as in the case of *o*-nitrophenol. In each system, the product was identified to be iodide or ammonia, respectively, while hypoiodite and hydroxylamine were not detected in the reaction mixture.

Based on the results described above, the hydroxylase reaction for 1-substituted analogs of salicylate is summarized as follows:

$$C_6H_4(OH)X + 2\,NADH + 2\,H^+ + O_2 \longrightarrow C_6H_4(OH)_2 + HX + H_2O + 2\,NAD^+ \quad (1)$$

where X denotes NO_2, NH_2 or I, respectively.

Reaction catalyzed by a chemically modified salicylate hydroxylase

We have postulated the reaction pathway for the salicylate hydroxylase-catalysed hydroxylation as summarized in Eqn. 2 [1, 2, 5], where E, F, S and DH_2 denote protein and flavin moieties of the enzyme, salicylate and NADH + H^+, respectively.

$$EF \xrightarrow{+S} EFS \xrightarrow{+DH_2} EFH_2S \xrightarrow{+O_2} EF + \text{catechol} + CO_2 + H_2O \quad (2)$$

When the enzyme was subjected to modification with glyoxal [13], the resulting "modified enzyme" catalyzed an NADH oxidation reaction which appeared to be dependent on the presence, but not on the consumption, of salicylate.

The bulk of the modified enzyme was prepared by the treatment of 50 μM salicylate hydroxylase in 300 mM sodium carbonate buffer, pH 9.2, with 120 mM glyoxal at 0°C for 3 h. The mixture was then passed through a Sephadex G-25 column. The enzyme fractions were combined and solid ammonium sulfate (500 mg/ml) was added. The precipitate was dissolved in a small amount of 500 mM potassium phosphate buffer, pH 8.0, containing 500 nmol of FAD and dialyzed against 33 mM phosphate buffer, pH 7.0. The preparation was homogeneous on polyacrylamide disc electrophoresis either at pH 9.4 or at pH 8.0.

The optimum pH of the native enzyme for the salicylate hydroxylation activity or for the NADH oxidation activity has been shown to be between 7.5 and 8 [14], while NADH oxidation activity of the modified enzyme had

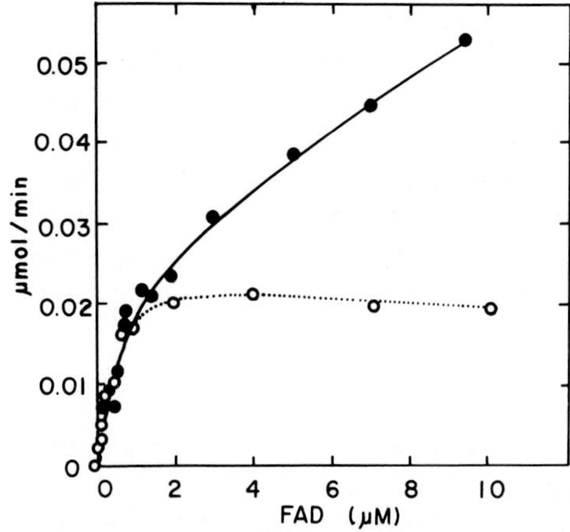

Fig. 1. Effect of FAD on NADH oxidation and catechol formation catalyzed by the modified salicylate hydroxylase. The reaction mixture (1 ml) contained 30 mM potassium phosphate buffer (pH 7.0), 120 μM NADH, 40 μM salicylate, the apoenzyme (5.4 μg) and varying amounts of FAD as indicated. The rates of catechol formation (○----○) and NADH consumption (●——●) were followed by the metapyrocatechase method [2] and at 340 nm absorbance, respectively. The ordinate indicates the rate of either catechol formation or NADH consumption.

a wider optimum pH range, down to 5.2. The modified enzyme contained a minor activity of salicylate hydroxylation. Under assay conditions at pH 5.2, NADH oxidation "stimulated" with salicylate was approx. 10 times greater than that coupled with salicylate hydroxylation. However, it is not at present clear whether the salicylate hydroxylation activity is ascribed to contamination of the native enzyme.

The absorbance spectrum of the modified enzyme had peaks at 275, 375 and 448 nm and exhibited a marked shoulder around 480 nm upon addition of salicylate. These characteristics were very similar to those of the native enzyme [1, 2]. From the data of the spectral titration curve, the dissociation constant of salicylate for the modified enzyme was calculated to be 4.5 μM which was slightly greater than the value for the native enzyme (3.5 μM). The K_m value for NADH was 5.6 μM for the modified enzyme (for the native enzyme 2.6 μM).

Benzoate is reported to stimulate NADH oxidation activity of a salicylate hydroxylase at high benzoate (K_m : 2 mM) and NADH (K_m :700 μM) concentrations [15]. Similar effects of benzoate were also observed either in our native or in the modified enzyme. The K_m value of 135 μM for the native enzyme coincided with that for the modified enzyme (139 μM).

The NADH oxidation activity of the modified enzyme appeared to be a diaphorase activity toward unbound FAD in the assay medium since the rate of NADH oxidation increased with concentration of FAD although the rate of catechol formation reached maximum at a FAD concentration of 2 μM (Fig. 1); while, when FAD was added to the native enzyme, the rate of NADH oxidation, whether coupled with the hydroxylation or stimulated by the presence of benzoate, was essentially unchanged between FAD concentrations of 2 μM and 10 μM.

These results suggest that the modification seems to be restricted basically on the "oxygen-binding site". Salicylate, in its "physiological" concentration, affects either the formation of the enzyme-salicylate complex or the stimulation of electron transfer from NADH to the FAD moiety. Although the nature of the chemical modification of the protein moiety remains to be studied, the flavin prosthetic group of the enzyme was not modified, since the characteristics of the native enzyme were not changed upon recombination of the native apoenzyme with the flavin isolated from the modified enzyme, and vice versa.

Riboflavin derivative of lactate oxygenase

Lactate oxygenase was reversibly resolved into FMN and apoenzyme moieties by an acid-ammonium sulfate precipitation method. The apoenzyme was specifically reactivated by addition of FMN or even riboflavin. However, the flavin derivatives such as lumiflavin, roseoflavin (7-methyl-8-dimethylamino-10-ribitylisoalloxazine) and FAD were ineffective.

The holo-riboflavin enzyme was prepared by the same procedure for the holo-FMN enzyme [7], except FMN was replaced by riboflavin. The overall recovery after three-fold recrystallization was approx. 30%. The crystals of the riboflavin enzyme were square transparent plates and their appearance was quite similar to that of the native enzyme.

Flavin analyses performed on the reconstituted enzyme confirmed that riboflavin was the sole flavin component. The absorbance band positions and the general profiles of the CD spectra were almost identical with those observed in the native enzyme. The anaerobic addition of L-lactate to the oxidized enzyme produced an instantaneous bleaching of the visible color, and no stable semiquinoid form of riboflavin could be detected. The riboflavin enzyme had a molecular weight of 340 000 as determined by gel filtration. The sedimentation coefficient of the enzyme was 12.5 S. These values were in good agreement with those assigned previously to the native enzyme [7]. When analyzed by the Ampholine electrofocusing technique, a considerably lower isoelectric point (less than pH 3) was estimated for both native and riboflavin enzyme.

The basic function of lactate oxygenase was not changed by the riboflavin substitution. A maximal turnover number of 1250 mol of O_2 per min per mol of riboflavin was obtained at the optimal pH at 6.0. This value was about half that for the native or FMN enzyme (3000). The K_m value for L-lactate was 50 mM for the riboflavin enzyme (for the native enzyme 34 mM). As characterized in the native enzyme [16], the riboflavin enzyme also acted as an oxidase when L-malate was used as substrate, producing oxalacetate with the stoichiometric consumption of molecular oxygen.

These results described above indicate that although the phosphate group of FMN is not essential for the function of lactate oxygenase, the ribityl side group has significant influence on the activation of the apoenzyme as well as on the interaction of the flavin with the protein moiety.

Acknowledgements

We are greatly indebted to Dr. S. Otani, Koshien University, for the gift of roseoflavin and to Dr. A. Yoshimura, Osaka University for lumiflavin. Experiments reported in this paper have been performed in collaboration with H. Nakashima, S. Takamiya, K. Onishi, T. Gomi and S. Nishimura in our laboratory. This work was supported by the Scientific Research Fund B-447051 from the Ministry of Education of Japan.

References

1 Katagiri, M., Takemori, S., Suzuki, K. and Yasuda, H. (1966) J. Biol. Chem. 241, 5675–5677
2 Takemori, S., Yasuda, H., Mihara, K., Suzuki, H. and Katagiri, M. (1969) Biochim. Biophys. Acta 191, 58–68
3 Takemori, S., Yasuda, H., Mihara, K., Suzuki, H. and Katagiri, M. (1969) Biochim. Biophys. Acta 191, 69–76
4 Suzuki, K., Takemori, S. and Katagiri, M. (1969) Biochim. Biophys. Acta 191, 77–85
5 Takemori, S., Nakamura, M., Suzuki, K., Katagiri, M. and Nakamura, T. (1972) Biochim. Biophys. Acta 284, 382–393
6 Takemori, S., Hon-nami, K., Kawahara, F. and Katagiri, M. (1974) Biochim. Biophys. Acta 342, 137–144
7 Takemori, S., Nakazawa, K., Nakai, Y., Suzuki, K. and Katagiri, M. (1968) J. Biol. Chem. 243, 313–319
8 Takemori, S., Nakai, Y., Nakazawa, K., Katagiri, M. and Nakamura, T. (1973) Arch. Biochem. Biophys. 154, 137–146
9 Takemori, S., Tajima, H., Kawahara, F., Nakai, Y. and Katagiri, M. (1974) Arch. Biochem. Biophys. 160, 289–303
10 Katagiri, M., Yamamoto, S. and Hayaishi, O. (1962) J. Biol. Chem. 237, PC2413–2414
11 Hayaishi, O. and Sutton, W.B. (1957) J. Am. Chem. Soc. 79, 4809–4810
12 Egami, F., Iida, K., Doke, T. and Taniguchi, S. (1954) Bull. Chem. Soc. Japan 27, 619–623
13 Nakaya, K., Horimishi, H. and Shibata, T. (1967) J. Biochem. Tokyo 61, 345–351
14 Yamamoto, S., Katagiri, M., Maeno, H. and Hayaishi, O. (1965) J. Biol. Chem. 240, 3408–3413
15 White-Stevens, R.H. and Kamin, H. (1972) J. Biol. Chem. 247, 2358–2370
16 Takemori, S. and Katagiri, M. (1975) in Reactivity of Flavins (Yagi, K., ed.), pp. 103–117, University of Tokyo Press, Tokyo

Part V

Mechanisms of proton abstraction

Flavins and Flavoproteins, edited by T.P. Singer

CHAPTER 19

THE STRUCTURE OF THE COVALENT FLAVIN ADDUCT FORMED BETWEEN LACTATE 2-MONOOXYGENASE* AND THE SUICIDE SUBSTRATE 2-HYDROXY-3-BUTYNOATE

A. SCHONBRUNN[a], R.H. ABELES[a], C. WALSH[b], S. GHISLA[c], H. OGATA[c] and V. MASSEY[c]

[a]*Graduate Department of Biochemistry, Brandeis University, Waltham, Mass. 02154,* [b]*Departments of Chemistry and Biology, Massachusetts Institute of Technology, Cambridge, Mass. 02139 and* [c]*Department of Biological Chemistry, The University of Michigan, Ann Arbor, Mich. 48104 (U.S.A.)*

Introduction

L-Lactate 2-monooxygenase* (EC 1.13.12.4, L-lactate: oxygen 2-oxidoreductase (decarboxylating)) from *Mycobacterium smegmatis* catalyzes the oxidative decaroxylation of L-lactate.

$$H_3C\text{—}\overset{\text{OH}}{\overset{|}{C}H}\text{—}COO^{(-)} + O_2 \rightarrow H_3C\text{—}COO^{(-)} + H_2O + CO_2$$

The substrate analogue DL-2-hydroxy-3-butynoate (Structure I) acts as both a substrate (i.e. is oxidized) and an irreversible inactivator of the enzyme [1].

$$HC{\equiv}C\text{—}\overset{\text{OH}}{\overset{|}{C}H}\text{—}COO^{(-)} \qquad \text{(I)}$$

Inactivation results in covalent modification of the enzyme-bound flavin, but not the apoprotein [1]. The structure of the flavin inhibitor adduct has not been elucidated, although preliminary isotope experiments have shown that in forming the adduct, the C-2 hydrogen of the inactivator is lost [1]. The C-4 hydrogen is present in the adduct but no longer in an acetylenic linkage [1].

We now report experiments which establish the structure of the flavin-2-hydroxy-3-butynoate adduct.

* Formerly lactate oxidase (EC 1.1.3.2).

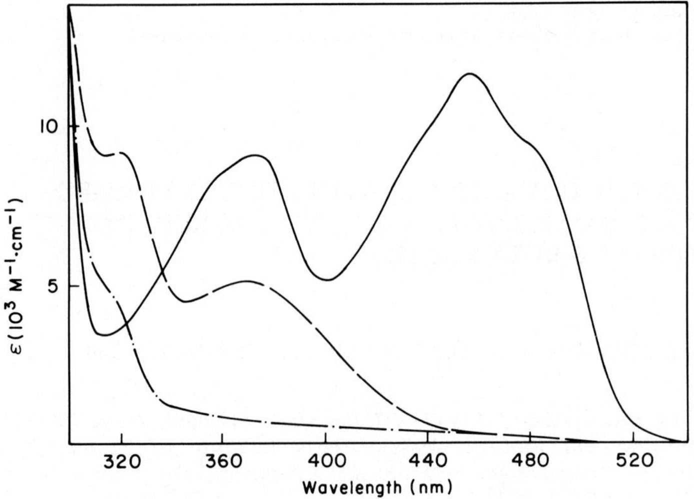

Fig. 1. Spectral changes observed on aerobic inactivation of lactate 2-monooxygenase with 2-hydroxy-3-butynoate before and after reduction of inactivated enzyme with $NaBH_4$. (——) Spectrum of 0.8 ml of 1.8 mg/ml lactate 2-monooxygenas in 0.01 M NaAc pH 8. (----) Spectrum 15 min after addition of 50 μl (1.23 μmol) DL-2-hydroxy-3-butynoate. (—·—·—) A few mg $NaBH_4$ were added to inactivated enzyme 20 min after the addition of inactivator. Reduction was allowed to proceed at room temperature for 25 min at which time 50 μl acetone was added to react with excess $NaBH_4$. The Spectrum was taken 5 min after the addition of acetone.

Spectral properties of the adduct

The absorbance spectrum of the inactivated enzyme is different from the spectrum of oxidized or reduced lactate 2-monooxygenase and is characterized by absorbance bands at 320 nm and 368 nm (Fig. 1) and strong fluorescence with an emission maximum at 502 nm. When these spectral properties are compared to those of model compounds, the flavin-inactivator adduct formed on the enzyme most closely resembles 4a, 5-dihydroflavin derivatives [2]. In particular, 4a, 5-dihydroflavin derivatives carrying a C-3 bridge between the C-4a and N-5 positions (Structures II and III) have characteristic spectra in solution which are very similar to the spectrum of inactivated lactate 2-monooxygenase [2, 3].

II (structure) III (structure)

When the N_{10}-ribitolphosphate derivative of II (C_{4a}, N_5-propano-4a, 5-dihydroFMN) is bound to apolactate 2-monooxygenase, the complex obtained has a spectrum similar to that of the inactivated holoenzyme (Fig. 2). Thus, the spectral properties of inactivated lactate 2-monooxygenase suggest that the modified flavin formed on the enzyme is a derivative of the

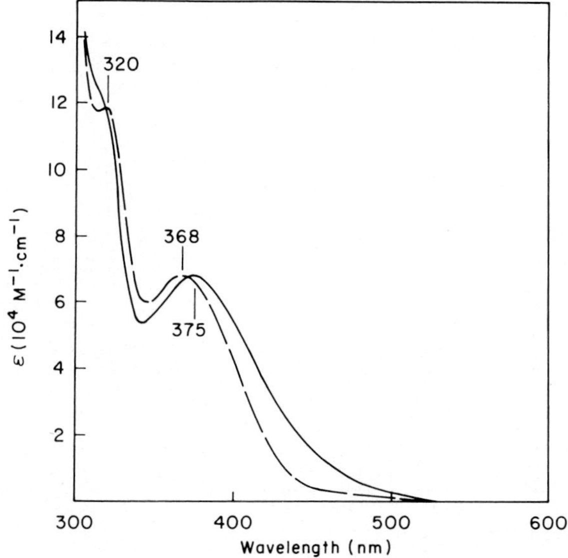

Fig. 2. Spectral comparison of 2-hydroxy-3-butynoate inactivated lactate oxidase (——) and 4a 5-propano-4a, 5-dihydro FMN-reconstituted apolactate 2-monooxygenase (- - - -).

4a, 5-dihydroflavin chromophore. However, identification of enzyme-bound adducts based on absorbance spectra alone is inadequate and can be misleading. Therefore, isotope substitution and Fourier Transform NMR experiments were carried out to determine the structure of the adduct formed between the flavin coenzyme of lactate 2-monooxygenase and the "suicide" substrate 2-hydroxy-3-butynoate.

Isolation of the flavin-inhibitor adduct

We attempted to resolve the flavin-inhibitor adduct from the enzyme so that it could be used for further structural studies. Experiments with 2-hydroxy-3-[1-^{14}C]butynoate established that all of the adduct could be released from the enzyme by treatment with 80% methanol. When inactivated enzyme is denatured in this manner, the released, modified flavin consists of a mixture of compounds. It was found that these adduct species decomposed further after release from the enzyme.

We found that $NaBH_4$ reduction of the enzyme-bound adduct led to one major stable species which could be isolated and identified. Reduction of inactivated lactate 2-monooxygenase with $NaBH_4$ results in the loss of the characteristic 368 nm absorbance of the adduct (Fig. 1) and its fluorescence and provides additional evidence for the initial formation of a 4a-mono or 4a, 5-disubstituted 4a, 5-dihydroflavin adduct. 4a, 5-Dihydroflavins, which contain C_{10a}—N_1 double bond, are known to be reduced rapidly by borohydride to give a 4a-5,10a-1-tetrahydroflavin, a nonfluorescent species with a maximum wavelength of absorbance close to 300 nm, typical of substituted phenylene diamines [4].

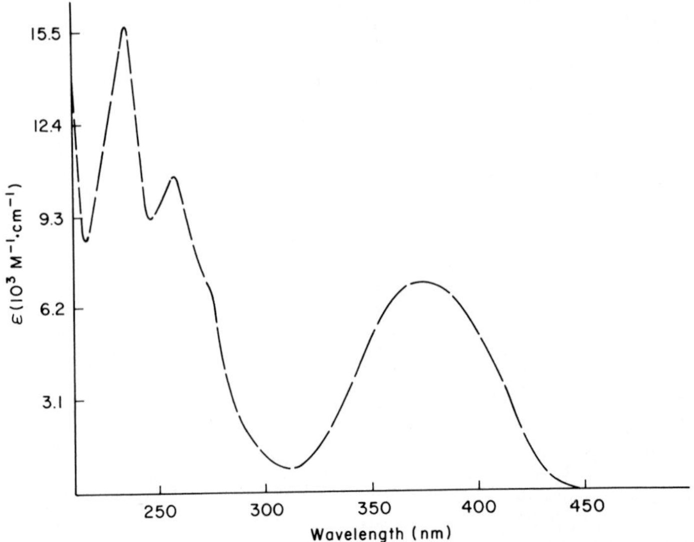

Fig. 3. Spectrum of the major turquoise-fluorescent adduct species eluted by water from silica gel thin-layer chromatography described in the legend to Table I.

The subsequent release of this borohydride-reduced adduct by methanol precipitation of the apoprotein yielded a stable species. A typical spectrum of this adduct species after release from the apoprotein is shown in Fig. 3. The spectrum of the released adduct differs from its spectrum while bound to the enzyme (cf. Figs. 1 and 3). Most notably, a seven-fold increase in the absorbance at 370 nm has occurred. Furthermore, while the enzyme-bound, borohydride-reduced adduct is nonfluorescent, the released adduct is strongly fluorescent with a maximum emission at 465 nm. This change in spectral properties suggests that a structural modification of the coenzyme complex has occurred after release from the enzyme.

Characterization of the flavin-inhibitor adduct

To characterize further the nature of the borohydride-reduced adduct, lactate oxidase was inactivated anaerobically with a 50-fold excess of DL-2-hydroxy-3-[1-^{14}C, 4-^{3}H] butynoate, reacted with $NaBH_4$ on the enzyme and the reduced flavin-inhibitor adduct isolated by the purification procedure described in Table I. Thin-layer chromatography of this adduct showed that greater than 90% of the radioactivity was located in a single peak. Several fluorescent species were observed on thin-layer chromatography. One spot, which fluoresced a bright turquoise, was by far the most intensely fluorescent and coincided with the radioactive peak. Therefore, the isolated adduct represents one major species. When this major spot was eluted from the thin-layer chromatogram, it was always found to have the absorbance spectrum shown in Fig. 3, and an intense fluorescent spectrum with an emission maximum at 465 nm.

When the purified, turquoise-fluorescent adduct was rechromatographed on silica gel plates, it always gave a single spot and was able to be stored at 4°C, protected from light, for several months. Thus, the procedure described in Table I was adopted in all subsequent experiments for the preparation and purification of the major stabilized adduct species released from inactivated lactate 2-monooxygenase.

The results shown in Table I demonstrate that the $^{14}C/^{3}H$ ratio remains unchanged throughout the purification procedure. Thus, both the gamma hydrogen and the carboxyl carbon of the inactivator are still present in the isolated, purified, turquoise-fluorescent species in the same ratio as in the original adduct formed on the protein.

Experiments with appropriately labelled radioactive compounds showed that (a) hydrogen from borohydride is incorporated into the purified,

TABLE I

INACTIVATION OF LACTATE 2-MONOOXYGENASE 2-HYDROXY-3-[1-^{14}C, 4-^{3}H] BUTYNOATE

One ml lactate 2-monooxygenase (7 mg, 56 units) in 20 mM Tris · Cl pH 8 was incubated anaerobically with 7.2 µmol DL-2-hydroxy-3-[1-^{14}C,4-^{3}H] butynoate ($^{3}H = 3.75 \cdot 10^{5}$ dpm/µmol, $^{14}C = 3.32 \cdot 10^{5}$ dpm/µmol) for 30 min before reduction with $NaBH_4$ for 30 min. After destroying excess $NaBH_4$ by pyruvate addition, the labelled, inactive holoenzyme was subjected to gel filtration. The protein contained 1 mol of both ^{3}H and ^{14}C per mol flavin. The holoenzyme was denatured with 80% MeOH to release quantitatively the labelled coenzyme adduct which was purified on Eastman precoated silica gel thin-layer chromatography plates with *n*BuOH:EtOH:0.01 M NaAc pH 5.4 (1:1:1) and then by electrophoresis on silica gel-coated glass plates at pH 7 in 0.02 M $NaPO_4$ at 400 V for 9 h.

	$^{14}C/^{3}H$
2-Hydroxy-3-[1-^{14}C, 4-^{3}H] butynoate	0.88
Protein off Sephadex G25	0.97
Concentrated adduct released from protein	0.92
Turquoise-fluorescent adduct	
after thin-layer chromatography	0.72
after thin-layer electrophoresis	0.93

TABLE II

INACTIVATION OF 2-^{14}C-LABELED FMN-LACTATE 2-MONOOXYGENASE WITH 2-HYDROXY-3-[4-^{3}H] BUTYNOATE

1.32 µmol of lactate 2-monooxygenase (spec.act. = 43.4 µmol/min/mg) was resolved into coenzyme and apoenzyme by the procedure of Choong et al. [5]. The apoenzyme retained less than 0.1% of the original activity. The apoenzyme was reconstituted with 3.79 µmol 2-^{14}C-labeled FMN (spec.act. = $1.19 \cdot 10^{5}$ dpm/µmol) prepared by enzymatic phosphorylation of commerical [2-^{14}C] riboflavin with partially purified flavokinase. Denatured apoprotein, not reconstituted with ^{14}C-labeled FMN was removed by centrifugation. 0.57 mol of ^{14}C-labeled FMN was bound per mol FMN resolved from lactate 2-monooxygenase. The ^{14}C-labeled FMN lactate 2-monooxygenase was then inactivated anaerobically with a 50-fold excess of DL-2-hydroxy-3-[4-^{3}H] butynoate (spec.act. $4.13 \cdot 10^{5}$ dpm/µmol), reduced with $NaBH_4$ and purified according to the procedure in the legend to Table I.

	$^{14}C/^{3}H$
Spec.act. ^{14}C-labeled FMN/spec. act. 2-hydroxy-3-[4-^{3}H] butynoate	0.29
Protein off Sephadex G25	0.27
Turquoise-fluorescent adduct after thin-layer chromatography	0.007

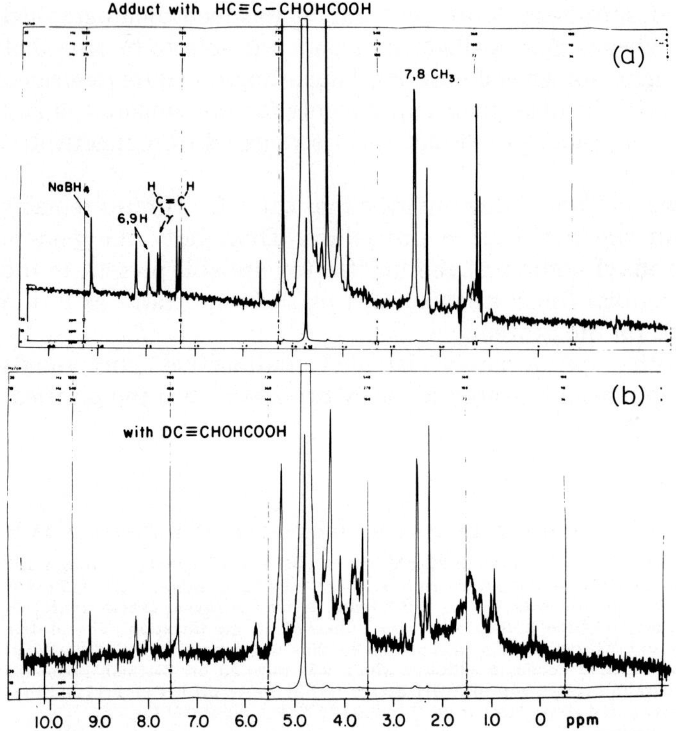

Fig. 4. Fourier Transform NMR spectra of the borohydride-stabilized adduct. PMR spectra were recorded with a Bruker WH90 Fourier Transform NMR spectrometer at a sweep width of 1200 cps. Samples of adduct, purified according to the procedure in the legend to Table I, were prepared for NMR by repeated lyophilization from 99.8% ^2H_2O to remove as much H_2O as possible. Spectra were run in 100.0% ^2H_2O at room temperature. Chemical shifts were measured in ppm using the absorbance peak of the solvent (H^2HO) as an internal standard. (A) Spectrum of adduct isolated from 2-hydroxy-3-butynoate-inactivated lactate monooxygenase. (B) Spectrum of adduct isolated from 2-hydroxy-3-[4-^{2}H] butynoate-inactivated lactate monooxygenase.

stabilized adduct*, (b) non-exchangeable hydrogen from water is incorporated* and (c) the pyrimidine nucleus of the flavin ring remains intact as long as the adduct is on the enzyme. However, ring opening occurs after removal from the apoprotein (Table II).

NMR studies of the borohydride-stabilized adduct

A large-scale preparation of the borohydride-stabilized adduct was undertaken in an effort to obtain further information about its structure by NMR spectroscopy. Lactate 2-monooxygenase (2 μmol) was inactivated anaerobically with DL-2-hydroxy-3-butynoate and the adduct was purified by the procedure described in the legend to Table I. The Fourier Transform

* Due to isotope effects of unknown magnitude, the number of atoms of hydrogen incorporated per mol adduct cannot be determined.

NMR is shown in Fig. 4A. Chemical shifts were assigned using the absorbance peak of the solvent as an internal standard (H^2HO = 4.75 ppm). The two singlets at 8.0 δ and 8.2 δ were assigned to the two hydrogens on the benzene ring respectively and were used to determine the relative areas of the other peaks in the spectrum**. The large singlet at 2.5 δ integrating to six protons corresponds to the 7, 8-methyl groups. The signals due to the hydrogens on the ribitol side chain are probably between 4.0 δ and 4.5 δ. Peaks observed between 0 δ and 2 δ were not reproducible and presumably represent contaminants introduced during the purification procedure.

The remaining signals were identified, as shown in Fig. 4A by appropriate isotope substitution. In one experiment, inactivated lactate 2-monooxygenase was reduced with NaB^2H_4, and the flavin-inhibitor adduct was purified according to the standard procedure. The NMR spectrum of the purified adduct is identical to that in Fig. 4A, except that the singlet integrating to one proton at 9.2 δ is absent.

In a separate experiment, lactate 2-monooxygenase was inactivated with 2-hydroxy-3-[4-^{2}H] butynoate, reduced with $NaBH_4$ on the enzyme, and the deuterated adduct was purified. The Fourier Transform NMR is shown in Fig. 4B. Comparison with Fig. 4A shows that the doublet at 7.8 δ is now missing and that the doublet at 7.5 δ has coalesced into a singlet. The coupling constant between the two doublets is 4.5 cps. Thus, the signal at 7.8 δ must be due to the C-4 hydrogen of 2-hydroxy-3-butynoate and the signal at 7.5 δ must be due to a hydrogen on an adjacent carbon atom, most probably on C-3 of 2-hydroxy-3-butynoate. This hydrogen on C-3 presumably comes from water and accounts for the previously discussed observation that at least one non-exchangeable tritium moiety is incorporated into the adduct when the inactivation reaction is carried out in ^3H_2O. Since each of these hydrogens gives rise to a doublet, they must be isolated from other hydrogens in the molecule. The chemical shifts suggest either aromatic hydrogens or olefinic hydrogens attached to a highly electron-deficient group. Furthermore, since these two hydrogens give rise to signals very close together, they must be in very similar chemical environments. The low value of the coupling constant between the two doublets in the NMR spectrum of the adduct (J = 4.5 cps) suggests that this group is part of a 5- or 6-membered ring, i.e., that there are two points of attachment between the inactivator and the flavin.

These results lead us to propose the sequence of reactions in Scheme I, and Structure V as the most resonable structure for the isolated borohydride-stabilized adduct. To verify this structure, the model compound 5, 7, 8-trimethylpyrroloquinoxalinium (VI) was synthesized by the procedure of Cheeseman and Tuck [6].

** The numbering system for flavins will continue to be followed even though the isoalloxazine ring is no longer intact in the adduct.

Sequence of Reactions in Isolation of Adduct

Ribitol-Ⓟ

$HC\equiv C-CH(OH)-COO^{(-)}$

(IV) Enzyme Bound Flavin–Inactivator Adduct

$NaBH_4$

Enzyme Bound Reduced Adduct

MeOH

Apoprotein Precipitate

(V) Stable Adduct Released from Enzyme

VI

VII

The chemical shifts of the protons of this model and the isolated adduct are compared in Table III. The chemical shifts agree closely and clearly eliminate Structure VII as a possibility for the structure of the isolated adduct. The H_1 and H_3 protons of the model can be distinguished readily by their chemical shifts. In Structure VII, the beta and gamma protons from 2-hydroxy-3-butynoate would be expected to have chemical shifts similar to H_2 and H_1 of the model, while in Structure V they would have signals similar to H_2 and H_3. Structure VII is clearly eliminated on this basis since in the adduct the beta and gamma protons give rise to signals close together, as do H_3 and H_2 in the model. Furthermore, the coupling of the model shows values of $J_{1-2} = 2.5$ cps and $J_{2-3} = 4.4$ cps. The coupling constant observed between the beta and gamma hydrogens of the adduct, $J = 4.5$ cps, is in much better agreement with their assignments in Structure V than in Structure VII. Therefore, synthesis of the model compound VI confirms that the proposed structure of the adduct, Structure V, is correct.

The sequence of reactions postulated in Scheme I will account for the introduction by borohydride of a hydrogen with an unusually downfield chemical shift and the loss of C-2 of the isoalloxazine ring. Borohydride reduction of the initially formed flavin-2-hydroxy-3-butynoate adduct is proposed to occur at C_{10a}-N_1. The flavin ring remains intact as long as the adduct is bound to the enzyme, but on release of the adduct into solution, breakdown of the carbon ring occurs as a consequence of the formation of an imine between N-10 and C-10a. Imine protons have chemical shifts consistent with that of the signal attributed to the hydrogen introduced from $NaBH_4$.

The presence of a positive charge on N-10 also accounts for the unusually high chemical shifts of H-9 and H-6 and the 7- and 8-position methyl groups on the benzene ring. The change from their normal chemical shifts in reduced flavins is comparable to the difference in the chemical shifts of these hydrogens in N_5-monoalkyl and N_5-dialkyl-1,5-dihydroflavins [7]. Furthermore, such a sequence of reactions leading to the introduction of a double bond at C_{10a}-N_1 conjugated with the benzene ring, also explains the large increase in absorbance at 370 nm when the borohydride-reduced adduct is released from the enzyme.

Discussion

The results presented provide evidence that Structure V (Scheme I) is the structure for the borohydride-stabilized adduct released from lactate 2-monooxygenase after inactivation with 2-hydroxy-3-butynoate. Therefore,

TABLE III

COMPARISON OF NMR ASSIGNMENTS AND CHEMICAL SHIFT DATA FOR THE ISOLATED ADDUCT (V) AND MODEL COMPOUND (VI)

Fourier Transform NMR in D_2O

	Model (VI)	Isolated Adduct (V)
	H_3C, H_3C, CH_3, N, N+, 1, 2, 3, 4, 6, 9	H_3C, H_3C, Ribitol Ⓟ, N, N+, ^-OOC
Hydrogen	Chemical Shift (ppm)	
H_1	8.4	—
H_2	7.2	7.5
H_3	7.6	7.8
H_4	9.0	9.2
H_6	7.9	8.2
H_9	7.7	8.0
CH_3-7,8	2.4	2.5
J1-2	2.50 cps	—
J2-3	4.38 cps	4.5

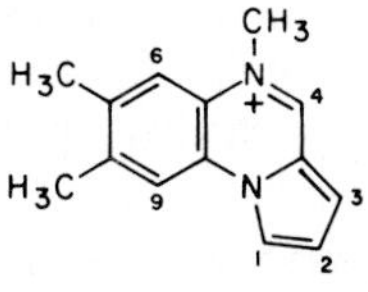

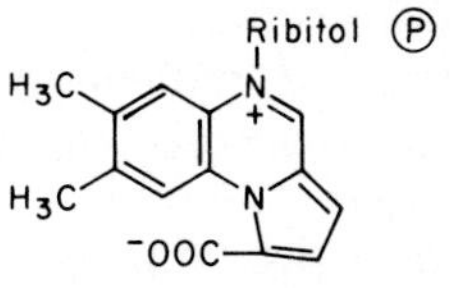

Adduct Formation from Oxidized FMN and 2-Hydroxy Butynoic Acid

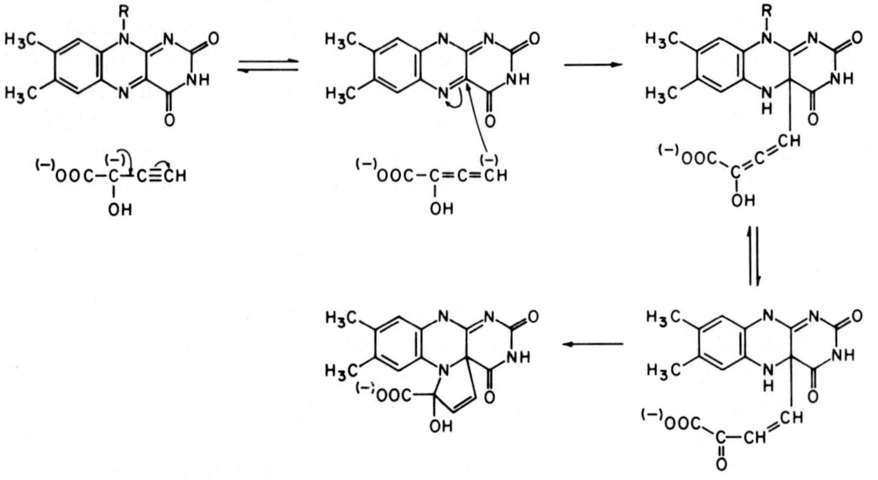

the most likely structure for the initial adduct formed on the enzyme between the flavin cofactor and 2-hydroxy-3-butynoate is Structure IV (Scheme I).

The flavin-inactivator adduct can be envisaged to be formed by one of several mechanisms. It is unlikely that an allene of 2-hydroxy-3-butynoate, formed by rearrangement after alpha-proton abstraction, is an intermediate, as in the case of inactivation of β-hydroxy-decanoyl-thioester dehydrase by an acetylenic substrate analogue [8]. Nucleophilic attack on an allene intermediate would occur on C-3 of the inactivator, nor at C-2 and C-4, the observed attachment sites. However, one possible mechanism for adduct formation involves formation of an allene anion which initiates nucleophilic attack at C-4a of the flavin ring (Scheme II). Alternatively, an allene radical may initiate attack on the flavin semiquinone formed after partial oxidation of the inactivator had occurred. No evidence has been presented to date to show the involvement of radical intermediates in flavoenzyme reactions, however, small concentrations of interacting radical pairs might escape detection.

It is also possible that 2-hydroxy-3-butynoate is first oxidized on the enzyme to the keto-acid. Alpha-keto-butynoate may react with the reduced flavin by Michael addition to an $\alpha\beta$-unsaturated system (Scheme III). A reduced flavin species is formed during the reaction of 2-hydroxy-3-butynoate with lactate oxidase indicating that this mechanism is possible [9]. The basic distinction between the two mechanisms is that in one case (Scheme I) inactivation occurs prior to complete reduction, while in the other case (Scheme II), after reduction of the flavin. Since E_{ox}-S and E_{red}-P are in equilibrium, it is diffucult to distinguish between the two alternatives by kinetic measurements. However, vinylglycolic acid, is oxidized to 2-keto-3-butenoic acid by lactate oxidase (Walsh, C.T. and Shonbrun, A., unpublished

Adduct Formation from Reduced FMN and 2-Keto Butynoic Acid

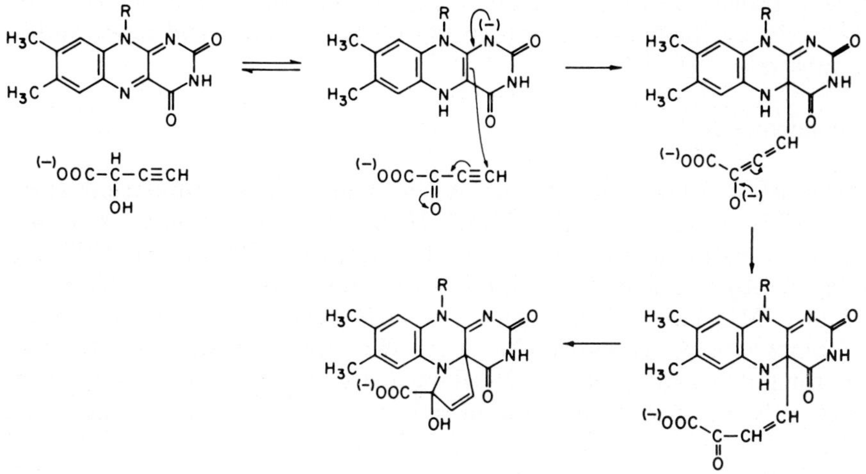

results) could undergo a Michael addition analogous to 2-keto-3-butynoic acid. This enzyme, as well as a number of other flavoenzymes [10, 11] which are also inactivated by 2-hydroxy-3-butynoate, are not inactivated by vinylglycolic acid. These observations tend to argue against the mechanism of Scheme III, but certainly cannot exclude it.

Regardless of the mechanism by which the adduct is formed, the structure of the adduct provides information concerning the orientation of the inhibitor on the enzyme. Thus, the structure of the flavin-inhibitor adduct shows that the α-carbon of the substrate must be located near N-5 and not C-4a of the isoalloxazine ring during the enzyme-catalyzed oxidation, and that the introduction of electrons into the flavin ring probably occurs through N-5. However, the isolation of an adduct between 2-hydroxy-3-butynoate and lactate 2-monooxygenase does not necessarily imply that a covalent adduct is formed as a real intermediate in the oxidation reaction with normal substrates. The proposal that interaction between substrate and flavin occurs at N-5 is consistent with results obtained with lactate 2-monooxygenase reconstituted with the coenzyme analogue deaza-FMN [12], as well as the experiments which showed that nitroethane forms an N-5 adduct with the flavin coenzyme of D-amino acid oxidase [13].

It is particularly noteworthy that the structure of the adduct formed on lactate 2-monooxygenase with 2-hydroxy-3-butynoate is different from the adduct formed on the flavin enzyme, monoamine oxidase with *N*, *N*-dimethyl propargylamine [14]. In the latter, the terminal carbon of the inactivator is attached to N-5. This observation implies that the substrates for these two enzymes are not situated in the same position relative to the flavin coenzyme. Thus, it is possible that alpha-hydroxy- and alpha-amino acid oxidases function by a different mechanism than amine oxidases. This may occur because alpha-proton abstraction and formation of an alpha-carbanion is far more favorable for alpha-hydroxy- and alpha-amino acids than for primary amines. In fact, elimination reactions with β-chlorosubstrates, indicative of transient carbanion formation, have to date been observed to be catalyzed only by flavoenzymes which use alpha-hydroxy- and alpha-amino acids as substrates [15, 16].

Acknowledgements

This work was supported in part by National Science Foundation Grant No. GB 31952X (RHA) and U.S. Public Health Service Training Grant No. GM 212 (AS); National Institutes of Health Grant no. GM 20011 (CW) and Helen Hay Whitney Fellowship (CW); National Institutes of Health Gran No. GM 11106 (VM). This is publication No. 1071 from the Graduate Department of Biochemistry, Brandeis University. We wish to thank Dr. Alfred G. Redfield for putting his excellent NMR facilities at our disposal and for his generosity with his advice and time in helping to run the spectra displayed in Fig. 4.

References

1 Walsh, C.T., Schonbrunn, A., Lockridge, O., Massey, V. and Abeles, R.H. (1972) J. Biol. Chem. 247, 6004–6006
2 Ghisla, S., Massey, V., Lhoste, J.M. and Mayhew, S.G. (1974) Biochemistry 13, 589–597
3 Zeller, E.A., Gärtner, B. and Hemmerich, P. (1972) Z. Naturforsch, Teil. B 27, 1050–1052
4 Dudley, K.H. and Hemmerich, P. (1967) J. Org. Chem. 32, 3049–3054
5 Choong, Y.S., Shepherd, M.G. and Sullivan, P.A. (1975) Biochem. J. 145, 37–45
6 Cheeseman, G.W.H. and Tuck, B. (1966) J. Chem. Soc. C, 852–855
7 Ghisla, S., Hartmann, I., Hemmerich, P. and Muller, F. (1973) Justus Leibings Ann. Chem. 1388–1415
8 Bloch, K. (1969) Acc. Chem. Res. 2, 193–202
9 Massey, V., Ghisla, S., Ballou, D.P., Walsh, C.T., Cheung, Y.T. and Abeles, R.H. (1976) in Flavins and Flavoproteins, Proceedings of the 9th Symposium (Singer, T.G., ed.), Chapt. 20, pp. 199–212, Elsevier, Amsterdam
10 Walsh, C.T., Abeles, R.H. and Kaback, H.R. (1972) J. Biol. Chem. 247, 7858–7863
11 Cromartie, T. and Walsh, C.T. (1975) Biochemistry in press
12 Averill, B.A., Shonbrunn, A., Abeles, R.H., Weinstock, L.T., Cheng, C.C., Fisher, J., Spencer, R. and Walsh, C.T. (1975) J. Biol. Chem. 250, 1603–1605
13 Porter, D.J.T., Voet, J.G. and Bright, H.J. (1973) J. Biol. Chem. 248, 4400–4416
14 Maycock, A.L., Abeles, R.H., Salach, J.I. and Singer, T.P. (1976) in Flavins and Flavoproteins, Proceedings of the 9th Symposium (Singer, T.P., ed.), Chapt. 22, pp. 218–224, Elsevier, Amsterdam
15 Walsh, C.T., Schonbrunn, A. and Abeles, R.H. (1971) J. Biol. Chem. 246, 6855–6866
16 Walsh, C.T., Lockridge, O., Massey, V. and Abeles, R.H. (1973) J. Biol. Chem. 248, 7049–7054

Flavins and Flavoproteins, edited by T.P. Singer

CHAPTER 20

RAPID REACTION STUDIES ON DEHYDROGENATION AND ELIMINATION REACTIONS OF D-AMINO ACID OXIDASE AND LACTATE 2-MONOOXYGENASE*

V. MASSEY[a], S. GHISLA[a], D.P. BALLOU[a], C.T. WALSH[b], Y.T. CHEUNG[b] and R.H. ABELES[c]

[a] *Department of Biological Chemistry, The University of Michigan, Ann Arbor, Mich. 48104,* [b] *Departments of Chemistry and Biology, Massachusetts Institute of Technology, Cambridge, Mass. 02139 and* [c] *Department of Biochemistry, Brandeis University, Waltham, Mass. 02154 (U.S.A.)*

Introduction

Recent studies have demonstrated that proton abstraction from the α-carbon atom of the substrate by an enzyme base is an early step in the catalytic reactions of D-amino acid oxidase (EC 1.4.3.3, D-amino acid: oxygen oxidoreductase (deaminating)) [1–3], L-amino acid oxidase (EC 1.4.3.2, L-amino acid: oxygen oxidoreductase (deaminating)) [1] and lactate 2-monooxygenase (EC 1.13.12.4, L-lactate: oxygen 2-oxidoreductase (decarboxylating)) [4, 5]. The first conclusive evidence for such a phenomenon came from the discovery that these enzymes could catalyze anaerobically an elimination reaction from β-chloro-substituted substrates. Under aerobic conditions these compounds were found to behave also as normal oxidase substrates, being converted by the enzyme to the corresponding chloro-substituted product. With α-amino β-chlorobutyrate and D-amino acid oxidase, only the elimination reaction was found to occur, even in the presence of oxygen-saturated solutions [6]. The present communication presents the results of further kinetic studies on this system.

We have also extended our previous studies on the reaction of lactate 2-monooxygenase with β-chlorolactate, and report here the results of rapid-reaction studies with this substrate, as well as with the normal oxidase substrate, L-lactate. This enzyme has also been found to undergo an irreversible modification of the flavin prosthetic group when reacting with the acetylenic "suicide" substrate, 2-hydroxy-3-butynoate [7]. The previous paper has dealt with the elucidation of the structure of this modified flavin [8]; we report here some rapid reaction studies dealing with intermediates in this reaction.

* Formerly lactate oxidase (EC 1.1.3.2).

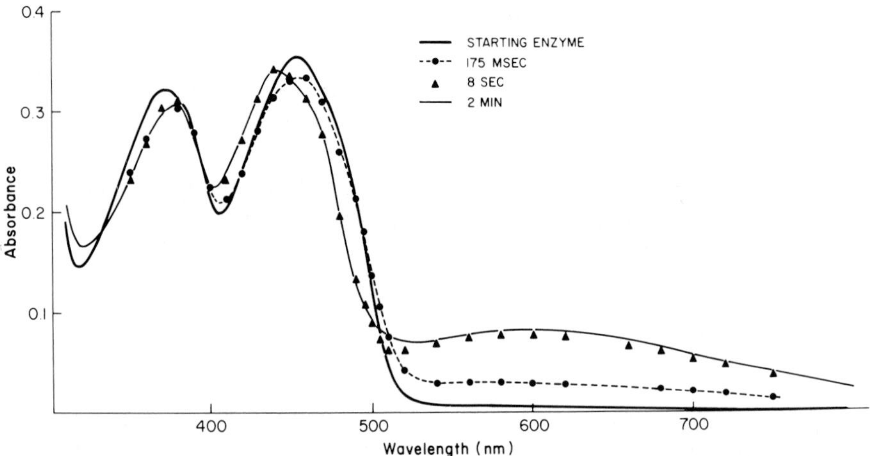

Fig. 1. Changes in absorbance of D-amino acid oxidase (DAAO) with α-amino β-chlorobutyrate. Conditions, $2.86 \cdot 10^{-5}$ M enzyme plus $6.67 \cdot 10^{-3}$ M DL-substrate, in 0.1 M pyrophosphate, pH 8.5, 25°C, determined with a Gibson–Milnes stopped-flow spectrophotometer. The spectrum at 175 ms is that of intermediate I; that at 8 s–2 min is that of intermediate II. The enzyme remains in the form of II during most of the catalytic turnover (from Ref. 6).

D-Amino acid oxidase

A previous rapid reaction study [6] had revealed the existence of two spectroscopically identifiable intermediates in the elimination reaction with D-α-amino β-chlorobutyrate. The spectral properties of these intermediates are shown in Fig. 1. The first intermediate (I) is characterized by a weak long-wavelength transition and only minor changes in the region 320–500 nm. The second intermediate (II) has much stronger absorbance at long wavelengths, with a maximum at 600 nm. It is also characterized by a shift of the 455 nm maximum of the free oxidized enzyme to 440 nm. The rate of production of I is dependent on substrate concentration, displaying saturation kinetics. It is also dependent on whether the α-carbon is substituted with hydrogen or deuterium. With the protio-substrate, the limiting rate at pH 8.5, 25°C is 2000 min^{-1}; with the deutero-substrate the limiting rate is 400 min^{-1}. In distinction the rate of conversion of intermediate I to intermediate II was found to be independent of substrate concentration, as well as whether H or ^2H forms of the substrate were used. The observed rate was 40 min^{-1} [6]. It was evident that neither of these steps was rate limiting in the catalytic cycle, since the turnover numbers were respectively 10 min^{-1} and 6 min^{-1} for the α-protio- and α-deutero-substrates.

In further efforts to elucidate the mechanism we have determined the rate of Cl^- release in the early stages of the reaction. Fig. 2 shows the results using α-H- and α-^2H- labelled forms of the substrate. It can be seen that the rate of Cl^- release in a rapid "burst" preceding steady state turnover is the same as that for the formation of intermediate I, and displaying the same five-fold deuterium isotope effect. Thus, the step in catalytic

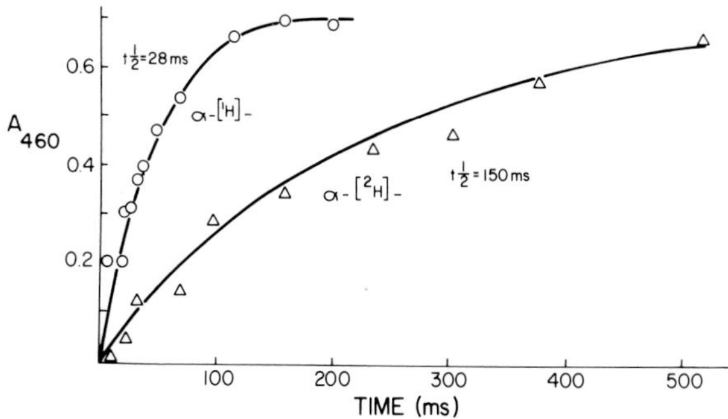

Fig. 2. Rate of Cl^- release during the "burst" phase of the reaction of D-amino acid oxidase ($1.42 \cdot 10^{-4}$ M E • FAD) with $1.25 \cdot 10^{-2}$ M D-β-chloro-α-aminobutyrate. The enzymes and α-H and α-^{2}H-labelled substrates were prepared in chloride-free 0.02 M pyrophosphate acetate buffer pH 8.5 and reacted at 25°C for the times shown before quenching the reaction with 0.1 M (final concentration) HNO_3. The rapid-quenching apparatus of Ballou and Palmer [18] was used. The quenched solutions were protected from light, centrifuged, and Cl^- determined in the supernatant fluid by the colorimetric method of Iwasaki et al. [19].

turnover involving Cl^- release occurs considerably faster than overall turnover. A puzzling aspect of these results however is that the stoichiometry of Cl^- release in the "burst" phase varied between 3–5 mol per mol enzyme-bound flavin in various experiments. A possible explanation will be considered later.

In the case of normal oxidase substrates with D-amino acid oxidase, the rate limiting step in catalysis has been found to be the dissociation of imino acid product from reoxidized enzyme [9]. The possibility that such a step was rate limiting in the present case was tested using D-α-aminobutyrate as oxidase substrate, since the same product, D-α-iminobutyrate is formed as in the Cl^- elimination reaction with α-amino-β-chlorobutyrate as substrate. The results of this study are summarized in Fig. 3. It had already been shown by Yagi and coworkers [10] that under anaerobic conditions the reaction of D-amino acid oxidase with D-α-aminobutyrate leads to the rapid production of a rather stable long-wavelength absorbing intermediate, with an absorbance maximum around 550 nm. Under aerobic conditions, as in Fig. 3, this same species is produced transiently in the first turnover, but in the steady state of turnover the enzyme exists largely in the oxidized form, returning to the anaerobic intermediate as the O_2 is consumed. Such behavior is typical for normal oxidase substrates of the enzyme [9, 11, 3]. Analysis of the absorbance changes with time by the method previously described [12] yield the Lineweaver–Burk plots of turnover number shown in the inset. From these data the V is calculated to be 125 min^{-1}. Hence, it can be concluded that the release of α-iminobutyrate from oxidized enzyme while probably the rate limiting step in the oxidation of α-aminobutyrate is not the rate-limiting step in turnover with α-amino-β-chlorobutyrate.

However, it may be that α-iminobutyrate is not the only product released during the α-amino-β-chlorobutyrate reaction. Our previous

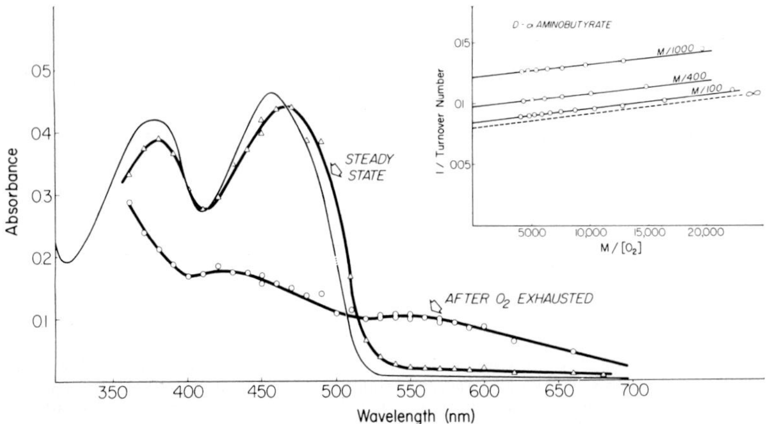

Fig. 3. Spectra of D-amino acid oxidase during aerobic turnover with D-α-aminobutyrate. Enzyme, $2.05 \cdot 10^{-5}$ M with respect to E • FAD was reacted in air-saturated solution with the concentrations of D-α-aminobutyrate shown, using the stopped flow spectrophotometer. After an initial burst reaction, (lasting between 3–20 ms for decreasing concentrations of the substrate) in which there was a considerable decrease in absorbance in the 400–500 nm region, and an increase in the 520–700 nm region, the steady state spectrum was reached within 100 ms. The spectrum then changed over a period of 7–12 s to yield that shown by the open circles. (The form of the turnover traces was similar to that shown previously for other substrates of the enzyme, cf. Refs. 9 and 11 for details.) The inset shows the result of analyses of such plots to yield catalytic turnover numbers. Conditions, 0.1 M pyrophosphate, pH 8.5, 25° C. The absorbances shown are for the 2 cm light path of the stopped flow spectrophotometer.

E FAD + CH_3-CH(Cl)-CH(NH_2)-COOH ⇌ ES ⇌ (2000 min^{-1}) {$EH^{\oplus}$ FAD / CH_3-CH(Cl)-$\overset{\ominus}{C}$(NH_2)-COOH}

fast ? $-Cl^{\ominus}$

O_2 (blocked) ← {$EH^{\oplus}$ $FADH_2$ / CH_3-CH(Cl)-C(=NH)-COOH} --- fast, $Cl^{\ominus}$ ---→ {$EH^{\oplus}$ FAD / CH_3-C(H)=C(NH_2)-COOH} **I**

H^+ ⇅ (40 min^{-1}) H^+

E FAD + CH_3-CH_2-C(=NH)-COOH ⇌ (125 min^{-1}) {E FAD / CH_3-CH_2-C(=NH)-COOH} ⇌ (10 min^{-1}) {$EH^{\oplus}$ FAD / CH_3-C(H)-$\overset{\ominus}{C}$(=NH)-COOH} **II**

Fig. 4. A possible scheme to account for the events occurring in the catalytic elimination reaction. For the sake of simplicity no attempt has been made to account for the changes in ionization state of the substrate during the course of catalysis.

experiments [6] on this reaction in ^2H_2O indicated 0.5 atom ^2H incorporated into the ketobutyrate product in a non-chiral fashion as analyzed by the ORD method, suggesting that 50% of the time the enamino acid is released into the solvent and protonated randomly at either face. That enzyme-catalyzed conversion of the enamino acid intermediate to α-iminobutyrate can occur some of the time, and is indeed a stereospecific process, will be detailed elsewhere (Cheung, Y.T. and Walsh, C.T., unpublished results).

A possible reaction scheme to account for the results is presented in Fig. 4. The kinetic isotope effect for the formation of intermediate I indicates a rate limiting abstraction of the α-proton from the substrate in this step. These results are consistent with rate limiting formation of an α-carbanion, followed by rapid Cl^- elimination, or a concerted elimination of HCl. If a process other than a concerted elimination is involved, elimination of Cl^- could involve several intermediates. One possibility is that a reductive elimination occurs after formation of reduced flavin.

In either case, the resulting intermediate I is envisaged as a complex between oxidized enzyme and the enamine form of the product. The low-intensity, charge-transfer band of I is very similar to that of similar complexes of oxidized enzyme with pyrrole 2-carboxylate and indole 2-carboxylate [14, 13]. These compounds should possess similar π electron distribution around the α-carbon atom to that of the proposed enamine. A protonated enamine in complex I should not give rise to long-wavelength, charge-transfer absorbance, since this cation would lack the donor properties required for a charge-transfer interaction with the flavin. To account for the spectral properties of intermediate II, we suggest that the enzyme-enamine complex I releases one proton. One of the possible isoelectronic species thus formed is represented in Fig. 4 as the complex of (protonated) enzyme and the β-carbanion of the α-imino acid. That the interconversion of complexes I and II could involve a slow proton uptake and release is made more feasible by the fact that the proton abstracted from the α-position of the substrate is also exchanged only slowly with the solvent [6]. Furthermore, with lactate 2-monooxygenase it has been shown that proton uptake and release is very slow, when the enzyme is complexed with dianionic acids such as oxalate or malonate [15]. Thermodynamic stabilization of a carbanionic species such as in complex II is conceivable; e.g. with lactate 2-monooxygenase, pK shifts of up to 5 units have been observed upon binding of transition state analogs [15].

The rate limiting step in the overall catalytic reaction is envisaged as the reprotonation of the β-carbanion by the protonated enzyme base which was responsible for the initial abstraction of the α-hydrogen atom. This interpretation would account satisfactorily for the observation that when α-^{3}H-labelled substrate is employed, ^{3}H is found in the β-carbon atom of the product [6]. It would also offer an explanation for the observed overall deuterium isotope effect on catalytic turnover. This step would result in the formation of the enzyme-α-iminobutyrate complex which dissociates comparatively rapidly ($k = 125\ \text{min}^{-1}$) to yield free enzyme available for the next cycle of catalysis. The oxidized enzyme-iminobutyrate complex does not have long wavelength absorbance, as can be determined readily by the addition of ketobutyrate + NH_4^+ to oxidized enzyme. Rather, this results in a slight perturbation of the absorbance spectrum, identical with that shown in Fig. 3 for the steady state in turnover of the enzyme with D-α-aminobutyrate.

As previously mentioned, there is also evidence that part of the time product is released in a form other than the imino acid. If product from either complex I or II was released into the medium, then medium proton incorporation at the β-carbon position would be expected. The possibility exists that rapid release of product at the level of complex I could account for the unusual stoichiometry of rapid Cl^- release observed in the experiments of Fig. 2. Thus, several turnovers might occur before the enzyme settled down into its steady state, in which most of the enzyme was in the form of complex II.

At present we have no convincing evidence whether the chloride elimination proceeds directly from the α-carbanion or whether it is preceeded by the latter first reducing the flavin to give a complex of reduced enzyme and β-chloro-α-iminobutyrate. The fact that no oxidase reaction is observed with this substrate could argue for the former. However, this phenomenon could equally well be explained if Cl^- elimination from the reduced enzyme β-chloro-α-iminobutyrate complex were much faster than reaction of the latter with O_2. A possible mechanism for such a reductive elimination is shown below:

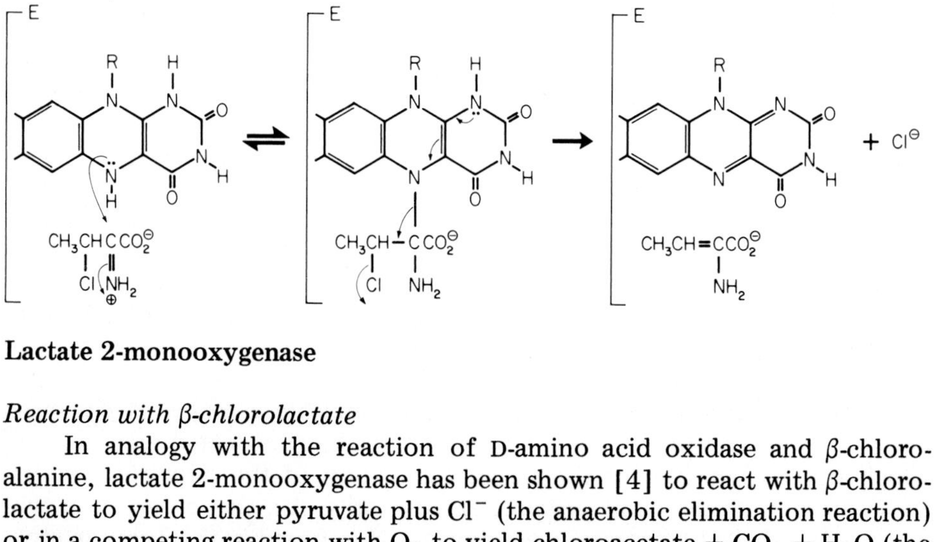

Lactate 2-monooxygenase

Reaction with β-chlorolactate

In analogy with the reaction of D-amino acid oxidase and β-chloroalanine, lactate 2-monooxygenase has been shown [4] to react with β-chlorolactate to yield either pyruvate plus Cl^- (the anaerobic elimination reaction) or in a competing reaction with O_2 to yield chloroacetate + CO_2 + H_2O (the normal oxygenase reaction).

The previous studies reported with this system used the racemic mixture. We have now reinvestigated the catalytic reactions using the resolved L-isomer. This was necessary since the D-forms of all hydroxy acids so far tested have been found to be reasonably good competitve inhibitors ($K_i \approx 1\,mM$) of the reactions with the L-isomers. The kinetic behavior was similar in form to that reported previously [4] but higher V values are obtained. These are listed in Table I. The K_m values at pH 7.0, 25°C, were found to be 1.8 mM for the elimination reaction and 2.0 mM for the oxidase reaction.

Rapid reaction studies, carried out under anaerobic conditions, reveal

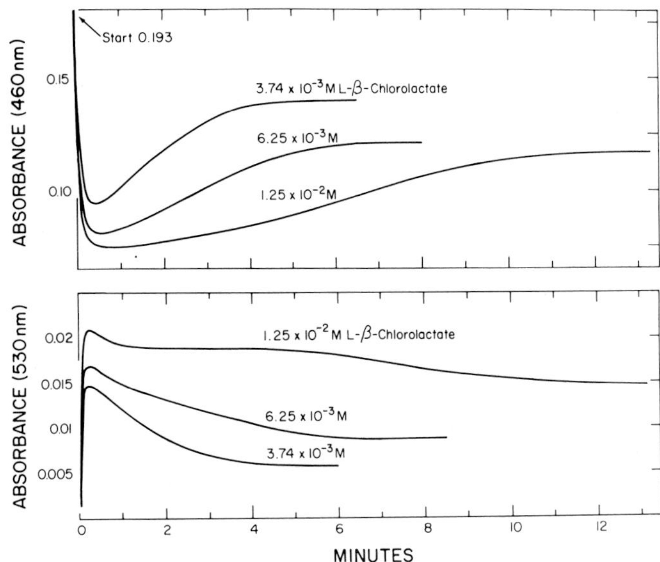

Fig. 5. Changes in absorbance of lactate 2-monooxygenase during the catalytic elimination reaction with L-β-chlorolactate as substrate. Enzyme, $1.72 \cdot 10^{-5}$ M with respect to E • FMN was reacted anaerobically with the concentrations of L-β-chlorolactate shown. Conditions, 0.02 M imidazole/acetate buffer, pH 7.0, 25°C. The experiments were performed with the stopped flow spectrophotometer; the absorbances shown are corrected to 1 cm light path.

that the enzyme becomes substantially reduced during turnover and returns partially to the oxidized state as the β-chlorolactate is exhausted (Fig. 5). The extent of reoxidation is dependent on the initial concentration of substrate. Parallel changes occur at 530 nm, with increased absorbance remaining at the end of turnover. These latter changes can probably be attributed to the accumulation of pyruvate (the product of the anaerobic elimination reaction) which forms a charge transfer complex with reduced enzyme [16]. If the keto acid-trapping agent, hydroxylamine (0.1 M) is included in the reaction mixture the enzyme becomes slowly converted to the fully reduced form instead of returning partially to the oxidized state.

While in the steady state of the enzyme during turnover, the flavin is

TABLE I

RATES OF REACTIONS OF LACTATE OXIDASE WITH β-Cl-LACTATE AT DIFFERENT OXYGEN CONCENTRATIONS

O_2 concentration		V		
		O_2 uptake	Pyruvate production	Sum
%	M	(min^{-1})	(min^{-1})	
100	$1.2 \cdot 10^{-3}$	385	79	464
21	$2.5 \cdot 10^{-4}$	173	209	382
10	$1.2 \cdot 10^{-4}$	100	238	338
0	0	0	286	286
Extrapolated for $C_{O_2} = \infty$		540	(0)	540

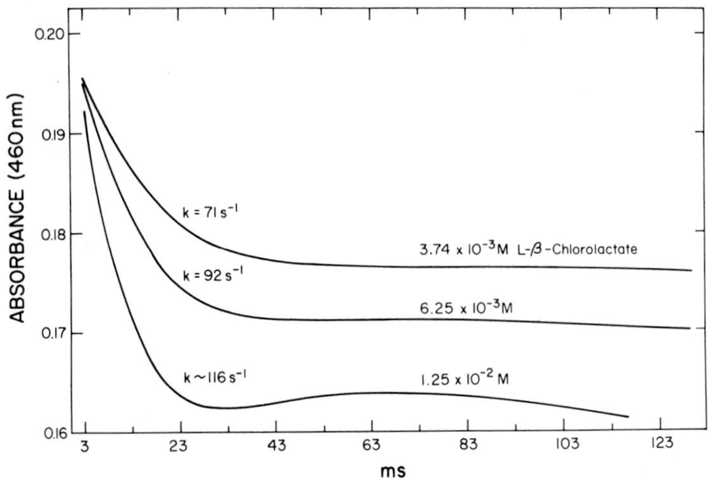

Fig. 6. Early phase of the reactions shown in Fig. 5.

substantially reduced, as described above, the rate of approach to the steady state level is appreciably slower than turnover and is almost independent of the concentration of β-chlorolactate. The observed rate is 15 min^{-1}. When the reaction is monitored over the time range of a single turnover, the results shown in Fig. 6 are obtained. As the concentration of β-chlorolactate is increased, the extent of the decrease in absorbance at 460 nm is greater, followed by a partial return before falling to the steady state levels shown in Fig. 5.

Clearly, the interpretation of such complex behavior must be made with caution. However, the results are compatible with a reductive elimina- pathway such as that outlined in Fig. 4 for D-amino acid oxidase. In the case of lactate 2-monooxygenase, the rate limiting step in the catalytic elimination reaction must also be some step occurring after Cl^- elimination and reoxidation of the enzyme. During the first turnover, i.e., before the rate limiting step becomes operational, the enzyme would appear to be first reduced and then reoxidized. The slow accumulation of reduced enzyme during the course of many catalytic turnovers could be ascribed to the reversible dissociation of the reduced enzyme-chloropyruvate complex. This equilibrium (displaced by NH_2OH) would also account for the fact that when catalytic turnover ceases, some of the enzyme remains in the reduced form.

Reduction of lactate 2-monooxygenase by L-lactate and turnover with O_2

The finding that proton abstraction from the α-carbon atom of β-chlorosubstrates by an enzyme base is an early step in the catalytic reactions of several enzymes lead us to examine further the interaction of the normal substrate of this enzyme, L-lactate. Previous rapid reaction studies had established that the enzyme is rapidly converted by L-lactate to a

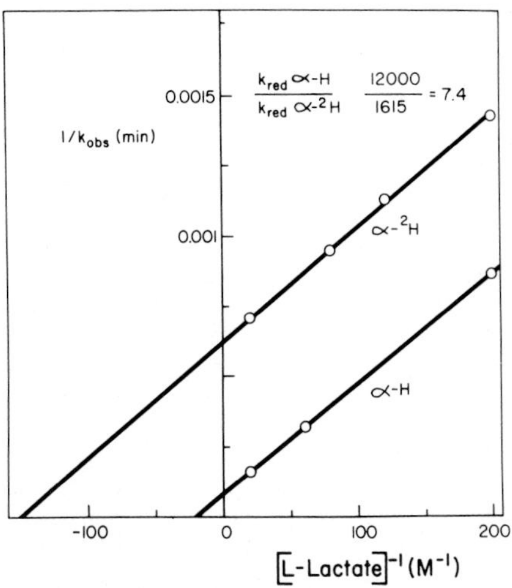

Fig. 7. Deuterium isotope effect on the observed rate of conversion of lactate 2-monooxygenase and L-lactate to the reduced enzyme-pyruvate complex. Lactate 2-monooxygenase, $1.61 \cdot 10^{-5}$ M with respect to enzyme-bound FMN was reacted under anaerobic conditions with the concentrations shown of the L-forms of α-H and [α-^{2}H] lactate. Conditions, 0.01 M imidazole, pH 7.0, 25°C.

reduced enzyme-pyruvate complex, and that it is this species which reacts with O_2 to yield the products, acetate, CO_2 and H_2O [16]. Fig. 7 shows the dramatic effect on the rate of reduction of the enzyme when α-deuterolactate is used instead of the normal protium form. A deuterium isotope effect of 7.4 fold is observed. The fact that the extrapolated negative intercept on the horizontal axis is also affected is inconsistent with a simple two-step process such as:

$$\text{EFMN} + \text{lactate} \rightleftharpoons \text{EFMN lactate} \rightleftharpoons \text{EFMNH}_2 \cdot \text{pyruvate}$$

and requires the intervention of another step. An obvious and attractive possibility is proton abstraction by the enzyme to form an α-carbanion intermediate:

$$\text{EFMN} + \text{lactate} \underset{k_{-1}}{\overset{k_1}{\rightleftharpoons}} \text{EFMN lactate} \underset{k_{-2}}{\overset{k_2}{\rightleftharpoons}} \begin{matrix}\text{EH}^+\text{FMN} \\ \text{CH}_3\text{—}\bar{\text{C}}\text{—COO}^- \\ \text{OH}\end{matrix} \underset{k_{-3}}{\overset{k_3}{\rightleftharpoons}} \left\{\begin{matrix}\text{EH}^+\text{FMNH}_2 \\ \cdot\ \text{pyruvate}\end{matrix}\right.$$

Computer simulations of this multiphasic equilibrium in which only k_2 and k_{-2} are varied approach the experimental results; however, no exact fit has so far been obtained (Frehland, H., Ghisla, S. and Massey, V., unpublished).

This large deuterium isotope effect on reduction is reflected still in the catalytic turnover number, where a 3.6 isotope effect on V (∞ L-lactate, ∞ O_2) is observed (Fig. 8). The effect on turnover number is what would be predicted from the previously reported rapid reaction analysis [16]. V is given by the expression $(k_{red} \cdot k_{ox})/(k_{red} + k_{ox})$ where k_{red} and k_{ox} are the

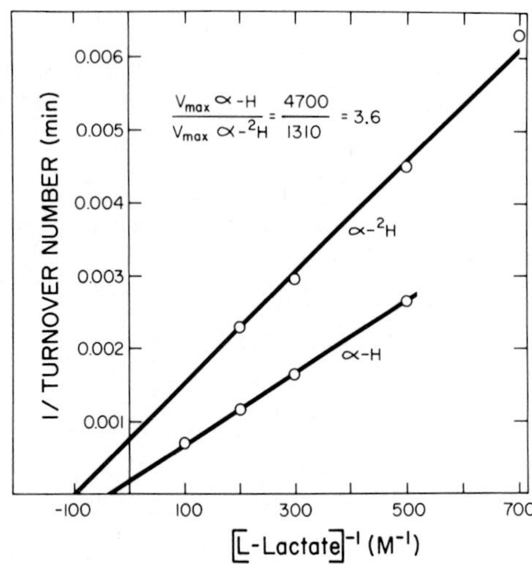

Fig. 8. Deuterium isotope effect on the catalytic turnover number of lactate 2-monooxygenase in its reaction with L-lactate and O_2. At each concentration of substrate shown, turnover analysis was performed as described in Ref. 15, Figs. 8 and 9. The experimental points shown in this figure are the extrapolated ∞ O_2 concentration values obtained from such plots. Hence the intercept values in the above graph represent true V values (∞ O_2, ∞ L-lactate).

limiting rates in the reductive and oxidative half reactions of catalysis. The value of k_{ox} determined previously is 11 000 min^{-1} [16], and should be independent of whether α-^{1}H- or α-^{2}H-labelled substrate were employed. The predicted values of V for the α-^{2}H-labelled and α-H forms of the substrate are 5700 min^{-1} and 1400 min^{-1}. These are in satisfactory agreement with the observed vaues, shown in Fig. 8. It should be noted that Lederer [17] has observed a 5-fold isotope effect on turnover with the yeast hemoflavoprotein, L-lactate dehydrogenase, employing α-H- and DL-[α-^{2}H] lactate. Comparatively small deuterium isotope effects on the rate of anaerobic reduction have also been observed with some substrates of D-amino acid oxidase [3, 10].

Reaction of lactate 2-monooxygenase with 2-hydroxy-3-butynoate

Previous studies have shown that the acetylenic hydroxy acid, 2-hydroxy-3-butynoate is an oxidase substrate for lactate 2-monooxygenase, but that a competing reaction occurs in which the flavin is covalently modified, resulting in loss of catalytic activity [7]. Under anaerobic conditions, complete loss of activity is obtained with equimolar amounts of enzyme-bound flavin and the L form of hydroxybutynoate. In the presence of O_2, as many as 100 catalytic turnovers can occur before the enzyme is fully inactivated [7].

We have, therefore, carried out some rapid reaction studies in an attempt to define further the nature of this suicide reaction. Unfortunately, the observed results are complicated somewhat by the fact that we have had to use the racemic substrate. As mentioned before, the D-isomers of hydroxy acids are good competitive inhibitors and so lead to biphasic changes in

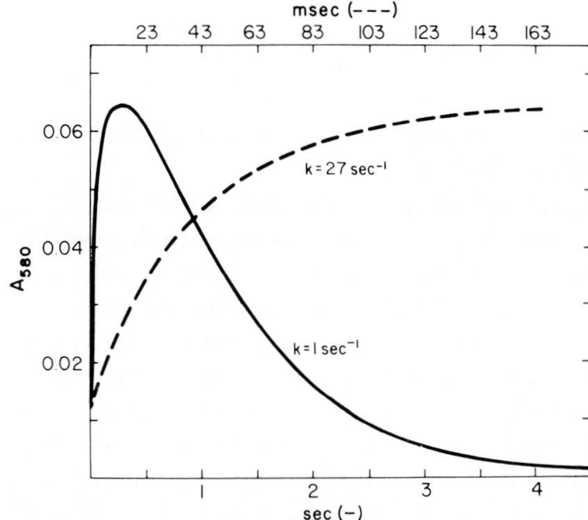

Fig. 9. Reaction of $1.77 \cdot 10^{-5}$ M lactate 2-monooxygenase under anaerobic conditions with $2 \cdot 10^{-3}$ M DL-2-hydroxy-3-butynoate. The absorbance changes shown are for the 2 cm light path of the stopped flow spectrophotometer. Conditions, 0.01 M imidazole, pH 7.0, 25°C.

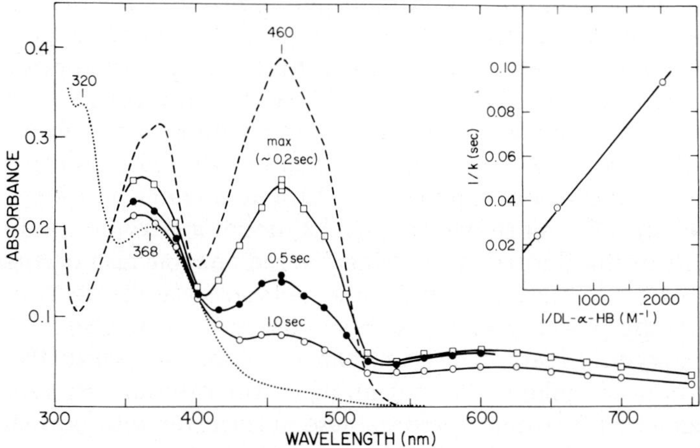

Fig. 10. Spectral changes during the reaction of lactate 2-monooxygenase with $5 \cdot 10^{-3}$ M DL-2-hydroxy-3-butynoate. The other conditions were the same as those for Fig. 9. The dashed line shows the spectrum of the startingenzyme; the dotted line the spectrum of the covalently modified flavin bound to the enzyme. Inset; the substrate concentration dependence of the rapid phase of the reaction, measured at 580 nm.

enzyme absorbance on reaction with the L form of the substrate. This is due to the formation of abortive complexes with the D form, with slow displacement determined by the off constant of this equilibrium and the concentration of the reactive L-isomer.

$$\text{E} + \text{D-hydroxy acid} \underset{k_{off}}{\overset{k_{on}}{\rightleftharpoons}} \text{E} \cdot \text{D-hydroxy acid}$$

$$E + \text{L-hydroxy acid} \underset{k_{-1}}{\overset{k_1}{\rightleftharpoons}} E \cdot \text{L-hydroxy acid} \overset{k_2}{\rightleftharpoons} EH_2 \cdot \text{keto acid}$$

(For an example of the effects of such competitive inhibitors cf. the results obtained in the presence of phosphate, Ref. 15.)

Figs. 9 and 10 show the spectral changes occurring under anaerobic conditions when lactate 2-monooxygenase is mixed with DL-α-hydroxybutynoate. There is a biphasic increase in absorbance at long wavelengths followed by a first order decay. At wavelengths below 500 nm there is a decrease in absorbance which occurs in three phases, the first corresponding in rate to the rapid increase at 580 nm and the last with the slow decrease in absorbance at 580 nm. The second phase is presumably due to the dissociation of D-isomer from its abortive complex. Only the first phase is markedly dependent in rate on the concentration of substrate; its dependence is shown in the inset of Fig. 10. The decay phase with a rate constant of $1\,s^{-1}$ (Fig. 9) is independent of substrate concentration. The long wavelength-absorbing intermediate rapidly produced has all of the characteristics of a charge-transfer complex in which the flavin is in the reduced form; by analogy with the similar spectra observed with reduced lactate 2-monooxygenase and pyruvate [16] and reduced D-amino acid oxidase and imino acids [9, 14, 13] it would appear likely that this intermediate is a complex of reduced enzyme and α-ketobutynoate. Under anaerobic conditions this intermediate appears to decay simply to the covalently modified flavin, with a rate constant of $1\,s^{-1}$. This conclusion is strengthened by companion stopped-flow absorbance and fluorescence studies, Figs. 11 and 12. The inactivated enzyme is highly fluorescent, with excitation maxima at 320 and 368 nm and emission maximum at 505 nm [7]. As shown in Fig. 12, under anaerobic conditions, the development of the fluorescence of inactivated enzyme also occurs with a rate constant of $1\,s^{-1}$. Figs. 11 and 12 also demonstrate the delay in inactivation caused by the presence of O_2, due to competing catalytic oxygenation of the hydroxybutynoate. The absorbance records shown in Fig. 11, with a rapidly produced spike at the beginning of the reaction, are very similar to those obtained with D-amino acid oxidase in turnover with normal substrates [9, 11]. Such results are indicative of a step after reoxidation of the intermediate being rate limiting in catalytic turnover; in the case of D-amino acid oxidase this is due to product dissociation being rate limiting [9].

While it is possible that other routes may be operational in the inactivation process (see below) the results are most simply interpreted as follows:

$$\text{EFMN} + \alpha\text{-hydroxybutynoate} \rightleftharpoons \text{ES} \rightleftharpoons \left\{ \begin{array}{l} \text{EFMNH}_2 \\ \alpha\text{-ketobutynoate} \end{array} \right.$$

$$\left\{ \begin{array}{l} \text{EFMNH}_2 \\ \alpha\text{-ketobutynoate} \end{array} \right. \begin{array}{l} \xrightarrow{O_2} \text{EFMN + products} \\ \longrightarrow \text{inactivated enzyme} \end{array}$$

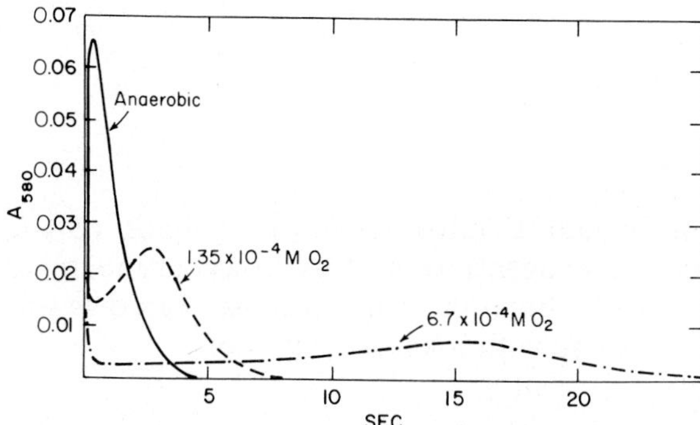

Fig. 11. Effect of O_2 on the reaction of lactate 2-monooxygenase with $2 \cdot 10^{-3}$ M DL-2-hydroxy-3-butynoate. Conditions, as in Fig. 9. Note the "spike" in 580 nm absorbance at the beginning of the reactions in the presence of O_2.

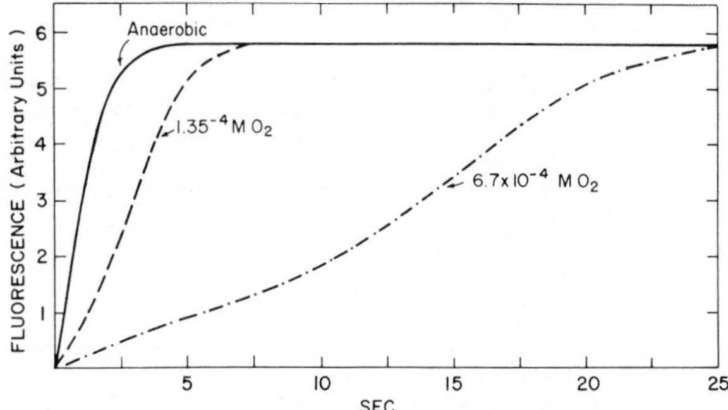

Fig. 12. Companion experiment to that of Fig. 11 except that the fluorescence of the inactivated enzyme was measured. The excitation wavelength used was 400 nm; emitted fluorescence was isolated from exciting light with the aid of a Corning CS372 filter placed between the reaction cell and the photomultiplier. This filter transmits light only of wavelengths greater than 430 nm.

There would certainly appear to be little doubt that the catalytic turnover pathway proceeds as written above, in the same way as already established for normal substrates [15]. The scheme, as written, is abbreviated, and undoubtedly involves the participation of an α-carbanion form of the substrate preceeding reduction:

$$\left\{\begin{matrix} \text{EFMN} \\ \text{HC}\equiv\text{C}-\underset{\text{OH}}{\overset{\text{H}}{\text{C}}}-\text{COO}^- \end{matrix}\right. \underset{k_{-2}}{\overset{k_2}{\rightleftharpoons}} \left\{\begin{matrix} \text{EH}^+\text{FMN} \\ \text{HC}\equiv\text{C}-\underset{\text{OH}}{\bar{\text{C}}}-\text{COO}^- \end{matrix}\right. \underset{k_{-3}}{\overset{k_3}{\rightleftharpoons}} \left\{\begin{matrix} \text{EH}^+\text{FMNH}^- \\ \text{HC}\equiv\text{C}-\underset{\text{O}}{\overset{}{\underset{\|}{\text{C}}}}-\text{COO}^- \end{matrix}\right.$$

The possibility cannot be ruled out that inactivation occurs from the level of the oxidized enzyme-carbanion complex in a reaction competing with reduction.

$$\left\{\begin{array}{l} EH^{+}FMN \\ HC{\equiv}C{-}\underset{OH}{\overset{=}{C}}{-}COO^{-} \end{array}\right. \xrightarrow{k_4} \begin{array}{l}\text{inactivated}\\ \text{enzyme}\end{array}$$

However, such a reaction would require some dependence of the observed rate of inactivation on the concentration of hydroxybutynoate, as well as a substantial value for k_{-3}. We hope in future studies to resolve this problem by reaction of reduced enzyme with α-ketobutynoate.

References

1 Walsh, C.T., Schonbrunn, A. and Abeles, R.H. (1971) J. Bio. Chem. 246, 6855–6866
2 Porter, D.J.T., Voet, J.G. and Bright, H.T. (1973) J. Biol. Chem. 248, 4400–4416
3 Yagi, K., Nishikimi, M., Takai, A. and Ohishi, N. (1973) Biochim. Biophys. Acta 321, 64–71
4 Walsh, C.T., Lockridge, O., Massey, V. and Abeles, R.H. (1973) J. Biol. Chem. 248, 7049–7054
5 Ghisla, S. and Massey, V. (1975) J. Biol. Chem. 250, 577–584
6 Walsh, C.T., Krodel, E., Massey, V. and Abeles, R.H. (1973) J. Biol. Chem. 248, 1946–1955
7 Walsh, C.T., Schonbrunn, A., Lockridge, O., Massey, V. and Abeles, R.H. (1972) J. Biol. Chem. 247, 6004–6006
8 Schonbrunn, A., Abeles, R.H., Walsh, C., Ghisla, S., Ogata, H. and Massey, V. (1976) Flavins and Flavoproteins, Proceedings of the 5th International Symposium (Singer, T.P., ed.), Chapt. 19, pp. 187–198, Elsevier, Amsterdam
9 Massey, V. and Gibson, Q.H. (1964) Fed. Proc. 23, 18–29
10 Yagi, K., Nishikimi, M., Ohishi, N. and Takai, A. (1971) in Flavins and Flavoproteins (Kamin, H., ed.), pp. 239–260, University Park Press, Baltimore
11 Massey, V., Curti, B. and Ganther, H. (1966) J. Biol. Chem. 241, 2347–2357
12 Swoboda, B.E.P., Gibson, Q.H. and Massey, V. (1964) J. Biol. Chem. 239, 3927–3934
13 Massey, V. and Ghisla, S. (1974) Ann. N.Y. Acad. Sci. 227, 446–465
14 Massey, V. and Ganther, H. (1965) Biochemistry 4, 1161–1173
15 Ghisla, S. and Massey, M. (1976) Flavins and Flavoproteins, Proceedings of the 5th International Symposium (Singer, T.P., ed.), Chapt. 21, pp. 213–217, Elsevier, Amsterdam
16 Lockridge, O., Massey, V. and Sullivan, P.A. (1972) J. Biol. Chem. 247, 8097–8106
17 Lederer, F. (1974) Eur. J. Biochem. 46, 393–399
18 Ballou, D.P. and Palmer, G. (1974) Anal. Chem. 46, 1248–1253
19 Iwasaki, I., Utsumi, S. and Ozawa, T. (1952) Bull. Chem. Soc. Jap. 25, 226

Flavins and Flavoproteins, edited by T.P. Singer

CHAPTER 21

STUDIES ON LACTATE 2-MONOOXYGENASE*. PROTON UPTAKE AND RELEASE DURING THE BINDING OF THE TRANSITION STATE ANALOGUE OXALATE

SANDRO GHISLA** and VINCENT MASSEY

Department of Biological Chemistry, The University of Michigan, Ann Arbor, Mich. 48104 (U.S.A.)

The reaction mechanism of the flavoenzyme lactate 2-monooxygenase (EC 1.13.12.4, L-lactate:oxygen 2-oxidoreductase (decarboxylating)) has been the object of several detailed investigations which lead to the postulation of the following sequence of reactions for the oxidation of substrate [1, 2].

$$E_{ox} + CH_3{-}\underset{OH}{\overset{H}{C}}{-}COO^- \rightleftharpoons E_{ox}{\sim}CH_3{-}\underset{OH}{\overset{H}{C}}{-}COO^- \rightleftharpoons E_{ox}\,H^+{\sim}CH_3{-}\underset{OH}{\overset{-}{C}}{-}COO^-$$

$$\updownarrow$$

$$CH_3{-}COO^- + CO_2 + H_2O \qquad E_{red} + CH_3{-}\underset{O}{\overset{}{\underset{\|}{C}}}{-}COO^- \rightleftharpoons E_{red}{\sim}CH_3{-}\underset{O}{\underset{\|}{C}}{-}COO^-$$

The proposed sequence of reactions in the reductive half of the pathway is based mainly on the observation that substrates containing a leaving group (e.g. Cl) in position β undergo an enzyme-catalyzed elimination reaction to yield the corresponding α-keto acids and halide ion [2] with the enzyme remaining in the oxidized state. Thus, an early step in the catalytic cycle would correspond to abstraction of the proton in position α of the substrate to form a (transient) carbanion, which would subsequently reduce the enzyme falvin. The formation of such carbanions, although stabilization through delocalization of the negative charge to the carboxylic function is possible, would be expected to be thermodynamically unfavorable and kinetically slow:

---

* Formerly lactate oxidase (EC 1.1.3.2)

** Present address: Fachbereich Biologie der Universität Konstanz, Postfach 733, D755 Konstanz, G.F.R.

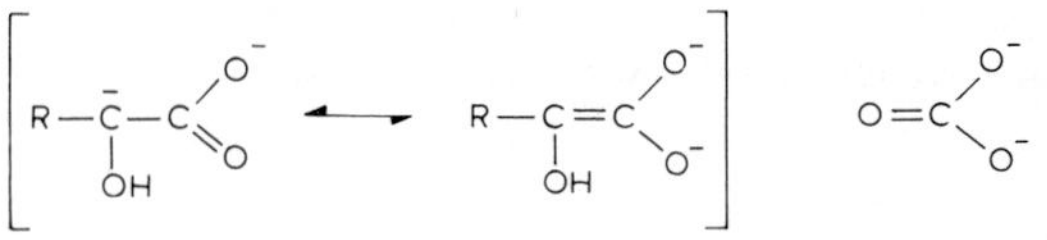

On the other hand, it should be noted that the pK for the second dissociation of the isoelectronic carbonate ion is ≈ 10. If carbanion intermediates do indeed occur in the catalytic cycle, the enzyme active site would require the presence of a base involved in the abstraction of the α proton. Its function should also consist of lowering the energy of the dianionic transition state by providing appropriately distributed counterions. This general concept led to the definition of transition state analogues and their successful use in the investigation of enzymatic reaction mechanisms [3].

Oxalate has been recently shown to be a potent inhibitor of lactate 2-monooxygenase ($K_d \approx 1.6 \cdot 10^{-5}$ M at pH 7.0) [4]. Although it is not strictly isosteric with the proposed carbanion intermediate, it is considered to be a reasonable analogue:

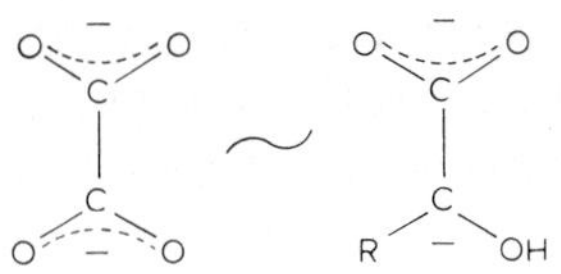

Oxalate binds to the enzyme in two kinetically separate steps: the first is fast, reversible and second order in oxalate and enzyme, leading to an intermediate complex I ($K_d = k_{-1}/k_1 \approx 10^{-2}$ M), which is converted to a complex II at a much slower rate [4]. The second step, however, shows clear saturation with increasing oxalate concentration; it is accompanied by major changes in the electronic absorbance spectrum of the oxidized enzyme flavin. Thus, the rates for the interconversion of complexes I and II were estimated as $k_{-2}/k_2 \approx 0.06\ (\text{min}^{-1})/40\ (\text{min}^{-1}) \approx 1.5 \cdot 10^{-3}$ at pH 7, yielding an overall K_d of $\approx 1.5 \cdot 10^{-5}$ M. In view of the possibility of oxalate being a transition state analogue, the hypothesis was put forward that the second slow process involves the uptake of one proton by the protein. This would parallel the abstraction of the α-proton from the substrate by the protein and would result in a thermodynamic stabilization of the enzyme-dianion complex II as compared to complex I

$$\text{E---B} + C_2O_4^{2-} \underset{k_{-1}}{\overset{k_1}{\rightleftharpoons}} \underset{\text{Complex I}}{\text{E---B} \approx C_2O_4^{2-}} \underset{k_{-2},\ \downarrow H^+}{\overset{H^+\downarrow,\ k_2}{\rightleftharpoons}} \underset{\text{Complex II}}{\text{E---BH}^+ \approx C_2O_4^{2-}}$$

In an experimental verification of this postulate, the rate of conversion of complex I into complex II was measured in weakly buffered solution and found closely similar to the rate of decrease of the H^+ concentration in the solvent (Fig. 1). (The internal pH indicator used, phenol red, was found to

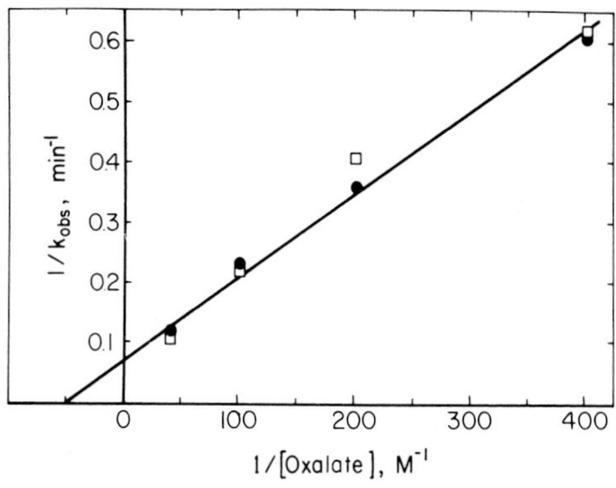

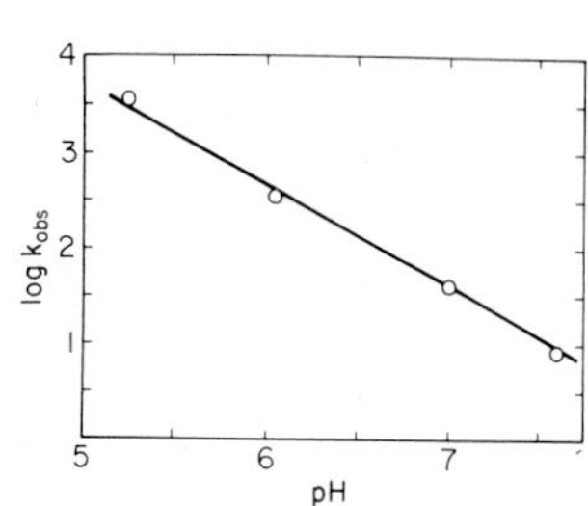

Fig. 1. Rates of formation of complex II and of H^+ uptake measured as function of oxalate concentration, in stopped-flow experiments. A solution of lactate 2-monooxygenase ($3.2 \cdot 10^{-5}$ M final concentration) in 10^{-2} M NaCl, and 10^{-4} M imidazole HCl containing phenol red ($8 \cdot 10^{-6}$ M final concentration) and catalytic amounts of carbonic anhydrase was equilibrated with CO_2-free nitrogen until the pH of the mixture was constant. The oxalate solutions at the concentrations shown, in unbuffered 10^{-2} M NaCl, were freed of CO_2 in the same way and both mixtures were reacted at 24.5°C. The conversion of complex I to complex II (●——●) was measured by the absorbance decrease at 365 nm (isosbestic point between neutral and anionic phenol red (cf. also Ref. 4). The uptake of H^+ was measured at 560 nm (□——□) where the enzyme or its complexes do not absorb. The pH of the solution after mixing was estimated from the initial absorbance at 560 nm and was 7.6. (The total pH change occurring during the reactions was ≈ 0.15. The absorbance changes at 560 nm and 365 nm obeyed first-order decay curves for at least three half-lives.) For the determination of the stoichiometry of proton uptake, solutions of exactly the same composition were titrated in the range pH 6.8–8.2 with known amounts of HCl and NaOH to yield a standard curve, which was then compared to the changes obtained in the stopped-flow measurements or in static experiments.

Fig. 2. Dependence of the rate of conversion of complex I to complex II (k_2) on H^+ concentration. Lactate 2-monooxygenase, $2 \cdot 10^{-5}$ M in the appropriate buffer was reacted with varying concentrations of oxalate in the same buffer at pH 5.25, 6.1 and 7.6, 25°C, and the rate of absorbance decrease at 505 nm (due to conversion of complex I to complex II) measured in the stopped-flow spectrophotometer. The rate at pH 7.0 is taken from Ref. 4. A universal buffer system was employed, consisting of 10^{-2} M piperazine/10^{-2} M Tris/10^{-2} M imidazole/$4 \cdot 10^{-2}$ M Cl^-. The desired pH values were obtained by titrating with HCl and adjusting the final chloride concentration to $4 \cdot 10^{-2}$ M with KCl.

have no measurable effect on the behaviour of the system.) Furthermore, the stoichiometry of proton uptake (cf. legend to Fig. 1) was 1 ± 0.15. If the above equation holds, the rate of conversion of complex I to complex II should also be linearly dependent on H^+ concentration. This was found to be the case in the range pH 5.2–7.6 (Fig. 2), when k_2 was measured in experiments similar to those of Fig. 1. Thus, k_2 is shown to be a second-order rate constant, with the value of $4 \cdot 10^8$ $M^{-1} \cdot min^{-1}$. From such plots, the K_d for the formation of complex I (k_{-1}/k_1) was estimated as $\approx 2 \cdot 10^{-2}$ M and found to be practically independent of pH (Fig. 3), as would be predicted if protons were involved only in the second step. It should be pointed out that most mono anions investigated so far are competitive inhibitors and bind to lactate 2-monooxygenase with K_d values typically around 10^{-2} M [1, 4]. The overall dissocation constant K_d was also measured spectrophotometrically as a function of pH and plotted according to Dixon [5]. In the range of pH 5–7, the slope of the plot $-\log K_d$/pH is unity, and it shows a break at ≈ 4.7, which would correspond to the pK for the protonation of uncomplexed enzyme. When this plot is extrapolated to higher pH, its intercept with the curve representing the pH-independent

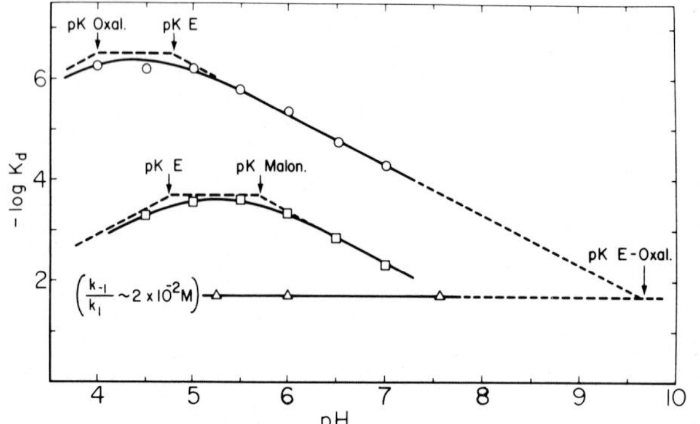

Fig. 3. Dixon plot of the dissocation constant of the lactate 2-monooxygenase/oxalate (○——○) and malonate (□——□) complexes as a function of pH. The experiments were carried out at 25°C, employing enzyme at a concentration of $2 \cdot 10^{-5}$ M. The K_d was obtained spectrophotometrically, as described previously [4]. The universal buffer system described for Fig. 2 was employed. The lower horizontal line (△——△) represents the values of k_{-1}/k_1 obtained from stopped-flow experiments such as shown in Fig. 1.

binding of oxalate to form complex I (k_{-1}/k_1) shows a pK of ≈9.7, which should thus correspond to the ionization of the oxalate-complexed enzyme. This compares with the kinetically determined pK value for the interconversion of complexes I and II, $-\log(k_{-2}/k_2) \approx 0.06/4 \cdot 10^8 [H^+] = 9.7$, (cf. Ref. 4, for the value of k_{-2}). Thus, the statically determined values for K_d (Fig. 3) agree reasonably well with the values obtained from stopped-flow measurements, ($K_d = (k_{-1} \cdot k_{-2})/(k_1 \cdot k_2[H^+])$). The next homologue of oxalate, malonate, exhibits a similar pH dependence of its overall K_d (Fig. 3). This reflects a pK of ≈4.7 for the uncomplexed enzyme in addition to the pK of 5.7 for the second ionization constant of malonate. Tartronate ($^{-}OOC{-}\underset{OH}{\overset{H}{C}}{-}COO^{-}$) was found to be a poor substrate of the enzyme and to bind in an analogous pH-dependent way (not shown, K_d at e.g. pH 7 ≈ $4 \cdot 10^{-4}$ M). These findings indicate that the active center of lactate 2-monooxygenase has a high affinity for dianionic small molecules and that the magnitude of this effect is considerably dependent on the separating distance between the two negative charges. This is taken as support of the concept that a dianionic species lies on the pathway of substrate oxidation. It should be stated that the results are compatible also with the occurrence of a transient dianion in which the second negative charge is located at the oxygen function in position α of the substrate $\left(CH_3{-}\underset{O_{-}}{\overset{H}{C}}{-}COO^{-}\right)$. This species would constitute an intermediate preceding hydride abstraction from the α position. On the other hand, it is difficult to rationalize how such an intermediate species could be the precursor leading to the elimination reactions reported for β-chlorolactate (cf. Ref. 2). Furthermore, reduced lactate 2-monooxygenase and β-chloropyruvate react to yield Cl^- elimination, and this again seems to preclude a mechanism involving the addition of "hydride" to the carbonyl function of pyruvate. (Ghisla, S. and Massey, V. unpublished results.)

The large pH shift of the enzyme function observed upon binding of oxalate is rather unusual and would correspond to a stabilization of ≈ 5 kcal/mol. Such a stabilization could presumably be even higher in the case of the normal substrate carbanion-enzyme complex and could thus shift the pK of the α-H in lactate to values compatible with catalysis. Large shifts of protein-base pK values have been observed, e.g. for lysine [6]. However, the magnitude of the pK shift which would be required to accommodate a pK of 4.7 would make a histidyl residue a more likely possibility. A pK of 9.7 seems unlikely for a carboxylate function. On the other hand, a solvent-shielded base has been implicated by Walsh et al. [2] in the abstraction of the α-H from substrate and in its reincorporation in position β after elimination has occurred. From the stoichiometry of ^{3}H incorporation in position β of eliminating substrates it was suggested that the base is monofunctional. The sum of the evidence presently available suggests that the residue in question may be histidine.

References

1 Lockridge, O., Massey, V. and Sullivan, P.A. (1972) J. Biol. Chem. 247, 8097–8106
2 Walsh, C., Lockridge, O., Massey, V. and Abeles, R.H. (1973) J. Biol. Chem. 248, 7049–7054
3 Lienhard, G.W. (1973) Science 180, 149–154
4 Ghisla, S. and Massey, V. (1975) J. Biol. Chem. 250, 577–584
5 Dixon, M. and Webb, E.C. (1964) Enzymes, 2nd Edn., pp. 116–145
6 Schmidt, D.E. and Westheimer, F.H. (1971) Biochemistry 10, 1249–1253

Flavins and Flavoproteins, edited by T.P. Singer

CHAPTER 22

STRUCTURE OF THE ADDUCT DERIVED FROM THE INHIBITION OF MONOAMINE OXIDASE BY 3-DIMETHYLAMINO-1-PROPYNE

ALAN L. MAYCOCK,[a,*] ROBERT H. ABELES,[a] JAMES I. SALACH[b,c] and THOMAS P. SINGER[b,c]

[a] *Graduate Department of Biochemistry, Brandeis University, Waltham, Mass. 02154,* [b] *Molecular Biology Division, Veterans Administration Hospital, San Francisco, Calif. 94121 and* [c] *Department of Biochemistry and Biophysics, University of California, San Francisco, Calif. 94143 (U.S.A.)*

Inactivation of monoamine oxidase (EC 1.4.3.4, amine : oxygen oxidoreductase (deaminating) (flavin-containing)) by appropriate acetylenic amines is known to occur with the formation of covalent adducts [1, 2]. Because of our general interest in the use of acetylenic suicide inactivators as probes of enzyme mechanisms [3–7], we decided to attempt to elucidate the structure of an adduct derived from monoamine oxidase. (Suicide inactivators are substrate analogs that require enzymatic conversion to highly reactive species which then react irreversibly with the enzyme to inactivate it.) We hoped this study would provide information helpful for understanding the mechanisms by which flavin catalysis can occur, and in particular we felt the structure of the adduct might be relevant to structures of covalent intermediates which have been postulated to occur during catalysis by flavin enzymes [8–11]. Herein we report our determination of a plausible structure for the covalent adduct formed when monoamine oxidase is inactivated with the suicide inactivator, 3-dimethylamino-1-propyne (Structure I).

Treatment of bovine liver monoamine oxidase [12] with I causes a rapid ($t_{1/2}$ = 10 min at 3.3 mM),

$$H{-}C{\equiv}C{-}CH_2{-}N(CH_3)_2 \qquad \text{(Structure I)}$$

complete, and first-order loss of enzyme activity which can be protected against by the presence of substrate (benzylamine). (As a competitive inhibitor of benzylamine oxidation, I has a K_i of approximately 1.2 mM). None of the lost enzyme activity is recovered by incubating the inactivated enzyme with substrate, by dialyzing it extensively, or by excluding it on Sephadex. The enzyme undergoes significant irreversible spectral changes during inactivation (Fig. 1). Loss of the long wavelength (450–500 nm) band is accompanied by development of a new band at 410 nm. From the

* Present address: Merck, Sharp and Dohme Research Laboratories, Rahway, N.J. 07065, U.S.A.

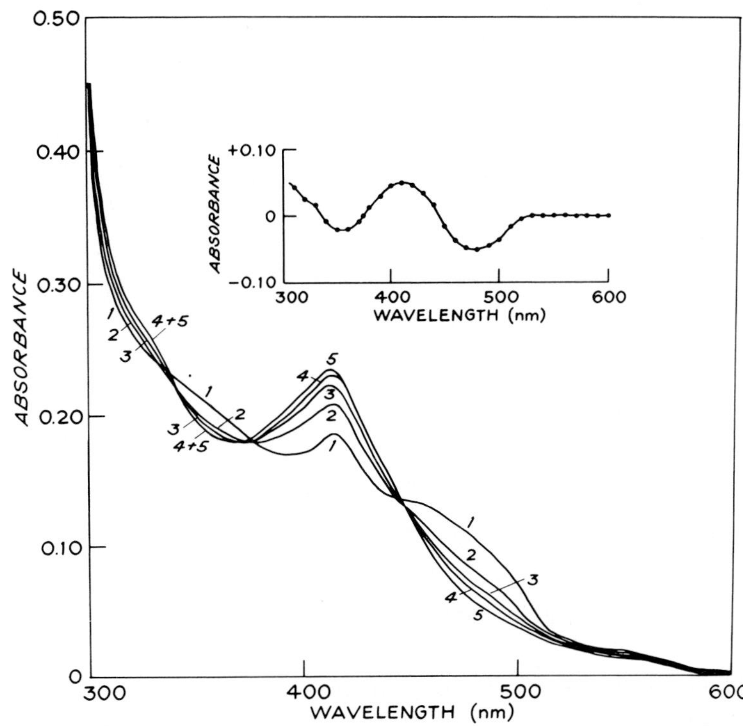

Fig. 1. Spectral changes accompanying inhibition of monoamine oxidase by I: monoamine oxidase, 0.38 mg, equivalent to 2.3 nmol cysteinyl flavin was incubated with 2.4 μmol of I (5.23 mM final) in 0.455 ml 50 mM NaP_i, pH 7.2, containing 1.1% (w/v) Triton X-100 at 22°C. Spectra were taken at intervals of 10 min against a reference containing everything except enzyme. 1: Native enzyme; 2,3,4,5: 10,20,30 and 40 min after addition of inhibitor. After 40 min the enzyme was 88% inhibited. Inset: difference spectrum of native enzyme vs. inhibited enzyme after 50 min incubation.

known cysteinyl flavin content [13] of starting enzyme, the molar absorbance (ϵ) of this new band was shown to be 28 800.

When monoamine oxidase is inactivated with either $HC{\equiv}C{-}CH_2{-}N(^{14}CH_3)_2$ ($^{14}CH_3$-inactivator) or $HC{\equiv}C{-}^{14}CH_2{-}N(CH_3)_2$ ($^{14}CH_2$-inactivator) radioactivity becomes associated with the protein as shown by the retention of counts in protein fractions recovered from dialysis or from exclusion on Sephadex. Based on cysteinyl flavin content, 2.3–2.8 equivalents of $^{14}CH_2$-inactivator become bound to protein during inactivation. However, only one equivalent is actually bound to flavin; the rest is due to nonspecific labelling of protein which is removed during subsequent isolation and purification of the flavin-inactivator adduct. The general characteristics of the inactivation process are similar to those reported [1, 2] for the inactivation of hog kidney monoamine oxidase by pargyline.

The adduct was purified from tryptic-chymotryptic digests of the inactivated enzyme by a procedure similar to that utilized to obtain the cysteinyl flavin pentapeptide from native monoamine oxidase [12]. Hence the adduct should be at the phosphorylation level of FMN and contain a

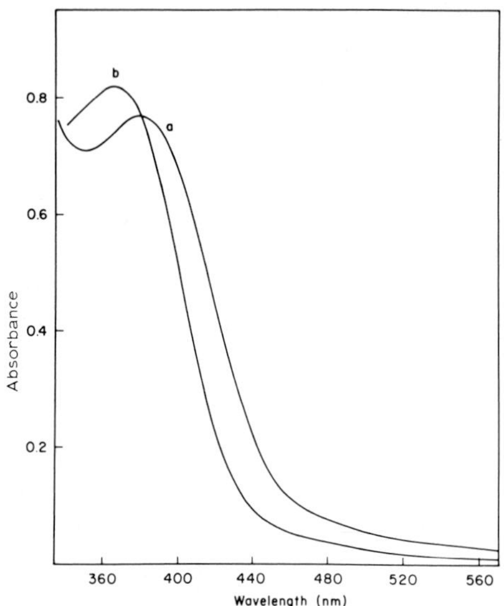

Fig. 2. The pH dependence of the optical spectrum of purified adduct. The sample (66 nmol) was in 2.85 ml water with (a) NaOH added (pH approximately 8.5); (b) concentrated HCl added (pH approximately 2).

short peptide linked to the 8a position through a thioether bond. When subjected to electrophoresis, the highly purified adduct migrates toward the anode (pH 6.5) or cathode (pH 1.6) as a yellow, single band of radioactivity. It is nonfluorescent but has intense visible absorbance (Fig. 2), $\lambda_{max}^{pH\ 8}$ 380 nm (ϵ 34 000) and $\lambda_{max}^{pH\ 2}$ 365 nm (ϵ 36 000), shifted somewhat from the band of the original enzyme-inactivator adduct. (The molar absorbance values are calculated from the ^{14}C content of the sample by assuming a 1:1 ratio of inactivator to flavin.) The satisfactory agreement of the molar absorbance (calculated from the ^{14}C content) of the pure flavin peptide adduct with that (calculated from cysteinyl flavin content) of the enzyme adduct is consistent with a 1:1 ratio of inactivator to flavin, which in turn shows that most nonspecifically incorporated radioactivity has been removed. The pH dependence of the spectrum shows that the adduct has a pK of 4 to 5.

During adduct pruification, there was a slow loss of radioactivity preferentially from the $^{14}CH_3$-inactivator adduct under moderately basic conditions. The only reasonable way to account for this preferential loss is by the expulsion of dimethylamine. Chemical studies were carried out with a partially purified sample (known to be contaminated with only small amounts of nonspecifically bound radioactivity) of adduct from $^{14}CH_2$-inactivator. It had never been subjected to basic conditions, and therefore still contained the dimethylamino group.

The partially purified adduct is stable in acid, since no spectral changes are observed and essentially all of the radioactivity can still be adsorbed on charcoal after treatment with acid. However, it is labile in strong base, decomposing to give oxidized flavin (67% based on the ^{14}C content of the adduct) and releasing the radioactivity in an unidentified fragment. The oxidized flavin is identified by the appearance of a band at 450–500 nm which can be bleached by dithionite then completely regenerated with oxygen. The molar absorbance of the flavin is assumed to be $12\,000\ M^{-1} \cdot cm^{-1}$ at 450 nm [14]. The acid stability rules out any ketal-type structures that upon hydrolysis would expel radioactive fragments or result in spectral changes. The fact that oxidized flavin can be produced shows that the isoalloxazine moiety remains intact in the adduct.

Ozonolysis of the adduct produces [^{14}C] glyoxal in yields up to 65%. This observation means the adduct must contain one of the following part structures, where D and E would have to be heteroatoms.

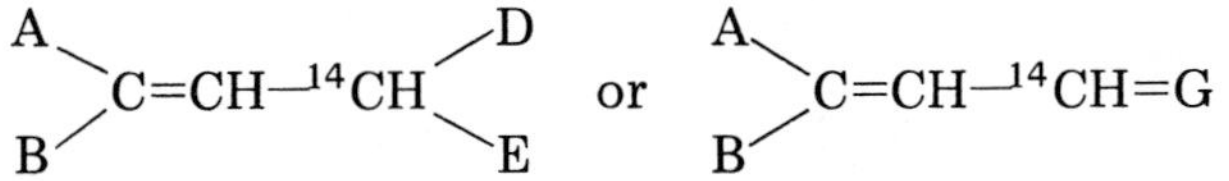

$$(A)(B)C{=}CH{-}^{14}CH(D)(E) \quad \text{or} \quad (A)(B)C{=}CH{-}^{14}CH{=}G$$

Since alkyl substituted propargylamines, $R{-}C{\equiv}C{-}CH_2{-}NR'R''$, also function as enzyme inhibitors [15], the hydrogen (R=H) of I must be retained during adduct formation. Therefore, substituent A or B is hydrogen. It has been suggested that any biological reactions involving covalent substrate-flavin intermediates will probably take place at N-5 or C-4a of the flavin nucleus. Assuming that as a substrate analog, the inactivator is most likely to interact with these same positions, we consider C-4a and N-5 as the most reasonable sites for attachment of the flavin to the inactivator.

We believe the following C-4a and/or N-5 adduct structures are the only

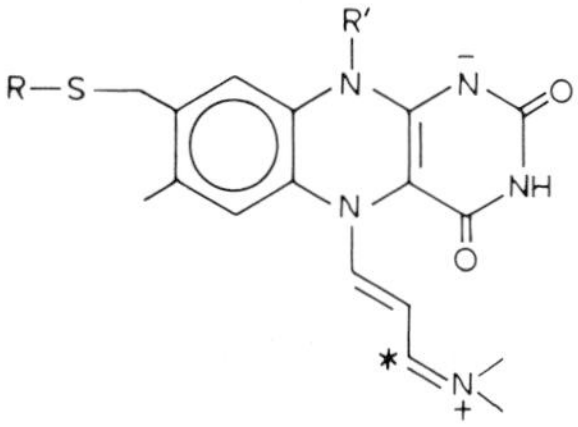

(II)

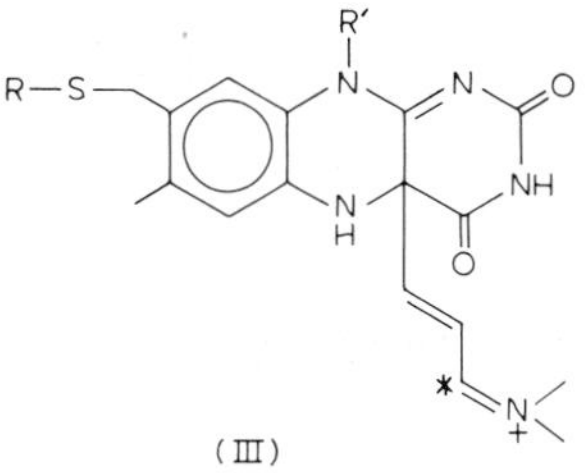

(III)

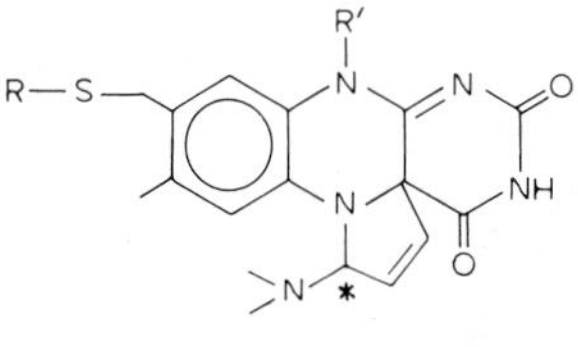

(IV)

ones consistent with the chemical facts. A comparison of the spectral and pK properties of alkyl-substituted dihydroflavins with those of the adduct are presented in Table I. None of the substituted flavins has a molar absorbance exceeding 10 000, whereas the adduct exhibits much more intense absorption. Only II has a structural feature which would account for this abnormally high ϵ. Although C-4a-substituted dihydroflavins ordinarily exhibit no pK between 2 and 10, N-5-substituted ones have pK values near that of the adduct.

The chemical and physical properties of a number of model compounds were compared to those of the enzymatically derived adduct. Two of them, V (prepared as previously reported [20, 21]) and VI* have properties consistent with those of the adduct:

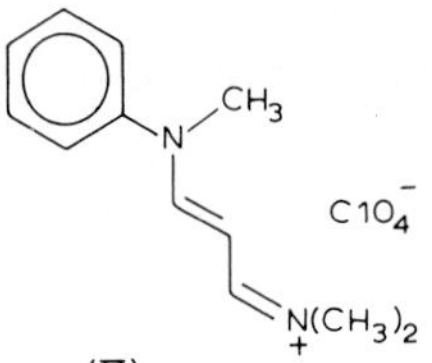

(V) 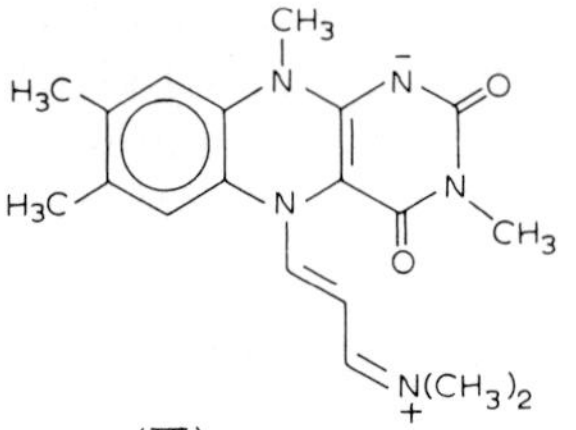 (VI)

acid stability, base lability, very intense optical spectra, glyoxal production upon ozonolysis and slow reduction by sodium borohydride. In addition, compound VI exhibits a pK of approximately 5 and its optical spectrum shows the same pH dependence as that of the adduct. Pertinent data are collected in Table II.

The properties of other model compounds differ in certain important respects from those of the adduct. For instance, compound VII,* has a

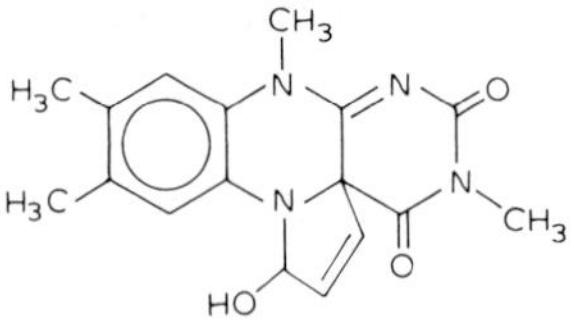

(VII) 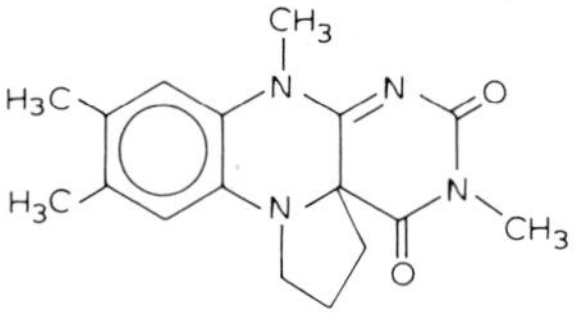 (VIII)

"normal" reduced flavin spectrum as discussed above and is labile in both

TABLE I

SPECTRAL AND ACID-BASE PROPERTIES OF SUBSTITUTED DIHYDROFLAVINS COMPARED TO THOSE OF THE ADDUCT
See Refs. 16–19.

Substitution	$\lambda_{max}^{H_2O}$ (nm)	$\epsilon (M^{-1} \cdot cm^{-1})$	pK_a
N-5	296–355	5000–8000	6–7
C-4a	360–370	5000–9000	none, 2–10
N-5, C-4a	325–360	5000–10 000	
adduct	365 (pH < 3)	36 000	4–5
	380 (pH > 7)	34 000	

* Compounds VI and VII were prepared by Dr. Hemmerich and coworkers [22]. The structure of VI was elucidated independently by Hemmerich and Gärtner [23] and by Maycock [24].

acid and base. Also, in contrast to the adduct and compound V, reduced flavins VIII, provided by Dr. S. Ghisla [25], and VII containing a C-10a, N-1 double bond undergo rapid reduction by sodium borohydride.

We believe these data constitute strong evidence that the adduct formed when bovine liver monoamine oxidase is inactivated by H—C≡C—CH_2—$N(CH_3)_2$ has the following structure:

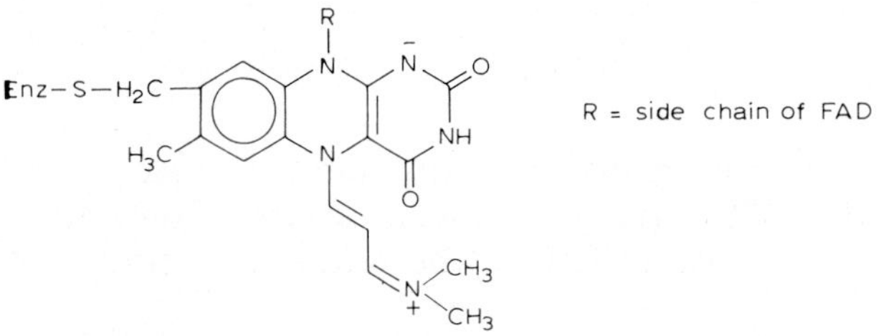

R = side chain of FAD

Although the mechanism by which adduct formation occurs is not known, we can envision at least three reasonable possibilities.

1. The enzyme catalyzes abstraction of the α-proton from I to form an allenyl carbanion which adds to N-5 of oxidized flavin.

2. The enzyme catalyzes transfer of (formally) the α-proton and an electron from I to the flavin to give the respective radicals which collapse to product.

3. The enzyme catalyzes oxidation of I to the "normal" product, H—C≡C—CH=$N^+(CH_3)_2$, which then undergoes Michael addition of the reduced flavin.

The usual oxidation catalyzed by monoamine oxidase occurs at C-1 of the substrate amine. If substrate-flavin covalent intermediates are catalytically important in such processes, intermediate formation could in principle take place at either C-1 or N of the substrate. In our adduct, however, the covalent bond is through C-3, a position which could not be available for formation of an intermediate during normal substrate oxidation. Hence the structure of the monoamine oxidase adduct described here leaves unresolved any questions concerning the importance of covalent intermediates in flavoprotein catalysis.

TABLE II

PROPERTIES OF COMPOUNDS V AND VI COMPARED TO THOSE OF THE ADDUCT

		V	VI	Adduct
Optical spectrum				
	pH 3		375 nm (ϵ 26 000)	365 (ϵ 36 000)
	pH 7	327 nm (ϵ 45 000)	390 nm (ϵ 24 000)	380 (ϵ 34 000)
pK_a		—	5.3	4–5
H^+		stable pH 0.6	stable pH 1.3	stable pH 1.3
HO^-		decomposes 0.1 M NaOH	decomposes pH 11.6	decomposes pH 12.6
BH_4^-		slow reduction	slow reduction	slow reduction
O_3		glyoxal + dimethylformamide + ?	glyoxal + dimethylformamide + ?	glyoxal + ?

A covalent adduct is also formed when lactate 2-monooxygenase is treated with the appropriate acetylenic substrate analog. In this case, the inactivator becomes attached to btoh N-5 and C-4a of the flavin, and the carbon atom undergoing oxidation becomes attached to N-5. A discussion of the inactivation of lactate oxidase and of the possible implications of these different modes of adduct formation are given in another chapter of this volume [26].

Acknowledgments

This research was supported by grants from the National Institutes of Health (HL 16251 and GM 12633) and the National Science Foundation (GB 36570X). A.L.M. was a Special Fellow of the National Institutes of Health (5FOZ GM 40063-02).

References

1 Hellerman, L. and Erwin, E.V. (1968) J. Biol. Chem. 243, 5234–5243
2 Chuang, H.Y.K., Patek, D.R. and Hellerman, L. (1974) J. Biol. Chem. 249, 2381–2384
3 Bloch, K. (1969) Acc. Chem. Res. 2, 193–202
4 Walsh, C.T., Schonbrunn, A., Lockridge, O., Massey, V. and Abeles, R.H. (1972) J. Biol. Chem. 247, 6004–6006
5 Walsh, C.T., Abeles, R.H. and Kaback, H.R. (1972) J. Biol. Chem. 247, 7858–7863
6 Abeles, R.H. and Walsh, C.T. (1973) J. Am. Chem. Soc. 95, 6124–6125
7 Hevey, R.C., Babson, J., Maycock, A.L. and Abeles, R.H. (1973) J. Am. Chem. Soc. 95, 6125–6127
8 Massey, V., Müller, F., Feldberg, R., Schuman, M., Sullivan, P., Howell, L.G., Mayhew, S.G., Mathews, R.G. and Foust, G.P. (1969) J. Biol. Chem. 244, 3999–4006
9 Hamilton, G.A. (1971) Progr. Bioorg. Chem. 1, 83–157
10 Sun, M. and Song, P.S. (1973) Biochemistry 12, 4663–4669
11 Clerin, D. and Bruice, T.C. (1974) J. Am. Chem. Soc. 96, 5571–5573
12 Kearney, E.B., Salach, J.I., Walker, W.H., Seng, R.L., Kenney, W., Zeszotek, E. and Singer, T.P. (1971) Eur. J. Biochem. 24, 321–327
13 Walker, W.H., Kearney, E.B., Seng, R.L. and Singer, T.P. (1971) Eur. J. Biochem. 24, 328–331
14 Ghisla, S. and Hemmerich, P. (1971) FEBS Lett. 16, 229–232
15 Swett, L.R., Martin, W.B., Taylor, J.D., Everett, G.M., Wykes, A.A. and Gladish, Y.C. (1963) Ann. N.Y. Acad. Sci. 107, 891–898
16 Ghisla, S., Hartmann, U., Hemmerich, P. and Müller, F. (1973) Liebligs Ann. Chem. 1388–1415
17 Porter, D.J., Voet, J.G. and Bright, H.J. (1973) J. Biol. Chem. 248, 4400–4416
18 Hevesi, L. and Bruice, T.C. (1973) Biochemistry 12, 290–297
19 Brustlein, M. and Hemmerich, P. (1968) FEBS Lett. 1, 335–338
20 Makin, S.A., Ishmael, A.A., Yastrebov, V.V. and Petrov, K.I. (1971) Zh. Org. Khim. 7, 2120–2124
21 Arnold, Z. and Holy, A. (1963) Coll. Czech. Chem. Commun. 28, 2040–2046
22 Zeller, E.A., Gärtner, B. and Hemmerich, P. (1972) Z. Naturforsch. 27b, 1050–1052
23 Gärtner, B. and Hemmerich, P. (1975) Angew. Chem. Internat. Edit. 14, 110–111
24 Maycock, A.L. (1975) J. Am. Chem. Soc. 97, 2270
25 Ghisla, S. and Massey, V. (1974) Biochemistry 13, 589–597
26 Schonbrunn, A., Abeles, R.H., Walsh, C., Ghisla, S., Ogata, H. and Massey, V. (1976) Flavins and Flavoproteins, Proceedings of the 5th International Symposium (Singer, T.P., ed.), Chapt. 19, pp. 187–198, Elsevier, Amsterdam

Flavins and Flavoproteins, edited by T.P. Singer

CHAPTER 23

FLAVOPROTEIN OXIDASE MECHANISMS

DAVID J.T. PORTER and HAROLD J. BRIGHT

Department of Biochemistry, School of Medicine, University of Pennsylvania, Philadelphia, Pa. 19174 (U.S.A.)

Introduction

As the accompanying reports will show, good progress has been made since the Fourth Symposium towards an understanding of the molecular events which comprise flavoprotein catalysis. However, there has been a significant move away from wholly physiological systems, whose kinetic mechanisms are well described but whose chemistry has proved intractable, toward experiments in which analogues of the substrate or of the coenzyme, or both, are utilized and to model systems utilizing flavin derivatives and substrate analogues in which various solvent species substitute for the apoenzyme. The burden of this kind of experimentation is intensified because, in addition to providing necessary and sufficient evidence for the mechanism of the artificial system, the investigator must further establish what, if any, relationship this mechanism has to the chemical pathway dictated by the holoenzyme in the physiological system. Until such time as both of these criteria are fulfilled, our understanding of flavoprotein catalysis will remain highly speculative.

We shall describe here a number of studies with D-amino acid oxidase from hog kidney and glucose oxidase from *Aspergillus niger* which involve both physiological substrates and substrate analogues. We are concerned, especially, with the interpretation of the mechanism of reactions involving substrate analogues because of the influence such interpretations have on attitudes toward the reactions of physiological substrates. We shall also describe some new aspects of the reactions of physiological substrates of D-amino acid oxidase.

Studies of D-amino acid oxidase

A reexamination of the reactions of β-chloroalanine

The α-β elimination of HCl from β-chloro-substituted substrates catalyzed by flavoprotein oxidases has been taken generally to mean that abstraction of the substrate α-hydrogen as a proton is an obligatory enzymic event in oxidation as well as elimination [1, 2, 3]. Our own work with nitroalkane carbanions is consistent with this notion [4]. However, two important aspects of the α-β elimination reactions have been largely overlooked. First, the kinetic mechanism of the β-chloroalanine reactions catalyzed by D-amino acid oxidase, as initially proposed [1], can not be correct [5]. Second, although simple α-β elimination reactions may well involve proton abstraction, the extrapolation of this principle to flavoprotein reactions would be invalid if oxidation of the substrate α-carbon were to precede Cl^- elimination. We have used double-mix stopped flow spectrophotometry to clarify the first issue and shall briefly discuss some aspects of the second issue.

When D-amino acid oxidase is mixed anaerobically with β-chloroalanine in a single-mix flow experiment, the enzyme retains most of its oxidized character but exhibits a long wavelength absorbance tail in which, at 550 nm, three kinetic events can be monitored [5]. Phase I consists of a rapid, substrate saturable, absorbance burst with a limiting value of $\sim 300\,s^{-1}$. At zero substrate, the kinetics of phase I extrapolate to a value of $\sim 90\,s^{-1}$. Such behavior is typical of the sequence $A \rightarrow B \rightarrow C$ when $A \rightarrow B$ is substrate-dependent and we assign the difference between these values (approx. $200\,s^{-1}$) to k_2 in Scheme 1 and the value of $90\,s^{-1}$ to k_3. Kinetic isotope effects of 4- and 2.5-fold on k_2 and k_3, respectively, are obtained with β-chloro-[α-^{2}H]alanine. Phase II consists of a transient absorbance decrease and is followed by phase III, a substrate saturable absorbance increase with a limiting value of $\sim 2\,s^{-1}$ and a value at zero substrate of $\sim 0.8\,s^{-1}$. We assign (Scheme 1) by reasoning similar to that given above, the difference between these values ($\sim 1.2\,s^{-1}$) to $k_{5,app}$ ($k_{5,app} < k_5$, see later) and the zero substrate value to k_6. Steps k_5 and k_6 must form a separate loop because the latter is substantially smaller than the maximum turnover number for elimination ($\phi_0^{-1} = 5\,s^{-1}$). Neither $k_{5,app}$ nor k_6 show a deuterium kinetic isotope effect.

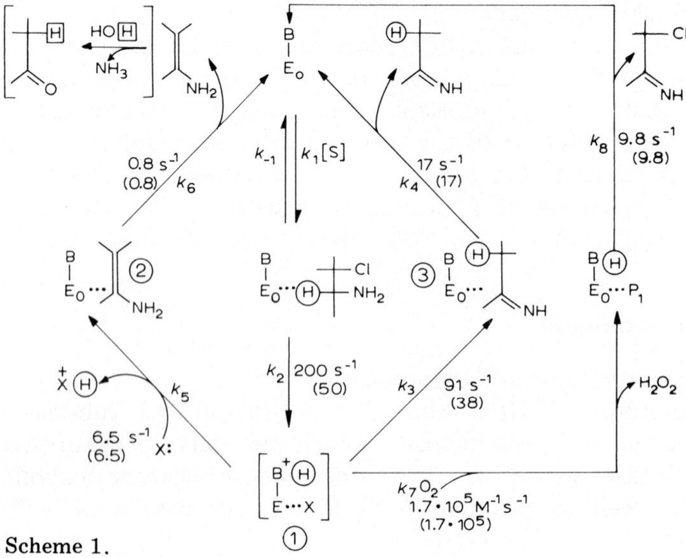

Scheme 1.

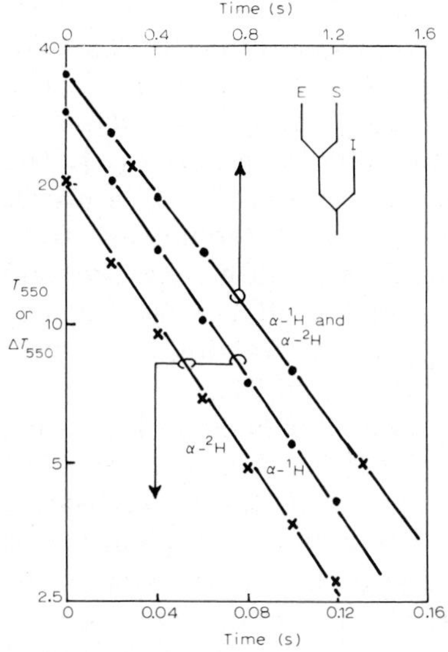

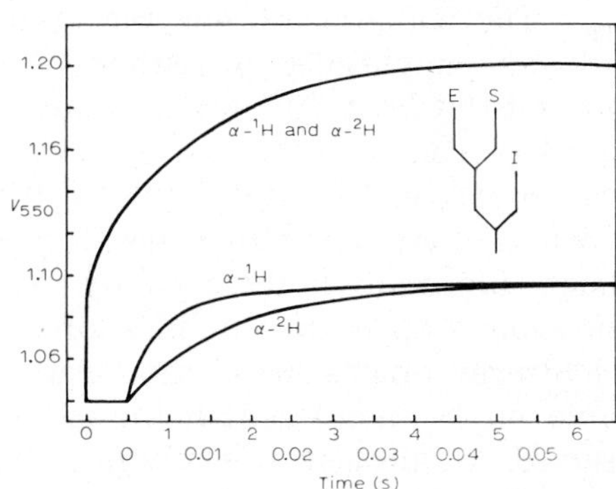

Fig. 1. Results from anaerobic double-mix flow experiments in which D-amino acid oxidase (E) is first mixed with β-chloroalanine (S, either α-^{1}H or α-^{2}H as shown) and then mixed again with anthranilate (I). The bottom two graphs (T_{550} on ordinate log scale) represent the rate of $E_0 \cdots$ I formation after the second mix while the top graph (ΔT_{550} on ordinate log scale) represents the decrease in total $E_0 \cdots$ I observed as the time between the two mixes is increased (see text and Scheme 1). Concentrations after the second mix: E, $2 \cdot 10^{-5}$ M; S, 10^{-2} M; I, $2.5 \cdot 10^{-2}$ M; potassium pyrophosphate (pH 8.3), 0.1 M. Temperature 25°C. The transmittance (T) values (arbitrary units) are small and do not require conversion to absorbance units.

Fig. 2. Reproduction of oscilloscope traces from anaerobic double-mix flow experiments in which D-amino acid oxidase (E) is first mixed with β-chloroalanine (S, either α-^{1}H or α-^{2}H as shown) and then mixed again with benzoate (I). The top trace shows the entire experiment (top time scale) while the bottom two traces show the initial phase at a faster sweep (bottom time scale). Concentrations after the second mix: E, $2 \cdot 10^{-5}$ M; S, 10^{-2} M; I, $2.5 \cdot 10^{-2}$ M. Other conditions as in Fig. 1.

In the first type of double-mix experiment, E_0 is mixed with S to generate reaction mixtures of pre-determined lifetimes. Anthranilate (I) is then added in the second mix to trap E_0 rapidly and essentially irreversibly as the complex $E_0 \cdots$ I after the first turnover. The rate of formation of $E_0 \cdots$ I monitored at 550 nm is uniphasic, not associated with a deuterium kinetic isotope effect and has a value of 16.5 s^{-1} which is independent of the time of the second mix over the range 0.02–2.0 s (see Fig. 1). If $k_4 > k_6$ and if $E_0 \cdots$ I and complex 2 absorb similarly at 550 nm, then $k_4 = 16.5\ \text{s}^{-1}$ (see Scheme 1) because, when I is added, most of the enzyme will exist as complexes 2 and 3. The total absorbance amplitude at 550 nm, however, decreases as the time of mixing of I is extended and corresponds to a process occuring at 1.5 s^{-1} (see Fig. 1). This represents the trapping of enzyme as complex 2 (see Scheme 1) in step k_5. No kinetic isotope effect is associated with k_5 and its value is in good agreement with that of phase III from the single-mix flow results [5].

In the second type of double-mix experiment, E_0 is trapped as its benzoate complex [6]. $E_0 \cdots$ I in this case has no absorbance at 550 nm and the rate of conversion of complexes 1 and 2 to 3 and $E_0 \cdots$ I can be monitored. As shown in Fig. 2, the traces are highly biphasic, the fast phase corresponding to $91\ s^{-1}$ and associated with an isotope effect of 2.4-fold. The slow phase (Fig. 2) has no isotope effect and has a value of $0.77\ s^{-1}$. The slow and fast phases are assigned to k_3 and k_6, respectively, and have their exact counterparts in the single mix experiments (phases I and III) [5]. The time dependence of the amplitude of the slow phase (see Fig. 2) confirms $k_{5,app}$. The amount of enzyme trapped as species 2 (computed from absorbance amplitudes in either of the double-mix flow experiments) amounts to at least 50% when steady state is achieved after about 2 s. The value of k_5 can be computed by dividing $k_{5,app}$ by the fraction of enzyme present as species 1 using $\phi_1^{-1}(= k_1 k_2/(k_{-1} + k_2))$, k_3 and k_4.

We have assigned structures to species 1, 2 and 3 in Scheme 1 by using chemical information reported by Walsh et al [2]. Species 3 in the fast elimination loop is the complex of E_0 and imino acid, the latter bearing the hydrogen originating at the α-carbon [2]. The kinetic isotope effect of 2.4-fold on k_3 indicates that the latter is dominated by the transfer of the conserved α-hydrogen, possibly in a tautomerization reaction. The value of k_4 is similar to the rate of the identical process involved in the oxidation of alanine ($9.8\ s^{-1}$, see below) as would be predicted. The slow loop (k_5 and k_6) is unique in flavoprotein oxidase reactions because of the generation of the enamine through α-β elimination of HCl. From studies in ^2H_2O, which showed that 0.5 atom solvent deuterium was incorporated (racemically in the case of β-chloroaminobutyrate) into the β-position of the α-keto acid product [2], it seems likely that at least one-half of the total product flux occurs through k_5 and k_6. We propose that k_5 results in a net deprotonation by solvent species of the branch point complex I to give an enzyme-bound enamine whose dissociation rate (k_6) is conspicuously small. A direct test of this feature of Scheme 1 is very difficult because of the instability of such enamines. However, we have found that the enamine phenyl-CH=C(—NH_2)—CO_2^- forms a complex with E_0 which has an electronic spectrum very similar to that of $E_0 \cdots$ anthranilate [6] and to that observed after 2 s of turnover with β-chloroalanine. Furthermore, this enamine is very tightly bound by E_0 because $4.5 \cdot 10^{-5}$ M E_0 was completely saturated by $5.5 \cdot 10^{-5}$ M enamine. These results support, qualitatively at least, the mechanism of Scheme 1.

We have added an oxidative pathway ($k_7[O_2]$ and k_8) to Scheme 1 which is entirely analogous to that for alanine (see below). The value of k_7 is given by Eqn. 1.

$$k_7 = \frac{\phi_2^{-1}(k_4 k_6 + k_4 k_5 + k_3 k_6)}{k_4 k_6} \qquad (1)$$

whereas k_8 is computed from $\phi_0^{-1} = k_2 k_8/(k_2 + k_8)$.

Scheme I explains the observations made by Walsh et al. [1] in their original studies as well as other curious features of the β-chloroalanine reaction. However, our interpretation is substantially different from that of Walsh et al. [1]. A cardinal feature of their interpretation was that a branch-point species was formed in a rate-determining step common to both oxidative and elimination pathways and that this species, on the basis of α-deuterium kinetic isotope effects of 1.7 to 1.9-fold on each pathway, was an enzyme-bound substrate carbanion or carbanion-flavin adduct. Scheme 1 shows that this interpretation is incorrect. It is entirely fortuitous that the maximum turnover numbers of the oxidative and elimination pathways, given by $k_2 k_8/(k_2 + k_8)$ and (approximately) by $k_3 k_4 k_6/(k_4 k_5 + k_3 k_6)$, respectively, are almost identical in magnitude. The calculated maximum turnover number for elimination is $6.8\,s^{-1}$, which compares very well with $\phi_0^{-1} = 5\,s^{-1}$. Similarly, the observed kinetic isotope effect on ϕ_0^{-1} for elimination, namely 2.1-fold, agrees well with the calculated value of 2-fold. However, this originates almost entirely from k_3 and not k_2 as originally supposed [1]. The ϕ_0^{-1} term for oxidation is $6\,s^{-1}$ and we find no detectable kinetic isotope effect. This agrees with the prediction of Scheme 1 because ϕ_0^{-1} is determined almost completely by k_8. Scheme 1 also explains why ϕ_2^{-1} for oxidation of β-chloroalanine is ten times smaller than that for alanine. Eqn. 1 shows that the bimolecular rate constant k_5 is related to ϕ_2^{-1} by a term which accounts for the small mole fraction of enzyme actually present in the steady state as the oxygen-reactive complex 1 (Scheme 1). Further, Scheme 1 predicts that the double reciprocal turnover plots of the oxidative reaction with β-chloroalanine as variable substrate should, unlike those for alanine (see below), converge and meet on the substrate axis. This is observed, and results from the fact that the elimination pathway, having an almost identical maximum turnover number, provides the required reversal of the pathway between species 1 and E_0. Finally, we should note that Scheme 1 accounts for the complex pattern seen in anaerobic single-mix flow experiments [5] at 550 nm. Phase I is the rapid formation of the absorbing complex I. This decays in phase II to the non-absorbing species 3. However, superimposed on this decay is the slow build-up of the highly absorbing enamine complex 2 in phase 3. After 2 s, the turnover absorbance trace is dominated by complex 2.

Thus far we have not addressed what is perhaps the most crucial question of all, namely, is there in fact any evidence to support the hypothesis that the oxidative and elimination pathways share any intermediates beyond E_0 or $E_0 \cdots S$? Although we believe that none of the published evidence shows that k_2 of Scheme 1 is common to both pathways, we can offer two independent lines of evidence to support such an hypothesis. First, the ϕ_1^{-1} values for oxidation and elimination ($\phi_1^{-1} = k_1 k_2/(k_{-1} + k_2)$ for Scheme 1) are quite similar in magnitude ($2 \cdot 10^4$ and $0.7 \cdot 10^4\,M^{-1}\,s^{-1}$, respectively) and each shows an α-deuterium kinetic isotope effect of 1.8-fold. Second, the rate and amplitude of phase I in single-mix experiments [5] increases and decreases, respectively, as the O_2 concentration is raised. This suggests that the branchpoint complex 1 of Scheme 1 is capable of reacting with O_2. Although one would wish for stronger evidence that a common α-hydrogen abstraction step (k_2) is shared by the oxidation and elimination pathways, in the absence of contrary evidence, we accept Scheme 1 as the best working hypothesis for the β-chloroalanine reactions.

Finally, we emphasize that even if a common C—H cleavage process (k_2) is assumed for elimination and oxidation, this does not lead inexorably to the conclusion that the α-hydrogen is abstracted as a proton. This issue depends critically on the structure of complex 1 which, within the bounds of present knowledge, can only be broadly defined as a rapidly interconverting mixture of four isomers of identical empirical formula ($\overset{+}{\text{HE}}\text{—}\overset{-}{\text{FADH}} \cdots \rangle\text{C=NH}$, $\text{E—}\overset{\bullet}{\text{FADH}} \cdots \overset{\bullet}{\text{C}}\text{—NH}_2$, $\overset{+}{\text{HE}}\text{—FAD} \cdots {}^{-}\text{C—NH}_2$ and $\overset{+}{\text{HE}}\text{—}\overset{-}{\text{FAD}}\text{—C—NH}_2$). Evidence for complexes involving substrate carbanions and covalent adducts has come from studies of nitroalkane oxidation [4] while Bruice favors biradical complexes in analogous model reactions [7]. If we argue that oxidation of the substrate is essentially completed in step k_2, N-5 of the reduced flavin might adventitiously attack the α-amino acid product in analogy with carbinolamine formation [7]. The resulting anionic adduct could then eliminate Cl^- ($\overset{+}{\text{HE}}\text{—}\overset{-}{\text{FAD}}\text{—C(NH}_2\text{)—CH}_2\text{—Cl}$) to form the enamine. Since this argument is independent of the mode of C—H bond scission in k_2 (atom, hydride or proton abstraction), we feel that the α-β elimination pathway by itself does not provide necessary and sufficient evidence for the hypothesis of α-proton abstraction.

The reversibility of flavin reduction and the locus of O_2 reactivity

The kinetic mechanism of the D-amino acid oxidase reaction conforms to Eqn. 2 [8–12].

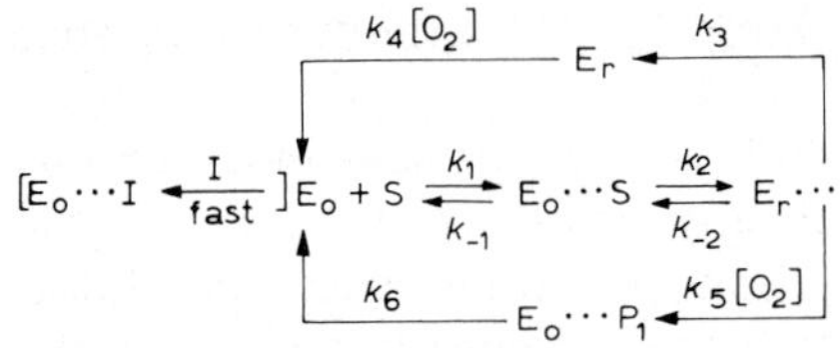

$$
\begin{array}{c}
E_r \xleftarrow{k_3} E_r \cdots P_1 \text{ (upper loop)}, \quad E_r \xrightarrow{k_4[O_2]} E_o \\
[E_o \cdots I] \xleftarrow[\text{fast}]{I} E_o + S \underset{k_{-1}}{\overset{k_1}{\rightleftharpoons}} E_o \cdots S \underset{k_{-2}}{\overset{k_2}{\rightleftharpoons}} E_r \cdots P_1 \\
E_r \cdots P_1 \xrightarrow{k_5[O_2]} E_o \cdots P_1 \xrightarrow{k_6} E_o \text{ (lower loop)}
\end{array} \qquad (2)
$$

All non-basic amino acid substrates appear to traverse the lower loop because $k_5[O_2] \gg k_3$. However, the reactivity of $E_r \cdots P_1$ with O_2 has only been measured indirectly [8, 9, 13]. Furthermore, the lower loop constitutes a sequential mechanism whereas double-reciprocal turnover patterns are conspicuously parallel, rather than intersecting [8, 12]. We have addressed both of these questions through the use of double-mix stopped flow spectrophotometry.

When enzyme is mixed with alanine and O_2 to generate $E_r \cdots P_1$, and then mixed again after a predetermined delay with benzoate (I), the kinetics of disappearance of $E_r \cdots P_1$ can be monitored at 550 nm (see Fig. 3) according to Eqn. 3 (where $k_2 > k_{-2}$).

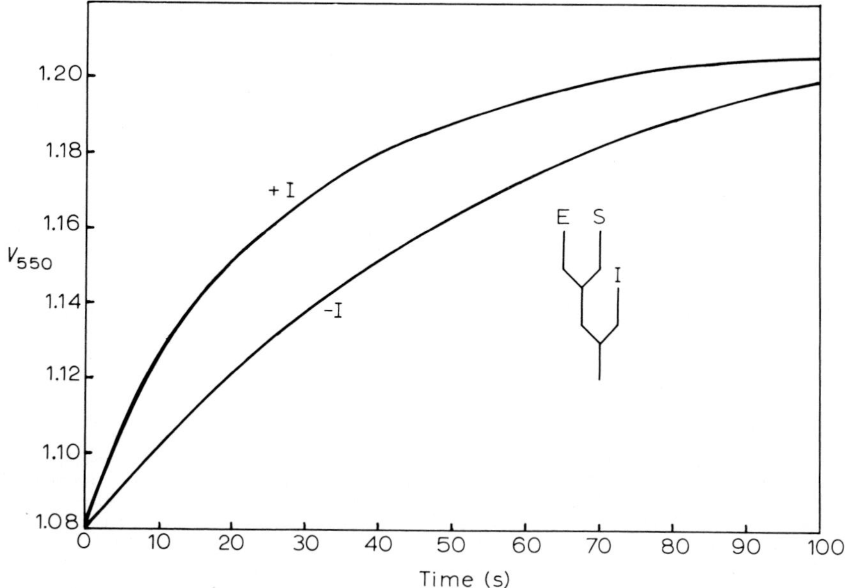

Fig. 3. Reproduction of oscilloscope traces from anaerobic double-mix flow experiments in which D-amino acid oxidase (E) is first mixed with D-alanine (S) and then mixed again with benzoate (I). The last two terms of Eqn. 3 are given by the upper trace, while k_3 is obtained from the lower trace (see text). When O_2 as well as I is admitted in the second mix (corresponding to all three terms of Eqn. 3) much faster sweeps are required. Conditions after the second mix: E, $2 \cdot 10^{-5}$ M; S, $5 \cdot 10^{-3}$ M; I, $5 \cdot 10^{-3}$ M. Other conditions as in Fig. 1.

$$k_{obs} = k_5[O_2] + k_3 + \frac{k_{-1}k_{-2}}{k_{-1} + k_{-2} + k_2} \tag{3}$$

Plots of k_{obs} versus $[O_2]$ yield k_5 directly ($1.8 \cdot 10^5$ M^{-1} s^{-1}) as well as the sum (0.039 s^{-1}) of the last two terms of Eqn. 3. The value of k_3 (0.018 s^{-1}) is determined in separate single-mix anaerobic experiments, giving a value for the last term of Eqn. 3 of 0.021 s^{-1}.

The major conclusions from these studies are the following. First, $E_r \cdots P_1$ reacts directly with O_2 at a rate which is in excellent agreement with other less direct estimates [8, 9, 13]. Second, the overall rate of conversion of $E_r \cdots P_1$ to E_0 is 0.021 s^{-1}. Since this value is much smaller than the value of $k_5[O_2]$ in routine turnover experiments, the sequence $E_0 \rightarrow E_0 \cdots P_1$ is effectively irreversible and the overall mechanism, though formally sequential, will yield parallel line turnover patterns as argued previously [14].

The irreversibility of steps 1 and 2 in the anaerobic alanine half-reaction (Eqn. 2) is also apparent from studies with α-^{2}H-alanine. Double reciprocal plots of k_{obs} for $E_0 \cdots P_1$ formation at 550 nm versus substrate concentration show a kinetic isotope effect on k_2 of 2-fold but no slope effect. This result is to be expected if k_{-1} or k_{-2} is effectively zero.

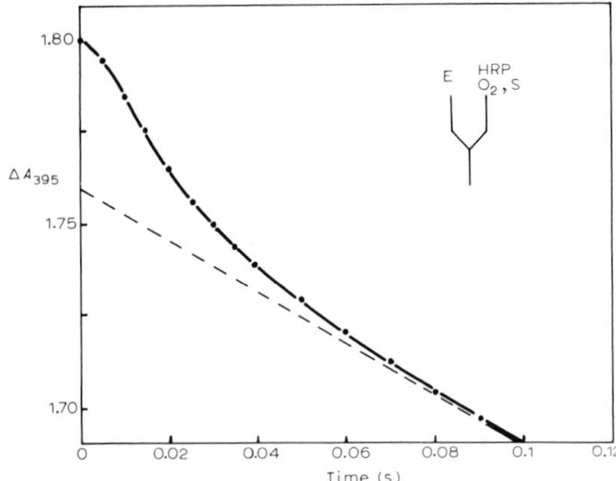

Fig. 4. Results of stopped flow experiment in which H_2O_2 is trapped as complex 1 of horseradish peroxidase (HRP) according to the conditions of Eqn. 4. This experiment actually measures the kinetics of disappearance of free HRP at a wavelength (395 nm) which is isosbestic for complexes I and II of HRP (Chance, B. (1949) Arch. Biochem. Biophys., 22, 224). The dashed line represents the steady state rate of disappearance of free HRP computed independently from the conditions of Eqn. 4. Concentrations after mixing: E, 10^{-6} M; S, 10^{-2} M; O_2, $7.2 \cdot 10^{-4}$ M; HRP, $1.2 \cdot 10^{-5}$ M. Other conditions as in Fig. 1.

The locus of H_2O_2 release

The scheme for alanine turnover (lower loop of Eqn. 2) reduces to Eqn. 4 when the alanine and O_2 concentrations are 10^{-2} M and $7.2 \cdot 10^{-4}$ M, respectively.

$$E_o \xrightarrow{10^3\ s^{-1}} E_r \cdots p \xrightarrow[125\ s^{-1}]{A} E_o \cdots p \xrightarrow[9.8\ s^{-1}]{B} E_o \qquad (4)$$

One may ask whether H_2O_2 is released at point A or B by trapping the H_2O_2 with horse radish peroxidase and measuring the kinetics of formation of complex 1 of the latter at 395 nm. Separately, the bimolecular rate constant for formation of complex 1 of horse radish peroxidase was determined to be $3 \cdot 10^7\ M^{-1}\ s^{-1}$. Computer simulation of Eqn. 4 clearly shows, as expected, that there will be a burst or a lag in complex 1 formation if H_2O_2 is released at A or B, respectively, in Eqn. 4. As shown in Fig. 4 the experimental results correspond unequivocally to H_2O_2 release at position A (step k_5 of Eqn. 2). We infer therefore that step k_6 of Eqn. 2 most probably represents the release of the imino acid (P_1) from $E_0 \cdots P_1$ as is shown.

Substrate stoichiometry and locus of proton release in the reductive half-reaction

The purple, or long wavelength-absorbing species $E_r \cdots P_1$ is formed anaerobically according to steps 1 and 2 of Eqn. 2. Although $E_0 \cdots S$ and $E_r \cdots P_1$ are commonly assumed to contain one equivalent of substrate and product (α-imino acid) per mol of oxidized and reduced flavin, respectively,

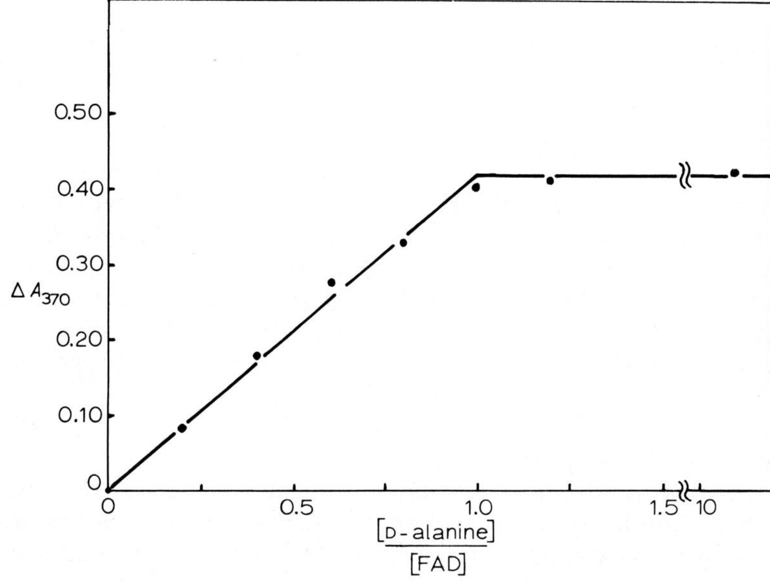

Fig. 5. Results of anaerobic stopped flow experiments at 370 nm which monitor the absorbance of $E_r \cdots P_1$ (Eqn. 2) as a function of the ratio of D-alanine to enzyme-bound FAD. The line is drawn assuming that $E_r \cdots P_1$ formation is irreversible and that one equivalent of D-alanine per mol enzyme-bound FAD is required to form $E_r \cdots P_1$. Concentrations after mixing: D-amino acid oxidase, $5 \cdot 10^{-5}$ M (as enzyme-bound FAD); D-alanine, as shown. Other conditions as in Fig. 1.

the evidence for this assumption [15] is not as strong as it might be. Because this stoichiometry is of fundamental importance in proposals for the mechanism of flavin reduction, we have redetermined it by a direct spectrophotometric method. We have also determined the stoichiometry of proton release in steps 1, 2 and 3 of Eqn. 2.

The rationale of the experiments to determine substrate stoichiometry is to mix E_0 and alanine anaerobically and then monitor the total absorbance increment due to $E_r \cdots P_1$ (Eqn. 2) at a wavelength which is isobestic for E_r and $E_r \cdots P_1$. The latter condition is necessary to accomodate the slow conversion of $E_r \cdots P_1$ to E_r via k_3 ($k_3 = 0.018\,s^{-1}$, see above). We have already shown (see above) that the conversion of E_0 to $E_r \cdots$ P1 is irreversible in the case of alanine. Fig. 5 shows very clearly that the formation of $E_r \cdots P_1$ requires only one substrate equivalent per mol of flavin.

The locus and stoichiometry of proton production in the anaerobic reductive half-reaction was investigated with D-valine because k_3 for this substrate (see Eqn. 2) is much smaller than the solvent-catalyzed hydrolysis of free P_1 (α-imino acid). The pK_a value [16] of E—$FADH_2$ is 7.1 and the argument of Eqn. 5 shows that the overall conversion of amino acid to α-keto acid and ammonia will release slightly less than one equivalent H^+ at pH 8.

$$\text{E-FAD} + \text{R}-\overset{\overset{\displaystyle NH_3^+}{|}}{\underset{\underset{\displaystyle H}{|}}{C}}-CO_2^- \longrightarrow \text{E-FAD}_2^- + \text{R}-\overset{\overset{\displaystyle O}{\|}}{C}-CO_2^- + NH_4^+ \tag{5}$$

$$pK_a \approx 9.5 \qquad \text{E-FAD}_2^- \rightleftharpoons \text{E-FADH}^- + H^+ \;\; (pK_a = 7.1) \qquad pK_a \sim 9$$

By using phenol red as buffer and pH indicator in the valine reductive half-reaction, we could show that this proton is actually released when substrate binds to E_0 in step k_1 of Eqn. 2. Because there was no proton release associated with the binding of either benzoate or L-leucine, and because the cleavage of the substrate C—H bond occurs in the k_2 step, we suggest that substrate binding involves deprotonation of the α-amino group of the substrate.

The mechanism of oxidation of nitroalkanes

We were able to establish a detailed chemical mechanism for the oxidation of nitroethane by D-amino acid oxidase through a combination of rapid kinetic and chemical trapping experiments [4]. We concluded that the carbanion of the model substrate attacked the flavin nucleus at N-5 to form a covalent adduct. Rapid elimination of NO_2^- from this adduct resulted in a N-5-imminium intermediate, which was then rapidly hydrated to form a carbinolamine. This, in turn, eliminated reduced flavin rapidly to form $E_r \cdots P$, a species formally identical in structure and O_2 reactivity to $E_r \cdots P_1$ in the alanine reaction (Eqn. 2). The deduction of this mechanism was predicated on the fact that CN^-, at a rate controlled by the initial attack of nitroethane carbanion on the flavin, completely inactivated the enzyme through the formation of enzyme-bound 5-cyanoethyl-1, 5-dihydroFAD. This was interpreted as resulting from the reaction of CN^-, in competition with hydration, with the N-5-imminium intermediate. We have now found, as would be predicted, that a variety of nucleophiles react rapidly with the imminium species derived from nitroethane anion.

The reactive nucleophiles fall into three classes (X, Y and Z) as shown in Scheme 2. Class X includes CN^-, H^- and the carbanions of nitromethane and nitroethane [4, 17]. The tetrahedral adducts substituted by X show the single-banded absorption spectra (λ_{max} in the range 330–340 nm) characteristic of such N-5 adducts [18]. The CN^- adduct was characterized previously [4], and we could show by tritium labeling that one atom of hydrogen was introduced as H^- from sodium borohydride. The class Y nucleophiles (H_2O and HH_3) interact very rapidly with the immonium species to form kinetically competent tetrahedral N-5 adducts which, in turn, form E—$FADH_2 \cdots P$ very rapidly. Thus, the implication of NH_3 as a class Y nucleophile could only be achieved by showing that NH_3 competitively protects against inactivation by CN^-. This effect of NH_3 was shown in control experiments not to be due to the trapping of CN^- as NH_4CN. The class Z nucleophiles are characterized by the fact that the oxygen-monitored turnover trace, corresponding to the oxidation of

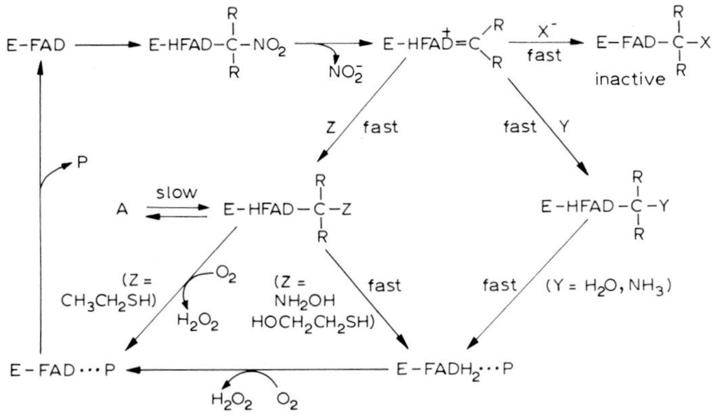

Scheme 2.

nitroethane anion, is decidedly biphasic in their presence. The initial rate of O_2 consumption is similar to that obtained in the absence of Z nucleophiles. However, the steady-state rate obtained after several minutes of turnover is much smaller than the initial rate. This effect is attributed to a relatively slow accumulation of a derivative of the tetrahedral adduct (species A in Scheme 2) as shown in Fig. 6. Species A is not formed irreversibly because addition of alanine results in a slow ($0.023\,s^{-1}$) recovery of almost total enzyme activity. The Z nucleophiles are of two types. Hydroxylamine and mercaptoethanol (see Scheme 2) are non-competitive inhibitors of nitropropane anion due to the formation of species A but appear otherwise to be similar in all other respects to the Y nucleophiles. Mercaptoethane, on the other hand, is different because the rate of inhibition, judged from oxygen-electrode traces, decreases as the O_2 concentration is raised. This suggests that the tetrahedral adduct formed with mercaptoethane can react directly with O_2 in a process which is competitive with the formation of species A.

In summary, these results substantiate the mechanism of nitroalkane oxidation deduced previously [14] because they show that a variety of nucleophiles are capable of reacting with the postulated N-5-immonium adduct. The results with NH_3 are of particular interest because the tetrahedral N-5 adduct in this case would be formally identical to that derived from a physiological amino acid substrate if the latter were oxidized by a mechanism involving α-proton abstraction and N-5 adduct formation.

Studies of glucose oxidase

The mechanism of oxidation of nitroalkanes

The oxidation of nitroalkanes by glucose oxidase [19] differs in certain important respects from the corresponding reactions of D-amino acid oxidase [4]. The reason for these differences is that, in part, the catalytic mechanism is free-radical in character. We shall briefly develop the evidence for this conclusion.

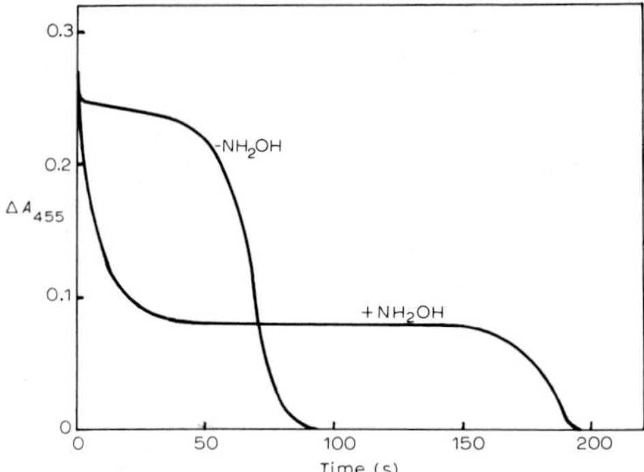

Fig. 6. Effect of hydroxylamine on stopped-flow monitored turnover of 2-nitropropane anion by D-amino acid oxidase. The experiments monitor oxidized enzyme species at 455 nm and clearly demonstrate the slow formation of a non-absorbing and inhibitory enzyme complex (species A in Scheme 2). Concentrations after mixing: D-amino acid oxidase, $1.2 \cdot 10^{-5}$ M; 2-nitropropane anion, [4] $1.3 \cdot 10^{-3}$ M; O_2, $1.2 \cdot 10^{-3}$ M; hydroxylamine, 0.1 M. Other conditions as in Fig. 1.

The stoichiometry of the overall reaction is non-integral and, as yet, incomplete (Eqn. 6). It is evident that a reduced electron acceptor on the right side of Eqn. 6

$$\bar{C}_2H_4NO_2 + O_2 \longrightarrow 1.09\ CH_3CHO + 0.94\ NO_2^- + 0.32\ H_2O_2 + 0.11\ NO_3^- + 0.07\ C_2H_4(NO_2)_2 \quad (6)$$

is as yet unidentified. Nevertheless, the non-integral character of the established products is highly suggestive of a free-radical mechanism.

Direct evidence for the kinetic mechanism comes from a variety of stopped-flow experiments and the scheme of Eqn. 7 has been established for nitroethane anion at pH 5.0.

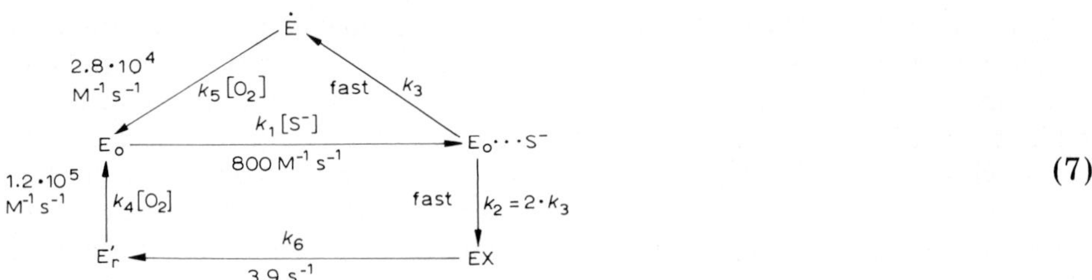

(7)

When E_0 is mixed anaerobically with nitroethane anion there is a rapid first order decrease in 450 nm absorbance (due to disappearance of E_0) and a concommitant increase in 550 nm absorbance. The kinetics at the two wavelengths are identical and define the bimolecular rate constant k_1. The value of k_1 agrees well with the coefficient ϕ_1^{-1} from oxygen-monitored turnover experiments, showing that k_1 is an obligatory step in turnover. The spectrum of the reduced enzyme from the anaerobic measurements clearly implicates [20] the semiquinone state of glucose oxidase ($\dot{E}$). This is substantiated by the line width and power saturation characteristics of the ESR spectrum and by the pH dependence of the optical spectrum [20–22]. However, at the end of the anaerobic experiment at pH 5.0 we calculate from the extinction coefficients that only 35% of the enzyme has been converted to the semiquinone while 65% is converted to EX, a reduced enzyme species having no significant absorbance at 450 nm and 550 nm. Since there is no kinetic evidence for the intermediate $E_0 \cdots S^-$ we conclude that steps k_2 and k_3 are very fast compared to $k_1[S^-]$ and that $E_0 \cdots S^-$ is thereby partitioned between $\dot{E}$ and EX at a total rate controlled by $k_1[S^-]$. A kinetically equivalent, but more complicated, scheme would have $\dot{E}$ and EX formed in entirely separate pathways. When the anaerobic reduced enzyme is mixed with O_2, the kinetics of appearance of E_0 and monitored at 450 nm are biphasic. The second, slower, phase is identical in rate to the uniphasic disappearance of $\dot{E}$ monitored at 550 nm. Thus, the reoxidative constant k_4 and k_5 can be assigned and evaluated as shown in Eqn. 7. The value of k_5 corresponded well to the value $(1.3 \cdot 10^4\ M^{-1}\ s^{-1})$ obtained from oxidation of $\dot{E}$ which had been obtained by irradiation in the presence of EDTA. The first order process, k_6, is evaluated from stopped-flow monitored turnover experiments.

These results are important because, for the first time, they provide direct evidence for a free radical mechanism in a flavoprotein oxidase reaction. The non-free radical pathway (k_2, k_6 and k_4) is difficult to evaluate chemically. The species EX has spectral properties which are consistent with those of a N-5 adduct. However, the oxidation of nitroethane anion by glucose oxidase is not inhibited by CN^-.

References

1 Walsh, C.T., Schonbrunn, A. and Abeles, R.H. (1971) J. Biol. Chem. 246, 6855
2 Walsh, C.T., Krodel, E., Massey, V. and Abeles, R.H. (1973) J. Biol. Chem. 248, 1946
3 Walsh, C.T., Lockridge, O., Massey, V. and Abeles, R.H. (1973) J. Biol. Chem. 248, 7049
4 Porter, D.J.T., Voet, J.G. and Bright, H.J. (1973) J. Biol. Chem. 248, 4400
5 Voet, J.G., Porter, D.J.T. and Bright, H.J. (1972) Z. Naturforsch. 27B, 1054
6 Nishikimi, M., Osamura, M. and Yagi, K. (1971) J. Biochem. (Tokyo) 70, 457
7 Bruice, T.C., in Progress in Bioorganic Chemistry (Kaiser, E.T. and Kezdy, E.J. eds.). Vol. 4, John Wiley and Sons, New York, in the press
8 Massey, V. and Gibson, Q.H. (1964) Fed. Proc. Fed. Amer. Soc. Exp. Biol. 23, 18
9 Nakamura, T., Nakamura, S. and Ogura, Y. (1963) J. Biochem. (Tokyo) 54, 512
10 Nakamura, T., Yashimura, Y. and Ogura, Y. (1965) J. Biochem. (Tokyo) 57, 554
11 Yagi, K (1971) Advan. Enzymol. 34, 41
12 Massey, V., Curti, B. and Ganther, H. (1966) J. Biol. Chem. 241, 2347
13 Rao, N.A., Nishikimi, M. and Yagi, K. (1972) Biochim. Biophys. Acta 276, 350
14 Palmer, G. and Massey, V. (1968) in Biological Oxidations (Singer, T.P., ed.) p. 236, John Wiley and Sons, New York
15 Yagi, K. and Ozawa, T. (1964) Biochim. Biophys. Acta 81, 29
16 Brunori, M., Rotilio, G., Antonini, E., Curti, B., Branzoli, U. and Massey, V. (1971) J. Biol. Chem. 246, 3140
17 Porter, D.J.T., Voet, J.G. and Bright, H.J. (1971) J. Biol. Chem. 247, 1971
18 Hemmerich, P., Ghisla, S., Hartmann, U. and Müller, F. (1971) in Flavins and Flavoproteins (Kamin. H., ed.) p. 83, University Park Press, Baltimore, Maryland
19 Porter, D.J.T., Voet, J.G. and Bright, H.J. (1972) Z. Naturforsch. 27B, 1052
20 Palmer, G. and Massey, V. (1966) Biochemistry 5, 3181
21 Müller, F., Brustlein, M., Hemmerich, P., Massey, V. and Walker, W.H. (1972) Eur. J. Biochem. 25, 573
22 Palmer, G., Müller, F. and Massey, V. (1971) Flavins and Flavoproteins (Kamin, H., ed.), p. 123, University Park Press, Baltimore, Maryland

Flavins and Flavoproteins, edited by T.P. Singer

CHAPTER 24

AFFINITY LABELLING OF BAKERS' YEAST L-LACTATE DEHYDROGENASE (CYTOCHROME b_2) BY BROMOPYRUVATE

FLORENCE LEDERER and CHRISTINE MULET

Centre de Génétique Moléculaire du C.N.R.S., Gif-sur-Yvette (France)

Introduction

The flavohemoprotein L-(+)-lactate dehydrogenase (or cytochrome b_2, EC 1.1.2.3) catalyzes the oxidation of lactate to pyruvate. Studies involving stopped flow techniques and E.P.R. spectroscopy combined with rapid freezing have recently led to the proposal of a kinetic scheme where the various steps of electron transfer from lactate to flavin to heme are detailed [1].

Another recent study concerned the chemical mechanism of lactate oxidation itself [2]. It led to the tentative conclusion that the initial step in the reaction is the same as that catalyzed by oxidase-type flavoproteins (D-amino acid oxidase and lactate oxidase for example), namely α-proton removal [3–7]. Studies in progress in our laboratory are trying to identify the side-chain of the active center which effects this reaction.

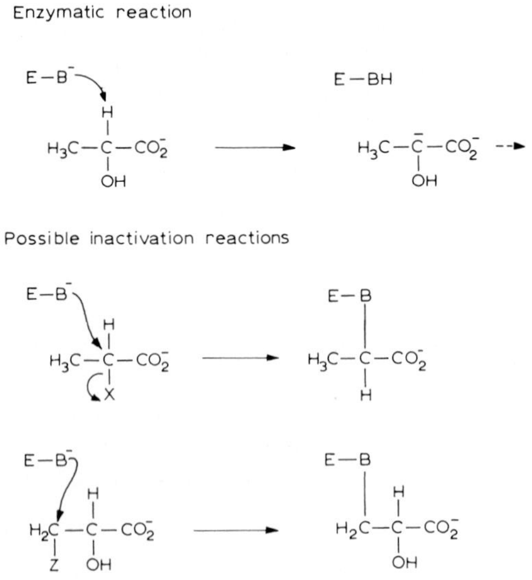

Enzymatic reaction

$E{-}B^- + H_3C{-}C(H)(OH){-}CO_2^- \longrightarrow E{-}BH + H_3C{-}\bar{C}(OH){-}CO_2^- \dashrightarrow$

Possible inactivation reactions

$E{-}B^- + H_3C{-}C(H)(X){-}CO_2^- \longrightarrow H_3C{-}C(B{-}E)(H){-}CO_2^-$

$E{-}B^- + H_2C(Z){-}C(H)(OH){-}CO_2^- \longrightarrow E{-}B{-}H_2C{-}C(H)(OH){-}CO_2^-$

Scheme I.

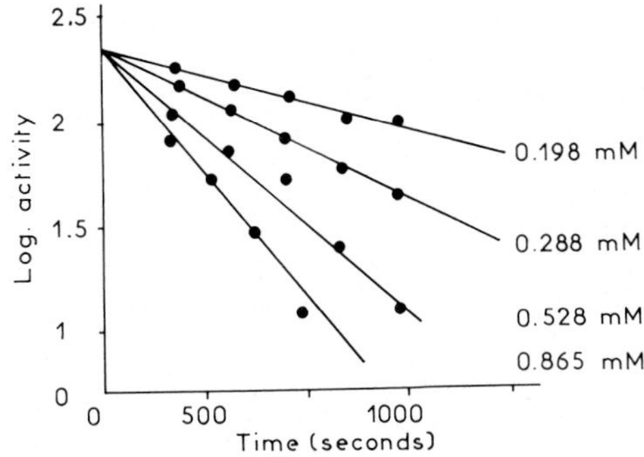

Fig. 1. Time course of inactivation. Incubations were carried out at 30°C in a total volume of 110 µl containing 2 µM enzyme and the indicated concentrations of bromopyruvate, in 0.1 M phosphate, pH 7, 1 mM EDTA. At various time intervals, 5 to 10 µl aliquots were withdrawn and immediately assayed in a spectrophotometer cuvette containing 10 mM L-(+)-lactate and 1 mM $K_4[Fe(CN)_6]$ in phosphate buffer.

As represented in Scheme I, one can expect to form a stable covalent bond between the nucleophile and an inhibitor by using a suitable substrate analog which would carry a strongly electron-withdrawing substituent in a position α or β to the carboxyl group. The inhibitor we used, bromopyruvate, is actually a product analog rather than a substrate analog.

Material and Methods

Cleaved cytochrome b_2 was prepared by the crystallization method of Appleby and Morton [8, 9]. Calbiochem bromopyruvic acid was recrystallized from chloroform or benzene/hexane. The radioactive reagent was synthesized from [2-^{14}C] pyruvate sodium salt (The Radiochemical Centre, Amersham) diluted with unlabelled pyruvic acid, according to Meloche [10], and recrystallized from benzene/hexane.

Incubations were conducted as described under the figures. Amino acid analyses were carried out on a Beckman 120C analyzer equipped for high sensitivity recording. Radioactive samples were counted in a Nuclear Chicago Mark I scintillation counter using Bray's scintillation fluid. Enzyme concentrations were expressed relative to heme ($\epsilon_{423}^{red} = 183\ mM^{-1} \cdot cm^{-1}$).

Results

Kinetic study of the inactivation

When cytochrome b_2 was incubated in the presence of bromopyruvate, the inactivation followed pseudo first-order kinetics (Fig. 1). When the derived k_{exp} were plotted versus the inhibitor concentration in a double reciprocal plot, a straight line was obtained, showing a finite intersection with the ordinate (Fig. 2). This indicated that bromopyruvate forms a reversible complex with the enzyme before covalent bond formation, according to the equation

$$E + I \underset{k_{-1}}{\overset{k_1}{\rightleftharpoons}} EI \xrightarrow{k_2} E - I$$

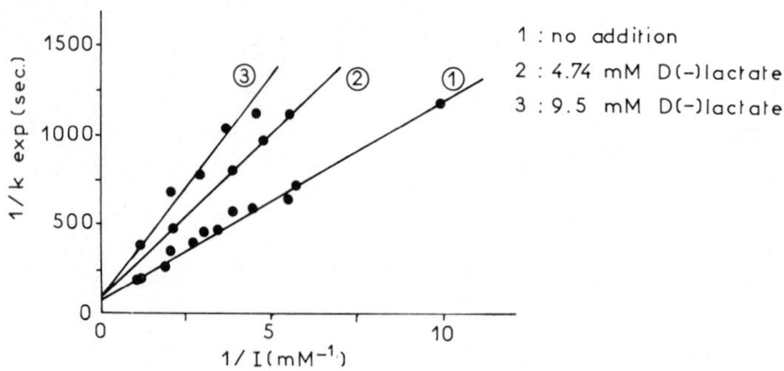

Fig. 2. Dependence of the inactivation rate on bromopyruvate and D-(—)-lactate concentration. k_{exp} is the slope of first order plots such as those represented in Fig. 1. Incubations in the presence of D-(—)-lactate were carried out exactly as described above.

From such experiments, the constants k_2 and $K_{inact} = \frac{k_{-1} + k_2}{k_1}$ were found to be $10.6 \cdot 10^{-3}$ s^{-1} and 1.3 mM, respectively.

D-(—)-Lactate acted as a competitive inhibitor of the inactivation by bromopyruvate. As shown in Fig. 2, inactivations carried out in the presence of 4.74 and 9.5 mM D-(—)-lactate gave K_i values of 6.5 and 7 mM, in good agreement with literature values of 6 to 7 mM obtained in competition experiments with L-(+)-lactate [11–13].

According then to these criteria, bromopyruvate behaves as a valid affinity label. As the reagent can also act as an unspecific alkylating agent, it is important to note that neither 10 mM 2-bromopropionate nor 10 mM 3-bromopropionate nor 5 mM iodoacetate inactivated enzyme under conditions similar to those used with bromopyruvate. Another argument in favor of the specificity of the reagent is the following: incubation with 4 mM bromopyruvate in the presence of 10 mM L-(+)-lactate leads to no inactivation. Dithionite is similarly effective in protecting the activity. It follows that the reduced enzyme is insensitive to bromopyruvate, in other words the looked for nucleophile is not active in the reduced enzyme. Several explanations are possible which will be considered in the conclusion.

Radioactive labelling stoichiometry and nature of labelled groups

When the enzyme was incubated with bromo-[2-^{14}C]pyruvate for 30 min in the presence of 10 mM L-(+)-lactate, about 1.1 groups were found to be labelled, without loss of activity, as just stated. Hence the enzyme was routinely prelabelled in the presence of lactate; after removal of lactate and excess nonradioactive reagent, inactivation was carried out with bromo-[2-^{14}C]pyruvate. A typical experiment is described in Fig. 3, showing the incorporation of a total of 3 groups, which clearly cannot be all essential for the activity. The protein was hydrolysed after borohydride reduction and the effluent of an analyzer column counted for radioactivity. A single peak was found which corresponded to *S*-carboxyhydroxyethylcysteine.

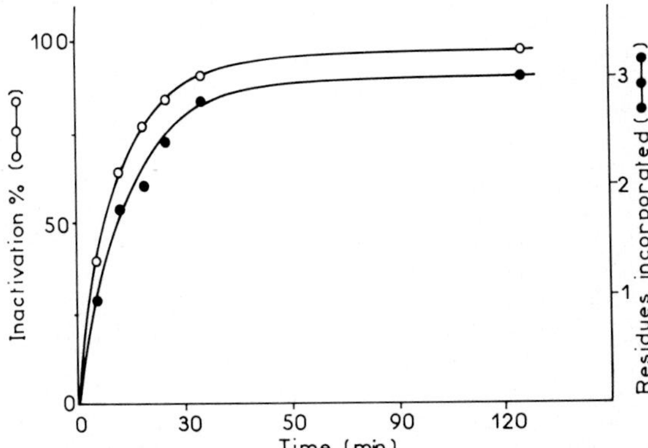

Fig. 3. Incorporation as a function of inactivation. 40 μM enzyme (3.5 ml) was incubated at 30°C for 30 mn with 4 mM cold bromopyruvate in the presence of 10 mM L-(+)-lactate. The enzyme was then precipitated by ammonium sulfate, redissolved in 0.1 M phosphate buffer, pH 7, 1 mM EDTA, 10 mM L-(+)-lactate. After a second precipitation, the precipitate was washed 2 or 3 times with a 70% saturated ammonium sulfate solution in buffer without lactate, and finally redissolved in 3.5 ml of 0.1 M phosphate buffer pH 7, 1 mM EDTA. 60 μl of 251 mM bromo-[2-^{14}C] pyruvate was added at zero time (incubation at 0°C). At various time intervals, aliquots were withdrawn and the reaction quenched with mercaptoethanol. Residual enzymatic activity was measured after suitable dilution, and radioactivity determined after exhaustive dialysis.

Assuming that only one of these residues could be essential for the activity, we tried to selectively modify the reactivity of the others. As all of them seemed to be essentially inert toward iodoacetate, the influence of sulfite was tested. Sulfite reacts with cytochrome b_2 to give a reduced-type flavin spectrum (Lederer, F., to be published). Such an effect has been attributed to the formation of a reversible covalent complex by addition of sulfite at the N-5 position of the flavin nucleus [14, 15]. We hoped that sulfite might exert a protecting effect in the immediate vicinity of the flavin nucleus. Labelling in the presence of sulfite (Fig. 4, top) seemed to indicate that some of the non-essential reactive groups were indeed labelled comparatively faster (under the conditions used, inactivation was slowed down compared to the control without sulfite; the progress of inactivation as a function of time suggests that the process could be limited by sulfite dissociations. Accordingly the enzyme was inactivated with bromo-[2-^{14}C] pyruvate after a 30 min prelabelling with nonradioactive reagent in the presence of sulfite. Fig. 4 (bottom) depicts the result of a very preliminary experiment, suggesting that under these conditions inactivation indeed depends on the labelling of essentially one group. Amino acid analysis again showed this group to be a cysteine.

Conclusions

The results reported in this paper, some of which must be considered as preliminary (Fig. 4), suggest the following conclusions:

(1) Kinetics and labelling attempts with various reagents point to bromopyruvate as a specific affinity label for the active site of cytochrome

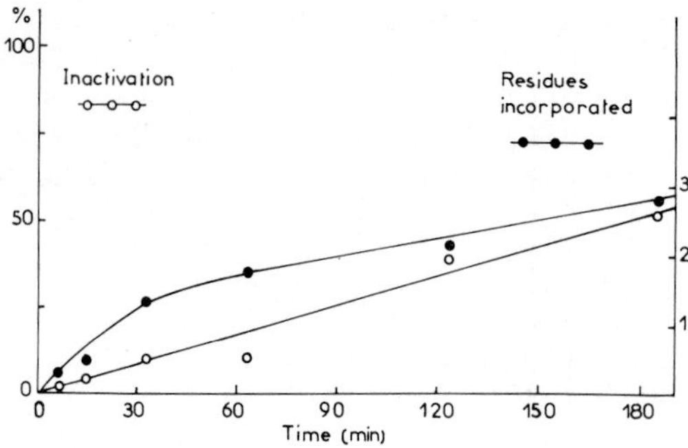

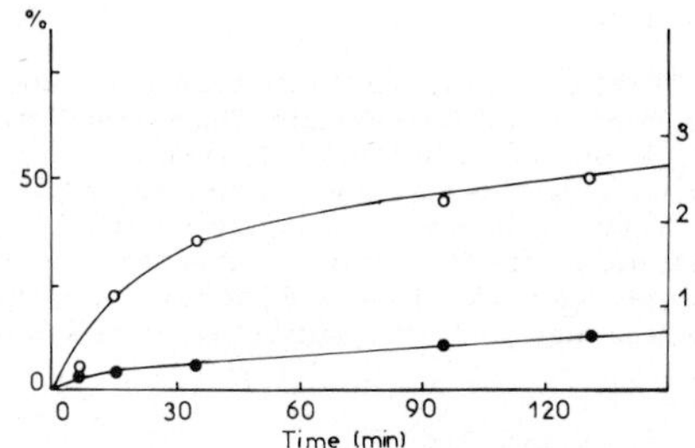

Fig. 4. Influence of sulfite on labelling. Top: labelling in the presence of sulfite. 0.135 μmol of enzyme was pretreated with non radioactive bromopyruvate in the presence of lactate as described under Fig. 3. Inactivation was carried out at 0°C in 3.2 ml of 0.1 M phosphate buffer, 1 mM EDTA, pH 7, containing 32 μM enzyme, 0.4 mM sodium sulfite and 4.7 mM bromo-[2-^{14}C] pyruvate. Residual activity and incorporation were measured as described under Fig. 3. Bottom: labelling after previous treatment with bromopyruvate in the presence of sulfite. 0.11 μmol of enzyme was pretreated in the presence of lactate, then incubated for 30 min in the presence of 0.4 mM sulfite and 5 mM bromopyruvate in a total volume of 3 ml at 0°C. Excess reagents was removed by two precipitations with ammonium sulfate; inactivation was then carried out at 0°C in a total volume of 3 ml at enzyme and bromo-[2-^{14}C] pyruvate concentrations of 29 μM and 5 mM, respectively.

b_2. A cysteine group is labelled which is essential for the activity; this group could be the base which abstracts the substrate α-proton. In addition the reagent specifically alkylates a number of cysteine residues which appear to be unnecessary for activity.

(2) The reduced enzyme is not inactivated by bromopyruvate. An attractive explanation is that whenever the flavin becomes reduced, protonation of the nucleophile occurs (or in our case of the system of interactions responsible for activation of the supposedly essential SH group). In view, however, of the fact that some non-essential cysteines are also unreactive in the reduced enzyme, one cannot discard the hypothesis of a structural change which would also modify the accessibility of the active site. Similarly, the effect of sulfite can be ascribed to either of two effects: steric protection at the active site, or a differential conformational effect (although the flavin shows a reduced-like spectrum, the heme is still oxidized). It is interesting to note that simulation studies led Capeillère-Blandin et al. [1] to postulate the occurence of a conformational change during reduction of cytochrome b_2 by lactate.

Future experiments will be directed toward ascertaining some of the results presented in the last section of this paper (Fig. 4), finding out whether the other molecular form of cytochrome b_2 [16] reacts in the same way, and localizing the various reactive cysteine residues in the supposedly polyglobular enzyme [16, 17].

References

1 Capeillère-Blandin, C., Bray, R.C., Iwatsubo, M. and Labeyrie, F. (1975) Eur. J. Biochem. 54, 549–566; see also this volume
2 Lederer, F. (1974) Eur. J. Biochem. 46, 393–399
3 Walsh, C.T., Schonbrunn, A. and Abeles, R.H. (1971) J. Biol. Chem. 246, 6855–6866
4 Walsh, C.T., Krodel, E., Massey, V. and Abeles, R.H. (1973) J. Biol. Chem. 248, 1946–1955
5 Walsh, C.T., Lockridge, O., Massey, V. and Abeles, R.H. (1973) J. Biol. Chem. 248, 7049–7054
6 Walsh, C.T., Schonbrunn, A., Lockridge, O., Massey, V. and Abeles, R.H. (1972) J. Biol. Chem. 247, 6004–6006
7 Porter, D.J.T., Voet, J.G. and Bright, H.J. (1973) J. Biol. Chem. 248, 4400–4416
8 Appleby, C.A. and Morton, R.K. (1959) Biochem. J. 71, 492–499
9 Spyridakis, A., Naslin, L. and Labeyrie, F. (1971) Biochimie, 53, 195–207
10 Meloche, H.P. (1967) Biochemistry 6, 2273–2280
11 Somlo, M. and Slonimski, P. (1966) Bull. Soc. Chim. Biol. 48, 1221–1249
12 Baudras, A. (1965) Bull. Soc. Chim. Biol. 47, 1177–1201
13 Hinkson, J.N. and Mahler, H.R., (1963) Biochemistry 2, 216–220
14 Massey, V., Müller, F., Feldberg, R., Schuman, M., Sullivan, P.A., Howell, L.G., Mayhew, S.G., Matthews, R.G. and Foust, G.P. (1969) J. Biol. Chem. 244, 3999–4006
15 Müller, F. and Massey, V. (1969) J. Biol. Chem. 244, 4007–4016
16 Jacq, C. and Lederer, F. (1974) Eur. J. Biochem. 41, 311–320
17 Labeyrie, F. and Baudras, A. (1972) Eur. J. Biochem. 25, 33–40

Flavins and Flavoproteins, edited by T.P. Singer

CHAPTER 25

SOME PHYSICAL ORGANIC STUDIES DEALING WITH FLAVIN CATALYSIS

THOMAS C. BRUICE

Department of Chemistry, University of California, Santa Barbara, Calif. 93106 (*U.S.A.*)

Summary

A discussion of the anticipated reactions of flavins, based on the constituent parts of the 1,5-dihydroisoalloxazine nucleus, is presented. The importance of the hitherto ignored role of N(5)-N(10) orbital splitting is discussed. The conclusion is reached that flavin redox reactions involve both $1e^-$ and $2e^-$ processes in competition and that the nature of a particular redox reaction will depend primarily upon the free energy of formation of radical species from the substrate and the nucleophilicyt (polarizability) of the nucleophilic entity of the substrate. Examples of $1e^-$ redox reactions involve carbonium ion, quinone, ninhydrin, chloral and nitroxide reduction whereas the oxidation of $2RS^-$ to RSSR represents the only known $2e^-$ (covalent adduct) redox reaction of flavins. Carbanion oxidation may represent either $1e^-$ or $2e^-$ processes. Carbonyl group reduction is considered to represent $1e^- + 1e^-$ or alternatively $1e^- + 1H\cdot$ transfer from dihydroflavin. The possibilities of this reaction are considered in detail. Plausible mechanisms for oxygen insertion and bacterial chemiluminescence are presented for consideration.

Though an appreciation of the chemical nature of an entity as complex as the isoalloxazine structure can only be appreciated through experimentation and discussed in terms of extended molecular orbital calculations it is useful, nonetheless, to consider the anticipated chemistry of flavins on the basis of the constituent parts of the isoalloxazine structure. The 1,5-dihydroisoalloxazine structure may be dissected as shown (because of tenets of microscopic reversibility a discussion of the chemistry of the dihydro-form serves the mechanistic purposes of discussing the flavoquinone structure as well).

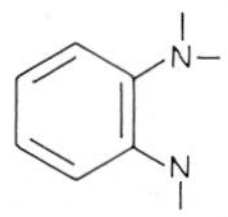

I 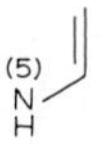 II 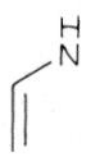 III 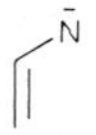 IV 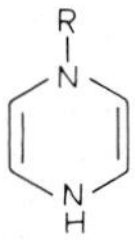 V

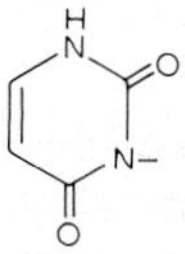

VI 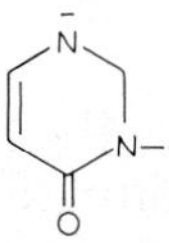 VII

Free radical reactions of *o*-phenylene diamines (I) are well appreciated. The characteristic reactions of enamines (II, III, IV) are their behavior as nucleophiles toward the proton [1, 2], and electrophilic centers in general (Eqn. 1), and their sensitivity to oxidants and susceptibility to N-alkylation [3]. The

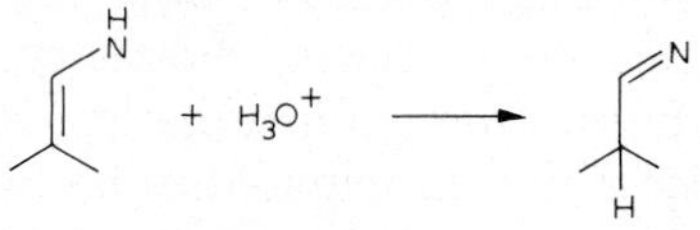

$$+\ H_3O^+ \longrightarrow \tag{1}$$

enamine-anion (IV) is anticipated to be a particularly effective nucleophilic entity leading to the prediction of nucleophilic $2\,e^-$ transfer reactions at the 4a- and 5-positions (Eqn. 2). The 1,4-dihydropyrazine structure (V)

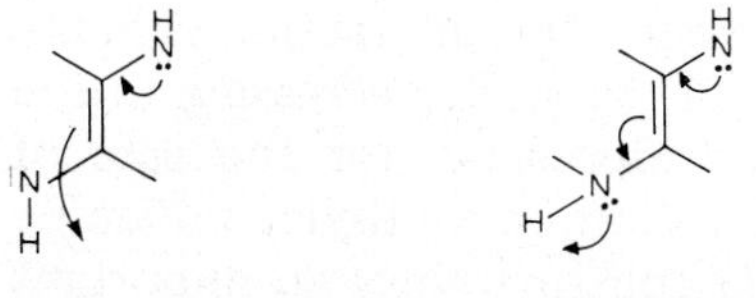

(2)

possesses unique chemical properties. 1,4-dihydropyrazines are known only in the form of their 1,4-dialkyl or diaryl derivatives and as such undergo ready sigmatropic rearrangements. An example is provided in the (1,3) sigmatropic shift with suprafacil allylic utilization of Eqn. 3 [4]. Hemmerich

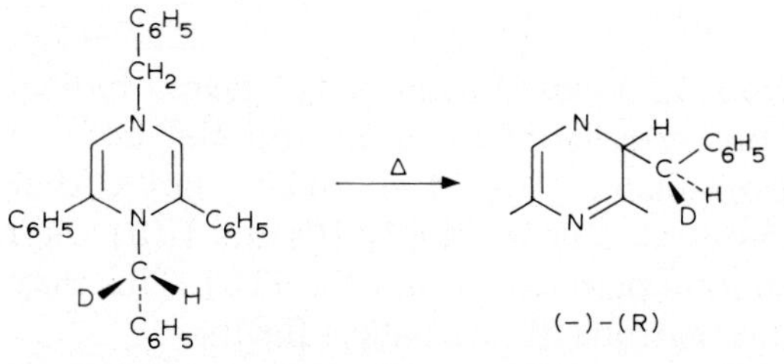

$$\xrightarrow{\Delta} \tag{3}$$

(−)-(R)

and coworkers provide examples of like rearrnagements involving N(5)-alkylated flavins [5, 6]. The 1,4-dihydropyrazines also possess potential anti-aromaticity (VIII), homo-aromaticity (IX) [7, 8], and N(5) vs. N(10) orbital–orbital splitting (X) [9–11]. The anti-aromatic state, which involves

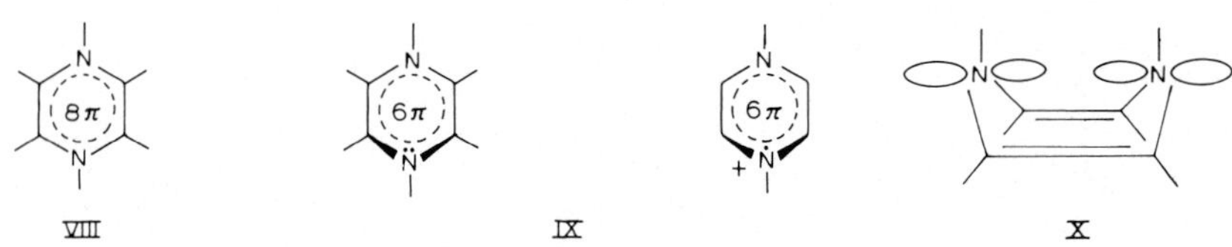

the four π electrons of the two C—C double bonds plus the unshared electron pairs of the two nitrogens would destabilize the dihydropyrazine ring. To obviate this destabilization 2 π electrons are transferred to separate antibonding orbitals to provide diradical character. This feature would provide a means for direct covalent bond formation with triplet states as in the case of ground state O_2 and lower $\Delta G^{\ddagger}$ for dihydropyrazine radical formation. Anti-aromaticity should come into play only when the dihydropyrazine ring is planar. The planarity is critical depending upon the out-of-plane bending angle (α) by $(\cos \alpha)^2$. Since 1,5-dihydroflavins normally possess a butterfly conformation, VIII is probably of little significance in flavin chemistry, but could contribute to the enzymology of flavins where conformation is controlled by the apoenzyme. Homo-aromaticity is not anticipated to be of significance in the chemistry of dihydropyrazines since both neutral and radical cation forms should receive equal stabilizing contributions from this feature.

Orbital–orbital interaction in ethylene diamines occurs both through space and through σ-bonds [9] and results in the splitting of the nitrogen lone-pair orbitals so that one exists at high energy. The interaction is maximal when the orbitals are equatorial, equatorial (as in X) decreasing in the order a,e and a,a. Orbital–orbital splitting is responsible for the ease of formation and stability ($t_{1/2}$ s) of the radical cation of triethylene diamine (XI) [12]. Without doubt, this orbital–orbital splitting plays an important role in the ease of formation and stability of flavin radicals.

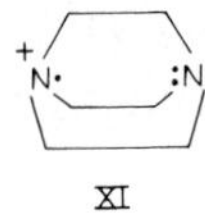

Both orbital–orbital interaction and anti-aromaticity favor radical formation. The former requires a butterfly conformation and the latter a planar conformation of the dihydropyrazine ring. It should be noted that in simple N(5)-alkyl substituted 1,5-dihydroflavins the N(10) and N(5) alkyl groups are axial so that the nitrogen lone pairs are equatorial [13]. This may well explain the stability of the radical cations of N(5)-alkyl flavins.

Lastly, one must take into consideration the uracil portion of (VI and VII) of the isoalloxazine structure. Uracils are known to undergo hydroxylation at both the 1- and 5-positions (for the latter a display of its enamine nature) [14]. These positions are equivalent to the 1- and 4a-positions of the isoalloxazine ring.

What the problems are

The foregoing considerations should lead one to anticipate that flavins should undergo reactions by both $1\,e^-$ transfer and by covalent bond formation which may, through a subsequent elimination or nucleophilic displacement, result in overall $2\,e^-$ transfer. In short, flavins should enter into oxidation reduction reactions which involve both radical and covalent pathways. The problem of mechanism is simply a decision as to which type mechanism is involved in each particular redox reactions. Because flavin radicals are rather stable, the reaction pathway (radical or covalent) should depend to a considerable degree upon: (1) the standard free energy (ΔG°) of formation of radical species of the substrate; and (2) the polarizability of the nucleophilic entity of the substrate.

Examples of 1 e⁻ transfer reactions not involving metal ions

One electron transfer from dihydroflavin or its anion (FlH_2 and FlH^- p$K_a \approx 7.0$) to substrate to yield a radical pair (4a) might be anticipated to be followed by either $1\,e^-$ (4b) or $H\cdot$ (4c) transfer to complete a $2\,e^-$ reduction process (Eqn. 4c). Both $1\,e^-$ and $H\cdot$ transfer reactions are

$$FlH^- + S \rightleftarrows FlH + S\cdot \tag{4a}$$

$$FlH\cdot + S\cdot \rightarrow Fl_{ox} + S^{(-)} + H^+ \tag{4b}$$

$$FlH\cdot + S\cdot \rightarrow Fl_{ox} + SH \tag{4c}$$

ubiquitous and diarylamines are particularly good sources of $H\cdot$ [15]. The mechanism of (5) would appear likely.

N—CH₃ O⁻ S· H S O SH O

(5)

The reaction of malachite green with $FlH_2 + FlH^-$ has been found to be second order in malachite green ($v = 2.6\,[FlH_2]\,[MG]^2\,s^{-1}$ at pH 5) [16] in accordance with a reaction involving consecutive $1\,e^-$ transfer.

$$FlH^- + \text{—}\overset{|}{\underset{|}{C}}{}^+ \rightleftarrows FlH\cdot + \text{—}\overset{|}{\underset{|}{C}}\cdot$$

$$FlH\cdot + \text{—}\overset{|}{\underset{|}{C}}{}^+ \rightleftarrows Fl_{ox} + \text{—}\overset{|}{\underset{|}{C}}\cdot + H^+ \tag{6}$$

$$2\text{—}\overset{|}{\underset{|}{C}}\cdot \rightarrow \text{Product}$$

In Eqn. 6 the rate of the reaction of $FlH\cdot$ with $\text{—}\overset{|}{\underset{|}{C}}{}^+$ exceeds that for $H\cdot$ transfer from $FlH\cdot$ to $\text{—}\overset{|}{\underset{|}{C}}\cdot$.

N(5)-methyl-1,5-dihydroflumiflavin (FlMeH + $FlMe^-$; XII) is oxidized to the corresponding flavinium radical by benzoquinone at a rate exceeding that for the comproportionation reaction of Fl^+Me_{ox} + (FlMeH + $FlMe^-$) → 2 FlMe· [16]. The radical, FlMe·, undergoes further reaction with benzoquinone to yield $FlMe_{ox}$ (Eqn. 7). Naphthoquinone reacts with FlMeH

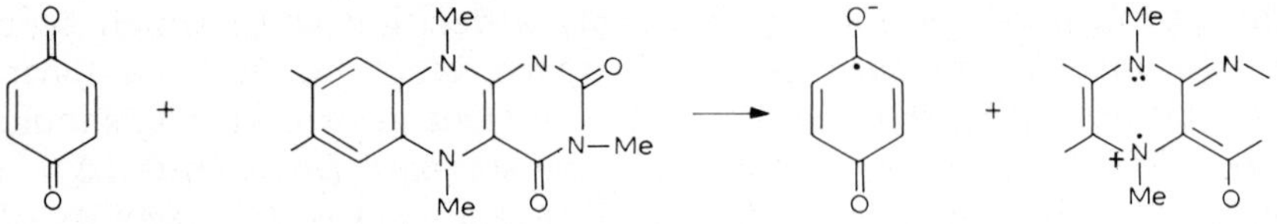

(7)

XII

and $FlMe^-$ with the same rate constant ($4 \cdot 10^4\ M^{-1} \cdot s^{-1}$) to yield FlMe· which is unreactive with naphthoquinone. These results suggest that quinone reductions by FlMeH and $FlMe^-$ follow consecutive 1 e^- transfer reactions (Eqn. 8). For FlH_2 + FlH^- reduction of quinones no radical intermediates

$$FlH^- + Q \underset{k_{-1}}{\overset{k_1}{\rightleftharpoons}} FlH\cdot + Q\cdot \rightarrow Fl_{ox} + HQ \tag{8}$$

may be detected [17]. This would be in accord with the intermediate flavin radical being present at steady state. The stability of the FlMe· radical far exceeds that of FlH· allowing the former's detection. Triketohydrindane hydrate (ninhydrin) is known to form radical species in solution [18, 19]. Ninhydrin reacts with both FlMeH and $FlMe^-$ to yield one FlMe· species for each ninhydrin species (Eqn. 9). This reaction must represent a 1 e^- transfer

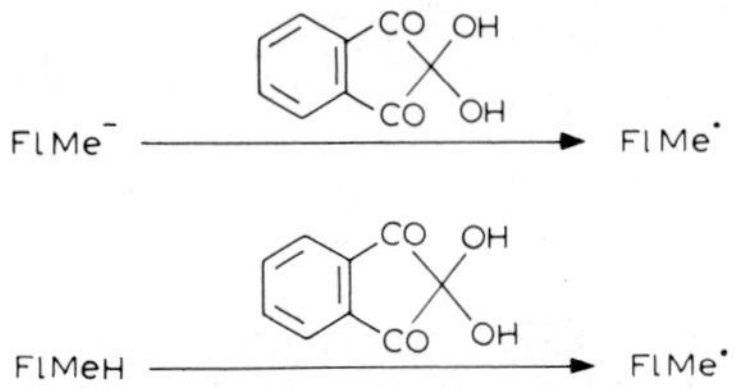

$FlMe^-$ → $FlMe^\cdot$

FlMeH → $FlMe^\cdot$ (9)

from dihydroflavin to ninhydrin for if FlMe· arose via a 2 e^- transfer followed by comproportionation of product $FlMe_{ox}$ and starting $FlMe^-$ and/or FlMeH it would have required one ninhydrin molecule to yield two FlMe· species. The aldehyde chloral is reduced by FlMeH ($k_v = 1.5 \cdot 10^{-5}$ $M^{-1} \cdot min^{-1}$) and $FlMe^-$ ($k_v = 1.05 \cdot 10^{-3}\ M^{-1} \cdot min^{-1}$) yielding FlMe·. That this reaction is also radical in nature is shown by the fact that the 2 e^- reduction product of chloral (2′,2′,2′-trichloroethanol) is not formed.

Reaction of dihydroflavins with O_2

The reaction of FlH_2 and FlH^- with O_2 provides both H_2O_2 and O_2^-. The reaction has been postulated to involve both 4a- and 10a-covalent adducts [20]. That the reaction is not initiated by formation of a 10a-adduct is shown by the observation that (in the presence of superoxide dismutase and Fl_{ox} species) the reaction constants for O_2 oxidation of XIII and XIV are comparable (Williams, R. and Bruice, T.C., unpublished results). Approach to the 10a-position of XIV is sterically hindered so that if 10a

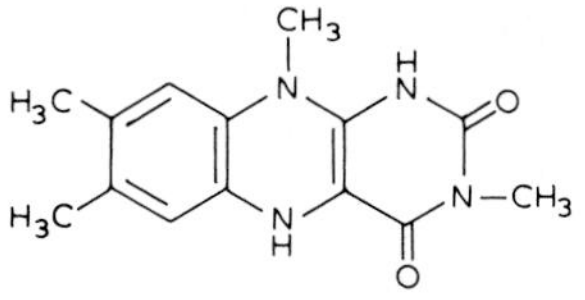

XIII

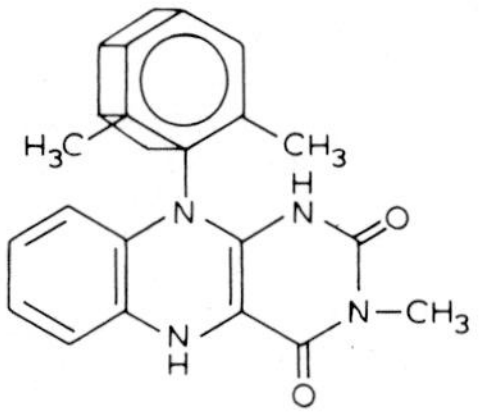

XIV

attack were predominant the rate of O_2 reaction with XIII should greatly exceed the rate of reaction with XIV. The extent of steric hindrance at the 10a-position of XIV is attested to by the fact that the HO^- catalyzed hydrolysis of its oxidized form is reversible (Eqn. 10) [21]. Compound XV is an imine of a urea which might normally be anticipated to fall apart in water.

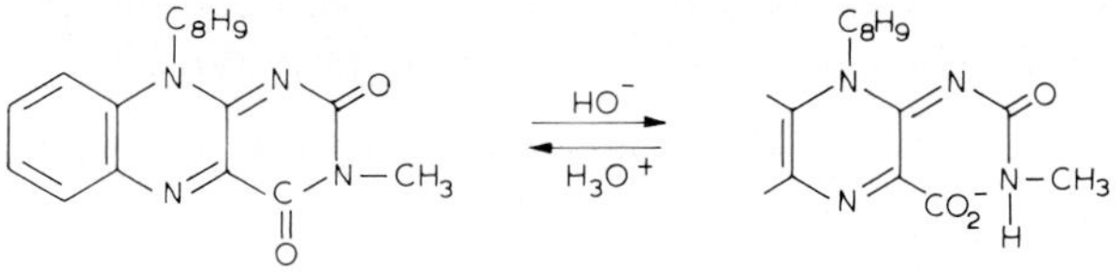

XIV a $\underset{H_3O^+}{\overset{HO^-}{\rightleftharpoons}}$ XV (10)

Hemmerich has considered the 6,8 [22] and 9a-positions (p. 9) at various times, as those which support O_2 addition. That the 6- and 8- positions are not involved as a general phenomenae may be shown by the fact that reaction of XVI with HO_2^- yields the 8-hydroxy flavin XVII while cyclic reduction of XVI and airation does not yield XVII (Eqn. 11) [23].

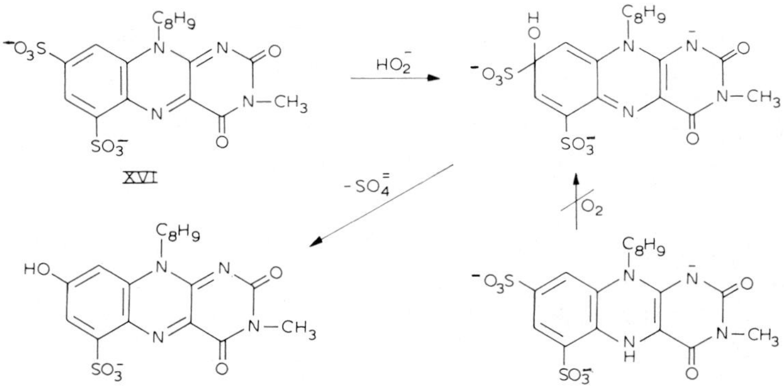

XVI $\xrightarrow{HO_2^-}$; $-SO_4^=$; $\not\uparrow O_2$; XVII (11)

A logical sequence for the direct formation of H_2O_2 and O_2^- on reaction of reduced flavin with O_2 is shown in Eqn. 12. The involvement of an

$$(12)$$

oxetane structure has been suggested by Orf and Dolphin [24]. The $O_2^{\overline{\cdot}}$ and H_2O_2 species then go on to react with FlH_2 and FlH^- etc. In order to shed light upon this complex sequence of events we have chosen to employ nitroxides as "monodentate" models for the "bidentate" oxygen triplet state (Chan, T.W. and Bruice, T.C., unpublished results). The analogy in structure between a nitroxide and triplet oxygen is obvious: ($\geq$N—O·, ·O—O·). In our investigations to date, the kinetics of reaction of a number of nitroxide molecules with flavins, substituted isoalloxazines and XII have been investigated. The reaction of nitroxides with ($FlH_2 + FlH^-$) to provide Fl_{ox} as product follows a $FlH_{2(total)} \xrightarrow{k_1'} X \xrightarrow{k_2'} Fl_{ox}$ kinetic path where in k_1' and

XVII + XIII → XVIII → X → ... → + 2 >N—OH (13a)

XVII + XIII → XVIII

XVIII → >N—OH + ... (13b)

2 ... → Fl_{ox} + XIII

k_2' are apparent first-order rate constants at constant nitroxide concentration ($k_1' = k_1$ [nitroxide]; $k_2' = k_2$ [nitroxide]). These results are in accord with the mechanisms of Eqn. 13 (written for FlH_2 but pertaining also to FlH^-). The rate of reaction of nitroxide with XII, yielding the blue radical FlMe·, has been found to be first order in nitroxide and first order in XII and independent of pH (pH 4.50 to 8.90). The reaction of Eqn. 14 serves as an additional example of a 1 e^- transfer from a dihydroflavin. Studies of the reaction of nitroxides with dihydroflavins is continuing and is being extended to flavoprotein oxidases.

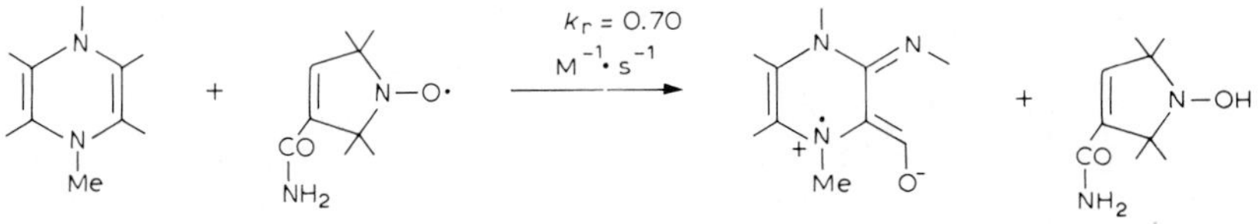

k_r = 0.70 $M^{-1} \cdot s^{-1}$ (14)

XII

The means by which oxygen is inserted into aromatic compounds by external flavomonooxygenases is an intriguing problem. Both Hamilton [25] and Hemmerich (p. 9) have proposed oxene mechanisms which involve the opening of the 4a (Hamilton) and 9a (Hemmerich) positions of the isoalloxazine ring on each turnover of enzyme. I would doubt that an enzymatic reaction would involve such disruptions of the ring system of a cofactor and suggest the more simplistic mechanism of Eqn. 15. The $-CO_2OH$ moiety could be generated from either an aspartic acid or glutamic

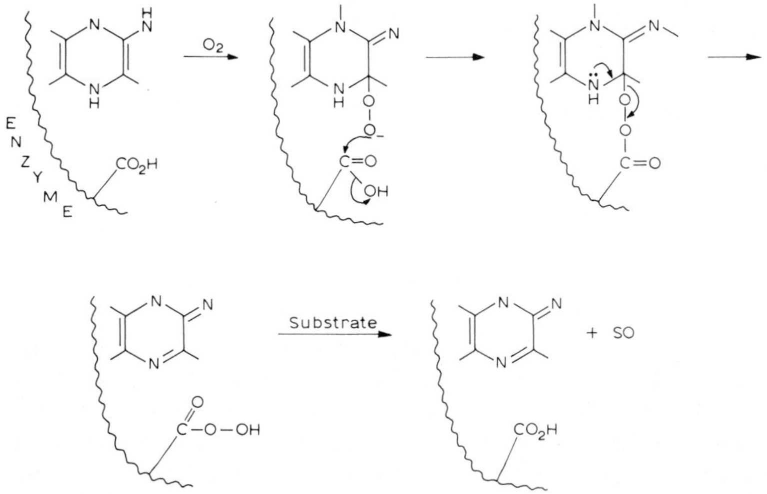

(15)

acid residue and it is well known that peracids are excellent oxene reagents [26]. This mechanism predicts that ^{18}O of $^{18}O_2$ should be incorporated into the carboxyl side chain of the enzyme after turnover.

A reasonable mechanism, and one which does not involve opening of the isoalloxazine ring, may also be written for the mechanism of light emission by bacterial lucipherase. The sequence of reactions of Eqn. 16 takes into account the probable intermediary of a 4a-peroxyadduct of FlH_2 [29]

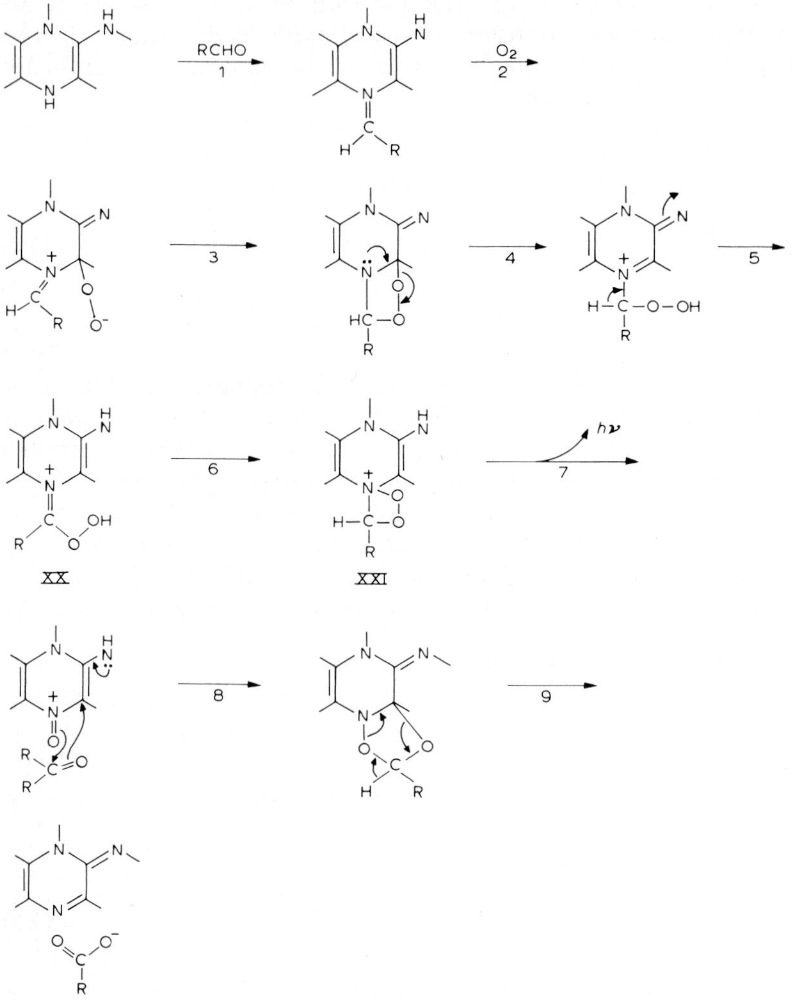

(16)

as well as the oxidation of aldehyde substrate to a carboxylic acid. For Eqn. 16 steps 1 to 3 involve simply the formation of a 4a-peroxy adduct of an imine of FlH_2. Step 3 should be highly exothermic being akin to intramolecular (a rate enhancing factor) formation of carbinolamine but involving the great nucleophilicity (α-effect) of a peroxyanion [30]. Step 5 has precedence in the base catalyzed elimination of CH_2O from $FlMe^{+}_{ox}$ (see p. 257). The product of step 5 (XX) is an imidate of a peracid and as such would be a powerful oxene (source of the HO^+ moiety) due to the resonance contribution of the $=\overset{+}{O}{-}OH$ canonical form. The intermediate XX could enter

(17)

into an oxidation or fragmentation producing light or in its cyclic form (XXI) undergo homolytic fragmentation to provide light via an excited

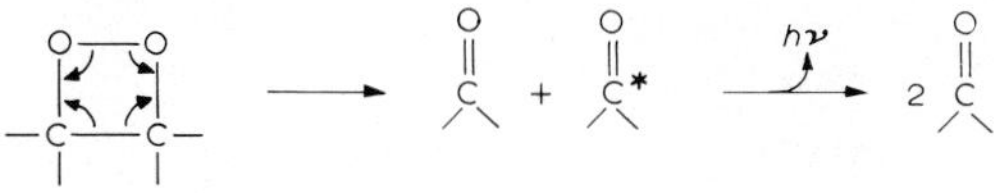

(18)

aldehyde moiety as is the case for dioxetane decomposition (Eqn. 18) [31–34]. Steps 8 and 9 involve the oxidation of aldehyde to carboxylic acid by the N-oxide of Fl_{ox} (Eqn. 19). In Eqn. 19, step a involves simple nucleophilic attack upon a carbonyl group, steb b has precedence [6] in the pseudo

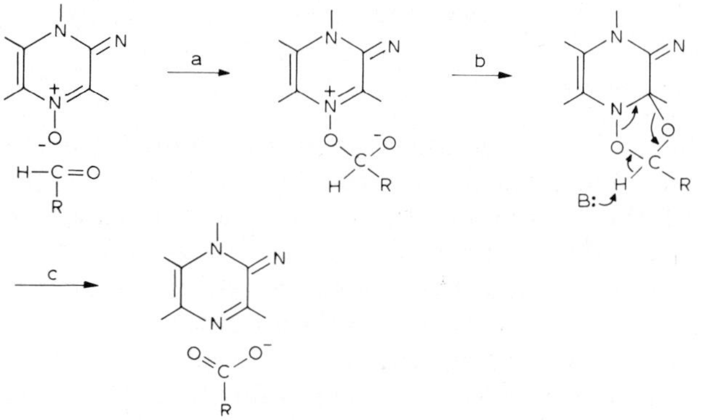

(19)

base equilibria of $FlMe_{ox}^{+} + HO^{-} \rightleftarrows FlMeOH$ while step c should be exothermic. I believe the possibilities of the N-oxide of oxidized flavin being an oxidant are intriguing and we are presently involved in its investigation. The mechanism of Eqn. 16 has precedence in studies of McCapra [35] and Rauhut [36] (Eqn. 20).

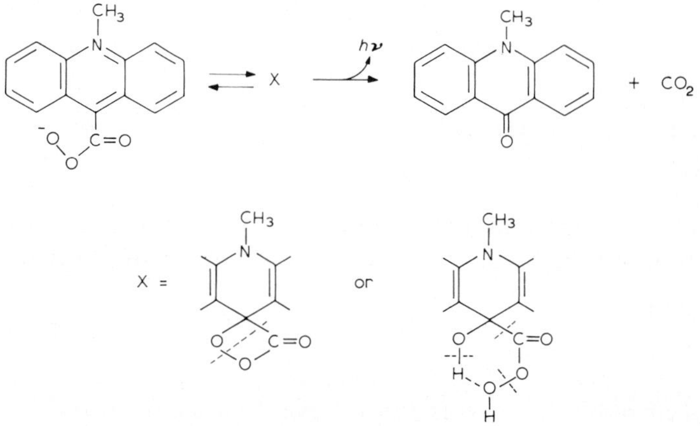

(20)

Covalent adducts

A means for differentiating between rate determining N(5)- and 4a-additions has been provided via kinetic and thermodynamic studies of SO_3^{2-} addition to XVI [37]. Addition to the N(5)-position involves

XVI $\underset{k_{gb}[SO_3^{2-}]}{\overset{k_{ga}[HSO_3^-][SO_3^{2-}]}{\rightleftharpoons}}$ (4a-adduct, SO_3 at 4a, N(5)H) — λ_{max} 304 nm shoulder 360-365 nm; Singlets 112 Hz (3H), 137 Hz (3H)

$k_n[SO_3^{2-}]$ ⇅ k_{-n} ; K_e ⇅ $4.12 \cdot 10^{-2}$ (21)

(N(5)-adduct anion, SO_3 at N(5)) $\underset{-H^+}{\overset{+H^+}{\rightleftharpoons}}$ (N(5)-adduct, protonated) — $pK_a = 3.21$ — λ_{max} 307 nm; Singlets 115 Hz (6H)

unassisted nucleophilic attack while 4a-addition of SO_3^{2-} required general-acid catalysis (by HSO_3^-) and in accord with the tenets of microscopic reversibility, dissociation of the 4a-adduct requires general-base catalysis. Attack of a nucleophile upon the 4a-position results in a great increase ($\approx 10^{17}$) in the basicity of the nitrogen at the 5-position. Protonation in the transition state by a general-acid is expected to greatly lower the free energy content of this state and, therefore, increase the rate constant. The lack of requirement for general acid catalysis of N(5) addition is due to the fact that the 5-adduct is stable as its anion (Eqn. 21).

The only known flavin redox reaction known to involve 4a-addition involves the oxidation of thiols by Fl_{ox} (Eqn. 22) (ref. 38 and Loechler, E. and Hollocher, T., unpublished results). From a steady state assumption in 4a-adduct:

$$RS^- + Fl_{ox} \underset{k_{gb}[B]}{\overset{k_{ga}[BH]}{\rightleftharpoons}} \text{(4a-SR adduct)} \xrightarrow{k_n[RS^-]} RSSR + FlH_2 \qquad (22)$$

$$\frac{d[RSSR]}{dt} = \frac{k_{ga}k_b[RS^-]^2[Fl_o][BH]}{k_{gb}[B] + k_n[RS^-]} \qquad (23)$$

Rate determining adduct formation should conform to the expression of (24) while rate controlling attack of RS^- upon the adduct would provide

$$v = k_{rate}[RS^-][Fl_o][BH]$$
$$k_{rate} = k_{ga} \qquad (24)$$

Eqn. 25. The kinetics of monothiol oxidation follows Eqn. 25 while dithiol

$$v = k_{rate}[RS^-][RSH][Fl_o]$$
$$k_{rate} = \frac{k_{ga}k_n K_{a_{RSH}}}{k_{gb} K_{a_{BH}}} \qquad (25)$$

oxidation follows the rate low of Eqn. 24. It is most reasonable that k_n for a dithiol, as lipoic acid, should exceed k_n for a monothiol (intramolecular vs. bimolecular attack).

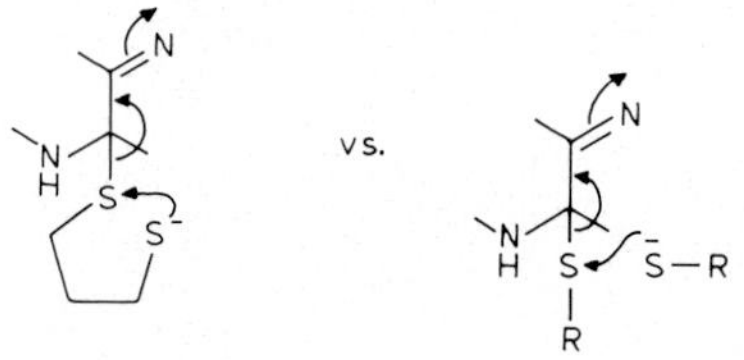

(26)

The reactions of Eqns. 27, 28 and 29 have been shown to involve carbanion species as immediate substrates. For Eqns. 27 [39] and 28 (ref. 40 and Taulane, J. and Brucie, T.C. unpublished results) the reactions may

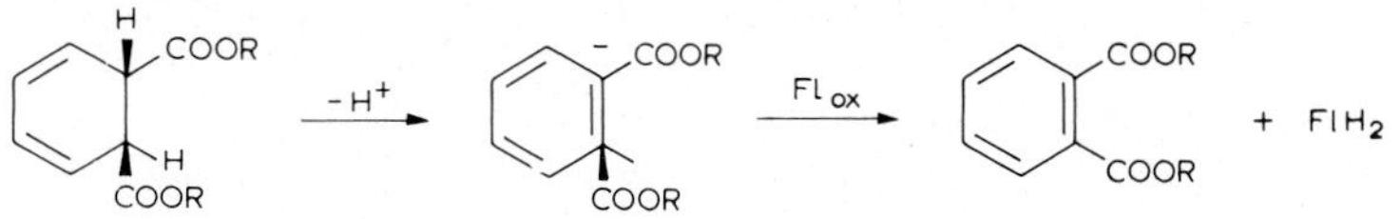

(27)

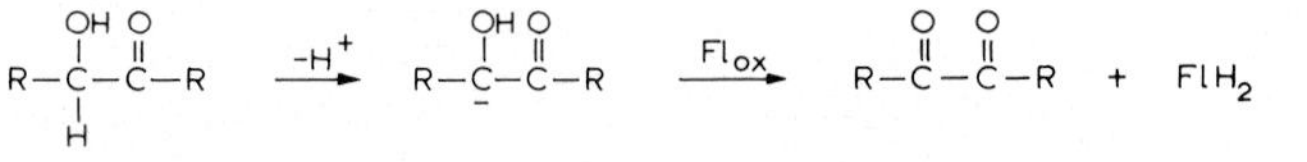

$$R{-}C(OH)(H){-}C(=O){-}R \xrightarrow{-H^+} R{-}\bar{C}(OH){-}C(=O){-}R \xrightarrow{Fl_{ox}} R{-}C(=O){-}C(=O){-}R + FlH_2 \qquad (28)$$

$$R{-}CH(R){-}NO_2 \xrightarrow{-H^+} R{-}\bar{C}(R){-}NO_2 \xrightarrow{Fl_{ox}} R{-}C(=O){-}R + NO_2^- + FlH_2 \qquad (29)$$

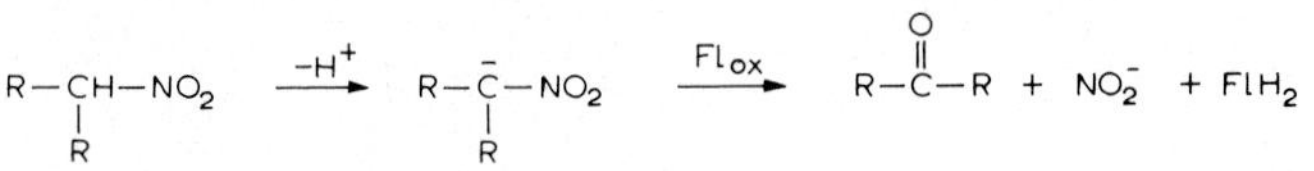

be carried out under conditions where the rate determining step is carbanion formation or oxidation by Fl_{ox}. The reaction [38] of Eqn. 29 was studied under condtiions where carbanion and carbon acid were allowed to equilibrate. In no case was general-acid catalysis or apparent specific acid catalysis required. Formation of a 4a-adduct does not seem likely for these reactions. Porter et al. [41] have established a N(5)-adduct in the oxidation of nitroalkene anion by D-amino acid oxidase (p. 225). This result may be interpreted to favor nitroalkane carbanion attack upon the N(5)-position. However, it must be recalled that carbanions, and in particular nitroalkane anions, undergo alkylation by free-radical mechanisms [43–46] so that it is of equal possibility in carbanion oxidations by Fl_{ox} to have a radical mechanism (Eqn. 30), which may or may not involve covalent adducts.

$$-\bar{C}- + Fl_{ox} \rightleftarrows -\dot{C}\cdot\cdot Fl^- \rightleftarrows \text{Adduct} \rightarrow \text{Product} + FlH_2 \qquad (30)$$

Once formed, 4a- and 5-carbon adducts at either the FlH_2 or Fl_{ox} level of oxidation solvolyze to products. Decomposition of XXII and XXIII occur via acid catalyzed hydrolysis and in the case of XXII but not XXIII general-base catalyzed hydrolysis [47]. The fact that general-base catalysis is seen

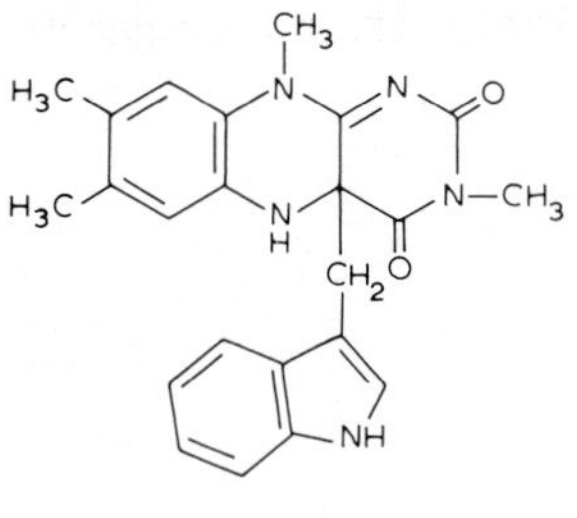

XXII

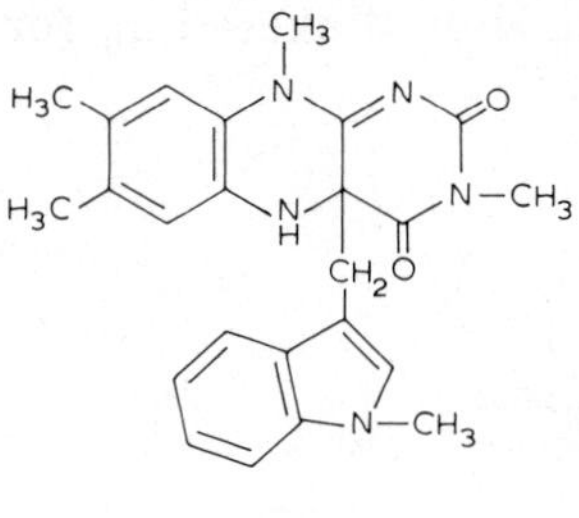

XXIII

only with XXII indicates an E-2 or E_1cB elimination reaction (Eqn. 31). To fit the log k_{obs} vs. pH profile it is required that the 4a- and 5-adducts are in

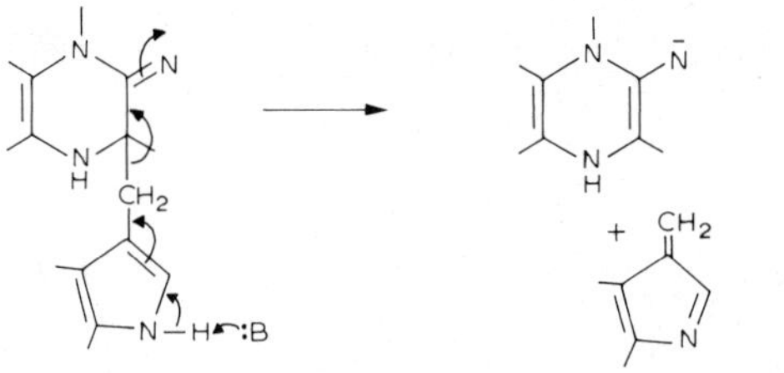

(31)

Brønsted β = + 0.5
k_{rate} for HO^- = 112 $M^{-1} \cdot min^{-1}$

equilibrium (Eqn. 32) providing the 5-isomer of XXII (i.e., XXII-5) and XXIII (i.e., XXIII-5). Compound XXII-5 but not XXIII-5 undergoes general-base catalyzed elimination. This establishes that base catalysis of elimination is not restricted to the 4a-adduct but is shared by the 5-adducts as well.

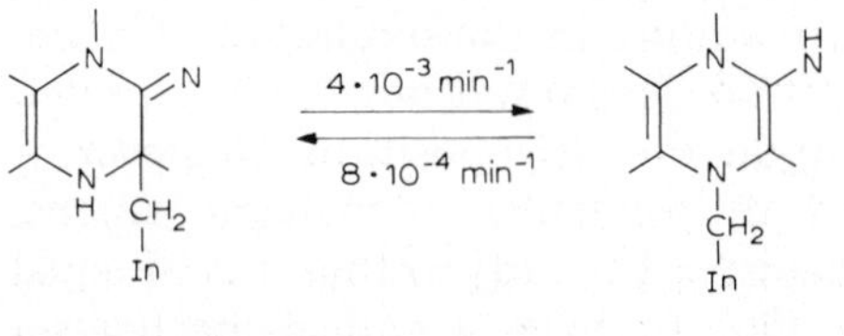

$4 \cdot 10^{-3}\ min^{-1}$
$8 \cdot 10^{-4}\ min^{-1}$

(32)

Compounds XXII, XXIII, XXIII-5 and XXIII-5 are readily oxidized to Fl_{ox}. Elimination of an alkyl substituent on N(5) of Fl_{ox} is a facile general-base catalyzed reaction (Eqn. 33) [48].

(33)

Brønsted β = ≈ 0.55
k_{gb} for B: = H_2O is $6.2 \cdot 10^{-6}$ $M^{-1} \cdot s^{-1}$
k_{gb} for B: = HO^- is ≈ 10^5 $M^{-1} \cdot s^{-1}$

The reduction of carbonyl groups and the oxidation of alcohols

In arriving at a general mechanism for the reaction, reduced flavin + >C=O ⇄ oxidized flavin + alcohol, the following features must be taken into consideration. The reaction of FlH_2 + FlH^- with aldehydes and ketones (CH_2O, CH_3—CO—CO_2Et, CH_3COCO_2H etc.) is characterized by an initial burst production of Fl_{ox} followed by a much slower formation of Fl_{ox} [49]. If in the second phase of the reaction additional carbonyl compound is added there does not ensue a second burst of Fl_{ox} formation. This observation establishes that all FlH^- and FlH_2 have reacted. In addition when borohydride is added following the initial burst and the solution oxygenated, the blue FlR· radical is obtained. This result establishes that the remaining dihydroflavin is present as an N(5) imine or carbinolamine. These results are in accord with the kinetic scheme of Eqn. 34. The rate constant for the initial burst is provided by $(k_1 + k_3)$ while the rate determining step for the

$$\mathrm{C{=}O} + \mathrm{FlH_2} \underset{k_2}{\overset{k_1}{\rightleftharpoons}} \mathrm{Adduct(s)} \quad \xrightarrow{k_3 \ (\text{from } \mathrm{FlH_2})} \quad \mathrm{Fl_{ox}} + \mathrm{HC{-}OH} \tag{34}$$

second phase is k_2. The burst reactions associated with reduction of CH_2O and CH_3COCO_2H by FlH_2 and FlH^- are specific acid catalyzed so that k_1 and k_3 represent acid catalyzed pathways. In the reverse reaction (i.e., H—C(OH)< + Fl_{ox} → >C=O + FlH_2) the carbon acid must be the direct substrate for CH_3OH oxidation while it is known (Eqn. 28) that for the oxidation of α-ketols by Fl_{ox} the carbanion [i.e., R—$\bar{C}$—(OH)—CO—R] is the substrate. Other considerations which should be taken into account are the observed $1e^-$ transfer reactions from 1,5-dihydro-5-methyllumiflavin to quinones, ninhydrin and chloral (p. 248) and the fact that the values of the calculated standard (standard state 1M) free energies of formation of the radical species obtained by $1e^-$ transfer from FlH_2 and FlH^- to $\gt C{=}O$ and $\gt C{=}\overset{+}{O}H$ (where $\gt C{=}O$ is H_2CO, $CH_3COCONH_2$ etc.) are less positive than the experimental $\Delta G^\ddagger$ values for the reduction of $\gt C{=}O$ by FlH^- and FlH_2 [50]. These two considerations (i.e., FlH^- and FlH_2 undergo $1e^-$ transfer to carbonyl compounds, and ΔG° of formation of radical pairs from FlH_2 and FlH^- by $1e^-$ transfer to aldehydes and ketones is less than $\Delta G^\ddagger_{\text{exptl.}}$) require that radical mechanisms be given serious consideration.

Any mechanistic scheme for carbonyl reduction through radical intermediates must take into account: (1) acid catalysis is required for certain reductions; and (2) when resonantly stabilized the product is a carbanion otherwise carbon acid is the direct product. The following scheme is offered (written for FlH^- but applying to FlH_2) [51]. The standard free energies

$$\begin{array}{ccc} \text{(A)}\quad FlH^- + \gt C{=}O & \underset{-H^+}{\overset{+H^+}{\rightleftharpoons}} & FlH^- + \gt C{=}\overset{+}{O}H \quad \text{(C)} \\ \updownarrow & & \updownarrow \\ \text{(B)}\quad FlH\cdot + \cdot\overset{|}{\underset{|}{C}}{-}\bar{O} & \underset{-H^+}{\overset{+H^+}{\rightleftharpoons}} & FlH\cdot + \cdot\overset{|}{\underset{|}{C}}{-}OH \quad \text{(D)} \end{array} \qquad (35)$$

$\Delta G_{A\to C}$, $\Delta G_{A\to B}$ and $\Delta G_{A\to D}$ may be calculated from the pK_a of $\gt C{=}\overset{+}{O}H$, the pH and the values of E° for formation of flavin radical plus $\cdot\overset{|}{\underset{|}{C}}{-}O^-$ and flavin radical plus $\cdot\overset{|}{\underset{|}{C}}{-}OH$. This provides the free energy levels of states B, C and D relative to A. Starting with A the reactants will proceed to that state with the lowest free energy content (i.e., B or C). Since $\Delta G_{A\to B}$ generally exceeds $\Delta G_{A\to D}$ the following cases are most probable:

$\Delta G^\circ_{A\to C} > \Delta G^\circ_{A\to B} > \Delta G^\circ_{A\to D}$ Case I

$\Delta G^\circ_{A\to B} > \Delta G^\circ_{A\to C} > \Delta G^\circ_{A\to D}$ Case II

$\Delta G^\circ_{A\to B} \simeq \Delta G_{A\to C} > \Delta G^\circ_{A\to D}$ Case III

For the formation of products from state D the following mechanism would then be in accord with the requirement that the reactants proceed along that path of minimum free energy: for Case I, Eqn. 36 for Case II, Eqn. 37; and

$$FlH^- + \gt C{=}O \rightleftarrows FlH\cdot\cdot\overset{|}{\underset{|}{C}}{-}O^- \underset{-H^+}{\overset{+H^+}{\rightleftharpoons}} FlH\cdot\cdot\overset{|}{\underset{|}{C}}{-}OH \xrightarrow{b} Fl_{ox} + HC{-}OH \;;\quad \xrightarrow{a} Fl_{ox} + (-)\overset{|}{\underset{|}{C}}{-}OH + H^+ \qquad (36)$$

for Case III the concerted $1e^- + H^+$ mechanism of Eqn. 38. The terminal step of 36a, 37a and 38a represent $1e^-$ transfer to yield carbanion while the terminal step of Eqns. 36b 37b and 38b are suggested to be $H\cdot$ transfer to provide carbon acid. The reduction of $H_2C{=}O$ may be fit quantitatively [50] to the mechanism of Eqn. 37b with electron transfer from reduced

$$\text{FlH}^- + \rangle\text{C=O} \underset{-\text{H}^+}{\overset{+\text{H}^+}{\rightleftharpoons}} \text{FlH}^+ + \rangle\text{C=}\overset{+}{\text{O}}\text{H} \rightleftarrows \text{FlH}\cdot\cdot\overset{|}{\underset{|}{\text{C}}}\text{—OH} \quad \begin{cases} \xrightarrow{b} \text{Fl}_{ox} + \text{H—}\overset{|}{\underset{|}{\text{C}}}\text{—OH} \\ \xrightarrow{a} \text{Fl}_{ox} + (-)\overset{|}{\underset{|}{\text{C}}}\text{—OH} + \text{H}^+ \end{cases} \tag{37}$$

flavin to $\rangle\text{C=}\overset{+}{\text{O}}\text{H}$ rate determining. Products may also arise directly from

$$\text{FlH}^- + \rangle\text{C=O} \underset{k_{gb}[\text{A}]}{\overset{k_{ga}[\text{AH}]}{\rightleftharpoons}} \text{FlH}\cdot\cdot\overset{|}{\underset{|}{\text{C}}}\text{—OH} \quad \begin{cases} \xrightarrow{b} \text{Fl}_{ox} + \text{H}\overset{|}{\underset{|}{\text{C}}}\text{—OH} \\ \xrightarrow{a} \text{Fl}_{ox} + (-)\overset{|}{\underset{|}{\text{C}}}\text{—OH} + \text{H}^+ \end{cases} \tag{38}$$

state B by the mechanism of Eqn. 39. In Eqn. 39 general acid proton transfer

$$\text{FlH}^- + \rangle\text{C=O} \rightleftarrows \text{FlH}\cdot\cdot\overset{|}{\underset{|}{\text{C}}}\text{—O}^- \quad \begin{cases} \underset{k_{gb}[\text{A}]}{\overset{(b)\; k_{ga}[\text{HA}]}{\rightleftharpoons}} \text{Fl}_{ox} + \text{H—}\overset{|}{\underset{|}{\text{C}}}\text{—OH} \\ \underset{(a)\; k_{gb}[\text{A}]}{\overset{k'_{ga}[\text{HA}]}{\rightleftharpoons}} \text{Fl}_{ox} + (-)\overset{|}{\underset{|}{\text{C}}}\text{—OH} + \text{H}^+ \end{cases} \tag{39}$$

to O^- should be required because the $\text{p}K_a$ of $\cdot\overset{|}{\underset{|}{\text{C}}}\text{—OH}$ greatly exceeds that of $\text{H—}\overset{|}{\underset{|}{\text{C}}}\text{—OH}$ (same reasoning as for 4a-addition see p. 254, Jenck's libido rule [52]).

Acknowledgement

This work was supported by a grant from the National Science Foundation.

References

1 Coward, J.K. and Bruice, T.C. (1969) J. Am. Chem. Soc. 91, 5339
2 Sollenberger, P.Y. and Martin, B.M. (1970) J. Am. Chem. Soc. 92, 4261
3 Cook, A.G. (1969) Enamines: Synthesis, Structure and Reactions, Marcel Dekker, New York, N.Y.
4 Lown, J.W., Ahktar, M.H. and McDaniel, R.S. (1974) J. Org. Chem. 39, 1999
5 Brüstlein, M., Knappe, W.-R. and Hemmerich, P. (1971) Angew. Chem. Int. edn. 10, 804
6 Ghisla, S., Hartmann, U., Hemmerich, P. and Müller, F. (1973) Ann. Chem. 1388
7 Breslow, R. (1973) Acc. Chem. Res. 6, 393
8 Streitwieser, A. (1961) Molecular Orbital Theory for Organic Chemists p. 275, Wiley, New York, N.Y.
9 Hoffman, R. (1971) Acc. Chem. Res. 4, 1
10 Hoffman, R., Imanura, A. and Hehre, J.W.J. (1968) J. Am. Chem. Soc. 90, 1499
11 Heilbronner, E. and Muszhat, A.K. (1970) J. Am. Chem. Soc. 92, 3878
12 Razuvaev, G.A., Abukumov, G.A. and Pestunovich, V.A. (1964) J. Struct. Chem. 5, 274
13 Kierkegard, P., Norrestam, R., Werner, P., Csöregh, I., Glehn, M., Karlsson, R., Meijonmarck, M., Röhnquist, O., Stensland, B., Tillberg, O. and Torbjörnsson, L. (1971) Flavins and Flavoproteins (Kamin, H., ed.), p. 1, University Press, Baltimore, Maryland
14 Tam, J.N.S., Vaughan, G.N., Mertes, M.P., Rork, G.S. and Pitman, I.H. J. Am. Chem. Soc., in the press

15 Poutsma, M.L. (1973) Free Radicals (Kochi, J.K., ed.), vol. II, p. 122, Wiley-Interscience, New York
16 Bruice, T.C. and Yano, Y. (1975) J. Am. Chem. Soc. 97, 5263
17 Gibian, M.J. and Rynd, J.A. (1967) Biochem. Biophys. Res. Commun. 34, 594
18 Russell, G.A. and Young, M.C. (1966) J. Am. Chem. Soc. 88, 2007
19 Lagercrantz, C. and Yhland, M. (1963) Acta Chem. Scand. 17, 277
20 Massey, V., Palmer, G. and Ballou, D. (1973) Oxidases and Related Systems (King, T.E., Mason, H.S. and Morrison, M., eds.), p. 25, University Park Press, Baltimore, Maryland
21 Smith, S.B. and Bruice, T.C. (1975) J. Am. Chem. Soc. 97, 2877
22 Hemmerich, P. and Müller, F. (1973) Ann. New York Acad. Sci. 212, 13
23 Smith, S.B., Brüstlein, M. and Bruice, T.C. (1974) J. Am. Chem. Soc. 96, 3696
24 Orf, H.W. and Dolphin, D. (1974) Proc. Natl. Acad. Sci. U.S. 71, 2646
25 Hamilton, G.A. (1971) Progress in Bioorganic Chemistry (Kaiser, E.T. and Kezdy, F.J., eds.), vol. I, p. 83, Wiley-Interscience, New York
26 van Douran, B.L., Berkersley, I. and Lefar, M. (1964) J. Org. Chem. 29, 686
27 Boyland, E. and Sims, P. (1964) Biochem. J. 90, 391
28 Jerina, D.M., Daly, J.W. and Witkop, B. (1971) Biochemistry 10, 366
29 Hastings, J.W., Balny, C., LePeuch, C. and Douznu, P. (1974) Proc. Natl. Acad. Sci. U.S. 70, 3468
30 Bruice, T.C. and Benkovic, S.J. (1966) Bioorganic Mechanisms vol. I, Chapt. 1, W.A. Benjamin, New York, N.Y.
31 Kopecky, K.R. and Mumford, C. (1969) Can. J. Chem. 47, 709
32 Bartlett, P.D. and Schaap, A.P. (1970) J. Am. Chem. Soc. 92, 3223
33 Mazur, S. and Foote, C.S. (1970) J. Am. Chem. Soc. 92, 3225
34 Wieringa, J.D.H., Strating, J., Wynberg, H. and Adams, W. (1972) Tetrahedron Lett. 169
35 McCapra, F., Richardson, D.G. and Chang, Y.C. (1965) Photochem. Photobiol. 4, 1111
36 Rauhut, M.M., Sheehan, D., Clarke, R.A., Roberts, B.G. and Sensel, A.M. (1965) J. Org. Chem. 30, 3587
37 Bruice, T.C., Hevesi, L. and Shinkai, S. (1973) Biochemistry 12, 2083
38 Yoko, I. and Bruice, T.C. (1975) J. Am. Chem. Soc. 97, 450
39 Main, L., Kasperek, G.J. and Bruice, T.C. (1972) Biochemistry 11, 3991
40 Shinkai, S., Kunitake, T. and Bruice, T.C. (1974) J. Am. Chem. Soc. 96, 7140
41 Porter, D.J.T., Voet, J.G. and Bright, H.J. (1972) J. Biol. Chem. 247, 1951
42 Porter, D.J.T., Voet, J.G. and Bright, H.J. (1973) J. Biol. Chem. 248, 4400
43 Kerber, R.C., Urry, G.W. and Kornblum, N. (1965) J. Am. Chem. Soc. 87, 4520
44 Kornblum, N., Kerber, R.C. and Urry, G.W. (1964) J. Am. Chem. Soc. 86, 3904
45 Russell, G.A. and Dannen, W.C. (1966) J. Am. Chem. Soc. 88, 5663
46 Kerber, R.C. (1966) J. Am. Chem. Soc. 88, 5660, 5662
47 Clerin, D.C. and Bruice, T.C. (1974) J. Am. Chem. Soc. 96, 5571
48 Kemal, C. and Bruice, T.C. J. Am. Chem. Soc., in the press
49 Shinkai, S. and Bruice, T.C. (1973) J. Am. Chem. Soc. 95, 7526
50 Williams, R.F., Shinkai, S. and Bruice, T.C. (1975) Proc. Natl. Acad. Sci. U.S. 72, 000
51 Bruice, T.C. (1975) Models and Flavin Catalysis in Progress in Bioorganic Chemistry (Kaiser, E.T. and Kezdy, F.J., eds.), vol. 5, Wiley-Interscience, New York, in the press
52 Jencks, W.P. (1972) Chem. Rev. 72, 705

Flavins and Flavoproteins, edited by T.P. Singer

CHAPTER 26

FLAVIN-NICOTINAMIDE CHARGE-TRANSFER CC THEIR FUNCTION IN OXIDOREDUCTION

GUNTER BLANKENHORN

Fachbereich Biologie der Universität, Konstanz (G.F.R.)

Introduction

NAD(P)H-dependent oxidoreduction of flavoenzymes occurs in many, functionally quite different, biological processes (Table I). Starting with the early experiments by Haas [1] on old yellow enzyme, a still growing body of information became available demonstrating the existence of long wavelength-absorbing species arising from the interaction between reduced flavoproteins and $NAD(P)^+$ [2–10]. However, only some of these complexes have been shown to be catalytical intermediates [7], since others are formed too slowly to participate in catalysis [2]. Investigation of flavoenzyme kinetics with the stopped-flow method revealed the existence of yet another long wavelength-absorbing, transient species which was correctly interpreted as an oxidized flavoprotein-NADH complex by Strittmatter [3] and

TABLE I

$Nic^+_{ox} \cdot Fl_{red}H^-$ CHARGE-TRANSFER COMPLEXES: EXAMPLES OF LOW MOLECULAR WEIGHT AND PROTEIN-BOUND SPECIES

CT, Charge-transfer complex.

Donor	Acceptor	$\lambda_{max\,CT}$ (nm)	Ref.
Lumiflavin H^-	NMN^+	505	
$FMNH^-$	NAD^+	527	23
$Fl_{red}H^-(CH_2)_3 \cdot Nic^+_{ox}$		600	22
Lipoyl dehydrogenase	NAD^+	700	2
p-OH-benzoate hydroxylase	$NADP^+$	750	6
NADH-cytochrome b_5 reductase	NAD^+	600	3
NADH-rubredoxin-reductase	NAD^+	700	8
Electron transferase	NAD^+	750	9
NADH peroxidase	NAD^+	700	10

Abbreviations: Nic^+_{ox}, oxidized nicotinamide; $Nic_{red}H$, reduced nicotinamide; Fl_{ox}, oxidized flavin; $Fl_{red}H_2$, reduced flavin; NMNH, *N*-methyl-1, 4-dihydro-nicotinamide.

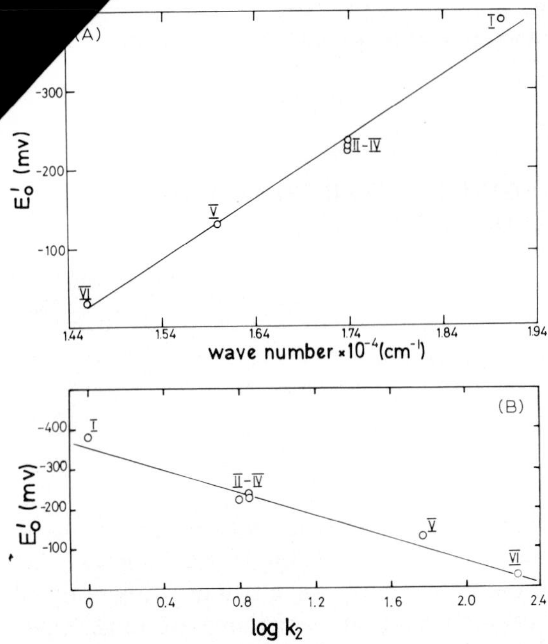

Fig. 1. (A) Plot of $h\nu_{CT}$ in NMNH • flavin charge-transfer (CT) complexes versus the flavin oxidation-reduction potential (pH 8.0). (B) Plot of log k_2 for flavin reduction versus the flavin oxidiation-reduction potential (pH 8.0). Numbers refer to those listed in Table II [35].

postulated to be of the charge-transfer type by Massey and Ghisla [34]. In the following years, analogous species were described for several other NADPH-dependent flavoenzymes by Massey's group and in each case reported so far, their formation and disappearance were sufficiently fast that they could be catalytic intermediates preceding oxidoreduction [6].

Since Singer and Kearney described nonenzymic oxidation of NADH by riboflavin [11], numerous model studies have been published aiming at a molecular interpretation of NAD(P)H dehydrogenation by flavoenzymes and low molecular weight flavins [12–23].

In contrast to the long wavelength-absorbing, stable complexes between $Fl_{red}H_2$ and Nic^+_{ox} which have been well characterized as charge-transfer complexes [24], long wavelength-absorbing complexes between $Nic_{red}H$ and Fl_{ox} have only recently been described in model systems and their nature has not as yet been established.

We have now been able to characterize complexes between *N*-methyl-1,4-dihydronicotinamide (NMNH) and flavins and elucidate the factors governing oxidoreduction.

Intermolecular flavin-nicotinamide charge-transfer complexes

Anaerobic reduction of flavins by NMNH, as investigated with the stopped-flow technique, revealed that prior to flavin reduction a long wavelength-absorbing intermediate was formed. The kinetics of its disappearance were the same as for flavin reduction, showing saturation at high NMNH concentrations. The charge-transfer character [25] of the long wavelength-absorbing intermediate was demonstrated by the linear relationship between the flavin oxidation-reduction potential, a substitute for electron affinity of flavin, and the energy of the long wavelength transition (Fig. 1A). Dissociation constants of Fl_{ox} · NMNH and $Fl_{red}H^-$ · NMN^+ complexes are in the order of 0.1 M (ΔG between 1 and 1.5 kcal/mol). Both species represent π-π donor-acceptor complexes showing a charge-transfer transition. It has to be pointed out that the charge-transfer interaction for this type of complex is not expected to contribute significantly to complex stability [26, 34]. Steric factors, however, should have a pronounced effect on complex stability. The validity of these predictions has been demonstrated for NMNH · Fl_{ox} charge-transfer complexes (Table II) [35].

Comparison with flavoprotein-nicotinamide charge-transfer complexes

Flavin-nicotinamide biscoenzymes linked covalently by two, three or four methylene groups through positions N-10 of the flavin and N-1 of the nicotinamide form long wavelength-absorbing, intramolecular complexes of the type described in the previous section. With three methylene groups the two molecule halves are held together in a most favourable conformation for charge-transfer interaction. The transition energies for this compound are minimal, whereas the transition intensities are maximal (Fig. 2; Tables I and III) [22].

Although optical properties of the long wavelength-absorbing $Nic_{red}H$ · Fl_{ox} complexes appear to be very similar to those of the Nic^+_{ox} · $Fl_{red}H^-$ complexes, there is one important difference: at physiological pH the former complexes are uncharged whereas the latter are present as ion pairs. Therefore, changes in polarity of the environment should shift the energy of the charge-transfer transition in different directions. Kosower [27] has shown that the energy of this transition in pyridinium iodides decreases dramatically with decreasing solvent polarity. This large effect was interpreted

TABLE II

ABSORBANCE MAXIMA AND DISSOCIATION CONSTANTS OF FLAVIN ANALOGUE · NMNH COMPLEXES [35]

First-order rate constants of flavin-analogue reduction versus the oxidation-reduction potential. CT, charge-transfer complex.

Flavin analogue	$\lambda_{max\,CT}$ (nm)	K_d (M)	k_2 (s^{-1})	Redox potential (pH 8, mV)
I 10-Methyl-deazaisoalloxazine	525	0.17	1.0	−380*
II Lumiflavin	575	0.10	7.0	−231
III FMN	575	0.13	7.0	−225
IV FAD	575	0.25	6.25	−223
V *S*-Methyl-2-thiolumiflavin	625	0.09	59	−127
VI 8-Nor-8-formyl-3-methyllumiflavin	685	0.16	192	−30

* This value does not represent a true oxidation-reduction potential, but is the midpoint of the potentiometric reduction wave.

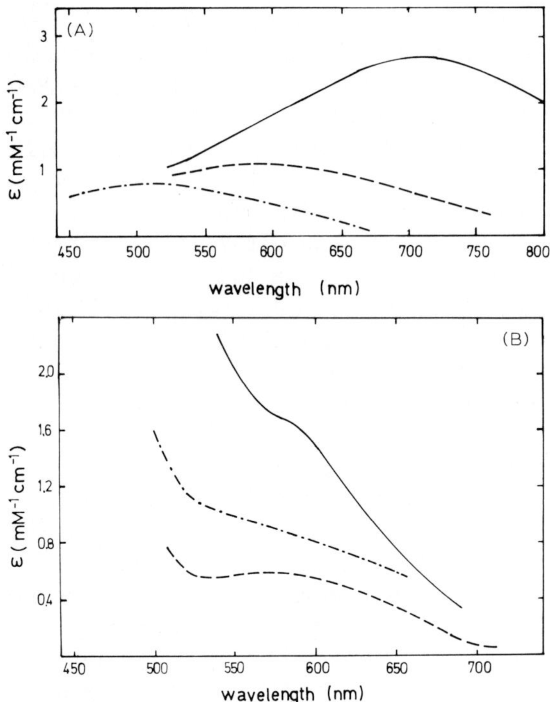

Fig. 2. Long wavelength-absorbing flavin-nicotinamide complexes. (A) $Nic^+_{ox} \cdot Fl_{red}H^-$; —·—·—, $NMN^+ \cdot$ dihydrolumiflavin; — — —, $Nic^+_{ox}(CH_2)_3 \cdot Fl_{red}H^-$ [22]; ——, NAD^+-reduced lipoyl dehydrogenase [2]. (B) $Nic_{red}H \cdot Fl_{ox}$; —·—·—, NMNH · lumiflavin; — — —, $Nic_{red}H(CH_2)_3 \cdot Fl_{ox}$ [22]; ——, NADPH · old yellow enzyme [6].

as a consequence of solvent stabilization of the ground-state ion pair and a lack of such stabilization in the much less polar excited state. For the $Nic^+_{ox} \cdot Fl_{red}H^-$ ion pair, the energy of the long wavelength transition should also be maximal in a highly polar environment and decrease with decreasing polarity of the environment. For the uncharged $Nic_{red}H \cdot Fl_{ox}$ the energy

TABLE III

$Nic_{red}H \cdot Fl_{ox}$ CHARGE-TRANSFER COMPLEXES: EXAMPLES OF LOW MOLECULAR WEIGHT AND PROTEIN-BOUND SPECIES AND FIRST-ORDER RATE CONSTANTS FOR FLAVIN REDUCTION

CT, charge-transfer complex.

Donor	Acceptor	λ_{max} CT (nm)	$k(s^{-1})$	Ref.
NMNH	Lumiflavin	575	7	20
$Nic_{red}H(CH_2)_3 \cdot Fl_{ox}$		575	345	22
NADPH	Old yellow enzyme	575	1.3	6
NADPH	*p*-OH-benzoate hydroxylase	600	57	6
NADPH	Thioredoxin reductase	570	44	6
NADPH	Ferredoxin-NADP reductase	620	50	6
NADH	NADH-cytochrome b_5 reductase		69	3
NADPH	NADPH-sulfite reductase		190	29

of the charge-transfer transition should be less solvent dependent [25]. The data presented in Tables I and III demonstrate that the energy of $Nic^{+}_{ox} \cdot Fl_{red}H^{-}$ charge-transfer complexes in proteins and model systems differ by as much as $6 \cdot 10^{3}$ cm^{-1}, whereas the transition energies for $Nic_{red}H \cdot Fl_{ox}$ charge-transfer complexes remain remarkably constant. Thus, polarity of the environment is reflected in the energies of the respective charge-transfer transitions but other factors are also important. Apart from electronic interactions between protein amino acids and the charge-transfer complex, slight differences in complex geometry and thereby in π-π overlap might be the most prominent ones. As a consequence, one cannot expect a correlation between charge-transfer transition energies of low molecular weight charge-transfer complexes and their protein-bound counterparts; nor can lack of correlation in transition energies for the same chromophores bound to different enzymes be taken as evidence against their charge-transfer character [28].

The mechanism of flavin-nicotinamide oxidoreduction

1. Formation of charge-transfer complexes

The formation of complexes between low molecular weight flavins and nicotinamides should be close to the diffusion-controlled limit. Second-order rate constants can be expected in the order of 10^{8}–10^{9} M^{-1} s^{-1}. Since the K_{d} of the complexes is in the order of 0.1 M, first-order rate constants for the reverse reaction can be expected to be in the order of 10^{7}–10^{8} s^{-1}. It is not surprising that the intermolecular flavin-nicotinamide charge-transfer complexes have only recently been observed. Their equilibrium concentration is very low and the equilibrium is established very rapidly, thus preventing their accumulation during oxidoreduction. A covalent link between flavin and nicotinamide or a protein environment overcomes these unfavourable facts and the respective charge-transfer complexes are more readily observed.

Two kinetically indistinguishable mechanistic possibilities have to be considered concerning the function of flavin-nicotinamide charge-transfer complexes. (a) They represent true precursors to oxidoreduction. (b) They are seen as products of a side reaction that is not on the pathway of the oxidation-reduction reaction. The experimental evidence favours the first alternative: for intramolecular models it has been demonstrated that a complex geometry allowing optimal charge-transfer interaction coincides with the highest oxidation-reduction rates [22]. Furthermore, the 50-fold rate increase observed for the intramolecular model suggests that its complex geometry more closely resembles the geometry of the transition state than the complex geometry of the intermolecular model.

2. Reaction rates

These should be primarily controlled by two factors: (a) complex geometry, which will determine the efficiency of the π-π overlap [22] and

(b) the difference in oxidation-reduction potential between donor and acceptor in the complex. This has been demonstrated by the linear free-energy relationship in the model system of NMNH-flavins (Fig. 1B). Oxido-reduction rates in an intramolecular model system ($345\,s^{-1}$) are higher than in most flavoenzymes (Table III). The only enzyme reported to have a higher turnover number, to the knowledge of the author, is mammalian NADH-dehydrogenase [30]. Thus, one may conclude that in most NAD(P)H-dependent flavoenzymes, proper orientation between protein-bound flavin and nicotinamide sufficiently accounts for the catalytic function of the enzyme oxidation-reduction step.

3. Stereospecificity of hydrogen transfer

Stereospecific transfer has to be expected if flavin and nicotinamide planes become fixed relative to each other. For several transhydrogenases [31, 32] and NADH-cytochrome b_5 reductase [3], direct and stereospecific hydrogen transfer has been observed. For NADH-cytochrome b_5 reductase it has been demonstrated that flavin-nicotinamide charge-transfer complexes are probably catalytic intermediates. Thus, stereospecificity of hydrogen transfer in NAD(P)H-dependent flavoenzymes may well arise from direct hydrogen transfer in the respective flavin-nicotinamide charge-transfer complexes.

4. Mechanism of hydrogen transfer

The exact mechanism of rate-limiting conversion of the $Nic_{red}H \cdot Fl_{ox}$ charge-transfer complex to the $Nic^{+}_{ox} \cdot Fl_{red}H^{-}$ charge-transfer complex (H^{+}—$H^{\cdot}$—H^{-} transfer) cannot be determined. In particular, it is not clear whether the charge-transfer interaction in the complexes is merely "accidental" and unrelated to the mechanism of hydrogen transfer or whether charge-transfer interaction is functionally significant for transfer of oxidation-reduction equivalents. In the first case, $2e^{-}$ transfer, either by H^{-} transfer or by group transfer [28] should be favoured. In the second case, however, which this author finds more attractive, $1e^{-}$ transfer would appear to be the logical possibility. In either case, one would expect direct hydrogen transfer (no exchange with solvent protons). The potential acceptor position in oxidized flavin, N-5 versus C-4a, has received much attention recently [19, 33]. In π-π charge-transfer interactions, this question probably is not very meaningful although the geometry of the transition state most probably does involve a conformation where the flavin N-5 = C-4a azomethine and dihydronicotinamide C-4 are in close proximity [22].

References

1 Haas, E. (1937) Biochem. Z. 290, 291–292
2 Massey, V. and Palmer, G. (1962) J. Biol. Chem. 237, 2347–2358
3 Strittmatter, P. (1965) J. Biol. Chem. 240, 4481–4487
4 Massey, V. and Williams, C.H. (1965) J. Biol. Chem. 240, 4470–4480
5 Zanetti, G. and Willaims, C.H. (1967) J. Biol. Chem. 242, 5232–5236

6 Massey, V., Matthews, R.G., Foust, G.P., Howell, L.G., Williams, C.H., Zanetti, G. and Ronchi, S. (1970) in Pyridine Nucleotide Dehydrogenases (Sund, H., ed.), pp. 393–411, Springer Verlag, Berlin
7 Strickland, S. and Massey, V. (1973) J. Biol. Chem. 248, 2953–2962
8 Ueda, T. and Coon, M.J. (1972) J. Biol. Chem. 247, 5010–5016
9 Whitfield, C.D. and Mayhew, S.G. (1974) J. Biol. Chem. 249, 2801–2810
10 Dolin, M.I. (1975) J. Biol. Chem. 250, 310–317
11 Singer, T.P. and Kearney, E.B. (1950) J. Biol. Chem. 183, 409–429
12 Sulter, C.H. and Metzler, D.E. (1960) Biochim. Biophys. Acta 44, 23–33
13 Radda, G.K. and Calvin, M. (1964) Biochemistry 3, 384–393
14 Fox, J.L. and Tollin, G. (1966) Biochemistry 5, 3873–3878
15 Gascoigne, J.M. and Radda, G.K. (1967) Biochim. Biophys. Acta 131, 502–507
16 Pappas, S.P., Pappas, B.C. and Merchant, K.A. (1967) Biochemistry 6, 3264–3269
17 Wu, F.Y., MacKenzie, R.E. and McCormick, D.B. (1970) Biochemistry 9, 2219–2224
18 Bruice, T.C., Main, L., Smith, S. and Bruice, P.Y. (1971) J. Am. Chem. Soc. 93, 7325–7328
19 Brüstlein, M. and Bruice, T.C. (1972) J. Am. Chem. Soc. 94, 6548–6549
20 Porter, D.J.T., Blankenhorn, G. and Ingraham, L.L. (1973) Biochem. Biophys. Res. Commun. 52, 447–452
21 Proffitt, R.T., Ingraham, L.L. and Blankenhorn, G. (1974) Biochim. Biophys. Acta 362, 534–548
22 Blankenhorn, G. (1975) Eur. J. Biochem. 50, 351–356
23 Grabe, B. (1958) Biochim. Biophys. Acta 30, 560–569
24 Sakurai, T. and Hosoya, H. (1966) Biochim. Biophys. Acta 112, 459–468
25 Mulliken, R.S. and Person, W.B. (1962) in Molecular Complexes, John Wiley and Sons Inc., New York
26 Suzuki, H. (1967) in Electronic Absorption Spectra and Geometry of Organic Molecules, p. 93, Academic Press, New York
27 Kosower, E.M. (1958) J. Am. Chem. Soc. 80, 3253–3260
28 Hemmerich, P. and Schumann-Jörns, M. (1973) in Enzymes: Structure and Function (Veeger, C., Drenth, J. and Oosterbaan, R.A., eds.), FEBS Symposia 29, pp. 95–117, North Holland, Amsterdam
29 Faeder, E.J., Davis, P.S. and Siegel, E.M. (1974) J. Biol. Chem. 249, 599–609
30 Cremona, T. and Kearney, E.B. (1964) J. Biol. Chem. 239, 2328–2334
31 Louie, D.D. and Kaplan, N. (1970) J. Biol. Chem. 245, 5691–5698
32 van den Broek, H.W.J. and Veeger, C. (1971) Eur. J. Biochem. 24, 72–82
33 Sun, M. and Song, P.S. (1973) Biochemistry 12, 4663–6449
34 Massey, V. and Ghisla, S. (1974) Ann. N.Y. Acad. Sci. 227, 446–465
35 Blankenhorn, G. (1975) Biochemistry 14, 3172–3176

Part VI

Covalently bound flavins and new flavins

Flavins and Flavoproteins, edited by T.P. Singer

CHAPTER 27

RECENT ADVANCES, PUZZLES, AND PARADOXES IN THE CHEMISTRY OF COVALENTLY BOUND FLAVINS

T.P. SINGER, D.E. EDMONDSON and W.C. KENNEY

Department of Biochemistry and Biophysics, University of California, San Francisco, Calif. 94143 and the Division of Molecular Biology, Veterans Adminstration Hosptial, San Francisco, Calif. 94121 (U.S.A.)

At the time of the first Symposium on Flavins and Flavoproteins in 1965 a single enzyme, succinate dehydrogenase, was known to contain covalently bound flavin and its structure was not yet established. By the time of the third Symposium in 1969, three or four enzymes of this class were recognized and although the complete structure of none of these flavins was known, the fact that in the flavin component of succinate dehydrogenase the substitution is at the 8α position was firmly established. By the time of the fourth Symposium in 1972 nine enzymes of this category had been discovered and the three types of chemical structures shown in Fig. 1 were known to exist in covalently bound flavin. Since then there has been a plethora of new findings in the field, many of which remain unpublished. We shall touch on some of these in the present talk. At least ten enzymes and four types of chemical structures are now known for covalently bound

(1) $N-H_2C$ (8α) ... $H_3N^{\oplus}-CH-COO^{\ominus}$

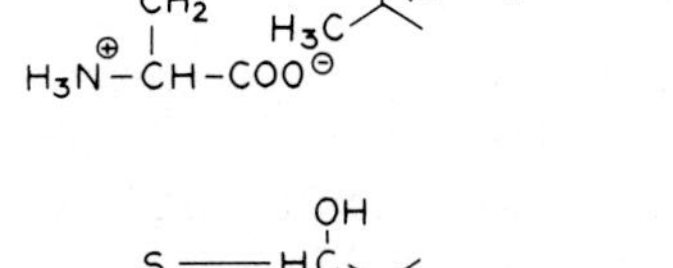

(2) $S-H_2C$ ([FAD]) ... $H_3N^{\oplus}-CH-COO^{\ominus}$

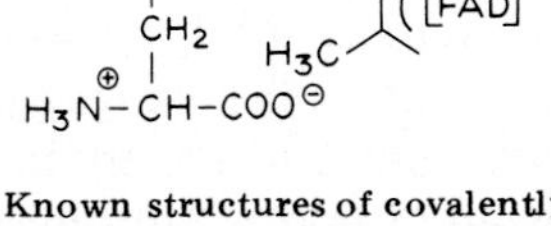

(3) $S-HC(OH)$ ([FAD]) ... $H_3N^{\oplus}-CH-COO^{\ominus}$

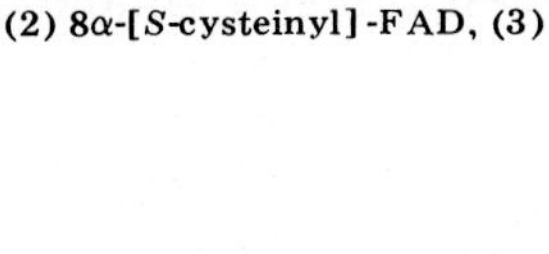

Fig. 1. Known structures of covalently-bound flavins occurring in flavoenzymes. (1) 8α-[*N*(3)-histidyl]-flavin, (2) 8α-[*S*-cysteinyl]-FAD, (3) 8α-[*S*-cysteinylthiohemiacetal]-FAD.

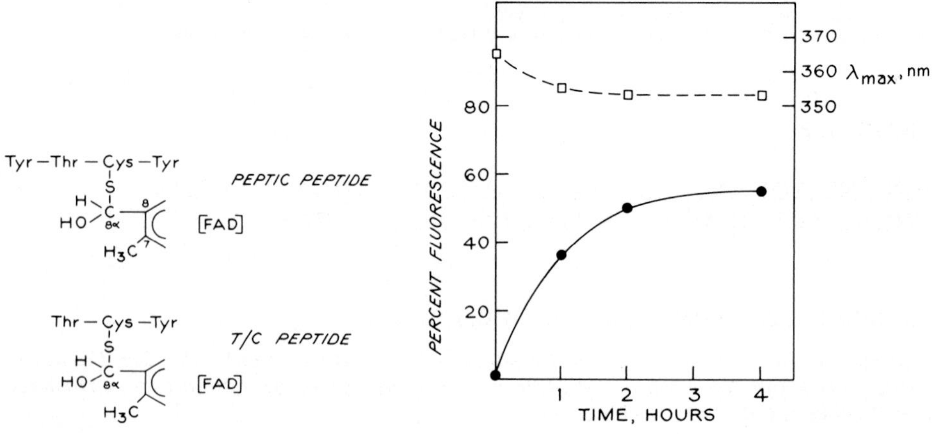

Fig. 2. Proposed structures and amino acid sequences of flavin peptides from *Chromatium* cytochrome *c*-552. T/C is tryptic-chymotryptic. Taken from Ref. 4.

Fig. 3. Time course of performic acid oxidation of the *Chromatium* peptic flavin peptide. ●——●, fluorescence at pH 3.4 relative to an equimolar solution of riboflavin; □---□, position of second fluorescence excitation maximum. Taken from Ref. 4.

flavins and much progress has been made in study of their chemistry, biological properties, and synthesis [1]. Because progress has been so rapid, some interesting, puzzling problems have been left by the wayside. This paper will briefly summarize progress in the field since the last Symposium and will attempt to point out some of the interesting and perplexing problems which remain for future research to resolve.

Among the three compounds shown in Fig. 1, the first two were firmly established by 1972 and the third one, the flavin thiohemiacetal, strongly suggested to be the structure of the flavin of *Chromatium* cytochrome *c*-552 but at that time not yet completely proven [2]. Unravelling of this structure proved to be more difficult than was the case with either of the previously known types of covalently bound flavins but by now the evidence seems to be convincing [3, 4]. It may be seen that in this adduct the flavin is at the oxidation level of 8-formylriboflavin. As expected from this, flavin peptides of *Chromatium* cytochrome *c*-552 yield on acid hydrolysis authentic 8-formylriboflavin, which has been identified by various chromatographic procedures and by the fact that on $TiCl_3$ reduction a blue color with a characteristic spectrum is formed: this is a specific test for 8-formylriboflavin [5].

The tryptic-chymotryptic and peptic peptides of the cytochrome differ by only a single tyrosine residue present in the latter (Fig. 2). Nevertheless, this is sufficient to cause major differences in stability and chemical reactivity [4]. The peptic peptide is much more stable to storage, less fluorescent and requires oxidation with performic acid at 40°C to yield the same fluorescence as does the tryptic-chymotryptic one at 0°C (Fig. 3). This is not due to resistance of the sulfur to oxidation but to an internal quenching caused by the tyrosine-flavin interaction, for if a sample of the

β-CYCLOPIAZONIC ACID — β-CYCLOPIAZONATE OXIDOCYCLASE → α-CYCLOPIAZONIC ACID

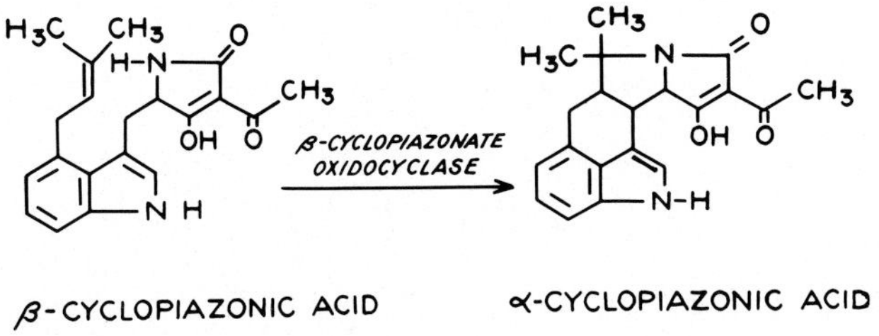

Fig. 4. Reaction catalyzed by β-cyclopiazonate oxidocyclase.

THIAMINE → THIAMINE ACETIC ACID

L-GULONO-γ-LACTONE + O_2 → 2-KETO-L-GULONO-γ-LACTONE + H_2O_2 — SPONTANEOUS ISOMERIZATION → L-ASCORBIC ACID

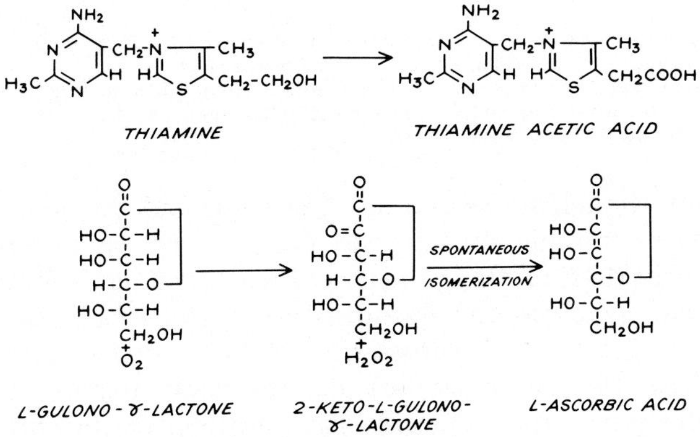

Fig. 5. Top: reaction catalyzed by thiamine dehydrogenase. Bottom: reaction catalyzed by L-gulono-γ-lactone oxidase.

peptic peptide is oxidized with performic acid at 0°C, then hydrolyzed with aminopeptidase, the fluorescence comes up immediately to the level obtained at 40°C. The fact that the NH_2-terminal tyrosine interacts with the flavin is also evident in comparison of the CD spectra of the two *Chromatium* peptides [4].

During the past two years a fourth type of covalently bound flavin has been discovered. It is a component of three enzymes, isolated from very different sources and catalyzing quite unrelated reactions: β-cyclopizonate oxidocyclase [6] (β-CPOC, Fig. 4) from *Penicillium cyclopium*, thiamine dehydrogenase from an unidentified soil bacterium [7], and gulono-γ-lactone oxidase from rat liver microsomes [8] (Fig. 5). Although the presence of covalently bound flavin in β-cyclopiazonate oxidocyclase and in thiamine dehydrogenase was reported some time ago [6, 7], identification of the flavin was undertaken only some two years ago in our laboratory. Although the description of this novel type of 8α-substituted flavin below is based on our studies with these two enzymes the properties of the flavin very recently isolated from gulono-γ-lactone oxidase [8] seem to coincide with it, so that it appears that all these enzymes contain the same type of covalently linked flavin.

The flavin components of both β-cyclopiazonate oxidocyclase and

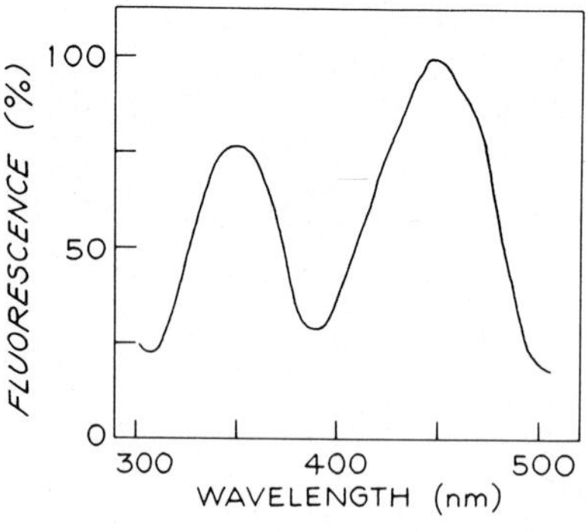

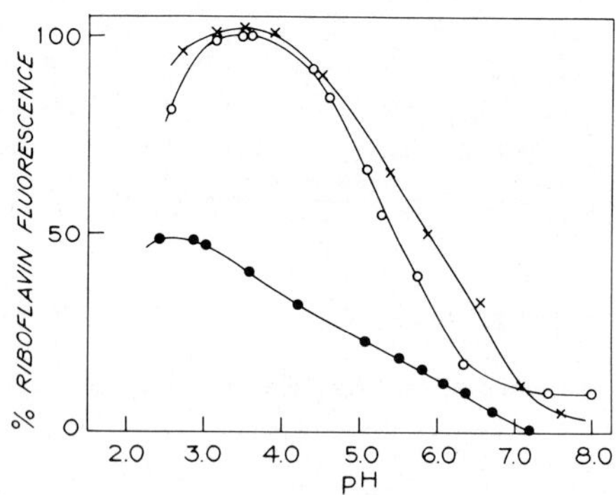

Fig. 6. Corrected fluorescence excitation spectrum of purified tryptic-chymotryptic flavin peptide (FMN level) from thiamine dehydrogenase in 1 mM citrate, pH 3.65. Taken from Ref. 9.

Fig. 7. pH vs. Fluorescence profile of tryptic-chymotryptic flavin peptides from thiamine dehydrogenase. ●——●, at FAD level; X——X, after treatment with nucleotide pyrophosphatase; ○——○, after hydrolysis in 6 M HCl at 95°C for 16 h. Taken from Ref. 9.

thiamine dehydrogenase are dinucleotides and are 8α-substituted. Evidence for this is seen in the hypsochromic shift of the fluorescence excitation spectrum of a pure flavin peptide from thiamine dehydrogenase from 372 to 345 nm (Fig. 6). Additional evidence for 8α-substitution from EPR and ENDOR data will be shown later. Fig. 7 shows the pH fluorescence curve of this peptide at the FAD level (solid circles) and the great increase in fluorescence yield on hydrolysis with nucleotide pyrophosphatase or with 6 M HCl at 95°C [9]. A noteworthy feature of this curve is the fluorescence quenching between pH 3.2 and 7.0. While in succinate dehydrogenase the pK of fluorescence quenching is 4.5 to 4.6, in the thiamine dehydrogenase flavin peptide, at the FMN level, it is 5.8, over 1 pH unit higher, while in the aminoacyl flavin produced by acid hydrolysis the pK is 5.1 ± 0.1, at least 0.5 pH unit higher than in N(3)-histidylriboflavin.

A pure tryptic-chymotryptic flavin peptide has also been isolated from β-cyclopiazonate oxidocyclase, in collaboration with Dr. Steenkamp [10]. Fig. 8 shows its fluorescence excitation spectrum at pH 3.4, where quenching by the adenylate part is minimal, before and after performic acid oxidation at 0°C. The increase in fluorescence is nearly 3-fold. The pH-fluorescence profile of the peptide at the FMN level before and after performic acid oxidation (Fig. 9) shows a pK_a of quenching of 5.8 and even after performic oxidation is only 60% of that of riboflavin. On the other hand, after hydrolysis in 6 M HCl at 95°C the fluorescence at pH 3.4 equals that of riboflavin and now shows a pK of 5.0, as in thiamine dehydrogenase (Fig. 10).

These data clearly suggest attachment of the flavin to an imidazole N, as in succinate dehydrogenase, although the compound cannot be simple histidyl flavin as in that enzyme because of the different pK_a values. Table I demonstrates that one mole of histidine is liberated by drastic acid hydrolysis. Since the Pauly test is negative before and positive after such cleavage, linkage of the flavin is to an imidazole N.

Several questions arise from these data. One question is, why the

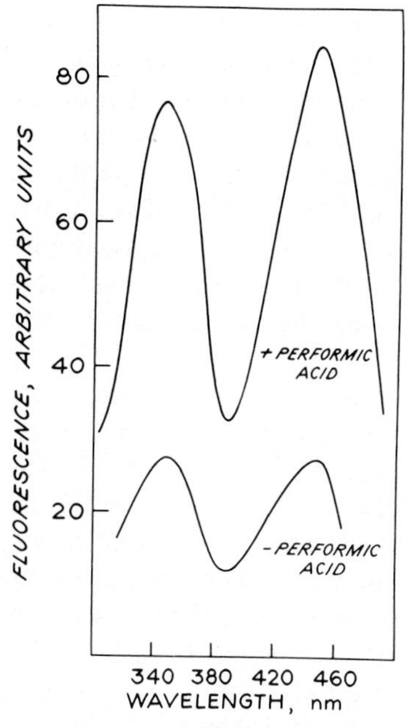

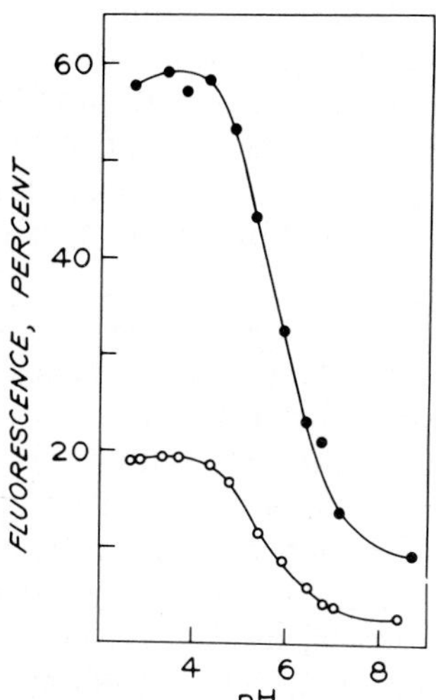

Fig. 8. Fluorescence excitation spectra of tryptic-chymotryptic flavin peptide of β-cyclopiazonate oxidocyclase at pH 3.4 before and after oxidation with performic acid at 0°C. Taken from Ref. 10.
Fig. 9. pH vs. Fluorescence profile of tryptic-chymotryptic flavin peptide (FMN level) from β-cyclopiazonate oxidocyclase. ○——○, before performic acid oxidation; ●——●, after performic acid oxidation at 0°C. Taken from Ref. 10.

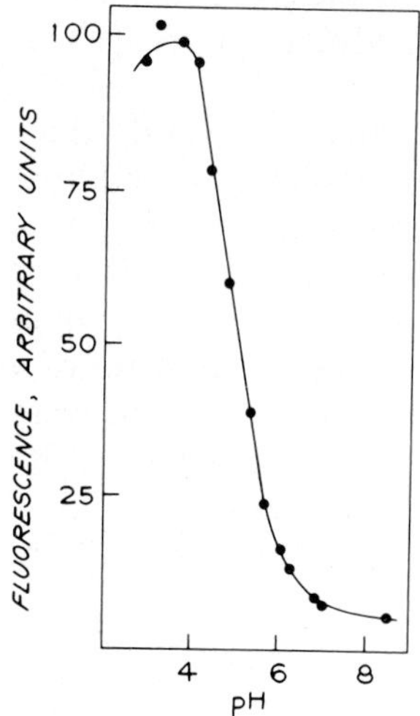

Fig. 10. pH vs. Fluorescence profile of amino acyl flavin from β-cyclopiazonate oxidocyclase. Taken from Ref. 10.

fluorescence is quenched in the β-cyclopiazonate oxidocyclase flavin peptide (FMN level) and why does performic acid overcome this quenching if the flavin is linked to a histidine, not to a sulfur group, since in the flavin peptide from succinate dehydrogenase at pH 3.2 the fluorescence is not quenched and performic acid has no effect on the fluorescence. The answer is in the amino acid composition of the peptide: it contains, besides 1 mol of histidine, 1 mol each of proline, valine and tryptophan and 2 mol of aspartate and threonine. Among these tryptophan is the only one likely to cause fluorescence quenching on interaction with the flavin. It is also known to be destroyed on oxidation with performic acid; hence the disappearance of the fluorescence quenching on oxidation of the flavin peptide. It is noteworthy that flavin also quenches the fluorescence of the tryptophan in the peptide by 90 to 95%.

A more difficult question is how does this compound differ from the histidyl flavin first described in succinate dehydrogenase and since then identified in several other enzymes? The differences we have to account for are quite marked: the p*K* values are sufficiently different that artificial mixtures of 8α-[*N*(3)-histidyl]-riboflavin and of the β-cyclopiazonate oxidocyclase flavin may be readily separated by high voltage electrophoresis at suitable pH. Moreover, the β-cyclopiazonate oxidocyclase flavin is much more labile to storage; unlike *N*(3)-histidylriboflavin, it is easily reduced by borohydride (Fig. 11). On reoxidation the original compound is not recovered but a new one, possibly dihydroflavin, is formed. A further difference from 8α-[*N*(3)-histidyl]-riboflavin is that the β-cyclopiazonate oxidocyclase flavin yields on hydrolysis 8-formylriboflavin. It should be emphasized that histidylriboflavin isolated from β-cyclopiazonate oxidocyclase and thiamine dehydrogenase also differed in these respects from the synthetic compound which was thought to be 8α-[*N*(1)-histidyl]-riboflavin and which had been obtained by acid treatment of 8α-[*N*(3)-histidyl]-riboflavin [11].

We have proposed several possible structures for this new histidylriboflavin, all of which turned out to be wrong. Since the flavin was histidine-linked but seemed to differ from both 8α-[*N*(3)-]- and 8α[*N*(1)-histidyl]-riboflavins, we first speculated that possibly the linkage is to C_2 or C_4 of the imidazole ring [9]. This possibility was ruled out when Dr. A. Schonbrunn at Brandeis University analyzed the NMR spectrum of a sample we had sent her and found that both the C_2 and the C_4 protons were present. Hence, linkage must be to an imidazole nitrogen, after all. (The same conclusion may be reached from the negative Pauly reaction, Table I). The same NMR data also seemed to suggest a single proton at the 8α carbon.

TABLE I

EVIDENCE FOR HISTIDINE ATTACHMENT IN β-CYCLOPIAZONATE OXIDOCYCLASE AND THIAMINE DEHYDROGENASE

Property	Result
Pauly test of amino acyl flavin	Negative
Same after hydrolysis in 6 M HCl, 125°C, 15 h	Positive
Histidine: flavin after 6 M HCl, 125°C, 15 h	1:1

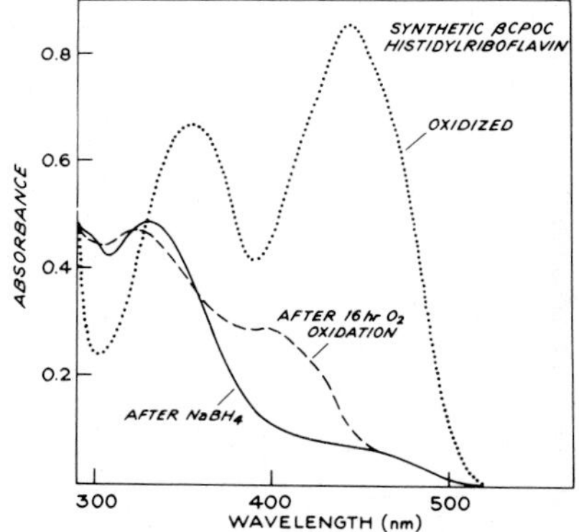

Fig. 11. Effect of sodium borohydride treatment on the absorbance spectral properties of synthetic cyclopiazonate oxidocyclase (β-CPOC) histidylriboflavin in 0.05 M sodium phosphate; pH 7.0.

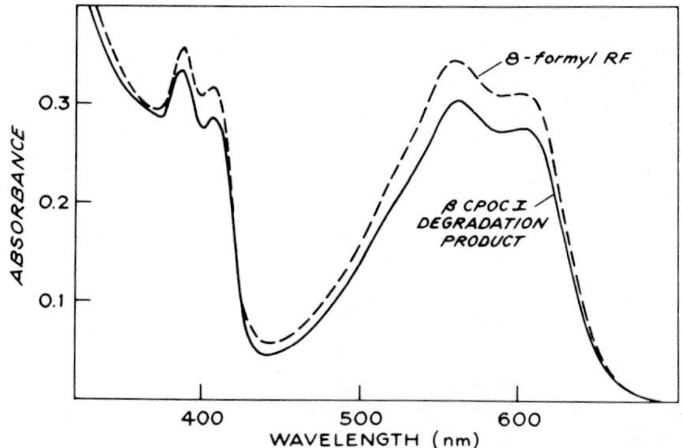

Fig. 12. Absorbance spectral properties of cation flavohydroquinone forms of 8-formylriboflavin (8-formyl RF) and the flavin degradation product of synthetic cyclopiazonate oxidocyclase (β-CPOC) histidylflavin. The flavins were reduced by an excess of $TiCl_3$ in 6 M HCl.

This gave rise to the second speculation, namely that the compound is an adduct of 8-formylriboflavin and histidine, i.e., that the 8α carbon is at the oxidation level of carbonyl, as in *Chromatium* flavin [1].

At first quite a bit of evidence accrued for this idea. Thus, storage of the β-cyclopiazonate oxidocyclase peptide yielded substantial amounts of a flavin which was chromatographically indistinguishable from 8-formylriboflavin. Moreover, on reduction with $TiCl_3$ the same characteristic blue compound was obtained as from synthetic 8-formylriboflavin (Fig. 12).

The idea of histidylhydroxyflavin was comforable, because if explained everything known about the new flavin and, besides, provided a comfortable symmetry in nature: 2 covalent flavin adducts with histidine, two with cysteine, one of each a derivative of 8α-hydroxyriboflavin, the other of 8-formylriboflavin. Critical examination of this idea called for optimally resolved NMR spectra as a basis for confirming the preliminary data which had suggested the presence of a single proton in the 8α position. This, in

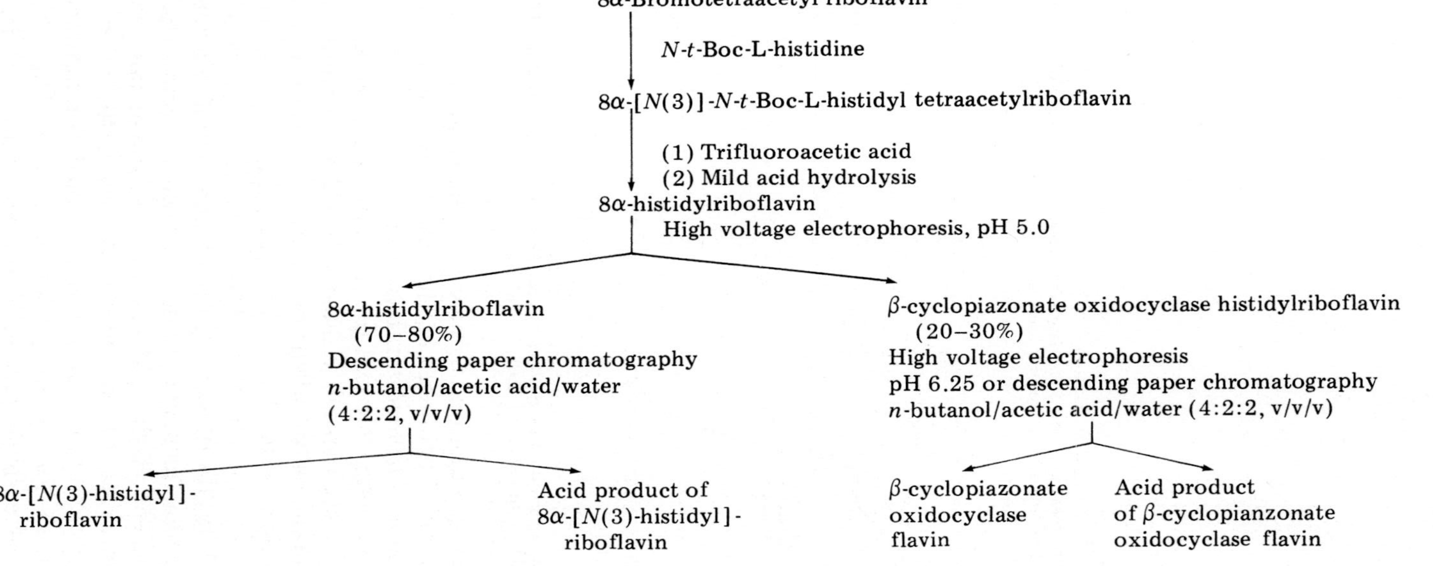

Scheme 1. Isolation of synthetic β-cyclopiazonate oxidocyclase histidylflavin.

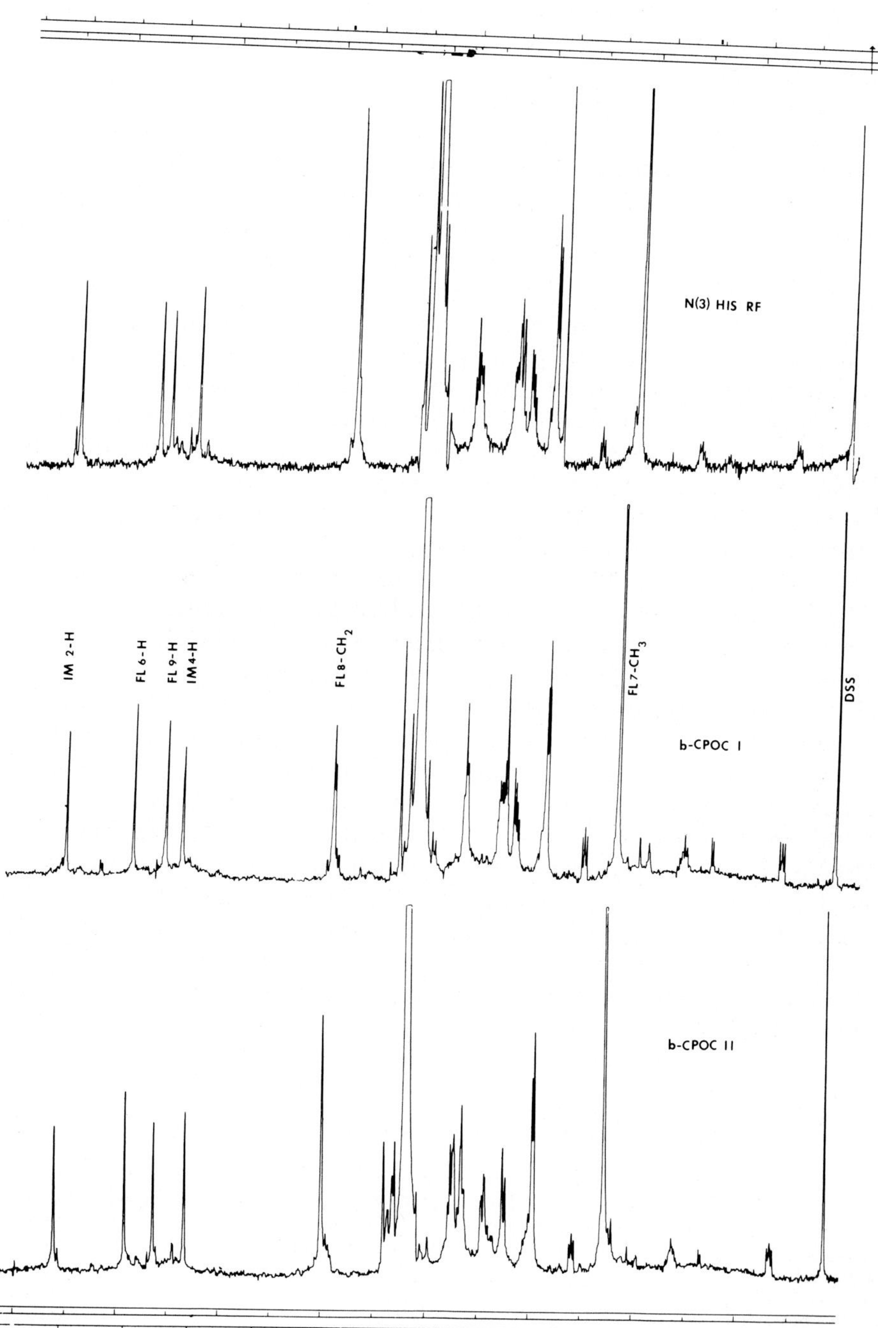

Fig. 13. Proton magnetic resonance spectra of 8α-[*N*(3)-histidyl]-riboflavin (top), synthetic cyclopiazonate oxidocyclase (β-CPOC) histidylriboflavin (center), and the acid product of synthetic cyclopiazonate oxidocyclase histidylriboflavin (bottom). The samples (4 μmol) were dissolved in ^2H_2O and the spectra taken on a 300 MHz Fourier-transform spectrometer in collaboration with Dr. J. Schoolery at Varian Instruments.

turn, required μmol quantities of the β-cyclopiazonate oxidocyclase flavin. This could not be obtained by isolation from natural sources and, therefore, had to be synthesized.

Fortunately, a histidylriboflavin obtained as a by-product of the chemical synthesis of 8α-[*N*(3)-histidyl]-riboflavin is indistinguishable from the β-cyclopiazonate oxidocyclase type of histidylriboflavin which is isolated from acid hydrolysates of the flavin peptides from either β-cyclopiazonate oxidocyclase or thiamine dehydrogenase by any of the criteria we have discussed. This compound is denoted as "β-cyclopiazonate oxidocyclase I" in Scheme 1 and in Fig. 13, to distinguish it from an acid modification product, which bears the same relation to the β-cyclopiazonate oxidocyclase flavin as does the acid product of 8α-[*N*(3)-histidyl]-riboflavin to its parent compound (Scheme 1). In both cases the desired (i.e., natural) compound may be separated from the acid modification product by paper chromatography or by high voltage electrophoresis. The "acid products" of either type histidylriboflavin arise during acid hydrolysis of the protecting groups in the synthesis [11] (Scheme 1) but may also be produced by acid treatment of the pure 8α-[*N*(3)-histidyl]-riboflavin of β-cyclopiazonate flavin. Since some 20 to 30% of histidylriboflavin produced on condensation of 8α-bromotetraacetylriboflavin with *N-t*-butoxycarbonyl-L-histidine is of the β-cyclopiazonate oxidocyclase type flavin, this method yields the desired quantities of this new flavin for chemical characterization.

The NMR spectra of 8α-[*N*(3)-histidyl]-riboflavin, of the β-cyclopiazonate oxidocyclase flavin obtained synthetically, and of its acid modification product in the 300 MHz Fourier transform spectrometer is given in Fig. 13. These beautifully resolved spectra show that the protons at C_2 and C_4 of the imidazole ring are present in the β-cyclopiazonate oxidocyclase flavin and that the 7-CH_3 and 8-CH_2 protons of the flavin are shifted downfield 20 and 15 cycles/s, respectively, as compared with 8α-[*N*(3)-histidyl]-riboflavin. Most importantly, mechanical integration of the area corresponding to the 8α position clearly shows the presence of two protons (Table II). This seems to rule out the speculation that β-cyclopiazonate oxidocyclase is an adduct of 8-formylriboflavin and histidine.

TABLE II

INTEGRATION OF FLAVIN AND IMIDAZOLE PROTONS OF SYNTHETIC HISTIDYL RIBOFLAVINS

Flavin	Fl-7-CH_3	Fl-8-CH_2	Fl(6)-H	Fl(9)-H	IM(4)-H	IM(2)-H
8α-[*N*(3)-histidyl]-riboflavin	3.0	2.0	0.8	0.6	0.6	0.9
Synthetic β-CPOC* I histidylriboflavin	3.0	1.8	0.8	0.9	0.7	0.9
Synthetic β-CPOC II histidylriboflavin (acid product of β-CPOC I)	3.0	2.1	0.7	0.6	0.7	0.8

* Cyclopiazonate oxidocyclase.

At this junction it became necessary to consider whether the β-cyclopiazonate oxidocyclase flavin differs from 8α-[N(3)-histidyl]-riboflavin in the flavin portion. There were two bits of evidence to rule this out. First, the synthetic compound made from authentic 8α-bromotetraacetyl riboflavin is absolutely identical with the natural one. Second, on methylation of β-cyclopiazonate oxidocyclase flavin, followed by Zn reduction, riboflavin was obtained in 80 to 90% yield.

Having eliminated all obvious possibilities, one of the assumptions we had made along the way must have been wrong. The first assumption was that because of the hypsochromic shift of the second absorbance band of the β-cyclopiazonate oxidocyclase flavin and by analogy with the three classes of known covalently bound flavins the substitution is on the 8α-CH_3 group.

What if the 7-CH_3 of riboflavin were not as inert as is generally supposed to be the case, so that the bromotetraacetylriboflavin used in the synthesis contained 20–30% of 7-bromotetraacetylriboflavin? On condensation with histidine we could get an adduct analogous to succinate dehydrogenase flavin but with different properties; probably greater lability, for instance. None of the data we had ruled out this possibility; even the NMR data could be explained to fit the picture by the observed downfield shift of the -CH_2- and CH_3- resonances relative to those of 8α-[N-(3)-histidyl]-riboflavin. To be sure, the EPR hyperfine spectra of the cation radicals of the β-cyclopiazonate oxidocyclase flavin and of histidylriboflavin

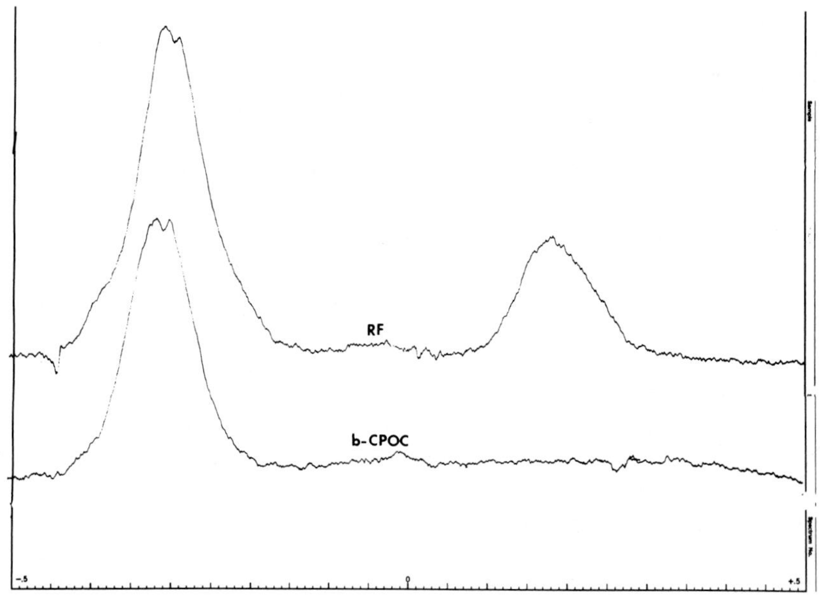

Fig. 14. ENDOR spectrum of riboflavin and of synthetic cyclopiazonate oxidocyclase (β-CPOC) histidylriboflavin in their cationic semiquinone form. The flavins were reduced with $TiCl_3$ under an argon atmosphere in 6 M HCl. The spectra were taken on a Varian +-700 system in collaboration with Dr. J. Hyde at Varian Instruments.

are identical and differ from the spectrum of riboflavin, suggesting that appreciable spin density occurs in the position of the flavin which is substituted. Moreover, we have emphasized [12] that the spin density of the 7 carbon is very low. But what if the presence of the bulky histidyl residue changed this spin density?

It became necessary to study the effect of substitution of the 7α and 8α carbons in the EPR spectra. Replacement of the 7-methyl group with an ethyl group in no way altered the EPR spectrum of riboflavin, but substitution of an ethyl for the methyl group in the 8 position caused an appreciable difference, confirming the fact that the 8 position is the one of high spin density. Since a formyl group would not be expected to hinder rotation any more than insertion of a methyl group does, if β-cyclopiazonate oxidocyclase were indeed 7-substituted, the formylriboflavin derived from it would be expected to show at least as great a difference from 8-formylriboflavin in the EPR spectra as do the 7α- and 8α-CH_3 derivatives. This was not the case: the two formyl-flavins showed identical EPR spectra. The clinching evidence that the peptide was attached to the 8α-CH_3 group came from ENDOR spectra kindly performed by Dr. Hyde of Varian (Fig. 14), showing that the β-cyclopiazonate oxidocyclase flavin lacks the 8-CH_3 ENDOR signal.

The remaining assumption was that the structure of the model compounds used in the characterization of the β-cyclopiazonate oxidocyclase flavin was correct. The two model compounds used were 8α-[N(3)-histidyl]-riboflavin and its acid modification product (cf. Scheme 1); the latter had been assigned the structure 8α-[N(1)-histidyl]-riboflavin in the literature [11]. While the evidence for the structure of the former, the flavin moiety of succinate dehydrogenase, was conclusive, the structure of the latter had not been firmly established. In fact, it has always been a paradox why acid treatment would cause a unidirectional shift from the less hindered to the more hindered isomer [11]. It was on the basis of the different pK_a values of the pH vs. fluorescence curves and different mobilities in electrophoresis of the β-cyclopiazonate oxidocyclase flavin and of the presumed 8α-[N(1)-histidyl]-riboflavin that the N(1)- of the imidazole was eliminated as the site of substitution in the β-cyclopiazonate oxidocyclase flavin [1, 9, 10]. If the assignment of the structure of the acid-modified 8α-[N(3)-histidyl]-riboflavin were wrong, the new flavin could be, after all, the N(1)-isomer of succinate dehydrogenase flavin.

The first clue that acid modification of 8α-[N(3)-histidyl]-riboflavin does not give the N(1)-isomer came from studies of the reductive cleavage of the two compounds. We found that the Zn cleavage of histidylriboflavins [11] may be greatly improved in regard to the recovery of the flavin if the histidylriboflavins are first methylated with CH_3I. With this modification the yield of riboflavin from reductive cleavage of 8α-[N(3)-histidyl]-riboflavin increases from less than 20% to over 80%. When the same procedure is applied to what has been called 8α-[N(1)-histidyl]-riboflavin (the acid modification product), no riboflavin is obtained but instead a compound arises with the same absorbance spectra and fluorescence as riboflavin, but readily speared from the latter by its higher mobility on thin-layer chromatography, using *n*-butanol/acetic acid/water (4:1:5, v/v/v) as the solvent. Conceivably, then, what has been called 8α-[*N*(1)-histidyl]-riboflavin in the literature is actually an *N*(3)-derivative, modified only in the ribityl position. This would explain why the NMR spectra of 8α-[*N*(3)-histidyl]-riboflavin and of its acid product differ in the ribityl region [11] and why on reductive cleavage the latter yields a product with the same spectrum as riboflavin but differing in mobility on thin-layer chromatography. Proof will obviously require the demonstration that on methylation and Zn reduction the acid-modified product yields 1-methylhistidine, as does the succinate dehydrogenase flavin [11], while β-cyclopiazonate oxidocyclase yields 3-methylhistidine. All this remains for future work to establish, as does the development of a rationale for the puzzling acid lability of the riboflavin part of histidylriboflavins, as compared with either free riboflavin or other 8α-substituted flavins, as well as the precise assignment of the structure of the acid modification products.

It also remains for future work to explain why the β-cyclopiazonate oxidocyclase flavin, if it is indeed the *N*(1)-isomer of the succinate dehydrogenase flavin, yields formylriboflavin on hydrolysis, although it is an adduct of 8α-hydroxyriboflavin, and why it is easily reduced by BH_4^- but not reoxidized by O_2.

Space does not permit one to discuss in detail the many remaining problems in this field but at least two should be briefly mentioned. After several years' search, the pathway of the biosynthesis of covalently bound flavins remains completely undefined. Yet, there must be at least one or two enzymatic steps involved. The very fact that among the 4 known types of covalently bound flavins 3 are adducts of 8α-hydroxyriboflavin and one of 8-formylriboflavin strongly suggests that an enzymic oxidation of the 8α-CH_3 group must precede the condensation. If the β-cyclopiazonate oxidocyclase type flavin were conclusively proven to be 8α-[*N*(1)-histidyl]-riboflavin, we would have three well established instances for an adduct of riboflavin with *N*(3)- of histidine (succinate dehydrogenase, D-hydrocynicotine oxidase, and sarcosine dehydrogenase) and three for an adduct with *N*(1)- of histidine (β-cyclopiazonate oxidocyclase, thiamine dehydrogenase, and L-gulono-γ-lactone oxidase). While this strict specificity of the condensation step could be determined by the tertiary structure of the apoenzyme, causing complete steric hindrance toward reaction of one of the two imidazole nitrogens, the authors would be more comfortable with the involvement of enzymes in the condensation as the basis of this specificity of attachment.

Perhaps the most important question which remains to be answered is the advantage that covalent linkage of flavin to proteins imparts to an enzyme. If we look over the list of 10 enzymes known to contain flavin in covalent linkage, no similarity is apparent in the reactions catalyzed, nor does intracellular location provide a clue. At a time when succinate dehydrogenase was the only enzyme of this class known, it was considered possible that covalent linkage may raise the redox potential enough to facilitate the oxidation of the high potential succinate/fumarate couple by FAD. Having now measured the redox potentials of several covalently linked flavins and of their peptides [13], we know that the potential is indeed raised in each case over that of the corresponding free flavin but only by some 25–30 mV, which would not be expected to facilitate redox reactions to a major extent. The reason for the existence of covalently linked flavins thus remains another puzzle of long standing in this field.

Note added in proof

The structure of the fourth type of covalently bound flavin (β-cyclopiazonate oxidocyclase flavin) has now been established as 8α-[N(1)-histidyl]-riboflavin by the isolation of 3-methylhistidine, following methylation and acid hydrolysis [14]. The acid modification of both N(3)- and N(1)-histidylriboflavins, as well as of riboflavin itself, has been shown to involve cyclization of the ribityl chain to the 2′,5′-anhydroflavin [15].

Acknowledgements

This work was supported by grants from the National Institutes of Health (1 P01HL16251) and from the National Science Foundation (GB 36570X and GB 41881)

References

1 Singer, T.P. and Edmonson, D.E. (1974) FEBS Lett. 42, 1–14
2 Kenney, W.C., Walker, W.H., Kearney, E.B., Seng, R., Singer, T.P., Cronin, J.R. and Hendriks, R. (1972) Z. Naturforsch. 27b, 1069–1071
3 Walker, W.H., Kenney, W.C., Edmondson, D.E. and Singer, T.P. (1974) Eur. J. Biochem. 48, 439–448
4 Kenney, W.C., Edmonson, D.E. and Singer, T.P. (1974) Eur. J. Biochem. 48, 449–453
5 Edmondson, D.E. (1974) Biochemistry 13, 2817–2821
6 Schabort, J.C. and Potgieter, D.J.J. (1971) Biochim. Biophys. Acta 250, 329–345
7 Neal, R.A. (1970) J. Biol. Chem. 245, 2599–2604
8 Nakagawa, H., Asano, A. and Sato, R. (1975) J. Biochem. 77, 221–232
9 Kenney, W.C., Edmondson, D.E. and Singer, T.P. (1974) Biochem. Biophys. Res. Commun. 57, 106–111
10 Kenney, W.C., Edmondson, D.E., Singer, T.P., Steenkamp, D.J. and Schabort, J.C. (1974) FEBS Lett. 41, 111–114
11 Walker, W.H., Singer, T.P., Ghisla, S. and Hemmerich, P. (1972) Eur. J. Biochem. 26, 279–289
12 Salach, J., Walker, W.H., Singer, T.P., Ehrenberg, A., Hemmerich, P., Ghisla, S. and Hartmann, U. (1972) Eur. J. Biochem. 26, 267–278
13 Edmondson, D.E. and Singer, T.P. (1973) J. Biol. Chem. 284, 8144–8149
14 Edmondson, D.E. and Kenney, W.C. (1976) Biochem. Biophys. Res. Commun. 68, 242–248
15 Edmondson, D.E. (1976) Biochemistry, to be published

Flavins and Flavoproteins, edited by T.P. Singer

CHAPTER 28

FLASH PHOTOLYSIS STUDIES OF 8α-SUBSTITUTED FLAVINS

DALE E. EDMONDSON[a], FRANK RIZZUTO[b] and GORDON TOLLIN[b]

[a] *Department of Biochemistry and Biophysics, University of California, San Francisco, Calif. 94143, Molecular Biology Division, Veterans Administration Hospital, 4150 Clement Street, San Francisco, Calif. 94121 and* [b] *Department of Chemistry, University of Arizona, Tucson, Ariz. 85721 (U.S.A.)*

Summary

Flash photolysis techniques have been used to study the effect of 8α substitution (histidine, morpholine, formyl) on flavin triplet state formation and on the reactivity of neutral and anionic flavin semiquinones. Only the 8-formyl derivative gave a measureable triplet yield. 8α-substituted neutral flavin semiquinones dismuted at a rate 2–3 times slower than the corresponding unsubstituted form, although the anionic semiquinones dismuted at approximately the same rate.

In the presence of oxygen, the kinetics of semiquinone decay changed from second order to pseudo-first order upon raising the pH, thus showing anion-radical oxidation as seen previously with unsubstituted flavins. The p*K* values for the ionization of the neutral 8α-flavin semiquinones are 1–1.4 units lower than the unsubstituted form. The anionic 8α-flavin semiquinones react with oxygen at a rate 4–5 times slower than does riboflavin.

Ferricyanide oxidizes both neutral and anionic 8α-substituted flavin semiquinones at rates which differ by at most a factor of five and which range from seven times smaller to three times larger than riboflavin.

These studies show that 8α substitution by histidine or morpholine, or by oxidation of the 8α-methyl group to an aldehyde, influences the reactivity of the flavin ring system in its anionic and neutral semiquinone forms. The altered p*K* values relative to riboflavin probably reflect the electron-withdrawing effect of the 8α-substituents on the flavin ring system.

Introduction

The anionic and neutral semiquinone forms of flavin coenzymes have been extensively studied over the past years by a variety of physical and chemical techniques. These include electron spin resonance spectroscopy [1], potentiometric titrations [2], pulse radiolysis [3] and flash photlysis

[4]. The results of these studies have defined the properties of these two forms of flavin semiquinone which exist under physiological conditions in a variety of flavoenzymes [5], with respect to spectral properties, sites of spin density, pK values and reactivity with a variety of oxidants.

In contrast, very little is known about the properties of 8α-substituted flavin semiquinones. This information is clearly of interest in view of the ever increasing number of enzymes (at present, there are 10 [6]) known to contain covalently bound 8α-substituted flavins.

As a first step towards this goal, this study assesses the effect of 8α substitution or modification on triplet state formation and on the reactivity of neutral and anionic flavin semiquinones with various oxidants. Furthermore, the pK values for ionization of the neutral semiquinones are determined by the pH dependence of semiquinone oxidation by oxygen.

Methods

8α-(N-3)-Histidylriboflavin and its acid-modification product* were synthesized and isolated as outlined by Walker et al. [7], while 8-formylriboflavin was synthesized as described by McCormick [8]. This latter compound was more than 90% in its intramolecular hemiacetal form as judged by thin-layer chromatography and CD spectra [9]. 8α-Morpholinolumiflavin was synthesized according to the procedure of Salach et al. [10] while lumiflavin was synthesized according to Guzzo and Tollin [11]. All other chemicals were of reagent grade and all solutions were made up in deionized water that had been distilled from a permanganate solution.

The flash apparatus and degassing procedures were as described earlier [12]. All flavin solutions were 10 μM in 0.1 M sodium formate (pH 3.3) or 0.1 M sodium pyrophosphate (pH 9.0). For experiments requiring intermediate pH values, 0.1 M sodium phosphate buffers were used. Flavin semiquinone concentrations were estimated using an $\epsilon_{M560nm} = 4700$ for the neutral species and an $\epsilon_{M510nm} = 1560$ for the anionic form [3]. All slopes and intercepts of linear plots were estimated by linear least-squares regression analysis of the data points.

Results and Discussion

Inasmuch as the photochemical production of flavin semiquinones occurs via the interaction of EDTA (or some other hydrogen donor) with the triplet state of the flavin [12], it is of interest to compare the effect of 8α substitution on both triplet yields and lifetimes. Absorbance due to the triplet is followed at 680 nm [12, 13] where both ground state and flavin semiquinone have little or no absorbance.

Fig. 1 shows that only one of the 8α-substituted flavins tested (8-formylriboflavin) shows any triplet absorbance upon flashing which is measurable after the duration of the flash (20–30 μs). No triplet absorbance was observed with either 8α-(N-3)-histidylriboflavins or with 8α-morpholinolumiflavin at either pH 3.3 or 9.0. The failure to observe a measurable

---

* This compound was thought previously to be 8α-(N-1)-histidylriboflavin [7] but it now seems to be an 8α-(N-3)-histidylriboflavin modified in the ribityl chain (cf. this volume, Singer et al., Chapt. 27).

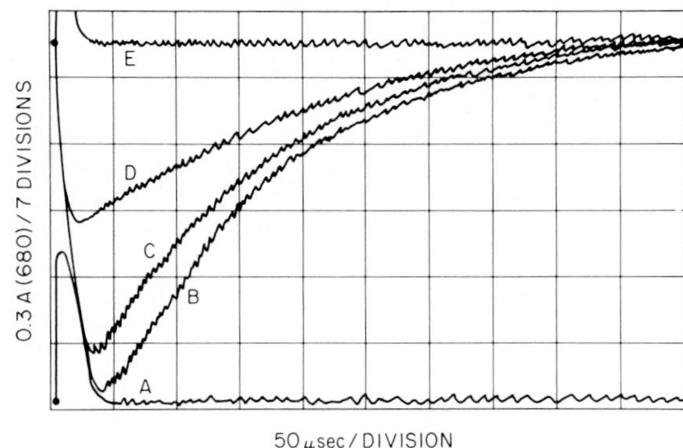

Fig. 1. Flash-induced transients absorbing at 680 nm of various flavin analogs. Flavin concentrations were 10 μM in 0.1 M sodium formate (pH 3.3) under anaerobic conditions. Curve A, flash decay in absence of flavin; Curve B, lumiflavin; Curve C, riboflavin; Curve D, 8-formylriboflavin; Curve E, 8α-(N-3)-histidylriboflavin.

triplet concentration is probably due to a faster decay than we could measure, since substantial flavin semiquinone is produced when EDTA is present (Figs. 2 and 3). Analysis of the triplet decay curves of 8-formylriboflavin and riboflavin show that the lifetimes for the two analogs are quite similar. The difference in yield is at present unexplained, although it could result from a smaller extinction coefficient of the triplet state of 8-formylriboflavin than for riboflavin. It is known that the molar extinction coefficient of the ground state at 445 nm is 9000 for 8-formylriboflavin [9] as compared to a value of 12 200 for riboflavin.

In the absence of EDTA, some neutral semiquinone is observed at pH 3.3 for all analogs tested and is due to either intra- or intermolecular hydrogen transfer [12, 13]. In the presence of EDTA, as shown in Fig. 2, substantial neutral flavin semiquinone is observed, which under anaerobic conditions decays by a second-order process of disproportionation. The neutral semiquinone yield was of similar magnitude for all analogs tested, thus suggesting a similar quantum yield of hydrogen abstraction from EDTA.

At pH 9.0, little or no transient absorbance at 560 nm is observed while substantial absorbance at 510 nm, characteristic of the anionic semiquinone, does occur (Fig. 3). Although 8-formylriboflavin gives only one-third the 510 nm absorbance of riboflavin (Fig. 3), the semiquinone yields are actually quite similar since the former dismutates at twice the rate of the latter (Table I). The 8α-histidylriboflavin analogs give substantially less (one-sixth) anionic semiquinone than does riboflavin (Fig. 3). This is not a result of differences in rate of disproportionation (Table I) but probably due to differing triplet yields at pH 9.0. It is known that the lumiflavin triplet has a p*K* of 9.8 [14] with decreasing yield as the pH is increased. This value could be different for the 8α-histidyl analogs. Also, at this pH, the quantum yield of fluorescence for 8α-histidylriboflavin is less than 10% that of riboflavin [7]. Thus, the singlet state population available for intersystem crossing to the triplet state is substantially reduced.

Table I shows that 8α substitution or modification slows the rate of

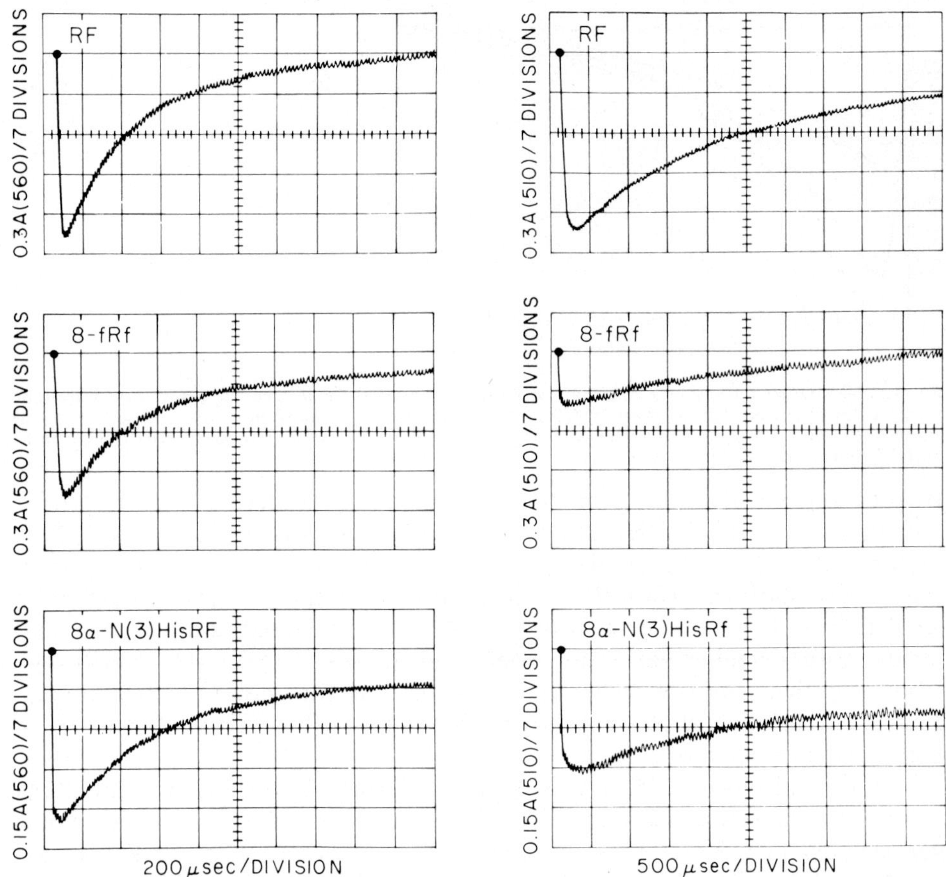

Fig. 2. Flash-induced transients absorbing at 560 nm in 0.1 M sodium formate (pH 3.3) in the presence of 0.05 M EDTA under anaerobic conditions. Rf, riboflavin; 8fRf, 8-formylriboflavin; 8α-N(3)HisRF, 8α-(N-3)-histidylriboflavin.

Fig. 3. Flash-induced transients absorbing at 510 nm in 0.1 M sodium pyrophosphate (pH 9.0) in the presence of 0.05 M EDTA under anaerobic conditions.

TABLE I

SECOND-ORDER RATE CONSTANTS FOR FLAVIN RADICAL DISPROPORTIONATION

Flavin	pH 3.3 $10^{-9} \cdot k$ $(M^{-1} \cdot s^{-1})$	pH 9.0 $10^{-9} \cdot k$ $(M^{-1} \cdot s^{-1})$
Riboflavin	2.5	0.75
8α-(N-3)-Histidylriboflavin	1.5	0.78
Acid-modified 8α-(N-3)-Histidylriboflavin	0.77	0.64
8-Formylriboflavin	1.3	1.4
Lumiflavin	4.3	0.91
8α-Morpholinolumiflavin	1.7	1.6

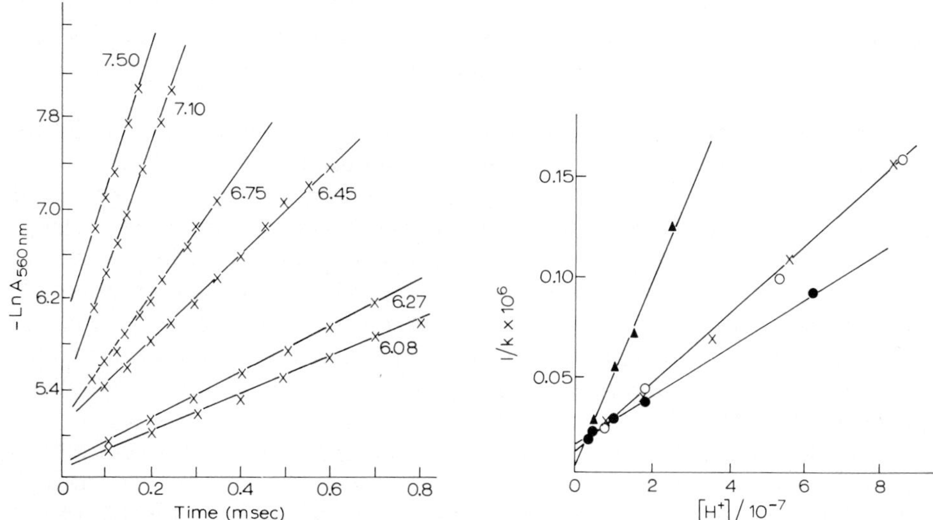

Fig. 4. First-order plots of decay of 8α-(N-3)-histidylriboflavin neutral semiquinone in 0.1 M sodium phosphate buffers at indicated pH values in the presence of 250 μM oxygen.

Fig. 5. Plot of $1/k$(apparent) versus $[H^+]$ for oxidation of flavin semiquinones by oxygen in 0.1 M sodium phosphate. ▲, riboflavin; ○, acid-modified 8α-(N-3)-histidylriboflavin; X, 8α-(N-3)-histidylriboflavin; ●, 8-formylriboflavin.

disproportionation of the neutral semiquinone form by a factor of 2–3 relative to the unsubstituted analog. In the case of the anionic semiquinone form, little difference in rate is observed with the 8α-histidylflavins; however, a 2-fold enhancement is seen in the case of 8-formylriboflavin and 8α-morpholinolumiflavin.

Under aerobic conditions, at pH 3.3, all of the flavin radicals decay by a second order process, as shown earlier for riboflavin and lumiflavin [4]. This observation demonstrates that neutral 8α-substituted flavin semiquinones dismutate more rapidly than the rate of oxidation by oxygen. If, as is the case with riboflavin, the anion semiquinone reacts with oxygen faster than it disproportionates, raising the pH under aerobic conditions should result in a pseudo-first order rate of decay. This rate should increase as the $[H^+]$ is lowered since the concentration of anion semiquinone increases. This behavior is observed with all 8α-substituted flavins and is demonstrated for 8α-(N-3)-histidylriboflavin in Fig. 4.

To determine the pK for neutral flavin semiquinone ionization as well as the second order rate constant for anion semiquinone oxidation by oxygen, the experimental data such as those shown in Fig. 4 were analyzed by the method of Vaish and Tollin [4]. Plots of the reciprocal of the apparent second order rate constants versus $[H^+]$ are linear as shown in Fig. 5. The intercepts give the reciprocal of the second order rate constant for anion semiquinone oxidation by oxygen (k) and the slope is the reciprocal of k and the equilibrium constant for neutral semiquinone ionization (K) [4]. The data of Vaish and Tollin [4] for lumiflavin and riboflavin are plotted for comparison with the 8α-flavin analogs. Although there are differences in the intercepts, the major differences are seen in the slopes of the plots, showing that 8α substitution has a substantial effect on the p*K* for neutral semiquinone ionization.

The p*K* values in Table II demonstrate that, irrespective of the nature of the 8α modification, the p*K* for neutral semiquinone ionization is lowered by 1–1.5 pH units. Previous results from oxidation/reduction-potential studies and from sulfite affinities [15] show that 8α substitution, again irrespective of the nature of the substituent, lowers the electron density of the isoalloxazine ring. This property is also apparent in the semiquinone p*K* since a lowering of electron density by an electron-withdrawing effect of the 8α substituent would be expected to stabilize the anionic semiquinone form, thus lowering the p*K* for its formation relative to the value of 8.3–8.4 for riboflavin [2, 3, 4, 16].

The slower rate of reaction of the 8α-substituted anionic semiquinones with oxygen may be a result of differing redox potentials for the oxidized to semiquinone conversion relative to riboflavin. The accepted potential for the $O_2/O_2^{\overline{\cdot}}$ conversion is −0.33 V which is independent of pH below the p*K* of 4.45 for $HO_2^{\overline{\cdot}}$ [17]. At high pH values, the oxidized to semiquinone redox couple for riboflavin (−0.325 V, pH = 10.00 [2]) approaches that of $O_2/O_2^{\overline{\cdot}}$. Although the oxidized to semiquinone potentials remain to be determined for the 8α-substituted flavins, previous work [15] has shown the two-electron redox potentials for oxidized to hydroquinone conversion to be 30–40 mV higher than that of riboflavin.

TABLE II

IONIZATION CONSTANTS OF THE NEUTRAL FLAVIN RADICAL ANALOGS AND SECOND-ORDER RATE CONSTANTS FOR ANION RADICAL OXIDATION BY OXYGEN

Flavin	p*K*	$10^{-8} \cdot k$ ($M^{-1} \cdot s^{-1}$)
Riboflavin	8.28	4.0
8α-(N-3)-Histidylriboflavin	7.23	0.96
Acid-modified 8α-(N-3)-Histidylriboflavin	7.20	0.98
8-Formylriboflavin	6.94	0.70

TABLE III

SECOND-ORDER RATE CONSTANTS FOR FLAVIN RADICAL OXIDATION BY FERRICYANIDE

Flavin	pH 3.3 $10^{-8} \cdot k$ ($M^{-1} \cdot s^{-1}$)	pH 9.0 $10^{-8} \cdot k$ ($M^{-1} \cdot s^{-1}$)
Riboflavin	3.9	4.2
8α-(N-3)-Histidylriboflavin	3.9	6.7
Acid-modified 8α-(N-3)-Histidylriboflavin	2.8	2.0
8-Formylriboflavin	0.65	2.7
Lumiflavin	5.2	4.6
8α-Morpholinolumiflavin	7.1	14.0

In contrast to the oxygen reaction, only small differences are observed in the oxidation of the neutral and anionic flavin radicals by potassium ferricyanide (Table III). The differences found in the case of 8-formylriboflavin and 8α-morpholinolumiflavin are not readily explainable at present although they may reflect changes in the one-electron redox potentials as compared with the other analogs.

The above results demonstrate that 8α substitution affects the reactivity of the flavin in its semiquinone form as well as its oxidized form [15] by decreasing the electron density of the isoalloxazine ring system. This information will be of importance in future studies on the mechanism of action of enzymes containing covalently-bound 8α-flavins.

Acknowledgements

Work supported in part by a National Institutes of Health Research Grant (AM 15057) to Gordon Tollin, by a National Science Foundation Research Grant (GB-41881) to Dale E. Edmondson and by program project grant 1 P01 HL 16251 from the National Institutes of Health to Dr. T.P. Singer.

References

1 Müller, F., Hemmerich, P., Ehrenberg, A., Palmer, G. and Massey, V. (1970) Eur. J. Biochem. 14, 185
2 Draper, R.D. and Ingraham, L.L. (1968) Arch. Biochem. Biophys. 125, 802
3 Land, E.J. and Swallow, A.J. (1969) Biochemistry 8, 2117
4 Vaish, S.P. and Tollin, G. (1971) J. Bioenerg. 2, 61
5 Massey, V. and Palmer, G. (1966) Biochemistry 5, 3181
6 Singer, T.P. and Kenney, W.C. (1974) Vitamins and Hormones, p. 1, Academic Press, New York
7 Walker, W.H., Singer, T.P., Ghisla, S. and Hemmerich, P. (1972) Eur. J. Biochem. 26, 279
8 McCormick, D.B. (1970) J. Heterocyclic Chem. 7, 447
9 Edmondson, D.E. (1974) Biochemistry 13, 2817
10 Salach, J., Walker, W.H., Singer, T.P., Ehrenberg, A., Hemmerich, P., Ghisla, S. and Hartmann, U. (1972) Eur. J. Biochem. 26, 267
11 Guzzo, A.V. and Tollin, G. (1964) Arch. Biochem. Biophys. 106, 380
12 Vaish, S.P. and Tollin, G. (1970) J. Bioenerg. 1, 181
13 Knowles, A. and Roe, E.M.F. (1968) Photochem. Photobiol. 7, 421
14 Schreiner, S., Steiner, U. and Kramer, H.E.A. (1975) Photochem. Photobiol. 21, 81
15 Edmondson, D.E. and Singer, T.P. (1973) J. Biol. Chem. 248, 8144
16 Ehrenberg, A., Müller, F. and Hemmerich, P. (1967) Eur. J. Biochem. 2, 286
17 Wood, P.M. (1974) FEBS Lett. 44, 22

Flavins and Flavoproteins, edited by T.P. Singer

CHAPTER 29

THE FUNCTION AND SOME MOLECULAR FEATURES OF CYTOCHROME *c*-553 DERIVED FROM *CHLOROBIUM THIOSULFATOPHILUM**

TATEO YAMANAKA and AKIRA KUSAI**

Department of Biology, Faculty of Science, Osaka University, Toyonaka, Osaka 560 (Japan)

Summary

Cytochrome *c*-553 (*Chlorobium thiosulfatophilum*) is a sulphide-cytochrome *c* reductase and cytochrome *c*-555 seems to act as the electron acceptor for the enzyme in vivo. The enzymatic activity is stronly inhibited by cyanide, and the absorbance spectrum of ferricytochrome *c*-553 is greatly affected by the reagent. Cytochrome *c*-552 (*Chromatium vinosum*) also shows the sulphide-cytochrome *c* reductase activity and the spectral change induced by cyanide.

Cytochrome *c*-553 is composed of two subunits: cytochrome and flavoprotein moieties. The molecular weights of these are 11 000 and 47 000 respectively. The cytochrome moiety molecule seems to contain only one cysteine residue. Cytochrome *c*-552 seems also to be composed of two subunits, one with a molecular weight of 20 000 and the other with a molecular weight of 46 000.

Introduction

The green sulphur bacterium *Chlorobium thiosulfatophilum* acquires energy and reducing power necessary for life processes by photosynthetic oxidation of inorganic sulphur compounds such as Na_2S and $Na_2S_2O_3$. The bacterium possesses three kinds of *c*-type cytochromes, cytochromes *c*-551, *c*-553 and *c*-555 [1]. Among them, cytochrome *c*-553 has been known to possess the covalently-bound flavin as well as haem *c* [2]. We have found that the cytochrome is a sulphide-cytochrome *c* reductase [3]. It catalyses reduction with sulphide of *c*-type cytochromes including cytochrome *c*-555.

---

* In the previous paper of this series (Biochim. Biophys. Acta (1973) 325, 304–314), the name of this author was misprinted as *K.* Kusai.

** Present address: Sankyo Co. Ltd., Tokyo, Japan.

The cytochrome seems to combine firmly with cyanide in its oxidized form, probably at the flavin moiety, and its enzymatic activity is strongly inhibited by the reagent.

Recently, we have succeeded in splitting the cytochrome molecule into cytochrome and flavoprotein moieties and determining some of their properties. The cytochrome molecule seems to be composed of one each of the two subunits.

Purification of cytochrome *c*-553

Chlorobium thiosulfatophilum (strain NCIB 8346) was cultivated in Larsen's medium for 5 days as described previously [4]. Cytochromes *c*-551 and *c*-553 were purified by the method of Meyer et al. [1], and cytochrome *c*-555 by the method as described previously [4].

Enzymatic properties of cytochrome *c*-553

Cytochrome *c*-553 was much more rapidly reduced by sulphide than other *C*-type cytochromes, at a concentration as low as 1 μM, and it catalyzed the reduction by the reagent of other *C*-type cytochromes, including cytochrome *c*-555 (*C. thiosulfatophilum*) [3] (Table I). As Fig. 1 shows, the reduction rate of yeast cytochrome *c* catalyzed by cytochrome *c*-553 was proportional to the concentration of this cytochrome when the concentration was less than about 50 nM. The catalytic activity of cytochrome *c*-553 was lost by heating at 80°C for 2 min and was inhibited strongly by cyanide, even at a concentration of 0.1 μM (Table II). Atebrin also inhibited the catalytic activity of cytochrome *c*-553, but EDTA, *o*-phenanthroline, azide and CO were not inhibitory. These facts appear to indicate that cytochrome *c*-553 is a sulphide-cytochrome *c* reductase. Between the two *c*-type

TABLE I

REDUCTION RATES OF VARIOUS CYTOCHROMES *c* BY Na_2S AND THE EFFECT OF ADDED CYTOCHROME *c*-553

The reaction mixture contained 0.1 M Tris HCl buffer pH 7.4, about 10 μM Na_2S and 50 μM each *C*-type cytochrome with (+) or without (—) 34 nm cytochrome *c*-553, in a total volume of 1.0 ml. The reaction (at 20°C) was started by adding Na_2S and followed spectrophotometrically by the increase in absorbance at the α peak of each cytochrome

Cytochrome	Addition of cytochrome *c*-553	Cytochrome reduced (nmol/min)
Chlorobium thiosulfatophilum		
c-551	—	16.2
	+	16.2
c-553		> 190
c-555	—	1.40
	+	14.0
Yeast cytochrome *c*	—	7.28
	+	21.8

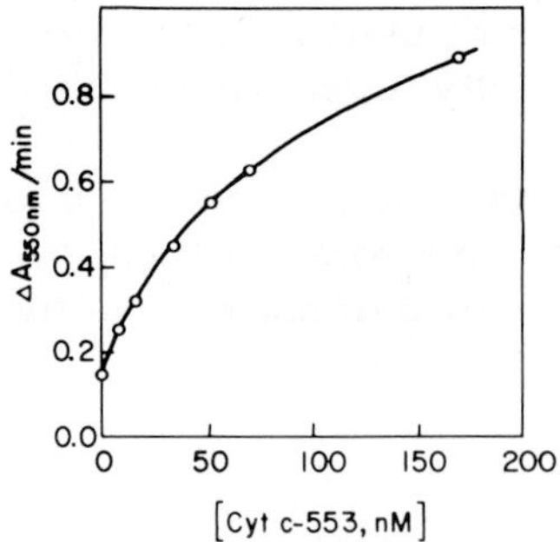

Fig. 1. Dependence of the reduction rate of yeast cytochrome *c*, with Na_2S, on the cytochrome *c*-553 concentration. The reaction mixture contained 0.1 M Tris HCl buffer, pH 7.4, 50 μM yeast cytochrome *c*, about 10 μM Na_2S and cytochrome *c*-553 at the various concentrations indicated, in a total volume of 1.0 ml.

cytochromes of *C. thiosulfatophilum* other than cytochrome *c*-553, cytochrome *c*-555 seems to act as the electron acceptor for the enzyme in vivo. Thus, the reduction of cytochrome *c*-551 by sulphide was not accelerated on addition of cytochrome *c*-553, while that of cytochrome *c*-555 was much accelerated. From the results shown in Table I, the molecular activity (mol cytochrome *c* reduced/mol cytochrome *c*-553) per min was calculated to be 370 with cytochrome *c*-555 as the electron acceptor under the present experimental conditions. The reaction product in the oxidation of sulphide catalyzed by cytochrome *c*-553 seems to be elementary sulphur, as judged from the stoichiometry in the reaction. That sulphite is not produced in the reaction has been confirmed by using the sulphite-cytochrome *c* reductase of *Thiobacillus novellus* (EC 1.8.2.1) [5].

As cytochromes *c* are generally non-enzymatically reducible by sulphide, one may think that the sulphide-cytochrome *c* reductase activity of

TABLE II

EFFECT OF INHIBITORS ON THE REDUCTION OF YEAST CYTOCHROME *c* WITH Na_2S CATALYSED BY CYTOCHROME *c*-553

The reaction mixture contained 0.1 M Tris HCl buffer pH 7.4, 50 μM yeast cytochrome *c*, about 10 μM Na_2S, 35 nM cytochrome *c*-553 and various compounds at the concentration as indicated, in a total volume of 1.0 ml. The reaction (20°C) was followed spectrophotometrically by the increase in the absorbance at 550 nm.

Compound added (μM)	Cytochrome *c* reduced (nmol/min)	Inhibition (%)
None	10.7	0
Cytochrome *c*-553 heated (80°C, 2 min)	0.0	100
Atebrin (100)	4.3	60
KCN (1.0)	2.1	80
KCN (0.1)	5.4	50
EDTA (100)	10.7	0
o-Phenanthroline (100)	10.7	0
NaN_3 (10 000)	10.0	7
CO (1 atm)	10.7	0

cytochrome *c*-553 is not of physiological significance. However, as seen from Table I, the non-enzymatic reduction rate of cytochrome *c*-555 with sulphide was very small as compared with yeast cytochrome *c*. Further, *C. thiosulfatophilum* has to take up electrons into the cells from sulphide present in the outside of the cells. Even in the case of *Chromatium*, electrons from sulphide will be transferred to the compartment other than the place where the compound is present. In this respect, the biological reduction of cytochromes *c* with sulphide could be different from the chemical reduction which occurs when the proteins are mixed with sulphide in a solution. Further, elementary sulphur, which is produced by oxidation of sulphide, is finally oxidized to sulphate by the organism. At present, we know nothing about the anaerobic oxidation of elementary sulphur in the organism. However, it could be possible that the oxidizing enzyme acts on only one of the solid states of elementary sulphur; elementary sulphur which is produced by action of cytochrome *c*-553 is oxidized by the oxidizing enzyme, while the substance of the other solid state may not be attacked by this enzyme. Indeed, the cells of *C. thiosulfatophilum* cultivated in the medium where all the thiosulphate was replaced by sulphide seemed to contain more cytochrome *c*-553 than those cultivated with thiosulphate as the major electron donor to the organism.

Cytochrome *c*-552 (*Chromatium vinosum*) [2] is also a flavocytochrome *c* similar to *Chlorobium* cytochrome *c*-553. By our preliminary experiments, cytochrome *c*-552 also acted as a sulphide-cytochrome *c* reductase and its catalytic activity was inhibited by cyanide. However, *Chromatium* cytochrome *c*-552 appeared to be less sensitive to cyanide than the *Chlorobium* cytochrome (Yamanaka, T. and Fukumori, Y., unpublished results).

Effect of cyanide on the absorbance spectrum of cytochrome *c*-553

As Fig. 2 shows, the absorbance spectrum of ferricytochrome *c*-553 was changed considerably on addition of cyanide [3]. The peaks or shoulders at 450 and 480 nm attributable to the flavin disappeared, while the absorbance between 600 and 700 nm increased. The absorbance in this wavelength region in ferrocytochrome *c*-553 was not affected by the presence of cyanide. Fig. 3 shows the difference spectra, ferricytochrome *c*-553 minus ferricytochrome *c*-553 + KCN. The spectral change was almost complete at a concentration of cyanide equimolar to the cytochrome. This implies that the spectral change was not caused by reduction of the flavin on addition of cyanide. This has been verified by the facts that the spectral change remained even after (1) addition of $K_3Fe(CN)_6$ to the solution of the cyanide-treated cytochrome, (2) overnight dialysis against a large amount of an appropriate buffer, (3) passage of the cyanide-treated cytochrome through a Sephadex G-25 column and (4) isoelectric focusing. The decrease in the absorbance at 450 nm on addition of cyanide ran parallel to the decrease in the sulphide-cytochrome *c* reductase activity (Fig. 4). The

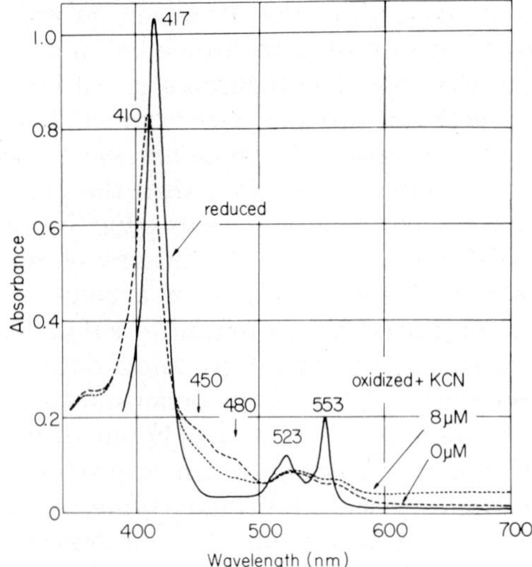

Fig. 2. Effect of cyanide on the absorbance spectrum of cytochrome *c*-553. Cytochrome *c*-553 was dissolved in 0.1 M Tris HCl buffer pH 7.4, (6.7 μM) and reduced with $Na_2S_2O_4$.

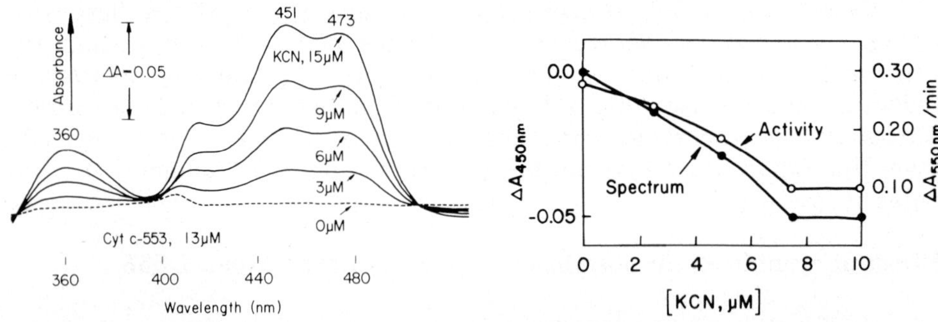

Fig. 3. Effect of cyanide on the absorbance spectrum of cytochrome *c*-553 as shown by difference spectra. Cytochrome *c*-553 was dissolved in 0.1 M Tris HCl buffer pH 7.4, (13 μM) and the difference spectra, ferricytochrome *c*-553 minus ferricytochrome *c*-553 + KCN, determined. Addition of KCN more than 15 μM did not cause further change in the spectrum.

Fig. 4. Relationship between the spectral change at 450 nm and the decrease in the enzymatic activity when cytochrome *c*-553 was treated with cyanide at various concentrations. To cytochrome *c*-553, dissolved in 0.1 M Tris HCl pH 7.4 (7 μM) was added cyanide at the concentration as indicated and the spectral change determined. Then, the enzymatic activity was measured by adding 5.0 μl of the above solution to 1.0 ml of the reaction mixture, as described in the legend to Fig. 1.

reduction rate of cytochrome *c*-553 itself by sulphide was also much decreased on addition of cyanide.

When $HgCl_2$ (e.g. 1 mM for 34 μM cytochrome *c*-553) was added to the cyanide-treated cytochrome *c*-553 and the resulting mixture passed through a Sephadex G-25 column, the cytochrome preparation thus obtained showed the same absorbance spectrum and enzymatic activity as the intact cytochrome. Cytochrome *c*-553 combined firmly with cyanide only in the oxidized state. Thus, when the cyanide-treated cytochrome was

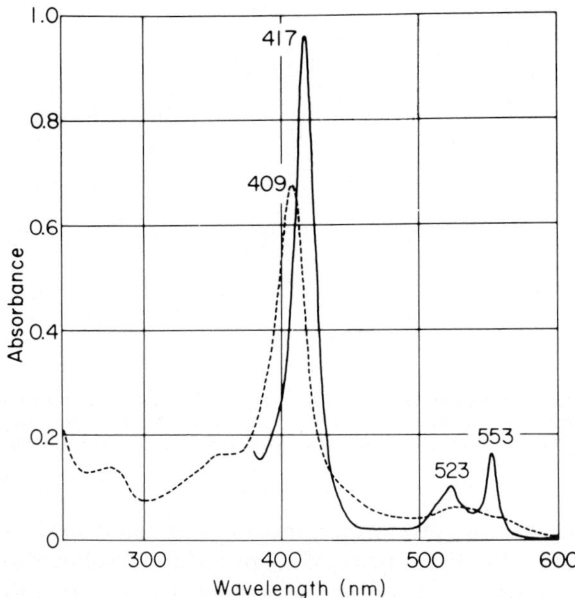

Fig. 5. Absorbance spectra of the cytochrome moiety of *Chlorobium* cytochrome *c*-553. The subunit was dissolved in 0.2 M phosphate buffer pH 7.0 (5.8 μM). ----, oxidized form; ——, reducted with $Na_2S_2O_4$.

first reduced and then passed through a Sephadex G-25 column, the ferricytochrome obtained by autoxidation showed the same absorbance spectrum as the intact cytochrome. Cytochrome *c*-553 showed a fluorescence spectrum with a peak at 450 nm at pH 8.5. The peak did not shift on addition of KCN.

Split of cytochrome *c*-553 into two moieties

When cytochrome *c*-553 precipitated by 2% trichloroacetic acid was dissolved in 0.1 M Tris HCl buffer, pH 8.5 and the turbid solution obtained centrifuged, the resulting supernatant contained a cytochrome, while the precipitate was yellow in color, suggesting that it contained flavin. The precipitate was insoluble at neutral pH, but soluble at pH 11. As the cytochrome moiety was still contaminated with the intact cytochrome *c*-553, it was again treated with 2% trichloroacetic acid, the resulting precipitate was dissolved in 0.1 M Tris HCl buffer pH 8.5, and the cytochrome moiety was separated from the insoluble material by centrifugation. The flavoprotein moiety that dissolved at an alkaline pH was again precipitated by gradual lowering of the pH of the solution. By repeating these procedures two or three times we obtained the flavoprotein moiety. Table III shows that the ratio of the cytochrome moiety to the flavoprotein moiety is unity.

As shown in Fig. 5, the cytochrome moiety displayed the absorbance spectrum of a typical *c*-type cytochrome. There was no peak or shoulder around 450 and 480 nm in the oxidized form. The millimolar extinction

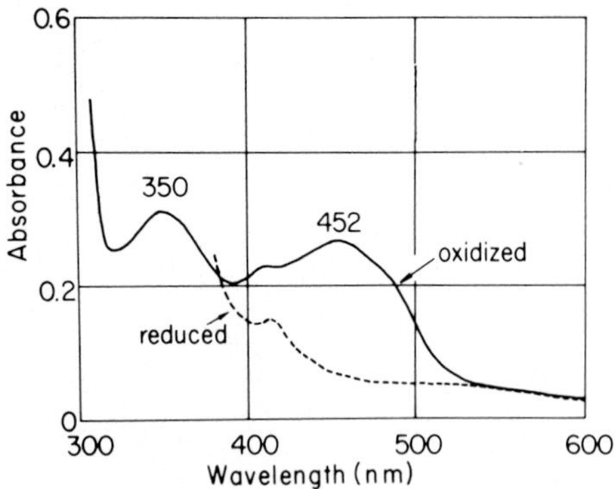

Fig. 6. Absorbance spectra of the flavoprotein moiety of *Chlorobium* c-553. The subunit was suspended in 0.1 M Tris HCl buffer and dissolved by gradual addition of 0.1 M NaOH. The final pH of the solution was 11.7. The concentration of the flavoprotein subunit was 24 μM, assuming ϵ_{mM} at 452 nm to be 11.

coefficient (ϵ_{mM}) of the α peak of the cytochrome moiety was calculated to be 28 on the basis of $\epsilon_{mM} = 29.1$ for the α peak of its pyridine ferrohaemochrome [6]. The flavoprotein moiety showed the absorbance peaks at 350 and 452 nm in the oxidized form, while there appeared a small peak at 408 nm on reduction, indicating that a small amount of the cytochrome moiety still remained in the preparation. The absorbance spectra of neither the cytochrome nor the flavoprotein moieties were affected by cyanide.

When cytochrome *c*-553 was subjected to sodium dodecyl sulfate-polyacrylamide gel electrophoresis [7], three bands were usually observed; their molecular weights were 11 000, 47 000 and 58 000 (Fig. 7). The two bands with molecular weights of 11 000 and 47 000 were identified as the cytochrome and the flavoprotein moieties by separate electrophoreses of each of these components (Fig. 8). The band with the largest molecular weight appeared usually when the intact cytochrome *c*-553 was subjected to electrophoresis. This band seems to be the original cytochrome *c*-553 which has remained unsplit as judged from its molecular weight. The molecular weight of cytochrome *c*-553 determined by electrophoresis in the present study is a little higher than that (50 000 ± 2000) determined

TABLE III

RELATION BETWEEN THE CYTOCHROME AND FLAVOPROTEIN MOIETIES

The quantitative relation between the cytochrome and flavoprotein moieties were obtained by splitting cytochrome *c*-553 into its two subunits. The amount of cytochrome moiety was calculated on the basis of $\epsilon_{mM} = 28$ at the α peak and that of the flavoprotein moiety was determined assuming its ϵ_{mM} at 452 nm to be 11.

Experiment	Cytochrome moiety (I) (nmol)	Flavoprotein moiety (II) (nmol)	I/II
1	159	145	1.1
2	218	228	0.96

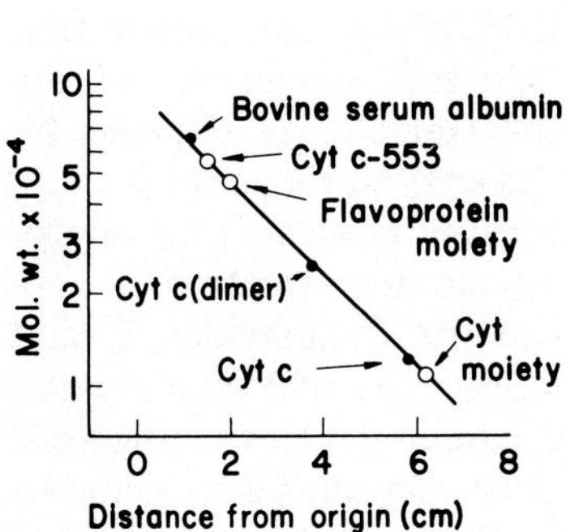

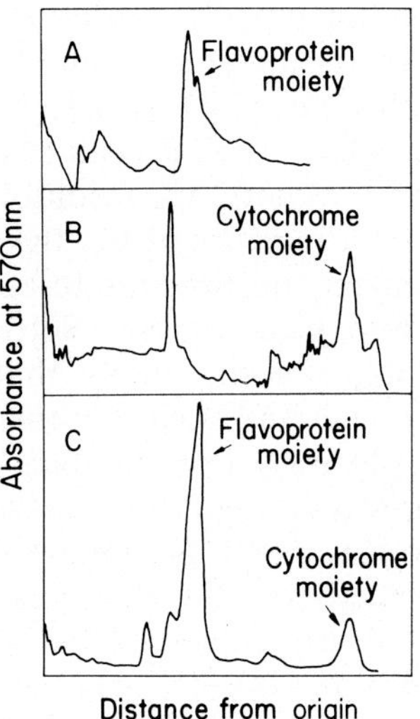

Fig. 7. Molecular weight estimations of cytochrome *c*-553 and of its two subunits by dodecyl sulfate-polyacrylamide gel electrophoresis. The concentrations of the gel, dodecyl sulfate and β-mercaptoethanol were 7.5, 2 and 0.5%, respectively. After electrophoresis the protein was stained with Coomassie blue. When the intact cytochrome *c*-553 was subjected to electrophoresis, a faint band appeared which corresponded to a molecular weight of 58 000, in addition to the bands of the cytochrome and flavoprotein moieties.

Fig. 8. Electrophoretic profile of cytochrome *c*-553 and of its subunits. The experimental conditions were the same as for Fig. 7. A, the flavoprotein moiety; B, the cytochrome moiety; C, cytochrome *c*-553.

by Meyer et al. [1]. Since the cytochrome *c*-553 molecule has been split into two moieties not only by treatment with trichloroacetic acid but also by the dodecyl sulfate-polyacrylamide gel electrophoresis, it seems likely that the cytochrome molecule is composed of one each of these two subunits.

Even when ferricytochrome *c*-553 was split by trichloroacetic acid, the cytochrome moiety was obtained in highly reduced state, although it was gradually oxidized in air. However, its autoxidation rate seemed to be slower than that of the intact cytochrome *c*-553. The reduced form of the cytochrome moiety was oxidized fairly rapidly by *Pseudomonas* cytochrome oxidase (EC 1.9.3.2) (= nitrate reductase) [8], while it did not react with bovine cytochrome *c* oxidase (EC 1.9.3.1) [9].

The amino acid compositions of cytochrome *c*-553 and its two components are shown in Table IV. The cytochrome *c*-553 molecule was composed of 534 amino acid residues of which only 6 were cysteine residues. It seems very interesting that there was only one cysteine residue in the cytochrome moiety. The amount of the residue was estimated as cysteic acid after oxidation by performic acid [10].

Even when the cytochrome subunit was oxidized by performic acid following removal of haem *c* with Ag_2SO_4 [12], the quantity of cysteic acid estimated was less than 1 mol per 11 000 g of the protein. Therefore, the haem in cytochrome *c*-553 could be bound to the protein through one thioether linkage. However, as the pyridine ferrohaemochrome of the cytochrome shows an α peak at 550 nm, the structure of the haem or its linkage to the protein moiety may be different from that in cytochrome *c*-558 (*Euglena gracilis*) and cytochrome *c*-557 (*Crithidia oncopelti*) [13] whose pyridine ferrohaeomochromes have their α peak at 553 nm.

The flavoprotein moiety was insoluble at neutral and soluble at alkaline pH. When the pH of a mixture of the flavoprotein and cytochrome moieties was lowered to neutral, no precipitate appeared. However, the preparation thus obtained showed no enzymatic activity or spectral change on cyanide treatment. Therefore, it seems likely that the spectral change induced by cyanide in cytochrome *c*-553 is closely related to its enzymatic activity.

Attempts to split cytochrome *c*-552 (*Chromatium vinosum*) into cytochrome and flavoprotein moieties by trichloroacetic acid treatment were unsuccessful. However, from preliminary experiments (Yamanaka, T, and Sasaki, K., unpublished results), it seems to be composed of two subunits, since it gave two bands on dodecyl sulfate-polyacrylamide gel electrophoresis; mol. wts. 20 000 and 46 000. The smaller seemed to contain haem as determined by the benzidine test [14]. Therefore, the *Chromatium* cytochrome *c*-552 molecule seems to be composed of one molecule each of a dihaem

TABLE IV

AMINO ACID COMPOSITIONS OF CYTOCHROME *c*-553 AND OF THE CYTOCHROME AND FLAVOPROTEIN MOIETIES DERIVED FROM IT

Numbers of amino acid residues were calculated on the basis of the molecular weights of each component as determined by dodecyl sulfate-polyacrylamide gel electrophoresis and expressed as the nearest integers. Tryptophan residues were not estimated.

Amino acid residue	Cytochrome *c*-553	Cytochrome moiety	Flavoprotein moiety	
			Experimental[a]	Calculated[b]
Lys	36	5	28	31
His	12	2	10	10
Arg	17	4	10	13
Asp	52	11	39	41
Thr	28	6	22	22
Ser	37	8	25	29
Glu	47	10	33	37
Pro	36	5	32	31
Gly	51	8	52	43
Ala	46	7	44	39
Cys	6	1	(5)	5
Val	39	6	40	33
Met	16	3	10	13
Ile	36	6	26	30
Leu	42	7	33	35
Tyr	16	2	13	14
Phe	16	2	15	14
Total	533	93	437	440
mol. wt.[c]	57 948	10 760	46 419	47 188

[a] Figures determined by direct analysis.
[b] Figures obtained by subtracting values for the cytochrome moiety from those for cytochrome *c*-553.
[c] Values including haem and/or flavin. Flavin is assumed to be FAD from the results obtained with *Chromatium* cytochrome *c*-552 [11].

cytochrome and a flavoprotein. Bartsch et al. [2] have reported that the cytochrome *c*-552 molecule contains two haem *c* molecules and that its molecular weight is 72 000.

Acknowledgements

The authors thank Professor H. Matsubara for his encouragement and valuable discussion during the course of this work. We are also grateful to Dr. K. Wada for his technical aid in amino acid analysis and to Dr. T. Yamamoto (present address: Pollution Prevention Technology Laboratory, Asahi Chemical Industry Co. Ltd., Kurashiki, Okayama, Japan) for his technical aid in the performance of the dodecyl sulfate-polyacrylamide gel electrophoresis.

References

1 Meyer, T.E., Bartsch, R.G., Cusanovich, M.A. and Mathewson, J.H. (1968) Biochim. Biophys. Acta 153, 854–861

2 Bartsch, R.G., Meyer, T.E. and Robinson, A.B. (1968) in Structure and Function of Cytochromes (Okunuki, K., Kamen, M.D. and Sekuzu, I., eds.), pp. 443–451, University of Tokyo Press, Tokyo

3 Kusai, A. and Yamanaka, T. (1973) Biochim. Biophys. Acta 325, 304–314

4 Yamanaka, T. and Okunuki, K. (1968) J. Biochem. Tokyo 63, 341–346

5 Charles, A.M. and Suzuki, I. (1966) Biochim. Biophys. Acta 128, 522–534

6 Drabkin, D.L. (1942) J. Biol. Chem. 146, 605–617

7 Wada, H. and Snell, E.E. (1972) Anal. Biochem. 46, 548–556

8 Yamanaka, T. (1966) in The Biochemistry of Copper (Peisach, J., Aisen, P. and Blumberg, W.E., eds.), pp. 275–292, Academic Press, New York

9 Okunuki, K. (1966) in Comprehensive Biochemistry (Florkin, M. and Stotz, E.H., eds.), vol. 14, pp. 232–308, Elsevier, Amsterdam

10 Hirs, C.H.W. (1967) Methods Enzymol. 11, 59–62

11 Walker, W.H., Kenney, W.C., Edmondson, D.E., Singer, T.P., Cronin, J.R. and Hendricks, R. (1974) Eur. J. Biochem. 48, 439–448

12 Paul, K.G. (1950) Acta Chem. Scand. 4, 239–244

13 Pettigrew, G.W. (1972) FEBS Lett. 22, 64–66

14 Connelly, J.L., Morrison, M. and Stotz, E. (1958) J. Biol. Chem. 233, 743–747

Flavins and Flavoproteins, edited by T.P. Singer

CHAPTER 30

BIOSYNTHESIS OF SUCCINATE DEHYDROGENASE CONTAINING COVALENTLY BOUND FLAVIN ANALOGS

S. GROSSMAN,* J. GOLDENBERG, F.B. KEARNEY, G. OESTREICHER and T.P. SINGER

Department of Biochemistry and Biophysics, University of California, San Francisco, Calif. 94143 and the Division of Molecular Biology, Veterans Administration Hospital, San Francisco, Calif. 94121 (U.S.A.)

Introduction

In flavoproteins in which the flavin is covalently bound to the protein, as in succinate dehydrogenase, the effect of chemical modification of the prosthetic group on catalytic activity cannot be tested by recombination of an apoenzyme with various flavin nucleotides. The most apparent approach to a study of this sort, until more is known about the mode of synthesis of the covalent flavin-protein bond, is to supply analogs of riboflavin for growth purposes, and allow the organism to make analog-bound enzymes. This approach tests the effect of the chemical modification on biosynthesis of covalently bound flavins as well as on the catalytic characteristics of the modified enzyme, if the analog is indeed incorporated. The availability of several mutants of *Saccharomyces cerevisiae* [1], blocked in the synthesis of riboflavin, provided a system for this study, since these mutants must be supplied riboflavin for growth, or, as shown here, an acceptable riboflavin analog.

A number of riboflavin analogs, substituted in various positions, were tested for their ability to support growth of the mutants, and of these, 7-methylriboflavin, 8-methylriboflavin, and deazaflavin appeared promising for this investigation. The use of 7-methylriboflavin, and 8-methylriboflavin in supporting growth of weanling rats was studied by Lambooy and his colleagues [2, 3] who reported reduced succinate dehydrogenase activity in the tissues of these animals; they did not, however, study the incorporation of the analogs into covalently bound flavins.

* On leave of absence from Bar Ilan University, Israel.

Materials and Methods

The yeast used was the riboflavinless mutant HK 645 [1]; it was grown as in previous work with wild type yeast [4], except that riboflavin was added at 5 mg/l, the medium and cultures were kept dark, and in aerobic growth the glucose concentration was 0.8%. When analogs were substituted for riboflavin, the yeast extract was extracted with benzyl alcohol to remove riboflavin and a supplement of biotin was added (1 mg/l); analogs were suppled at 5 mg/l. Analytical methods and assays were as in previous work [5]. Submitochondrial particles for assay were prepared with the aid of the Nossal shaker, and for isolation of the enzyme, with a blendor, both as already described [4]. Flavin peptides were isolated with the aid of apoflavodoxin from *Azotobacter vinelandii*, as described elsewhere in this volume [6]. Corrected fluorescence spectra were determined with a Perkin-Elmer fluorescence spectrophotometer (MPF 3).

Results and Discussion

The growth characteristics of the mutant yeast on 7-methylriboflavin and 8-methylriboflavin are illustrated in Fig. 1. The yeast grow well on these two analogs, but not on deazaflavin when it is present as the only flavin; deazaflavin, however, supports optimal growth if a low level of riboflavin is also supplied, although this amount of riboflavin by itself is insufficient for good growth. Deazaflavin was shown to enter the yeast cell and the phosphorylated (coenzyme) forms were recovered from the cells after breakage.

Cells grown on the analogs had normal respiratory capacity (Q_{O_2}) (Table I), and the succinate dehydrogenase activity and covalently bound flavin content of submitochondrial particle preparations from the yeasts grown on the 7- and 8-methyl analogs were not very different from those of the riboflavin-grown yeast. Anomalous results were observed with deazaflavin; these will be discussed later. When the membrane preparations were analyzed for covalently bound flavin, pH vs. fluorescence curves were obtained as illustrated in Fig. 2; the p*K* of fluorescnce quenching was essentially the same as in riboflavin-grown cells and may be taken as indicative of a histidyl flavin bond [7].

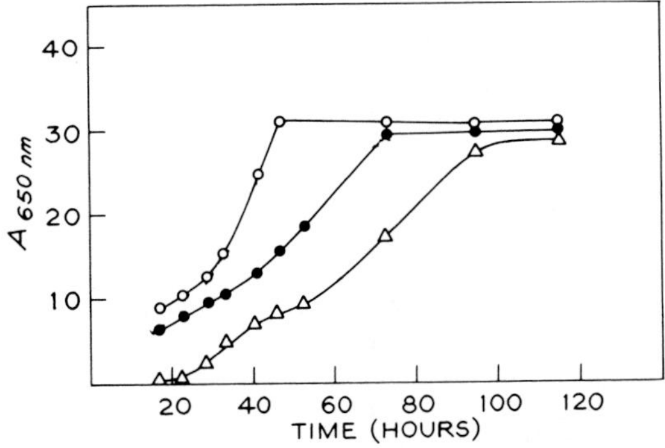

Fig. 1. Aerobic growth of the yeast mutant HK-645 on riboflavin (●—●), 7-methylriboflavin (△—△), and 8-methylriboflavin (○—○).

TABLE I

INCORPORATION OF RIBOFLAVIN AND ANALOGS INTO SUCCINATE DEHYDROGENASE DURING AEROBIC GROWTH OF YEAT MUTANT

Source of flavin	Flavin concn. in medium (mg/l)	Q_{O_2} of cells on glucose	Enzyme activity in ETP* (μmol succinate/min/mg)	Bound flavin in ETP* (pmol/mg)		Turnover no.** at 30°C per total bound flavin
				Riboflavin	Deazaflavin	
Riboflavin	5	88	0.30	21		14 300
8-CH_3-riboflavin	5	102	0.34	48		7 080
7-CH_3-riboflavin	5	90	0.36	36		10 000
Riboflavin +	1					
Deazaflavin	5	92	0.74	13	0.19	56 000

* Submitochondrial particles.

** In phenazine methosulfate assay.

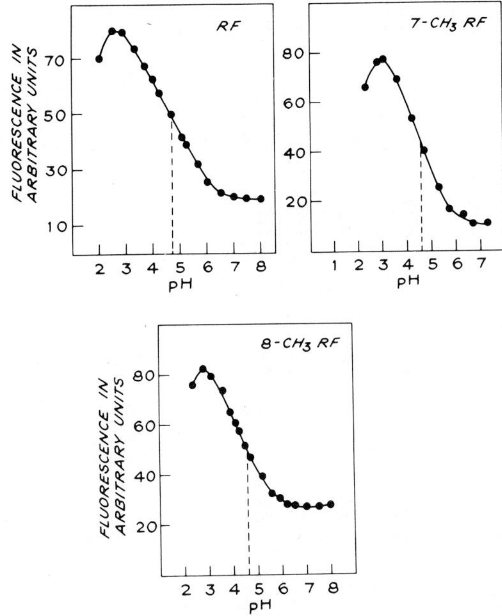

Fig. 2. pH vs. fluorescence profiles of flavin peptides from HK-645 grown aerobically on riboflavin (RF), 7-methylriboflavin (7-CH_3RF), and 8-methylriboflavin (8-CH_3RF). Following proteolytic digestion of the enzyme preparations, samples were treated with trichloracetic acid (5% final concentration), incubated to hydrolyze FAD to FMN, and the fluorescence of the supernatant liquid was measured at various pH values, obtained on sequential additions of NaOH; measurements are corrected for dilution.

Positive demonstration that succinate dehydrogenase indeed contained the modified flavin required isolation of the flavin peptides and cleavage to yield the analogs. This procedure could be carried out only with the 7-methyl analog, since insufficient 8-methylriboflavin was available. The enzyme was extracted from the membrane with sodium perchlorate and partially purified (Table II). The flavin peptide was then prepared from the enzyme by tryptic-chymotryptic digestion, hydrolysed to the FMN level and treated with apoflavodoxin to bind the flavin peptide (Scheme 1). The flavodoxin fraction was then separated from the free flavin peptide on Sephadex G-50, and the process was repeated with the free flavin peptide. The flavin was dissociated from flavodoxin by the addition of acid, absorbed on Florisil and eluted. The flavin peptide so isolated was then methylated with methyl iodide, and subjected to reduction with Zn, whereupon 7-methylriboflavin was obtained in excellent yield and was identified by thin layer chromatography (Fig. 3). In Fig. 4 the fluorescence properties are illustrated for the 7-methylriboflavin peptide (solid line) and for the free 7-methylriboflavin obtained after cleavage with Zn (dashed line). Also evident is the hypsochromic shift in the excitation spectrum of the peptide-bound flavin, in comparison with the normal riboflavin-like excitation spectrum of the free flavin after cleavage of the covalent flavin-peptide bond. These data are consistent with complete substitution of the analog for riboflavin in the 8α-histidyl flavin structure.

TABLE II

PURIFICATION OF SUCCINATE DEHYDROGENASE FROM COMMERCIAL YEAST

The turnover number of the membrane bound enzyme from both mammalian and yeast cells is twice that of the soluble enzyme. This has been traditionally assumed to be due to the operation of two sites for phenazine methosulfate in the membrane and only one in the soluble enzyme. This is why specific activity is expressed for both "2 sites" and "1 site."

Step	Total protein (mg)	Total activity (units	Specific activity (μmol succinate/min/mg)		Activity yield (%)	Histidyl flavin		
			2 sites	1 site		Total (nmol)	Per mg (nmol/mg)	Yield (%)
Submitochondrial particles	8 600	14 750	1.72	0.86	(100)	940	0.11	(100)
Perchlorate extract	1 740					750	0.44	80
0–30.5% $(NH_4)_2SO_4$ pellet	1 650	960		0.64	13	165	0.10	18
30.5–50.5% $(NH_4)_2SO_4$ pellet	280	3 900		14.0	52	392	1.40	42
Sephadex, peak 1	59	925		15.7	12	117	2.0	12
Sephadex, peak 11	149	2 670		18.0	35	328	2.2	35

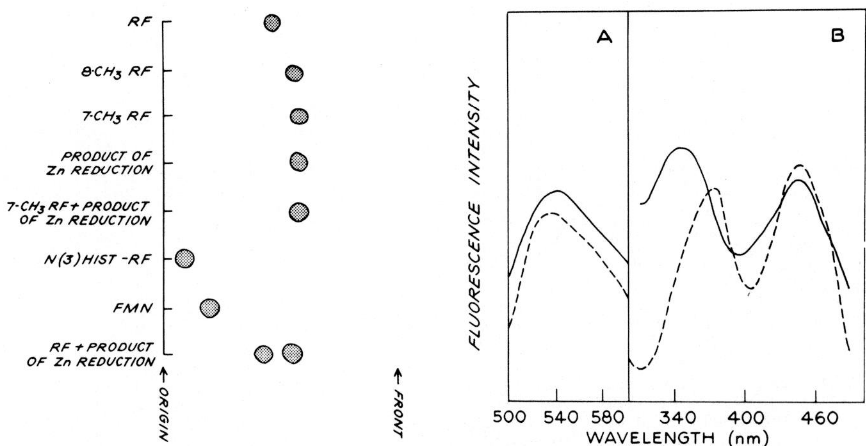

Fig. 3. Demonstration of 7-methylriboflavin in histidyl flavin fraction by reductive cleavage with Zn. Thin-layer chromatography was carried out on a cellulose plate with *n*-butanol/acetic acid/water (12 : 3 : 5, v/v/v) as the developing solvent system.

Fig. 4. Fluorescence spectra of 7-methylriboflavin peptide before and after reductive cleavage with Zn. (A) Emission spectra of the 7-methylriboflavin peptide (——) and the free 7-methylriboflavin obtained after reduction with Zn (----), determined at pH 3.4 with exciting light at 445 nm. (B) Excitation spectra of same.

Succinate dehydrogenase
↓ Trypsin/chymotrypsin
Florisil
↓
Treatment with flavodoxin
↓
Sephadex G-50 → unbound fraction treated again with flavodoxin
↓
Acidification
↓
Florisil
↓
Methylation
↓
Zn reduction

Scheme 1.

Similar studies with deazaflavin were more complicated to undertake since the cells required some riboflavin for growth in addition to deazaflavin. The flavin peptide fraction from such cells would thus be expected to contain normal riboflavin peptides as well as possible deazaflavin peptides. It was necessary, therefore, to analyze for the two types of flavin differentially in mixtures. This could be done by taking advantage of their different fluorescence emission characteristics, or, as shown in Fig. 5, by making use of the much higher affinity of deazaflavin for sulfite. The flavin peptide fraction from cells grown on riboflavin alone shows the same emission spectrum in the presence and absence of sulfite (right side of figure) whereas the fluorescence of free deazaflavin is nearly completely quenched by sulfite (left side). The left side of the figure illustrates also that in the covalently bound

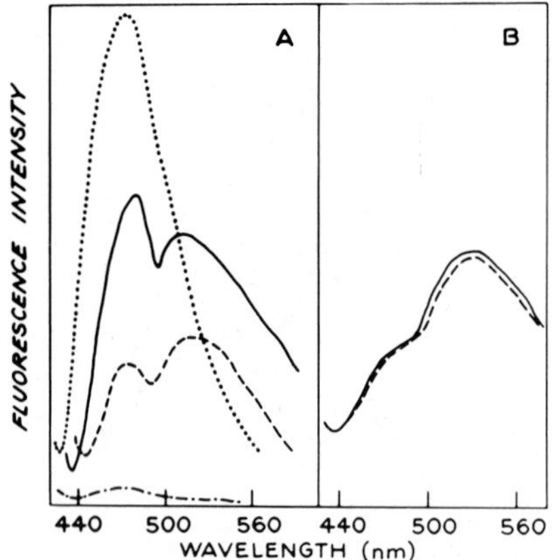

Fig. 5. Demonstration of deazaflavin in flavin peptide of yeast. (A) Fluorescence emission spectra of deazaflavin (····); deazaflavin plus 50 mM sodium sulfite (-··-·); flavin peptides from submitochondrial particles prepared from HK-645 grown anaerobically on riboflavin, then oxygenated in the presence of deazaflavin, before (——) and after the addition of 50 mM sodium sulfite (----). (B) Flavin peptides from submitochondrial particles prepared from yeast grown as in (A), but oxygenated in the presence of riboflavin instead of deazaflavin, before (——) and after the addition of 50 mM sodium sulfite (----). All spectra were taken in 100 mM phosphate buffer, pH 7.0, with excitation wavelength of 400 nm.

flavin fraction containing both types of flavin sulfite quenches the 470 nm emission band (that due to deazaflavin) but not the 540 nm band of riboflavin. It is of interest that no quenching of fluorescence above pH 3.2 is observable in the deazaflavin type of fluorescence in these preparations. The lack of this characteristic feature of the succinate dehydrogenase-flavin bond made it difficult to be certain that deazaflavin was actually incorporated into the active site via a histidyl linkage, rather than bound elsewhere, possibly in a non-functional capacity. It was possible, however, to obtain membrane preparations in which 15 to 30% of the covalently bound flavin present was deazaflavin, by growing the yeast anaerobically on riboflavin, and then exposing the yeast to deazaflavin, and to oxygen to induce mitochondrial development (Table III). The activity of succinate dehydrogenase in these preparations was very much higher than normal. If a turnover number is calculated from this catalytic activity and the total bound flavin is 2 to 3 times that of preparations from normal riboflavin-grown cells. If either the deazaflavin content alone, or that of riboflavin is used as the basis of the calculation, the turnover number is, of course, very much higher still. Thus, the presence of deazaflavin during growth gives rise to anomalously high succinate dehydrogenase activity. To eliminate the possibility that this might be the result of a non-enzymatic carrier function of deazaflavin, non-specifically bound in the membrane fraction, soluble succinate dehydrogenase was prepared from such particles. Table IV shows its characteristics: deazaflavin remained with the succinate dehydrogenase activity during solubilization and partial purification, and both the specific activity and turnover number were still abnormally high. As with the enzyme from wild type yeast (and from animal tissues) the turnover number here too is lowered to half when the

TABLE III

DEVELOPMENT OF RESPIRATION AND OF SUCCINATE DEHYDROGENASE ON EXPOSURE OF ANAEROBIC CELLS OF MUTANT TO O_2

Stage	Q_{O_2}	Enzyme activity in submitochondrial particles (μmol/min/mg)	Bound flavin (pmol/mg) in submitochondrial particles		Turnover no. at 30°C per mol of total covalently bound flavin
			Riboflavin	Deazaflavin	
Anaerobic cells	3	0 to 0.03	0 to 2		
Oxygenated in presence of riboflavin (5 mg/l)	68	0.14	7.9		17 500
Oxygenated in presence of deazaflavin (5 mg/l)	96	0.21	5	0.9	37 200

enzyme is extracted free of the membrane environment; this suggests that the higher turnover number is a real characteristic of succinate dehydrogenase from deazaflavin-grown yeast. No evidence is presently available, however, to show that deazaflavin is actually reduced by succinate and reoxidized by phenazine methosulfate in the catalytic reaction. The extremely limited amount of enzyme which could be isolated so far from the mutant yeast has precluded the demonstration of tritium transfer from labelled succinate to deazaflavin. Moreover, it is difficult to visualize a catalytic function for deazaflavin in view of its low redox potential in comparison with that of the succinate-fumarate couple. The possibility must be considered, therefore, that the effect of deazaflavin is exerted in an indirect fashion on the riboflavin containing form of the enzyme.

TABLE IV

COMPARISON OF PURIFIED SUCCINATE DEHYDROGENASE CONTAINING RIBOFLAVIN (WILD TYPE CELLS) AND RIBOFLAVIN + DEAZAFLAVIN (MUTANT)

Cell type	Turnover number of purified succinate dehydrogenase	% Deazaflavin in covalently bound flavin	K_m for succinate (30°C)(mM)
Wild (contains riboflavin only)	8 000	0	0.5
Mutant (grown on riboflavin + deazaflavin)	24 000	18	0.44

It will be of great interest to isolate flavin peptides from deazaflavin-grown yeast, and to determine with assurance that deazaflavin is covalently bound to protein, as the data now indicate, and to determine to what amino acid it may be bound, whether it is at the active center or elsewhere and whether both types of flavin are present on the same enzyme molecule or one or the other alone.

Acknowledgements

We are most grateful to Dr. J.P. Lambooy for the gift of 7-methylriboflavin and 8-methylriboflavin, and to Dr. A. Bacher for the yeast strain used in this work. This work was supported in part by grant No.30078 from the National Science Foundation, and by the National Institutes of Health (1 PO 1 HL 16251).

References

1 Oltmanns, O., Bacher, A., Lingens, F. and Zimmerman, F.K. (1969) Mol. Gen. Genetics 105, 306–313

2 Kim, Y.S. and Lambooy, J.P. (1967) Arch. Biochem. Biophys. 122, 644–647

3 Dombrowski, J.J. and Lambooy, J.P. (1973) Arch. Biochem. Biophys. 159, 378–382

4 Singer, T.P., Rocca, E. and Kearney, E.B. (1966) in Flavins and Flavoproteins (Slater, E.C., ed.), pp. 391–426, Elsevier, Amsterdam

5 Grossman, S., Cobley, J., Hogue, P.K., Kearney, E.B. and Singer, T.P. (1973) Arch. Biochem. Biophys. 158, 744–753

6 Oestreicher, G., Edmondson, D.E. and Singer, T.P. (1976) Flavins and Flavoproteins, Proceedings of the 5th International Symposium (Singer, T.P., ed.), Chapt. 47, pp. 447–452, Elsevier, Amsterdam

7 Walker, W.H., Singer, T.P., Ghisla, S. and Hemmerich, P. (1972) Eur. J. Biochem. 26, 279–289

Flavins and Flavoproteins, edited by T.P. Singer

CHAPTER 31

THE REACTION OF FLAVOCYTOCHROMES *c* OF THE PHOTOTROPHIC SULFUR BACTERIA WITH THIOSULFATE, SULFITE, CYANIDE AND MERCAPTANS

T.E. MEYER and R.G. BARTSCH

Department of Chemistry, University of California, San Diego, La Jolla, Calif. 92093 (U.S.A.)

Summary

The flavocytochromes *c* from *Chromatium* and *Chlorobium* react with thiosulfate, sulfite, cyanide, GSH, and CoASH with resultant loss of the near ultraviolet and visible absorbance of flavin. New absorbance bands are developed at 310–320 nm and at 650–700 nm. It is postulated that the new ultraviolet band indicates the presence of an adduct between the flavin and the added ligand and that the far red band is the consequence of a charge-transfer complex between the liganded flavin and a protein component, possibly a tyrosine residue located adjacent to the covalently bound 8-α-cysteinyl flavin thiohemiacetal.

Introduction

Flavocytochromes *c* constitute a small class of electron transport proteins which contain both covalently bound heme and flavin [1]. These cytochromes have been isolated only from two species of phototrophic sulfur bacteria which metabolize thiosulfate in addition to sulfide, and they are absent from extracts of all other bacteria tested, including the non-sulfur purple bacteria, the blue-green bacteria and two species of phototrophic sulfur bacteria which metabolize sulfide only. *Chromatium vinosum* strain D flavocytochrome *c* has two *c*-type hemes and one FAD [2] linked at the 8-α position via a thiohemiacetal bond to a cysteine residue [3] in a molecule of approx. 72 000 mol. wt. [1]. The protein can be dissociated into dissimilar diheme- and flavin-containing subunits [1] of 21 and 45 kdaltons [4] respectively. The *Chlorobium thiosulfatophilum* (strains PM, L and Tassajara) flavocytochrome *c* has one heme and one flavin in a molecule of approx. 50 000 mol. wt. and this protein can also be dissociated into heme and flavin subunits (ref. 1 and Chapt. 29, this volume). The redox potentials ($E_{m,7}$) of the heme portion of the cytochromes are 10 and 100 mV for

Chromatium and *Chlorobium*, respectively [1], and the redox potential of the flavin [5] of the *Chromatium* protein is close to that measured for the heme. Both flavocytochromes *c* are relatively autoxidizable and are isolated and handled in the oxidized form.

Because it was thought possible that the presence of flavocytochrome *c* was related to thiosulfate metabolism by the bacteria, the reaction between thiosulfate and the cytochromes was tested. Thiosulfate was found to cause loss of the flavin absorbance at pH 5.5 without reduction of the heme [6]. It was also noted that there was transient appearance of 650 nm absorbance during reduction of both heme and flavin with β-mercaptoethanol or dithionite. These observations led to a general survey of compounds which might react with the flavin.

An early study of the phenomena was presented by Meyer [6], limited kinetic studies were made by Vorkink [5] and Yamanaka and Kusai (ref. 7 and Chapt. 29, this volume) observed it while studying catalysis of sulfide oxidation by the *Chlorobium* flavocytochrome *c* and its inhibition by cyanide.

Methods and Materials

Chr. vinosum D and *Chl. thiosulfatophilum* Tassajara flavocytochromes *c* were prepared as described [1]. The samples used for the experiments were those with the highest observed flavin/heme absorbance ratio, A_{475}/A_{525} nm = 1.30 for *Chromatium* and 1.50 for *Chlorobium* flavocytochromes *c*. Reactions were routinely carried out in open cuvettes of 1 ml vol. and 1.0 cm optical path. Buffers of 0.1 M concentration were routinely used: sodium acetate, pH 5; potassium phosphate, pH 6 and 7; Tris·chloride, pH 8; sodium glycinate, pH 9; sodium carbonate, pH 10. For routine observation of the reactions, absolute spectra were recorded using cytochrome concentrations similar to those used for Fig. 1, and the extent of flavin bleaching was monitored at 475 nm.

Results

The flavin of *Chlorobium* and *Chromatium* flavocytochromes *c* reacted with cyanide, sulfite, thiosulfate, GSH and CoASH resulting in bleaching of the flavin absorbance at 450–475 nm with concomitant formation of a new, broad absorbance band centered between 650 and 700 nm, approx. 250 nm in width at half height. Fig. 1 presents difference spectra for the reaction between the two cytochromes and sulfite, thiosulfate and cyanide. From these it is clear that essentially no change in heme absorbance occurs but that the flavin bands disappear. From comparison of the ligand-induced spectra, it is apparent that the long wavelength absorbance-band maxima shift position with various combinations of pH, cytochrome and ligand. The reactions in all cases were multiphasic over a time span varying from seconds to hours, thus precluding true equilibrium measurements. The initial reactions which resulted in flavin bleaching together with the appearance of the long wavelength absorbance band were in some cases stoichiometric and rapid. Reversibility in the strict sense was not determined; nevertheless the original

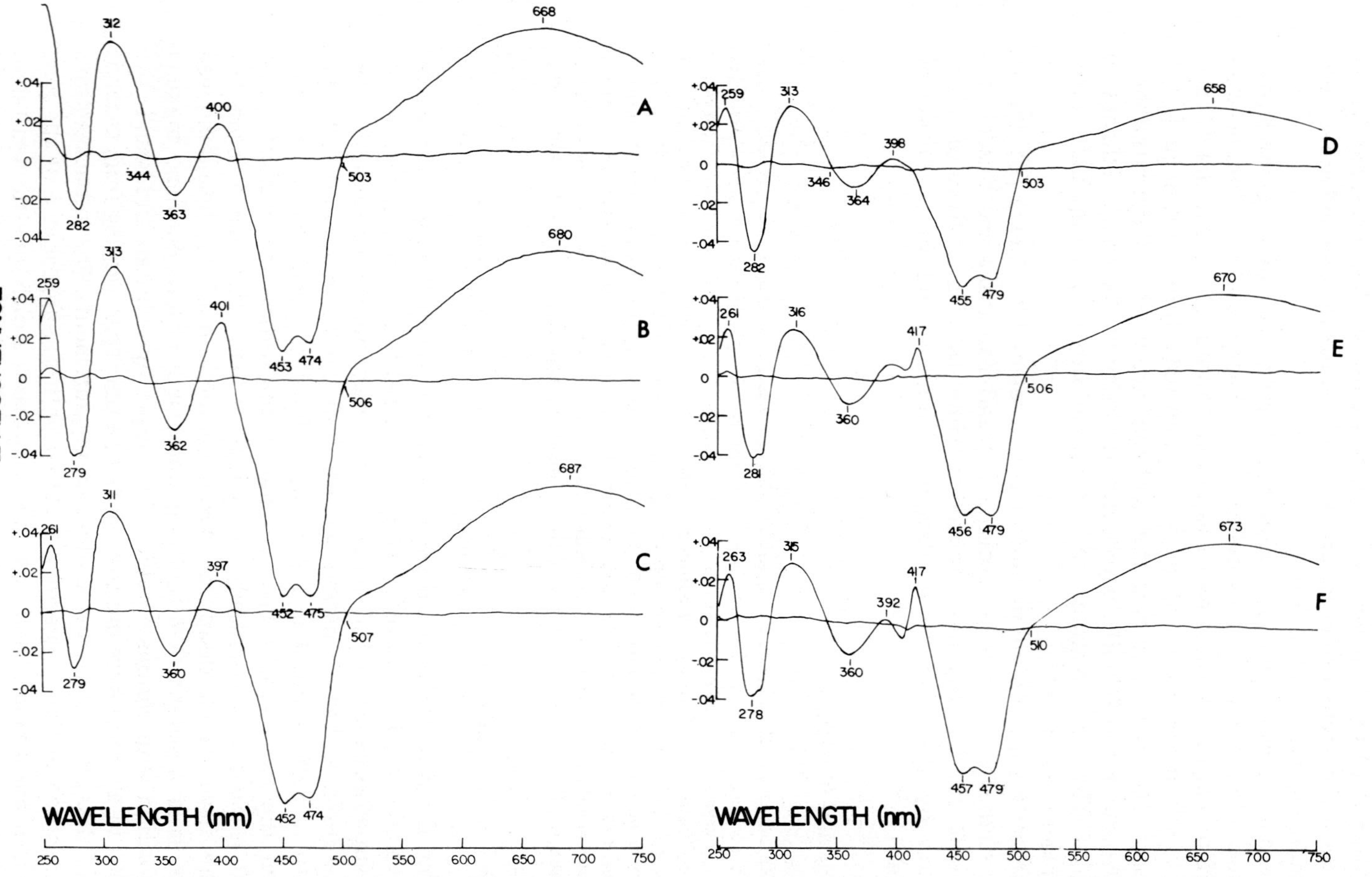

Fig. 1. Difference spectra of flavocytochromes *c* reaction products. A, *Chlorobium* flavocytochrome *c* plus thiosulfate at pH 5. B, C, *Chlorobium* flavocytochrome *c* plus cyanide at pH 5 and 10, respectively. D, *Chromatium* flavocytochrome *c* plus sulfite at pH 5. E, F, *Chromatium* flavocytochrome *c* plus cyanide at pH 5 and 10. The concentration of the *Chlorobium* cyotchrome was approx. 10 μM and of the *Chromatium* cytochrome approx. 7 μM. The reference cell for each spectrum contained the appropriate unreacted cytochrome solution. The ligand concentrations were as follows: [$Na_2S_2O_3$] 330 μM, [Na_2SO_3] [NaCN] 170 μM.

absorbance spectrum could be restored by adding GSSG or mercuriphenyl sulfonate and in some instances by dialyzing the sulfite or cyanide complexes formed at pH 5.0. Subsequent addition of cyanide to the reaction mixtures would again generate the spectral changes. We have not determined if the added reactants are oxidized or otherwise changed, nor if the flavocytochromes *c* become altered in some way not indicated in the spectra.

GSH was more reactive with the *Chlorobium* than with the *Chromatium* cytochrome and the reaction proceeded most rapidly at alkaline pH. Unlike the reaction with other compounds, the *Chlorobium* flavocytochrome *c* heme became reduced above pH 7 and most rapidly at pH 10. The flavin bleach and the new long wavelength absorbance band persisted through overnight incubation at pH 5, but the original absorbance returned within 16 h at pH 10, presumably due to air oxidation of both GSH and reduced cytochrome. With the lower-redox-potential *Chromatium* cytochrome, only slight reduction of the heme occurred at pH 7 and 8. It is probable that the GSH anion was the effective reactant, inasmuch as the reaction rates were very slow at low pH. At pH 5.5, CoASH mimicked the effect of GSH by reacting with the flavin without reducing the heme. Smaller analogs of glutathione, such as sulfide, cysteine and β-mercaptoethanol, reduced the heme rapidly at both high and low pH, but the transient appearance of the long wavelength absorbance band could be observed before reduction was complete.

An essentially stoichiometric reaction occurred between cyanide and the *Chromatium* flavocytochrome *c*, although attempts to determine the equilibrium dissociation constant through titration gave variable results, suggesting an alternative sink for the cyanide. The reaction with *Chlorobium* flavocytochrome *c* was less reproducible since one set of experiments performed gave an approximately stoichiometric reaction with cyanide, whereas other experiments required an approx. 100-fold excess of cyanide and gave little or no pH effect. The reaction products were stable to overnight incubation or longer if minimal cyanide was added but there was indication of a cyanide- and acid-dependent recoloring. The reactions with cyanide are potentially interesting in that they may be sensitive to subtle differences in protein preparations and reaction conditions.

With the *Chlorobium* cytochrome, an essentially stoichiometric reaction occurred with thiosulfate ion but the *Chromatium* cytochrome was much less reactive, requiring as much as 1000-fold more thiosulfate for an equivalent reaction. For both cytochromes, the extent of flavin bleaching within the first few minutes of reaction decreased as the pH was increased from 5 to 10. On prolonged incubation the bleaching continued at alkaline pH, was stable at pH 5, but recoloring occurred at neutral pH, with the result that the net absorbance change was least at intermediate pH values.

An approximately stoichiometric and rapid reaction was observed between sulfite and the *Chromatium* cytochrome, whereas the reaction with the *Chlorobium* cytochrome required a 100-fold higher sulfite concentration for complete reaction. The latter reaction proceeded relatively slowly and the products were stable to prolonged incubation.

In contrast to the *Chlorobium* reaction, there were three distinct kinetic phases observed in the *Chromatium* sulfite reaction. In the first phase (using an approx. 100-fold excess of sulfite), the flavin absorbance bleached within the 10–15 s mixing time at all pH values tested. There followed a second, recoloring phase which required 10 min at pH 5 and 6, but only 30 s at pH 9 and 10, to reach a pseudo-equilibrium state. The third phase, which occurred over a period of hours, resulted in further recoloring between pH 5–7 and rebleaching at higher pH. The extent of the bleaching at the end of the second, recoloring phase and the time required to reach this plateau or pseudo-equilibrium were highly dependent on pH with approximate transition pK values of 6.5 and 7, respectively. Within the first few hours of reaction the extent of flavin bleaching and of increase in long wavelength absorbance remained proportional, but after prolonged incubation the intensity of the latter band was decreased relative to flavin bleaching.

Discussion

There is no precendent for the reaction between a flavoprotein and the variety of reagents reported here, which results in the bleaching of flavin absorbance with concomitant formation of a broad absorbance band at 650–700 nm. However, a possible explanation for the overall phenomenon may be synthesized through separate consideration of the flavin bleaching and the formation of long wavelength absorbance.

The loss of flavin absorbance in general could be due to (1) destruction, (2) reduction or (3) a reversible addition reaction. The flavin in the flavocytochromes *c* was not irreversibly destroyed by sulfite, thiosulfate, cyanide or mercaptans, because the visible absorbance was recovered by various treatments and rebleaching occured upon addition of excess cyanide.

It is conceivable that the flavin in flavocytochrome *c* may be reduced by these various reactants. If this were so, the heme would also be reduced because its oxidation-reduction potential is not lower than that of the flavin. However, the heme was not reduced by sulfite, cyanide or thiosulfate and anaerobiosis had no effect on the reaction with flavin at pH 5.

The reaction between flavocytochrome *c* and GSH was exceptional in that while the heme was reduced at high pH, the long wavelength absorbance band formed, persisting until GSH was presumed to become oxidized by air. There was no heme reduction at pH 5 and the product of reaction with flavin was stable to overnight incubation in air. Therefore, reduction of the flavin may be an explanation for loss of 475 nm absorbance only in the case of reaction with small mercaptans such as cysteine and mercaptoethanol.

The best explanation for reversible bleaching of flavin in flavocytochrome *c* induced by sulfite, thiosulfate, cyanide and mercaptans is the formation of a flavin adduct. Sulfite, but not cyanide, is known to add reversibly to the N-5 position of free flavins [8] and to flavin bound to certain flavoproteins [9]. The flavin-sulfite adduct is equivalent in structure to reduced flavin and has neither visible nor infrared absorbance, but does have a peak at 320 nm [8]. This is approximately the same as that observed in difference spectra for the reaction of sulfite with flavocytochrome *c*. The formation of adducts with the other reagents may be restricted to the peculiar bound flavins of these cytochromes.

Charge transfer would appear to be a possible explanation for the long wavelength absorbance in the flavocytochromes *c*-nucleophile reaction products. For example, the reduced FAD in lipoyl dehydrogenase forms a charge-transfer complex with oxidized NAD [10], and has a broad band at 720 nm with a width, at half-height, of approx. 280 nm and an ϵ_{mM} of approx. 3. The long wavelength bands resulting from reaction of flavocytochromes *c* with sulfite, cyanide, thiosulfate and mercaptans have maxima between 650 and 700 nm (varies with protein, nucleophile and pH), $\epsilon_{mM} \approx 6$, and width at half-height approx. 250 nm. A probable explanation is that the flavin adduct may form a charge-transfer complex with a peptide component, possibly the tyrosine which quenches flavin fluorescence in the flavin pentapeptide obtained from *Chromatium* flavocytochrome *c* [3]. It is not known whether this peptide can form a complex with any of the ligands reactive with the intact cytochrome.

It would be of interest to test the reactivity of other flavoproteins with the ligands which react with the flavocytochromes *c* to determine the generality of the reaction with flavoproteins and to gain thereby possible insight into the biochemical significance of the reactions.

Acknowledgements

This research was supported by grants from the National Science Foundation (BMS72-02409) and the National Institutes of Health (GM-18528) to Professor Martin D. Kamen.

References

1 Bartsch, R.G., Meyer, T.E. and Robinson, A.B. (1968) in Structure and Function of Cytochromes (Okunuki, K., Kamen, M.D. and Sekuzu, I., eds.), pp. 443–451, University Park Press, Baltimore
2 Hendriks, R. and Cronin, J.R. (1971) Biochem. Biophys. Res. Commun. 44, 313–318
3 Walker, W.H., Kenney, W.C., Edmondson, D.E., Singer, T.P., Cronin, J.R. and Hendriks, R. (1974) Eur. J. Biochem. 48, 439–453
4 Kennel, S.J. (1971) Ph.D. thesis, University of California, San Diego
5 Vorkink, W.P. (1972) Ph.D. thesis, University of Arizona, Tucson
6 Meyer, T.E. (1970) Ph.D. thesis, University of California, San Diego
7 Kusai, K. and Yamanaka, T. (1973) Biochim. Biophys. Acta 325, 304–314
8 Müller, F. and Massey, V. (1969) J. Biol. Chem. 244, 4007–4016
9 Massey, V., Müller, F., Feldberg, R., Schuman, M., Sullivan, P.A., Howell, L.G., Mayhew, S.G., Matthews, R.G. and Foust, G.P. (1969) J. Biol. Chem. 244, 3999–4006
10 Massey, V. and Palmer, G. (1962) J. Biol. Chem. 237, 2347–2358

Flavins and Flavoproteins, edited by T.P. Singer

CHAPTER 32

COVALENTLY BOUND FAD IN 6-HYDROXY-D-NICOTINE OXIDASE. LACK OF RIBOFLAVIN INCORPORATION INTO ITS PRECURSOR PROTEIN

MARGARETE BRÜHMÜLLER, ANGELIKA SCHIMZ, LUITGARD MESSMER and KARL DECKER

Biochemisches Institut an der Medizinischen Fakultät der Universität Freiburg, D-7800 Freiburg (G.F.R.)

Summary

6-Hydroxy-D-nicotine oxidase containing covalently bound FAD becomes labelled when induced in the presence of [2-^{14}C]riboflavin. A protein with the immunological properties of a precursor protein was highly purified. Its molecular weight of 53 000 corresponds to that of the active enzyme. This D-enzyme precursor, however, when isolated from cells grown in the presence of [2-^{14}C]riboflavin lacks the flavin moiety.

Introduction

6-Hydroxy-D-nicotine oxidase (D-enzyme) from *Arthrobacter oxidans* [1] is induced in the presence of DL-nicotine (fully induced cells) although its induction kinetics are different from that of the other enzymes of the nicotine degradation pathway, including 6-hydroxy-L-nicotine oxidase [2]. Growth on 0.5% glucose and 0.2% DL-nicotine as sole nitrogen source, specifically prevents the formation of active D-enzyme (L-induced cells) [3]. These L-induced cells, however, contain a protein which forms a precipitin band with antiserum prepared against the purified D-enzyme (anti-D-antiserum); this enzymatically inactive protein is converted in vivo to the active D-enzyme [3]. Data will be presented which strongly indicate that the D-enzyme precursor lacks the flavin moiety and may be identical with or closely related to the apoprotein of 6-hydroxy-D-nicotine oxidase.

Methods

Anti-D-antiserum was prepared as described previously [3]. The Ouchterlony double-diffusion technique was used for immunological assays. The yeast extract required as growth supplement was freed of riboflavin by illumination of its alkaline solution.

Incorporation experiments were performed in the presence of 30 μCi of [2-^{14}C] riboflavin (26 Ci/mol) per liter of medium. Radioactive flavin covalently bound to the protein was determined in acid-precipitable protein after removal of the soluble flavin and in the purified protein after sodium dodecyl sulfate disc electrophoresis. The purification of a small quantity of [2-^{14}C]riboflavin-labelled protein was conducted essentially as described for 6-hydroxy-D-nicotine oxidase [1]. The molecular weight was determined by sodium dodecyl sulfate disc electrophoresis.

Results

Incorporation of [2-^{14}C]riboflavin into protein

When *A. oxidans* was grown in the presence of labelled riboflavin radioactivity was incorporated into the acid-precipitable protein (Table I). The specific radioactivity of the protein, however, was different in cells grown in the presence and in the absence of DL-nicotine, respectively. This difference between induced and non-induced cells corresponded to the amount of active 6-hydroxy-D-nicotine oxidase formed in the presence of DL-nicotine. 3 h-induced cells [3] which show L- but not D-enzyme activity incorporated only as much radioactivity per mg of protein as did repressed cells; L-induced cells, too, incorporated much less label than fully induced bacteria (Table I). Both types of cells lacking active D-enzyme were shown to contain a D-enzyme precursor [3]. The selectivity of [2-^{14}C]riboflavin incorporation was substantiated by Ouchterlony double-diffusion tests with anti-D-antiserum. With extracts of fully-induced cells a radioactive D-enzyme precipitin line was obtained, whereas the lines elicited by the D-enzyme precursor in extracts from 3 h-induced and L-induced cells were unlabelled (Fig. 1).

Purification of the D-enzyme precursor of cells grown on [2-^{14}C]riboflavin

The precursor was purified from extracts of L-induced cells which do not show 6-hydroxy-D-nicotine oxidase activity. Using the immunoprecipitation elicited by this protein with anti-D-antiserum [3] as an assay it was found that the D-enzyme precursor followed sharply the purification pattern of the active D-enzyme. The molecular weight of the purified precursor protein was found to be 53 000 corresponding to that of the active D-enzyme.

The absence of flavin in the D-enzyme precursor was verified after its

TABLE I

INCORPORATION OF RADIOACTIVITY FROM [2-^{14}C] RIBOFLAVIN

Acid-precipitable protein of the cell extracts was solubilized in 0.1 ml tissue solubilizer (NCS) and the bound radioactivity was counted in a Liquid Scintillation Counter.

Type of cells	cpm/mg protein	Δcpm/mg protein	Δcpm/m units L-activity
Non-induced	2300	—	
3 h-induced	2300	0	—
L-induced	4000	1700	3.4
Fully induced	7500	5200	10.4

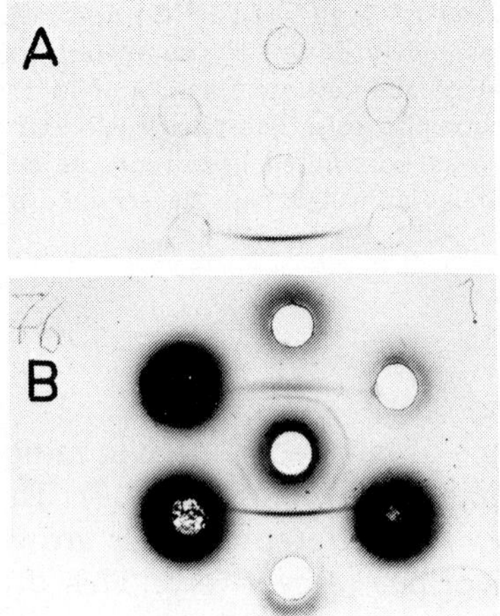

Fig. 1. Immunoprecipitation of differently induced cells grown in the presence of [2-^{14}C]riboflavin. The center well contained anti D-antiserum; the peripheral wells (read clockwise from top) contained extract of 1, unlabelled fully induced cells; 3 and 5, unlabelled 3 h-induced cells; 2, labelled L-induced cells; 4, labelled fully induced cells; 6, labelled 3 h-induced cells. (A) represents the autoradiogram of the stained Ouchterlony plate (B) developed 5 days after exposure.

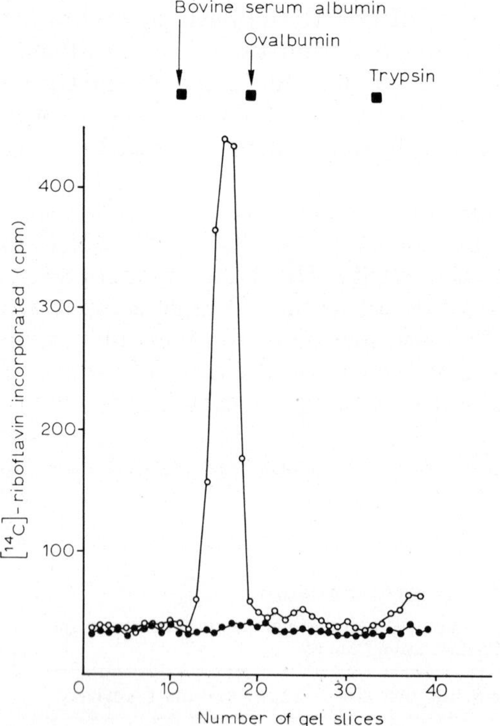

Fig. 2. Sodium dodecyl sulfate disc electrophoresis of 6-hydroxy-D-nicotine oxidase and its precursor purified from [2-^{14}C]riboflavin-grown cells, 1.5 mm gel slices were eluted with 1 mM Tris buffer, pH 7.5, and the eluates counted. o—o, 6-hydroxy-D-nicotine oxidase from fully-induced cells, ●—●, precursor protein from L-induced cells.

purification from L-induced bacteria grown in the presence of [2-^{14}C]riboflavin. Active D-enzyme of flavin-labelled, fully induced cells was purified concomitantly. Both purified fractions were submitted to 8 M urea/1% sodium dodecyl sulfate treatment followed by dialysis and sodium dodecyl sulfate disc electrophoresis (Fig. 2). While the area of the D-enzyme was heavily labelled, the corresponding region from L-induced cells did not contain any radioactivity.

Discussion

These experiments are connected with the process of covalent attachment of the flavin moiety to the protein of 6-hydroxy-D-nicotine oxidase. The incorporation of radioactivity into the active D-enzyme from [2-^{14}C]-riboflavin added to a fully inducing medium indicates that riboflavin or, more likely, one of its derivatives is attached to the protein. The quantity of labelled flavin covalently incorporated by the protein of fully induced bacteria as compared to that of repressed or 3 h-induced cells accounts satisfactorily for the amount required by the newly synthesized D-enzyme. Whether the substantial incorporation of labelled riboflavin into non-induced cells is due to the presence of another protein with covalently bound flavin or merely to a high background cannot be decided yet.

The D-enzyme precursor protein demonstrated immunologically in cells with L- but without 6-hydroxy-D-nicotine oxidase activity does not contain the flavin moiety. This was shown by the lack of riboflavin-labelling of the respective precipitin line and, more clearly, after purification of this protein from cells grown in the presence of [2-^{14}C]riboflavin. The correspondence of the molecular weight of the D-enzyme precursor with that of the active D-enzyme and the absence of the cofactor suggest that this precursor may be the apoprotein of 6-hydroxy-D-nicotine oxidase. Its availability from L-induced cells in reasonable quantities uncontaminated with active D-enzyme make studies on the mechanism of covalent FAD attachment feasible.

The induction of the D-enzyme precursor [3] coincides with that of L- rather than 6-hydroxy-D-nicotine oxidase. This induction behaviour together with the other properties of the precursor indicate that the retarded appearance of the active D-enzyme relative to the other nicotine-degrading enzymes of *A. oxidans* is due to a delayed induction of an enzyme system involved in the covalent attachment of the flavin moiety to the polypeptide chain.

Acknowledgement

This work was supported by grants from the Deutsche Forschungsgemeinschaft, Bonn-Bad Godesberg, G.F.R., through Sonderforschungsbereich 46.

References

1 Brühmüller, M., Möhler, H. and Decker, K. (1972) Eur. J. Biochem. 29, 143–151
2 Gloger, M. and Decker, K. (1969) Z. Naturforsch. 246, 1016–1025
3 Brühmüller, M., Schimz, A., Messmer, L. and Decker, K. (1975) J. Biol. Chem. 250, 7747–7751

Flavins and Flavoproteins, edited by T.P. Singer

CHAPTER 33

STUDIES ON ROSEOFLAVIN: ISOLATION, PHYSICAL, CHEMICAL AND BIOLOGICAL PROPERTIES

SHOHEI OTANI

Department of Nutrition, Koshien University, Momijigaoka 10, Takarazuka 665 (Japan)

Summary

A new pigment has been extracted, as a dark red needle-shaped crystal, from the culture broth of *Streptomyces* strain No. 768 isolated from a soil sample obtained in the Philippines. It was identified as a new flavin compound and named Roseoflavin (Rof) after its color [1]. Its physical, chemical and biological properties were investigated.

Isolation and Purification

Streptomyces strain No. 768 was cultured aerobically at 37°C for 80–120 hours in a culture medium consisting of starch, soybean meal, meat extracts and salts. Roseoflavin in the filtered broth was adsorbed onto diatomaceous earth (Celite FC), and eluted with 5% hot pyridine. After concentration and cooling of the eluate, crude crystals were obtained. The crude roseoflavin was dissolved in a mixture of *n*-butanol, dimethylformamide and water (4:1:1) and applied to a cellulose column which had been treated previously with aqueous *n*-butanol and eluted with the same solvent. The red-colored fraction of the eluate was collected and evaporated in vacuo. On cooling, the deep red crystals of roseoflavin were isolated. By recryastallisation with hot water pure roseoflavin was obtained.

Physical and chemical properties

Rof has a melting point of 276–278°C, $[\alpha]_D^{22} = -315°$ (0.1 M NaOH). Elemental analysis gave the following data: C 52.22, H 5.46, N 16.81%; no halogen and sulphur were found. The molecular weight was estimated to be 405 from the ion peak in the mass spectrum. These values coincide well with those calculated as $C_{18}H_{23}N_5O_6$. Roseoflavin is easily soluble in dimethylsulfoxide; soluble in water, pyridine, methanol, acetone and *n*-butanol; insoluble in benzene, *n*-hexane, chloroform and carbon tetrachloride. The absorbance maxima (nm) and molar extinction coefficients ($\times 10^{-3}$, in brackets) of the ultraviolet/visible absorbance spectrum of an aqueous solution of roseoflavin are as follows: 223(14.0), 258(40.0), 314(6.77) and 505(32.8).

Solutions of roseoflavin, both in water and in organic solvents, were photosensitive and turned yellow and became fluorescent on exposure to sunlight. Determination of redox potential gave the value of E_0 (pH = 7.01, t = 25°C) = −222 mV (determined by Professor Yamano).

The chemical structure of roseoflavin was studied by Professor Matsui and it proved to be 7-methyl-8-dimethylamino-D-ribitylisoalloxazine [2].

Biological properties

Roseoflavin was tested for riboflavin activity in a culture of *Lactobacillus casei* ATCC 7469, which is absolutely heterotrophic to riboflavin, and in the nutrition of rats.

Effect on Lactobacillus casei

The medium, the composition of which is shown in Table I, was used as the basal medium in the culture of *Lactobacillus.* To this basal medium, riboflavin and roseoflavin were added in different ratios and the three series of such media were prepared for the incubation at 37°C for 72, 96 and 120 h.

TABLE I

BASAL MEDIUM USED FOR THE INCUBATION OF *LACTOBACILLUS CASEI*

The pH was adjusted to 6.8; quantities expressed as mg, unless otherwise stated.

Casamino acids (DIFCO)	4.5 g	Adenine	10
L-Cystine	100	Guanine	10
DL-Tryptophan	100	Xanthine	10
Thiamine/HCl	1.0	Uracil	10
Pyridoxine	1.0	Glucose	20 g
Calcium pantothenate	1.0	Sodium acetate	20 g
Nicotinic acid	1.0	K_2HPO_4	500
p-Aminobenzoic acid	0.2	KH_2PO_4	500
Biotin	0.01	$MgSO_4 \cdot 7H_2O$	200
Folic acid	0.01	$FeSO_4 \cdot 7H_2O$	10
		$MnSO_4 \cdot 4H_2O$	10
		NaCl	10
		H_2O	1000 ml

The growth of *Lactobacillus* was estimated by titration of lactic acid formed in the medium with 0.1 M NaOH solution.

In the inhibition test, 0.3 μg of riboflavin was added to 10 ml of the basal medium and to this the increasing amounts of roseoflavin were added. As shown in Fig. 1, with the addition of a small amount of roseoflavin to the medium, promotion of the growth of *Lactobacillus* was observed, but on further addition growth was inhibited. Concentrations of 9, 10 and 12 μg to 10 ml of the medium stopped all growth of the *Lactobacillus* when observed at 72, 96 and 120 h of incubation, respectively.

The ratio of riboflavin to roseoflavin at half-maximum growth, in other words the values of the inhibition index were 1:14, 1:19 and 1:30 after 72, 96 and 120 h incubation, respectively.

However, when growth was completely inhibited by roseoflavin, addition of riboflavin resulted in removal of inhibition. Ten μg of roseoflavin

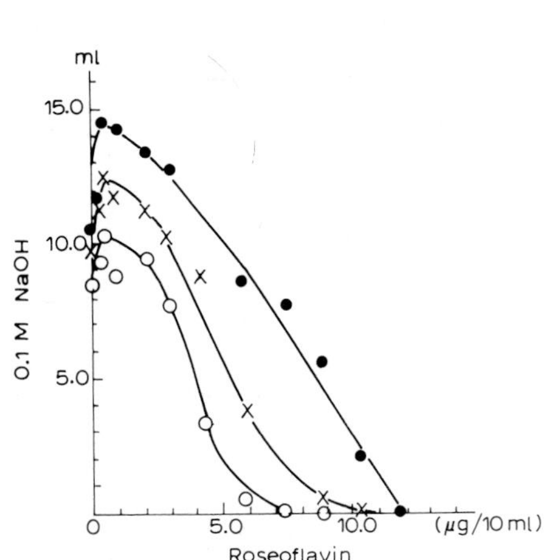

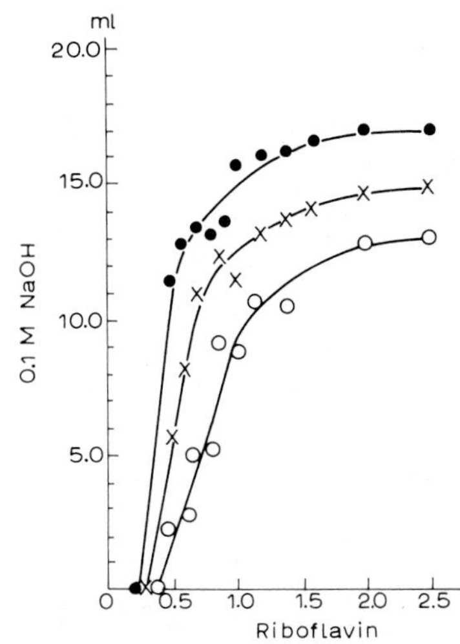

Fig. 1. Inhibitory effect of roseoflavin in the culture of *Lactobacillus casei* ATCC 7469. 0.3 μg of roseoflavin in 10 ml of the basal medium. Riboflavin added at concentrations shown. ○——○, after 72 h; X——X, after 96 h; ●——●, after 120 h.

Fig. 2. Deinhibition effect of riboflavin in the culture of *Lactobacillus casei* ATCC 7469. 10 μg roseoflavin in 10 ml basal medium. Riboflavin added at concentrations shown. ○——○, after 72 h; X——X, after 96 h; ●——●, after 120 h.

were added to 10 ml of the basal medium and to this medium increasing amounts of riboflavin were added, and three series of such media were prepared for incubation at 37°C for 72, 96 and 120 h. As shown in Fig. 2, maximum growth of *Lactobacillus* was observed with 2.0 μg of riboflavin to 10 ml of the medium in each series of experiments. The ratios of riboflavin to roseoflavin at half maximum growth were 1:12, 1:20 and 1:23 at 72, 96 and 120 h incubation respectively.

Effect on rat

Male weanling *Donryu* rats fed with riboflavin-free basal diet were daily given a mixture of riboflavin and roseoflavin in 0.5 ml of 10% gum acacia solution by stomach tube. As a positive control rats were given 40 μg of riboflavin daily and as a negative control rats were fed basal diet only.

As shown in Table II, the adminstration of roseoflavin affected the growth-promoting effect of riboflavin and little difference was found, after seven weeks, in body-weight gain between the negative controls and the rats given 40 μg of riboflavin plus 120 μg of roseoflavin. The ratio of riboflavin to roseoflavin at half maximum weight gain was 40 μg:100 μg, in other words, 1:2.5.

In recovery tests, 21 rats fed with basal diet were given 40 μg of roseoflavin daily for two weeks and then they were divided into three groups. The rats of each group were given daily riboflavin and roseoflavin in different ratios and growth was observed for another four weeks.

As shown in Table III, in the group given 44 μg of riboflavin and 40 μg of roseoflavin, the weekly body-weight gain was almost the same as the

TABLE II

GROWTH RESPONSE OF RATS OF RIBOFLAVIN AND ROSEOFLAVIN

Group	Riboflavin (μg) per day	Roseoflavin (μg) per day	No. of rats	No. of deaths	Average weekly body-weight gain (g)							
					Week:1	2	3	4	5	6	7	Total average (g)
I	0	0	6	0	8.1	8.7	3.7	7.4	5.3	3.3	6.9	6.2
II	40	100	6	0	20.4	13.9	15.2	15.4	14.6	12.3	20.3	16.0
III	40	120	6	0	14.0	5.3	0.4	5.9	8.2	8.2	9.0	7.3
IV	40	0	6	0	20.2	17.4	25.5	31.8	25.6	20.4	23.7	23.5
II − I/IV − I × 100: 100					104.7	84.5	67.6	54.7	52.6	52.6	79.8	56.6

TABLE III

GROWTH RESPONSE OF RATS OF RIBOFLAVIN AND ROSEOFLAVIN

Group	Riboflavin (μg) per day	Roseoflavin (μg) per day	No. of rats	No. of deaths	Average weekly body-weight gain (g)						
					Week:1	2	3	4	5	6	Total average (g)
I	0	0	7	0	3.7	−1.8	0.9	3.0	4.2	0	1.7
II	0	40	21	1	8.1	1.8	—	—	—	—	
II_a	16	40	7	0	—	—	7.2	15.2	19.1	14.0	13.9
II_b	20	40	7	0	—	—	8.1	21.0	21.3	13.7	16.0
II_c	44	40	6	0	—	—	15.2	40.0	34.3	32.7	30.6
III	40	0	7	0	19.2	17.8	31.0	35.1	37.8	33.6	29.8
II_b − I/II_c − I × 100:					—	—	50.3	48.6	56.8	41.9	49.5

positive controls. In the group given 20 μg of riboflavin and 40 μg of roseoflavin, the ratio of riboflavin to roseoflavin was 1:2, the body-weight gain was almost half the positive control.

From the results mentioned above it appears that roseoflavin is a naturally occuring analogue of riboflavin, which has an antagonistic effect on the growth-promoting activity of riboflavin.

References

1 Otani, S., Takatsu, M., Nakano, M., Kasai, S., Miura, R. and Matsui, K. (1974) J. Antibiot. 27, 88
2 Kasai, S., Miura, R. and Matsui, K. (1976) Bull. Chem. Soc. Japan, in press

Flavins and Flavoproteins, edited by T.P. Singer

CHAPTER 34

PHOTOLYSIS PRODUCTS OF ROSEOFLAVIN

KUNIO MATSUI and SABU KASAI

Division of Biology, Research Institute for Atomic Energy, Osaka City University, Sumiyoshi-ku, Osaka 558 (Japan)

The known photolysis products of flavins are classified to two groups: one is the group formed by degradation of the ribityl chain [1–5], the other by hydroxylation of the isoalloxazine ring [6–8]. We identified two of the photolysis products of roseoflavin, 7-methyl-8-dimethylamino-10-D-ribityl-isoalloxazine [9] (I); one was a compound of the former group, 7-methyl-8-dimethylamino-alloxazine (II), the other was a new type of photoproduct, 7-methyl-8-methylamino-10-D-ribityl-isoalloxazine (III). A probable product of lumiflavin type, 7,10-dimethyl-8-dimethylamino-isoalloxazine (IV), was not found in the photolysis mixtures.

Compound I was prepared according to Kasai et al. [10]. For thin-layer chromatography, Merck's precoated silica gel plate 60F, the solvent system A (isoamyl alcohol/methyl ethyl ketone/acetic acid/water, 40:40:7:13, by vol.) and B (chloroform methanol, 19:1, by vol.) were used.

Isolation and synthesis of II

I (0.4 g), suspended in 20 mM HCl (0.6 l) was exposed to sunlight. After complete dissolution (10–20 days), the newly formed fine precipitate was collected and recrystallized from pyridine (pyridine complex, orange yellow). Yield, 60 mg. It was converted to the orange red hydrate by treatment with water.

2-Dimethylamino-4-amino-toluene [10] (9.7 g) in methanol (30 ml) was added to a refluxing mixture of violuric acid (10 g), methanol (150 ml), and acetic acid (20 ml), and the reflux was continued for 5 h [11]. The crystals formed were filtered and washed with methanol. Yield, 12.8 g. The product was dissolved in 1 M NaOH at 70–80°C and recrystallized by adding 0.1 vol of acetic acid. The orange red crystals were washed with water and air dried. When they were recrystallized from pyridine, the pyridine complex was obtained.

Isolation and synthesis of III

I (0.5 g), suspended in 50 mM NH_3 (1 l) was exposed to sunlight. After complete dissolution (40–60 days), the solution was evaporated to dryness

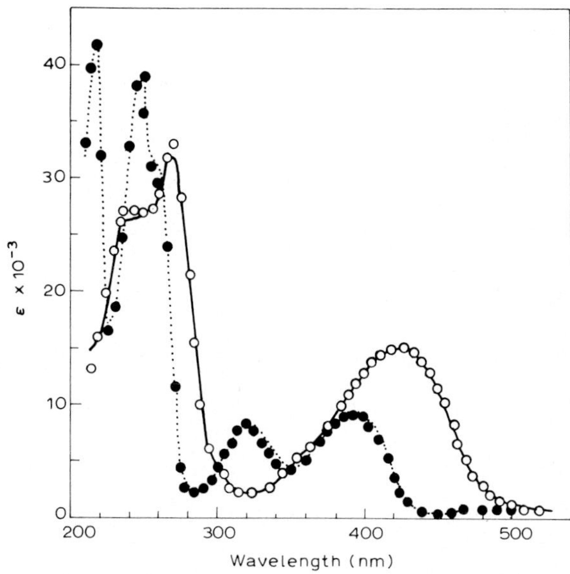

Fig. 1. Absorption spectra of II. ○ circle, photoproduct in 0.1 M NaOH; ———, synthesized sample in 0.1 M NaOH; ●, photoproduct in 6 M HCl; ----- synthesized sample in 6 M HCl.

and the residue was treated with pyridine (20 ml) and acetic anhydride (2.3 ml). After stirring overnight, the mixture was evaporated to dryness and treated with water (40 ml). The suspension was extracted four times with 30 ml aliquots of chloroform. The dried extracts were evaporated with $MgSO_4$ and the residue dissolved in 5% methanol in chloroform (v/v, 10 ml) was charged to a dry column of silica gel (1.8 x 16 cm). The materials were eluted with the same solvent (160 ml) and then with 10% methanol in chloroform (v/v, 200 ml) by suction. Being monitored with thin-layer chromatography (solvent B), the fractions, containing main green-yellow fluorescent substance (R_F about 0.4), were combined and evaporated. The residue was recrystallized from ethanol (orange red crystals). Yield, 0.1 g. The isolated acetate of III (93 mg) was dissolved in 1 M NaOH (1 ml) and after 10 min, the solution was neutralized with acetic acid (0.1 ml). The precipitates were washed with water and dried. Yield, 65 mg.

A mixture of 2-methylamino-4-amino-toluene [12] (9 g), D-ribose (9 g), and methanol (20 ml) was refluxed for 2 h. The reaction mixture was transferred to an autoclave and reduced with H_2 in the presence of Raney nickle prepared from 50% alloy (24 g). After reduction (120 atm, 100°C, 1 h, with stirring), the catalyst and solvent were removed. The syrup was mixed with hot methanolic solution of violuric acid (9.3 g in 150 ml), and the mixture was refluxed for 2 h. The crystals were filtered, washed with methanol and dried. Yield, 16 g (62%). The product (3 g) was acetylated as described above (anhydrous pyridine, 20 ml; acetic anhydride, 5 ml). The reaction mixture was evaporated, and the residue was dissolved in chloroform. After washing with water, the chloroform

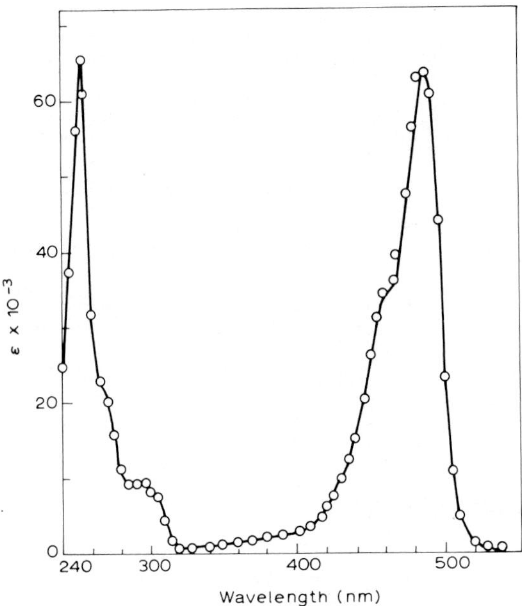

Fig. 2. Absorption spectra of III acetate in chloroform. o, photoproduct; ——, synthesized sample.

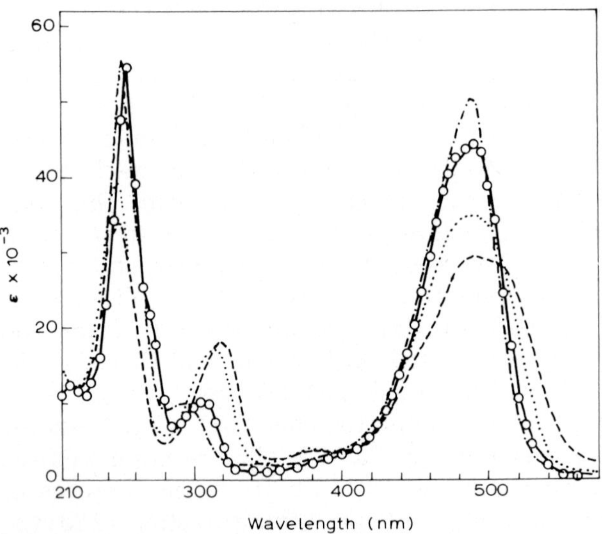

Fig. 3. Absorption spectra of III (free form). o, photoproduct in water; ——, synthesized sample at pH 7.01; ------ at pH 0.85 (cation); — — — in 11 M HCl (divalent cation); — · — · — at pH 12.39 (anion).

solution was evaporated, and the residue was recrystallized from ethanol. Yield, 3.1 g (79%).

A pure preparation of free III was obtained by hydrolysis of the acetate with NaOH as described above. pK_a, 11.1, 1.7 and < 0.

Synthesis of IV

2-Dimethylamino-4-methylamino-toluene (*V*). To a mixture of 2-dimethylamino-4-amino-toluene [10] (17.3 g), dioxane (100 ml) and 30% (w/v, 30 ml) was added dimethyl sulfate (9 ml) gradually under stirring overnight at room temperature. Monitoring with gas chromatography, methylation was repeated until all of the amine was methylated. After reaction, water was added to the reaction mixture, and the product was extracted with ether, and purified by distillation. b.p., 142°C/25 mmHg. Yield, 11.8 g. Found: C, 73.48; H, 9.57; N, 16.33%. Calculated for

TABLE I

PROPERTIES OF II AND III

Substance	m.p. (°C)	Elemental analysis (%) C	H	N	R_F	$[\alpha]_D$ (°)
II Pyridine complex						
Photoproduct	> 300[a]	61.87	5.17	23.57	0.72[c]	
Synthesized	> 300[a]	62.23	5.23	23.98	0.72[c]	
Calculated for $C_{13}H_{13}N_5O_2$—C_5H_5N		61.70	5.18	23.99		
II Hydrate						
Photoproduct	> 300[a]	51.70	5.18	21.69		
Synthesized	> 300[a]	51.60	5.38	23.44		
Calculated for $C_{13}H_{13}N_5O_2$–I · $5H_2O$		52.34	5.41	23.48		
III Acetate						
Photoproduct	267–271[b]	53.17	5.16	12.30	0.40[d]	+ 242[e]
Synthesized	268–272[b]	53.02	5.22	12.64	0.40[d]	+ 234[e]
Calculated for $C_{25}H_{29}N_5O_{10}$		53.66	5.22	12.52		
III (free)						
Photoproduct	290–293[b]	47.59	5.85	16.07	0.16[c]	− 112[f]
Synthesized	290–295[b]	47.13	5.73	15.83	0.16[c]	− 107[f]
Calculated for $C_{17}H_{21}N_5O_6$ · $2H_2O$		47.77	5.90	16.39		

[a] Sublimate. [b] Decomposition. [c] Solvent A. [d] Solvent B. [e] Chloroform. [f] 0.1 M NaOH.

TABLE II

δ VALUES OF ASSIGNABLE PROTONS OF II HYDRATE AND III TETRAACETATE

Substance	δ (ppm) CO—CH_3				C—CH_3	NCH_3		$N(CH_3)_2$	Aryl H	
II Hydrate[a]										
Photoproduct					2.92			3.72	8.47	8.67
Synthesized					2.90			3.72	8.47	8.68
III Acetate[b]										
Photoproduct	1.83	2.07	2.12	2.28	2.28	3.15	3.20		6.37	7.60
Synthesized	1.83	2.07	2.12	2.28	2.28	3.13	3.18		6.37	7.58

[a] In CF_3CO_2H. [b] In C^2HCl_3.

$C_{10}H_{16}N_2$: C, 73.12; H, 9.82; N, 17.06%. NMR(C^2HCl_3) δ: 2.20_s(C—CH_3), 2.61_s(NH), 2.65_s($N(CH_3)_2$), 2.83_s(NCH_3), 6.40_s(3—H).

V was added to a hot methanolic solution of violuric acid (5 g in 100 ml) in methanol (5.5 g in 20 ml), and the mixture was refluxed for 2 h. The crystals were filtered and washed with methanol. The product was dissolved in 1 M NaOH and recrystallized by adding 0.1 vol of acetic acid. The brown red crystals were washed with ethanol. Yield, 5.8 g (64%). Found: C, 58.55; H, 5.32; N, 24.54%. Calculated for $C_{14}H_{15}N_5O_2$: C, 58.93; H, 5.30; N, 24.55%. NMR(CF_3CO_2H), δ: 2.77_s(C—CH_3), 3.62_s($N(CH_3)_2$), 4.23_s(NCH_3), 7.05_s, 7.98_s(6,9—H).

The properties of photoproducts of I and synthesized samples are shown in Table I, II and Figs. 1–3. The product II coincided with the synthesized sample in all properties tested. In mass spectra they also coincided each other. The acetate and free form of product III coincided with those of the synthesized in properties including specific rotation. The absorption spectra of III in 0.1 M NaOH and 6 M HCl also coincided with those of the synthesized, and the mass spectrum of acetate of product III also with the synthesized acetate (not shown). Thus, II and III were identified as 7-methyl-8-dimethylamino-alloxazine and 7-methyl-8-methylamino-10-D-ribityl-isoalloxazine, respectively.

The photochemical formation of III is a new type of photolysis reaction of flavin. The fate of methyl group detached from dimethylamino group is unknown. However, as the reaction mixture smelled of an aldehyde, the methyl group might be oxidized to HCHO. The mechanism of the reaction was not studied. The dimerization reaction of riboflavin [13] may give a hint to elucidate the mechanism. The photolysis reaction of I in aqueous solution was very slow. But the analogous reaction in organic solvent was faster; the tetraacetate of I was rapidly photolyzed in chloroform. Though the product was not finally identified, was obtained evidences suggesting that the main product was III tetraacetate. Dr. S. Otani also observed rapid photolysis of I in pyridine (personal communication). The photolysis reaction of I would be affected by protons.

The other probable photoproduct of lumiflavin type (IV) was not detected thin-layer chromatographically in photolysis mixtures of I. Even if IV is formed, it may be photolyzed to demethylated compound, 7,10-dimethyl-8-methylamino-isoalloxazine, the occurrence of which is not established.

References

1 Karrer, P., Salomon, H., Schöpp, K., Schlittler, E. and Fritzsche, H. (1934) Helv. Chim. Acta 17, 1010
2 Ghisla, S. and Mayhew, S.G. (1973) J. Biol. Chem. 248, 6568
3 Kuhn, R., Rudy, H. and Wagner-Jauregg, T. (1933) Bericht 66, 1950
4 Smith, E.C. and Metzler, D.E. (1963) J. Am. Chem. Soc. 85, 3285
5 Cairns, W.L. and Metzler, D.E. (1971) J. Am. Chem. Soc. 93, 2772
6 Nijenhuis, Te. B. and Berends, W. (1967) Photochem. Photobiol. 6, 491
7 Schöllnhammer, G. and Hemmerich, P. (1974) Eur. J. Biochem. 44, 561
8 Mayhew, S.G., Whitfield, C.D. and Ghisla, S. (1974) Eur. J. Biochem. 44, 579

9 Otani, S., Takatsu, M., Nakano, M., Kasai, S., Miura, R. and Matsui, K. (1974) J. Antibiotics 27, 88
10 Kasai, S., Miura, R. and Matsui, K. (1975) Bull. Chem. Soc. Japan 48, 2877
11 Piloty, O. (1904) Liebig's Ann. Chem. 33, 22
12 Gnehm, R. and Blumer, E. (1899) Lieberg's Ann. Chem. 304, 87
13 Hemmerich, P., Prijs, B. and Erlenmeyer, H. (1959) Helv. Chim. Acta 42, 2164

Flavins and Flavoproteins, edited by T.P. Singer

CHAPTER 35

STUDIES ON THE ACTIVE CENTERS OF FLAVOPROTEINS: BINDING OF 8-HYDROXY-FAD AND 8-HYDROXY-FMN TO APOPROTEINS

SANDRO GHISLA[a]*, VINCENT MASSEY[a] and STEPHEN G. MAYHEW[b]

[a] *Department of Biological Chemistry, The University of Michigan, Ann Arbor, Mich. 48104 U.S.A. and* [b] *Department of Biochemistry, Agricultural University, Wageningen (The Netherlands)*

Summary

8-OH-FAD and 8-OH-FMN were bound to a series of apoflavoproteins and the spectra of the complexes and their pH dependencies compared with those of model 8-OH-flavins with functional substituents in the isoalloxazine nucleus. Apoflavodoxins, and also apo-old yellow enzyme, shift the p*K* for the 8-OH group to higher values, indicating that they probably contain a negative charge near their flavin-binding sites; the spectral characteristics of the bound flavin indicates that protonation in the enzyme complexes occurs at position C_8—O rather than at N_1—C_2=O. In contrast, the flavin-binding sites of lactate oxidase, D-amino-acid oxidase and electron-transferring flavoprotein probably contain a positive charge which stabilizes the anionic form of the 8-OH-flavin. Pentafluorophenol and 8-OH-FMN-old yellow enzyme form a charge-transfer complex in which the p*K* of the C-8 hydroxyl group is shifted to higher values. Although phenolate anion is bound to both ionic states of the flavin, charge transfer occurs only with the neutral flavin.

Structure and properties of 8-hydroxy-flavins

The presence of an orange prosthetic group in an NADH-dehydrogenase from *Peptostreptococcus elsdenii* was reported some years ago by Mayhew and Massey [1]. This enzyme was later shown to be a modified form of electron-transferring flavoprotein [2] and the structure of the chromophore was established as the 8-hydroxy analog of FAD [3]. Compared with unmodified flavins at neutral pH, the visible absorbance of 8-OH-flavins in red shifted and about 4 times as intense; it also depends on pH, reflecting a p*K* of ≈ 4.8 (Fig. 1). The chromophore is fluorescent, with emission maxima in the range 505–528 nm which similarly reflect the p*K* of deprotonation at the

* Present address: Fachbereich Biologie der Universität, Postfach 733, D-775 Konstanz, G.F.R.

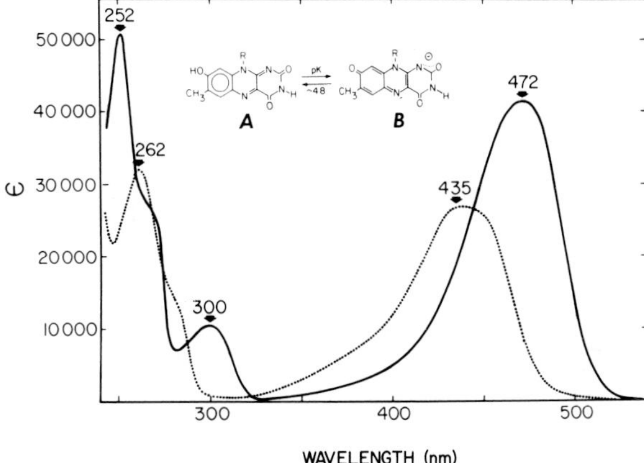

Fig. 1. Absorbance spectra of the neutral (······) and anionic (——) forms of 8-OH-FMN recorded at pH 3 and 8, respectively.

8-hydroxyl (p$K \approx 4.8$), the pK of deprotonation of position N_3—H (pK = 11.5) and also the pK of the adenine moiety in the FAD derivative (Ghisla, S. and Mayhew, S.G., in preparation). The neutral form of the flavin can be represented by two tautomeric structural formulae, a "phenolic" one (cf. A, Fig. 1) and a "paraquinoid" structure (cf. B, with a proton at N_1—C_2 = O, Fig. 1). Deprotonation can lead to electronic distribution approximated by either A or B with the negative charge located predominantly either at C_8—O or at N_1—C_2=O. Structures A and B should have markedly different electronic spectra, as indeed is the case (Fig. 1.)

To assign the observed spectral types to either A or B, models were synthesized with substituents at N-1, N-3, C_8—O and N-10 of the isoalloxazine nucleus. Blocking of N-1 and N-10 leads to spectra of type B with absorbance maxima at ≈ 475 nm ($\epsilon \approx 25\,000$) and a characteristic band at 318 nm ($\epsilon \approx 17\,400$) and with fluorescence emission maximal at 557 nm for the neutral molecules. Substitution at C_8—O leads to spectra very similar to that of neutral orange flavin (A, Fig. 1) with an absorbance maximum at 436 nm ($\epsilon \approx 22\,600$) and shoulders at 380 and 280 nm.

Binding of 8-hydroxy-flavins to apoflavoproteins

The favorable spectral properties and pK of 8-OH-flavins suggested that these flavins might be sensitive probes for the active centres of flavoenzymes. Thus, hydrogen bridges directed towards N_1—C_2=O or C_8—O might stabilize structures similar to A or B. Charged protein residues near these regions of the coenzyme could have analogous effects on the distribution of the negative charge of the 8-OH-flavin anion and influence its spectrum. Furthermore, the net charge at the coenzyme-binding site could induce pK shifts in the bound 8-OH-flavin. Similar considerations apply for the 6-OH derivatives of FAD and FMN (pK = 7.1), which are characterized by a broad band at 600 nm [4].

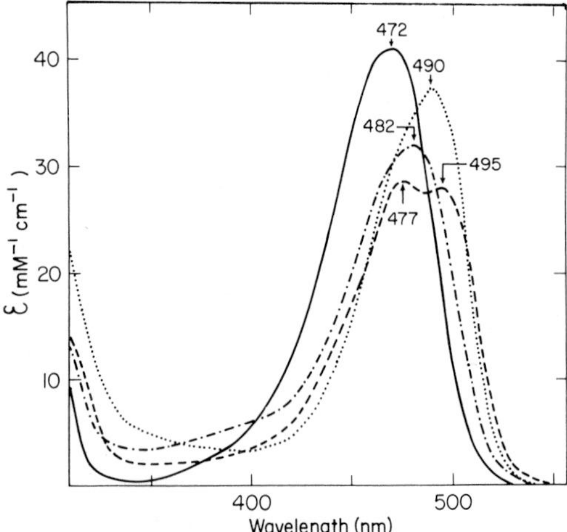

Fig. 2. Absorbance spectra of complexes of the anion of 8-OH-FMN with apoflavodoxins. Spectra were recorded at pH 8–9 and are shown for 8-OH-FMN free in solution (——) and bound to apoflavodoxins from *P elsdenii* (-·-·-·-), *D. vulgaris* (---) and *A. vinelandii* (······).

(1) Apoflavodoxins

The spectral changes which accompany the binding of 8-OH-FMN to apoflavodoxin from *P. elsdenii* at pH 8 (Fig. 2) are analogous to those observed with the natural coenzyme (5); the absorbance maximum is shifted to longer wavelength and decreased in intensity, and the fluorescence is >99% quenched. The spectrum of the complex shows a pH dependence similar to that of the free flavin, but the p*K* is shifted to 6.1 (Fig. 3). The spectra of the neutral complexes with apoflavodoxins from *Desulfovibrio vulgaris* and *Azotobacter vinelandii* (Shethna flavoprotein [6]) are similar to that with *P. elsdenii* apoflavodoxin. However, the spectra of the anions are markedly different as shown in Fig. 2. The resolution of the spectrum in the case of the complex with *D. vulgaris* apoflavodoxin possibly reflects a more hydrophobic flavin environment provided by the tyrosine and tryptophan residues known to sandwich the flavin in this protein [7]. The p*K* values for the 8-OH-FMN complexes with *D. vulgaris* and *A. vinelandii* apoflavodoxins are shifted to higher pH, but the shifts are not as marked as in the case of the complex with *P. elsdenii* apoflavodoxin. These p*K* shifts indicate that the flavin-binding sites in all three apoproteins are negatively charged (glutamic and aspartic acid residues are known to be close to the pyrimidine subnucleus in the cyrstal structure of flavodoxin from *Clostridium* MP [8]). The shapes of the spectra of the neutral complexes indicate that protonation occurs at position 8-O of the flavin.

*(2) Apo-*D*-amino acid oxidase and apolactate oxidase*

Very different behavior is observed upon binding 8-OH-FAD to apo-D-amino acid oxidase. The modified flavin is bound more tightly (K_d = 10^{-7}M) than FAD ($K_d = 2 \cdot 10^{-7}$M [9]), the absorbance spectrum of the

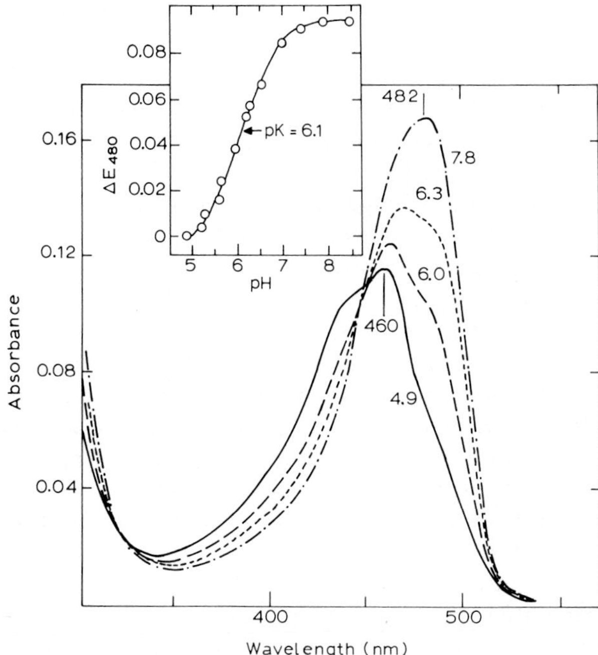

Fig. 3. Effect of pH on the absorbance spectrum of 8-OH-FMN-apoflavodoxin. A solution of $4.5 \cdot 10^{-6}$ M 8-OH-FMN and $7.3 \cdot 10^{-6}$ M apoflavodoxin (*P. elsdenii*) in 2 ml 0.05 M NaCl and 0.01 M potassium phosphate buffer pH 7.8 was titrated first to pH 4.9 by adding increments of solid citric acid and then to pH 8.5 by addition of 2 M NaOH. Absorbance spectra and pH were recorded after each addition. The inset shows a plot of the change in absorbance at 480 nm versus pH.

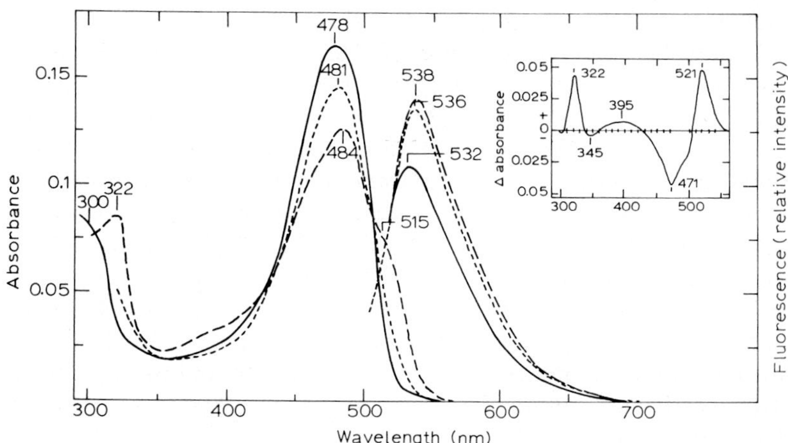

Fig. 4. Spectral properties of 8-OH-FAD bound to apo-D-amino acid oxidase. A solution of 8-OH-FAD ($9 \cdot 10^{-6}$ M) in 0.1 M pyrophosphate buffer, pH 8.5 (——) was treated with increments of apo-D-amino acid oxidase ($3.3 \cdot 10^{-4}$ M) in the same buffer. ----, after addition of apoprotein to a final concentration of $2 \cdot 10^{-5}$ M; ······, after the subsequent addition of sodium benzoate to $2 \cdot 10^{-3}$ M. The inset shows the difference spectrum between 8-OH-FAD-apo-D-amino acid oxidase and its complex with benzoate.

complex is resolved, with shoulders at 515 and 390 nm, and has a second maximum at 322 nm (Fig. 4), and the fluorescence is increased and shifted to 538 nm. The absorbance spectrum indicates that the flavin structure is of the B type and suggests that there may be a positive charge near N-1. Binding of benzoate to native D-amino acid oxidase increases the resolution of the absorbance spectrum [10] but binding of the inhibitor to the complex with 8-OH-FAD has the opposite effect; in particular the band at 320 nm is lost (Fig. 4). This suggests that benzoate neutralizes a positive charge in the protein which is probably responsible for stabilization of the anion form of the 8-OH-flavin. This same positively charged group is probably responsible for substrate binding, since benzoate is a competitive inhibitor.

The spectrum of the complex of apolactate oxidase from *Mycobacterium smegmatic* and 8-OH-FMN (not shown) is very similar to that of the D-amino acid oxidase complex, with marked absorbance bands at 323 and 492 nm and shoulders at 400, 470 and 520 nm. The complex is also strongly fluorescent with a well-resolved emission spectrum showing maxima at 550 and 585 nm. Addition of the strong inhibitor oxalate [11] does not result in any spectral changes, indicating that a positive charge is probably already neutralized by the flavin anion. Consistent with this interpretation, lowering the pH to 4 fails to cause detectable spectral changes, thus indicating a considerable lowering of the p*K*. The artificial complexes with apo-D-amino acid oxidase and apolactate oxidase do not have catalytic activity with the natural substrates. Binding of 8-OH-FAD to apoelectron-transferring flavoprotein has been shown to give a spectrum similar to that observed with D-amino acid oxidase and lactate oxidase [2].

(3) Apo-old yellow enzyme

8-OH-FMN is bound very tightly to the apoprotein of the old yellow enzyme; the K_d at 25°C, pH 7 is less than 10^{-8} M. The modified enzyme, in contrast to the native enzyme, is strongly fluorescent, with an emission maximum at 543 nm and a marked shoulder at 575 nm. The fluorescence intensity is increased by a factor of appox. 1.5 over that of the free coenzyme. However, in contrast to the free coenzyme, which is fluorescent in both its neutral and anionic forms, the complex with old yellow enzyme is fluorescent only in the anionic form. As shown in Fig. 5 (inset) the p*K* for the enzyme-bound flavin so determined is 6.2. The same p*K* value is also obtained from changes in absorbance spectrum as a function of pH. Fig. 5 shows the spectra of the protein-bound chromophore in its neutral and anion forms. The former is unusual among those so far studied in exhibiting a well-resolved spectrum with maxima at 458 and 478 nm. The anion is similar to that found with D-amino acid oxidase and lactate oxidase.

It was of particular interest to examine the effects resulting from interaction of phenols with the modified old yellow enzyme. Native old yellow enzyme has recently been isolated in a green form [12] which was shown to be due to liganding with a small molecule metabolite. The naturally occuring ligand was identified as *p*-hydroxybenzaldehyde (Matthews, R.G. and

Massey, R.G. unpublished). Subsequent work has shown that a large number of phenols can bind to old yellow enzyme to produce long wavelength absorbance bands. These have now been characterized unequivocally as charge-transfer absorbance bands, with the oxidized flavin as acceptor and the phenolate anion as charge-transfer donor. This characterization was achieved by the correlation of the energy of the transition (i.e., wavelength maximum) with the electron-donating power of the phenolate and the redox properties of the flavin acceptor. For the latter a large number of modified flavins has been bound to the apoenzyme (Abramovitz, A. and Massey, V., unpublished). In the case of 8-OH-FMN bound to old yellow enzyme, the effect of one such phenolate, pentafluorophenol, is shown in Fig. 5. The resulting complex has long wavelength absorption, considerably decreased intensity in the 430–500 m region and is devoid of fluorescence. If at any phenolate concentration the pH is increased, the long wavelength absorbance decreases, the spectrum approaches that of the anion in the absence of phenolate, and fluorescence reappears. The latter, both in terms of its excitation and emission characteristics, is identified as that of the anion. The pH dependence of the fluorescence in the presence of 1.8 mM pentafluorophenol is shown in Fig. 5; at this concentration the apparent pK is shifted to 9.2. However, this is only an apparent pK, since the mid-point of such a titration curve depends on the phenolate concentration, being shifted to higher pH values the higher the phenolate concentration. These data are consistent with

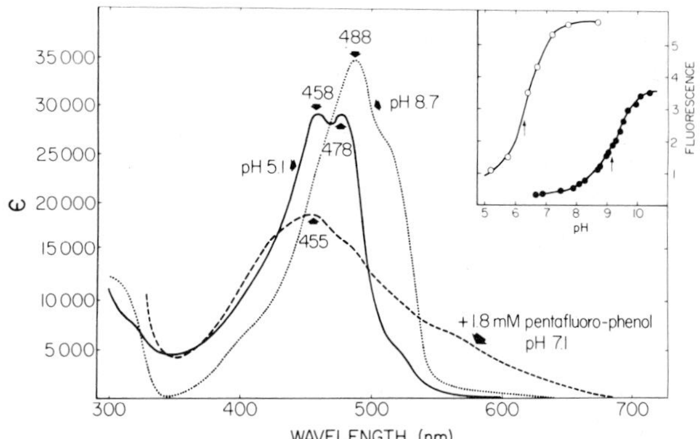

Fig. 5. Effect of pH on the complex of old yellow enzyme apoprotein and 8-OH-FMN. 8-OH-FMN ($2.3 \cdot 10^{-6}$ M) in 0.25 M acetate pH 5 was titrated with apo-old yellow enzyme and the spectrum recorded. The pH was adjusted by the addition of solid K_2HPO_4 and finally K_2CO_3. After each addition the spectrum was recorded, as well as the characteristic fluorescence of the anionic form of the modified enzyme (excitation λ_{488}, emission $\lambda_{543\,nm}$). In a similar experiment apo-old yellow enzyme was complexed with 8-OH-FMN at pH 7.1 (in 0.1 M phosphate buffer) then titrated with pentafluorophenol. The pH was adjusted as above, and absorbance and fluorescence monitored after each addition. In both experiments the absorbance and fluorescence changes were reversible on lowering the pH again (by the addition of solid KH_2PO_4 and citric acid).

the following model:

$$\begin{array}{ccc} \text{E—8—OH—Fl} & \underset{}{\overset{K_A}{\rightleftharpoons}} & \text{E—8—OH—Fl}^- + \text{H}^+ \\ K_B \updownarrow + \text{phenolate} & & K_D \updownarrow + \text{phenolate} \\ \text{E—8—OH—Fl} & \overset{K_C}{\rightleftharpoons} & \text{E—8—OH—Fl}^- + \text{H}^+ \\ \text{phenolate} & & \text{phenolate} \end{array}$$

A good fit to the data is given with the following values for the dissociation constants; $K_A = 6 \cdot 10^{-7}$ M, $K_B = 7 \cdot 10^{-5}$ M, $K_D = 2.9 \cdot 10^{-3}$ M. With these values the value of K_C (i.e., the apparent pK) is defined by the phenolate concentration, as observed. Of the two enzyme-phenolate complexes, only the neutral flavin-phenolate complex exhibits charge-transfer interaction. Thus, the flavin-anion form of the enzyme, although able to bind phenolate anion more weakly than the neutral species, does not give a charge-transfer interaction. This result is what would be expected since the negative charge of the flavin would render it unacceptable as a charge-transfer acceptor.

Acknowledgements

This work was supported by N.I.H. Research Grant GM-11106 to V.M. and a grant from the Schweizerische Naturforschende Gesellschaft to S.G.

References

1 Mayhew, S.G. and Massey, V. (1971) Biochim. Biophys. Acta 235, 303–310
2 Whitfield, C.D. and Mayhew, S.G. (1974) J. Biol. Chem. 2811–2815
3 Ghisla, S. and Mayhew, S.G. (1973) J. Biol. Chem. 248, 6568–6570
4 Mayhew, S.G., Whitfield, C.D., Ghisla, S. and Shuman-Jorns, M. (1974) Eur. J. Biochem. 44, 579–591
5 Mayhew, S.G. (1971) Biochim. Biophys. Acta 235, 289–302
6 Tollin, G. and Edmondson, D.E. (1971) in Flavins and Flavoproteins (Kamin, H., ed.), pp. 153–170, University Park Press, Baltimore
7 Watenpaugh, K.D., Seiker, L.C. and Jensen, L.H. (1973) Proc. Natl. Acad. Sci. U.S. 70, 3857
8 Burnett, R.M., Darling, G.D., Kendall, D.S., Le Quesene, M.E., Mayhew, S.G., Smith, W.W. and Ludwig, M.L. (1974) J. Biol. Chem. 249, 4383–4392
9 Massey, V. and Curti, B. (1966) J. Biol. Chem. 241, 3417–3423
10 Massey, V. and Ganther, H. (1965) Biochemistry 4, 1161–1173
11 Ghisla, S. and Massey, V. (1975) J. Biol. Chem. 250, 577–584
12 Matthews, R.G. and Massey, V. (1969) J. Biol. Chem. 244, 1779

Part VII

Deazaflavin and other model flavins

Flavins and Flavoproteins, edited by T.P. Singer

CHAPTER 36

STRUCTURE AND REACTIVITY OF 5-DEAZAFLAVINS

H. FENNER, H.H. ROESSLER, H.J. DUCHSTEIN and P. HEMMERICH

Fachbereich Pharmazie-Freie Universität Berlin (Germany) and Fachbereich Biologie, Universität Konstanz (G.F.R.)

Introduction

With respect to a better understanding of flavoprotein active site structure and function, especially the biological significance of the N-5 position in the flavin nucleus, it seemed to be a promising approach to modify the flavin system by replacement of the N-5 by isovalent groups. The synthesis of pyrimidobenzothiazines, which can be regarded as sulfur-analogous dihydroflavins, and their ability to form stable radicals was reported by us in 1970 [1, 2]. O'Brien et al. [3] synthesized a 5-deazariboflavin; the first studies on the biochemical properties of this compound have been published by Tollin et al. [4]. Independently, an easier synthesis of 5-deazadihydroflavins and their chemical and spectral properties was studied in our group [5, 6]. Bruestlein and Bruice demonstrated the reduction of 5-deazalumiflavin by NADH [7]. Recent studies have established the importance of deazaflavins: deazariboflavin of deaza-FMN have been combined with the apoproteins of bacterial luciferase [8] and *N*-methylglutamate synthetase [9]. In these studies the modified enzymes were found to be active, albeit at lower rates than the native enzymes. The 5-thiaflavins could be of some interest for checking the hypothesis, that the mechanism of action of phenothiazines should be a phenothiazine-flavin interaction, proposed by Yagi et al. [10] and other authors.

Considering all these data, we have set out to bring a little more light into the problem: whether deazaflavins are flavin models — in what respect and to what extent.

Changes of the flavin redox system induced by modification of N-5

The reaction of the 5-deaza-lumiflavin with NADH, yielding the 1,5-dihydrodeazaflavin [7], which is identical with the product obtained by dithionite or borohydride reduction as well as with the dihydrodeazaflavin synthesized directly [11], was interpreted as evidence for a direct hydride transfer from NADH to the deazaflavin. The authors, therefore, claimed the deazaflavin to be a chemical model for the biological cofactor. What would be the alternative for these results: a carbanion mechanism, implying proton abstraction from NADH within an NADH-deaza-Fl_{ox} "pre-equilibrium complex" (cf. [12]). In the absence of any stronger base, however, the deaza-Fl_{ox} would be the acceptor, and since the acceptor site is a carbon, proton transfer from NADH to deaza-Fl_{ox} will occur without exchange with the environment. Hence, there is no alternative to this result, irrespective of what kind of redox mechanism is preferred.

On the whole, one might expect that the replacement of N-5 in the flavin nucleus by carbon might suffice to transform a flavin into a nicotinamide analog rather than a flavin analog.

Regarding the "redox shuttles" of flavoprotein subclasses, as discussed by Hemmerich [13], the modification of the flavin nucleus by replacement of N-5 by carbon yields a redox system with a $2e^-$ input-output function, in which the quinoid redox state is a flavin-analogous system, while the reduced form is non-hydroquinoid. Any $1e^-$ transfer would require 5-deazaflavin-radicals, but it is obvious, that the replacement of N-5 in this system eliminates radical stability (cf. Scheme 1). A radical, obtained by hydrogen abstraction from the C-5 in the 1,5-dihydro system is hypthetical, because this species is thermodynamically and kinetically unstable even with normal flavins. The $\dot{F}l^-$-analogous system will undergo irreversible protonation at N-1, yielding the same unstable species. The radical, formally comparable with the blue $H\dot{F}l$ radical, is no semi-quinone system, because any tricyclic resonance is blocked, therefore it is "$H\dot{F}l$-non-analog" and will undergo irreversible disproportionation very rapidly. All these hypothetical species demonstrate the importance of the N-5 in the flavin nucleus, regarding the requirements for spin delocalization and stabilisation of semiquinone radicals. Hence, deazaflavin is a "non-radical" flavin and can, therefore, mimick only "transhydrogenase" functions of flavoproteins.

Even with normal flavin thermodyn. and kinetically unstable

Deaza Fl_{ox}

$-e^-$ $+e^-$

$\dot{F}l^-$ analog but liable to irrevers. protonation

Tricyclic resonance blocked
$H\dot{F}l$-non-analog:
disproportionation irrevers.

Scheme 1. Hypothetical "5-Deazaflavin radicals". (cf. Note added in proof).

In the 5-thiaflavin series the hydroquinone state is stabilized by the insertion of sulfur in position 5, while the corresponding quinone level (the thiazonium cation) is too acidic and undergoes covalent hydration at

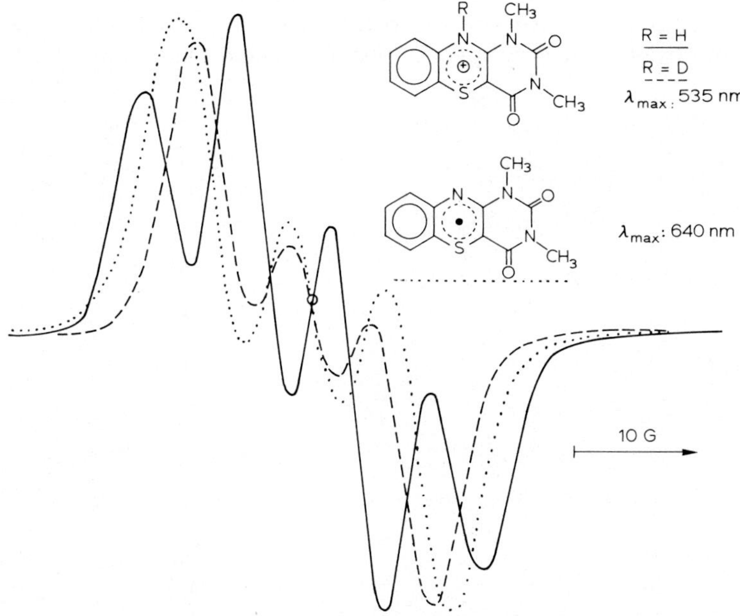

Fig. 1. ESR spectra of thiaflavin radicals.

any pH. Formation of the 4a-pseudobase or the sulfoxide is dependent on the pH.

The radicals of 5-thiaflavins, however, are found to be extensively stabilized. Hence, the 5-thiaflavin radicals are analogs or models for 5-HFl radicals. (Fig. 1) the 5-thiaflavins, therefore, can mimick only the e^- transferase flavoproteins, e.g. flavodoxins, with respect to their "output", but not with their "input" function. The spin density in the thiaflavin radicals is similar to the flavin radicals, since N-1 does not couple, while S-5 does (as can be shown by the ESR spectra of S_5—CH_3 radicals), as well as N-10.

TABLE I

SUMMARY OF THE DIFFERENCES BETWEEN FLAVINS, DEAZAFLAVINS AND THIAFLAVINS

	Flavins	Deazaflavins	Thiaflavins
Radical			
Chemically	Reversible disproportionation	Irreversible disproportionation	Stable
Protein bound	Stable	Unstable	?
Oxidized	Quinoid	Quinoid (flavin analog)	Non-quinoid (sulfoxide)
Reduced	Hydroquinoid	Non-hydroquinoid (NADH analog)	Hydroquinoid (flavin analog)
Redox shuttles			
$2e^-$	+ (Dehydrogenase)	+	—
Upper $1e^-$	+ (Artificial?)	—	—
Lower $1e^-$	+ (e^--transferase)	—	+

In general the steric shape of the various 5-deazaflavins is altered only to a minor extent compared with the flavin nucleus, but their behavior as redox systems is modified in a very specific way (cf. Table I and Scheme 2).

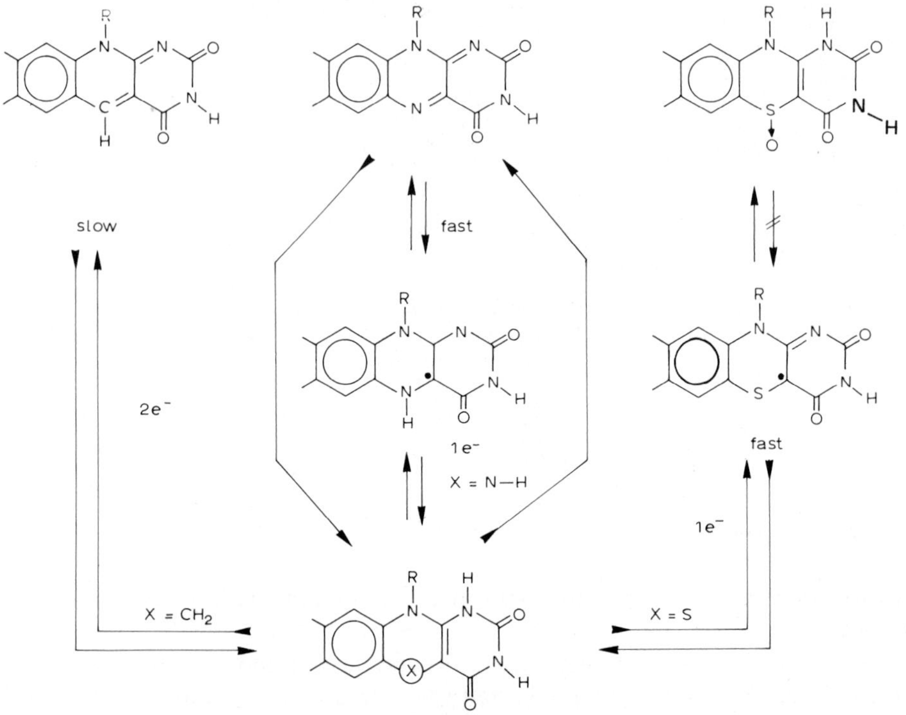

Scheme 2. Changes in the redox system of flavins induced by replacement of N-5.

Chemical properties of reduced deazaflavins

The reduction of lumiflavin-analogous deazaflavins with dithionite or borohydride yields 1,5-dihydro-deazaflavins, identical with the product obtained by NADH reduction [7]. The NMR-data given in the literature, are different in the chemical shift of the N-1 proton: from our results it is obvious that this proton shows a singlet $\delta = 10.55$ ppm in C^2HCl_3. This chemical shift excludes a 4a,5-dihydro structure, perhaps because there is a very rapid 4a,1-prototropy. These NMR studies fit the thesis that deazaflavins are nicotinamide models rather than flavin models. As shown in Fig. 2, there is only a small influence on the spectral behavior, if N-1 is deprotonated. The pK_a value of this species if 7.8. The assumption, that deazaflavins are reduced by several substrates in a light reaction [4] could be shown by us to be not correct: in the photo-reaction with mandelate, $O{-}CH_3$-mandelate, oxalate, EDTA and other substrates photo-addition could be demonstrated, even in those cases where the corresponding reactions with flavins do not yield adducts.

From NMR data and ultraviolet spectra (cf. Fig. 2) it can be shown that no 4a-adducts but 5-adducts are formed in all cases.

Scheme 3. 2e⁻ oxidoreduction of deazaflavin.

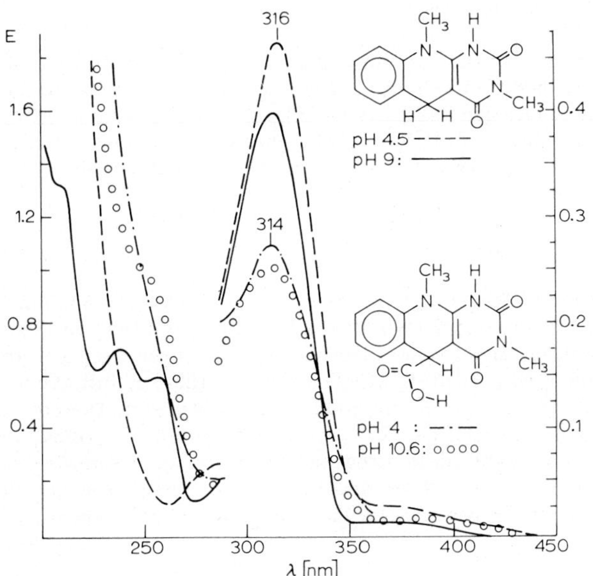

Fig. 2. Ultraviolet spectra of 1,5-dihydro-deazaflavin obtained by BH_4^- reduction at pH 4.5 and pH 9 and ultraviolet spectra of the deazaflavin-oxalate adduct at pH 4 and pH 10.6. (cf. Note added in proof).

Chemical properties of 5-thiaflavins

The stable redox state of 5-thiaflavin is a dihydroflavin model. We have studied the significance of the 4a, S-5 and N-10 positions as nucleophilic centers in alkylation reactions [14]. As Ghisla et al. [15] have reported, alkylation reactions in the dihydroflavin series can be interpreted as electrophilic attacks at the different nucleophilic centers, depending on the structure of N-5: in the case of the N_5-acetyldihydroflavin the alkylation reactions are different from those found with the N_5—H system. Under the same conditions the 5-thiaflavins yield alkylation products similar to those described by Ghisla et al. [15].

Note added in proof

The preliminary results presented in the present Proceedings Volume have been extensively magnified in 1975. This requires the following indispensable corrections of the interpretations given above: (1) The "oxalate photo-adduct" interpreted as 5-carboxylic acid in Fig. 2 is not the acid, as shown by a comparison with material now obtained by direct synthesis. The authentic carboxylic acid is, however, found to undergo an extremely rapid photoreaction with deazaflavin, much faster than the initial reaction of deazaflavin with oxalate. The product of this reaction is a mixture of diastereomeric deazaflavin 5,5′-covalent dimers. The low extinction spectra of Fig. 2 belong to this mixture. The full sequence of the deazaflavin-oxalate photoreaction must, therefore, read:

$$dFl^*_{ox} + (COO)_2^{2-} \xrightarrow[slow]{} {}^{-}dFl_{red}\text{-}5\text{-}COO^- + CO_2,$$

$$^{-}dFl_{red}\text{-}5\text{-}COO^- + dFl^*_{ox} \xrightarrow[fast]{} {}^{-}dFl\text{-}5,5'\text{-}dFl^- + CO_2$$

The chromophors of the 5-carboxylate and the 5,5′-dimer are, of course, very similar. The dimer is readily characterized by NMR (Duchstein, Knappe and Hemmerich, to be published) Spectra and reactivities of reduced deazaflavin ($dFl_{red}H_2$, Fig. 2, above) differ characteristically from those of the dimer $(dFlH)_2$. (2) The enforced reaction of deazaflavin with e^- has been checked by radiolysis (Goldberg, Pecht and Hemmerich, to be published). The sequence is as follows:

$$dFl_{ox} + e^-_{aq.} \rightarrow \dot{d}Fl^- \underset{pK \approx 6}{\overset{+H^+}{\rightleftharpoons}} 1\text{-}Hd\dot{F}l \rightarrow (dFlH)_2$$

These shortlived radical species do not differ strongly in their absorption ($\lambda_{max} \approx 530$ nm), they are correctly represented by the two upper formulae in Scheme 1, and they undergo irreversible dimerisation within milliseconds to yield the same dimer as obtained before by "photocomproportionation" of $dFl_{ox} + HdFl_{red}COOH$. Hence, there is full experimental support for the thermodynamic as well as kinetic instability and, therefore, biological irrelevance of deazaflavin radicals, i.e. for deazaflavin being a nicotinamide rather than flavin model, which retains the (de)hydrogenase, but not the e^--transferase properties of flavoproteins. Furthermore, deazaflavin dependent (photo)dehydrogenation is shown to involve group rather than hydrogen transfer in a notorious $2e^-$-reaction.

References

1 Fenner, H. (1970) Arzneimittel-Forsch. (Drug Res.) 20, 1815–1818
2 Fenner, H. (1970) Tetrahedron Lett. 617–620
3 O'Brien, D.E., Weinstock, L.T. and Cheng, C.C. (1970) J. Heterocycl. Chem. 7, 99–105
4 Fenner, H. and Michaelis, P. (1972) Arch. Pharmaz. 305, 751–756
6 Fenner, H., Michaelis, P., Ghisla, S. and Hemmerich, P., to be published
7 Bruestlein, M. and Bruice, T.C. (1972) J. Am. Chem. Soc. 94, 6548–6549
8 Fisher, J. and Walsh, C. (1974) J. Am. Chem. Soc. 96, 4345–4346
9 Schuman-Jorns, M. and Hersh, L.B. (1974) J. Am. Chem. Soc. 96, 4012–4014
10 Yagi, K., Ozawa, T. and Nagatsu, T. (1959) Nature, 184, 982
11 Fenner, H. and Bauch, W., to be published
12 Blankenhorn, G. (1976) Flavins and Flavoproteins, Proceedings of the 5th International Symposium (Singer, T.P., ed.), Chapt. 26, pp. 261–267, Elsevier, Amsterdam
13 Hemmerich, P. and Wessiak, A. (1976) Flavins and Flavoproteins, Proceedings of the 5th International Symposium (Singer, T.P., ed.), Chapt. 1, pp. 9–22, Elsevier, Amsterdam
14 Fenner, H., Roessler, H.H., Ghisla, S. and Hemmerich, P. Liebigs Ann. Chem. in the press
15 Jefcoate, C.R., Ghisla, S. and Hemmerich, P. (1971) J. Chem. Soc. (C) 1689–1694

Flavins and Flavoproteins, edited by T.P. Singer

CHAPTER 37

5-DEAZAFLAVIN COENZYMES AS PROBES OF FLAVOENZYME MECHANISMS

ROB SPENCER, JED FISHER, RICHARD LAURA and CHRISTOPHER WALSH

Departments of Chemistry and Biology, Massachusetts Institute of Technology, Cambridge, Mass. 02139 (U.S.A.)

Introduction

Flavin coenzymes are obligate cofactors for a wide variety of enzyme-catalyzed oxidation processes such as the dehydrogenation of acyl CoAs, the oxidation of amines, amino acids and hydroxy acids, and the oxidation of nicotinamide coenzymes. Flavoprotein oxidases use molecular oxygen to reoxidize the reduced bound coenzyme while many flavoprotein dehydrogenases are membrane bound and are vital elements of electron transport chains. Evaluation of the chemical behavior of the flavin coenzymes in these catalyses is necessary to understand flavoenzyme reaction mechanisms.

Recent efforts in our laboratory, among others, have probed the utility of 5-deazaflavin analogs, in which N-5 of the flavin isoalloxazine ring structure is replaced with a carbon (C-5), for providing mechanistic insight into flavoenzyme mechanisms. In particular, we feel that the deaza analogs may provide unique information on some of the following unresolved aspects of flavoenzyme catalysis: (1) whether N-5 is necessary for enzyme-catalyzed coenzyme reduction; (2) whether N-5 is necessary for catalytic reoxidation of reduced coenzyme with O_2; (3) whether direct hydrogen transfer occurs from substrate to deazaflavin; (4) whether covalent adducts can be detected during catalysis; (5) whether the stereochemistry of hydrogen transfer to and from deazaflavins can be determined.

In this paper we shall discuss some studies on the chemistry of deazaflavins relevant to their enzymatic function and then summarize some information on coenzymatic function of 5-deazariboflavin, 5-deazariboflavin 5′-phosphate, and 5-deazariboflavin adenine dinucleotide when reconstituted with appropriate apoenzymes.

Materials and Methods

10-Methyldeazaisoalloxazine and 5-deazariboflavin were prepared by the general synthetic procedure developed by Cheng and coworkers [1]. Enzymatic synthesis of

---

Abbreviations: dRF, 5-deazariboflavin; dFMN, 5-deazariboflavin 5′-phosphate; dFAD, 5-deazariboflavin adenine dinucleotide; $dRFH_2$, 1,5-dihydrodeazariboflavin; RF, riboflavin.

dFMN and dFAD was achieved with the partially purified enzymes from *Brevibacterium ammoniagenes* [2]. Chiral, isotopically labelled NADH was prepared and isolated by the procedure of Kaplan [3].

The *Beneckea harveyi* NAD(P)H:flavin oxidoreductase was purified by modification of the procedure developed for isolation of the luciferase [4]. Brewer's bottom yeast NADPH:(acceptor) oxidoreductase (EC 1.6.99.1, trivial name old yellow enzyme) was isolated by the affinity method of Abramovitz and Massey (personal communication), and resolved into the apoenzyme by their procedure. The *Mycobacterium smegmatis* L-lactate:oxygen 2-oxidoreductase (decarboxylating) (EC 1.13.12.4, trivial name lactate 2-monooxygenase*) was purified and resolved by the procedure of Sullivan [5]. D-amino acid:oxygen oxidoreductase (deaminating) (EC 1.4.3.3, trivial name D-amino acid oxidase) from hog kidney was prepared by the method of Brumby and Massey [6] and resolved by high-salt dialysis [7]; the *Aspergillus niger* β-D-glucose:oxygen 1-oxidoreductase (EC 1.1.3.4, trivial name glucose oxidase) was purchased from Sigma and resolved by acid ammonium sulfate precipitation [8].

The Fourier transform PMR spectra of the deazariboflavins were obtained on an interfaced 60 MHz Perkin-Elmer R20B spectrometer, using a 10 μs pulse width followed by acquisition of the free induction decay for a resolution of 1.0 Hz. The solvent used for these samples consisted of 0.67 [$^2H_6$] dimethylsulfoxide 0.33 C^2HCl_3 with chloroform or tetramethylsilane as internal reference.

Results

Chemistry of deazaflavins and dihydrodeazaflavins

Our current interest in the deazaflavins revolves around their ability to undergo enzyme-catalyzed reduction and oxidation. Since any mechanistic interpretation of a reaction, enzymatic or otherwise, requires knowledge of the structures and reactivities of both reactants and products, we have undertaken a brief examination of the redox properties of the deazaisoalloxazine system and of the structure of the two-electron-reduced, dihydro species.

The numerous similarities in reactivity of isoalloxazines and deazaisoalloxazines confirms the correctness of the structure (Fig. 1) proposed by Cheng and colleagues [1] for the condensation product of barbituric acid and *N*-ribityl-2-aminobenzaldehyde. Both chromophores are reduced by sodium borohydride, by hydrogen/platinum, and by sodium dithionite. Reduction in the deazaflavin system proceeds with loss of blue fluorescence and the long-wavelength (λ_{max} = 396 nm) absorption; the 338 nm band is shifted to 320 nm with no loss in intensity. Enzymatic reduction of free deazariboflavin by the *B. harveyi* oxidoreductase using NADH as reductant yields a dihydro species with an identical ultraviolet spectrum (Fig. 2). Indeed, the dihydrodeazaflavins produced by any of the chemical reductions all undergo oxidoreductase-catalyzed oxidation with concomitant reduction of NAD^+.

Crucial to an evaluation of the coenzymatic properties of deazaisoalloxazines is the structural identification of the dihydro compound generated.

* Formerly EC 1.1.32, lactate oxidase.

Fig. 1. The structure of the 5-deazariboflavins in oxidized (left) and reduced forms. dRF, R = H; dFMN, R = PO_3^{2-}; dFAD, R = PO_2^--adenylate. The flavin-numbering scheme is indicated on the oxidized form.

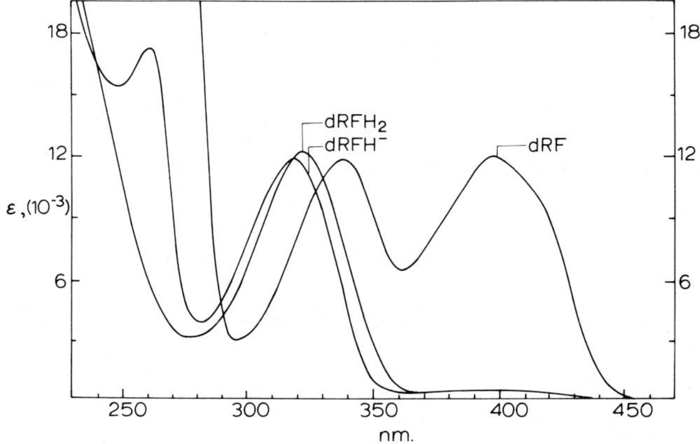

Fig. 2. The absorption spectra of oxidized deazariboflavin and the enzymatically reduced (pH 5.7) and reduced anion (pH 8.7) forms.

With a simple 10-methyldeazaisoalloxazine, Brustlein and Bruice [9] have observed that non-enzymatic reduction by NADH proceeds with direct hydrogen transfer to yield the 1,5-dihydrodeazaisoalloxazine, on the basis of its PMR spectrum. Since there are no references to the spectral properties of these dihydroquinolines [10], it remained to be demonstrated that the generalized structure of the 1,5-dihydrodeazaflavins produced by the methods of the preceding paragraph was the 1,5-dihydroquinoline of Fig. 1.

Consistent with this dihydro structure is our observation that dihydrodeazariboflavin preparations possess a pK_a of 7.2 ± 0.1 ($dRFH^-$; λ_{max} 262 nm, $\epsilon = 17\,500\ M^{-1} \cdot cm^{-1}$). The appearance of a pK_a near neutrality in isoalloxazines is diagnostic of saturation at the 10a-1 carbon–nitrogen bond [11] as would also occur in the 1,5-dihydrodeazaisoalloxazines.

However, PMR analysis offers the more direct corroboration of the dihydroquinoline structure and allows the establishment of C-5 as the position of hydrogen transfer on reduction. To compare our findings with those of Brustlein and Bruice, the 10-methyldeazaisoalloxazine was reduced with sodium borohydride, giving a dihydroflavin whose PMR spectrum was identical to that obtained previously from the non-enzymatic reduction and, as stated previously, is consistent only with the 1,5-dihydrodeazaisoalloxazine

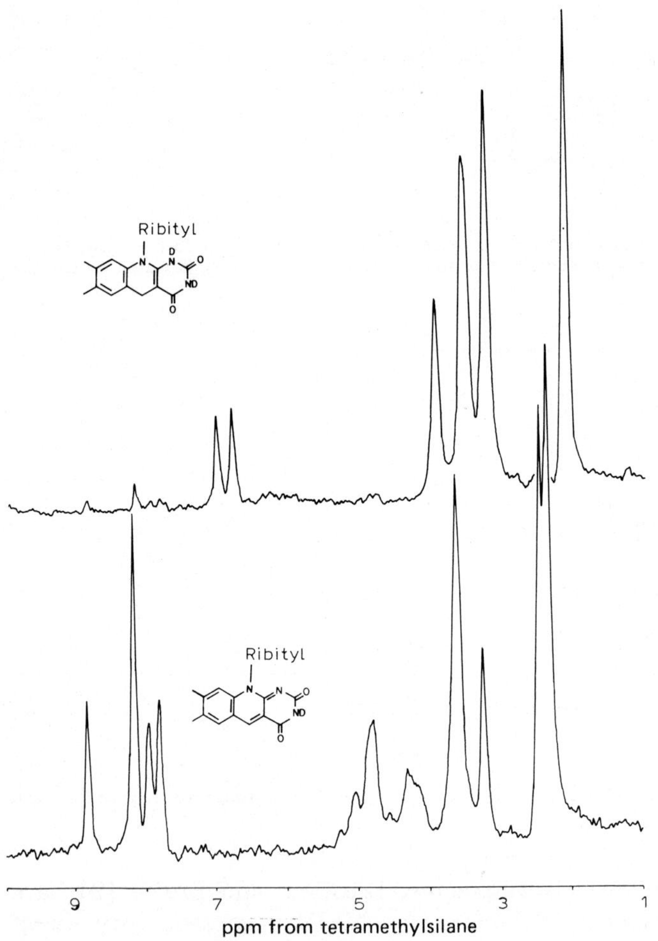

Fig. 3. The PMR spectra of deazariboflavin. Above: reduced with sodium borohydride. Below: oxidized form.

structure. Repetition of this reduction with sodium borodeuteride gives a marked diminution in the two proton singlet at δ 3.57 (C-5 hydrogens) confirming direct hydrogen (deuterium) transfer to C-5.

We have similarly examined the PMR spectrum of 5-deazariboflavin and its borohydride reduction product and have also observed a 1,5-dihydrodeazaisoalloxazine product (Fig. 3). On reduction there is increased shielding of the aromatic and methyl protons with collapse of the methyl doublet to a singlet. A concomitant change in the chemical shift and line shape of the ribityl protons is also seen on reduction, supporting a significant electronic and conformational change. On borodeuteride reduction the δ 3.27 singlet is somewhat reduced but explicit identification of hydrogen transfer to C-5 is obscured by adventitious water protons absorbing in the same region. However, we reasoned that reoxidation of the borodeuteride-reduced dihydrodeazariboflavin should give oxidized deazariboflavin containing deuterium, specifically at C-5, easily detectable by PMR analysis.

Since autoxidation of the dihydrodeazariboflavin is slow (see below) we have employed catalytic amounts of riboflavin to hasten reoxidation. The RFH_2 thus generated is in turn rapidly reoxidized by molecular oxygen.

$$dRFH_2 + RF \rightarrow dRF + RFH_2 \xrightarrow[O_2\,(fast)]{} RF$$

In oxygenated buffer at pH 8.6 the reaction between $dRFH_2$ and RF exhibits pseudo-first order kinetics for the first three half lives, yielding a bimolecular rate constant of $4.6 \cdot 10^4\ M^{-1} \cdot min^{-1}$. A pH rate profile unambiguously implicates the dihydrodeazaflavin anion as the reactive redox species, the rate increasing through the pK_a to a plateau between pH 8.0–9.5. When the borodeuteride-reduced dihydrodeazariboflavin or *N*-methyldihydrodeaza-isoalloxazine samples are oxidized by riboflavin, an identical kinetic isotope effect of 2.0 ± 0.1 shows that hydrogen transfer from C-5 is now partially rate limiting. With no isotope effect on oxidation, the oxidized dRF should have 0.5 atoms ^2H and 0.5 atoms ^1H at C-5; a two-fold isotope effect predicts only 0.33 atoms ^1H at C-5. The actual PMR spectrum obtained is identical to normal dRF except that absorption due to the C-5 proton is nearly abolished, consistent with the above prediction. Integration indicates less than 0.2 atoms ^1H at C-5 relative to the one hydrogen signals at C-6 and C-9.

The retention of more than 0.8 atoms ^2H at C-5 on oxidation suggests a product isotope effect differing from the observed rate effect, a phenomenon previously reported for dihydronicotinamide model systems [12, 13], interpreted as requiring the partially rate-determining formation of an intermediate prior to electron transfer [14]. While borodeuteride presumably reduces randomly at either face of dRF, one expects that enzymatic reduction might be chiral, only to one face. Monodeutero-$dRFH_2$ can be prepared with the oxidoreductase and 4-R-[^2H]NADH. When this sample is reoxidized by riboflavin the rate isotope effect is 3.0 ± 0.1. The difference in rate isotope effects between racemic and (presumably chiral) C-5 monodeutero-$dRFH_2$ may reflect some preferential alignment in a productive RF-$dRFH^-$ intermediate in the oxidation process.

Indication that the oxidoreductase also removes a hydrogen species from C-5 of $dRFH_2$ on reoxidation with NAD^+ comes from two experiments. The monodeutero-$dRFH_2$ species are enzymatically reoxidized 1.6 ± 0.2 fold more slowly than the protio species. Further, if 5-^3H-labelled $dRFH_2$, generated by NaB^3H_4 reduction of dRF, is incubated with enzyme and NAD^+ to isotopic equilibrium, 40% of the original tritium is transferred to the NADH formed, with 60% retained in the dRF coproduct.

$$[5\text{-}^3H]dRFH_2 + NAD^+ \overset{enzyme}{\rightleftharpoons} 4\text{-R-}[^3H]NADH + [5\text{-}^3H]dRF$$

If the ^3H-labelled NADH is isolated and incubated with acetaldehyde and dehydrogenase, all of the tritium is transferred to the ethanol pro-, showing that tritium was localized specifically at the 4-R position of NADH and establishing that tritium transfer by the oxidoreductase in the above experiment is completely chiral with respect to the nicotinamide. The 40:60 split between NADH and dRF also suggests chiral recognition at C-5 of the dihydrodeazariboflavin.

Finally, the sensitivity of 5-dihydrodeazaflavins to oxidation by molecular oxygen deserves comment. Not surprisingly, loss of the non-bonded nitrogen lone pair in going from dihydroisoalloxazine to dihydrodeazaisoalloxazine dramatically lowers oxygen reactivity. Whereas simple 1,4-dihydropyrazines, a classical antiaromatic system, are extremely unstable to one-electron oxidation by oxygen and to substituent migration [15, 16], 1,4-dihydropyridines with electron-withdrawing substituents (e.g. NADH) require few precautions in handling and are only very slowly decomposed by oxygen [17]. With respect to the dihydrodeazaflavins, Tollin and co-workers [18] have reported a half time of 3 h for oxygen oxidation of $dRFH_2$ prepared from dithionite. Dihydrodeazariboflavin prepared by hydrogen/platinum reduction behaves similarly ($t_{1/2}$ = 3–5 h) while material prepared enzymatically or from borohydride reduction has much slower kinetics ($t_{1/2}$ = 30 h). As with the riboflavin oxidation, the reactive species is the dihydrodeazariboflavin anion. The lack of stability of the $dRFH_2$ produced by these first two methods may well be due to contaminants arising from the reducing agents themselves: dithionite-produced NADH (containing sulfite) is significantly less stable [19] and there is some possibility that a metal contaminant associated with the platinum/asbestos catalyst could mediate electron transfer between the reduced deazaflavin and oxygen. Also, we have observed that prolonged catalytic hydrogenation and light/EDTA photoreduction can lead to irreversible decomposition of the deazaflavin. For these reasons, borohydride represents the preferred method of chemical access to the dihydrodeazaflavins. Since light itself catalyzes the oxygen oxidation [18], routine handling of these compounds requires degassed solutions and darkened conditions. As will be detailed in the following sections on the enzymatic reactions of deazaflavins, we have not observed any enzymatic catalysis of this reaction by the flavoprotein oxidases.

Enzymatic behavior of deazaflavins

NAD(P)H:flavin oxidoreductase

We have recently reported that the *B. harveyi* NAD(P):flavin oxidoreductase will reduce 5-deazariboflavin with direct hydrogen transfer from NADH [20]. This enzyme is unusual in using flavins as substrates rather than tightly bound coenzymes, in accord with its physiological role of supplying luciferase with $FMNH_2$. This allows use of catalytic quantities of enzyme in making a detailed comparison of flavins and deazaflavins. The oxidoreductase tolerates a broad range of dihydronicotinamide:flavin pairs with varying degrees of efficiency, with the most reactive pair being NADH and riboflavin. Deazariboflavin behaves as a competitive inhibitor ($K_i = 8 \cdot 10^{-6}$ M) to riboflavin reduction, and at higher enzyme concentrations can be observed to undergo reduction itself at a rate 0.65% that of riboflavin. Since our experiments with the dihydrodeazaflavins indicate complete loss of absorption at 396 nm, monitoring the course of the reaction at this wavelength provides a quantitative measure of the reaction's progress. The reduction of deazariboflavin by NADH proceeds to an equilibrium position ($K_{eq} = 1.7 \pm 0.1$) that can be manipulated by incremental addition of the reactants. Confirmation that an equilibrium position is reached is indicated by the effect of adding either riboflavin, or ethanol and alcohol dehydrogenase after the equilibrium has stabilized. As a result of the low K_m for riboflavin ($8.2 \cdot 10^{-7}$ M) and the large amount of enzyme required to effect deazariboflavin reduction, the presence of trace amounts of riboflavin will rapidly and irreversibly consume the NADH to NAD^+ and thus drive the reaction towards deazariboflavin oxidation. On the other hand, ethanol and alcohol dehydrogenase provide a regenerating system for NADH and allow the reaction to proceed slowly towards the complete reduction of deazariboflavin. The back reaction between NAD^+ and $dRFH_2$ is independent of the presence of oxygen. The equilibrium value (the ratio of the forward and reverse rate constants) is sensitive to substituents on the reactants with the NADH:dFMN pair having a lower K_{eq} (1.0 ± 0.1). The fact that the reaction between the deazaflavin and the dihydronicotinamide shows an equilibrium constant close to unity indicates that the $dRFH^-$:dRF redox couple is essentially identical to that of the NADH:NAD^+ couple ($E_0' = -0.320$ V) [21]. Given the structural similarities between the reduced forms of each, this observation is not surprising. The large difference in redox potential that must then exist between deazaisoalloxazines and isoalloxazines (old yellow enzyme, $E_0' = -0.122$ V; riboflavin, $E_0' = -0.208$ V) [21] leads to the prediction that the preferred mode of interaction of deazaflavins with many normal flavoenzyme substrates will be in a reverse manner, since these substrates have potentials more negative than flavins but greater than dihydronicotinamides. Thus, the reduction of keto acids and imino acids by $dRFH^-$ will be more favorable than the corresponding substrate oxidative process.

As a further mechanistic probe of the reaction, we have examined the kinetic isotope effects for the direct hydrogen transfer from NADH to the flavins. Preparation of [4-R-^{2}H]NADH and incubation with oxidoreductase and riboflavin indicated stereospecific transfer of this hydrogen by the complete (> 90%) volatilization of the counts. This was confirmed by the kinetics observed with [4-R-^{2}H]NADH, which exhibited an unambiguous primary effect equal to 4.8 ± 0.1, measured both aerobically at 340 nm (NADH oxidation) and anaerobically at 445 nm (riboflavin reduction). A small, possibly secondary isotope effect of 1.2 ± 0.1 by [4-S-^{2}H]NADH was seen for riboflavin reduction.

Use of deazariboflavin has no effect on the NADH stereospecificity; a kinetic isotope effect of 3.3 ± 0.1 is observed only for the [4-R-^{3}H]NADH, while the 4-S isomer gave a rate identical to the protio control. dFMN, which reacts at 0.70 times the rate of dRF, has kinetic isotope effects identical with the deazariboflavin. Even though a large difference in absolute rates exists between riboflavin and deazariboflavin, hydrogen transfer remains rate determining for both. With this data, we may eliminate the possibility that substrate binding of the dRF to the active site, or product release are the rate-determining steps of catalysis. The lower turnover number with deazariboflavin may be due in part to the less favorable redox potential or it may be that the active site is not designed to maximize the proper donor/acceptor orbital overlap required in the productive dRF · NADH intermolecular complex.

Other enzymes utilizing dFMN: lactate 2-monooxygenase and NADPH dehydrogenase (old yellow enzyme)

Both of these enzymes contain FMN as a tightly bound coenzyme and are capable of resolution into apoenzymes. The behavior of dFMN bound to lactate 2-monooxygenase has been detailed elsewhere [22].

Addition of dFMN to old yellow enzyme apoenzyme allows enzyme-catalyzed reduction of the coenzyme analog by NADPH. We have not as yet determined that the NADPH:dFMN interaction here parallels that of the *B. harveyi* oxidoreductase; work on this enzyme is hampered by the poor binding of the dFMN relative to normal FMN which precludes gel filtration of thr reconstituted enzyme. Experiments of Massey indicate that charge-transfer agents for normal enzyme tighten the binding of dFMN and produce charge-transfer complexes with this coenzyme analog also (personal communication).

Neither $dFMNH_2$-lactate 2-monooxygenase nor $dFMNH_2$-old yellow enzyme accelerate the sluggish reoxidation of the reduced deazaflavin by molecular oxygen despite the fact that the normal holoenzymes have oxidase activity.

Jorns and Hersh [23] have established conclusively that dFMN can function analogously to FMN in many catalytic aspects in the *N*-methylglutamate synthetase reaction, confirming that deazaflavins are versatile analogs of flavins. Again direct hydrogen transfer was seen in each half reaction between substrate and coenzyme.

Enzymes-utilizing dFAD: D-amino acid oxidase

D-amino acid oxidase from hog kidney catalyzes the oxidation of D-amino acids by molecular oxygen to form hydrogen peroxide and the imino acid, which then rapidly hydrolyzes non-enzymatically to the keto acid and ammonium ion. The FAD coenzyme is readily removed by high salt dialysis; the apoprotein thus obtained can be loaded with dFAD and dialyzed against buffer to yield about a 60% reconstitution of the soluble protein, which remains stable through gel filtration.

D-Alanine will reduce approximately half the dFAD enzyme with a half time of from 20 to 40 min (depending on temperature and ionic strength) under saturating substrate conditions. Direct tritium transfer to deazaflavin can be demonstrated on incubation with DL-[α-^{3}H]alanine. Reaction of reduced $dFADH_2$ enzyme with pyruvate and ammonium ion gives rapid reoxidation of the deazaflavin; use of [^{14}C]pyruvate yields isolable [^{14}C]alanine.

Addition of benzoate to the dFAD enzyme gives spectral changes mimicking those seen with the normal flavoenzyme and benzoate; with decreased absorbance at $\lambda_{max} = 340$ nm, a red shift of the long wavelength peak from 403 to 407 nm, and appearance of a distinct shoulder at 425–435 nm. Hersh confirms our observations with benzoate and dFAD reconstituted enzyme, and also finds direct tritium transfer and reversability with several amino and keto acids, respectively (see Chapt. 38, this volume).

Glucose oxidase

The normal reaction catalyzed by this enzyme is the FAD- and oxygen-dependent oxidation of the C-1 position of glucose to form the δ-gluconolactone. The tightly bound FAD can be removed by repeated acid/ammonium sulfate precipitation, although in our hands with appreciable loss of reconstitutable activity. Reconstitution by dFAD gives incomplete loading of the soluble protein (estimated at 80% by the 280/402 nm ratio) with the coenzyme analog binding sufficiently strongly to allow dialysis or gel filtration of the reconstituted enzyme without dissociation. This enzyme fits into the now expected pattern for deazaflavin-reconstituted apoenzymes, with direct tritium transfer from the C-1 atom of glucose to the dFAD upon reduction. The rate of deazacoenzyme reduction (20°C, 0.1 M sodium acetate pH 5.7) is relatively rapid, complete in less than 5 s under saturating glucose conditions. The reduction levels off at an $A_{402\,nm}$ value corresponding to approximately 40–60% reduction of the total bound dFAD, suggesting that only this fraction of the active sites are catalytically competent. Again, no evidence for a covalent dFAD-glucose adduct is obtained after reduction of the bound coenzyme by [^{14}C]glucose and gel filtration. The final $A_{402\,nm}$ value is unaffected by further additions of substrate or product indicating that an equilibrium position is not reached. The inability of gluconolactone to oxidize the bound $dFADH_2$ represents an exception to the enzymes thus far examined.

When [1-^{3}H]glucose is used with dFAD-glucose oxidase and the reaction mixture subjected to gel filtration, tritium is found to run with the protein peak, allowing the calculation of a product isotope selection of 14.

In a preliminary effort to relate the chirality of tritium transfer to dFAD bound to glucose oxidase and tritium transfer catalyzed by the *B. harveyi* oxidoreductase, we have removed the [5-^{3}H]$dFADH_2$ from glucose oxidase by heat treatment. Consecutive phosphodiesterase and alkaline phosphatase treatment produced [5-^{3}H]$dRFH_2$ which was isolated by gel filtration on Biogel P2. Subsequent incubation with excess NAD^+ and oxidoreductase to achieve complete reoxidation resulted in transfer of about 25% of the radioacitivity to the NADH formed and retention of 75% of the tritium as dRF. The basis for incomplete chirality is currently unclear but this initial experiment may suggest that the glucose oxidase and oxidoreductase transfer hydrogen to different faces of the deazaflavin ring system.

Discussion

The ability of 5-deazaisoalloxazines to substitute for the normal isoalloxazines has now been examined with several flavoenzymes. Without exception, the deazaflavin is bound to the active site and is rendered capable of reduction by the specific enzyme substrate. Furthermore, reoxidation of the dihydrodeazaflavin by the oxidized substrate has also been demonstrated with several enzymes, with the only exception so far being glucose oxidase. The sluggish oxidation of reduced deazaflavins by molecular oxygen, enzymatically and nonenzymatically, confirms beyond doubt that the unbonded electron pair found at N-5 is essential for this process to occur rapidly.

We have established that the chemical and enzymatic reduction of 5-deazariboflavins both proceed to produce the 1,5-dihydrodeazaflavin, the structural analogue of 1,5-dihydroriboflavin. This reduction of the deazaflavin results in hydrogen transfer to the non-exchangeable locus of C-5. In all cases thus far examined, enzyme-catalyzed substrate oxidation results in direct hydrogen transfer to this carbon. In experiments to probe for covalent flavin-substrate adducts, none have been detected with lactate 2-monooxygenase [22], D-amino acid oxidase, glucose oxidase and *N*-methylglutamate synthetase [23].

The slow reduction of deazaflavins compared to the normal flavins by the specific apoenzymes is explicable by the large redox potential difference. The deazaflavin couple is considerably more negative and in accord with this, we have observed faster rates of oxidation of the dihydrodeazaflavins by the oxidized substrates. Our examination of the *B. harveyi* oxidoreductase suggests that the enzyme displays similar affinities for riboflavin and deazariboflavin, that the rate-determining step of catalysis is identical and that the stereochemistry of hydrogen transfer to and from the nicotinamide is the same. These data implicate a similar mechanism for electron transfer with the flavin and deazaflavin coenzymes. This conclusion has also been attained by Jorns and Hersh [23] on *N*-methylglutamate synthetase and by both laboratories with D-amino acid oxidase supporting the view that the insights from deazaflavin experiments pertain to normal flavin mechanisms.

We have investigated two classes of flavoenzyme reactions with the coenzyme analogues: those using dihydronicotinamide substrates (NAD(P)H:flavin oxidoreductase, old yellow enzyme) and those involving carbonyl compounds (lactate 2-monooxygenase, D-amino acid oxidase and glucose oxidase). Since distinct mechanisms have often been proposed for these categories we may consider separately what mechanistic insight the deazaflavins may offer.

The nature of electron transfer from dihydronicotinamides in enzymatic catalysis has long been enigmatic. Numerous model and enzymatic studies have been interpreted as hydride transfer from the dihydropyridine to aromatic systems [24], carbonyls [25–27] and isoalloxazines [28, 29]. An alternative scheme involving intramolecular transfer of two electrons and a proton concerted with decomposition of a dihydronicotinamide-isoalloxazine covalent adduct has been presented by Hamilton [30]. However, neither we nor others have found evidence for such intermediates. While the repeated observation of pH-independent, direct hydrogen transfer in non-enzymatic dihydronicotinamide reductions would appear to eliminate hydrogen transfer as a proton and thus favor hydride transfer, there remains a viable alternative to both. The concept of dihydronicotinamide reductions proceeding by an initial one-electron transfer followed by hydrogen-atom abstraction has been mentioned on numerous occasions (e.g. refs. 31–33) but in recent years has rarely received serious consideration. But, as Klinman [32] has pointed out, all the evidence advanced for hydride transfer could equally well be applied to hydrogen atom transfer mechanisms. Kosower, on the basis of the ability of several substituted pyridiniums to undergo one-electron reduction to stable, neutral, free radicals [34, 35], has consistently advanced a homolytic process involving initial charge-transfer complexation with the subsequent hydrogen atom migration completing the net two-electron transfer [36]. Hydrogen atom transfer could well account for the direct hydrogen transfer seen for the reversible enzymatic reaction between NADH and dRF, and for the direct hydrogen transfer now inferred for the isoalloxazines. One might envision that the active site environment necessary to induce NADH · RF complexing would be identical to that required for NADH · dRF complexing and subsequent electron transfer.

The reaction of dihydroflavins with carbonyls may, or may not, represent a different mechanistic class. An impressive body of evidence has accumulated requiring the formation of α-carbanions in the isoalloxazine-dependent oxidation of hydroxy acids and amino acids, for both the enzymatic [37, 38] and non-enzymatic pathways [39–42]. Whether α-carbanionic-substrate species collapse to covalent flavin-substrate adducts has been debated. A flavin N-5 adduct has been trapped during D-amino acid oxidase action on preformed nitroalkane anions [38]. Non-enzymatically with deazaflavins, these anions form transient adducts at low pH that rapidly decompose to return oxidized deazaflavin. If carbanionic species occur during catalysis by deazaflavin-reconstituted enzymes, then one must explain the direct hydrogen transfer to C-5 of the coenzyme analog. It may be that the substrate-derived proton is fully shielded from solvent and is transferred from the initial enzyme base to the C-5 position on reduction of the deazaflavin. Sequestered, substrate-derived protons are implicated in reactions of D-amino acid oxidase [43], lactate 2-monooxygenase [37], and *N*-methylglutamate synthetase [23] containing the normal flavin coenzymes. On the other hand, the enzymatic reduction of deazaflavins with these enzymes may proceed via hydride transfer or two one-electron steps (an electron and a hydrogen atom to C-5 of the deazaflavin) as noted above for the nicotinamide-utilizing enzymes.

As before, the available information may be interpreted by an alternative process. Kosower [44] has examined the energetics of one-electron transfer between dihydropyridines and carbonyls and found them favorable. The structural similarities between dihydrodeazaflavins and dihydronicotinamides, and the closeness of their redox potential might allow identical arguments to be made here. Bruice and coworkers [45] have done similar calculations in more detail for isoalloxazines and carbonyls and have reached the same conclusion. That one-electron transfers represent the actual mechanism for oxidation and reduction of carbonyls is supported by two experiments from his laboratories: the carbinolamine formed in solution between lumiflavin and ethylpyruvate is nonproductive for electron transfer [46]; and incubation of simple carbonyls with N-5-methyllumiflavin results in formation of spectrally detectable flavin semiquinones [45]. This latter observation is consistent only with hydrogen atom transfer from the N-5 position as the final step for carbonyl reduction. Indeed, it remains very likely that a fundamental mechanistic homology exists between the enzymatic and non-enzymatic electron transfers amongst dihydropyridines, deazaisoalloxazines, isoalloxazines and carbonyls.

Acknowledgements

We wish to thank the National Institutes of Health (GM 20011) and the Petroleum Research Fund (administered by the American Chemical Society) for research support. R.S. is an N.S.F. predoctoral fellow.

References

1 O'Brien, D., Weinstock, L. and Cheng, C. (1970) J. Heterocycl. Chem. 7, 99–105
2 Nakamura, N. and Tanaka, M. (1968) J. Agric. Chem. Soc. Japan 42, 281–287
3 Oppenheimer, N., Arnold, L. and Kaplan, N. (1971) Proc. Natl. Acad. Sci. U.S. 68, 3200–3205
4 Gunsalus-Miguel, A., Meighen, E., Nicoli, M., Nealson, K. and Hastings, J. (1972) J. Biol. Chem. 247, 398–404
5 Choong, Y., Sheperd, M. and Sullivan, P. (1975) Biochem. J. 145, 37–45
6 Brumby, P. and Massey, V. (1968) in Biochemical Preparations (Lands, W., ed.), Vol. 12, pp. 29–41, Wiley, New York
7 Massey, V. and Curti, B. (1966) J. Biol. Chem. 241, 3417–3423
8 Swoboda, B. (1969) Biochim. Biophys. Acta 175, 365–379
9 Brustlein, M. and Bruice, T. (1972) J. Am. Chem. Soc. 94, 6548–6549
10 Fenner, H. and Michaelis, P. (1972) Arch. Pharmacol. 305, 751–757
11 Ghisla, S., Hartmann, U., Hemmerich, P. and Muller, F. (1973) Liebig's Ann. Chem. 1388–1415
12 Steffens, J. and Chipman, D. (1971) J. Am. Chem. Soc. 93, 6694–6696
13 Creighton, D., Hajdu, J., Mooser, G. and Sigman, D. (1973) J. Am. Chem. Soc. 95, 6855–6857
14 Bruice, T., Main, L., Smith, S. and Bruice, P. (1971) J. Am. Chem. Soc. 93, 7526–7528
15 Lown, J., Akhtar, M. and McDaniel, R. (1974) J. Org. Chem. 39, 1998–2006
16 Tauscher, L., Ghisla, S. and Hemmerich, P. (1973) Helv. Chim. Acta 56, 630–644
17 Fowler, F. (1972) J. Org. Chem. 37, 1321–1323
18 Edmondson, D., Barman, B. and Tollin, G. (1972) Biochemistry 11, 1133–1138
19 Rafter, G. and Colowick, S. (1957) in Methods in Enzymology (Colowick, S. and Kaplan, N., eds.), Vol. 3, pp. 887–890, Academic Press, New York
20 Fisher, J. and Walsh, C. (1974) J. Am. Chem. Soc. 96, 4345–4346

21 Loach, P. (1968) in Handbook of Biochemistry (Sober, H., ed.), pp. J31–J33, Chemical Rubber Company, Cleveland
22 Averill, B., Schonbrunn, A., Abeles, R., Weinstock, L., Cheng, C., Fisher, J., Spencer, R. and Walsh, C. (1975) J. Biol. Chem. 250, 1603–1605
23 Jorns, M. and Hersh, L. (1974) J. Am. Chem. Soc. 96, 4012–4014
24 Kurz, L. and Frieden, C. (1975) J. Am. Chem. Soc. 97, 677–679
25 Dunn, M. and Hutchison, J. (1973) Biochemistry 12, 4882–4892
26 Shore, J., Gutfreund, H., Brooks, R., Santiago, D. and Santiago, P. (1974) Biochemistry 13, 4185–4191
27 Sund, H. (1968) in Biological Oxidations (Singer, T., ed.), pp. 621–624, Wiley Interscience, New York
28 Suelter, C. and Metzler, D. (1960) Biochim. Biophys. Acta 44, 23–33
29 Wu, F., MacKenzie, R. and McCormick, D. (1970) Biochemistry 9, 2219–2224
30 Hamilton, G. (1971) in Progress in Bioorganic Chemistry (Kezdy, F. and Kaiser, E., eds.), Vol. 1, pp. 83–157, Wiley Interscience, New York
31 Bruice, T. and Benkovic, S. (1966) in Bioorganic Mechanisms, Vol. 2, pp. 301–349, W. Benjamin, New York
32 Klinman, J. (1972) J. Biol. Chem. 247, 7977–7987
33 Kosower, E., Teuerstein, A. and Swallow, A. (1973) J. Am. Chem. Soc. 95, 6127–6128
34 Itoh, M. and Kosower, E. (1968) J. Am. Chem. Soc. 90, 1843–1849
35 Badock, R., Hudson, P., Katritzky, A. and Soti, F. (1974) J. Chem. Soc. Perkin I, 1422–1432
36 Kosower, E. (1966) in Flavins and Flavoproteins (Slater,E., ed.), pp. 1–11. Elsevier, Amsterdam
37 Walsh, C., Lockridge, O., Massey, V. and Abeles, R. (1973) J. Biol. Chem. 248, 7049–7054
38 Porter, D., Voet, J. and Bright, H. (1973) J. Biol. Chem. 248, 4400–4416
39 Brown, L. and Hamilton, G. (1970) J. Am. Chem. Soc. 92, 7225–7227
40 Main, L., Kasperek, G. and Bruice, T. (1972) Biochemistry 11, 3991–4000
41 Shinkai, S., Kunitake, T. and Bruice, T. (1974) J. Am. Chem. Soc. 96, 7140–7141
42 Zeller, E., Gartner, B. and Hemmerich, P. (1972) Z. Naturforsch. 27, 1051–1052
43 Walsh, C., Krodel, E., Massey, V. and Abeles, R. (1973) J. Biol. Chem. 248, 1946–1955
44 Kosower, E. (1965) in Progress in Physical Organic Chemistry (Cohen, S., Streitwieser, A. and Taft, R., eds.), pp. 81–163, Wiley Interscience, New York
45 Williams, R., Shinkai, S. and Bruice, T. (1975) Proc. Natl. Acad. Sci. U.S. 72, 1763–1767
46 Shinkai, S. and Brucie, T. (1974) J. Am. Chem. Soc. 95, 7526–7528

Flavins and Flavoproteins, edited by T.P. Singer

CHAPTER 38

USE OF DEAZAFLAVINS TO STUDY HYDROGEN TRANSFER IN THE GLYCOLATE OXIDASE AND D-AMINO ACID OXIDASE REACTIONS

LOUIS B. HERSH and MARILYN SCHUMAN JORNS

The University of Texas, Health Science Center at Dallas, Southwestern Medical School, 5323 Harry Hines Boulevard, Dallas, Texas 75235 (U.S.A.)

In recent years deazaflavins have been used as model compounds to study both non-enzymic [1, 2] and enzymic flavin reactions [3–6]. The similarity in chemical reactivity between normal flavins and deazaflavins has been described [1, 2, 4, 7–9]. We have previously demonstrated that 5-deaza-FMN can replace FMN as a coenzyme in the *N*-methylglutamate synthase reaction. Studies utilizing [α-^{3}H]glutamate and the deaza-FMN-containing enzyme led to the demonstration of reversible hydrogen transfer from the substrate to flavin as part of the catalytic cycle [3, 4]. Fisher and Walsh [5] have been able to demonstrate hydrogen transfer from NADH to 5-deazariboflavin, when the flavin analogue is used as a substrate in the NADH-FMN oxidoreductase reaction, while Averill et al. [6] have demonstrated hydrogen transfer from L-lactate to deaza FMN in the lactate oxidase reaction.

In order to determine whether hydrogen transfer from substrate to deazaflavin is a general property of oxidases we have studied this phenomenon in pig liver glycolate oxidase and hog kidney D-amino acid oxidase. In both cases, it has been demonstrated that hydrogen transfer from substrate to flavin does occur, leading to dihydroflavin as the product of the reaction.

Results

Glycolate oxidase

Apo-glycolate oxidase (free of 6-hydroxy-FMN) prepared by dialysis of pig liver holoenzyme [10] against 3 M potassium bromide in 0.1 M sodium phosphate buffer, pH 6.3, containing 0.3 mM disodium EDTA, was reconstituted with deaza-FMN. The reconstituted enzyme, containing 2 mol of bound deaza-FMN/mol of enzyme (the same as native enzyme [10]), exhibits no catalytic activity as measured by the glycolate-dependent reduction of 2,6-dichlorophenol indophenol [10]. However, the addition of glycolate to the deaza-FMN-enzyme under anaerobic conditions results in reduction of the bound coenzyme as evidenced by a decrease in absorbance at 400 nm, and a general increase in absorbance below 345 nm which is the

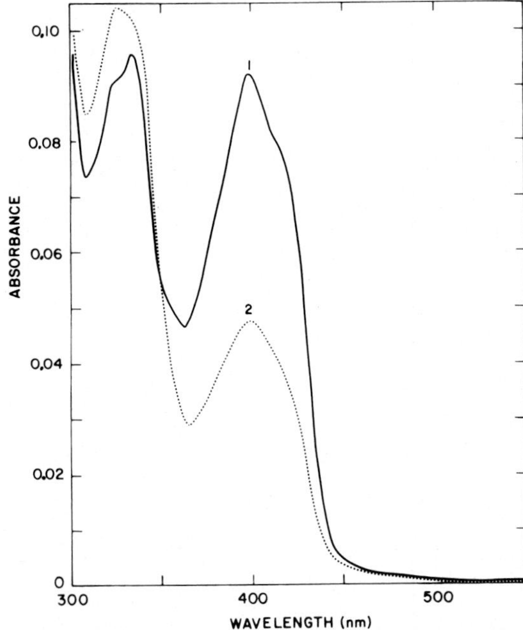

Fig. 1. Reduction of deaza-FMN glycolate oxidase by glycolate. Curve 1, oxidized enzyme in 0.1 M sodium phosphate buffer, pH 7.0, containing 0.3 mM disodium EDTA; Curve 2, 10 min after the anaerobic addition of 10 mM glycolate at 25° C.

isosbestic point of the reaction, Fig. 1. At 10 mM glycolate, an apparent pseudo-first-order rate constant of 0.5 min^{-1} was observed. This value is only an approximation, for as shown below, this reaction represents an approach to equilibrium, rather than full reduction of enzyme-bound flavin. If the reduced flavin is released into solution it re-oxidizes with a half-time of approximately 180 min, suggesting 1,5-dihydrodeaza-FMN as the product of glycolate reduction [7].

The extent of reduction of the bound deazaflavin is dependent on the glycolate concentration while the addition of glyoxylate, to glycolate-reduced enzyme causes rapid reoxidation. Reduction also occurs in the presence of L-lactate, with reoxidation again occurring upon the addition of glyoxylate. Fig. 2 shows that the steady state level of reduced enzyme is a function of the glycolate to glyoxylate ratio. Thus, although deaza-FMN-containing glycolate oxidase does not catalyze oxidation of hydroxy acids via electron transfer from substrate to an electron acceptor, the enzyme is capable of catalyzing anaerobic hydrogen transfer via the reaction sequence shown in Eqns. 1 and 2.

$$R\text{—}\underset{\substack{|\\ XH}}{\overset{\substack{H\\ |}}{C}}\text{—}COO^- + E\text{—}dFl \rightleftharpoons R\text{—}\overset{\substack{X\\ \|}}{C}\text{—}COO^- + E\text{—}dFlH_2, \quad \text{where } X{=}O, \quad dFl{=}\text{deaza-FMN} \qquad (1)$$

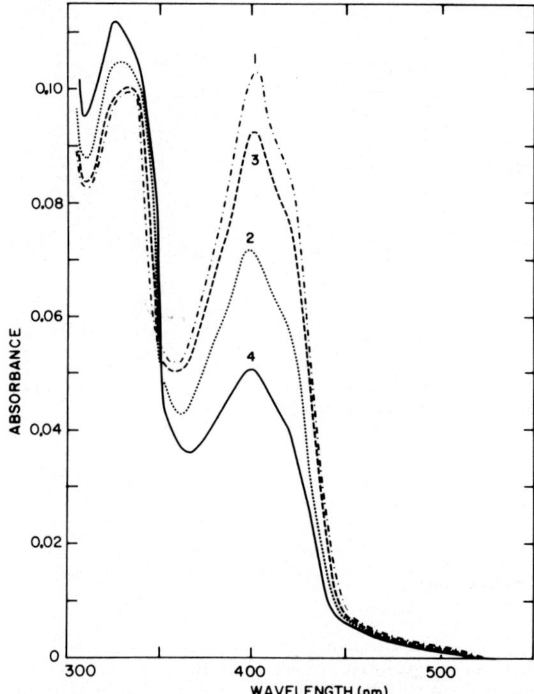

Fig. 2. Effect of glycolate and glyoxylate on the extent of reduction of deaza-FMN glycolate oxidase. Curve 1, oxidized enzyme; Curve 2, 20 min after anaerobic addition of 1 mM glycolate; Curve 3, 5 min after anaerobic addition of 50 μM glyoxylate. No further spectral changes were noted after an additional 40 min incubation at 25°C; Curve 4, 10 min after increasing the glycolate concentration to 25 mM.

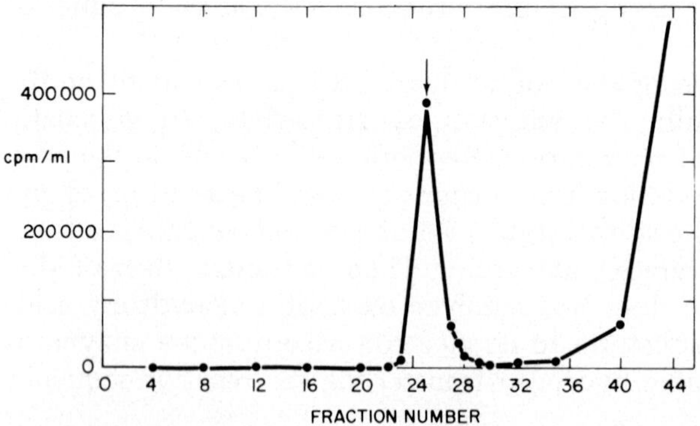

Fig. 3. Isolation of reduced deaza-FMN glycolate oxidase following reduction by [α-^{3}H]glycolate. Deaza-FMN glycolate oxidase was reduced for 60 min at 25°C by 1 mM [α-^{3}H]glycolate (specific activity = $3.1 \cdot 10^5$ cpm/nmol) as described in Fig. 1, and separated from glycolate by anaerobic chromatography on a Sephadex G-25 column (1 × 57 cm). The arrow indicates the position of the protein peak. The isolated enzyme contained 3.2 nmol of tritium corresponds to 0.5 mol of tritium/mol of deaza-$FMNH_2$ produced.

$$R'-\overset{\overset{\displaystyle X}{\|}}{C}-COO^- + dFlH_2 \rightleftharpoons R'-\underset{\underset{\displaystyle XH}{|}}{\overset{\overset{\displaystyle H}{|}}{C}}-COO^- + E-dFl. \qquad (2)$$

From the data in Fig. 2 and experiments not shown, it was possible to measure the equilibrium constant for the reactions described by Eqns. 1 and 2 with glycolate and glyoxylate as substrates. The equilibrium constant of $3 \cdot 10^{-3}$ (K_{eq} = [E—dFMNH$_2$] [glyoxylate]/[E—dFMN] [glycolate]) shows that the equilibrium position of the reaction strongly favors the oxidized form of the enzyme. A similar result was obtained with deaza-FMN *N*-methylglutamate synthase, in which the equilibrium of that reaction strongly favors oxidized deaza-FMN as compared to an equilbrium constant of 1.0 for the native enzyme [4].

When deaza-FMN-containing glycolate oxidase is anaerobically reduced with [α-^{3}H] glycolate and then separated from tritiated glycolate by anaerobic chromatography on a Sephadex G-25 column, tritium is found to migrate with the protein, Fig. 3. The amount of tritium co-chromatographing with the enzyme corresponds to 0.5 mol of bound tritium/mol of deaza-FMNH$_2$ formed, suggesting the kinetic isotope discrimination against tritium is 2.0. The same experiment conducted with [1-^{14}C] glycolate results in no incorporation of radioactivity into the enzyme.

When isolated tritium-labeled enzyme is denatured by acid, tritium is

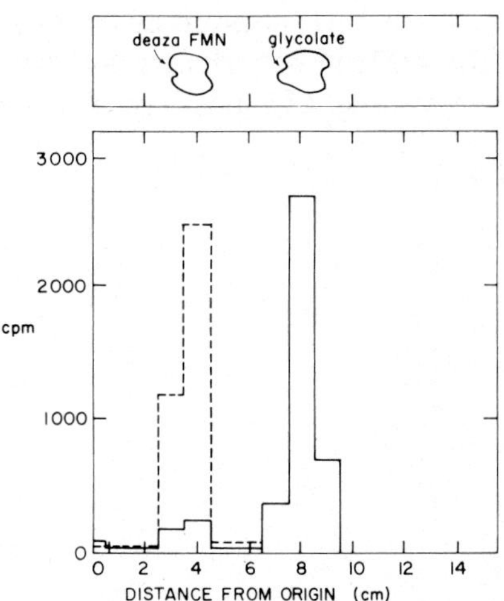

Fig. 4. Identification of deaza-[^{3}H] FMN and [α-^{3}H] glycolate. Tritiated deaza-FMN glycolate oxidase prepared and isolated as described in Fig. 4 except that the column fractions were incubated for 5 min in the presence or absence of 0.1 mM glyoxylate prior to acid denaturation. After removal of denatured protein by centrifugation and extraction of trichloroacetic acid into ether, the samples were chromatographed on Whatmann 3 MM paper using *n*-butanol/acetic acid/water (12 : 3 : 5 v/v) as solvent. Deaza-FMN was located by fluorescence and glycolate by spraying with bromphenol blue. The chromatogram was cut into 1 cm strips, eluated with water, and the radioactive content determined by liquid scintillation counting. After incubation with glyoxylate, 90% of the radioactivity is present as [^{3}H] glycolate (solid line) whereas all of the radioactivity in the untreated sample (dashed line) is present as deaza-[^{3}H] FMN.

released into solution. About 20% of this radioactivity is lost upon repeated evaporation, while the remaining tritium co-chromatographs with authentic deaza-FMN, Fig. 4. When glyoxylate is added to the isolated enzyme prior to denaturation, no tritium is lost upon evaporation, and tritiated glycolate is identified as the product, Fig. 4.

These results indicate that reduction of the enzyme by glycolate leads to the transfer of the α hydrogen from the substrate to the enzyme-bound flavin. Upon denaturation and non-enzymic reoxidation of free deaza-FMN, approximately 20% of this tritium is lost since there are two hydrogens at the 5 position of deaza-$FMNH_2$ one of which is lost upon reoxidation. When glyoxylate is added to the reduced enzyme, reoxidation of reduced deaza-FMN involves the stereospecific transfer of hydrogen from the reduced flavin to the substrate. Thus, in the case of glycolate oxidase, reduction and reoxidation of bound flavin involves hydrogen transfer between substrates and flavin.

D-*Amino acid oxidase*

D-Amino acid oxidase and apo-D-amino acid oxidase were prepared by published procedures [11, 12]. The apoenzyme was reconstituted with deaza-FAD, prepared enzymatically from deaza-FMN by the action of FAD pyrophosphorylase, EC 2.7.7.2. The reconstituted enzyme contained 1.8 mol of deaza-FAD/mol enzyme (0.9 mol/45 500 g) [12].

The deaza-FAD-containing enzyme exhibits less than 0.05% the activity of native enzyme as measured by the production of pyruvate from D-alanine. Characteristic of native D-amino acid oxidase, deaza-FAD-containing enzyme binds benzoate as evidenced by a shift of the 400 nm absorbance band to 405 nm and a broadening of the shoulder at aroung 415 nm, Fig. 5. The

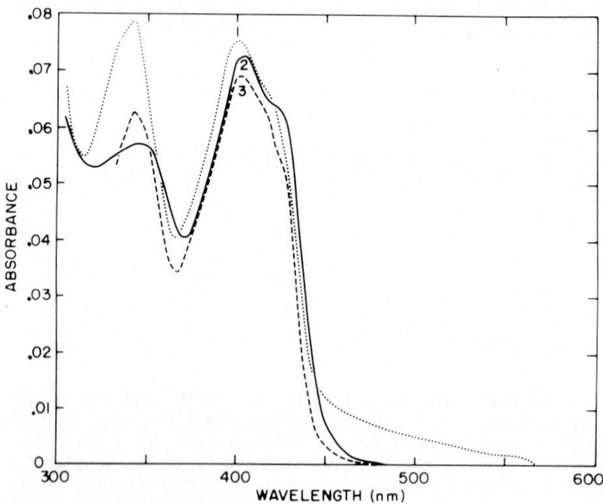

Fig. 5. Complex formation with deaza-FAD-D-amino acid oxidase. Curve 1, deaza-FAD D-amino acid oxidase in 0.1 M sodium pyrophosphate buffer, pH 8.5. After the addition of 1 mM sodium benzoate (Curve 2) or the addition of 1 mM anthranilate (Curve 3) to oxidized enzyme.

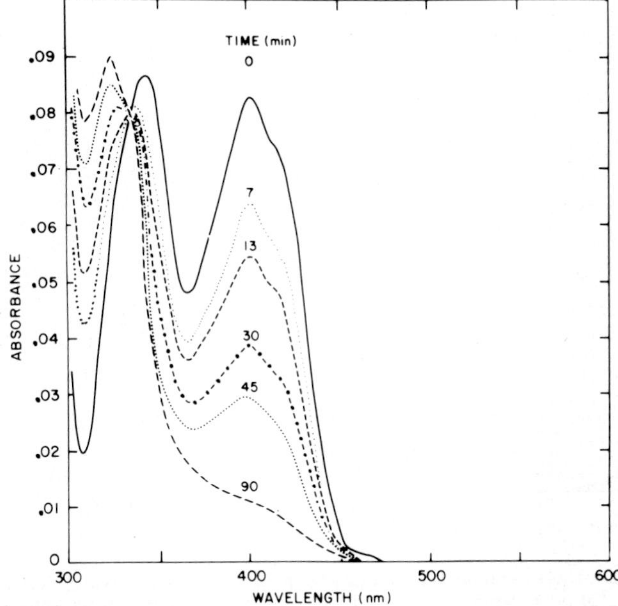

Fig. 6. Spectral changes accompanying alanine reduction of deaza-FAD-D-amino acid oxidase. To deaza-FAD-D-amino acid oxidase ($7 \cdot 10^{-6}$ M with respect to deaza-FAD) in 0.1 ml of 0.1 M sodium pyrophosphate buffer, pH 8.5, was added 50 μl of 0.5 M DL-alanine. Spectra were recorded at the time periods indicated at 20°C.

TABLE I

REDUCTION OF DEAZA-FAD-D-AMINO ACID OXIDASE BY VARIOUS SUBSTRATES

Reaction mixtures contained 100 mM sodium pyrophosphate buffer, pH 8.5, 0.1 mM dithioerythritol, 0.06 mM deaza-FAD D-amino acid oxidase, and substrates as indicated in a final volume of 1.0 ml. Reduction was monitored by the decrease in absorbance at 400 nm at 20°C. Rate constants were determined from pseudo-first-order rate plots.

Substrate	$k(\text{min}^{-1})$
DL-Alanine	0.039
DL-Methionine	0.023
DL-Phenylalanine	0.011
DL-Proline	0.20*
DL-Valine	0.012
DL-Glutamate	$<$0.0005
DL-Aspartate	$<$0.0005
DL-[α-^{2}H] Alanine	0.50

* Estimated from initial rates. Only 33% of the bound flavin was reduced.

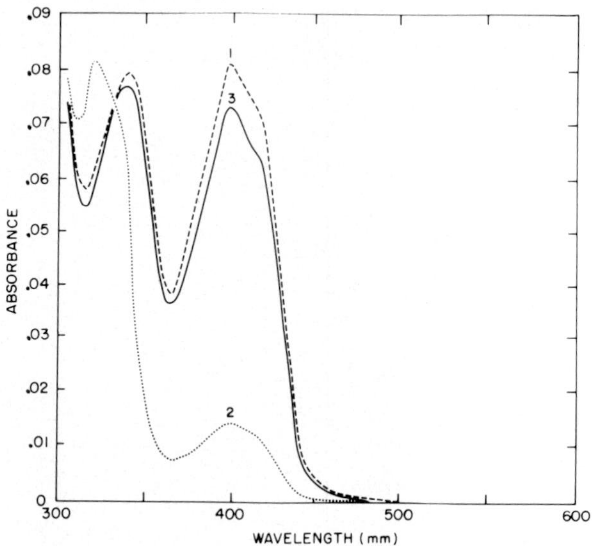

Fig. 7. Reoxidation of reduced deaza-FAD-D-amino acid oxidase by pyruvate-imine. Curve 1, deaza-FAD-D-amino acid oxidase ($7 \cdot 10^{-6}$ M with respect to deaza-FAD) in 0.1 M sodium pyrophosphate buffer, pH 8.5. Curve 2, 75 min after the addition of 25 mM DL-alanine at 20° C. Curve 3, 10 min after the addition of 50 mM pyruvate and 50 mM ammonium chloride. No further spectral changes were observed after an additional 20 min.

enzyme also binds 2-aminobenzoate with the production of increased absorbance in the region of 425 to 580 nm, Fig. 5.

As observed with glycolate oxidase, deaza-FAD D-amino acid oxidase can be reduced by substrates for the native enzyme. Table I lists the pseudo-first-order rate constants for reduction of the enzyme by a number of DL-amino acids. With the exception of proline, the extent of reduction of the bound chromophore proceeds to at least 85% completion. The lack of reactivity of both aspartate and glutamate, is consistent with the known substrate specificity of the native enzyme [13]. As shown in Fig. 6, the spectral changes accompanying reduction, a decrease in absorbance at 400 nm, an increase in absorbance at 322 nm, and an isosbestic point at 333 nm, are very similar to that observed upon substrate reduction of deaza-FMN bound to *N*-methylglutamate synthetase [4] and glycolate oxidase.

The rate of reduction by DL-alanine compared to the rate of reduction by DL-[α-^{2}H] alanine shows a deuterium isotope effect of 1.67, Table I, indicating the importance of carbon-hydrogen bond breakage in the rate-determining step of the reaction.

Evidence that reduction of deaza-FAD by alanine is reversible is shown by the reoxidation of the reduced flavin upon addition of pyruvate plus ammonia (Fig. 7). No re-oxidation by pyruvate alone or by ammonia alone is observed. Reoxidation is also observed if 20 mM β-phyenylpyruvate or 20 mM α-ketobutyrate, but not 20 mM α-ketoglutarate, is substituted for pyruvate. The rate of re-oxidation of reduced deaza-FAD-amino acid oxidase, isolated by Sephadex G-25 chromatography, in the presence of 20 mM

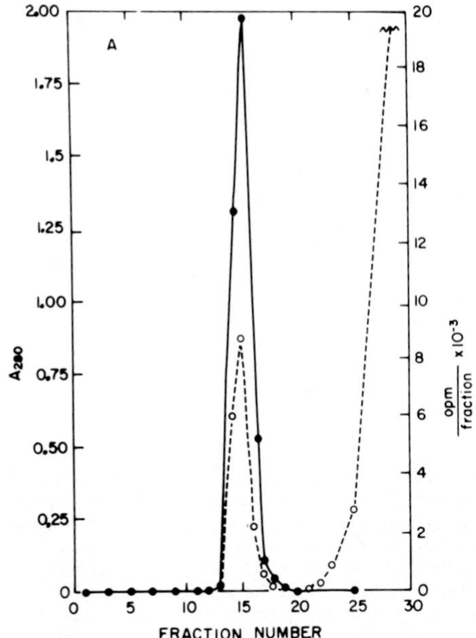

Fig. 8. Incorporation of tritium from DL-[α-^{3}H]alanine into deaza-FAD D-amino acid oxidase. (A). A 1 ml reaction mixture containing 40 nmol of deaza-FAD D-amino acid oxidase in 0.1 M sodium pyrophosphate buffer, pH 8.5 was reacted for 60 min at 20°C with 25 mM DL[α-^{3}H]alanine, specific activity $1.1 \cdot 10^6$ cpm/μmol. A total of 34.7 nmol of reduced deaza-FAD were produced during this time. The reaction mixture was chromatographed on a Sephadex G-25 column (1.2 × 50 cm) equilibrated with 1 mM sodium pyrophosphate buffer, pH 8.5. Fractions were assayed for protein (●) and tritium content (○). The total readioactivity in the protein-containing fractions corresponds to 17.2 nmol of tritium incorporated.

pyruvate and 20 mM ammonia is 0.50 min^{-1}, considerably faster than reduction by alanine; $k' = 0.039$ min.

These results suggest that like deaza-FMN-containing glycolate oxidase, deaza-FAD-containing D-amino acid oxidase catalyzes a transhydrogenation reaction according to Eqns. 1 and 2, where X = NH and dFl = deaza-FAD.

In order to determine whether reduction of the enzyme by alanine results in hydrogen transfer from substrate to flavin, deaza-FAD D-amino acid oxidase was reduced with DL-[α-^{3}H] alanine, and then chromatographed on a Sephadex G-25 column. As shown in Fig. 8 tritium was found to co-chromatograph with the enzyme. Denaturation of the enzyme by heat or acid released the tritium into solution, and after repeated evaporations 65% of the tritium remained. The tritiated compound was identified as deaza-FAD by thin-layer chromatography, Fig. 9. No tritium was incorporated into the enzyme when the same experiment was conducted using apoenzyme in place of deaza-FAD enzyme. If DL-[U-^{14}C] alanine was used, no radioactivity co-chromatographed with the enzyme. When the isolated tritiated enzyme was treated with 20 mM pyruvate/20 mM ammonia prior to denaturation, radioactivity was no longer observed in the flavin, but instead co-chromatographed with alanine, Fig. 9.

These results thus show that deaza-FAD-containing D-amino acid

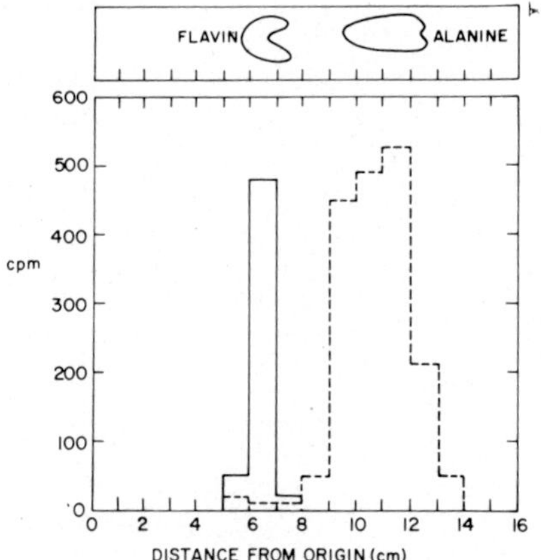

Fig. 9. Identification of tritium in deaza-FAD-D-amino acid oxidase as tritiated deaza-FMN. The tritiated protein fractions obtained in the experiment shown in Fig. 8 were pooled, and two aliquots were withdrawn. One was boiled for 1 min and then subjected to repeated evaporation: to the other was added pyruvate and ammonia to give a final concentration of 25 mM each. This fraction was incubated for 30 min, boiled and then evaporated as described above. After dissolving the fractions, aliquots were spotted on a Silica Gel F-254 (Merck). After development of the plates in 5% Na_2HPO_4 1 cm blocks were scraped from the plates and the tritium content determined by liquid scintillation counting. Deaza-FAD was identified by its fluorescence, and alanine by ninhydrin reagent.—, elution of radioactivity from the untreated sample; - - - -, enzyme pre-treated with ammonia plus pyruvate.

oxidase, like deaza-FMN-containing glycolate oxidase, catalyzes hydrogen transfer from substrate to flavin, and that this hydrogen transfer involves the stereospecific incorporation of hydrogen into the flavin.

Discussion

It is clear from these studies and from those of Averill et al. [6] that deazaflavin-containing oxidases are not active in catalyzing substrate oxidation with oxygen as the electron acceptor. These enzymes are, however, capable of catalyzing reversible half-reactions leading to oxidized substrate and reduced flavin. These observations are consistent with the known chemical reactivity of deazaflavins. Brustlein and Bruice [1] showed that the rate of reduction of 3,10-dimethyl-5 deazaisoalloxazine by NADH or reduced *N*-propyl nicotinamide proceeds at 3.5–7% the rate with 3,10-dimethyl isoalloxazine. Edmondson et al. [7] have shown that the rate of reoxidation of reduced deaza-FMN by oxygen is less than 10^{-4} times as fast as the rate of reoxidation of reduced FMN.

It is clear from our studies on deaza-FMN glyclate oxidase and deaza-FAD D-amino acid oxidase as well as from other studies using deazaflavin-containing enzymes [3, 4, 6] that hydrogen transfer from substrate to flavin is an integral part of the reaction sequence in going from oxidized deazaflavin to reduced deazaflavin. Evidence has been presented by Walsh et al. [14–16] and by others [17–20] that one of the initial steps in flavoprotein oxidase reactions is the abstraction of the substrate α hydrogen as proton. Since the pK for protonation of oxidized flavin is on the order of zero [21] it is unlikely that flavin itself could serve as a general base to abstract a proton. It thus seems likely that hydrogen transfer from substrate to flavin is mediated by a basic residue on the protein. At the present time, it is not clear whether the subsequent transfer of this proton plus two electrons to flavin involves adduct formation between substrate carbanion and flavin [22, 23], or is mediated through a non-covalent carbanion-flavin complex [24].

Acknowledgements

This work was supported in part by grant No. I-391 from The Robert A. Welch Foundation, Houston, Texas and grant No. AM13443 from the National Institutes of Health. L.B.H. acknowledges the receipt of a Research Career Development Award from the National Institute of General Medical Science, No. GM70239.

References

1 Brustlein, M. and Bruice, T.C. (1972) J. Am. Chem. Soc. 94, 6548–6549
2 Shinkai, S. and Bruice, T.C. (1973) J. Am. Chem. Soc. 95 7526–7528
3 Jorns, M.S. and Hersh, L.B. (1974) J. Am. Chem. Soc. 96, 4012–4014
4 Jorns, M.S. and Hersh, L.B. (1975) J. Biol. Chem. 250, 3620–3628
5 Fisher, J. and Walsh, C. (1974) J. Am. Chem. Soc. 96, 4345–4346
6 Averill, B.A., Schonbrunn, A., Abeles, R.H., Weinstock, L.T., Cheng, C.C., Fisher, J., Spencer, R. and Walsh, C. (1975) J. Biol. Chem. 250, 1603–1605
7 Edmondson, D.E., Barman, B. and Tollen, G. (1972) Biochemistry 11, 7526–7528
8 Sun, M. and Song, P.S. (1973) Biochemistry 12, 4663–4669
9 Bruice, T.C., in Progress in Bioorganic Chemistry (Karser, T. and Keady, F., eds.), Vol. 5, Wiley-Interscience, N.Y., in the press
10 Schuman, M. and Massey, V. (1971) Biochim. Biophys, Acta 227, 500–520
11 Brumby, P.E. and Massey, V.E. (1968) Biochem. Prep. 12, 29–41
12 Massey, V. and Curti, B. (1966) J. Biol. Chem. 241, 3417–3423
13 Dixon, M. and Kleppe, K. (1965) Biochim. Biophys. Acta 96, 368–382
14 Walsh, C.T., Schonbrunn, A. and Abeles, R.H. (1971) J. Biol. Chem. 246, 6855–6866
15 Walsh, C.T., Krodel, E., Massey, V. and Abeles, R.H. (1973) J. Biol. Chem. 248, 1946–1955
16 Walsh, C.T., Lockridge, O., Massey, V. and Abeles, R.H. (1973) J. Biol. Chem. 248, 7049–7054
17 Neims, A.H., DeLuca, D.C. and Hellerman, L. (1966) Biochemistry 5, 203–213
18 Yagi, K., Nishikimi, M., Takai, A. and Ohishi, N. (1974) J. Biochem. (Tokyo) 76, 451–454
19 Yagi, K., Nishikimi, M., Takai, A. and Ohishi, N. (1974) Biochim. Biophys. Acta 341, 256–264
20 Yagi, K., Nishikimi, M., Takai, A. and Ohishi, N. (1973) Biochim. Biophys. Acta 321, 64–71
21 Hemmerich, P., Veeger, C. and Wood, H.C.S. (1965) Ang. Chem. 4, 671–687
22 Porter, D.J.T., Voet, J.G. and Bright, H.J. (1973) J. Biol. Chem. 248, 4400–4416
23 Walsh, C.T., Schonbrunn, A., Lockridge, O., Massey, V. and Abeles, R.H. (1972) J. Biol. Chem. 247, 6004–6006
24 Bruice, T.C., Main, I., Smith, S. and Bruice, P.Y. (1971) J. Am. Chem. Soc. 93, 7327–7328

Flavins and Flavoproteins, edited by T.P. Singer

CHAPTER 39

USE OF DEAZA-FMN TO STUDY HYDROGEN TRANSFER IN THE *N*-METHYLGLUTAMATE SYNTHASE REACTION

MARILYN SCHUMAN JORNS and LOUIS B. HERSH

The University of Texas, Southwestern Medical School, Department of Biochemistry, 5323 Harry Hines Boulevard, Dallas, Texas 75235 (U.S.A.)

Introduction

The flavoenzyme, *N*-methylglutamate synthase [1, 2] catalyzes the reversible formation of *N*-methyl-L-glutamate from L-glutamate and methylamine via the two-step mechanism shown in Reactions 1 and 2 [3];

$$\text{E–FMN} + \text{L-glutamate} \rightleftharpoons \text{E}\begin{matrix} \diagup \text{FMN reduced} \\ \diagdown \text{glutaryl} \end{matrix} + \text{ammonia} \qquad (1)$$

$$\text{E}\begin{matrix} \diagup \text{FMN reduced} \\ \diagdown \text{glutaryl} \end{matrix} + \text{methylamine} \rightleftharpoons \text{E–FMN} + N\text{-methyl-L-glutamate} \qquad (2)$$

It has been shown that the α hydrogen abstracted from glutamate during formation of the reduced enzyme intermediate (Reaction 1) is not exchangeable with solvent, and is used in the reformation of the α carbon-hydrogen bond, (Reaction 2) [3]. The α substrate hydrogen may be transferred to a basic amino acid residue in the protein or to the flavin prosthetic group during formation of the reduced enzyme intermediate. Hydrogen attached to weakly basic nitrogens in free reduced flavins exchanges rapidly with solvent, preventing direct study of substrate-flavin hydrogen transfer during catalysis. This difficulty has been avoided by use of 5-deaza-FMN (the N-5 nitrogen of FMN is replaced by CH [4, 5]), a flavin analogue which, although less reactive, has been shown to exhibit qualitatively similar chemical properties as compared with normal flavin [5–9]. Substitution of deaza-FMN for FMN in *N*-methylglutamate synthase thus permits a unique opportunity to directly study the involvement of flavin in enzymic hydrogen transfer.

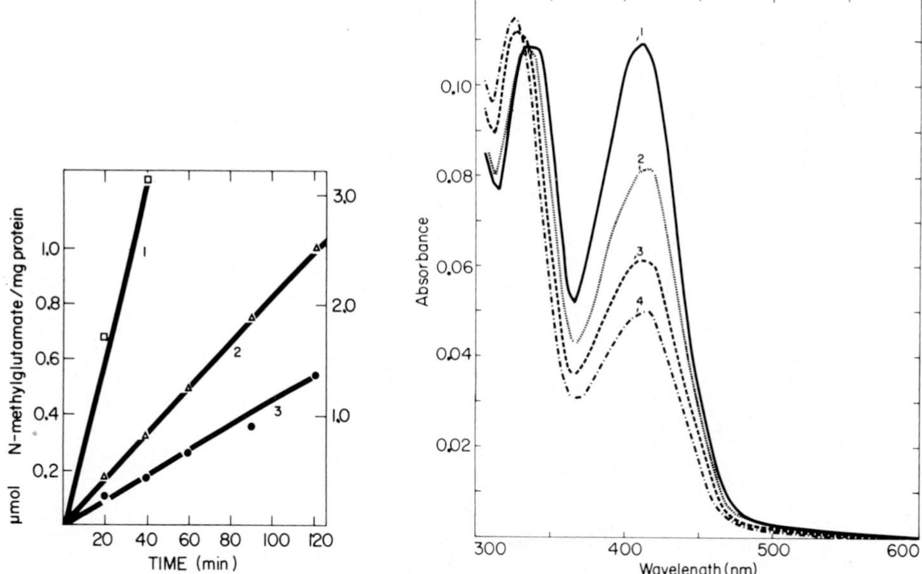

Fig. 1. Time course of *N*-methylglutamate formation from glutamate and methylamine in the presence of FMN-reconstituted (1) and deaza-FMN-reconstituted (2) enzymes and apoprotein (3). Line 1 is plotted according to the scale on the right and lines 2 and 3 are plotted according to the scale on the left. Enzyme acitivity was measured at 30°C as previously described [2]. Specific activity values (μmol *N*-methylglutamate/min/mg protein $\times 10^3$) of native, FMN-reconstituted and deaza-FMN-reconstituted enzymes are 115.0, 79.4, and 3.8, respectively. The latter two values are corrected for the blank reaction observed with apoprotein alone.

Fig. 2. Reduction of deaza-FMN enzyme by L-glutamate. Curve 1, oxidized enzyme in 110 mM potassium tricine buffer, pH 8.3, containing 16.7 mM potassium chloride, 1.1 mM dithioerythritol, 1.4 mM magnesium chloride and 0.3 mM magnesium disodium EDTA. Curves 2–4 were recorded 1.5, 11.5, and 34 min, respectively, after the addition of 22.2 mM L-glutamate at 8°C.

Results

The binding of either FMN of deaza-FMN, as evidenced by the quenching of fluorescence of the free flavin derivatives, follows pseudo-first-order kinetics in the presence of excess apoprotein ($1.5 \cdot 10^{-4}$ M in flavin binding sites). Similar rate constants are obtained for the binding of FMN ($k_{obs} = 3.5$ min^{-1}) and deaza-FMN ($k_{obs} = 2.9$ min^{-1}). Preparations of enzyme reconstituted with FMN and deaza-FMN contain 50–70% of the flavin content observed for native enzyme. The catalytic efficiency of deaza-FMN-reconstituted enzyme is 3.3% of native enzyme whereas the FMN-reconstituted enzyme exhibits a specific activity equal to 69% of the native enzyme (Fig. 1).

Flavin reduction, observed during the reaction of FMN enzyme with amino acid substrates (Reaction 1), is also observed with deaza-FMN enzyme upon addition of L-glutamate or *N*-methyl-L-glutamate (Fig. 2), but not D-glutamate, indicating that in Reaction 1 deaza-FMN enzyme exhibits the same specificity as native enzyme. The spectral course of substrate reduction of deaza-FMN enzyme is consistent with the direct conversion of oxidized

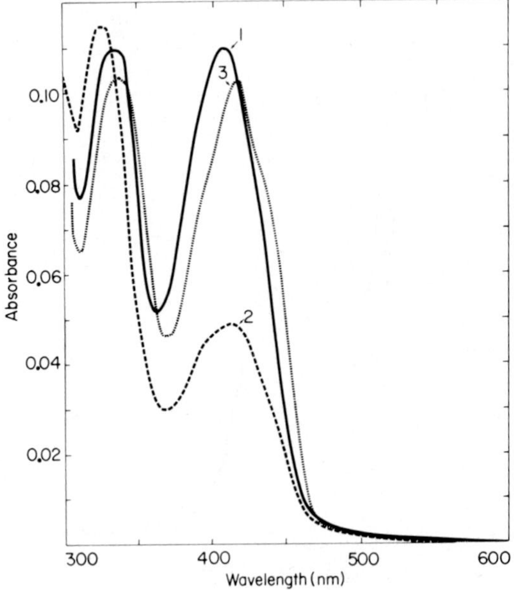

Fig. 3. Reoxidation of substrate-reduced deaza-FMN enzyme by ammonia. Curve 1, oxidized enzyme in the same buffer as described in Fig. 2. Curve 2, 37 min after the addition of 22.2 mM L-glutamate at 8°C. Curve 3, after the addition of 22.2 mM ammonium sulfate to the reduced enzyme, represents the spectrum of the enzyme in the steady state.

deaza-FMN to 1,5-dihydrodeaza-FMN since a similar isosbestic reaction and hypsochromic shift of the 340 nm band of the oxidized chromophore also occurs during dithionite reduction of free deazariboflavin and partially reduced samples of deazariboflavin show an absorbance maximum at 320 nm [10]. At saturating concentrations of glutamate, the reduction of deaza-FMN-reconstituted enzyme ($k = 0.17\,\text{min}^{-1}$) proceeds at approximately 1% the rate observed for FMN-reconstituted enzyme ($k = 20.8\,\text{min}^{-1}$).

Free reduced deaza-FMN, released from substrate-reduced enzyme by acid denaturation, is slowly reoxidized ($t_{1/2} = 3$ h), similar to results obtained for the reoxidation of 1,5-dihydrodeazaflavins [5]. Rapid reoxidation of enzyme-bound reduced deaza-FMN occurs upon addition of either ammonia or methylamine (Fig. 3). Reoxidation in the presence of saturating glutamate with 22.2 mM ammonium sulfate is too fast to measure on the Cary spectrophotometer and at the same concentration of methylamine the reaction is complete within 1 min. These results show that deazaflavin reduction is slower than reoxidation since the $t_{1/2}$ for reduction in the presence of saturating glutamate is 4.0 min.

The addition of a 2-fold excess of ammonia with respect to glutamate causes nearly complete flavin reoxidation, as judged by the increase in absorbance at 410 nm (Fig. 3). When the glutamate concentration was held constant and the ammonia concentration varied to give partial flavin reoxidation, the equilibrium was found to strongly favor oxidized flavin.

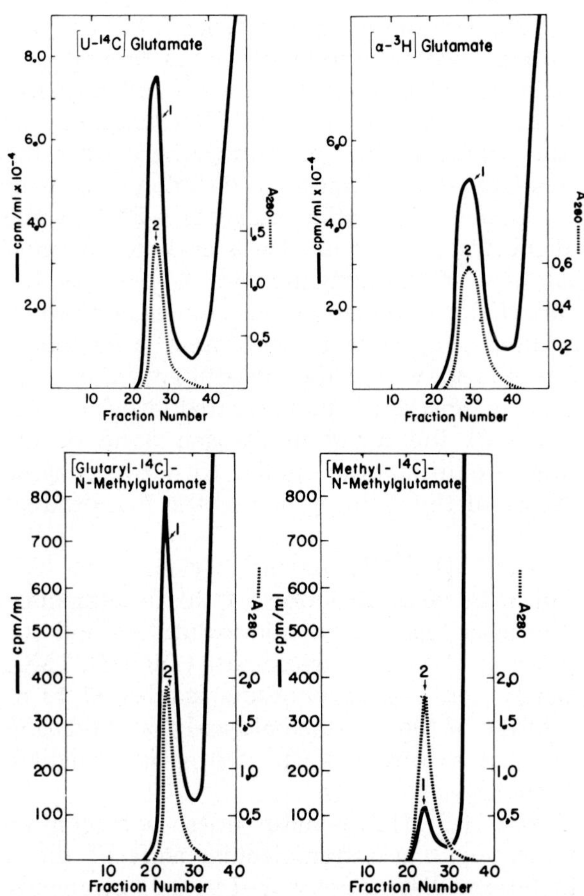

Fig. 4. Isolation of deaza-FMN glutaryl enzyme intermediate. Enzyme was incubated for 60 min at 10°C in 110 mM tricine buffer, pH 8.3, 88 mM potassium chloride, 1.1 mM dithioerythritol, 1 mM magnesium disodium EDTA, 5 mM magnesium chloride and the indicated substrate at 1 mM concentration and then chromatographed on a Sephadex G-25 column, as previously described [11]. The amount of labeling observed corresponds to 0.8, 0.7, 0.5, and 0.08 mol of substrate per mol of reduced deaza-FMN formed during reaction with [U-^{14}C]glutamate, [α-^{3}H]glutamate, *N*-[*glutaryl*-^{14}C]methylglutamate and *N*-[*Me*-^{14}C]methylglutamate, respectively. The deaza-$FMNH_2$ formed was determined by the decrease in absorbance at 410 nm ($\epsilon_{\text{deaza-FMN}} - \epsilon_{\text{deaza-FMNH}_2} = 9400\ M^{-1} \cdot cm^{-1}$).

This can be contrasted with the equilibrium constant of 1 obtained for Reaction 1 with native enzyme [3].

A perturbation of the original oxidized enzyme spectrum (Fig. 3, curve 1) identical to that observed after reoxidation (Fig. 3, curve 3), is also observed with non-reducing amino acid substrate analogues, such as D-glutamate, α-ketoglutarate or α-L-hydroxyglutarate but not with amines, suggesting that the spectrum obtained after reoxidation with amine substrate is due to an L-glutamate-oxidized enzyme complex. This complex would be expected to be the predominant enzyme species under the conditions of Fig. 3 since the steady-state would favor an oxidized form of the enzyme and glutamate is present at saturating levels.

The reduction-reoxidation reactions observed with deaza-FMN enzyme suggested that a deaza-FMN glutaryl enzyme intermediate was formed and, as with native enzyme [3], could be isolated by Sephadex G-25 chormatography. Incubation of deaza-FMN enzyme with [U-^{14}C]glutamate or [α-^{3}H]glutamate results in the incorporation of radioactivity into the enzyme (Fig. 4) which is approximately stoichiometric (70–90% yields) with the amount of deaza-$FMNH_2$ produced, indicating that deaza-FMN is the sole oxidant of the amino acid substrate. Radioactivity is also incorporated into the enzyme when *N*-[*glutaryl*-^{14}C]methylglutamate is used as the reductant, but virtually no radioactivity chromatographs with enzyme when *N*-[*Me*-^{14}C]methylglutamate is used (Fig. 4) or if apoprotein is incubated with [U-^{14}C]glutamate. These results show that the glutaryl group, but not the amino group, of the amino acid substrate is incorporated into the intermediate, indicating that cleavage of the α-carbon-nitrogen bond of the substrate occurs during intermediate formation, similar to native enzyme [3] and consistent with the release of the amino group as the first product of the overall reaction.

Intermediate prepared with [U-^{14}C]glutamate yields α-[^{14}C]-ketoglutarate when denaturated with acid. Reaction of this intermediate with methylamine results in the disappearance of α-ketoglutarate and the formation of stoichiometric amounts of *N*-methylglutamate (Table I). These results show that the catalytically active glutaryl residue is released as α-ketoglutarate upon acid denaturation of the intermediate and that oxidation at the α carbon of the amino acid substrate occurs during intermediate formation similar to native enzyme [3].

Intermediate prepared from [α-^{3}H]glutamate does not contain radioactive α-ketoglutarate, indicating that intramolecular shift of the α hydrogen does not occur. If the intermediate is denatured by acid or phenol, 23% of the tritium is converted to $^{3}H_2O$ during air reoxidation and 48% of the tritium chromatographs with deaza-FMN during paper and thin layer chromatography [11], indicating that greater than 70% of the tritium in the isolated intermediate is present as deaza-[^{3}H] $FMNH_2$. (α-Hydroxy[^{3}H]-glutarate accounts for rest of the tritium label, as discussed later.) If the

TABLE I

REACTION PRODUCTS OF DEAZA-FMN ENZYME INTERMEDIATE FORMED WITH [U-^{14}C] GLUTAMATE

Enzyme intermediate, prepared as described in the legend to Fig. 4, was incubated for 2.5 min at 30°C in 10 mM sodium pyrophosphate buffer, pH 8.5, 1 mM magnesium disodium EDTA, 5 mM 2-mercaptoethanol, 2.5 mM dithioerythritol and, where indicated, 120 mM methylamine. The intermediate was then denatured with 5% trichloroacetic acid and aliquots were separately analysed [10] for α-ketoglutarate and *N*-methylglutamate.

Treatment	Products (cpm)	
	α-Ketoglutarate	*N*-Methylglutamate
Acid denaturation	2440	0
Methylamine	180	2230

intermediate is incubated with methylamine prior to denaturation, 85% of the tritium, initially present as deaza-$FMNH_2$, is incorporated into *N*-methylglutamate (Table II). Intermediate prepared with FMN enzyme and [α-^{3}H]glutamate yields ^3H_2O when denatured with acid and *N*-[^{3}H]-methylglutamate when reacted with methylamine [3, 10]. The results with deaza-FMN enzyme indicate that the α hydrogen of glutamate is transferred to deaza-$FMNH_2$ during intermediate formation and that this hydrogen is utilized for the reformation of the α-C—H bond during reaction of this intermediate with methylamine.

The pseudo-first-order rate constant for formation of the deaza-FMN enzyme intermediate with glutamate is equal to the rate of the corresponding overall reaction and the rate of reaction of the isolated intermediate with methylamine is considerably greater than the rate of the corresponding

TABLE II

REACTION OF ^{3}H-LABELED DEAZA-FMN ENZYME INTERMEDIATE WITH METHYLAMINE

Treatment[a]	Radioactivity incorporated[b] (cpm)	
	Deaza-$FMNH_2$	Amino acids
Unincubated	46449	—
Incubated 3.5 min with 100 mM CH_3NH_2	6861	39423[c]
Incubated 3.5 min without CH_3NH_2	40330	6184[d]

[a] Intermediate, prepared as described in the legend to Fig. 4, was incubated where indicated at 30°C and then denatured with 0.25 M formic acid.
[b] Values determined by the fraction procedure previously described [11].
[c] Identified as *N*-methylglutamate by paper and silica gel chromatography [11].
[d] The low incorporation of ^{3}H into amino acid, identified as glutamate, is presumably due to trace contamination by ammonia.

TABLE III

RATES OF DEAZA-FMN ENZYME INTERMEDIATE FORMATION AND REACTION

	Rate of intermediate formation or reaction (min^{-1})	Rate of overall reaction (min^{-1})
Intermediate formation[a]	0.17 ± 0.02	0.18 ± 0.03
Intermediate reaction[b]	≈13	0.21 ± 0.03

[a] The rate of the overall reaction was measured at 10°C by determining the amount of radioactive *N*-methylglutamate formed during incubation of deaza-FMN enzyme in 100 mM potassium tricine buffer, pH 8.3, 1 mM dithioerythritol, 20 mM L-glutamate, varying concentrations of [^{14}C] methylamine and potassium chloride to maintain an ionic strength of 200 mM. Intermediate formation, measured under the same conditions except that methylamine was not present, was monitored by the decrease in absorbance at 410 nm. This method is equivalent to measuring the rate of formation of α-ketoglutarate as an estimate of intermediate formation (Fig. 5).
[b] The rate of the overall reaction was determined as described above except the glutamate concentration was varied at a constant [^{14}C] methylamine concentration (100 mM). The rate of intermediate (prepared with [U-^{14}C] glutamate as described in the legend to Fig. 4) reaction was measured under the same conditions except unlabeled methylamine was used and glutamate was not present. Sufficient points for a pseudo-first-order plot were not obtained since 87 and 100 percent of the reaction occurred at 10 and 20 s, respectively. Thus, the half-time of this reaction was estimated to be on the order of 3 s, assuming pseudo-first-order kinetics.

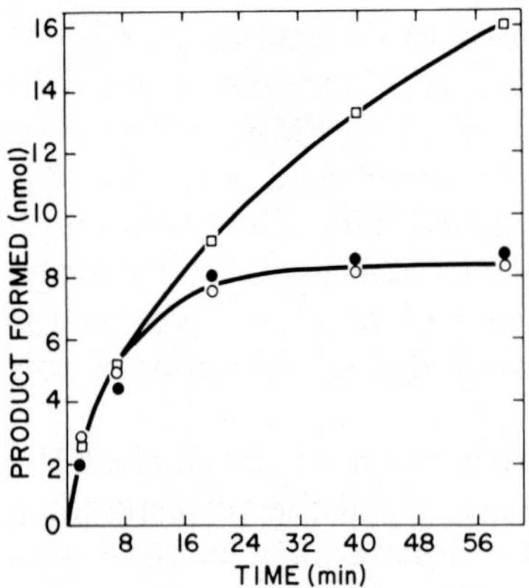

Fig. 5. Formation of reduced deaza-FMN (○), α-ketoglutarate (●) and methylamine (□) during reaction of deaza-FMN enzyme with *N*-methylglutamate. Enzyme was incubated at 10°C in 110 mM potassium tricine buffer, pH 8.3, containing 89 mM potassium chloride, 5 mM magnesium chloride, 1 mM magnesium disodium EDTA and 22.2 mM *N*-methylglutamate. Samples were withdrawn at various times, the enzyme was denatured by addition of 3.3% trichloroacetic acid and the amounts of α-ketoglutarate and methylamine that had been generated were determined with glutamate dehydrogenase and primary amine dehydrogenase [10]. The reduced deaza-FMN formed was determined from the decrease in absorbance at 410 nm of a parallel reaction mixture.

overall reaction (Table III). These results indicate that the deaza-FMN enzyme intermediate exhibits kinetic properties expected for a true catalytic reaction intermediate. The rapid rate of intermediate reaction, as measured by the rate of amino acid product formation, is consistent with the previously described rapid rate of reoxidation of reduced deaza-FMN observed upon addition of methylamine or ammonia to substrate-reduced deaza-FMN enzyme.

Although most of the radioactivity isolated from deaza-FMN enzyme intermediates prepared with [U-^{14}C]- and [α-^{3}H]glutamate was identified as α-ketoglutarate and deaza-FMNH$_2$, respectively, α-hydroxyglutarate accounts for 20–25% of the radioactivity present in both cases [10, 11]. Since tritium is transferred from [α-^{3}H]glutamate to deaza-FMNH$_2$, the formation of α-hydroxy[^{3}H]glutarate could occur via a slow oxidation of deaza-[^{3}H]FMNH$_2$ by either enzyme-bound (unlabeled) α-ketoglutarate or an intermediate at the same oxidation level as α-ketoglutarate [3], as suggested by the following observations: (1) a slow disappearance of α-ketoglutarate is observed when isolated intermediate is incubated in the absence of methylamine, although, unlike the rapid reaction observed in the presence of methylamine (Tables I and III), no amino acid product is formed, and free α-ketoglutarate is stable under these conditions [10]; (2) the amount of α-ketoglutarate formed during reaction of deaza-FMN enzyme with *N*-methylglutamate is stoichiometric with the amount of deaza-FMN reduced whereas the amount of methylamine, initially stoichiometric with reduced flavin, continues to increase after flavin reduction is apparently complete (Fig. 5). These results are consistent with the slow formation of α-hydroxyglutarate via Reactions 3 and 4.

$$\text{E—deaza-FMN} + N\text{-methylglutamate} \rightleftharpoons \text{E}\begin{matrix}\diagup \text{deaza-FMNH}_2 \\ \diagdown \text{glutaryl}\end{matrix} + \text{methylamine} \qquad (3)$$

$$\text{E}\begin{matrix}\diagup \text{deaza-FMNH}_2 \\ \diagdown \text{glutaryl}\end{matrix} \rightleftharpoons \text{E—deaza-FMN} + \alpha\text{-hydroxyglutarate} \qquad (4)$$

The decay of the enzyme intermediate via Reaction 4 is consistent with the slow disappearance of α-ketoglutarate observed during incubation of the isolated intermediate in the absence of methylamine, whereas the recycling of oxidized enzyme produced in this reaction accounts for the slow formation of methylamine in the absence of net flavin reduction.

Discussion

A number of criteria have been presented in this paper which strongly suggest that deaza-FMN *N*-methylglutamate synthase catalyzes the conversion of glutamate and methylamine into *N*-methylglutamate and ammonia via the same two-step mechanism previously elucidated for native enzyme [3]. The results show that the α hydrogen of glutamate is transferred to deaza-FMNH$_2$ and that this same hydrogen is utilized for the reformation of the α-C—H bond during reoxidation with amine substrate. A similar hydrogen transfer appears likely with native enzyme since other aspects of catalysis are similar even though the relative reaction rates are different. The tritium in deaza-FMNH$_2$, released by denaturation, is nonexchangeable with solvent and less than 50% is lost during air oxidation, indicating that the α substrate hydrogen is probably attached to C-5 in deaza-FMNH$_2$. Since deaza-FMNH$_2$ has two potential hydrogens at C-5, either of which could be removed during reaction of the intermediate with methylamine, the observed incorporation of 85% of the tritium into *N*-methylglutamate indicates stereospecific enzymic reoxidation of deaza-FMNH$_2$.

Studies by Bruice and coworkers [7, 8] demonstrated non-enzymic hydrogen transfer to deazaflavin while direct enzymic transfer of substrate hydrogen to deazaflavin has recently been observed with several flavoprotein oxidases [12, 13] and with NADH:FMN oxidoreductase [14], consistent with earlier indirect evidence for hydrogen transfer to flavin in enzymic reactions [15, 16].

Reoxidation of the reduced glutaryl enzyme intermediate by amine substrate (Reaction 2), is similar to the reaction observed when reduced (deaza)FAD D-amino acid oxidase is reoxidized with α-keto acid and ammonia [13, 17] since these reactions probably proceed via hydrogen transfer from reduced (deaza)-flavin to an α-imino acid. Formation of α-hydroxyglutarate (Reaction 4) is analogous to the reduction of α-keto acids by the reduced forms of L-lactate oxidase [12] and glycolate oxidase [13]. Hydrogen transfer from reduced deazaflavin to the α-hydroxy acid product has also been demonstrated in the case of glycolate oxidase [13] and in similar model reaction studies [8]. Although non-enzymic reduction of carbonyl compounds is observed for both reduced flavins [18] and deazaflavins [8], the enzymic reactions are only observed with the deazaflavin enzymes.

Although our data clearly demonstrate enzymic hydrogen transfer from substrate to flavin, the mechanism of this hydrogen transfer is not presently understood. Theoretical calculations indicate that the 5-position of flavins and deazaflavins is the most electrophilic and, therefore, most likely to receive a transferred hydride ion [9]. However, the observed hydrogen transfer reaction could also proceed via a proton transfer mechanism and recent studies have provided evidence for carbanion formation in the reactions catalyzed by several flavoprotein oxidases [19–24] which also involve transfer of substrate hydrogen to flavin.

Acknowledgments

This work was supported in part by USPHS Grant AM 13443 and Grant I-391 from The Robert A. Welch Foundation. Louis B. Hersh is the recipient of Research Career Development Award GM 70239 from The National Institute of General Medical Sciences.

References

1 Shaw, W.V., Tsai, L. and Stadtman, E.R. (1966) J. Biol. Chem. 241, 935–945
2 Pollock, R.J. and Hersh, L.B. (1971) J. Biol. Chem. 246, 4737–4743
3 Pollock, R.J. and Hersh, L.B. (1973) J. Biol. Chem. 248, 6724–6733
4 O'Brien, D.E., Weinstock, L.T. and Cheng, C.C. (1970) J. Heterocyclic Chem. 7, 99–105
5 Edmondson, D.E., Barman, B. and Tollin, G. (1972) Biochemistry 11, 1133–1144
6 Müller, F. and Massey, V. (1969) J. Biol. Chem. 244, 4007–4016
7 Brüstlein, M. and Bruice, T.C. (1972) J. Am. Chem. Soc. 94, 6548–6549
8 Shinkai, S. and Bruice, T.C. (1973) J. Am. Chem. Soc. 95, 7526–7528
9 Sun, M. and Song, P.S. (1973) Biochemistry 12, 4663–4669
10 Jorns, M.S. and Hersh, L.B. (1975) J. Biol. Chem. 250, 3620–3628
11 Jorns, M.S. and Hersh, L.B. (1974) J. Am. Chem. Soc. 96, 4012–4014
12 Averill, B.A., Schonbrunn, A., Abeles, R.H., Weinstock, L.T., Cheng, C.C., Fisher, J., Spencer, R. and Walsh, C.T. (1975) J. Biol. Chem. 250, 1603–1605
13 Hersh, L.B. and Jorns, M.S. (1976) Flavins and Flavoproteins, Proceedings of the 5th International Symposium (Singer, T.P., ed.), Chapt. 38, pp. 362–371, Elsevier, Amsterdam
14 Fisher, J. and Walsh, C. (1974) J. Am. Chem. Soc. 96, 4345–4346
15 Louie, D.D. and Kaplan, N.O. (1970) J. Biol. Chem. 245, 5691–5698
16 Drysdale, G.R., Spiegel, M.J. and Strittmatter, P. (1961) J. Biol. Chem. 236, 2323–2328
17 Radhakrishnan, A.N. and Meister, A. (1958) J. Biol. Chem. 233, 444–450
18 Blankenhorn, G., Ghisla, S. and Hemmerich, P. (1972) Z. Naturforsch. 27b, 1038–1040
19 Walsh, C.T., Schonbrunn, A. and Abeles, R.H. (1971) J. Biol. Chem. 246, 6855–6866
20 Walsh, C.T., Krodel, E., Massey, V. and Abeles, R.H. (1973) J. Biol. Chem. 1946–1955
21 Walsh, C.T., Lockridge, O., Massey, V. and Abeles, R.H. (1973) J. Biol. Chem. 248, 7049–7054
22 Porter, D.J.T., Voet, J.G. and Bright, H.J. (1973) J. Biol. Chem. 248, 4400–4416
23 Page, D.S. and VanEtten, R.L. (1971) Biochim. Biophys, Acta 227, 16–31
24 Page, D.S. and VanEtten, R.L. (1971) Bioorg. Chem. 1, 361–373

Flavins and Flavoproteins, edited by T.P. Singer

CHAPTER 40

NATURE OF THE N-5 REACTIVITY IN ENZYMATIC AND PHOTOCHEMICAL OXIDATIONS BY FLAVINS

PILL-SOON SONG,[a] JUNG DO CHOI,[a] ROBERT D. FUGATE[a] and KUNIO YAGI[b]

[a] *Department of Chemistry, Texas Tech University, Lubbock, Texas 79409 (U.S.A.)* and [b] *Institute of Biochemistry, Faculty of Medicine, University of Nagoya, Nagoya 466 (Japan)*

Summary

1. There is no incorporation of ^3H into the C-5 position from ^3H_2O upon photoreduction of deazariboflavin with NADH. Thus, the site of hydrogen (or hydride) transfer is identical in both thermal and photochemical oxidations of NADH by deazariboflavin. The same hydrogen transfer site (N-5) is probably involved in the oxidation of NADH by flavins. The importance of the frontier orbital at C-5 and N-5 has been elucidated by spectroscopic characterization of flavins.

2. The addition of hydrogen (or hydride) and other nucleophiles (e.g. carbanion intermediates) to C-5 of deazariboflavin and N-5 of flavins can be explained by the reactivity of the 5 position which can accommodate the incoming electron pair of the reagent in its frontier orbital.

3. Flavoenzyme-carbanion adducts and D-amino acid oxidase's "purple complex" are likely to involve the N-5 site, via 2Pσ-type overlap of the N-5 frontier orbital with the filled carbanion orbital lobe.

Introduction

In both photochemical and enzymatic oxidation-reduction reactions involving flavins, the N-5 position plays an important role [1]. The reactivity behavior of the N-5 in flavins is particularly intriguing in contrast to that of common pyridinyl nitrogen. Thus, it is well known that the pyridinyl nitrogen (with nonbonding orbital electrons) acts as a nucleophile (Lewis base) in forming a Lewis acid-base adduct with an electrophile (Lewis acid; e.g., trimethyl boron and sulfur trioxide) [2]. On the other hand, there is apparently no reported evidence for the nucleophilicity of the nitrogen-5 in forming Lewis acid-base complexes. Instead, the nitrogen-5 attacks nucleophiles such as sulfite anion [3] and phosphine [4]. This unique reactivity appears to be of importance in enzymatic and photochemical oxidation

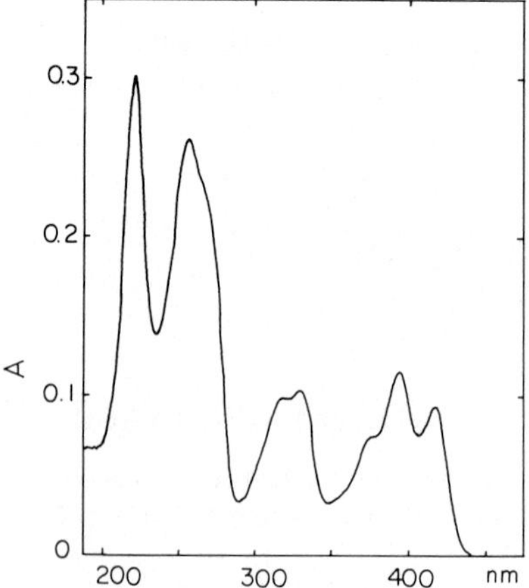

Fig. 1. The absorbance spectrum of deazariboflavin ($\approx 1 \cdot 10^{-5}$ M) in ethanol at 77°K.

reactions [5–8]. In addition, it has been shown that nitrogen-5 is crucial in stabilizing flavosemiquinone. [9].

In our earlier paper [10], we proposed that the N-5 electrophilicity of flavins is of orbital origin and used the photooxidation of NADH by flavins for elucidating the electronic mechanism of the flavin-catalyzed reaction. However, the site of the photochemical hydrogen transfer in flavin from NADH was not experimentally established. In the present paper, we describe results of the NADH hydrogen transfer experiments and further elucidate the nature of the N-5 reactivity of flavins using 5-deazariboflavin (5-deaza-7,8-dimethyl-10-D-ribityl-isoalloxazine) as a model flavin.

We also examine the possibility of the "purple complex" on D-amino acid oxidase [D-amino acid:O_2 oxidoreductase (deaminating), EC 1.4.3.3] being an ES adduct formed at N-5, as such an adduct is relevant to the discussion of the N-5 reactivity in an enzyme reaction.

Materials and Methods

Materials

5-Deaza-isoalloxazine and 5-deazariboflavin were gifts from Professors T.C. Bruice and G. Tollin, respectively. β-NADH, Sigma grade III, 98%, was used without further purification. ^3H_2O (1 mCi/2 g) was obtained from New England Nuclear. Scintanalyzed grade PPO (2,5-diphenyloxazole) employed for autoradiography-photofluorography was from Fisher Scientific Co. Both cellulose thin-layer chromatography sheets and X-ray film (single coated, blue sensitivity for photofluorography, SB54) were from Eastman Kodak. D-Amino acid oxidase prepared according to Yagi et al. [11] ($8.64 \cdot 10^{-4}$ M, from hog kidney) was mixed with lithium pyruvate (0.1 M), D-alanine ($5 \cdot 10^{-2}$ M) and $(NH_4)_2SO_4$ ($5 \cdot 10^{-2}$ M) in 0.017 M pyrophosphate buffer, pH 8.3 [12]. The "purple complex" formed was immediately subjected to absorbance (on Cary 118C spectrophotometer) and CD measurements.

Photolysis

Photolysis of 5-deazariboflavin solution ($1 \cdot 10^{-3}$ M) in the presence of excess NADH (0.2 M) in ^3H_2O was carried out in a 10 mm light pass cuvette at 405 nm under

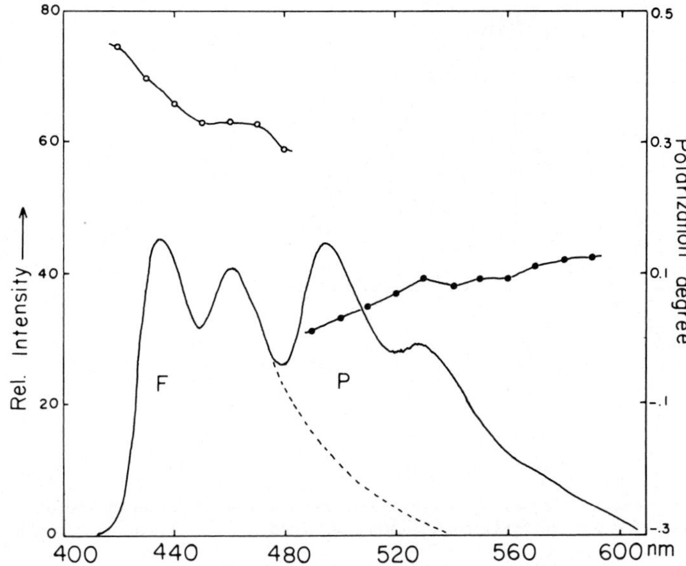

Fig. 2. The fluorescence (F) and phosphorescence (P) spectra of deazariboflavin ($\approx 3 \cdot 10^{-6}$ M) in ethanol at 77°K. The band-widths for emission and excitation (400 nm) were 2 and 7 nm, respectively. Fluorescence (○) and phosphorescence (●) polarizations are also shown.

anaerobic conditions for 2 h. The anaerobic condition was achieved by flushing the serum rubber-stoppered cuvette with prepurified nitrogen for about 20 min. The monochromatic source was provided by a Bausch and Lomb high intensity grating monochromator (150 W Xe lamp). The photolytic reduction of 5-deazariboflavin by NADH was carried out to 60–70% completion, as followed by an absorbance decrease on a Cary 118C spectrophotometer.

Thin-layer chromatography and autoradiography

The solvent ^3H_2O of the photolyzed solution was evaporated below 30°C at 0.1 mm Hg in the dark. The residue was dissolved with 0.5 ml water in order to "wash out" exchangeable ^3H in the reduced 5-deazariboflavin ($DRFH_2$ or DRFHT). About one-third of the aliquot was spotted on a cellulose plate for thin-layer chromatography and was developed with the top layer of *n*-butanol/acetic acid/water (4:1:5, v/v). The developed thin-layer chromatographic plate was treated with 15 ml of 7% PPO ether solution [13]. The thin-layer chromatogram was then exposed to the X-ray film at −80°C for more than two weeks [14]. The remaining aliquot of the photolyzed solution was also chromatographed similarly after aerobic photo-oxidation of the reduced DRF ($DRFH_2$ or DRFHT) by irradiating it at 360 nm. The film was developed, the thin-layer chromatogram was examined under an ultraviolet lamp, and spots were cut out for liquid scintillation counting.

Luminescence measurements

Fluorescence spectra, polarization, quantum yields, and the phosphorescence lifetime of 5-deazariboflavin were measured as described previously [10, 15].

Circular Dichroic (CD) spectrum

The CD spectrum of the "purple complex" was recorded on a JASCO-J20 CD-ORD spectrometer which was modified to enhance the signal-to-noise ratio sufficiently by replacing the Pockel cell and circuitry with a Morvue photoelastic modulator (Model PEM-3) and lock-in amplifier/phase detector (Model PAR 121).

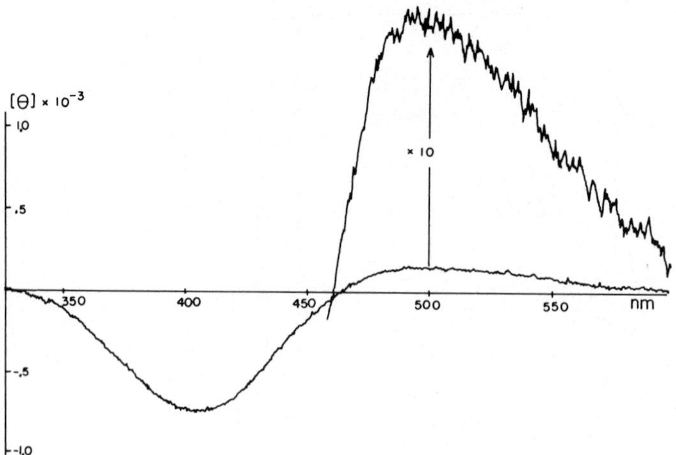

Fig. 3. The CD spectrum of the purple complex of D-amino acid oxidase and substrates in pyrophosphate buffer (0.017 M, pH 8.3) at room temperature. CD curve obtained at 10-fold sensitivity is also shown.

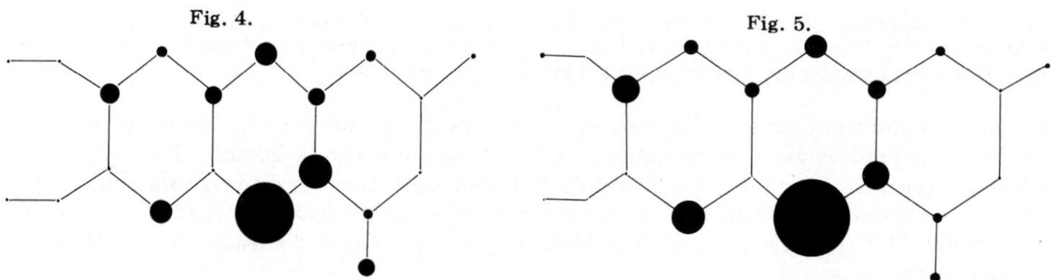

Fig. 4. The frontier orbital density distribution in riboflavin calculated by the frontier electron method [29].
Fig. 5. The frontier orbital density distribution in deazariboflavin.

Results

The anaerobic photolysis solution yielded four spots on the thin-layer chromatogram corresponding to R_F values of 0 (origin, $DRFH_2$), 0.42 (DRF), 0.05 (NADH), and 0.94 (NAD^+). These spots were identified with reference spots. The 5-deazariboflavin spot was enhanced at the expense of $DRFH_2$ when the anaerobic photolysis solution of the spot eluate was reoxidized by air upon irradiation at 360 nm, showing conversion of $DRFH_2$ to DRF. There was no ^3H incorporation into the $DRFH_2$, as judged by the thin-layer chromatography-autoradiography after about two weeks exposure and liquid scintillation counting (88 cpm for $DRFH_2$ spot, 33 cpm for reoxidized DRF spot, and 95 cpm for unphotolyzed DRF spot). It should be noted that the autoradiography of the PPO-coated thin-layer chromatogram employed is sensitive enough to detect 2–3 nCi $^3H/cm^2/24$ h ($\approx 4 \cdot 10^3$ dpm) [14].

The present results indicate that there is no non-exchangeable ^3H incorporation during the photoreduction of DRF to $DRFH_2$.

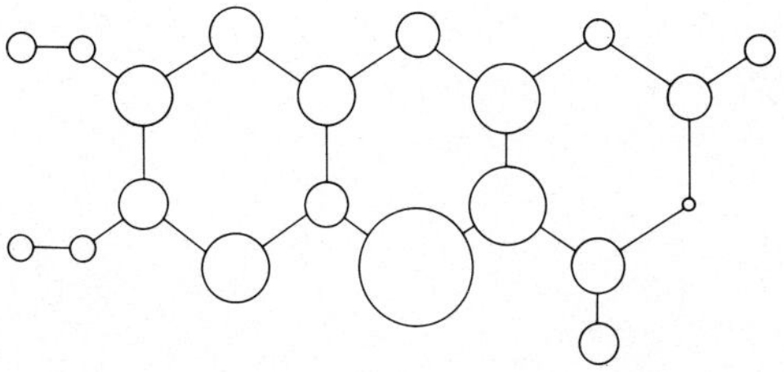

Fig. 6. The nucleophilic superdelocalizability distribution of riboflavin calculated by the perturbation method [30].

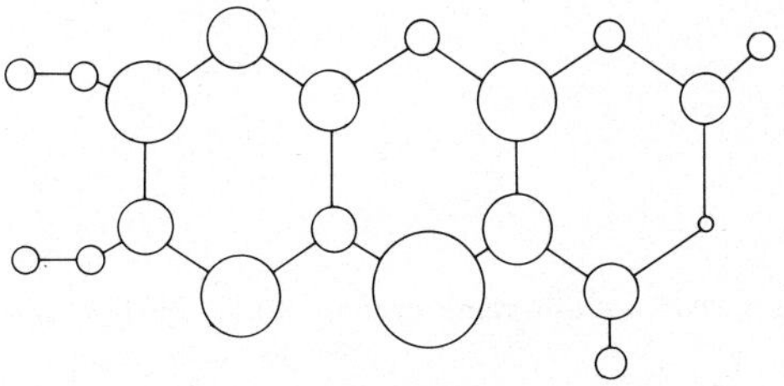

Fig. 7.The nucleophilic superdelocalizability distribution in deazariboflavin.

In order to elucidate the electronic reactivity of 5-deazariboflavin, its spectroscopic characterization was carried out. Fig. 1 shows the absorbance spectrum of 5-deazariboflavin at 77°K, showing a close similarity to the absorbance spectra of riboflavin and deazaflavin [10, 15] although C-5 deaza and 7,8-dimethyl substiuations exert a blue shift of the main absorbance band. A strong fluorescence ($\Phi_F = 0.49$) of 5-deazariboflavin and high fluorescence polarization are consistent with the S_1 state being $^1(\pi, \pi^*)$ type (Fig. 2). The phosphorescence ($\Phi_P = 0.33$) shows a long lifetime (1.6 s) and its polarization is close to zero, suggesting that the lowest triplet is of $^3(\pi, \pi^*)$ type, as has been assigned for N-5 flavins and deazaflavins [10, 15]. The electronic structural similarity between the triplet states of N-5 flavins and C-5 deazaflavin has also been suggested in a recent optical detection of magnetic resonance (ODMR) experiment [16].

The CD spectrum of the "purple complex" is shown in Fig. 3, revealing a positive ellipticity for the long wavelength band. The positive ellipticity band of the purple complex was resolved on a sensitive CD-ORD spectrometer, in contrast to the previous report [17].

Discussion

We have discussed previously that the photooxidation of NADH by flavins proceeds via the $^3(\pi, \pi^*)$ state of the flavin [10]. The fact that there is no incorporation of ^3H at C-5 in the photoreduction of DRF to $DRFH_2$ by NADH is consistent with the direct hydrogen transfer to C-5. It is

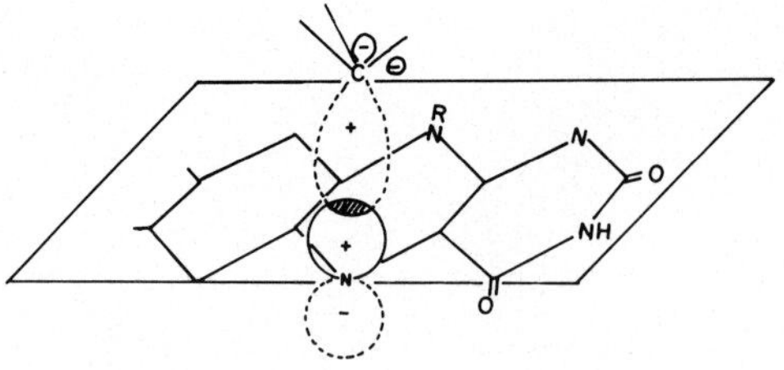

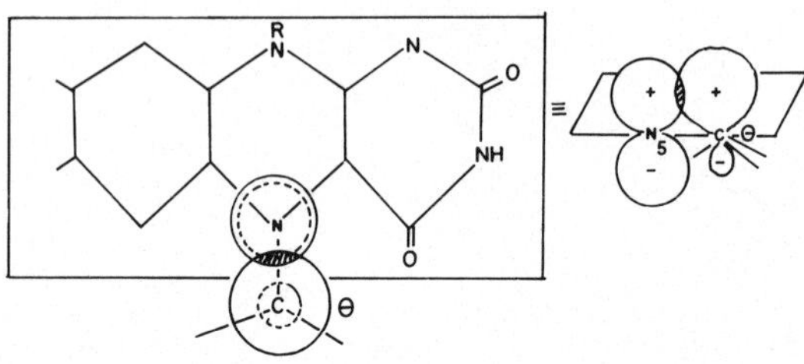

Fig. 8. Orbital overlap for a carbanion-flavin adduct. Top: out-of-plane overlap (2Pσ); bottom: in-plane overlap (2Pπ).

noteworthy that the thermal [6] (also possibly enzymatic) and photochemical reactions operate via essentially the identical mechanism of hydrogen transfer.

Although the hydrogen transfer from NADH to flavins is not clearly discernible in terms of either a two-step-one-electron (i.e. radical mechanism) or a one-step-two-electron (i.e. hydride mechanism) process, the hydrogen transfer may well proceed via the latter mechanism in the case of NADH-deazariboflavin or deazaflavin pair. This is based on the observation that the semiquinone intermediates are not detected in the photoreduction of deazariboflavin [9, 18].

We note, therefore, that the hydrogen transfer to C-5 (and to N-5 by analogy) reveals the common reactivity pattern shown by the addition of nucleophiles (e.g., sulfite ion) to N-5, in spite of the non-bonding electron pair on the nitrogen. This unique reactivity of flavins can be understood in terms of the participation of the frontier orbital at N-5 (and C-5) in accommodating the incoming nucleophiles [10].

Fig. 4 shows the relative size of the frontier orbital at each position in riboflavin. It is apparent that N-5 can best accommodate the incoming nucleophiles. This is also true for deazariboflavin (Fig. 5). In fact, the frontier orbital at C-5 is significantly larger than at C-4a in deazariboflavin and is slightly larger than N-5 in riboflavin. It is, thus, interesting to note that the photochemical hydride transfer from NADH to C-5 is somewhat larger (by about 30%) than to N-5 [10].

Since the thermal and photochemical hydrogen transfer and nucleophilic addition occur at C-5 and N-5 and to a first approximation the frontier orbital contributes predominantly to the make-up of the π^* orbital, the

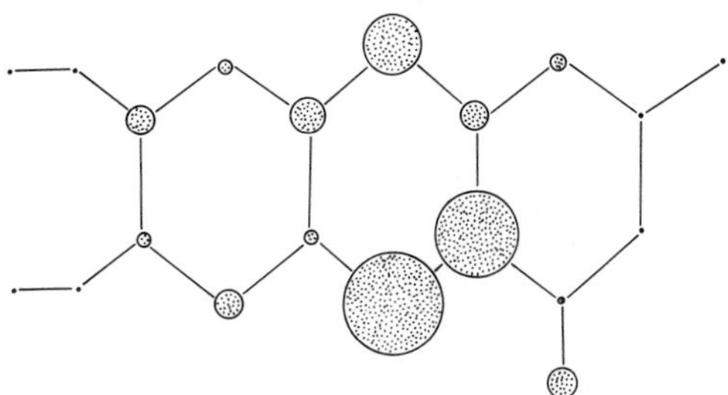

Fig. 9. The frontier electron density distribution in RFH_2 calculated by the frontier electron theory [29].

frontier orbital contribution to the electrophilicity of C-5 and N-5 in flavins is experimentally justified. Furthermore, spectroscopic results discussed in this paper (Fig. 2) confirm our previous proposal that the reactive orbital for hydrogen transfer from NADH to flavin is π^* [10], because the excited triplet states of deazaflavin, deazariboflavin and riboflavin abstract hydrogen from NADH efficiently, independent of C-5 or N-5 substitution. In other words, we can describe the triplet states of riboflavin and deazariboflavin as follows

$$^3\Psi_{RF}(\pi,\pi^*) = {}^3\Psi^\circ(\pi,\pi^*) + \sum_i \lambda_i {}^3\Psi_i^\circ(n,\pi^*) \simeq {}^3\Psi^\circ(\pi,\pi^*) + \lambda_{N\text{-}5}\,{}^3\Psi_{N\text{-}5}(n,\pi^*)$$

$$^3\Psi_{DRF}(\pi,\pi^*) = {}^3\Psi^\circ(\pi,\pi^*) + \sum_i \lambda_i {}^3\Psi_i^\circ(n,\pi^*) \simeq {}^3\Psi^\circ(\pi,\pi^*)$$

where summation is over nitrogen and oxygen n,π^* centers. The suggestion that the triplet state of deazariboflavin is of higher purity (π,π^*) character than that of riboflavin is supported by the above mentioned spectroscopic results and the photoreactivity for NADH oxidation. Thus, the frontier orbital-based reactivity of C-5 and N-5 is spectroscopically consistent.

To substantiate the frontier orbital-based reactivity of C-5 and N-5, we also present Fig. 6 which illustrates the super-delocalizability for addition of a nucleophile at each position as an approximate perturbation (second order perturbation theory) measure of the activation energy. Again, N-5 as well as C-5 (Fig. 7) are predicted to be most reactive.

Figs. 3–6 are qualitatively useful in discussing the reaction mechanisms for flavoenzymes, if N-5 is involved in the reaction scheme. In the case of D-amino acid oxidase, the α-proton is liberated in concert with the formation of the purple complex [19–21], with the resulting substrate carbanion being stabilized via interactions with the flavin ring at N-5, analogous to the intermediate carbanion adduct proposed by Porter et al. [5]. Two modes of the adduct formation are possible, as illustrated in Fig. 8. The in-plane adduct can be ruled out, since the presence of N-5 nonbonding electron pair would

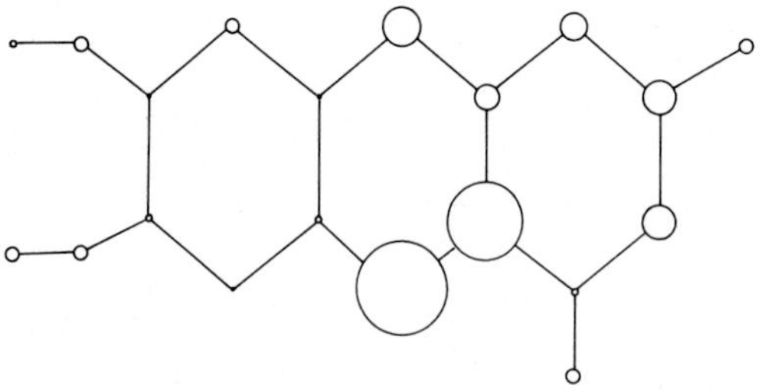

Fig. 10. The electrophilic superdelocalizability distribution in RFH_2 calculated by the perturbation method [30].

prevent the filled carbanion orbital lobe (sp^3) from overlapping with the frontier orbital at N-5*.

The carbanion-flavin adducts [5, 22] and "purple complex" may well involve the out-of-plane adduct depicted in Fig. 8, since both of these complexes show a broad, long wavelength absorption band (> 500 nm) which can arise from the charge transfer transition (from filled carbanion orbital to flavin antibonding molecular orbital)**. We examine this possibility further for the "purple complex" which is characterized by its charge transfer absorbance band [17, 23–25] and a positive ellipticity (Fig. 3).

The calculated transition energies in Table I are qualitatively consistent with the absorbance and CD spectra ($\lambda_{max} \approx 540$–$550$ nm, $[\theta] = 116$ degree $cm^2/dmol$; 430 nm, $[\theta] = -702.4$ degree $cm^2/dmol$; 340 nm, $[\theta] \approx 0°$) and polarization directions of the two long wavelength bands are in agreement with the CD spectrum***. The crystallographic and NMR structure of the

TABLE I

Predicted transition energies for an out-of-plane carbanion-N-5-flavin adduct (distance between N-5 and C^- is assumed to be 3 Å, Fig. 8) calculated by the SCF MO CI Pariser-Parr-Pople method. The one-center repulsion integral (ionization potential minus electron affinity) for the carbanion was taken as 10 eV and the 2Pσ-resonance core integral (− 0.5 eV) between N-5 and C^- was evaluated by the published method [26]. Calculated energies are subject to variation by these parameters (± 30 nm).

Transition	Energy (nm)	Oscillator strength	Approximate polarization*
$S_1 \leftarrow S_0$	519	0.09	o.p., inclined to short axis
$S_2 \leftarrow S_0$	485	0.49	i.p., short axis
$S_3 \leftarrow S_0$	434	0.59	i.p., long axis
$S_4 \leftarrow S_0$	413	0.39	o.p., inclined to short axis
$S_5 \leftarrow S_0$	367	0.32	i.p., intermediate axis

* i.p. and o.p. stand for in-plane and out-of-plane polarization components, respectively.

* The long-wavelength charge transfer band of an in-plane complex is predicted to be intense, while that of an out-of-plane complex is weak (vide infra, Table I).

** The ground and charge transfer (S_1) states of the carbanion-N-5 adduct (Fig. 8) show a considerable charge transfer character with 0.86 and 1.21 electrons transferred from the filled carbanion orbital to the flavin ring, respectively.

*** The long-wavelength CD band can be attributed to two electronic transitions with the same sign for their rotational strength. This is consistent with the predicted transitions at 519 nm and 485 nm in Table I.

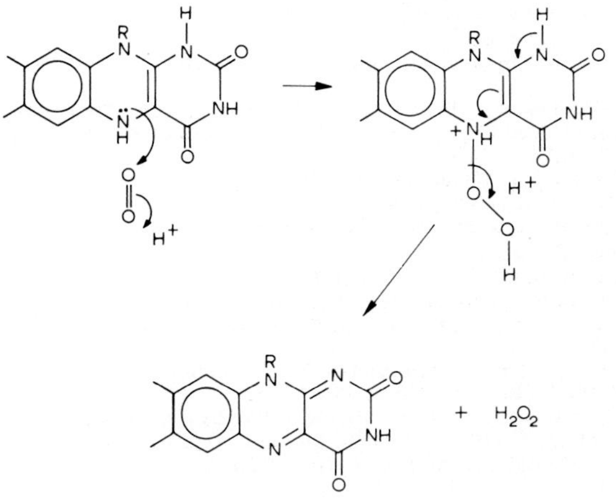

purple complex is being investigated to ascertain the spectroscopic assignment.

Finally, the reactivity of reduced flavins is briefly discussed with emphasis on the role of N-5. Fig. 9 shows that N-5 and C-4a are the most susceptible positions for attack by an electrophile (e.g., oxygen). Fig. 10 shows the same trend in terms of the perturbation energy, i.e., superdelocalizability for electrophilic attack. The proposed electrophilic addition of oxygen at C-4a in the reduced flavin model [1, 27] and flavoenzyme reactions [28] is supported by the calculated reactivity maps (Figs. 9 and 10). It is also tempting to speculate that singlet oxygen is likely to attack N-5 (Scheme 1) because of the highest nucleophilicity of this position. However, there is no experimental evidence for N-5 peroxide adduct in this mechanism either as an alternative or a concommitant route for reoxidation of the reduced flavin by oxygen. The hydrogen transfer from N-5 of reduced flavins to carbonyl carbon is consistent with the reactivity in Figs. 9 and 10 [1].

Acknowledgements

This work was supported by the Robert A. Welch Foundation (D-182) and National Science Foundation (GB 21266). Assistance of C.-N. Ou with autoradiography experiments is greatly appreciated.

References

1 Bruice, T.C. (1975) Progr. Bioorg. Chem. 4, in the press
2 Gould, E.S. (1959) Mechanism and Structure in Organic Chemistry, p. 115, Holt, New York
3 Massey, V., Müller, F., Feldberg, R., Sullivan, M., Howell, L.G., Mayhew, S.G., Matthews, R.G. and Foust, G.P. (1969) J. Biol. Chem. 244, 3999
4 Müller, F. (1972) Z. Naturforsch 27b, 1023
5 Porter, D.J.T., Voet, J.G. and Bright, H.J. (1973) J. Biol. Chem. 248, 4400
6 Brüstlein, M. and Bruice, T.C. (1972) J. Am. Chem. Soc. 94, 6548
7 Moore, W.M. and Baylor, Jr., C. (1969) J. Am. Chem. Soc. 91, 7170

8 Haas, W. and Hemmerich, P. (1972) Z. Naturforsch 27b, 1035
9 Edmonson, D.E., Barman, B. and Tollin, G. (1972) Biochemistry 11, 1133
10 Sun, M. and Song, P.S. (1973) Biochemistry 12, 4663
11 Yagi, K., Naoi, M., Harada, M., Okamura, K., Hidaka, H., Ozawa, T. and Kotaki, A. (1967) J. Biochem. (Tokyo) 61, 580
12 Yagi, K. (1970) Methods in Enzymology (Colowick, S.P. and Kaplan, N.O., eds.), Vol. 17B, p. 608
13 Horowitz, J., Ou, C.-N., Ishaq, M., Ofengand, J. and Bierbaum, J. (1974) J. Mol. Biol. 88, 301
14 Randerath, K. (1970) Anal. Biochem. 34, 188
15 Sun, M., Moore, T.A. and Song, P.S. (1972) J. Am. Chem. Soc. 94, 1730
16 Moore, T.A. and Kwiram, A.L. (1974) Biochemistry 13, 5403
17 Kotaki, A., Sugiura, N. and Yagi, K. (1968) Biochim. Biophys. Acta 151, 686
18 Fox, J.L. and Tollin, G. (1966) Biochemistry 5, 3865
19 Yagi, K., Nishikimi, M., Ohishi, N. and Takai, A. (1970) FEBS Lett. 6, 22
20 Yagi, K., Nishikimi, M., Takai, A. and Ohishi, N. (1973) Biochim. Biophys. Acta 321, 64
21 Yagi, K., Nishikimi, M., Ohishi, N. and Takai, A. (1971) Flavins and Flavoproteins (Kamin, H., ed.), p. 239, University Park Press, Baltimore, Md.
22 Porter, D.J.T., Voet, J.G. and Bright, H.J. (1972) Biochim. Biophys, Res. Commun. 49, 257
23 Yagi, K., Okuda, J. and Okamura, K. (1965) J. Biochim. (Tokyo) 58, 300
24 Yagi, K., Okuda, J., Naoi, M., Sugiura, N. and Kotai, A. (1967) Biochim. Biophys. Acta 146, 77
25 Yagi, K., Sugiura, N., Okamura, K. and Kotaki, A. (1968) Biochim. Biophys. Acta 151, 343
26 Nagata, C., Fujita, H. and Imamura, A. (1967) Bull. Chem. Soc. Japan 40, 2564
27 Hamilton, G. (1971) Progr. Bioorg. Chem. 1, 83
28 Massey, V., Palmer, G. and Ballou, D. (1973) Oxidases and Related Systems (King, T.E., Mason, H.S. and Morrison, M., eds.), p. 25, University Park Press, Baltimore, Maryland
29 Fukui, K., Yonezawa, T., Nagata, C. and Shingu, H. (1954) J. Chem. Phys. 22, 1433
30 Fukui, K., Yonezawa, T. and Nagata, C. (1957) J. Chem. Phys. 27, 1247

Part VIII

Flavodoxin

Flavins and Flavoproteins, edited by T.P. Singer

CHAPTER 41

THE STRUCTURE OF *CLOSTRIDIUM MP* FLAVODOXIN AS A FUNCTION OF OXIDATION STATE: SOME COMPARISONS OF THE FMN-BINDING SITES IN OXIDIZED, SEMIQUINONE AND REDUCED FORMS

MARTHA L. LUDWIG, ROGER M. BURNETT*, GEORGE D. DARLING, STEVEN R. JORDAN, DOUGLAS S. KENDALL** and WARD W. SMITH

Biophysics Research Division, Institute of Science and Technology, and Department of Biological Chemistry, University of Michigan, Ann Arbor, Mich. 48105 (U.S.A.)

Introduction

The crystal structure analyses of two flavodoxins, one from *Clostridium MP*, studied in our laboratory [1, 2] and the other from *Desulfovibrio vulgaris*, described by Watenpaugh et al. [3, 4] have afforded a detailed view of the interactions between oxidized FMN and apoprotein in these electron carriers. As might have been anticipated, the tight binding of FMN arises from a complicated set of hydrogen-bonding interactions and van der Waals contacts. However, each of these flavodoxins provides a somewhat different environment for the isoalloxazine ring. In *Cl. MP* flavodoxin, the residues adjoining the flavin ring are Met 56 and Trp 90; in the *D. vulgaris* structure the corresponding side chains are Trp and Tyr, respectively. Such species-dependent differences in the flavin-binding site are probably responsible for the distinctive spectral properties of these two flavodoxins, but they also seem to preclude an identical pathway of electron transfer if the redox reactions utilize side chains in contact with the isoalloxazine ring.

From the beginning of our structural analyses it has been our intention to determine the conformation of the holoprotein with FMN in each of its three oxidation states. In this chapter we present some structural comparisons of the oxidized, semiquinone and reduced states, emphasizing the features which we believe may be related to the chemical properties, particularly the oxidation-reduction potentials, of *Cl. MP* flavodoxin.

Methods

(a) Structure determination

Because of the large intensity changes accompanying the conversion of the oxidized

---

* Present addresses: Biozentrum der Universität Basel, Cell Biology Klingelbergstrasse, 70, CH-4056, Basel, Switzerland and ** National Center for Atmospheric Research, Box 1470, Boulder, Colo. 80302, U.S.A.

crystals to the radical state [5], each of these structures has been determined independently, using isomorphous replacement with the heavy atoms Sm^{3+} and Au^{+} (Table I) [1, 2]. Refinement has been conducted by a combination of difference Fourier [6, 7] and real space procedures [8, 9]. The current R values are included in Table I. The details of the fully reduced structure have been derived from difference maps employing phases calculated from coordinates for the semiquinone form, a procedure which assumes that the semiquinone and reduced structures are closely isomorphous, i.e. that the majority of the atoms are identically situated in the two unit cells. The isomorphism of crystals of reduced and radical forms of *Cl. MP* flavodoxin is evidenced by the value of $R_I = (\Sigma |I_{red} - I_{sq}|)/(\Sigma I_{red} + I_{sq}) = 0.08$. ($|F|$, structure factor amplitude; α, phase angle in radians; I, intensity. Subscripts obs or o refer to observed quantities, calc or c, to quantities calculated from coordinates, ox, sq and red refer to the oxidation states of the flavin. $\vec{F}$ is equivalent to $|F| \exp i\alpha$.) In all difference maps, the $\vec{F}_c$ values of reflections in the resolution range 5–15 Å were corrected for the effect of disordered solvent according to a procedure of Moews and Kretsinger [10], in which it is assumed that the protein and solvent contributions have opposite phases.

(b) Characterization of the oxidation states of crystalline Cl. MP flavodoxin

Crystals have routinely been grown in a nitrogen atmosphere from solutions of the semiquinone form, in 2.3–2.6 M ammonium sulfate/0.1 M Tris, pH 6.8 [5]. For intensity measurements on the other oxidation states, they are allowed to oxidize in air-saturated buffered ammonium sulfate (pH 6.8) or are reduced with dithionite, the final pH being 7.2–7.5. We have inspected diffraction patterns of crystals grown from solutions of oxidized or of predominantly reduced flavodoxin and patterns from crystals in which the redox reactions were performed in the crystalline state. X-ray photographs of the centric zone at approx. 1.85 Å resolution indicate that the intensities of the oxidized crystals are the same whether oxidation took place before or after crystallization. In similar fashion, photographs of the fully reduced species at 2.7 Å resolution give no evidence for structural changes which can occur only if the protein is in solution during reduction.

The oxidation state of the flavin in crystals designated semiquinone or reduced has been established by X-ray and EPR measurements. The cell parameters and selected intensities have been monitored throughout data collection from semiquinone crystals. Measurements have been terminated when the changes correspond to 15% oxidation. X-ray intensities for the reduced state have been obtained from four crystals. For two of these crystals, including 28% of the approx. 11 000 unique reflections, measurements were concluded when the surface of the crystals began to change color. For the remaining 72% of the reduced data, the extent of oxidation after completion of the intensity measurements was determined by integration of the EPR spectra of the crystals. (Spectra recorded by R.H. Sands and J.A. Fee.) The fraction of reduced molecules in these two crystals was found to be no less than 77 and 88%, respectively.

TABLE I

CRYSTALLOGRAPHIC DATA FOR FLAVODOXIN FROM *CL. MP* (SPACE GROUP $P3_121$)

	Oxidized	Semiquinone	Reduced
Cell dimensions			
a	61.56 Å	61.63 Å	61.68 Å
c	70.36 Å	70.98 Å	71.05 Å
Resolution	1.9 Å	1.8 Å	2.0 Å
$\langle m \rangle$*	0.808	0.725	
R values**	0.271	0.285	

* Mean figure of merit for isomorphous replacement phasing.

** $R = \dfrac{\Sigma \, ||F|_{obs} - |F|_{calc}|}{\Sigma \, |F_{obs}|}$ for data with $I > 2\sigma$, 1.8 or 1.9 Å to 5 Å

Oxidized *Cl. MP* flavodoxin

The chain of 138 residues [11] is folded as shown in Fig. 1. At the present stage of refinement of the structure, only 65 of the atoms determined by sequence analysis are unobserved in the electron density. The earlier hydrogen-bonding schemes [2] have been reassessed employing distances and angles from the latest refinements, and a number of solvent positions have been assigned.

Fig. 2 shows our current view of some of the interactions between FMN and the protein, and includes several solvent molecules bound in the vicinity of the active center. The essential picture of FMN-protein binding remains as reported previously [2]; the phosphate group resides in a neutral environment comprising backbone amides and side-chain hydroxyl groups (Fig. 2). The only possible countercharge is solvent W-1, which could be NH_4^+ rather than water. Ribityl oxygens form hydrogen bonds to both solvent and protein, the latter including Asn 11, Ser 87, and Ala 55. Most of the van der Waals contacts made by the isoalloxazine ring are with atoms of Met 56 and Trp 90 (Fig. 8 of Ref. 2); the indole ring of the latter residue lies on the solvent side of the flavin ring, inclined to the flavin at an angle of the order of 20°.

The flavin ring atoms N-1, O-2, N-3 and O-4 may all participate in hydrogen bonding to the protein, as their distances from donor or acceptor protein neighbors are less than 3.3 Å. The angles for the Glu 59 COO^-: N-3 and Gly 89 NH: N-1 interactions are quite acceptable, those for interaction of Gly 91 NH with O-2 and Glu 59 NH with O-4 are somewhat less

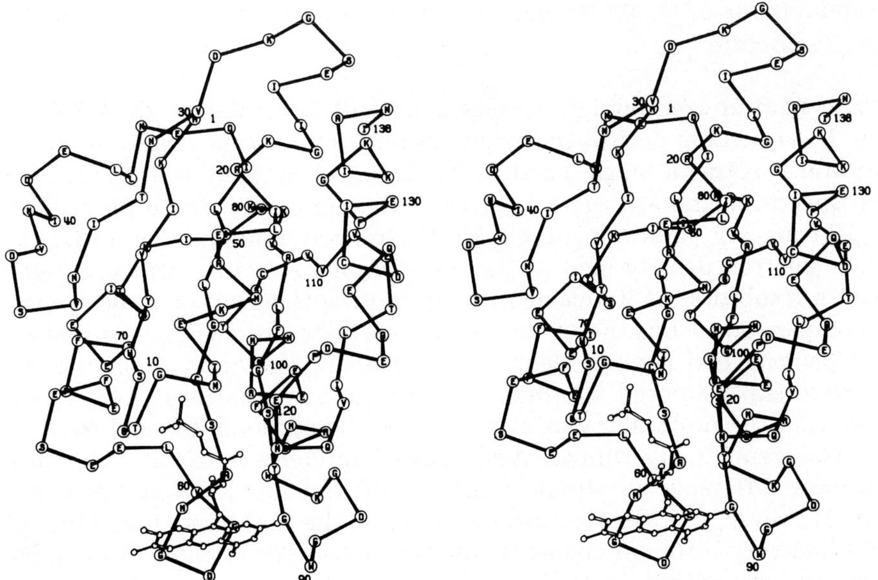

Fig. 1. A stereoview of the backbone and FMN atoms of oxidized flavodoxin from *Cl.MP*. Residues are identified by the one-letter code. The viewing angle is down a direction approximately 120° from the $\vec{c}$ axis; the vertical direction is parallel to $\vec{b}$. (Reproduced from Atlas of Protein Sequence and Structure, by permission of the editors.)

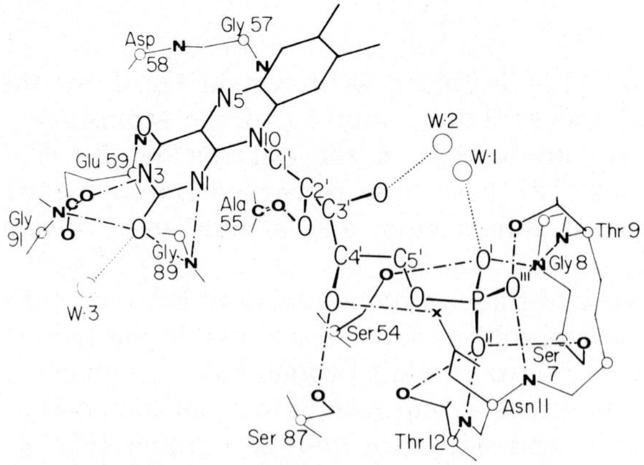

Fig. 2. A drawing of the FMN-binding site in oxidized *Cl.MP* flavodoxin, showing some of the FMN-protein interactions. (Trp 90 and Met 56 omitted for clarity.) The three solvent atoms are in positions where they may form hydrogen bonds to O^{I}, O-3′ and O-2, respectively. ····, neighbors at distances less than 3.3 Å. The probabilities of hydrogen bond formation between these pairs are considered in the text.

satisfactory, while Gly 89 NH and O-2 are very poorly oriented for hydrogen bonding. The flavin N-5 does not appear to be involved in a hydrogen bond. In the conformation depicted in Fig. 3a, the nearest potential donors to N-5 are Asp 58 NH (4.2 Å) and Gly 57 NH (3.7 Å). The conformation of residues 56–59 is considered in the next section.

The semiquinone of *Cl. MP* flavodoxin: comparisons with the oxidized structure

The neutral semiquinone species is present in crystalline *Cl. MP* flavodoxin. The isotropic optical spectrum, reconstructed from polarized optical absorption spectra of single crystals by Eaton et al. [12] is a typical blue radical spectrum; the λ_{max} and λ_{min} values coincide with those observed in the spectrum of solutions of *Cl. MP* flavodoxin semiquinone [12, 13].

An electron density map of the semiquinone structure has been calculated at a resolution of 2 Å using isomorphous replacement phasing. Preliminary comparison of this map with the oxidized structure suggested that the crystal packing had altered slightly as a result of reduction. Once the two maps were superimposed by small rotations and translations [14], it was obvious that the molecular conformations of the semiquinone and oxidized states were remarkably similar. After partial refinement of each structure, the average difference in atomic positions of 954 non-FMN atoms is only 0.36 Å. However, the density corresponding to the backbone at residues 57 and 58 had very different shapes in the two maps. The possible biochemical importance of changes in the flavin-protein interactions prompted us to scrutinize the conformation of the FMN and the surrounding residues, particularly the region including residues 56–59, with more care.

Contours in maps of the oxidized structure, determined by isomorphous replacement, indicated that the carbonyl group of Gly 57 points "down", with the oxygen away from the FMN ring, whereas in the semiquinone map, the density conforms better to a peptide unit whose oxygen points "up", toward N-5. The conformation first suggested by the maps of the oxidized form of flavodoxin is quite close to that shown in Fig. 3a and somewhat resembles a Type II 3_{10} bend [15] in which the oxygen of the second residue, Gly 57, and the β-carbon of the third residue are *gauche* to one another. A Type II arrangement is relatively unstable whenever the third residue, Asp 58 in this case, is not glycine. On the other hand, the semiquinone density seemed to correspond approximately to a Type I 3_{10} bend, with O–57 and the β-carbon of Asp 58 pointing in opposite directions. The differences in the maps thus appeared to require an inversion of the peptide unit, in which the peptide planes rotates approx. 180° about a line connecting the α-carbons of residues 57 and 58.

To test our interpretations, we first utilized real space refinement [8] with isomorphous replacement maps. Since the refinement is not expected to invert a peptide unit when an incorrect conformation is introduced [16, 17], the computations were performed twice, beginning with models having the carbonyl group up or down. With the semiquinone map, the oxygen up conformation was clearly a better fit to the observed density. The oxygen down conformation was preferred over oxygen up in the oxidized structure, as measured by the electron occupancies. In the down conformation the poorest fit was for N-58, 1.3 electrons (occupancies for well-placed atoms in the polypeptide backbone range from 3.0 to 6.0 electrons), whereas for O-57 up, both O-57 and N-57 refined to positions corresponding to less than 0.3 electrons. The bends formed by residues 56–59 deviate considerably from the models proposed by Venkatachalam [15]. The distances between N-59 and O-56 are appropriate for 3_{10} hydrogen bonding, being within 3.0 Å, however, the angles are not so favorable as in idealized models of the 3_{10} bend. In both oxidation states, but particularly in the semiquinone structure, NH-59 seems better oriented for hydrogen bonding with O-4 of the flavin ring than with the Met 56 carbonyl group.

We have sought to verify the conformation of this chemically crucial region by computing electron densities which are independent of heavy atom phasing, in a fashion which avoids any a priori assumptions about the atomic arrangement of residues 56–59. A Fourier synthesis employing measured amplitudes with phases computed from known portions of the structure reveals the positions of atoms omitted from the phase calculation [18–20]. Syntheses using coefficients $|F|_{\text{obs}} \exp i\alpha_{\text{calc}}$, $(|F|_{\text{obs}} - |F|_{\text{calc}}) \exp i\alpha_{\text{calc}}$, or $(2|F|_{\text{obs}} - |F|_{\text{calc}}) \exp i\alpha_{\text{calc}}$ [7] where $\vec{F}_c$ does not include the backbone atoms 56–59 should therefore yield an unbiased view of the region. To increase the accuracy of the images, we have refined both the oxidized and semiquinone structures (Table I).

Fig. 3a and b represents the densities of the backbone atoms of residues 56–59 in syntheses with coefficients $(2|F|_0 - |F|_c) \exp i\alpha_c$. The map of the semiquinone form confirms the assignment of the O-57 up conformation

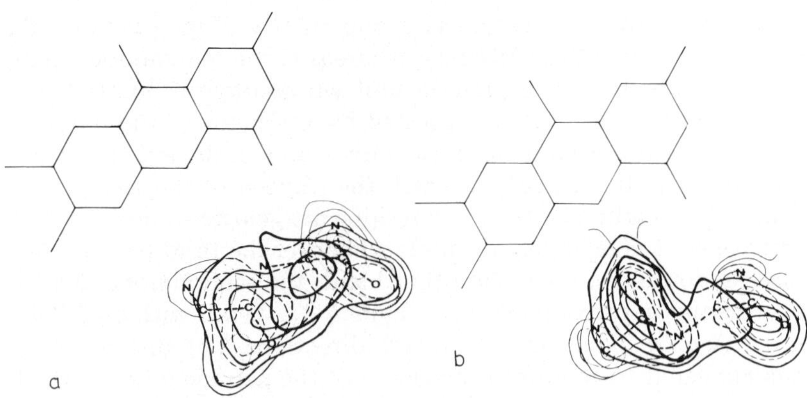

Fig. 3. Drawings of the electron density of the polypeptide chain from N-57 to N-59, with accompanying interpretations. The position of an idealized isoalloxazine, with all ring bonds of equal length, is drawn for reference, and represents an averaged fit to the actual atomic coordinates of the flavin ring. For N-57 to N-59 the atomic positions are those obtained by real space refinement using the isomorphous replacement electron density. Sections perpendicular to b, equivalent to a depth of 3.0 Å, have been contoured. (a) $(2|F_0| - |F|_c) \exp i\alpha_c$ map of oxidized flavodoxin, 1.9 Å resolution. Model refined in the O-57 down conformation; (b) $(2|F|_0 - |F|_c) \exp i\alpha_c$ map of flavodoxin semiquinone, 1.8 Å resolution, O-57 up.

and suggests a connection between O-57 and N-5 of the FMN ring; these atoms are separated by 2.8 Å, and are nicely oriented for hydrogen bonding. The density of oxidized flavodoxin (Fig. 3a) has much the same shape as observed in the isomorphous replacement maps, and still gives no evidence for such a close approach of any protein atom to N-5. However, from Fig. 3a it is evident that the model does not fit the contours as satisfactorily as in the interpretation of the semiquinone structure. Density "below" the assigned position of O-57 has persisted in all the maps of oxidized flavodoxin. This density is too close to the α-carbon of Gly 57 to represent a solvent molecule. One other alternative has been explored. Insertion of a *cis* peptide bond improves significantly the visual fit of the model to the map of Fig. 3a. Apart from the expected *cis* peptides at proline residues [21, 22], other *cis* peptides have rarely been documented in protein structures; one may occur in carboxypeptidase A [23]. We have been cautious therefore about accepting the *cis* conformation as the unique interpretation of the density. The correct choice may become obvious as refinement proceeds.

Reduced *Cl. MP* flavodoxin

The series of crystal structure analyses by Kierkegaard and co-workers [24–26] demonstrated that several N-1- or N-5-substituted 1,5-dihydroisoalloxazines adopt a butterfly conformation. Spectroscopic measurements [27, 28] suggest that in solution at room temperature, 1,5-dihydroflavins are also nonplanar, although they oscillate [29] between the two possible bent conformations. However, the fluorescence of several reduced flavoproteins implies that the motion of the reduced flavin ring may be restricted upon binding to an apoprotein [28], and the absorbance at wavelengths above

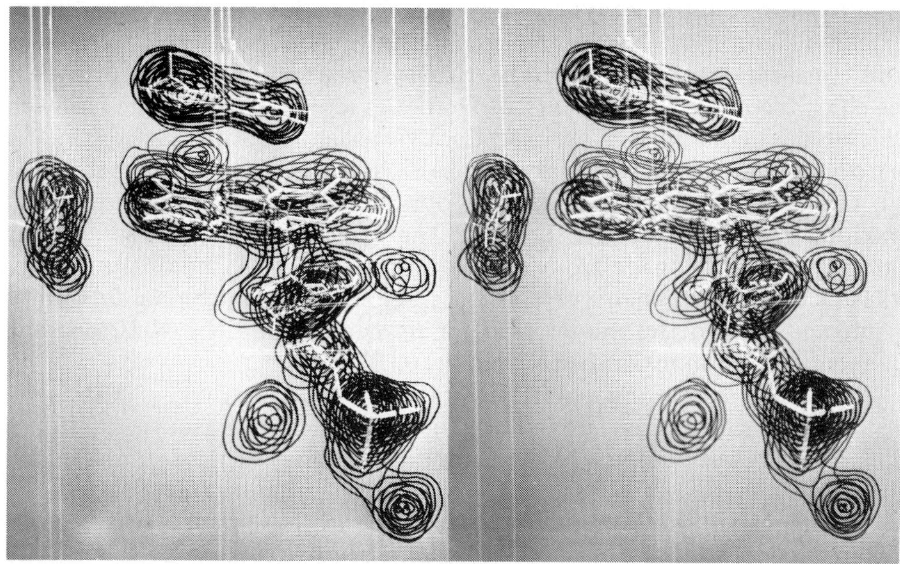

Fig. 4. A stereoview of the electron density of reduced *Cl. MP* flavodoxin in FMN-binding region, with model superposed. The map was calculated with coefficients $(2|F|_{red} - |F|_{calc\,sq})$ exp i$\alpha_{calc\,sq}$, deleting isoalloxazine atoms from $|F|_c$ and α_c. The ribityl phosphate chain extends to the lower right; the interaction of a phosphate oxygen with Ser 7 OH, of O-4′ with Ser 87 OH, and of O-3′ with solvent can also be seen. At the upper left is the side chain of Glu 59; Met 56 (behind the ribityl side chain) and Trp 90 have also been included. Density corresponding to the position of O-57 appears near N-5.

400 nm in reduced *Cl. MP* and *P. elsdenii* flavodoxins may denote a nearly planar conformation for the reduced flavin in these proteins [28]. Our first objective in examining the crystal structure of fully reduced *Cl. MP* flavodoxin has therefore been to ascertain the shape of the dihydroisoalloxazine ring.

To calculate images of fully reduced *Cl. MP* flavodoxin, we have relied primarily on syntheses with coefficients $(2|F|_{red} - |F|_{calc,sq})$ exp i$\alpha_{calc,sq}$. Such maps allow one to obtain an unprejudiced view of a local region of the electron density if the atoms in that portion of the structure are not included in $|F_c|$ and α_c. Use of $\vec{F}_{calc,sq}$ appears justified by the test described in Methods.

In Fig. 4, we show part of a map computed to determine the conformation of the reduced isoalloxazine ring. In this case, all ring atoms, but not C-1^1, were deleted from the calculation of $\vec{F}_{sq}$. The flavin ring is seen to be slightly bent, with its extreme ends displaced away from the ribityl side chain. As a first approximation to a description of the ring conformation, a model permitting bending only about the N-5/N-10 line, with the atoms on either side of this line coplanar, has been oriented in the density by real space refinement [8]. The bending angle was found to be 8.6°. Although this deviation from planarity is considered significant (Table II), the angle is much smaller than the 30–35.5° found for most substituted dihydroflavins [24–26], and is comparable with the angle of 9.1° in

5-diethyl-3,7,8,10-tetramethyl-1,5-dihydroisoalloxazine [30]. (To estimate the bending angle of the protein-bound flavin, both oxygens and methyl groups were retained as part of the ring system, and therefore to make a strict comparison with the results of Werner et al. [30], their angle has been recalculated, including the peripheral oxygens and methyl carbons.) The sense of the bending can be described as follows: in a view facing the N-1, N-10, C-9 edge of the flavin, with the pyrimidine ring to the left, the "V" formed by the two planes is upright. The assignment of C-1′ as axial or equatorial cannot be made from the map of Fig. 4. The shape of the density (Fig. 4) also suggests a slight warp of the ring, but when additional flexibility was introduced into the model used for refinement, the N-5/N-10 bending angle increased by an insignificant amount (0.6°).

A second map using coefficients $(2|F|_0 - |F|_c)\exp i\,\alpha_{calc}$, this time with the flavin ring atoms retained in $\vec{F}_c$ and the backbone atoms of residues 56–59 deleted, was computed to examine the arrangement of atoms in the bend adjacent to the FMN ring. O-57 is observed in approximately the position it occupies in the semiquinone structure, where it points toward N-5. In the reduced form, the N-5/O-57 distance is estimated to be 2.8 Å. The overlap for hydrogen bonding between N-5 and O-57 is better if the hydrogen is equatorial rather than axial. It seems probable, from other difference maps, that as in the semiquinone and oxidized forms, a hydrogen bond can be formed between N-1 and Gly 89 NH. We assume that N-1 is ionized in our reduced structure [31].

Discussion

(a) Comparison of the oxidized and semiquinone forms: the conformation of residues 56–59

The conformations of the oxidized and semiquinone forms have proven to be rather similar, as deduced earlier from the combined results of X-ray and NMR studies [5, 32]. (A thorough comparison of all regions of the molecule awaits further refinement of the semiquinone structure.) The large changes in diffracted intensities, which occur when crystals of oxidized *Cl. MP* flavodoxin are reduced, have now been shown to result from a combination of small conformation differences with a 'rigid-body' motion of the molecule in the unit cell. This latter motion undoubtedly dominates the X-ray changes [33]. We do not understand how the repacking is initiated, but it could be associated with alterations in the isoalloxazine ring and its environment, since Trp 90 and the dimethylbenzene end of the FMN ring are close to Glu 25 of a symmetry-related molecule. Upon one-electron reduction of the oxidized crystals, both Trp 90 and the dimethylbenzene portion of the FMN ring are slightly displaced relative to the remainder of the protein molecule.

The electron densities corresponding to backbone atoms 56–59 display distinguishing features in maps of the oxidized and semiquinone species and their interpretations yield different conformations for the protein in these oxidation states. In flavodoxin semiquinone, the peptide oxygen of residue 57 is in a position to form a strong hydrogen bond to the protonated flavin atom N-5. This interaction, first observed in isomorphous replacement maps at 2.0 Å resolution, has been corroborated at 1.8 Å resolution from independent views of the density obtained by syntheses with coefficients $(2|F|_{0sq} - |F|_{c\,sq})\exp i\alpha_{c\,sq}$. O-57 occupies a different position in the oxidized molecule. Although interpretations of the oxidized maps have proven more difficult, the structure provides no evidence for a hydrogen-bonding interaction between N-5 of oxidized FMN and the protein.

Rearrangement of the 57–58 peptide unit, as suggested by the crystal structures (Fig. 3), has several implications for the chemical behavior of *Cl. MP* flavodoxin. Formation of a hydrogen bond to N-5 in the semiquinone would be expected to increase the pK of the neutral species, in accord with spectral measurements on flavodoxins. The presence of two carboxylate residues, Asp 58 and Glu 59, in the vicinity of the flavin ring, may further suppress deprotonation of N-5. Moreover, the addition of a strong interaction between the flavin radical and the apoprotein is in the correct direction to account for preferential binding of the radical, as opposed to the oxidized, species of FMN. Stronger binding of FMN semiquinone is demanded by the observed redox potentials (E_1(pH 7) = − 0.399 V; E_2(pH 7) = − 0.092 V [13, 34]). From our examination of the oxidized structure, we conclude tentatively that residues 56–59 are in a relatively less stable conformation when the prosthetic group is oxidized. If so, the conformation change would also contribute to the greater affinity for FMN radical. A more thorough analysis of the relative energies [35] of the oxidized and semiquinone conformation is clearly warranted. The rate of isomerization could limit the overall rate of interconversion of oxidized and semiquinone forms; e.g. the known barriers for *cis-trans* conversions [37, 38] would require a rate less than $10^2\ s^{-1}$. Comparison with reported second-order rate constants [39] suggests that further experiments ought to be performed in an effort to obtain kinetic evidence for an isomerization step.

The differences we find in *Cl. MP* flavodoxin may not be universal for all flavodoxins. In the *D. vulgaris* molecule, the residues corresponding to

TABLE II

BENDING ANGLES FOR THE ISOALLOXAZINE RING IN FLAVODOXIN FROM *CL. MP*

Determined by real-space refinement [8] of a model permitting bending along the N-5/N-10 line. Atoms in each wing were kept coplanar. Refinement was begun with an idealized flat flavin [12]. Overall accuracy is not better than ± 2°.

	Isomorphous Replacement Map	$(2|F|_0 - |F_c|)\exp i\alpha_c^*$ Map
Oxidized	1.5°	1.0°
Semiquinone	3.3°	2.6°**
Reduced		8.6°

* Flavin ring atoms were not included in $|F|_c$ or α_c.
** From a map including data to 2.0 Å resolution.

our 56–59 are 60–63; they do not adopt the same conformation in both species of flavodoxin [36].

(b) The isoalloxazine ring conformation in Cl. MP flavodoxin

To a good approximation, distortions of the isoalloxazine ring from planarity can be ascribed solely to bending along a line connecting N—5 with N—10. Even in model oxidized flavins, the mean planes through the benzene and pyrimidine rings do not coincide perfectly; bending angles of 2.9° [40] and as large as 4.1° [41] have been found. All of the torsion angles we have determined for the ring in oxidized flavodoxin (Table II) fall within the range observed in model flavins.

For the semiquinone oxidation state, no models have been available from crystal-structure analyses, but spectral data for certain flavodoxins [12, 42] have lent support to the notion that the neutral semiquinone ring is essentially planar. At 3.25 Å resolution, we could detect no deviation of the ring from planarity in flavodoxin semiquinone by visual fitting of the electron density. The bending angle determined from a map at 2 Å resolution was 3.3°, with the same "sense" as we now find in the fully reduced structure, but was not considered significantly different from the angles typical of so-called "planar" oxidized flavins. After refinement only minimal bending can be detected (Table II).

Ring bending is increased after complete reduction, although the changes associated with reduction of the radical species are very small indeed. The "equilibrium" torsion angles for 1,5-dihydroisoalloxazines without alkyl substituents at N-1 or N-5 have not been measured; one might surmise that they lie somewhere between the value of 9° for the N-5 disubstituted model [30] and the 30° observed in 9-Br-1,3,7,8,10-pentamethyl-1,5-dihydroisoalloxazine [24, 29]. A dihydroflavin with a torsion angle of 8–9° may therefore possess a relatively unstable conformation. Unfortunately, we cannot assess accurately the magnitude of destabilization. The free energies of activation for *N*- and coupled *N*- and ring inversion of 1,5-dihydroisoalloxazines have been measured in acetone and estimated for aqueous solutions; in water they are about 13 kcal/mol and 18 kcal/mol, respectively [29]. Thus, "freezing" of the transition state for inversion by attachment to the apoprotein would be achieved at a considerable cost in conformational energy. However, we do not know how closely our observed conformation resembles the transition state.

One might inquire whether the image in Fig. 4 really represents a "fixed" conformation or instead, the average position of a ring undergoing rapid vibration. Spectroscopic rather than X-ray methods are the procedures of choice to address this question. We can only comment that the density in Fig. 4 does not appear smeared out; the peaks have magnitudes comparable with those of the oxidized FMN bound to flavodoxin and with the densities of atoms in the protein backbone itself.

The difference in ΔG of binding to apoflavodoxin for FMN$^{\cdot}$ and dihydro-FMN amounts to 5.3 kcal/mol, with the reduced FMN · apoprotein complex being less stable [13, 14]. Previously we had proposed that the reduced FMN ring might be constrained in a high energy (nearly flat) conformation by the apoprotein and that "coenzyme strain" was responsible, at least in part, for the decrease in K_a and low E_1 potential. The results reported here seem consistent qualitatively with this view.

Acknowledgement

This research has been supported by P.H.S. grants GM 16429, a Career Development Award (GM 1611) to M.L.L., and Training Grants GM 1355 and GM 00187.

References

1 Andersen, R.D., Apgar, P.A., Burnett, R.M., Darling, G.D., LeQuesne, M.E., Mayhew, S.G. and Ludwig, M.L. (1972) Proc. Natl. Acad. Sci. U.S. 69, 3189–3191
2 Burnett, R.M., Darling, G.D., Kendall, D.S., LeQuesne, M.E., Mayhew, S.G., Smith, W.W. and Ludwig, M.L. (1974) J. Biol. Chem. 249 4383–4392
3 Watenpaugh, K.D., Sieker, L.C., Jensen, L.H., LeGall, J. and Dubourdieu, M. (1972) Proc. Natl. Acad. Sci. U.S. 69, 3185–3188
4 Watenpaugh, K.D., Sieker, L.C. and Jensen, L.H. (1973) Proc. Natl. Acad. Sci. U.S. 70, 3857–3860
5 Ludwig, M.L., Andersen, R.D., Mayhew, S.G. and Massey, V. (1969) J. Biol. Chem. 244, 6047–6048
6 Watenpaugh, K.D., Sieker, L.C., Herriott, J.R. and Jensen, L.H. (1973) Acta Crystallogr. B29. 943–956
7 Freer, S.T., Alden, R.A., Carter, Jr., C.W. and Kraut, J. (1975) J. Biol. Chem. 250, 46–54
8 Diamond, R. (1971) Acta Crystallogr. A27, 436–452
9 Huber, R., Kukla, D., Bode, W., Schwager, P., Bartels, K., Deisenhofer, J. and Steigemann, W. (1974) J. Mol. Biol. 89, 73–101
10 Moews, P.C. and Kretsinger, R.H. (1975) J. Mol. Biol. 91, 201–228
11 Tanaka, M., Haniu, M., Yasunobu, K.T. and Mayhew, S.G. (1974) J. Biol. Chem. 249, 4393–4396
12 Eaton, W.A., Hofrichter, J., Makinen, M.W., Andersen, R.D. and Ludwig, M.L. (1975) Biochemistry, 14, 2146–2151
13 Mayhew, S.G. (1971) Biochim. Biophys. Acta 235, 276–288
14 Cox, J.M. (1967) J. Mol. Biol. 28, 151–155
15 Venkatachalam, C.M. (1968) Biopolymers 6, 1425–1436
16 Diamond, R. (1974) J. Mol. Biol. 82, 371–391
17 Ruhlmann, A., Kukla, D., Schwager, P., Bartels, K. and Huber, R. (1973) J. Mol. Biol. 77, 417–436
18 Stout, G.H. and Jensen, L.H. (1968) in X-ray Structure Determination, a Practical Guide, Chap. 15, Macmillan, New York
19 Luzzati, V. (1953) Acta Crystallogr. 6, 142–152
20 Ramachandran, G.N. and Srinivasan, R. (1970) in Fourier Methods in Crystallography, Chap. 6, Wiley-Interscience, New York
21 Ramachandran, G.H. and Sasisekharan (1968) Adv. Prot. Chem. 23, 284–437
22 Wyckoff, H.W., Tsernoglou, D., Hanson, A.W., Knox, J.R., Lee, B. and Richards, F.M. (1970) J. Biol. Chem. 245, 305–328
23 Hartsuck, J.A. and Lipscomb, W.N. (1971) in The Enzymes (Boyer, P.D., ed.), Vol. 3, pp. 1–56, Academic Press, New York
24 Norrestam, R. and von Glehn, M. (1972) Acta Crystallogr. B28, 434–446
25 Werner, P.-E. and Ronnquist, O. (1970) Acta Chem. Scand. 24, 997–1009
26 Norrestam, R., Kierkegaard, P., Stensland, B. and Torbjornsson, L. (1969) Chem, Commun. 1227, 1250–1251
27 Dudley, K.H., Ehrenberg, A., Hemmerich, P. and Muller, F. (1964) Helv. Chim. Acta 47, 1354–1383
28 Ghisla, S., Massey, V., Lhoste, J.-M. and Mayhew, S.G. (1974) Biochemistry 13, 589–597
29 Tauscher, L., Ghisla, S. and Hemmerich, P. (1973) Helv. Chim. Acta 56, 630–644
30 Werner, P.-E., Linnros, B. and Leijonmarck, M. (1971) Acta Chem. Scand. 25, 1297–1312
31 Mayhew, S.G., Foust, G.P. and Massey, V. (1969) J. Biol. Chem. 244, 803–810
32 James, T.L., Ludwig, M.L. and Cohn, M. (1973) Proc. Natl. Acad. Sci. U.S. 70, 3292–3295
33 Crick, F.H.C. and Magdoff, B.S. (1956) Acta Crystallogr. 9, 901–908
34 Mayhew, S.G. (1971) Biochim. Biophys. Acta 235, 289–302
35 Levitt, M. (1974) J. Mol. Biol. 82, 393–420

36 Watenpaugh, K.D., Sieker, L.C. and Jensen, L.H. (1976) Flavins and Flavoproteins, Proceedings of the 5th International Symposium (Singer, T.P., ed.), Chapt. 42, pp. 405–410, Elsevier, Amsterdam
37 Rogers, M.T. and Woodbrey, J.C. (1962) J. Phys. Chem. 66, 540–546
38 Bushweller, C.H., O'Neill, J.W., Halford, M.H. and Bisset, F.H. (1971) J. Am. Chem. Soc. 93, 1471–1473
39 Mayhew, S.G. and Massey, V. (1973) Biochim. Biophys. Acta 315, 181–190
40 Wang, M. and Fritchie, Jr., C.J. (1973) Acta Crystallogr. B29 2040–2045
41 Tanaka, M. Ashida, T., Sasada, Y. and Kakudo, M. (1969) Bull. Chem. Soc. Japan 42, 1546–1554
42 Eriksson, L.E.G. and Ehrenberg, A. (1973) Biochim. Biophys. Acta 293, 57–66

Flavins and Flavoproteins, edited by T.P. Singer

CHAPTER 42

A CRYSTALLOGRAPHIC STRUCTURAL STUDY OF THE OXIDATION STATES OF *DESULFOVIBRIO VULGARIS* FLAVODOXIN

K.D. WATENPAUGH, L.C. SIEKER and L.H. JENSEN

Department of Biological Structure, University of Washington, School of Medicine, Seattle, Wash. 98195 (U.S.A.)

Summary

Crystallographic studies have been carried out to compare the reduced states of *Desulfovibrio vulgaris* flavodoxin to the oxidized state determined to 2.0 Å resolution. Differences in the diffraction data of the three oxidation states indicate sizable differences between their crystallographic structure but not necessarily conformational changes. The flavin remains nearly planar in the semiquinone and the aromatic groups around it show little movement. Three possible models of conformational changes between the oxidized and semiquinone states are discussed. They are based on the observation that the primary differences observed in the structures appear to be near residues 62 and 64 of the protein.

Introduction

Flavodoxins are a class of small proteins involved in electron transport in a variety of organisms. They contain a single FMN group and no metal atoms, but normally appear to replace functionally the iron-sulfur protein ferredoxin. Detailed three-dimensional structures have been determined for two bacterial flavodoxins (*Clostridium MP* [1, 2] and *Desulfovibrio vulgaris* [3, 4]). Although both of these flavodoxins have similar tertiary structure, the binding between the flavin group and the protein is quite different, and while they have similar oxidation-reduction properties, they must have significant differences in the electron transport mechanisms.

Detailed discussions of the similarities and differences in the environment of the FMN in the oxidized form have previously been published [5, 6]. Ludwig et al. discuss in Chapter 41 various aspects of the different oxidation states of the *Cl. MP* flavodoxin deduced from their X-ray crystallographic studies. This paper presents the possible conformational states associated with the change in the oxidation states of *D. vulgaris* flavodoxin from a similar study.

Methods of studying the oxidation states and problems

The methods used to determine the structure to 2.0 Å resolution using the anomalous scattering of the samarium heavy atom derivation have been discussed elsewhere [7]. We are now extending the structural study to the various oxidation states. Crystals are reduced to the blue semiquinone in ammonium sulphate solutions at pH 7.5 under an inert atmosphere using sodium dithionite. The reduced single crystals are then inserted in glass capillary tubes for collecting the X-ray diffraction data. Crystals can also be reduced to the pale yellow fully reduced form in the same manner at pH 9.0. Unfortunately, we have not been able to keep the crystals in the particular oxidation states during the data-collection process. Data collection even to moderate resolution requires several days. While it is easy to monitor the reoxidation by collecting a standard set of reflections at intervals, it is difficult to interpret X-ray diffraction data on a sample that is changing conformation during the collection process. A comparison of the data in the different oxidation states shows that there are substantial changes in the crystals (at least from a crystallographic viewpoint) when going from the oxidized form to the semiquinone and finally to the fully reduced form. It should be pointed out, however, that these large intensity changes may be attributed to a number of factors. They could be due to relatively large conformational changes in a part of the molecule, many small changes distributed over a large portion of the molecule or even relatively small orientational or translational changes of the entire molecule without any change in the conformation. Intensity changes occur from any absolute shift of the atomic

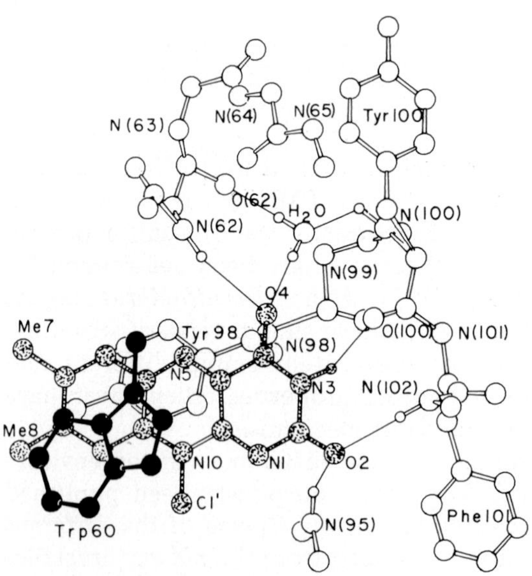

Fig. 1. Environment of the flavin in the oxidized state of *D. vulgaris* flavodoxin. Flavin and aromatic groups are idealized but the polypeptide chain is not idealized.

positions in the crystal unit cell whether conformational changes occur or not. Unfortunately, intensity changes are also accompanied by changes in the relative phase angles associated with them. Therefore, applying phases determined for one oxidation state to data of a different oxidation state will at the very least introduce "noise" in the calculated electron density maps.

While we do have three data sets on reduced forms of the crystals, only one set consisting of data to a resolution of approx. 2.5 Å on the semiquinone appears to consist primarily of one oxidation state and to warrant serious examination. Unfortunately, at this resolution it is difficult to interpret unequivocally shifts in the molecule, particularly with phases determined from the oxidized form. Both an electron density map and a difference electron density map between the oxidized and semiquinone forms have been calculated using the phases determined for the oxidized form.

Possible conformational differences between the oxidation states of *D. vulgaris* flavodoxin

Fig. 1 shows the flavin environment determined from the 2.0 Å resolution map of the oxidized form. The flavin is between the two aromatic residues tryptophan 60 and tyrosine 98. The tryptophan is nonparallel with the flavin while the tyrosine is nearly parallel. The two methyl groups on the benzenoid ring of the flavin are on the surface of the protein while the pyrimidine ring is deeply buried and strongly hydrogen bonded to the protein. When compared to the *Cl. MP* flavodoxin, this region and the hydrogen bonding are quite different. O-2 of the flavin hydrogen binds to

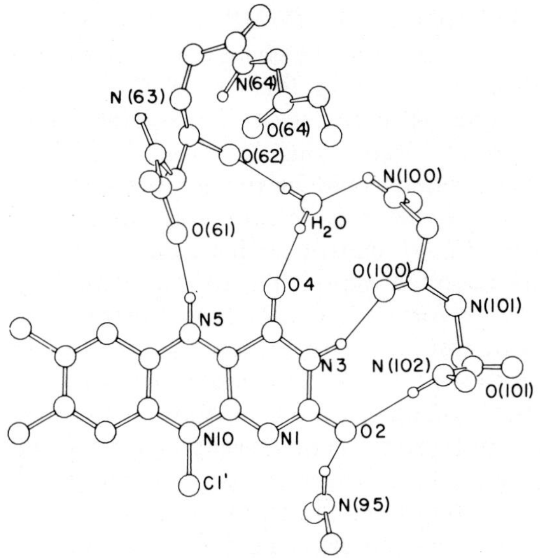

Fig. 2. First possible model of conformation of semiquinone. Peptide 62 is rotated to form a H bond between CO-61 and NH-5 of flavin.

NH-95 and NH-102 of the polypeptide chain, NH-3 bind to CO-100 and O-4 to a water molecule and weakly to NH-62. The water molecule in turn binds to CO-62 and NH-100. Fig. 1 also shows the position of two other aromatic groups near the flavin which could influence various physico-chemical properties such as circular dichrosism. Tyrosine 100 is roughly in the same plane as the flavin and phenylalanine 101 is in a plane roughly halfway between the flavin and tyrosine 98.

Examination of the difference map between the oxidized and semiquinone indicates that there are many small changes over the entire semiquinone molecule relative to the oxidized, perhaps from some changes in rotational and translational orientation. There do not appear to be large conformational changes and, unfortunately, the small conformational changes are hard to separate from the overall movement of the molecule. Certain facts are suggested by the data. The flavin group appears to remain nearly planar on reduction to the semiquinone form and the orientation of the tyrosine (Tyr 98) and the tryptophan (Trp 60) residues neighboring the flavin remain approximately the same. Some of the largest shifts appear to be in the region of residue 62. The NH of this residue as shown in Fig. 1 is close to N-5 of the flavin, although it is probably weakly hydrogen bonded to O-4. On reduction of the flavin, a hydrogen would bond to N-5, producing an unfavorable hydrogen-hydrogen interaction. Consequently, some conformational change must occur to relieve it.

There are several alternatives to accomplish this. One is to move the peptide chain and the flavin group away from each other without changing the conformation of the chain. This is not suggested by examination of the difference map. Peptide 61 could also rotate in several ways to form more favorable interaction with no other changes in the structure. Ludwig et al. (Chapt. 41, this symposium) have shown how the *Cl. MP* appears to have done this. It should be pointed out, however, that the chain conformation in the *D. vulgaris* flavodoxin is quite different from the *Cl. MP* one in this region and than an exactly analogous change cannot occur. Three alternatives suggest themselves. If the peptide group rotates roughly about the C_α^{61} to C_α^{62} vector by about 150°, a hydrogen bond could be formed between the carbonyl of residue 61 and NH-5 of the flavin as is shown in Fig. 2. This would be similar to that proposed for *Cl. MP* except in that case it converts a 3_{10} bend from type II to the more favorable type I. The second possibility is that the peptide rotates approx. 90° to form a hydrogen bond between the carbonyl of residue 61 and the NH of residue 64 to form a 3_{10} type I bend but losing a hydrogen bond to the flavin (Fig. 3). This model would place the hydrogen of NH-62 close to the hydrogen of C-6 on the flavin unless there is additional adjustment. This configuration could easily be that of the apoprotein. The third possibility (Fig. 4) is to rotate the peptide about 60° in the direction opposite that of the second model to form a hydrogen bond between NH-62 and CO-64. This third model appears to best fit the difference map but the first model (Fig. 2) with the hydrogen bond between NH-5 of the flavin and CO-61 cannot be ruled out on the basis of the data. The first

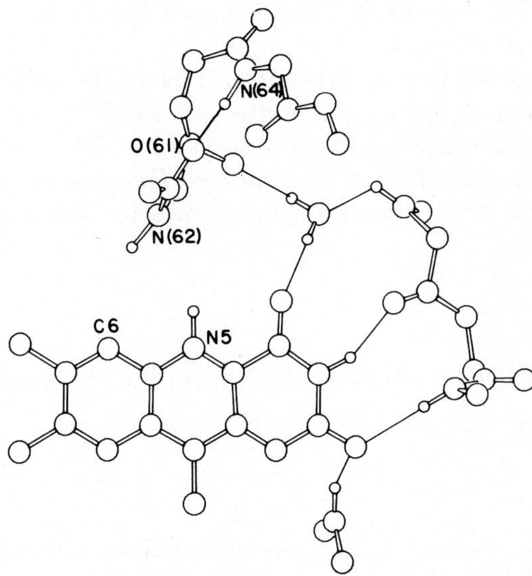

Fig. 3. Second possible model of conformation of semiquinone. Peptide 62 is rotated to form 3_{10} bond with H bond between CO-61 and NH-64.

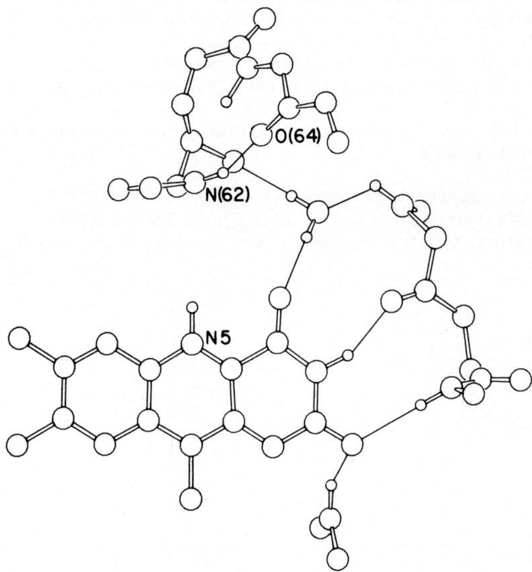

Fig. 4. Third possible model of conformation of semiquinone. Peptide 62 is rotated to form a H bond between NH-62 and CO-64.

model would also permit a situation more nearly in accord with that proposed in the *Cl. MP* flavodoxin case.

A comparison of the flavodoxin sequences [9, 10, 11] indicates a thus far conserved glycine residue at residue 61 of *D. vulgaris* flavodoxin. This would be important if the peptide 61 does in fact rotate when the flavin is converted from one oxidation state to another.

Finally, the large changes observed in the diffraction data in the fully reduced form cannot as yet be explained. They may be due just to additional small atomic shifts over the entire molecule like those observed for the semiquinone form. A comparison of data collected on the oxidized flavodoxin at pH 9 and pH 7.5 shows little or no conformation change due to this pH change.

Acknowledgement

This work was supported by USPHS Grant AM-03288 from the National Institutes of Health.

References

1 Anderson, R.D., Apgar, P.A., Burnett, R.M., Darling, G.D., LeQuesne, M.E., Mayhew, S.G. and Ludwig, M.L. (1972) Proc. Natl. Acad. Sci. U.S. 69, 3189

2 Burnett, R.M., Darling, G.D., Kendall, D.S., LeQuesne, M.E., Mayhew, S.G., Smith, W.W. and Ludwig, M.L. (1974) J. Biol. Chem. 249, 4383

3 Watenpaugh, K.D., Sieker, L.C., Jensen, L.H., LeGall, J. and Dubourdieu, M. (1972) Proc. Natl. Acad. Sci. U.S. 69, 3185

4 Watenpaugh, K.D., Sieker, L.C. and Jensen, L.H. (1973) Proc. Natl. Acad. Sci. U.S. 70, 3857

5 Ludwig, M.L., Burnett, R.M., Darling, G.D., Jordan, S.R., Kendall, D.S. and Smith, W.W. (1975) 4th Steenbock Symposium on Structure and Conformation of Nucleic Acids and Protein-Nucleic Acid Interactions (Sundaralingam, M. and Rao, S.T., eds.), Madison, Wisconsin, 1974, p. 407

6 Watenpaugh, K.D., Sieker, L.C. and Jensen, L.H. (1975) 4th Steenbock Symposium on Structure and Conformation of Nucleic Acid and Protein-Nucleic Acid Interactions (Sundaralingam, M. and Rao, S.T., eds.), Madison, Wisconsin, 1974, p. 431

7 Watenpaugh, K.D., Sieker, L.C. and Jensen, L.H. (1975) Inter-Congress Conference on Anomalous Scattering (Ramaseshan, S., ed.), Madrid, 1974, p. 393

8 Venkatachalam, C.M. (1968) Biopolymers 6, 1425

9 Dubourdieu, M., LeGall, J. and Fox, J.L. (1973) Biochem. Biophys. Res. Commun. 28, 1418

10 Tanaka, M., Haniu, M., Yasunobu, K.T. and Mayhew, S.G. (1974) J. Biol. Chem. 249, 4393

11 Tanaka, M., Haniu, M., Yasunobu, K.T., Mayhew, S.G. and Massey, V. (1974) J. Biol. Chem. 249, 4397

Flavins and Flavoproteins, edited by T.P. Singer

CHAPTER 43

THE EFFECTS OF pH ON DITHIONITE-REDUCED FLAVODOXIN FROM *PEPTOSTREPTOCOCCUS ELSDENII* AND THE USE OF APOFLAVODOXIN TO DETERMINE AND PURIFY FMN

STEPHEN G. MAYHEW, EEF J.F. VAN AREM, MARIJKE J.J. STRATING and JOHANNES H. WASSINK

Department of Biochemistry, Agricultural University, De Dreijen 11, Wageningen, (The Netherlands)

Summary

After reduction at pH 8 with an equimolar amount of sodium dithionite, flavodoxin from *Peptostreptococcus elsdenii* remains fully reduced indefinitely under anaerobic conditions. However, when the pH is decreased to 5 by adding acid anaerobically, the flavin is reoxidised to the semiquinone. Part of the flavin is oxidised during the time of mixing and measurement (approx. 30 s) and the remainder over a period of several hours. The extent of the rapid reaction depends on pH (50% of the flavin is converted to the semiquinone at pH 5) and the concentration of the oxidation products of dithionite, and it is increased by addition of SO_3^{2-} (but not by SO_4^{2-} or $S_2O_3^{2-}$). A chromophore which absorbs maximally at about 315 nm is generated during the reaction. The rapid reaction is reversed by a subsequent increase of pH, but the reversibility decreases during incubation at low pH. It is proposed that the rapid reoxidation at low pH is due to electron transfer to an oxidation product of dithionite, and that a redox equilibrium is established between the flavin hydroquinone/semiquinone couple and the dithionite system.

A method has been developed for the determination of FMN by fluorescence titration with apoflavodoxin from *P. elsdenii*. Such titrations reveal a major impurity in FMN that has been purified from a commerical product by chromatography on DEAE-cellulose. This impurity, and similar impurities in FMN derivatives, can be removed by affinity chromatography with apoflavodoxin as the immobilised ligand.

Introduction

The reaction of dithionite with flavodoxins is of considerable practical interest because dithionite is the only convenient reagent for the full chemical reduction of these low-potential electron carriers. The rate and extent of reduction varies with the source of the protein [1–7]. Flavodoxin from *Peptostreptococcus elsdenii* is readily reduced at pH values of 7 or greater and full reduction by equimolar dithionite occurs within a few minutes [1, 7]. Kinetic studies at pH 8.5 have suggested that this flavodoxin is reduced by a dissociation product of dithionite, such as SO_2^-, and that full reduction of the oxidised protein involves two sequential one-electron transfers; a relatively slow reduction to the semiquinone, followed by a more rapid reaction to form the hydroquinone [7]. Comproportionation between oxidised and fully reduced flavodoxin can occur [1] but is not significant on reduction by excess dithionite unless the ionic strength is high [7]. Reduction at pH values below about 6.5 is not stoichiometric, the extent of reduction of the semiquinone to the hydroquinone depending on pH and the concentration of dithionite in excess. This concentration effect suggested that at low pH a pH-dependent redox equilibrium is established between dithionite (and its oxidation product(s)) and the semiquinone/hydroquinone couple of flavodoxin. Further evidence for such a redox equilibrium has now been obtained from studies on the effects of pH on flavodoxin that has been reduced stoichiometrically with dithionite.

Studies on the specificity of flavin binding by *P. elsdenii* apoflavodoxin have indicated that although flavins modified in the isoalloxazine nucleus are bound to the apoprotein, an N-10 side chain on the flavin with 5 carbon atoms and a terminal phosphate group are essential for binding [8, 9]. This high specificity has been exploited in methods for the determination of FMN (and FAD after enzymic hydrolysis) by fluorescence titration with apoflavodoxin [24], and for the purification of FMN and its analogues by affinity chromatography with apoflavodoxin immobilised on Sepharose 4B [25].

Results and Discussion

Effect of pH on dithionite-reduced flavodoxin

After titration to full reduction at pH 8.1 with a stoichiometric amount of dithionite, flavodoxin from *P. elsdenii* remains fully reduced indefinitely under anaerobic conditions. However, when the pH of the solution of the reduced protein is decreased to 4.9 by the anaerobic addition of acid, the flavin is partially converted to a blue intermediate which in the long wavelength region has spectral properties identical with those of flavodoxin semiquinone [1] (Fig. 1). The reaction is biphasic; about 50% of the flavin is oxidised before measurements can be made in a static spectrophotometer, and the remainder over a period of several hours. The rapid reaction is almost completely reversible, the flavin returning to its fully reduced state when the pH is returned to 8.1 (Fig. 2). The much slower phase of the reaction is not reversible and has not been investigated further.

The spectrum in the region 300–360 nm at pH 4.9 is higher than expected for a mixture of semiquinone and hydroquinone (Fig. 1), and the

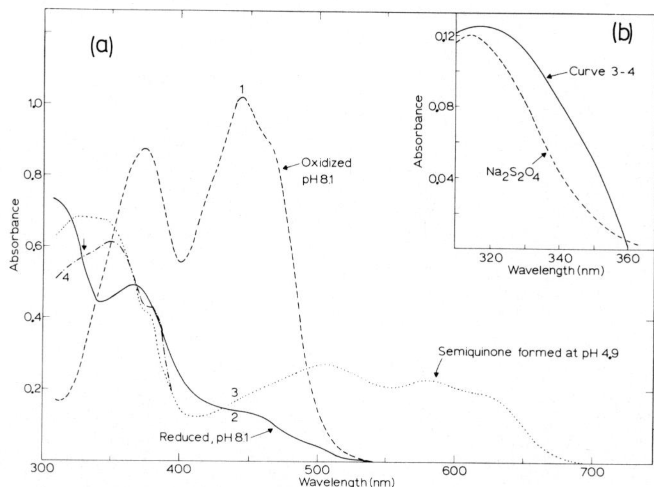

Fig. 1. Effect of pH on dithionite-reduced flavodoxin. The experiments were carried out at 22°C in a 3 ml anaerobic cuvette equipped with a silicon rubber stopper which contained a glass electrode, a saturated KCl bridge in acrylamide gel, gas inlet and outlet ports for flushing the cell with N_2 (purified by passage over BASF catalyst at 80°C) and a port through which additions to the cell could be made with a Hamilton (type 87001) gas-tight syringe. The salt bridge was connected to a saturated calomel electrode. (a) A solution containing 98 μM flavodoxin, 20 mM potassium phosphate buffer, pH 8.1, 0.3 M NaCl and 10 μl octanol (to prevent frothing) in a final volume of 3 ml was made anaerobic (curve 1) and the flavin then titrated carefully to full reduction with standard $Na_2S_2O_4$ dissolved in 0.1 M Tris/HCl buffer, pH 8.6 (curve 2). The pH of the solution was then adjusted to pH 4.9 by adding 18 μl 3 M HCl and a spectrum recorded immediately (curve 3). (b) A mixture similar to that of (a) but at pH 9.7 and containing in addition 0.1 M EDTA was made anaerobic (curve 1) and illuminated for 15 h with light from a 60 W tungsten lamp. The pH of the solution of partially reduced flavodoxin was then adjusted to pH 4.9 (curve 4).

isosbestic point at 329 nm [1] between these two redox states of the flavin is lost. Several observations suggest that this new absorbance is due to a chromophore free in solution, and that it is produced by electron transfer from flavodoxin to an oxidation product of dithionite:

1. After partial photoreduction of flavodoxin with EDTA at pH 9.7 [10], a pH change to pH 5 causes a small increase of absorbance at 580 nm, due to comproportionation, but the resulting spectrum passes through the 329 nm isosbestic point (Fig. 1, curve 4). It is unlikely therefore that the increase of absorbance at 329 nm with dithionite-reduced flavodoxin is due to a pH-dependent change in the flavin absorbance (by protonation of the hydroquinone [1] for example).

2. Dithionite-reduced FMN remains fully reduced when the pH is changed from 8.1 to 4.9; further, there are no increases in absorbance at short wavelength, indicating that the increase observed with flavodoxin is not due to a pH-dependent absorbance by an oxidised product of dithionite.

3. When flavodoxin is partly photoreduced and then taken to full reduction by adding dithionite (about 0.25 mol per mol of flavin) only 15% of the flavin is rapidly reoxidised after a pH change to 4.9, and the increase at 329 nm is correspondingly lower.

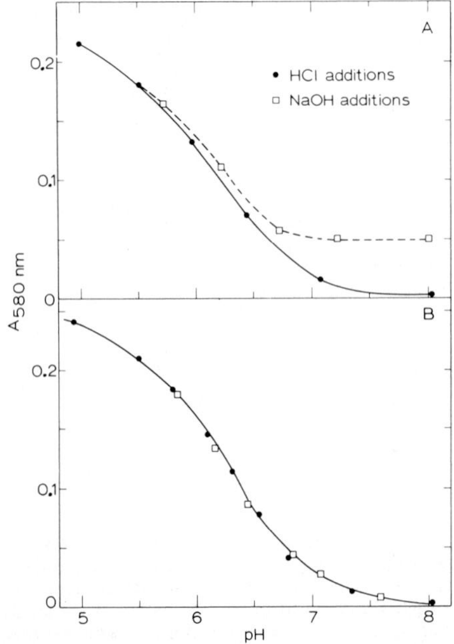

Fig. 2. Formation of semiquinone by pH titration of dithionite-reduced flavodoxin. (A) Flavodoxin (0.1 mM) was fully reduced with dithionite as described in Fig. 1 and then the pH was changed stepwise first to pH 5 by adding increments of 3 M HCl, and then to pH 8 by adding increments of 3 M NaOH. The absorbance at 580 nm due to flavodoxin semiquinone was recorded after each addition. (B) Flavodoxin (98 μM) was fully reduced with dithionite as described in Fig. 1. More dithionite was then added to give a solution 0.1 mM in unreacted dithionite. The excess dithionite was then oxidised by adding increments of O_2-saturated H_2O until the absorbance at 315 nm reached a minimum and absorbance at 580 nm reappeared; this small amount of semiquinone (about 4 μM) was reduced again by titration with dithionite. Acid and base titrations were then carried out as described above.

It should be noted that the maximum at about 315 nm in the difference spectrum between dithionite- and photo-reduced flavodoxin at pH 4.9 (Fig. 1, inset) and its extinction coefficient (5000 $M^{-1} \cdot cm^{-1}$ per 2 reducing equivalents), calculated from the amount of flavodoxin oxidised, are similar to those of dithionite (see references in Ref. 7).

Most of the flavodoxin semiquinone generated during the pH change is converted back to the hydroquinone when the pH is returned to 8.1 (Fig. 2A). However, the reversibility of the reaction decreases with increasing times of incubation at low pH, suggesting that like the dithionite ion (see references in Ref. 7) the reduced electron acceptor is unstable in acid solution; the new absorbance at 329 nm also decays slowly at low pH. The electron acceptor from flavodoxin is stable at high pH since the pH-dependent redox equilibrium can be established even after storage of reduced flavodoxin for 16 h at pH 8. An excess of the oxidation product(s) of dithionite, produced by treating flavodoxin with 2 mol of dithionite per mol of protein at pH 8.5 and then oxidising away the excess by careful titration with O_2-saturated water or anaerobic K_3FeCN_6, causes a small increase in the amount of semiquinone formed at low pH, and unexpectedly enhances the reversibility

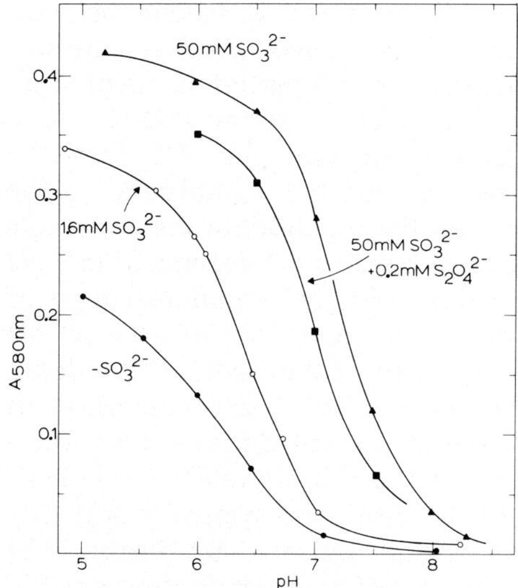

Fig. 3. Effect of sulphite on the pH-dependent oxidation of dithionite-reduced flavodoxin. The experimental conditions were as described in Figs. 1 and 2A except that the cuvette also contained Na_2SO_3; ●——●, minus Na_2SO_3; ○——○, plus 1.6 mM Na_2SO_3; ▲——▲, plus 50 mM Na_2SO_3; ■——■ plus 50 mM Na_2SO_3 and excess $Na_2S_2O_4$ (0.2 mM).

of the reaction (Fig. 2B). No similar effects are observed when dithionite is oxidized with air before it is added to flavodoxin, which probably indicates that the dithionite oxidation products formed with O_2 in the presence and absence of flavodoxin are different.

According to the reaction mechanism proposed earlier [7], the oxidation product of the reaction of dithionite with flavodoxin is sulphur dioxide, and by hydration this would give rise to sulphite. It was thought likely therefore that sulphite might influence the pH-dependent equilibrium between flavodoxin and the dithionite system. Fig. 3 shows that added sulphite shifts the equilibrium to higher pH values and increases the amount of semiquinone that is formed at low pH; in the presence of 50 mM sulphite, the semiquinone formed at pH 5.2 is about 95% of the theoretical maximum. However, sulphite appears to increase the acid lability of the reduced product so that full reversibility of the reaction is more difficult. Nevertheless, as might be anticipated from Fig. 3, fully reduced flavodoxin at pH 7 can be partially converted to the semiquinone by addition of sulphite, and then reduced again simply by increasing the pH to 9. In the presence of sulphite at pH 7, the protein is not fully reduced by a 2-fold molar excess of dithionite (Fig. 3). Sulphate (33 mM) and thiosulphate (33 mM) were also tested in the system but they had no effect.

These experiments show that electrons can be transferred from fully reduced flavodoxin to the oxidation product(s) of dithionite. It is clear that the reaction could complicate measurements of electron transfer between dithionite-reduced flavodoxins and other electron acceptors, especially when such experiments are carried out in a pH region known to favour the equilibrium [11, 23]. The electron acceptor and its reduced product have not been identified. However, the spectral and stability properties of the product are similar to those of dithionite, and it is possible therefore that the equilibrium is described, at least in part, by the equation [7]: $FH_2^{\cdot} + SO_2^- \rightleftharpoons FlH_2^- + SO_2$. Sulphite might be expected to shift this equilibrium to the left, as was observed. The effects of pH on the equilibrium are explicable if the redox potential of the dithionite system remains pH dependent in a pH region where the potential (E_1) of the semiquinone/hydroquinone couple of flavodoxin changes from pH dependent ($\Delta E_1/\Delta \text{pH} = -59$ mV) to pH independent (pK = 5.8) [1]. At pH 5, where half of the flavin of flavodoxin is oxidised to the semiquinone in the presence of 1 mol equivalent of dithionite oxidation product(s), the potential of the dithionite acceptor-donor is probably similar to E_1 (−323 mV); at pH 8 where E_1 is −375 mV and independent of pH, the potential of the dithionite system is evidently more reducing (but possibly not as reducing as indicated by Burton [12] (−471 mV)). As far as we know, the effect of pH on the redox potential of dithionite has not been measured.

If this interpretation of our observations is correct, then the failure of dithionite to reduce fully flavodoxin from *Azotobacter vinelandii* at pH 7 [2] is not surprising; the redox potential (E_1) of this flavodoxin is −495 mV at pH 8.2 [11] and is therefore 120 mV more negative than E_1 of *P.elsdenii* flavodoxin. Titration of *A.vinelandii* flavodoxin with dithionite at pH values between 7 and 8 would indicate whether this protein forms a similar redox equilibrium with the dithionite system, or whether, as has been claimed [2, 11], reduction is independent of the concentration of dithionite in excess and controlled only by pH and the ionisation of the hydroquinone. Since the redox potentials of this flavodoxin are known [11], such studies might lead to a measurement of the redox potential of dithionite at pH values above 7.

Determination of FMN and FAD with apoflavodoxin

The apoprotein of *P.elsdenii* flavodoxin is unusually specific for flavin deriviatives which contain an N-10 side chain with 5 carbon atoms and a 5′ phosphate group [8, 9]. It binds FMN tightly ($K_d \approx 4 \cdot 10^{-10}$ M) and causes almost complete quenching of the flavin fluorescence. Fluorimetric titration with this apoflavodoxin therefore provides a convenient and sensitive assay for riboflavin 5′-phosphate. The assay requires only a measurement of the amount of standard apoprotein to cause the maximum decrease in flavin fluorescence. Consequently, it offers some advantages over other fluorimetric methods for FMN which involve at least a partial purification of the flavin and comparison of its fluorescence with that of authentic FMN (Ref. 13 and references contained therein). For example, FMN can be determined in the presence of similarly fluorescent compounds such as FAD or riboflavin, and in the presence of materials which decrease the observed fluorescence (Fig. 4). FAD can be similarly determined in mixtures from the difference

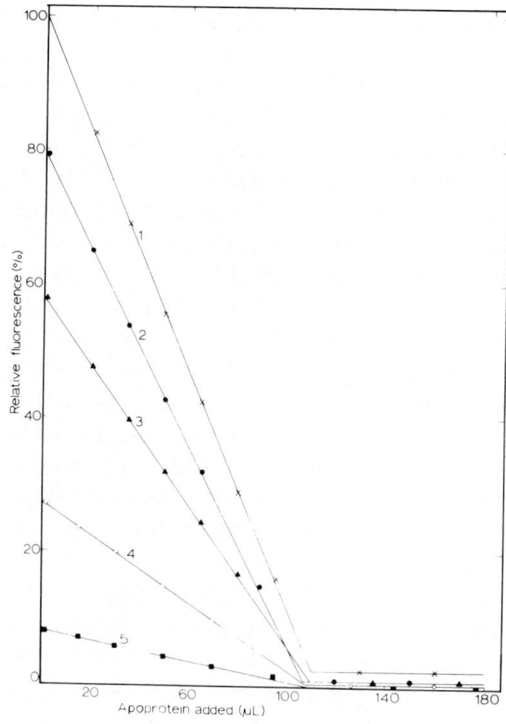

Fig. 4. Fluorimetric titration of FMN with apoflavodoxin in the presence of compounds which decrease flavin fluorescence. Increments of apoflavodoxin (18.2 μM) were added to solutions of 1 μM FMN in 2 ml 10 mM sodium acetate buffer, pH 6, and 0.2 M NaCl and the fluorescence (excitation 445 nm, emission 525 nm) was measured after each addition. The following compounds were added to the assay mixture: curve 1, no additions; curve 2, 10 μM cytochrome *c*; curve 3, 10 mM adenosine; curve 4, 50 mM KI; curve 5, 0.1 mM cytochrome *c*.

between the endpoints of two titrations [24]; an initial titration to determine FMN, and a second titration following treatment of the test sample with phosphodiesterase (*Naja naja* venom from Sigma) to hydrolyse FAD to FMN and AMP.

As mentioned previously [14], titrations of this kind show that the commercial preparations of FMN which we have available are very impure and contain much fluorescent material which is not bound to the apoprotein (Table I, Fig. 5). Treatment on DEAE-cellulose [15] removes some of these impurities [8] but at least 12% of the resulting preparation of FMN (Sigma F 2253) fails to bind to the apoprotein. Hemmerich [14] has proposed that the impurities which co-chromatograph with FMN include the 4′-phosphate isomer, presumably formed during chemical phosphorylation of riboflavin. A more satisfactory preparation of FMN for routine use (2% residual fluorescence with apoflavodoxin) can be obtained by enzymic hydrolysis of FAD following purification of commercial preparations of this flavin on DEAE-cellulose [15]. However, our most pure preparations are readily made by affinity chromatography of FMN with apoflavodoxin as the immobilised ligand [25].

Properties of immobilised flavodoxin and use of the immobilised apoprotein in the purification of FMN

Flavodoxin from *P. elsdenii* has been covalently coupled to CNBr-activated Sepharose 4B (Ref. 16 and Strating and Mayhew [25]). By carrying out the coupling reaction at pH 10.5 with a high concentration of flavodoxin (12 mg/ml), the resulting preparation of packed gel contains at least 0.2 mM immobilised protein. Like the free protein [11], the bound protein is reduced by dithionite, it stabilises the blue (neutral) semiquinone of FMN, and it replaces ferredoxin in hydrogen production from dithionite or pyruvate in the presence of ferredoxin-free extracts of *P. elsdenii* [10]. The catalytic activities are low (13% and 3.5% of the activity of soluble flavodoxin in the dithionite and pyruvate oxidation reactions, respectively) but indicate that the bound protein is accessible for electron transfer to and from the larger protein molecules hydrogenase [17] and pyruvate dehydrogenase [18]. Free flavodoxin forms a 1:1 complex with ferredoxin-NADP$^+$ reductase from spinach [19]. Immobilised flavodoxin also binds this protein. The complex is dissociated at high ionic strength and chromatography of partially purified ferredoxin-NADP$^+$ reductase on a column of the electron carrier leads to substantial purification (Strating and Mayhew [25]). However, it is not yet clear whether there is any specificity in this interaction since positively charged proteins such as cytochrome *c* and ribonuclease are also strongly bound; evidently the negatively charged flavodoxin behaves as an ion exchanger for these proteins.

The immobilised apoprotein has proved to be very useful as a ligand for the purification of FMN and FMN analogues. Although such flavins can be purified by combination with soluble apoflavodoxin, followed by dialysis or

TABLE I

DETERMINATION OF FMN IN COMMERICAL AND PURIFIED SAMPLES OF FLAVIN

Fluorimetric titrations were performed with standard apoflavodoxin as described in Fig. 4. The theoretical concentration of flavin was 1 μM and was based on the absorbance at 445 nm (FMN) or 450 nm (FAD). Where indicated, FAD was hydrolysed with *Naja naja* venom, chromatography was performed on DEAE-cellulose [15], and affinity chromatography was carried out on a column of immobilised apoflavodoxin. The residual fluorescence is the fluorescence remaining at the end of a titration in the presence of excess apoflavodoxin.

Flavin	Source	Pretreatment	FMN determined (μM)	Residual fluorescence (% initial)
FAD	Boehringer 15403	Hydrolysis	0.90	4.9
	Sigma F 6625	Hydrolysis	0.88	13.3
		Hydrolysis then chromatography	0.90	7.0
		Chromatography then hydrolysis	0.96	2.0
FMN	Boehringer 15405	None	0.68	32.5
	Sigma F 6750	None	0.73	25.0
	F 2253	None	0.75	20.5
		Chromatography	0.88	13.0
		Affinity chromatography	1.00	0.7

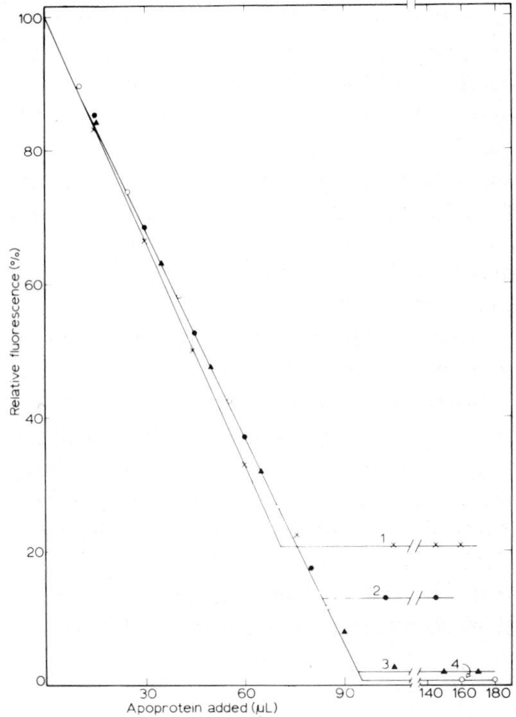

Fig. 5. Fluorimetric titration of FMN samples of different purity. The experimental conditions were as described in Fig. 4. The flavin concentrations were nominally 1 μM and were determined spectrophotometrically. Curve 1, unpurified FMN; curve 2, FMN after purification on DEAE-cellulose [15]; curve 3, FMN obtained by enzymic hydrolysis of FAD purified by chromatography on DEAE-cellulose; curve 4, FMN after affinity chromatography on immobilised apoflavodoxin. The FMN (F 2253) and FAD (F 6625) were from Sigma.

gel filtration to remove unbound flavin impurities and extraction of the bound flavin from the protein [14], this procedure is time consuming and it is not suitable for FMN derivatives which are bound only weakly. Binding of FMN to the immobilised apoflavodoxin is tight ($K_d \approx 1.5 \cdot 10^{-9}$ M in 0.025 M sodium acetate buffer, pH 6.5, and 0.1 M NaCl at 25°C), the bound flavin is reversibly dissociated by acid, and the specificity of flavin binding is similar to that of the apoprotein in solution. Therefore, FMN and its analogues can be purified rapidly by adding the crude flavin to a column of the immobilised apoprotein, washing the column to remove impurities, and then eluting the bound flavin with 5% trichloroacetic acid (or 0.1 M HCl or 0.1 M acetic acid). The flavin-binding capacity of the column decreases by about 10% during the first acid cycle but further losses in many subsequent cycles over several months are small.

As judged by the fluorescence-quenching test with apoflavodoxin, the purity of a number of FMN derivatives is greatly improved by treatment of the flavin with immobilised apoflavodoxin (Table II). The purified flavins suffer no detectable deterioration during several months of storage in frozen solution at −20°C or as a freeze-dried powder at −20°C or 4°C; after storage of the dry powder at 25°C for 1 month, 1.5% of the purified FMN failed to bind to apoflavodoxin. The spectroscopic and thin-layer chromatographic properties of the flavin fraction of FMN (12%) which is not bound to immobilised apoflavodoxin have so far proved to be indistinguishable from those of the flavin, which is bound.

The presence of residual impurity in electrophoretically or chromatographically purified FMN was first noted 20 years ago by Theorell and Nygaard [20, 21]; they observed a rest fluorescence of 11% when the purified flavin was treated with apo-old yellow enzyme. This impurity has been largely ignored by later workers, undoubtedly because a convenient method for its removal was not known. Affinity chromatography with immobilised apoflavodoxin may rectify this situation by permitting the preparation of purified riboflavin 5′-phosphate on a large scale.

Although *P. elsdenii* apoflavodoxin was used in this work, apoflavodoxins from *Clostridium pasteurianum* and *Clostridium MP* are similarly specific [8]. In contrast, apoflavodoxins from *A. vinelandii* and *Desulfovibrio vulgaris* are less suitable for the determination and purification of FMN since they also bind FAD and riboflavin [22]. The large quantities of flavodoxin required for preparation of the immobilised apoprotein can be readily obtained from *P. elsdenii*; this organism contains almost 10 mg of flavodoxin per g of dry cell material [10].

TABLE II

PURIFICATION OF FMN AND FMN ANALOGUES BY AFFINITY CHROMATOGRAPHY WITH IMMOBILISED APOFLAVODOXIN

Affinity chromatography was carried out on a column of immobilised apoflavodoxin. The column material was prepared by mixing flavodoxin with activated Sepharose 4B (200 mg CNBr per ml gel) (reference 16 and Strating and Mayhew [25]) in 0.1 M sodium carbonate buffer pH 10.5, at 4°C, to give a final protein concentration of 12 mg/ml. After 16 h with gentle stirring, the gel was made into a column, washed with 10 mM potassium phosphate pH 7, plus 0.3 mM EDTA to remove unreacted protein, and then washed with 5% trichloroacetic acid plus 0.3 mM EDTA to remove FMN from the bound protein. As soon as all of the flavin was removed, the column was washed with a mixture of 0.1 M potassium phosphate buffer pH 7, 0.3 mM EDTA and 0.1 mM dithiothreitol (added as a precaution when the immobilised protein is to be stored in the apoprotein form). The impure flavin derivative was added to the column in 0.1 M potassium phosphate buffer pH 7, 0.3 mM EDTA. After about 1 h the column was washed with 0.01 M phosphate buffer, pH 7, plus 0.3 mM EDTA to remove unbound flavin, and the bound flavin then eluted with 5% trichloroacetic acid as described above. The acid was removed from the solution of flavin by extraction (4 times) with a 10-fold excess of ether, and the solution finally adjusted to pH 7 with solid Na_2CO_3. DEAE-cellulose chromatography was performed according to Massey and Swoboda [15]. After purification, each flavin was subjected to fluorimetric titration with apoflavodoxin as described in Fig. 4. The results are expressed as residual fluorescence i.e. the fluorescence remaining at the end of the titration in the presence of a 2-fold molar excess of apoprotein.

Treatment	FMN derivative				
	FMN(Sigma)	3-CH_3-	3-CH_2COO^--	iso-	8-OH-
None	20.5	30	31	15	25
DEAE-cellulose chromatography	13	23	21	7	—
Affinity chromatography	0.7	1	3	0.6	0.6

Acknowledgements

We thank W. Smith and M.L. Ludwig for their gift of ferredoxin-NADP$^+$ reductase, P. Hemmerich for his gift of FMN analogues, J. Visser for helpful advice and B.J. Sachteleben for drawing the figures.

References

1 Mayhew, S.G., Foust, G.P. and Massey, V. (1969) J. Biol. Chem. 244, 803–810
2 Edmondson, D.E. and Tollin, G. (1971) Biochemistry 10, 133–145
3 Bothe, H., Hemmerich, P. and Sund, H. (1971) in Flavins and Flavoproteins (Kamin, H., ed.), pp. 211–226, University Park Press, Baltimore
4 Cusanovich, M.A. and Edmondson, D.E. (1971) Biochem. Biophys. Res. Commun. 45, 327–336
5 Mayhew, S.G. (1971) Biochim. Biophys. Acta 235, 276–288.
6 Van Lin, B. and Bothe, H. (1972) Arch. Mikrobiol. 82, 155–172
7 Mayhew, S.G. and Massey, V. (1973) Biochim. Biophys. Acta 315, 181–190
8 Mayhew, S.G. (1971) Biochim. Biophys. Acta 235, 289–302
9 Mayhew, S.G. and Ludwig, M.L. (1975) in The Enzymes (Boyer, P.D., ed.), Vol. 12, 57–118
10 Mayhew, S.G. and Massey, V. (1969) J. Biol. Chem. 244, 794–802
11 Barman, B.G. and Tollin, G. (1972) Biochemistry 11, 4755–4759
12 Burton, K. (1957) Ergeb. Physiol. 49, 275–298
13 Faeder, E.J. and Siegel, L.M. (1973) Anal. Biochem. 53, 332–336
14 Mayhew, S.G. (1971) in Flavins and Flavoproteins (Kamin, H., ed.), pp. 185–205 and following discussion pp. 205–209, University Park Press, Baltimore
15 Massey, V. and Swoboda, B.E.P. (1963) Biochem. Z. 338, 474
16 Cuatrecasas, P. and Anfinsen, C.B. (1971) in Methods in Enzymology (Jakoby, W.B., ed.), Vol. 22, pp. 345–378, Academic Press, New York
17 Chen, J.-S. and Mortensen, L.E. (1974) Biochim. Biophys. Acta 371, 283–298
18 Uyeda, K. and Rabinowitz, J.C. (1971) J. Biol. Chem. 246, 3111–3119
19 Foust, G.P., Mayhew, S.G. and Massey, V. (1969) J. Biol. Chem. 244, 964–970
20 Theorell, H. and Nygaard, A.P. (1954) Acta Chem. Scand. 8, 877–887
21 Theorell, H. and Nygaard, A.P. (1954) Acta Chem. Scand. 8, 1649–1658
22 D'Anna, Jr., J.A. and Tollin, G. (1972) Biochemistry 11, 1073–1080
23 Yates, M.G. (1972) FEBS Lett. 27 63–67
24 Wassink, J.H. and Mayhew, S.G. (1975) Anal. Biochem. 68, 609–616
25 Mayhew, S.G. and Strating, M.J.J. (1975) Eur. J. Biochem. 59, 539–544

Flavins and Flavoproteins, edited by T.P. Singer

CHAPTER 44

STUDIES ON THE MECHANISM OF ELECTRON TRANSFER IN FLAVODOXINS

KIYOSHI SHIGA and GORDON TOLLIN

Department of Chemistry, University of Arizona, Tucson, Ariz. 85721 (U.S.A.)

Summary

One consequence of the recent X-ray crystallographic studies of the flavodoxins is that detailed questions can now be asked concerning the mode of entry and exit of electrons in these proteins. A clear possibility which is suggested by the structural information, and supported by our knowledge of flavin chemistry, is that the dimethylbenzene ring plays an important role in this process. We have carried out some experiments designed to test this hypothesis. Our approach has been to modify chemically this portion of the flavin structure, to bind such modified coenzymes to apoflavodoxin and to investigate the redox properties of the analog proteins. The present studies involve the 7,8-dichloro analogs of FMN and riboflavin. These bind quite well to apoflavodoxins, with affinity constants and spectral properties closely similar to those of the normal flavins. The redox properties are modified as follows. Photoreduction of the dichloro-FMN flavodoxins proceeds much faster (300–1000 times) than with the normal proteins. Similarly, O_2 oxidation of the semiquinone forms is greatly enhanced in rate (50–300 times). On the other hand, ferricyanide and cytochrome *c* oxidations of the semiquinone forms of the dichloro analog proteins are greatly reduced in rate (30–100 times). Experiments with free dichloro-flavins show that chlorine substitution produces only a small perturbation of reaction rates. Thus, these large effects with the flavodoxins must be a consequence of the flavin-protein interaction. The results of these studies are consistent with the hypothesis that semiquinone oxidation in the flavodoxins by ferricyanide and cytochrome *c* involves intimately the dimethylbenzene portion of the flavin ring system.

Introduction

It is well known that many of the chemical properties of free flavins are substantially modified as a consequence of the flavin-protein interactions which occur in flavoenzymes. This is not, however, well understood at the molecular level. One way to approach this problem is to compare the properties of free flavins with those of flavin bound to apoenzyme. Recently, we have studied the kinetics of the reactions of the semiquinone forms of free flavins with oxygen, ferricyanide and cytochrome *c* [1]. X-ray crystallographic investigations of two flavodoxins [2, 3] have now produced a detailed picture of the flavin environment in these flavoenzymes. Thus, it was felt that similar kinetic studies with the flavodoxins would allow structure/function correlations to be made. Furthermore, the X-ray work has suggested that the dimethylbenzene portion of the flavin ring system might be intimately involved in the mechanism of electron transfer, and this possibility is consistent with our knowledge of flavin chemistry [1]. In view of this, we have also investigated the effects of replacing the 7- and 8-position methyl groups in the flavin structure by chlorine on the redox properties of the flavodoxins.

The flavodoxins chosen for these studies were from *Azotobacter vinelandii* (*A.v.*), *Clostridium pasteurianum* (*C.p.*) and *Desulfovibrio vulgaris* (*D.v.*). The *A.v.* and *D.v.* proteins are known to be similar in many respects [4, 5], and the X-ray structure work has been performed on the *D.v.* and *Clostridium MP* flavoproteins [2, 3] (we shall assume that the *C.p.* and *C.MP* proteins are identical structurally). The results of the present experiments are interpreted in the light of X-ray crystallography as being consistent with the hypothesis that flavin-semiquinone oxidation in the flavodoxins by ferricyanide and cytochrome *c* intimately involves the relatively exposed dimethylbenzene ring of the flavin.

Experimental section

The flavodoxins from *Azotobacter vinelandii* and *Clostridium pasteurianum* were purified and apoproteins were obtained by previously published methods [6–9]. The *Desulfovibrio vulgaris* flavodoxin was a gift from Drs. M. Dubourdieu and J. LeGall. Protein concentrations were determined spectrophotometrically with a Coleman-Hitachi Model 124 spectrophotometer using the absorption coefficients already published [6–10]. Riboflavin and equine heart cytochrome *c* were obtained from Calbiochem and used without further purification. 7,8-Dichlororiboflavin and 7,8-dichloro-FMN were synthesized according to published procedures [11, 12] and purified by DEAE-cellulose chromatography [13]. Both flavins gave single spots on thin-layer chromatography using Eastman 6065 cellulose plates and *n*-butanol acetic acid water (2 :1:1 v/v) as solvent. The absorption spectra were similar to that previously obtained [14] for 7,8-dichloro-10-ethylisoalloxazine. We have determined the absorption coefficient of the dichloroflavins to be $1.11 \cdot 10^4\ M^{-1} \cdot cm^{-1}$ at 442 nm.

The neutral semiquinones of the analog flavoproteins were made by photo-irradiation in the presence of 0.05 M EDTA [15] under anaerobic conditions and the rate of semiquinone appearance (the rate of photo-reduction) was measured spectrophotometrically. In the case of the analog proteins, excess apoprotein (2–4 fold) was added to minimize the free flavin concentration. On the basis of the known association constant for riboflavin and *A.v.* apoflavodoxin [8, 16] and control experiments, it was determined that little or no contribution to the rate was made by free flavin under the conditions of our experiments. The concentrations of the semiquinone forms of the flavodoxins were determined using published absorbance values [6, 7, 10]. In the case of the analog protein complexes, i.e. Rib-*A.v.*, Cl_2Rib-*A.v.*, Cl_2FMN-*A.v.*, and Cl_2FMN-*C.p.*, values of ϵ_{580} = 5132, 5029, 4926 and $\epsilon_{575} = 4576\ M^{-1} \cdot cm^{-1}$ respectively were used. These are based on the assumption that the ratio of the absorption peak at about 445 nm of the oxidized flavin-protein complex and that at about 580 nm of the semiquinone form is the same as the ratio in the normal protein complex. Although this approximation may not be totally correct, it is adequate for the present study. We have determined the absorption coefficients of the oxidized Rib-, Cl_2Rib-, Cl_2FMN-*A.v.* and Cl_2FMN-*C.p.* proteins to be $\epsilon_{450} = 1.00 \cdot 10^4$, $\epsilon_{448} = 9.80 \cdot 10^3$, $\epsilon_{446} = 9.60 \cdot 10^3$ and $\epsilon_{445} = 1.04 \cdot 10^4\ M^{-1} \cdot cm^{-1}$ respectively.

The reaction rates between oxygen and semiquinones were measured by monitoring spectrophotometrically the decrease of the concentration of radical under air-saturated conditions as previously described [17]. The results were analyzed by obtaining pseudo-first order rate constants and then calculating second order rate constants using an oxygen concentration of $2 \cdot 10^{-4}$ M (at 35°C).

The reaction rates between semiquinones and ferricyanide or cytochrome *c* were measured in the Gibson-Durrum apparatus or in the Coleman-Hitachi spectrophotometer under anaerobic conditions by monitoring the decrease of semiquinone in the presence of excess ferricyanide or cytochrome *c*. Second order rate constants were obtained by dividing the pseudo-first order rate constants by the concentration of oxidant. It should be noted that in these oxidation reaction measurements, only a part of the oxidized protein was converted into the semiquinone form by photo-irradiation, and therefore the solutions contained mixtures of oxidized and semiquinone forms. However, no fully reduced protein was present.

CD spectra were measured with a Cary Model 60 spectropolarimeter equipped with a Model 6001 circular dichroism attachment. The binding of dichloroflavin to apoflavodoxin was measured fluorimetrically as previously described [8].

Results

1. Photochemical production of semiquinones of flavoproteins and their stability to dithionite reduction

The fact that light irradiation of flavoproteins in the presence of EDTA under anaerobic conditions produces semiquinone is well known [15]. In the case of the native *A.v.*, *D.v.* and *C.p.* flavodoxins, results similar to previously published data [6, 7, 13] were obtained. With the Rib-, Cl_2Rib-, Cl_2FMN-*A.v.* and Cl_2FMN-*C.p.* proteins, a normal neutral semiquinone was obtained. A typical experiment is shown in Fig. 1. As with the normal proteins, the absorption around 445 nm decreased and that around 580 nm increased upon photo-irradiation. It was noted that there were isosbestic points during this conversion from oxidized to semiquinone form, which shows clearly that the solution contained only oxidized and semiquinone forms, and not the fully reduced form. When air was mixed into the semiquinone solutions, the spectra changed to the original form keeping the isosbestic points.

When the solution contains only the oxidized and semiquinone forms, the following formula applies:

$$\Delta A_{580} = \frac{\epsilon_{580}^{semiq}}{\epsilon_{450}^{ox} - \epsilon_{450}^{semiq}} \, |\Delta A_{450}|$$

In this equation, ΔA_{450} and ΔA_{580} are the changes in absorption during

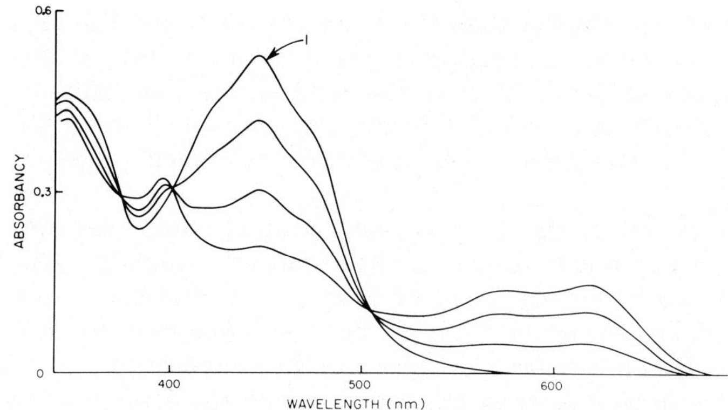

Fig. 1. Photoreduction of Cl_2FMN-*A.v.* flavodoxin in 0.025 M phosphate buffer pH 7.0, 25°C containing 0.05 M EDTA. Curve 1 is fully oxidized. Other curves represent intermediate stages of reduction. Concentration of $Cl_2FMN = 5.5 \cdot 10^{-5}$ M; concentration of apoprotein $= 9.5 \cdot 10^{-5}$ M.

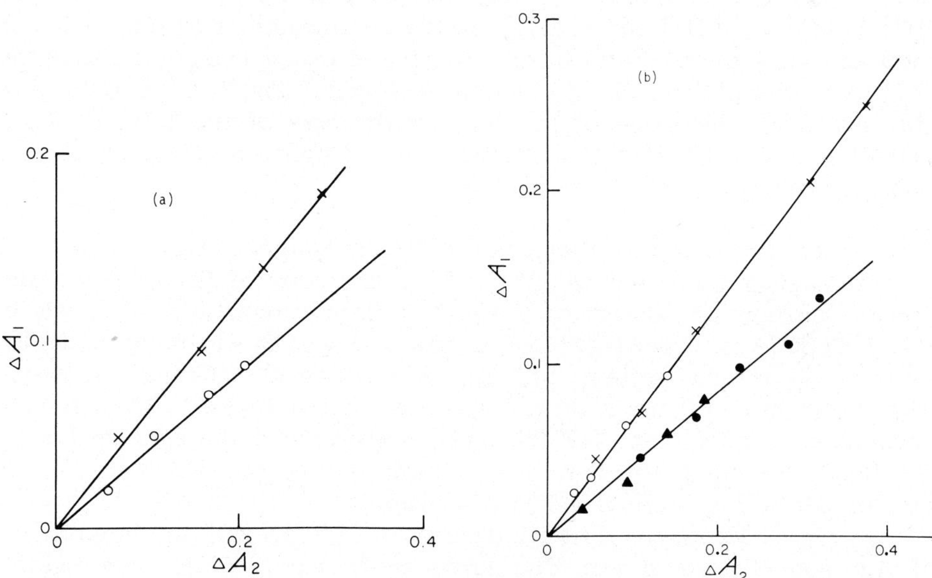

Fig. 2. Plot of changes in absorption at two different wave lengths during photochemical reduction of flavoproteins. Experimental conditions same in Fig. 1. (a) (○) Cl_2FMN-*C.p.*, $\lambda_1 = 575$ nm and $\lambda_2 = 443$ nm. (X) *C.p.*, $\lambda_1 = 575$ nm and $\lambda_2 = 443$ nm. (b) (X) Rib-*A.v.*, $\lambda_1 = 580$ nm, $\lambda_2 = 450$ nm. (●) Cl_2FMN-*A.v.*; $\lambda_1 = 580$ nm, $\lambda_2 = 448$ nm. (○) *A.v.*, $\lambda_1 = 580$ nm, $\lambda_2 = 450$ nm. (▲) Cl_2Rib-*A.v.*, $\lambda_1 = 580$ nm, $\lambda_2 = 448$ nm.

photochemical reduction at these wavelengths. ϵ_{580}^{semiq} and ϵ_{450}^{semiq} are the absorption coefficients of the semiquinone forms at these wavelengths. ϵ_{450}^{ox} is the absorption coefficient of oxidized form at this wavelength. In Fig. 2, plots of this relation are shown for several flavodoxins. The agreement is again consistent with the lack of full reduction. As Fig. 2 shows, the slopes of the plots for the Cl_2FMN and Cl_2Rib protein complexes, both for the

A.v. and *C.p.* flavodoxins, are smaller than those for the FMN and Rib protein complexes. This could be due to smaller values of the absorption coefficients of the semiquinones of the dichloro protein complexes. The values of the slopes of the plots for the *A.v.* and *C.p.* flavodoxins were 0.67 and 0.60, respectively. These can be compared with previously published values of 0.68 [6] and 0.55 [7].

As mentioned above, all of the dichloroflavin protein complexes produce typical neutral semiquinones during photochemical reduction. This might indicate that the flavin moiety in all of these analog proteins is in a similar environment. Such an assumption is further supported by CD spectroscopy and flavin-protein titration, as shown in the next section.

It has been shown previously that the reduction of the semiquinone forms of flavodoxins by excess (1 mg/ml) dithionite is controlled only by pH and not by the concentration of dithionite [7, 17]. Even large excesses of dithionite will not reduce the semiquinone to the fully reduced form when the pH is low. Apparent pK values for this interconversion were obtained (pK = 7 for *A.v.* and pK = 6.6 for *D.v.*) by plotting $[PFH_2]/[PFH\cdot]$ vs. $[H^+]^{-1}$, where $[PFH_2]$ and $[PFH\cdot]$ are the concentrations of fully reduced form and semiquinone form in the presence of excess (1 mg/ml) dithionite [17]. A similar plot for *C.p.* flavodoxin is shown in Fig. 3. A pK of 6.2 was obtained from the slope of the line. In the case of the Rib-, Cl_2Rib-, Cl_2FMN-*A.v.* and Cl_2FMN-*C.p.* proteins, no resistance to dithionite reduction was observed from pH 7 to pH 5.3.

2. CD spectroscopy and titration of Cl_2FMN and apoflavodoxins

As already published [4, 18], CD-spectroscopy of flavoproteins can provide information concerning the protein-flavin interaction. As shown in Fig. 4, Cl_2FMN-*A.v.* and Cl_2FMN-*C.p.* have CD spectra which are similar to those of the normal proteins [4, 18], although small differences do exist. Also shown in this figure is the CD spectrum of free Cl_2FMN. These results demonstrate clearly that Cl_2FMN binds specifically to the apoflavodoxins and that these complexes have protein-flavin interactions which are similar but not identical to those of the native flavoprotein.

In Fig. 5, fluorescence titration curves of Cl_2FMN with the apoproteins of *A.v.* and *C.p.* are shown. The curves show that Cl_2FMN binds tightly ($K_a > 10^8\ M^{-1}$) to the apoproteins to give a 1:1 complex. This is consistent with a specific mode of protein-flavin binding.

3. Rate of semiquinone appearance by photo-irradiation and of reduction of oxidized form by dithionite

As stated in Section 1, semiquinone forms of flavodoxins can be made by photo-irradiation with EDTA under anaerobic conditions. A first-order plot of semiquinone appearance vs. irradiation time is a straight line. The half time of photoreduction can be obtained from these plots and is given in Table I. The rate of *A.v.* photoreduction is much slower than for the other flavodoxins, as well as for the analog proteins. Similarly, the Cl_2FMN analog

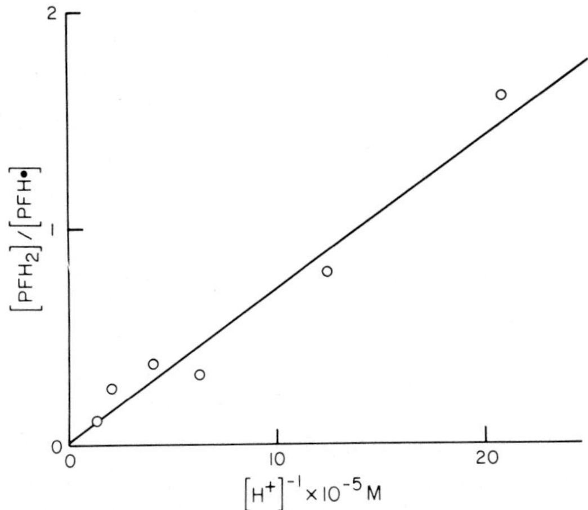

Fig. 3. Relation between the ratio of the concentration of fully reduced form to that of semiquinone of *C.p.* flavodoxin and the reciprocal of $[H^+]$ in the presence of excess dithionite (1 mg/ml). Buffer was 0.05 M phosphate acetate at each pH; temperature was 25°C and the concentration of *C.p.* flavodoxin was $6.5 \cdot 10^{-5}$ M.

of *C.p.* flavodoxin photoreduces considerably faster than the normal protein. Note also that, with free flavin, chlorine substitution acts to slow down the photoreduction rate. Thus, removal of the ribityl phosphate or replacement of the ring methyl groups by chlorine must lead to changes in the nature of the flavin-protein interaction.

4. The reaction of semiquinone with oxygen

The reaction rates of semiquinone forms with oxygen were determined under conditions of air saturation by measuring the decrease of semiquinone spectrophotometrically. Second order rate constants were obtained by dividing the pseudo-first order constants by the concentration of oxygen ($2 \cdot 10^{-4}$ M, at 35°C) and are shown in Table I. Again, the rate of the *A.v.* reaction is much slower than those of the *D.v.* and *C.p.* flavodoxins. The rates of the Cl_2FMN-, Cl_2Rib- and Rib-*A.v.* complexes are faster than that of the native protein, and an identical pattern is observed upon chlorine substitution with *C.p.* flavodoxin. Although the rates of reaction of the neutral semiquinones of free flavins with oxygen are too slow to be measured, chlorine substitution is essentially without effect on the rate of the anion radical reaction.

5. The reaction of semiquinone with ferricyanide and cytochrome c

Semiquinone forms of the flavodoxins were reacted with excess ferricyanide and cytochrome *c* and the reaction was followed spectrophotometrically, measuring the decrease of semiquinone. To determine that the reaction was indeed pseudo-first order, and that no complications such as complex formation existed, we studied the dependence of the reaction rate on the oxidant concentration. Fig. 6a and b show the relation between

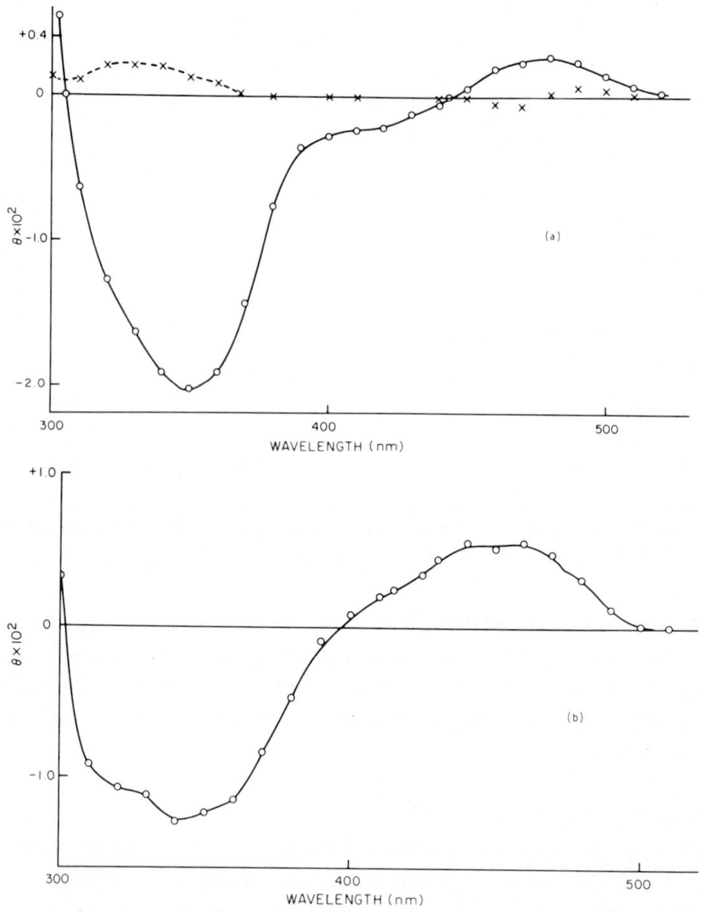

Fig. 4. CD spectra of Cl_2FMN-*A.v.*, Cl_2FMN-*C.p.* and free Cl_2FMN. Spectra were measured in 0.025 M pH 7.0 phosphate buffer at room temperature. (a) (○) Cl_2FMN-*A.v.* ($5.3 \cdot 10^{-5}$ M flavin and $6.9 \cdot 10^{-5}$ M protein) (X) Cl_2FMN ($5.3 \cdot 10^{-5}$ M). (b) Cl_2FMN-*C.p.* ($5.4 \cdot 10^{-5}$ M flavin and $6.6 \cdot 10^{-5}$ M protein).

the reciprocal of the half-fime and the concentration of ferricyanide or cytochrome *c* for the reactions of *C.p.* and *A.v.* flavodoxin semiquinones. The linearity of the plots is good evidence for the absence of complicating factors, at least over the concentration range investigated. In Table I, rate constants for the various reactions are given.

In the case of the reaction with ferricyanide, it should be noted that the rates of the *A.v.* and *D.v.* reactions are identical within experimental error, but that for *C.p.* is about ten times higher. Note also that the rate of Cl_2FMN-*A.v.* is considerably slower than that of the normal protein, whereas that for Rib-*A.v.* is significantly faster. A similar effect of chlorine substitution is observed for the *C.p.* flavodoxin.

In the cytochrome *c* reaction, the rates of the three native flavodoxins are all quite similar to each other, although the rate of the *C.p.* reaction is

TABLE I

RATES OF VARIOUS FLAVODOXIN REDOX REACTIONS

Estimated errors of the rates are as follows: photoreduction ≈ 30%; O_2 oxidation ≈ 20%; dithionite reduction, ferricyanide oxidation and cytochrome oxidation ≈ 15%. Fld represents Flavodoxin.

Flavin derivative	Photoreduction to $FH\cdot$ ($\tau_{1/2}$)	Reaction			
		$FH\cdot + O_2$		$Fe(CN)_6^{-3} + FH\cdot$	Cytochrome c + $FH\cdot$
		$\tau_{1/2}$	k $(M^{-1}\cdot s^{-1})$	k $(M^{-1}\cdot s^{-1})$	k $(M^{-1}\cdot s^{-1})$
Azotobacter Fld. (FMN)	457 min	700 min	$8.3\cdot10^{-2}$	$8.3\cdot10^{3}$	$1.1\cdot10^{5}$
Azotobacter Fld. (Rb)	25 s	2 min	29	$2.0\cdot10^{4}$	—
Azotobacter Fld. (Cl_2Rb)	33 s	2 min	29	—	—
Azotobacter Fld. (Cl_2FMN)	24 s	7 min	8.3	$1.4\cdot10^{2}$	$1.2\cdot10^{3}$
C. pasteurianum Fld. (FMN)	46 min	18 min	3.2	$1.1\cdot10^{5}$	$2.3\cdot10^{5}$
C. pasteurianum Fld. (Cl_2FMN)	9 s	22 s	160	$3.0\cdot10^{3}$	$1.9\cdot10^{3}$
D. vulgaris Fld. (FMN)	78 min	10 min	5.8	$8.3\cdot10^{3}$	$1.7\cdot10^{5}$
Lf	< 1 s	≈ 10^{-3} s	$< 10^{4}$* $3\cdot10^{8}(F^{-}\cdot)$* $< 10^{4}$**	$6.3\cdot10^{8}$**	$< 6\cdot10^{6}$*
Cl_2Lf	≈ 1 s	—	$3\cdot10^{8}(F^{-}\cdot)$**	$2.1\cdot10^{8}$**	—

* Vaish, S.P. and Tollin, G. [1].
** Tollin, G. and Rizzuto, F. [21].

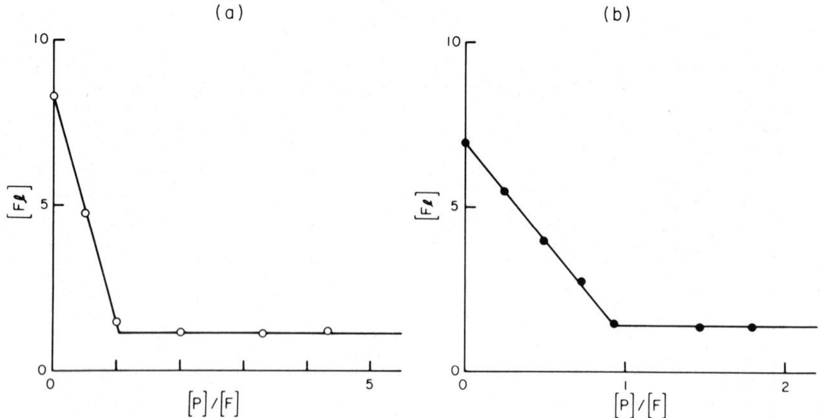

Fig. 5. Fluorescence titration curves of Cl_2FMN with apoproteins of *A.v.*, or *C.p.* flavodoxins. The ordinate is the fluorescence intensity in arbitrary units and the abscissa is the ratio of the concentration of protein to flavin. Excitation and emission were at 360 and 543 nm, respectively, in pH 7.0, 0.025 M phosphate buffer at 25°C. (a) *A.v.*, flavin concentration = $4.6 \cdot 10^{-7}$ M. (b) *C.p.*, flavin concentration = $6.7 \cdot 10^{-7}$ M.

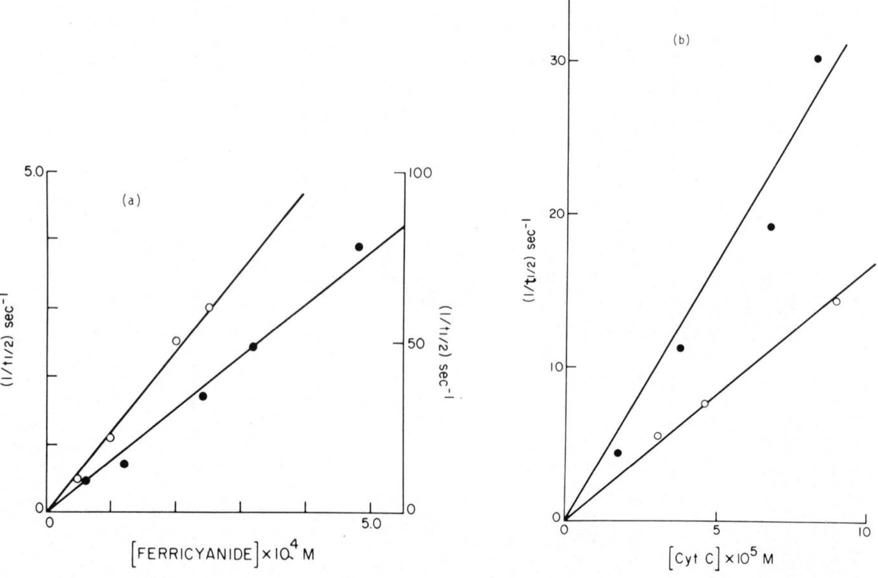

Fig. 6. The relation between the reciprocal of the half time of the reaction of ferricyanide or cytochrome *c* with semiquinones of *A.v.*, or *C.p.* flavodoxins and the concentration of ferricyanide or cytochrome *c*. Reactions run in 0.025 M phosphate buffer, pH 7.0, containing 0.025 M EDTA at 20°C. (a) (○) *A.v.* flavodoxin plus ferricyanide. Ordinate is left. (●) *C.p.* flavodoxin plus ferricyanide. Ordinate is right. (b) (○) *A.v.* flavodoxin plus cytochrome *c*. (●) *C.p.* flavodoxin plus cytochrome *c*.

slightly higher than the others. Again, chlorine substitution greatly reduces the reaction rates.

It is important to point out that, in the free flavins, replacement of the 7,8-methyl groups by chlorine causes a much smaller decrease in the rate of ferricyanide oxidation of the semiquinone than is observed with the flavoproteins. Thus, again, protein-flavin interactions are playing an important role in determining relative reaction rates.

Discussion

Photoreduction of flavin with EDTA undoubtedly involves the region of the molecule near the N-1, N-5 double bond system. Thus, it is not surprising that the flavoproteins are reduced more slowly than free flavin, considering the fact that this portion of the coenzyme is deeply buried within the protein cavity [2, 3, 17]. Furthermore, the generally large extent of the flavin-protein interaction, and in particular the proximity of aromatic amino acid side chains, should lead to a high level of quenching of flavin-excited states. This, of course, is shown quite clearly by the essentially complete quenching of flavin fluorescence upon binding.

In view of the above considerations, it is quite interesting that replacement of FMN by riboflavin or dichloroflavins greatly increases the photoreduction rate. However, it is rather difficult to rationalize this, inasmuch as it is not clear which factors are crucial in determining this rate. It is known that, for free flavin, it is the lowest triplet state which is photoreactive [19]. Thus, one important determinant may be the extent of population of this state. Chlorine substitution would be expected to enhance intersystem crossing into the triplet state, and also to reduce the triplet state lifetime. It is conceivable that the former property could account for the rate enhancement in the flavoproteins, whereas the latter property leads to a slower rate of photoreduction in the free flavins. An increase in the triplet state population might be more significant in a situation, such as exists in the flavoproteins, in which there is extensive quenching at the singlet level and hence a very low concentration of triplet states. In the case of riboflavin, there is evidence [20] which indicates that the presence of a ribityl phosphate group causes changes in the conformation of the apoprotein upon flavin binding. Thus, one could attribute the faster rate of riboflavin analog protein photoreduction to as yet unknown differences in the nature of the flavin-protein interaction.

Another possibility which needs to be noted, and which relates to the other reactions as well, is that replacement of methyl groups by larger chlorine atoms could in principle cause steric interference between the flavin ring and the protein and thus lead to modified flavin-protein interactions. However, the X-ray results (refs. 2, 3 and personal communication from Ludwig, M.L. and Watenpaugh, K.) indicate that this is unlikely inasmuch as the 7,8-methyl groups are quite far from any protein atoms.

It is clear that additional studies are needed to resolve this problem. For example, structural and spectroscopic studies of riboflavin and dichloroflavin analog proteins would be highly informative.

It is interesting to note that for photoreduction, the *D.v.* flavodoxin has a much shorter half time than does the *A.v.* flavodoxin. This might be a consequence of whatever differences exist between these two proteins in the nature of the flavin environment. Obviously, structural studies of the *A.v.* protein would be most informative.

Effects similar to those found for photoreduction are observed in the case of the reaction of semiquinone with oxygen. The reaction of free flavin semiquinone with oxygen is known to proceed via the anionic form [1]. The rates of reaction of all of the protein semiquinones with oxygen are much slower than that of free flavin at pH 7. This is not unreasonable when we consider that oxygen probably reacts at or near the N-5 position of flavin and that this position is deeply buried within the protein cavity. Furthermore, a second factor in the low reactivity of the flavoproteins is the stabilization of the neutral semiquinone by the protein. Thus, the p*K* for semiquinone ionization in *A.v.* flavodoxin is increased by at least three pH units [17].

The faster rates of reaction of the semiquinones of Rib-, Cl_2Rib- and Cl_2FMN-*A.v.* with oxygen may also be rationalized in terms of protein conformation differences and semiquinone p*K* shifts. Since the riboflavin and dichlororiboflavin analog proteins react with oxygen at similar rates, replacement of CH_3 by Cl at the 7,8 position of flavin probably does not directly change the reactivity to oxygen. This is consistent with the results obtained with free flavins [21]. However, chlorine substitution does shift the p*K* of the flavin semiquinone to lower values (Tollin, G. and Rizzuto, F., unpublished results). Thus, if one assumes that the radical p*K* values for both the riboflavin and dichloroflavin protein semiquinones are lower than those of the native proteins, the oxygen reactivity results are easily rationalized in terms of a higher equilibrium concentration of the anionic semiquinone. In the riboflavin case, this would be due to differences in protein conformation [20] and in the Cl_2FMN case to the electronegativity of the chlorine atoms. It is not immediately clear, however, why there is not a further rate increase in going from the Rib- to the Cl_2Rib-*A.v.* protein, unless accessibility to the reactive site on the flavin becomes rate limiting in the riboflavin analogs.

As was the case with photochemical reduction, the rate of reaction of the *A.v.* semiquinone with oxygen is much slower than that of the *D.v.* and *C.p.* flavodoxins. This difference must again reflect the nature of the protein-flavin interactions.

The most striking results obtained in these studies are the totally opposite patterns observed for ferricyanide and cytochrome *c* oxidation of the protein semiquinones. Thus, all three native proteins react at equal or nearly equal rates with these oxidants, chlorine substitution causes a large decrease in rates of reaction, and removal of the ribityl phosphate group (in the case of ferricyanide oxidation of the *A.v.* analog protein) causes only a small increase in reaction rate. These results can be easily rationalized if one assumes that the reactions with these oxidants directly involve the dimethylbenzene portion of the flavin ring. Inasmuch as this is relatively exposed in the native protein [2, 3, 17], differences in protein-flavin interactions caused by a change in the apoprotein in the various flavodoxins, or by removal of the ribityl phosphate group in going from FMN to riboflavin, might be expected to have only minor consequences in this region, thus accounting for the rather small rate effects noted. On the other hand, the highly electronegative chlorine atoms would act to decrease the unpaired electron density in the benzene ring of the flavin and thus make it less reactive to an oxidant which interacts with this region. However, chlorine substitution in free flavins has only a relatively small effect on the rate of ferricyanide oxidation of the semiquinone [21]. Thus, some effect on the flavin-protein interaction must also be occurring in the chloroflavin analogs of the flavodoxins. It is also significant that the rate of semiquinone oxidation by cytochrome *c* is hardly affected by incorporating the flavin into the protein. Since the cytochrome is such a large molecule, it would most probably be unable to penetrate very deeply into the flavin-binding site, and thus would react with the most exposed region of the flavoprotein.

It is clear that, although further studies are desirable, the results of the present work are consistent with the information obtained from X-ray crystallography of the flavodoxins, and are at least suggestive that protein semiquinone oxidation by ferricyanide and cytochrome *c* involves the dimethylbenzene portion of the flavin ring system.

Acknowledgements

This work was supported by an N.I.H. Research Grant (AM 15057).

References

1 Vaish, S.P. and Tollin, G. (1971) Bioenergetics 2, 61–72
2 Watenpaugh, K.D., Sieker, L.C. and Jensen, L.H. (1973) Proc. Natl. Acad. Sci. U.S. 70, 3857–3860
3 Burnett, R.M., Darling, G.D., Kendall, D.S., LeQuesne, M.E., Mayhew, S.G., Smith, W.W. and Ludwig, M.L. (1974) J. Biol. Chem. 249, 4383–4392
4 D'Anna, J.A. and Tollin, G. (1972) Biochemistry 11, 1073–1080
5 MacKnight, M.L., Gray, W.R. and Tollin, G. (1974) Biochem. Biophys. Res. Commun. 59, 630–637
6 Hinkson, J.W. and Bulen, W.A. (1967) J. Biol. Chem. 242, 3345–3351
7 Mayhew, S.G. (1971) Biochim. Biophys. Acta 235, 276–288
8 Edmondson, D.E. and Tollin, G. (1971) Biochemistry 10, 124–132
9 Mayhew, S.T. (1971) Biochim. Biophys. Acta 235, 289–302
10 Dubourdieu, M. and Legall, J. (1970) Biochem. Biophys. Res. Commun. 38, 965–972
11 Lambooy, J.P., Scala, R.A. and Haley, E.E. (1961) J. Nutr. 74, 466–472
12 Flexser, L.A. and Farkas, W.G. (1952) U.S. Patent 2, 610–611
13 Massey, V. and Swohoda, B.E.P. (1963) Biochem. Z. 338, 474–484
14 Guzzo, A.V. and Tollin, G. (1963) Arch. Biochem. Biophys. 103, 231–243
15 Massey, V. and Palmer, G. (1966) Biochemistry 5, 3181–3189
16 Shiga, K. and Tollin, G. (1974) Biochemistry 13, 3268–3273
17 Edmondson, D.E. and Tollin, G. (1971) Biochemistry 10, 133–145
18 Edmondson, D.E. and Tollin, G. (1971) Biochemistry 10, 113–123
19 Vaish, S.P. and Tollin, G. (1970) Bioenergetics 1, 181–192
20 Barman, B.G. and Tollin, G. (1972) Biochemistry 11, 4755–4759
21 Tollin, G. and Rizzuto, F. (1975) Reactivity of Flavins (Yagi, K., ed.), p. 25, Univ. Tokyo Press

Flavins and Flavoproteins, edited by T.P. Singer

CHAPTER 45

PROTON MAGNETIC RESONANCE OF *DESULFOVIBRIO VULGARIS* AND *DESULFOVIBRIO GIGAS* FLAVODOXINS

VINCENT FAVAUDON[a], JEAN LE GALL[b] and JEAN-MARC LHOSTE[c,*]

[a] *Laboratoire de Biophysique, Muséum National d'Histoire Naturelle, Paris;* [b] *Laboratoire de Chimie Bactérienne, C.N.R.S., Marseille;* [c] *Fondation Curie-Institut du Radium, Section de Biologie, 91405-ORSAY (France)*

Summary

The structure of two flavodoxins from sulfate reducing bacteria appears to behave similarly upon oxidoreduction. The flavoquinone forms are more related to the deflavoproteins than to the flavohydroquinone or semiquinone forms which exhibit close structural resemblance. Most resonances characteristic of the native structures do not involve the FMN prosthetic group, either directly or indirectly. Two resonances assigned to the isoalloxazine ring indicate that this group is in a very similar environment in both proteins.

The proton magnetic resonance spectra of two flavodoxins from sulfate-reducing bacteria, *Desulfovibrio vulgaris* and *Desulfovibrio gigas*, have been recorded at 100 MHz in the fully oxidized and reduced forms as well as in the pure semiquinone form of these FMN containing electron transfering proteins.

The PMR spectra of the oxidized flavodoxins [1, 2], recorded in the continuous wave or Fourier transform modes on a Varian XL 100 Spectrometer, have been obtained at 34°C using proteins lyophilized once in ^2H_2O (Fig. 1). In such conditions, a large amount of the peptidic protons appearing in the low field range are not exchanged in 99.7% ^2H_2O, 0.06 M phosphate buffer (p^{2}H 8.4). In the aliphatic range the methyl peak around one ppm (Phe, Val, Leu) and the resonances at 2.0 to 2.2 ppm (Met, Pro) are more intense in the *D. gigas* flavodoxin, reflecting the differences in aminoacid composition [2, 3]. Similarly the phenylalanine peak at 7.1 ppm dominates the aromatic region in the *D. vulgaris* flavodoxin. The high field part of the spectra exhibits resonances characteristic of the native proteins and shifted up to −0.20 ppm by the ring current of aromatic groups. They should cor-

* To whom correspondence should be addressed.

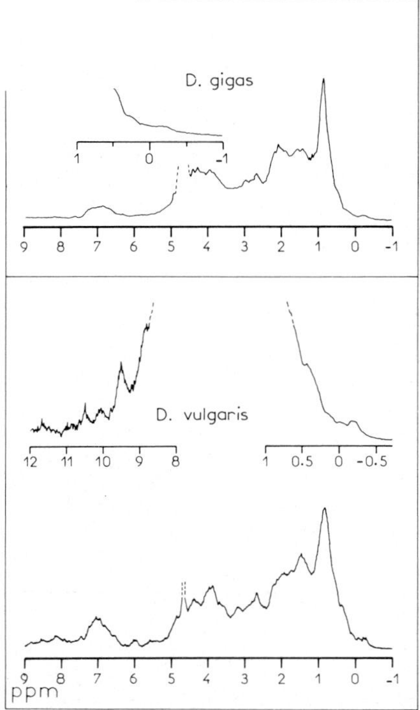

Fig. 1. 100 MHz proton magnetic resonance spectra of the fully oxidized form of the flavodoxins from *D. gigas* (top), $3.5 \cdot 10^{-4}$ M, and from *D. vulgaris* (bottom), $7 \cdot 10^{-4}$ M, in phosphate buffer, p^2H 8.4. The lower spectrum was recorded using a single scan in the CW mode (the expansion of the high field range is an accumulation of 160 scans). The upper spectrum was obtained by block accumulation in the FT mode using 10 000 transients and an acquisition time of one second. The chemical shifts are referenced to internal trimethylsilyltetradeuteriopropionate.

respond to at least five different methyl groups located in the vicinity of the flavin ring or of aromatic aminoacids.

Reduction by dithionite of the flavodoxins was carried out under argon directly in the 10 mm sample tube. At optimum reduction, adjusted by optical titration, the 100% semiquinones were stable for at least 1 h. Two cycles of reduction and reoxidation were carried out for each sample (Fig. 2 and 3). The high field part of the spectra is not very sensitive to the redox state with small modifications of the lineshape but no evident disappearance of resonances. The low field part is more variable as already noted for two other flavodoxins, from *Clostridium MP* and *Streptopeptococcus elsdenii*, by James et al. [4]. The extent of deuteration of the exchangeable protons in the *D. gigas* flavodoxin is already important after a single lyophilisation in ^2H_2O and is completed at 34°C after reduction, leaving an isolated resonance at 8.21 ppm. On the contrary, a number of protons comparable to the number of aromatic protons appears below 7.5 ppm down to 11.6 ppm in the *D. vulgaris* flavodoxin spectrum at 34°C and their exchange is not favored by the reduction. One third of these protons exchanges rapidly upon heating at 45°C and a resonance appears better resolved at 8.20 ppm. The

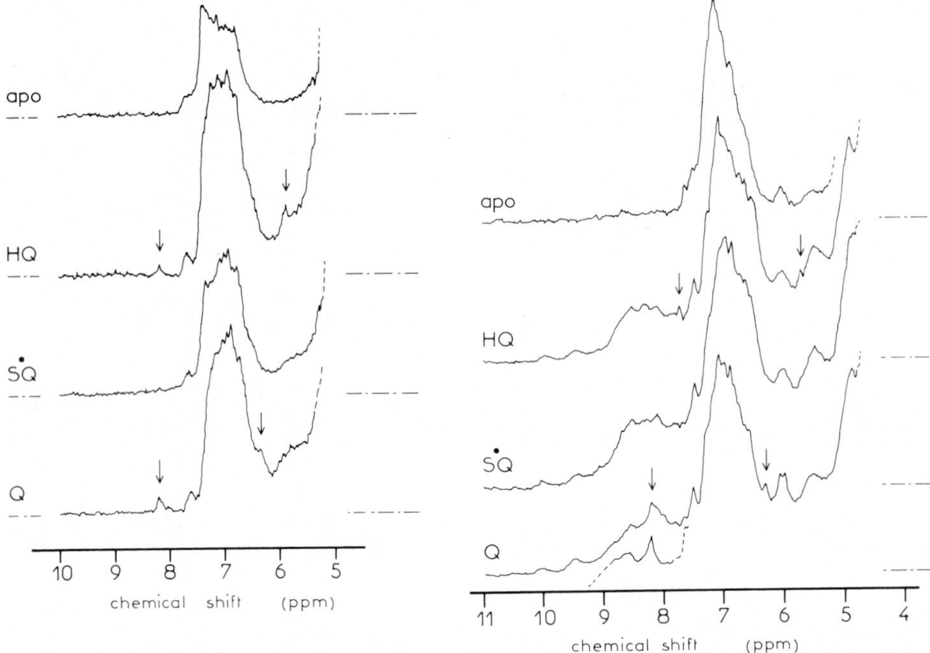

Fig. 2. The low field part of the 100 MHz PMR spectra of the flavodoxin from *D. gigas*, $2.3 \cdot 10^{-4}$ M, in phosphate buffer, p^2H 8.4, at 34°C after complete deuteration of exchangeable protons in the flavoquinone (Q) (bottom), flavosemiquinone (S˙Q) and flavohydroquinone (HQ) states and of the deflavoprotein, $2 \cdot 10^{-4}$ M, (apo) (top). The spectra, referenced to internal trimethylsilyltetradeuteriopropionate, are expanded from complete FT spectra using 10 000 transients and an acquisition time of one second. Single resonances marked by arrows are discussed in the text.

Fig. 3. The low field part of the 100 MHz PMR spectra of the flavodoxin from *D. vulgaris*, $7 \cdot 10^{-4}$ M, in phosphate buffer, p^2H 8.4, at 34°C in flavoquinone (Q) (bottom, with part of the spectrum obtained after 30 min deuterium exchange at 45°C), flavosemiquinone (S˙Q) and flavohydroquinone (HQ) states and of the deflavoprotein, $5 \cdot 10^{-4}$ M (apo) (top). The conditions are the same as in Fig. 2, using 5000 transients.

complete exchange requires the denaturation of the protein with irreversible damage.

These low field resonances should correspond mainly, if not exclusively, to peptidic NH protons buried within the protein. Such protons are expected to participate in maintaining the secondary and tertiary structures of the proteins and it is of major interest that they exhibit more similar spectra in the partly and fully reduced states of the *D. vulgaris* flavodoxin than in the oxidized form. These differences are alike those in the lineshape of the high field region and support the idea that structural changes in flavodoxins upon oxidoreduction concern mainly the "non physiological" flavoquinone state. The very low resonance at 11.6 ppm (Fig. 1) can be assigned to the NH-3 proton of the isoalloxazine prosthetic group of *D. vulgaris* flavodoxin. Its chemical shift confirms that this hydrogen is strongly bound to the protein, as revealed by X-ray crystallography [5, 6], since it appears 0.2 and 1.6 ppm downfield from the corresponding resonance of tetraacetylriboflavin in $[^2H_6]Me_2SO$ and $[^2H_6]$acetone respectively. It disappears irreversibly upon reduction. Thus, a conformational change modifies its binding or opens the structure to the solvent in order to permit exchange though the other exchangeable protons, such as that corresponding to the 10.48 ppm resonance and tentatively assigned to the indolic NH of one of the tryptophane residues, remain inaccessible to water. The absence of magnetic saturation of the water peak in the semiquinone solutions is in favor of a direct contact of the flavin ring with water, at least in the semiquinone state of both proteins, as already suggested by ENDOR for the *P. elsdenii* flavodoxin [7]. We could show by nuclear Overhauser effect that a water molecule is bound to N-5 of tetraacetylriboflavin in $[^2H_6]Me_2SO$, but there is no evidence that water can penetrate so far in the native proteins.

Two resonances corresponding to single protons are well resolved in both proteins in the oxidized state at 8.20 and 6.29 ppm (*D. vulgaris*) and at 8.22 and 6.33 ppm (*D. gigas*). These resonances shift to 7.74 and 5.73 ppm and to 8.13 and 5.82 ppm respectively in the fully reduced state (Fig. 2 and 3). They disappear in the semiquinone form as well as in the freshly prepared [8] apoproteins. On the other hand, all resonances characteristic of the native structure either in the high field region, +0.8 to −0.2 ppm, or in the region of 5.0 to 6.5 ppm are still present in the semiquinones and in the apoproteins. Therefore, apart from the two single resonances mentioned above, all resonances characteristic of the native structure of the flavodoxins belong to protein protons and result from the three-dimensional structure without involving the FMN group. They cannot belong to this group or involve ring current effects of the isoalloxazine ring, as proposed for related resonances in the *Cl. MP* and *P. elsdenii* flavodoxins [4], since they are still present in the apoproteins. As a matter of fact, both proteins are rich in phenylalanine and tyrosine residues which may induce such shifts. The native structure of the apoproteins is relatively unstable in our experimental conditions as indicated by the complete deuteration of the exchangeable protons and a rapid evolution of the PMR spectra corresponding to partial denaturation.

The PMR spectra in the different forms of the flavodoxins show that both proteins have related structures, especially in the vicinity of the FMN prosthetic group. The two single resonances characteristic of the fully reduced and fully oxidized states are clearly related to the FMN group since the *D. gigas* flavodoxin has no histidine residue which could exhibit low field resonances. One can refer to the X-ray analysis of the structure of *D. vulgaris* flavodoxin, already resolved at 2 Å [6], in order to identify the protons responsible for these resonances. Using the tables of Johnson and Bovey [9] for the benzenoid rings of Tyr-98 and Trp-60 which surround the flavin ring and the calculations of Giessner-Prettre and Pullman [10] for the ring current induced shifts due to a reduced or an oxidized isoalloxazine altogether with the crystallographic data, it appears that none of the protons of these neighbor aromatic groups can exhibit a shift due to reciprocal ring current effects larger than 0.2 ppm apart for the resonance of the H-9 proton of the isoalloxazine ring which must be shifted upfield by more than one ppm because it is positioned just atop the ring of Trp-60. We could therefore assign the resonances at 6.29 or 6.33 ppm to this H-9 proton and those at 8.20 or 8.22 ppm to the H-6 proton of the oxidized flavin. These protons appear poorly resolved at 7.95 ppm in free FMN in similar conditions. This assignment is further confirmed by the relative narrowness of the resonances (7–10 Hz), since the corresponding hydrogen atoms interact weakly with the protein [6]. The two methyl groups of the flavin, which appear at 2.48 and 2.57 ppm in free FMN solutions, have not been clearly identified within the large absorption of the protein in this range.

The ring current effect of Trp-60 upon the H-9 proton is much lower in the reduced flavodoxins since upfield shifts larger than one ppm are observed upon reduction of free FMN for the two methine protons. Furthermore, the resonances attributed to H-6 are broadened in the reduced state and little shifted indicating a much stronger interaction with the protein. These phenomena reflect large changes in the protein-flavin interactions similar in both flavodoxins.

These resonances just disappear upon reduction to the semiquinone state without noticeable broadening as observed for some resonances in the *Cl. MP* and *P. elsdenii* flavodoxins [4]. As a matter of fact, our investigations have been carried out using protein concentrations one order of magnitude lower and slow electron transfer was expected.

References

1 Le Gall, J. and Hatchikian, E.C. (1967) C. R. Acad. Sci. Paris 264, 2580–2583

2 Dubourdieu, M. and Le Gall, J. (1970) Biochem. Biophys. Res. Commun. 38, 965–972

3 Dubourdieu, M., Le Gall, J. and Fox, J.L. (1973) Biochem. Biophys. Res. Commun. 52, 1418–1425

4 James, T.L., Ludwig, M.L. and Cohn, M. (1973) Proc. Natl. Acad. Sci. U.S. 70, 3292–3295

5 Watenpaugh, K.D., Sieker, L.C., Jensen, L.H., Le Gall, J. and Dubourdieu, M. (1972) Proc. Natl. Acad. Sci. U.S. 69, 3185–3188

6 Watenpaugh, K.D., Sieker, L.C. and Jensen, L.H. (1973) Proc. Natl. Acad. Sci. U.S. 70, 3857–3860

7 Fritz, J., Müller, F. and Mayhew, S.G. (1973) Helv. Chim. Acta 56, 2250–2254

8 Hinkson, J.W. (1968) Biochemistry 7, 2666–2672

9 Johnson, C.E. and Bovey, F.A. (1958) J. Chem. Phys. 29, 1012–1014

10 Giessner-Prettre, C. and Pullman, B. (1970) J. Theoret. Biol. 27, 87–96

Flavins and Flavoproteins, edited by T.P. Singer

CHAPTER 46

EVOLUTION OF THE FLAVODOXINS

J. LAWRENCE FOX

Department of Zoology, The University of Texas, Austin, Texas 78712 (U.S.A.)

Flavoenzymes are present in nearly all fundamental electron transferring systems throughout the biological world. To date, the only flavoenzymes for which we have extensive structural information are the flavodoxins. The flavodoxins are a low molecular weight class of FMN requiring proteins found in a variety of microorganisms [1]. Flavodoxin circular dichroism spectra indicate that the FMN environment is similar to that found for higher molecular weight flavoenzyme dehydrogenases in eukaryotes [2]. Similarly, they can be classified as dehydrogenases by their "blue" semiquinone [3]. The circular dichroism spectra, apoenzyme fluorescence maxima and flavin binding constants for flavodoxins from different species of organisms subdivide the flavodoxins into two classes labelled the "pasterianum" and "rubrum" types [4].

Flavodoxins function as intermediate electron carriers in a variety of metabolic pathways. They are functionally equivalent to the ferredoxins, very low molecular weight non-heme iron–sulfur proteins [5]. These two redox proteins possess the same, very negative redox potential (roughly −0.40 V) [6] and their presence appears to be mutually exclusive in a given organism. In some organisms the flavodoxins are constitutive enzymes, e.g. *Desulfovibrio vulgaris* [7], *Azotobacter vinelandii* [8], *Synechococcus lividus* [9], while in other organisms the ferredoxins are constitutive and the flavodoxins are only induced under iron starvation conditions, e.g. *Desulfovibrio gigas* [7], *Clostridium pasteurianum* [10], *Clostridium MP* [11], and *Chlorella fusca* [12]. By analogy with the observations made for ferredoxins and other proteins, two distinct molecular weight ranges of apoenzyme are found [13]. The non-photosynthetic bacteria possess apoenzymes of roughly 15 000 molecular weight and are labelled the bacterial type while the photosynthetic bacteria and eukaryotes possess apoenzymes with molecular weights roughly 50% larger, i.e. 22 000 which are labelled the plant type [11].

Complete amino acid sequences have been published for three apoenzymes [14–17] and three more have had partial sequences reported [1, 18, 19]. In addition, two bacterial type proteins have been subjected to X-ray diffraction crystallographic analysis and the three-dimensional solutions have been published [20–23]. Fortunately, the two proteins for which

TABLE I

FLAVODOXIN AMINO ACID SEQUENCES

D.v.	Met	Pro	Lys	Ala	Leu	Ile	Val	Tyr	Gly	Ser	Thr	Thr	Gly	Asn	Thr	Glu	Tyr	Thr	Ala	Glu	Thr	Ile	Ala	Arg	Glu	
C.MP	Met	—	Lys	—	—	Ile	Val	Tyr	Trp	Ser	Gly	Thr	Gly	Asn	Thr	Glu	Lys	Met	Ala	Glu	Leu	Ile	Ala	Lys	Gly	
P.e.	Met	—	—	Val	Glu	Ile	Val	Tyr	Trp	Ser	Gly	Thr	Gly	Asn	Thr	Glu	Ala	Met	Ala	Asn	Glu	Ile	Glu	Ala	Ala	25
C.p.	Met	—	Lys	Val	Asn	Ile	Ile	Tyr	Trp	Ser	Gly	Thr	Gly	Asn	Thr	Glu	Ala	Met	Ala	Lys	Leu	Ile	Ala	Glu	Gly	
A.v.		Ala	Lys	Ile	Gly	Leu	Phe	Phe	Gly	Ser	Asn	Thr	Gly	Lys	Thr	Arg	Lys	Val	Ala	Lys	Ser	Ile	Lys	Lys	Arg	
R.r			Gly	Thr	Thr	Val	Ile	Tyr	Gly	Ser	Asp	Gly	Gly	Thr	Thr	Glu	Gly	Val	Ala	Lys	Arg	Ile	Ala	(A/S)	Pro	
D.v.	Leu	Ala	Asn	Ala	Gly	Tyr	Glu	Val	Asp	Ser	Arg	Asp	Ala	Ala	Ser	Val	Glu	Ala	Gly	Gly	Leu	Phe	Glu	Gly	Phe	
C.MP	Ile	Ile	Glu	Ser	Gly	Lys	Asp	Val	Asn	Thr	Ile	Asn	Val	Ser	Asp	Val	Asn	Ile	Asp	Glu	Leu	Leu	Asn	Glu	—	
P.e.	Val	Lys	Ala	Ala	Gly	Ala	Asp	Val	Glu	Ser	Val	Arg	Phe	Glu	Asp	Thr	Asn	Val	Asp	Asp	Val	Ala	Ser	Lys	—	50
C.p.	Ala	Gln	Glu	Lys	Gly	Ala	Glu	Val	Lys	Leu	Leu	Asp	Val	Ser	Asp	Ala	Lys	Glu	Asp	Asp	Val	Lys	Glu	Ala	—	
A.v.	Phe	Asx	Asx	(Glx)	Thr	Met	(Ser)	Asp	Ala	Leu	Asx	Val	(Asx)	Arg	Val	Ser	Ala	Glu	Asx	Phe	Ala					
R.r.	Leu	Glx	Ala	Lys	Val	Val	Asx	Ile	Lys	(Val)	Ala	Thr	(Thr)													
D.v.	Asp	Leu	Val	Leu	Leu	Gly	CyS	Ser	Thr	Trp	Gly	Asp	Asp	Ser	Ile	Glu	Leu	Glu	Asp	Asp	Phe	Ile	Pro	Leu	Phe	
C.MP	Asp	Ile	Leu	Ile	Leu	Gly	CyS	Ser	Ala	Met	Gly	Asp	Glu	Val	Leu	Glu	—	Glu	Ser	Glu	Phe	Glu	Pro	—	Phe	
P.e.	Asp	Val	Ile	Leu	Leu	Gly	CyS	Pro	Ala	Met	Gly	Ser	Glu	Glu	Leu	Glu	—	Asp	Ser	Val	Val	Glu	Pro	—	Phe	75
C.p.	Asp	Val	Val	Ala	Phe	Gly	Ser	Pro	Ser	Met	Gly	Ser	Glu	Val	Ser	Glu	Pro	Glu	—	—	—	Glu	Met	—	Phe	
D.v.	Asn	Ser	Leu	Glx	Glx	Thr	Gly	Ala	Glx	Gly	Arg	Lys	Val	Ala	CyS	Phe	Gly	CyS	Gly	Asp	Ser	Ser	Tyr	Glu	Tyr	
C.MP	Ile	Glu	Glu	Ile	Ser	Thr	Lys	Ile	Ser	Gly	Lys	Lys	Val	Ala	Leu	Phe	Gly	—	—	—	—	Ser	Tyr	Gly	Trp	
P.e.	Phe	Thr	Asp	Leu	Ala	Pro	Lys	Leu	Lys	Gly	Lys	Lys	Val	Gly	Leu	Phe	Gly	—	—	—	—	Ser	Tyr	Gly	Trp	100
C.p.	Leu	Asp	Val	Val	Ser	Ser	Ile	Val	Thr	Gly	Lys	Lys	Val	Ala	Leu	Phe	Gly	—	—	—	—	Ser	Tyr	Gly	Trp	
D.v.	—	Phe	CyS	Gly	Ala	Val	Asp	Ala	Ile	Glu	Glu	Lys	Leu	Lys	Asn	Leu	Gly	Ala	Glx	Ile	Val	Glx	Asx	Gly	Leu	
C.MP	Gly	Asp	Gly	Lys	Trp	Met	Arg	Asp	Phe	Glu	Glu	Arg	Met	Asn	Gly	Tyr	Gly	CyS	Val	Val	Val	Glu	Thr	Pro	Leu	
P.e.	Gly	Ser	Gly	Glu	Trp	Met	Asp	Ala	Trp	Lys	Gln	Arg	Thr	Glu	Asp	Thr	Gly	Ala	Thr	Val	Ile	Gly	Thr	Ala	Ile	125
C.p.																										
D.v.	Arg	Ile	Asp	Gly	Asp	Pro	Arg	Ala	Ala	Arg	Asp	Asp	Ile	Val	Gly	Trp	Ala	His	Asn	Val	—	Arg	Gly	Ala	Ile	
C.MP	Ile	Val	Gin	Asn	Glu	Pro	Asp	Glu	Ala	Glu	Gln	Asp	CyS	Ile	Glu	Phe	Gly	Lys	Lys	Ile	—	—	Ala	Asn	Ile	
P.e.	Val	—	Asn	Glu	Met	Pro	Asp	Asn	Ala	Pro	Glu	—	CyS	Lys	Glu	Leu	Gly	Glu	Ala	Ala	—	—	Ala	Lys	Ala	150
C.p.																Leu	Gly	Lys	Asn	Leu	Gly	Arg	Glu	Leu	Val	

Abbreviation	Full name	Reference
D.v.	*Desulfovibrio vulgaris*	16
C.MP	*Clostridium MP*	17
P.e.	*Peptostreptococcus elsdenii*	14, 15
C.p.	*Clostridium pasteurianum*	1, 18
A.v.	*Azotobacter vinelandii*	19 and private communication
R.r.	*Rhodospirillum rubrum*	19

β-pleated sheet regions: D.v. 3–10, 32–38, 52–58, 87–93, 118–126 (Watenpaugh, private communication, 1975)
C.MP 1–6, 30–35, 48–55, 80–89, 108–119. [22, 23]

α-helix regions: D.v. 14–30, 71–83, 102–115, 133–148 (Watenpaugh, private communication, 1975)
C.MP 10–27, 66–74, 93–107, 124–138. [22, 23]

the X-ray diffraction studies were conducted are members of the two flavodoxin sub-classes described before. Since Drs. Ludwig and Watenpaugh have already discussed many of the interpretations of the three dimensional structure of these proteins, I will direct my remarks to the relationships of the primary level of structure for these proteins.

We have conducted our studies with computer programs developed by Dr. Walter Fitch [24] for analyzing the minimal number of nucleotide replacements necessary to transform one protein sequence into another. Table I lists the published flavodoxin sequences. They have been aligned according to minimal nucleic acid base changes and greatest homology, rather than by structural considerations. This is really only important at the C- and N-termini. Overall, the homology of these proteins ranges between 30 and 40% identity between each other.

Particular regions of the proteins are more homologous than others. The N-terminus possesses the highest level of homology among the four bacterial type proteins. This region serves as the FMN phosphate binding site using Ser(10), Thr(12), Asn(14), and Thr(15) of the *D. vibrio* protein [20, 21]. These positions are not rigorously homologous in the plant type proteins (for which only partial sequences are presently available) [19]. Five regions of the bacterial type apoenzymes form β-pleated sheet structures [20, 21]. These general regions of the protein are the most homologous possessing 40–50% identity levels, depending in part upon whether the region is defined from the *D. vibrio* [20, 21] or the *C. MP* [22, 23] protein. In a given protein, the degree of identity for both the α-helical regions and the rest of the protein are about the same and roughly one-half the level of identity observed for the β-pleated sheet regions, i.e. 20–30%. There is somewhat of a gradient of identity which decreases as one moves from the N- to the C-terminus.

The evolutionary relationships can be arrived at by two procedures

TABLE II

HOMOLOGIES AND MINIMUM BASE CHANGES/CODON

For abbreviation list, see Table I.

Ist 38 residues

Homologies

	D.v.	C.MP	P.e.	C.p.	A.v.	R.r
D.v.						
C.MP	45%					
P.e.	42%	47%				
C.p.	45%	58%	50%			
A.v.	24%	24%	16%	24%		
R.r.	26%	24%	24%	34%	21%	

MBC/C

	D.v.	C.MP	P.e.	C.p.	A.v.	R.r.
D.v.						
C.MP	0.60					
P.e	0.84	0.60				
C.p.	0.79	0.47	0.55			
A.v.	1.03	0.89	1.08	1.00		
R.r.	0.97	0.89	0.84	0.79	0.95	

Total protein

Homologies

	D.v.	C.MP	P.e.
D.v.			
C.MP	31%		
P.e.	24%	40%	

MBC/C

	D.v.	C.MP	P.e.
D.v.			
C.MP	0.92		
P.e.	1.03	0.66	

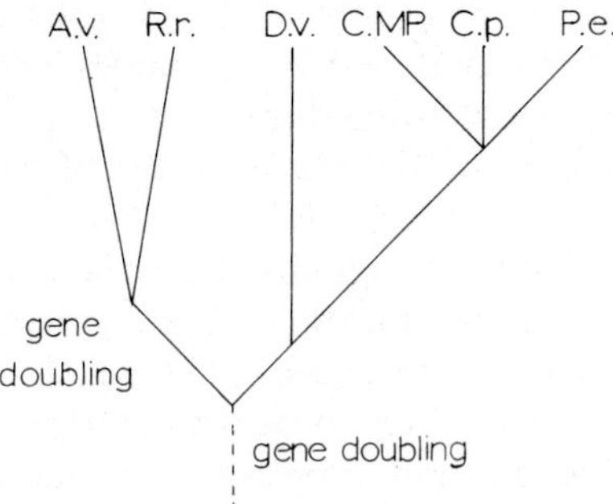

Fig. 1. Evolutionary tree for flavodoxins. For abbreviations, see Table I.

which complement each other. Either the number of homologous positions can be tabulated or the minimum base changes per codon (MBC/C) can be determined. Both of these are shown in Table II. From this the chart shown in Fig. 1 was deduced. Several lines of evidence argue for *D. vibrio* being of ancient age [25].

Very little evidence has been noted to date for homology between the flavodoxins and the ferredoxins. This analysis has been badly handicapped by the absence of two complete sequences for these proteins from the same organism. One mild suggestion is shown in Table III. The amino acid sequence of the flavodoxin from *D. vibrio* [16] and the ferredoxin from *D. gigas* [26] (Dubourdieu and Fox, unpublished), two seemingly closely related species which possess flavodoxins and ferredoxins of similar amino acid compostion, but for which only one protein each has been sequenced, are compared. This leaves the question of convergent or divergent evolution unresolved [16].

The study of ferredoxins has indicated an interesting conclusion. The bacterial type proteins apparently possess evidence of internal gene duplication from a more primitive fragment, roughly one-half the length of the contemporary protein [27–30]. The plant type ferredoxins then demonstrate a second duplication to establish their 50% longer protein. For this reason we examined the sequences of the bacterial flavodoxins shown in Table I. Analysis of the *Peptostreptococcus elsdenii* protein indicated a statistically significant homology for two regions of the protein in which the groups which sandwich the isoalloxazine ring are located. This is illustrated in Table IV with the analogous regions of the other three bacterial type proteins. While this argument does not hold for the *D. vibrio* protein, many of the residues are single base changes or commonly found changes. This picture is consistent with allowing *D. vibrio* a longer period of evolution in which to permit more mutational events.

Lastly, I would like to discuss some findings in Dr. Rossmann's laboratory [31, 32]. Dr. Rossmann noted from slides at a talk given by Dr. Jensen that the flavodoxin flavin binding region was structurally homologous in three dimensional space with the pyridine nucleotide binding regions of other

TABLE III

FLAVODOXIN-FERREDOXIN HOMOLGY

MR is the minimum number of base changes necessary to transform one amino acid sequence into the other sequence.

	84	85	86	87	88	89	90	91	92	93	94	95	96	97	98	99	100
D.v. flavodoxin	Glx	Gly	Arg	Lys	Val	Ala	CyS	Phe	Gly	CyS	Gly	Asx	Ser	Ser	Tyr	Glu	Tyr
MR=	1	2	1	2	1	2	0	2	1	0	1	1	1	1	1	0	1
D.g. ferredoxin	Pro	Ile	Gln	Val	Asp	Asn	CyS	Met	Ala	CyS	Gln	Ala	CyS	Ile	Asn	Glu	CyS
	1	2	3	4	5	6	7	8	9	10	11	12	13	14	15	16	17

	101	102	103	104	105	106	107	108	109	110	111	112	113	114	115	116	117
D.v. flavodoxin	Phe	CyS	Gly	Ala	Val	Asp	Ala	Ile	Glu	Glu	Lys	Leu	Lys	Asn	Leu	Gly	Ala
MR=	2	2	1	1	1	1	2	2	0	1	2	2	0	2	1	2	2
D.g. ferredoxin	Pro	Val	Asp	Val	Phe	Gln	Met	Asp	Glu	Gln	Gly	Asp	Lys	Ala	Val	Asn	Ile
	18	19	20	21	22	23	24	25	26	27	28	29	30	31	32	33	34

TABLE IV

FLAVODOXIN INTERNAL GENE DUPLICATION

The numbers are the residue locations within each flavodoxin. The arrows indicate the flavin binding residues. For abbreviation list, see Table I.

											↓										
	47	48	49	50	51	52	53	54	55	56	57	58	59	60	61	62	63	64	65	66	67
P.e.	Lys	Asp	Val	Ile	Leu	Leu	Gly	CyS	Pro	Ala	Met	Gly	Ser	Glu	Glu	Leu	Glu	Asp	Ser	Val	Val
	Lys	Lys	Val	Gly	Leu	Phe	Gly	Ser	Tyr	Gly	Trp	Gly	Ser	Gly	Glu	Trp	Met	Asp	Ala	Trp	Lys
	81	82	83	84	85	86	87	88	89	90	91	92	93	94	95	96	97	98	99	100	101
											↕										
	46	47	48	49	50	51	52	53	54	55	56	57	58	59	60	61	62	63	64	65	66
C.MP	Glu	Asp	Ile	Leu	Ile	Leu	Gly	CyS	Ser	Ala	Met	Gly	Asp	Glu	Val	Leu	Glu	Glu	Ser	Glu	Phe
	Lys	Lys	Val	Ala	Leu	Phe	Gly	Ser	Tyr	Gly	Trp	Gly	Asp	Gly	Lys	Trp	Met	Arg	Asp	Phe	Glu
	80	81	82	83	84	85	86	87	88	89	90	91	92	93	94	95	96	97	98	99	100
											↕										
	50	51	52	53	54	55	56	57	58	59	60	61	62	63	64	65	66	67	68	69	70
D.v.	Phe	Asp	Leu	Val	Leu	Leu	Gly	CyS	Ser	Thr	Trp	Gly	Asp	Asp	Ser	Ile	Glu	Leu	Gln	Asp	Asp
	Val	Ala	CyS	Phe	Gly	CyS	Gly	Asp	Ser	Ser	Tyr	Glu	Tyr	Phe	CyS	Gly	Ala	Val	Asp	Ala	Ile
	88	89	90	91	92	93	94	95	96	97	98	99	100	101	102	103	104	105	106	107	108
											↕										
	48	49	50	51	52	53	54	55	56	57	58	59	60	61	62	63	64	65	66	67	68
C.P.	Ala	Asp	Val	Val	Ala	Phe	Gly	Ser	Pro	Ser	Met	Gly	Ser	Glu	Val	Ser	Glu	Pro	Glu	Glu	Met
	Lys	Lys	Val	Ala	Leu	Phe	Gly	Ser	Tyr	Gly	Trp. . .										
	80	81	82	83	84	85	86	87	88	89	90										
											↑										

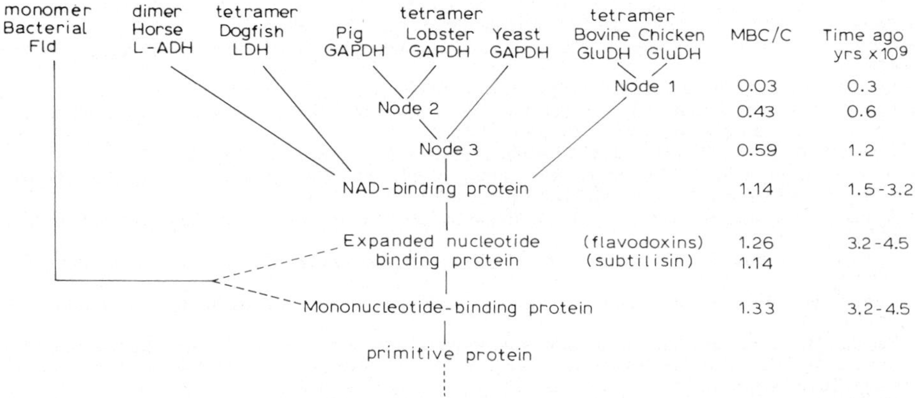

Fig. 2. Evolutionary development of the nucleotide binding protein (after Rossmann [32]). Fld, flavodoxin; L-ADH, liver alcohol dehydrogenase; LDH, lactate dehydrogenase; GAPDH, glyceraldehyde-3-phosphate dehydrogenase; GluDH, glutamate dehydrogenase. MBC/C minimum base changes per codon. Nodes 1–3 are correlated with fossil data and the earlier times result from extrapolations of the divergence times based on the MBC/C values.

dehydrogenases while the primary structures showed no discernable homology. The relationships are summarized in Fig. 2. A correlation of the minimum base changes per codon to transform the sequence for one nucleotide binding site into another with fossil records for the more recent evolutionary changes leads Rossmann et al. [32] to date the appearance of flavodoxins at the earliest dates for which living forms are thought to have been found on the earth. An evolutionary selection process at the structural, but not sequence, level as noted by Rossmann raises a whole host of new questions. Evolutionary data at the sequence level have only begun to allow our inspection, so interpretation of structural data will be a long way off into the future.

Acknowledgements

This research was supported by grants from the National Science Foundation and the Biomedical Sciences Study Group at the University of Texas. I wish to thank Drs. J.R. Brown, K. Kobayashi, K. Watenpaugh, and M. Dubourdieu for discussions of this research. Special thanks are given to Dr. W. Fitch for providing us with copies of his programs and Mr. M. Teitell for assistance in converting these programs to our facilities here and in performing some of the calculations.

References

1 Fox, J.L., Smith, S.S. and Brown, J.R.(1972) Z. Naturforsch. 27b, 1096–1100
2 Edmondson, D.E. and Tollin, G. (1971) Biochemistry 10, 113–124
3 Massey, V. and Palmer, G. (1966) Biochemistry 5, 3181–3188
4 D'Anna, J.A., Jr. and Tollin, G. (1972) Biochemistry 11, 1073–1080
5 Steicher, S.L. and Vallentine, R.C. (1973) Annu. Rev. Biochem. 42, 279–302
6 Mayhew, S.G. (1971) Biochim. Biophys. Acta 235, 276–288

7 Dubourdieu, M. and LeGall, J. (1970) Biochem. Biophys. Res. Commun. 38, 965–972
8 Edmondson, D.E. and Tollin, G. (1971) Biochemistry 10, 124–132
9 Crespi, H.L., Smith, U., Gajda, L., Tisue, T. and Ameraal, R.M. (1972) Biochim. Biophys. Acta 256 611–618
10 Knight, Jr., E. and Hardy, R.W.F. (1966) J. Biol. Chem. 241, 2752–2756
11 Ludwig, M.L., Andersen, R.D., Apgar, P.A., Burnett, R.M., LeQuesne, M.E. and Mayhew, S.G. (1971) Cold Spring Harbor Symp. 36, 369–380
12 Zumft, W.G. and Spiller, H. (1971) Biochem. Biophys. Res. Commun. 45, 112–118
13 Orme-Johnson, W.M. (1973) Annu. Rev. Biochem. 42, 159–204. Lovenberg, W. (1973) The Iron Sulfur Proteins 1 and 2, Academic Press, New York
14 Tanaka, M., Haniu, M., Yasunobu, K.T., Mayhew, S.G. and Massey, V. (1973) J. Biol. Chem. 248, 4354–4366
15 Tanaka, M., Haniu, M., Yasunobu, K.T., Mayhew, S.G. and Massey, V. (1974) J. Biol. Chem. 249, 4397
16 Dubourdieu, M., LeGall, J. and Fox, J.L. (1973) Biochem. Biophys. Res. Commun. 52, 1418–1425
17 Tanaka, M., Haniu, M., Yasunobu, K.T. and Mayhew, S.G. (1974) J. Biol. Chem. 249, 4393–4396
18 Tanaka, M., Haniu, M., Matsueda, G., Yasunobu, K.T., Mayhew, S.G. and Massey, V. (1971) Biochem. 10, 3041–3046
19 MacKnight, M.L., Gray, W.R. and Tollin, G. (1974) Biochem. Biophys. Res. Commun. 59, 630–637
20 Watenpaugh, K.D., Sieker, L.C., Jensen, L.H., LeGall, J. and Dubourdieu, M. (1972) Proc. Natl. Acad. Sci. U.S. 69, 3185–3188
21 Watenpaugh, K.D., Sieker, L.C. and Jensen, L.H. (1973) Proc. Natl. Acad. Sci. U.S. 70, 3857–3860
22 Andersen, R.D., Apgar, P.A., Burnett, R.M., Darling, G.D., LeQuesne, M.E., Mayhew, S.G. and Ludwig, M.L. (1972) Proc. Natl. Acad. Sci. U.S. 69, 3189–3191
23 Burnett, R.M., Darling, G.D., Kendall, D.S., LeQuesne, M.E., Mayhew, S.G., Smith, W.W. and Ludwig, M.L. (1974) J. Biol. Chem. 249, 4383–4392
24 Fitch, W.M. (1970) J. Mol. Biol. 49, 1–14
25 Peck, Jr., H.D. (1974) Symp. Soc. Gen. Microbiol. (London) Evolution in the Microbial World 24, 241–262
26 Travis, J., Newman, D.J., LeGall, J. and Peck, Jr., H.D. (1971) Biochem. Biophys. Res. Commun. 45, 452–458
27 Eck, R.V. and Dayhoff, M.O. (1966) Science 152, 363–366
28 Jukes, T.H. (1966) Molecules and Evolution, Columbia University Press, New York
29 Fitch, W.M. (1973) Annu. Rev. Genet. 7, 343–380
30 Dayhoff, M.O. Ed. (1972) Atlas of Protein Sequence and Structure 5, National Biomedical Research Foundation, Silver Spring, Maryland
31 Liljas, A. and Rossmann, M.G. (1974) Annu. Rev. Biochem. 43, 475–507
32 Rossmann, M.G., Moras, D. and Olsen, K.W. (1974) Nature 250, 194–199

Flavins and Flavoproteins, edited by T.P. Singer

CHAPTER 47

BINDING OF 8α-SUBSTITUTED FLAVINS AND FLAVIN PEPTIDES TO *AZOTOBACTER* FLAVODOXIN

GUILLERMO OESTREICHER*, DALE E. EDMONDSON and THOMAS P. SINGER

Molecular Biology Division, Veterans Administration Hospital, 4150 Clement St., San Francisco, Calif. 94121 and Department of Biochemistry and Biophysics, University of California, San Francisco, Calif. 94143 (U.S.A.)

Summary

The binding of 8α-substituted flavins and flavin peptides to the apoenzyme of *Azotobacter* flavodoxin has been investigated using Sephadex column chromatography and by fluorescence spectroscopy. Substantial purification of flavin peptides in 60–75% yield may be achieved by gel filtration of apoflavodoxin-treated crude proteolytic digests of the enzymes: beef heart succinate dehydrogenase, liver monoamine oxidase and bacterial thiamine dehydrogenase. In the case of succinate dehydrogenase, a pronase treatment, following tryptic-chymotryptic digestion, increased the yield of bound flavin peptide from 40% to 75%.

The 8α-flavin peptides tested (FMN level) are bound 10 to 20-fold less tightly to the apoprotein than FMN, while synthetic 8α-substituted riboflavin analogs are bound 2 to 28-fold more tightly than riboflavin itself. The reconstituted flavin peptide-apoflavodoxin complexes are reduced to the neutral semiquinone form by the anaerobic addition of one electron-equivalent of dithionite. With the exception of the 8α-cysteinylflavin-pentapeptide complex, a second reducing equivalent yields the flavin hydroquinone.

These data show that extensive substitution in the 8α position of the flavin ring does not prevent binding of the flavin to the apoprotein and are in agreement with X-ray crystallographic data on flavodoxin, which show the benzenoid portion of the flavin ring to be exposed to the solvent.

---

* Predoctoral Fellow of the Organization of American States, present address: Departmento de Bioquimica, Facultad de Medicina, Sede Norte, Universidad de Chile, Casilla 6679, Santiago 4, Chile.

Introduction

The binding properties of 8α-substituted flavins and flavin peptides to apoflavoenzymes are of interest from two standpoints. First, it provides information regarding the steric limitations about the benzenoid portion of the flavin molecule in its binding site on the enzyme. Secondly, it may provide an attractive alternative method for the isolation of flavin peptides from proteolytic digests of enzymes containing covalently bound flavin since the methods at hand are tedious and result in low yields of flavin peptides.

With this in mind, we present here a study of the binding of various 8α-substituted flavins with *Azotobacter* apoflavodoxin. This particular flavoprotein was chosen because it is readily purified in reasonable quantities, it has a relatively low specificity in binding of flavin analogs [1], and because available X-ray crystallographic data on other flavodoxins [2, 3] show the benzenoid portion of the bound flavin (FMN) to be exposed to the solvent.

Methods

Azotobacter flavodoxin was purified from *Azotobacter vinelandii*, strain "OP" (the culture was a gift from Dr. D. Yoch, University of California, Berkeley) according to the procedure of Hinkson and Bulen [4]. The apoprotein was prepared as described by Edmondson and Tollin [1].

Synthetic 8α-histidyflavins were prepared as according to Walker et al. [5], while 8α-cysteinylriboflavin was synthesized by the procedure of Ghisla and Hemmerich [6]. Phosphorylation of the riboflavin analogs to the FMN level was according to the procedure of Flexer and Farkas [7].

Preparations of the following enzymes containing covalently bound flavin were obtained according to published procedures: succinate dehydrogenase [8], monoamine oxidase [9] and thiamine dehydrogenase [10]. Tryptic-chymotryptic digests of these enzymes were prepared, following trichloroacetic acid precipitation, in 0.1 M Tris/HCl, pH 8.0, at 38°C for 4 h. The dinucleotides were hydrolyzed to the FMN level either by overnight incubation in 5% trichloroacetic acid at 38°C or enzymatically, using snake venom phosphodiesterase.

Binding of the purified 8α-flavins and flavin peptides to apoflavodoxin was measured fluorometrically. Since the flavin fluorescence yield of these analogs is low at pH 7, binding was measured by monitoring the quenching of protein fluorescence due to tryptophan using a Perkin-Elmer MPF-3 spectrofluorometer.

Results and Discussion

When a tryptic-chymotryptic digest of succinate dehydrogenase (at the FMN level) was incubated with a 2-fold molar excess of apoflavodoxin and subjected to Sephadex G-50 column chromatography, only 38–40% of the total flavin appeared with the flavodoxin in the excluded peak (Fig. 1A). Treatment of the tryptic-chymotryptic digest with pronase prior to the addition of apoflavodoxin resulted in the binding of 80% of the total flavin to the apoprotein (Fig. 1B). These results suggest that a substantial fraction of the flavin peptide ($\approx$ 40%) was in a longer form than the pentapeptide and that the presence of a long peptide chain hinders binding to the apoflavodoxin.

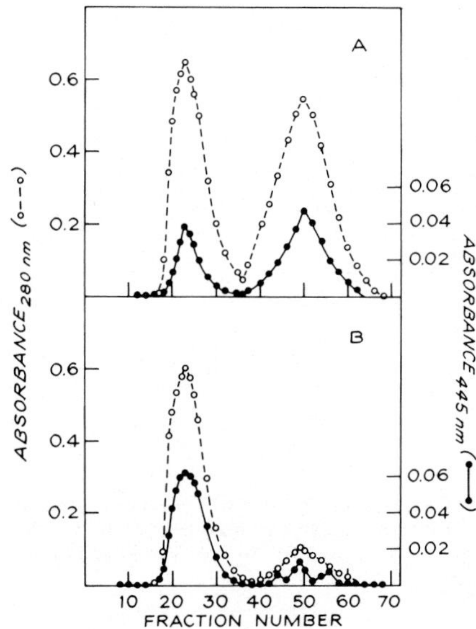

Fig. 1. Sephadex G-50 Column (1 × 50 cm) elution profile of apoflavodoxin-treated proteolytic digest of succinate dehydrogenase. The molar ratio of apoflavodoxin to bound flavin was 2 in each case. A. tryptic-chymotryptic digest. B. tryptic-chymotryptic digest treated with pronase for 24 h pH 7.4 at 38°C. The elution buffer was 25 mM sodium phosphate pH 7.0.

Less likely, non-flavin peptides could interfere with the binding of the flavin peptide and the increased combination on pronase treatment might be the result of hydrolysis of such interfering peptides.

The data in Table I show that when the method was applied to tryptic-chymotryptic digests of monoamine oxidase and of thiamine dehydrogenase (FMN level), 60–70% of the total covalently-bound flavin was bound to apoflavodoxin. For comparison it is shown that riboflavin and 8α-(N-3)-histidylriboflavin were bound to an extent of 92 and 86% respectively under

TABLE I

BINDING OF 8α-SUBSTITUTED FLAVIN PEPTIDES FROM PROTEOLYTIC DIGESTS OF VARIOUS ENZYMES TO APOFLAVODOXIN AS DETERMINED BY SEPHADEX G-50 CHROMATOGRAPHY

The molar ratio of flavin to apoflavodoxin was 1:2. All flavin peptides were at the FMN level.

Flavin	Flavin concentration (μM)	% Bound to apoflavodoxin
Riboflavin	85	92
8α-(N-3)-histidylriboflavin	100	86
Flavin in trypsin-chymotrypsin digest of beef heart succinate dehydrogenase	100	38
Same but digested with pronase after trypsin-chymotrypsin	82	75
Flavin in trypsin-chymotrypsin digest	80	69
of monoamine oxidase	80	69
Flavin in trypsin-chymotrypsin digest of thiamine dehydrogenase	200	63

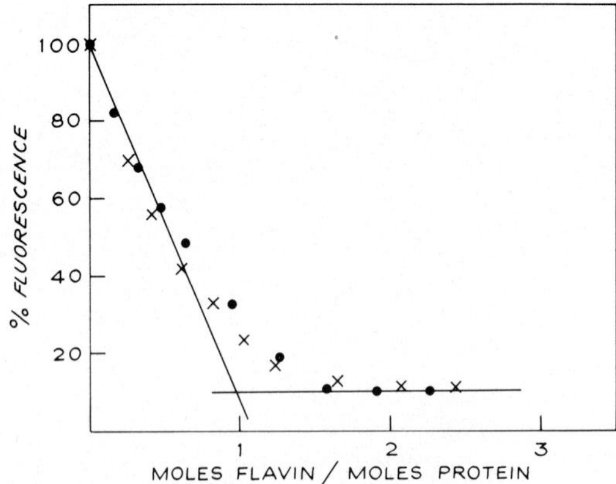

Fig. 2. Tryptophan fluorescence quenching upon binding of 8α-(N-3)-histidyl FMN (●) or of purified tryptic-chymotryptic flavin pentapeptide (FMN level) (X) from succinate dehydrogenase at 25°C. The excitation wavelength is 280 nm and the emission wavelength monitored is 330 nm. The protein concentration is $3 \cdot 10^{-7}$ M in 25 mM sodium phosphate, pH 7.0.

the same conditions. These results indicate that if the K_D were of 10^{-6} M or less, the flavin peptide would be bound under the conditions of Sephadex chromatography.

Although substantial purification of flavin peptides is achieved by this single step, thin-layer chromatography of the bound flavin peptide (after acid liberation from apoflavodoxin) showed some ninhydrin-positive material not associated with flavin in all cases tested. This is not surprising, since apoflavodoxin, being an acidic protein [1], may bind peptides non-specifically. In our experience, however, a single thin-layer chromatographic step usually suffices to remove such non-flavin impurities.

To quantitate the binding affinities of various 8α-flavins and flavin peptides to apoflavodoxin, dissociation constants (K_D) were determined

TABLE II

DISSOCIATION CONSTANTS OF VARIOUS 8α-SUBSTITUTED FLAVIN AND FLAVIN PEPTIDE-APOFLAVODOXIN COMPLEXES

Flavin	$10^7 \times K_D$ (M)
Riboflavin	28.4
8α-(N-1)-Histidylriboflavin	4.1
8α-(N-3)-Histidylriboflavin	1.0
8α-(N-3)-Histidyl FMN	0.15
Tryptic flavin peptide from succinate dehydrogenase (FMN level)	2.9
Tryptic-chymotryptic flavin peptide from succinate dehydrogenase (FMN level)	0.14
Tryptic-chymotryptic, then pronase flavin peptide from succinate dehydrogenase (FMN level)	0.42
Tryptic-chymotryptic flavin peptide from monoamine oxidase (FMN level)	1.22
8α-Cysteinylriboflavin	14.1

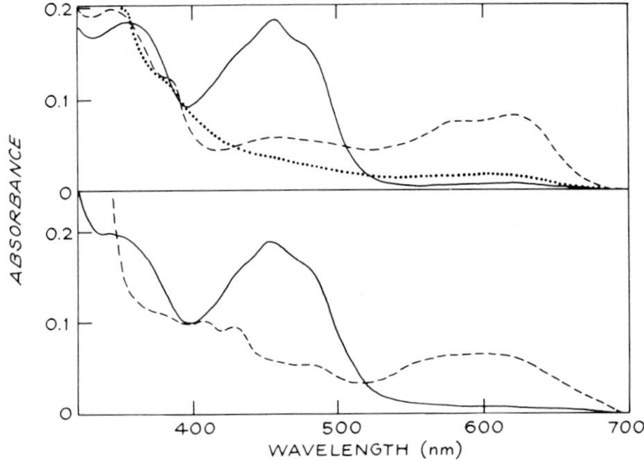

Fig. 3. Absorbance spectra of the (top) oxidized (——), semiquinone (---) and hydroquinone (···) forms of 8α-(N-3)-histidyl FMN-apoflavodoxin complex and (bottom) of the oxidized (——) and semiquinone (---) forms of monoamine oxidase-flavin peptide-apoflavodoxin complex under anaerobic conditions in 25 mM sodium phosphate pH 7.0.

fluorometrically by monitoring the quenching of tryptophan fluorescence on flavin binding. This method was preferable to monitoring flavin fluorescence quenching, since flavin fluorescence is characterized by low quantum yields under the conditions of binding. The titration curve in Fig. 2 shows that synthetic 8α-(N-3)-histidyl FMN and the flavin pentapeptide from beef heart succinate dehydrogenase (FMN level) are bound with nearly equal affinities and with a stoichiometry of 1 mol flavin per mol of protein.

Table II shows the relative binding affinities of a number of 8α-substituted flavins and flavin peptides to apoflavodoxin. Both the N-1 and N-3 isomers of 8α-histidylriboflavin are bound to apoflavodoxin more tightly than riboflavin itself by approximately one order of magnitude. This indicates an energetically favorable interaction of the 8α-histidyl substituent with the protein, which increases the binding affinity. This behavior is not seen in comparing the binding of 8α-(N-3)-histidyl FMN or the succinate dehydrogenase flavin peptides with FMN [1] (Table II). The lack of correlation between the binding of riboflavin and FMN analogs to *Azotobacter* apoflavodoxin has been observed previously [1] and is due to a phosphate-induced conformational change about the flavin binding site.

The largest flavin peptide isolated to date is the tryptic peptide of succinate dehydrogenase, which contains 23 residues [11]. Table II shows that this flavin peptide is bound to apoflavodoxin at least as tightly as the 8α-histidylriboflavin analogs and further shows that the flavin peptide, which is not bound by apoflavodoxin in tryptic-chymotryptic succinate dehydrogenase digests but becomes bound on further treatment with pronase (see Fig. 1 and Table I), is not the tryptic flavin peptide. (The authors are grateful to Dr. W.C. Kenney of this department for the gift of this material).

Reduction of the apoflavodoxin complexes of 8α-(N-3)-histidyl FMN, flavin pentapeptide from succinate dehydrogenase (FMN level) and flavin pentapeptide from monoamine oxidase (FMN level) by one electron-equivalent of dithionite results in substantial yield of neutral flavin semiquinone, as judged by the resulting absorbance spectra (Fig. 3). While the apoflavodoxin complexes of the 8α-histidyl FMN analogs are reduced to their hydroquinone forms upon addition of a second electron-equivalent of dithionite, the apoflavodoxin-monoamine oxidase flavin peptide complex is not further reduced, even in the presence of 4 to 5 electron equivalents. The reason for this behavior requires further investigation. However, studies with the flavodoxin holoprotein [12] show this behavior to be dependent on the ionization of the protein-bound flavin hydroquinone.

Acknowledgements

Work supported by a program project grant 1 PO 1 HL 16251 from the National Institutes of Health to Dr. T.P. Singer and by a National Science Foundation Grant (GB 41881) to Dr. D.E. Edmondson.

References

1 Edmondson, D.E. and Tollin, G. (1971) Biochemistry 10, 124
2 Watenpaugh, K.D., Sieker, L.C. and Jensen, L.H. (1973) Proc. Natl. Acad. Sci. U.S. 70, 3857
3 Burnett, R.M., Darling, G.D., Kendall, D.S., LeQuesne, M.E., Mayhew, S.G., Smith, W.W. and Ludwig, M. (1974) J. Biol. Chem. 249, 4383
4 Hinkson, J.W. and Bulen, W.A. (1967) J. Biol. Chem. 242, 3345
5 Walker, W.H., Singer, T.P., Ghisla, S. and Hemmerich, P. (1972) Eur, J. Biochem. 26, 279
6 Ghisla, S. and Hemmerich, P. (1971) FEBS Lett. 16, 229
7 Flexser, L.A. and Farkas, W.G. (1952) U.S. Patent 2, 610, 179; Chem Abstr, 47, 8781
8 Salach, J., Walker, W.H., Singer, T.P., Ehrenberg, A., Hemmerich, P., Ghisla, S. and Hartmann, U. (1972) Eur. J. Biochem. 26, 267
9 Kearney, E.B., Salach, J.I., Walker, W.H., Seng, R.L., Kenney, W., Zeszotek, E. and Singer, T.P. (1971) Eur. J. Biochem. 24, 321
10 Neal, R.A. (1970) J. Biol. Chem. 245, 2599
11 Kenney, W.C., Walker, W.H. and Singer, T.P. (1972) J. Biol. Chem. 247, 4510
12 Edmondson, D.E. and Tollin, G. (1971) Biochemistry 10, 133

Part IX

Lipoyl dehydrogenase and glutathione reductase

Flavins and Flavoproteins, edited by T.P. Singer

CHAPTER 48

ACTIVE CENTER SEQUENCE HOMOLOGY AND MECHANISTIC HOMOLOGY BETWEEN GLUTATHIONE REDUCTASE AND LIPOAMIDE DEHYDROGENASE

CHARLES H. WILLIAMS, Jr., L. DAVID ARSCOTT and ERIC T. JONES

Veterans Administration Hospital and The Department of Biological Chemistry, The University of Michigan, Ann Arbor, Mich. 48105 (U.S.A.)

Summary

The sequence of amino acid residues around the active-center cystine in yeast glutathione reductase is presented and compared with the structures of the analogous regions of lipoamide dehydrogenase and thioredoxin reductase. There is a high degree of homology in this region between glutathione reductase and lipoamide dehydrogenase. These two enzymes have been known to be mechanistically very similar.

The 2-electron-reduced form of glutathione reductase, known to be the catalytically important intermediate, forms complexes with both NADPH and NADP. The spectral characteristics of these complexes have been established for the enzyme from *Escherichia coli* and are very similar to analogous complexes formed by pig heart lipoamide dehydrogenase with NAD and NADH. Thus, complex formation with the reduced pyridine nucleotide results in decreased absorbance in the 450 nm region and increased absorbance at wavelengths above 500 nm, while complex formation with the oxidized pyridine nucleotide results in decreased absorbance in the 450 nm region and in the 530 nm region, with increased absorbance at wavelengths greater than 560 nm.

The complex between two-electron-reduced lipoamide dehydrogenase and NAD is quite stable. The analogous complex between glutathione reductase and NADP, however, slowly forms another species. This species has spectral characteristics similar to those of the flavin semiquinone anion and its formation is associated with the production of an ESR signal.

Three well characterized flavoproteins catalyze electron transfer between pyridine nucleotides and disulfides: lipoamide dehydrogenase, glutathione reductase and thioredoxin reductase. In spite of the chemical similarity of the reactions catalyzed, these enzymes exhibit a high degree of substrate specificity in most species. Each of the three enzymes is composed of two identical or near identical polypeptide chains and two molecules of FAD, tightly bound in an apolar environment [1–6]. Each polypeptide chain contains a single cystine residue which shows redox activity acting with the FAD in catalysis [1, 7–10]. The two active centers on a single protein molecule appear to be kinetically independent. The minimum requirements, then, for each active center would be: an FAD-binding site, a specific disulfide substrate-binding site, a pyridine nucleotide-binding site and finally the disulfide of the redox-active cystine residue.

Recent reports from our laboratory [4, 6, 11, 12] have focused on the amino acid sequences around the two halves of the redox-active cystine residue in lipoamide dehydrogenase and thioredoxin reductase. We are now able to add the sequence in this region of the final member of this trio, glutathione reductase [13]. It was found that the halves of the redox-active cystine residue were quite close to one another in the primary structures, as shown in Fig. 1. Models of these peptides indicate that the close juxtaposition forces a flat conformation upon this section of the polypeptide chain and we were tempted to speculate that this flat area might accommodate the planar isoalloxazine ring.

Since there is no homology around the active-site cystine residue between lipoamide dehydrogenase and thioredoxin reductase, these structures must have been evolved convergently. The same sequence for this region of thioredoxin reductase was obtained by Thelander [14]. In contrast, the high degree of homology in this region of lipoamide dehydrogenase from a prokaryote and a eukaryote may indicate that this enzyme has undergone divergent evolution from a common ancestor, but such a conclusion must

Pig heart lipoamide dehydrogenase:

Glu-Lys-Asx-Glu-Thr-Leu-Gly-Gly-Thr-Cys-Leu-Asn-Val-Gly-Cys-Ile-Pro-Ser-Lys-Ala-Leu (S–S bridge between the two Cys)

Yeast glutathione reductase:

Lys-Ala-Gly-Lys-Ala-Leu-Gly-Gly-Thr-Cys-Val-Asn-Val-Gly-Cys-Val-Pro-Lys-Val-Val-Met (S–S bridge between the two Cys)

Escherichia coli lipoamide dehydrogenase:

Tyr-Asn-Thr-Leu-Gly-Gly-Val-Cys-Leu-Asn-Val-Gly-Cys-Ile-Pro-Ser-Lys (S–S bridge between the two Cys)

Escherichia coli thioredoxin reductase:

Ala-Cys-Ala-Thr-Cys-Asp-Gly-Phe (S–S bridge between the two Cys)

Fig. 1. Sequences of amino acid residues around the redox active cystine residue in pig heart and *E. coli* lipoamide dehydrogenase, yeast glutathione reductase and *E. coli* thioredoxin reductase.

remain tentative with so short a sequence from a polypeptide chain of almost 500 residues. The lipoamide dehydrogenase sequences shown in Fig. 1 combine our data with that of Brown and Perham [15, 16] who have extended our sequences toward the amino terminus. They have also corrected two errors in our sequence of this region in the pig heart enzyme [12] and we have confirmed their findings.

The recently determined sequence in this region of glutathione reductase is also shown in Fig. 1. It is homologous to a very high degree with the lipoamide dehydrogenases. If we consider the 15 residue sequences, 14 of which are homologous between the two lipoamide dehydrogenases, beginning with the Thr residue in the *Escherichia coli* lipoamide dehydrogenase sequence and running to the end, it can be seen that 10 of the positions in glutathione reductase are homologous and two of the five changes are chemically conservative, namely a Val for a Leu and a Val for an Ile. In addition, both enzymes have a lysine residue on the carboxyl-terminal side of the disulfide. Again, the sequence is much too short to allow speculation on the divergent evolution of these catalytically distinct enzymes from a common ancestor. However, the important point is the homology in this region known to be crucial in catalysis, thus reinforcing the long recognized mechanistic similarity between these two flavoproteins. It seems reasonable to assume that the region of homology contains some of the determinants necessary to maintain the close juxtaposition of the two half cystines during rapid reduction and reoxidation as well as the determinants vital in the interaction of the disulfide with the FAD. On the other hand, the region of homology must not contain any groups functional in the binding of the disulfide or dithiol substrate since the substrates are very different and the degree of specificity is very high. A parallel can be drawn between this group and the aldolases, where the closely related aldolases A and B show extensive homology around the active-site lysine residue but do not shown homology with transaldolase around the analogous lysine. Such is the case with glutathione reductase and lipoamide dehydrogenase, where there is no homology with thioredoxin reductase around the active-center cystine residue.

Glutathione reductase and lipoamide dehydrogenase, containing as they do two redox-active groups, can accept four electrons upon reduction with strong reagents such as dithionite. The anaerobic titration of *E. coli* glutathione reductase with dithionite is shown in Fig. 2 and is essentially the same as similar titrations of lipoamide dehydrogenase. The titration has two phases; in the phase up to two electrons per active center (curves 1–4) there is a modest decrease in the 450 nm region and an increase in the absorbance at longer wavelengths. The set of four isosbestic points in this phase indicate that two species are involved, the oxidized enzyme and a species which we will refer to as EH_2 without any commitment as to the redox states of the two active groups on the enzyme. In the second phase of the titration, the absorbance decreases in both regions and the final spectrum is that of reduced FAD. The classical studies of Massey, Veeger and Gibson [7, 8] have shown that in catalysis, these enzymes cycle between the oxidized and EH_2 states.

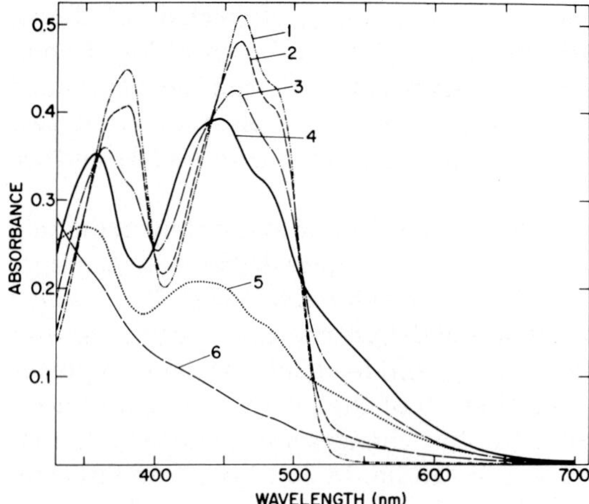

Fig. 2. Titration of *E. coli* glutathione reductase with dithionite. Curve 1, 44.2 μM oxidized enzyme in 0.05 M phosphate buffer, 0.15 mM EDTA, pH 7.6 following anaerobiosis. The dithionite was 1.95 mM (determined by titration with lumiflavin acetic acid) in 0.1 M pyrophosphate buffer, pH 8.4. Curve 2, immediately after the addition of 0.24 equivalents (FAD) of dithionite; curve 3, immediately after 0.71 equivalents; curve 4, 40 min after 1.19 equivalents; curve 5, 12.7 h after 2.38 equivalents; curve 6, 29.5 h after 3.57 equivalents. The spectra were taken when no further changes with time were observed.

A possible mechanism for lipoamide dehydrogenase is shown in Fig. 3; it is essentially a simplification of the mechanism of Massey and Veeger [8]. Catalysis in the physiological direction is clockwise. The enzyme is reduced by dihydrolipoamide through a dithiol-disulfide interchange and lipoamide dissociates to give free EH_2. In this picture we have indicated a charge-transfer interaction between a thiolate anion and the FAD which is an interpretation of Massey and Ghisla [17] based on the spectrum of EH_2. The reaction is then completed in the reoxidation of the enzyme by NAD. This is pictured as a simple binary-complex mechanism but it is far from simple, as the debate of the past 10 or so years attests. It should be emphasized that lipoamide dehydrogenase is reversible over a wide pH range.

An essentially identical mechanism is shown for glutathione reductase in Fig. 4. Again, physiological catalysis is clockwise but, physiologically, glutathione reductase oxidizes rather than reduces pyridine nucleotide. Unlike lipoamide dehydrogenase, it is reversible only at high pH with the aid of a low potential reagent, dithiotheitol. Reduction of these enzymes is very rapid and in the stopped flow, no intermediates have been observed between species I and IV (Fig. 3) and species I and II (Fig. 4). However, it seems reasonable that pyridine nucleotide intially reduces the FAD which then exchanges electrons with the disulfide as suggested by Searls et al. [18]. This is the case with thioredoxin reductase but with this enzyme, subsequent charge transfer between the thiolate and the FAD is not observed [9].

Given that the simple binary-complex formulation is an inadequate description of the mechanism of these flavoproteins, it is essential to quantitate the relative importance of all species of the enzymes which are either part of the catalytic cycle or are on significant side paths from that cycle.

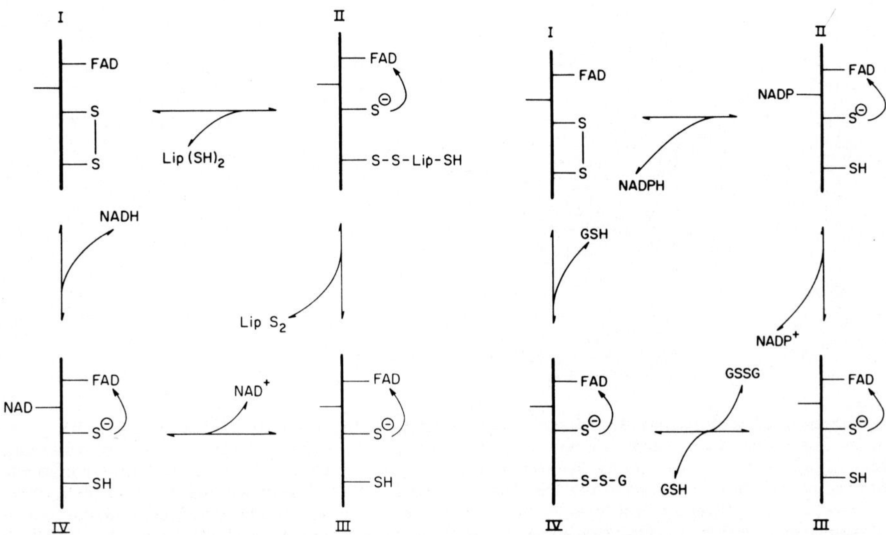

Fig. 3. A working hypothesis for the enzyme and enzyme-substrate intermediates in catalysis by lipoamide dehydrogenase.
Fig. 4. A working hypothesis for the enzyme and enzyme-substrate intermediates in catalysis by glutathione reductase.

Veeger and Massey [19] were the first to point out the perturbation by NAD of the spectrum of lipoamide dehydrogenase at the EH_2 redox level and Bulger and Brandt [20] have demonstrated a perturbation by NADPH of the spectrum of glutathione reductase at the EH_2 redox level. These perturbations are indications of complex formation and Bulger and Brandt determined a dissociation constant of 2 μM for the complex of yeast glutathione reductase at the EH_2 level with NADPH. A complex of EH_2 with reduced pyridine nucleotide would either constitute a side path from species III (Figs. 3 and 4) or, if its presence did not prevent the binding of the oxidizing substrate, it could constitute an alternate pathway of catalysis. Both these complexes can be demonstrated in both enzymes and their spectral characteristics are essentially the same in both enzymes.

In demonstrating such complexes it is desirable to generate EH_2 in the absence of pyridine nucleotide. As we have pointed out previously, this can be done very conveniently with glutathione reductase by the addition of a slight excess of borohydride, allowing sufficient time for the hydrolysis of the excess before proceeding with the pyridine nucleotide titration. Fig. 5 shows such an experiment with the *E. coli* glutathione reductase. Curve 2, the highest in the 450 nm region and the lowest in the 600 nm region, is the spectrum of free EH_2. Curves 3–5 show the progressive changes upon the addition of NADPH: increasing absorbance at long wavelengths and a modest decrease in the 450 nm region. The K_D calculated from this titration is 11 μM or some 5-fold higher than that found by Bulger and Brandt for the yeast enzyme. This again illustrates what has been found so often in the past, that the differences between species and between the two enzymes are

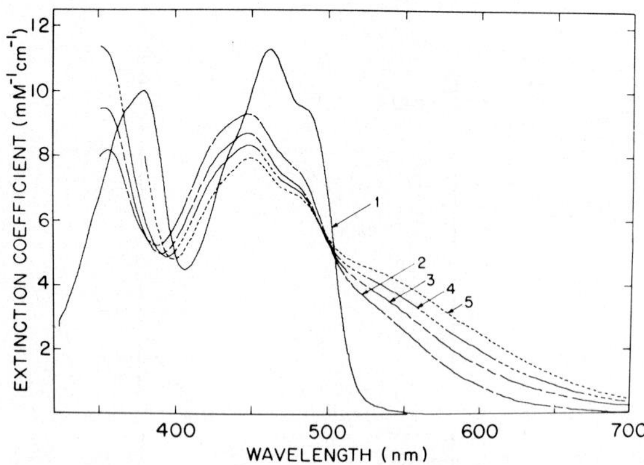

Fig. 5. Titration of 2-electron-reduced *E. coli* glutathione reductase (EH_2) with NADPH. EH_2 was formed by the addition of a slight excess (1.9 equivalents in 15 μl 0.2 M NaOH) of borohydride and sufficient time was allowed for the hydrolysis of the excess before beginning the titration; borohydride does not reduce this enzyme beyond the EH_2 level. Curve 1, 39.9 μM oxidized enzyme in 0.10 M phosphate buffer, pH 7.0, after anaerobiosis; curve 2, after the addition of the borohydride and the hydrolysis of the excess; curve 3, after the addition of 0.45 equivalents of NADPH; curve 4, after 0.90 equivalents; curve 5, after 2.75 equivalents. The spectral changes were immediate. All spectra were corrected for dilution.

quantitative rather than qualitative. An analogous complex can be formed with lipoamide dehydrogenase and its spectrum is shown in the accompanying paper [21]. Such rapidly formed, tight complexes between EH_2 and reduced pyridine nucleotide give spectral confirmation to such kinetic studies as those of Reed [22] with lipoamide dehydrogenase in which the product-inhibition patterns indicated dead-end inhibition.

Direct evidence for the formation of an EH_2·NADP complex with the *E. coli* glutathione reductase is given in Fig. 6. Indirect evidence for such a complex in the yeast enzyme has been given [20]. Again EH_2 was formed with borohydride. The spectral changes are quite different from those observed in the formation of the EH_2·NADPH complex. Whereas the complex with reduced pyridine nucleotide has higher extinction than that of free EH_2 at all wavelengths greater than 500 nm, the complex with oxidized pyridine nucleotide shown here has decreased extinction at 530 nm and increased extinction above 560 nm. A moderate decrease in absorbance in the 450 nm region is observed.

Similar spectral changes are observed with lipoamide dehydrogenase when free EH_2 is titrated with NAD as shown in the accompanying paper [21].

Titration of the oxidized glutathione reductase directly with NADPH, as shown in Fig. 7, is much more complex but can be interpreted in the light of the spectra of free EH_2 and its complexes. In the first phase, when less than two electron equivalents have been added (curves 2 and 3), a set of four isosbestic points at 505, 434, 400 and 363 nm would lead one to believe that two species are involved. However, it must be pointed out that in the dithionite experiment, where the two species in the first phase were fairly unambiguous, the isosbestic point was at 444 nm not at 434. We believe that

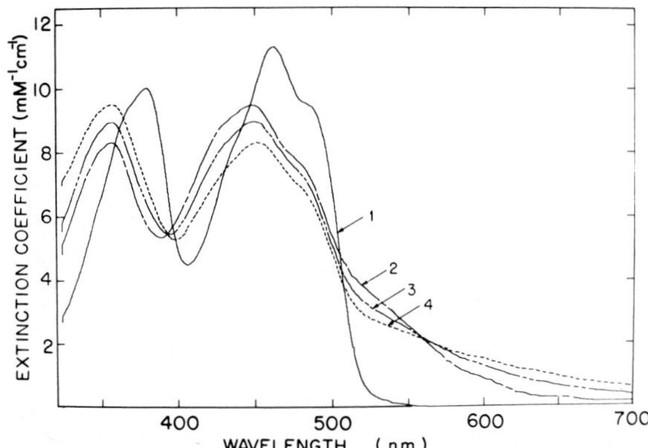

Fig. 6. Titration of 2-electron-reduced *E. coli* glutathione reductase (EH_2) with NADP. EH_2 was formed as in Fig. 5. Curve 1, 38.6 μM oxidized enzyme in 0.14 M phosphate buffer, pH 7.1, after anaerobiosis; curve 2, after the addition of the borohydride and the hydrolysis of the excess; curve 3, after the addition of 1.72 equivalents of NADP; curve 4, afte 10.36 equivalents. The spectral changes were immediate. All spectra were corrected for dilution.

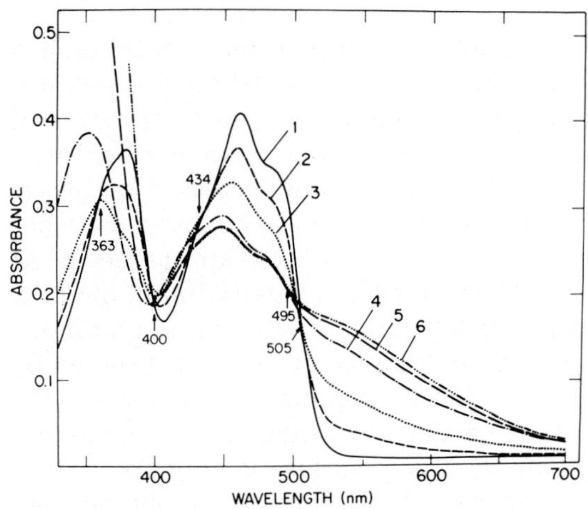

Fig. 7. Titration of *E. coli* glutathione reductase with NADPH. Curve 1, 35.4 μM oxidized enzyme in 0.10 M phosphate buffer, 0.3 mM EDTA, pH 7.6, after anaerobiosis; curve 2, after the addition of 0.40 equivalents of NADPH; curve 3, after 0.80 equivalents; curve 4, after 1.80 equivalents; curve 5, after 4.4 equivalents; curve 6, after 9.8 equivalents. The spectral changes were immediate. The spectra have not been corrected for dilution (up to 2%).

this indicates that when less than two electron equivalents have been added, the oxidized enzyme is being converted to a mixture of free EH_2 and its complexes. It does not seem likely that the EH_2·NADPH complex makes a large contribution to this mixture in light of the efficiency with which NADPH reduces the enzyme. In the latter phase of this experiment the EH_2·NADPH complex is clearly being formed (curves 4, 5 and 6).

Bulger and Brandt [20] gave preliminary data on still another reaction involving the 2-electron-reduced glutathione reductase. Its quite slow rate of

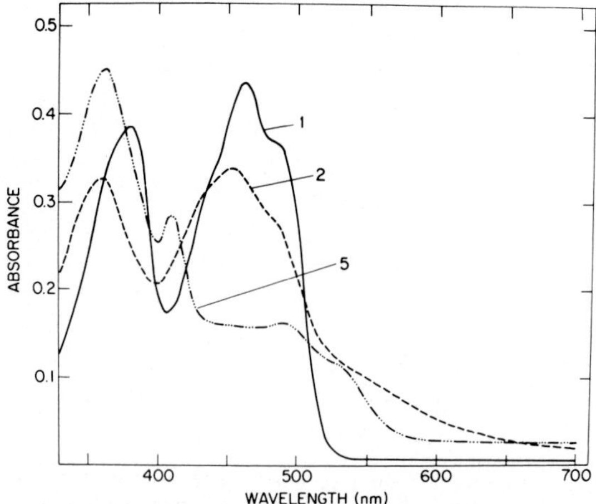

Fig. 8. Conversion of 2-electron-reduced *E. coli* glutathione reductase to another species. Curve 1, 37.7 μM oxidized enzyme in 0.10 M phosphate buffer, 0.3 mM EDTA, pH 7.6, after anaerobiosis; curve 2, immediately after the addition of 1.0 equivalent of NADPH; curve 5, 24 h after the addition of 10 equivalents of NADP; the NADP was added 22 h after the NADPH.

formation makes it of unlikely catalytic interest. Since a similar reaction had been observed earlier by Massey et al. [23] in lipoamide dehydrogenase, it is another point of similarity between the two. Fig. 8 shows the formation of this species from EH_2. The rate and extent of this reaction are increased by a high concentration of NADP and by high pH, and are decreased by excess NADPH. The conversion of EH_2 to this species appears to be incomplete but the extent of conversion can be correlated roughly with the appearance of an ESR signal. (We are grateful to Dr. G. Palmer for his collaboration in the ESR measurements.) The optical spectrum is like that of a flavin semiquinone anion having peaks at 365, 410 and 490 nm. The last of these is obscured by the remaining EH_2. Again the difference between glutathione reductase and lipoamide dehydrogenase is quantitative rather than qualitative. The extent of formation of this species in lipoamide dehydrogenase is only about 10% and is very slow [23]. The spectral features of the anion semiquinone were therefore not obvious.

A vast body of evidence has attested to the remarkable similarity between glutathione reductase and lipoamide dehydrogenase. The homology at the active sites and the virtually identical spectral characteristics of the EH_2·pyridine nucleotide complexes further demonstrate the basic similarity.

Acknowledgements

This work was supported in part by Grant GM 21444 from the National Institute of General Medical Sciences, National Institutes of Health, United States Public Health Service. This work was aided in part by United States Public Health Service Grant AM 12734 (Equipment Grant) to the Department of Biological Chemistry.

References

1 Massey, V. and Williams, Jr., C.H. (1965) J. Biol. Chem. 240, 4470–4480
2 Mavis, R.D. and Stellwagen, E. (1968). J. Biol. Chem. 243, 809–814
3 Massey, V., Hofmann, T. and Palmer, G. (1962) J. Biol. Chem. 237, 3820–3828
4 Burleigh, Jr., B.D. and Williams, Jr., C.H. (1972) J. Biol. Chem. 247, 2077–2082
5 Thelander, L. (1967) J. Biol. Chem. 242, 852–859
6 Ronchi, S. and Williams, Jr., C.H. (1972) J. Biol. Chem. 247, 2083–2086
7 Massey, V., Gibson, Q.H. and Veeger, C. (1960) Biochem. J. 77, 341–351
8 Massey, V. and Veeger, C. (1961) Biochim. Biophys. Acta 48, 33–47
9 Zanetti, G. and Williams, Jr., C.H. (1967) J. Biol. Chem. 242, 5232–5236
10 Thelander, L. (1968) Eur. J. Biochem. 4, 407–422
11 Williams, Jr., C.H., Burleigh, Jr., B.D., Ronchi, S., Arscott, L.D. and Jones, E.T. (1971) in Flavins and Flavoproteins (Kamin, H., ed.), pp. 295–311, University Park Press, Baltimore
12 Williams, Jr., C.H. and Arscott, L.D. (1972) Z. Naturforsch. 27b, 1078–1080
13 Jones, E.T. and Williams, Jr., C.H. (1975) J. Biol. Chem. 250, 3779–3784
14 Thelander, L. (1970) J. Biol. Chem. 245, 6026–6029
15 Brown, J.P. and Perham, R.N. (1972) FEBS Lett. 26, 221–224
16 Brown, J.P. and Perham, R.N. (1974) Biochem. J. 138, 505–512
17 Massey, V. and Ghisla, S. (1974) Ann. N. Y. Acad. Sci. 227, 446–465
18 Searls, R.L., Peters, J.M. and Sanadi, D.R. (1961) J. Biol. Chem. 236 2317–2322
19 Veeger, C. and Massey, V. (1963) Biochim. Biophys. Acta 67, 679–681
20 Bulger, J.E. and Brandt, K.G. (1971) J. Biol. Chem. 246, 5578–5587
21 Matthews, R.G., Wilkinson, K.D., Ballou, D.P. and Williams, Jr., C.H. (1976) Flavins and Flavoproteins, Proceedings of the 5th International Symposium (Singer, T.P., ed.), Chapt. 49, pp. 464–472, Elsevier, Amsterdam
22 Reed, J. (1973) J. Biol. Chem. 248, 4834–4839
23 Massey, V., Palmer, G., Williams, Jr., C.H., Swoboda, B.E.P. and Sands, R.H. (1966) in Flavins and Flavoproteins (Slater, E.C., ed.), pp. 133–152, Elsevier Publishing Company, Amsterdam

Flavins and Flavoproteins, edited by T.P. Singer

CHAPTER 49

THE ROLE OF NAD AS AN ACTIVATOR IN THE NADH-LIPOAMIDE REDUCTASE REACTION OF PIG HEART LIPOAMIDE DEHYDROGENASE

ROWENA G. MATTHEWS, KEITH D. WILKINSON, DAVID P. BALLOU and CHARLES H. WILLIAMS, Jr.

Veterans Administration Hospital and The Department of Biological Chemistry, The University of Michigan, Ann Arbor, Mich. 48105 (U.S.A.)

Summary

The role of NAD as an activator in the NADH-lipoamide reductase reaction catalyzed by pig heart lipoamide dehydrogenase has been studied.

1. NAD functions both as an activator (apparent $K_A = 5\ \mu M$) and as a product inhibitor ($K_{ip} = 280\ \mu M$). The activation is maximal at pH 6.3. The effect of increasing the NAD concentration in the range 0–67 μM is to decrease the apparent Michaelis constant for oxidized lipoamide while leaving V unchanged.

2. Stopped-flow studies at pH 6.3 show that 65 μM NAD does not affect the kinetics of the reaction of oxidized lipoamide with 2-electron-reduced enzyme produced by dithionite titration. Extrapolation of this stopped-flow data to infinite lipoamide concentration gives a limiting velocity of 50 000 per min. This is in agreement with $V = 50\,000$ per min extrapolated from steady state kinetic studies of the NADH-lipoamide reductase activity at pH 6.3.

3. 4-electron-reduced enzyme produced by stoichiometric titration with dithionite is rapidly oxidized to 2-electron-reduced enzyme by NAD. The rate of reoxidation is 38 000 per min for 65 μM NAD and 17 μM enzyme at pH 6.3.

4. NADH functions both as a substrate and as a competitive inhibitor when oxidized lipoamide is the varied substrate in the NADH-lipoamide reductase reaction at pH 6.3.

5. A model for the mechanism of lipoamide dehydrogenase is proposed to explain these results.

In this paper we shall examine the role of NAD as an activator in the NADH-lipoamide reductase reaction catalyzed by pig heart lipoamide dehydrogenase and discuss the implications of our findings for the kinetic mechanism of lipoamide dehydrogenase.

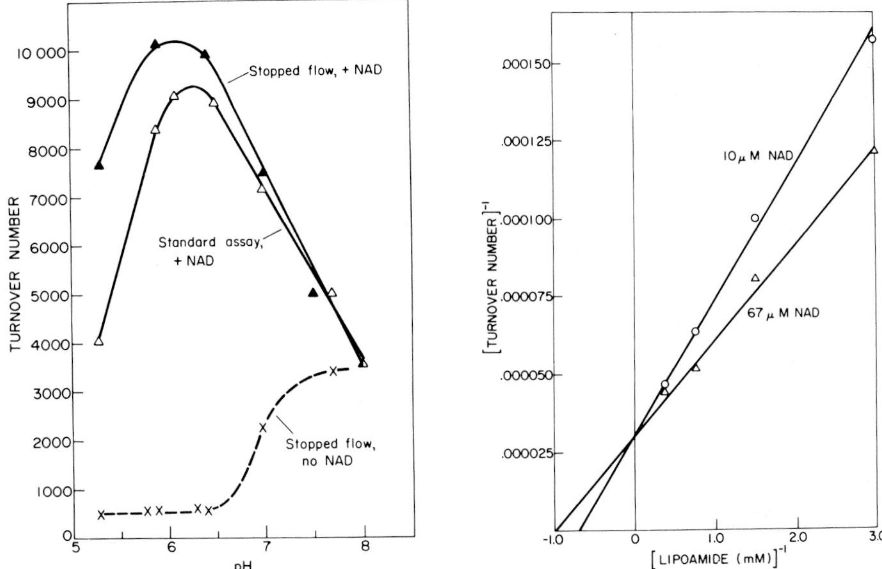

Fig. 1. Comparison of the pH-activity profiles for NADH-lipoamide reductase activity in the presence or absence of added NAD. Turnover numbers were determined from the changes in absorbance at 340 nm. ▲——▲, 1.7 nM enzvme in 1 mM EDTA pH 7.0, containing 0.13% bovine serum albumin, 1.34 mM oxidized lipoamide and 67 μM NAD, was reacted in the stopped-flow apparatus with an equal volume of 100 μM NADH and 67 μM NAD in 0.1 M phosphate of the desired pH; X------X, conditions as above except that no NAD was added to the reaction mixtures; △——△, turnover numbers obtained for enzyme activity measured spectrophotometrically at 340 nm in 3-ml volumes containing 0.05 M phosphate buffer at the indicated pH values; 2 mg bovine serum albumin; 0.3 μmol NADH; 2 μmol DL-lipoamide; 1.023 pmol lipoamide dehydrogenase and 0.2 μmol NAD. NADH followed by enzyme were added to start the reaction.

Fig. 2. The effect of NAD activation on the NADH-lipoamide reductase reaction at pH 6.3. Assays were performed spectrophotometrically at 340 nm, 25°C, in a total volume of 3 ml containing 0.05 M phosphate pH 6.3; 1 mM EDTA; 2 mg bovine serum albumin; 0.15 μmol NADH; 1.023 pmol lipoamide dehydrogenase; 0.03 or 0.20 μmol NAD; and 0.99–8.04 μmol DL-lipoamide. All assays were initiated by the addition of enzyme to the reaction mixture. Each point represents the average of rates determined from four assays.

Massey and Veeger [1] observed a distinct lag phase in the NADH-lipoate reductase reaction if NAD was not added to the reaction mixture. The rate of the reaction accelerated as NAD was produced, but was negligible if NADase was present in the assay mixture. Addition of NAD to the starting mixture abolished the lag phase.

Searls and Sanadi [2] showed that the lag phase was most pronounced at low pH and that quite linear initial rates were observed at pH 7 and above. The effect of NAD on the pH-activity profile for NADH-lipoate reductase activity was described by Voetberg et al. [3]. We have obtained a very similar pH-activity profile for the NADH-lipoamide reductase reaction, which is shown in Fig. 1. When lipoamide is the substrate, the maximum stimulation exerted by NAD is at pH 6.3, which is also the optimum for the NADH-lipoamide reductase reaction. At higher pH the turnover numbers measured in the stopped flow approach one another and the observed time courses for NADH oxidation become more and more linear. The turnover

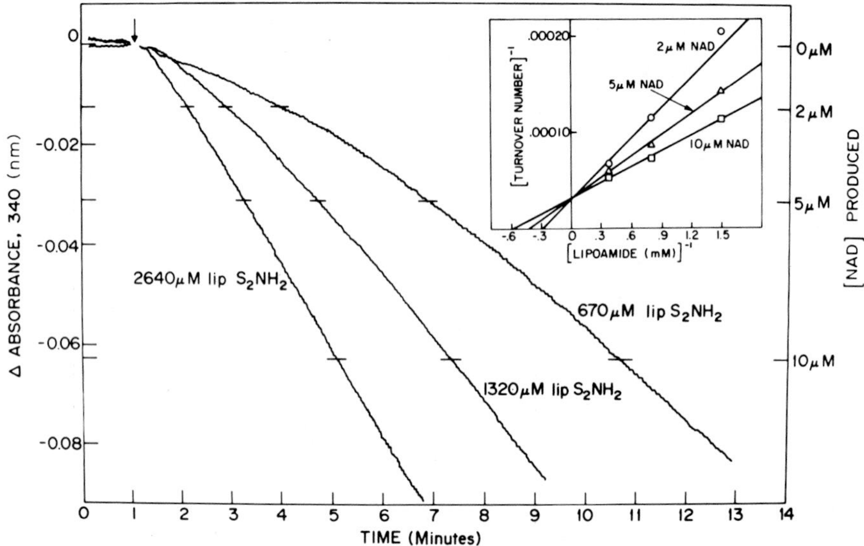

Fig. 3. Time courses of the NADH-lipoamide reductase reaction in the absence of added NAD, as a function of lipoamide concentration. The assays were performed under the conditions described in Fig. 2, except that no NAD was added to the reaction mixture. The lipoamide added varied from 2 to 8 μmol for the three assays. The horizontal lines intersecting each time course correspond to the production of 2, 5 and 10 μM NAD respectively. Tangents to each time course at these NAD concentrations were used to determine the catalytic rates. Control time courses in the presence of 67 μM NAD were used to determine the decrease in rate due to oxidation of NADH at these NAD concentrations. (At 10 μM NAD, the remaining NADH is 40 μM and the rate in the presence of 67 μM NAD is decreased 6–12%). Rates measured in the absence of added NAD were then corrected for effects due to decreases in NADH concentration, and used to construct the Lineweaver–Burk plot shown in the inset.

numbers determined from stopped-flow kinetics were calculated from rates measured during the first 2 s of reaction. Under these conditions, a linear rate of NADH oxidation is reached within 200 ms of mixing, when 67 μM NAD is present. In the absence of NAD, no oxidation of NADH could be detected in the first 200 ms after mixing and the data points represent upper limits for initial rates at these pH values.

Since NADH-lipoamide reductase assays are performed at concentrations of lipoamide which are below saturation, we were interested in examining the effect of NAD on the apparent affinity for lipoamide. Fig. 2 shows a Lineweaver–Burk plot of initial rates where lipoamide is the varied substrate. The effect of increasing concentrations of NAD is an increase in the apparent affinity of the enzyme for lipoamide, with no observable change in the intercept velocity. Increasing the NAD concentration above 67 μM produces no further activation, and below 10 μM NAD, the initial rates are too non-linear to obtain accurate initial slopes.

The effect of lower concentrations of NAD on NADH-lipoamide reductase activity was examined in another way. Fig. 3 shows time courses of NADH oxidation at three different lipoamide concentrations. No NAD was added to the reaction mixtures so that NAD produced was stoichiometric with NADH oxidized. The slopes of these curves were measured at the points

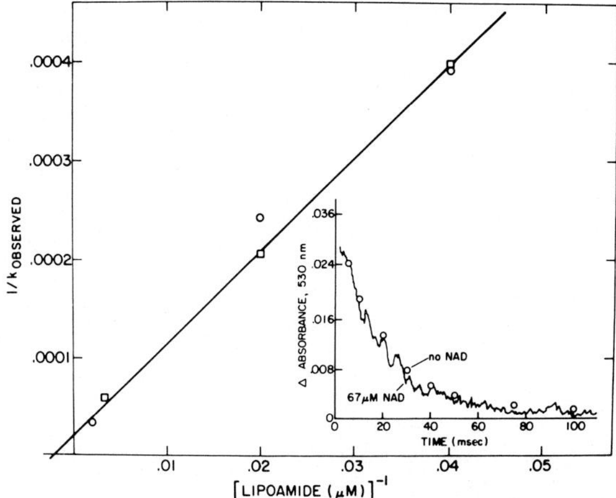

Fig. 4. The rate of reaction of anaerobic 2-electron-reduced enzyme (EH_2) with oxidized DL-lipoamide. EH_2 was produced by the addition of one equivalent of dithionite per flavin. The dithionite concentration was determined by titration with lumiflavin acetic acid. The circles represent reaction rates determined in the stopped flow apparatus under anaerobic conditions with 15 μM EH_2 and 25–500 μM oxidized DL-lipoamide (concentrations after mixing). The squares represent reaction rates determined in the same way, but in the presence of 65 μM NAD. The data shown by the squares at 50 μM and 294 μM lipoamide concentrations were obtained by mixing EH_2 and lipoamide; 65 μM NAD were present in both syringes. The inset compares the data obtained when EH_2 was reacted with 25 μM lipoamide in the presence (trace) or absence (circles) of NAD. A solution containing 50 μM lipoamide in 0.1 M phosphate buffer, pH 6.3, with or without 130 μM NAD, was mixed with an equal volume of enzyme solution in the same buffer.

where the NAD concentration had reached 2, 5 and 10 μM, and the rates were corrected for the corresponding decreases in NADH concentration (see legend to Fig. 3). A Lineweaver–Burk plot of the corrected rates is shown as an inset. The plot confirms the indication that the primary effect of NAD is on the apparent affinity for lipoamide and that the intercept velocity is unchanged within the limits of experimental error.

In the concentration range 2–67 μM, NAD as an activator has a linear effect on the slopes of reciprocal plots, which means that within this concentration range, NAD is reacting predominantly with only one enzyme species [4]. The apparent K_A for NAD is 5 μM, but this is not a simple binding constant.

If we examine the reaction of 2-electron-reduced enzyme (EH_2) with oxidized lipoamide

$$EH_2 + LipS_2NH_2 \rightleftharpoons EH_2{-}LipS_2NH_2 \rightarrow E + Lip(SH)_2NH_2 \qquad (1)$$

we can see that NAD could exert the observed effect on the apparent affinity of EH_2 for lipoamide in either of two ways: (a) by increasing the affinity of EH_2 for lipoamide, or (b) by increasing the concentration of EH_2 in the steady state. Measurements of the reoxidation of EH_2 by oxidized lipoamide can be performed in the stopped-flow apparatus and such measurements allow us to distinguish between these two alternatives. Fig. 4 shows experiments performed to evaluate the effect of NAD on the rate of

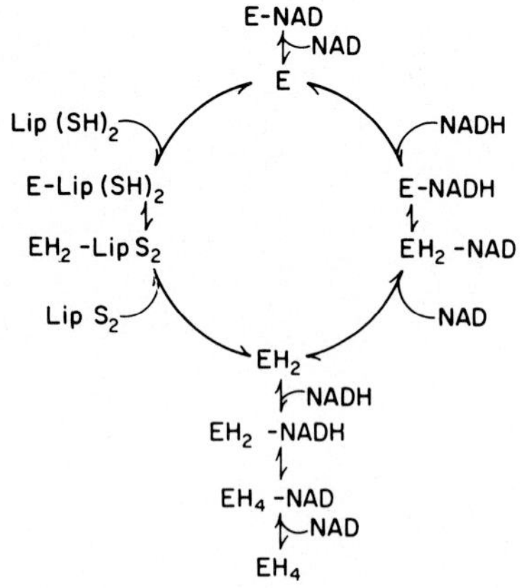

E-NAD
NAD
E
Lip $(SH)_2$
NADH
E-Lip $(SH)_2$
E-NADH
EH_2-Lip S_2
EH_2-NAD
Lip S_2
NAD
EH_2
NADH
EH_2-NADH
EH_4-NAD
NAD
EH_4

Fig. 5. Scheme for the proposed reaction mechanism of lipoamide dehydrogenase.

oxidation of EH_2 by lipoamide. Within the limits of the experimental technique, we were unable to observe an effect of 65 μM NAD. Thus, we conclude that the activation effect of NAD cannot be explained by alterations of the affinity of EH_2 for lipoamide. Extrapolation of these stopped flow data to infinite lipoamide concentration yields a limiting velocity of 50 000 per min.

Fig. 5 shows the mechanism with which we propose to explain how NAD alters the steady state concentrations of EH_2. This mechanism is very similar to the classical lipoamide dehydrogenase mechanism of Massey and Veeger [1] but incorporates the suggestion made by Sanadi (in the discussion following ref. 5) that NAD functions as an activator by reacting with 4-electron-reduced enzyme (EH_4) to produce EH_2. Massey and Gibson [5] observed that EH_4 was produced in substantial amounts during steady state turnover in the absence of NAD, and that addition of NAD led to rapid disappearance of the EH_4.

We have made predictions on the basis of this model and then subjected them to experimental investigation. (1) NADH should be a competitive inhibitor with respect to lipoamide even in the presence of NAD. (2) NAD must react rapidly with EH_4 to produce 2-electron-reduced enzyme and the overall equilibrium must lie in the direction of 2-electron-reduced enzyme at 67 μM NAD.

Massey et al. [6] demonstrated that at pH 7.6 and 25°C, the enzyme showed parallel line kinetics for the transfer of electrons from reduced lipoamide to NAD, and that the rate-limiting step was the reduction of enzyme by reduced lipoamide. Reed [7] has studied steady state kinetics of the transfer of electrons from reduced lipoamide to NAD, and from NADH to oxidized lipoamide at pH 8, 37°C, using the rat liver enzyme. She concluded that the kinetic and isotope-exchange data are compatible with a parallel-line mechanism with dead-end inhibition by both NADH and lipoamide. However, no data have been published for the kinetics of transfer of electrons

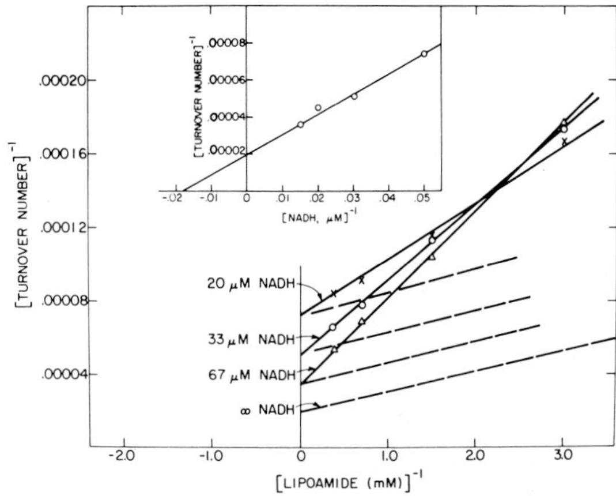

Fig. 6. Lineweaver–Burk plots of the NADH-lipoamide reductase reaction at pH 6.3. The assays were performed as in Fig. 2, except that the NAD concentration was fixed at 67 μM, and the NADH concentration was varied from 20 to 67 μM. The dashed lines at finite NADH concentrations indicate the expected pattern in the absence of competitive inhibition by either NADH or NAD, and the dashed line at inifinite NADH represents the rate of reoxidation of EH_2 by oxidized lipoamide determined in Fig. 4. The inset shows a secondary plot of intercepts vs. reciprocal NADH concentration.

from NADH to lipoamide in the pH region where NAD is required as an activator. Fig. 6 shows Lineweaver–Burk plots of the data obtained at pH 6.3. In this series of reciprocal plots, the intercepts are decreasing and the slopes increasing with increasing NADH concentration. Garces and Cleland [8] have obtained very similar results for an enzyme reaction in which the fixed substrate also functions as a competitive inhibitor with respect to the varied substrate. NADH as the fixed substrate in a reaction, which proceeds by a ping-pong kinetic mechanism, should affect the intercepts of reciprocal plots but not the slopes. As a competitive inhibitor, NADH should affect the slopes of the reciprocal plots but not the intercepts. Secondary plots of intercepts vs. reciprocal NADH concentration should yield V for the catalytic reaction. The inset shows the secondary plot. V obtained in this way is 50 000 per min in agreement with the limiting velocity extrapolated from stopped-flow measurements of the reoxidation of EH_2 by oxidized lipoamide. The dashed line at inifinite NADH concentration represents the rate of reoxidation of EH_2 by oxidized lipoamide as determined in Fig. 4. This rate of reoxidation, in the presence or absence of NAD, is sufficiently rapid to account for turnover, not only at infinite lipoamide concentrations but also at finite lipoamide concentrations, and should be rate limiting in the absence of NADH inhibition. It can be seen that all the reciprocal plots at finite NADH concentrations show steeper slopes than this, and inhibition is very marked even though all measurements were made in the presence of 67 μM NAD. Although these data lead us to conclude that NADH is a competitive inhibitor with respect to lipoamide in this reaction, we are unable to distinguish between fully and partially competitive inhibition. Indeed, a secondary plot of the slopes of reciprocal plots vs. the NADH concentration was concave

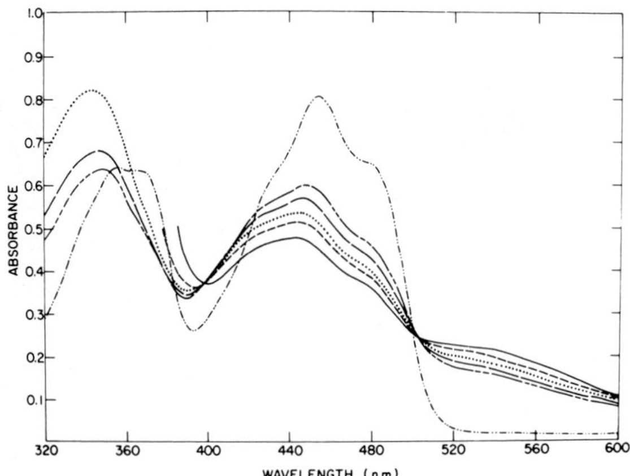

Fig. 7. Anaerobic titration of the enzyme with NADH. —···—, 2 ml of 71 μM enzyme in 0.05 M phosphate pH 7.5, 25°C, after anaerobiosis; — --, in presence of 1 equivalent of NADH; ——, 1.25 equivalents; · · · · · ·, 1.75 equivalents; ----, 2.75 equivalents; ——, 6 equivalents NADH. All spectra are corrected for dilution.

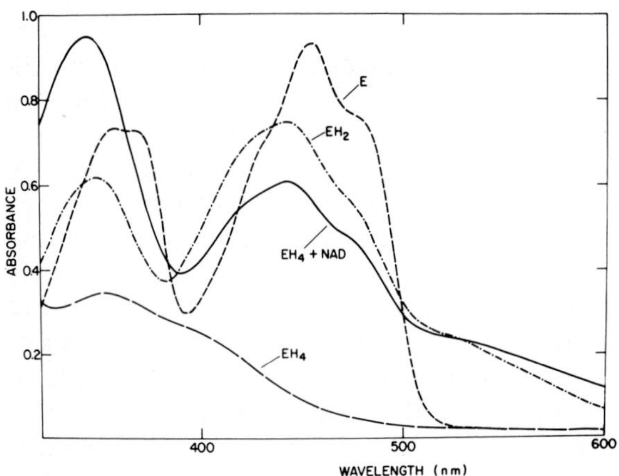

Fig. 8. Anaerobic reaction of 4-electron-reduced enzyme (EH_4) with NAD. — — —, 2 ml of 81 μM enzyme in 0.05 M phosphate buffer pH 6.3, 25°C, following anaerobiosis; — · —, after addition of 1.04 equivalents of dithionite; — —, after addition of 2.08 equivalents of dithionite; ——, immediately after tipping in 2.47 equivalents of NAD (200 μM).

downwards, which leads us to suspect that NADH is only partially competitive with respect to lipoamide.

Fig. 7 shows a spectral titration of oxidized enzyme with NADH. This experiment was done at pH 7.5, but very similar results are obtained from titrations at pH 6.3. Addition of one equivalent of NADH produces a spectrum which differs somewhat from free EH_2 (Fig. 8), it is higher at 340 nm and lower at 440 nm. As we add NADH the absorbance falls at 440 and rises at wavelengths above 503 nm. After addition of six equivalents of NADH, no further increases at long wavelengths are observed on addition

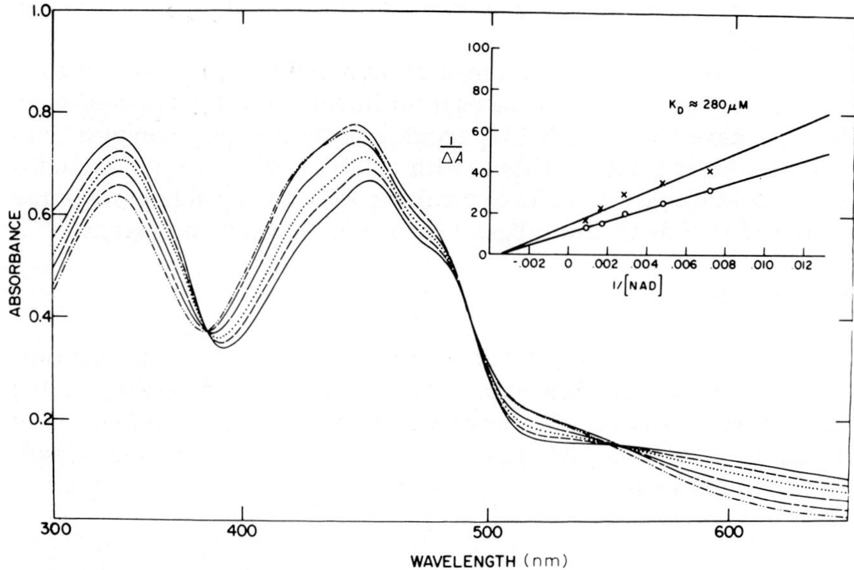

Fig. 9. Anaerobic titration of 2-electron-reduced enzyme with NAD. — · · —, 2 ml 86 μM enzyme in 0.05 M phosphate pH 6.3, 25°C, following addition of 1 equivalent dithionite; — -- —, in presence of 0.84 equivalents NAD; — —, 1.6 equivalents NAD; · · · · · ·, 4.2 equivalents NAD; — — —, 7.0 equivalents NAD; and ——, 13.5 equivalents NAD. The enzyme was titrated with 48 mM NAD. The inset shows the data at 340 nm (X) and 650 nm (o) plotted by the method of Benesi and Hildebrand [12].

of NADH. These spectral changes are compatible with the formation of an $EH_2 \cdot NADH$ complex, as had previously been suggested by data of Veeger (in the discussion of ref. 9). The spectral changes are quite similar to those observed by Bulger and Brandt [10] on titration of 2-electron-reduced glutathione reductase with NADPH.

Our second prediction was that NAD should react rapidly with EH_4 to produce EH_2 and the overall equilibrium should favor 2-electron-reduced enzyme at 67 μM NAD. Fig. 8 shows spectral changes which result when NAD is added to 4-electron-reduced enzyme. The products appear to be a mixture of free EH_2 and $EH_2 \cdot NADH$. From the rise in absorbance at 340 nm it is evident that NADH has been produced. We have measured the rate of oxidation of EH_4 by NAD in the stopped flow using 65 μM NAD and 17 μM enzyme, and the observed rate constant is 38 000 per min.

Veeger and Massey [11] have described a complex between EH_2 and NAD. We have determined the dissociation constant for the $EH_2 \cdot NAD$ complex by spectrophotometric titration of EH_2 with NAD and the results are shown in Fig. 9. K for the dissociation is 280 μM. Under these conditions, no reoxidation of the enzyme is seen. Such weak binding of NAD to EH_2 means that in the presence of 67 μM NAD the ratio of EH_2 to $EH_2 \cdot NAD$ will be 4 :1. Hence our stopped-flow measurements of the rate of reaction of EH_2 with lipoamide in the presence of 65 μM NAD were performed under conditions where the majority of the enzyme was present as free EH_2. We have not studied the rate of reaction of $EH_2 \cdot NAD$ with oxidized lipoamide.

Thus, we have obtained experimental verification of the two predictions made on the basis of the mechanism for lipoamide dehydrogenase catalysis. While we have shown that EH_2 reoxidizes via a binary complex with oxidized lipoamide at rates consistent with catalysis, we have not excluded the possibility that ternary complexes involving EH_2, oxidized lipoamide and either NAD or NADH as the third ligand are also kinetically competent.

Acknowledgements

This work was supported in part by Grant GM 21444 from the National Institute of General Medical Sciences, National Institutes of Health, United States Public Health Service. This work was aided in part by United States Public Health Service Grant AM 12734 (Equipment Grant) to the Department of Biological Chemistry.

References

1 Massey, V. and Veeger, C. (1961) Biochim. Biophys. Acta 48, 33–47
2 Searls, R.L. and Sanadi, D.R. (1961) J. Biol. Chem. 236, 580–583
3 van Muiswinkel-Voetberg, H., Visser, J. and Veeger, C. (1973) Eur. J. Biochem. 33, 265–270
4 Cleland, W.W. (1963) Biochim. Biophys. Acta 67, 188–196
5 Massey, V. and Gibson, Q.H. (1963) in Proceedings of the Fifth International Congress of Biochemistry, Vol. 5, pp. 157–173, Pergamon Press, Oxford
6 Massey, V., Gibson, Q.H. and Veeger, C. (1960) Biochem. J. 77, 341–351
7 Reed, J.K. (1973) J. Biol. Chem. 248, 4834–4839
8 Garces, E. and Cleland, W.W. (1969) Biochemistry 8, 633–640
9 Massey, V., Palmer, G., Williams, Jr., C.H., Swoboda, B.E.P. and Sands, R.H. (1966) in Flavins and Flavoproteins (Slater, E.C., ed.), B.B.A. Library, Vol. 8, Elsevier, Amsterdam
10 Bulger, J.E. and Brandt, K.G. (1971) J. Biol. Chem. 246, 5578–5587
11 Veeger, C. and Massey, V. (1963) Biochim. Biophys Acta 67, 679–681
12 Benesi, H.A. and Hildebrand, T.A. (1949) J. Am. Chem. Soc. 71, 2703–2707

Flavins and Flavoproteins, edited by T.P. Singer

CHAPTER 50

REACTION MECHANISM OF THE COMPLEXED AND UNCOMPLEXED LIPOAMIDE DEHYDROGENASES IN PIG HEART 2-OXOGLUTARATE DEHYDROGENASE MULTIENZYME COMPLEX

MASAHIKO KOIKE and KICHIKO KOIKE

Department of Pathological Biochemistry, Atomic Disease Institute, Nagasaki University School of Medicine, 12-4, Sakamoto-cho, Nagasaki-shi 852 (Japan)

Summary

The kinetic mechanisms of the complexed and uncomplexed lipoamide dehydrogenases of the pig heart 2-oxoglutarate dehydrogenase complex were studied. Initial velocity patterns and product inhibition data were consistent with a bi-bi ping-pong mechanism for the action of this enzyme.

Lipoamide dehydrogenase (reduced nicotinamide adenine dinucleotide: lipoamide oxidoreductase, EC 1.6.4.3) is an essential component of the two species of α-keto acid dehydrogenase multienzyme complexes of pig heart mitochondria which catalyze the CoA- and NAD-linked oxidative decarboxylation of pyruvate [1, 2] or of 2-oxoglutarate [3] (Eqn. 1).

$$\text{R—COCOOH} + \text{CoA—SH} + \text{NAD}^+ \longrightarrow \text{R—CO—S—CoA} + \text{CO}_2 + \text{NADH} + \text{H}^+ \qquad (1)$$

A comparison of enzymatic activities, thiol contents, electrophoretic patterns after incubation with NADH and arsenite, and optical rotatory dispersion and circular dichroism spectra of the lipoamide dehydrogenase isolated from each of the two types of complexes indicated only conformational differences around the active site involving thiol groups and FAD [4]. Although the two forms of lipoamide dehydrogenase appear unique, they are interchangeable with respect to both function and the ability to form the multienzyme complexes.

Lipoamide dehydrogenase is a flavoprotein, whose physiological function appears to be the oxidation of protein-bound dihydrolipoic acid with NAD^+ serving as the electron acceptor of a coordinated sequence of reactions (see Fig. 1). From the kinetic and product inhibition studies of the 2-oxoglutarate dehydrogenase complex, a series of parallel lines were obtained from plots of $1/v$ versus 2-oxoglutarate concentrations for various

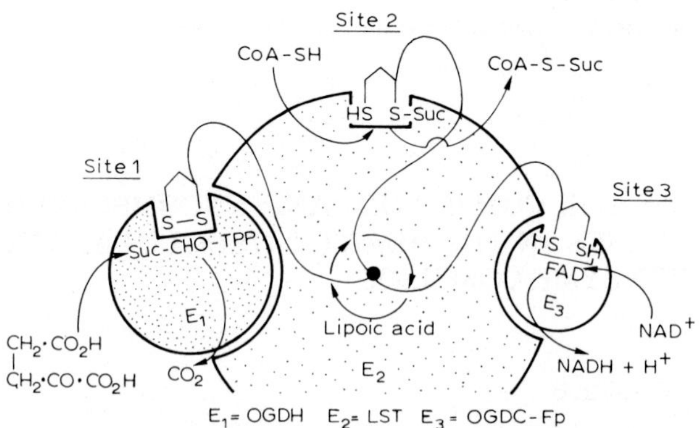

Fig. 1. Schematic representation of the reaction sequence of oxidative decarboxylation of 2-oxoglutarate (Eqn. 1). The scheme depicts a rotation of the lipoyllysyl residue from succinic semialdehyde-thiamine-*PP* (Suc-CHO-TPP) at Site 1 (E_1) to Site 2 (E_2) for the transfer of the succinyl group to CoA and to Site 3 (E_3) for the oxidation of FAD. The lipoyllysyl residue is covalently bound to the core enzyme (E_2). The abbreviations used are: E_1 = OGDH, 2-oxoglutarate dehydrogenase; E_2 = LST, lipoate succinyltransferase; E_3 = OGDC-Fp, lipoamide dehydrogenase.

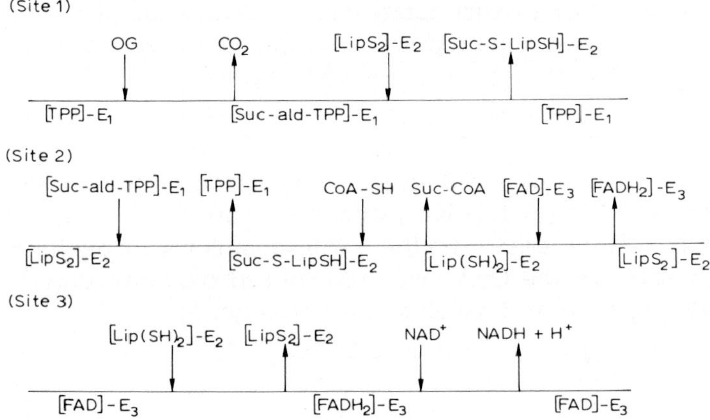

Fig. 2. Diagram of the overall oxidative decarboxylation reaction mechanism catalyzed by the 2-oxoglutarate dehydrogenase complex. The abbreviations used are: Suc-ald-TPP, succinic semialdehyde-thiamine-*PP*; $LipS_2$, lipoic acid; Suc-S-LipSH, *S*-succinyl dihydrolipoic acid; and $Lip(SH)_2$, dihydrolipoic acid.

levels of CoA and a level of NAD^+ near its Michaelis constant value [5]. These results were taken as evidence in favor of the three-site ping-pong mechanism (Fig. 2) originally proposed by Cleland [6] for the mechanism of action of the complex. Recently, similar observations were made with the bovine kidney pyruvate dehydrogenase multienzyme complex [7]. From the diagram of Site 3 in Fig. 2, a bi-bi ping-pong mechanism would be predicted for the lipoamide dehydrogenase. This paper reports on initial velocity and product inhibition studies of the complexed and uncomplexed lipoamide dehydrogenases of the 2-oxoglutarate dehydrogenase complex.

Materials and Methods

The kinetic mechanisms of lipoamide dehydrogenase in the complexed (E_1-E_2-E_3) and uncomplexed (E_3) forms were studied. Initial velocity experiments were performed by measuring the oxidation of NADH by lipoamide (Eqn. 2). The assay mixture contained in a total

$$\text{lipoamide} + \text{NADH} + \text{H}^+ \rightleftharpoons \text{dihydrolipoamide} + \text{NAD}^+ \tag{2}$$

volume of 2.0 ml, 100 μmol of potassium phosphate buffer (pH 7.5), 0.1 μmol of thiamine-*PP* and the amounts of the reactant specified in the text. Lipoamide was added in 0.05 ml of 95% ethanol. After equilibration at 25°C, the reaction was started by addition of 2 μg of 2-oxoglutarate dehydrogenase complex (complexed form) or 0.5 μg of isolated lipoamide dehydrogenase (uncomplexed form). The rate of oxidation of NADH was followed at 340 nm with a recording spectrophotometer. Initial velocities are expressed as μmol of NADH oxidized per min. The 2-oxoglutarate dehydrogenase complex (complexed form) and lipoamide dehydrogenase (uncomplexed form) were prepared essentially as described in a previous paper [3]. The sources of other materials and description of methods are described in a previous paper [5].

Result and Discussion

Initial velocity experiments

The kinetics obtained from initial velocity studies of the forward reaction (Eqn. 2) with the uncomplexed lipoamide dehydrogenase gave a series of parallel lines, as shown in Fig. 3A and B. Similar results were obtained with the complexed form of lipoamide dehydrogenase. The data are consistent with a bi-bi ping-pong mechanism for the two forms of lipoamide dehydrogenase and fit the general equation (Eqn. 3) derived by Cleland [8] for a bi-bi ping-pong mechanism in the absence of products; where K_a and K_b

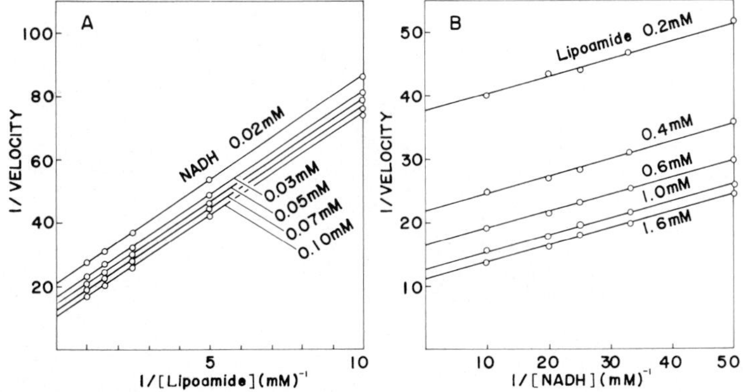

Fig. 3. Plot of the reciprocal of initial velocities (v) versus the reciprocal of lipoamide (A) or NADH (B) concentrations. A, the NADH concentrations were held constant as indicated. B, the lipoamide concentrations were held constant as indicated. Assay conditions were as described under Materials and Methods. Velocity is expressed as micromoles of NADH oxidized per min.

$$v = \frac{V}{1 + \frac{K_a}{[A]} + \frac{K_b}{[B]}} \tag{3}$$

are Michaelis constants for lipoamide (A) and NADH (B), respectively. Michaelis constants for reactants were estimated to be 0.85 mM for lipoamide and 0.024 mM for NADH.

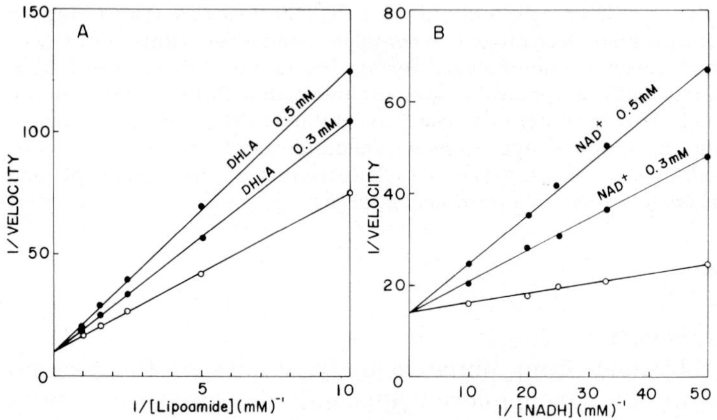

Fig. 4. (A) Product inhibition by dihydrolipoamide (DHLA). Lipoamide was the varied substrate. NADH was held constant at 0.1 mM. Dihydrolipoamide added as indicated. (B) Product inhibition by NAD^+. NADH was the varied substrate. Lipoamide was held constant at 1.0 mM. NAD^+ added as indicated. Other conditions were as in Fig. 3.

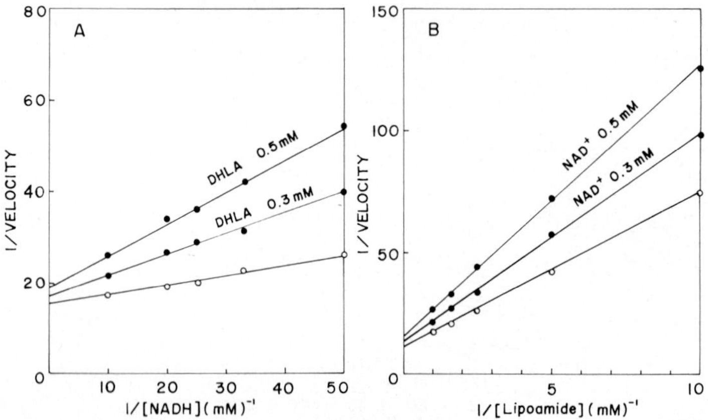

Fig. 5. (A) Product inhibition by dihydrolipoamide (DHLA). NADH was the varied substrate. Lipoamide was held constant at 1.0 mM. Dihydrolipoamide added as indicated. (B) Product inhibition by NAD^+. Lipoamide was the varied substrate. NADH was held constant at 0.1 mM. NAD^+ added as indicated.

Product inhibition experiments

The rate equation of a bi-bi ping-pong mechanism (Eqn. 3) predicts each product to be competitive with respect to substrates. In agreement with these predictions dihydrolipoamide and NAD^+ were competitive with respect to lipoamide and NADH, respectively, as shown in Fig. 4A and B. Dihydrolipoamide and NAD^+ were noncompetitive with respect to NADH and lipoamide, respectively, as shown in Fig. 5A and B. Similar product inhibitions were obtained with either the complexed or the uncomplexed lipoamide dehydrogenase. That no interaction occurs between E_1 and E_3 or E_2 and E_3 (see Fig. 1) in the complex was indicated by the observation that 2-oxoglutarate or succinyl-CoA had no effect. However, CoA inhibited the complexed lipoamide dehydrogenase noncompetitively with respect to NADH when the concentration of the fixed substrate (lipoamide) was near its Michaelis constant. CoA did not inhibit the complexed lipoamide dehydrogenase with respect to lipoamide in the complex. These observations with the 2-oxoglutarate dehydrogenase complex suggest that the combination of CoA with E_2 in some way hinders the combination of NADH.

Using the uncomplexed lipoamide dehydrogenase from rat liver mitochondria, Reed [9] reported that their results were consistent with a bi-bi ping-pong mechanism originally proposed by Massey et al. [10]. In addition, the formation of kinetically significant, abortive complexes between the enzyme and NADH or lipoamide were observed.

Acknowledgements

The authors wish to thank Dr. W.W. Cleland, Department of Biochemistry, University of Wisconsin for his criticism. We also thank Dr. Kenneth H. Tachiki, Section of Neurobiology, Institute of Psychiatric Research, Indiana University Medical School for his criticism and kind help in preparation of this manuscript. This investigation was supported in part by the grants from the Ministry of Education of Japan, the Vitamin B Research Committee and U.S. Army R & D Group (Far East).

References

1 Hayakawa, T., Hirashima, M., Ide, S., Hamada, M., Okabe, K. and Koike, M. (1966) J. Biol. Chem. 241, 4694–4699
2 Hayakawa, T., Kanzaki, T., Kitamura, T., Fukuyoshi, Y., Sakurai, Y., Koike, K., Suematsu, T. and Koike, M. (1969) J. Biol. Chem. 244, 3660–3670
3 Tanaka, N., Koike, K., Hamada, M., Otsuka, K.-I., Suematsu, T. and Koike, M. (1972) J. Biol. Chem. 247, 4043–4049
4 Sakurai, Y., Fukuyoshi, Y., Hamada, M., Hayakawa, T. and Koike, M. (1970) J. Biol. Chem. 245, 4453–4462
5 Hamada, M., Koike, K., Nakaula, Y., Hiraoka, T., Koike, M. and Hashimoto, T. (1975) J. Biochem. Tokyo 77, 1047–1056
6 Cleland, W.W. (1973) J. Biol. Chem. 248, 8353–8355
7 Tsai, C., Burgett, M.W. and Reed, L.J. (1973) J. Biol. Chem. 248, 8348–8352
8 Cleland, W.W. (1967) Annu. Rev. Biochem. 36, 77–112
9 Reed, J.K. (1973) J. Biol. Chem. 248, 4834–4839
10 Massey, V., Gibson, Q.H. and Veeger, C. (1960) Biochem. J. 77, 341–351

Flavins and flavoproteins, edited by T.P. Singer

CHAPTER 51

KINETIC STUDIES OF THE EFFECT OF ADENOSINE DIPHOSPHORIBOSE ON PIG HEART LIPOAMIDE DEHYDROGENASE

KARL G. BRANDT and JOHN K. LUTTON

Department of Biochemistry, Purdue University, West Lafayette, Ind. 47907 (U.S.A.)

Summary

Steady-state kinetic studies of the pyridine nucleotide transhydrogenase reaction catalysed by lipoamide dehydrogenase show that adenosine diphosphoribose inhibits the transhydrogenase reaction by binding to both the oxidized and reduced forms of flavoenzyme. In contrast, adenosine diphosphoribose activates the NADH-lipoate reductase activity of the enzyme. Stopped-flow kinetic studies show that adenosine diphosphoribose, when bound to the enzyme, inhibits reduction of the enzyme by NADH, but facilitates reoxidation of the reduced enzyme by lipoic acid. It is concluded that the active site of lipoamide dehydrogenase has separate binding sites for pyridine nucleotide and lipoic acid.

Pig heart lipoamide dehydrogenase (EC 1.6.4.3) catalyzes both an NADH-lipoate reductase reaction (Eqn. 1) and a pyridine nucleotide transhydrogenase reaction (Eqn. 2) [1, 2]. Steady-state kinetic [3]

$$NADH + Lip(S_2) \rightleftharpoons NAD^+ + Lip(SH)_2 \quad (1)$$

$$NADH + thio\text{-}NAD^+ \rightleftharpoons NAD^+ + thio\text{-}NADH \quad (2)$$

and static spectroscopic [3, 4] studies suggest that the mechanism of the reaction in Eqn. 1 involves a ternary complex of enzyme with both substrates. Formation of a ternary complex requires separate binding sites for lipoic acid and pyridine nucleotide at the active site of the enzyme. A test for the existence of such separate sites would be to show that pyridine nucleotide bound at the active site does not prevent reoxidation of the reduced flavoenzyme by lipoic acid. This paper presents a kinetic study of the effect of adenosine diphosphoribose (ADPR) on the lipoate reductase

---

Abbreviations: $Lip(S_2)$ and $Lip(SH)_2$, oxidized and reduced lipoic acid; thio-NAD^+ and thio-NADH, oxidized and reduced thionicotinamide-adenine dinucleotide; ADPR, adenosine diphosphoribose.

and transhydrogenase activities of lipoamide dehydrogenase. ADPR, an analog of NAD lacking the nicotinamide ring, cannot undergo electron transfer reactions. The results indicate that ADPR binds at the active site of the enzyme without preventing reoxidation of the reduced enzyme by lipoic acid.

Materials and Methods

Pig heart lipoamide dehydrogenase, D,L-lipoic acid, and thio-NAD^+ were purchased from Sigma Chemical Co. NAD^+, NADH, and ADPR were obtained from Calbiochem. Enzyme-bound FAD concentration was determined at 455 nm using an absorption of $11.3\ mM^{-1} \cdot cm^{-1}$ [5].

Lipoate reductase activity was measured by observing the oxidation of NADH at 340 nm. The transhydrogenase activity was measured by observing the formation of thio-NADH at 395 nm. Velocities were computed in units of μM/min using absorption values of $6.22\ mM^{-1} \cdot cm^{-1}$ for NADH and $11.3\ mM^{-1} \cdot cm^{-1}$ for thio-NADH [6]. All experiments were performed in 55 mM sodium phosphate, pH 6.3, containing 1 mM EDTA. Stopped-flow kinetic studies were performed on a Durrum-Gibson stopped-flow spectrophotometer. Pseudo-first-order rate constants were evaluated from oscilloscope traces as described previously [7].

Results and Discussion

The pyridine nucleotide transhydrogenase activity of lipoamide dehydrogenase involves the interaction of only pyridine nucleotides with the enzyme. The steady-state kinetics of the transhydrogenase reaction and its inhibition by ADPR were studied in order to characterize the interactions of ADPR with the enzyme. Fig. 1 shows a Lineweaver-Burk plot of the initial

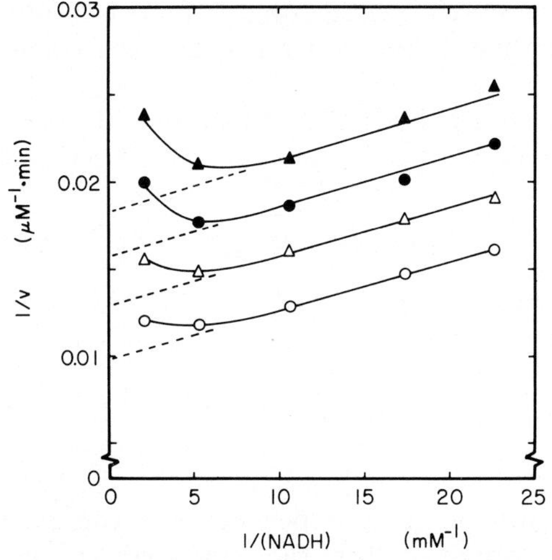

Fig. 1. Lineweaver-Burk plot of initial velocity data for the pyridine nucleotide transhydrogenase activity of lipoamide dehydrogenase at 25°C. Assay mixtures contained 24 nM enzyme-bound FAD, the indicated concentrations of NADH, and thio-NAD^+ concentrations of: ○, 224 μM; △, 112 μM; ●, 74 μM; ▲, 56 μM.

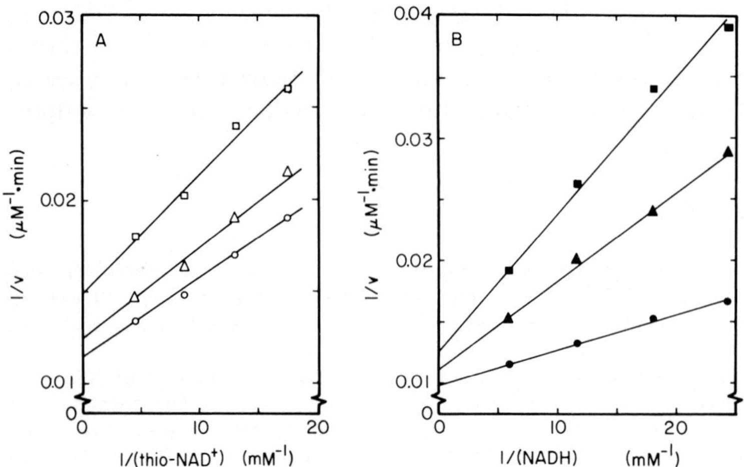

Fig. 2. Lineweaver-Burk plots of inhibition by ADPR of the pyridine nucleotide transhydrogenase activity of lipoamide dehydrogenase at 25°C. (A) Thio-NAD$^+$ as the variable substrate at a fixed concentration of 193 μM NADH. The concentrations of ADPR were: ○, zero; △, 0.19 mM; □, 0.38 mM. (B) NADH as the variable substrate at a fixed concentration of 224 μM thio-NAD$^+$. The concentrations of ADPR were: ●. zero; ▲, 0.39 mM; ■, 0.78 mM.

velocity with NADH as the variable substrate and thio-NAD$^+$ as the changing fixed substrate. The pattern of parallel lines at low NADH concentration indicates a ping-pong mechanism; the substrate inhibition by NADH at high NADH concentration is also typical of ping-pong mechanisms [8], resulting from the binding of substrate to the improper stable enzyme form. Fig. 2 shows that ADPR inhibits the transhydrogenase reaction non-competitively with respect to both NADH and thio-NAD$^+$, indicating [9] that ADPR competes with each substrate for binding at the active site on both stable forms of the enzyme.

The results in Figs. 1 and 2, neglecting the NADH substrate inhibition, are consistent with the mechanism in Eqns. 3–6. E_o and E_r are oxidized and reduced

$$E_o + NADH \rightleftharpoons E_o \cdot NADH \rightleftharpoons E_r \cdot NAD^+ \rightleftharpoons E_r + NAD^+ \quad (3)$$

$$E_r + \text{thio-NAD}^+ \rightleftharpoons E_r \cdot \text{thio-NAD}^+ \rightleftharpoons E_o \cdot \text{thio-NADH} \rightleftharpoons E_o + \text{thio-NADH} \quad (4)$$

$$E_o + ADPR \rightleftharpoons E_o \cdot ADPR \quad (5)$$

$$E_r + ADPR \rightleftharpoons E_r \cdot ADPR \quad (6)$$

forms of the enzyme. From Fig. 1 and replicate experiments, a turnover number of $6200 \pm 670\ \text{min}^{-1}$ was obtained for the transhydrogenase reaction with Michaelis constants for NADH of $35 \pm 2\ \mu$M and for thio-NAD$^+$ of $74 \pm 10\ \mu$M. Analysis of the data in Fig. 2 and a duplicate experiment yields values of the dissociation constants for $E_o \cdot$ ADPR and $E_r \cdot$ ADPR of $K_{i1} = 0.24 \pm 0.03$ mM and $K_{i2} = 0.61 \pm 0.06$ mM, respectively.

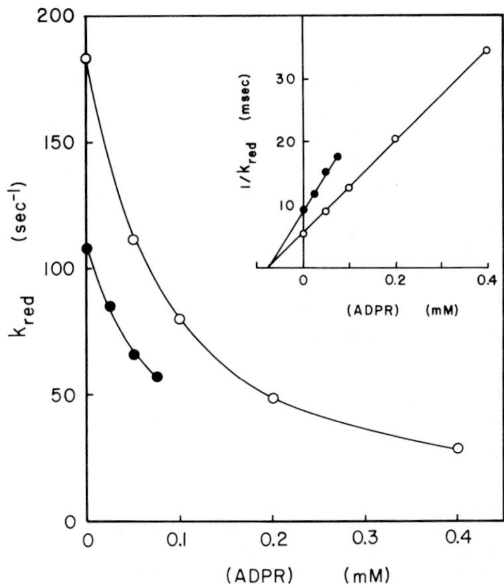

Fig. 3. The effect of ADPR on the apparent first order rate constant, k_{red}, for reduction of the enzyme-bound FAD of lipoamide dehydrogenase at 5°C. The initial concentrations of reactants, after mixing in the stopped-flow apparatus, were 3.9 μM enzyme-bound FAD, the indicated concentration of ADPR, and: ○, 15.6 μM NADH; ●, 9.2 μM NADH. Inset: Plot of the reciprocal of k_{red} vs. ADPR concentration.

K_{i1} represents the dissociation constant of the $E_o \cdot$ ADPR complex. It can be compared with the dissociation constant reported [4] for the analogous $E_o \cdot NAD^+$ complex. From difference spectral studies at 25°C, a value of $K_d = 0.20–0.25$ mM was reported [4] for NAD^+ bound at the active site. The pyridine nucleotide binding site at the active site of lipoamide dehydrogenase appears to have comparable affinities for NAD^+ and ADPR.

To provide further evidence that the inhibition of the transhydrogenase reaction resulted from the binding of ADPR at the active site of the enzyme, stopped-flow kinetic studies of the effect of ADPR on the reduction of the enzyme by NADH were performed. These studies were done at 5°C; reduction of the enzyme by NADH at 25°C is too fast to be observed [5]. Because the reduction is still rapid at 5°C, low ratios of NADH to enzyme-bound FAD were used to slow down the reaction. The reaction was followed by observing the absorbance increase at 530 nm characteristic of formation of the 2-electron reduced enzyme [5]. Fig. 3 shows that ADPR inhibits reduction of the enzyme. The observation that k_{red}, the apparent first order rate constant for reduction, decreases toward zero with increasing ADPR concentration, and does not level off at an intermediate value, indicates competitive inhibition. The inset of Fig. 3 shows a graph of $1/k_{red}$ vs. ADPR concentration. For competitive inhibition, the abscissa coordinate at the intersection of the two lines equals $-K_{i1}$, yielding $K_{i1} = 0.08$ mM at 5°C for the $E_o \cdot$ ADPR complex. This value of K_{i1} is smaller than the value obtained at 25°C. A similar decrease has been reported [4] for the dissociation constant of the analogous active site $E_o \cdot NAD^+$ complex at 0°C relative to 25°C.

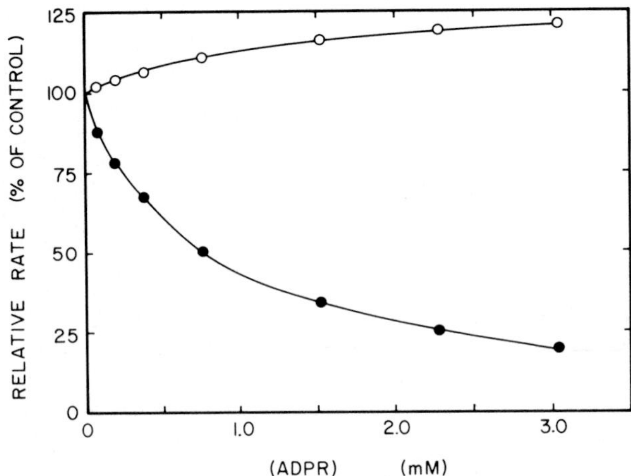

Fig. 4. Comparison of the effect of ADPR concentration on the lipoate reductase activity and the transhydrogenase activity of lipoamide dehydrogenase at 25°C. Both sets of data were obtained using the same enzyme preparation. The assay mixture for the lipoate reductase activity (○) contained 1.0 mM lipoic acid, 110 μM NADH, 100 μM NAD^+, 200 nM enzyme-bound FAD, and the indicated ADPR concentrations. The assay mixture for the transhydrogenase activity (●) contained 110 μM NADH, 211 μM thio-NAD^+, 10 nM enzyme-bound FAD, and the indicated ADPR concentrations.

From the data in Figs. 1–3, we conclude that ADPR binds at the active site of lipoamide dehydrogenase on both the oxidized and reduced forms of the enzyme. When the effect of ADPR on the lipoate reductase activity (Eqn. 1) of lipoamide dehydrogenase was investigated, however, inhibition was not observed. Fig. 4 compares the effect of ADPR concentration on the transhydrogenase and lipoate reductase activities of the enzyme. Whereas the transhydrogenase activity is inhibited by ADPR, the lipoate reductase activity is activated by ADPR. It is known [1] that NAD^+ is necessary in the assay mixture to prevent conversion of the enzyme to an inactive, 4-electron reduced form. The activation by ADPR shown in Fig. 4 is observed in the presence of NAD^+, and thus is not simply due to ADPR preventing over-reduction by NADH. Activation by ADPR is also observed in the absence of NAD^+.

The rate-limiting step in the lipoate reductase reaction is the reoxidation of the 2-electron reduced enzyme by lipoic acid [5]. The effect of ADPR on this step was studied by stopped-flow methods. Lipoamide dehydrogenase was reduced anaerobically with a 3-fold molar ratio of NADH to enzyme-bound FAD, and then mixed in the stopped-flow apparatus with solutions containing lipoic acid and ADPR. Oxidation of the 2-electron reduced enzyme was followed by observing the absorbance decrease at 530 nm [5]. With 4.5 μM enzyme-bound FAD and 90 μM lipoic acid at 25°C, the pseudo-first order rate constant for reoxidation, k_{ox}, increased with ADPR concentration: no ADPR, $k_{ox} = 0.44\ s^{-1}$; 0.5 mM ADPR, $k_{ox} = 0.51\ s^{-1}$; 1.0 mM ADPR, $k_{ox} = 0.55\ s^{-1}$; 2.0 mM ADPR, $k_{ox} = 0.61\ s^{-1}$. The increase in k_{ox} with ADPR concentration parallels the effect of ADPR on the steady-state rate of the lipoate reductase reaction, indicating that the presence of ADPR bound to the enzyme facilitates electron transfer from the 2-electron reduced enzyme to lipoic acid.

Lipoamide dehydrogenase has two different binding sites for NAD^+ on each of its two subunits [3, 4]. Of these two types of site, the lower affinity site is thought to represent the binding site at the active site [3, 4]. Inhibition of the transhydrogenase activity of lipoamide dehydrogenase by ADPR is characterized by inhibition constants comparable to the affinity of the active site for NAD^+. Furthermore, the inhibition by ADPR gives the pattern predicted for ADPR binding to the pyridine nucleotide binding site at the active site in competition with NADH and thio-NAD^+. We conclude that ADPR binds to the pyridine nucleotide binding site at the active site on both the oxidized and reduced forms of the enzyme. Thus, it is significant that, under conditions where ADPR inhibits the transhydrogenase activity, inhibition of the lipoate reductase activity is not observed. The lack of inhibition of the lipoate reductase activity by ADPR must mean that ADPR bound at the active site of the enzyme prevents neither the binding of lipoic acid to the 2-electron reduced enzyme nor the reoxidation of the reduced enzyme by the bound lipoic acid. This conclusion is supported by the stopped-flow kinetic studies of the effect of ADPR on k_{ox}. Therefore, it must be possible for both lipoic acid and ADPR to be bound simultaneously at the active site of the 2-electron reduced enzyme, to produce a kinetically competent ternary complex. We conclude that the active site of the 2-electron reduced enzyme has separate binding sites for pyridine nucleotide and lipoic acid.

The active site of lipoamide dehydrogenase contains both FAD and a redox-active protein disulfide [1]. Furthermore, the isoalloxazine ring of flavins can undergo a conformational change upon reduction [10]. It may be proposed that electron transfer from bound NADH to bound lipoic acid within the active site of the enzyme effectively involves electron transport between topographically distinct substrate binding sites, perhaps mediated by a conformational change in the isoalloxazine ring and/or the enzyme: for example, from bound NADH to FAD to protein disulfide to bound lipoic acid.

Acknowledgement

This work was supported by U.S. Public Health Service Research Grant No. GM 19458. This is Journal Paper No. 5843 from the Purdue University Agricultural Experiment Station, W. Lafayette, Ind. 47907.

References

1 Massey, V. and Veeger, C. (1961) Biochim. Biophys. Acta 48, 33–47
2 Sakurai, Y., Fukuyoshi, Y., Hamada, M., Hayakawa, T. and Koike, M. (1970) J. Biol. Chem. 245, 4453–4462
3 Visser, J., Voetberg, H. and Veeger, C. (1970) in Pyridine Nucleotide-Dependent Reductases (Sund, H., ed.), pp. 359–370, Springer-Verlag, New York

4 van Muiswinkel-Voetberg, H. and Veeger, C. (1973) Eur. J. Biochem. 33, 285–291
5 Massey, V., Gibson, Q.H. and Veeger, C. (1960) Biochem. J. 77, 341–351
6 Keister, D.L. and Yike, N.J. (1967) Biochemistry 6, 3847–3857
7 Bulger, J.E. and Brandt, K.G. (1971) J. Biol. Chem. 246, 5570–5577
8 Garces, E. and Cleland, W.W. (1969) Biochemistry 8, 633–640
9 Mourad, N. and Park, R.E. (1966) J. Biol. Chem. 241, 271–278
10 Kierkegaard, P., Norrestam, R., Werner, P.E., Csoregh, I., Glehn, M.V., Karlsson, R., Leijonmarck, M., Ronnquist, O., Stensland, B., Tillberg, O. and Torbjornsson, L. (1971) in Flavins and Flavoproteins (Kamin, H., ed.), pp. 1–20, University Park Press, Baltimore

Flavins and Flavoproteins, edited by T.P. Singer

CHAPTER 52

THE KINETIC MECHANISM OF GLUTATHIONE REDUCTASE: A BRANCHING REACTION SCHEME APPLICABLE TO MANY FLAVOPROTEINS

BENGT MANNERVIK

Department of Biochemistry, Arrhenius Laboratory, University of Stockholm, Stockholm (Sweden)

Introduction

Glutatione reductase, like many other flavoenzymes, catalyzes the transfer of chemical groups (in the present case "reducing equivalents") between substrate molecules of greatly dissimilar molecular structure. The group transfer is mediated by the flavin, and the donor and acceptor substrates, consequently, do not have to be present simultanously on the enzyme to make possible the reaction. Distinct binding sites probably exist for donor and acceptor substrates to account for the high specificity with respect to the two kinds of structurally different substrates. Thus, the question arises whether donor and acceptor substrates do in fact bind

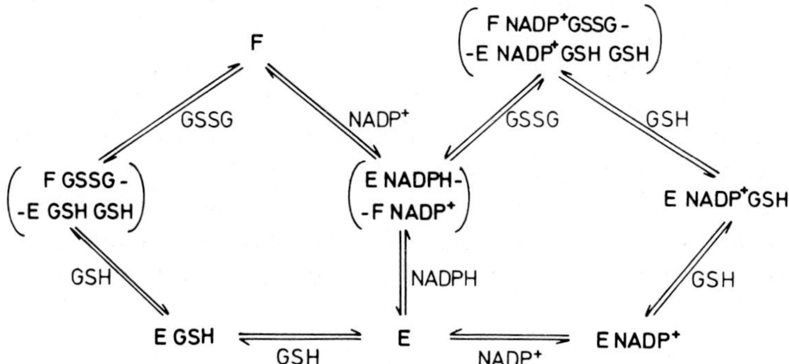

Fig. 1. Reaction scheme for the branching mechanism proposed for glutathione reductase. The left loop (counter-clockwise in the forward direction) corresponds to a ping-pong mechanism, and the right loop (clockwise) corresponds to a sequential mechanism. These loops describe the limiting kinetic patterns, which obtain at infinitely low and high concentrations of GSSG, respectively. E and F denote oxidized and reduced forms of the free enzyme. The substrates and products are written beside the arrows, which indicate the direction of addition of the particular reactant (i.e. products are indicated beside the arrows of steps directed backwards in the scheme). Isomerization steps, which are necessary for the catalysis (e.g. E • NADPH → F • NADP$^+$) but without effect on the algebraic form of the steady state rate equation are written within parentheses and not explicitly taken into account.

simultaneously to the enzyme or whether the acceptor substrate binds after the release of the product formed from the donor substrate. Although the fundamental chemical mechanism, a double-displacement reaction, is the same in both cases, the kinetic mechanisms of the two extreme cases are basically different. This paper presents a reaction scheme (Fig. 1) [1, 2] in which glutathione reductase operates by both the ping-pong and the sequential kinetic patterns. The concentration of the acceptor substrate, glutathione disulfide, governs the flux of reactants through the alternative reaction pathways: at low concentrations the ping-pong pattern prevails, whereas at high concentrations the sequential pathway dominates.

Theory

Wong and Hanes have shown that it is possible to predict the algebraic form of the initial-rate equation with respect to a particular reactant, R, by inspection of the reaction scheme considered for the mechanism [3, 4]. It was found [3] that the general form of the rate law was

$$v = \frac{\sum_{i=0}^{n} a_i [\mathrm{R}]^i}{\sum_{i=0}^{n} b_i [\mathrm{R}]^i} \tag{1}$$

where v is the initial velocity under steady-state conditions, a_i and b_i are coefficients (which may be functions of other variables than the particular reactant, R, considered), and n is equal to or less than the number of enzyme forms in the scheme which react with R. When R is a substrate, a_0 is always zero. The number n can be used to define the degree of a mechanism with respect to R [3]. Thus, it can be predicted that the reaction scheme presented in Fig. 1 will yield a rate law which is first degree in [NADPH] and second degree in [GSSG]. In the present paper the dependence of the initial velocity on the concentrations of the substrates and products will be compared with the predictions made on the basis of the scheme in Fig. 1. The comparison of alternative rate laws has been based on nonlinear regression analysis and statistical criteria as previously described [5, 6, 7], but the quantitative data are omitted from the present paper owing to space limitations. A detailed analysis of the kinetics of yeast glutathione reductase which also takes into account the experimental error structure has recently been presented [8, 9].

Examination of the kinetic model

The experimental data which will be used to test the reaction scheme in Fig. 1 were obtained with yeast glutathione reductase. Experimental details and parts of the extensive nonlinear regression analysis have previously been published [1, 9]. Although the present analysis is restricted to yeast glutathione reductase, it is clear that the kinetic data obtained with the enzyme from other sources, including rat liver [10], are compatible with the same reaction scheme (Fig. 1).

The rate equation of the forward reaction in the absence of products is, according to the scheme in Fig. 1:

$$v = \frac{V_1[\text{GSSG}][\text{NADPH}] + V_2[\text{GSSG}]^2[\text{NADPH}]}{K_1[\text{GSSG}] + K_2[\text{NADPH}] + [\text{GSSG}][\text{NADPH}] + K_3[\text{GSSG}]^2 + K_4[\text{GSSG}]^2[\text{NADPH}]} \quad (2)$$

where V_1, V_2, and K_i ($i = 1$–4) are constants. It is clear that v is first degree in [NADPH] and second degree in [GSSG] and consequently expected to give linear v versus v/[NADPH] and nonlinear v versus v/[GSSG] plots. These results have been obtained [9]. Another important property of Eqn. 2 is that it lacks a constant term in the denominator and thus degenerates to the rate law of a simple ping-pong mechanism at low [GSSG], which causes terms containing $[\text{GSSG}]^2$ to vanish. This feature of Eqn. 2 is an explanation of the finding of parallel lines in a double reciprocal plot [11] or a single point of intersection on the x-axis in a v versus v/[NADPH] plot [9]. The latter method of plotting is preferred because nonlinearities indicating second or higher degree rate laws are more easily discovered in these plots [12].

The rate equation of the reverse reaction is expected to be second degree in $[\text{NADP}^+]$ and fourth degree in [GSH]. Fig. 2 and previously

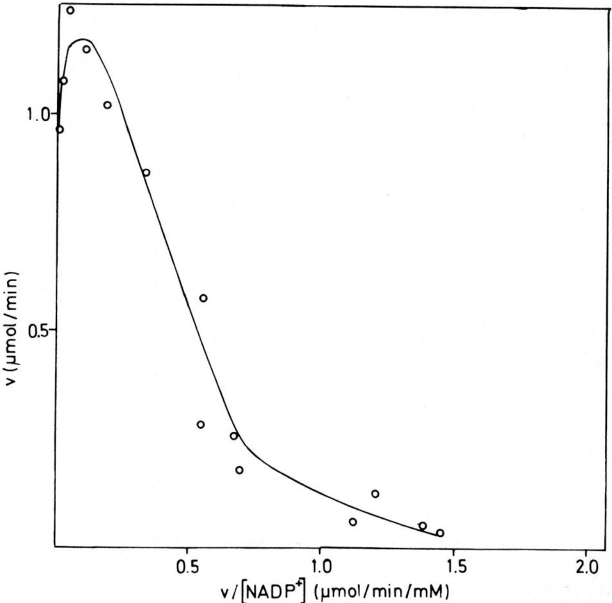

Fig. 2. Effect on initial velocity of varying $[\text{NADP}^+]$ in the reverse reaction. The measurements were made at a constant GSH concentration (6.0 mM) in a reaction system (30°C) containing 0.1 M sodium phosphate pH 7.6, 0.5 mM EDTA, 40 mM dithioerythritol, 0.5 mg human serum albumin, and 12 μg yeast glutathione reductase (Boehringer). Each point is the mean of four replicates.

published data [2] show nonlinearities which are in agreement with the predictions. At this point it can be concluded that the finding of nonlinearities which necessitate second degree terms of [GSSG] and [NADP$^+$] in the numerator of the corresponding rate laws of the forward [2] and reverse reactions (a finding which has been confirmed by statistical criteria) exclude the simple ping-pong or sequential mechanisms from consideration. These alternatives, which have been discussed in previous symposia [13, 14], are in fact both included as basic structures which are linked by a common step in the scheme in Fig. 1.

Further support for the branching mechanism is provided by the nonlinear inhibition of the forward reaction caused by NADP$^+$ [1]. The experimental data are consistent with the predicted rate law ([GSSG] is constant):

$$v = \frac{V_1\,[\mathrm{NADPH}] + V_2\,[\mathrm{NADPH}]\,[\mathrm{NADP^+}]}{K_1 + [\mathrm{NADPH}] + K_2\,[\mathrm{NADPH}]\,[\mathrm{NADP^+}] + K_3\,[\mathrm{NADP^+}] + K_4\,[\mathrm{NADP^+}]^2} \qquad (3)$$

The [NADPH] [NADP$^+$] and [NADP$^+$]^2 terms will vanish at high GSSG concentrations and Eqn. 3 thus degenerates to the case of a simple competitive inhibition as found previously [2]. The corresponding (more complex) rate equation applicable for variable [GSSG] and [NADP$^+$] at constant [NADPH] also adequately describes the pertinent experimental data. The effect of GSH on the forward reaction is expected to give a rate law second

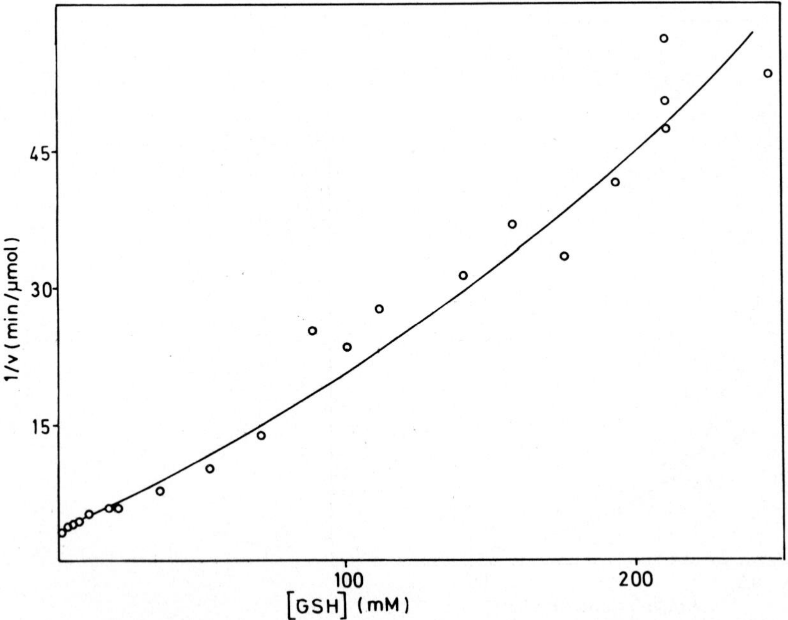

Fig. 3. Inhibition of the forward reaction by GSH, added as a product inhibitor (Dixon plot). The measurements were made at constant concentrations of NADPH (26 µM) and GSSG (53 mM). The medium was as described in the legend to Fig. 2 except for the amount of enzyme which was 1.2 µg and the absence of dithioerythritol.

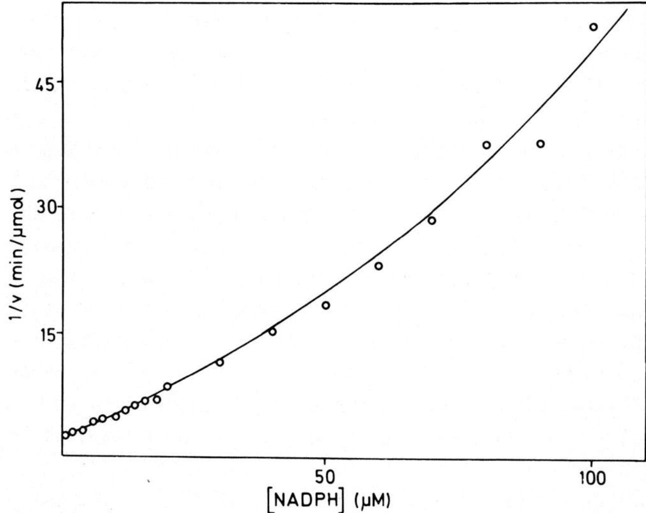

Fig. 4. Inhibition of the reverse reaction by NADPH, added as a product inhibitor (Dixon plot). The measurements were made at constant concentrations of $NADP^+$ (0.23 mM) and GSH (2.2 mM) in the system described in the legend to Fig. 2.

degree in [GSH] in the numerator and fourth degree in the denominator in agreement with Fig. 3.

The reverse reaction was found to be inhibited nonlinearly (parabolically as a first approximation) by NADPH (Fig. 4). The basic scheme of Fig. 1 can easily accommodate this finding if it is assumed that the reduced form of the enzyme (F) can bind NADPH. This assumption is in good agreement with the experimental data of Bulger and Brandt [15].

Thus, a large body of steady state kinetic data is in agreement with the branching reaction scheme depicted in Fig. 1. The ping-pong loop of the scheme has also direct support from experiments in which the flavin of the enzyme is reduced by NADPH and in a separate step reoxidized by GSSG [14]. Furthermore, preliminary experiments in our laboratory demonstrate isotope exchange between ^{14}C labelled $NADP^+$ and NADPH in the absence of GSSG and GSH, which is expected when a ping-pong branch is operative. However, although Fig. 1 presents the simplest reaction scheme compatible with the experimental data, it is possible that more complex models have to be considered in the future. In particular, the tendency of higher-degree dependence of v on [NADPH] at high [GSSG] (which was not considered significant [9]) might indicate that a third alternative loop, in which GSSG is added as the first substrate to the oxidized form of the enzyme (E), has to be included.

It seems logical to assume that other flavoenzymes, such as lipoamide dehydrogenase, DT diaphorase, pyridine nucleotide transhydrogenase and many others, also operate by branching mechanisms which unite a ping-pong and a sequential reaction pattern. The following criteria may help to identify such mechanisms.

(a) When the concentrations of two substrates are varied, the steady state kinetics will asymptotically approach a ping-pong pattern as the concentrations are decreased [16]. This results from the absence of a constant term in the denominator of the complete-rate equation.

(b) The rate equation will be second or higher degree in the concentration of acceptor substrate (cf. ref. 3). The same dependence on donor substrate obtains if either acceptor or donor can add as the first reactant to the enzyme. Nonlinearities are expected in v versus $v/[\text{substrate}]$ and double reciprocal plots if the concentration ranges examined are sufficiently wide.

(c) In contrast to the ping-pong pattern of (a), the product-inhibition patterns will differ from those of a classical ping-pong mechanism. For example, the product resulting from the donor substrate will not be competitive with the acceptor substrate. The reciprocal-rate equation is expected to be nonlinear with respect to the concentration of the product used as an inhibitor.

(d) Under favourable conditions the partial reactions of the ping-pong loop can be studied by chemical or spectroscopic methods. Such experiments give direct evidence for the existence of this branch and may yield quantitative data to support its kinetic competence.

(e) Isotope exchange is expected in the presence of two suitable reactants owing to the ping-pong loop. This exchange may be blocked by adding excess of two other reactants under equilibrium conditions (cf. ref. 17). For example, the exchange between ^{14}C-labelled $NADP^+$ and NADPH will be inhibited by excess of GSSG and GSH if the branching scheme of Fig. 1 is applicable. A pure sequential mechanism requires the presence of all reactants to make possible any isotope exchange.

Acknowledgements

The excellent technical assistance by Miss Kerstin Jacobsson is gratefully acknowledged. This investigation was supported by the Swedish Cancer Society.

References

1 Mannervik, B. (1973) Biochem. Biophys. Res. Commun. 53, 1151–1158
2 Mannervik, B. (1974) in Glutathione (Flohé, L., Behöhr, H.C., Sies, H., Waller, H.D. and Wendel, A., eds.), pp. 114–120, Georg Thieme Verlag, Stuttgart
3 Wong, J.T. and Hanes, C.S. (1962) Can. J. Biochem. Physiol. 40, 763–804
4 Wong, J.T. and Hanes, C.S. (1973) Acta Biol. Med. Ger. 31, 507–514
5 Bartfai, T. and Mannervik, B. (1972) FEBS Lett. 26, 252–256
6 Bartfai, T. and Mannervik, B. (1972) in Analysis and Simulation of Biochemical Systems (Hemker, H.C. and Hess, B., eds.), pp. 197–209, North-Holland, Amsterdam
7 Mannervik, B. and Bartfai, T. (1973) Acta Biol. Med. Ger. 31, 203–215
8 Mannervik, B. and Bartfai, T. (1974) Abstr. Commun. Meet. Fed. Eur. Biochem. Soc. 9, 335
9 Mannervik, B. (1975) Biosystems 7, 101–119
10 Cariberg, I. and Mannervik, B. (1975) J. Biol. Chem. 250, 5475–5480
11 Massey, V. and Williams, Jr., C.H. (1965) J. Biol. Chem. 240, 4470–4480
12 Mannervik, B. (1975) Anal. Biochem. 63, 12–16
13 Massey, V., Veeger, C. and Mannervik, B. (1970) in Pyridine Nucleotide-Dependent Dehydrogenases (Sund, H., ed.), (discussion comments) pp. 370–373, Springer, Berlin

14 Williams, Jr., C.H. and Veeger, C. (1971) in Flavins and Flavoproteins (Kamin, H., ed.), (discussion comments) pp. 289–294, University Park Press, Baltimore
15 Bulger, J.E. and Brandt, K.G. (1971) J. Biol. Chem. 246, 5578–5587
16 Mannervik, B. and Askelöf, P. (1975) FEBS Lett. 56, 218–221
17 Wong, J.T. and Hanes, C.S. (1964) Nature 203, 492–494

Flavins and Flavoproteins, edited by T.P. Singer

CHAPTER 53

STUDIES ON GSSG- AND CoASSG REDUCTASES FROM RAT LIVER AND YEAST

RAUL N. ONDARZA*

Facultad de Medicina, Universidad Nacional Autónoma de México, México 20, D.F. (Mexico)

Introduction

In 1960 we were able to isolate an adenine-free acid-soluble nucleotide from rat liver [1] which was finally characterized as a mixed disulfide compound formed by Coenzyme A and glutathione moieties linked by a disulfide bridge [2]; this compound corresponds to the one isolated independently from bovine liver by Chang and Wilken [3]. In fact the existence of such a compound in yeast was suggested by Kornberg and Stadtman in their report on the isolation of Coenzyme A [4]. Their suggestion was that Coenzyme A and glutathione may be present largely as the mixed disulfide derivative.

In order to gain insight into the probable role of this compound in the liver cell and in yeast, the presence of a new enzyme activity was investigated by analogy to GSSG reductase [5] which might metabolize this complex in the presence of a hydrogen coenzyme donor according to the following reaction:

$$\text{CoASSG} \xrightarrow[\text{Enzyme}]{\text{NADPH H} + \text{NAD}} \text{CoASH} + \text{GSH}$$

Preliminary studies on CoASSG reductase from rat liver and yeast [5]

A new disulfide reductase activity from normal rate liver extracts has been established. Since the enzyme preparation was not pure, there was the possibility that the CoASSG-reducing activity was due to its GSSG reductase. Further study was carried out to the commercially available "purified" GSSG reductase of yeast. With this preparation it was clearly shown that by increasing the amount of yeast enzyme (4.158 units) up to 50 μg, as much as 0.137 units of CoASSG reducing activity could be measured.

---

* At present commissioned by the U.N.A.M. to the Consejo Nacional de Ciencia y Tecnologia. Preparation of CoASSG: (For information about the assay method, preparation and properties of CoASSG see ref. 6).

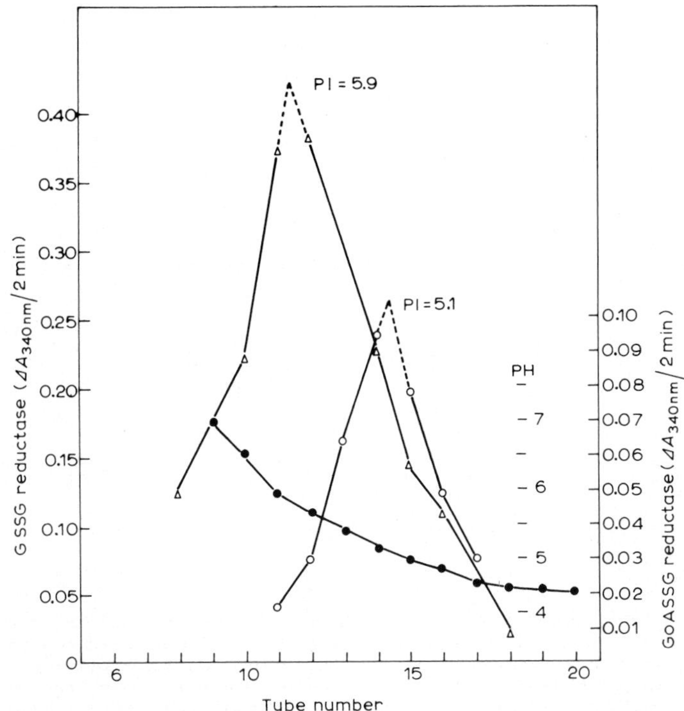

Fig. 1. Isoelectric focusing separation of CoASSG reductase from yeast GSSG reductase. A sample of 200 μl (1 mg) of commercially purified GSSG reductase, (dialyzed against 4 l of water, for 3 h) in a concentration gradient from 6 to 2% of carrier ampholtyes, within a pH range of 3 to 6, was applied to an electrophoresis columb. For evaluation of CoASSG reductase (measured in 0.05 M phosphate buffer with 1 mM EDTA, at pH 5.5) 100 μl were used, and 20 μl for GSSG reductase (0.1 M phosphate buffer with 1 mM EDTA, at pH 7.0). Each point represents the initial velocity expressed as a change in absorbance at 340 nm per 2 min. (Taken from ref. 8.)

The above finding indicated that: (a) the purified yeast enzyme had some contaminating CoASSG-reducing activity, or (b) that the GSSG reductase enzyme might not be as specific as originally described by Pihl et al. [7], who studied the specificity from rat liver and found that the enzyme fails to reduce mixed disulfides containing glutathione.

Characterization of a NADPH-dependent CoASSG reductase from yeast [*8*]

In order to give an answer to one of the above questions, enzyme extracts were prepared from the yeast *Saccharomyces cerevisae* ATCC-1946, cultivated and grown in our laboratory.

According to the data obtained from polyacrylamide column gel electrophoresis and isoelectric-focusing, the yeast GSSG reductase could be partially separated from the new enzyme activity (Fig. 1). The enzyme has the characteristics which appear for the purpose of comparison together with data for GSSG reductase and CoASSG-GSH transhydrogenase in Table I.

It was assumed that the CoASSG reductase reaction is GSH independent since purified CoASSG is free of GSH. Moreover, when GSH is added in

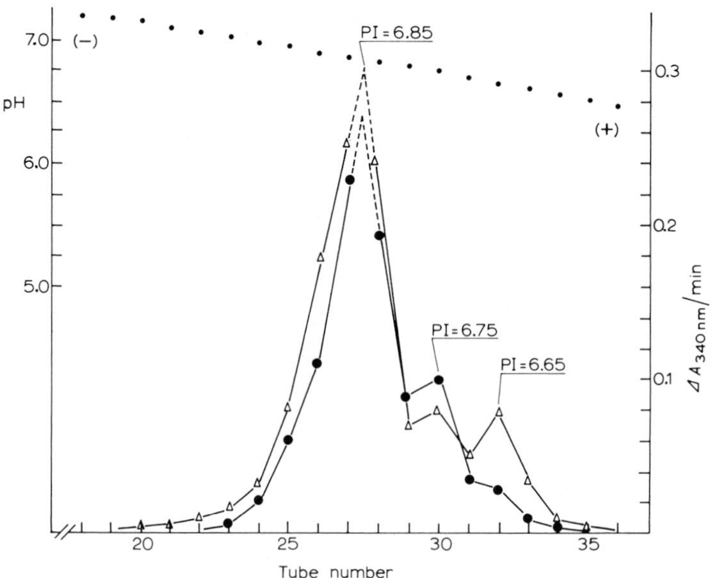

Fig. 2. Treatment of rat liver GSSG and CoASSG reductase (30 mg of protein obtained from the phosphate gel column step) by means of isoelectric-focusing with an electrophoresis column (LKB 8100 of 110 ml of capacity) in a concentration gradient from 6 to 2% of carrier ampholytes within a pH range of 5 to 8. The initial charge of the column was 450 V, 6 mA. The experiment was carried out at 5°C for 48 h. Fractions of 1.0 ml were collected. The enzymatic activities were determined with 20 μl GSSG reductase and 100 μl for CoASSG reductase under optimal conditions. △—△, GSSG reductase; ●—●, CoASSG reductase. (Taken from ref. 12.)

competing amounts to the reaction mixture, the initial velocity is not affected. On the contrary, if GSSG is added, the reaction is inhibited probably by competing for the active site of the new enzyme. These experiments were done by measuring the CoASH liberated [10].

TABLE I

COMPARISON OF DATA OBTAINED FOR YEAST GSSG REDUCTASE, CoASSG REDUCTASE AND CoASSG-GSH, TRANSHYDROGENASE

Properties	GSSG reductase activity			CoASSG reductase activity		CoASSG-GSH trans-hydro-genase
	Partially purified yeast	Purified yeast (Boeh-ringer)	Ref. 20	Partially purified yeast	Purified yeast (Boeh-ringer)	
$M_r \times 10^{-3}$	108	110	56.5*	108	110	12
isoelectric point	—	5.9	—	—	5.1	—
pH optimum	7.0	7.0	7.0	5.5	5.5	8.2
$K_m \times 10^5$(M)	4	6.41	8.5	20	19.8	4.5
Cofactors	NADPH	NADPH	NADPH	NADPH	NADPH	GSH
Reversibility	irreversible	irreversible	irreversible	irreversible	irreversible	reversible
Phosphate ion effect	activation	activation	activation	inhibition	inhibition	activation

* Based on one flavin per mol of enzyme. (Taken from ref. 8.)

There was no reducing acitivity under optimal conditions for the purified yeast extract on CoASSCys, GSSCys or CysSSCys.

A striking inhibition effect was found on CoASSG reductase by phosphate ion which is less accentuated under substrate saturation conditions. However, GSSG reductase can be stimulated by this ion or by ionic strength as has also been demonstrated by Woodin and Segel [11].

Characterization of a NADPH-dependent CoASSG and GSSG reductases from rat liver [12]

In contrast to the enzyme obtained from yeast during the purification procedure using different types of column chromatography, the GSSG- and CoASSG-reducing enzymes from rat liver extracts eluted together and kept a similar ratio of activity throughout the different steps of purification. Also, the new reducing enzyme in rat liver was found to purify together with GSSG reductase by isoelectric-focusing technique.

The CoASSG-reducing enzyme has an optimum pH of 5.75 and the GSSG reductase of 7.0. The apparent Michaelis constant for CoASSG-reducing enzyme from liver at a fixed value of NADPH (110 nmol) at pH 5.75, was calculated by the Lineweaver–Burk method, the value obtained being $2.3 \cdot 10^{-4}$ M.

A molecular weight of 42 500 for GSSG and CoASSG reductases was found by gel filtration in Sephadex G-100 by the method of Andrews [13]. Mize et al. [14] have reported on the basis of sedimentation velocity and diffusion techniques, a molecular weight of 44 000 for GSSG reductase from rat liver which is consistent with our results.

The two reducing enzymes showed an identical isoelectric point of 6.85 in a concentration gradient from 6 to 2% of carrier ampholites within a pH range of 3 to 10. The activity of GSSG reductase in this step was of 15.87 units and for CoASSG reductase 1.3 units. To increase the accuracy in the possible separation, a pH gradient between 5 and 8 was chosen (Fig. 2). The results indicate that even within this pH range both reducing enzymes remained in a single peak and the ratio of activities was constant between the two extremes of the peak with the same pI of 6.85. The same result has been obtained in several experiments.

It is important to point out, that when using this technique in a pH range between 5 and 8 in some preparations, two additional disulfide reducing activities appeared at a pI of 6.75 and 6.65 (see Fig. 2).

The specificity of the GSSG and CoASSG reductases was measured with several disulfides (Table II, Fraction A) as well as their dependence on the thermostable protein cofactor reported by Tietze [15] to be necessary for an NADPH-dependent non-specific disulfide reductase activity also present in rat liver extracts (Table II, Fraction B). This non-specific reductase activity could be easily separated and distinguished from the GSSG and CoASSG reducing enzymes by ion-exchange column chromatography and isoelectric focusing.

The enzymatic preparation containing GSSG and CoASSG reducing activities reduced GSSG, CoASSG and dithionitrobenzoic acid in that order of effectiveness. There was no effect of the protein cofactor in either of these activities (see Table II).

On the active site of the CoASSG reductase from yeast and rat liver [*16*]

Taking into consideration that the CoASSG reductase catalyses a reaction similar to those of lipoamide dehydrogenase (EC 1.6.4.3), GSSG reductase (EC 1.6.4.2) and thioredoxin reductase, which are well-known pyridine nucleotide-dependent flavoproteins and that their active center and mechanism of action have been studied [17–20]; it was decided to see if the new reductase has the particular characteristics of this group of enzymes.

Since arsenite is known to react with an NADPH-dependent reducible SS bridge, forming covalent bonds, such as in GSSG reductase and lipoamide dehydrogenase, with resulting inhibition of enzyme activity, the effect of this compound was assayed on CoASSG reductase activity from yeast and rat liver. Arsenite was found to inhibit at 10^{-1}, 10^{-2} and 10^{-3} M only when it was preincubated with the NADPH-dependent reduced enzyme, for a minimum period of 3 min.

It is worthwhile mentioning that there is a partial protection of the inhibition by arsenite if at the same time the enzyme is incubated in the presence of 5-nitrofurans at 10^{-2} M even in the presence of NADPH.

If the enzyme is not previously reduced with NADPH, no inhibiting effect is observed. The inhibitory effect of arsenite is not reversed by dialysis, and longer time of incubation (1h) produces total inactivation of yeast and rat liver preparations.

TABLE II

RAT LIVER ENZYME SPECIFICITY AND EFFECT OF A THERMOSTABLE PROTEIN COFACTOR ON THE REDUCTION OF SEVERAL DISULFIDE COMPOUNDS

The GSSG and CoASSG reducing activities (Fraction A) passed first through a DEAE-cellulose column and was then further purified by filtration on a column of Sephadex G-150. The non-specific reductase was prepared according to Tietze [15] (Fraction B). In each experiment, where indicated, 140 μg of thermostable protein cofactor were added. Incubating conditions: 100 μg of enzyme protein; 110 nmol NADPH; 350 nmol of substrate (except for CoASSG where 125 nmol were used) and 100 mM Tris · HCl (pH 7.5) with 1 mM EDTA in a final volume of 1 ml. The buffer used for CoASSG reducing acitivity was of sodium phosphate (pH 5.75). (Taken from ref. 12.)

Substrate	Enzyme preparation (m units)			
	Fraction A		Fraction B	
	No cofactor	Cofactor	No cofactor	Cofactor
GSSG	1280.00	1280.0	12.8	29.00
CoASSG	96.6	100.0	0.64	1.28
Dithionitrobenzoic acid	59.0	53.2	1500.00	1154.00
Cystine	1.28	1.28	13.58	41.8
Cystamine	2.58	2.58	3.22	58.06
Pantethine	1.93	1.93	6.44	25.8
Insulin	1.28	1.28	7.1	71.0

TABLE III

INHIBITION OF CoASSG REDUCTASE AND GSSG REDUCTASE FROM YEAST AND RAT LIVER BY 5-NITROFURAN DERIVATIVES (10^{-4} M)

The percent inhibition was calculated from the normal activity for each reductase. For details and conditions of the experiment consult ref. 16. (Taken from ref. 16.)

Nitrofuran derivative	Yeast		Rat liver	
	CoASSG reductase (inhibition, %)	GSSG reductase (inhibition, %)	CoASSG reductase (inhibition, %)	GSSG reductase (inhibition, %)
Micofur ($M_r = 156$)	41	67	34	85
Furaspor ($M_r = 157$)	58	48	19	75
Furacin ($M_r = 198.14$)	86	72	77	85
Furoxone ($M_r = 225.16$)	74	45	40	97
Furadantin ($M_r = 238.16$)	64	65	55	82

On the other hand, it has been shown by Buzard and Kopko [21] that 5-nitorfuran derivatives can inhibit the GSSG reductase from rat liver. In this sense, we utilized 5-nitrofuran derivatives as inhibitors in order to see a difference in sensitivity and to gain insight on the active site of both enzymes from the two sources*. The inhibition was reversed by dialysis and is independent of NADPH. The results appear in Table III.

Inhibition of GSSG reductase by CoASSG [22]

It was found that the commercial purified yeast GSSG reductase can be largely inhibited with CoASSG if the enzyme is previously incubated in the presence of NADPH. The inhibition can be partially eliminated after a 60 min dialysis. Furthermore, the presence of NADPH is very important; if it is absent during the incubation period, no inhibition can be demonstrated upon addition of GSSG followed by NADPH. These studies have been extended to rat liver and yeast extracts prepared in our laboratory which display a proportionally higher level of CoASSG-reducing enzyme. In both cases the inhibition effect by CoASSG is less accentuated.

Discussion

According to the results, the CoASSG is a disulfide oxido-reductase and must be a flavoprotein. A disulfide bridge is present as shown by the fact that the enzyme is inhibited by arsenite provided that it has been previously reduced with NADPH. Both CoASSG and GSSG reductases from yeast and liver, seem to be intimately associated within the cell constituting one enzymatic unit. In liver, the new reducing activity corresponds to the same protein entity of GSSG reductase. In yeast, the CoASSG reductase can be partially separated from the GSSG reductase. In such a case, it would be necessary to prove whether the two originally described polypeptide chains for GSSG reductase are being separated. In liver, the presence of two additional reducing activities detected with a pI of 6.75 and 6.65, opens the possibility that there are two isoenzymes respectively, for CoASSG and GSSG reductases.

In yeast, there is a discrepancy with our results from those of Mannervick [23] who has not been able to separate the two reducing activities. This might be due to the different conditions of separation.

Although it is as yet unknown whether 5-nitrofuran derivatives are specific inhibitors of disulfide oxidoreductases, we would like to propose that they could inhibit the electron transport in two sites of the enzyme — one could be between the NADPH and FAD and the other between FAD and the disulfide bridge of the active center of the enzyme. The above

* The 5-nitrofuran derivatives utilized were: furadantin (N-5-nitro-2-furfurylidene-1-amino hydantoin), furoxone (N-5-nitro-2-furfurylidene-3-amino-2-oxazolidone), micofur (5-nitro-2-furfuraldehyde oxime), furaspor (5-nitro furfuryl methyl ether) and furacin (5-nitro-2-furfuraldehyde semicarbazone); at final concentration of 10^{-4} M in the reaction mixtures.

suggestion is possible taking into consideration the electronic structure of 5-nitrofurans. Moreover, it was shown that they can protect the arsenite inhibition effect.

It was observed that the GSSG reductase from yeast and rat liver, is largely inhibited by 5-nitrofurans, independent of their molecular weight. On the other hand, the CoASSG-reductase from both sources is also inhibited, but to a much lesser degree when the molecular weight of the 5-nitrofuran derivative is increased. This finding indicates that the CoASSG almost completely occupies the active site of the enzyme and does not permit the accomodation of larger nitrofuran inhibitors.

References

1 Ondarza, R.N. and Aubanel, M. (1960) Biochim. Biophys. Acta 44, 381–383
2 Ondarza, R.N. (1965) Biochim. Biophys. Acta 107, 112–119
3 Chang, S.H. and Wilken, D.R. (1965) J. Biol. Chem. Vol. 240, 3136–3139
4 Kornberg, A. and Stadtman, E.R. (1970) Methods in Enzymology (Colowick, S.P. and Kaplan, N.O., eds.), Vol. 3, p. 907, Academic Press, N.Y.
5 Ondarza, R.N. and Martinez, J. (1966) Biochim. Biophys. Acta 113, 409–411
6 Ondarza, R.N. (1970) Methods in Enzymology (Colowick, S.P. and Kaplan, N.O. eds.), Vol. XVIII, Academic Press, N.Y.
7 Pihl, A., Eldjarn, L. and Bremer, J. (1957) J. Biol. Chem. 227, 339–345
8 Ondarza, R.N., Abney, R. and Lopez-Colomé, A.M. (1969) Biochim. Biophys. Acta 191, 239–248
9 Racker, E. (1955) Methods in Enzymology (Colowick, S.P. and Kaplan, N.O., eds.), Vol. 2, p. 722, Academic Press, N.Y.
10 Michal, G. and Bergmeyer, H.U. (1963) Methods in Enzymatic Analysis (Bergmeye, H.U., ed.), p. 512, Academic Press, N.Y.
11 Woodin, T.S. and Segel, I.H. (1968) Biochim. Biophys. Acta 164, 64–77
12 Ondarza, R.N., Escamilla, E., Gutiérrez, J. and De la Chica, G. (1974) Biochim. Biophys. Acta 341, 162–171
13 Andrews, P. (1964) Biochem. J. 91, 222–233
14 Mize, C.E., Thompson, T.E. and Langdon, R.G. (1962) J. Biol. Chem. 237, 1596–1600
15 Tietze, F. (1970) Biochim. Biophys. Acta 220, 449–462
16 Ondarza, R.N. and Abney, R. (1970) FEBS Lett., Vol. 7, 227–230
17 Searle, R.L., Peters, J.M. and Sanodi, D.R. (1961) J. Biol. Chem. 236, 2317–2322
18 Massey, V. and Williams, C.H. (1965) J. Biol. Chem. 240, 4470–4480
19 Moore, E.C., Reichard, P. and Thelander, L. (1964) J. Biol. Chem. 239, 3445–3452
20 Jones, E.T. and Williams, Jr., C.H. (1975) J. Biol. Chem. in the press
21 Buzard, J.A. and Kopko, F. (1963) J. Biol. Chem. 238, 464
22 Ondarza, R.N. (1967) National Cancer Inst., Monogr. No. 27, 81–88
23 Eriksson, S., Guthenberg, C. and Mannervick, B. (1974) FEBS Lett. 39, 296–300

Flavins and Flavoproteins, edited by T.P. Singer

CHAPTER 54

FLUORESCENCE STUDIES ON LIPOAMIDE DEHYDROGENASES, PYRUVATE DEHYDROGENASE COMPLEXES AND TRANSHYDROGENASE

C. VEEGER, A.J.W.G. VISSER, J. KRUL, H.J. GRANDE, R.A. de ABREU and A. de KOK

Department of Biochemistry, Agricultural University, Wageningen (The Netherlands)

Introducti

Recent studies [1] have shown that the two flavin molecules of the dimeric lipoamide dehydrogenase (E.C. 1.6.4.3, NADH:lipoamide oxidoreductase) from pig heart are bound in different environments to the two peptides. This conclusion was derived by measuring the non-exponential fluorescence decay curve by the single-photon-counting technique. To eliminate the possibility that this phenomenon is induced by the isolation procedure, we have extended our studies to the pyruvate dehydrogenase complexes of pig heart and *Azotobacter vinelandii*, as well as the purified enzyme from the latter complex. These studies were performed by determination of the fluorescence lifetime of the flavins by phase- and modulation fluorimetry. It was assumed that the relation between the lifetime values determined from either the phase shift $\langle\tau_p\rangle$ or from the degree of modulation $\langle\tau_m\rangle$ is $\langle\tau_m\rangle = \langle\tau_p\rangle$ for emittors with equal lifetimes and $\langle\tau_m\rangle > \langle\tau_p\rangle$ in the case of non-homogeneous fluorescence decay [2, 3].

In addition, it has been shown from polarization studies [4] that the degree of depolarization of fluorescence emission, obtained by excitation in the second absorption band of lipoamide dehydrogenase, is larger than the value observed with 3-methyllumiflavin in a solid matrix. Therefore, we have studied the fluorescence properties of the FAD-containing flavoprotein $NAD(P)^+$ transhydrogenase (E.C. 1.6.1.1, NADPH:NAD^+ oxidoreductase) from *A. vinelandii*.

Evidence will be presented from fluorescence-lifetime studies that a ternary complex exists between oxidized transhydrogenase, phosphate and $NADP^+$. Furthermore, these studies show that the difference between the FAD moieties in pig heart lipoamide dehydrogenase is not an artifact; the flavins have different reactivities towards NADH.

Studies on lipoamide dehydrogenase

Table I shows that a large difference exists between the fluorescence properties of this flavoprotein in the free as well as in the complexed form, depending on the source from which it is isolated. The fact that in the free *A. vinelandii* enzyme $\langle\tau_p\rangle \simeq \langle\tau_m\rangle$ indicates that the two flavins have identical emission characteristics and thus are located in more or less similar environments. The results suggest that in contrast with the free enzyme the two flavins bound to the complex are probably not entirely identical. The difference between $\langle\tau_p\rangle$ and $\langle\tau_m\rangle$ values is certainly outside the experimental error of the method (approx. 50 ps) at a modulation frequency of 60 MHz. On the other hand in the pig heart enzyme, free and bound to the complex, the two flavins are bound to different environments. The results suggest that binding within the complex induces changes in the microenvironments of the FADs. Thus, an isolation artifact can be excluded and the effects are due to differences in intrinsic binding. The lifetime of the long-living emittor indicates that the flavin is at 4°C in an environment enabling it to emit with a lifetime comparable with that of free FMN.

In agreement with the single-photon-counting experiments [1], these results indicate that at 22°C the difference between the two flavin moieties become less, while at 37°C hardly any difference exists. The decrease in lifetime of the long-living species at higher temperatures is much larger than is expected from the temperature dependence of the lifetime of free FMN, while the lifetime of the short-living species seems to become relatively longer and to change little [1].

Knowing the absolute lifetime values it is possible to calculate from $\langle\tau_p\rangle$ and $\langle\tau_m\rangle$ the ratio of the quantum yields of the two species and to compare it with the ratio of the actual lifetimes (c.f. ref. 1). The value of 10 at 3.5°C (2.5 at 22°C) is larger than can be expected from a pure, dynamic, quenching process. Thus, it is our conclusion that the short-living species forms a dark complex at 3.5°C; at higher temperatures the long-living species also forms such a dark complex, while the contribution of the dark complex of the short-living species becomes less.

The observation that at temperatures less than 37°C differences exist between the two FAD moieties may account for the irregular kinetic

TABLE I

FLUORESCENCE LIFETIMES OF LIPOAMIDE DEHYDROGENASE FROM PIG HEART AND *A. VINELANDII* FREE AND BOUND TO PYRUVATE DEHYDROGENASE COMPLEX (PDC)

Source	Form of enzyme	3.5°C		22°C		37°C	
		$\langle\tau_p\rangle$ (ns)	$\langle\tau_m\rangle$	$\langle\tau_p\rangle$ (ns)	$\langle\tau_m\rangle$	$\langle\tau_p\rangle$ (ns)	$\langle\tau_m\rangle$
Pig heart	PDC	3.2	3.9	2.1	2.8	—	—
	free	3.6	4.7	2.5	2.9	1.7	1.9
A. vinelandii	PDC	3.0	3.2	1.8	2.1	—	—
	free	3.1	3.1	2.1	2.2	—	—

behaviour of the enzyme under these conditions [5] and the negative cooperativity of NAD^+-binding to the oxidized enzyme [6]. In accordance with this idea is the regular kinetic pattern as observed at 37°C [7]. Furthermore, this difference could also provide an explanation for the observed biphasic reduction of the enzyme by reduced lipoamide, for which a different explanation was given [8, 9]. The question may arise whether this difference is due to the presence of isoenzymes, since an equilibrium between two forms was excluded [1]. Dr. J. Visser in our laboratory succeeded recently in isolating one of the isoenzymes of lipoamide dehydrogenase from the pig heart pyruvate dehydrogenase complex by means of isoelectric focussing. This isoenzyme also has FAD moieties non-identically bound. Although it is possible that this isoenzyme is a hybrid form, we are inclined to disregard this view, since no evidence has been obtained until now for non-identical peptide chains.

To obtain evidence for a difference in behaviour between the two FADs of the pig heart enzyme, lifetimes were measured after reduction with reduced lipoamide and NADH. Table II shows that after reduction with 2 mol of reduced lipoamide $\langle\tau_p\rangle$ and $\langle\tau_m\rangle$ change drastically; the FADs become more closely similar, as judged from the 15% residual fluorescence and assuming random distribution of the reducing equivalents. To eliminate reoxidation by traces of O_2, the experiments were performed anaerobically in the presence of glucose, glucose oxidase and catalase. Taking into account the different forms of the enzyme (Fig. 1), it is logical to assume that this residual fluorescence is due to the $FAD{<}^{SH}_{SH}$-form of the enzyme. Either the interaction with the vicinal SH groups or with other groups of the protein decreases the fluorescence lifetimes of the flavins. Addition of AsO_3^{3-} increases the flavin fluorescence, while the difference between the FADs becomes larger. From the data it can be calculated that the equilibrium constant between the $FAD{<}^{SH}_{SH}$- and $FAD(H_2)\begin{matrix}\text{-S}\\ |\\ \text{-S}\end{matrix}$ forms of the enzyme lies in

TABLE II

$\langle\tau_p\rangle$ AND $\langle\tau_m\rangle$ VALUES OF PIG HEART LIPOAMIDE DEHYDROGENASE UNDER DIFFERENT CONDITIONS

	4°C		I (%)
	$\langle\tau_p\rangle$ (ns)	$\langle\tau_m\rangle$	
Oxidized	3.6	4.7	100
After addition of reduced lipoamide:			
0.9 mol/mol FAD	3.0	3.5	18
2.0 mol/mol FAD	2.9	3.3	15
After addition of 1.1 mol NADH/mol FAD + 10 mM AsO_3^{3-}:			
1 min after mixing	2.0	2.7	35
2.5 h after mixing	2.8	3.8	75

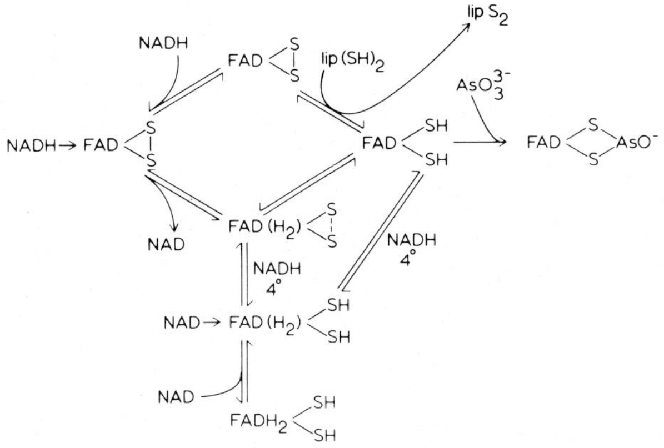

Fig. 1. Different reduced species of pig heart lipoamide dehydrogenase (c.f. refs. 5 and 8).

the direction of the latter species with $K_{eq} = 5\text{–}7$, assuming random distribution of the reducing equivalents over the two peptides.

It cannot be excluded however that a specific form of the two FADs remains in a more oxidized state. Evidence for this possibility comes from titration with NADH at 4°C. Although at this temperature the difference between the FADs is largest and addition of NADH leads to rapid formation of two-equivalent reduced forms, in a slower secondary reaction a partial redistribution of reducing equivalents occurs with the subsequent formation of the oxidized and four-equivalent reduced forms (c.f. refs. 8 and 9). The initial drop in fluorescence intensity, $\langle\tau_p\rangle$ and $\langle\tau_m\rangle$ (measured, due to technical difficulties such as proper temperature control, about 1–1.25 min after mixing) is followed by a slow increase in the values of these three parameters (Fig. 2) (half time about 2 min). This indicates that the initial drop in fluorescence intensity has been initially larger than the value measured; for instance, the 40% quenching measured after addition of 0.29 mol NADH/FAD. But even from the value measured, it can be seen that this low concentration of NADH is able to quench the flavin fluroescence more strongly than expected for random distribution of the reducing equivalents. This can only be explained by preferential reaction with the FAD with the longest lifetime (largest quantum yield). At higher ratios, this tendency seems to disappear. Under the latter conditions the residual (15%) fluorescence has the same lifetime values as observed by reduction with reduced lipoamide. The changes in fluorescence intensity, $\langle\tau_p\rangle$ and $\langle\tau_m\rangle$, indicate that even at 0.29 mol of NADH/FAD in addition to the $\mathrm{FAD}\begin{smallmatrix}\text{-S}\\ |\\ \text{-S}\end{smallmatrix}$- and the $\mathrm{FAD(H_2)}\begin{smallmatrix}\text{-S}\\ |\\ \text{-S}\end{smallmatrix}$ forms, a substantial amount of the $\mathrm{FAD}\begin{smallmatrix}\text{SH}\\ \text{SH}\end{smallmatrix}$ form is present (about 10–15% of the FAD). At 0.58 mol NADH/FAD about 50% of the enzyme is in the $\mathrm{FAD(H_2)}\begin{smallmatrix}\text{-S}\\ |\\ \text{-S}\end{smallmatrix}$ form and 20–25% in the $\mathrm{FAD}\begin{smallmatrix}\text{SH}\\ \text{SH}\end{smallmatrix}$ form, again indicating preferential

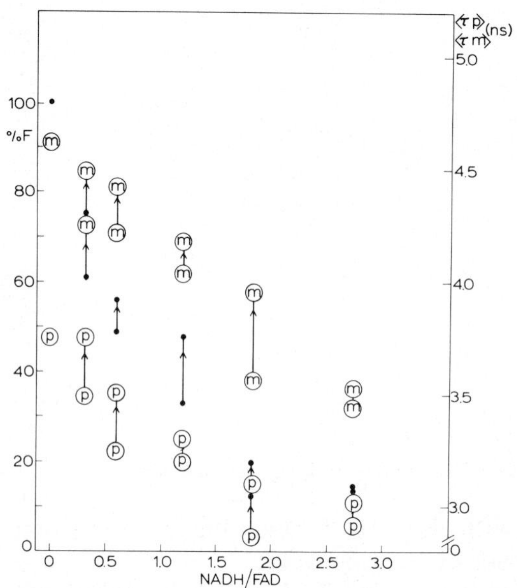

Fig. 2. Anaerobic fluorimetric titration of pig heart lipoamide dehydrogenase with NADH. To individual spectrophotometric Thunberg cuvettes (containing 8 μM enzyme flavin in 0.1 M phosphate buffer pH 7.6, 0.3 mM EDTA, 1 mM glucose, traces of glucose oxidase and catalase), NADH, of which the concentration was determined enzymatically, was added in the amounts indicated. Within 60–75 s, $\langle\tau_p\rangle$, $\langle\tau_m\rangle$ and the emission were measured. These parameters were followed until they were constant (about 15 min). m , $\langle\tau_m\rangle$, p , $\langle\tau_p\rangle$; •, emission intensity. The arrows indicate the changes in the parameter values with time and their final values. λ_{exc} = 440 nm, emission was passed through a Schott-Jena GG 495 nm filter. $T = 4^\circ$C.

reduction of the FAD with the longest lifetime. Fig. 3 shows that even at 7 mol NADH/FAD, residual, polarized, flavin fluorescence is present with $\langle\tau_p\rangle$ and $\langle\tau_m\rangle$ identical with the values obtained after reduction with reduced lipoamide, again supporting these postulates. The titration curve shows that reaction of the enzyme with 2 mol NADH/FAD completely quenches the FAD fluorescence.

In summary, these studies show that large differences exist between the lipoamide dehydrogenases from pig heart and *A. vinelandii*. The two non-identical FADs of the pig heart enzyme seem to have different reactivities with respect to NADH.

Studies on transhydrogenase

Oxidized FAD-containing transhydrogenase shows a pronounced flavin fluorescence with an emission maximum at 511 nm when it is dissolved in Tris/HCl buffer. The quantum yield is about 25% of that of FMN. The maximum fluorescence polarization is high (p = 0.43), but not exceptionally high for a molecule of this size. The addition of 10 mM phosphate causes a small decrease in the degree of polarization. The fluorescence polarization spectrum (Fig. 4) shows a similar fine structure in the first excitation band as reported by Visser et al. [4] for pig heart lipoamide dehydrogenase and 3-methyllumi-

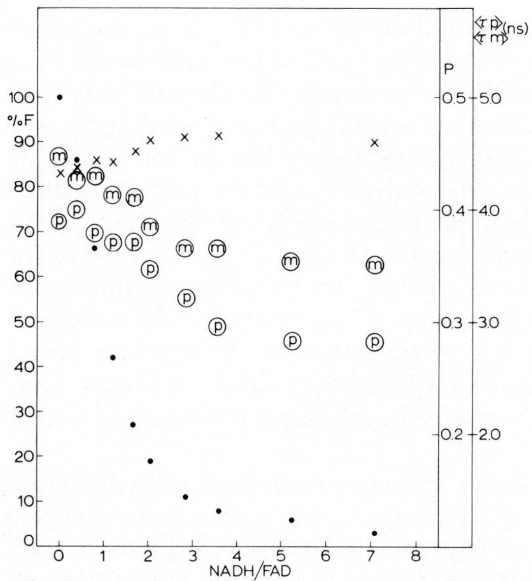

Fig. 3. Anaerobic fluorimetric titration of pig heart lipoamide dehydrogenase with NADH. To one spectrophotometric Thunberg cuvette (containing 8.8 μM enzyme flavin in 30 mM phosphate buffer pH 7.6, 0.3 mM EDTA, 1 mM glucose, traces of glucose oxidase and catalase), successive additions of NADH were made and equilibrium established before the next amount of pyridine nucleotide was added (constant values of fluorescence intensity, $\langle\tau_p\rangle$ and $\langle\tau_m\rangle$). λ_{exc} = 440 nm, emission passed through a Schott-Jena GG 495 nm filter. $T = 4^\circ$C. ⓜ, $\langle\tau_m\rangle$; ⓟ, $\langle\tau_p\rangle$; ●, emission intensity; X, degree of polarization.

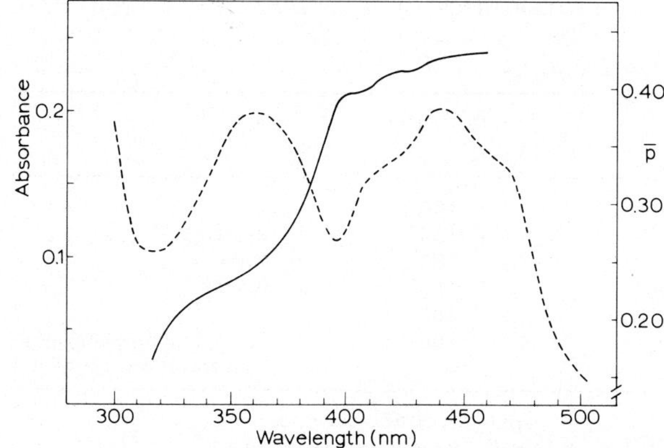

Fig. 4. Fluorescence polarization spectrum of *A. vinelandii* transhydrogenase. The enzyme was dissolved in 50 mM Tris/HCl pH 7.6, $T = 22^\circ$C. The measurements were performed with a polarization spectrofluorimeter designed by Weber and Bablouzian [12]. A pair of Schott Jena GG 495 filters was used.

flavin in polymethylmethacrylate. The degree of polarization of the fluorescence emission excited in the second flavin absorption band is much lower than the values reported for lipoamide dehydrogenase and 3-methyllumiflavin [4]. Calculation of the angle between the S_1- and S_2 excitation vectors gives a value of 37–39° for transhydrogenase (c.f. 30–31° for lipoamide dehydrogenase and 26–27° for 3-methyllumiflavin [4]). It seems hardly justified to attribute these differences to distortions in the isoalloxazine ring upon binding to the protein, although the maxima of the absorption spectra of the two enzymes are at totally different wavelengths (transhydrogenase in Tris/HCl buffer 442 nm, lipoamide dehydrogenase 455 nm). On the other hand, Harbury et al. [10, 11] have demonstrated that H-bridge formation and polarity of environment have a marked influence on the positions of the absorption bands. Thus, it is more reasonable to assume that binding to the protein enlarges the contribution of a n–π^* transition or of the vibronic coupling, thus leading to a larger depolarization of the S_2-excited fluorescence emission. Similarly, the polarization fine structure can be due to vibronic coupling. However, proper explanation of these phenomena awaits more detailed insight into the exact molecular physical parameters contributing to the absorption spectrum.

Taking the quantum yield of fluorescence emission of transhydrogenase as measured in Tris/HCl as 100%, the addition of 5 mM phosphate quenches the fluorescence about 35%. Similarly, the addition of 0.1 mM $NADP^+$ quenches the fluorescence about 20%, but the combination phosphate plus $NADP^+$ quenches the fluorescence about 70%; the final level of quenching is dependent on the phosphate concentration. Similarly, the dissociation constant for $NADP^+$ decreases about an order of magnitude (Table III). 2′-AMP, which itself has no influence on the fluorescence in Tris/HCl, completely reverses the quenching effect of phosphate and lowers the affinity of the

TABLE III

SPECTRAL AND FLUORESCENCE CHARACTERISTICS OF *A. VINELANDII* TRANSHYDROGENASE

The reference enzyme was dissolved in Tris HCl pH 7.6.

Fluorescence emission species	% yield	K_d
E_{ox}	100	—
$E_{ox} + P_i \rightarrow E_{ox} - P_i$	65–55	2–10 mM
$E_{ox} + NADP^+ \rightarrow E_{ox} - NADP^+$	80	7 μM
$E_{ox} - P_i + NADP^+ \rightarrow E_{ox} - P_i - NADP^+$	30	0.4 μM
E_{ox} + 2′-AMP $\rightarrow E_{ox}$ − 2′-AMP	100	—
$E_{ox} + P_i$ + 2′-AMP $\rightarrow E_0$ − 2′-AMP + P_i	≈100	depending on phosphate concentration: 10–50 μM

Absorption spectrum species	Spectral characteristics in difference spectrum		
	positive maxima (nm)	negative minima (nm)	isosbestic points (nm)
$E_{ox} - P_i + NADP^+ \rightarrow E_{ox} - P_i - NADP^+$	464, 434, 412	483, 453, 383	469, 460, 448, 398
$E_{ox} + P_i \rightarrow E_{ox} - P_i$	464, 434, 412	483, 453, 383	469, 460, 448, 398
$E_{ox} - P_i$ + 2′-AMP $\rightarrow E_{ox}$ − 2′-AMP + P_i	483, 453, 383	464, 434, 412	469, 460, 448, 398

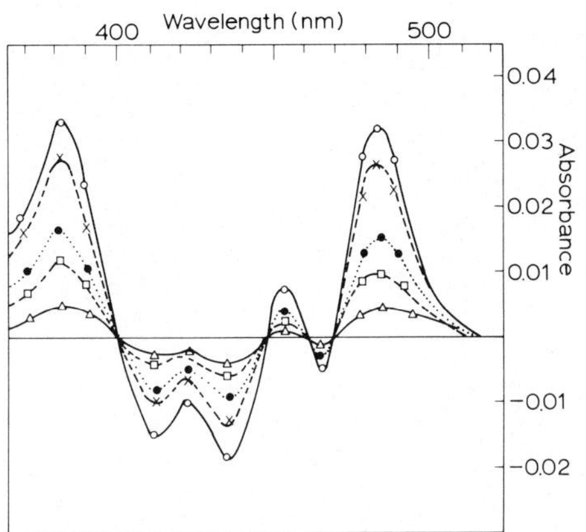

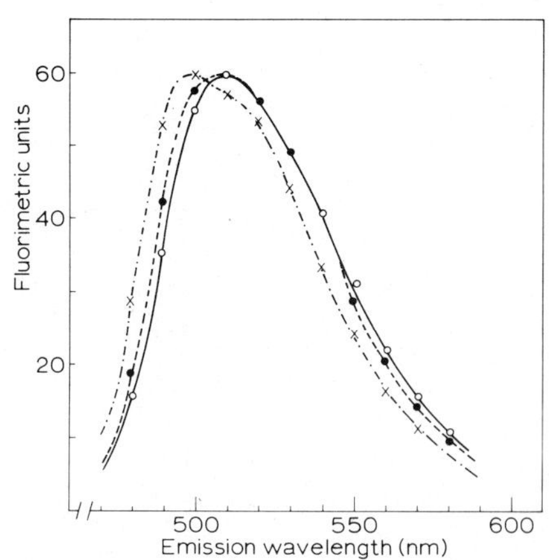

Fig. 5. Spectral titration of *A. vinelandii* transhydrogenase with 2′-AMP. Enzyme (36 μM enzyme flavin) was dissolved in 50 mM Tris/HCl pH 7.6, containing 10 mM potassium phosphate. The experiment was carried out in tandem cells with the following 2′-AMP concentrations: △—△, 35 μM; □– – –□, 70 μM; ●· · · ·●, 100 μM; X– – –X, 270 μM; ○—○, 1.1 mM.

Fig. 6. Effect of $NADP^+$, in the presence or absence of phosphate, on the flavin fluorescence emission maximum of *A. vinelandii* transhydrogenase. The spectra of normalized to the same maximum emission. Excitation wavelength 440 nm, $T = 22°C$. ○—○, enzyme FAD in Tris/HCl pH 7.6; ●– – –●, in the presence of either 0.1 mM $NADP^+$ or 10 mM potassium phosphate alone; X–·–X, in the presence of 0.7 mM $NADP^+$ plus 10 mM potassium phosphate.

enzyme for $NADP^+$ accordingly, but it cannot reverse the quenching of fluorescence by $NADP^+$. Although both phosphate and $NADP^+$ or their combination induce similar difference spectra (Table III), their effect on the absorption spectrum is rather small. Phosphate or $NADP^+$ alone shift the maximum of the first absorption band from 442 to 440 nm, but their combination induces a further shift to about 438 nm. In addition, the bands on both sides of the maximum become sharper (c.f. ref. 13). 2′-AMP reverses the spectral shift induced by phosphate (Table III and Fig. 5).

Either phosphate or $NADP^+$ alone hardly induce a shift in the fluorescence emission spectrum of the enzyme in Tris/HCl (Fig. 6). Neither do these compounds have a marked influence on the fluorescence lifetime (Table IV). Again, their combination induces a large shift to the blue of the

TABLE IV

FLUORESCENCE LIFETIMES OF TRANSHYDROGENASE FAD

22°C, 5 μM enzyme flavin in Tris/HCl pH 7.6, Excitation wavelength 440 nm, emission filter Schott-Jena OG 530.

	τ_p (ns)
Enzyme (in 50 mM Tris/HCl pH 7.6)	2.3
E + 10 mM P_i	2.1
+ 1.1 μM $NADP^+$	2.1
+ 3.8 μM $NADP^+$	1.9
+ 7.6 μM $NADP^+$	1.9
+ 50 μM $NADP^+$	1.6
+ 130 μM $NADP^+$	1.6
+ 740 μM NAD^+	1.6

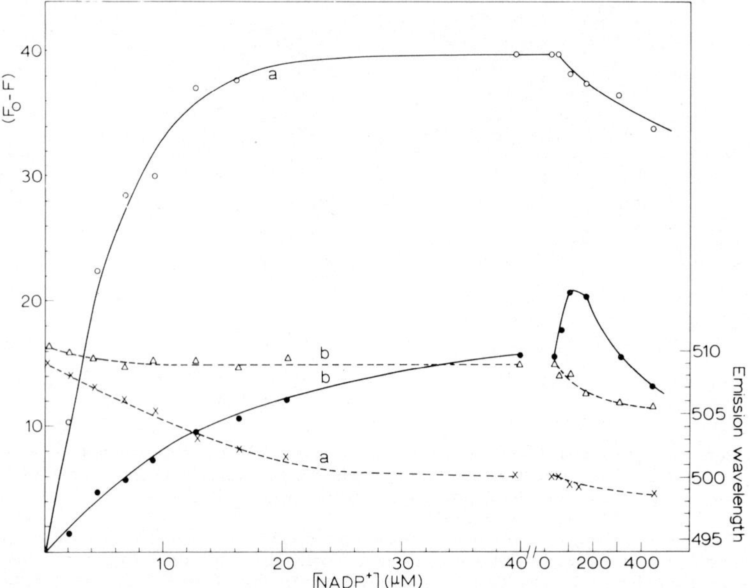

Fig. 7. Effect of $NADP^+$ on the flavin fluorescence spectrum of *A. vinelandii* transhydrogenase. 10 μM enzyme flavin in 50 mM Tris/HCl pH 7.6, $T = 22°C$. Excitation wavelength 440 nm, ○—○, effect on the emission intensity in the presence of 10 mM potassium phosphate (the fluorescence intensity after addition of phosphate was taken as 100); X---X, shift of the fluorescence maximum in the presence of phosphate; ●—●, effect on the emission intensity in the absence of phosphate; △---△, shift of the fluorescence maximum in the absence of phosphate.

emission maximum, while a significant decrease in fluorescence lifetime is observed. The emission maximum (511 nm in Tris/HCl) shifts by the presence of either phosphate or 0.1 mM $NADP^+$ alone to 509 nm; the shape of the emission band is hardly affected under these conditions. However, the combination of phosphate and $NADP^+$ also affects the shape of the emission band. To the red side of the emission maximum at 498 nm, a shoulder at about 510 nm is visible. Measurements of τ_p at different emission wavelengths showed that within experimental error the fluorescence lifetimes are identical. A titration curve with $NADP^+$ is given in Fig. 7. The results confirm earlier spectral studies [13] showing that two $NADP^+$-binding sites are present. The high-affinity site leading in the presence as well in the absence of phosphate to maximum fluorescence quenching; the low-affinity site shows a small increase in fluorescence intensity as compared with the maximum quenching effect. Similarly, it can be shown that two phosphate-binding sites are present, quenching the emission 35% (K_d about 2 mM) and 45% (K_d about 10 mM) respectively. 2′-AMP seems to affect mainly the high-affinity, phosphate-binding site.

The molecular physical interpretation for these phenomena is rather speculative at this moment. Clearly, either phosphate or $NADP^+$ each form a dark complex with the enzyme FAD, but their combination probably forms a fluorescent complex (exiplex) as well. The totally different shape of the emission band and the shift in emission maximum support this view. Any interpretation is however hampered by the two phosphate-binding sites with about equal affinities. This makes a proper selection of experimental conditions rather difficult, i.e. conditions saturating one site specifically are difficult to achieve. It is possible that low-affinity phosphate binding occurs at the high-affinity phosphate binding occurs at the high-affinity $NADP^+$-binding site and vice versa. The shift in emission maximum 509 → 505 nm in the absence of phosphate at $[NADP^+] > 0.1$ mM suggests this, as compared with hardly any shift in the presence of phosphate. It is difficult however to account for the additional quenching by phosphate upon saturating the low-affinity phosphate site, while at $[NADP^+] > 0.1$ mM an increase in fluorescence is observed.

In summary the studies show that a ternary complex exists between phosphate, $NADP^+$ and oxidized transhydrogenase; phosphate promoting the affinity for $NADP^+$. The kinetic significance of this complex awaits further study.

Acknowledgements

We acknowledge gratefully the support from the Netherlands Foundation for Chemical Research (S.O.N.) with financial aid from the Netherlands Organization for the Advancement of Pure Research (Z.W.O.).

References

1 Wahl, P., Auchet, J.C., Visser, A.J.W.G. and Veeger, C. (1974) Eur. J. Biochem. 50, 413–418
2 Spencer, R.D. and Weber, G. (1969) Ann. N.Y. Acad. Sci. 158, 361–376
3 Spencer, R.D., Vaugham, W.M. and Weber, G. (1969) Molecular Luminescence (Lim, E.C., ed.), pp. 607–629, Benjamin, New York
4 Visser, A.J.W.G., Grande, H.J., Müller, F. and Veeger, C. (1974) Eur. J. Biochem. 45, 99–107
5 Visser, J., Voetberg, H. and Veeger, C. (1970) Pyridine Nucleotide-dependent Dehydrogenases (Sund, H., ed.), pp. 359–373, Springer Verlag, Berlin
6 van Muiswinkel-Voetberg, H. and Veeger, C. (1973) Eur. J. Biochem. 33, 285–291
7 Reed, J.K. (1973) J. Biol. Chem. 248, 4834–4839
8 Massey, V. and Veeger, C. (1961) Biochim. Biophys. Acta 48, 33–47
9 Massey, V., Gibson, Q.H. and Veeger, C. (1960) Biochem. J. 77, 341–351
10 Harbury, H. and Foley, K.A. (1958) Proc. Natl. Acad. Sci. U.S. 44, 662–668
11 Harbury, H., la Noue, K.F., Loach, P.A. and Amick, R.M. (1959) Proc. Natl. Acad. Sci. U.S. 45, 1708–1717
12 Weber, G. and Bablouzian, B. (1966) J. Biol. Chem. 241, 2258–2261
13 van den Broek, H.W.J. and Veeger, C. (1971) Eur. J. Biochem. 24, 55–62

Part X

Complex flavoproteins

Flavins and Flavoproteins, edited by T.P. Singer

CHAPTER 55

STUDIES ON THE MECHANISM OF ACTION OF NADH DEHYDROGENASE FROM *AZOTOBACTER VINELANDII*

D.V. DERVARTANIAN

Department of Biochemistry, Boyd Graduate Studies Research Center, University of Georgia, Athens, Ga. 30602 (U.S.A.)

Summary

1. NADH dehydrogenase isolated from *Azotobacter vinelandii* is of low molecular weight and has functional FMN and iron-sulfur centers with high reactivity with the electron acceptors menadione and ferricyanide. Enzymatic reduction with NADH results in the appearance of iron-sulfur centers (1 + 2) as detected by low temperature EPR spectroscopy (15°K).

2. NADPH has been found to be a substrate of NADH dehydrogenase. A constant ratio of NADPH to NADH oxidation activities with electron acceptors has been observed in cell-free extract, phosphorylating particles and at various stages of purification up to the final purified NADH dehydrogenase. NADPH oxidation relative to that of NADH oxidation is approx. 10%. NADPH when added to NADH dehydrogenase causes a similar degree of reduction as monitored by absorption spectroscopy and in the same time interval as does NADH.

3. The site of dicumarol inhibition has been studied by examining its interaction with membrane-bound and purified NADH dehydrogenase. This inhibition site has been conserved in purified NADH dehydrogenase. In the latter enzyme a high degree of inhibition (97%) by dicumarol of NADH or NADPH oxidation occurs with menadione as acceptor, while slightly less inhibition (84%) is observed and at higher dicumarol concentrations when ferricyanide is the acceptor.

4. A direct interaction of dicumarol with purified NADH dehydrogenase has been found consisting of strong perturbations in the absorption spectrum; specifically a negative decrease of absorption in the 400–480 nm region and a positive increase in the 485–520 nm region, with maximum increase of approx. 20% at 508 nm. Spectral titration of enzyme with varying concentrations of dicumarol has allowed determination of the dissociation constant for the enzyme-dicumarol complex as 75 μM. This value is in good agreement with the kinetically determined inhibitor constants.

5. It is proposed that the interaction of dicumarol with NADH

dehydrogenase results in a charge-transfer complex with enzyme-bound FMN as electron acceptor and dicumarol as electron donor. The uncoupling effect of dicumarol in oxidative phosphorylation may be attributable in part to a competitive molecular charge-transfer complex involving isoalloxazine of enzyme-bound FMN and dicumarol.

Introduction

A low molecular weight NADH dehydrogenase has been isolated previously [1] which has functional iron-sulfur centers as determined by the appearance of substrate-reducible, g-1.94-type, reduced-iron signals plus high reactivity with the electron acceptors menadione and ferricyanide. The latter acceptor is known [2] to react with complex ironflavoproteins via the iron-sulfur centers. During the purification of NADH dehydrogenase from *Azotobacter vinelandii*, no unusual changes in Michaelis-Menten constants or electron-acceptor reactivities were observed [1].

It was of interest to consider an investigation of NADPH as a substrate of the enzyme and also to study the nature of the reaction of dicumarol with NADH dehydrogenase since this inhibitor is a known uncoupler of oxidative phosphorylation and a potent inhibitor of NADH menadione reductase activity. As far as it is known, low molecular weight NADH dehydrogenases from heart or bacterial sources do not generally react with dicumarol.

TABLE I

PROPERTIES OF NADH DEHYDROGENASE ISOLATED FROM *AZOTOBACTER VINELANDII*

Mol. wt.		56500
Sedimentation coefficient ($s_{20,w}$)		4.64
Iron content (gatoms/mol)		4
Labile sulfide content (gatoms/mol)		4
Catalytic-center activities (mol substrate/mol FMN at 25°C at infinite substrate concentration)		
Substrate	Acceptor	Catalytic-center activity
NADH	Menadione	16510
NADPH	Menadione	1572
NADH	Ferricyanide	6260
NADPH	Ferricyanide	2268
Absorption spectra		
Oxidized	Maximum at 414 nm, shoulders at 336 nm, 375 nm and 470 nm.	
Reduced	Bleaching of spectrum with maximum decrease of absorption at 465 nm of 48%	
EPR spectra		
Oxidized 77°K	No signals	
15°K	Asymmetric signal at $g = 2.00$	
NADH-reduced		
77°K	Broad, unreasolvable signal	
15°K	Center 1: $g = 2.02, 1.92$	
	Center 2: $g = 2.04, 1.93$	

Methods

NADH dehydrogenase was isolated from *A. vinelandii* OP (kindly provided by Dr. P. Wilson) by a procedure briefly summarized as follows: 25 g washed, previously frozen, log-phase cells are homogenized in 25 mM potassium phosphate buffer (pH 7.4) and passed twice at 0–4°C through a French pressure cell at 12 000 lb/inch^2. The supernatant after centrifugation at 5000 × *g* is the cell-free extract. Henceforth, the procedure is performed anaerobically (in the presence of H_2 and/or with H_2-saturated buffers) due to the extreme lability of the enzyme. The following steps occur in succession: reaction with RNAase and DNAase at 25°C for 10 min to degrade nucleic acids, heat treatment for 10 min at 55°C to dissociate NADH dehydrogenase from the respiratory chain, centrifugation for 2 h at 40 000 × *g* until a clear supernatant is obtained. To this solution is added 1 M potassium phosphate buffer (pH 7.4) to make the supernatant 50 mM with respect to phosphate. Ammonium sulfate fractionation is carried out and the 20–48% pellet cut is taken up in a minimum volume of 50 mM phosphate buffer and dialyzed for 6–8 h, centrifuged for 30 min at 40 000 × *g* to remove denatured protein, and subjected to Sephadex G-100 chromatography. The most active fractions, as determined by NADH (NADPH) menadione [3] or ferricyanide [4] reductase activities and by absorption spectra (Cary models 14 and 15 or Aminco DW-2), are pooled and concentrated under H_2 (Sartorius vacuum membrane apparatus). The concentrated enzyme is stored in liquid nitrogen under anaerobic conditions in tightly-capped storage tubes.

Fresh enzyme (after concentration) shows in the ultracentrifuge (sedimentation-velocity run) a single, nearly symmetrical peak. The enzyme, after 2–4 h ageing under anaerobic conditions at 0°C, has lost at least 60–75% of its initial enzymatic activity and in the ultracentrifuge reveals a broadened peak indicating heterogenous species. Simultaneously, free FMN can be detected, indicating the presence of inactive apoprotein molecules.

Table I summarizes properties of purified NADH dehydrogenase. It is necessary to isolate the enzyme from whole cells from which derived phosphorylating particles show a P/NADH ratio of at least 0.7. Lower P/NADH ratios result in NADH dehydrogenase preparations of low specific activities, especially with respect to ferricyanide reductase activities with either NADH or NADPH.

Kinetic inhibitor constants and spectral dissociation constants were determined as previously described [5] with the purified enzyme. Inhibitors were reacted with enzyme for 3 min at 25°C in the absence of substrates and then the kinetic or spectral runs were carried out. For determination of the spectral dissociation constant [5] for the enzyme-dicumarol complex, a 1 ml cuvette of 1 cm light-path was used. Each separate cell contained fresh enzyme plus a different inhibitor concentration.

Results

NADPH as a substrate of NADH dehydrogenase

Table II reveals that significant NADPH oxidation activity is present in cell-free extracts with the electron acceptor, oxygen, menadione and ferricyanide. NADPH oxidase like NADH oxidase activity is subject to the same respiratory chain inhibitors, including for example cyanide, antimycin, rotenone and dicumarol. The ratio of NADH to NADPH oxidation is fairly constant in cell-free extract, phosphorylating particles (prepared as previously described, ref. 6) and at various stages of NADH dehydrogenase purification with the various electron acceptors. The rate of NADPH oxidation with respect to NADH oxidase activity is approx. 10%. NADPH oxidation is an integral property of the purified NADH dehydrogenase. Hatefi and Hanstein

TABLE II

NADPH AND DICUMAROL REACTIVITY IN *AZOTOBACTER VINELANDII*

Fraction	Specific activity*				NADH-menadione / NADPH-menadione	% Inhibition at 0.2 mM Dicumarol**
	Substrate: NADH Acceptor: O_2	NADPH O_2	NADH Menadione	NADPH Menadione		
Cell-free extract	1.43	0.12	1.41	0.15	9.4	76.9
Phosphorylating particles	3.32	0.27	2.06	0.17	12.1	82.2
Ammonium sulfate cut (20–48%), after dialysis	none	none	12.9	1.28	10.1	83.8
Purified NADH dehydrogenase	none	none	63.0	6.01	10.5	83.2

*μmol substrate/min/mg protein at 25°C.
**NADH as substrate and menadione as acceptor.

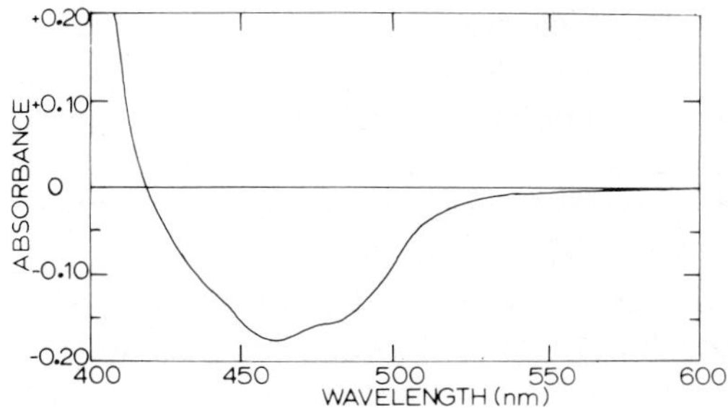

Fig. 1. The reaction of NADPH with NADH dehydrogenase (2.4 mg/ml in 100 mM potassium phosphate buffer, pH 7.4). ——, difference spectrum of NADPH (0.2 mM) reduced minus oxidized.

[7] reported that NADPH oxidase activity was approx. 5% of NADH oxidase activity in bovine heart phosphorylating particles (ETP).

A direct reduction of NADH dehydrogenase by NADPH* could be observed by monitoring the spectral changes occurring in the absorption spectrum. Fig. 1 shows the difference spectrum (NADPH reduced minus oxidized) after addition of NADPH to the enzyme. A similar degree of reduction (approx. 40% at 465 nm) was noted in the same time interval as when NADH was added to the enzyme. The subsequent addition of sodium dithionite (not shown) resulted in a further decrease of absorbance similar in magnitude to that previously reported [1] when dithionite was added to NADH-reduced enzyme. These results differ from those reported by Hatefi and Hanstein [7] who found that addition of NADPH to Complex I resulted in reduction of centers 2 and 3 plus 4 but not of center 1.

The effect of dicumarol or rotenone on NADH or NADPH oxidation

Dicumarol and rotenone are well-known mammalian and bacterial respiratory inhibitors which operate in the region of membrane-bound NADH dehydrogenase. The inhibition by rotenone of NADH oxidation has been studied in detail by Singer and coworkers. These latter workers concluded [8, 9], on the basis of inhibitor binding and EPR studies, that the site of action of rotenone was on the oxygen side of the iron-sulfur centers since in the presence of rotenone, NADH was still able to cause reduction of all known iron-sulfur centers. In addition to being a potent inhibitor of NADH dehydrogenase, dicumarol is also an uncoupler of oxidative phosphorylation and thus has additional significance. In general, higher concentrations of either rotenone or dicumarol are required in the bacterial respiratory systems to achieve a similar degree of inhibition as in the mammalian system. With *A. vinelandii* phosphorylating particles, 79% inhibition of NADH oxidase activity is found at 15 μmol of rotenone per mg protein whereas 91% inhibition is observed at 0.2 mM dicumarol. The inhibition of NADH (or NADPH) oxidation by dicumarol when menadione is the electron acceptor is

---

* Preliminary experiments indicate that NADPH causes the appearance of iron-sulfur centers 1 + 2 in the same intensity and time as does NADH.

slightly lower at comparable dicumarol concentrations. Table II also shows that the inhibition at 0.2 mM dicumarol with NADH or NADPH as substrate and menadione as acceptor is reasonably constant in cell-free extract, phosphorylating particles and at various stages of purification involved in the isolation of NADH dehydrogenase. It appears then that the site of dicumarol inhibition is maintained in the purified NADH dehydrogenase.

Both NADH and NADPH oxidation catalyzed by purified NADH dehydrogenase are strongly inhibited by dicumarol (maximum inhibition, 97%) when menadione is the electron acceptor. When ferricyanide is the acceptor, approx. 2–3-fold higher concentrations of dicumarol are required to achieve a slightly lower degree of inhibition (maximum inhibition, 84%). The kinetically determined inhibitor constants and Michaelis-Menten constants with both substrates and acceptors are listed in Table III. The kinetic plots of dicumarol inhibition indicate competitive-type inhibition. Rotenone (16 μmol per mg purified NADH dehydrogenase) had no inhibitory effect when either ferricyanide or menadione were the electron acceptors and NADH was the substrate.

When NADH dehydrogenase was titrated with varying dicumarol concentrations, distinct and reproducible perturbations in the absorption spectrum of NADH dehydrogenase could be observed. The maximum spectral effect is seen in Fig. 2. The difference spectrum (dicumarol-treated minus oxidized) shows a decrease of absorption in the 400–480 nm region (maximum decrease of 5.0% at 450 nm) and a positive increase of absorption in the 485–520 nm region (maximum increase of 19.6% at 508 nm). This spectral change is stable for at least 30 min and undergoes no further absorbance changes. The respiratory chain succinate dehydrogenase [5] is comparable in many properties with NADH dehydrogenase (both are multi-iron-sulfur flavoproteins) and undergoes similar spectral changes on the addition of fumarate or other competitive inhibitors to the oxidized enzyme. A number of other flavoproteins likewise show similar spectral changes on the addition of their inhibitors [10, 11]. The reaction of dicumarol, a strong competitive inhibitor of NADH (NADPH) oxidation, with NADH dehydrogenase results in the appearance of a spectrally detectable, enzyme-dicumarol complex. From the titration of NADH dehydrogenase with varying amounts of dicumarol, a dissociation constant of 75 μm was found. This value is intermediate in value (Table III) between the kinetically determined inhibitor

TABLE III

KINETIC CONSTANTS OF NADH DEHYDROGENASE

Substrate	Acceptor	K_m(μM)	K_I(dicumarol)(μM)	K_D^*
NADH	Menadione	140	45	
NADPH	Menadione	260	48	75 μM
NADH	Ferricyanide	280	130	
NADPH	Ferricyanide	160	110	

*Dissociation constant of enzyme-dicumarol complex determined spectrophotometrically as in Methods section.

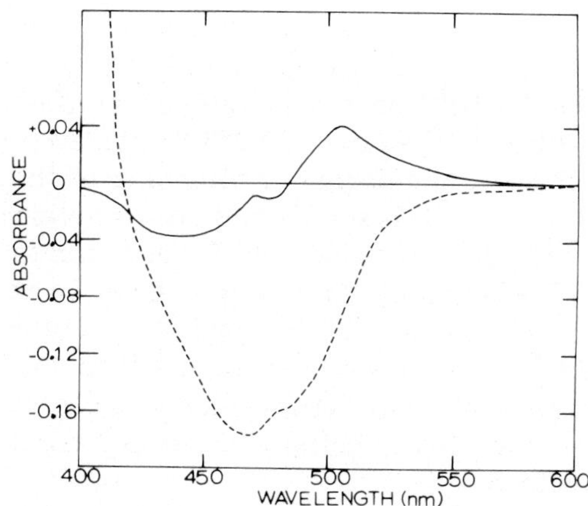

Fig. 2. The reaction of dicumarol with NADH dehydrogenase (5.1 mg/ml in the buffer above). ——, difference spectrum of dicumarol (0.1 mM) reacted minus oxidized; – – –, NADH (0.2 mM) reduced minus oxidized.

constants for NADH or NADPH oxidation by menadione or ferricyanide and may be considered a good agreement. Rotenone, which exhibited no inhibitory effect on NADH oxidation when ferricyanide or menadione were acceptors, had no effect on the absorption spectrum at 15 μmol per mg protein. The addition of rotenone to NADH-reduced enzyme also caused no spectral changes nor did it diminish the magnitude of the NADH-induced absorbance changes.

The dicumarol-enzyme complex appears to be partially reversible based on the abolishment of the spectral features of the complex and appearance of the typical NADH-reduced difference spectrum (Fig. 2, lower spectrum) on the subsequent addition of excess NADH. The difference spectrum shown (NADH-reduced minus oxidized) is of the same type as shown by the direct reaction of NADH with oxidized NADH dehydrogenase. However, the magnitude of spectral response is 60% less than that found when NADH is added in the absence of dicumarol. The addition of dicumarol (0.4 mM) to NADH-reduced enzyme caused no spectral changes attributable to the enzyme-dicumarol complex, indicating that prior reduction with substrate offers protection against the inhibitor action of dicumarol.

Since it is possible that rotenone or dicumarol could cause spectral alterations in the FMN moiety in the absence of bound protein, the effects of these compounds on free FMN were tested under identical conditions and concentrations. No spectral changes were noted with either compound when added to FMN. A small decrease of absorbance in the 375–500 nm region of FMN was noted when ethanol (2.5%) was added. Ethanol is used to dissolve rotenone. Likewise, a similar small decrease of absorbance (about 3–5% at 450 nm) in the 375–550 nm region was noted on addition of 2.5% ethanol to oxidized NADH dehydrogenase.

Discussion

This study has established that NADPH is a substrate of purified azotobacter NADH dehydrogenase. Although the rate of NADPH oxidation is approx. 10% of that found with NADH, the degree of bleaching of the absorption spectrum by either NADH or NADPH is similar and occurs within the same time interval. Beinert et al. [12] reported that NADPH addition to the Singer high-molecular weight NADH dehydrogenase derived from heart mitochondria caused the appearance of the $g = 1.94$ signal (iron-sulfur center 1) although more slowly than NADH. It was also reported in this paper that the NADH/NADPH ratio was 3500 with ferricyanide as acceptor and that NADPH qualitatively caused the same spectral changes as NADH but at far slower rates than observed with NADH.

The nature of the spectral interaction of dicumarol with NADH dehydrogenase is not known with certainty. Based on the model studes of Harbury and coworkers [13, 14], it was suggested earlier [6] that the enzyme-inhibitor spectral complex in the case of succinate dehydrogenase was due in part to a charge-transfer complex with the inhibitor acting as electron donor and the FMN moiety as the electron-accepting molecule. Other model flavin and flavoprotein studies [11] indicate that very similar and reversible spectral changes are detectable by temperature-difference spectroscopy. These spectral changes were attributed by Müller et al. [11] in part to vibronic transitions in two electronic absorption bands. These authors considered that the spectral changes represented temperature-dependent conformational changes occurring within the flavoproteins. Rigers and Brandth [15] observed that even the addition of Cl^- to *Aspergillus niger* glucose oxidase resulted in the same type of spectral difference spectrum as found in this study.

Harbury and Foley [13] showed that the addition of a large excess of benzoate to 3-methylriboflavin resulted in a decrease of absorption in the 400–475 nm region and a positive increase of absorption in the 480–500 nm region. No spectra were taken above 500 nm. In fact, these spectral changes are very similar to those found when dicumarol is added to NADH dehydrogenase. The tight binding of dicumarol at the active site permits interaction with the FMN moiety, while in the model system a mass action shift in equilibrium caused by the high concentration of reactant permits a similar type of spectral perturbation. One possible explanation, therefore, for the spectral enzyme-dicumarol complex is that it represents a molecular charge-transfer complex where the closs interaction of dicumarol with FMN has permitted partial charge transfer. In this regard, dicumarol is a possible electron donor by virtue of a strategically situated methylene group.

The mechanism of dicumarol inhibition at the NADH oxidation site of the respiratory chain may differ significantly from that of rotenone. This study establishes an interaction of dicumarol with enzyme-bound FMN resulting in some type of competitive enzyme-inhibitor complex. Rotenone, however, reacts at a site(s) on the oxygen side of the known iron-sulfur centers. It will be necessary to study the direct reaction of dicumarol with the *Azotobacter* respiratory chain (by monitoring absorbance changes at 508 nm) to determine whether this interaction also occurs in the intact respiratory system. Such studies are currently underway.

Harbury et al. [14] suggested that an uncoupler of oxidative phosphorylation might act with isoalloxazine via a competitive charge-transfer interaction. The experimental finding of a dicumarol · FMN (enzyme-bound) complex raises the question whether such a concept may be a feasible explanation for at least part of the uncoupling effect of dicumarol.

Acknowledgements

Thanks are due to Mrs. Pamela Hartsell for the excellent technical assistance and to Mr. Vance Morgan for part of the *Azotobacter* cells. This study was supported in part by Research Grant 1-RO1-GM-18895 and Research Career Development Award 1-KO4-GM-70010 from the National Institute for General Medical Science, U.S. Public Health Service.

References

1 DerVartanian, D.V. (1972) Z. Naturforsch. 27b, 1082–1084
2 Komai, H., Massey, V. and Palmer, G. (1969) J. Biol. Chem. 244, 1692–1700
3 Hatefi, Y. and Stempel, K.E. (1969) J. Biol. Chem. 244, 2350–2357
4 Smith, S.T., Rajagopalan, K.V. and Handler, P. (1967) J. Biol. Chem. 242, 4108–4117
5 DerVartanian, D.V. and Veeger, C. (1964) Biochim. Biophys. Acta 92, 233–247
6 DerVartanian, D.V., Iburg, L.K. and Morgan, T.V. (1973) Biochim. Biophys. Acta 305, 173–178
7 Hatefi, Y. and Hanstein, W.G. (1973) Biochemistry 12, 3515–3522
8 Horgan, D.J., Singer, T.P. and Casida, J.E. (1968) J. Biol. Chem. 243, 834–843
9 Gutman, M., Singer, T.P. and Beinert, H. (1972) Biochemistry 11, 556–562
10 Veeger, C., DerVartanian, D.V., Kalse, J.F., DeKok, A. and Koster, J.F. (1966) in Flavins and Flavoproteins, (Slater, E.C., ed.), pp. 242–262, Elsevier, Amsterdam
11 Müller, F., Mayhew, S.G. and Massey, V. (1973) Biochemistry 12, 4654–4662
12 Beinert, H., Palmer, G., Cremona, T. and Singer, T.P. (1965) J. Biol. Chem. 240, 475–480
13 Harbury, H. and Foley, K.A. (1958) Proc. Natl. Acad. Sci. U.S. 44, 662–668
14 Harbury, H., LaNoue, K.F., Loach, P.A. and Amick, R.M. (1959) Proc. Natl. Acad. Sci. U.S. 45, 1708–1717
15 Rogers, M.J. and Brandt, K.G. (1971) Biochemistry 10, 4630–4641

Flavins and Flavoproteins, edited by T.P. Singer

CHAPTER 56

ROLE OF FLAVIN IN REDUCTIVE ACTIVATION OF SUCCINATE DEHYDROGENASE

BRIAN A.C. ACKRELL, EDNA B. KEARNEY and DALE EDMONDSON

Department of Biochemistry and Biophysics, University of California, San Francisco, Calif. 94143 and the Division of Molecular Biology, Veterans Administration Hospital, San Francisco, Calif. 94121 (U.S.A.)

Summary

It is known that succinate dehydrogenase, deactivated by combination with oxaloacetate, can be activated by reducing agents. The question of which group in the enzyme must be reduced for activation to occur was studied in anaerobic titrations with dithionite and in poised potential experiments.

In titrations with dithionite, 6–7 reducing equivalents bring about complete reduction of the chromophoric groups. The first 2 electrons apparently go to Fe—S centers and the second two are accepted by the flavin moiety, as judged from spectral changes and from loss of fluorescence of the enzyme. Activation of the enzyme occurs with the incorporation of the third and fourth electrons. In poised potential experiments activation occurred concurrently with the uptake of 2 H^+ and 2 e^- by a group with a half-reduced potential of -140 mV at pH 7.4; these data confirm the involvement of the flavin in reductive activation of the enzyme. The remaining electrons required for full reduction of the enzyme in the anaerobic titration experiments are taken up by the Fe—S center(s) which is of very low redox potential. This center is not involved in the activation-deactivation cycle, and is also not reducible by succinate.

Binding of oxaloacetate to a fully activated enzyme causes not only the expected deactivation, but also loss of the flavin fluorescence of the enzyme. This may indicate an interaction between the covalently bound flavin and the oxaloacetate binding site.

Abbreviations: ETP, non-phosphorylating sub-mitochondrial particle preparation from beef heart; Complex II, succinate-ubiquinone reductase; Hepes, *N*-2-hydroxyethylpiperazine-*N*′-2-ethanesulfonic acid; Mes, 2-(*N*-morpholino) ethanesulfonic acid; PMS, phenazine methosulfate.

The activation-deactivation cycle of succinate dehydrogenase has recently been related to the binding of oxaloacetate to the enzyme [1]. Oxaloacetate binds in almost irreversible fashion to succinate dehydrogenase and the activity of the enzyme cannot be expressed when oxaloacetate is so bound. Thus the process of activation must involve dissociation of this complex.

In previous work [1, 2] we have studied the activation of succinate dehydrogenase, by incubation of the enzyme with substrate, substrate analogues, and simple inorganic anions. The enzyme may also be activated, however, by reduction [3, 4] and preliminary experiments showed that reductive activation differs from that by the other methods studied in that it occurs readily even at 0°C. This suggested that a different mechanism might be involved. Two questions had to be asked therefore. The first concerned the component in the enzyme molecule that was reduced concurrently with activation and the other concerned the mechanism of activation itself. Did reductive activation represent a reduction of bound oxaloacetate to the more easily dissociable form, malate, or did it reflect a difference in the binding affinities of the oxidized and reduced forms of the enzyme for oxaloacetate? Since succinate dehydrogenase contains flavin and non-heme iron chromophores characterized by different half-reduction potentials, our experimental approach in answering these questions was to correlate increases in the level of activation of the enzyme during dithionite titrations with changes in its visible absorption spectrum and fluorescence properties, and also to measure the level of activation of the enzyme at different poised potentials during oxidation-reduction titrations.

Correlating activation with the reduction of a specific redox component is complicated because the contribution of the flavin and non-heme iron clusters to the optical spectra are not clearly distinguished, and because the binding of oxaloacetate, as well as causing deactivation of the enzyme, also induces spectral changes. This latter point is illustrated in Fig. 1, which shows the difference spectrum of the oxaloacetate-bound and oxaloacetate-free forms of the enzyme. The spectral changes are in agreement with the observation of DerVartanian and Veeger [5] with a different type of succinate dehydrogenase preparation. The two isosbestic points at 410 nm and 500 nm allow absorbance measurements without interference from spectral changes due to the enzyme-oxaloacetate complex and permit a direct comparison of the reduction characteristics of the oxaloacetate-bound and oxaloacetate-free forms of the enzyme. Such an experiment is shown in Fig. 2, where the bleaching of both forms of the enzyme at these isosbestic wavelengths is plotted against the number of electrons added per enzyme molecule in anaerobic titrations with dithionite. In agreement with EPR data [6], full reduction of the enzyme is effected by the incorporation of 6 to 7 electrons. This electron stoichiometry is that expected from the flavin and non-heme iron content of this particular preparation of enzyme (8.3 gatoms of iron and 1 mol flavin per mol enzyme), if one accepts that each pair of iron atoms requires one electron for reduction [7] and the flavin moiety is

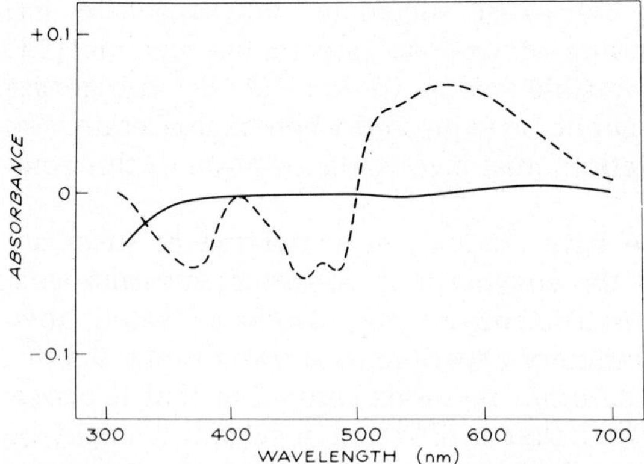

Fig. 1. Oxaloacetate-induced difference spectrum of succinate dehydrogenase. The enzyme was an oxaloacetate-free type 3-NS preparation, dissolved in 50 mM Hepes/NaOH, pH 7.5, at 25°C (2.1 mg of protein/ml; 14.4 μM enzyme). A baseline was taken of enzyme vs. enzyme (——), following which oxaloacetate was added to the sample cuvette (final concentration = 106 μM), an equal volume of buffer was added to the reference cuvette, and the spectrum was again recorded (---).

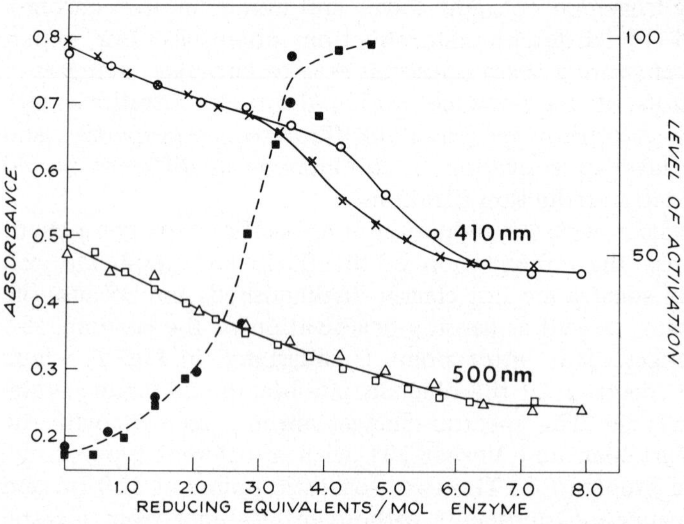

Fig. 2. Titration of active and deactivated succinate dehydrogenase with dithionite: absorbance changes and activation. Both oxaloacetate-free and oxaloacetate-bound enzyme were titrated with sodium dithionite anaerobically in 50 mM Hepes/NaOH, pH 7.5, at 25°C, and at a protein concentration of 2.1 mg/ml (16.8 μM enzyme). The oxaloacetate-bound enzyme contained 0.63 mol of oxaloacetate per mole of enzyme. The absorbance of the oxaloacetate-bound enzyme at 410 nm and 500 nm is given by (○) and (△), respectively, that of the oxaloacetate-free enzyme by (X) and (□), and the level of actiavtion of the oxaloacetate-bound enzyme by (●). In another experiment the level of activation is plotted (■) during a similar titration of type 3-NS enzyme in the presence of excess oxaloacetate (7.7 mol/mol enzyme; 1.52 mg of protein/ml, 10.4 μM enzyme).

reduced to the hydroquinone form. Full reduction of the enzyme requires 6–7 electrons whether or not it is bound with oxaloacetate, which suggests that oxaloacetate is not reduced to malate during the activation process. This point was confirmed directly by precipitating a sample of the

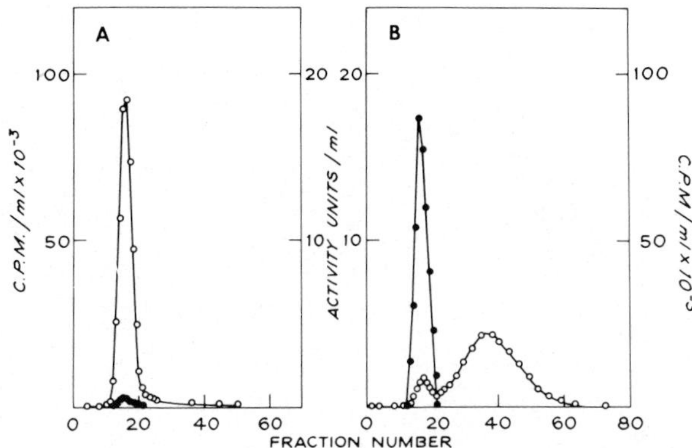

Fig. 3. Chromatography of [^{14}C] oxaloacetate-bound succinate dehydrogenase in the oxidized and reduced state. 1 ml aliquots of type 1 enzyme, containing 6.48 mg of protein, 26.6 nmol of enzyme, and 24 nmol of uniformly labelled [^{14}C] oxaloacetate, were applied to Sephadex G-50 columns, 0.9 × 20 cm, equilibrated at 4°C with oxygen-free 50 mM Hepes, pH 7.8, in the absence (A) and in the presence (B) of 3 mM sodium dithionite. The collected fractions were monitored for enzyme activity (●) and radioactivity (○). The specific activity of the oxaloacetate was 5348 cpm/nmol.

oxaloacetate-bound enzyme anaerobically with perchloric acid after it had been fully reduced with dithionite. All the oxaloacetate was still there, and no malate was detectable, although even a 10% conversion of oxaloacetate to malate could have been detected. It appeared likely that the mechanism would involve dissociation of oxaloacetate from the enzyme as a result of a change in affinity of the reduced enzyme for oxaloacetate. This possibility was tested in the experiment illustrated in Fig. 3, in which aliquots of the enzyme bound with [^{14}C] oxaloacetate were passed through anaerobic columns of Sephadex G-50 in the presence and absence of 3 mM dithionite. In the absence of dithionite (Fig. 3A) all the counts eluted with the deactivated enzyme, demonstrating essentially irreversible binding of oxaloacetate to the oxidized form of the enzyme. In the presence of dithionite (Fig. 3B), however, only 10% of the counts eluted with the enzyme, which was at this point reduced and thereby also activated. These data show that reduction of the enzyme decreases its affinity for oxaloacetate by at least an order of magnitude.

The number of reducing equivalents needed to cause dissociation of the bound oxaloacetate from the enzyme is shown in Fig. 2, which gives the activation profile for the oxaloacetate-bound enzyme during the course of reductive titration. A 10-fold excess of free oxaloacetate did not influence the electron stoichiometry. Thus, whereas complete reduction of the enzyme requires incorporation of 6–7 electrons, major activation takes place with the uptake of the 3rd and 4th electrons. Although significant bleaching of the enzyme is brought about by the first two electrons to enter the molecule, this is not associated with extensive activation.

The nature of the component, or components of the enzyme, titrated with the third and fourth electrons, whose oxidized or reduced state influences oxaloacetate binding, was studied in experiments where the effect of

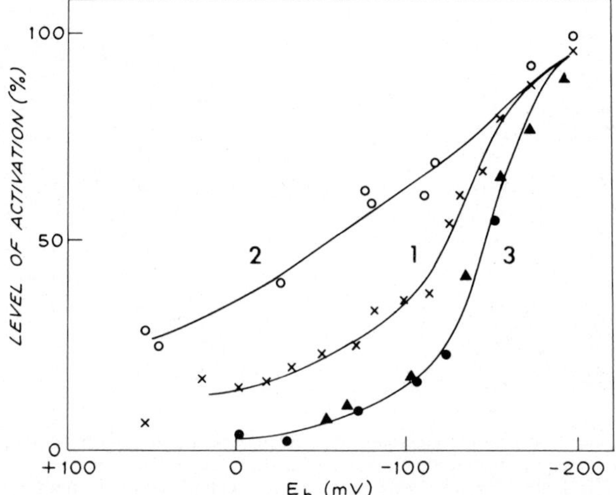

Fig. 4. Potentiometric titration of succinate dehydrogenase. Deactivated type 1 enzyme (2.15 mg of protein/ml, 9.0 μM enzyme) was titrated at 25°C in 50 mM Hepes/NaOH, pH 7.5, in the presence of 60 μM concentrations of the redox mediators listed as follows, with their respective midpoint potentials given in parentheses: PMS (+80 mV), juglone (+30 mV), duroquinone (+5 mV), indigo tetrasulfonate (—46 mV), indigo disulfonate (—125 mV), and anthraquinone-2-sulfonate (—225 mV). Aliquots of the reaction mixture were taken anaerobically for assay of enzyme activity during the titrations, which were carried out on the same sample in the following sequence: (curve 1) (X), plus dithionite; (curve 2) (○), plus ferricyanide; (curve 3), plus additional oxaloacetate to give 67 μM and then dithionite again (▲); plus ferricyanide once again (●).

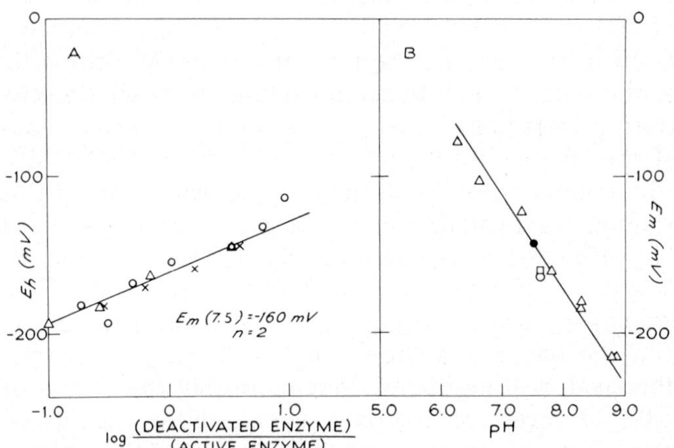

Fig. 5. (A) Relation of redox potential and the activation level of Complex II in the presence of excess oxalacetate. The enzyme (2.3 mg of protein/ml, 8 μM enzyme) was titrated at pH 7.5 with dithionite in the presence of redox mediators as in Fig. 4. Activity was monitored during reductive titration with dithionite (X) in the presence of 67 μM oxaloacetate, during reoxidation with ferricyanide (○), and during reduction with dithionite (△) in the presence of 202 μM oxaloacetate. (B) pH dependence of the half-reduction potential in the presence of 67 μM oxaloacetate. ETP (5.76 mg protein/ml) were titrated as in Fig. 4, but in a medium containing, in addition to the mediators, 200 mM sucrose, antimycin A (1 μg/mg protein), and one of the following buffers at 50 mM concentration: Mes/NaOH (pH 6.27), Hepes/NaOH (pH 6.6, 7.3, 7.8), glycylglycine/NaOH (pH 8.3) and Hepes/Tris (pH 8.81). Single points are also included for type 1 enzyme (○), Complex II (□), and for type 3-NS enzyme (●).

poising the oxidation-reduction potential on the level of activation was measured.

Fig. 4 shows that the reductive activation of an oxaloacetate-bound soluble preparation of the enzyme at pH 7.5 is a sigmoidal function of the potential of the system relative to the hydrogen electrode (curve 1). Reoxidation of the system with ferricyanide (curve 2) did not re-establish the original level of activation, showing that the system was not at equilibrium. A fully reversible system was achieved only when an excess of oxaloacetate was present (curve 3).

Under conditions where a fully reversible system pertains, the half-reduction potential of the component of the enzyme influencing activation may be obtained from a plot according to the Nernst equation, where oxidized component is equated with the deactivated fraction of the enzyme and reduced component with the active fraction. Such a plot for a reduction and reoxidation titration of Complex II, carried out in the presence of 67 μM oxaloacetate, is shown in Fig. 5A. An additional reduction titration was done in the presence of 200 μM oxaloacetate. The fact that the data from these titrations can be expressed within experimental error as a single linear plot strongly implies that the activation of the enzyme stems from the reduction of only one of its several redox components. The half-reduction potential was -160 mV at pH 7.5 with an n value of 2 electrons. Similar values were obtained with other types of enzyme preparations.

Fig. 5B shows the pH dependence on the half-reduction potential for activation of the membrane-bound enzyme in sub-mitochondrial particles (ETP)*. Single values for Complex II and two different types of soluble enzyme preparations are also included for comparative purposes. The slope of the line, equivalent to 0.056 V/pH unit, indicates that 2 protons as well as 2 electrons are required for reduction of the component in the key steps of the activation process. These results are consistent with the reduction of the flavin to the hydroquinone form. It is proposed, therefore, that the half-reduction potential of -160 mV is that of the FAD/$FADH_2$ couple of succinate dehydrogenase under these experimental conditions. At pH 7.0, this value becomes -120 mV by interpolation of the data in Fig. 5B.

The half-reduction potential of the couple could be corrected for the effect of oxaloacetate according to the equation

$$E_m\,(\text{observed}) = E_m + \frac{RT}{nF}\log\frac{1 + K_R\,[\text{OAA}]}{1 + K_O\,[\text{OAA}]}$$

where K_R and K_O are the equilibrium constants for the binding of oxaloacetate to the reduced and oxidized forms of the enzyme, respectively. Although the equilibrium constants are not precisely known, it is evident that the value of -120 mV will become approximately 30 mV more positive for each order of magnitude that the affinity of oxaloacetate for the oxidized

---

* See footnote p. 522.

form of the enzyme exceeds that for the reduced form. On the basis of the data presented earlier, this figure of −120 mV may therefore be corrected upwards to at least −90 mV.

This value, ascribed here to the flavin couple of the enzyme, apparently reflects the influence of the protein environment, as it is some 60–70 mV more positive than that of either synthetic histidyl-flavin [8] or of a histidyl-flavin peptide from sarcosine dehydrogenase [9], compounds which themselves reproduce the main point of attachment of the flavin to the enzyme, i.e. the covalent bond between a histidine residue of the protein and the 8-methyl group of riboflavin.

The value of −90 mV is more negative than either of the values of +65 mV and −5 mV recently ascribed [10, 11] to two of the non-heme iron clusters present in the enzyme [6]. Thus reduction of the flavin would occur subsequent to reduction of these same non-heme iron groups, in agreement with spectrophotometric observations (Fig. 2) and EPR data [6].

Spectrophotometric data from anaerobic titration experiments may be expressed for emphasis of the flavin contribution if the ratio of absorbance changes occurring at 450 and 550 nm is plotted as a function of electron input into the enzyme, as has been done by Olson and co-workers with xanthine oxidase [12]. With succinate dehydrogenase this approach can only be used with the oxaloacetate-free enzyme, since 450 and 550 nm are not isosbestic wavelengths (Fig. 1). This is permissible, however, since oxaloacetate does not change the stoichiometry of the titration (Fig. 2). Fig. 6A shows that the ratio $\Delta A_{450\,nm}/\Delta A_{550\,nm}$ is about 2.0 early in the titration, when, according to EPR data [6], only non-heme iron groups are being

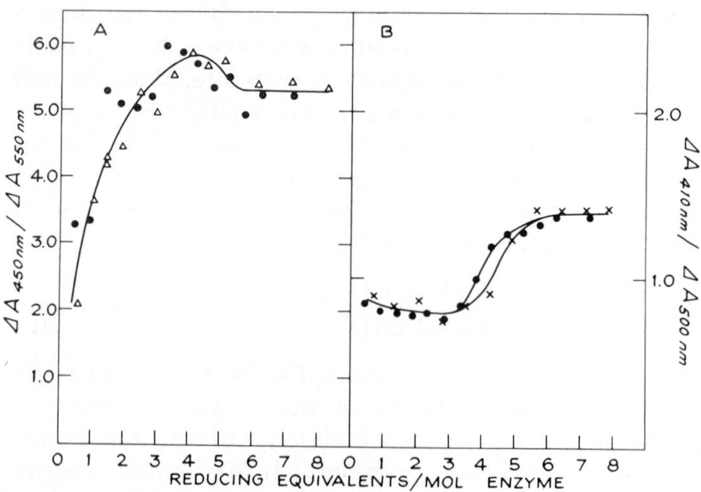

Fig. 6. (A) Ratio of absorbance changes at 450 nm and 550 nm during reductive titration. Data for the type 3-NS enzyme (●) were taken from the experiment of Fig. 2. Data for type 1 enzyme (3.4 mg of protein/ml, 12.7 μM enzyme, ratio of non-heme iron to flavin, 5.4:1) were similarly obtained (△). (B) Ratio of absorbance changes at 410 nm and 500 nm during reductive titration. Data are taken from Fig. 2; X, ratio for oxaloacetate-bound enzyme; ●, ratio for oxaloacetate-free enzyme.

titrated. The ratio reaches a maximum of 6.0 exactly coincident with the incorporation of four electrons. This is consistent with reduction of the flavin, which contributes maximally to the absorbance loss at 450 nm, but not at all at 550 nm. Four electrons is, as mentioned above, the number required to elicit full activation of the oxaloacetate-bound enzyme.

Conversely, when spectrophotometric changes are plotted so as to minimize the flavin contribution, a correlation with activation is not seen. The course of the titration plotted in Fig. 2 suggests a different degree of change in absorbance at 410 nm in the early and late phases of the titration, while loss of absorbance at 500 nm follows a smoother curve. In Fig. 6B, the ratio of absorbance loss at these two isosbestic wavelengths is plotted against the number of reducing equivalents added. The absorbance is almost entirely due to non-heme iron groups, since the contribution of the flavin, calculated from extinction coefficients derived from the absorbance spectrum of a histidyl flavin peptide, is only 15% of the total absorbance at 410 nm, and 8% at 500 nm. It is evident that the non-heme iron groups fall into at least two types in terms of chromophoric properties. Those with the more positive half-reduction potentials show an average ratio of absorbance changes at the two wavelengths of approximately 0.8, whereas the ratio rises to 1.5 when the iron moieties of lower half-reduction potential than the flavin are also reduced. It is also clear that the higher potential iron groups can be reduced without activation of the enzyme, whereas reduction of the lower potential iron groups is still incomplete when full activation has been achieved by the incorporation of four electrons by the enzyme.

The final piece of evidence to suggest that the reduction of flavin is synonomous with activation came from fluorometric data. We have confirmed the observation of Veeger [13] that soluble preparations of succinate dehydrogenase exhibit flavin fluorescence. The fluorescence is about 1% that expected from the same concentration of free riboflavin and it is quenched by dithionite. In anaerobic titrations of the enzyme with dithionite, complete quenching was achieved by the incorporation of four electrons.

Interestingly, fluorescence is also quenched by very low concentrations of oxaloacetate; this supports the assumption that the observed fluorescence is due to succinate dehydrogenase rather than to a trace impurity in the preparation. When oxaloacetate is added in substoichiometric increments to the enzyme, both deactivation and concurrent and proportional quenching of fluorescence occur (Table I).

Just how reduction of the flavin moiety can influence the dissociation of oxaloacetate remains to be investigated. Oxaloacetate is presumed to be bound in a thiohemiacetal linkage to a sulfhydryl at the active center of the enzyme [14]. As this is a relatively labile bond, we may assume it is stabilized by the protein environment. The change in structure of the flavin which may occur on its reduction could induce a configurational change in the protein, particularly where a covalent bond holds the flavin in position, and this in turn could alter the stabilizing influence of the oxidized protein structure on the thiohemiacetal, with resulting dissociation of the oxaloacetate.

Materials and Methods

ETP was isolated from beef heart mitochondria as previously described [15]. Soluble succinate dehydrogenase was prepared either from an acetone poweder of ETP according to Coles et al. [16] (type 1 enzyme) or by perchlorate extraction of Complex II [17], but in the absence of succinate (type 3-NS enzyme) [16].

Type 3-NS enzyme is fully activated as prepared. Type 1 enzyme was activated by incubation at 20°C with 500 mM $NaNO_3$/100 mM semicarbazide [1]. The oxaloacetate dissociated thereby was removed by passing the reaction mixture through a Sephadex G-50 column equilibrated with the same activating mixture. The activating anion mixture was then removed by passage through Sephadex G-50 equilibrated with 50 mM Hepes/NaOH, pH 7.8; preparations of type 3-NS enzyme were rendered free of extraneous salts in similar fashion.

Deactivated enzyme preparations were obtained by incubating enzyme with oxaloacetate at 20°C until essentially no catalytic activity could be detected; unbound oxaloacetate was then removed by Sephadex chromatography, or in the case of particulate preparations, by repeating washing and centrifugation. Where used, radioactive oxaloacetate was generated from L-[^{14}C]aspartate by glutamate-oxaloacetate transaminase [18].

Fully activated enzyme is defined as that obtained by incubation of the enzyme with succinate prior to assay [2]. The deactivated fraction of the enzyme is equated with the difference between the activities observed in the catalytic assay at 15°C, with and without this prior treatment. Determinations of activity and protein were done as in previous work [2]. The concentration of enzyme was determined by analysis for histidyl flavin [19]. Non-heme iron was measured by the method of Brumby and Massey [20] and oxaloacetate and malate were measured fluorometrically, according to Williamson and Corkey [21].

All anaerobic experiments were performed in an atmosphere of helium that had been purified over hot copper and saturated with water in a gas train made of glass and butyl rubber tubing. Dithionite solutions, prepared anaerobically and standardized by anaerobic titration of a solution of 3-methyllumiflavin, were stored in a gas-tight Hamilton syringe. Anaerobic titrations of succinate dehydrogenase with the standardized dithionite solution were performed in a glass cuvette, similar to that described by Burleigh et al. [22]. Potentiometric titrations were carried out at 25°C in an apparatus essentially like that described by Dutton [23] under a continuous stream of water-saturated helium. Aliquots of enzyme taken for measurement of activity during both types of titrations were withdrawn from the respective reaction vessels under a stream of helium into a helium-flushed gas-tight syringe. A small amount was first ejected from the syringe and the requisite aliquot was then injected from the syringe directly into a prepared cuvette containing the otherwise complete reaction mixture for assay. This precaution was necessary

TABLE I

RELATION OF FLUORESCENCE QUENCHING AND ACTIVITY LOSS ON TITRATION OF SUCCINATE DEHYDROGENASE WITH OXALOACETATE

Oxaloacetate (μM)	Quenching of fluorescence* (%)	Level of activation (%)
—	0	100
2.1	38.4	69.3
3.2	56.8	56.2
4.2	66.0	34.3
5.2	77.2	23.9
6.3	92.5	9.0
7.4	100	0

* Calculated as percentage of maximal change induced by oxaloacetate.

because, after reduction with dithionite, the enzyme was very rapidly inactivated on exposure to air.

Absorbance spectra were taken in a Cary 14 spectrophotometer thermostated at 25°C and corrected fluorescence spectra were determined with a Perkin-Elmer fluorescence spectrophotometer (MPF 3).

Acknowledgements

We wish to express our appreciation of the excellent technical assistance of Miss Maria Mayr. This investigation was supported by the National Institutes of Health (1 PO 1 HL16251) and the National Science Foundation (GB 36570X).

References

1 Ackrell, B.A.C., Kearney, E.B. and Mayr, M. (1974) J. Biol. Chem. 249, 2021–2027
2 Kearney, E.B., Ackrell, B.A.C., Mayr, M. and Singer, T.P. (1974) J. Biol. Chem. 249, 2016–2020
3 Klaasse, A.D.M. and Slater, E.C. (1972) Z. Naturforsch. 27b, 1077–1078
4 Salach, J.I. and Singer, T.P. (1974) J. Biol. Chem. 249, 3765–3767
5 DerVartanian, D.V. and Veeger, C. (1964) Biochim. Biophys. Acta 92, 233–247
6 Beinert, H., Ackrell, B.A.C., Kearney, E.B. and Singer, T.P. (1974) Biochem. Biophys. Res. Commun. 58, 564–571
7 DerVartanian, D.V., Veeger, C., Orme-Johnson, W.H. and Beinert, H. (1969) Biochim. Biophys. Acta 191, 22–27
8 Edmondson, D.E. and Singer, T.P. (1973) J. Biol. Chem. 248, 8144–8149
9 Patek, D.R. and Frisell, W.R. (1972) Arch. Biochem. Biophys. 150, 347–354
10 Ohnishi, T., Winter, D.B., Lim, J. and King, T.E. (1974) Biochem. Biophys. Res. Commun. 53, 231–237
11 Ohnishi, T., Winter, D.B., Lim, J. and King, T.E. (1974) Biochem. Biophys. Res. Commun. 61, 1017–1025
12 Olson, J.S., Ballou, D.P., Palmer, G. and Massey, V. (1974) J. Biol. Chem. 249, 4363–4382
13 Veeger, C. (1971) Flavins and Flavoproteins (Kamin, H., ed.), p. 644, University Park Press, Baltimore, Md.
14 Vinogradov, A.D., Winter, D.B. and King, T.E. (1972) Biochem. Biophys. Res. Commun. 49, 441–444
15 Ringler, R.L., Minakami, S. and Singer, T.P. (1963) J. Biol. Chem. 238, 801–810
16 Coles, C.J., Tisdale, H.D., Kenney, W.C. and Singer, T.P. (1972) Physiol. Chem. Phys. 4, 301–316
17 Baginsky, M.L. and Hatefi, Y. (1971) Biochemistry 10, 2517–2524
18 Kenney, W.C. (1975) J. Biol. Chem. 250, 3089–3094
19 Singer, T.P., Salach, J.I., Hemmerich, P. and Ehrenberg, A. (1971) in Methods of Enzymology (McCormick, D.B. and Wright, L.D., eds.), Vol. 18B, pp. 416–427, Academic Press, New York
20 Brumby, P.E. and Massey, V. (1967) in Methods of Enzymology (Estabrook, R.W. and Pullman, M.E., eds.), Vol. 10, pp. 463–474, Academic Press, New York
21 Williamson, J.R. and Corkey, B.E. (1969) in Methods of Enzymology (Lowenstein, J.M., ed.), Vol. 13, pp. 434–513, Academic Press, New York
22 Burleigh, B.D., Jr., Foust, G.P. and Williams, Jr., C.M. (1969) Anal. Biochem. 27, 536–544
23 Dutton, P.L. (1971) Biochim. Biophys. Acta 226, 63–80

Flavins and Flavoproteins, edited by T.P. Singer

CHAPTER 57

RELATIONSHIP OF THE ACTIVE SITE TO THE FLAVIN SUBUNIT IN SUCCINATE DEHYDROGENASE

WILLIAM C. KENNEY and PATRICK C. MOWERY

Department of Biochemistry and Biophysics, University of California, San Francisco, Calif. 94143 and Division of Molecular Biology, Veterans Administration Hospital, San Francisco, Calif. 94121 (U.S.A.)

Succinate dehydrogenase is composed of two subunits, one of mol. wt. 70 000, containing FAD in covalent linkage to a histidyl residue of the polypeptide chain, the other subunit of mol. wt. 30 000 [1–3]. The amino acid sequence of the flavin site of this enzyme has recently been determined [4]; however, no cysteinyl residue was found in the flavin peptide.

This enzyme was one of the first shown to be reversibly inhibited by sulfhydryl reagents and this inhibition was prevented by substrates or competitive inhibitors [5–8]. These observations suggested that sulfhydryl residues are essential for catalytic activity. Thus, in addition to the FAD moiety, a thiol group must be present at the active site of succinate dehydrogenase. The flavin peptide contains no thiol group, and hence, no substrate site. We set out therefore to determine which subunit contains the essential sulfhydryl residue and to ascertain its relationship to the flavin site. As will be shown below, identification of the substrate site and its localization in the flavin subunit required covalent labeling of this thiol group with suitable alkylating agents because substrates and competitive inhibitors do not form a covalent adduct stable enough for isolation.

Vinogradov et al [9] have postulated that the tightly-bound oxalacetate in the deactivated form of the enzyme forms a covalent thiohemiacetal bond with a cysteinyl residue of the enzyme. This structure is somehow stabilized in the native state of the enzyme. On denaturation, however, oxaloacetate is readily dissociated and thus no stable adduct may be obtained. Since it has not been possible to make covalent adducts that survive denaturation between the enzyme and substrates or competitive inhibitors, we have resorted to the use of *N*-ethylmaleimide to ascertain the location of the active site cysteinyl residue. As will be shown, this has been accomplished by differential labelling of the enzyme with *N*-[^{14}C]ethylmaleimide in the absence or presence of malonate, since substrates and competitive inhibitors prevent the inhibition of succinate dehydrogenase by *N*-ethylmaleimide [10].

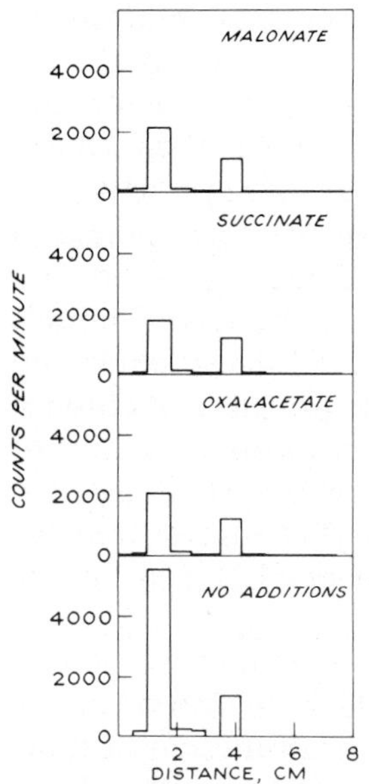

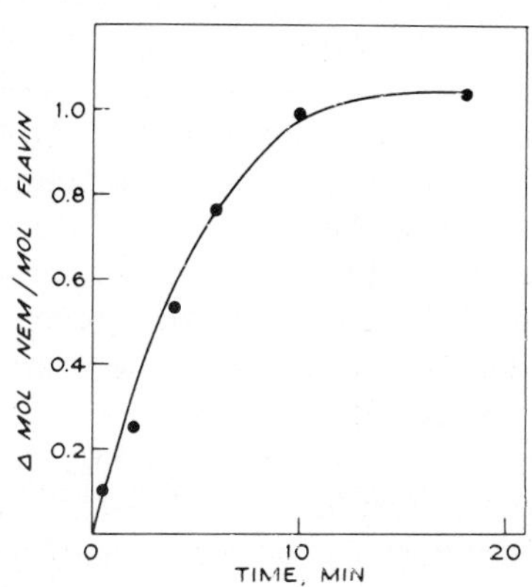

Fig. 1. Incorporation of N-[^{14}C]ethylmaleimide into subunits of succinate dehydrogenase. The enzyme (1.1 μM) was incubated in 50 mM HEPES, pH 8.0, at 15°C with N-[^{14}C]ethylmaleimide (25 μM) in the absence or presence of malonate (0.32 mM), succinate (16 mM), or oxalacetate (0.09 mM) until 70% inhibition had occurred in the sample containing no protective agent. To 2.0 ml of each reaction mixture was added 10 mmol dithiothreitol to stop the reaction. The enzyme was then denatured with 5% trichloroacetic acid, washed and an aliquot subjected to dodecyl sulfate-polyacrylamide gel electrophoresis which results in the resolution of the 70 000 (1.5 cm) and 30 000 (4 cm) mol. wt. subunits. The gels were then sliced, solubilized in H_2O_2, and counted for ^{14}C incorporation.

Fig. 2. Net incorporation of N-[^{14}C]ethylmaleimide into succinate dehydrogenase and presence of malonate. Succinate dehydrogenase was incubated with N-[^{14}C]ethylmaleimide in the absence or presence of malonate as described in Fig. 1. Aliquots were removed and the reaction stopped with dithiothreitol at the designated times. The enzyme was precipitated with 5% trichloroacetic acid. The pellet was washed with 5% and then 1% trichloroacetic acid, and subsequently solubilized in 10 mM phosphate, 0.1% dodecyl sulfate pH 7.0. Aliquots of these solutions were counted for radioactivity. Incorporation obtained in the presence of malonate was subtracted from that obtained in the absence of malonate to obtain the net incorporation responsible for loss of activity. NEM, N-ethylmaleimide.

Inactivation of succinate dehydrogenase by excess N-ethylmaleimide essentially goes to completion following pseudo-first order kinetics. On detailed examination, following the inactivation over an extended period of time, it is possible to describe the inhibition as the sum of two pseudo-first order reactions [11]. For the purpose of this discussion though, the relationship of N-ethylmaleimide incorporated to the loss in activity during this initial rapid phase is of immediate interest.

Although there are a large number of cysteinyl residues in the enzyme, most do not react under the conditions used in this study. Even in the presence of protective agents, however, a small amount of nonessential thiol groups is alkylated. By differential labelling of succinate dehydrogenase with N-[^{14}C]ethylmaleimide in the presence or absence of substrates or competitive inhibitors, it is possible to ascertain the net incorporation of N-ethylmaleimide responsible for the inactivation of the enzyme and to determine whether the thiol group reacting is located on the flavoprotein subunit. The results of this experiment are shown in Fig. 1. It is seen that the level of incorporation of ^{14}C into the 30 000-dalton subunit is the same whether or not malonate, succinate or oxalacetate were present in the incubation medium. The incorporation into the 70 000-dalton subunit was significantly decreased, however, if malonate, succinate or oxalacetate were present. The additional incorporation in the absence of protective agents corresponds to the titration of an extra thiol group in the enzyme. Thus, it may be concluded that the thiol group responsible for activity is on the flavin subunit of succinate dehydrogenase. Felberg and Hollocher [12] have shown by differential labelling with *N*-[^{14}C]ethylmaleimide in the presence or absence of malonate that a crucial cysteine is located on the flavoprotein subunit of succinate dehydrogenase. They concluded, however, that this thiol group is not at the catalytically active site. More recent results [11] suggest that the cysteinyl residue characterized by Felberg and Hollocher [12] may be the same as that characterized in this report.

In the experiments summarized in Fig. 1, a single time point was taken. Fig. 2 shows the net incorporation of *N*-[^{14}C]ethylmaleimide into succinate dehydrogenase in the absence or presence of malonate as a function of time. In this figure the difference in incorporation is plotted. It is seen that 1 mol of *N*-ethylmaleimide is incorporated per mol of flavin, representing the titration of the active-site thiol group.

It is possible, once again, to follow the incorporation of radioactivity into both the 70 000 and 30 000 mol. wt. subunits by use of dodecyl sulfate-polyacrylamide gel electrophoresis. The results of these studies are shown in Fig. 3. The left hand portion shows that the uptake of *N*-[^{14}C]ethylmaleimide into the 30 000 mol. wt. subunit is the same with or without malonate. In this experiment, a plateau of radioactivity is reached, which corresponds to approximately 1 mol of *N*-ethylmaleimide incorporated per mol of enzyme. The right half of this figure shows the time course of incorporation into the 70 000 mol. wt. subunit. There is a dramatic decrease in incorporation of radioactivity when malonate is present as a protective agent. The difference in incorporation in the absence or presence of malonate is plotted in the middle curve, represented by the triangles. The plateau reached again corresponds to about 1 mol of *N*-[^{14}C]ethylmaleimide incorporated per mol of flavin. Since under the experimental conditions we used, activity loss seems to be restricted to alkylation of a single cysteinyl residue at the active site, it is possible to compare the rate of alkylation of this cysteinyl residue with the rate of loss of activity. Fig. 4 shows that the kinetics of titration of this sulfhydryl is essentially identical to the kinetics of inactivation of the enzyme. Since the two rates are identical, in spite of the large number of thiol groups present in the enzyme, it is possible to demonstrate the absolute requirement for a cysteinyl residue in the catalytic activity of succinate dehydrogenase.

One last point we would like to make is that oxalacetate seems to bind in the tightly bound form with the same sulfhydryl group as the one whose alkylation is prevented by succinate or malonate. The evidence for this is

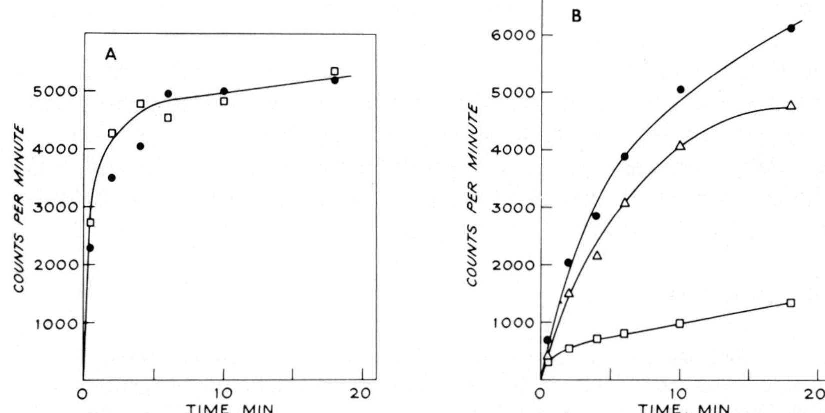

Fig. 3. Time course of incorporation of N-[^{14}C]ethylmaleimide into succinate dehydrogenase subunits. Aliquots of the protein prepared as described in Fig. 2 were subjected to dodecyl sulfate-polyacrylamide electrophoresis. The portions of the gel containing the 30 000 and 70 000 mol. wt. subunits were solublized in H_2O_2 and analyzed for ^{14}C incorporation. A, 30 000 mol. wt. subunit; B, 70 000 mol. wt. subunit. □, incubation with malonate; ●, no malonate present; △, see legend to Fig. 4.

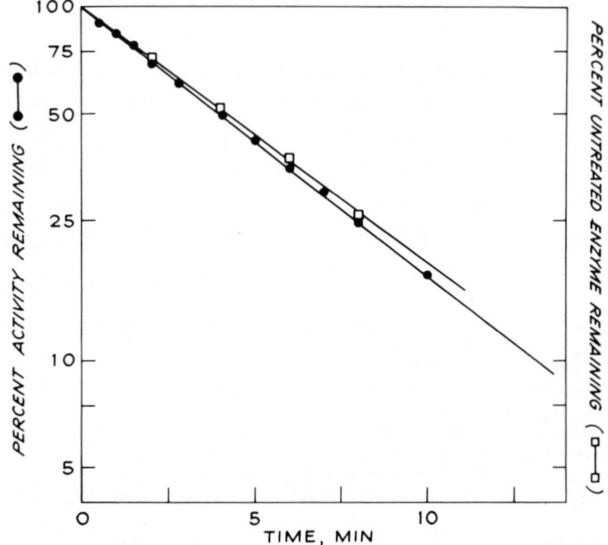

Fig. 4. Kinetics of inactivation of succinate dehydrogenase with N-ethylmaleimide and of N-[^{14}C] ethylmaleimide alkylation of an essential cysteinyl residue. The enzyme was incubated with N-ethylmaleimide as described in Fig. 1. The percent untreated enzyme remaining was determined from Fig. 3 (triangles) assuming the maximum level obtained to represent the titration of the active site thiol group.

that first, as already shown in Fig. 1, oxalacetate prevents alkylation of this sulfhydryl group and second, prior alkylation of succinate dehydrogenase by N-ethylmaleimide prevents the binding of [^{14}C]oxalacetate [11]. The possibility that the tight binding of oxalacetate occurs on the flavin subunit has been suggested previously by Winter and King [13] on the basis of inconclusive data, whereas the present data seem to be unambiguous on this point.

The results presented show that the cysteinyl residue, essential for catalytic activity and protected by substrates or competitive inhibitors from inhibition by sulfydryl reagents, is located on the 70 000 mol. wt. subunit of succinate dehydrogenase which also contains the covalently bound FAD moiety. The proximity of this thiol residue to the flavin must be brought about by the tertiary structure of the enzyme.

Experimental procedure

Succinate dehydrogenase was prepared by the method of Davis and Hatefi [2] for the experiment documented in Fig. 1. In subsequent experiments, the enzyme was prepared in the absence of succinate [14] using a single extraction with 0.89 M sodium perchlorate. Sodium dodecyl sulfate-polyacrylamide gel electrophoresis was performed according to the procedure of Weber and Osborn [15]. Radioactivity was determined with a Beckman LS-150 liquid scintillation counter and Aquasol (New England Nuclear) as the scintillation medium. Other methods have been previously documented [11].

Acknowledgements

This investigation was supported by a Grant-in-Aid from the American Heart Association (73674) and with funds contributed in part by the Bay Area Heart Association, and by a grant from the National Heart and Lung Institute (1 PO 1HL 16251). The excellent technical assistance of Mr. Richard Seng is gratefully acknowledged.

References

1 Coles, C.J., Tisdale, H.D., Kenney, W.C. and Singer, T.P. (1972) Physiol. Chem. Phys. 4, 301
2 Davis, K.A. and Hatefi, Y. (1971) Biochemistry 10, 2509
3 Righetti, P. and Cerletti, P. (1971) FEBS Lett. 13, 181
4 Kenney, W.C., Walker, W.H. and Singer, T.P. (1972) J. Biol. Chem. 247, 4510
5 Hopkins, F.G. and Morgan, E. (1938) Biochem. J. 32, 611
6 Hopkins, F.G., Morgan, E. and Lutwok-Mann, C. (1938) Biochem. J. 32, 1829
7 Barron, E.S.G. and Singer, T.P. (1945) J. Biol. Chem. 157, 221
8 Singer, T.P., Kearney, E.B. and Bernath, P. (1956) J. Biol. Chem. 223, 599
9 Vinogradov, A.D., Zimakova, N.I. and Solntseva, T.I. (1971) Dokl. Akad. Nauk. SSR Ser. Biol. 201, 359
10 Kenney, W.C. (1973) Fed. Proc. 32, 595
11 Kenney, W.C. (1975) J. Biol. Chem. 250, 3089
12Felberg, N.T. and Hollocher, T.C. (1972) J. Biol. Chem. 247, 4539
13 Winter, D.B. and King, T.E. (1974) Biochem. Biophys. Res. Commun. 56, 290
14 Coles, C.J., Tisdale, H.D., Kenney, W.C. and Singer, T.P. (1974) J. Biol. Chem. 249, 381
15 Weber, K. and Osborn, M.J. (1969) J. Biol. Chem. 244, 4409

Flavins and Flavoproteins, edited by T.P. Singer

CHAPTER 58

REDUCTIVE ACTIVATION OF SUCCINATE DEHYDROGENASE: EQUILIBRIUM AND KINETICS STUDIES*

MENACHEM GUTMAN and NITZA SILMAN

Department of Biochemistry, The George S. Wise Center for Life Sciences, Tel-Aviv University, Tel-Aviv (Israel)

It is now 20 years since the first indication for the activation of succinate dehydrogenase was observed [1]. During this period the family of activators expanded from just substrate and its analogues [2], and now includes reduced ubiquinone [3], ATP and ITP [4], various anions [5], and reductive conditions [6–8]. Many mechanisms have proposed to describe the activation of the enzyme but a detailed, quantitative, model had to wait for the discovery of the negative modifier of the enzyme, oxaloacetate [8], which converts it to its nonactive form.

There is no doubt that activation is accompanied with dissociation of the succinate dehydrogenase oxaloacetate complex [9] but a quantitative description which can account for activator, oxaloacetate and active enzyme concentrations is still not at hand.

An attempt at quantitative analysis of the mechanism of activation of succinate dehydrogenase has been carried in our laboratory for some years, combining both equilibrium and kinetic analysis of the activation [15].

These studies indicated that at least with some activators (succinate, malonate and Br^-) the mechanism of activation consists of an equilibrium between two stable forms: free enzyme in its active form and a nonactive enzyme-oxaloacetate complex. The activators in this case are ligands which bind to the active enzyme, thus stabilizing this species with respect to nonactive oxaloacetate complex. Such a model predicts that at equilibrium the fraction of the active enzyme will be a function of the activator: oxaloacetate ratio, and not a function of the absolute concentration of these effectors. Typical results of such experiments are shown in Fig. 1. In Fig. 1A the enzyme was equilibrated with constant oxaloacetate concentrations

* This publication is dedicated to the memory of Mrs. Nitza Silman-Movshovitz, who died before submission of the manuscript.
Abbreviations and symbols: $[\overline{OAA}]$, concentration of oxaloacetate at mid-activation potentials; ETP_H phosphorylating sub-mitochondrial particles; E'_{ma}, mid-activation potential.

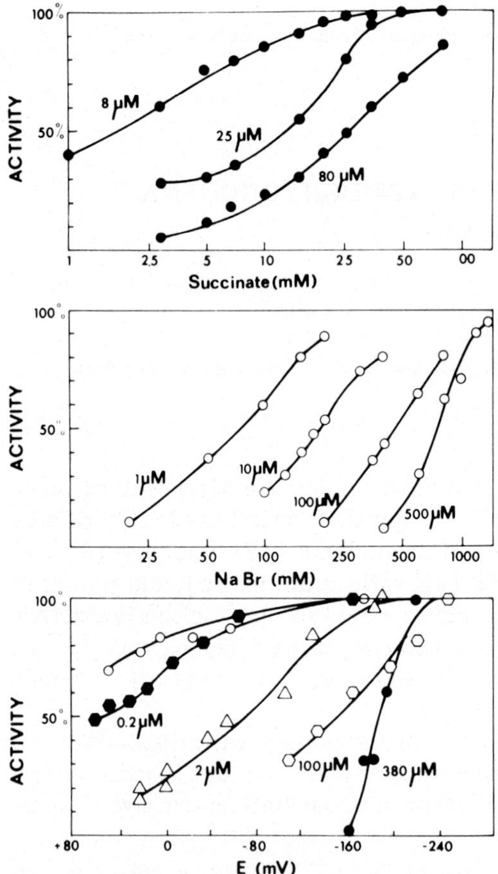

Fig. 1. Equilibrium level of active succinate dehydrogenase in presence of oxaloacetate and activators (A) (top) ETP_H in 0.18 M sucrose, 50 mM Tris acetate, 5 mM $MgSO_4$ (STM buffer) pH 7.2, 30°C were equilibrated in presence of the indicated concentration of oxaloacetate, against variable succinate concentrations. Samples were withdrawn and assayed at 13°C with phenazine methasulfate/dichlorophenol indophenol (20 mM succinate, 30 mM potassium phosphate, 2 mM KCN [3]) (100% = 0.32 μmol/min mg). (B) ETP_H in 0.18 M sucrose 100 mM HEPES p*K* 6.2, 30°C were equilibrated with the indicated oxaloacetate concentrations against variable NaBr concentrations, samples were assayed as before. (100% = 0.34 μmol/min mg (13°C)) (C) ETP_H were suspended in STM buffer at 1 mg/ml in the presence of 30 μM of the following redox mediators: 2 hydroxy-1,4-naphthoquinone, menadione, duroquinone, phenanthraquinone, anthraquinone 2-sulfonate, 2-hydroxy 3-methyl naphthoquinone, tetra-methylparaphenylendiamine, phenazine methasulfate, dichlorophenol indophenol. The solution was equilibrated in a potentiometric titration vessel with helium saturated with water. Anaerobiosis was accelerated by addition of neutralized solution of ascorbate (0.5 M). The redox potential was adjusted to the desired values by addition of either a solution of $Na_2S_2O_4$ or 10 mM $K_3Fe(CN)_6$. The titration was carried out at room temperature. The measurement of the succinate dehydrogenase activity carried as before was started 5–10 s after the removal of the sample. The activity measured is related to the potential measured at the time of sampling. During the titrations, samples were taken and activated by succinate to account for possible inactivation of the enzyme. No inactivation of the enzyme was observed even after 60 min titration. (100% = 0.254 μmol/min mg (13°C)).

and variable concentrations of succinate. The higher the oxaloacetate, the higher the succinate concentration required to keep the enzyme active. In Fig. 1B the activator employed was Br^- and the same relationship was observed. A detailed analysis of these systems will be published elsewhere [15].

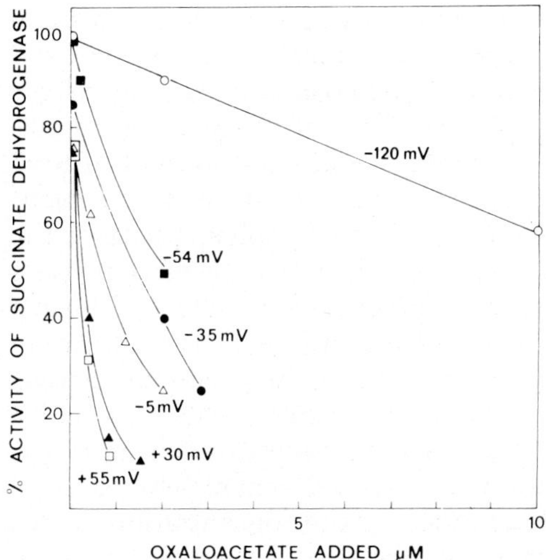

Fig. 2. Titration of the succinate dehydrogenase activity by oxaloacetate under constant redox potential. The ETP_H were activated by malonate, washed and brought to the desired potential with $Na_2S_2O_4$ and 10 mM $K_3Fe(CN)_6$ as in Fig. 1C. Graded amounts of oxaloacetate were added and the activity was measured 3–4 min after each addition of oxaloacetate. (100% activity was 0.25 μmol/min/mg (13°C)).

In the present communication we shall demonstrate that reductive activation of succinate dehydrogenase is also subjected to quantitative analysis similar to that found with other activators. The resemblance between reductive activation and Br^- or succinate activation is shown in Fig. 1C, where activation was achieved by reductive conditions. ETP_H particles were equilibrated with oxaloacetate and redox mediators under a helium atmosphere; samples were removed in gas-tight microsyringes and assayed within 10 s. In the assay mixture oxaloacetate was diluted to 4 μM or less, and in the presence of 20 mM succinate no further deactivation could occur (see Fig. 1A). The low temperature of the assay (13°C) prevented activation in situ. Thus this technique can measure accurately the fraction of the active enzyme at the time of sampling. The similarity of this sytem to those depicted above (Figs. 1A and B) indicates that reducing potential can substitute for succinate of Br^- in maintaining succinate dehydrogenase active in presence of oxaloacetate. The same proportionality observed for oxaloacetate and activator is maintained in this case; the higher the concentration of oxaloacetate, the more negative the potential which must be applied to keep the enzyme active.

The effect of redox potential on deactivation of succinate dehydrogenase by oxaloacetate is also documented in Fig. 2, where the active fraction of succinate dehydrogenase is related to the oxaloacetate concentration at various poise potentials. It is evident that the higher the poise potential, the lower the oxaloacetate concentration need be to deactivate the enzyme.

The identity of the nonactive form of the enzyme was investigated by correlating its quantity with the amount of [U-^{14}C]oxaloacetate bound to the enzyme. ETP_H particles were equilibrated at various redox potentials with 3 μM [U-^{14}C]oxaloacetate. When equilibrium was reached, the particles were diluted 1:10 in ice-cold buffer containing 10 mM semicarbazide and spun down. The enzymic acitivity and [^{14}C]oxaloacetate content of the pellet were measured. A constant ratio of [^{14}C]oxaloacetate/nonactive enzyme was obtained, corresponding to 1 mol of oxaloacetate/mol of non-active enzyme*. Thus the ratio of active/nonactive enzyme is equivalent to the ratio of free enzyme to enzyme-oxaloacetate complex, whatever the redox potential. As reductive activation proceeds at potentials more positive than the redox potential of the oxaloacetate-malate couple (−160 mV pH 7.2 [11]) see Figs. 1C and 5), it is concluded that in reductive activation not the ligand but the enzyme is being reduced. Apparently reductive activation of succinate dehydrogenase is an equilibrium system, where the active enzyme is favored by reduction, in analogy to a high activator concentration shown in Fig. 1A and B. Upon reduction the reduced enzyme-oxaloacetate complex will dissociate, to a degree determined by the concentration of free oxaloacetate, and this ligand-free enzyme will assume its active conformation. A quantitative description of the mechanism is given by the following equations (SDHA and SDHNA), succinate dehydrogenase represent the active and non-active forms, respectively):

$$\mathrm{SDH_{ox}(A)} + ne^- \overset{K_1}{\rightleftharpoons} \mathrm{SDH_{red}(A)} \qquad \text{(I)}$$

$$\mathrm{SDH_{ox}(A)} + \text{oxaloacetate} \overset{K_2}{\rightleftharpoons} \mathrm{SDH_{ox}(NA)}\text{-oxaloacetate} \qquad \text{(II)}$$

$$\mathrm{SDH_{red}(A)} + \text{oxaloacetate} \overset{K_3}{\rightleftharpoons} \mathrm{SDH_{red}(NA)}\text{-oxaloacetate} \qquad \text{(III)}$$

$$\mathrm{SDH_{ox}(NA)}\text{-oxaloacetate} + ne^- \overset{K_4}{\rightleftharpoons} \mathrm{SDH_{red}(NA)}\text{-oxaloacetate} \qquad \text{(IV)}$$

These four reactions can be written in a cyclic form with the appropriate equilibrium constants presented, so as to describe a closed circle in the clockwise direction. At equilibrium the product of these constants as written must equal one (OAA, oxaloacetate).

$$\begin{array}{ccc} \mathrm{SDH_{ox}(A)} + \mathrm{OAA} + ne^- & \xrightarrow{K_1} & \mathrm{SDH_{red}(A)} + \mathrm{OAA} \\ \uparrow {\scriptstyle 1/K_2} & & \downarrow {\scriptstyle K_3} \\ \mathrm{SDH_{ox}(NA)\text{-}OAA} + ne^- & \xrightarrow{1/K_4} & \mathrm{SDH_{red}(NA)\text{-}OAA} \end{array}$$

$$\frac{K_1 K_3}{K_2 K_4} = 1 \quad \text{and} \quad K_4 = K_1 \frac{K_3}{K_2}$$

The relative instability of the reduced complex with respect to the oxidized one ($K_2 > K_3$) leads to the conclusion that ($K_1 > K_4$). Consequently the

* The concentration of succinate dehydrogenase was calculated from V and the published turnover number for this type of particle [10].

oxidized complex will be reduced at a lower poise potential than the free, oxidized enzyme, in accord with the experimental results (Fig. 1C).

This analysis can be taken through a further step. We measured at a series of oxaloacetate concentrations the mid-activation potential, where at equilibrium 50% of the succinate dehydrogenase was in its active form. The relationship between the two parameters ($[\overline{OAA}]$ and E'_{ma}) was derived as follows: the active enzyme is equated with the free forms of the enzyme ($E_A = SDH_{red} + SDH_{ox}$) and the nonactive fraction with the two complexes ($E_{NA} = SDH_{ox}\text{-}OAA + SDH_{red}\text{-}OAA$). The ratio of E_A/E_{NA} is now of function of the following parameters: oxaloacetate concentration, the association constants for the oxidized and reduced complexes (K_2, K_3), the poise redox potential, and redox potentials $E_{m(1)}$ and $E_{m(4)}$ for Reactions I and IV, respectively. The relationship between concentration $[\overline{OAA}]$ and poise potential (E'_{ma}) at point of midactivation can be shown to be

$$[\overline{OAA}] = \frac{1}{K_2} \times \frac{1}{(1 + e^{-\frac{n}{60}\cdot\Delta E'})} + \frac{1}{K_3} \times \frac{1}{(1 + e^{\frac{n}{60}\cdot\Delta E'})}$$

Where $\Delta E' = E'_{ma} - E_{m(4)}$. As will be shown below $K_2/K_3 > 100$ thus we can neglect the first term. This leads to

$$\log \frac{1}{[\overline{OAA}]} = \log K_3 + \log (1 + e^{\frac{n}{60}\cdot\Delta E'})$$

At Potentials where $e^{\frac{n}{60}\cdot\Delta E'} \gg 1$, a linear function of $\log 1/[\overline{OAA}]$ vs. E_h i.e. E'_{ma} is expected with slope of ($n/60 \cdot \Delta E'$). This function is shown in Fig. 3 where

$$K_{eq} = \frac{E_{NA}}{E_N} \cdot \frac{1}{[\overline{OAA}]} = \frac{1}{[\overline{OAA}]}$$

is drawn with respect to the poise potential. The slope of the line yields $n =$ 6. It is of interest to point out that full reduction of the $g = 1.94$ signal of succinate dehydrogenase, calls for addition of 6 equivalents per mol of flavin [12], suggesting that in the course of reductive activation the whole enzyme must be reduced in order to attain the reduced conformation, the one with lower affinity to oxaloacetate. The highest value of K_{eq}, measured at $+ 55$ mV, is $6 \cdot 10^6\ M^{-1}$, and the lowest value is $6 \cdot 10^4\ M^{-1}$, thus $K_2/K_3 \geqslant 100$. The curvature of the line below $- 100$ mV might represent the region where the exponent diminishes in magnitude and deviation from linearity is expected. Presently neither the absolute midpotentials for Eqns. I and IV nor the values of K_2 and K_3 can be evaluated. Anyhow these results clearly demonstrate that the reducible component in reductive activation is not oxalacetate.

The relative instability of $SDH_{red}\text{-}OAA$ is reflected by its ability to exchange its oxaloacetate with the solution. There were some contradictory reports in the literature [13, 14] where such reactions take place. Presently it is evident that such exchange can take place, given that the complex is

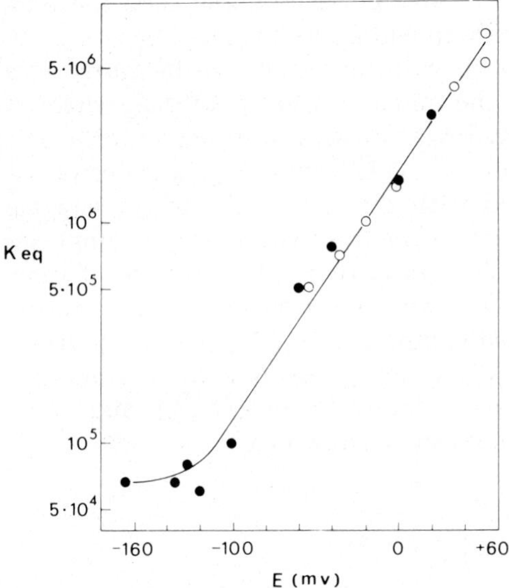

Fig. 3. Correlation between the apparent equilibrium constant for deactivation and the poise potential. The results at low potential were corrected for the quilibrium concentrations of oxaloacetate according to the Nernst equation. ●, results calculated from experiments where oxaloacetate was constant and the potential was varied as in Fig. 1C. ○, results calculated from titration at constant potential as in Fig. 2.

destabilized either by reduction [6] or by anions [15].This exchange is demonstrated in Fig. 4. The lower the poise potential, the faster the exchange, but owing to the high oxaloacetate concentration (1mM), no activation took place. This finding is compatible with two mechanisms; the first assumes that dissociation and activation occurs in a single step.

$$SDH_{red}(NA)\text{-}OAA \rightleftharpoons SDH_{red}(A) + OAA$$

Where the level of active enzyme is kept very low by the high concentration of oxaloacetate. The second model assumes that dissociation of oxaloacetate and activation are consecutive reactions.

$$SDH_{red}(NA)\text{-}OAA \underset{k_{-1}}{\overset{k_1}{\rightleftharpoons}} SDH_{red}(NA) + \text{oxaloacetate} \qquad \text{(V)}$$

$$SDH_{red}(NA) \xrightarrow{k_2} SDH_{red}(A) \qquad \text{(VI)}$$

As before, exchange is a first order reaction, unaffected by oxaloacetate concentrations, but the high ligand concentration shifts the equilibrium of Reaction V to the left so that the concentratio of the activatable species is partially nil.

According to the first model, the rate of exchange and activation must be identical, as the two parameters reflect the same step in the reaction, while the second model (V and VI) can account for the exchange being

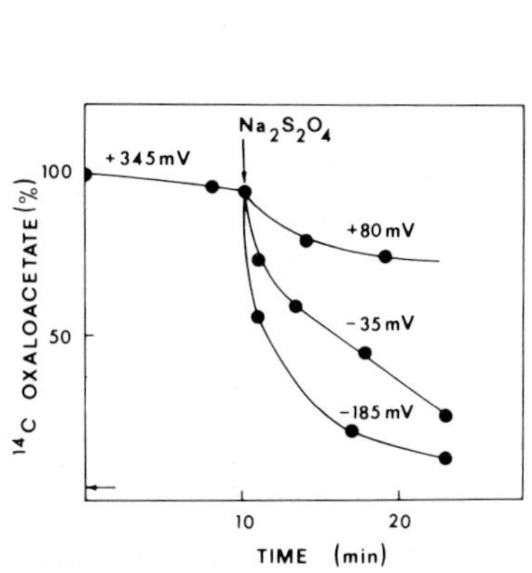

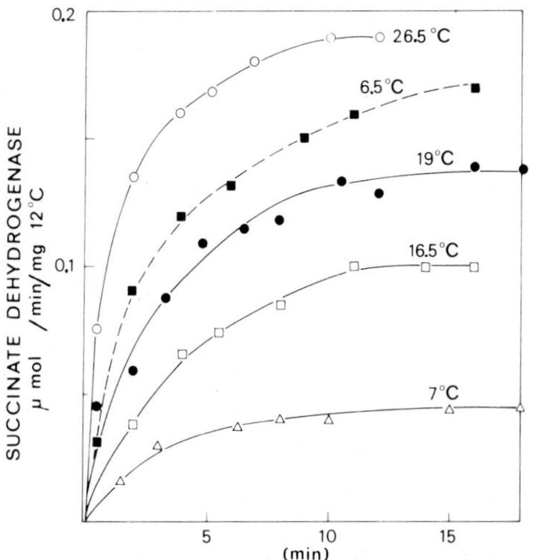

Fig. 4. Exchange of [^{14}C]oxaloacetate bound to succinate dehydrogenase with unlabelled oxaloacetate at different poise potentials. 40 mg ETP$_H$ suspended in STM buffer at 1 mg/ml were activated by NADH (250 μM), and antimycin A, (1 nmol/mg) in the presence of KCN (1 mM) and washed by centrifugation. The enzyme was deactivated by 36 nmol of L-[U-^{14}C]malate (82 Ci/mmol) in the presence of 100 μM phenazine methosulfate. The enzyme was completely deactivated after 3 min. The particles were washed twice with STM buffer to remove the spurious radioactivity, then equilibrated against 1 mM oxaloacetate in the potentiometric titration vessel, at + 345 mV. At the indicated time, the potential was lowered by the addition of a predetermined volume of freshly prepared solution of $Na_2S_2O_4$ in order to obtain the desired redox potential. At intervals, samples of 1.5 mg protein were withdrawn and injected into 4.5 ml of ice cold STM buffer containing 10 mM semicarbazide. After 10 min at 0°C, the particles were centrifuged and the radioactivity of samples containing 1 mg protein was measured. The activity of the enzyme was monitored throughout the experiment and was found in all cases to be less than 3% active. To correct for possible non-specific absorption of [^{14}C]oxaloacetate on the protein, a sample at zero time was denatured with 5% trichloroacetic acid before the washing with semicarbazide. The counts measured for this sample is indicated by the arrow on the ordinate. The mediators used in the reactions were those detailed in Fig. 1C plus Indigo tetrasulfonate, Indigo trisulfonate, Indigo carmine all at 30 μM. (100% = 0.285 μmol/min/mg (13°C)).

Fig. 5. The time course of reductive activation of succinate dehydrogenase in the presence of 1 μM oxaloacetate. ETP$_H$ were washed in STM buffer and resuspended to 15 mg/ml in the same buffer containing 1 μM oxaloacetate. These particles were further diluted in the titration vessel to 2 mg/ml by STM containing 1 μM oxaloacetate and 30 μM of all mediators listed in Fig. 4. The potential was adjusted to + 60 mV by addition of neutralized ascorbate (0.1 M) and samples were removed for determination of succinate dehydrogenase activity. The potential was lowered by addition of predetermined volume of freshly prepared solution of $Na_2S_2O_4$ to achieve the desired potential and increase in activity was followed with time. In parallel, samples were removed and activated by succinate for determination of 100% activity. ——, reductive activation of − 85 ± 5 mV; – – –, reductive activation at − 245 ± mV. The temperature of the reaction was controlled by a water jacket around the titration vessel and was measured during the reaction by a thermocouple.

faster than the activation. The rate of exchange (measured in Fig. 4) and of reductive activation (see below, Fig. 5) were measured under comparable conditions (− 185 mV, 25°C). The exchange was twice as fast as the activation (1.2 min^{-1} and 0.6 min^{-1}, respectively). Consequently the second mechanism should be adopted. This model is very similar to that derived for activation of succinate dehydrogenase by Br^-, where exchanges of oxaloacetate also take place [15].

The kinetics of reductive activation were measured using a potential jump technique. The enzyme was equilibrated with 1 mM oxaloacetate, a concentration where at + 55 mV the enzyme is not active but at − 90 mV

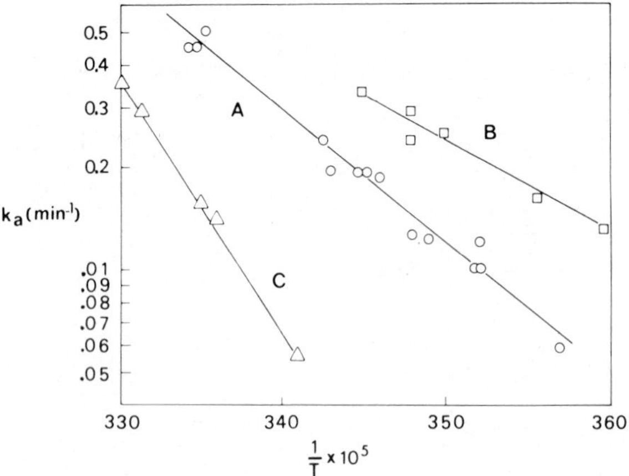

Fig. 6. Arrhenius plot for the rate constant of reductive activation. The results of experiments carried out as described in Fig. 6: A(○—○) reductive activation at − 85 ± mV; B(□—□) reductive activation at − 260 ± 20 mV; C(△—△) activation by reduced ubiquinone, data for this line was taken from ref. 3.

it is fully active. The poise potential was changed very rapidly from + 55 mV to − 90 mV and the increase in enzyme activity was followed with time. These conditions allow to measure full activation of succinate dehydrogenase, at a potential well above the midpotential of oxaloacetate-malate. Thus during the course of the reaction there is no change in oxaloacetate concentration.

Fig. 5 depicts the time course of reductive activation at various temperatures. Apparently both the rate and the final level are functions of the temperature: at 25°C all of the enzyme is activated but at lower temperatures this level is decreased. The enthalpy change calculated for the reaction is $\Delta H = 26.5$ kcal/mol. The effect of temperature on the rate of activation is shown in Fig. 6 line A. The activation energy is 18 kcal/mol, very close to that reported by Salach and Singer [8]. In other experiments the redox potential was lowered to − 260 mV, well below the mid-potential of oxaloacetate-malate couple. Thus we probably reduced both enzyme and ligand. Under such conditions Reaction V becomes irreversible (due to reduction of the oxaloacetate), and the kinetics of the reaction will approach with time the kinetics of Reaction VI.

As shown by the dotted line in Fig. 5 at − 260 mV full activation takes place at temperature where at − 90 mV activation is only partial, and the rate of activation is much faster.

As the rate of exchange is faster than rate of activation (see above) we can treat SDH_{red}(NA) as a steady state intermediate in the course of activation. For such sequential reactions given in Eqns. V and VI it can be shown that $K_{obs} = (k_1 \cdot k_2)/(k_{-1} + k_2)$. The rate of association of oxaloacetate with the enzyme (k_{-1}) is much faster than the rate of activation, thus given that $k_{-1} \gg k_2$, we end up with $k_{obs} = k_2 \times (k_1/k_{-1})$. In a similar way the observed activation energy for such reactions will be $E_{a(obs)} = E_{a(1)} + E_{a(2)} - E_{a(-1)}$. At − 260 mV, where only Reaction VI is observed, the rate of activation is faster than at -90 mV. This might indicate that the slower rate at -90 mV is due to the contribution of k_1/k_{-1} to k_{obs} and $k_{-1} > k_1$. The activation energy at -260 mV, 12.5 kcal/mol, is also smaller than that measured at -90 mV, probably representing the contribution of $E_{a(1)} - E_{a(-1)}$ to the value measured at -90 mV ($E_{a(1)} > E_{a(-1)}$).

As shown in Fig. 6, line C, activation by reduced ubiquinone is much slower and is associated with a much higher energy of activation (E_a = 31.7 kcal/mol). Furthermore, in presence of free oxaloacetate, reduction of membranal quinones cannot activate succinate dehydrogenase [6]. This leaves no doubt that activation by reduced ubiquinone is not via reduction but involves a modulation of succinate dehydrogenase. The reductive activation itself is a complex reaction in which the binding of oxaloacetate is first destabilized by reduction of the enzyme. The resulting reduced enzyme-oxaloacetate complex dissociates and the free form of the nonactive enzyme is the species which undergoes the conformational change to yield active succinate dehydrogenase.

References

1 Kearney, E.B., Singer, T.P. and Zastrow, N. (1955) Arch. Biochim. Biophys. 55, 579–581
2 Kearney, E.B. (1957) J. Biol. Chem. 299, 363–375
3 Gutman, M., Kearney, E.B. and Singer, T.P. (1971) Biochemistry 10, 2726–2733
4 Gutman, M., Kearney, E.B. and Singer, T.P. (1971) Biochemistry 10, 4763–4770
5 Kearney, E.B., Ackrell, B.A.C., Mayer, M. and Singer, T.P. (1974) J. Biol. Chem. 249, 2010–2020
6 Gutman, M. and Silman, N. (1975) Mol. Cell. Biochem. 7, 177–185
7 Klaase, A.D.M. and Slater, E.C. (1972) F. Naturforsch. 27b, 1077
8 Salach, J. and Singer, T.P. (1974) J. Biol. Chem. 249, 3765–3769
9 Ackrell, B.A.C., Kearney, E.B. and Mayer, M. (1974) J. Biol. Chem. 249, 2021–2027
10 Singer, T.P. (1966) Comprehensive Biochemistry 14, 127–198
11 Barton, K. and Wilson, T.M. (1953) Biochem. J. 54, 86–94
12 Beinert, H., Ackrell, B.A.C., Kearney, E.B. and Singer, T.P. (1974) Biochim. Biophys. Acta 58, 564–571
13 Priegnitz, A., Brzhevska, O.N. and Wojtzak, L. (1973) Biochim. Biophys. Res. Commun. 51, 1034–1041
14 Winter, D.B. and King, J.E. (1974) Biochim. Biophys. Res. Commun. 56, 290–295
15 Gutman, M. Biochemistry, in the press

Flavins and Flavoproteins, edited by T.P. Singer

CHAPTER 59

THE INTERACTION OF RHODANESE AND SUCCINATE DEHYDROGENASE

FRANCO BONOMI, SILVIA PAGANI, CARLO CANNELLA and PAOLO CERLETTI

Department of General Biochemistry of the University of Milan, and Department of Biological Chemistry of the University of Rome (Italy)

Summary

Rhodanese shifts the association equilibrium of aged succinate dehydrogenase preparations towards higher molecular weight aggregates. It appears to transfer its sulfane sulfur to the flavoprotein. More radioactivity is incorporated where treatment with [^{35}S]rhodanese results in a larger increase of the reconstitutive capacity of succinate dehydrogenase.

Introduction

Succinate dehydrogenase (EC 1.3.99.1, succinate: (acceptor) oxidoreductase), as other complex flavoproteins, contains an iron-sulfur structure which in the isolated flavoprotein is quite labile.

The role of this structure in the catalytic process has received much attention, but little is known so far about the physiological processes of sulfur transfer to the flavoprotein or to other iron-sulfur proteins. It has been found that adrenodoxin is reconstituted by treating the apoprotein with 3-mercaptopyruvate and 3-mercaptopyruvate sulfurtransferase (EC 2.8.1.2) [1] and that rhodanese (EC 2.8.1.1, thiosulfate: cyanide sulfurtransferase), an enzyme transferring sulfane sulfur, and thiosulfate, its substrate, substitute for inorganic sulfide in the reaction medium for restoring the iron-sulfur center of ferredoxin [2].

In previous work [3] we investigated the effects of rhodanese on beef heart succinate dehydrogenase by measuring the changes in labile sulfur content and in the reconstitutive capacity, i.e. changes in the ability of the purified flavoprotein to restore a functioning oxidase system in submitochondrial particles deactivated by alkali. Circumstantial evidence [4–7] indicates that the reconstitutive capacity is a measure of the integrity of the iron-sulfur structure in isolated succinate dehydrogenase.

We found that rhodanese protects the labile sulfide content and the reconstitutive capacity of aged preparations. Thiosulfate and inorganic sulfide do not influence the effect of rhodanese. Sulfur-free rhodanese is inactive, but it competes with rhodanese, decreasing the effect of the latter. When the mixture containing succinate dehydrogenase and rhodanese was centrifuged on a density gradient, the peak of the flavoprotein sedimented faster and that of rhodanese, measured from the activity in thiocyanate production, appeared to have a smaller area [3].

These results showed an interaction between rhodanese and succinate dehydrogenase requiring the presence of the sulfane sulfur on the sulfurtransferase. The data from ultracentrifugation were interpreted as indicating either that rhodanese forms a stable complex with the flavoprotein or that it shifts the association equilibrium of succinate dehydrogenase towards higher molecular weight aggregates. The reduced activity of rhodanese could alternatively be attributed to formation of an inactive complex with succinate dehydrogenase or to inactivation of the enzyme in its interaction with the flavoprotein.

Further work to elucidate these problems is briefly reported here.

Materials and Methods

Enzyme preparations and assays were those described previously [3]. Succinate dehydrogenase was aged enough before treatment with rhodanese to inactivate it partially to reconstitution.

Sodium dodecyl sulfate electrophoresis on acrylamide gel was done according to Dunker and Rueckert [8]. After the run was completed, proteins were stained with Coomassie blue and the gel scanned at 600 nm with a Beckman recording spectrophotometer.

[^{35}S]Rhodanese was prepared by treating sulfur-free rhodanese with 1 mM $Na_2{}^{35}SSO_3$ (purchased from the Radiochemical Centre, Amersham) having a specific activity of 6.5–13.7 $Ci \cdot M^{-1}$. The enzyme was then crystallized in the presence of labeled thiosulfate. The specific radioactivity of the preparation agreed with the content of 1.35 atoms sulfane sulfur per mol enzyme, measured independently [9].

The apparent molecular weight of the proteins separated in gel chromatography was determined by comparison to known standards as described by Andrews [10].

Results

Succinate dehydrogenase was treated with rhodanese and the mixture was analyzed by gel chromatography. Sufficiently small fractions were collected so as to resolve partially the broad peak of the flavoprotein: the elution pattern showed increased areas of peaks with smaller retention volumes and a new peak to appear, as compared to the untreated preparation (Fig. 1).

The tubes corresponding to each peak in the preparation treated with rhodanese were separately pooled and analyzed by acrylamide gel electrophoresis in sodium dodecyl sulfate. In each peak, the two protomers of succinate dehydrogenase and minor protein impurities were evidenced but none contained rhodanese, except the one with a retention volume corresponding

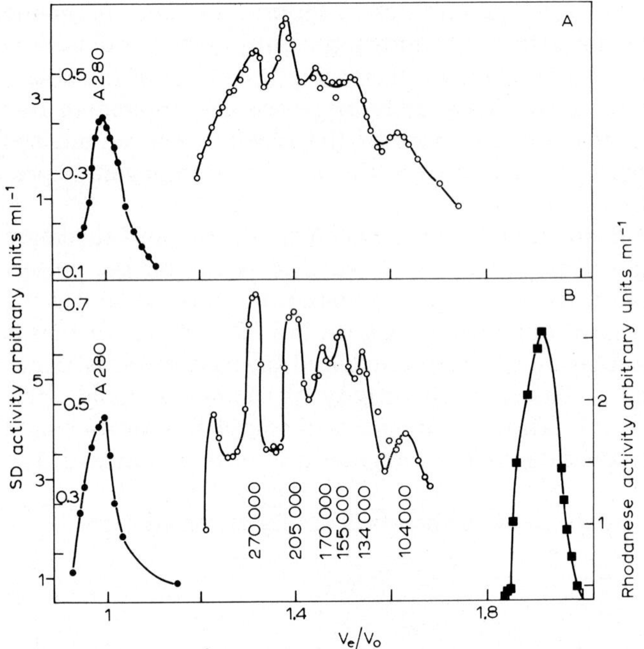

Fig. 1. Gel chromatography of succinate dehydrogenase and rhodanese. Purified succinate dehydrogenase (SD) (3.96 mg) was incubated for 1 h at 0°C in 50 mM phosphate buffer pH 7.6 and 20 mM succinate without (A) and with (B) rhodanese (0.396 mg). The mixture was then chromatographed on a column (2 X 93 cm) of Sephadex G-200 fine. Fractions of 0.6 ml were collected. Blue dextran was used to check the void volume of the column. ○, succinate dehydrogenase activity; ■, rhodanese activity, ●, blue dextran. The apparent molecular weight corresponding to succinate dehydrogenase peaks is given in the figures. Abscissa: retention volumes. Ordinate: absorbance at 280 nm (blue dextran) and enzyme activities, arbitrary units · ml^{-1}.

to an apparent molecular weight of 155 000, where the enzyme was less than 2% of the total protein. The fractions containing rhodanese had no appreciable activity in thiocyanate formation, whereas similar amounts of the enzyme assayed in the absence of succinate dehydrogenase could easily be detected.

Sulfur transfer to succinate dehydrogenase was investigated by using thiosulfate labelled in the outer sulfur and rhodanese labelled with ^{35}S. This enzyme was chromatographed before use on a column of Sephadex G-25 to remove any low molecular weight contaminant.

Succinate dehydrogenase was treated with [^{35}S]rhodanese and the mixture analyzed by gel chromatography. As shown in Fig. 2, radioactivity appeared in the peak of succinate dehydrogenase. The balance of incorporation is given in Table I, first line. Analysis of the peak of succinate dehydrogenase by dodecyl sulphate electrophoresis on acrylamide gel showed that rhodanese was less than 0.5% of the total proteins (Fig. 3).

When succinate dehydrogenase was treated under similar condition with labelled thiosulfate having the same radioactivity as the [^{35}S]rhodanese preparation, no radioactivity was found in the peak of the flavoprotein after gel filtration.

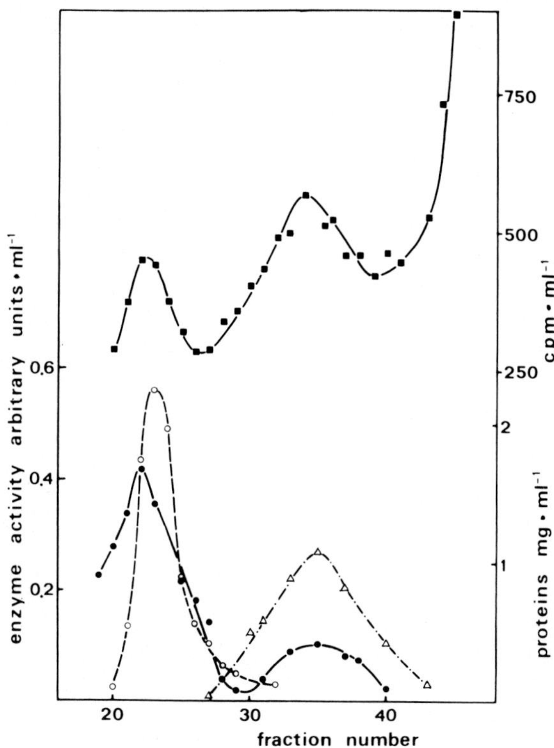

Fig. 2. Gel chromatography of succinate dehydrogenase treated with [^{35}S] rhodanese. Succinate dehydrogenase (5.37 mg) was incubateo for 1 h at 0°C in 50 mM phosphate buffer pH 7.6 and 20 mM succinate with [^{35}S] rhodanese (0.537 mg, 197 541 cpm • mg^{-1}). The mixture was then chromatographed on a column (2 × 18 cm) of Sephadex G-75 fine. Fractions of 1.2 ml were collected and in each the protein content, enzyme activity and radioactivity were assayed. ○, succinate dehydrogenase activity; △, rhodanese activity; ●, proteins; ■, radioactivity. Abscissa: fraction number. Ordinate: enzymic activities arbitrary units • ml^{-1}, radioactivity cpm • ml^{-1}, protein mg • ml^{-1}.

TABLE I

DISTRIBUTION OF RADIOACTIVITY IN [^{35}S] RHODANESE AND IN SUCCINATE DEHYDROGENASE

Crystalline [^{35}S] rhodanese, passed through Sephadex G-25, was incubated for 1 h at 0°C in 50 mM phosphate buffer pH 7.6 and 20 mM succinate without or with succinate dehydrogenase (hystidyl FAD: rhodanese 3:1, mol/mol). It was then chromatographed on a column (2 × 18 cm) on Sephadex G-75 fine. Fractions of 1.2 ml were collected and in each the protein content, enzyme activity and radioactivity were assayed. The reconstitutive capacity of succinate dehydrogenase was measured before gel chromatography.

Initial [^{35}S] rhodanese (cpm)	Radioactivity recovered			Increase of reconstitutive capacity (% of undamaged succinoxidase activity
	in succinate dehydrogenase (cpm)	in rhodanese (cpm)	in retained compounds (% of amount applied)	
86 500	7200	13 000	76.7	19
88 000	3940	46 400	42.8	0.5
112 800	—	76 700	32.0	—

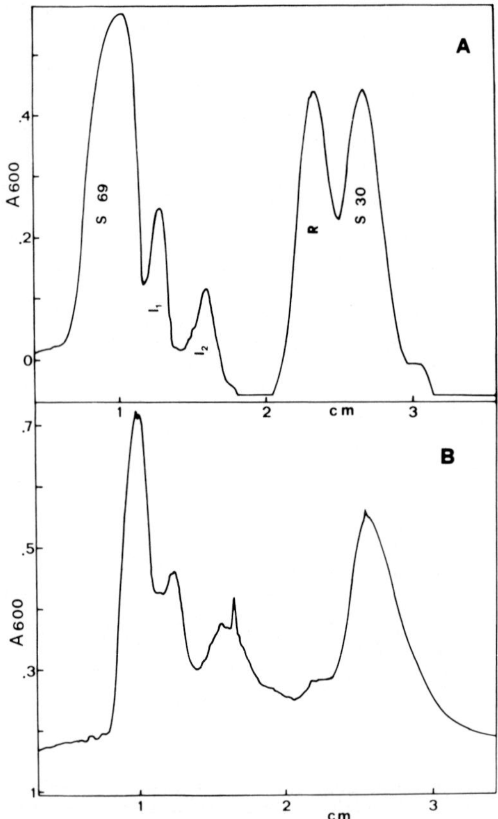

Fig. 3. Sodium dodecyl sulfate electrophoresis of succinate dehydrogenase treated with [^{35}S]rhodanese. The reaction mixture of the experiment shown in Fig. 2 was analysed by electrophoresis on a column of acrylamide gel in sodium dodecyl sulfate before (A) and after (B) fractionation by gel filtration. In this latter case the analysis was performed on the pooled fractions corresponding to the peak of succinate dehydrogenase (tubes 22–24). After completion of the run, the proteins were stained with Coomassie blue and the absorbance at 600 nm was recorded. The recorder tracing is shown. Migration pattern: S 69 and S 30 are the succinate dehydrogenase subunits with mol. wts. 69 000 and 30 000 respectively, R is rhodanese, I_1 and I_2 are impurities. Abscissa: cm from start. Ordinate: absorbance at 600 nm.

As shown in Table I, more radioactivity was incorporated into succinate dehydrogenase and the specific radioactivity of the sulfurtransferase decreased most when treatment with rhodanese resulted in a larger increase of the reconstitutive capacity of the flavoprotein. Considerable radioactivity was released with low molecular weight compounds. Some was found with them also when [^{35}S]rhodanese was incubated in the absence of succinate dehydrogenase.

Discussion

The results reported indicate that rhodanese does not form a stable complex with succinate dehydrogenase: it probably shifts the association equilibrium of the flavoprotein towards higher molecular weight aggregates.

The small amount of rhodanese associated with succinate dehydrogenase is inactive in the catalysis of thiocyanate formation. We had previously observed external thiosulfate not to affect the action of rhodanese on the flavoprotein [3]. These data suggest that the active site of the sulfurtransferase may be engaged in the interaction with the flavoprotein.

Rhodanese appears to transfer its sulfane sulfur to the flavoprotein. In our experiments, the molar ratio of rhodanese to succinate dehyrodgenase was 1:3 since in previous work optimal effects were obtained using this proportion [3]. All the sulfur incorporated into the flavoprotein may therefore originate from the sulfrutransferase. It has been found that sulfur, from iron-sulfur systems degraded under oxidizing conditions, is bound to the protein molecule as a trisulfide [11]. Trisulfides are sulfur donors to rhodanese [12] and in the present case, trisulfide sulfur formed from labile sulfide in deactivated molecules of succinate dehydrogenase might be transferred to rhodanese and be further exchanged with the flavoprotein. Although in our experiments anaerobiosis was strictly controlled, some access of air may have occured during the manipulations. Moreover, limited internal oxidation might be supported by Fe^{3+} in the molecule of succinate dehydrogenase. On the other hand, considering the scheme in the reverse direction, rhodanese might transfer sulfur to the flavoprotein forming a persulfide or a trisulfide.

The nature of the sulfur incorporated into the flavoprotein is still under investigation. The parallel increase of sulfur incorporation and restored reconstitutive capacity observed in these experiments confirms that the two phenomena are related.

Sulfane sulfur has a finite dissociability from the protein and gradually leaches away from rhodanese [13]. This may explain the radioactivity found with low molecular weight compounds after incubating [^{35}S]rhodanese in the absence of succinate dehydrogenase. The enhanced release in the presence of the flavoprotein indicates that sulfur is labilized in the interaction. Persulfide groups in proteins are generally very labile [14]: if a persulfide is formed on the flavoprotein it may contribute to the release.

Acknowledgements

This work was supported by grants of the Italian National Research Council (C.N.R.).

References

1 Taniguchi, T. and Kimura, T. (1974) Biochim. Biophys. Acta 364, 284–295
2 Finazzi Agrò, A., Cannella, C., Graziani, M.T. and Cavallini, D. (1971) FEBS Lett. 16, 172–174
3 Pagani, S., Cannella, C., Cerletti, P. and Pecci, L. (1975) FEBS Lett. 51, 112–115
4 Baginsky, M.L. and Hatefi, Y. (1969) J. Biol. Chem. 244, 5313–5319
5 Hanstein, W.G., Davis, K.A., Ghalambor, M.A. and Hatefi, Y. (1971) Biochemistry 10, 2517–2524
6 King, T.E. (1963) J. Biol. Chem. 238, 4037–4051
7 Kimura, T. and Hauber, J. (1963) Biochem. Biophys. Res. Commun. 13, 169–174.
8 Dunker, A.K. and Ruckert, R.R. (1969) J. Biol. Chem. 244, 5074–5080
9 Cannella, C., Pecci, L., Pensa, B., Costa, M. and Cavallini, D. (1974) FEBS Lett. 49, 22–24
10 Andrews, P. (1065) Biochem. J. 96, 595–605

11 Petering, D., Fee, J.A. and Palmer, G. (1971) J. Biol. Chem. 246, 634–653
12 Szczepkowski, T.W. and Wood, J.L. (1967) Biochim. Biophys. Acta 139, 469–478
13 Volini, M. and Wang, S.F. (1973) J. Biol. Chem. 248, 7392–7395
14 Cavallini, D., Federici, G. and Barboni, E. (1970) Eur. J. Biochem. 14, 169–174

Flavins and Flavoproteins, edited by T.P. Singer

CHAPTER 60

COMPARATIVE STUDIES BY ELECTRON PARAMAGNETIC RESONANCE SPECTROSCOPY ON THE MECHANISMS OF ACTION OF MOLYBEDNUM-IRON/SULPHUR-FLAVIN HYDROXYLASES

R.C. BRAY[a], M.J. BARBER[a], M.P. COUGHLAN[b], H. DALTON[a], E.M. FIELDEN[c] and D.J. LOWE[a]

[a]*University of Sussex, Brighton, Sussex (U.K.),* [b]*University College, Galway (Republic of Ireland) and* [c]*Institute of Cancer Research, Sutton (U.K.)*

Summary

Detailed EPR studies have been carried out comparing turkey liver xanthine dehydrogenase and *Veillonella alcalescens* [*Micrococcus lactilyticus*] xanthine dehydrogenase with milk xanthine oxidase. The studies have included rapid-freezing experiments and quantitative titrations with xanthine. Conclusions are:

1. All the enzymes give the 'blue' FADH• radical. Amounts of this are much higher in the dehydrogenases than in the oxidases. For turkey enzyme, maximum conversion to the radical is about 50%, with xanthine failing to reduce the enzyme completely to the $FADH_2$ form. Low conversion to $FADH_2$ presumably explains the low oxidase activity.

2. The environments of molybdenum in the three enzymes are extremely similar. The turkey enzyme gives Very Rapid, Rapid, Slow and Inhibited signals, all very like those from the milk enzyme. Similar, though less detailed, results have been obtained from the *Veillonella* enzyme. In all three enzymes, molybdenum cycles between the 6- and the 5- and 4-valent forms during turnover.

3. Work with the enzymes is complicated by the presence in ordinary preparations of several non-functional forms, which are readily distinguishable by their EPR properties. Three non-functional forms give Mo(V) EPR signals designated Resting I, Resting II and Slow. Redox behaviour of these species is quite different from that of Mo in the active enzymes.

4. Pulse radiolysis work has provided evidence that the Slow signal from Mo in the desulpho form of milk xanthine oxidase is due to Mo(V) not Mo(III).

5. The *Veillonella* enzyme contains Fe/S in an environment like that reported for this centre in aldehyde oxidase. In the turkey enzyme there are two sorts of Fe/S centres, quite like the two centres of the milk enzyme. In the turkey, in contrast to the milk enzyme, the redox potentials of the two Fe/S centres are close together.

Introduction

Xanthine oxidase, xanthine dehydrogenases and aldehyde oxidase all contain molybdenum and iron-sulphur centres as well as flavin in their molecules. The name 'molybdenum hydroxylases' has recently [1] been proposed for these enzymes. Milk xanthine oxidase is by far the most studied of these and its mechanism of action is quite reasonably well understood, largely as a result of studies by electron paramagnetic resonance spectroscopy (EPR) [1, 2].

Of the other enzymes of the group, only rabbit liver aldehyde oxidase has been at all extensively studied by EPR [3, 4]. However, this work, carried out some while ago, revealed little similarity between this enzyme and milk xanthine oxidase, apart from the fact that various molybdenum, flavin and iron-sulphur EPR signals were observed.

Limited EPR work on chicken liver xanthine dehydrogenase has been reported (see e.g., ref. 5). In the case of the enzyme from *Veillonella alcalescens* (*Micrococcus lactilyticus*), the only EPR data available [6] seem to suggest [4, 6] that the enzyme may operate in a manner entirely different from that of milk xanthine oxidase.

We now report detailed EPR studies on two molybdenum hydroxylases, namely the xanthine dehydrogenases from turkey liver and from *V. alcalescens*. The turkey enzyme was prepared by the method of Cleere and Coughlan [19] and that from *Veillonella* by a method based on that of Smith et al. [9]. The work will be published in greater detail elsehwhere [7, 8].

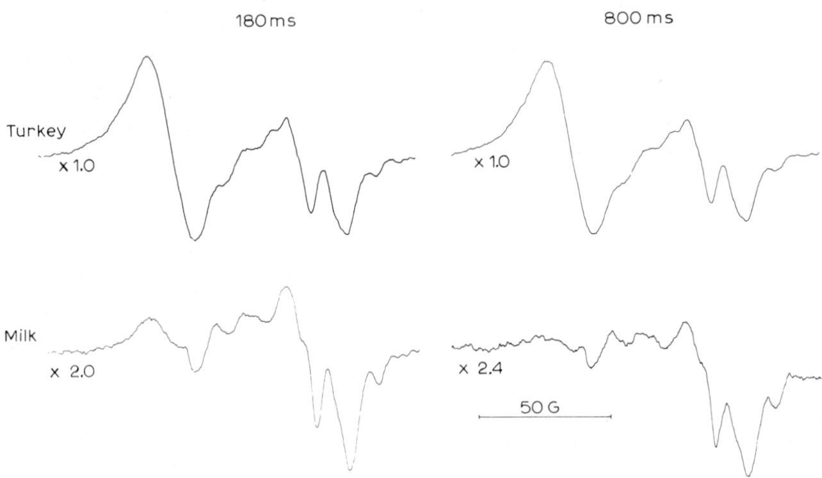

Fig. 1. Comparison of turkey liver xanthine dehydrogenase and milk xanthine oxidase in rapid-freezing experiments. Samples of the two enzymes were reduced anaerobically with excess xanthine at pH 8.2 for 180 ms or for 800 ms. All samples correspond to the same concentration of functional active sites (18 μM). Relative instrument gain settings are shown on the figure. EPR conditions were: 120°K; microwave power 5 mW; frequency 9.1–9.3 GHz; modulation 2.5 G. (Reproduced from ref. 8.)

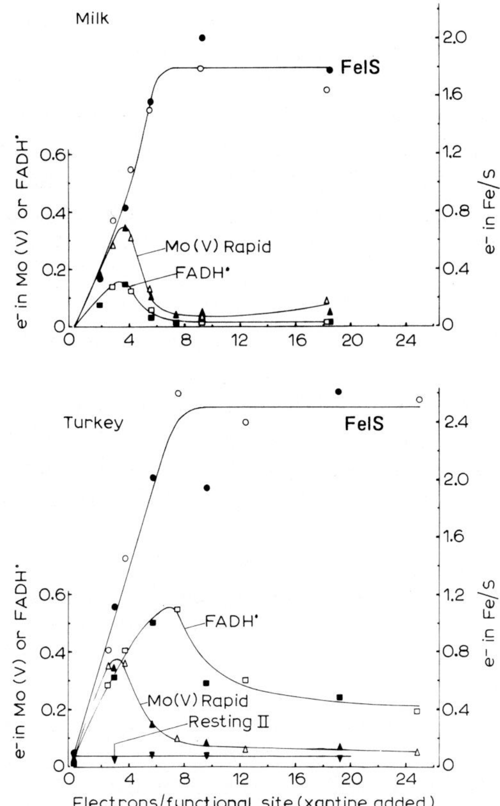

Fig. 2. Titration of milk xanthine oxidase and of turkey liver xanthine dehydrogenase with xanthine. EPR signal intensities are expressed as electrons per functional active site and concentrations of xanthine are given in the same units. Under the conditions of the experiments (1.5 min reaction time), only functional enzyme was titrated. The symbols used are: total iron-sulphur, circles; Mo (V) Rapid, triangles; FADH$^{\bullet}$, squares; Mo (V) Resting II, inverted triangles. Each point on the graph corresponds to the average from two separate samples. At each xanthine level a fresh enzyme sample was used. For both enzymes, two separate experiments, each employing a different batch of enzyme, are illustrated (full and open symbols). Both samples of the milk enzyme contained 80% of functional enzyme [18]; those of the turkey enzyme contained, respectively, 23% (full symbols) and 39% (open symbols) of functional enzyme [11]. (Reproduced from ref. 8.)

FADH$^{\bullet}$ signals

Since this volume is devoted to flavins and flavoproteins, FADH$^{\bullet}$ signals from the enzymes will be discussed first. As shown in Fig. 1, FADH$^{\bullet}$ semiquinone EPR signals were readily observable from the turkey enzyme in rapid-freezing experiments, after reduction with xanthine. The signals were considerably stronger than the ones obtained from the milk enzyme under similar conditions. In these and other experiments, the FADH$^{\bullet}$ signal had a line width of about 19 gauss, for both the turkey and the *Veillonella* enzymes, so that the radicals must be of the 'blue' type, as in the milk enzyme [20].

Fig. 2 compares quantitative reductive titration studies on the turkey and milk enzymes. Although the turkey enzyme is normally contaminated with non-functional forms (see below), only functional enzyme was reduced under the conditions of the titration experiments. Thus, data obtained using two samples of the turkey enzyme, differing almost two fold in their activities, fall on the same curves (Fig. 2). In the titration, the behaviour of the turkey enzyme, with regard to the intensities of the molybdenum and iron-sulphur signals, was virtually indistinguishable throughout from that of the milk enzyme (details of the forms of these signals are considered below). On the other hand, there was a striking difference in the behaviour of the radical signal between the two enzymes. In the case of the milk enzyme, production of the flavin semiquinone never exceeded 15%, whereas with the turkey enzyme it reached about 50%. In further contrast to the behaviour of the milk enzyme, the intensity of the turkey FADH$^{\bullet}$ signal remained quite high in the presence of excess xanthine. This would imply (cf. ref. 2) that in the turkey enzyme the redox potential for the system FADH$^{\bullet}$/FADH$_2$ must be higher than that of xanthine/uric acid. Olson et al. [2] concluded that the reactivity of milk xanthine oxidase with oxygen as an oxidising substrate depends primarily on the reaction of this acceptor with FADH$_2$ in the reduced enzyme. In keeping with this, we conclude that the lack of reactivity of the turkey enzyme with oxygen is a consequence of the difficulty of fully reducing its flavin.

Fig. 3 shows a rapid-freezing experiment with the *Veillonella* enzyme. The FADH$^{\bullet}$ signal was present at a relatively long reaction time with excess xanthine, suggesting that in this enzyme, too, it is difficult to reduce flavin to FADH$_2$. Indeed, Smith et al. [9] had reached this conclusion from optical studies, though they did not relate it to the lack of oxidase activity.

Molybdenum EPR signals

Detailed examination of the molybdenum EPR signals from the turkey and the *Veillonella* enzymes leads to the conclusion that the metal in the active sites of these enzymes must be in an environment extremely similar to that in the milk enzyme, having the same ligand atoms arranged in a similar

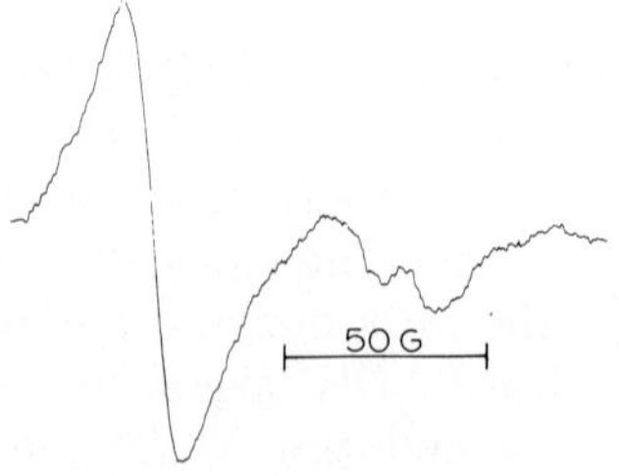

Fig. 3. EPR signals obtained on reducing *V. alcalescens* xanthine dehydrogenase with xanthine in a rapid-freezing experiment. The enzyme was reduced anaerobically at pH 7.8 for 450 ms with excess xanthine. EPR conditions were: 110°K; microwave power 5 mW; frequency 9.1 GHz; modulation 8 G.

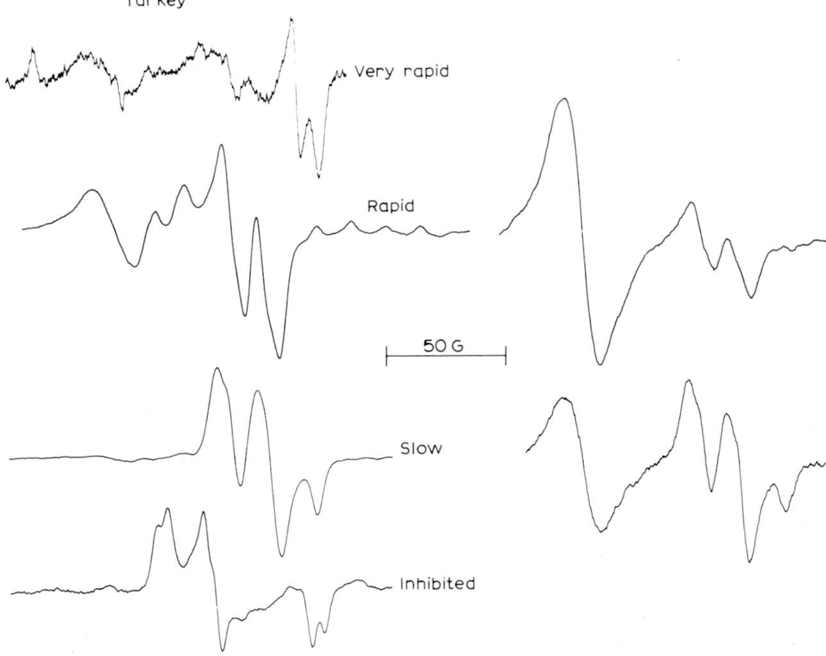

Fig. 4. Comparison of Very Rapid, Rapid, Slow and Inhibited EPR signals from turkey liver and *V. alcalescens* xanthine dehydrogenases. The Very Rapid signal was obtained by reduction with xanthine, at pH 9.9 for 16 ms. For the Rapid signal, the turkey and *Veillonella* enzymes were reduced with low concentrations of xanthine and dithionite, respectively. Extended reduction with dithionite was used to obtain the Slow signal and treatment with formaldehyde for the Inhibited. The pH was 8.2 for all signals except the Very Rapid. EPR conditions were: 120°K; microwave power 5 mW; frequency 9.1–9.3 GHz; modulation 2.5 G.

manner. As shown in Fig. 4, the turkey enzyme can, under suitable conditions, give rise to the characteristic molybdenum signals known as Very Rapid, Rapid, Slow and Inhibited (see refs. 1 or 10 for details of these signals from milk xanthine oxidase). The *Veillonella* enzyme gave two of these signals (Fig. 4), shortage of material preventing us looking for the other two. Detailed measurements of g values and proton splittings for all the observed signals showed that differences among the three enzymes were indeed very small.

For the turkey, as with the milk enzyme, only the Very Rapid and Rapid signals developed within the turnover time. The Rapid signal was the one followed in titration experiments (Fig. 2). Changes in its intensity on titration correspond to reduction of molybdenum(VI) of the enzyme by the substrate, first to Mo(V) and then to Mo(IV). The Slow signal (Fig. 4) is due to a non-functional form of these enzymes (the 'desulpho' form [1]). Existence of the desulpho form of the turkey enzyme has been shown chemically [11]. Development of an Inhibited signal (Fig. 4), on treatment of turkey enzyme with formaldehyde, is consistent with the sensitivity [12] of this enzyme to inhibition by methanol.

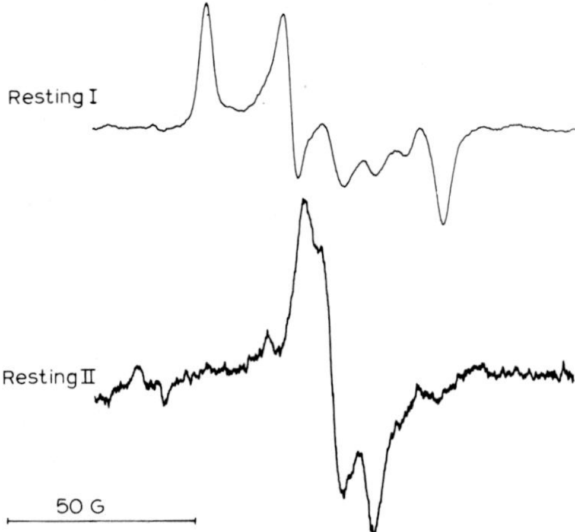

Fig. 5. Mo(V) EPR signals from two new non-functional species present in some preparations of turkey liver xanthine dehydrogenase. With Resting I, in the upper spectrum, a small amount of the Resting II species is also present, as well as traces of the Rapid signal. The lower spectrum is Resting II. Signals were recorded at 120°K, 9.2–9.3 GHz and 2.5 G modulation, with 5 mW microwave power for the upper spectrum and 2 mW in the lower.

Molybdenum EPR signals from additional non-functional forms of the enzymes

Fig. 5 shows two additional molybdenum signals which were obtained from certain preparations of the turkey enzyme. These signals have no analogy in ordinary samples of the milk enzyme. Neither signal shows any indications of proton splittings. Since they were given by untreated samples of the enzyme, they will be referred to as Resting I and Resting II. g values for Resting I and Resting II were, respectively, g_1 1.993, g_2 1.980, g_3 1.956, g_{av} 1.976; g_1 1.977, g_2 1.972, g_3 1.965, g_{av} 1.971. That these signals are due to non-functional forms is shown by a number of pieces of information. Firstly, the intensities of the signals were always small (e.g. 1% of the total molybdenum present) and varied considerably from one batch of enzyme to another. Furthermore, both signals were resistant to oxidation or reduction. Thus, Resting II was apparently completely stable, being resistant not only to xanthine (Fig. 2) but also to dithionite, phenazine methosulphate and ferricyanide, as well as to air. Resting I was relatively stable. On addition of xanthine it did not change (in rapid-freezing experiments) and it was not affected by air. However, it disappeared on treatment with dithionite.

Some preparations of the *Veillonella* enzyme gave a signal quite like Resting I of the turkey enzyme. Behaviour of such a sample on treatment with xanthine is illustrated in Fig. 6. The signal did not change on addition of ferricyanide (not illustrated) but decreased relatively slowly when xanthine was added. At the same time, as expected, the Rapid signal developed. On prolonged incubation with xanthine, the Rapid signal was replaced by Slow. Finally, on admission of air, the Rapid and Slow signals disappeared and the

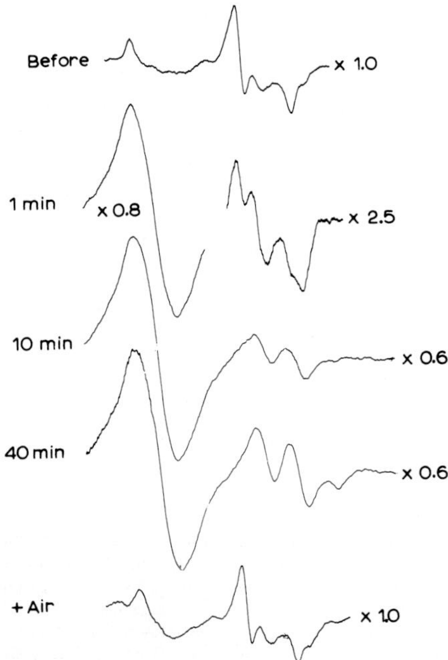

Fig. 6. Changes in the EPR spectrum of *Veillonella* xanthine dehydrogenase on addition of xanthine. Before addition of xanthine, the sample showed a strong Resting I signal. After treatment with xanthine (1 mM at pH 7.8) for 1 min, small amounts of this signal remained, together with some Rapid signal and a free-radical signal. After 10 min, the Resting I signal had disappeared, while after 40 min the Rapid signal had been replaced by the Slow. On reoxidation with air, free radical, Slow and Rapid signals all disappeared, while Resting I returned, at its original intensity. EPR conditions were: 110°K; microwave power 5 mW; frequency 9.1 GHz; modulation 5 G. (Reproduced from ref. 7.)

Active Mo(VI) $\underset{\text{Reoxidation}}{\overset{\text{Substrate}}{\rightleftharpoons}}$ Mo(V) (Rapid signal) $\underset{\text{Reoxidation}}{\overset{\text{Substrate}}{\rightleftharpoons}}$ Mo(IV)

Inactive Mo(VI) $\underset{\text{Reoxidation}}{\overset{\text{Reduced active enzyme}}{\rightleftharpoons}}$ Mo(V) (Slow signal)

Inactive (new form) Mo(V) (Resting I signal) $\underset{\text{Reoxidation}}{\overset{\text{Reduced active enzyme}}{\rightleftharpoons}}$ Mo(IV)

Scheme 1. Proposed oxidation-reduction reactions of molybdenum in active and in non-functional forms of xanthine dehydrogenase from *V. alcalescens*

Resting I signal returned. This behaviour is readily interpreted in terms of the reactions shown in Scheme I. Disappearance of the Resting I species is presumed to occur by a process analogous to reduction of the desulpho enzyme, that is, via interaction of an oxidised non-functional form with

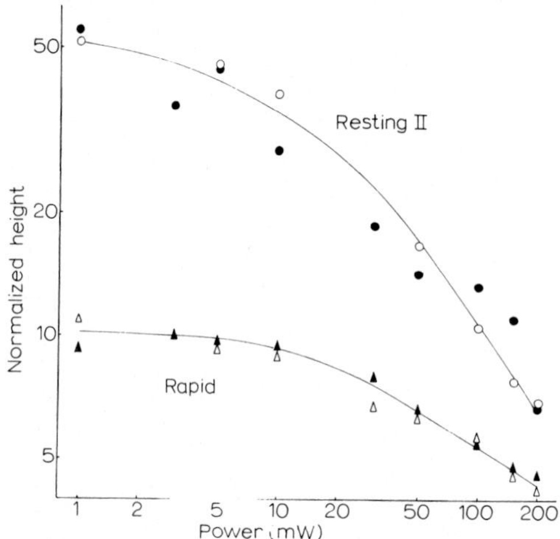

Fig. 7. Saturation behaviour of the Rapid and Resting II signals on samples of the turkey enzyme. Signal amplitudes in arbitrary units (normalized by dividing by the instrument gain setting and the square root of the microwave power) are plotted against microwave power, on a double logarithmic scale. Full symbols correspond to measurements on a sample reduced with xanthine and showing both signals simultaneously. Open circles correspond to measurements of Resting II on the same sample of enzyme, but prior to the addition of xanthine and hence without Rapid. Open triangles were obtained on another sample of enzyme showing a Rapid signal only. Measurements were made at 120°K, with 2.5 G modulation. (Reproduced from ref. 8.)

reduced, active, enzyme molecules (cf. ref. 13). Rajagopalan et al. [4] reported decreases rather than increases in intensities of the molybdenum signals on adding purine to the *Veilonella* enzyme. Examination of the spectra published by these workers [6] clearly shows the presence of Resting I. We conclude, therefore, that in their experiments, decrease of a relatively large Resting I signal must have overshadowed development of Rapid, leading to an observed net decrease.

Resting II of the turkey enzyme appears to be virtually identical with the resting signal reported for aldehyde oxidase by Rajagopalan et al. [3]. Furthermore, we have recently found that by special treatment, desulpho xanthine oxidase can be converted to a species with similar properties (Lowe, D.J., Sherwood, D. and Bray, R.C., unpublished). From saturation studies during titration with reducing substrates, Rajagopalan et al. [4] concluded that the resting signal of aldehyde oxidase could not be due to catalytically inactive enzyme molecules. For the turkey enzyme, despite the overlap of the signals, we were able to quantitate Rapid and Resting II in the presence of one another, and so to carry out separate saturation measurements on each. We found (Fig. 7) that the behaviour of Resting II, in a sample showing only this signal, was no different from that of Resting II in the same sample, after addition of xanthine, when a Rapid signal was also present. By contrast, Rajagopalan et al. [3] reported that their resting signal became progressively more difficult to saturate as substrate was added. The evidence we have presented above shows that signals of the Resting II type are due to non-functional molecules. Thus, we can only conclude that in the experiments on aldehyde oxidase, other centres in such molecules must have become reduced, so causing the observed changes in saturation.

Valency of molybdenum in the Slow signal

In Scheme I, the valency of molybdenum in the Slow signal is given as five. However, Massey and co-workers have repeatedly suggested (eg. refs. 2 and 14) that Mo(III) may be more likely. We have now succeeded (Bray, R.C., and Fielden, E.M. unpublished) in getting evidence on this point from pulse radiolysis experiments (cf. ref. 15). Desulpho milk xanthine oxidase (prepared by cyanide treatment [16]), was reduced with a relatively low concentration of the hydrated electron and frozen 30 ms later. The dose of radiation employed was such that there was about one electron per molybdenum atom and formation of the Slow signal was very clearly demonstrated (Fig. 8). Under such conditions, the statistical probability of three electrons reacting with a single half-enzyme molecule is low. Though the intensity of the Slow signal was itself low, it increased proportionately when the radiation dose was increased three fold. Therefore, if it is accepted (and there seems no reason to doubt this) that in the normal, oxidased state, molybdenum of desulpho xanthine oxidase is 6-valent, then reduction from the 6- to the 5-valent state is fully compatible with our results, whereas reduction to the 3-valent state is not. We conclude that the Slow-signal-giving species, like that in the other molybdenum signals, is Mo(V).

Iron-sulphur signals

Fig. 9 shows iron-sulphur signals from the turkey and from the *Veillonella* enzymes. The former gave two signals, very like the two from milk xanthine oxidase [17], whereas the latter gave a single signal, rather

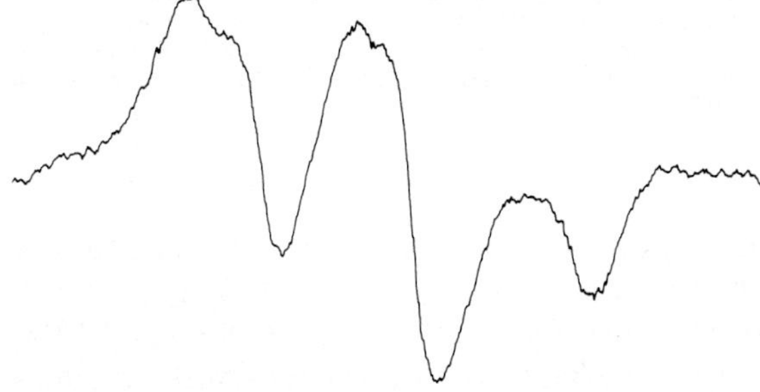

Fig. 8. Slow signal obtained in pulse radiolysis experiments on reduction of desulpho xanthine by low concentrations of the hydrated electron. Desulpho xanthine oxidase was prepared by cyanide treatment [16]. Pulse radiolysis of the flowing solution, prior to freezing for EPR, was according to Bray et al. [15]. *t*-Butanol (10 mM) was used as hydroxyl scavenger, the pH was 7.2 and the time between irradiation and freezing, 30 ms. The concentration of hydrated electrons generated (calculated from dosimetry with KCNS, with proper allowance for the geometry of the irradiation coil) corresponded to 1.0 e^-_{aq} per Mo. EPR conditions were: 120°K; 9.3 GHz; 5 mW microwave power and 3.2 G modulation.

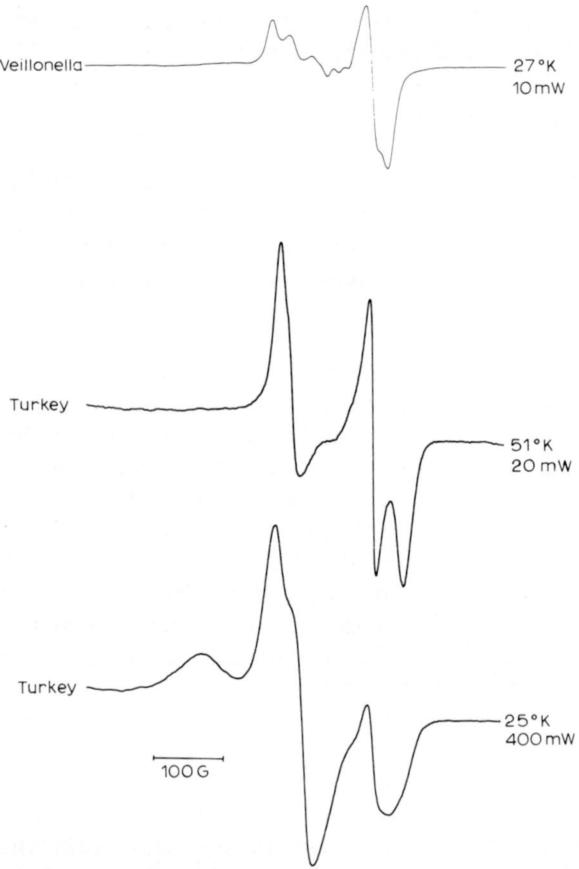

Fig. 9. EPR signals from the iron-sulphur centres of *Veillonella* and turkey xanthine dehydrogenases. The *Veillonella* enzyme was reduced at pH 7.8 with purine and the turkey enzyme with allopurinol at pH 8.2. Small amounts of molybdenum and radical signals are detectable in the *Veillonella* spectrum and a free radical signal is detectable in the turkey enzyme, at the higher temperature. Only one species of iron sulphur was detected in the *Veillonella* enzyme, whereas the turkey enzyme showed two signals, corresponding to Fe/S I and Fe/S II of the milk enzyme. The latter signal was observed at low temperature, only and its features were most readily apparent when the Fe/S I signal was saturated by employing high microwave powers.

like that from aldehyde oxidase [3]. g values were as follows: turkey enzyme: FeS/I, g_1 2.017, g_2 1.932, g_3 1.906, g_{av} 1.95; FeS/II, g_1 2.08, g_2 2.00, g_3 1.92, g_{av} 2.00; *Veillonella* enzyme, g_1 2.026, g_2 1.939, g_3 1.925, g_{av} 1.96. A most important difference between the turkey and milk enzymes is illustrated in Fig. 10. This shows additional data from the titration experiment illustrated in Fig. 2. Clearly, whereas for the milk enzyme FeS/II has a higher redox potential than does FeS/I [2, 17], this does not hold for the turkey enzyme, where the two systems must have very similar redox potentials.

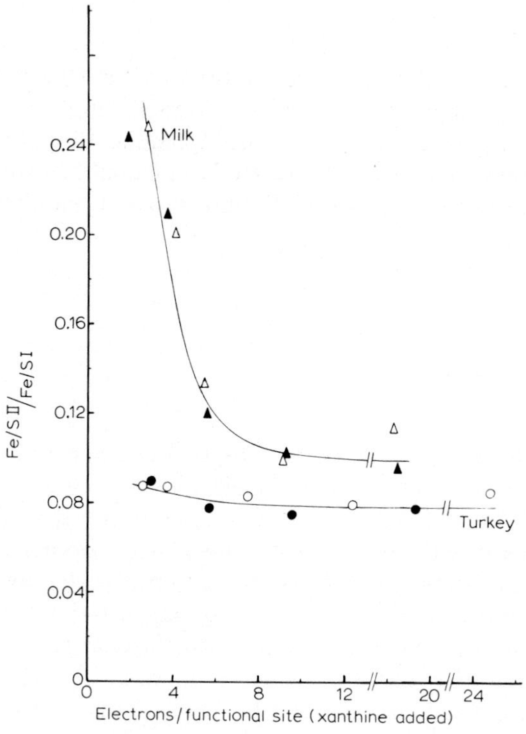

Fig. 10. Changes in the relative intensities of the Fe/S I and Fe/S II signals of the milk and turkey enzymes during titration with xanthine. The ratio of the amplitudes of the two signals (measured from the g_1 feature of Fe/S II and the g_2 feature of Fe/S I) are plotted as a function of the amount of xanthine added per functional site. The experiment is the one illustrated in Fig. 2 (see legend to Fig. 2 for explanation of symbols). EPR measurements were carried out at 21°K. (Reproduced from ref. 8.)

TABLE I (Addendum)

RELATIVE AMPLITUDES OF EPR SIGNALS OBTAINED ON REDUCTION OF XANTHINE OXIDASE AFTER VARIOUS PRETREATMENTS

Amplitudes or their ratios are given in arbitrary units. Enzyme was prepared by a modification of the method of Battelli et al [23]; treatment with chymotypsin (0.05 or 0.5 mg/ml) was for 45 min at 37°C and that with dithiothreitol (1.5 mg/ml) was for 15 min at 37°C; these reagents were not removed for the EPR. Reduction of the enzyme with xanthine (0.1 mM) or with purine (0.5 mM) was carried out anaerobically at pH 8.1 for 1 min at about 20°C. Untreated enzyme gave no signals prior to addition of substrates but dithiothreitol caused partial reduction (Fe/S I, 0.1; Mo/Fe—S I, 4; FADH•/Fe—S I, 4).

Pretreatment of enzyme	Reducing substrate	Signal amplitudes (and ratios)		
		Fe—S I	Mo(Rapid)/Fe—S I	FADH•/Fe—S I
None	Xanthine	0.7	2.0	0.8
Chymotrypsin	Xanthine	0.5	1.7	0.4
Dithiothreitol	Xanthine	1.0	1.0	2.9
None	Purine	0.5	5.2	0.7
Chymotrypsin	Purine	0.6	5.4	0
Dithiothreitol	Purine	0.9	1.6	2.1

Acknowledgements

Turkey xanthine dehydrogenase was kindly provided by Mr. W.F. Cleere. The work was supported by the Medical Research Council and was facilitated by a short-term followship from the European Molecular Biology Organization to M.P.C. Travel grants made available to M.P. Coughlan by the National Dairy Council of America and by the Wellcome Trust (London) are also gratefully acknowledged.

Addendum

(in collaboration with F. STIRPE, *University of Bologna, Italy*)

An important conclusion in the above work is that the reduced xanthine dehydrogenases do not react with oxygen because their flavin is present predominantly as FADH$^\bullet$ rather than as $FADH_2$. A clue to the origin of this behaviour is provided by the work of Stirpe and co-workers, who reported [21–23] on dehydrogenase/oxidase interconversions in some molybdenum hydroxylases, including milk xanthine oxidase [23]. They suggested [23] that a thiol group present in the dehydrogenase ('D') form of these enzymes prevents their flavin from reacting with oxygen, whereas in oxidase ('O') forms, this thiol has been either oxidized or removed by proteolysis. It was therefore of considerable interest to study 'D' and 'O' forms of the milk enzyme by EPR. Our data, presented below, are in agreement with and extend somewhat that of Waud and Rajagopalan (see this volume, Chapt. 61).

Milk xanthine oxidase was prepared without the use of proteolytic enzymes by a modification of the method of Battelli et al. [23] and showed only one component in the ultracentrifuge. Table I shows that treatment of this enzyme with a thiol or with a proteolytic enzyme profoundly influenced the yield of FADH$^\bullet$ obtained on reduction with substrates. Chymotrypin decreased FADH$^\bullet$ levels, whereas dithiothreitol increased them. Molybdenum and Fe/S signals also changed in intensity, but considerably less than did those from FADH$^\bullet$. Careful examination of the forms of the various signals in a considerable number of experiments and under a variety of EPR conditions failed to reveal any form changes caused by the reagents. Though rapid-freezing studies have not been carried out, the work, in conjunction with the known effects [23] of thiols and proteolytic enzymes on the catalytic properties of this enzyme, therefore strengthens the evidence that 'O' forms of molybdenum hydroxylases yield less flavin semiquinone than do the 'D' forms and also provides a direct indication that the proposed thiol group of the D-forms is associated with flavin.

References

1 Bray, R.C. (1975) in The Enzymes (Boyer, P.D., Ed.), 3rd edn., Vol. 12, pp. 299–419, Academic Press, New York

2 Olson, J.S., Ballou, D.P., Palmer, G. and Massey, V. (1974) J. Biol. Chem. 249, 4363–4382
3 Rajagopalan, K.V., Handler, P., Palmer, G. and Beinert, H. (1968) J. Biol. Chem. 243, 3784–3796
4 Rajagopalan, K.V., Handler, P., Palmer, G. and Beinert, H. (1968) J. Biol. Chem. 243, 3797–3806
5 Kanda, M. and Rajagopalan, K.V. (1972) J. Biol. Chem. 247, 2177–2182
6 Aleman, V., Smith, S.T., Rajagopalan, K.V. and Handler, P. (1966) in Flavins and Flavoproteins (Slater, E.C., ed.), pp. 99–114, Elsevier, Amsterdam
7 Dalton, H., Lowe, D.J., Pawlik, R.T. and Bray, R.C. (1976) Biochem. J. 153, 287–295
8 Barber, M.J., Bray, R.C., Lowe, D.J. and Coughlan, M.P. (1976) Biochem. J. 153, 297–307
9 Smith, S.T., Rajagopalan, K.V. and Handler, P. (1967) J. Biol. Chem. 242, 4108–4117
10 Bray, R.C. and Swann, J.C. (1972) Struct. Bonding 11, 107–144
11 Cleere, W.F., O'Regan, C. and Coughlan, M.P. (1974) Biochem. J. 143, 465–468
12 Cleere, W.F. and Coughlan, M.P. (1974) Biochem. J. 143, 331–340
13 Morrell, D.B. (1952) Biochem. J. 51, 657–666
14 Massey, V., Komai, H., Palmer, G. and Elion, G.B. (1970) Vitam. Horm. 28, 505–531
15 Bray, R.C., Lowe, D.J., Capeillère-Blandin, C. and Fielden, E.M. (1973) Biochem. Soc. Trans. 1, 1067–1072
16 Massey, V. and Edmondson, D. (1970) J. Biol. Chem. 245, 6595–6598
17 Lowe, D.J., Lynden-Bell, R. and Bray, R.C. (1972) Biochem. J. 130, 239–249
18 McGartoll, M.A., Pick, F.M., Swann, J.C. and Bray, R.C. (1970) Biochim. Biophys. Acta 212, 523–526
19 Cleere, W.F. and Coughlan, M.P. (1975) Comp. Biochem. Physiol. 50B, 311–322
20 Palmer, G., Müller, F. and Massey, V. (1971) in Flavins and Flavoproteins (Kamin. H., ed.), pp. 123–137, University Park press, Baltimore
21 Della Corte, E. and Stirpe, F. (1968) Biochem. J. 108, 349–351
22 Della Corte, E. and Stirpe, F. (1972) Biochem. J. 126, 739–745
23 Battelli, M.G., Lorenzoni, E. and Stirpe, F. (1973) Biochem. J. 131, 191–198

Flavins and Flavoproteins, edited by T.P. Singer

CHAPTER 61

RAT LIVER XANTHINE DEHYDROGENASE: STUDIES ON DEHYDROGENASE TO OXIDASE CONVERSION

WILLIAM R. WAUD and K.V. RAJAGOPALAN

Department of Biochemistry, Duke University Medical Center, Durham, N.C. 27710 (U.S.A.)

The xanthine oxidizing enzyme of rat liver has been isolated and described as an oxidase [1]. However, in crude preparations of rat liver, the enzyme appears to be an NAD^+-utilizing dehydrogenase (D form), which is readily converted into an oxidase (O form) by treatments such as incubation at 37°C or proteolysis with trypsin [2]. We have purified the D form of rat liver xanthine dehydrogenase and have studied the D form to O form conversion.

The purified D and O forms of the rat liver enzyme are indistinguishable from each other in their absorbance spectra, cofactor composition or electrophoretic behavior on 5% polyacrylamide gels. Samples of D and O forms displayed predominant, apparently identical protein subunits after electrophoresis on 7.5% sodium dodecyl sulfate-urea polyacrylamide gels in the presence of β-mercaptoethanol.

The relative activities of the D and O forms with various electron acceptors were determined. The low aerobic rate of oxidation of xanthine by the D form was stimulated almost 4-fold by NAD^+ and 6-fold by methylene blue. As expected of a dehydrogenase, NAD^+ was reduced to NADH as a concomitance of the stimulated xanthine oxidation. $NADP^+$ did not act as an electron acceptor. In contrast, the O form showed little or no stimulation of xanthine oxidation by NAD^+ or methylene blue and did not use NAD^+ as an electron acceptor. Further, while both forms effected reduction of cytochrome *c* and 2,6-dichlorobenzenone indophenol (Cl_2-Ind) when oxidizing xanthine, ability to interact with the above electron acceptors in the presence of NADH as the alternative substrate was a property of the D form alone.

Purified D and O forms of rat liver xanthine dehydrogenase were incubated with sufficient antiserum (directed against the O form) to precipitate all the enzymatic activity. The resuspended precipitates were assayed for various activities. Because of the measurable, albeit attenuated, activities exhibited by these antibody complexes, the use of antibodies for rapid isolation of dehydrogenase of maximum D/O ratio appeared to offer the best

Abbreviations: Cl_2-Ind, 2,6-dichlorobenzenone indophenol; Nbs_2, 5,5′-dithiobis-(2-nitrobenzoic acid); $NsBzO^-$, 2-nitro-5-thiobenzoate.

means for studying the native properties of rat liver xanthine dehydrogenase. For detailed comparative studies two types of the O form [2, 3] were utilized: (1) the O form obtained from liver extracts heated at 37°C for 30 min (heated O form) and (2) the O form prepared from liver extracts incubated with trypsin (trypsinized O form). Fresh rat livers were homogenized in 0.25 M sucrose or 0.1 M Tris · HCl, pH 8.1. Sucrose homogenates were used for obtaining antibody complexes of the D and heated O forms while Tris · HCl homogenates were used for trypsinized O antibody complex. The heated O form was generated by incubation of the sucrose homogenate at 37°C for 30 min. All homogenates were centrifuged at 100 000 × *g*. The 100 000 × *g* sucrose supernatants were incubated with sufficient antiserum. The Tris · HCl 100 000 × *g* supernatant was incubated at 37°C for 45 min after addition of 0.1 mg trypsin/ml [3]. The extract was recentrifuged at 100 000 × *g* and then incubated with sufficient antiserum.

The activities of the three antibody complexes with various electron acceptors were determined. Xanthine → NAD^+ (295 nm), xanthine → methylene blue and xanthine → Cl_2-Ind activities were essentially the same for all three antibody complexes. The other activities reflected the expected differences between D and O forms. Both xanthine → O_2 and xanthine → cytochrome *c* activities were lowest for the D form and highest for the trypsinized O form, with the heated O form displaying values approaching those of the trypsinized O form. The ability to utilize NAD^+ as an electron acceptor and NADH as an electron donor was greatest for the D form and least for the trypsinized O form with the heated O form showing intermediate values. The experimentally observed differences in activities of the three antibody complexes reflect even greater actual differences, in view of the differential inhibitions of purified D and O forms upon complexing with antibody. The effect of heat on the D form was similar to that produced by trypsinization with respect to the oxidation of xanthine but was different with respect to the NADH oxidizing activities. In fact some antibody complexes of the heated O form showed but marginal losses in NADH → methylene blue and NADH → Cl_2-Ind activities, while having singnificantly decreased xanthine → NAD^+ activity. Thus heating seems to have uncoupled the sequence of electron transfer reactions required for the manifestation of xanthine → NAD^+ activity with no effect on the interaction of xanthine with the enzyme, and with only a partial effect on the site of interaction with NADH.

Antibody complexes of xanthine dehydrogenase D form, heated O form and trypsinized O form were subjected to electrophoresis on 6% dodecyl sulfate-urea polyacrylamide gels in the presence of β-mercaptoethanol. The D form and the heated O form showed one major polypeptide band with a molecular weight of approximately 150 000. Trypsinized O form also showed one major band but with an approximate molecular weight of 125 000–130 000.

Antibody complexes of xanthine dehydrogenase D form, heated O form and trypsinized O form were examined by EPR spectroscopy at −100°C with either xanthine or NADH as the electron donor under

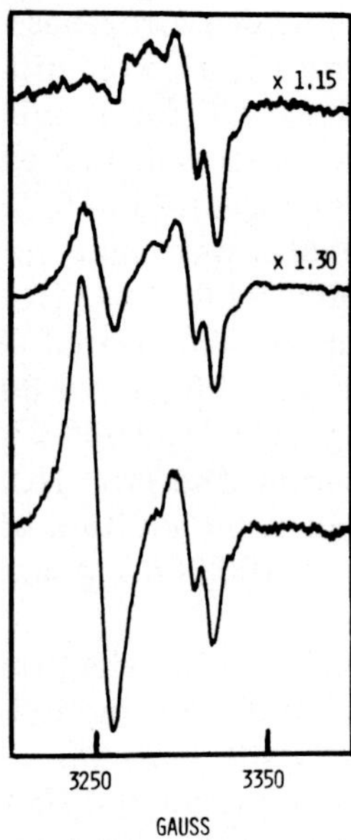

Fig. 1. Effect of trypsin and heat on the flavin and molybdenum EPR signals of rat liver xanthine dehydrogenase antibody complex. The antibody complexes were reduced anaerobically in the presence of 1.5 mM xanthine. Upper spectrum, trypsinized O form antibody complex. Middle spectrum, heated O form antibody complex. Lower spectrum, D form antibody complex.

anaerobic condtions. Only the free radical signal of flavin semiquinone and the $g = 1.97$ signal of molybdenum (V) were monitored in these experiments. Fig. 1 shows the spectra obtained from the three antibody complexes after reduction by xanthine for 80 s. The D form displayed an intense FADH$^{\cdot}$ signal in addition to the molybdenum (V) signal. The amplitudes of both signals generated by NADH were much higher than those seen on reduction with xanthine. The trypsinized O form differed markedly from the D form by the virtual absence of the FADH$^{\cdot}$ signal on reduction with xanthine. The amplitude of the molybdenum (V) signal was, on the other hand, somewhat larger than that seen with the D form. The heated O form showed intermediate behavior and displayed a small but measurable FADH$^{\cdot}$ signal in addition to the expected molybdenum signal upon reduction with xanthine. The amplitude of the FADH$^{\cdot}$ signal produced by reduction of the heated O form with NADH was the same as in the case of the D form. However, the intensity of the molybdenum (V) signal was considerably less. Further, while signals seen on reduction of the D form with NADH were maximal within 20 s after mixing, the signals of the heated O form attained their maximal intensities at a much slower rate. The FADH$^{\cdot}$ and molybdenum (V) signals of the NADH-reduced trypsinized O form also showed a slower rate of development and attained final intensities less than those seen with the D form.

Della Corte and Stirpe [4] showed that the stimulatory effect of NAD^+ on xanthine oxidation by rat liver extracts could be abolished with concomitant increase in O_2 reactivity by treatment of the extracts with 1 mM

5,5′-dithiobis-(2-nitrobenzoic acid) (Nbs_2). In the present studies isolated D form enzyme and antibody complex of the D form were both incubated in the presence of 1 mM Nbs_2 at 4°C. After 15 min the D/O ratios were reduced to 1.0–1.1. For the antibody complex xanthine → NAD^+ (340 nm) and NADH → Cl_2-Ind activities were lowered to 15% and 31%, respectively, of control. The effect of Nbs_2 on the D form antibody complex was further studied by titration with Nbs_2. Both xanthine → NAD^+ (340 nm) and NADH → Cl_2-Ind activities decreased gradually even as the amount of 2-nitro-5-thiobenzoate ($NsBzO^-$) released went up. After approximately 55% inhibition of the two activities there occurred no further release of $NsBzO^-$, even though with the time the inhibition increased. It appeared the the inhibition by Nbs_2 was in part time-dependent and that possibly only a few sulfhydryls needed to be modified in order to effect the inhibition. As a result, D form antibody complex was incubated with varying amounts of Nbs_2 for 18 h at 4°C. Xanthine → NAD^+ (340 nm) activity was inhibited 27% for 8 equiv. $NsBzO^-$ released per equiv. enzyme and 90% for 14 equiv. $NsBzO^-$ released per equiv. enzyme. Hence it appears that the inhibition by Nbs_2 is not due to modification of just one or two sulfhydryls nor is it due to modification of all the exposed sulfhydryls, but is the result of time-dependent changes consequent to modification of some intermediate number of sulfhydryl groups.

Because of the implication of sulfhydryl groups in the D to O conversion, the three antibody complexes, D form, heated O form and trypsinized O form, were reacted with excess Nbs_2 at 4°C. The results are given in Table I. Also included are values for purified D and O forms. D form antibody complex and heated O form antibody complex effected the release of approximately the same amount of $NsBzO^-$, about 28 mol per mol of enzyme, despite the large difference in D/O ratios. Purified D form, having a D/O ratio of 2.0, brought about the release of 9.1 mol $NsBzO^-$ per mol of enzyme. Trypsinized O form antibody complex produced only 7.5 moles $NsBzO^-$ while purified O form had the lowest value, 3.7.

A preparation of xanthine dehydrogenase O form was incubated at 37°C in the presence of 10 mM dithiothreitol for 30 min. After passage through a small Sephadex G-25 column to remove the thiol reagent, the

TABLE I

Nbs_2 REACTIVITY OF XANTHINE DEHYDROGENASE ANTIBODY COMPLEXES

The results are presented as mean ± S.E.

Antibody complex	D/O ratio before complexing	TNB released (mol/mol enzyme)
D form	7.4 ± 0.4	28.2 ± 2.1
D form*	2.0	9.1**
Heated O form	1.2 ± 0.05	27.2 ± 2.0
Trypsinized O form	1.1 ± 0.0	7.5 ± 0.9
O form*	1.0 ± 0.0	3.7 ± 0.1

* Uncomplexed enzyme.

** Single determination.

enzyme displayed a D/O ratio of 2.2 and effected the reduction of NAD^+ with xanthine as substrate. The converted enzyme was examined by EPR spectroscopy at −100°C with either xanthine or NADH as electron donor under anaerobic conditions. The treated enzyme displayed an intense FADH˙ signal in addition to the molybdenum (V) signal. The untreated enzyme, while giving rise to the molybdenum (V) signal, displayed no FADH˙ signal. With NADH as reductant the treated enzyme developed both FADH˙ and molybdenum (V) signals at a faster rate and to a higher level than did the untreated enzyme.

These results suggest the following description of the D to O conversion. Xanthine dehydrogenase in vivo has many free sulfhydryls. During purification most of these free sulfhydryls are oxidized to disulfides with concomitant loss of dehydrogenase activity. Rapid purification will allow one to see a form of the enzyme which has some dehydrogenase activity and an intermediate number of free sulfhydryls. Heating curde homogenates at 37°C merely alters the conformation of the enzyme, thereby abolishing its dehydrogenase activity, even though the free sulfhydryls have not had time to form disulfides. Trypsinization removes a peptide of approximately 20 000 molecular weight which is rich in sulfhydryl content and is essential for dehydrogenase activity.

Olson et al. [5] have recently postulated that reduced milk xanthine oxidase reacts with oxygen principally via fully reduced flavin. Also anaerobic reduction of the milk enzyme by xanthine generates no stable EPR signal ascribable to the semiquinone flavin. In contrast, chicken liver xanthine dehydrogenase when reduced by xanthine generates a large radical signal due to the semiquinone radical of FAD [6]. The rat liver enzyme, isolated rapidly as an antibody complex, also displays a stable semiquinone signal when reduced by xanthine. Treatments which convert the D form to an O form, i.e. heat, aging and trypsin lead to decreases in the flavin signal intensity. The amount of signal observable correlates well with the amount of residual dehydrogenase activity. It appears that the ability to maintain the semiquinone form of FAD after reduction by xanthine is essential for the dehydrogenase function of the enzyme. The chicken and rat enzymes stabilize the semiquinone and prevent it from being fully reduced. When NAD^+ interacts with the enzyme quite possibly the fully reduced flavin is now favored and the cofactor then reduced. Or equally possible is that the semiquinone form is the species that interacts and reduces NAD^+. Both enzymes, by maintaining the semiquinone form of FAD, can bind and interact with NAD^+.

In the rat enzyme the semiquinone form of FAD is stabilized through the interaction of the FAD with a cysteine-rich terminal segment of the polypeptide chain. When the enzyme is converted to an O form by heating or by trypsinization, the terminal segment no longer interacts with the FAD, with concomitant destabilization of the semiquinone form and preferential formation of fully reduced flavin. As a consequence of the facilitated reduction of the flavin to the level of $FADH_2$, the enzymes loses the ability to interact with NAD^+, and at the same time acquires the enhanced ability to transfer electrons to molecular oxygen.

Addendum

As demonstrated above, purified rat liver xanthine dehydrogenase O form is converted into the D form in the presence of dithiothreitol, as reported earlier by Della Corte and Stirpe [4]. Battelli et al. [7] reported similar conversion of milk xanthine oxidase into a dehydrogenase in the presence of dithioerythritol, but did not characterize the nature of the enzyme after removal of the thiol reagent. We present here preliminary studies on the EPR properties of the D and O forms of milk xanthine oxidase.

Milk xanthine oxidase was prepared without proteolytic treatment as described elsewhere [3]. The purified enzyme was incubated at 37°C in the presence of 10 mM dithiothreitol for 90 min, and passed through a Sephadex G-25 column to remove the thiol agent. The thiol-free enzyme displayed a D/O ratio of 2.5. As with the rat enzyme xanthine → O_2 and xanthine → cytochrome *c* activities were significantly lowered after thiol treatment, while the rate of reduction of NAD^+ with xanthine as substrate was elevated. Ability to utilize NADH as an electron donor was also greater for the treated enzyme.

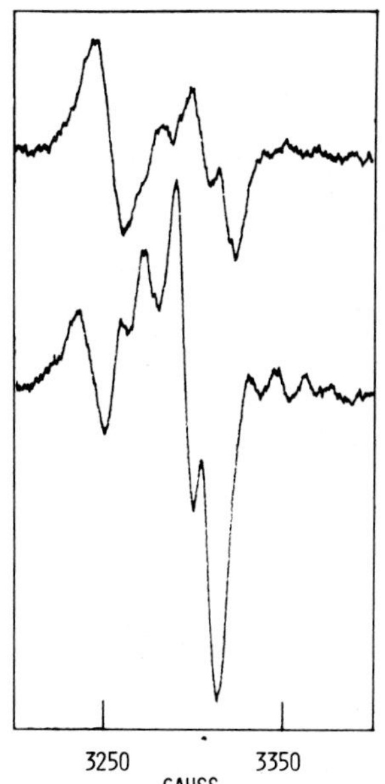

Fig. 2. Effect of dithiothreitol treatment on the flavin and molybdenum EPR signals of purified milk xanthine oxidase. Upper spectrum, treated enzyme. Lower spectrum, untreated enzyme. The concentration of the untreated enzyme was 3.88 times that of the treated enzyme. EPR conditions were the same as given in the preceding paper.

The converted enzyme (D/O = 2.6) was examined by EPR spectroscopy at −100°C with either xanthine or NADH as electron donor under anaerobic conditions. Fig. 2 shows the spectra obtained from treated and untreated enzyme after reduction by xanthine for 20 s. The treated enzyme displayed an intense FADH$^{\cdot}$ signal in addition to the molybdenum (V) signal. The untreated enzyme, while giving rise to the molybdenum (V) signal, produced litttle FADH$^{\cdot}$ signal. With NADH as reductant the treated enzyme developed both FADH$^{\cdot}$ and molybdenum (V) signals at a faster rate and at greater amplitude than did untreated enzyme.

Enzyme purified with proteolytic treatment [8], when subjected to treatment with dithiothreitol, displayed a D/O ratio of 1.0 after removal of the reagent and did not effect reduction of NAD^+. The treated and untreated enzyme were examined by EPR spectroscopy. With xanthine as reductant the molybdenum (V) signals developed at the same rate and to the same extent. There was, however, no FADH$^{\cdot}$ signal for either enzyme. With NADH as reductant, both enzymes developed molybdenum (V) and FADH$^{\cdot}$ signals at the same slow rate and to the same extent.

The results obtained with the milk enzyme support the contention in the preceding paper that the ability to maintain the semiquinone form of FAD after reduction by xanthine is essential for dehydrogenase function and further show that a decrease in subunit size by proteolysis leads to the formation of an irreversible oxidase.

References

1 Rowe, P.B. and Wyngaarden, J.B. (1966) J. Biol. Chem. 241, 5571
2 Stirpe, F. and Della Corte, E. (1969) J. Biol. Chem. 244, 3855
3 Della Corte, E. and Stirpe, F. (1968) FEBS Lett, 2, 83
4 Della Corte, E. and Stirpe, F. (1972) Biochem. J. 126, 739
5 Olson, J.S., Ballou, D.P., Palmer, G. and Massey, V. (1974) J. Biol. Chem. 249, 4350
6 Kanda, M. and Rajagopalan, K.V. (1972) J. Biol. Chem. 247, 2177
7 Battelli, M.G., Lorenzoni, E. and Stirpe, F. (1973) Biochem. J. 131, 191
8 Waud, W.R., Brady, F.O., Wiley, R.D. and Rajagopalan, K.V. Arch. Biochem. Biophys., in the press

Flavins and Flavoproteins, edited by T.P. Singer

CHAPTER 62

COMPARATIVE ASPECTS OF ALDEHYDE OXIDASE AND XANTHINE OXIDASE IN MAMMALIAN AS OPPOSED TO NON-MAMMALIAN TISSUES

MAUREEN J. SMYTH and EDWARD J. DUKE

Department of Zoology, University College, Stillorgan Road, Dublin 4 (Republic of Ireland)

Aldehyde oxidase (EC 1.2.3.1, aldehyde:oxygen oxidoreductase) and xanthine oxidase (EC 1.2.3.2, xanthine:oxygen oxidoreductase) are both flavoproteins which appear to have very similar structures and functions. Up to quite recently, most of the aldehyde and xanthine oxidases which had been purified were from mammalian sources. Generally, only one or other enzyme was purified from a given tissue, principally cow milk xanthine oxidase [1] and pig [2, 3] and rabbit [3, 4] liver aldehyde oxidase. In some early studies [5], it was stated that pig liver aldehyde oxidase is a form of xanthine oxidase. A more recent investigation of this enzyme shows, however, that at a certain stage in the purification xanthine oxidase activity is lost. From these data it is not clear whether in fact the latter activity is recoverable or has been destroyed.

The present report is concerned with an electrophoretic analysis of the aldehyde and xanthine oxidases of various crude mammalian liver tissues. It was hoped that such analysis might clarify the relationship between the two enzyme activities from a single mammalian tissue. These mammalian electrophoretic patterns are compared with those obtained from *Drosophila melanogaster* and *Locusta migratoria*.

Fresh liver samples of cow, pig and sheep were obtained from the Dublin City Abattoir. Mouse liver was taken from the Q-strain [7] and rabbit liver from New Zealand White animals. Crude extracts for electrophoresis were prepared in all cases by homogenizing the tissues in a Potter-Elvehjem hand homogenizer in 2 vol. (w/v) of phosphate buffer (0.01 M phosphate containing 1 mM EDTA and 0.1 mM dithiothreitol, pH 7.0). The homogenates were centrifuged at $30\,000 \times g$ for 20 min at 4°C. *Drosophila* and locust extracts were prepared in similar fashion using 4 vol. (w/v) of phosphate buffer.

Polyacrylamide gel electrophoresis was carried out in 5% gels as previously described [6]. Starch gel electrophoresis was carried out in phosphate gels at pH values from 6 to 8 (Lush, I., unpublished). The bridge buffer was 0.393 M sodium phosphate adjusted to the appropriate pH

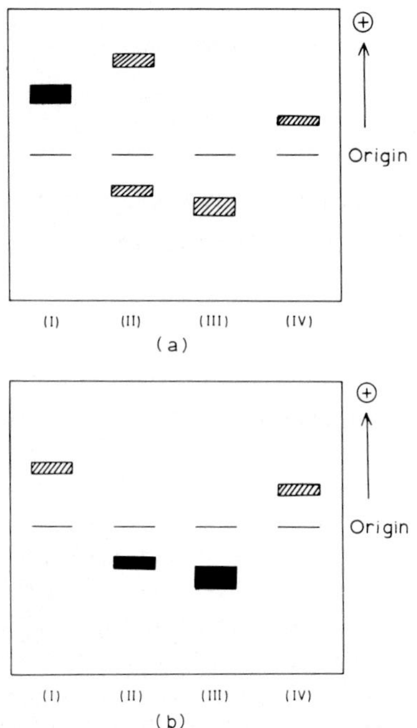

Fig. 1. Starch gel electrophoretograms of aldehyde oxidase and xanthine oxidase/dehydrogenase. Crude extracts were prepared by homogenization of mammalian livers in 2 vol. (w/v) of phosphate buffer and in 4 vol. (w/v) for *D. melanogaster*. Starch gel electrophoresis was carried out in phosphate buffer at pH 6.0 for 15 h at 4°C using 20 mA per gel of cross section 160 × 60 mm. (a) Gels were stained for aldehyde oxidase in 100 ml of 0.01 M phosphate pH 8.0 containing 30 mg nitroblue tetrazolium, 8 mg phenazine methosulphate and 0.1 ml salicylaldehyde. The slots contained: (i) *Drosophila* (ii) mouse liver (iii) cow liver and (iv) pig liver. (b) Gels are similar to those in (a) except that 14 mg hypoxanthine was substituted for salicylaldehyde in the staining mixture. Closed symbols represent strongly staining enzymes and cross-hatched symbols denote those of medium-staining intensities.

value. Gels were made in a 40-fold dilution of bridge buffer. Electrophoresis was allowed to proceed at 4°C for 15 h using 20 mA through a gel of cross-section 160 × 6 mm. After electrophoresis, polyacrylamide gels were stained for both xanthine [6] and aldehyde [8] oxidases. Starch gels were stained in 100 ml 0.01M sodium phosphate, pH 8.8, containing 30 mg nitro blue tetrazolium and 8 mg phenazine methosulphate, with 14 mg hypoxanthine or 0.1 ml salicylaldehyde for xanthine oxidase and aldehyde oxidase respectively. Staining for xanthine dehydrogenase was carried out by adding 20 mg NAD to the above.

Mouse liver was partially purified through ammonium sulphate precipitation and Sephadex G-200 column chromatography as previously described [6] except that the above-mentioned phosphate-homogenizing buffer was used in preference to Tris. Further purification through DEAE-cellulose, also incorporating phosphate buffer, was carried out [8].

Polyacrylamide gel electrophoresis of all mammalian liver samples reveals the presence of only a single enzyme type which stains equally with hypoxanthine or salicylaldehyde. This is also true of samples purified as far as Sephadex G-200. These results contrast sharply with those previously obtained with *D. melanogaster* and *L. migratoria* (ref. 8 and Hayden, T.J., unpublished) where clear separations of xanthine dehydrogenase and aldehyde oxidase were obtained. In *L. migratoria*, seven aldehyde oxidase isozymes, belonging to three distinct families, have been demonstrated. Fig. 1 shows starch gel (pH 6.0) electrophoretograms of aldehyde and xanthine oxidases from mouse, cow and pig liver. *D. melanogaster* extracts, stained for aldehyde oxidase and xanthine dehydrogenase, are included as a reference. Two distinct enzyme bands can be observed for mouse liver. The cathodal band stains with both salicylaldehyde and hypoxanthine whereas the anodal band stains only with salicylaldehyde. Only one enzyme can be observed for both pig and cow liver and in each case the staining is stronger with hypoxanthine than with salicylaldehyde. Under these conditions, no staining with either substrate could be seen for rabbit and sheep liver. *D. melanogaster* extracts on starch gel, in contrast to polyacrylamide gels, exhibit only one staining zone. When the same extracts are electrophoresed at pH 8.0, all enzyme bands migrate anodally but no improvement in resolution is evident. At this pH a single enzyme band, staining with both substrates, can now be seen from sheep liver. Surprisingly, no distinct aldehyde oxidase band was observed for rabbit liver. It is probable that N_1-methylnicotinamide is required as substrate rather than salicylaldehyde [3]. For comparison with the polyacrylamide data, the two distinct mouse enzymes were extracted from starch gel and re-electrophoresed on polyacrylamide gel. Both migrate to a similar position corresponding to that of the single band from crude extract. Each individual enzyme retains its unique staining characteristics in polyacrylamide. During purification of the mouse liver, the two enzymes remain unfractionated until eluted through DEAE-cellulose which separates both completely.

It is quite clear that mammalian aldehyde and xanthine oxidases are more closely related than, for example, those of insects. Under the conditions described in this report, only the mouse liver enzymes are electrophoretically distinct. In other mammalian tissues a single electrophoretic enzyme form was observed which oxidises hypoxanthine and salicylaldehyde, to a lesser extent. In insects, and in many other invertebrates (Hayden, T.J., unpublished) the two enzyme types are readily separable and, in fact, the aldehyde oxidase may be highly polymorphic. It is unclear at this stage whether these differences have physiological significance.

References

1 Rajagopalan, K.V. and Handler, P. (1964) J. Biol. Chem. 239, 1509–1514
2 Mahler, H.R., Mackler, B., Green, D.E. and Bock, R.M. (1954) J. Biol. Chem. 210, 465–480
3 Felsted, R.L., Chu, A.E. and Chaykin, S. (1973) J. Biol. Chem. 248, 2580–2587
4 Rajagopalan, K.V., Fridovich, I. and Handler, P. (1962) J. Biol. Chem. 237, 922–928
5 Igo, R.P. and Mackler, B. (1960) Biochim. Biophys. Acta 44, 310–314
6 Duke, E.J., Joyce, P. and Ryan, J.P. (1973) Biochem. J. 131, 187–190
7 McCarthy, J.C. (1967) Genet. Res. 10, 73–80
8 Hayden, T.J. and Duke, E.J. (1975) Isozymes (Markert, C.L., ed.), Vol. 2, pp. 501–517, Academic Press, New York

Flavins and Flavoproteins, edited by T.P. Singer

CHAPTER 63

ON THE INTERACTION BETWEEN SULPHIDE AND TURKEY LIVER XANTHINE DEHYDROGENASE

MICHAEL P. COUGHLAN and WILLIAM F. CLEERE

Department of Biochemistry, University College, Galway (Republic of Ireland)

Turkey liver xanthine dehydrogenase, as normally isolated, was recently shown [1, 2] to contain significant amounts of non-functional or desulpho enzyme. The desulpho forms of these enzymes lack the active-centre persulphide groups essential to activity. For a review of the work leading to the proposed existence of these groups in the milk enzyme see ref. 3. A second non-functional form of milk xanthine oxidase, i.e. the demolybdo enzyme (for review see ref. 4) does not appear to contaminate turkey enzyme preparations.

This has necessitated a reappraisal of some calculations in reports [5, 6] that went to press before these findings (though they appeared in print some what later). The turnover numbers using various electron acceptors, recalculated for fully functional enzyme ($AFR_{340}^{30°} = 325$), are shown in Table I. Xanthine-NAD$^+$ oxidoreductase activity is expressed as the $AFR_{340}^{30°}$ value, i.e. the activity/flavin ratio, obtained by dividing the change in E_{340}/min by the E_{450} of the enzyme used in the assay. Fully functional turkey enzyme has an AFR value of 325 (see ref. 2).

The enzyme shows a pattern of specificity similar to that exhibited by the enzyme from other avian species [7, 8]. The turnover number of the xanthine-NAD$^+$ oxidoreductase reaction (at 30°C, ∞ xanthine, ∞ NAD$^+$) catalyzed by fully functional enzyme is 2500 mol per mol of enzyme-bound flavin per min. This is equivalent to a turnover number of 1666 at 25°C and is similar to the values reported for the reaction catalyzed by xanthine dehydrogenase isolated from the livers of chickens fed a high-protein diet [9] and for the xanthine oxygen oxidoreductase reaction catalyzed by fully functional milk enzyme [10].

The available evidence suggests that the active centre of turkey liver xanthine dehydrogenase [1, 2], like those of milk xanthine oxidase [11] and of rabbit liver aldehyde oxidase [12], contains a persulphide group essential to activity with all substrates except NADH. Unequivocal proof of the existence of persulphide groups in these enzymes is not available. However, the sulphur released as thiocyanate on cyanide treatment [1, 11, 12] does not derive from the iron-sulphide chromophores [11] nor from cysteine [1].

Moreover, the difference spectrum between the native and cyanide-inactivated enzyme [1] shows a distinct peak around 325 nm, at which wavelength persulphide groups are known to absorb (see ref. 15).

The persulphide group may be lyzed during purification, aging or cyanide treatment of the enzyme and may be regenerated on treatment with sulphide (Fig. 1).

The effects of sulphide treatment vary from enzyme to enzyme and in the case of the turkey enzyme are rather complex. Thus, while sulphide effected significant reactivation of the cyanide-inactivated enzyme, similar treatment of the native enzyme resulted in the rapid loss of 50% of the xanthine-oxidizing activity (Fig. 1, Table II) and in bleaching of the visible absorbance spectrum of the enzyme (Fig. 2). It may be noted that the time required for completion of spectral change and the time for loss of sulphide-sensitive activity were the same.

Loss of half of the xanthine-oxidizing activity might be due to non-equivalence of the intramolecular electron-transfer chains such that sulphide disrupts the reactivity of one but not of the other. However, various observations in the past that appeared to provide evidence for non-equivalence were later shown to be consequent on the presence of non-functional enzyme in the preparations used [2, 13, 14]. Similar considerations can explain the findings reported in Fig. 1 and Table II. We know that the preparations used in these investigations contained 56% non-functional enzyme. Moreover, more recent data shows that the extent of loss of activity and of bleaching of the visible absorbance spectrum vary with the degree of functionality of the preparations used.

In an earlier report [1] we showed that sulphide interacted with the enzyme in more than one way. Several lines of evidence suggest that the

TABLE I

ELECTRON ACCEPTOR SPECIFICITY OF FULLY FUNCTIONAL TURKEY LIVER XANTHINE DEHYDROGENASE WITH XANTHINE OR NADH AS ELECTRON DONOR

Turnover numbers were calculated from the apparent V at infinite concentration of the acceptor and at a fixed concentration of xanthine (0.15 mM) or of NADH (0.05 mM).

Acceptor	Turnover number (mol/mol of flavin/min)	
	Xanthine	NADH
Dichlorophenolindophenol	2172	1130 (AFR=82)*
Phenazinemethosulphate	3341	—
NAD^+	1252	—
Methylene blue	138	1510
Trinitrobenzenesulphonate	83	320 (AFR=65)**
Ferricyande	36	22
O_2 (air)	43	1.5

* The turnover number obtained experimentally was 1130 with enzyme of $AFR_{340}^{30°} = 82$. Assuming that cyanide treatment of the turkey enzyme effects loss of all but 30% of the NADH-dichlorophenolindophenol activity, as is the case with the chicken liver enzyme (see ref. 7), one can calculate a turnover number of 2380 for fully functional enzyme.

** We do not know if the content of functional enzyme affects NADH-trinitrobenzenesulphonate activity.

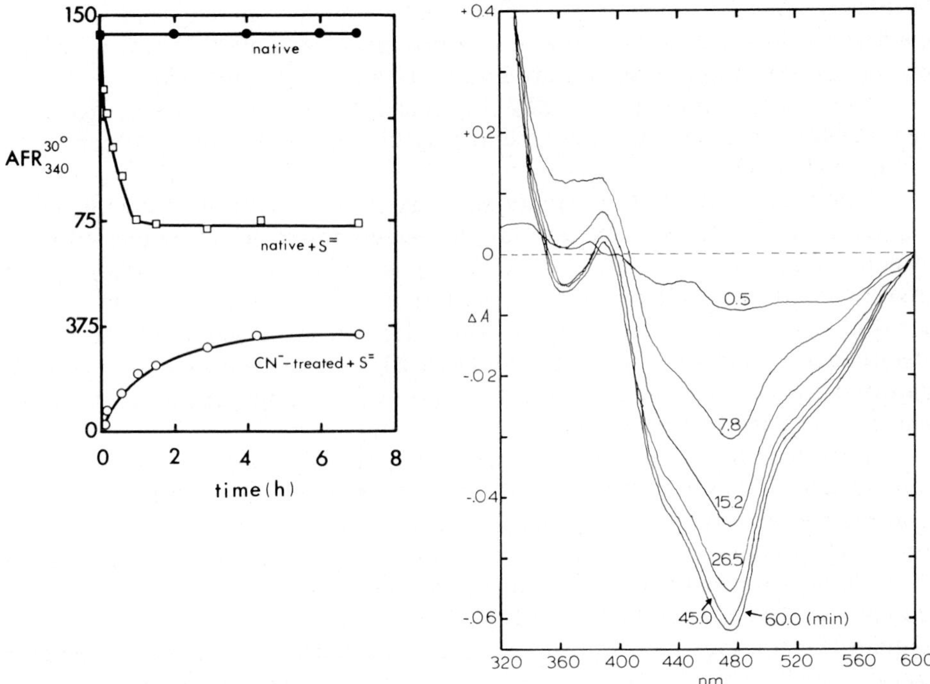

Fig. 1. The effect of sulphide treatment on the xanthine-NAD$^+$ oxidoreductase activity of native and of cyanide-inactivated turkey xanthine dehydrogenase (redrawn from ref. 1, courtesy of the Biochemical Journal). Native and cyanide-inactivated enzyme (from which excess of cyanide had been removed by gel filtration following inactivation) were incubated in stoppered tubes at 45°C in 0.05 M potassium phosphate buffer pH 7.8, containing 0.1 mM EDTA and 0.01 M sodium sulphide. Samples for assay of activity were withdrawn at intervals. The xanthine-NAD$^+$ oxidoreductase activities of the untreated enzyme (●), of the sulphide-treated enzyme (□) and of the sulphide-treated, cyanide-inactivated enzyme (○) are expressed as the $AFR^{30°}_{340}$ values.

Fig. 2. Difference spectra between sulphide-treated and native turkey xanthine dehydrogenase (from ref. 1, courtesy of the Biochemical Journal). Enzyme (A_{450} = 0.68) was incubated in capped cuvettes at 45°C in 0.05 M potassium phosphate buffer pH 7.8, containing 0.1 mM EDTA and 0.01 M sodium sulphide. Difference spectra against the same concentration of untreated enzyme were recorded thereafter at the times (min) indicated.

active-centre persulphide group (or its remains) at the molybdenum locus is one such site of interaction. Thus, (a) sulphide treatment reactivates non-functional enzyme, (b) the extent of the loss of xanthine-oxidizing activity on sulphide treatment is related to the content of functional enzyme and (c) the "sulphide difference spectrum" of the turkey enzyme is similar to the difference spectrum between anaerobic and aerobic cyanide-inactivated milk xanthine oxidase (see ref. 11). Sulphide treatment leads to decreases in visible absorbance (Fig. 2) and variably affects NADH diaphorase activities (Table II). Taken together, these findings suggested to us that one of the iron-sulphur chromophores was also disrupted by sulphide treatment [1].

How does one reconcile two seemingly disparate effects of sulphide treatment? viz. activation of cyanide-inactivated enzyme and inactivation of the native enzyme. Any explanation must take account of the differential loss of activities (Table II) and the fact the the extent of loss of xanthine-oxidizing activity and of spectral change may be correlated with the content of fully functional enzyme. Moreover, such explanations must allow for the gradual return towards the native activity and spectrum following the removal of sulphide [1].

A possible explanation for most of the above observations is outlined in Scheme 1. Sulphide treatment of the native enzyme reactivates the desulpho fraction present by regenerating the active centre persulphide groups. Concurrently, interaction of sulphide at a second site results in a large decrease in activity relative to untreated enzyme. This may be due to an effect on one of the Fe/S centres, either directly or through a conformational change in the enzyme. The effects of sulphide on the activities of functional molecules cannot be due to mere reduction of the enzyme since restoration of activity and of the oxidized absorbance spectrum is a very slow process. EPR studies may clarify the mechanism of this interaction.

It follows therefore, that sulphide treatment of any preparation of the turkey enzyme, containing functional and non-functional or desulpho forms, should effect a decrease (or an increase) in activity compared to that exhibited by the conformationally-altered enzyme. Moreover, the loss (or gain) of activity should be proportional to the content of fully functional enzyme initially present. If the above hypothesis is correct, one would expect the $AFR^{30°}_{340}$ values of different preparations to be the same following sulphide treatment regardless of the initial value. That this may in fact be the case is shown in Table III.

Reactivation of cyanide-inactivated or naturally occurring desulpho-enzyme by sulphide is postulated to result from the regeneration of the active site persulphide groups through the interaction of sulphide with a disulphide (Scheme 1 and see ref. 11). Treatment of disulphide-containing proteins with sulphide under alkaline conditions produces an absorbance in the range of 320–350 nm due to the formation of persulphide groups (see

TABLE II

EFFECTS OF SULPHIDE TREATMENT ON TURKEY LIVER XANTHINE AND NADH OXIDOREDUCTASE ACTIVITIES

Enzyme in stoppered tubes was incubated at 45°C for 1 h in 0.05 M phosphate buffer pH 7.8, containing 0.1 mM EDTA and 0.01 M sodium sulphide. After removal of excess of reagent by passage through Sephadex G-25, xanthine oxidoreductase and NADH oxidoreductase activities were measured and are expressed as a percentage of those of the untreated enzyme. Each value is the average of three separate experiments. Table reproduced from ref. 1, courtesy of the Biochemical Journal.

Electron acceptor	Oxidoreductase activity (% of control)	
	Xanthine	NADH
NAD^+	53.0	—
Phenazine methosulphate	47.6	—
Dichlorophenolindophenol	48.2	69.0
Trinitrobenzenesulphonate	45.4	63.9
O_2	51.7	97.9
Ferricyanide	51.3	98.3
Methylene blue	52.0	97.0

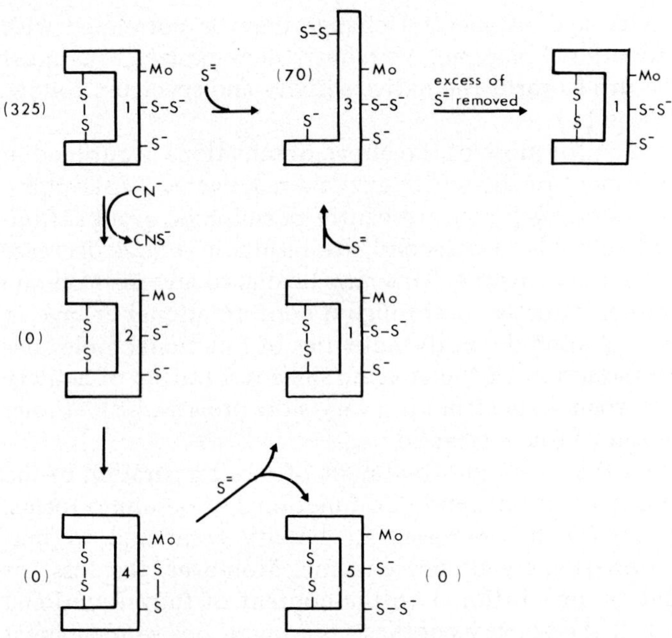

Scheme. 1. Possible reactions of sulphide with turkey liver xanthine dehydrogenase. Turkey liver xanthine dehydrogenase as normally isolated, is a mixture of functional (form 1) and non-functional (form 4) enzyme. Loss of sulphur from the active-centre persulphide groups gives inactive enzyme (form 2) and allows formation of a disulphide bond with a vicinal thiol to give the inactive form 4 (see ref. 11). Reaction between form 4 and sulphide produces equal amounts of inactive form 5 and the active form 1. Interaction of sulphide with a second site on form 1 affects one of the Fe/S centres directly, or indirectly via a conformational change, to give form 3. The latter may be due to cleavage of a disulphide bond by sulphide with the formation of a new persulphide and a new thiol group, (see ref. 15). The difference spectrum between sulphide-treated and native turkey xanthine dehydrogenase (Fig. 2) exhibits a significant positive absorbance in the region 300–350 nm consistent with the formation of persulphide groups. $^{35}S^{2-}$ treatment of cyanide-inactivated milk enzyme showed that in addition to reincorporation of sulphur to the active site, extraneous sulphur had been introduced into the protein, presumably through formation of persulphide by reaction with disulphides (see ref. 11). Form 3 exhibits a low level of xanthine-oxidizing activity (20% of that of fully functional enzyme). The new persulphide group (unlike the active-centre persulphide group which is stabilized by molybdenum [3]) is unstable, especially in the absence of excess of sulphide (see e.g. ref. 15). Thus, removal of excess sulphide by gel filtration would result in the gradual restoration of form 1. The values in parentheses beside each of the different forms of the enzyme represent the $AFR_{340}^{30°}$ value of that form.

TABLE III

THE EFFECT OF SULPHIDE TREATMENT ON THE $AFR_{340}^{30°}$ VALUE OF TURKEY LIVER XANTHINE DEHYDROGENASE

Conditions as for Table II.

Sample	$AFR_{340}^{30°}$ value	
	before treatment	after treatment
Native 1	143	70.1
2	138	72.7
3	114	69.5
CN-inactivated	1	35.0

e.g. ref. 15). It would, however, be predicted that only half of the resultant product would be functional. This may explain why the $AFR^{30°}_{340}$ of the cyanide-inactivated enzyme following sulphide treatment was 35 rather than 70 (Table III).

Loss of native conformation or direct attack on the iron-sulphur centres may explain the observed spectral changes (Fig. 2) when sulphide interacts with fully functional enzyme molecules. By the same token, the gradual return of the original absorbance following removal of excess of reagent may be explained by the restoration of the native conformation under these conditions.

Acknowledgements

M.P.C. thanks the National Dairy Council of America and the Wellcome Trust for travel grants. W.F.C. was in receipt of a maintenance grant from the Department of Education, Republic of Ireland. Our thanks are also due to Dr. K.V. Rajagopalan for his comments on this manuscript.

References

1 Cleere, W.F. and Coughlan, M.P. (1974) Biochem. J. 143, 331–340
2 Cleere, W.F., O'Regan, C. and Coughlan, M.P. (1974) Biochem. J. 143, 465–468
3 Massey, V. (1973) in Iron-Sulphur Proteins (Lovenberg, W., ed.), pp. 301–360, Academic Press, New York
4 Bray, R.C. and Swann, J.C. (1972) Struct. Bonding II, 107–144
5 Cleere, W.F. and Coughlan, M.P. (1975) Comp. Biochem. Physiol. 50B, 311–322
6 Cleere, W.F., Mulhern, J.F. and Coughlan, M.P. (1975) Comp. Biochem. Physiol 50B, 323–329
7 Rajagopalan, K.V. and Handler, P. (1967) J. Biol. Chem. 242, 4097–4107
8 Krenitsky, T.A., Tuttle, J.V., Cattau, Jr., E.L. and Wang, P. (1974) Comp. Biochem. Physiol. 49B, 687–703
9 Nishino, T. (1974) Biochim. Biophys. Acta, 341, 93–98
10 Olson, J.S., Ballou, D.P., Palmer, G.P. and Massey, V. (1974) J. Biol. Chem. 249, 4350–4362
11 Massey, V. and Edmondson, D. (1970) J. Biol. Chem. 245, 6595–6598
12 Branzoli, U. and Massey, V. (1974) J. Biol. Chem. 249 4346–4349
13 Komai, H. and Massey, V. (1969) in Flavins and Flavoproteins (Kamin, H., ed.), pp. 399–423, University Park Press, Baltimore
14 McGartoll, M.A., Pick, F.M., Swann, J.C. and Bray, R.C. (1970) Biochim. Biophys. Acta 212, 523–526
15 Cavallini, D., Federici, G. and Barboni, E. (1970) Eur. J. Biochem. 14, 169–174

Flavins and Flavoproteins, edited by T.P. Singer

CHAPTER 64

FORMATE DEHYDROGENASE: PREPARATION AND EXAMINATION OF FMN, DEFLAVO, FAD AND RIBOFLAVIN ENZYMES

THOMAS HÖPNER*, ULRICH MÜLLER, ULRICH RUSCHIG and PETER WILLNOW

Institut für Biologische Chemie der Universität, D-6900 Heidelberg, Neuenheimer Feld 501 (G.F.R.)

Summary

1. Formate dehydrogenase from *Pseudomonas oxalaticus* was isolated. The two forms "I" (M_r = 320 000) and "II" (M_r = 175 000) were characterized. I tentatively contains 2 FMN and 17–20 iron and labile sulphides. II tentatively contains 1 FMN and 8–10 iron and labile sulphides.

2. During anerobic incubation in the presence of ammonium sulphate, sodium sulphide and EDTA formate dehydrogenase slowly releases the FMN which is rapidly re-incorporated after admission of oxygen if not removed by dialysis. Deflavo formate dehydrogenase is inactive in catalyzing the formate : NAD oxidoreductase reaction. It is fully reactivated by FMN.

3. While the incorporation of FMN into the deflavo protein as observed by the alteration of the flavin fluorescence is a rapid process the reappearance of the catalytic activity takes place rather slowly.

4. Besides FMN deflavo formate dehydrogenase incorporates FAD and riboflavin, thereby regaining the formate and NADH oxidase activity which disappears during the removement of the flavin. FMN, deflavo, FAD, and riboflavin formate dehydrogenase were examined in view of further catalytic abilities.

Introduction

Formate dehydrogenase (Formate : NAD^+ oxidoreductase EC 1.2.1.2) from *Pseudomonas oxalaticus* has been inroduced by us [1] as an iron sulphur flavoprotein containing FMN. It catalyzes the oxidation of formate to CO_2 by NAD, oxygen and several non-physiological electron acceptors, and the oxidation of NADH by CO_2 [2], oxygen, and dyes. The enzyme has been recommended as a tool for formate assay [3].

* Present address: Universität, D-2900 Oldenburg, Postfach 943, G.F.R.

This report deals with the molecular properties, the reversible removement of the flavin moiety, and some of the catalytic abilities of the holo and the deflavo enzyme. In addition it is shown that the deflavo enzyme also binds FAD and riboflavin. Catalytic activities of the resulting flavoenzymes are compared to those of FMN and the deflavo enzyme.

Materials and methods

Isolation of formate dehydrogenase

The principal steps in the method used to isolate the enzyme can be briefly outlined as follows. *Pseudomonas oxalaticus*, grown aerobically in a formate pyruvate salt medium [4], was sonicated in 0.05 M histidine hydrochloride buffer, containing 5 mM sodium sulphide and 0.5 mM EDTA (pH 6). After centrifugation the pH was brought to 5.2 (0.4 M hydrochloric acid). After centrifugation 1.5% (v/v) of a 10% polyethylene imine (average $M_r = 1800$) solution was added to the supernatant. The precipitate was then centrifuged off. From the supernatant solution formate dehydrogenase was precipitated by ammonium sulphate (up to 1.9 M). The precipitate was eluted stepwise by ammonium sulphate solutions, pH 5.6.

Formate dehydrogenase dissolved between 1.35 and 0.80 M. After desalting by gel filtration (0.05 M histidine buffer, pH 5.6) the supernatant of an acid precipitation (pH 5.2, 1 M acetic acid) was fractionated with acetone. Combined active fractions were adsorbed to DEAE-Sephadex A 25 and eluted by a potassium chloride gradient in the histidine buffer. Active fractions were adsorbed to calcium phosphate gel and eluted by a phosphate gradient in the histidine buffer. Formate dehydrogenase was collected by ammonium sulphate precipitation and further purified by sucrose density gradient centrifugation. Results of this last step are discussed in the next section.

Exclusion of oxygen during the purification enhances the yield which even under this precaution does not exceed 5% in the 185-fold enrichment.

The enzyme is stable for some weeks when stored anaerobically at 0°C as a suspension in 1.2 M ammonium sulphate, pH 5.6, containing 5 mM sodium sulphide and 0.5 mM EDTA. Assays for activity were performed as given in ref. 3.

Results and Discussion

Examination of sucrose density gradient centrifugation fractions

Density gradient centrifugation resulted in the two active forms "I" and "II" already reported by us [1]. Fig. 1 shows the distribution of enzyme activity, iron, and labile sulphide in the fractions of the gradient. Specific acitivities, FMN, iron and labile sulphide content of the fractions based on the molecular weights 320 000 and 175 000 are summarized in Table I. Molecular weights were determined by sucrose density gradient centrifugation using the customary internal standards. Purity of the fractions was examined by disc (Fig. 2) and sodium dodecyl sulphate gel electrophoresis (Fig. 3). Both methods show the presence of an inactive contaminating protein* in the fractions 7–11 while the remaining fractions contain only disc electrophoresis bands visible both after staining for protein and activity (see legend of Fig. 2).

* This protein had not been separated off in three different disc electrophoresis systems used at the time of our first report [1].

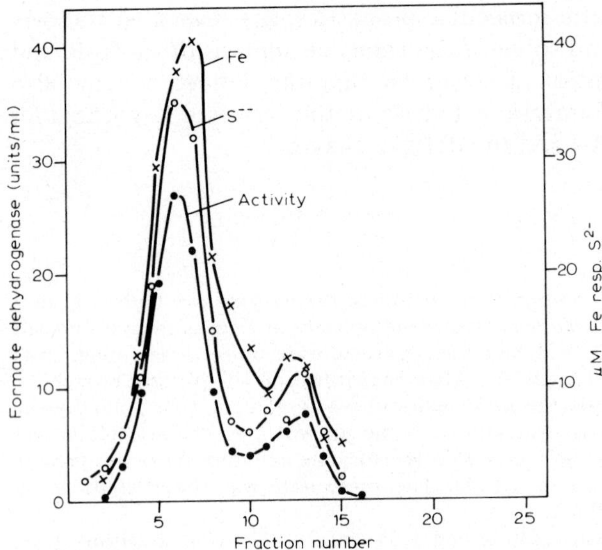

Fig. 1. Distribution of activity, iron and labile sulphide in a sucrose density gradient (5–20% sucrose in 0.05 M histidine buffer pH 5.6). Rotor SW 40 Ti (Beckman), 21.5 h, 40 000 rev/min. 0.45 ml formate dehydrogenase per tube (156 units/ml, 14 units/mg). Bottom at left.

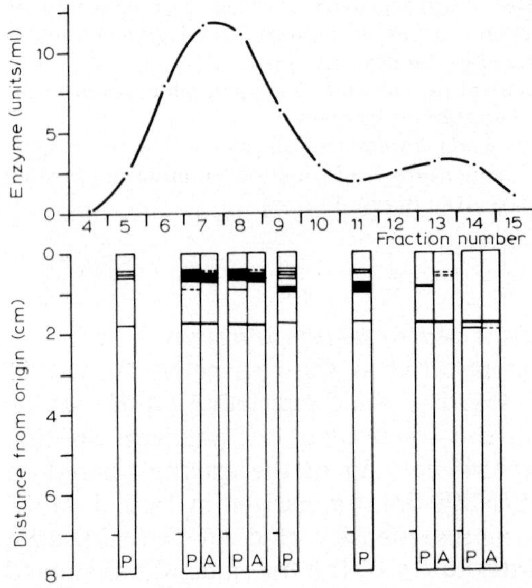

Fig. 2. Disc electrophoresis of sucrose density gradient fractions. System 6 of Maurer [6]. Stain for protein ("P") by amino black. Stain for activity ("A") by formate and nitroblue tetrazolium salt at pH 7.

The dodecyl sulphate electrophoresis results were exploited by means of standard proteins to obtain molecular weights of formate dehydrogenase subunits (Fig. 4). We found two subunits for formate dehydrogenase I with molecular weights of 100 000 and 59 000, respectively, while formate dehydrogenase II in addition showed two minor bands with molecular weights of 83 000 and 56 000

From the values of Table I and Fig. 4 formate dehydrogenases I and II are characterized as follows:

Formate dehydrogenase I (M_r = 320 000) contains 2 FMN, 17–20 iron and presumably about the same amount of labile sulphide per mol. Dodecyl sulphate electrophoresis shows two chains with molecular weights of 100 000 and 59 000. The FMN content of formate dehydrogenase I

TABLE I

Analyses of sucrose density gradient centrifugation fractions as in Fig. 1. Assay for FMN by extinction at 450 nm of the perchloric acid supernatant ($\epsilon = 10.7\ cm^{-1} \cdot nM^{-1}$ under the conditions of the supernatant). Identification of the FMN performed by paper chromatography, iron by atomic absorption spectroscopy, labile sulphide by the methylene blue method of Fogo and Popowsky [5]. Enzyme activity measured according to ref. 3 and protein by the biuret method.

	Fraction						
	3	4	5	6	12	13	14
mol FMN/mol	—	—	1.7	1.4	—	0.6	—
mol iron/mol	17	18	21	20	10	8	6
mol sulphide/mol	17	15	13	18	6	8	7
specific activity	29	42	42	44	32	33	26

TABLE II

Catalytic abilities of formate dehydrogenase I. Methods: 1a, optically (A_{340nm}). 1b, optically at the appropriate wavelength. 1c, Clark oxygen electrode with and without catalase. 2a, optically (A_{340nm}) and by reduction of $^{14}CO_2$ to [^{14}C]formate [2]. Formate dehydrogenase catalyzes the adjustment of the equilibrium of the NAD:formate oxidoreductase reaction from both the formate and the CO_2-direction reaching the final value calculated from E_0'(NAD/NADH) = − 0.32 V and E_0'(CO_2 plus bicarbonate/formate) = − 0.42 V. CO_2 was identified to be the "active species" by kinetic methods [10]. 2b as 1b. 2c, optically (A_{340nm}) and as 1c. All specific activities at 20°C at substrate saturation (for the dyes and CO_2 extrapolated theoretical saturations). Exception: oxygen at saturation concentration. Units/mg: μmol formate or NADH/min per mg.

No.	Electron donator (product)	Electron acceptor (product)	units/mg	Remarks
1a	formate (CO_2)	NAD	42	"Standard reaction"
1b	formate	Cl_2-Ind, methylen blue, methyl viologen, benzyl viologen, nitroblue tetrazolium salt	≈ 20	not dependent on NAD(H)
1c	formate	oxygen (H_2O_2)	2	Not dependent on NAD(H)
2a	NADH	CO_2 (formate)	4	Reverse reaction of 1a. See legend
2b	NADH	see 1b	≈ 20	
2c	NADH	oxygen (H_2O_2)	2	

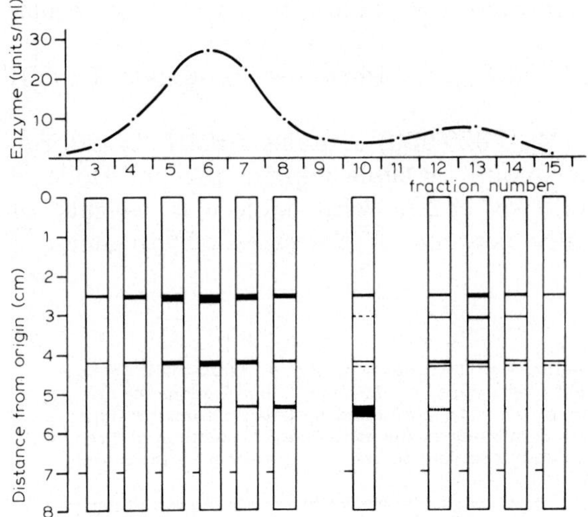

Fig. 3. Sodium dodecyl sulphate electrophoresis of sucrose density gradient fractions. 5% acrylamide, 0.2% dodecyl sulphate in 0.2 M sodium phosphate pH 7.0. Stained with coomassie blue. For further details see ref. 7.

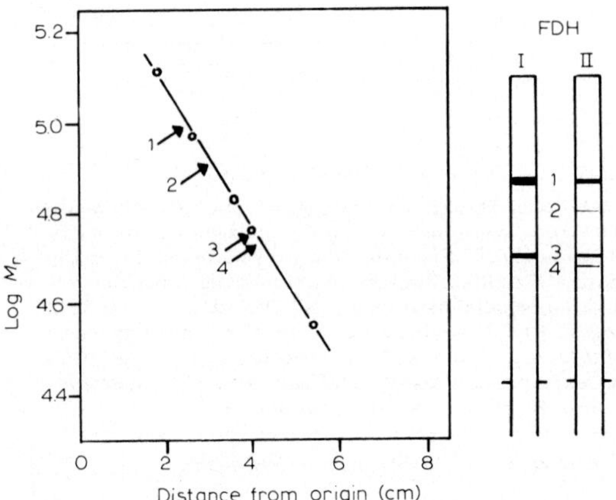

Fig. 4. Molecular weight determination of the subunits by sodium dodecyl sulphate electrophoresis. For conditions see legend to Fig. 3. Standard proteins [7]: pig muscle lactate dehydrogenase (M_r = 36 000), bovine liver catalase (58 000), bovine serum albumin (68 000), rabbit muscle phosphorylase *a* (93 000 [8]).

(2 mol/mol) can also be calculated from the data of the experiment described in Fig. 6 where a deflavo protein consisting of a formate dehydrogenase I/II mixture in a ratio visible in Figs. 2 and 3 was titrated with FMN.

Formate dehydrogenase II (M_r = 175 000) contains 1 FMN, 8–10 iron and about the same amount of labile sulphide per mol. Dodecyl sulphate electrophoresis shows the two chains found for formate dehydrogenase I and additionally two bands with much lower intensity (M_r = 83 000 and 56 000).

Formate dehydrogenases I and II always show multiple bands on disc-electrophoreses, but not in dodecyl sulphate electrophoreses. We often examined samples of the individual purification steps for the occurrence of formate dehydrogenases I and II by sucrose density gradient centrifugation. We sometimes observed that cell extracts after sonification contained no formate dehydrogenase II which appeared during the purification steps. That formate dehydrogenase I as isolated from sucrose gradient centrifugation may convert to II was already mentioned by us [1]. It has not been possible to reverse this conversion.

The turnover numbers of the two forms are in the range of 100 (mol/mol flavin per s, formate:NAD oxidoreductase reaction). We did not find significant differences in the catalytic abilities of the two forms. Apparently form I is a dimer of form II.

Preparation of deflavo formate dehydrogenase

The starting material for the deflavo formate dehydroganse preparation was a mixture of I and II as subjected to sucrose density gradient centrifugations.

The systematic and fundamental procedures developed by Kanda et al. [9] for the preparation of flavin-free metalloflavoproteins failed when applied to formate dehydrogenase as well as other usual methods. By chance we found the rather specific condition suited for formate dehydrogenase: a sample stored overnight anaerobically in the presence of 1.2 M ammonium sulphate, 5 mM sodium sulphide and 0.5 mM EDTA seemed to have lost its activity. Full activity appeared by testing it aerobically in the presence of FMN (note: nothing had been removed!) or after admission of a trace of air to the sample ("auto reactivation"). In this case the addition of FMN to the test cuvette was not necessary.

From this we deduced the deflavo enzyme preparation method: we dialyzed the enzyme for 10–20 h at 0°C against anaerobic histidine buffer, pH 5.6, containing 1.2 M ammonium sulphate, 5 mM sodium sulphide, 0.5 mM EDTA, and some charcoal. After this the "auto reactivation" after admission of air did not take place. We could not replace the sodium sulphide by dithiols.

The removal of the FMN proceeds very slowly. The "auto reactivation" apparently consists of a re-incorporation of the flavin after the admission of a trace of oxygen which has no measurable consequence on the oxidation-reduction potential of the solution. Free FMN is very slowly reduced under these conditions* and it is rapidly reoxidized even by that trace of oxygen. Therefore we consider that the enzyme looses reduced FMN and rapidly re-incorporates reoxidized FMN.

It would be expected that reduced substrates of formate dehydrogenase are more suitable reagents to achieve the flavin reduction than sodium sulphide. However, attempts to realize this have failed.

---

* See also De Sa, Chapter 79.

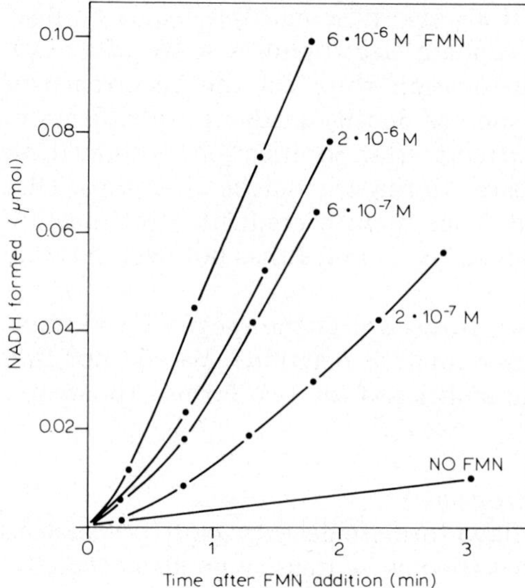

Fig. 5. Appearance of formate:NAD oxidoreductase activity upon addition of FMN (as indicated) to deflavo formate dehydrogenase. The deflavo protein contained residual traces of FMN.

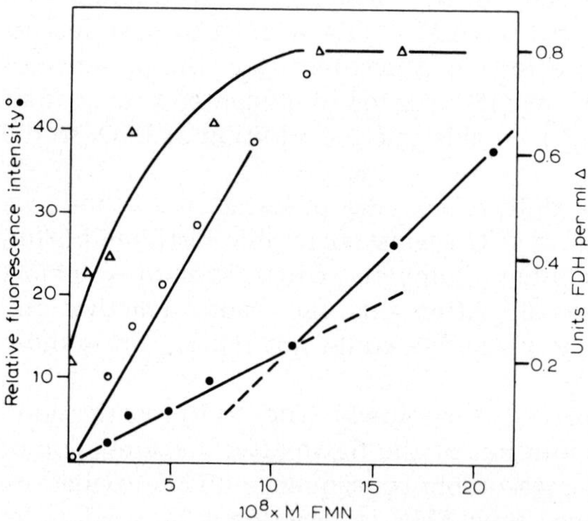

Fig. 6. Fluorescence titration of deflavo formate dehydrogenase with FMN. Behaviour of flavin fluorescence and formate:NAD oxidoreductase activity. In the fluorescence cuvette (2.0 ml, 0.05 M histidine buffer, pH 5.6, aerobic) 30 μg deflavo formate dehydrogenase (5.3 units/mg after saturation with FMN) were titrated. 5 min after each addition of FMN a sample was withdrawn and assayed for formate:NAD oxidoreductase activity. △—△, activity 5 min after addition of FMN as indicated; ●—●, relative fluorescence in the cuvette containing the protein; ○—○, fluorescence in a blank cuvette without protein. All values corrected for dilution. Notable is the speccific activity of the deflavo formate dehydrogenase used here: it is partly due to denaturated deflavo formate dehydrogenase and partly to the contaminating protein (see Figs. 2 and 3) which is colourless. The fluorescence visible below the saturation point is possibly due to FMN contaminations present in spite of the homogeneity in paper chromatography of the FMN (purchased from Boehringer, cf. Mayhew et al., Chapter 43).

Reactivation of deflavo formate dehydrogenase

The following results concern the formate:NAD oxido reductase reaction ("standard reaction"). The behaviour of the other activities is discussed later on.

Regaining of activity after addition of FMN is a rather slow process (Fig. 5). In contrast to this, the incorporation of FMN into the protein (observed by fluorescence titration) is a rapid process which is completely accomplished within the mixing time.

Fig. 6 shows a fluorescence titration of the deflavo enzyme followed by assays for activity. It is seen that full activity is gained when the protein is saturated with respect to FMN. This experiment showed best the difference in the velocities of the incorporation and the reactivation processes. While the final fluorescence value could be measured immediately after the flavin addition we had to wait for several minutes (especially after the addition of low flavin amounts) until the final enzyme activity was reached.

We conclude that the incorporation is only a first step of the whole reactivation process. To decide whether the subsequent steps involve a reoxidation of the enzyme corresponding to a reduction during the flavin removal procedure, we tried to influence the reactivation velocity by oxidizing or reducing conditions. There was no effect, but we observed a remarkable difference when we re-activated deflavo formate dehydrogenase by FMN or $FMNH_2$ under anaerobic conditions in the presence of the reactants of the standard reaction (Fig. 7). The reactivation proceeds more slowly and up to a lower final activity when $FMNH_2$ is added. This is reflected by the half maximal reaction velocities catalyzed by the deflavo enzyme in the presence of FMN or $FMNH_2$, respectively. The apparent values are $1 \cdot 10^{-7}$ M for FMN and $25 \cdot 10^{-7}$ M for $FMNH_2$. These values may argue for the high affinity of the deflavo protein to FMN and a lower one to $FMNH_2$.

Whether an observed enzyme catalyzed the reduction of free FMN, FAD and riboflavins by formate is due to an electron transfer from the reduced enzyme to the free flavin or to a displacement of enzyme bound reduced flavin by free flavin remains to be decided by an experiment with radioactive flavins.

Displacement of enzyme bound FAD and riboflavin by FMN

It can be easily demonstrated by fluorescence titration that the deflavo protein binds not only FMN but also FAD and riboflavin and in the same amounts. Displacement experiments show that the affinity of the protein to the flavins increases in the direction riboflavin → FAD → FMN. The experiments were followed by the increasing formate:NAD oxidoreductase activity which is peculiar only to the FMN-containing formate dehydrogenase.

When the deflavo protein is preincubated with FAD or riboflavin the activity due to subsequently added FMN is dependent on the ratio of the flavin concentrations. Fig. 8 shows this for a preincubation with FAD. Using riboflavin instead of FAD we observed higher activities for the same

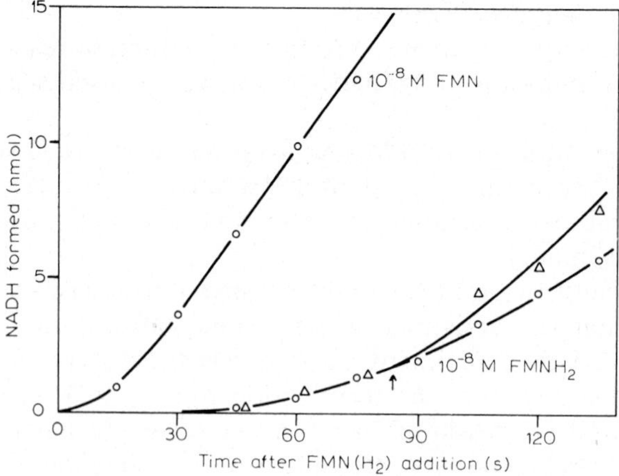

Fig. 7. Increase of formate: NAD oxidoreductase activity after addition of 8 μM FMN or $FMNH_2$ under anaerobic conditions. △—△, air admitted after 80 s.

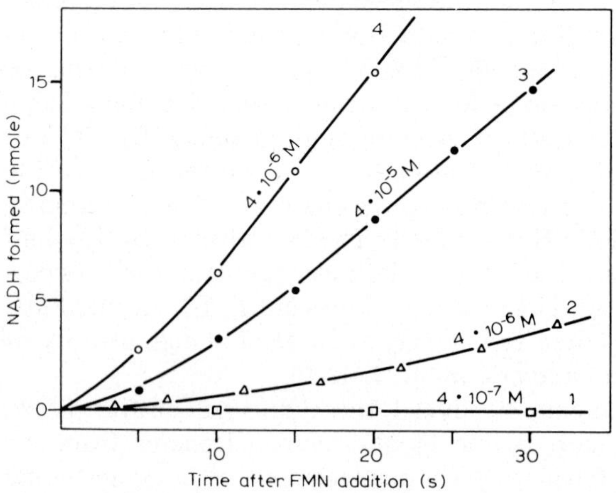

Fig. 8. Increase of formate: NAD oxidoreductase activity after FMN addition to the deflavo enzyme preincubated for 2 min with FAD as indicated. Curves 1–3 4 · 10^6 M FAD; curve 4, no preincubation with FAD, then FMN added as indicated.

FMN concentrations as in the Fig. 8 which were reached more rapidly than after preincubation with FAD.

The reverse experiment, i.e. the addition of FAD or riboflavin after preincubation with FMN leads essentially to no decrease of activity. Hence there is no displacement of the "right" by the "wrong" flavin.

Four enzyme species: FMN, deflavo, FAD and riboflavin formate dehydrogenase

With the incorporation of FAD and riboflavin into deflavo formate dehydroganse we gained two further species besides FMN and deflavo formate dehydrogenase. In this section we discuss some catalytic abilities of these four species. To facilitate the discussion we refer to the enumeration of the reactions as used in the Table II.

The specific demand of the formate:NAD oxidoreductase reaction 1a for FMN is unambiguous. Deflavo preparations are inactive with respect to this reaction and regain full activity after addition of FMN. A deflavo formate dehydrogenase is defined as a preparation inable of catalyzing the standard reaction 1a and capable of being fully reactivated by FMN with respect to the standard reaction.

Is FMN the linking site between NAD and the enzyme or between formate and the enzyme? To study this we turned to the reactions in which NAD(H) or formate are involved. In reaction 1b FMN behaves as in reaction 1a. A deflavo enzyme as defined above is inactive in reaction 1b as well as the FAD and riboflavin enzyme. It regains full activity after addition of FMN.

An intermediate result was achieved for reaction 1c. FMN formate dehydrogenase is active, deflavo formate dehydrogenase is inactive. But we observed a reactivation by FAD (up to 70% of the reaction velocity obtained with the FMN enzyme) and riboflavin (up to 30%).

Examining the reactions of the group 2 we found full activity of all four species for reaction 2b. Reaction 2c is not catalyzed by the deflavo formate dehydrogenase while FAD and riboflavin reactivated the enzyme up to 50% of the activity of the FMN enzyme. Hence the two oxidase reactions 1c and 2c are dependent on the presence of a flavin, regardless of the nature of the electron donator. While FMN is specifically necessary for the standard reaction this specificity is not peculiar to the oxidase reactions. Reaction 2a has not been examined.

The results show that the interaction of the enzyme with NAD(H) as well as with formate is dependent on the presence of the flavin.

We mention a result which is not in accordance with the conclusion drawn above: formate dehydrogenase catalyzes the oxidation of formate (specific activity 42 units/mg) and NADH (60 units/mg) by ferricyanide. We were surprised when we found that the deflavo enzyme as well as the FAD and the riboflavin enzyme catalyze these reactions with unchanged specific activity. The interaction of the deflavo enzyme at least with NAD(H) disagrees with the common experiences. The suspicion that a residual flavin content is responsible for the NAD ferricyanide reaction cannot be rejected substantially at the present stage of this work.

Acknowledgement

This work was supported by grants Ho 441-447 of the Deutsche Forschungsgemeinschaft. Ms. A. Trautwein and Ms. R. Reitberger are thanked for their skilful technical assistance.

References

1 Höpner, Th. and Trautwein, A. (1972) Z. Naturforsch. 27b, 1075–1076
2 Höpner, Th., Ruschig, U. and Müller, U. (1973) Hoppe-Seyler's Z. Physiol. Chem. 354, 216

3 Höpner, Th. and Knappe, J. (1974) Methoden der enzymatischen Analyse (Bergmeyer, H.U., ed.), pp. 1596–1600, Verlag Chemie, Weinheim
4 Höpner, Th. and Trautwein, A. (1971) Arch. Microbiol. 77, 26–35
5 Fogo, J.K. and Popowsky, N. (1949) Anal. Chem. 21, 732–734
6 Maurer, H.R. (1968) Disk-Elektrophorese, pp. 42, 43, W. d. Gruyter, Berlin
7 Weber, K., Pringle, J.R. and Osborn, M. (1972) Methods in Enzymology (Hirs, C.H.W. and Timascheff, S.N., eds.), Vol. 26, pp. 3–27 (especially pp. 6–11)
8 Seery, V.L., Fischer, E.H. and Teller, D.C. (1967) Biochemistry 6, 3315–3327
9 Kanda, M., Brady, F.O., Rajagopalan, K.V. and Handler, P. (1971) J. Biol. Chem. 247, 765–770
10 Ruschig, U. (1974) Dissertation, Universität Heidelberg

Flavins and Flavoproteins, edited by T.P. Singer

CHAPTER 65

RAT LIVER MITOCHONDRIAL MONOAMINE OXIDASE — AN IRON-REQUIRING FLAVOPROTEIN

M.B.H. YOUDIM

MRC Unit and University Department of Clinical Pharmacology, Radcliffe Infirmary, Woodstock Road, Oxford OX2 6HE (U.K.)

Summary

The purified enzyme has a spectrum characteristic of flavoprotein and there is a maximum reduction of absorption at 460 nm when either substrate (benzylamine) or sodium dithionite is added under anaerobic conditions. The native enzyme gives a fluorescence-emission peak at 520 nm when activated at 375 or 450 nm, which is also evidence for the presence of a flavin group. There is 1 mol of FAD per 120 000 of monoamine oxidase protein. Monoamine oxidase consists of two subunits of similar molecular weight (69 500). From the values of flavin content it would appear that only one of the subunits possesses a catalytic site. A study has been made of the binding of the active site-directed inhibitors [^{14}C]phenylethylhydrazine and [^{14}C]deprenil to native rat liver monoamine oxidase, its multiple forms and subunits. Inhibition of enzyme activity was time dependent and was accompanied by irreversible binding of the inhibitors to the enzyme. When fully inhibited, 1 mol of each inhibitor is bound per mol of enzyme in all cases and evidence suggests that the preparation consists of two types of monoamine oxidase. Recent studies indicated that [^{14}C]phenylethylhydrazine inhibits the enzyme by forming a stable adduct with the flavin component. The isolated [^{14}C]phenylethylhydrazine adduct has spectral properties resembling those of a flavin and was still associated with an amino acid or a peptide as shown by the presence of ninhydrin-positive material. Evidence for the functional role of iron in monoamine oxidase activity comes from studies using enzyme prepared from livers of rats made nutritionally deficient in iron. A significant decrease in the oxidase activity was observed in iron deficiency. These results would suggest that the metal may have some role as a prosthetic group or that iron is involved in the activity of an enzyme responsible for the incorporation of flavin into the apoprotein. Alternatively, iron may be necessary for the synthesis of monoamine oxidase apoprotein.

Introduction

Information regarding the nature and the function of cofactors in the monoamine oxidase (EC 1.4.3.4., amine:oxygen oxidoreductase (deaminating) (flavin-containing)) reaction; $R—CH_2—NH_2 \rightarrow R—CHO + H_2O_2$ is necessary if we are to understand the role of this ubiquitous enzyme in the metabolism of monoamine neurotransmitters (see ref. 1 for review).

The role of flavin in the monoamine oxidase reaction was first envisaged by Richter [2] because of the similarity of the oxidative reaction of this enzyme to D-amino acid oxidase, a known flavin adenine dinucleotide (FAD)-containing enzyme. This hypothesis was strengthened when Hawkins [3] and others [4–8] showed that rats fed a diet nutritionally deficient in riboflavin have significantly lowered liver monoamine oxidase activity.

Information regarding riboflavin requirements of monoamine oxidase was slow to develop because of the strong association of this enzyme with mitochondria [9, 10] and difficulties encountered in solubilizing and purifying it. Absorption spectra of highly purified monoamine oxidase [11–17] indicated the presence of a flavin.

Recently, the presence of flavin identified as covalently bound FAD in purified enzyme from beef liver [18–20], bovine kidney cortex [13, 21], rat liver [22–24] and pig liver [16, 17] have been confirmed. An exception is pig brain monoamine oxidase [25], whose FAD was reported to be non-covalently bound and could be removed relatively easily.

It has been suspected for some time that a metal may also be involved in the action of monoamine oxidase (see refs. 1, 26 and 27 for review) and thus there has been some interest in determining whether the enzyme contains a heavy metal. Gorkin [28] originally reported that monoamine oxidase could be inhibited reversibly with various chelating agents, and this has been confirmed by others using a variety of chelating agents [1, 7, 12, 13, 22, 29].

The original evidence presented by Gomes et al. [12] that the enzyme contains copper and is therefore a copper-containing protein has not received much support, for purified preparations of monoamine oxidase contain very little of the metal [11, 13, 17, 22, 25, 30] and copper deficiency developed in rats does not affect enzyme activity [7, 31]. However, iron has been reported to be present in rat liver [1] and pig liver monoamine oxidase preparations [17]. Support for the role of iron in full functional activity of monoamine oxidase has been derived from studies on iron deficiency in rats, for liver monoamine oxidase has been shown to have significantly lower activity [31] and iron deficiency leads to a great decrease in the rate at which the animals metabolise [^{14}C]pentylamine to $^{14}CO_2$ [8].

This paper concerns the nature and role of flavin and iron in the activity of monoamine oxidase.

Flavin content of monoamine oxidase

The absorption spectrum of purified rat liver monoamine oxidase possesses a shoulder at about 460 nm. The preparation also emits a fluorescence with a peak at 520 nm when activated at 375 or 460 nm, a characteristic of a flavin compound being present [7, 11]. Recent studies have established rat liver mitochondrial monoamine oxidase is an FAD-containing enzyme [7, 11, 22, 23, 24]. Present evidence suggests that there are 2 mol of FAD per 290 000 g enzyme, or 1 mol per minimum molecular weight of 150 000 [23, 24]. This is in close agreement with the finding of others for monoamine oxidase from different sources [13, 17, 18, 19, 21, 25].

In rat liver [11], the flavin is covalently bound to the protein, because treatment of the enzyme with trichloroacetic acid released only a small portion of the flavin. Only when the enzyme was digested with pronase for 4 h at 37°C (in the dark) was a substantial amount of flavin released. This observation is very similar to that of succinic dehydrogenase [32]. The flavin from the enzyme, when subjected to thin-layer chromatography, was still associated with an amino acid or a peptide, as shown by the presence of ninhydrin-positive material associated with the extracted flavin, but not with standard FAD or FMN. Similar conclusions have been arrived at for beef liver monoamine oxidase [19, 20, 33], bovine kidney cortex monoamine oxidase [13, 21] and enzyme derived from pig liver [17, 34].

Tipton [25] has made a study of pig brain mitochondrial monoamine oxidase and reported the presence of 1 mol of FAD per 120 000 g of protein. The spectrum properties of this enzyme was similar to those described by others. A fully inactive, FAD-free apoenzyme has been obtained by careful treatment with acid and reported to be partly reactivated by FAD. With some preparations of the enzyme it has been found difficult to reduce the activity below 15% of the original. The reason for these differences was not discussed.

These observations raise the question of whether in different locations the mode of attachment of the flavin moiety to the enzyme protein is different and whether or not the reactivated enzyme has similar properties to the native, undissociated enzyme.

Further evidence that monoamine oxidase is a flavin containing a dependent enzyme is provided in our studies on the effect of galactoflavin, an antimetabolite of riboflavin, on the activity of the rat brain and liver enzyme. Rats fed a riboflavin-deficient diet are known to have significantly decreased liver monoamine oxidase activity (see Introduction). In the recent experiments, rats given 2 mg/kg galactoflavin in their daily diet for 42 days showed a progressive decrease in monoamine oxidase activity. Within 14 days, liver monoamine oxidase activity declined by 33% but that of brain was unchanged. However, 42 days of treatment resulted in 80% and 31% decreases in liver- and brain-enzyme [24] activity. The analogue did not completely eliminate the activity so that the results obtained are not significantly different from simple riboflavin deficiency [7]. One explanation may be that galactoflavin could not deplete all the stores of flavin. In the light of the present discussion on the mode of attachment of flavin, it is interesting to note that galactoflavin had significantly greater inhibitory effect on the activity of the liver than of the brain enzyme. This differential effect may indicate that brain and liver monoamine oxidases are different or that flavin has limited access to the brain. Neither FAD nor FMN at concentrations as high 1 mM reactivated depressed enzyme activity in vitro. These results suggest that either monoamine oxidase apoenzyme synthesis has been lost owing to the deficiency of FAD or that the flavin coenzyme of monoamine oxidase must be covalently linked to the enzyme protein. Thus, there is a considerable amount of experimental evidence in favour of monoamine oxidase possessing a covalently linked flavin. The above experiments on liver and brain raise the question whether pig brain enzyme [25] also contains a covalently linked flavin compound of this kind. It would appear that the results to the contrary described by Tipton [25] may represent an artifact of the method used to solubilize the enzyme, considering that in order to release monoamine oxidase from the outer mitochondrial membrane of pig brain into a soluble form, Tipton [25] used a total of 5 h of ultrasonic disintegration at maximum output and treatment with acid. It seems conceivable

TABLE I

THE BINDING OF [^{14}C]PHENYLETHYLHYDRAZINE AND [^{14}C]DEPRENIL TO NATIVE MONOAMINE OXIDASE AND ISOLATED FLAVIN-PEPTIDES

Preparation	Monoamine oxidase protein (μg)	Total counts	^{14}C-labelled inhibitor (nmol)	mol ^{14}C-labelled inhibitor bound per 150 000 g monoamine oxidase protein
[^{14}C]Phenylethylhydrazine-treated monoamine oxidase	1800	$2.77 \cdot 10^4$	13.0 ± 0.58*	1.18
Flavin peptide isolated from [^{14}C]phenylethylhydrazine-treated monoamine oxidase	1800	$2.47 \cdot 10^4$	11.6 ± 1.10**	0.97
[^{14}C]Deprinil-treated monoamine oxidase	1800	$4.80 \cdot 10^3$	13.4 ± 0.37*	1.12
Flavin peptide isolated from [^{14}C]deprenil-treated monoamine oxidase	1800	$4.65 \cdot 10^3$	12.8 ± 0.81**	1.06

* For these studies, [^{14}C]phenylethylhydrazine and [^{14}C]deprenil were diluted with 0.05 M phosphate buffer pH 7.4 and 1 ml added to 4 ml enzyme preparation (spec. act. 5875, containing 1.8 mg of protein/ml) to give a final concentration of 100 μM [^{14}C]phenylethylhydrazine and 50 μM [^{14}C]deprenil. After incubation at 37°C for 40 min, samples were tested to make sure that the enzyme activity was fully inhibited, using kynuramine and phenylethylamine as substrates. The excess inhibitor was removed as described previously [43]. The inhibited enzyme preparation was precipitated with 1 ml 6% perchloric acid and after centrifugation and washing with 1 ml 1% trichloracetic acid the precipitate was dissolved in 1 ml of 1 M NaOH and radioactivity counted.

** Aliquots of native (untreated), [^{14}C]phenylethylhydrazine- and [^{14}C]deprenil-treated purified monoamine oxidase as described above were freed of excess inhibitors. The trichloroacetic acid-washed precipitate was suspended in 0.1 M phosphate buffer pH 7.4 containing 1.0 mg of pronase per ml and the mixture was incubated at 37°C (in the dark) with continuous shaking for 4 h. The flavin was isolated and purified as described previously [24] and the aliquot counted for its radioactivity. The radioactivity could not be resolved from the flavin by thin-layer chromatography [24].

that the free FAD was obtained by a cleavage of a covalent bond. It is of interest to note that the recent studies of Salach et al. [35] involving the preparation of purified mitochondrial monoamine oxidase from pig and rat brain suggests that for the most part brain monoamine oxidase also contains covalently linked flavin.

The mode of attachment of covalently bound flavin moiety to the beef liver enzyme protein has recently been elucidated by Kearney et al. [19] and by Walker et al. [20]. The pentapeptide riboflavin 5-phosphate of hepatic monoamine oxidase isolated contains 1 mol each of serine, tyrosine and cystein and 2 mol of glycine. The amino acid sequence being Ser—Gly—Gly—Cys—Tyr and the flavin is attached via the 8α-carbon of riboflavin as a thioether to the cysteine. In the experiments of Tipton [25] already described, it seems conceivable that the free FAD was obtained by a cleavage of a covalent bond at the level of thio-ether, but the reactivation of the apoenzyme by added FAD remains to be accounted for in the light of negative results obtained with the attempted reactivation of monoamine oxidase prepared from riboflavin-deficient liver and brain in vitro [24].

Binding of [^{14}C]phenylethylhydrazine and [^{14}C]deprenil to monoamine oxidase

Recent studies have shown that monoamine oxidase from both rat [36] and pig liver [37] consists of at least two subunits of similar molecular weight (about 70 000). From the published values of the flavin content of rat liver monoamine oxidase [24] it is apparent that only one of the subunits possesses a catalytic site [36, 37]. These earlier results have led us [43] to study the mechanism by which the active site-directed inhibitors phenylethylhydrazine and deprenil (phenylisopropylmethylpropinylamine hydrochloride) inhibit monoamine oxidase (see refs. 29, 38, and 40). The inhibition of enzyme activity by [^{14}C]phenylethylhydrazine was time dependent and hyperbolic; approx. 40 min incubation at 37°C being necessary for 95% inhibition at a drug concentration of 5 μM (Fig. 1). Similar results have been observed with 0.5 μM deprenil (Fig. 2). Furthermore, the inhibition was a function of the amount of ^{14}C-labelled inhibitors bound to the enzyme (Figs. 1 and 2). Presumably binding occurs at the active site, since the presence of the 1 mM substrate benzylamine or 0.5 mM kynuramine blocks not only the inhibitory action of [^{14}C]phenylethylhydrazine and [^{14}C]-deprenil but also their binding (Fig. 3) [43]. Non-specific binding of [^{14}C]-phenylethylhydrazine and [^{14}C]deprenil may be occurring; the addition of a large molar excess of the cold inhibitor followed by dialysis or gel filtration on Sephadex G-25 presumably displaces non-covalently bound ^{14}C-labelled inhibitors from the sites of attachments (Fig. 1). However, the amount of inhibitor bound is approximately the same whether the excess is removed by dialysis or by gel filtration through Sephadex G-25. Since the molecular weight of rat liver monoamine oxidase is 150 000 [11, 36], then each mol of enzyme binds 1.18 mol of [^{14}C]phenylethylhydrazine or 1.12 mol of

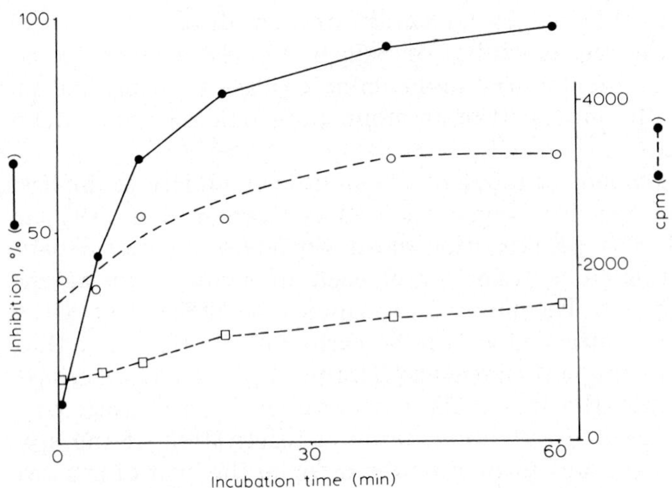

Fig. 1. The binding of [^{14}C]phenylethylhydrazine to purified rat liver mitochondrial monoamine oxidase. Purified enzyme (5 mg/ml) was incubated with 5 μM [^{14}C]phenylethylhydrazine final concentration (spec. act. 1.66 Ci/mol) and the residual activity (●——●) estimated using kynuramine as substrate. The binding of the radioactive inhibitor was estimated by dialysis of samples against excess 0.05 M phosphate buffer (pH 7.4) alone (○– – – –○) or against the same buffer containing 100 μM phenylethylhydrazine (□– – – –□) until the counts associated with the enzyme protein reached a constant low value. The non-covalent bound ^{14}C-labelled inhibitor could also be removed by the method described in footnote to Fig. 3. The radioactive material was prepared for counting as described earlier [43].

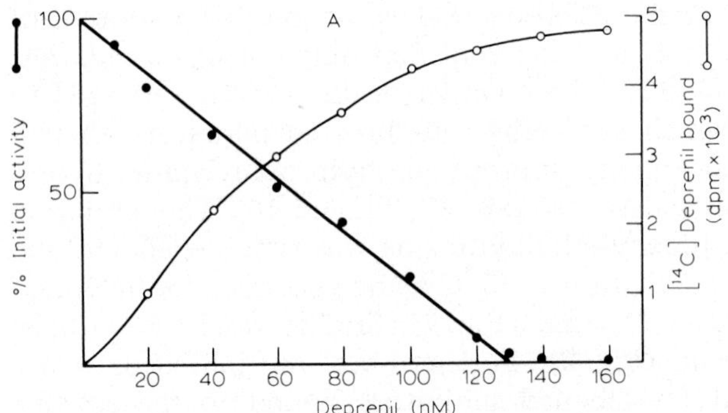

Fig. 2. Irreversible inhibition and binding of rat liver mitochondrial monoamine oxidase with deprenil. Percentage initial enzyme activity towards phenylethylamine and binding of [^{14}C]deprenil against increasing concentrations of inhibitor. Preincubation was carried out for 60 min until no further inhibition occurred at 25°C in 0.05 M phosphate buffer pH 7.4, in a total volume of 1 ml. Activity was determined after incubation by the addition of 100 μl of 1 mM substrate [47]. For the binding studies, enzyme preparations were treated with increasing concentrations of [^{14}C]deprenil (2.15 Ci/mol) for 60 min under the conditions described. The excess inhibitor was removed by precipitation of protein with 1 ml 6% perchloric acid, followed by centrifugation. The precipitate was washed twice with 1% trichloroacetic acid and finally dissolved in 1 ml 0.1 M NaOH and an aliquot was counted for its radioactivity.

[^{14}C]deprenil (Table I). These results are in close agreement with those reported for binding of [^{14}C]pargyline [21, 41] and [^{14}C]phenylhydrazine [42] with monoamine oxidase from pig liver [37] and bovine kidney cortex [21].

Previous work [36] has shown that when purified enzyme is treated with 1% sodium dodecyl sulphate and then subjected to dodecyl sulphate-polyacrylamide gel electrophoresis in the presence of 0.1 M 2-mercaptoethanol, the enzyme dissociates to give a single band of protein having a molecular weight of approx. 70 000. When [^{14}C]phenylethylhydrazine-treated monoamine oxidase is submitted to dodecyl sulphate-polyacrylamide gel electrophoresis in gels containing 0.1 M 2-mercaptoethanol, a single protein band having a molecular weight of 69 500 migrating from cathode to anode was visible [43]. This would suggest that each molecule of monoamine oxidase consists of at least 2 subunits of equal molecular weight. On the basis of these results and of the reported flavin content and the [^{14}C]phenylethylhydrazine- and [^{14}C]deprenil-binding capacity of the enzyme it is concluded that only one of the subunits should possess an active site. However, if the ^{14}C-labelled inhibitor-treated enzyme was subjected to dodecyl sulphate-polyacrylamide gel electrophoresis in the absence of 2-mercaptoethanol, two bands of protein were visible (Fig. 4) with molecuar weights of (A) 72 500 and (B) 75 000. Both protein bands bind the inhibitor [^{14}C]-phenylethylhydrazine to equal extents (approx. 1 mol inhibitor per 1 mol enzyme) and must both possess active sites. Moreover, the amount of inhibitor bound is similar to that bound to the native enzyme (approx. 1 mol

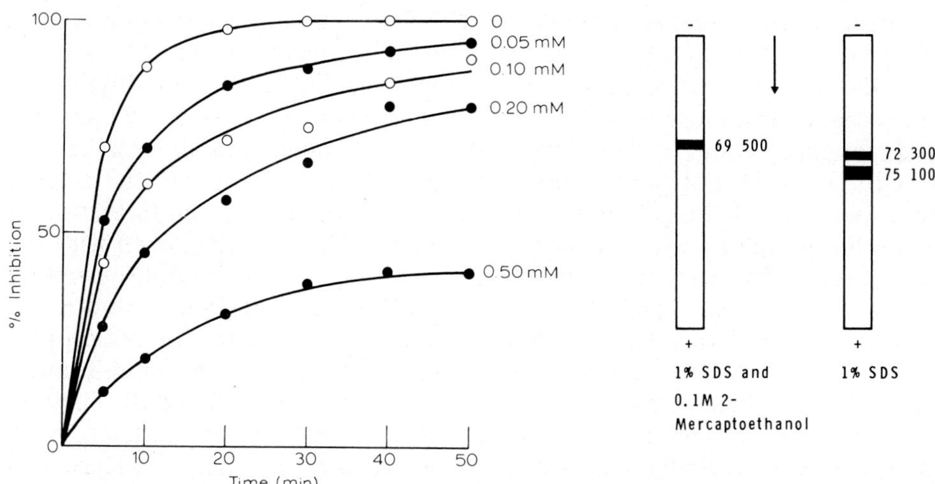

Fig. 3. The effect of various concentrations of substrate on the inhibition of monoamine oxidase by deprenil. Purified rat liver monoamine oxidase (spec. act. 5600) preparations in 0.05 M phosphate buffer pH 7.4 containing 10 μM deprenil and kynuramine, 0, 0.05, 0.10, 0.20 and 0.50 mM in a total volume of 1 ml were incubated at 37°C for various periods indicated before termination of activity.

Fig. 4. The preparation and binding of [^{14}C]phenylethylhydrazine to the subunits of purified monoamine oxidase. Monoamine oxidase with specific activity of 5665 was incubated with [^{14}C]phenylethylhydrazine (10 μM) for 60 min at 37°C. The excess of noncovalently bound [^{14}C]phenylethylhydrazine was removed by gel filtration on Sephadex G-25 equiliberated with 0.05 M phosphate buffer (pH 7.4) [43]. Sodium dodecyl sulphate (SDS) was added to the samples to give a final concentration of 1% (w/v) and in other samples, 0.1 M 2-mercaptoethanol was also present. After incubation at 37°C for 2 h, polyacrylamide gel electrophoresis was carried out in 5% gel containing 1% (w/v) dodecyl sulphate [36, 43]. The molecular weights of the bands of protein separated were determined [36]. The proteins extracted from the gels [36] were passed through a Sephadex G-25 column (2.5 × 10 cm) and the radioactivity and the protein contents of the eluates determined [43].

[^{14}C]phenylethylhydrazine per 150 000 g monoamine oxidase protein) [43]. Oreland et al. [37], who measured the binding of [^{14}C]pargyline, a non-hydrazine inhibitor, have reported very similar results and have suggested that the two protein bands separated are not artifacts [37] so that one can only conclude that each represents a different monoamine oxidase [44–46].

Interaction of [^{14}C]phenylethylhydrazine and [^{14}C]deprenil with flavin component of monoamine oxidase

It has been established that either 1 mol of [^{14}C]phenylethylhydrazine or 1 mol of [^{14}C]deprenil binds with 1 equiv. monoamine oxidase. In view of the apparent involvement of the oxidase flavin component when [^{14}C]pargyline or [^{14}C]phenylhydrazine was employed [21, 37, 41, 42], it was pertinent to examine the possible binding of [^{14}C]phenylethylhydrazine and [^{14}C]deprenil to the flavin component. Purified rat liver monoamine oxidase with a specific activity of 5875 was prepared [11] and fully inactivated with [^{14}C]phenylethylhydrazine (100 μM) or with 50 μM [^{14}C]deprenil to give 1.0 mol each of ^{14}C-labelled inhibitor bound per 150 000 g monoamine oxidase protein [47]. The flavin component, which is covalently bound [24] was isolated by subjecting the enzyme to pronase proteolysis, thin-layer chromatography and Sephadex G-25 gel filtration [24, 43]. The yellow pigment obtained contained more than 90% of the ^{14}C label originally bound to the protein (Table I). The radioactivity could not be resolved from the flavin peptide. The absorption spectrum properties of [^{14}C]phenylethylhydrazine aduct resembled those of a flavin although bleached (see refs 21, 24 and 42) and retained about 1.0 mol residues of [^{14}C]phenylethylhydrazine and [^{14}C]deprenil bound per 150 000 g protein (Table I). However, the absorption spectrum of the [^{14}C]deprenil adduct isolated showed the formation of a new chromophore which absorbed at 403 nm. These results, taken with the alteration of the enzyme spectrum shown in Fig. 5 and those discussed by Patek and Hellerman [42] and others [21, 59], suggest that inhibition of monoamine oxidase by hydrazines and deprenil involve the formation of a stable, "reduced-type", flavin-inhibitor adduct. Moreover, the inhibition by phenylethylhydrazine and deprenil is stoichiometric (Figs. 2 and 3). Furthermore, titration of enzyme was valid for preparations of homogenates, mitochondrial and partially purified enzymes of various purity, because end points that relate to enzyme activity and K_m values of the initial rate of irreversible loss of activity were found to be the same, regardless of enzyme homogeneity [47].

Relationship of metal content to enzyme activity

Metal-binding agents are known to inhibit the mitochondrial monoamine oxidase [1, 12, 13, 16, 28, 31]. These observations suggest the presence of metal as a cofactor. Nevertheless, the question of the involvement of a metal in monoamine oxidase has not been satisfactorily resolved.

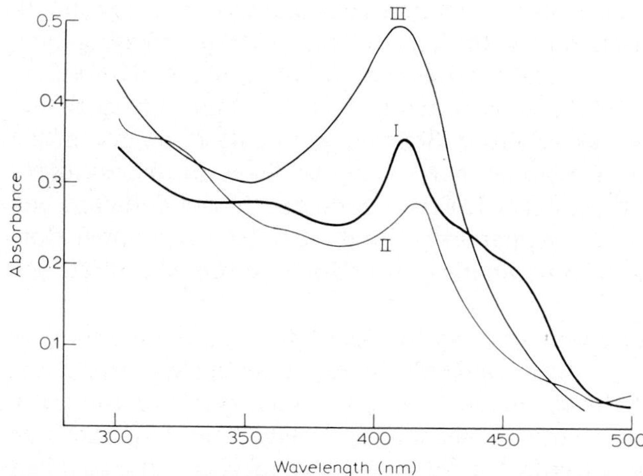

Fig. 5. Absorption spectra of purified rat liver mitochondrial monoamine oxidase before and after inactivation by phenylethylhydrazine and deprenil. Native monoamine oxidase (I) (1.35 mg/ml) in 0.05 M phosphate buffer (pH 7.4) was treated with either 1 mM phenylethylhydrazine (II) or 0.5 mM deprenil (III) for 60 min at 25°C before spectra were recorded.

The iron content of liver enzyme preparation was measured [48] at each step of purification: there was an increase in this value proportional to the increase in specific activity of monoamine oxidase. On the basis of the estimated minimal molecular weight of rat liver monoamine oxidase being 150 000 and the iron content 0.12%, the enzyme would contain about 2–3 atoms of iron per mol [11]. Comparable values have been reported for spermidine dehydrogenase of *Serratia marascens* [49] and for pig liver monoamine oxidase [17]. The analysed copper content would on the same basis be equivalent to less than 1 g atom of copper per 450 000 g of enzyme [11]. Similar results have been obtained for the enzyme of beef kidney cortex [13], pig brain [25] *Sarcina lulea* [30] and pig liver [17]. Copper concentrations of this order are very much lower than those reported for copper and pyridoxal containing soluble amine oxidases [50]. According to Nara et al. [12] and Nara and Yasunobu [51], the enzyme prepared from beef liver mitochondria is a copper-containing protein. However, the same authors have recently described the preparation of purified enzyme with much lower copper content than originally reported [33], indicating that copper is not associated with the enzyme.

In studies with metal-binding agents, a greater degree of inhibition was observed with the chelators which are known to chelate iron [1, 22, 31, 52, 53].

Evidence that iron is implicated in the activity of monoamine oxidase comes from studies on iron-deficient rats [8, 31] and iron-deficiency anaemia in man [54, 55]. Thus, iron deficiency causes a significant decrease in the activity of hepatic rat liver monoamine oxidase [31], and human platelet monoamine oxidase [54, 55] in vitro as well as in the initial rate of oxidation of *n*-[^{14}C]pentylamine to $^{14}CO_2$ in vivo [8]. The iron dependency of this phenomenon has been demonstrated for animals and humans by supplementing the diet of iron-deficient rats with ferric citrate [8] and treating iron-deficient human subjects with oral ferrous sulfate [55]. Animals attained the same rate of amine metabolism as controls by day 6 and in human subjects between 2 and 4 weeks. Copper deficiency in rats does not affect enzyme activity [31]. The nutritional lack of copper leads to deposition of excess iron in the liver of the rat [56]; this is reflected in the deficient biosynthesis of haemoglobin [57]. Apparently the dietary lack of copper does not affect the utilisation of iron in whatever function is subserved in relation to monoamine oxidase.

The nutritional requirement of iron in regard to monoamine oxidase activity is thus clear but not fully understood mechanistically. Iron could function in the binding of the enzyme at the outer membrane of the mitochondrion, although recent evidence seems to exclude this [8]. Another possibility is that it serves as a cofactor for an enzyme affecting the covalent linking of the flavin moiety the peptide chain [13, 17, 19, 20, 24]. It is conceivable that iron is actually a part of the monoamine oxidase molecule in the rat liver and human platelet, bound in some way by the enzyme and accessible to some, but not all, iron-binding chelating agents. Finally, iron

TABLE II

SOME PROPERTIES OF HUMAN PLATELET MONOAMINE OXIDASE FROM NORMAL AND IRON-DEFICIENT SUBJECTS

A. Specific activity is defined as nmol of deaminated product formed per mg protein in 30 min incubation at 37°C. The K_m is given at μM and was calculated from double-reciprocal plots. The figures in parenthesis are the maximum velocity relative to that of dopamine, which was taken as 100. B. Heat inactivations were performed in 0.05 M phosphate buffer pH 7.4 for 40 min and the results are expressed as percentage loss in activity compared to unheated preparation for particular substrate. C. Human platelet preparations in 0.05 M phosphate buffer (pH 7.4) were incubated at 37°C for 30 min with deprenil at final concentrations indicated until no further inhibition occurred. Enzyme activity was determined using kynuramine at substrate and the results are expressed as percentage inhibition.

	Normal	Iron deficient
A. Specific Activity		
Dopamine	5.20 ± 0.71 (100)	2.40 ± 0.55 (100)
K_m	100	85
5-Hydroxytryptamine	4.90 ± 0.41 (94)	2.75 ± 0.38 (114)
K_m	200	175
Phenylehtylamine	3.87 ± 0.6 (74)	2.46 ± 0.50 (100)
K_m	90	124
Kynuramine	6.80 ± 0.37 (130)	2.46 ± 0.50 (103)
K_m	48	53
B. Heat Inactivation at 45°C		
Dopamine	10	27
5-Hydroxytryptamine	17	40
Phenylethylamine	35	64
Tyramine	8	62
Kynuramine	32	81
C. Inactivation by Deprenil (μmol/l)		
0.05	40	90
0.10	58	97
1.00	70	99
10.00	85	100
100.00	95	100

may be necessary at some stage in the biosynthesis of the apoenzyme. Recent studies on platelet monoamine oxidase from iron-deficient subjects indicate that this may indeed be so [58]. It is apparent that although the K_m of platelet monoamine oxidase from iron-deficient subjects is similar to the K_m of the enzyme from normal subjects using the substrates dopamine, 5-hydroxytryptamine and kynuramine, the maximum velocity is significantly reduced (Table II). However, iron-deficient platelet monoamine oxidase is much more sensitive to heat inactivation and the action of the irreversible inhibitors clorgyline and deprenil, which bind stoichiometrically to the enzyme via the flavin cofactor. Thus, platelets from iron-deficient human subjects bind significantly less [^{14}C]deprenil per mg platelet protein than platelets from normal subjects. These results may imply that iron is necessary for the biosynthesis of monoamine oxidase protein (Table II) [58].

Conclusions

From the present and previous studies it is apparent that rat liver mitochondria monoamine oxidase contains a covalently bound FAD as a cofactor. Iron which is present in purified preparations has been implicated to play some role in its activity but its function as a possible second cofactor is obscure. This enzyme possesses 8 mol of sulfhydryl groups per g equivalent [1] which are capable of functioning reversibly in an oxidation reduction process [13]. Sulfhydryl groups appear to be essential for enzyme activity. However, contrary results have been reported for the enzyme from beef liver [59]. The role of iron and sulfhydryl groups in the activity and biological function of monoamine oxidase must await further clarification.

References

1 Youdim, M.B.H. (1975) in MTP International Review of Science (Blaschko, H., ed.), Vol. 5, pp. 169–211 Butterworth Ltd., London
2 Richter, D. (1937) Biochem. J. 31, 2022
3 Hawkins, J. (1952) Biochem. J. 51, 399
4 Wiseman-Distler, M.H. and Sourkes, T.L. (1961) Biochem. J. 78, 123
5 Wiseman-Distler, M.H. and Sourkes, T.L. (1963) Can. J. Biochem. Physiol. 41, 57
6 Sourkes, T.L. (1958) Rev. Cand. de Biol. 17, 328
7 Youdim, M.B.H. and Sourkes, T.L. (1965) Can. J. Biochem. 43, 1305
8 Symes, A.L., Missala, K. and Sourkes, T.L. (1971) Science 174, 153
9 Hawkins, J. (1952) Biochem. J. 50, 577
10 Schnaitman, C., Erwin, V.G. and Greenawalt, J.W. (1967) J. Cell. Biol. 34, 719
11 Youdim, M.B.H. and Sourkes, T.L. (1966) Can. J. Biochem. 44, 1397
12 Nara, S., Gomes, S.B. and Yasanobu, K.T. (1966) J. Biol. Chem. 241, 2774
13 Erwin, V.G. and Hellerman, L. (1967) J. Biol. Chem. 242, 4230
14 Tipton, K.F. (1958) Eur. J. Biochem. 4, 103
15 Murali, D.K. and Radha-Krishnan, A.M. (1970) Biochem. Biophys. Acta 206, 61
16 Hollunger, G. and Oreland, L. (1970) Arch. Biochem. Biophys. 139, 320
17 Oreland, L. (1971) Arch. Biochem. Biophys. 146, 410
18 Nara, S., Igaue, I., Gomes, B. and Yasanobu, K.T. (1966) Biochem. Biophys. Res. Commun. 23, 324
19 Kearney, E.B., Salach, J.I., Walker, W.H., Seng, R.L., Kenney, W., Zeszotek, E. and Singer, T.P. (1971) Eur. J. Biochem. 24, 321
20 Walker, W.H., Kearney, E.B., Seng, R.L. and Singer, T.P. (1971) Eur. J. Biochem. 24, 328
21 Chuang, H.Y.K., Patek, D.R. and Hellerman, L. (1974) J. Biol. Chem. 249, 2381
22 Sourkes, T.L. (1968) Adv. Pharmacol. 6A, 61

23 Youdim, M.B.H. and Sourkes, T.L. (1971) Biochem. J. 121, 20P
24 Youdim, M.B.H. and Sourkes, T.L. (1972) Adv. Biochem. Psychopharmacol. 5, 45
25 Tipton, K.F. (1968) Biochim. Biophys. Acta 159, 451
26 Tipton, K.F. (1975) in Handbook of Physiology (Blaschko, H., ed.), pp. 677–697 American Physiological Society Publishers, Washington D.C., in press
27 Lagnado, J. and Sourkes, T.L. (1956) Can. J. Biochem. Physiol. 34, 1095
28 Gorkin, V.Z. (1959) Biokhimiya 24, 758
29 Gorkin, V.Z. (1966) Pharmacol. Rev. 18, 115
30 Kumagai, H., Matsui, H., Ogata, H. and Yamada, H. (1969) Biochim. Biophys. Acta 171, 1
31 Symes, A.L., Sourkes, T.L., Youdim, M.B.H., Gregoriadis, G. and Birnbaum, H. (1969) Can. J. Biochem. 47, 999
32 Walker, W.H., Salach, J., Gutman, M., Singer, T.P., Hyde, J.S. and Ehrenberg, A. (1969) FEBS Lett. 5, 237
33 Yasanobu, K.T., Iagaue, I. and Gomes, B. (1968).Adv. Pharmacol. 6A, 43
34 Oreland, L., Kinemuchi, H. and Yoo, B.Y. (1973) Life Sci. 13. 1533
35 Salach, J.I. Minamiura, M., Yasunobu, K.T., Youdim, M.B.H., and Singer, T.P. (1976) Flavins and Flavoproteins, Proceedings of the 5th International Symposium (Singer, T.P., ed.), Chapt. 66, pp. 605—609, Elsevier, Amsterdam
36 Youdim, M.B.H. and Collins, G.G.S. (1971) Eur. J. Biochem. 18, 73
37 Oreland, L., Kinemuchi, H. and Stingbrand, T. (1973) Arch. Biochem. Biophys. 159, 854
38 Tipton, K.F. and Spires, I.P.C. (1971) Biochem. J. 125, 521
39 Tipton, K.F. (1972) Biochem. J. 128, 913
40 Clineschmidt, B.V. and Horita, A. (1969) Biochem. Pharmacol. 18, 1011
41 Hellerman, L. and Erwin, V.G. (1968) J. Biol. Chem. 243, 5234
42 Paket, D.R. and Hellerman, L. (1974) J. Biol. Chem. 249, 2373
43 Collins, G.G.S. and Youdim, M.B.H. (1975) Biochem. Pharmacol. 24, 703
44 Sandler, M. and Youdim, M.B.H. (1972) Pharmacol. Rev. 24, 331
45 Youdim, M.B.H. (1974) Adv. Biochem. Psychopharmacol. 10, 59
46 McCauley, R. and Racker, E. (1973) Mol. Cell Biochem. 1, 73
47 Youdim, M.B.H. (1976) Br. J. Pharmacol. 56, 375
48 Ramsay, W.N.M. (1954) Biochem. J. 54, xvii
49 Tabor, C.W. and Kellogg, P.D. (1971) J. Biol. Chem. 245, 5424
50 Blaschko, H. (1974) Rev. Physiol. Biochem. Pharmacol. 76, 84
51 Nara, S. and Yasanobu, K.T. (1965) in Biochemistry of Copper (Peisach, J., Aisen, P. and Blumberg, W.E., eds.), pp. 423–436, Academic Press, New York
52 Bernheim, F. (1940) J. Biol. Chem. 133, 485
53 Youdim, M.B.H. (1966) Ph.D. Thesis, McGill University
54 Callendar, S., Grahame-Smith, D.G., Woods, H.F. and Youdim, M.B.H. (1974) Brit. J. Pharmacol. 52, 447
55 Youdim, M.B.H., Woods, H.F., Mitchell, B., Grahame-Smith, D.G. and Callendar, S. (1975) Clin. Sci. Mol. Med. 48, 289
56 Elvehjem, C.A. and Sherman, W.C. (1932) J. Biol. Chem. 98, 309
57 Matrone, G. (1960) Fed. Proc. 19, 659
58 Youdim, M.B.H., Grahame-Smith, D.G. and Woods, H.F. (1976) Clin. Sci. Mol. Med., in the press
59 Gomes, B., Naguwa, G., Kloepfer, H.G. and Yasunobo, K.T. (1969) 132, 28

Flavins and Flavoproteins, edited by T.P. Singer

CHAPTER 66

IS THERE COVALENTLY-BOUND FLAVIN IN MONOAMINE OXIDASE FROM BRAIN?

J.I. SALACH,[a] M. MINAMIURA,[b] K.T. YASUNOBU[b] and M.B.H. YOUDIM[c]

[a] *Molecular Biology Division, Veterans Adminsitration Hospital, San Francisco, California 94121 and Department of Biochemistry and Biophysics, University of California, San Francisco, Calif. 94143,* [b] *Department of Biochemistry and Biophysics, University of Hawaii, Honolulu, Hawaii (U.S.A.) and* [c] *M.R.C. Clinical Pharmacology Unit, Radcliffe Infirmary, Oxford, (U.K.)*

Several years ago, after the prescence of covalently-bound flavin in mitochondrial monoamine oxidase from liver and kidney had been recognized [1–3], Tipton [4] reported that the enzyme isolated from pig brain contains, in contrast, flavin in non-covalent linkage. The evidence on which this conclusion rested was persuasive at first glance but puzzling on closer scrutiny. Thus, at the end of a rather long purification procedure trichloroacetic acid released free FAD from the enzyme. Second, again at the end of the procedure but not at earlier stages, acid ammonium sulfate produced a virtually inactive protein, the activity of which could be restored on adding back FAD, as in a typical reconstitution of a holoenzyme from its components. While the release of free FAD from a preparation which had not been proven to be completely homogenous is of no great moment, the apparent reversible resolution seemed more impressive. Tipton [4] apparently recognized that even this evidence had puzzling aspects, because the enzyme could not be resolved into FAD and apoenzyme at earlier stages by the same procedure, nor did trichloroacetic acid extraction yield free FAD prior to the last step, and because K_D, the dissociation constant for FAD determined from the reconstitution experiment was too high to be compatible with the resistance of the enzyme to dissociation during isolation. For all these reasons the conclusion that monoamine oxidase from mammalian brain differs from the liver and kidney enzymes in the linkage of the flavin to the protein needed further examination. The development in subsequent years of a method for analyzing for the covalently-bound cysteinyl flavin component of monoamine oxidase [5] in this laboratory afforded an opportunity to do so.

Attempts were made in all three of our laboratories to repeat these experiments following Tipton's purification procedure [6] without success, mostly because the enzyme was not extracted by us from brain mitochondria by the sonication method. Alternate methods of purification were developed, therefore. Isolation of the enzyme from either pig, beef, or rat brain proved to be uncommonly difficult because of the very high lipid content of the source material. Moreover, because we planned to follow the cysteinyl flavin content through purification, it was essential to remove, as much as possible, any other sources of covalently-bound flavin which would interfere with the analysis. This was even more difficult, because brain mitochondria contain far more succinate dehydrogenase than monoamine oxidase and the two cannot be separated in brain mitochondria by existing methods designed to separate inner and outer membranes.

It was decided, therefore, to work out alternative methods for purification of the enzyme from the central nervous system. The first procedure (procedure A), elaborated in Honolulu, yields an enzyme of high specific activity, although not as high as that isolated from liver, and which is apparently not far from homogeneous, since on electrophoresis in polyacrylamide gel in the presence of sodium dodecyl sulfate and mercaptoethanol a single protein band was detected. The procedure is summarized in Table I. The alternate procedure (B), developed in Oxford, involves the extraction of pig brain mitochondria with Triton X-100, precipitation with ammonium sulfate, and chromatographies on Sephadex G-200 and on DEAE-cellulose, yielding an approximately 50-fold purification from mitochondria. While this is less purified than the alternative procedure in Table I, it is short and thus minimizes inactivation. The advantage of this will be apparent from what follows.

Scheme 1 shows the various steps involved in the analytical procedure for cysteinyl flavin. As may be seen, cysteinyl flavin is determined by fluorescence after oxidation with cold performic acid, since the fluorescence is extremely low in the thioether stage but rises to about 70% of that given by riboflavin after oxidation to the sulfone. Histidyl flavin content is subtracted from the value obtained by taking advantage of the fact that in the case of

TABLE I

SUMMARY OF PURIFICATION OF MONOAMINE OXIDASE FROM PIG BRAIN

Protein was determined by the Lowry method. The enzyme was assayed in a volume of 3.0 ml, in the presence of benzylamine, potassium phosphate buffer, pH 7.4, and 0.5% Triton X-100 at 25°C. Specific activity is in nmol benzylamine oxidized/min/mg at 25°C. Specific activity at 30°C may be approx 30% higher.

Step	Total protein (mg)	Specific activity	Yield
1. Mitochondria	12 500	1.0	100
2. Extract (Triton-cholate)	3 900	3.5	108
3. Unabsorbed fraction from DEAE-cellulose treatment	2 077	6.0	100
4. Calcium phosphate gel eluate	1 000	8.4	68
5. DEAE-cellulose eluate	30	124	30
6. Hydroxylapatite eluate	17.5	356	22
7. Biogel A-1.5 m eluate	2.8	803	19
8. Biogel A-0.5 m eluate	0.81	1163	8

Heat denaturation, trichloroacetic acid precipitation

↓

Wash with acetone/HCl, alkaline $CHCl_3/CH_3OH$, precipitate with acetone/trichloroacetic acid, wash with trichloroacetic acid

↓

Digestion with trypsin/chymotrypsin or pronase

↓

Lyophilization, sublimation of NH_4HCO_3, oxidation of thioether with performic acid to sulfone

↓

Removal of performic and formic acids by lyophilization; acid hydrolysis of remaining pyrophosphate linkage

↓

Fluorescence measurements at pH 3.4 and pH 7.0 before and after dithionite, with and without internal standards

↓

Subtraction of histidyl flavin from total flavin gives cysteinyl flavin content

Scheme 1. Determination of cysteinyl flavin in monoamine oxidase.

histidyl flavin the fluorescence is 9 times as high at pH 3.4 as at pH 7, while with cysteinyl flavin the fluorescence does not vary with pH in this range. The method is satisfactory as long as the histidyl flavin content is not much greater than the cysteinyl flavin content. The problem in the case of monoamine oxidase from brain is the mitochondria from this source contain far more succinate dehydrogenase than they do monoamine oxidase, so that cysteinyl flavin content can be reliably estimated only after some initial purification steps, i.e., after most of the succinate dehydrogenase has been removed.

Cysteinyl flavin was found in all monoamine oxidase samples from brain that were examined, but reliable values could be obtained only from the most purified samples because of interference by succinate dehydrogenase. Thus for procedure A (Table I) the catalytic activity per cysteinyl flavin content (i.e., turnover number) remained reasonably constant from the second DEAE-cellulose eluate to the end of the procedure. The value obtained (360 ± 50) in these terminal steps was reasonably constant and in satisfactory agreement with the turnover number in the same assay which has been published for the highly purified enzyme from kidney [7], (Table II). In the latter preparation the concentration of monoamine oxidase had been calculated from flavin absorbance and from pargyline binding, indirect measures of the covalently bound flavin content.

As shown in Table III, the monoamine oxidase preparations obtained by the shorter procedure B appear to have a turnover number some three times higher, perhaps reflecting the possibility that procedure A, like that of Chuang et al. [7], although yielding a purer enzyme, is considerably longer and that this may incur inactivation of the enzyme. This higher turnover number agrees well with the value determined in outer membranes of beef liver and in the purified enzyme from liver mitochondria obtained by a short and rapid isolation method recently developed by the senior author.

The fact that cysteinyl flavin was present in both types of monoamine oxidase preparations from pig brain reported here and in samples of varying purity from beef and rat brain strongly suggest that at least one form of monoamine oxidase in the central nervous system does contain covalently-bound flavin. We cannot state unequivocally, however, that all monoamine oxidase in mammalian brain contains covalently-bound cysteinyl flavin, for two reasons. The first is the range of turnover numbers observed, with the short preparation agreeing with that of the liver enzyme and that isolated by the longer procedure agreeing with the value reported for the kidney enzyme. The second reason for leaving open the question of the possible existence of a form of monoamine oxidase in brain which contains flavin in non-covalent linkage is that all samples of the brain enzyme, including the most purified ones obtained by procedure A, contained some acid-extractable flavin. Unitl it is proven that this flavin originated from trace impurities, it seems best not to exclude completely the possibility that it originated from a second type of monoamine oxidase, which does not contain cysteinyl flavin.

TABLE II

TURNOVER NUMBERS OF EXTENSIVELY PURIFIED PREPARATIONS OF MONOAMINE OXIDASE

The turnover number for the kidney enzyme is recalculated from the published data of Chuang et al. [7] for 30°C and for standard assay conditions by the use of a factor of 1.8–1.9. Turnover number is based on total flavin and on pargyline bonding. The turnover number of the pig brain enzyme (preparation A, Table I) was determined under standard condtions but at 25°C; at 30°C the value would be slightly higher.

Source	Preparation	Turnover number at 30° (mol of benzylamine/min per mol of cysteinyl flavin)
Beef kidney	Chuang et al. [7]	approx 298
Pig brain	This paper, procedure A	approx 318

TABLE III

TURNOVER NUMBERS OF PARTIALLY PURIFIED PREPARATIONS OF MONOAMINE OXIDASE

The data for the liver preparations refer to an unpublished procedure of J.I. Salach. The pig brain preparation (procedure B) is that developed in Oxford.

Source	Preparation	Turnover number at 30°C (mol of benzylamine/min per mol of cysteinyl flavin)
Beef liver	Outer membrane and partially purified, soluble preparations	715 ± 320
Beef liver	Highly purified preparation	1,020
Pig brain	Procedure B (this paper)	860 ± 270

Acknowledgements

This investigation was supported by the National Science Foundation (GB 36570X) and the National Institutes of Health (MH 21539).

References

1 Nara, S., Igaue, I., Gomes, B. and Yasunobu, K.T. (1966) Biochem. Biophys. Res. Commun. 23 324–328
2 Igaue, I., Gomes, B. and Yasunobu, K.T. (1967) Biochem. Biophys. Res. Commun. 29, 562–570
3 Erwin, V.G. and Hellerman, L. (1967) J. Biol. Chem. 242, 4230–4238
4 Tipton, K.F. (1968) Biochim. Biophys. Acta 159, 451–459
5 Walker, W.H., Kearney, E.B., Seng, R.L. and Singer, T.P. (1971) Eur. J. Biochem. 24, 328–331
6 Tipton, K.F. (1968) Eur. J. Biochem. 4, 103–107
7 Chuang, H.K., Patek, D. and Hellerman, L. (1974) J. Biol. Chem. 249, 2381–2384

Flavins and Flavoproteins, edited by T.P. Singer

CHAPTER 67

ESCHERICHIA COLI SULFITE REDUCTASE: ROLE OF SIROHEME-SULFUR COMPLEXES IN THE MULTIELECTRON REDUCTION PROCESS

DAVID C. RUEGER and LEWIS M. SIEGEL

Department of Biochemistry, Duke University Medical School, and the Basic Science Division, Veterans Administration Hospital, Durham, N.C. 27705 (U.S.A.)

Escherichia coli sulfite reductase is a complex hemoflavoprotein [1] which catalyzes the stoichiometric reduction of SO_3^{2-} to H_2S [2], a six-electron transfer reaction, with either NADPH or the low potential dye reduced methyl viologen as electron donor. No sulfur compounds of oxidation state intermediate between SO_3^{2-} and H_2S appear to be released from the enzyme during the course of the reduction. The *E. coli* hemoflavoprotein, molecular weight 670 000, is composed of 8α and 4β polypeptide chains [3] and the following prosthetic groups (per molecule) [1]: 4 FAD, 4 Fe/S, a result which indicates that sulfite reductase contains four Fe_4S_4 unusual heme. When treated with 5 M urea and chromatographed on DEAE-cellulose, *E. coli* sulfite reductase can be separated into two functional "subenzymes": a flavoprotein of structure α_8, which contains both FMN and FAD (but negligible amounts of iron) and catalyzes all of the NADPH-diaphorase and transhydrogenase activities characteristic of the sulfite reductase holoenzyme, and a hemoprotein containing the Fe/S and heme (but no flavin), which exists in standard buffer (0.05 M potassium phosphate/0.1 mM EDTA, pH 7.7) as the monomeric β polypeptide and catalyzes the stoichiometric reduction of SO_3^{2-} to H_2S with reduced methyl viologen (but not NADPH) as electron donor [3]. The flavoprotein and hemoprotein have also been isolated from mutants of *Salmonella typhimurium* [4, 5] (an organism which produces a sulfite reductase hemoflavoprotein closely related in structure to the *E. coli* enzyme [4]) which are blocked in the ability to reduce SO_3^{2-} to H_2S in vivo. By combining the isolated flavoprotein and hemoprotein "subenzymes", NADPH-linked sulfite reductase activity can be reconstituted in vitro [3, 4].

EPR spectra of sulfite reductase (Fig. 1A) show the heme to be present largely in the high-spin ferriheme state [1] with *g* values of 6.63, 5.24 and 1.98. There is also a small amount of low-spin ferriheme. Upon treatment of the enzyme with strong reducing agents (Fig. 1D), the ferriheme signals disappear and are replaced by signals characteristic of reduced Fe/S centers,

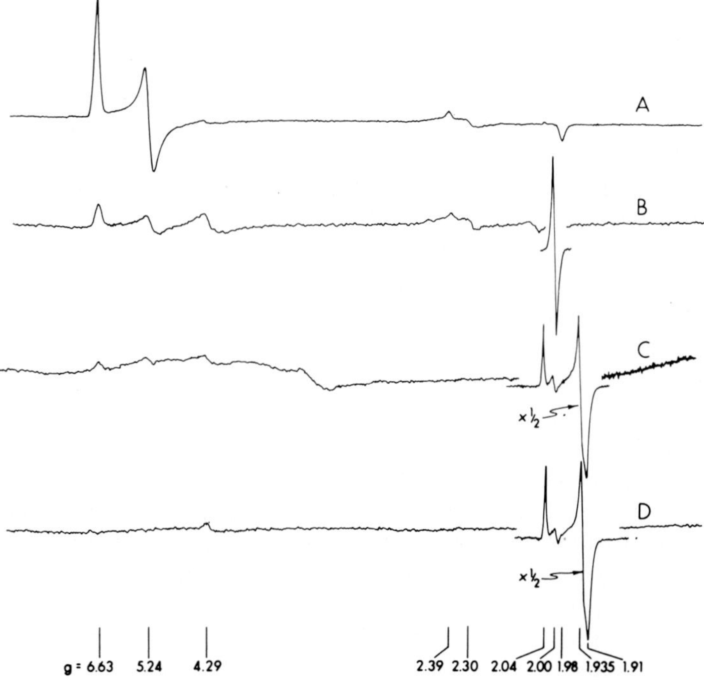

Fig. 1. EPR spectra of sulfite reductase hemoflavoprotein. All samples were in 0.1 M potassium phosphate buffer, pH 7.7. Spectra were recorded at 13°K at 3 mW (except for the inset to B where power was 0.1 mW). A: 25 μM (in heme) sulfite reductase. Instrument gain 250, time constant 0.01 s. B: 0.2 ml 16.7 μM sulfite reductase were made anaerobic and equilibrated with 0.25 atm CO/0.75 atm N_2. 0.1 ml of 1 mM NADPH were then added, and the solution incubated at 23°C for 1 h. After CO had been removed by evacuation, the solution was opened to air and 0.05 ml O_2-saturated buffer added. The solution was incubated at 25°C for 15 min, then evacuated and frozen. Instrument gain 1250, time constant 0.1 s. Intensity of $g = 6.63$ signal (corrected to gain and dilution), relative to A, 0.13. C: Solution B was thawed, made anaerobic with N_2, 0.09 ml of 100 mM NADPH added, and the sample then quickly frozen. Instrument settings for main spectrum as in B; for inset, gain 630, time constant 0.03 s. Intensity of $g = 6.63$ signal, relative to A, 0.06. Intensity of $g = 1.93$ signal, relative to D, 0.85. D: Solution C was thawed and 0.025 ml of 50 mM sodium dithionite added. The sample was incubated for 5 min at 25°C and then frozen. Instrument settings as in B.

with g values of 2.04, 1.935 and 1.91. There is one unpaired spin present per 4 Fe/S, a result which indicates that sulfite reductase contains four Fe_4S_4 centers, each capable of accepting a single electron. The hemoprotein "subenzyme" (β monomer) contains, then, a single Fe_4S_4 center and a single heme per molecule. In agreement with the presence of Fe_4S_4 clusters is the fact that each molecule of β polypeptide contains only 5 half-cystine residues [3], whereas a minimum of 8 half-cystines would be required to form two Fe_2S_2 centers.

With the use of agents which characteristically bind to the heme (e.g., CO or CN^-) [2, 6] and of enzyme from which the FMN (but not the FAD) has been specifically removed [2, 7], it has been possible to show that electron flow within the sulfite reductase hemoflavoprotein follows the minimum linear sequence: NADPH → FAD → FMN → Heme → SO_3^{2-}. Although the role of Fe/S in electron transfer has not been precisely defined, it is clear from the experiment shown in Fig. 1 that electron flow through the heme is not required in order to achieve reduction of Fe/S by NADPH. In this experiment , enzyme was reacted with CO in the presence of a small amount of NADPH in order to permit formation of the heme • CO complex. After a period of incubation sufficient to allow 87% conversion of high spin ferriheme to ferroheme • CO, the enzyme was exposed to O_2 to allow reoxidation of Fe/S and non-complexed heme. Fig. 1B shows that no reduced Fe/S was detected after this oxidation process, although some electrons remained on the enzyme as flavin radical ($g = 2.003$). When excess NADPH was added to an anaerobic solution of the reoxidized enzyme-CO complex, the spectrum of reduced Fe/S appeared (Fig. 1C). The reduction obtained with NADPH was 85% of the Fe/S reduction seen when dithionite was subsequently added to the NADPH + CO-enzyme solution (Fig. 1D). In view of the fact that there appears to be little influence of the hemoprotein subunits on the rates of the various NADPH-diaphorase activities which can be catalyzed by the flavoprotein itself [4], and the fact that no interaction between the flavin radical and the Fe/S centers of the hemoflavoprotein could be detected by EPR power saturation studies [8], it seems likely that the Fe/S centers do not function in catalysis "between" the FAD and FMN groups of the enzyme. The results, then, are consistent with the following electron transfer sequence: NADPH → FAD → FMN → Fe/S → Heme → SO_3^{2-}. A number of "branched" electron flow schemes are also consistent with the data, however.

Siroheme

Sulfite reductase is one of a small group of enzymes which can be termed "multielectron reductases", in that they catalyze the transfer of more than two electrons from appropriate donors to an enzyme-bound substrate without release of partially reduced intermediates from the enzyme. The "multielectron reductases" include cytochrome oxidase ($O_2 + 4H^+ + 4e^- \rightarrow 2H_2O$), nitrogenase ($N_2 + 8H^+ + 6e^- \rightarrow 2NH_4^+$), and nitrite reductase ($NO_2^- + 8H^+ + 6e^- \rightarrow NH_4^+ + 2H_2O$), in addition to sulfite reductase. Each of these enzymes occupies a crucial position in the cycling of a major element (O, N or S) through the biosphere.

We have found [9–13] that two of these "multielectron reductases", sulfite and nitrite reductase, possess a novel type of heme prosthetic group which we have termed "siroheme". In no case has the heme been found to be covalently bound to its enzyme. The enzymes studied have included the nitrite reductases of spinach and *Neurospora crassa* and four different bacterial sulfite reductases; in all cases the heme prosthetic groups were shown to be identical by a variety of chromatographic and spectroscopic criteria.

The structure of siroheme extracted from *E. coli* sulfite reductase has been studied by our laboratory, with the following results (most of which are reported in detail by Murphy, et al. [9]): (a) Siroheme is an iron tetrahydroporphyrin in which adjacent pyrrole rings are reduced. (b) Siroheme possesses eight carboxylic acid-containing side chains, which, on the basis of

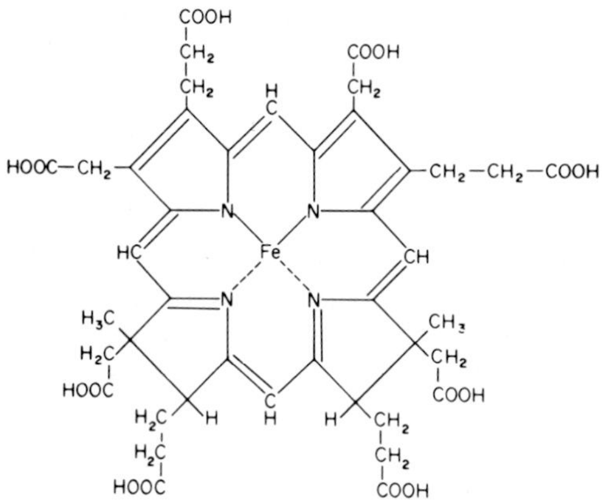

Fig. 2. Proposed structure of siroheme [9].

mass spectroscopy cleavage patterns, appear to be acetyl and propionyl groups. (c) Siroheme is methylated. An *E. coli* methionine auxotroph was grown on a minimal medium supplemented with [$^3H_3$]methionine and [U-^{14}C]glucose, and the sulfite reductase produced by the organism was purified. The siroheme was extracted and purified by chromatography on a column of Sephadex LH-20 equilibrated with pyridine [1, 9]. The extracted siroheme was found to contain 1.8 methyl groups per 42 total carbon atoms. Since the molecular formula of sirheme, obtained by mass spectrometry [9], is $FeC_{42}H_{44}N_4O_{16}$, it can be concluded that each siroheme molecule contains two methionine-derived methyl groups. The stability properties of siroheme [9] suggest that the methyl groups are added to the partially saturated ("reduced") pyrolle rings.

Murphy, et al. [9] have proposed the structure for siroheme shown in Fig. 2. This structure, although one of a number of possible isomers, is biologically attractive, since it can be derived, in principle, from uroporphyrinogen III by two successive methylation reactions followed by oxidation of the porphyrin nucleus and insertion of iron (the latter two steps need not involve enzymes specific for siroheme synthesis). Uroporphyringogen III is an intermediate common to the biosynthesis of all known heme, chlorophyll, and corrin compounds. If the structure proposed in Fig. 2 is correct, siroheme biosynthesis must represent a new branch of tetrapyrrole metabolism, since it is not likely that this structure could be derived metabolically from protoporphyrin IX, the common precursor of all known enzyme-associated heme and chlorophyll compounds other than siroheme. It has been pointed out that the structure proposed for siroheme bears some resemblance to the corrin series, in that both types of compound are highly carboxylated, contain reduced pyrrole rings, and are methylated. It is interesting to note that *Clostridia*, thought to be primitive organisms which do not contain protoheme-derived compounds, are able to produce both corrins and a sulfite reductase [14] with a heme-like absorption spectrum [16]. The possibility that siroheme may represent an "ancestral" type of heme compound has been discussed [10].

Siroheme is both widespread in nature (having been found in organisms ranging from anaerobic bateria to higher plants) and demonstrably involved in the catalytic function of both sulfite reductase [2] and nitrite reductase [13]. Spectroscopically identifiable complexes between each of these enzymes and its respective substrate have been found; the spectral shifts are such as to suggest that siroheme represents the site of substrate binding (see below). It is not clear as yet, however, what there is about the structure of siroheme that can make it particularly suited to function in these two "multielectron reduction" reactions.

Role of enzyme-sulfur complexes in sulfite reduction

When *E. coli* sulfite reducase reacts with NADPH, 0.1 mM or more, the ferriheme moiety is reduced [8] (following a lag to allow for flavin reduction and intramolecular electron transfer between the flavin and/or Fe/S groups) in a first order process with $k = 6 \pm 1\,s^{-1}$ at 25°C in standard buffer. The rates obtained by following the absorbance changes associated with heme reduction in the stopped flow apparatus or by following the loss in $g = 6$ EPR signal in a rapid-freeze system appear to be identical. The turnover number for NADPH-dependent SO_3^{2-} reduction under the same conditions is 2.5 SO_3^{2-} reduced $s^{-1} \cdot heme^{-1}$, or 15 electron $s^{-1} \cdot heme^{-1}$. Since the rate of ferriheme reduction is not very different from the overall rate of sulfite reduction, one would expect to see significant amounts of ferriheme during the process of enzyme turnover with NADPH and sulfite, if, indeed, the ferriheme species present in the enzyme as isolated participates in catalysis. In fact, when enzyme is mixed with NADPH, 0.1 mM or more, and SO_3^{2-}, equimolar to NADPH, in the rapid freeze apparatus, no $g = 6$ EPR signal can be detected in the enzyme during the turnover process. The $g = 6$ signal does not reappear following turnover under these conditions. The rate of reduction of the $g = 6$ signal by NADPH appears to be the same whether or not sulfite is present. These results suggest that "free" ferriheme may not play a role in the sulfite reduction process.

E. coli sulfite reductase catalyzes the NADPH-dependent reduction of NH_2OH to ammonia [2, 15]. This reaction is inhibited by CN^- and CO, and thus appears to require electron transfer through the heme. The *V* for NH_2OH reduction is 57 NH_2OH $s^{-1} \cdot heme^{-1}$ (115 electron $s^{-1} \cdot heme^{-1}$) at 25°C in standard buffer [2]. Thus, the rate of ferriheme reduction by NADPH is far too slow to account for the rate of the hydroxylamine reduction reaction. The rate of heme-dependent reduction reactions is clearly dependent on the substrate involved. It seems likely that complexes between sulfite reductase and reducible substrates, in which the electron flow rates to (and through) the heme are altered when compared to the "free" heme, play a major role in catalysis by this enzyme.

We have thus far detected two types of spectrophotometrically distinct species formed upon reaction of sulfite reductase with sulfite (or reduced sulfur compounds derived from sulfite). One species exhibits an α-band maximum at 595 nm and is present only when there is reducing agent available to the enzyme. The second species exhibits an α-band maximum at 582 nm and is present when reducing agent has been exhausted; it can also form slowly by reaction between enzyme and sulfite or sulfide in the absence of externally added reductants.

We have examined the nature of the "582 nm" species in some detail, using hemoprotein (and MVH*) rather than holoenzyme, in order to eliminate complications in the absorption spectra due to the presence of multiple flavin species in the hemoflavoprotein. Table I shows that a complex between hemoprotein and ^{35}S derived from $^{35}SO_3^{2-}$ is formed rapidly when hemoprotein is mixed with MVH and excess $^{35}SO_3^{2-}$. The ^{35}S remains bound to the hemoprotein following repeated passage of the protein through columns of Sephadex G-25. The amount of ^{35}S bound is a function of the concentration of $^{35}SO_3^{2-}$ present; in the presence of small amounts of MVH (2–3 MVH per hemoprotein), there is little reduction of $^{35}SO_3^{2-}$ to $H_2{}^{35}S$, and the amount of ^{35}S bound to hemoprotein (HP) appears to follow the reaction: $HP + {}^{35}SO_3^{2-} \rightleftharpoons HP \cdot {}^{35}S$ with a dissociation constant of $7 \cdot 10^{-6}$ M. The maximum amount of ^{35}S binding observed corresponds to 1.0 ± 0.1 ^{35}S per heme. The spectrum of the hemoprotein · ^{35}S complex is shown in Fig. 3. Maxima

TABLE I

BINDING OF SULFITE AND CYANIDE TO HEMOPROTEIN AND HEMOPROTEIN COMPLEXES

$Na^{35}SO_3$, $K^{14}CN$, and MVH (reduced with H_2/platinum asbestos) were present (0.1 mM). $K^{12}CN$ was added at 1 mM. Hemoprotein (HP) was 8 μM and hemoprotein complexes (indicated by brackets) were 4 μM. Incubations were performed anaerobically for 5 min at 23°C, except for experiment I, in which the incubation was extended to 16 h. Analyses for ^{35}S, ^{14}C and protein were performed on protein-containing fractions following passage of each incubation mixture through a column of Sephadex G-25 aerobically. Hemoprotein complexes with sulfite ($^{35}SO_3^{2-}$) and with cyanide ($^{14}CN^-$ or $^{12}CN^-$) were made by incubating components as described in experiments B and F, respectively, and isolating the complexes following passage through columns of Sephadex G-25. Hemoprotein-CO complex was made by a procedure like that described in Fig. 4.

Experiment	Components present in incubation mixture	Analysis of Hemoprotein after passage through Sephadex G-25	
		^{35}S/Heme	^{14}C/Heme
A	$HP + {}^{35}SO_3^{2-}$	0.02	—
B	$HP + {}^{35}SO_3^{2-}$ + MVH	0.98	—
C	$[HP \cdot {}^{35}SO_3^{2-}]$	0.97	—
D	$[HP \cdot {}^{35}SO_3^{2-}]$ + MVH	0.05	—
E	$HP + {}^{14}CN^-$	—	0.02
F	$HP + {}^{14}CN^-$ + MVH	—	1.01
G	$[HP \cdot {}^{14}CN^-]$	—	1.00
H	$[HP \cdot {}^{14}CN^-]$ + MVH	—	1.02
I	$[HP \cdot {}^{14}CN^-]$ + MVH + $^{12}CN^-$	—	0.98
J	$[HP \cdot {}^{12}CN^-]$ + MVH + $^{35}SO_3^{2-}$	0.08	—
K	[HP · CO] + MVH + $^{35}SO_3^{2-}$	0.09	—

* MVH, reduced methyl viologen.

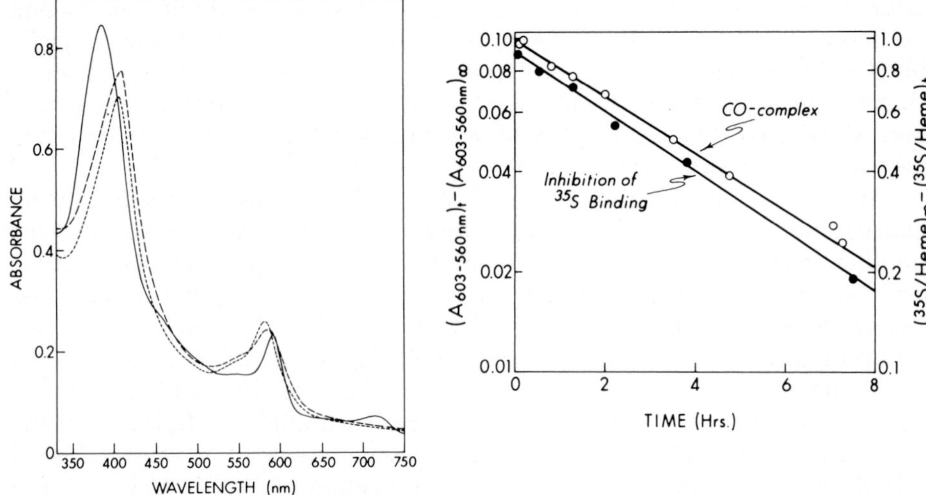

Fig. 3. Absorption spectra of hemoprotein complexes. All samples contained 12.8 μM (in heme) hemoprotein in standard buffer (see text) and were centrifuged for 10 min at 48 000 × g to remove any turbid material before spectra were recorded (25°C). Cuvettes 1 cm in path length and a Cary model 14 spectrophotometer were used. ———, Hemoprotein. · · · · · · ·, Hemoprotein-cyanide complex, formed by incubating hemoprotein anaerobically for 6 min with 0.26 mM MVH and 1 mM KCN, then opening the solution to air. - - - - - -, Hemoprotein-sulfite complex, formed by incubating hemoprotein anaerobically with 0.13 mM MVH and 1 mM Na_2SO_3 for 1 min, then opening the solution to air.

Fig. 4. Dissociation of CO from hemoprotein permits $^{35}SO_3^{2-}$ to bind. Hemoprotein · CO complex was formed by incubating 1.0 ml of an anaerobic solution containing 32 μM hemoprotein, 0.3 mM MVH, and 0.8 mM CO in standard buffer for 90 min at 23°C. The solution was then chilled to 4°C (to lessen the rate of CO dissociation), opened to air, and passed through a column of Sephadex G-25 to remove unbound CO and methyl viologen. Dissociation of CO from complex: 4 ml of 5.8 μM hemoprotein·CO complex in standard buffer were incubated anaerobically at 23°C and absorption spectra recorded at intervals in cuvettes of 1 cm light path with a Cary model 14 spectrophotometer equipped with 0–0.1, 0.1–0.2 absorbance units slide wire. The concentration of CO complex is proportional to the difference in absorbance between the wavelengths 603 nm and 560 nm [6]. ○—○, semilogarithmic plot of $A_{603-560\text{nm}}$ at time t minus $A_{603-560\text{nm}}$ at infinite time (21.5 h of incubation at 23°C) vs time. Loss of inhibition of ability to bind $^{35}SO_3^{2-}$: At intervals, 0.6 ml aliquots of the dissociating hemoprotein · CO complex were removed, made anerobic with argon, and 0.34 ml 0.3 mM $Na_2{}^{35}SO_3$ and 0.05 ml 1 mM MVH added. The solutions were incubated to 60 s, opened to air, and then passed through a column of Sephadex G-25. Hemoprotein-containing fractions were assayed for protein by the microbiuret method [1] and for ^{35}S. The amount of ^{35}S bound per heme was measured and found to be 1.08 after 21.5 h ($t = \infty$) of incubation of the hemoprotein · CO complex at 23°C. ●—●, semilogarithmic plot of ^{35}S bound per heme at $t = \infty$ minus ^{35}S bound per heme at time t.

are observed at 408 and 582 nm. The $g = 6$ EPR signal of the "free" hemoprotein is over 95% abolished in the hemoprotein · ^{35}S complex. No EPR signal attributable to the heme of the complex has been detected yet.

The data of Table I and Fig. 3 also show that if hemoprotein is treated with MVH and cyanide, a complex is formed which contains one mol of cyanide per mol of heme and exhibits absorption maxima at 404 and 581 nm. EPR spectra of the hemoprotein-cyanide complex show that the heme is in the low-spin ferriheme state, with g values at 2.39, 2.31 and 1.73. The ferriheme forms upon reoxidation of the ferroheme-cyanide complex which is originally formed when sulfite reductase is incubated with cyanide in the presence of reducing agents [1]. The presence of reductant during the process of formation of the complexes with SO_3^{2-} or cyanide seems to be necessary to permit rapid "accessibility" of these agents to the site on the enzyme to which they can tightly bind. Thus, complexes between hemoprotein and cyanide or sulfite can be formed simply by incubating the protein with either ligand in standard buffer and waiting for a sufficient period of time. In the presence of 1 mM ligand, the half-time for formation of the complex with sulfite is 2.7 h at 23°C; with KCN it is 8.5 h. The spectral properties of the complexes appear to be identical whether or not reducing agent was present during their formation.

Dissociation of cyanide from its complex with hemoprotein in standard buffer has not been detected (in the presence or absence of MVH). If cyanide is bound to the hemoprotein, $^{35}SO_3^{2-}$ cannot bind (Table I). Similarly, if the CO complex of the hemoprotein is exposed to $^{35}SO_3^{2-}$ and MVH, the sulfite cannot bind. Unlike cyanide, CO can dissociate from its complex with the siroheme moiety of the hemoprotein [6]. Fig. 4 shows that the dissociation of CO from the heme is coincident with appearance of the ability for the hemoprotein to bind $^{35}SO_3^{2-}$. It has been shown previously [2] that CO binding to and dissociation from sulfite reductase are correlated in time with the development and release of inhibition of sulfite reductase activity. These results strongly suggest that the sulfur moiety derived from sulfite in the binding experiments cited above, like CO and cyanide, binds to the heme moiety of sulfite reductase.

^{35}S can be released from its complex with the hemoprotein in any of the following ways: (a) It dissociates slowly from the hemprotein · ^{35}S complex, with a half-life of 20 h at 23°C in standard buffer. (b) It is released rapidly (in less than 2 min) from the hemoprotein · ^{35}S complex in the presence of 3.3 M urea. The ^{35}S released from the hemoprotein in urea can be reduced to $H_2{}^{35}S$ upon addition of NADPH and catalytic amounts of sulfite reductase hemoflavoprotein; this reduction is accompanied by the oxidation of 2.8 NADPH per ^{35}S originally bound to the hemoprotein. Thus, the ^{35}S appears to be released from the enzyme at the oxidation level of sulfite. (c) It is released rapidly from the ^{35}S · hemoprotein complex upon addition of one mol of MVH per mol of hemoprotein. Virtually no $H_2{}^{35}S$ is formed under these conditions. (d) The bound ^{35}S can be reduced in 90% yield to $H_2{}^{35}S$ upon addition of eight or more mol of MVH per mol of bound ^{35}S.

Bound ^{35}S is reduced in preference to exogenous SO_3^{2-}, as shown by the fact that if excess MVH (identical results were obtained with 10, 20 or 40 MVH per heme) is added to $4 \cdot 10^{-6}$ M hemoprotein · ^{35}S complex in the presence of 1 mM non-radioactive SO_3^{2-}, 25 ± 1% of the ^{35}S originally bound is converted to $H_2{}^{35}S$ (2% remains associated with the enzyme after passage through a column of Sephadex G-25, and the remainder is released from the enzyme as ^{35}S which is non-volatile at pH 7.7). The resulting hemoprotein exhibits a spectrum like that of Fig. 3, showing that non-radioactive sulfite has replaced the ^{35}S in complex with the enzyme. Thus, the rate of reduction of the bound ^{35}S to $H_2{}^{35}S$ compares favorably with the rate of its release from the enzyme in the presence of excess reductant. These results indicate that the "582 nm" species of hemprotein may be catalytically competent to serve as an intermediate in the sulfite reduction process.

In view of (a) the similarities in spectrum between the cyanide (ferriheme) and sulfite-derived complexes of the hemoprotein, and (b) the fact that the complexes can form in the absence of reducing agents, it seems probable that the "582 nm" species represents a complex between a sulfur compound and ferriheme. The results with 3.3 M urea release of the bound ^{35}S indicate that the sulfur compound, in the hemoprotein complex as normally isolated, is sulfite itself. It should be noted however, that a species with a spectrum very similar to that shown in Fig. 3 can be formed upon aerobic incubation of sulfite reductase with 5 mM HS^-. This sulfide-derived "582 nm" species could be made by treatment with small amounts of reductant followed by reoxidation of the latter in air, as well as in the absence of reductant. Thus, the "582 nm" type of spectrum may specify the oxidation state of the heme and the fact that it is ligated by a sulfur compound, rather than the precise nature of the sulfur ligand.

The "595 nm" species has been detected only in the presence of reducing agents. It is seen if either hemoprotein or holoenzyme is mixed with large amounts of sodium dithionite and spectra recorded rapidly. Since interaction of dithionite with the enzyme can give rise to sulfite, and this in turn can be reduced to sulfide by the enzyme with dithionite as reductant [1], the enzyme can catalyze conversion of the dithionite to sulfite and sulfide. During this turnover process the enzyme spectrum is converted to that of the "582 nm" species; the latter species persists following oxidation of all the dithionite. The "595 nm" species may also be observed if holoenzyme is mixed with 1 mM NADPH and 5 mM HS^- under anaerobic conditions. If this species is allowed to oxidize in air, it becomes converted to one with a spectrum like that of the "582 nm" species.

More pertinent to catalysis, however, is the fact that the "595 nm" species represents the dominant species present when sulfite reductase is undergoing turnover with NADPH and SO_3^{2-}. Fig. 5 shows that when enzyme reacts in the stopped-flow apparatus with NADPH and sulfite (equimolar to NADPH), the ferriheme is reduced to a species with an α-band maximum at 595 nm. (k for this reduction process is 6 s^{-1}.) In the absence of sulfite, reduction of enzyme with NADPH produces no significant shift in the α-band maximum of the enzyme (although there is a marked broadening of the entire α-band of the enzyme). The 595 nm species persists during most of turnover until, upon exhaustion of the NADPH, the α-band maximum gradually shifts to 582 nm (Fig. 5). If sulfite reductase is first reduced with 1 mM NADPH (to achieve over 95% reduction of the ferriheme) and then allowed to react with 1 mM SO_3^{2-} in the stopped-flow apparatus, the species formed within the mixing time of the apparatus (3 ms) exhibits its α-band maximum at 595 nm. As in the experiment of Fig. 5, the α-band maximum remains at 595 nm throughout turnover until the NADPH is exhausted, whereupon it "decays" to the "582 nm" species.

These results suggest that the "595 nm" species of sulfite reductase

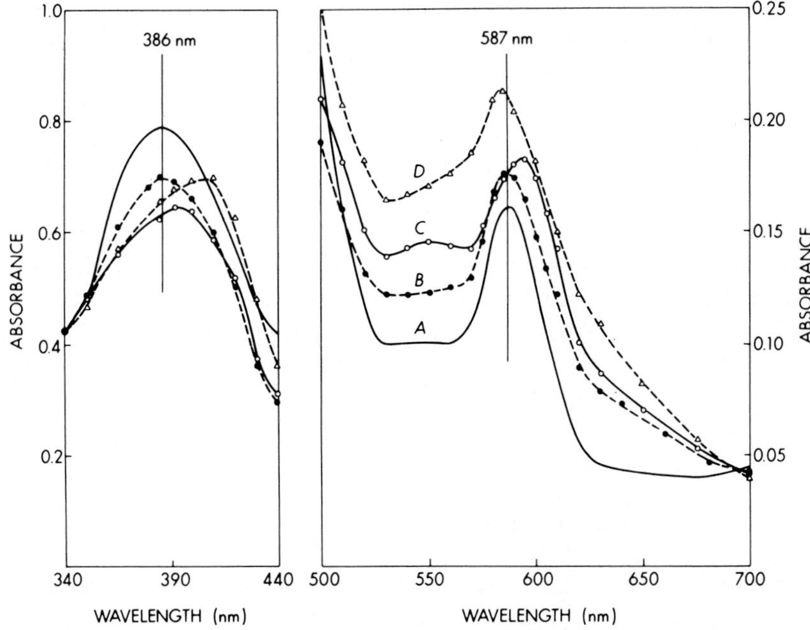

Fig. 5. Spectral characteristics of intermediates formed upon reaction of sulfite reductase hemoflavoprotein with sulfite and NADPH. All solutions were in standard buffer at 23°C. A (——): Absorbance spectrum of 2.5 μM (in heme) sulfite reductase recorded in cells of 1 cm path length with a Cary 14 spectrophotometer. B (●- - - -●): 2.5 μM sulfite reductase was mixed with an equal volume of 0.1 mM NADPH in a Durrum model 110 stopped-flow apparatus, and the spectrum shown was constructed by adding the absorbance changes observed between zero time and 0.6 s of reaction to the spectrum shown in A. Path length of the stopped-flow observation cell was 2 cm. C and D: 2.5 μM sulfite reductase was mixed with an equal volume of a solution containing 0.1 mM NADPH and 0.1 mM Na_2SO_3 in the stopped-flow apparatus, and the spectra shown were constructed by adding the absorbance changes observed between zero time and either 0.6 s (○——○) or 9 s (△- - - -△) of reaction to the spectrum shown in A. For spectra in the region 340–440 nm (left) the spectrum of the NADPH remaining in solution at the times indicated has been subtracted from each of the spectra shown. The concentration of NADPH was determined approximately by means of the residual absorbance at 340 nm of each solution minus that of the oxidized enzyme.

represents a complex of ferroheme with sulfite, sulfide, or sulfur compounds of intermediate oxidation state. If this were so, one should be able to reduce the "582 nm" species of the sulfite reductase hemoflavoprotein to the "595 nm" species with NADPH. Indeed, one can isolate the "582 nm" species of holoenzyme following incubation with NADPH and SO_3^{2-} and gel filtration [2]. When this species is mixed with 0.1 M NADPH in the stopped flow apparatus, the spectrum is reduced to that of the "595 nm" species with $k = 16\,s^{-1}$. Subsequently, the spectrum shifts to that of "free" heme (α-band maximum at 587 nm) and some $g = 6$ EPR signal can be detected in the resulting enzyme. The latter process, which must require dissociation of the H_2S formed, occurs with a $k = 1\,s^{-1}$. The small value of this latter rate constant suggests that sulfide dissociation from the heme may be rate-limiting in the sulfite reduction process.

It appears, then, that the heme moiety of sulfite reductase represents the binding site for sulfite, and that sulfur remains bound to the heme during the entire six-electron transfer process. Free ferroheme is a minor enzyme species during turnover, and free ferriheme cannot be detected. The nature of the sulfur-containing compounds which remain bound to the heme during the six-electron transfer process remains of course a question of paramount interest.

Acknowledgements

These studies were supported in part by research grant AM-13460 from the National Institutes of Health. Veterans Administration Project No. 7875-01. We are indebted to Professors W. Orme-Johnson who performed the EPR experiments reported in this paper, and to Prof. Quentin Gibson who performed some of the stopped-flow studies.

References

1 Siegel, L.M., Murphy, M.J. and Kamin, H. (1973) J. Biol. Chem. 248, 251
2 Siegel, L.M., Davis, P.S. and Kamin, H. (1974) J. Biol. Chem. 249, 1572
3 Siegel, L.M. and Davis, P.S. (1974) J. Biol. Chem. 249, 1587
4 Siegel, L.M., Kamin, H., Rueger, D.C., Presswood, R.P. and Gibson, Q.H. (1971) in Flavins and Flavoproteins (Kamin, H., ed.), p. 523, University Park Press, Baltimore, Md.
5 Siegel, L.M. and Kamin, H. (1971) Fed. Proc. 30, 1261
6 Murphy, M.J., Siegel, L.M. and Kamin, H. (1974) J. Biol. Chem. 249, 1610
7 Faeder, E.J., Davis, P.S. and Siegel, L.M. (1974) J. Biol. Chem. 249, 1599
8 Siegel, L.M. and Kamin, H. (1968) in Flavins and Flavoproteins (Yagi, K., ed.), p. 15, University Park Press, Baltimore, Md.
9 Murphy, M.J., Siegel, L.M., Kamin, H. and Rosenthal, D. (1973) J. Biol. Chem. 248, 2801
10 Murphy, M.J. and Siegel, L.M. (1973) J. Biol. Chem. 248, 6911
11 Murphy, M.J., Siegel, L.M., Kamin, H., Dervertanian, D.V., Lee, J.-P., LeGall, J. and Peck, Jr., H.D. (1973) Biochim. Biophys. Res. Commun. 54, 82
12 Murphy, M.J., Siegel, L.M., Tove, S.R. and Kamin, H. (1974) Proc. Natl. Acad. Sci. U.S. 71, 612
13 Vega, J.M., Garrett, R.H. and Siegel, L.M. (1975) J. Biol. Chem. 250, 7980
14 Laishley, E.J., Lin, P.M. and Peck, Jr., H.D. (1971) Can. J. Microbiol. 17, 889
15 Kemp, J.D., Atkinson, D.E., Ehret, A. and Lazzarini, R.A. (1963) J. Biol. Chem. 238, 3466
16 Peck, Jr., H.D. and LeGall, J., personal communication

Flavins and Flavoproteins, edited by T.P. Singer

CHAPTER 68

FLAVOCYTOCHROME b_2: STUDIES OF INTRAMOLECULAR ELECTRON EXCHANGE BETWEEN PROSTHETIC GROUPS

C. CAPEILLERE-BLANDIN, M. IWATSUBO, F. LABEYRIE and [a] R.C. BRAY

Centre de Génétique Moléculaire, CNRS, 91190 Gif-sur-Yvette (France) and [a] The School of Molecular Sciences, University of Sussex, Brighton (U.K.)

The kinetic study of flavocytochrome b_2 reduction was done with the hope of shedding some light on the characteristics of intramolecular electron transfers within a functional flavocytochrome assembly. In fact, flavocytochrome b_2 represents a complex between a flavoreductase and its specific acceptor, a cytochrome *b*, these two elements being covalently bound and assembled as a tetramer molecule. This can accept a total of 12 electrons which can then be redistributed, one by one, to cytochrome *c* molecules. Such information might indeed help in the understanding of the importance of the topology in the electron exchanges within the respiratory chain, where a number of such transfers, if not most of them, probably take place within stable complexes as intra-"super"-molecular events.

A functional role of flavocytochrome b_2 in aerobic yeast

Under aerobic conditions, the yeast *Saccharomyces cerevisiae* can grow as well on ethanol, glycerol and pyruvate as on D- or L-lactate, so that the role of a L-lactate cytochrome *c* reductase was not understood for a long time. The fact that it seemed to be present in mitochondria in relatively small amounts, compared with the other components of the respiratory chain [1–3], even in lactate-adapted yeast, contributed to deny its possible important role as a participant of a secondary respiratory chain.

Recently, a specific function for this enzyme (and for the D-lactate cytochrome *c* reductase simultaneously present in yeast mitochondria) was revealed by Pajot and Claisse [4] in our laboratory. They found that yeast can grow aerobically in the presence of antimycin A when, and only when, the respiring substrate is L- or D-lactate. The yield of (protein)/(lactate consumed) is 1/11 that found in a control growth without antimycin A. This ratio is in agreement with the assumption that the D- and L-lactate

---

Abbreviations and symbols: H_{ox}, oxidized haem; H_{red}, reduced haem; Fl_{ox}, oxidized flavin (quinone); Fl_{SQ}, Flavin radical (semiquinone); Fl_{red}, reduced flavin (hydroquinone); n, number of electrons per mol exchanged in an oxidoreduction reaction; E_m, midpoint oxidation-reduction potential (with reference to the standard hydrogen electrode); k_H, heme-reduction rate constant; k_{Fl}, flavin (Fl_{ox}), reduction rate constant.

dehydrogenases allow the reduction of cytochrome *c* by lactate, the electrons being afterwards transferred to the cytochrome oxidase and then to oxygen. The reaction should be coupled with a single phosphorylating site. The origin and the maintenance of such a system pose a number of interesting questions.

In fact, flavocytochrome b_2 content in yeast cells can be much higher than estimated previously: in the lactate-adapted aerobic *Hansenula anomala* yeast, it can approach the cytochrome *c* content. The amount of enzyme present in the best strains of (commercial) bakers yeast *S. cerevisiae* is 3 nmol/kg dry w. [5] instead of 15–20 nmol/kg for the lactate-adapted *Hansenula* yeast (Iwatsubo, M., unpublished results).

A gene fusion between the flavoreductase part and the cytochrome b part

As soon as the double flavocytochromic character of yeast L-lactate cytochrome *c* reductase (EC 1.1.2.3) was recognized [6, 7], the fact that it may result from the association between a flavoreductase and a cytochome *b* was considered and discussed [7].

However, it is only recently that some solid information has been obtained concerning such an association; several findings support, in fact, the assumption that the long polypeptide chains of approx. 60 000 daltons, which constituted each protomer [8, 9], result from a gene fusion between a gene coding for the flavoreductase part and a gene coding for the cytochrome *b* part.

The data relevant to this view are the following:

(a) When fully structured in the "native" conformation, the tetramer molecule of flavocytochrome b_2 presents a polyglobular aspect: along each chain there are three main loci, the trypsic splitting of which gives rise to three relatively stable globules, termed n, ϵ and β, with small fragments (Naslin et al. [10], Gervais et al. [11]). Under certain conditions, it is then possible to separate the cytochrome globule n from a flavoreductase keeping the lactate-binding site and formed by the association of ϵ and β globules in a 1:1 ratio [11].

(b) The cytochrome globule, termed the cytochrome b_2 core, was found and characterized a long time ago as being resistant to an exhaustive tryptic action. It presents a marked similarity, in all its spectral and physical properties [12–18] with liver microsome cytochrome b_5 studied in Strittmatters's laboratory. It constitutes the N-terminal part of the flavocytochrome b_2 chain (Guiard, Lederer and Jacq, 1975, submitted for publication). Recently, the determination of the amino acid sequence revealed in fact a considerable homology between the two molecules' fragments, cytochrome b_2 core and cytochrome b_5 core, leading to the well supported conclusion that they derive from a common ancestor (ref. 19 and Guiard and Lederer, to be published). These sequential homologies are in line with the possible conservation of the structural features described for cytochrome b_5, by Mathews et al. [20], in particular the β-sheet, the hydrophobic cores no.1 (heme site) and no.2 (tryptophan site), as discussed by Guiard [21]; this conservation is supported further by the analogies in the NMR spectra [18], in which are resolved a number of lines of methyl protons located near the heme and near the tryptophan.

(c) The correlation between the presence of a bond sensitive to proteases and located between two relatively stable globules, and the trace of a gene-fusion event has been reported and discussed in several cases [22–24]; this idea finds additional support in the case of flavocytochrome b_2.

It should be emphasized that the liver microsome flavoreductase cytochrome b_5 system has not undergone a gene fusion similar to the one that took place in the yeast flavocytochrome b_2 system: the cytochrome b_5 and the NADPH (cytochrome b_5) flavoreductase have been shown to be separate molecules, both being independently and tightly bound to the microsome membrane [25, 26]. However, it should be noted that it could be possible that not only the cytochrome parts of the two systems derive from a common ancestor, but also the flavoreductase parts. In effect, the total polypeptide lengths of the two systems are very similar and moreover, in both cases, the flavoreductase has byitself a bi-globular character (Gervais et al., unpublished results, and ref. 11).

Technological problems in the study of flavocytochrome b_2 reduction

In flavocytochome b_2, there is no problem for the independent observation of the heme group; the α band at 557 nm is quite convenient. For the flavin, the situation is very different since the absorbing bands are in the same region as the δ and γ bands of the heme, and with much lower ϵ_{max} values. An analysis of the isobestic points of the heme in the deflavoderivatives [27, 28] and the study of stopped-flow recordings at a series of wavelengths in the 436 to 480 nm range [29], showed that the only satisfactory means to observe purely flavin modifications is to work precisely at the heme isobestic point of 438.3 nm. The wavelength has to be adjusted precisely on the stopped-flow apparatus for each series of experiments. At this wavelength, the absorbance coefficient values expected for the red semiquinone flavin (as identified by its EPR signal bandwidth of 15 gauss [30]) is not significantly different (as discussed in ref. 28) from that of the fully reduced flavin, so that the changes in absorbance in the course of the reduction reflect directly the changes of the Fl_{ox}/Fl_{total} ratio. This would not hold if the wavelengths 470 or 480 nm were used, as in the former kinetic studies of Morton and Sturtevant [31] and Suzuki and Ogura [32], wavelengths where the red semiquinone has a higher absorbance than the fully reduced form [28].

Advantages of parallel EPR rapid-freezing and stopped-flow absorbance studies

Studies of the EPR signals of the two paramagnetic species, the oxidized heme and the semiquinone flavin (at powers and temperatures where these signals are not saturated), offer the possibility after a standardization process to measure quantitatively the amounts of each of them at different reaction times. The accuracy of EPR rapid-freezing data will evidently be markedly lower than that of absorbance stopped-flow continuous recordings. There can be a systematic error in the evaluation of the amount of semiquinone if the value taken into consideration for the "dilution factor" of the enzyme sample in isopentane, for the frozen sample under study, is not correct. The data presented here are calculated with a dilution factor of 1.7 as checked very recently by Barber and Bray (unpublished results) instead of 2.0 in the study presented in ref. 28. For the haem, such an error cannot occur since the initial 100% oxidized level is known. Once the amounts of the various forms H_{ox}, H_{total}, Fl_{ox}, Fl_{total} and Fl_{SQ} are known, it is possible to calculate Fl_{red} and to trace the whole reaction pattern.

Consideration of the flavocytochrome b_2 reduction processes

The oxidation reduction potentials (E_m) of the various systems involved in flavocytochrome b_2 are at pH 7: −190 mV for the lactate/pyruvate couple [33], −52 mV for the bound "reduced flavin/oxidized flavin" couple [28] at 21°C pH 7.05, +6 mV for the bound "reduced heme/oxidized heme" couple, under the same conditions [28], in close agreement with earlier data [34]. Therefore, the reduction of the enzyme prosthetic groups by lactate, once the reaction is completed, should be quantitative, the number of electrons accepted being equal to twice the initial L-lactate molar amount.

In the tetramer molecule of flavocytochrome b_2 there is one substrate-binding site per protomer [29], therefore, as depicted in Fig. 1, the total reduction of the enzyme requires two successive cycles of lactate binding, one of two lactate molecules per pair of protomers and the other of one lactate molecule per pair of protomers.

A priori, after the binding of lactate in the first cycle, binding which, at saturating lactate concentration, takes place very rapidly ($k_{app} > 1500\ s^{-1}$) [35] three successive events must occur:

(a) The first event will correspond to the transfer of an electron couple given by the lactate molecule to the associated flavin molecule in each active site. Such a direct transfer is predictable from chemical theory and is supported in particular by data presented in the following section. The fact that this step is the main limiting one in the overall process, at the catalytic level, was demonstrated by Morton and Sturtevant [31] since they found that in the presence of both lactate and ferricyanide, the heme and the flavin prosthetic groups remain quantitatively under the oxidized form. On the other hand, Lederer, [36] has found a marked isotopic effect, a reduction by a factor of 5 of the molecular activity, when α-[^{2}H] lactate is compared to the usual protonated form, supporting the idea that it is precisely the activation of the C—H bond of bound lactate which constitutes the much slower step in the whole enzymic process.

(b) The second event, once four reduced flavins and four oxidized hemes are present in the enzyme molecule considering the relative redox potentials of these two systems, should be a transfer of one electron from each flavin to one heme, producing simultaneously one flavin semiquinone and one reduced heme. This reaction, a priori, could be inter- and/or intramolecular, and reversible.

(c) The third event, taking place once the semiquinone is formed, has to be a dismutation reaction between couples of semiquinone flavin, the reaction required to regenerate oxidized flavin groups, the only form which is able to accept the additional electrons given by lactate in the second turn of the cycle. Massey has shown that in the case of L-amino acid oxidase, the semiquinone cannot be reduced by the substrate [37].

Such a series of expected events can be considered as oversimplified, but the results presented and discussed below show that it gives a satisfactory reactional scheme. The clearly established points which should be retained are the following:

There are two, unambiguously established, successive cycles of lactate to enzyme transfers, the first takes place with a rate constant in the 120 s^{-1} range at 24°C and corresponds to the delivery of one electron pair per protomer, all of them apparently in a synchronous manner. The electrons accepted in this process are distributed between heme and flavin groups by means of intramolecular processes.

The second turn of the cycle takes place with a rate constant in the 5 s^{-1} range, that is to say about 25 times slower than in the first turn, and corresponds to the delivery of the third and final electron per protomer. The assumption that the second turn is limited by pyruvate dissociation cannot hold since such a limiting process should also affect in the same proportion the turnover of the enzyme. In fact, the molar activity, 175 electrons per s per mol of monomers, is just 30% lower that the rate expected for a two-electrons transfer with the rate constant established for the first turn ($2 \times 120 = 240\ s^{-1}$)(data at 24°C).

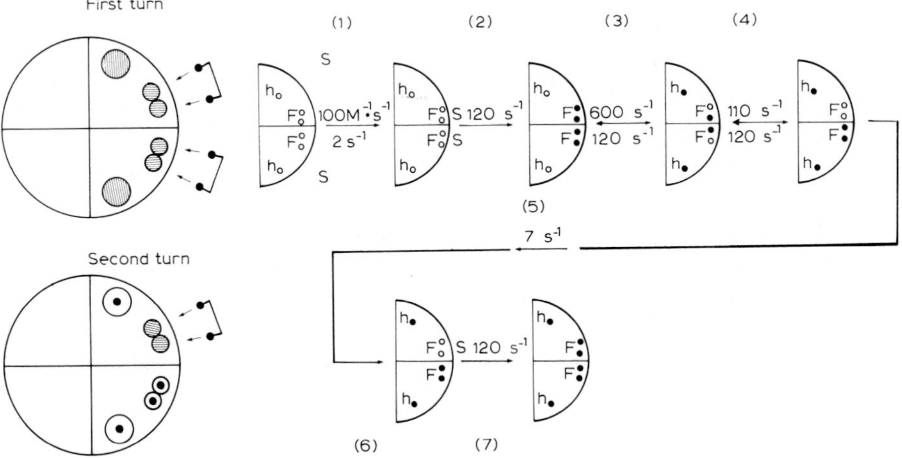

Fig. 1. Scheme of the two-turn mechanism for the reduction of flavocytochrome b_2 (three-electron acceptor) by its substrate (a two-electron donor): Best reactional scheme and rate constants. The rate constants indicated fit the experiments at 24°C with 15 mM lactate presented in Fig. 2 as well as experiments at 250 μM L-lactate [38]. For step 1 read 100 $mM^{-1} \cdot s^{-1}$ instead of 100 $M^{-1} \cdot s^{-1}$.

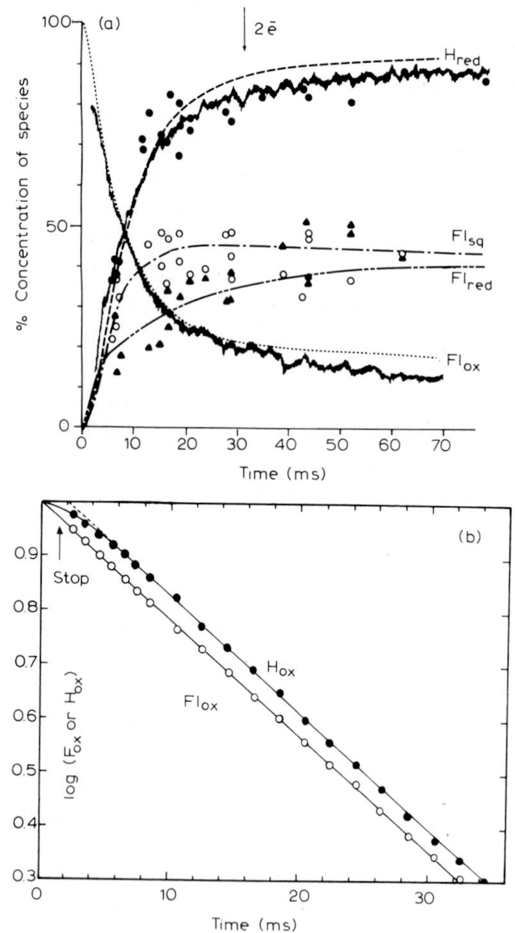

Fig. 2. Reaction pattern during flavocytochrome b_2 reduction. (a) Experiments at 24°C; along with typical stopped-flow absorbance recordings at two wavelengths at 557 nm for H_{red} ($\Delta_{ox}^{red} = 22\ mM^{-1} \cdot cm^{-1}$), at 438.3 nm for Fl_{ox} ($\Delta_{ox}^{red} = 8\ mM^{-1} \cdot cm^{-1}$), (3.9 μM flavocytochrome b_2, 15 mM L-lactate) are given EPR rapid-freezing results (solid circles for H_{red}, circles for Fl_{SQ}) (170 μM flavocytochrome b_2, 15 mM L-lactate). The black triangles correspond to Fl_{red} calculated by the relation $Fl_{total} - Fl_{ox} - Fl_{SQ}$. Dotted lines are simulated data with the scheme and rate constants given in Fig. 1. (b) Experiment at 5°C; first-order plots of the stopped-flow recordings.

In such a situation the reaction scheme can be taken as sequential monomolecular, reversible multistep, of the type shown in Fig. 1.

After a fast pseudo-first-order binding process [1], a second event occurs defined by k_2, slow and irreversible, then several faster reversible steps, then a much slower irreversible step, No. 5. In such a case the aspect of the time courses of the different transients, 2, 3, 4, will be very dependent of the relative magnitude of the various k values. In particular, if k_3 is not very high compared to k_2, we should expect a lag in the appearance of the transient 3, which in our case would be the reduced heme and the semiquinone flavin, and a burst in the appearance of 2, which would be the reduced flavin.

If on the contrary all the rate constants implied in steps 3 and 4, that is k_3, k_{-3}, k_4, k_{-4}, were several orders of magnitude higher than k_2 and k_5, we should expect a true equilibrium between the forms 2, 3 and 4, established after a certain time i.e.

$$(2):(3):(4) = \frac{1}{k_3 k_4}:\frac{1}{k_4 k_{-3}}:\frac{1}{k_{-3} k_{-4}}$$

In intermediate situations, there can be cases of "pseudo-equilibrium" conditions, where the relative proportions of the transients will remain constant for a certain time, but differ from the true equilibrium value.

Reactional time courses of the flavin and heme during reduction

In Fig. 2 (top) is presented the pattern established (at 24°C, pH 7.36), pooling the absorbance stopped-flow recordings at two wavelengths (H_{red} and Fl_{ox}) and the EPR rapid-freezing results at a series of time (H_{red} and Fl_{SQ}). An estimation of the Fl_{red} amount was also calculated by using the relationship $[Fl_{total}] - [Fl_{ox}] - [Fl_{SQ}]$; the pattern is a little different from that presented in ref. 28, the first one being established with a slightly erroneous "dilution factor" (2 instead of 1.7).

This pattern does not change with flavocytochrome b_2 concentrations over a large range; the EPR and absorbance results for H_{red} are in excellent agreement. The limiting processes are intramolecular when the experiments are carried out at saturating lactate concentrations.

The main features of the time courses are the following:

The reaction is biphasic for the species Fl_{ox}, H_{red} and Fl_{sq}; there is a second slow first-order phase which extends over more than 1 s, with a rate constant of $5 \pm 2\,s^{-1}$. Its extrapolation to time zero allows us to determine the plateau of the first phase, corresponding to about 85% of the total amplitude as well for Fl_{ox} as for H_{red}. The first phase is also first order, with the same rate constants for H_{red} and Fl_{ox}, no detectable lag for H_{red}, an increase of Fl_{SQ} up to 45% with an approximately constant ratio Fl_{SQ}/H_{red} and a burst of small amplitude ($>$ 20% before 6 ms) for Fl_{red}.

These various aspects are in line with Fl_{red} as the first product of the rate-limiting process followed by faster processes of electron distribution between the internal acceptor groups.

The first phase ends at 35 ms. If the number of electrons per protomers are summed up at this time, a total of 2 is found. It is therefore possible to conclude, as done previously, that the first phase corresponds to the first turnover of lactate binding, with the four sites of the tetramer reacting apparently synchronously, with a limiting process at the level of a two-electron transfer between bound lactate and flavin, while the second phase of the reaction corresponds to a second turn of lactate-flavin transfer, one per couple of protomers taking place with a rate constant much lower than in the first turn. It is difficult to understand why the lactate-flavin transfer is so different in the two rotations; a feasible explanation would be that the active sites have not the same conformation and the same reactivity before and after the first reduction turn.

Since all the interpretations are supported by the observed biphasicity of the absorbance stopped-flow recordings, it is very important to note that the slow phase in the experiments reported is certainly not due to some denatured enzyme. This fact was established by testing the turnover of all the samples studied under standard conditions, and by testing a series of enzyme samples more or less inactivated; when there is some inactivated enzyme, the reaction is triphasic and it is the slowest additional phase with a rate constant in the 0.04–0.1 s^{-1} range which corresponds to the reduction of this inactivated material by the active one. This third rate constant is that expected, considering the bimolecular rate constant determined in the reduction reaction of the deflavocytochrome b_2 by the reduced flavocytochrome b_2, i.e. $10^5 M^{-1} \cdot s^{-1}$ [29].

Study of the relationship between instantaneous amounts of the transients: equilibrium or pseudo equilibrium?

Considering the proportions of the transients involved in a sequential reversible multistep process as discussed in a previous section, two possible equilibria were examined:
The dismutation equilibrium between two flavins:

$$2\ Fl_{sq} \rightleftharpoons Fl_{red} + Fl_{ox} \qquad K_{FF} = \frac{(Fl_{sq})^2}{(Fl_{red})(Fl_{ox})}$$

The redox equilibrium between the heme and flavin systems:

$$H_{ox} + Fl_{red} \rightleftharpoons H_{red} + Fl_{sq} \qquad K_{FH} = \frac{(H_{ox})(Fl_{red})}{(Fl_{sq})(H_{red})}$$

The calculated K values were compared for true final equilibrium data K_e, at partial reduction, in the range (L-lactate)/(flavocytochrome b_2) $\leqslant 1.5$, and for transients time, K_k, in the course of the reduction by L-lactate at saturating concentrations. The results indicated that for the dismutation, K_e is equal to 0.9 ± 0.2 (S.D.) and $K_k = 2.6 \pm 0.9$ (S.D.) established after about 15 ms. For the heme-flavin reaction, $K_e = 0.11 \pm 0.02$ (S.D.) and $K_k = 0.21 \pm 0.07$ (S.D.) established within 15 ms.

This approach tends to show that for both the reactions considered, the values during the time courses are a little different from that of the final equilibrium, but there may be two possible explanations: either the situation is that of a pseudo equilibrium, or it is that of an equilibrium different from the "final" one, a situation which could be due to the difference in the enzyme state saturated by lactate, or free of lactate, or to a difference in conformations.

In any case, the interpretation is that the rate of electron exchange between the prosthetic groups within the molecule are relatively fast. In the following section, a best estimate of the rate constant involved will be given.

A simulation study to propose a best scheme and best rate constants

A logical reaction scheme shown in Fig. 1 was tested in a thorough simulation study by Capeillère-Blandin [38]. It was shown that it is not possible to find a simpler linear scheme, whatever the rate constants introduced, which gives a satisfactory fitting of the experimental results under various conditions. Similarly, it is not possible to suppress the reversiblity of the steps 2 and 3.

With the scheme considered, a convenient set of rate constants is indicated in Fig. 1. in fact, keeping a constant ratio between the rate constants k_3, k_{-3}, k_4 and k_{-4}, all of them can be increased up to a certain limit but not decreased, the values given being the minimal ones for the fitting. A difficult problem is to know whether or not a highest set of values can be defined.

An answer to this question can be deduced from a similar study, in absorbance stopped flow, but carried out at 5°C instead of 24°C (Fig. 2 at bottom). The first-order plots of the first reduction phase, presented in fine solid lines have the same slopes for the H_{red} and Fl_{ox} recordings, but the flavin one actually extrapolates to time zero, while the heme one extrapolates at a time significantly different from zero, that is 2.0 ± 0.2 ms. These results can be fitted precisely in simulations of a two-step linear scheme, with the "on" rate constants of 50 and 500 s^{-1} for steps 2 and 3; 400 s^{-1} for the second one gives a lag of 2.2 ms while 700 s^{-1} gives a lag of 1.4 ms. Previous studies have shown that this two-step scheme gives the same results as the scheme in Fig. 1 for the early time of the raction [38]. This result indicates that the rate constants proposed in Fig. 1 at 24°C are in a convenient range and are not markedly underestimated.

In conclusion, the results support the estimations that the transfer of one electron from reduced flavin to heme within the molecule takes place at 24°C with a k_+ value between 600 and 1200 s^{-1}, and a k_- value of about 150 s^{-1}, and that k values in the 100 s^{-1} range are involved in the intramolecular dismutation exchange of one electron between two flavins.

Studies on flavocytochrome b_2 of the "non-cleaved" or "physiological" type

The crystalline Appleby and Morton type of preparation [6] which was the object of the study reported above is in fact an enzyme modified as compared with that present in fresh crude extracts. Somlo and Slonimski [3] found that abrupt modifications of K_m and K_i values take place at a certain moment during the preparations. The modification, which also affects the turnover rate by a factor of about × 2 [39] was correlated with a splitting into two parts of the unique polypeptide chain of each protomer [8, 9, 39]; this splitting is due to contaminating proteases.

As the main study presented here was carried out with the "cleaved" form, S_X, of the *S. cerevisiae* flavocytochrome b_2, parallel stopped-flow studies were performed with two types of "non-cleaved" enzyme, that of *S. cerevisiae* (Iwatsubo, Jacq and Lederer, unpublished results), and that of the *H. anomala* yeast (Iwatsubo, unpublished results); termed respectively S_p and H_p (p for "physiological"). The kinetic characteristics of two of them are presented in Table I.

The comparison of the behaviors of cleaved and non-cleaved flavocytochrome b_2 indicates that it is only in the flavoreductase parts that important differences occur; it is also there that the more marked difference is found between the two non-cleaved forms H_p and S_p. The large variations in the rate constants of lactate to flavin transfer are directly reflected in variations of the molar activities and of the first-order rate constant of the first reduction phase; one should have expected such a correlation if in all the cases, the much slower process which controls the rates observed is always the same, that is the lactate to flavin transfer at saturating lactate concentrations.

The stopped-flow absorbance recordings of the heme and flavin time courses present, for the three enzyme species, both common and different aspects. In common is found the biphasicity which indicates that the two-turn mechanism is valid for all of them, the entry of the third electron being much slower than that of the first pair of electrons in each protomer. Also in common is the lag observed at the heme level.

The main apparent difference appears at the level of the first-order rate constant for heme, k_H, at 5°C for the H_p enzyme; there, k_H seems much smaller than k_F, while they are equal for S_x.

In fact, simulation studies have shown that the difference lies not at all at the k_3 level (flavin to heme transfer rate constant, in both cases equal to approx. 400–500 s^{-1}) but at the level of the lactate flavin transfer where the k_2 values are 300 s^{-1} for H_p and 50 s^{-1} for S_x.

There is one interesting point to note in the comparison of the molar activities (at saturating L-lactate and ferricyanide concentrations) with the

TABLE I

Under standard conditions at 30°C the molar activity (*TN*) of various samples were: 210 s^{-1} for S_x and 1250 s^{-1} for H_p.

Experimental stopped-flow studies	S_x (24°C)	S_x (15°C)	H_p (15°C)	S_x (5.5°C)	H_p (5.5°C)
k_{F1} (s^{-1})	120 ± 20	50	450	50	300
k_H (s^{-1})	125 ± 15	75	440 — 520	45	170
haem lag (ms)	<0.5	<0.5	1	2	2
TN (s^{-1})	175	115	520	70	225
$TN/k_{F1} + k_H$*	0.73	0.90	0.56	0.74	0.48
Simulation studies	**				***
k_{+2} (s^{-1}) L-lactate → flavin	120			50	300
k_{+3} (s^{-1}) Flavin → haem	600			500	400 — 500

* It corresponds approximately to the expression $TN/2k_2$ used in the text.
** Simulation of the pattern with the scheme and rate constants presented in Fig. 2.
*** Simulation by use of a simple two-step linear scheme which allows a good fitting of the initial reaction times (see Fig. 1).

reduction rate constants in phase I. If the first one is uniquely limited by the two-electron transfer with k_2, it should be equal to $2k_2$. A smaller value would indicate that another process is slow enough to introduce an additional limitation. The data presented in Table I show that this is actually the case; it is our assumption that the process in question is very probably at the pyruvate dissociation.

Comparison with previous published data

Since certain data and the conclusions of the work presented differ partly from previously reported studies, some comments have to be made to explain these differences. Rapid kinetic studies were carried out in three laboratories, those of Sturtevant, of Ogura and of Iwatsubo. The first and the last one were working with Appleby and Morton type I (with DNA) preparations [5] while a non-crystalline, unstable preparation, probably of the cleaved type was studied in Ogura's laboratory. Only kinetic data obtained directly from reduction stopped-flow recordings will be considered in the comparison; a series of earlier data [40] were obtained from the recordings of oxidation-rereduction pulses when small amounts of ferricyanide were added to lactate-flavocytochrome b_2 mixtures; the interpretations of such experiments in terms of oxidation and reduction rate constants requires certain assumptions concerning the distribution of the electrons between the heme and the flavin groups, so that even though certain conclusions in terms of mechanism might be correct, they were not founded [41].

The differences in the rate constants k_F and k_H, and in the molar activities at saturating lactate and ferricyanide concentrations, if the data presented above are compared to the other previous studies [27, 31, 32], are due to three factors: (1) in refs. 27–31, the biphasicity of the reaction was not observed so that an error in the final level resulted in an underestimation of the rate constants; (2) in refs. 31 and 32 the flavin was followed at 470 nm where both the heme and the flavin semiquinone contribute significantly to the absorbance changes, so that the flavin kinetic parameters were underestimated [28]; (3) in the three works, the absorbance coefficient used to calculate the flavocytochrome b_2 concentrations was overestimated by 27% [42]; consequently, the molar activities were overestimated by the same percentage.

As a result of these different errors, it was not possible to understand the mechanism of the reaction: however, in the discussion of their article, Suzuki and Ogura [32] postulated a scheme in which two turns of lactate binding were proposed, as well as the fact that a similar process, the lactate flavin transfer in turn I, limits also the reduction first phase and the molar activity.

Although the presence of the semiquinone flavin in significant amounts was demonstrated by Hiromi and Sturtevant [43] in partially reduced flavocytochrome b_2 at equilibrium, its kinetic behavior was first studied by Suzuki and Ogura [32]: they found that the accumulation of the semiquinone reaches a plateau in less than 100 ms, and that this plateau, if the signal height is corrected for a saturation effect, might correspond to 100% of the flavin under the semiquinone form. However, the precise correlation between the semiquinone time course and that of heme and flavin was not studied. Now, it is this precise correlation leading to the determination of the complete reaction pattern which has allowed, by means of the simulation studies, to afford estimations of the rate constants involved in the various intramolecular processes within flavocytochrome b_2.

Concluding remarks

One of the aims of the kinetic study of flavocytochrome b_2, to delineate approximated values for the rate constants involved in the internal electron exchanges between the prosthetic groups, seems actually to be fulfilled. The next step should be to elucidate the disposition of hemes and flavins in the molecules; a crystallographic answer to this problem is seriously desired.

Many efforts have been directed towards whether or not heterologous interactions between heme and flavin groups, or homologous interactions between heme-heme or flavin-flavin could be detected. In fact, all of the answers were ambiguous or negative. The studies were of the following types: spectral [42, 44] normal CD [45–47] and magnetic CD [17], fluorescence and phosphorescence [48] and EPR [15, 28]. The only fact possibly revealing an interaction was in a magnetic CD study by Risler et al. [17]. Indeed, a band with a high temperature coefficient, therefore due to a paramagnetic Fe^{3+} species, at the level of the δ band of the haem spectrum, was markedly altered when flavin is released. Another spectral effect, simultaneous small modifications of the heme and flavin spectrum afforded by the fixation of oxalate, was recently reported (Blazy et al., unpublished results and ref. 45).

Concerning the structures allowing the possibility of an intramolecular flavin-flavin dismutation coupling, it should be recalled that, due to the presence of three rectangular, 2-fold symmetry axes in the molecule [49, 50], the position of a flavin close and perpendicular to a symmetry axis will result in its proximity to its symmetric neighbour in another protomer, this design of flavin pairs being present twice in the whole tetramer.

If the reactive site of the flavin for the electron transfer in dismutation is at the level of the 7- and 8-methyl on the aromatic ring as proposed by Favaudon and Lhoste [51], we could imagine a close proximity of pairs of such methyl in two symmetric flavins. It should be underlined that a study of the paramagnetic signals of the semiquinone does not reveal any kind of spin-spin interactions between couples of semiquinone so that up to now a close proximity is purely hypothetical.

Acknowledgements

A. di Franco is gratefully acknowldeged for the communication of unpublished results concerning flavocytochrome b_2 titrations, used here, and for stimulating discussions in particular about the simulation study.

References

1 Somlo, M. (1962) Biochim. Biophys. Acta 65, 333–346
2 Somlo, M. (1964) Bull. Soc. Chim. Biol. 46, 1793–1828
3 Somlo, M. and Slonimski, P.P. (1966) Bull. Soc. Chim. Biol. 48, 1221–1249
4 Pajot, P. and Claisse, M. (1974) Eur. J. Biochem. 49, 275–285
5 Baudras, A. and Spyridakis, A. (1971) Biochimie 53, 943–955
6 Appleby, C.A. and Morton, R.K. (1954) Nature 173, 749–753
7 Morton, R.K. (1955) Soc. Biol. Chem. India, 177–185
8 Labeyrie, F. and Baudras, A. (1972) Eur. J. Biochem. 25, 33–40
9 Jacq, C. and Lederer, F. (1972) Eur. J. Biochem. 25, 41–48
10 Naslin, L., Spyridakis, A. and Labeyrie, F. (1973) Eur. J. Biochem. 34, 268–283
11 Gervais, M., Groudinsky, O. and Labeyrie, F. (1975) Abstr., 10th Meeting FEBS, Paris
12 Yamashita, J., Higashi, T., Yamanaka, T., Nozaki, M., Mizushima, H., Matsubara, H., Horio, T. and Okunuki, K. (1957) Nature 179, 959–960
13 Labeyrie, F., Groudinsky, O., Jacquot-Armand, Y. and Naslin, L. (1966) Biochim. Biophys. Acta 128, 492–503
14 Labeyrie, F., di Franco, A., Iwatsubo, M. and Baudras, A. (1967) Biochemistry 6, 1791–1797
15 Watari, H., Groudinsky, O. and Labeyrie, F. (1967) Biochim. Biophys. Acta 131, 592–594
16 Labeyrie, F. (1966) Bull. Soc. Fr. Physiol. Vég. 12, 17–28
17 Risler, J.L. and Groudinsky, O. (1973) Eur. J. Biochem. 35, 201–205
18 Keller, R., Groudinsky, O. and Wuthrich, K. (1973) Biochim. Biophys. Acta 328, 233–238
19 Guiard, B., Groudinsky, O. and Lederer, F. (1974) Proc. Natl. Acad. Sci. U.S. 71, 2539–2543
20 Mathews, F.S., Czerwinski, E.W. and Argos, P. (1975) in The Porphyrins, (Dolphin, D., ed.), Academic Press, New York, in press
21 Guiard, B. (1974) Thèse de 3ème Cycle, Orsay
22 Goldberg, M. (1969) J. Mol. Biol. 46, 441–446
23 Cassio, D. and Waller, J.P. (1971) Eur. J. Biochem. 20, 283–300
24 Véron, M., Falcoz-Kelly, F. and Cohen, G.N. (1972) Eur. J. Biochem. 28, 520–527
25 Rogers, M.J. and Strittmatter, P. (1974) J. Biol. Chem. 249, 5565–5569
26 Strittmatter, P., Rogers, M.J. and Spatz, L. (1972) J. Biol. Chem. 247, 7188–7194
27 Iwatsubo, M., Baudras, A., di Franco, A., Capeillère, C. and Labeyrie, F. (1968) in Flavins and Flavoproteins (Yagi, K., ed.), pp. 41–58, University Park Press, Baltimore
28 Capeillère-Blandin, C., Bray, R.C., Iwatsubo, M. and Labeyrie, F. (1975) Eur. J. Biochem. 54, 549–566
29 Capeillère-Blandin, C. (1974) Doctoral Thesis, University of Orsay, Paris-Sud No. A09989
30 Palmer, G., Muller, F. and Massey, V. (1971) in Flavins and Flavoproteins (Kamin, H., ed.), pp. 123–140, University Park Press, Baltimore
31 Morton, R.K. and Sturtevant, J. (1964) J. Biol. Chem. 239, 1614–1624
32 Suzuki, H. and Ogura, Y. (1969) J. Biochem. Tokyo 67, 277–289
33 Labeyrie, F., Naslin, L., Curdel, A. and Wurmser, R. (1960) Biochim. Biophys. Acta 41, 501–515
34 Baudras, A. (1965) Bull. Soc. Chim. Biol. 47, 1449–1466
35 Iwatsubo, M. and Capeillère, C. (1967) Biochim. Biophys. Acta 146, 349–366
36 Lederer, F. (1974) Eur. J. Biochem. 46, 393–399
37 Massey, V. and Curti, B. (1967) J. Biol. Chem. 242, 1259–
38 Capeillère-Blandin, C. (1975) Eur. J. Biochem. 56, 91–101
39 Jacq, C. and Lederer, F. (1974) Eur. J. Biochem. 41, 311–320
40 Nakamura, T. and Ogura, Y. (1966) J. Biochem. Tokyo 60, 77–86
41 Baudras, A., Capeillère-Blandin, C., Iwatsubo, M. and Labeyrie, F. (1972) in Structure and Function of Oxidation Reduction Enzymes (Akinson, A. and Ehrenberg, A., eds.), pp. 273–290, Pergamon Press, London
42 Pajot, P. and Groudinsky, O. (1970) Eur. J. Biochem. 12, 158–164
43 Hiromi, K. and Sturtevant, J.M. (1966) in Flavins and Flavoproteins (Slater, E.C., ed.), pp. 283–303, Elsevier, Amsterdam
44 Baudras, A. (1965) Bull, Soc. Chim. Biol. 47, 1143–1175
45 Iwatsubo, M. and di Franco, A. (1968) in Structure and Function of Cytochromes (Okunuki, K., Kamen, M.D. and Sekuzu, I., eds.), pp. 613–625, University Park Press, Baltimore
46 Iwatsubo, M. and Risler, J.L. (1969) Eur. J. Biochem. 9, 280–285
47 Sturtevant, J.M. and Tsong, T.Y. (1968) J. Biol. Chem. 243, 2359–2366

48 Risler, J.L. (1971) Biochemistry 10, 2664–2669
49 Monteilhet, C. and Risler, J.L. (1970) Eur. J. Biochem. 12, 165–169
50 Olive, J., Barbotin, J.N. and Risler, J.L. (1973) Int. J. Peptide Protein Res. 5, 219–228
51 Favaudon, V. and Lhoste, J.M. (1975 I and II) Biochemistry 14, 4731–4744

Part XI

Adrenodoxin, ferredoxin and cytochrome reductases and electron transferring flavoprotein

Flavins and Flavoproteins, edited by T.P. Singer

CHAPTER 69

ADRENODOXIN REDUCTASE: FLUORESCENCE PROPERTIES AND THE EFFECTS OF SALTS ON ITS CATALYTIC ACTIVITIES

TOKUJI KIMURA, JAU-WEN CHU and JEFF PARCELLS

Department of Chemistry, Wayne State University, Detroit, Mich. 48202 (U.S.A.)

Introduction

The electron transfer system which functions in steroid hydroxylation reactions in adrenal cortex mitochondria consists of an NADPH-dependent flavoprotein dehydrogenase (adrenodoxin reductase, EC 1.6.7.1, NADPH: ferredoxin oxidoreductase), an iron-sulfur protein (adrenodoxin) and a heme protein (cytochrome *P*-450).

In our previous reports [1, 2] some molecular and catalytic properties of this NADPH-dependent dehydrogenase system have been described. Briefly, these characteristics can be summarized as follows. Adrenodoxin reductase contains 1 mol of FAD along with 2% hexose (w/w) per mol of protein. The amino acid composition of this flavoprotein revealed a high content (46%) of hydrophobic residues. Its molecular weight was estimated at about 54 000. The oxidized protein had absorbance maxima at 270, 377 and 450 nm with small shoulders at 425 and 480 nm. The oxidation-reduction potential was found to be −0.274 volts at pH 7.0. The K_m values NADPH and NADH are 1.82 μM and 5.56 mM, respectively. Adrenodoxin reductase and adrenodoxin form a complex with a stoichiometric molar ratio of 1:1 [3]. The dissociation constant was calculated to be approx. 10^{-9} M in a low ion strength medium.

In this paper, we wish to report the fluorescence properties of this flavoprotein and the changes in the catalytic activity of adrenodoxin reductase-adrenodoxin complex with salt.

Results and Discussion

Fluorescence properties

As prepared by a method [2] essentially similar to the procedures of Omura et al. [4], adrenodoxin reductase had a high intensity fluorescence emission at 520 nm. This intensity was comparable to that of free FAD. In

Abbreviation: Cl_2Ind, 2, 6-dichlorophenolindophenol.

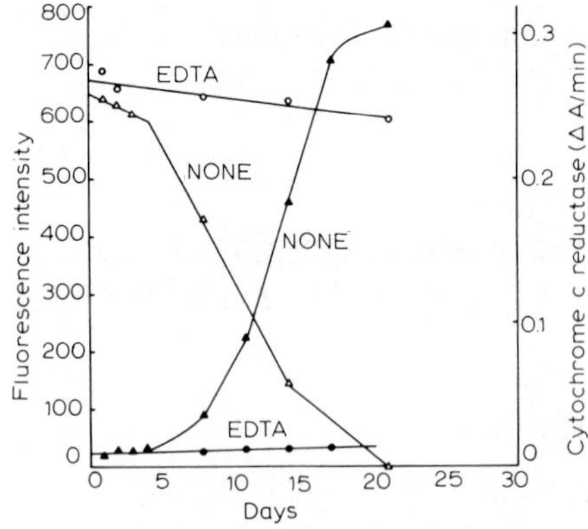

Fig. 1. Relationship between increase in fluorescence intensity and decrease in NADPH-cytochrome *c* reductase. The reaction mixture for the cytochrome *c* reduction contained: adrenodoxin reductase which was kept on at 5°C for an indicated period of time, $3.3 \cdot 10^{-8}$ M; cytochrome *c*, $6.7 \cdot 10^{-6}$ M; adrenodoxin, (without the heat treatment); $1.0 \cdot 10^{-6}$ M; and NADPH, $1.6 \cdot 10^{-5}$ M; in 3.0 ml of 0.01 M phosphate, pH 7.4. The assay temperature was 25°C. The mixture for the fluorescence measurements contained: adrenodoxin reductase, $1.0 \cdot 10^{-6}$ M in 3.0 ml of 0.05 M phosphate buffer, pH 7.4. The temperature of measurements was 5°C.

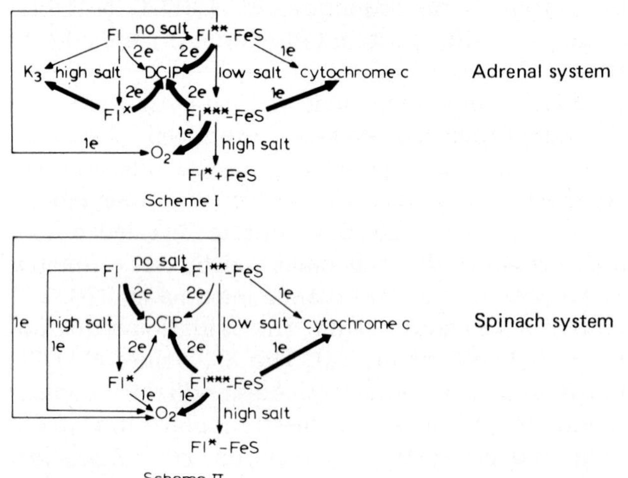

TABLE I

FLUORESCENCE INTENSITY OF ADRENODOXIN REDUCTASE

The flavin concentrations were adjusted to $2.25 \cdot 10^{-6}$ M. The excitation and emission wavelengths were 380 and 515 nm, respectively.

Sample	Emission Intensity	%
Free FAD	70	100
Free FMN	770	1100
Arenodoxin reductase	11	16
+ Mn (8 mM)*	385	550

*Incubated at 38°C for 2 h.

the literature, there are not many flavoproteins with such a high fluorescence intensity. One well-known example is lipoyl dehydrogenase [5]. We have, therefore, investigated further the fluorescence properties of this reductase.

When the purified enzyme preparation, using the previous method, was kept at 0°C for a prolonged period of time, the fluorescence intensity of the flavin gradually increased to reach an intensity comparable to that of FMN. Concomitantly, its enzymatic activity was lost (Fig. 1). We subsequently found that this process was strongly inhibited by the addition of EDTA. This result suggested that the purified preparation contains a catalytic amount of Mn^{2+}- or Mg^{2+}-activated phosphodiesterase which catalyzes the conversion of FAD to FMN. Therefore, the enzyme was then purified using solutions containing 5–10 mM EDTA, and the preparation was subjected to hydroxyapatite column chromatography in its last purification step. Most of the diesterase was removed from the protein. The resulting enzyme preparation, which was much more stable upon storage than the enzyme prepared without EDTA, had a flavin fluorescence corresponding to about 15–20% of free FAD. When this preparation was incubated at 38°C for 2 h in the prescence of Mn^{2+}, an increase in the fluorescence intensity was observed.

The purified flavoprotein, with the improved method, had a ratio of absorbance at 270 nm to that of 450 nm of about 7. The turnover number of the reductase was estimated to be 650/min at 22°C and pH 7.4, in the presence of 0.02 M phosphate buffer with an excess amount of adrenodoxin to the reductase and a fixed amount of NADPH.

Our value for the turnover number appears to be low in comparison with a recently reported value of 800/min by Foster and Wilson [6]. Yet, in any direct comparison, one has to take into consideration the concentration of reactants present in the assay mixture. As will be shown later, this activity is dependent on both the ionic strength of the medium and the temperature of the assay (the activation energy is 14.2 kcal/mol).

Binding of $NADP^+$ to adrenodoxin reductase

When adrenodoxin reductase was titrated with NADPH under anaerobic conditions, a flat absorbance band extending from 505 to 750 nm was observed. This is due to the formation of a complex between $NADP^+$ and the reductase. $NADP^+$ is a strong inhibitor for the reductase with an inhibition constant of 5.32 μM. In fact, the oxidized protein binds to $NADP^+$, producing a difference spectrum in the flavin region (Fig. 2). The stoichiometry of $NADP^+$ bound to the enzyme was approximately 1:1.

Shin [7] has reported that the fluorescence of spinach ferredoxin is enhanced by the addition of $NADP^+$. We have examined the effects of $NADP^+$ on the fluorescence emission of adrenodoxin reductase. Fig. 3 shows the titration of intensity versus $NADP^+$. In contrast to Shin's observations on spinach ferredoxin reductase, the flavin fluorescence intensity of adrenodoxin reductase neither increased nor dramatically decreased on addition of $NADP^+$.

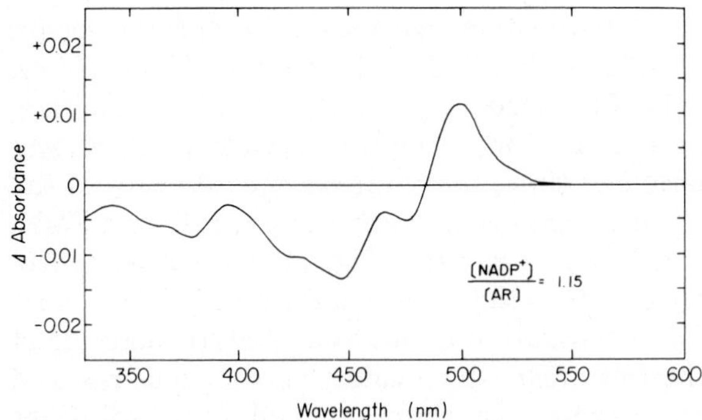

Fig. 2. Difference absorbance spectrum induced by the binding of $NADP^+$ to adrenodoxin reductase. The reaction mixture contained: adrenodoxin reductase, $2.54 \cdot 10^{-5}$ M, and $NADP^+$, $2.79 \cdot 10^{-5}$ M in 1.20 ml of 0.04 M phosphate buffer, pH 7.4. The difference spectrum between a mixture of adrenodoxin reductase and $NADP^+$ versus its individual components in separate compartments of a reference tandem cuvette was recorded at 12.5°C.

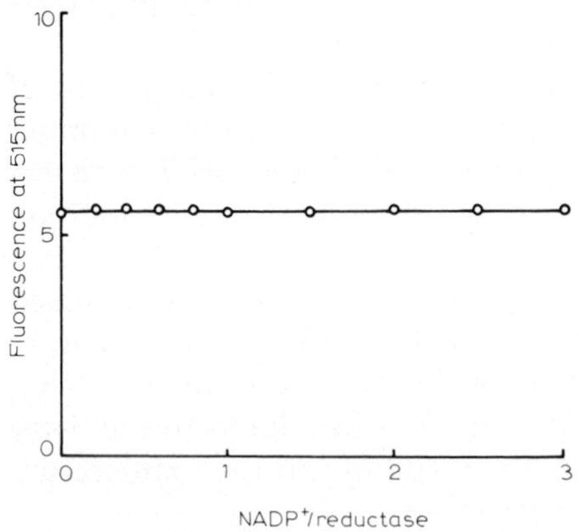

Fig. 3. Titration of fluorescence intensity with $NADP^+$. The reaction mixture contained adrenodoxin reductase, $1.9 \cdot 10^{-6}$ M in 3.0 ml of 0.025 M phosphate buffer, pH 7.4. The titration was carried out by adding $1.2 \cdot 10^{-3}$ M $NADP^+$. The excitation and emission wavelengths were 470 and 515 nm, respectively. Temperature was 22°C.

Effects of salt on flavoprotein catalysis

Studies on the effects of salt on flavoprotein catalysis have received considerable interest. In 1968, Kamin and Masters [8] were the first to report the effects of high concentrations of salt on the specificity of the lipase-solubilized, microsomal NADPH-cytochrome *c* reductase system. Their detailed results, which appeared in a later publication [9], showed that the reductase cannot reduce known microsomal cytochromes such as b_5 and *P*-450. However, in a medium of high ionic strength, it does in fact reduce cytochrome b_5. In addition, these NADPH-cytochrome *c* and menadione reductase activities were largely stimulated by the addition of salts. These findings were recently confirmed by Prough and Masters [10].

In 1971 Nakamura elaborated upon this with studies on the effects of

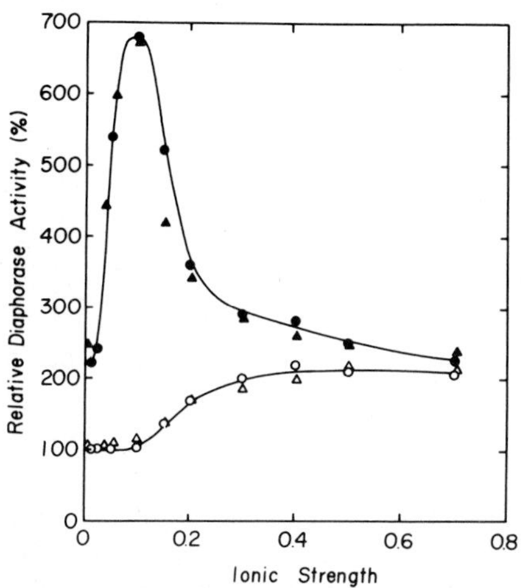

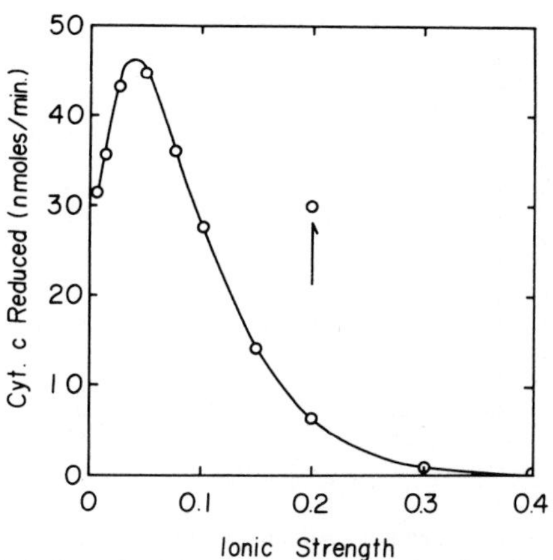

Fig. 4. Effects of salt concentrations on Cl_2Ind reductase activity by adrenodoxin reductase. The reaction mixture contained: adrenodoxin reductase, $1.18 \cdot 10^{-7}$ M; adrenodoxin (when present), $1.13 \cdot 10^{-6}$ M; NADPH, $1.37 \cdot 10^{-5}$ M; and Cl_2Ind, $1.93 \cdot 10^{-5}$ M in 3.0 ml of phosphate buffer, pH 7.4 (○ and ●), or Tris buffer pH 7.4 (△ and ▲). The assay temperature was 22°C. ●, ▲, with adrenodoxin; ○, △, without adrenodoxin.

Fig. 5. Effects of salt concentrations on cytochrome *c* reductase activity by adrenodoxin reductase and adrenodoxin. The reaction mixture contained: adrenodoxin reductase, $2.75 \cdot 10^{-7}$ M; adrenodoxin, $2.75 \cdot 10^{-6}$ M; NADPH, $1.2 \cdot 10^{-5}$ M; cytochrome *c* $1.98 \cdot 10^{-5}$ M in 3.0 ml of 0.01 M phosphate buffer, pH 7.4. The ionic strength was adjusted by the addition of NaCl. (○) indicated by the arrow represents the activity at ionic strength of 0.1 of the complex, which was preincubated at ionic strength of 0.2 and 10 min at 4°C. The assay temperature was 22°C.

salt on spinach ferredoxin reductase activity [11]. He found that the NADPH-Cl_2Ind reductase activity of the reductase alone was inhibited about 80% by increasing the concentration of NaCl from zero to 0.4 M, whereas the activity catalyzed by the reductase in the presence of ferredoxin increased by varying the NaCl concentrations from zero to 0.1 M. Higher concentrations of NaCl than 0.1 M had inhibitory effects on this activity. At zero concentration of NaCl, the Cl_2Ind-reductase activity was inhibited about 50% by the addition of ferredoxin. Ferredoxin acts as a partially competitive inhibitor with respect to NADPH.

In the case of adrenodoxin reductase, the effects of salt are similar to those of spinach ferredoxin reductase, but not identical. The Cl_2Ind reductase activity of adrenodoxin reductase alone increased two times by increasing the concentrations of NaCl or KCl. The activity by the reductase in the presence of adrenodoxin increased as the salt concentrations increased from an ionic strength of zero to 0.15 (Fig. 4). At the zero salt concentration, the Cl_2Ind reductase activity was stimulated two fold by the addition of adrenodoxin. Essentially similar effects were observed when NaCl was replaced by KCl or $MgCl_2$, indicating that the effects were not due to a specific salt, but by the ionic strength of the solution.

The effects of salt on cytochrome *c* reductase activity were then examined. Here, adrenodoxin reductase lacks activity toward cytochrome *c* reduction in the absence of adrenodoxin. Fig. 5 shows the results with

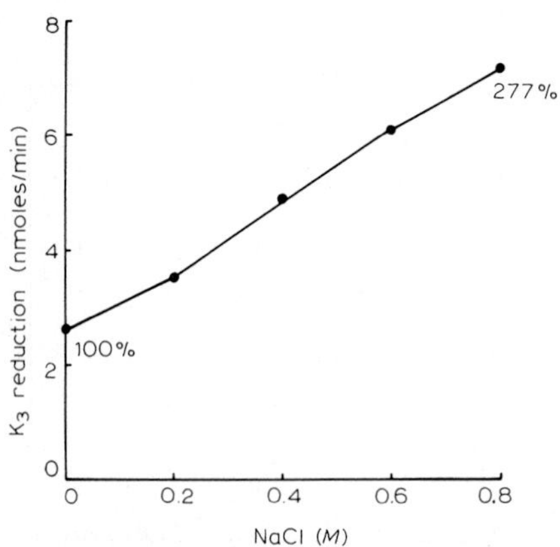

Fig. 6. Effects of salt concentrations on vitamin K_3 reductase activity by adrenodoxin reductase. The reaction mixture contained: adrenodoxin reductase, $3.4 \cdot 10^{-7}$ M; vitamin K_3 which was suspended in H_2O by sonication, $7.5 \cdot 10^{-5}$ M; cytochrome *c*, $1.3 \cdot 10^{-5}$ M; and NADPH, $8.8 \cdot 10^{-5}$ M in 3.0 ml of 0.006 M phosphate buffer, pH 7.2. The absorbance changes at 550 nm by cytochrome *c* were recorded at 22°C. For calculation, vitamin K_3 was assumed as a two-electron acceptor for this reaction. The salt concentrations were adjusted by adding NaCl.

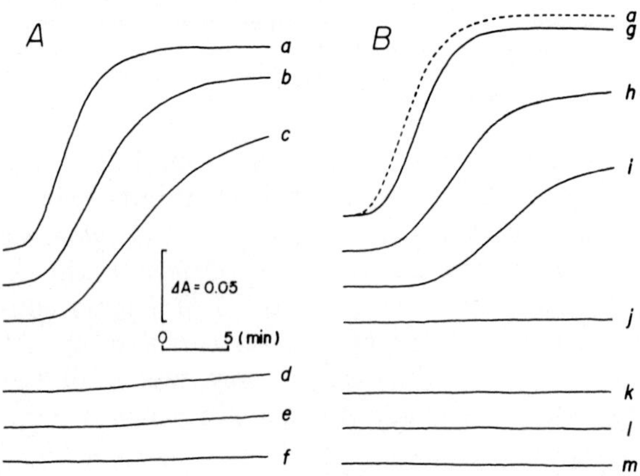

Fig. 7. Superoxide generation by adrenodoxin reductase and inhibition by superoxide dismutase. a: adrenodoxin reductase, $7.56 \cdot 10^{-7}$ M; adrenodoxin, $7.56 \cdot 10^{-7}$ M; b: adrenodoxin reductase, $4.4 \cdot 10^{-7}$ M; adrenodoxin, $4.4 \cdot 10^{-7}$ M; c: adrenodoxin reductase, $3.24 \cdot 10^{-7}$ M; adrenodoxin, $3.24 \cdot 10^{-7}$ M; d: adrenodoxin reductase, $7.56 \cdot 10^{-7}$ M; e: adrenodoxin reductase, $4.4 \cdot 10^{-7}$ M; f: adrenodoxin reductase, $3.24 \cdot 10^{-7}$ M; g: same as curve a; dismutase, 0.67 μg/ml; h: same as curve a; dismutase, 1.67 μg/ml; i: same as curve a; dismutase, 3.0 μg/ml; j: same as curve a; dismutase, 6.7 μg/ml; k: same as curve d; dismutase, 0.67 μg/ml; l: same as curve d; dismutase, 1.67 μg/ml; m: same as curve d; dismutase, 3.0 μg/ml. The concentrations for other components in the assay mixture are: epinephrin, $5.03 \cdot 10^{-5}$ M; NADPH, $15.8 \cdot 10^{-5}$ M; EDTA, $3.02 \cdot 10^{-5}$ M in 3.0 ml of 0.01 M phosphate buffer pH 7.4. Assay temperature was 25°C.

adrenodoxin reductase and adrenodoxin. The maximal activity was obtained at an ionic strength of 0.05. Additionally, antiserum against adrenodoxin inhibited the cytochrome *c* reductase activity completely, whereas it had neither inhibitory nor stimulatory effects on the Cl_2Ind reductase activity catalyzed by adrenodoxin reductase and adrenodoxin. Therefore, electrons

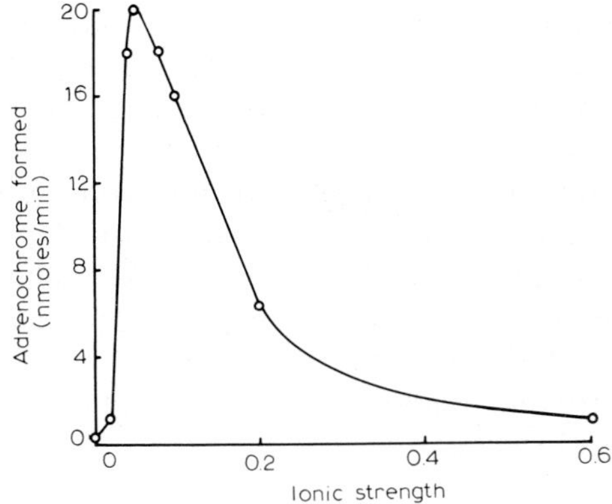

Fig. 8. Effects of salt concentrations on epinephrin cooxidation by adrenodoxin reductase and adrenodoxin. The reaction mixture contained: adrenodoxin reductase, $3.4 \cdot 10^{-7}$ M; adrenodoxin, $3.9 \cdot 10^{-7}$ M; epinephrin, $5.0 \cdot 10^{-5}$ M; EDTA, $2.5 \cdot 10^{-5}$ M; and NADPH, $17.4 \cdot 10^{-5}$ M in 3.0 ml of 0.006 M phosphate buffer, pH 7.2. The ionic strength was adjusted by the addition of NaCl. Assay temperature was 22°C.

from NADPH flow through the reductase toward Cl_2Ind even in the presence of adrenodoxin, and through the reductase and adrenodoxin to cytochrome c.

Next, we examined the effects of salt on the vitamin K_3 reductase activity of adrenodoxin reductase alone. Fig. 6 indicates that as the concentration of NaCl increases, the K_3 reductase activity without adrenodoxin increases. The profile was similar to that of the Cl_2Ind reductase activity with the reductase alone (Fig. 4, lower curve).

Superoxide generation and effects of salt

Fig. 7 shows the adrenodoxin reductase-mediated formation of adrenochrome from epinephrin, which is known to be initiated by superoxide [12]. Adrenodoxin reductase alone did not produce any significant amount of superoxide, but upon the addition of adrenodoxin a marked generation of superoxide was detected. This reaction was able to be inhibited by the addition of superoxide dismutase.

Fig. 8 indicates the effects of NaCl on the epinephrin co-oxidation catalyzed by adrenodoxin reductase and adrenodoxin. As can be seen clearly from the figure, the epinephrin co-oxidation during the course of NADPH oxidation is stimulated dramatically by a low concentration of NaCl. The maximal stimulation was seen at an ionic strength of 0.05. In the absence of adrenodoxin, the reductase alone did not generate any significant amount of O_2^- in a wide range of salt concentrations. The optimum ionic strength for the generation coincides with that for the cytochrome *c* reductase activity and it is slightly lower than that for the Cl_2Ind reductase activity by adrenodoxin reductase and adrenodoxin.

Nakamura [13] observed the effects of NaCl on sulfite co-oxidation catalyzed by spinach ferredoxin reductase during the course of NADPH oxidation by molecular oxygen. The rate of the sulfite co-oxidation reaction was dependent on both ferredoxin and salt. At 0.4 M NaCl, the rate of oxygen consumption increased five fold, relative to that in the absence of NaCl.

Table II shows the effects of superoxide dismutase on the NADPH-Cl_2Ind, cytochrome *c*, K_3 reductase and epinephrin co-oxidation mediated by adrenodoxin reductase. These results indicate strongly that the reaction mediated by O_2^- is less than 15% of the total reductase activity in these reactions. As illustrated in Fig. 8, at an ionic strength of 0.05 the adrenodoxin reductase-adrenodoxin system should generate a considerable amount of O_2^-. Yet, these oxygen radicals do not seem to be responsible for the cytochrome *c* reduction reaction. This can be inferred by both the pronounced lag period (about 5 min) for the O_2^- generation and the rate of their reaction being about 20-times slower than the cytochrome *c* reduction. In fact, the cytochrome *c* reduction is completed during this lag period.

Conclusions

From these results, the effects of salt on the flavoprotein-iron-sulfur protein systems can be summarized as follows (Schemes I and II):

(1) At high salt concentrations, the flavoprotein, per se, presumably changes it conformation. The activity mediated by the flavoprotein alone either increases in the case of adrenodoxin reductase, or decreases, as in the case of ferredoxin reductase.

(2) In the absence of salt, a tight complex between the two proteins occurs with a dissociation constant in the order of 10^{-9}–10^{-8} M. In the adrenal system, the Cl_2Ind reductase activity increases by the addition of adrenodoxin, while that of spinach ferredoxin reductase decreases by the addition of ferredoxin.

(3) At low salt concentrations, the reductase is able to form a complex with the respective ferredoxins with a dissociation constant of 10^{-6} M, and reductase activities largely increase. Here, at least two factors have to be considered: one is the direct effect of salt on the flavoprotein level; and the other is the influence of ferredoxin binding to the flavoprotein. The former effect is relatively small at these salt concentrations (0.5–0.15 M): the

TABLE II

EFFECTS OF SUPEROXIDE DISMUTASE ON ADRENODOXIN REDUCTASE ACTIVITIES

The reaction mixtures contained the concentrations indicated below. The other conditions were similar to those shown in Fig. 3, 4, 5 and 7. Adrenodoxin reductase (AR): for Cl_2Ind reductase, $6.7 \cdot 10^{-8}$ M; for cytochrome *c* and K_3 reductase, $3.3 \cdot 10^{-8}$ M, and for epinephrin cooxidation, $3.4 \cdot 10^{-7}$ M. Adrenodoxin (AD) (when present): for cytochrome *c* reductase, $3.3 \cdot 10^{-7}$ M and for epinephrin cooxidation, $5 \cdot 10^{-5}$ M. The assay temperature was 22°C.

Systems	NaCl(M)	Superoxide dismutase (mol/min/mol)		
		—	+	+/— (%)
NADPH $\xrightarrow{AR}$ Cl_2Ind	0.60	75	65	86
NADPH $\xrightarrow{AR-AD}$ cytochrome *c*	0.05	440	390	89
NADPH $\xrightarrow{AR}$ K_3	0.60	140	125	89
NADPH $\xrightarrow{AR-AD}$ epinephrin	0.05	20	0	0

Cl_2Ind reductase activity of the reductase alone is not affected by these amounts of NaCl (Fig. 4), the latter effect being predominant, causing the enhanced activity.

(4) At high concentrations, the complex tends to dissociate into its components. Accordingly, the enzyme activities mediated by the complex decrease to the level of the flavoprotein alone.

(5) With respect to the generation of superoxide, the adrenal system differs from the spinach system. Adrenodoxin reductase alone does not generate any superoxide, regardless of the ionic strength of the medium. When the reductase molecule binds to the adrenodoxin molecule, this complex generates superoxide at 0.05 ionic strength. Neither at zero nor at higher salt concentrations (more than 0.05 ionic strength) is the superoxide generation markedly produced.

In contrast, spinach ferredoxin reductase generates appreciable amounts of superoxide in both the presence and absence of ferredoxin. The rate of production increases as the ionic strength increases. The salt concentration for maximal rate stimulation was 0.4 M NaCl, which is eight times higher than that for the adrenal system.

These comparisons are made at $2-3 \cdot 10^{-5}$ M NADPH. The affinity of NADPH to the flavoprotein also depends on the salt concentrations. The K_m for NADPH remains unchanged up to an ionic strength of 0.15 and then increases with a further increase in ionic strength. In addition to this, an excess substrate inhibition was observed in the absence of NaCl, while this inhibition was not observed in a higher ionic strength buffer. These results suggest that the configuration of the NADPH-binding site is rather resistant to ionic strength. However, changes in the site may occur at high ionic strengths.

Finally, it is of interest to compare the properties of the flavoprotein ferredoxin complex with those of the microsomal NADPH-cytochrome c reductase. Table III summarizes this comparison. Microsomal NADPH-cytochrome c reductase has 1 mol of FAD and 1 mol of FMN per mol of

TABLE III

EFFECTS OF SALT ON ADRENODOXIN REDUCTASE, SPINACH FERREDOXIN REDUCTASE AND MICROSOMAL NADPH-CYTOCHROME *c* REDUCTASE

The percent of activity was calculated as the activity at optimum salt concentration/the activity without salt × 100. ND: not determined; — no or negligible activity.

Enzyme systems	Cl_2Ind (%)	Cytochrome *c* (%)	K_3 (%)	O_2^- generation (%)
Adrenodoxin reductase	200	—	> 280	—
+ adrenodoxin	350	150	—	6700
Ferredoxin reductase	20	—	ND	100
+ ferredoxin	450	450	ND	500
microsomal NADPH-cytochrome *c* reductase	ND	260*	1500*	stimulated**

*From Bilimoria and Kamin (9).
**From Coon et al. (15).

protein [14]. Our complexes have 1 mol of FAD and 1 mol of Fe_2S_2 center per mol of protein complex. Most cases showed a marked stimulation of their enzyme activities with increasing ionic strength, although there are salient differences between adrenodoxin reductase and spinach ferredoxin reductase in this respect. It is safe to say that salts change the specificity of flavoproteins to their electron acceptors, presumably due to the changes in protein conformation. The critical salt concentration appears to be around 0.10 M.

Acknowledgements

This study was supported by a Research Grant from the National Institutes of Health (AM-12713).

References

1 Kimura, T. (1968) in Functions of the Adrenal Cortex (McKerns, K.W., ed.), Vol. 2, pp. 993–1006, Appleton-Century-Crofts, New York
2 Chu, J.W. and Kimura, T. (1973) J. Biol. Chem. 248, 2089–94
3 Chu, J.W. and Kimura, T. (1973) J. Biol. Chem. 248, 5183–87
4 Omura, T., Sanders, E., Estabrook, R.W., Cooper, D.Y. and Rosenthal, O. (1966) Arch. Biochem. Biophys. 117, 660–673
5 Palmer, G. and Massey, V. (1968) in Biological Oxidations (Signer, T.P., ed.), pp. 263–300, Interscience, New York
6 Foster, R.P. and Wilson, L.D. (1974) Fed. Proc. 33, Abstr. 1291
7 Shin, M. (1973) Biochim. Biophys. Acta 292, 13–19
8 Kamin, H. and Masters, B.S.S. (1968) in Enzyme Oxidation of Toxicants (Hodgson, E., ed.), p. 5, North Carolina State University Press, Raleigh, N.C.
9 Bilimoria, M.H. and Kamin, H. (1973) Ann. N.Y. Acad. Sci. 212, 428–48
10 Prough, R.A. and Masters, B.S.S. (1974) Arch. Biochem. Biophys. 165, 263–67
11 Nakamura, S. and Kimura, T. (1971) J. Biol. Chem. 246, 6235–41
12 McCord, M. and Fridovich, I. (1969) J. Biol. Chem. 244, 6049–55
13 Nakamura, S. (1970) Biochem. Biophys. Res. Commun. 41, 177–83
14 Vermilion, J.L. and Coon, M.J. (1974) Biochem. Biophys. Res. Commun. 60, 1315–22
15 Coon, M.J., Van der Hoeven, T.A., Kaschnitz, R.M. and Stobel, H.W. (1973) Ann. N.Y. Acad. Sci. 212, 449–457

Flavins and Flavoproteins, edited by T.P. Singer

CHAPTER 70

ADRENODOXIN REDUCTASE: COMPLEX FORMATION WITH NADP(H)

JOHN DAVID LAMBETH and HENRY KAMIN

Department of Biochemistry, Duke University Medical Center, Durham, N.C. 27710 (U.S.A.)

Introduction

Adrenodoxin reductase, (NADPH: ferredoxin oxidoreductase, EC 1.6.7.1), from adrenocortical mitochondria, contains one FAD per molecule and no FMN [1–4]. It is part of this organelle's NADPH-dependent steroid hydroxylating system, which has been resolved into the flavoprotein, an iron-sulfur protein (adrenodoxin), and a heme-protein or proteins (cytochrome *P*-450) [5–7]. Chu and Kimura previously reported that anaerobic reduction of bovine adrenodoxin reductase with NADPH resulted in a spectrum with a broad absorbance band, 505 to 750 nm [1], suggestive of a charge-transfer complex between the flavoprotein and NADP(H). They also reported a marked catalytic specificity for NADPH over NADH as reductant [1]. Because the specificity of the hydroxylating system for NADPH as reductant resides with the flavoprotein, we decided to measure the spectra during titrations of adrenodoxin reductase using this and a number of other reductants, and to investigate the nature of spectral changes seen on reduction with NADPH.

Methods, Results and Discussion

Adrenodoxin reductase and adrenodoxin were purified from bovine adrenal cortex according to a modification of the method of Omura et al. [5], to a 272:450 nm absorbance ratio of under 10, suggesting a high degree of purity [3]. Titrations were performed at room temperature using a Foust apparatus [8], with 10 mM glucose and catalytic glucose oxidase to further insure anaerobiosis. EPR spectra were obtained at −100 and −170°C, at protein concentrations of 3–5 mg/ml.

We have confirmed that reduction of adrenodoxin reductase with NADPH yields a spectrum with low, broad long-wavelength absorbance (Fig. 1). The flavoprotein, reduced with 1.1 equivalents of NADPH, also shows residual absorbance at 450 nm and 377 nm. This spectrum is unlike that of semiquinone forms hitherto described [9, 10]. Oxidized and NADPH-reduced adrenodoxin reductase showed no EPR signals at $g = 2.00$. These

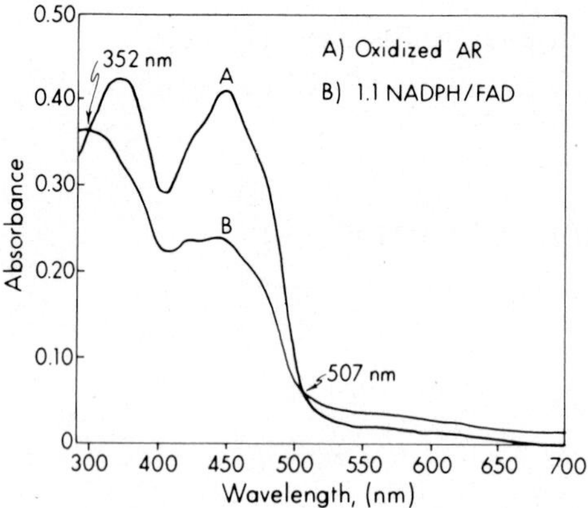

Fig. 1. Spectrum of adrenodoxin reductase reduced by NADPH. (A) Absorbance spectrum of oxidized adrenodoxin reductase; 84.6 nmol in 2.25 ml 10 mM potassium phosphate, pH 7.76. (B) Absorbance spectrum of adrenodoxin reductase following anaerobic addition of 93.1 nmol of NADPH. All spectra were recorded using a Cary-14 recording spectrophotometer.

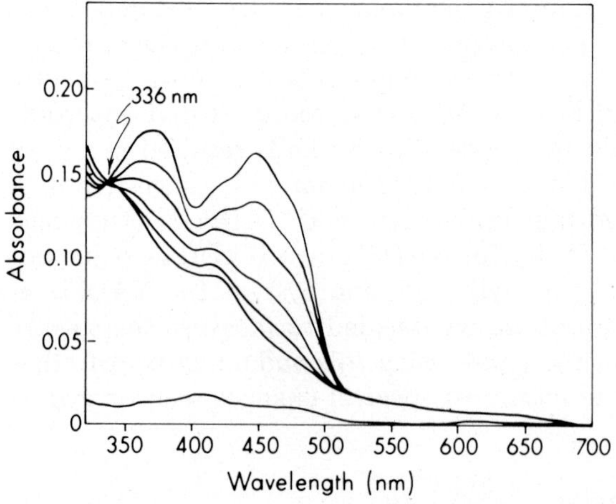

Fig. 2. Anaerobic reduction of adrenodoxin reductase with an NADPH generating system. Solution contained 30.3 nmol of adrenodoxin reductase and 3.0 nmol $NADP^+$ in 2.2 ml volume; also $2.2 \cdot 10^{-3}$ M isocitrate and 0.72 units isocitrate dehydrogenase. Absorbance spectra were recorded at 0,1,9,14,21 and 36 min and did not change thereafter.

data suggested that the spectrum observed represented a species other than a semiquinone.

We have observed that reduction of the flavoprotein with reducing agents alternate to stoichiometric NADPH yielded a spectrum without long wavelength absorbance or residual peaks at 450 or 377 nm, a spectrum more typical of fully reduced flavin. We will designate this spectral species as the "fully reduced" spectrum. Such a spectrum has been reported for the enzyme following dithionite reduction [1]. We confirm this report, and have also observed the fully reduced spectrum upon photo-reduction by light-EDTA, by a 3-fold excess of NADH, and by using an NADPH-generating system in place of equimolar NADPH.

The spectrum (Fig. 2) of flavoprotein reduced by an NADPH-generating system consisting of a catalytic amount of $NADP^+$, isocitrate, and isocitrate dehydrogenase showed no long-wavelength absorbance and no residual peaks at 450 or 377 nm. This suggested that the spectrum observed on reduction of adrenodoxin reductase with equimolar NADPH (Fig. 1B) was due to direct complex formation between the NADP(H) and flavoprotein, and did not arise as an inevitable mechanistic consequence of reduction of enzyme by this coenzyme.

Anaerobic titration of oxidized adrenodoxin reductase with NADPH (Fig. 3), showed decreasing absorbance at 450 nm and increasing absorbance at 700 nm, with a sharp break at 1 NADPH added per flavin, indicating a 1:1 stoichiometry. Titration of reduced adrenodoxin reductase with $NADP^+$ is indicated in Fig. 4. Adrenodoxin reductase was first reduced with 1.1 equivalents of dithionite to produce the fully reduced spectrum (Fig. 4, curve 2). The reduced flavoprotein was then titrated with $NADP^+$, producing a spectrum identical to that obtained by reduction of oxidized flavoprotein with NADPH (Fig. 4, curve 3). Again a 1:1 stoichiometry was observed (see inset of Fig. 4), at either 450 or 700 nm. These data indicate a 1:1 complex between NADP(H) and flavoprotein. The sharpness of the titration plots indicates that the complex has a low K_D, less than $10^{-8}\,M^{-1}$. The complex may be considered as containing two reducing equivalents (with the location of the hydrogens undefined), and is designated adrenodoxin reductase (red) $NADP^+$ or AR(red)$NADP^+$.

These interpretations were supported by the observation that *Neurospora* NADase [11] which cleaves $NADP^+$, converts the spectrum of the

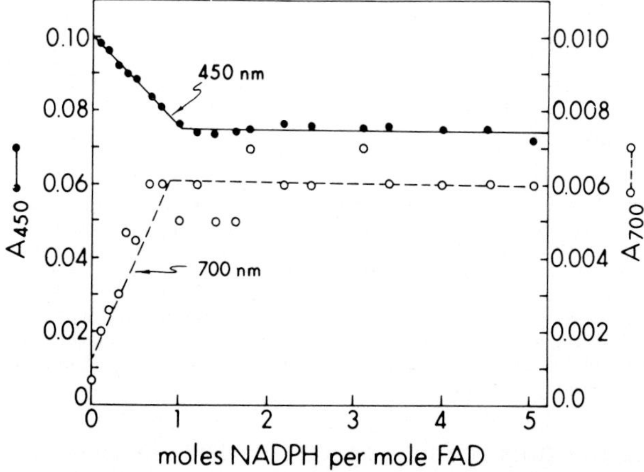

Fig. 3. Anaerobic titration of adrenodoxin reductase with NADPH. 19.7 nmol of adrenodoxin reductase in 2.1 ml volume was titrated with $1.93 \cdot 10^{-4}$ M NADPH. ●, absorbance at 450 nm after successive additions of NADPH; ○, absorbance at 700 nm, as above

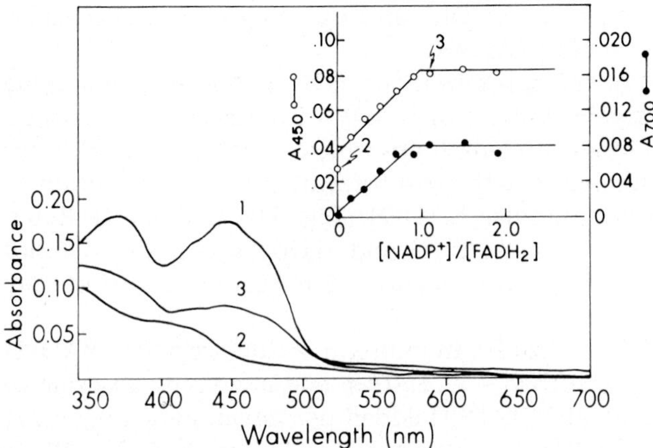

Fig. 4. Anaerobic titration of dithionite-reduced adrenodoxin reductase with $NADP^+$ (1) Spectrum of oxidized adrenodoxin reductase (36.6 nmol). (2) Spectrum of $Na_2S_2O_4$-reduced adrenodoxin reductase 40.2 nmol $Na_2S_2O_4$ added. (3) Spectrum obtained after addition of 40.2 nmol $NADP^+$ to (2). Inset: Titration curves showing absorbance at 450 and 700 nm after successive additions of $NADP^+$ to dithionite-reduced adrenodoxin reductase. The numbers "2" and "3" designate the experimental points at which the spectra shown were recorded.

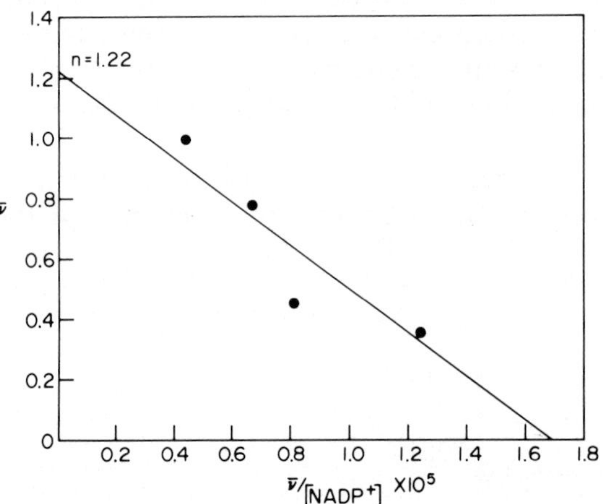

Fig. 5. Binding of $NADP^+$ to oxidized adrenodoxin reductase. Measurements were made using a 50 ml ultrafiltration assembly with a PM-30 ultrafilter. Each point represents a separate experiment in which a known amount of $NADP^+$ (59.3, 118.6, 243.3, and 474.4 nmol respectively) was added to 68.0 nmol of adrenodoxin reductase in 21 mls of 0.1 M potassium phosphate, pH 7.5, within the assembly. Several fractions of filtrate were then collected at 2.5 lb/inch^2 and with constant stirring, and the concentration of free $NADP^+$ was then calculated from the absorbance at 258 nm of the filtrate. Apparent binding due to a small retardation of movement of $NADP^+$ by the membrane was corrected for by repeating the experiment without adrenodoxin reductase, and subtracting this "apparent" binding in the absence of protein from that in the presence of protein.

complex into one resembling the fully-reduced spectrum. The cleavage products of $NADP^+$ (nicotinamide and 2′-monophospho-adenosine 5′-diphosphoribose) were shown in a separate experiment to cause no spectral changes when added to dithionite-reduced flavoprotein.

A spectrum representing a "tight" complex, such as that observed between reduced adrenodoxin reductase and $NADP^+$, was not observed for oxidized adrenodoxin reductase and $NADP^+$. Some binding does occur, but with a higher K_D than for adrenodoxin reductase (red) $NADP^+$. Ultrafiltration data [12], Fig. 5, was consistent with binding with an apparent K_D of $1.4 \cdot 10^{-5}$ M, and indicates approximately one binding site per flavin. It should be noted that $NADP^+$ has been observed to be a "mixed" type inhibitor with respect to NADPH, with a "K_I" of $5.32 \cdot 10^{-6}$ M [1].

Reduced adrenodoxin reductase was shown to yield an unusual spectrum when titrated with NADPH (Fig. 6). Adrenodoxin reductase was first reduced with a slight excess of dithionite, (Fig. 6, curve 2) and then titrated with NADPH. Spectral changes were observed over a broad wavelength range (Fig. 6, curve 3). Data from a titration, recording at 570 nm, is shown in the inset of Fig. 6. A break in the titration plot at 1 NADPH per $FADH_2$ can be seen. These data indicate that reduced enzyme forms a "tight" complex, not only with $NADP^+$, but with NADPH.

Anaerobic titration of oxidized adrenodoxin reductase with NADH yielded a curved titration plot, rather than the sharp break seen on titration with NADPH (Fig. 7). Data points for the titration are indicated by open circles. The theoretical titration plot, assuming complete reduction of flavoprotein (without detectable reversibility) by 2 reducing equivalents, is indicated by the dotted line, which was calculated on the basis of the 17% residual absorbance seen at 450 nm after full reduction by dithionite. At increasing concentrations of NADH, the absorbance at 450 nm becomes asymptotic to the "fully reduced" absorbance.

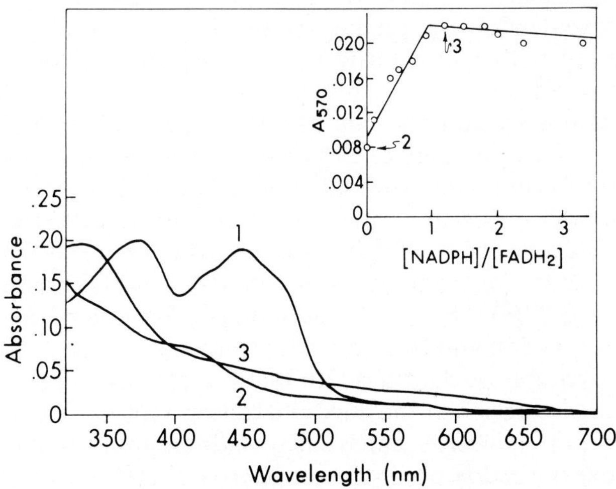

Fig. 6. Anaerobic titration of dithionite-reduced adrenodoxin reductase with NADPH. (1) Spectrum of oxidized adrenodoxin reductase. (2) Spectrum obtained after addition of 1.5 equivalents of dithionite to (1). (3) Spectrum obtained after addition of 1.2 equivalents of NADPH to reduce adrenodoxin reductase. Inset: Titration curve showing absorbance at 570 nm after successive additions of NADPH to dithionite-reduced adrenodoxin reductase. The numbers "2" and "3" designate the experimental points at which the complete spectra were recorded.

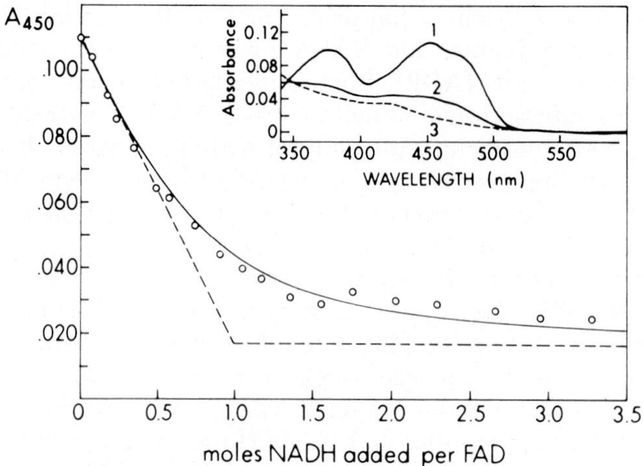

Fig. 7. Anaerobic titration of adrenodoxin reductase with NADH. 19.2 nmol of adrenodoxin reductase in 10 mM potassium phosphate, pH 7.5 was titrated with NADH. ○, data points after successive additions of NADH. ······, theoretical line for reduction of adrenodoxin reductase by 2 reducing equivalents assuming reaction to go to completion (see text). ——, calculated assuming an E_0 for reduced: oxidized adrenodoxin reductase of −0.295 V. Inset: (1) spectrum of oxidized adrenodoxin reductase. (2) 0.9 NADPH added per flavin. (3) Calculated spectrum of fully-reduced form.

This curved titration plot indicates an equilibrium between oxidized and reduced forms of flavoprotein and NAD. From this equilibrium, the reduction potential for adrenodoxin reductase could be calculated, assuming an E_0 for $NADH/NAD^+$ of −0.320 V. The solid curve shown in Fig. 7, is a theoretical line generated by assuming a difference in reduction potential between NADH and adrenodoxin reductase of 0.025 V, i.e., a reduction potential for reduced: oxidized adrenodoxin reductase of −0.295 V at pH 7.5. This is in good agreement with a previously reported E_0' of −0.274 V, obtained by different methods [1].

The data for the titrations of adrenodoxin reductase by NADPH and NADH are replotted in Fig. 8, as per cent of fully reduced spectrum. This shows that at 1 NADPH per flavin (2 electrons added per flavin), the formation of fully-reduced form was essentially complete. However, at 1 NADH per flavin, only about 75% of adrenodoxin reductase molecules were in the reduced form. If one considers the potentials for $NADPH/NADP^+$ and $NADH/NAD^+$ to be equal, −0.320 V then, without complex formation, one should expect the composition of the mixture of redox forms at equilibrium to be identical for both pyridine nucleotides. Reduction of adrenodoxin reductase by NADH does indeed result in the curved titration plot predicted by the previously reported [1] reduction potential for adrenodoxin reductase, and the assumption that the redox potential for $NADH/NAD^+$ is −0.32 V. Reduction of adrenodoxin reductase by NADH can be represented as the equilibrium (Eqn. 1) between oxidized and reduced forms of flavoprotein and NAD(H), (AR, adrenodoxin reductase).

$$NADH + H^+ + AR \rightleftarrows NAD^+ + ARH_2 \qquad (1)$$

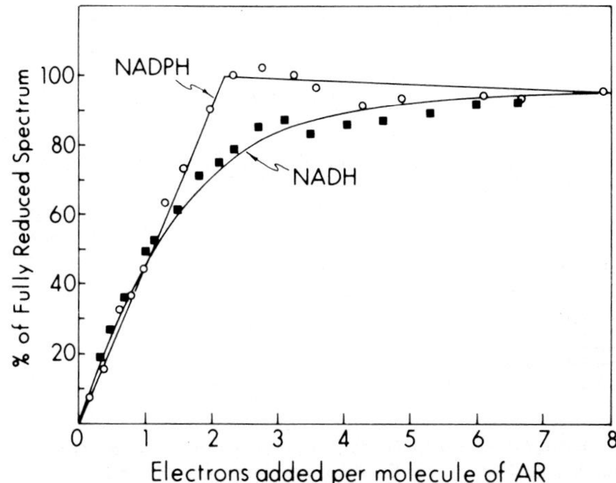

Fig. 8. Anaerobic titration of adrenodoxin reductase by NADH and NADPH. Data from Fig. 3 and Fig. 7 were plotted as per cent of fully reduced spectrum.

However, addition of one equivalent of NADPH to adrenodoxin reductase results in the complete rather than partial transfer of 2 reducing equivalents. This can be explained by complex formation between NADP(H) and reduced adrenodoxin reductase providing the thermodynamic driving force for full reduction of the flavoprotein under conditions wherein the reduction potentials of the flavin and the free pyridine nucleotides would otherwise dictate an equilibrium between oxidized and reduced forms.

The partial reactions for this process are indicated: Reduction by NADH (without complex formation) results in the equilibrium indicated in Eqn. 1. NADP(H) and adrenodoxin reductase can also be considered, for convenience, to first form an equilibrium mixture of oxidized and reduced forms, as dictated by their potentials (Eqn. 2). Complex formation between $NADP^+$ and reduced flavoprotein then removes the reduced flavoprotein from the equilibrium (Eqn. 3), driving the reaction towards complete formation of the adrenodoxin reductase (red) $NADP^+$ complex which contains 2 reducing equivalents (Sum, Eqn. 4).

$$NADPH + H^+ + AR \rightleftarrows NADP^+ + ARH_2 \quad (2)$$

$$NADP^+ + ARH_2 \rightarrow AR(red)NADP^+ \quad (3)$$

Sum of (2) plus (3):

$$NADPH + H^+ + AR \rightarrow AR(red)NADP^+ \quad (4)$$

This scheme predicts that the equilibrium produced by addition of equimolar NADH (Eqn. 1) may be driven towards full reduction of adrenodoxin reductase by the addition of $NADP^+$, by removal of reduced enzyme from the equilibrium (Eqn. 3). The sum of Eqns. 1 plus 3, Eqn. 5, predicts that 2 reducing equivalents from NADH will be utilized, in the presence of $NADP^+$, to form fully reduced complex.

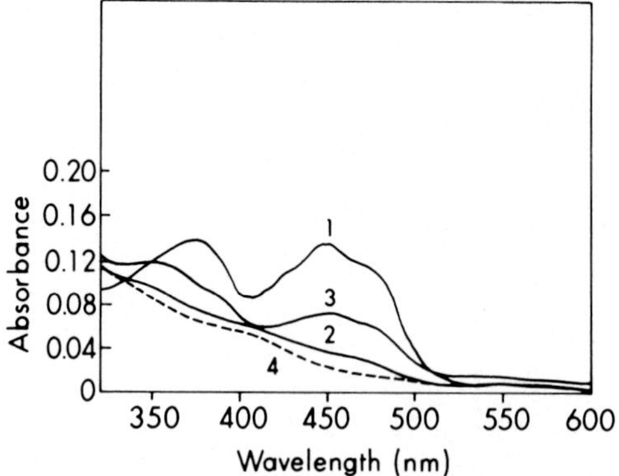

Fig. 9. Adrenodoxin reductase: anaerobic reduction by NADH, followed by addition of $NADP^+$. (1) Spectrum of oxidized adrenodoxin reductase, 22.5 nmol. (2) Spectrum of adrenodoxin reductase following addition of 21.1 nmol NADH to (1). (3) Spectrum of adrenodoxin reductase following addition of 48.6 nmol $NADP^+$ to (2). (4) Spectrum calculated for fully reduced adrenodoxin reductase.

$$NADH + H^+ + NADP^+ + AR \rightarrow NAD^+ + AR(red)NADP^+ \quad (5)$$

The results of such an experiment are indicated (Fig. 9). Addition of equimolar NADH to adrenodoxin reductase results in incompletely reduced adrenodoxin reductase (Fig. 9, curve 2). Subsequent addition of a slight excess of $NADP^+$ then results in the spectrum of fully reduced adrenodoxin reductase(red)$NADP^+$ (Fig. 9, curve 3). Thus, we have the paradoxical but predictable situation in which addition of an oxidized substrate drives a reaction toward complete reduction. Finally, we must caution that we do not yet know whether formation of this complex occurs during catalysis.

Acknowledgements

We wish to thank Dr. K.V. Rajagopalan and Dr. Jean Johnson for performing the EPR experiments. We also thank Mrs. Paula Leyva for her technical assitance. This work was supported by N.I.H. Grant GM-21226.

References

1 Kimura, T. and Chu, J.W. (1973) J. Biol. Chem. 248, 2089
2 Foster, R.P. and Wilson, L.D. (1974) Fed. Proc. 33, 1291
3 Suhara, K., Ikeda, Y., Takemori, S. and Katagiri, M. (1972) FEBS Lett. 28, 45
4 Masters, B.S.S., Prough, R.A. and Kamin, H. (1975) Biochemistry 14, in the press
5 Omura, T., Sanders, E., Estabrook, R.W., Cooper, D.Y. and Rosenthal, O. (1966) Arch. Biochem. Biophys. 117, 660
6 Kimura, T. and Suzuki, K. (1967) J. Biol. Chem. 242, 485
7 Suzki, K. and Kimura, T. (1965) Biochem. Biophys. Res. Commun. 19, 340
8 Foust, G.P., Burleigh, B.D., Mayhew, S.G., Williams, C.H. and Massey, V. (1969) Anal. Biochem. 27, 530
9 Kamin, H. (ed.), (1971) Flavins and Flavoproteins, University Park Press, Baltimore, Maryland
10 Massey, V., Palmer, G., Williams, C.H., Swoboda, B.E.P. and Sands, R.H. (1966) Flavins and Flavoproteins (Slater, E.C., ed.), p. 133, Elsevier, Amsterdam
11 Kaplan, O. (1955) Methods of Enzymology (Colowick, S.P. and Kaplan, N.O., eds.), Vol. II, p. 664, Academic Press, New York, New York
12 Steinhardt, J. and Reynolds, J.A. (1969) Multiple Equilibria in Proteins, pp. 50–52, Academic Press, New York

Flavins and Flavoproteins, edited by T.P. Singer

CHAPTER 71

PURIFICATION AND PROPERTIES OF ADRENAL FERREDOXIN:NADP$^+$ REDUCTASE (ADRENODOXIN REDUCTASE) FROM BOVINE ADRENOCORTICAL MITOCHONDRIA

KATSUKO SUHARA, YOSHINO IKEDA, SHIGEKI TAKEMORI and MASAYUKI KATAGIRI

Department of Chemistry, Faculty of Science, Kanazawa University, Ishikawa 920 (Japan)

Introduction

The steroid-hydroxylating system of adrenocortical mitochondria has been resolved into 3 components: a flavoprotein (adrenal ferredoxin-NADP$^+$ reductase, EC 1.6.7.1, NADPH: adrenal ferredoxin oxidoreductase), an Fe-S protein (adrenal ferredoxin) and a hemoprotein (cytochrome *P*-450) [1]. The purification of these components would be desirable to clarify the reaction mechanism of this electron transfer system. We have improved the procedure for preparing crystalline adrenal ferredoxin [2, 3]. Recently, cytochrome *P*-450 has been obtained in a pure state [4]. Adrenal ferredoxin-NDAP$^+$ reductase has been isolated from adrenal mitochondria of beef [1], pig [5] and rat [6]. Recently, Chu and Kimura [7] and our group [8] have independently purified the enzyme from bovine adrenocortical mitochondria.

The present paper deals with the more detailed investigation of this adrenal ferredoxin-NADP$^+$ reductase.

Materials and Methods

Mitochondrial pellets were prepared from bovine adrenal cortex as previously described [2]. Activity of adrenal ferredoxin-NADP$^+$ reductase was routinely assayed by measuring the rate of the adrenal ferredoxin-dependent reduction of cytochrome *c* [2]. The NADPH-diaphorase activity of the enzyme was assayed by measuring the rate of reduction of 2,6-dichlorophenolindophenol. The assay mixture (1 ml) contained 30 nmol of NADPH, 50 μmol of potassium phosphate pH 7.4, 35 nmol of the dye and the enzyme. Enzymic activity is expressed in katals (amount of activity that converts 1 mol of substrate per second) at 25°C and the specific activity was defined as the number of katals per kg of protein. The apoenzyme was prepared by the acid-ammonium sulfate precipitation [9]. The acid-liberated flavin was concentrated using a Florisil column and subjected to the flavin analysis.

Results and Discussion

Purification of adrenal ferredoxin-NADP$^+$ reductase

It has been found that adrenal ferredoxin is easily extracted into the sucrose solution during the isolation of mitochondria from bovine adrenal cortexes [2]. Adrenal ferredoxin-NADP$^+$ reductase was also partially extractable into this fraction: half of the total activity. However, purification of the enzyme from the sucrose extracts appeared to be impractical because the extracts had a low starting specific activity and the activity was rapidly lost during purification. The reductase activity was also found either in soluble or in residue fractions obtained after the mitochrondria had been sonicated. When the sonic residue fraction was dispersed with 1.5% Triton X-100, about 15% of the total activity was solubilized. The preparation could be partially purified. However, either the specific activity or the yield of the enzyme was low. Considering this, together with the fact that the sonic residue is rich in cytochrome *P*-450, the soluble fraction obtained after sonication was used for the purification of adrenal ferredoxin-NADP$^+$ reductase throughout the study. The fraction contained about 35% of the total reductase activity and seemed to be a good and the most economical source.

Method of purification

The following purification was perfomed at 4°C. Unless otherwise stated, potassium phosphate buffer (pH 7.4) was used as a solvent for the enzyme. A typical protocol is presented in Table I. Mitochondrial pellets (27 g) were suspended in about 900 ml of 100 mM buffer to approximately 30 mg protein per ml. 50-ml aliquots of the suspension were sonicated for 5 min at 0°C in a Quigley-Rochester sonic desmembrator (20 KHz). The sonicated samples were centrifuged at 30 000 × *g* for 90 min. The yellow-brown supernatant fraction was fractionated between 35 and 60% saturation of ammonium sulfate. The precipitate was dissolved in 10 mM buffer and dialyzed overnight against the same buffer. After FAD was added (25 μM to stabilize the holoenzyme) to the dialyzate, the enzyme was applied to a DEAE-cellulose column (2.8 × 30 cm) equilibrated with 10 mM buffer. The enzyme was washed onto the column with 100 ml of 10 mM buffer and finally eluted with 50 mM buffer. On adding FAD, the combined active fractions were fractionated between 35 and 55% saturation of ammonium sulfate. The precipitate was dissolved in 10 mM buffer and dialyzed overnight against the same buffer. The dialyzate was again applied to a DEAE-cellulose column (2.8 × 25 cm) equilibrated with

TABLE I

SUMMARY OF THE PURIFICATION OF ADRENAL FERREDOXIN-NADP$^+$ REDUCTASE

Fraction	Volume (ml)	Protein (mg)	Activity (μkat)	Specific activity (mkat/kg protein)	Yield (%)
Crude extract	680	5830	16.2	2.8	100
Ammonium sulfate	50	2000	8.3	4.2	52
1st DEAE-cellulose	10	170	4.8	28	30
2nd DEAE-cellulose	2.3	35	1.6	46	10
Sephadex G-100	1.5	19	1.2	63	7.2

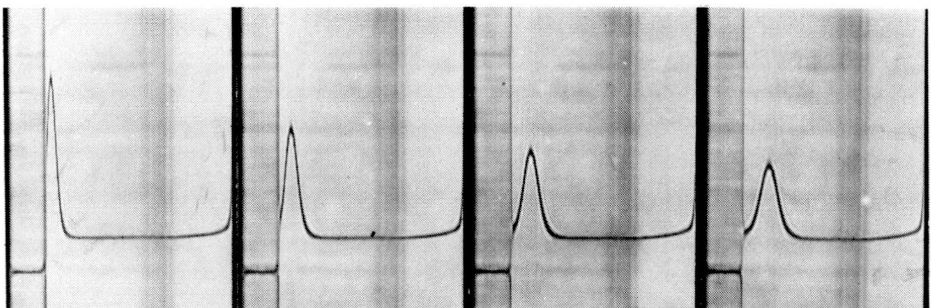

Fig. 1. Ultracentrifugation pattern of adrenal ferredoxin-NADP$^+$ reductase. The enzyme was a 7.5 mg per ml solution in 50 mM potassium phosphate buffer (pH 7.4). Photographs were taken at 8 min intervals after attaining a rotor speed of 59 780 rev/min.

10 mM buffer. After washing the column with 100 ml of 10 mM buffer, the enzyme was eluted with a linear gradient consisting of 200 ml of 10 mM buffer (the mixing chamber) and 200 ml of 50 mM buffer (the reservoir). The enzyme was usually separated into two fractions. The active major fractions were combined and FAD was added. The enzyme was fractionated between 40 and 55% saturation of ammonium sulfate. The precipitate was dissolved in 10 mM buffer. On adding FAD, the enzyme was applied to a Sephadex G-100 column (2.8 × 60 cm) equilibrated with 10 mM buffer containing 100 mM KCl and eluted with the same buffer. FAD was added to the combined active fractions and the enzyme was concentrated with ammonium sulfate (60% saturation) and dissolved in 50 mM buffer. After dialysis against 50 mM buffer, the enzyme could be stored for several weeks at − 80° C without substantial loss of activity. At 5° C, half of the original activity was lost in a week.

Properties of adrenal ferredoxin-NADP$^+$ reductase

When subjected to disc electrophoresis on polyacrylamide gel [10], both native and sodium dodecyl sulfate-treated preparations were essentially homogeneous. As shown in Fig. 1, the sedimentation pattern of the enzyme showed a single symmetrical peak, and the $s_{20,w}$ was estimated to be 3.9 S.

The enzyme catalyzed adrenal ferredoxin-dependent reduction of cytochrome *c* in the presence of NADPH. In the absence of adrenal ferredoxin, no reduction of cytochrome *c* occurred. When adrenal ferredoxin was present in excess, the activity of the enzyme was saturated and the specific activity was found to be 63 mkat per kg of protein. The enzyme had a rather broad pH optimum at 7.5–9.5. The K_m value for NADPH was $1.81 \cdot 10^{-6}$ M. The NADPH-diaphorase activity in the presence of 2,6-dichlorophenolindophenol was estimated to be 9 mkat per kg of protein under the conditions described in the Methods. The enzyme, together with adrenal ferredoxin and cytochrome *P*-450, could function in the reconstitution of the cholesterol side-chain cleavage system [4].

The oxidized enzyme shows an absorbance spectrum with maxima at 272, 378 and 450 nm and shoulders at 425 and 475 nm. Its absorbance ratio at 272/450 nm was 7.3. The enzyme was fully reduced upon addition of sodium dithionite. When NADPH was added anaerobically to the enzyme, the fully reduced species was not observed. The NADPH-reduced enzyme

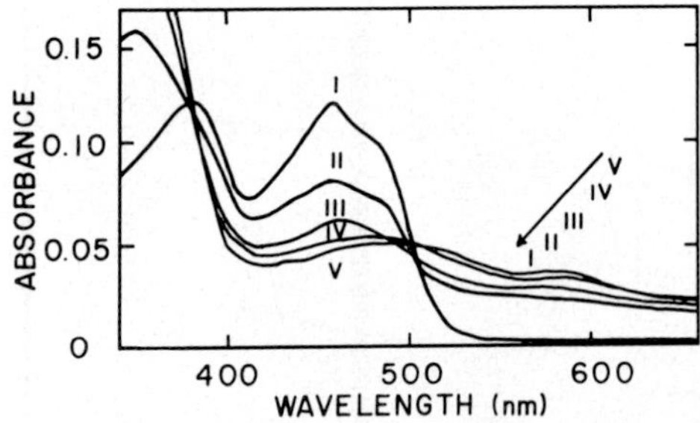

Fig. 2. Anaerobic reduction of adrenal ferredoxin-NADP$^+$ reductase by the NADP$^+$-glucose-6-phosphate dehydrogenase system. The assay system (3 ml) contained in Thunberg type cuvettes, 30 nmol of adrenal feredoxin-NADP$^+$ reductase, 150 μmol of potassium phosphate (pH 7.4), 3 μmol of $MgCl_2$, 4.5 μmol of glucose 6-phosphate, 260 nmol of NADP$^+$ and 4 units of glucose-6-phosphate dehydrogenase. Curve I, oxidized; Curves II–V, after incubation for 10, 30, 60 and 90 min, respectively.

showed a spectrum with a long wavelength absorbance band. Fig. 2 shows the reduction of the enzyme by NADPH which is continuously generated by the glucose-6-phosphate dehydrogenase system. Since the enzyme reacts wih adrenal ferredoxin to form a protein complex in the molecular ratio of 1:1 [11], adrenal ferredoxin was added to the system. However, even in this case, only the reduction rate was accelerated and the absorbance spectrum was not affected. A similar spectrum for a reduced flavoprotein was reported by Shin and Arnon for chloroplast ferredoxin-NADP$^+$ reductase [12].

The early findings [5, 13] that FAD appears to be the flavin component of the enzyme were confirmed from the following different lines of evidence*. (a) Paper chromatography of the acid-liberated flavin with three different solvent system (1-butanol/acetic acid/H_2O (4:1:5), pyridine/H_2O (2:1) and 5% Na_2HPO_4) revealed a single yellow spot with the same R_f value as authentic FAD. (b) The fluorescence of the enzyme flavin at 523 nm was identical with that of authentic FAD, when activated by light of 470, 379 and 273 nm. (c) The apoenzyme was reactivated by FAD, but not FMN. FMN did not affect the extent of activation by FAD. (d) The enzyme flavin could fully activate both apo-D-amino acid oxidase and apo-salicylate hydroxylase from *Pseudomonas putida* [14], which are specifically reactivated by FAD. These results indicated that the sole flavin component of the enzyme is FAD and excluded the presence of FMN as a co-component.

The molecular weight of the enzyme was estimated to be 54 000 by the gel filtration on a Sephadex G-100 column. The Stokes radius of the enzyme was measured with the use of a Sephadex G-100 column [15] and its value was calculated to be 3.29 nm. From the Stokes radius, a diffusion coefficient of $6.10 \cdot 10^{-7}$ $cm^2 \cdot s^{-1}$ was calculated [16]. High-speed sedimentation equilibrium experiments [17] yielded a molecular weight of 60 000. Assuming a molar extinction coefficient of $11.3 \cdot 10^3$ $M^{-1} \cdot cm^{-1}$ at 450 nm, the protein weight per FAD was estimated to be 55 000. Sodium dodecyl sulfate-gel electrophoresis [18] gave a minimum molecular weight of 50 000. These results suggest that the enzyme is a single-chain protein containing one FAD.

* The enzyme used for flavin analysis was prepared without addition of FAD during purification.

There are other enzymes which have a quite similar catalytic function to adrenal ferredoxin-$NADP^+$ reductase. The ferredoxin-NAD^+ reductase in the camphor methylene hydroxylase system from *P. putida* [19] and rubredoxin-NAD^+ reductase in the hydrocarbon ω-hydroxylase system from *Pseudomonas oleovorans* [20] have molecular weights of 43 500 and 55 000, respectively. Both contain one molecule of FAD in a single-chain protein. Thus, adrenal ferredoxin-$NADP^+$ reductase exhibits a similarity in the relationship between structure and catalytic function to these enzymes.

Acknowledgements

We gratefully acknowledge the help of Professor R. Sato, Institute for Protein Research, Osaka University, for supplying the adrenal glands. This work was supported by the Scientific Research Fund B-447051 from the Ministry of Education of Japan.

References

1 Omura, T., Sanders, E., Estabrook, R.W., Cooper, D.Y. and Rosenthal, O. (1966) Arch. Biochem. Biophys. 117, 660–673
2 Suhara, K., Takemori, S. and Katagiri, M. (1972) Biochim. Biophys. Acta 263, 272–278
3 Suhara, K., Kanayama, K., Takemori, S. and Katagiri, M. (1974) Biochim. Biophys. Acta 336, 309–317
4 Takemori, S., Suhara, K., Hashimoto, S., Hashimoto, M., Sato, H., Gomi, T. and Katagiri, M. (1975) Biochem. Biophys. Res. Commun. 63, 588–593
5 Kimura, T. and Suzuki, K. (1967) J. Biol. Chem. 242, 485–491
6 Nakamura, Y., Otsuka, H. and Tamaoki, B. (1966) Biochim. Biophys. Acta 122, 34–42
7 Chu, J.W. and Kimura, T. (1973) J. Biol. Chem. 248, 2089–2094
8 Suhara, K., Ikeda, Y., Takemori, S. and Katagiri, M. (1972) FEBS Lett. 28, 45–47
9 Warburg, O. and Christian, W. (1938) Biochem. Z. 298, 150–168
10 Davis, B.J. (1964) Ann. N.Y. Actad. Sci. 121, 404–427
11 Nakamura, S., Kimura, T. and Chu, J.W. (1972) FEBS Lett. 25, 249–252
12 Shin, M. and Arnon, D.I. (1965) J. Biol. Chem. 240, 1405–1411
13 Omura, T., Sanders, E., Cooper, D.Y. and Estabrook, R.W. (1967) in Methods in Enzymology (Colowick, S.P. and Kaplan, N.O., eds.), Vol. 10, pp. 362–367, Academic Press, New York
14 Katagiri, M., Yamamoto, S. and Hayaishi, O. (1962) J. Biol. Chem. 237, 2413–2414
15 Ackers, G.K. (1964) Biochemistry 3, 723–730
16 Gosting, L.J. (1956) in Advances in Protein Chemistry (Anson, M.L., Bailey, K. and Edsall, J.T., eds.), Vol. 11, pp. 449–522, Academic Press, New York
17 Yphantis, D.A. (1964) Biochemistry 3, 297–317
18 Weber, K. and Osborn, M. (1969) J. Biol. Chem. 244, 4406–4412
19 Tsai, R.L., Gunsalus, I.C. and Dus, K. (1971) Biochem. Biophys. Res. Commun. 54, 1300–1306
20 Ueda, T., Lode, E.T. and Coon, M.J. (1972) J. Biol. Chem. 247, 2109–2116

Flavins and Flavoproteins, edited by T.P. Singer

CHAPTER 72

SPECTRAL STUDIES ON THE STABLE AEROBIC AND ANAEROBIC HALF-REDUCED STATES OF LIVER MICROSOMAL NADPH-CYTOCHROME *c* REDUCTASE

BETTIE SUE SILER MASTERS, RUSSELL A. PROUGH and HENRY KAMIN

Department of Biochemistry, The University of Texas Health Science Center at Dallas, 5323 Harry Hines Boulevard, Dallas, Texas 75235 and Department of Biochemistry, Duke University Medical Center, Durham, N.C. 27710 (U.S.A.)

Introduction

The existence of NADPH-cytochrome *c* reductase (EC 1.6.2.4, NADPH: ferricytochrome oxidoreductase) in the endoplasmic reticulum of a variety of organs, including liver [1, 2], kidney cortex [3–5], lung [6], adrenal cortex [7, 8], spleen [9] and heart [10] has made the understanding of the mechanism of electron transfer through this flavoprotein to cytochrome *P*-450 for the mono-oxygenation of a variety of substrates of great interest. Studies demonstrating the immunochemical similarities of NADPH-cytochrome *c* reductases from various extrahepatic organs to the liver enzyme [3, 7, 9, 11] and the similarities in kinetic and physico-chemical properties between the liver and kidney microsomal flavoproteins [12] indicate strongly that only subtle differences exist among the flavoproteins derived from these various sources. Therefore, it is of interest to study in great detail the microsomal flavoprotein derived from its richest source, the liver, to understand its mode of action in other tissues as well.

Recent reports from Mason's laboratory [13, 14] regarding the flavin content, spectral properties and redox states of NADPH-cytochrome *c* reductase have necessitated a re-examination of this flavoprotein by our laboratories. We have confirmed the conclusions of Iyanagi and Mason [13] that NADPH-cytochrome *c* reductase contains equimolar concentrations of FAD and FMN [15, 16], as have the recent studies of Vermilion and Coon [17] on detergent-solubilized reductase. Masters et al. [18, 19] had previously reported that this flavoprotein contains 2 mol of flavin per mol of enzyme with a molecular weight of 69 000 for the pancreatic lipase-solubilized preparation.

Our studies, however, do not confirm the data of Iyanagi et al. [13, 14] stating that the O_2-stable, half-reduced state of NADPH-cytochrome *c* reductase reported by Masters et al. [18, 20] in 1965 is a one-electron-reduced intermediate. Instead, our data show that the "O_2-stable semiquinone" (which we have designated as "half-reduced flavoprotein" since we

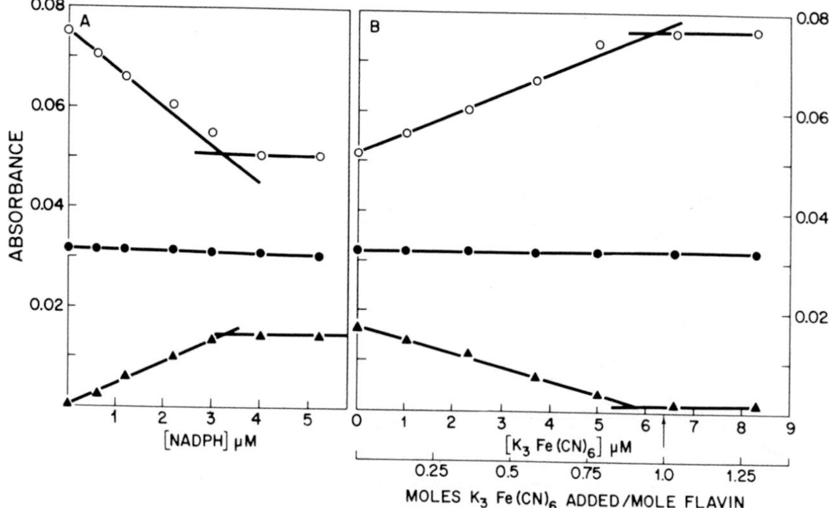

Fig. 1. Aerobic titration of porcine liver microsomal NADPH-cytochrome *c* reductase with NADPH and $K_3Fe(CN)_6$. Absorbance changes at 453 (○—○), 500 (●—●) and 585 nm (▲—▲) are plotted against either NADP (A) or $K_3Fe(CN)_6$ (B) added. The abscissa in B is also labeled as mol $K_3Fe(CN)_6$ added per mol flavin as the data of Fig. 7 of Iyanagi and Mason [13]. Pig liver microsomal NADPH-cytochrome *c* reductase was diluted in 0.1 M potassium phosphate pH 7.7, to a final concentration of 6.4 μM in a total volume of 3 ml and titrated with NADPH and then $K_3Fe(CN)_6$ sequentially under aerobic conditions at 25°C. The total volume change was less than 3.5% and was corrected for upon each addition. Reprinted with permission from the American Chemical Society (Biochemistry (1975) 14, 607–613).

cannot produce a free radical signal in the EPR which represents quantitatively more than 10% of the flavoprotein present) is two-electron-reduced intermediate.

Results

The aerobic titration of NADPH-cytochrome *c* reductase with NADPH is shown in Fig. 1A. It is obvious that during the stepwise addition of NADPH in the presence of air, no appreciable oxidase activity occurs when less than 0.5 mol of NADPH is added per mol of flavin, since the titration proceeds linearly toward this equivalence point before the curve breaks. In Fig. 1B, the back titration of this same air-stable enzyme intermediate with potassium ferricyanide requires 1 mol of oxidant per mol of flavin (as indicated on the lower abscissa) and not 0.5 mol of potassium ferricyanide as reported by Iyanagi et al. [13, 14]. This titration was performed on porcine liver microsomal flavoprotein, solubilized by the method of Masters et al. [23], but utilizing the conditions of Iyanagi and Mason [13] for the titration experiment.

To confirm that the oxidation-reduction state of the half-reduced flavoprotein was not dependent upon the species from which the reductase was derived, upon the presence or absence of EDTA, or upon aerobic or anaerobic conditions, a titration of rat liver microsomal reductase with

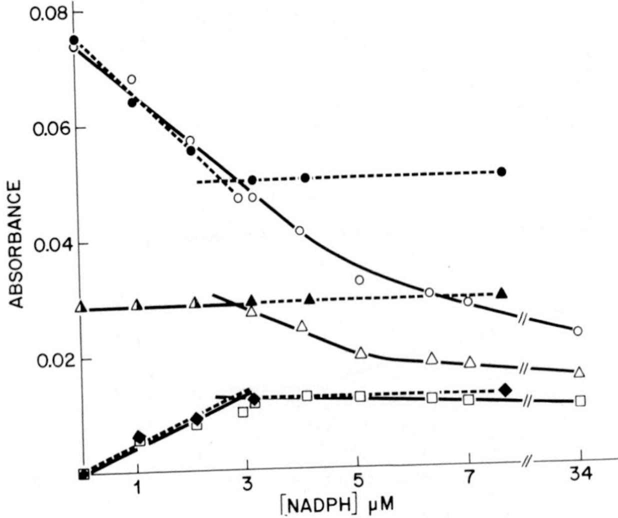

Fig. 2. Anaerobic and aerobic titrations of rat liver microsomal NADPH-cytochrome *c* reductase. Absorbance changes under aerobic conditions at 453 (●——●), 500 (▲——▲), and 585 nm (◆——◆) and under anaerobic conditions at 453 (○——○), 500 (△——△), and 585 nm (□——□) are plotted against NADPH added. Rat liver microsomal NADPH-cytochrome *c* reductase was dissolved in 0.05 M potassium phosphate pH 7.7, containing 0.1 mM EDTA and 0.02% NaN_3, to a final concentration of 6.4 μM in a total volume of 3 ml and titrated with NADPH either anaerobically or aerobically at 25°C. The total volume change was less than 3.5% and was corrected for upon each addition. Reprinted with permission form the American Chemical Society (Biochemistry (1975) 14, 607–613).

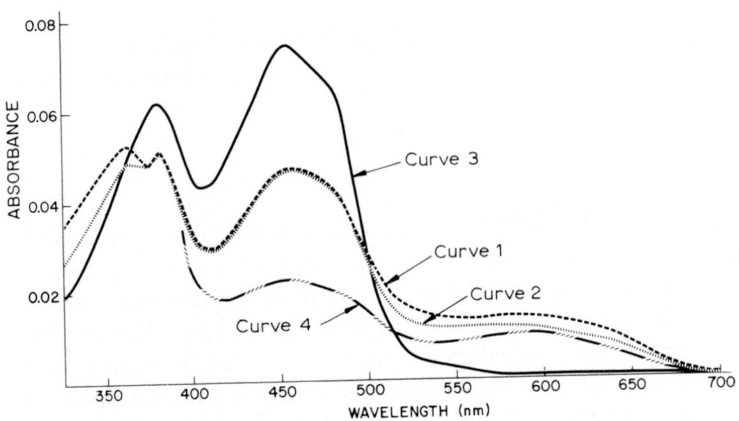

Fig. 3. Aerobic and anaerobic spectra of rat liver microsomal NADPH-cytochrome *c* reductase. NADPH was added aerobically (curve 1) and anaerobically (curve 2) to 6.3 μM oxidized flavoprotein (curve 3) dissolved in 0.05 M potassium phosphate pH 7.7, containing 0.1 mM EDTA and 0.02% NaN_3. Excess NADPH (34 μM final concentration) was added aerobically (curve 4). Reprinted with permission from the American Chemical Society (Biochemistry (1975) 14, 607–613).

NADPH was performed (Fig. 2). As seen in Fig. 1, the curve breaks in the aerobic titration upon the addition of 3.1 μM NADPH to 6.4 μM flavoprotein. As either the aerobic or anaerobic titrations proceed, the 585 nm absorbance changes, representing the half-reduced species, become maximal upon the

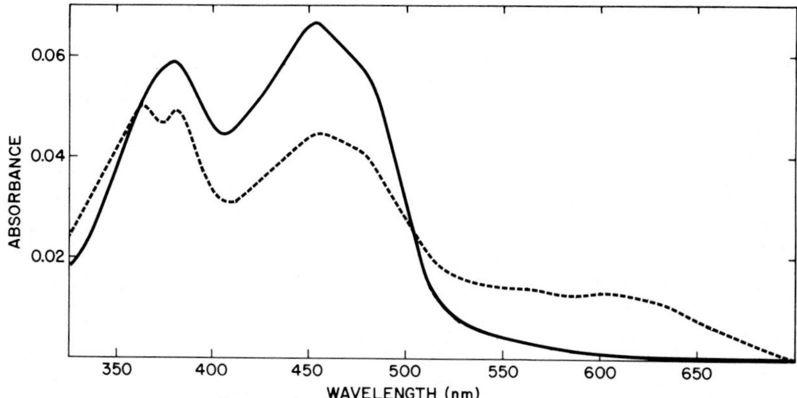

Fig. 4. Spectrum of NADPH-cytochrome *c* reductase produced after reoxidation of reduced cytochrome b_5 in 0.8 M KCl. The reference cuvette contained $3.4 \cdot 10^{-6}$ M cytochrome b_5 and the sample cuvette contained $3.4 \cdot 10^{-6}$ M cytochrome b_5 and $5.6 \cdot 10^{-6}$ M NADPH-cytochrome *c* reductase. The buffer was 0.8 KCl, 0.05 M potassium phosphate, $1 \cdot 10^{-4}$ M EDTA pH 7.7. The reaction was initiated with NADPH ($8.5 \cdot 10^{-6}$ M final concentration) and was monitored at 557 nm until the cytochrome b_5 was reoxidized prior to recording the final spectrum. (——) spectrum of reaction mixture before addition of NADPH; (----) spectrum of reaction mixture after reoxidation of cytochrome b_5. Reprinted with permission from Academic Press, Inc. (Arch. Biochem. Biophys. (1974) 165, 263–267).

addition of 2.9 μM NADPH, and the 500 nm absorbance, representing the fully reduced species, decreases in the anaerobic titration only after 3.1 μM NADPH has been added. Identical results were obtained with trypsin-solubilized rat liver microsomal flavoprotein, indicating that our titration results were not peculiar to the lipase-solubilized preparations.

For purposes of comparison, Fig. 3 presents the absorbance spectra of the oxidized flavoprotein, the aerobic and anaerobic half-reduced enzyme and the flavoprotein reduced with excess NADPH aerobically. It can be seen that the species exhibiting maximal absorbance at 453 nm is the oxidized state, at 585 nm, the half-reduced state, and at 500 nm, the oxidized and half-reduced states. The similarity between the aerobically and anaerobically produced half-reduced species is shown in Fig. 3; the differences in absorbance at 360 nm and above 500 nm in the anaerobic spectrum are simply due to slight over-reduction beyond the 0.5 mol NADPH per mol enzyme-bound flavin.

In previous studies, Master et al. [18, 20] reported that the spectral species resulting from the reoxidation of the reduced form of NADPH-cytochrome *c* reductase by either one- or two-electron acceptors during catalytic turnover was the half-reduced state of the flavoprotein. Fig. 4 shows the oxidized spectrum of 5.6 μM NADPH-cytochrome *c* reductase in 0.05 M potassium phosphate buffer pH 7.7, containing 0.8 M KCl and 0.1 mM EDTA. The sample and reference cuvettes contained 3.4 μM cytochrome b_5 pipetted very accurately. NADPH (8.5 μM) was then added to the sample cuvette and the spectrum of the enzyme was recorded after the cytochrome b_5 had been reduced and reoxidized. The resulting spectrum is identical to that of the half-reduced flavoprotein shown in Fig. 3, indicating that

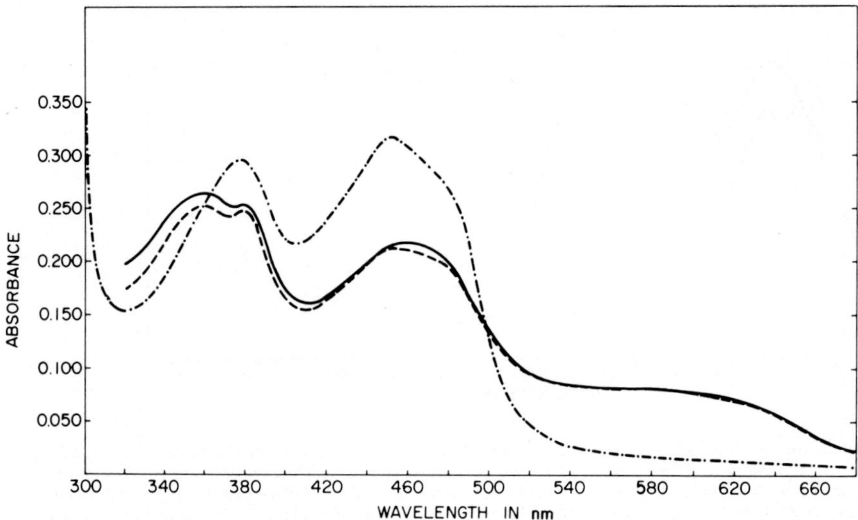

Fig. 5. Treatment of oxygen-stable, half-reduced porcine liver microsomal NADPH-cytochrome *c* reductase with charcoal. The spectra are shown for oxidized (· — ·) and half-reduced (——) flavoprotein, and half-reduced flavoprotein passed through charcoal (----). To obtain the oxidized spectrum, NADPH-cytochrome *c* reductase was dissolved to a final concentration of 28.1 μM in 3 ml of 0.05 M potassium phosphate pH 7.7, containing 0.1 mM EDTA. For the air-stable, half-reduced spectrum, 2 mol of NADPH per mol of flavin were added aerobically and the flavoprotein was allowed to reoxidize to the half-reduced state. The resulting enzyme solution was then filtered through an acid-washed charcoal-celite (1:2) column and the spectrum was recorded on a Cary Model 14 recording spectrophotometer. All spectra were corrected for dilution during the various procedures. Such a procedure results in the removal of 99.5% of the $NADP^+$ formed. Reprinted with permission from the American Chemical Society (Biochemistry (1975) 14, 607–613).

cytochrome b_5 reoxidizes the enzyme to the same redox state obtained with cytochrome *c*, potassium ferricyanide and 2, 6-dichlorophenolindophenol as electron acceptors. It is interesting to note that the high ionic strength required for the reduction of cytochrome b_5 by this flavoprotein [21] does not alter the spectrum nor does it affect the stoichiometry obtained in the titration experiments.

To establish that enzyme-bound pyridine nucleotide was not contributing to the absorbance spectrum of the O_2-stable, half-reduced flavoprotein, the enzyme was passed through acid-washed charcoal to remove the $NADP^+$ formed during reduction to the half-reduced state. The spectrum, shown in Fig. 5, resulting from such charcoal treatment is identical to that of the O_2-stable, half-reduced species before charcoal treatment. (Using $NADP^3H$, the recovery of at least 99.5% of the pyridine nucleotide added to the enzyme was effected after elution of the charcoal with ammonia/water/ethanol (1:49:50.)

The aerobic titration of such a charcoal-treated preparation resulted in the curves shown in Fig. 6. It can be seen that 1 mol of potassium ferricyanide per mol of flavin is required to reoxidize the flavoprotein to the fully oxidized form. The stoichiometry obtained in this experiment indicates that the two-electron-reduced intermediate does not involve pyridine

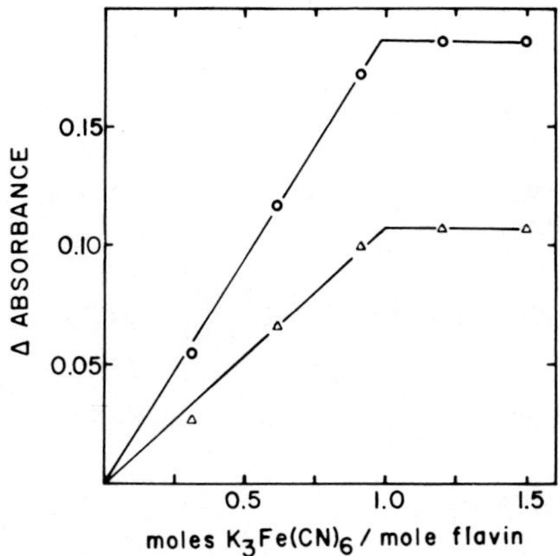

Fig. 6. Titration of charcoal-treated, half-reduced NADPH-cytochrome *c* reductase with $K_3Fe(CN)_6$. The reductase was reduced with 1 mol NADPH per mol flavin and allowed to reoxidize in air until no further change in absorbance at 453 nm was observed. 1 mol of this stable, half-reduced intermediate was then filtered through 20 ml of acid-washed charcoal-celite (1:2), 0.95 ml of eluate was collected by vacuum. The enzyme solution containing 54 μM flavin was then back titrated with increments of $K_3Fe(CN)_6$ until the absorbance at 453 nm remained constant. Plots of the absorbance changes at 453 (○——○) and 585 nm (△——△) are shown in the figure. Reprinted with permission from the American Chemical Society (Biochemistry (1975) 14, 607–613).

nucleotide bound to the enzyme which could serve as an additional electron sink. The concentration of flavoprotein utilized in this titration was approx. ten times that titrated in the experiments reported in Fig. 1 and 2, thus precluding the possibility that low concentrations of flavoprotein (due to slower NADPH oxidation rates) may yield different stoichiometry and preventing loss of enzyme upon charcoal filtration.

Discussion

In considering the early steady-state kinetics data of Williams and Kamin [1] and Masters et al. [18, 20], one is faced with the dilemma of explaining the facility with which this flavoprotein reduces both one- and two-electron acceptors with approximately equal turnover numbers, when expressed in terms of electron equivalents per min per mol of flavin. The stoichiometry of reduction of cytochrome *c* when equimolar concentrations of NADPH and flavoprotein were mixed in the stopped-flow indicated that only 1 mol of cytochrome *c* was reduced per mol of NADPH, i.e., two electron equivalents were retained by the enzyme-bound flavin. On the other hand, as the enzyme was diluted towards catalytic concentrations, approx. 2 mol of cytochrome *c* were reduced by 1 mol of NADPH. These data can be explained by the retention of two electron equivalents (from 1 mol of NADPH) by the half-reduced form of the enzyme which is the functionally oxidized state of the flavin during catalytic turnover.

The data presented in this paper support the earlier experiments of

STOICHIOMETRIC [ENZYME]

$$\mathrm{NADPH} + \mathrm{H}^+ + \left[\begin{matrix}\mathrm{FMN}\\ \mathrm{FAD}\end{matrix}\right. \longrightarrow \mathrm{NADP}^+ + \left[\begin{matrix}\mathrm{FMNH}\\ \mathrm{FADH}\end{matrix}\right.$$

$$\left[\begin{matrix}\mathrm{FMNH}\\ \mathrm{FADH}\end{matrix}\right. + \mathrm{NADPH} + \mathrm{H}^+ \longrightarrow \left[\begin{matrix}\mathrm{FMNH_2}\\ \mathrm{FADH_2}\end{matrix}\right. + \mathrm{NADP}^+$$

$$2\ \text{Cytochrome}\ c\ (\mathrm{Fe}^{+3}) + \left[\begin{matrix}\mathrm{FMNH_2}\\ \mathrm{FADH_2}\end{matrix}\right. \longrightarrow 2\ \text{Cytochrome}\ c\ (\mathrm{Fe}^{+2}) + \left[\begin{matrix}\mathrm{FMNH}\\ \mathrm{FADH}\end{matrix}\right. + 2\mathrm{H}^+$$

$$2\ \mathrm{NADPH} + \left[\begin{matrix}\mathrm{FMN}\\ \mathrm{FAD}\end{matrix}\right. + 2\ \text{Cytochrome}\ c\ (\mathrm{Fe}^{+3}) \longrightarrow 2\ \mathrm{NADP}^+ + \left[\begin{matrix}\mathrm{FMNH}\\ \mathrm{FADH}\end{matrix}\right. + 2\ \text{Cytochrome}\ c\ (\mathrm{Fe}^{+2})$$

CATALYTIC [ENZYME]

$$\mathrm{NADPH} + \mathrm{H}^+ + \left[\begin{matrix}\mathrm{FMNH}\\ \mathrm{FADH}\end{matrix}\right. \longrightarrow \mathrm{NADP}^+ + \left[\begin{matrix}\mathrm{FMNH_2}\\ \mathrm{FADH_2}\end{matrix}\right.$$

$$2\ \text{Cytochrome}\ c\ (\mathrm{Fe}^{+3}) + \left[\begin{matrix}\mathrm{FMNH_2}\\ \mathrm{FADH_2}\end{matrix}\right. \longrightarrow 2\ \text{Cytochrome}\ c\ (\mathrm{Fe}^{+2}) + \left[\begin{matrix}\mathrm{FMNH}\\ \mathrm{FADH}\end{matrix}\right. + 2\mathrm{H}^+$$

$$\mathrm{NADPH} + \mathrm{H}^+ + 2\ \text{Cytochrome}\ c\ (\mathrm{Fe}^{+3}) \longrightarrow \mathrm{NADP}^+ + 2\ \text{Cytochrome}\ c\ (\mathrm{Fe}^{+2}) + 2\mathrm{H}^+$$

Masters et al. [18, 20] indicating that the O_2-stable, half-reduced flavoprotein contains two electron equivalents in contrast to the recent reports of Iyanagi et al. [13, 14]. It would appear that the titration experiments of Iyanagi et al. were performed on trypsin-solubilized enzyme since the preparation of a detergent-solubilized flavoprotein of sufficient purity to exhibit a flavoprotein spectrum has only recently been achieved by Vermilion and Coon [17] and by Dignam et al. [22]. In our hands, although the trypsin- and lipase-solubilized enzymes are kinetically identical and appear to have similar physical properties, these preparations are not immunochemically identical and the trypsin-solubilized preparations always give multiple bands on disc-gel electrophoresis which stain for NADPH-neotetrazolium diaphorase activity. We have performed these titrations on a large number of preparations of NADPH-cytochrome *c* reductase with both the trypsin- and the lipase-solubilized preparations yielding similar results, but the trypsin-solubilized enzyme was less stable to storage. In our studies, we have been careful to include only those preparations which are 80–90% reducible by NADPH (indicating enzyme-bound flavin), while it has not been possible to assess the kinetic properties of the preparations used by Iyanagi et al. [13, 14].

Finally, although we understand a great deal about its mechanism of action with artificial electron acceptors, little is known about NADPH-cytochrome *c* reductase as it functions in its role as the ultimate physiological reductant of cytochrome *P*-450. These studies, however, must be performed with preparations which are capable of reducing cytochrome *P*-450, a capacity which the proteolytically prepared reductases of Iyanagi et al. [13, 14] and Masters et al. [18, 20] do not possess. Future experiments will be directed toward this end.

Acknowledgments

This study was supported by American Cancer Society Grant BC-153 (R.A.P.) and by U.S. Public Health Service Grant No. HL13619 and Grant No. I-453 from The Robert A. Welch Foundation, Houston, Texas (B.S.S.M.).

References

1 Williams, Jr., C.H. and Kamin, H. (1962) J. Biol. Chem. 237, 587–595
2 Phillips, A.H. and Langdon, R.G. (1962) J. Biol. Chem. 237, 2652–2660
3 Wada, F., Shibata, G.M. and Sakamoto, Y. (1968) Biochim. Biophys. Acta 162, 518–524
4 Ichihara, K., Kusunose, E. and Kusunose, M. (1971) Biochim. Biophys. Acta 239, 178–189
5 Sasame, H.A., Thorgeirsson, S.S., Mitchell, J.R. and Gillette, J.R. (1974) Biochem. Biophys. Res. Commun. 14, 35–46
6 Matsubara, T., Prough, R.A., Burke, M.D. and Estabrook, R.W. (1974) Cancer Res. 34, 2196–2203
7 Masters, B.S.S., Taylor, W.E., Isaacson, E.L., Baron, J., Harkins, J.B., Nelson, E.B. and Bryan, G.T. (1973) Ann. N.Y. Acad. Sci. 212, 76–88
8 Masters, B.S.S., Baron, J., Taylor, W.E., Isaacson, E.L. and LoSpalluto, J.J. (1971) J. Biol. Chem. 246, 4143–4150
9 Schacter, B.A., Nelson, E.B., Marver, H.S. and Masters, B.S.S. (1972) J. Biol. Chem. 247, 3601–3607
10 Imai, K., Omura, T. and Sato, R. (1966) J. Biochem. Tokyo 60, 274–285
11 Masters, B.S.S., Nelson, E.B., Schacter, B.A., Baron, J. and Isaacson, E.L. (1973) Drug Metab. Disposition 1, 121–128
12 Fan, L.L. and Masters, B.S.S. (1974) Arch. Biochem. Biophys. 165, 665–671
13 Iyanagi, T. and Mason, H.S. (1973) Biochemistry 12, 2297–2308
14 Iyanagi, T., Makino, N. and Mason, H.S. (1974) Biochemistry 13, 1701–1710
15 Prough, R.A., Masters, B.S. and Kamin, H. (1974) Fed. Proc. 33, 982
16 Masters, B.S.S., Prough, R.A. and Kamin, H. (1975) Biochemistry 14, 607–613
17 Vermilion, J. and Coon, M.J. (1974) Biochem. Biophys. Res. Commun. 60, 1315–1322
18 Masters, B.S.S., Kamin, H., Gibson, Q.H. and Williams, Jr., C.H. (1965) J. Biol. Chem. 240, 921–931
19 Masters, B.S.S. and Ziegler, D.M. (1971) Arch. Biochem. Biophys. 145, 358–364
20 Masters, B.S.S., Bilimoria, M.H., Kamin, H. and Gibson, Q.H. (1965) J. Biol. Chem. 240, 4081–4088
21 Prough, R.A. and Masters, B.S.S. (1974) Arch. Biochem. Biophys. 165, 263–267
22 Dignam, J.D., Saine, S. and Strobel, H.W. (1975) Fed. Proc. 34, 2162
23 Masters, B.S.S., Williams, C.H. and Kamin, H. (1967) Methods in Enzymology (Estabrook, R.W. and Pullman, M.E., eds.), Vol. X, pp. 565–573, Academic Press, New York

Flavins and Flavoproteins, edited by T.P. Singer

CHAPTER 73

KINETIC AND SPECTRAL STUDIES ON THE REDUCTION OF LIVER MICROSOMAL NADPH-CYTOCHROME *c* REDUCTASE BY NADH

RUSSELL A. PROUGH and BETTIE SUE S. MASTERS

Department of Biochemistry, The University of Texas Health Science Center at Dallas, Southwestern Medical School, 5323 Harry Hines Boulevard, Dallas, Texas 75235 (U.S.A.)

Summary

The liver microsomal flavoprotein, NADPH-cytochrome *c* reductase, is the initial electron acceptor in the electron transport system involved in cytochrome *P*-450-mediated, mixed-function oxidase reactions. NADPH is the preferred reductant of the microsomal oxidation reactions and in the reduction of cytochrome *c* by purified enzyme. However, NADH has been shown to support both reactions, albeit at much reduced rates. Present experiments indicate that the steady-state levels of fully reduced flavin determine the catalytic rate produced by the two pyridine nucleotides, thus partially explaining the differences in *V* obtained in the reduction of cytochrome *c*. Titration of the flavoprotein with NADH under anaerobic conditions yields a stoichiometry of reduction which is identical with that obtained with NADPH. Transient-state studies on flavin reduction have shown that the maximal rate of flavin reduction is identical for either NADH or NADPH. Steady-state kinetic experiments on cytochrome *c* reduction show identical turnover numbers for the reactions with NADH and NADPH. These data are consistent with a single, common, active site for NADH and NADPH but with different dissociation constants for the two pyridine nucleotides.

Introduction

The liver microsomal flavoprotein, NADPH-cytochrome *c* reductase (EC 1.6.2.4, NADPH: ferricytochrome oxidoreductase), is the initial electron acceptor in the electron transport system involved in cytochrome *P*-450-mediated, mixed-function oxidase reactions [1, 2]. NADPH is the preferred reductant of the microsomal hydroxylase reactions and in the reduction of cytochrome *c* by the purified enzyme, but NADH in physiological concentrations has been shown to support both reactions, albeit at much reduced rates. Although NADH serves as a poor reductant in these hydroxylase reactions, Cohen and Estabrook [3] have established that low concentrations of NADPH make possible the efficient use of NADH for mixed-function oxidations. This phenomenon not only results in an increase in the rate of hydroxylation reactions over that of NADPH but also in an increase in the extent of the reaction. The possibility of a complex interaction between the two reduced pyridine nucleotides at the flavoprotein level prompted the study of the reduction of NADPH-cytochrome *c* reductase by NADH.

An earlier report by Ichikawa and Yamano [4] on the steady-state reduction of cytochrome *c* by NADH and NADPH catalyzed by purified rabbit liver NADPH-cytochrome *c* reductase suggested that the maximal velocities (V) of cytochrome *c* reduction were not identical for the two pyridine nucleotides. Their turnover numbers for cytochrome *c* reduction were approx. 425 and 160 mol cytochrome *c* reduced per min per mol flavoprotein for NADPH and NADH, respectively, which do not agree with the published values of 1250–1350 mol cytochrome *c* reduced per min per mol flavin [5–7] for NADPH. Ichikawa and Yamano [4] reported that the K_m for NADH was approx. 1 mM.

This report will establish that NADPH- and NADH-dependent cytochrome *c* reduction and flavin reduction have identical turnover numbers, that the stoichiometry of reduction of the flavoprotein by either reduced pyridine nucleotide is 1 mol NAD(P)H per mol flavoprotein, and that the rate-limiting step in NADH reduction of NADPH-cytochrome *c* reductase is the formation of an NADH-enzyme binary complex.

Materials and Methods

NADPH-cytochrome *c* reductase was prepared as described previously [6]. All experiments were performed at 25°C in 0.05 M potassium phosphate (pH 7.7), $1 \cdot 10^{-4}$ M EDTA and 0.02% sodium azide unless otherwise stated in the figure legends. All steady-state kinetic and anaerobic spectral titrations were performed as described by Masters et al. [7]. Transient-state kinetic measurements were obtained with an Aminco-Morrow stopped-flow spectrophotometer at 500 nm, using the $\Delta\epsilon_{500\,nm}$ for F_{ox}—F_{ox}(FH—FH) → FH_2—FH_2 of $-3.55 \cdot 10^3\,M^{-1} \cdot cm^{-1}$. NADH and NADPH were obtained from P.-L. Biochemicals Inc. and the NADH used was the chromatopure™ grade.

Results

Steady-state measurements of cytochrome *c* reduction by NADH or NADPH catalyzed by NADPH-cytochrome *c* reductase indicated that the kinetic mechanism was bi-bi, ping-pong, in agreement with earlier reports [4, 5]. The maximal velocities and apparent Michaelis constants are shown in Table I. The experiments involving NADH-dependent reduction were complicated by a chemical reduction of cytochrome *c* with NADH and all measurements were corrected by subtracting the chemical rate from the total enzymic rate. The maximal rates of cytochrome *c* reduction by either reduced pyridine nucleotide were identical and agreed with the previously reported values [5]. The apparent K_m for NADH was 10^3–10^4-times larger than the K_m for NADPH. Inhibitors of NADPH-cytochrome *c* reductase, such as $NADP^+$ and 2′-AMP, were equally effective as inhibitors of the reduction of cytochrome *c* by either reduced pyridine nucleotide (Table II). In agreement with Ichikawa and Yamano [4], $NADP^+$ and 2′-AMP are competitive inhibitors of cytochrome *c* reduction.

Anaerobic titration of NADPH-cytochrome *c* reductase with NADH or NADPH at either 0.5 or 2.2 equiv. of reduced pyridine nucleotide yielded nearly identical spectra (Fig. 1). Higher levels of NADH than NADPH were required to reduce completely the flavin, but in the presence of NADase nearly stoichiometric concentrations of NADH or NADPH could fully reduce the flavin. Fig. 2 shows the anaerobic titration of the flavoprotein reductase. The 500 and 585 nm breakpoints occurred at 0.5 mol NADH per mol of flavin. The spectrum obtained at this point is identical to the spectrum of the air-stable, half-reduced form of the flavoprotein which we have shown to contain two electron equivalents per two flavins per mol enzyme upon the anaerobic addition of 0.5 mol reduced pyridine nucleotide per mol flavin [7]. Complete NADH reduction of the flavoprotein occurred at approx. 1.0 mol NADH per mol flavin.

Preliminary transient-state kinetic measurements of flavin reduction at 500 nm were performed using NADH as a reductant. The maximal rate of flavin reduction (T.N.) by NADH was found to be $2.43 \cdot 10^3$ equivalents per

TABLE I

KINETIC CONSTANTS FOR NADH- OR NADPH-DEPENDENT CYTOCHROME *c* REDUCTION CATALYZED BY NADPH-CYTOCHROME *c* REDUCTASE

These experiments were performed at 25°C in 50 mM potassium phosphate (pH 7.7), $1 \cdot 10^{-4}$ M EDTA and 0.02% sodium azide. The enzyme concentrations were $1.0 \cdot 10^{-8}$ M and $5.0 \cdot 10^{-8}$ M for the reductions by NADPH and NADH, respectively. Cytochrome *c* concentrations were varied from 6 to 45 μM. The NADPH and NADH concentrations were varied from 4 to 100 μM and 3.5 to 50 mM, respectively.

Reductant	V (mol cytochrome *c* reduced per min per mol flavin)	NAD(P)H K_m (M)	Cytochrome *c* K_m (M)
NADPH	1220	$4.3 \cdot 10^{-6}$	$6.6 \cdot 10^{-6}$
NADH	1250	$1.0 \cdot 10^{-1}$	$3.3 \cdot 10^{-5}$

TABLE II

INHIBITION OF CYTOCHROME *c* REDUCTION BY $NADP^+$ AND 2′-AMP

The reaction mixture consisted of 18 μM NADPH, $2.55 \cdot 10^{-7}$ M NADPH-cytochrome *c* reductase or 15 mM NADH, $2.75 \cdot 10^{-6}$ M NADPH-cytochrome *c* reductase in a solution of 40 μM cytochrome *c*, 50 mM potassium phosphate (pH 7.7), $1 \cdot 10^{-4}$ M EDTA, 0.02% sodium azide at 25°C. The NADH concentration utilized in the experiment was sub-optimal and the maximal rate was not obtained. However, sufficient rates of cytochrome *c* reduction were obtained to demonstrate the inhibition.

Inhibitor	mol cytochrome *c* reduced/min/mol flavin	
	NADPH	NADH
—	1200	108
$NADP^+$ (450 μM)	112	3
2′-AMP (1 mM)	355	9

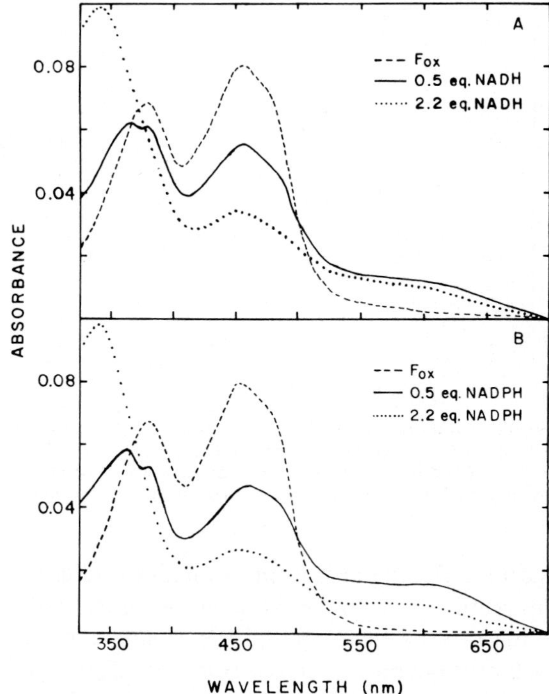

Fig. 1. Anaerobic spectra of NADH- or NADPH-reduced NADPH-cytochrome *c* reductase. NADPH-cytochrome *c* reductase, 6.8 μM, in 50 mM potassium phosphate (pH 7.7), $1 \cdot 10^{-4}$ M EDTA, 0.02% sodium azide was reduced by (A) 3.4 (——) and 15.0 μM (····) NADH or (B) 3.4 (——) and 15 μM (····) NADPH. The reduction was performed under an anaerobic atmosphere of argon. The spectra of oxidized enzyme are designated by the dashed lines (----).

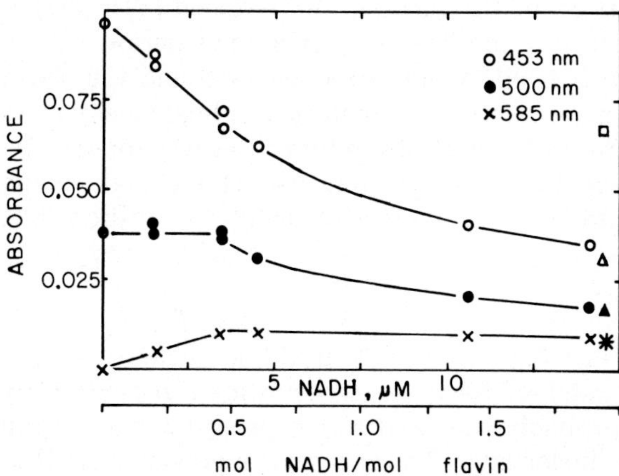

Fig. 2. Anaerobic titration of NADPH-cytochrome *c* reductase by NADH. NADPH-cytochrome *c* reductase, 8.3 μM, in 50 mM potassium phosphate (pH 7.7), $1 \cdot 10^{-4}$ M EDTA, 0.02% sodium azide was reduced by NADH. The reduction was performed under an anaerobic atmosphere of Argon. The absorbances measured were 453 (○, △), 500 (●, ▲), and 585 nm (×, ✻) for the anaerobic titration by NADH or NADPH, respectively. The absorbance at 453 nm of the air-stable, half-reduced form of the flavoprotein is shown (□).

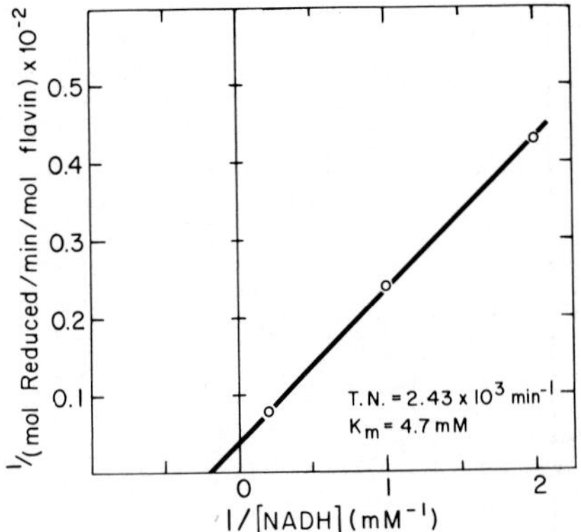

Fig. 3. Double-reciprocal plot of transient-state reduction of NADPH-cytochrome *c* reductase by NADH. NADPH-cytochrome *c* reductase, 20 μM, in 50 mM potassium phosphate (pH 7.7), $1 \cdot 10^{-4}$ M EDTA, 0.02% sodium azide was mixed with NADH and monitored at 500 nm to measure flavin reduction using an Aminco-Morrow, stopped-flow spectrophotometer. The instrument dead-time was determined to be 3 ms.

min per mol flavin or 1200 mol reduced flavin per min per mol flavin (Fig. 3), which is consistent with the maximal rate of reduction using NADPH reported by Masters et al. [8]. This rate of flavin reduction is sufficient to account for the maximal rate of cytochrome *c* reduction of 1250 min^{-1} obtained with NADH. The apparent K_m was found to be greater than 5 mM.

The possibility that the high apparent K_m for NADH may result from a small contamination of NADPH in the NADH is difficult to ascertain. However, the stoichiometric reduction by NADH supports the role of NADH as a reductant for the enzyme. Furthermore, 1 mM NADH yields a steady-state level of reduced flavin, such that approx. two thirds of the flavin is fully reduced (FH_2—FH_2) and one third of the flavin is half reduced (FH—FH) as measured at 453 nm under aerobic conditions over a time period during which a 10% contamination of NADPH would be consumed and the flavin reoxidized (Fig. 4). The fact that NADH alone maintains a high steady-state level of the catalytic form of the enzyme (fully reduced flavin) supports the conclusion of the anaerobic spectral titration that NADH will reduce the flavoprotein, i.e., NADPH contamination is most probably not responsible.

Discussion

These results suggest the existence of a single pyridine nucleotide binding site and the differences in rates of flavin or cytochrome *c* reduction by either NADH or NADPH are probably due to differences in the dissociation constants of the two pyridine nucleotides. The reactions involving NADH as a reductant under most conditions are limited by the steady-state concentration of an enzyme · NADH binary complex (Michaelis complex). Since the binding of NADPH occurs in the dead-time of fluorescence stopped-flow measurements, the rate-limiting step for NADPH-mediated reduction cannot be assigned due to the similarity in rate constants for reduction and

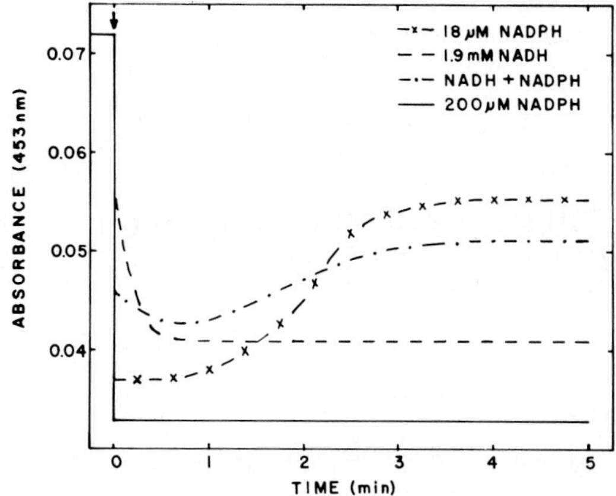

Fig. 4. Time course of reduction and reoxidation of NADPH-cytochrome *c* reductase by NADH and NADPH. NADPH-cytochrome *c* reductase, 6.2 μM, in 50 mM potassium phosphate (pH 7.7), $1 \cdot 10^{-4}$ M EDTA, 0.02% sodium azide was mixed with 18 μM NADPH (—X—), 1.9 mM NADH (--), 200 μM NADPH (——), or 18 μM NADPH and 1.9 mM NADH (-·-). The rate of flavin reduction and reoxidation were monitored with time at 453 nm.

reoxidation of flavin. Future studies utilizing flavin analogues to alter these rate-determining steps may resolve this question.

Acknowledgements

This study was supported by American Cancer Society Grant BC-153 (R.A.P.) and by U.S. Public Health Service Grant No. Hl13619 and Grant No. I-453 from The Robert A. Welch Foundation, Houston, Texas (B.S.S.M.)

References

1 Orrenius, S., Berg, A. and Ernster, L. (1969) Eur. J. Biochem. 11, 193–200
2 Masters, B.S.S., Baron, J., Taylor, W.E., Isaacson, E.L. and LoSpalluto, J.J. (1971) J. Biol. Chem. 246, 4143–4150
3 Cohen, B.S. and Estabrook, R.W. (1971) Arch. Biochem. Biophys. 143, 46–53
4 Ichikawa, Y. and Yamano, T. (1969) J. Biochem. Tokyo 66, 351–360
5 Masters, B.S.S., Kamin, H., Gibson, Q.H. and Williams, Jr., C.H. (1965) J. Biol. Chem. 240, 921–931
6 Prough, R.A. and Masters, B.S.S. (1973) Ann. N.Y. Acad. Sci. 212, 89–93
7 Masters, B.S.S., Prough, R.A. and Kamin, H. (1975) Biochemistry 14, 607–613
8 Masters, B.S.S., Bilimoria, M.H., Kamin, H. and Gibson, Q.H. (1965) J. Biol. Chem. 240, 4081–4088

Flavins and Flavoproteins, edited by T.P. Singer

CHAPTER 74

PROPERTIES OF HIGHLY PURIFIED DETERGENT-SOLUBILIZED NADPH-CYTOCHROME *P*-450 REDUCTASE FROM LIVER MICROSOMES

JANICE L. VERMILION and MINOR J. COON

Department of Biological Chemistry, Medical School, The University of Michigan, Ann Arbor, Mich. 48104 (U.S.A.)

The flavoprotein, NADPH-cytochrome *P*-450 reductase, is an essential component of the liver microsomal electron transfer system which catalyzes the NADPH-dependent hydroxylation of a wide variety of both naturally occurring and non-physiological compounds. The terminal oxidase, cytochrome *P*-450, is responsible for the broad substrate specificity of the system and is now known to exist in several forms [1, 2]. After detergent solubilization of the protein components from the microsomes, a third component, phosphatidylcholine, was shown to be necessary for successful reconstitution of the NADPH-dependent hydroxylation reactions characteristic of the membrane-bound system [3, 4]. Recent work in this laboratory has centered on the purification and characterization of the individual proteins of this electron transport chain from mammalian liver.

For a number of years the microsomal flavoprotein, NADPH-cytochrome *c* reductase, first extensively studied by Williams and Kamin [5] and by Phillips and Langdon [6], was considered to be the catalyst which transfers electrons from NADPH to cytochrome *P*-450. Purified preparations of this enzyme solubilized by proteolytic procedures have never been shown to support substrate hydroxylation by cytochrome *P*-450. On the other hand, antibodies prepared against the purified cytochrome *c* reductase cause inhibition of NADPH-dependent drug hydroxylation in intact microsomes [7, 8], thereby providing evidence that this protein is somehow involved in drug hydroxylation reactions. It has been well established that purification of certain membrane-bound proteins after proteolytic solubilization yields smaller proteins whose physical and catalytic properties may differ significantly from those of the native form [9, 10]. For this reason, a number of investigators have undertaken purification of the intact flavoprotein after detergent solubilization [11–17].

We have recently described an NADPH-cytochrome *P*-450 reductase highly purified from phenobarbital-induced rat liver microsomes after

Abbreviation: Cl_2Ind, 2,6-dichlorophenolindophenol.

solubilization with deoxycholate and chromatography in the presence of deoxycholate or the nonionic detergent, Renex 690 [18]. The following observations provide further evidence for the identity of microsomal NADPH-cytochrome *P*-450 reductase and NADPH-cytochrome *c* reductase. When NADPH oxidation in a reconstituted enzyme system containing phospholipid, purified cytochrome *P*-450 and benzphetamine as substrate is used as an indirect measure of cytochrome *P*-450 reduction, this activity and cytochrome *c* reductase activity copurify. The minimal molecular weight of NADPH-cytochrome *P*-450 reductase determined by sodium dodecyl sulfate-gel electrophoresis is 79 000, the same as that of the polypeptide precipitated from rat liver microsomes by antibodies against bromelain-treated NADPH-cytochrome *c* reductase [19]. The detergent-solubilized reductase contains FMN and FAD in approximately equimolar quantities, as first demonstrated for NADPH-cytochrome *c* reductase by Iyanagi and Mason [11]. In addition, the purified NADPH-cytochrome *P*-450 reductase has spectral properties similar to those of NADPH-cytochrome *c* reductase. As shown in Fig. 1, the reductase exists in an oxygen-stable semiquinone state after air oxidation of the NADPH-reduced enzyme.

Since the enzyme contains both FMN and FAD, an understanding of the role of each flavin and the nature of their interaction in catalysis becomes of central importance to an understanding of the mechanism of electron transfer. We have approached this problem by studying the catalytic properties of enzyme from which FMN has been removed.

Selective removal of FMN may be accomplished by irradiation of the

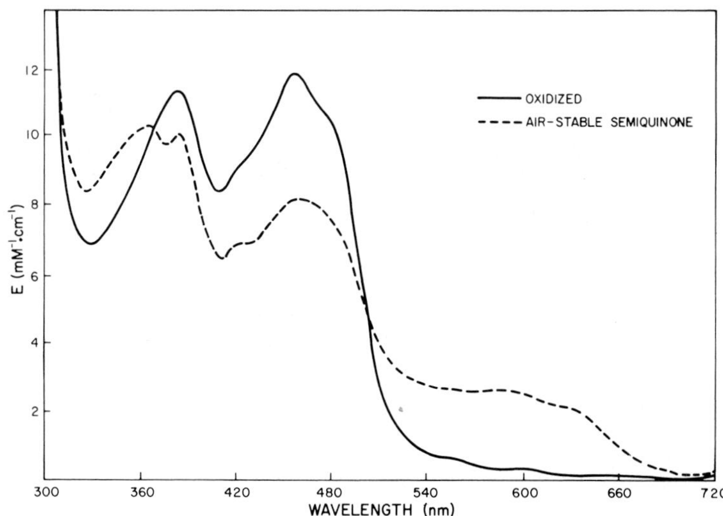

Fig. 1. The oxidized spectrum of NADPH-cytochrome *P*-450 reductase in 0.1 M potassium phosphate buffer pH 7.7, containing 0.1 mM EDTA and 10% glycerol, is shown by a solid line. The shoulder at 418 nm is due to a trace of heme remaining in the preparation. The spectrum of the air-stable semiquinone (dashed line) was obtained after the addition of 1 mol of NADPH per mol of flavin under aerobic conditions.

enzyme with visible light in the presence of ammonium sulfate, essentially as described by Baggott and Langdon [20], or by dialysis against 2 M KBr at pH 8.5. The data in Table I summarize the various activities of the native enzyme and FMN-depleted enzyme prepared by KBr treatment. It should be noted that the native enzyme shows some stimulation of activity towards several electron acceptors when preincubated in the presence of FMN, presumably because a small amount of this flavin is lost from the enzyme during purification. In contrast to the results with cytochrome *P*-450, cytochrome *c* and 2,6-dichlorophenolindophenol (Cl_2Ind), FMN did not stimulate ferricyanide reduction. Upon KBr treatment, very little loss of FAD occurred, but only 30% of the theoretical amount of FMN was still present. The treated enzyme had decreased activity towards cytochrome *P*-450, cytochrome *c*, and Cl_2Ind; the residual activity corresponded fairly well to the amount of FMN still remaining. Ferricyanide activity, on the other hand, was not altered significantly by treatment. Preincubation of the depleted enzyme with FMN restored cytochrome *c* activity to nearly the maximal value. Slightly less, but still significant, restoration of activity towards cytochrome *P*-450 and Cl_2Ind was also demonstrated.

That participation of FMN is apparently not required for ferricyanide reduction by the enzyme was also confirmed in studies with depleted enzyme prepared by irradiation in the presence of ammonium sulfate. Table II summarizes the data regarding reduction of cytochrome *c* and ferricyanide by such a preparation. Activities before and after dialysis to remove ammonium sulfate are presented since, as also reported by Baggott and Langdon [20], dialysis of the depleted enzyme invariably results in some loss of FMN-

TABLE I

EFFECT OF FMN REMOVAL BY KBr TREATMENT OF REDUCTASE

Enzyme, 1.2 μM in total flavin, was dialyzed for 36 h at 4°C against three changes of 0.05 M pyrophosphate buffer pH 8.5, containing 2 M KBr (purchased from Matheson, Coleman and Bell), 0.1 mM EDTA and 10% glycerol. Potassium bromide was then removed by dialysis against 0.05 M potassium phosphate buffer pH 7.6, also containing EDTA and glycerol. Cytochrome *P*-450 reductase activity was measured under conditions similar to those described earlier [12] using rabbit cytochrome *P*-450_{LM_2} purified by a published procedure [1, 21] and benzphetamine as substrate. Activity towards other acceptors was measured at 30° in 0.3 M potassium phosphate buffer pH 7.7, containing 0.1 mM EDTA, at the following concentrations of acceptors: cytochrome *c*, $4 \cdot 10^{-5}$ M; Cl_2Ind, $5 \cdot 10^{-5}$ M; and ferricyanide, $5 \cdot 10^{-4}$ M. Where indicated, FMN was added to the assay mixture at a final concentration of 4.5 μM. All assay mixtures were preincubated 4 min prior to initiation of the reaction with NADPH. The rates were corrected for the non-enzymatic reaction where necessary. FMN and FAD were purified by the methods of Mayhew [22] and Massey and Swoboda [23], respectively. Flavin was determined by the fluorimetric procedure of Faeder and Siegel [24].

Reductase preparation	Flavin content[a]		Activities toward acceptors[b]			
	FMN	FAD	*P*-450	Cytochrome *c*	Cl_2Ind	Ferricyanide
Native	84	100	81	81	89	102
+ FMN			100	100	100	100
KBr-treated	30	93	22	24	28	106
+ FMN			74	97	81	103

[a] % relative to FAD content of native enzyme.
[b] % relative to activities of FMN-supplemented native enzyme.

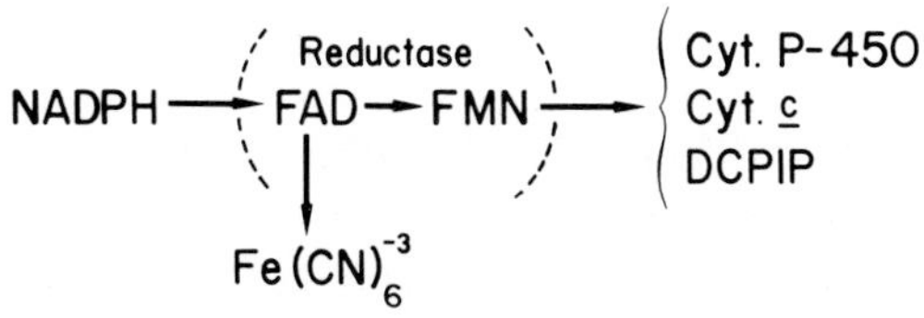

Reductase

NADPH → FAD → FMN → Cyt. P-450 / Cyt. c / DCPIP

FAD → $Fe(CN)_6^{-3}$

Fig. 2. Postulated pathway for electron flow for NADPH-cytochrome *P*-450 reductase. DCPIP, 2,6-dichlorophenolindophenol.

dependent activity. The treated enzyme before dialysis showed very little activity towards cytochrome *c*, but the ability to reduce ferricyanide was essentially unchanged. Preincubation with FMN restored activity towards cytochrome *c* to 80% of the maximal value but was without significant effect on ferricyanide activity. Analysis of the dialyzed sample for flavin confirmed the selective removal of FMN, but compared to the native enzyme only 83% of the starting FAD remained. If loss of FAD occurs on dialysis, this may account for the failure to recover all of the FMN-dependent activity after removal of ammonium sulfate. In other experiments, we have found that the loss of FAD always results in a loss of activity towards all four acceptors studied, and the activity cannot be recovered by preincubation with FMN or a mixture of both flavins.

We have concluded from these experiments that FMN is necessary for the transfer of electrons to cytochrome *P*-450, cytochrome *c*, and Cl_2Ind, but not to ferricyanide. The ability of the FMN-depleted enzyme to reduce ferricyanide suggests that electrons may enter FAD without the participation of FMN. The possibility may be considered that the two flavins have distinct roles in overall catalysis by the enzyme, as shown in Fig. 2. Electrons entering FAD presumably must be transferred via FMN to acceptors such as cytochrome *P*-450, cytochrome *c* and Cl_2Ind. Ferricyanide, on the other hand, is capable of accepting electrons from enzyme containing only FAD; however, at the present time it is not possible to exclude the participation of FMN in the reaction of the native enzyme with ferricyanide.

TABLE II

EFFECT OF FMN REMOVAL BY IRRADIATION OF REDUCTASE IN PRESENCE OF AMMONIUM SULFATE

Enzyme, 1.9 μM in total flavin, in 0.04 M pyrophosphate buffer pH 8.5, containing 40% ammonium sulfate, 0.01 mM EDTA and 10% glycerol, was placed at a distance of 6 inches from the center of a circular array of twelve 15-watt fluorescent lights and irradiated for 8.5 h. The temperature was maintained at about 10°C by means of an ice-water bath.

Additions	Activities before dialysis[a]		Activities after dialysis[a]		Flavin content after dialysis[b]	
	Cytochrome *c*	Ferricyanide	Cytochrome *c*	Ferricyanide	FMN	FAD
None	7	96	6	77	8	83
FMN	81	103	60	78		

[a]% relative to activity of FMN-supplemented native enzyme.
[b]% relative to FAD content of native enzyme.

Acknowledgements

This research was supported by Grant BMS71-01195 from the National Science Foundation and Grant AM-10339 from the United States Public Health Service. J.L.V. is a Predoctoral Trainee, United States Public Health Service, Grant GM-00187.

References

1 Haugen, D.A., van der Hoeven, T.A. and Coon, M.J. (1975) J. Biol. Chem. 250, 3567–3570
2 Ryan, D., Lu, A.Y.H., West, S. and Levin, W. (1975) J. Biol. Chem. 250, 2157–2163
3 Lu, A.Y.H., Junk, K.W. and Coon, M.J. (1969) J. Biol. Chem. 244, 3714–3721
4 Strobel, H.W., Lu, A.Y.H., Heidema, J. and Coon, M.J. (1970) J. Biol. Chem. 245, 4851–4854
5 Williams, Jr., C.H. and Kamin, H. (1962) J. Biol. Chem. 237, 587–595
6 Phillips, A.H. and Langdon, R.G. (1962) J. Biol. Chem. 237, 2652–2660
7 Omura, T. (1969) in Microsomes and Drug Oxidations (Gillette, J.R., Conney, A.H., Cosmides, G.J., Estabrook, R.W., Fouts, J.R. and Mannering, G.J., eds.), pp. 160–162, Academic Press, New York
8 Masters, B.S.S., Baron, J., Taylor, W.E., Isaacson, E.L. and LoSpalluto, J. (1971) J. Biol. Chem. 246, 4143–4150
9 Ito, A. and Sato, R. (1968) J. Biol. Chem. 243, 4922–4923.
10 Spatz, L. and Strittmatter, P. (1973) J. Biol. Chem. 248, 793–799
11 Iyanagi, T. and Mason, H.S. (1973) Biochemistry 12, 2297–2308
12 van der Hoeven, T.A. and Coon, M.J. (1974) J. Biol. Chem. 249, 6302–6310
13 Levin, W., Ryan, D., West, S. and Lu, A.Y.H. (1974) J. Biol. Chem. 249, 1747–1753
14 Ichihara, K., Kusunose, E. and Kusunose, M. (1973) Eur. J. Biochem. 38, 463–472
15 Satake, H., Imai, Y. and Sato, R. (1972) Annu. Meet. Jap. Biochem. Soc. Abstr.
16 Golf, S.W., Graef, V. and Staudinger, H. (1974) Hoppe-Seyler's Z. Physiol. Chem. 355, 1063–1069
17 Dignam, J.D., Saine, S. and Strobel, H.W. (1975) Fed. Proc. 34, 601
18 Vermilion, J.L. and Coon, M.J. (1974) Biochem. Biophys. Res. Commun. 60, 1315–1321
19 Welton, A.F., Pederson, T.C., Buege, J.A. and Aust, S.D. (1973) Biochem. Biophys. Res. Commun. 54, 161–167
20 Baggott, J.P. and Langdon, R.G. (1970) J. Biol. Chem. 245, 5888–5896
21 van der Hoeven, T.A., Haugen, D.A. and Coon, M.J. (1974) Biochem. Biophys. Res. Commun. 60, 569–575
22 Mayhew, S.G. (1971) in Flavins and Flavoproteins (Third International Symposium) (Kamin, H., ed.), pp. 185–209, University Park Press, Baltimore
23 Massey, V. and Swoboda, B.E.P. (1963) Biochem. Z. 338, 474–484
24 Faeder, E.J. and Siegel, L.M. (1973) Anal. Biochem. 53, 332–336

Flavins and Flavoproteins, edited by T.P. Singer

CHAPTER 75

FLUORESCENCE AND ABSORBANCE PROPERTIES OF ELECTRON TRANSFER FLAVOPROTEIN FROM PIG LIVER MITOCHONDRIA

CAROLE L. HALL and HENRY KAMIN

Department of Biochemistry, Duke University Medical Center, Durham. N.C. 27710 (U.S.A.)

Introduction

The electron-transferring flavoprotein (ETF) of the fatty acyl coenzyme A dehydrogenase system (EC 1.3.99.3, acyl-CoA: (acceptor) oxidoreductase) in mitochondria accepts electrons from another flavoprotein, one of the three reduced acyl dehydrogenases [1, 2]. ETF can donate electrons to one- and two-electron acceptors, such as indophenol, quinones and ferricyanide [1]. Its natural electron acceptor in the electron transport chain is still uncertain but may be coenzyme Q [3]. We have been studying these enzymes, primarily from pig liver, in the hope of elucidating more fully their properties and the nature of this flavoprotein-flavoprotein interaction.

Results

We have shown [4] that ETF has a molecular weight of 58 000, with two subunits containing one FAD each. The spectrum of ETF (Fig. 1) shows peaks at 370 and 438 nm and shoulders at 460 and 270 nm. ETF is fluorescent, with a polarized emission at 495 nm 3.5 times as intense as that of its own free flavin at pH 7 [4]. This enzyme has been described as rather unstable [2]. We have also been made quite aware that ETF is easily "lost" [4]. We have observed losses of yellow color and greenish fluorescence under various conditions of storage and handling, which correlated with losses in enzyme activity. Since fluorescence of bound flavin is relatively rare (*Peptostreptococcus elsdenii* flavoprotein [5] and lipoyl dehydrogenase [6] are fluorescent), we hoped that changes in fluorescence could give information on changes in flavin binding. We therefore examined the relationship between optical properties of the ETF flavin and enzyme activity.

Table I presents the results of an experiment [4] which shows the

---

Abbreviation: G-AD, general acyl-CoA dehydrogenase [4], ETF, electron-transferring flavoprotein, see Enzyme Nomenclature (1973) Recommendations (1972) IUPAC-IUB, p. 34, Elsevier, Amsterdam.

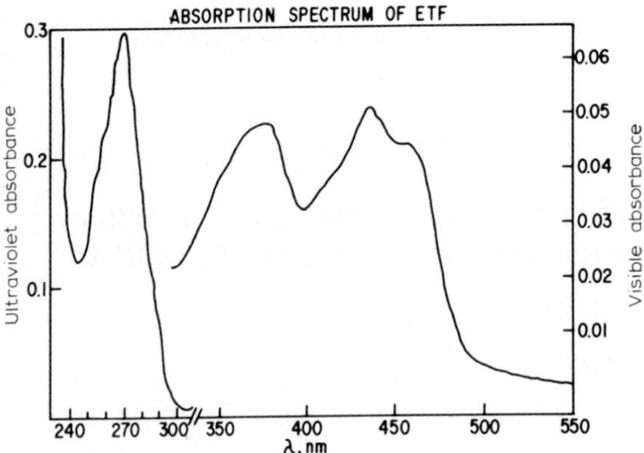

Fig. 1. Absorbance spectrum of ETF, see ref. 4, Fig. 4.

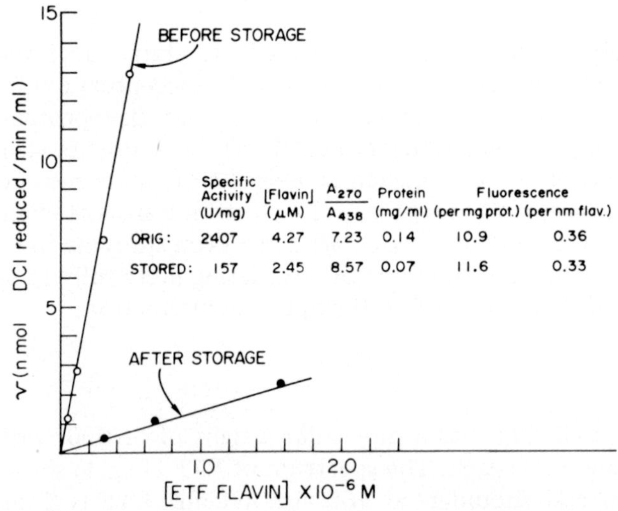

	Specific Activity (U/mg)	[Flavin] (μM)	A_{270}/A_{438}	Protein (mg/ml)	Fluorescence (per mg prot.)	Fluorescence (per nm flav.)
ORIG:	2407	4.27	7.23	0.14	10.9	0.36
STORED:	157	2.45	8.57	0.07	11.6	0.33

Fig. 2. Diluted ETF solution was passed through a millipore filter into a sterile vacutainer and assayed with $1.3 \cdot 10^{-7}$ M G-AD flavin, purity = 24 nmol flavin/mg protein. Specific activity in the inset table was calculated on the basis of dehydrogenase protein only, velocities taken at $4.85 \cdot 10^{-7}$ M ETF flavin. Indophenol was added to an absorbance of about 0.6 and 20 mM potassium phosphate pH 7.6 were added in a 1.2 ml final volume, 20 μM in octanoyl CoA. ETF used was a fraction from pH 8.1 $(NH_4)_2SO_4$ precipitation (see ref. 4).

relationship between the purity of ETF and its spectral properties. Tube 18, which represents a sharp breakthrough of the flavoprotein peak, has the highest flavin (A_{438}) to protein (microbiuret) ratio, the lowest A_{270}/A_{438} ratio, the highest fluorescence per mg protein and per nm flavin, and the highest index of polarization. The table also demonstrates the losses of ETF required for achieving the slight final purification. The changes in optical properties during storage, associated with changes in enzyme activity, are shown in Fig. 2 and its inset table. The activity, flavin content, fluorescence and A_{270}/A_{438} ratio of a sample of ETF were measured before and after

TABLE I

GEL FILTRATION OF ETF ON SEPHADEX G-75[a]

Fraction	Protein[b] (mg/ml)	Flavin[c] (nmol/ml)	$\frac{A_{270\,nm}}{A_{438\,nm}}$	nmol flavin[c]/mg protein[b]	Minimum mol. wt.	Fluorescence[d] intensity/mg protein	Polarization[e] (490 nm)	Fluorescence/mol flavin
Before gel filtration	1.46	38.6	6.4	26.4	37 870	8.3	0.23	0.31
Tube 18	0.16	6.27	5.94	38.7	25 840	14.3	0.26	0.36
19	0.14	4.27	6.81	31.0	32 300	10.7	0.25	0.35
20	0.08	2.09	8.26	25.8	38 730	9.0	0.24	0.34
21	0.05	1.45	9.31	26.9	37 120	8.8	0.22	0.30

Total flavin applied = 46.36 nmol; flavin recovered (tubes 18–21) = 13.4 nmol = 29%. Fraction 17 had no significant fluorescence when observed under a hand-held ultraviolet source and was not used. Total protein applied = 1.75 mg; protein recovered (tubes 18–21) = 0.41 mg = 23%.

a Eluted with 2.5 mM K_2HPO_4. ETF sample was pH 8.1 $(NH_4)_2SO_4$ 80% precipitate [4].

b Determined by the microbiuret method.

c A_{438}, using an E of $11.3 \cdot 10^6$ $cm^{-1} \cdot M^{-1}$

d Arbitrary units: exciting λ = 380 nm, analyzing λ = 490 nm; slits = 20 nm for exciter and 5 nm for analyzer. Determined on Farrand spectrofluorimeter.

e $P = \frac{\parallel \text{intensity} - \perp \text{intensity}}{\parallel \text{intensity} + \perp \text{intensity}}$ exciting λ = 450 nm, analyzing λ and slits as in footnote d.

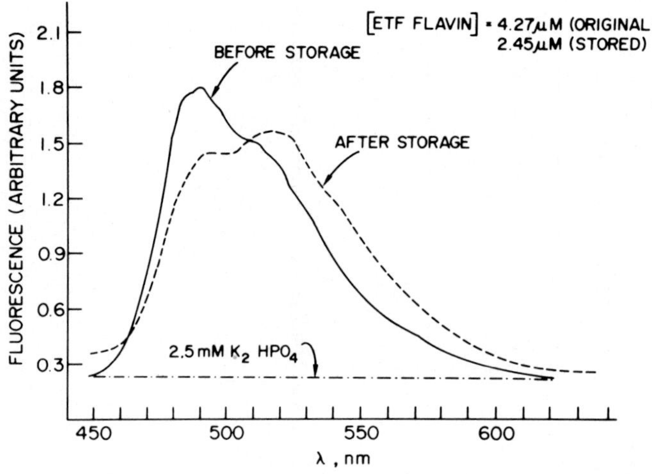

Fig. 3. The ETF used was a sample of the same lot used for the experiment described in Fig. 2, which had been passed through Sephadex G-75. The excitation wavelength was 375 nm and the emission was scanned from 425 to 650 nm in the Farrand Mark I Spectrofluorimeter.

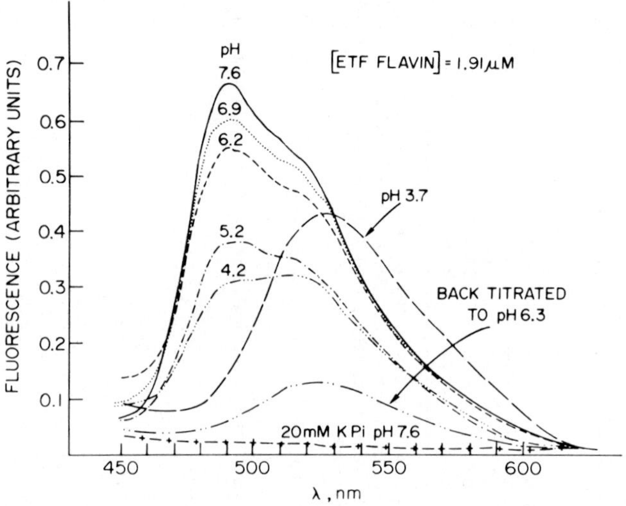

Fig. 4. See text, and legend to Fig. 3.

(sterile) storage for one day at room temperature. Although both the specific fluorescence (per mg protein and per nmol flavin) and the A_{270}/A_{438} ratio were essentially unchanged before and after storage, (see inset of Fig. 2) the specific activity, based upon either the protein content of the dehydrogenase or of the ETF flavin content, was now only 6% of the activity of the original sample. This can be explained by the data of Fig. 3, (normalized for loss of protein) which show a marked change in the shape of the fluorescence emission spectrum. However, ETF is relatively stable at $-70°C$ by optical and catalytic criteria.

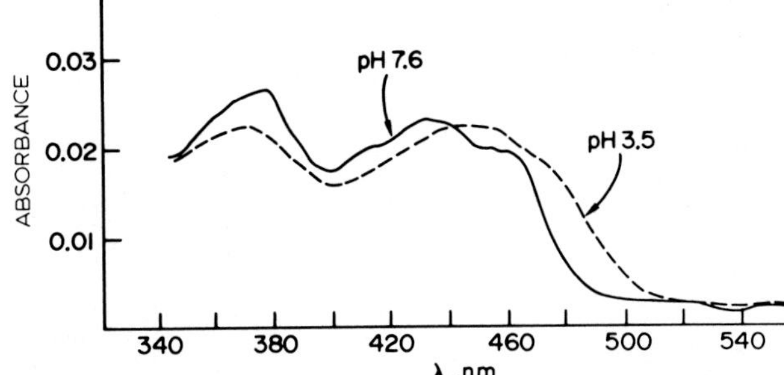

Fig. 5. See text. Experimental conditions were identical to those in Fig. 4.

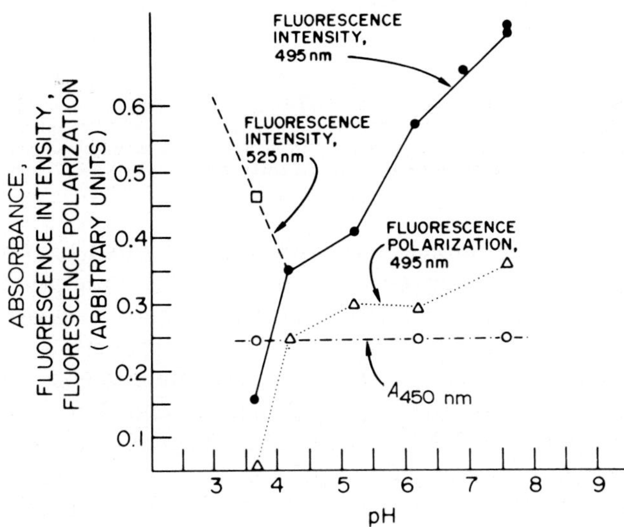

Fig. 6. See text.

Fig. 4 shows the effect of titration with acid on the fluorescence emission spectrum of ETF. The 495 nm emission decreased with decreasing pH until about pH 4. Between there and pH 3.7, the fluorescence changed abruptly, to a spectrum with the 525 nm emission characteristic of free flavin. The absorbance spectrum (Fig. 5) shows little change in the 450 nm absorbance, but the spectrum has shifted from that characterisitc of ETF to that characteristic of FAD. These data suggest that at this low pH, free FAD has been released, and if the solution is back-titrated to pH 6.3 with 1 M dibasic potassium phosphate, the fluorescence returns to the low level characteristic of FAD (Fig. 4, open circles). These data indicate that the events between pH 7 and 4 are qualitatively different from those between pH 4 and 3.5. Fig. 6 displays the previous data as changes in fluorescence intensity, fluorescence polarization and absorbances plotted against pH,

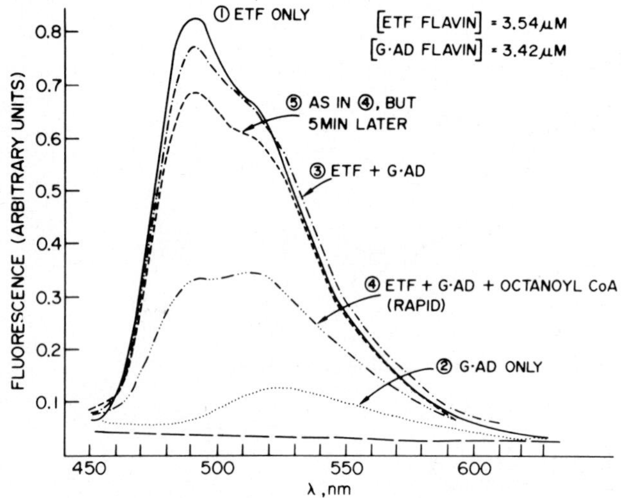

Fig. 7. See text. The ETF used was similar to that used in Figs. 2 and 3, the A_{270}/A_{438} was 5.6. The G-AD used was from $(NH_4)_2SO_4$ fractionation after preparative electrophoresis on G-200 (see ref. 4), purity = 25 nmol/mg protein prior to passing through Sephadex G-150 in 20 mM potassium phosphate pH 7.6. Fluorescence was measured on a Farrand Mark I Spectrofluorimeter. Exciter slits were 20 mm, and analyzer slits were 5 mm.

and shows the striking decrease in fluorescence intensity compared to the much smaller changes in polarization between pH 7.6 and 4. When the pH is lowered to 3.75, both intensity and polarization of fluorescence decrease markedly; the flavin has apparently dissociated and has the properties of free FAD. Other experiments have shown that the ETF fluorescence at 495 nm is essentially unchanged between pH 7.6 and 9 and that the fluorescence of the free flavin at pH 2.75 at its emission maximum has nearly the same intensity as that of native ETF at pH 7.6.

Although this curve has the appearance "formally" of a possibly biphasic titration curve, it would be an oversimplification to ascribe these data to the effects of dissociation of one or even two specific groups: the fluorescence of the flavin can be a complex function not only of the dissociation of its own protons but also of the properties of multiple groups in the protein to which it is bound. But the data probably reflect at least two processes, one (pH 9 to 4.2) where fluorescence intensity changes with little change in polarization, and one in which both fluorescence intensity at 495 nm and polarization decline rapidly with decreasing pH (pH 4.2 to 3.7). The latter probably reflects dissociation of the flavin.

Fig. 7 shows changes in the emission spectrum of ETF when ETF is reacted with G-AD and octanoyl CoA. In this experiment ETF (curve 1) and G-AD (curve 2) were equimolar (approximately 3.5 μM) with respect to flavin. Little change occurs upon mixing (Curve 3), but addition of octanoyl CoA (2.5 μM) results in a rapid drop in the fluorescence intensity (Curve 4). After 5 min, when the ETF may be presumed to have been largely reoxidized by air, most of the original emission spectrum returns (Curve 5).

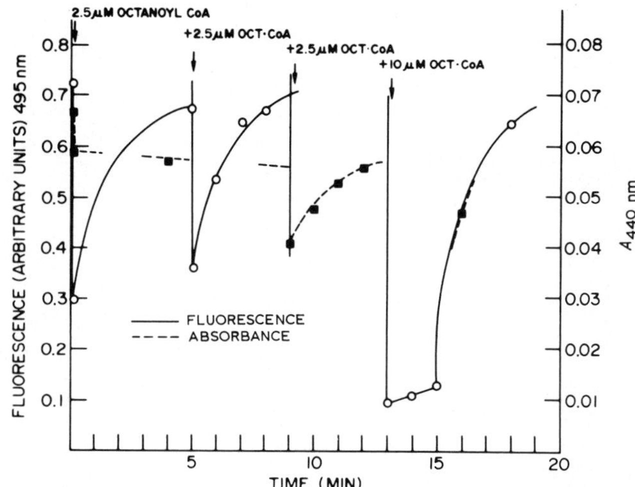

Fig. 8. See text. The solid line represents fluorescence and the dashed line represents absorbance.

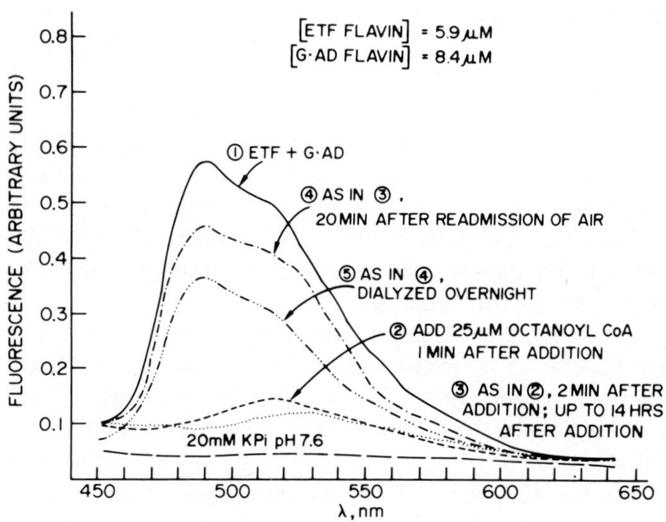

Fig. 9. ETF and G-AD were fractions similar to those described in Figs. 1–4. G-AD had been passed through G-150 Sephadex equilibrated to 20 mM potassium phosphate pH 7.6. Purity of ETF; $A_{270}/A_{438} = 5.3$; purity of G-AD: $A_{270}/A_{48} = 7.8$. A fluorescence cuvette fitted with a Thunberg type tube for addition of substrate by means of ground-glass fittings and a ground-glass stopcock was used, and evacuated by means of a vacuum pump.

The results of four repeated additions of octanoyl CoA to this system are shown in Fig. 8, 3 of them followed by fluorescence at 495 nm and one by absorbance at 440 nm. With each addition of substrate the extent of the return of emission by reoxidation is a little less. With the addition of excess substrate, a lag is seen. Fig. 9 shows the result of an anaerobic experiment, adding 30% excess of G-AD to ETF (Curve 1), and then a 2-fold molar excess of octanoyl CoA. The fluorescence decreases essentially to zero

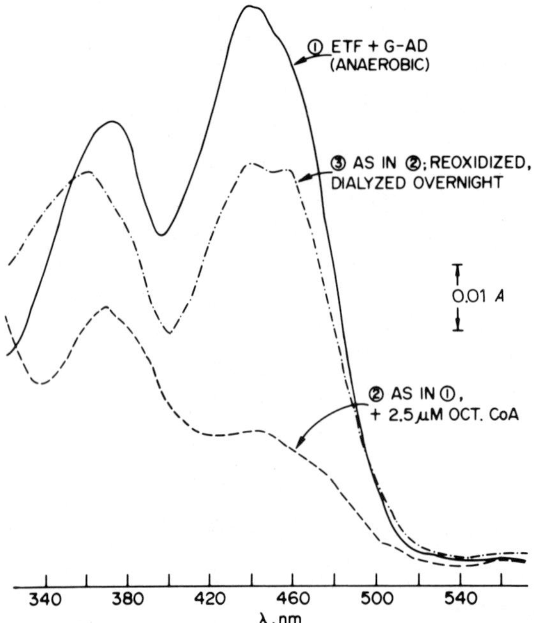

Fig. 10. The conditions were the same as in Fig. 9. See text.

within 1 min (Curve 2), with some further drop and some peak shift within another minute (Curve 3). Fluorescence remained at this level for 14 h. When air was readmitted, the reoxidation phase required 20 min to approach completion (Curve 4), with a lag of several minutes which probably reflected a steady state during which the excess substrate was dehydrogenated. The emission returned to about 80% of the original.

Some corresponding absorbance changes are shown in Fig. 10. The spectrum of the mixture (Curve 1) is dominated by the characteristic absorbance of the dehydrogenase rather than by that of ETF, because of the excess of dehydrogenase. Upon addition of substrate, only about 25% of the absorbance remains, and the resulting spectrum (Curve 2) resembles that of G-AD plus octanoyl CoA in the absence of ETF, which suggests that most of the ETF flavin absorbance had been bleached. Crane and Beinert had shown that with catalytic quantities of general acyl dehydrogenase about 90% of the ETF flavin is bleached (1). After reoxidation and dialysis the spectrum (Curve 3) shows the features of oxidized ETF and is no longer dominated by the G-AD features. This suggests that some of the dehydrogenase may still be reduced, even after prolonged exposure to oxygen.

Discussion

The experiments described suggest that the flavin of ETF is held in a highly specific configuration. This configuration can "decay" to a form (Fig. 3) which, while fluorescent and polarized is catalytically less effective; this change is accompanied by a shift in the fluorescence emission spectrum. Additional "mishandling" can lead to complete liberation of FAD, with decreased fluorescence, polarization and activity (Figs. 4, 5 and 6). Changes occurring during catalysis (Figs. 7, 8 and 9), while leading to marked diminution of fluorescence and absorbance, do not result in the peak shifts seen, for example, during titration with acid (Figs. 4, 5 and 6). Addition of substrate to G-AD/ETF mixtures causes decrease in absorbance as well as fluorescence, indicating reduction of the ETF flavin. No EPR experiments have yet been performed.

Since the increased fluorescence of FAD at pH 3 compared to that at pH 7 has been ascribed to unfolding of the molecule so that the adenine ring does not quench the isoalloxazine ring fluorescence [6], the flavin of ETF may be held in the extended form in the native enzyme. The pH-fluorescence curves shown (Figs. 4–6) suggest that the mode of attachment of the ETF flavin to the protein is pH dependent. Lowering of pH may permit the configuration of the flavin to become altered, without becoming completely detached from the protein, as shown by the retention of polarization. Breakage of critical ligands does, however, result in detachment of the flavin at pH 3.2, and the optical properties of the ETF flavin approach those of free FAD. This process, in our hands, has thus far been irreversible.

Increases in the resolution of absorbance spectra, as well as blue shifts and increased fluorescence emission yields of free flavins, have been ascribed to increased hydrophobicity of the medium [6]. Thus, the flavin moiety of flavoproteins like ETF, which show such spectral characteristics, may also be located in a hydrophobic milieu. The fluorescence of ETF may also be due in part to its low levels of aromatic amino acids [4], which may be factors in the quenching of fluorescence [6]. It has been suggested that the fluorescent flavin of lipoyl dehydrogenase is held in open configuration by a hydrogen bond to the protein. The similarities in the optical characteristics between lipoyl dehydrogenase and ETF may suggest a similar binding of flavin in ETF protein.

Acknowledgements

We wish to thank Paula Leyva and Marie Hall for expert technical assistance. These studies were supported by U.S.P.H.S. Grants AM-04662 and HSAA-5-504-RR-06148 and National Science Foundation Grant GB-7905.

References

1 Crane, F.L. and Beinert, H. (1956) J. Biol. Chem. 218, 717
2 Beinert, H. and Lee, J.R. (1963) Methods in Enzymology (Colowick, S.P. and Kaplan, N.O., eds.), Vol. 6, p. 424, Academic Press, New York
3 Beinert, H. (1963) The Enzymes (Lardy, P.D. and Myrback, H., eds.), 2nd edn., Vol. 7, p. 447, Academic Press, New York
4 Hall, C.L. and Kamin, H. (1975) J. Biol. Chem. 250, 3476
5 Whitfield, C.D. and Mayhew, S.G. (1974) J. Biol. Chem. 249, 2801
6 Palmer, G. and Massey, V. (1968) Biological Oxidations (Singer, T.P. ed.), p. 263, Interscience, New York

Part XII

Structure and function of flavoprotein oxidases

Flavins and Flavoproteins, edited by T.P. Singer

CHAPTER 76

STRUCTURE AND STABILITY OF GLUCOSE OXIDASE: EFFECT OF PERIODATE OXIDATION

SATOSHI NAKAMURA, SUEKO HAYASHI and HIDEYO HASUMI

Department of Biochemistry, Kitasato University School of Medicine, Sagamihara, Kanagawa 228 (Japan)

Summary

For the investigation of the molecular structure and of the roles of the carbohydrate moiety of glucose oxidase, physicochemical and immunological experiments were performed both before and after periodate oxidation. Hydrodynamic parameters indicated that the enzyme was a globular protein with a value of 1.21 for the frictional ratio and of 43 Å for the Stokes radius. The periodate oxidation decreased the carbohydrate content to about 40% of its original amount. No significant alteration was brought about in catalytic, immunological or physicochemical properties by this treatment. The native enzyme was quite resistant to sodium dodecyl sulfate and fairly stable against urea or heat treatment. Although the periodate-oxidized enzyme was also stable against heat, it showed a diminished stability in the presence of dodecyl sulfate and urea. It was found by kinetic analyses that the entropy of activation in the inactivation processes was greatly decreased by the denaturing agents, especially in the case of the periodate-oxidized enzyme, while the free energy of activation remained constant. It is deduced that the carbohydrate moiety of the enzyme increases the stability of the protein moiety, but it does not participate in the catalytic activity, in the immunological reactivity, or in maintaining the conformations of the enzyme molecule.

Introduction

Glucose oxidase (EC 1.1.3.4, β-D-glucose: oxygen 1-oxidoreductase) from fungal sources is a flavoprotein with a molecular weight of around 150 000 and it is known to contain two mol of FAD per mol of enzyme [1–5]. It has been shown that the enzyme contains 11–16% carbohydrate, depending on its source [5–7]. We reported in a previous communication [8] that glucose oxidase isolated from *Aspergillus niger* (Kyowa Hakko

Kogyo Co.) possesses some 11% of carbohydrate by weight and that this enzyme has a relatively small amount of "ordered" structure such as α-helix and β-pleated sheet conformations, suggesting that the main part of the polypeptide chains might be of an "unordered" conformation.

To obtain further insight into the molecular structure of the enzyme, physicochemical properties of the native enzyme are the first to be established.

This paper is concerned with the gross structure and the properties of the enzyme before and after periodate oxidation, and also with possible roles of the carbohydrate moiety of the enzyme.

Materials and Methods

Crude samples of *A. niger* glucose oxidase were obtained from Kyowa Hakko Kogyo Co., Tokyo and purified as described elsewhere [5]. The enzyme activity was measured by the use of a Clark oxygen electrode. Total carbohydrate content was estimated by colorimetric methods using mannose as the standard. Qualitative and quantitative analyses of neutral carbohydrate components were carried out by the use of a Shimadzu Gas-Chromatograph, GC-4BM. Basic carbohydrate components and amino acid residues were determined by the use of a Hitachi 034 Liquid Chromatograph. Sedimentation experiments were performed with a Beckman Model E Ultracentrifuge at 20°C. Circular dichroism (CD) spectra were measured at room temperature with a Jasco spectropolarimeter, J-20. Viscosity and diffusion measurements were made in Ostwald capillary viscometers and in a Tiselius electrophoresis apparatus, respectively. Periodate oxidation was done in a 0.1 M sodium metaperiodate solution at 0°C in the dark as described [9]. Antisera were prepared by the immunization of rabbits with purified glucose oxidase. Reagents used were of the purest grade commercially available.

TABLE I

CARBOHYDRATE COMPOSITIONS OF GLUCOSE OXIDASE

Method	Total Sugar (%)	Component Sugar (residues/mol FAD)				
		Man	Gal	Glc	Hexosamine	(GlcN)
Phenol/sulfuric acid	12.4					
	12.2					
Orcinol/sulfuric acid	10.8					
	12.4					
Park-Johnson	13.2					
Gas-chromatorgaphy		58.7	4.4	1.0		
		53.0	4.4	1.3		
		51.5		1.2		
				0.9		
Amino acid analyzer						11.1
Elson-Morgan					8.9	
					10.2	
Average	12.2	55	4	1	10	

Measurements were done in a 0.05 M acetate buffer pH 5.5, at 25°C unless otherwise specified.

Results

Carbohydrate composition

The carbohydrate composition of the purified glucose oxidase was determined by various methods. The results are given in Table I, together with the methods adopted for analyses. The total carbohydrate content was about 12%, which agreed with our previous results [8], but was slightly lower than those reported by several groups who had studied the *Aspergillus* enzyme [5–7]. However, our preliminary observations on the enzyme samples from different manufacturers (Boehringer, Mannheim and Sigma, St. Louis) indicated that the carbohydrate content of the enzyme was dependent on the production lot, probably because of possible differences in culture conditions and in the methods of purification. For this reason, it is required that a single lot of sample from a single company be used when carry out quantitative carbohydrate analyses. Hence, the investigations were done with a sample of Lot No. 93005 from Kyowa Hakko Kogyo Co. throughout the present study. In accordance with other reports [5, 7], the enzyme contained mannose as the main carbohydrate component, with glucose and galactose as minor neutral carbohydrate components. Glucosamine was the only hexosamine identified and no sialic acid was detected. In our previous communication [8], the presence of a small amount of galactosamine was described, but the hexosamine was not detected this time by gas-chromatographic method, by the amino acid analyzer method or by a paper chromatographic method (pyridine/ethanol/acetic acid/water, 5:5:1:3, v/v); so that the previous result was not reproducible. However, when the analyses were carried out with the amino acid analyzer, an unidentified peak at a position close to that expected for galactosamine was sometimes observed.

Preparation of periodate-oxidized enzyme. Some of the possible roles of the carbohydrate moiety of a glycoenzyme are: (1) to participate in the catalytic activities of the enzyme; (2) to maintain the three-dimensional

TABLE II

EFFECT OF PERIODATE-OXIDATION ON THE CARBOHYDRATE CONTENT

Incubation time (h)	Man* (Residues/mol FAD)	Gal*	Glc*	Hexosamine**	(GlcN)***
0	51.5	4.4	0.9	8.9 (10.2)	11.1
0.5	24.5	1.5	trace	9.1	11.4
1.0	23.7	trace	trace	8.8	10.9
2.0	ND	ND	ND	8.2	ND
3.0	16.4	trace	trace	7.8	ND
5.0	18.5	trace	trace	8.0	9.5
20.0	16.6	trace	trace	7.4	ND

*Gas-chromatography. **Elson-Morgan method. ***Amino acid analyser. ND: not determined.

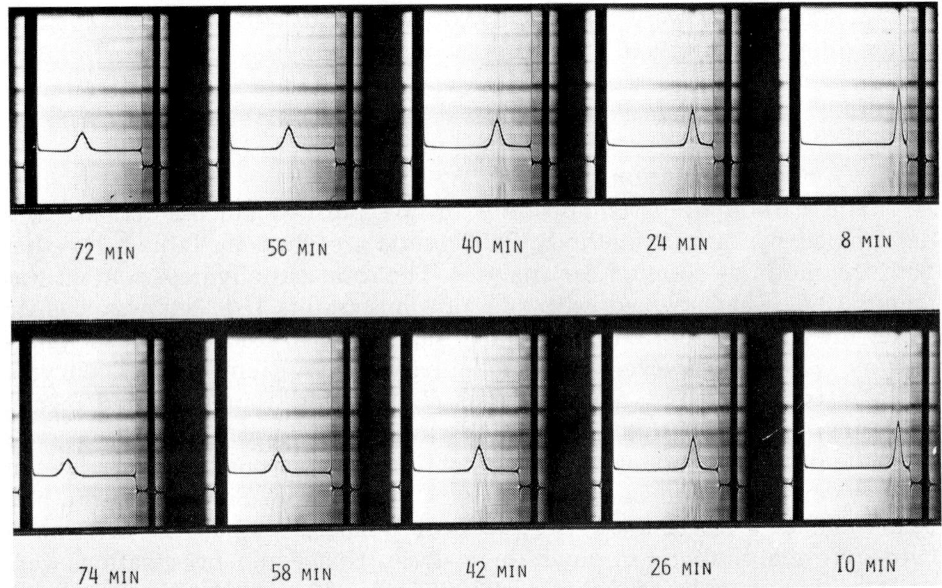

Fig. 1. Sedimentation patterns of the native (upper) and the periodate-oxidized (lower) enzymes. Enzyme concentrations: the native enzyme, 5.6 mg/ml; the periodate-oxidized enzyme, 5.7 mg/ml.

structure of the enzyme protein; (3) to protect the enzyme from proteolysis and (4) to facilitate the transport of the enzyme through cellular membranes [10].

For the investigation of these possibilities, one of the approaches is to compare the properties of the native enzyme with those of the enzyme from which the carbohydrate moiety is removed. For this purpose, glucose oxidase was subjected to the periodate treatment.

Upon incubation of the native glucose oxidase with 0.1 M sodium metaperiodate in the dark, it was found that the amount of the carbohydrate was decreased to approx. 40% of the original content in about 3 h, but no further decrease was seen even after a 20 h incubation (Table II). Quantitative analyses revealed that the main carbohydrate component, mannose, showed the greatest decrease during the treatment. It is noteworthy that the amount of glucosamine was not significantly changed by the treatment, suggesting that the component is deeply buried in the framework of the polypeptide chains. Since the carbohydrate content was not appreciably changed by an incubation of more than 5 h, the enzyme sample treated with the periodate for 5 h was used as the "periodate-oxidized" enzyme hereafter throughout the present study.

Catalytic activity. No significant alteration in the catalytic properties resulted from the treatment, since the apparent Michaelis constants for glucose and for molecular oxygen of the periodate-oxidized enzyme were 28 and 0.18 mM, while those of the native enzyme were 26 and 0.20 mM, respectively.

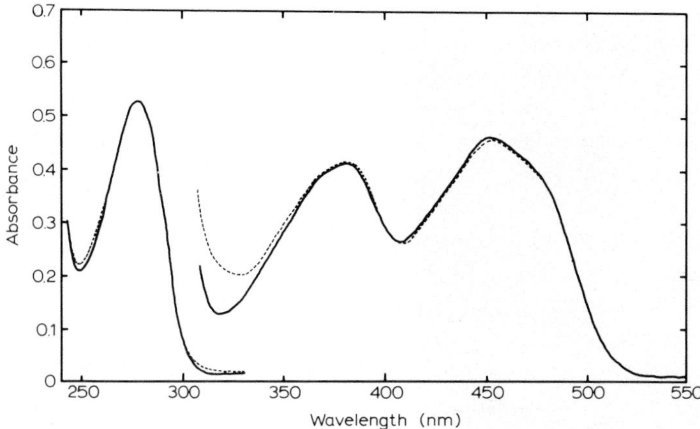

Fig. 2. Absorbance spectra of the native (———) and the periodate-oxidized (----) enzymes. The spectra in the ultraviolet region were measured with 10-fold diluted samples of those used for the measurements in the visible region.

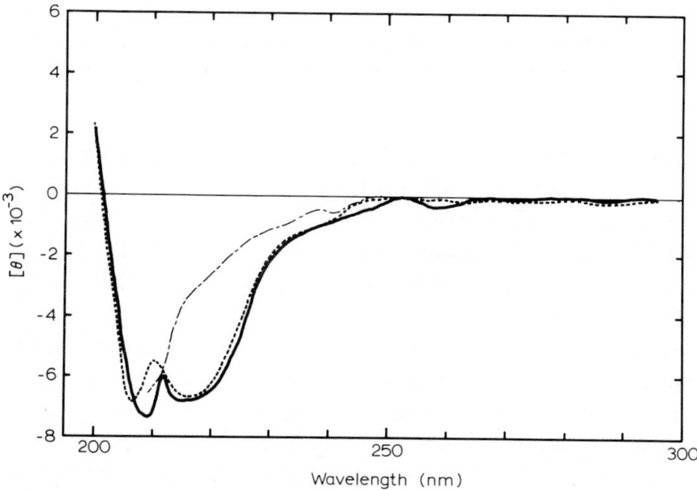

Fig. 3. CD spectra of the native enzyme (———), of the periodate-oxidized enzyme (------) and of the urea-denatured enzyme (—·—). The spectra were measured with a 1-mm cell at room temperature. The denatured enzyme was prepared by incubation of the enzyme with 6 M urea for 10 h. The mean residue weight of amino acids was taken as 109 in the calculation of molar ellipticity [θ].

Sedimentation behavior and molecular weight. The enzyme samples before and after the periodate treatment were subjected to ultracentrifugation. As shown in Fig. 1, both of the samples were homogeneous by this method. The sedimentation coefficient, $s^0_{20,w}$, of the native enzyme was 8.06 S and that of the periodate-oxidized enzyme was 8,37 S. It is thus known that the periodate oxidation did not cause any aggregation or dissociation into subunits of this enzyme. It is interesting to note that the periodate-oxidized enzyme had a slightly higher sedimentation coefficient

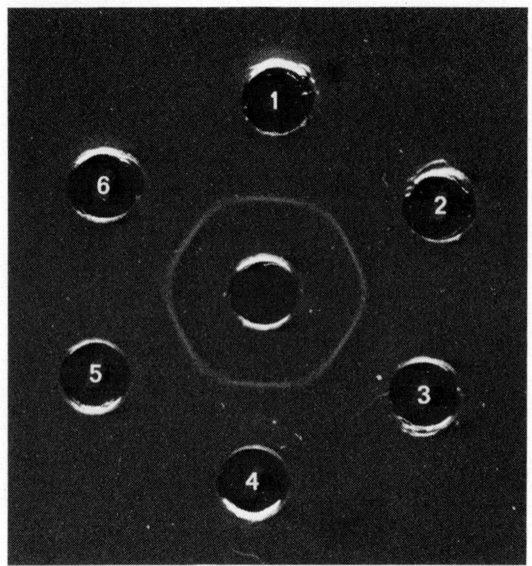

Fig. 4. Agar immunodouble diffusion experiments with the antiserum against the native enzyme. Center well: antiserum, 0.75 mg; (1) (3) (5): the native enzyme, 4.3 μg; (2) (4) (6): the periodate-oxidized enzyme, 4.2 μg.

than the native enzyme, implying that the periodate-oxidized enzyme molecule is somewhat more compact with a higher density than the native enzyme molecule. The molecular weight determined by the meniscus-depletion method [11], using the values for the partial specific volume (see below), was 153 000 for the native enzyme and 150 000 for the periodate-oxidized enzyme.

Absorbance spectrum. Fig. 2 shows that the periodate treatment caused no significant alteration in the absorbance spectrum of glucose oxidase, indicating that the FAD moiety as well as the largest part of the protein moiety remained unattacked by the reagent. Only a slight increase in absorbance was observed in the near-ultraviolet region 300–350 nm. Preliminary analyses in our laboratory indicated that some of arginyl residues might be destroyed during the incubation. Further studies of this problem are now underway.

Circular dichroism spectrum. CD spectra of the native and the periodate-oxidized enzymes are shown in Fig. 3. As reported previously [8], this enzyme protein has a limited degree of ordered structure, such as α-helix and/or β-pleated sheet conformations. Assuming that the CD behavior is mainly due to the α-helix conformation, the helical contents of the native and the periodate-oxidized enzymes were calculated to be about 15% [12, 13]. The periodate oxidation, therefore, did not affect the secondary structure of the protein moiety of the enzyme.

Immunological reactivity. As shown in Fig. 4, the antibody formed against the native enzyme cross-reacted with the periodate-oxidized enzyme.

Immunological association constants measured by the quantitative precipitin method were $3.3 \cdot 10^6$ and $3.7 \cdot 10^6$ M^{-1} for the native and the periodate-oxidized enzymes, respectively. We may conclude therefore that the periodate treatment did not change the immunological properties of the glucose oxidase and that the antigenic site of this enzyme is mainly located on the protein moiety rather than on the carbohydrate moiety.

Partial specific volume. The partial specific volumes of the native and the periodate-oxidized enzymes were determined picnometrically to be 0.727 and 0.721 ml/g, respectively. The values were a little larger than those reported earlier for the *Aspergillus* enzyme [4, 5], but distinctly smaller than those for the *Penicillium* enzyme [2, 4, 5] in accordance with our former results [5].

Intrinsic viscosity. The intrinsic viscosity of the native enzyme was 4.57 ml/g and that of the periodate-oxidized enzyme was 4.63 ml/g. Both values were nearly identical and well within the range of those for globular proteins. The value obtained for the native enzyme under high ionic-strength conditions (in 0.1M acetate buffer with 0.3M NaCl) was 4.6 ml/g. These facts may exclude the possibility that the protein moiety of the enzyme possesses quite a randomly ordered structure as might be suggested from the results of CD measurements. Rather than this, the enzyme has a fairly rigid structure as a globular protein with a frictional ratio of 1.21 (see below).

Diffusion. Diffusion was measured by the free boundary method using a Tiselius electrophoresis apparatus. The values for $D_{20,w}$ of the native enzyme and the periodate-oxidized enzyme were $4.94 \cdot 10^{-7}$ and $4.74 \cdot 10^{-7}$ cm^2/s, respectively. The diffusion coefficient value measured under high ionic strength conditions with the native enzyme was virtually indentical ($4.98 \cdot 10^{-7}$ cm^2/s).

Table III summarizes the results of the hydrodynamic measurements, the values of the frictional ratios and the Stokes radii calculated from these hydrodynamic parameters.

Thermostability. As reported previously [8], glucose oxidase was very stable against denaturing agents, such as sodium dodecyl sulfate and urea. It retained full activity after a 30-h incubation in a 1% sodium dodecyl sulfate

TABLE III

PHYSICOCHEMICAL PROPERTIES OF THE NATIVE AND THE PERIODATE-OXIDIZED GLUCOSE OXIDASES

Physicochemical parameters	Native enzyme	Periodate-oxidized enzyme
Sedimentation coefficient (S)	8.06	8.37
Partial specific volume (ml/g)	0.727	0.721
Intrinsic viscosity (ml/g)	4.57	4.63
Diffusion coefficient (cm^2/s)	$4.94 \cdot 10^{-7}$	$4.74 \cdot 10^{-7}$
Frictional ratio	1.21	1.20
Axial ratio* (b/a, $a \geqslant b$)	1/4.40	1/4.25
Stokes radius (Å)	42.7	44.5

*Prolate ellipsoid is assumed.

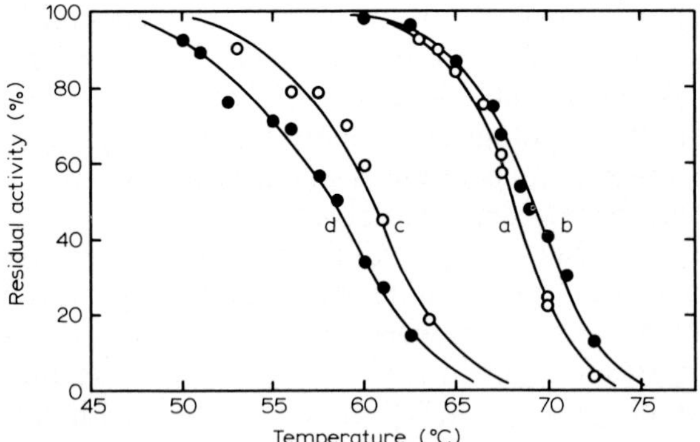

Fig. 5. Thermal stability of glucose oxidase with and without 1% sodium dodecyl sulfate. The enzyme was incubated for 5 min at each temperature, and an aliquot was taken for the assay of the enzymatic activity. Curves a (—dodecyl sulfate) and c (+ dodecyl sulfat) are for the native enzyme, and curves b (—dodecyl sulfate) and d (+ dodecyl sulfate) are for the periodate-oxidized enzyme. Assay conditions were: glucose oxidase, $8 \cdot 10^{-9}$ M (as FAD); D-glucose, 57 mM.

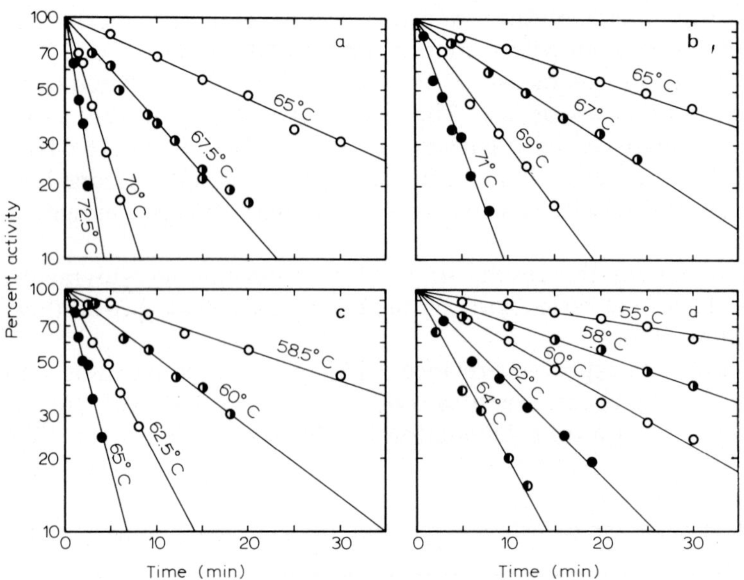

Fig. 6. Kinetic patterns of the heat-inactivation processes. (a): The native enzyme in the absence of denaturing agents. (b): The periodate-oxidized enzyme in the absence of denaturing agents. (c): The native enzyme in the presence of 1.5 M urea. (d): The periodate-oxidized enzyme in the presence of 1.5 M urea.

solution at 30°C pH 5.5. The activity was also fully restored after a 5-min contact with 7 M urea. As shown in Fig. 5, the enzyme was fairly stable at an elevated temperature. The stability of the enzyme, however, was reduced to an appreciable degree in the presence of dodecyl sulfate. An interesting fact is that the periodate-oxidized enzyme was as stable as the native enzyme when no denaturing agent was added. Kinetic analyses of the heat-inactivation

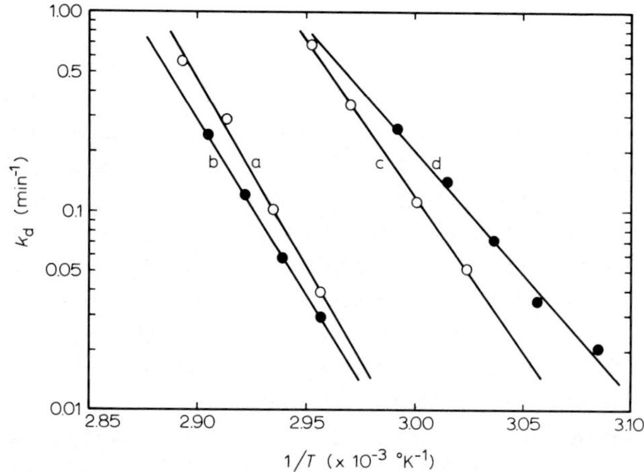

Fig. 7. Arrhenius plot for thermal inactivation of glucose oxidase with and without 1% sodium dodecyl sulfate. Curves a (—dodecyl sulfate) and c (+ dodecyl sulfate) are for the native enzyme, and curves b (—dodecyl sulfate) and d (+ dodecyl sulfate) are for the periodate-oxidized enzyme.

processes with both the native and the periodate-oxidized enzymes were performed in the absence and in the presence of the denaturing agents. Fig. 6 demonstrates that the inactiviation processes follow the first-order reactions, and the temperature dependence of the rate constant, k_d, is shown in the form of the Arrhenius plot (Fig. 7). Activation parameters for the heat-inactivation kinetics were calculated by the use of the activation energy obtained from the figure, and are presented in Table IV. An important observation is that a dramatic change was found in the activation enthalpy ($\Delta H^{\ddagger}$), particularly in the case of the periodate-oxidized enzyme when the denaturing agents were present. Since the free energy of activation ($\Delta G^{\ddagger}$) was found nearly constant in all cases, the entropy of activation ($\Delta S^{\ddagger}$) was varied in accordance with the change in the enthalpy.

TABLE IV

ACTIVATION PARAMETERS FOR HEAT-INACTIVATION PROCESSES OF GLUCOSE OXIDASE IN THE PRESENCE AND IN THE ABSENCE OF DENATURING AGENTS

Activation parameters were calculated from $\Delta H^{\ddagger} = E^* - RT$, where E^* is the experimental activation energy; $\Delta G^{\ddagger} = -RT \cdot \ln[(h/kT)k_d]$, where k_d is the experimental rate constant for the heat inactivation, R, h and k are the gas constant, Planck's constant and Boltzmann's constant, respectively; and $\Delta S^{\ddagger} = (\Delta H^{\ddagger} - \Delta G^{\ddagger})/T$. Values listed here are the average of four to five calculations on the basis of the experimental data in Fig. 7.

	Conditions	$\Delta H^{\ddagger}$ (kcal/mol)	$\Delta G^{\ddagger}$ (kcal/mol)	$\Delta S^{\ddagger}$ (cal/degree/mol)
Native enzyme	Control	87.3	24.1	184.8
	+1% SDS*	74.8	23.1	153.0
	+1.5 M urea	77.8	23.1	161.2
	+3.0 M urea	66.1	23.3	130.5
Periodate-oxidized enzyme	Control	81.0	24.5	165.6
	+1% SDS*	53.2	23.7	89.5
	+1.5 M urea	59.8	24.3	106.8
	+3.0 M urea	50.0	23.5	81.2

*Sodium dodecyl sulfate

Discussion

The purposes of the present study are: (1) to investigate the gross structure of the glucose oxidase molecule, since it seemingly has a randomly extended conformation in the main part of the polypeptide chains as judged by its CD spectrum in the far-ultraviolet region; (2) to examine the effects of the periodate treatment on the various properties of the enzyme, since this method is usually adopted to delete the carbohydrate moiety, without necessary consideration of possible alterations in the structure and properties of the protein moiety; and (3) to know the possible roles of the carbohydrate moiety of the glycoenzyme by comparing the physicochemical properties of the periodate-oxidized enzyme with those of the native enzyme.

From the hydrodynamic parameters obtained in the present study, the size and shape of the glucose oxidase molecule can be depicted. The molecule has a value of about 43 Å for the Stokes radius with a frictional ratio of 1.21, which may correspond to an axial ratio (b/a, $a \geqslant b$) of 1/4.40, if a prolate ellipsoidal form is assumed. The size and shape of the periodate-oxidized enzyme is essentially the same as those of the native enzyme. These findings may enable us to conclude that glucose oxidase is a typical globular protein with a rigid structure, and is not a protein of disordered or denatured configuration. In favor of this interpretation, no alteration in the hydrodynamic properties was brought forth by the addition of neutral salt, which may give rise to such an alteration by weakening the intra-peptide ionic interactions as in the case of a disordered polypeptide.

As described in the text, the periodate treatment did not affect the properties of the enzyme, as for instance by the catalytic properties, immunological properties, the secondary and tertiary structures and the stability against heating. Accordingly, it is reasonable to conclude that the protein moiety of the enzyme remains virtually intact during the periodate treatment, although a minor modification was detected in the absorbance spectrum (Fig. 2). The only remarkable alteration in the periodate-oxidized enzyme in comparison with the native enzyme was a decreased thermal stability in the presence of the denaturing agents. Therefore, a protective effect of the carbohydrate moiety must be postulated, which does increase the stability of the protein moiety when the denaturing agents are present. The protective effect is quantitatively reflected by the values of the entropy of activation in the inactivation processes (Table IV). Since a decrease in this value, in general, might well be ascribed to a less-ordered state of the protein structure, as compared with a standard state, the notable decrease observed with the periodate-oxidized enzyme in the presence of the denaturing agents implies that the protein is forced by the denaturing agents to take on a somewhat disordered conformation. Hence, the periodate-oxidized enzyme is much more susceptible to heat treatment than the native enzyme in the presence of the denaturing agents.

Acknowledgments

We want to thank Professor S. Horie for his interest in this work and his encouragement. We are indebted to Mr. T. Kondo, Department of Molecular Biology, for ultracentrifugal analyses; to Professor G. Matsumura, Showa University, and to Dr. Y. Kawai, Ochanomizu University, for carbohydrate analyses. Thanks are also due to Dr. T. Oshima, Mitsubishi-Kasei Institute of Life Sciences, for his facilitating the use of the Jasco Spectropolarimeter.

References

1 Cecil, R. and Ogston, A.G. (1948) Biochem. J. 42, 229
2 Kusai, K. (1960) Annu. Rep. Sci. Works, Osaka Univ. 8, 43–74
3 Pazur, J.H. and Kleppe, K. (1964) Biochemistry 3, 578–583
4 Bodmann, O. and Walter, M. (1965) Biochim. Biophys. Acta 110, 496–506
5 Nakamura, S. and Fujiki, S. (1968) J. Biochem. Tokyo 63, 51–58
6 Pazur, J.H., Kleppe, K. and Cepute, A. (1965) Arch. Biochem. Biophys. 111, 351–357
7 Swoboda, B.E.P. and Massey, V. (1965) J. Biol. Chem. 24, 2209–2215
8 Nakamura, S. and Hayashi, S. (1974) FEBS Lett. 41, 327–330
9 Yasuda, Y., Takahashi, N. and Murachi, T. (1971) Biochemistry 10, 2624–2630
10 Pazur, J.H. and Anderson, Jr., N.N. (1972) in Advances in Carbohydrate Chemistry (Tipson, R.S. and Horton, D., eds.), Vol. 27, pp. 301–341, Academic Press, New York
11 Yphantis, D.Y. (1964) Biochemistry 3, 297–317
12 Townend, R., Kumonosinski, T.F., Timasheff, S.N., Fassman, G.D. and Davidson, B. (1966) Biochem. Biophys. Res. Commun. 23, 163–169
13 Cassim, J.Y. and Yang, J.T. (1967) Biochem. Biophys. Res. Commun. 26, 58–64

Flavins and Flavoproteins, edited by T.P. Singer

CHAPTER 77

D-AMINO ACID OXIDASE PRIMARY STRUCTURE: CYANOGEN BROMIDE FRAGMENTS AND SEQUENCES OF THE CYSTEINYL PEPTIDES

SEVERINO RONCHI, LORENZO MINCHIOTTI, BRUNO CURTI and [a] JOHN BRIDGEN

Institute of Biological Chemistry, Faculty of Medicine and Pharmacy, The University of Pavia, 27100 Pavia (Italy) and [a] M.R.C. Laboratory of Molecular Biology, University Postgraduate Medical School, Cambridge (U.K.)

Introduction

Much work has been carried out in the past years concerning the catalytical mechanism and the kinetic properties of D-amino acid oxidase from pig kidney (E.C. 1.4.3.3, D-amino-acid: oxygen oxidoreductase (deaminating)) [1, 2, 3]. However, very little data is available concerning the primary structure of this enzyme and many discrepancies exist in the literature about the molecular weight and amino acid composition of this flavoprotein [4]. Although many specific reagents have been tested with the aim of labelling essential amino acid residues in the active center of the enzyme, no unequivocal results have so far been obtained, except for some indications concerning the role of cysteine residues in the binding of the coenzyme [5] and the kinetic evidence for the role of positively charged groups in the catalytic mechanism [1].

The scarce structural data on D-amino acid oxidase, compared with a considerable amount of information produced by kinetic studies, prompted us to investigate the primary structure of this flavoprotein, with the aim of obtaining a better insight into its structural properties and to facilitate the identification of the active center of the enzyme, once a suitable labelling agent was obtained.

In the present study we report data concerning the CNBr digestion of D-amino acid oxidase and the identification of the amino acid sequences around the five cysteinyl residues of the enzyme.

Methods

D-Amino acid oxidase, purified from pig kidney according to our procedure [6], was carboxymethylated with iodo[$^{14}C_2$]acetic acid as described by Butler et al. [7]. All subsequent experiments were carried out on the carboxymethylated freeze-dried protein. CNBr, tryptic and chymotryptic digestion were substantially performed according to the procedures of Gross [8], Gibbons et al. [9] and Smith [10], respectively. CNBr fragments were separated by gel chromatography on a Sephadex G-75 Superfine column and further purified by rechromatography on the same column or by high-voltage electrophoresis. Tryptic and chymotryptic peptides were separated using ion-exchange chromatography on PA 35 resin and high-voltage electroohoresis at pH 6.5 and 2.2 followed by paper chromatography. Sequential degradation of the whole protein or the largest CNBr fragments was performed by the automatic Edman procedure [11] whereas small peptides were sequenced by the manual dansyl-Edman degradation [12]. Protein and peptides samples were hydrolyzed and analyzed for their amino acid content by using a Beckman 120 B amino acid analyzer equipped with a high-sensitivity cuvette, according to the procedure of Spackman [13].

Results and Discussion

The amino acid composition of D-amino acid oxidase, purified according to our procedure, is reported in Table I, column 1. Five methionine and five cysteine residues are present per FAD monomer enzyme (mol. wt. of 39 600). Methionine is the N-terminal amino acid, whereas leucine is the C-terminal amino acid.

As a first approach to the elucidation of the primary structure we have carried out a CNBr digestion of the [^{14}C] carboxymethylated protein. The elution profile of the digested D-amino acid oxidase, following chromatography on Sephadex G-75 Superfine, is shown in Fig. 1. Three distinct peaks were found, along with a fourth which consisted mainly of homoserine lactone.

The first two peaks, after a further purification by rechromatography on the same Sephadex column, were found to contain two pure peptide fragments, namely CNBr 1 and CNBr 2. The third peak of Fig. 1, following high-voltage electrophoresis, resolved into two peptide fragments, chromatographically pure, namely CNBr 3a and CNBr 3b. Their amino acid composition has been reported in Table I. CNBr 1, CNBr 2, CNBr 3a and CNBr 3b fragments contain respectively 217, 84, 23 and 14 amino acid residues. For peptides CNBr 3a and CNBr 3b it was possible to determine the whole sequence by the dansyl-Edman degradation. There is a good correlation between the amino acid content of the whole protein (Table I, column 1) and the sum of the amino acid content of the single CNBr fragments.

The results of the CNBr digestion however, did not show the appearance of the fifth expected peptide, considering that the enzyme contains five methionines, one being N-terminal. To elucidate this point and to produce a tentative alignment of the CNBr peptides, we carried out two different sets of analyses. The first was an analysis of the N-terminal portion of the protein both by the method of Weiner et al. [14] for a limited number of amino acid residues and by using the automatic Edman procedure which allowed us to obtain the sequence of the first 42 residues, as shown in Table II. The second analysis was a tryptic digestion of the protein, followed by a close examination of the methionine-containing peptides. Three peptides, namely TM 1, TM 2 and TM 3 were obtained and the sequences of TM 2 and TM 3 were determined by the dansyl-Edman degradation. From its amino acid

TABLE I

AMINO ACID COMPOSITION OF D-AMINO ACID OXIDASE AND ITS CYANOGEN BROMIDE FRAGMENTS

Amino acid	Composition (residues/mol FAD)									
	Enzyme subunit		Fragment CNBr 1		Fragment CNBr 2		Fragment CNBr 3a		Fragment CNBr 3b	
	Analysis	Integer	Analysis	Integer	Analysis	Integer	Analysis	Sequence	Analysis	Sequence
Carboxymethylcysteine	5.0	5	4.0	4	0.9	1	—	—	—	—
Aspartic acid	32.2	32	19.0	19	7.9	8	2.9	3	1.0	1
Threonine*	22.6	23	12.9	13	6.6	7	1.0	1	1.0	1
Serine*	13.6	14	5.3	6	5.4	6	1.0	1	—	—
Glutamic acid	36.2	36	26.5	26	8.3	8	1.1	1	1.1	1
Proline	23.4	23	11.4	11	6.7	7	2.9	3	1.1	1
Glycine	33.2	33	24.1	24	5.2	5	2.0	2	1.0	1
Alanine	17.9	18	8.3	9	7.6	8	1.1	1	—	—
Valine*	24.8	25	13.7	14	7.6	8	2.0	2	0.9	1
Methionine	5.0	5	0.8	1	0.8	1	0.9	1	0.8	1
Isoleucine*	16.8	17	13.5	14	3.6	4	—	—	—	—
Leucine	36.2	36	22.7	23	6.8	7	2.0	2	2.8	3
Tyrosine	13.9	14	8.1	8	3.6	4	1.6	2	—	—
Phenylalanine	14.9	15	9.9	10	3.1	3	1.0	1	0.9	1
Lysine	11.9	12	8.3	9	1.1	1	0.9	1	1.0	1
Histidine	7.5	8	4.9	5	2.1	2	—	—	—	—
Arginine	21.2	21	14.4	15	2.9	3	1.0	1	1.8	2
Tryptophan	8.0	8	—	6	—	1	—	1	—	—
Total residues		345		217		84		23		14

* Extrapolated values

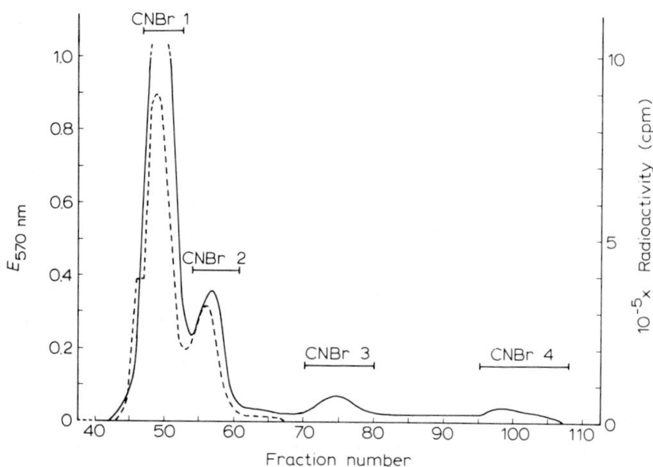

Fig. 1. Gel filtration on a Sephadex G-75 Superfine column of CNBr fragments from [^{14}C]carboxymethylated D-amino acid oxidase. ——, E_{570} - - -, radioactivity.

composition, TM 1 was easily positioned in the N-terminal region of the protein, whereas TM 2 overlapped the C-terminal portion of CNBr 3a and the N-terminal portion of CNBr 3b. TM 3 overlapped the C-terminal portion of CNBr 3b and the N-terminal portion of CNBr 1. We were unable to purify other methionyl peptides from the soluble portion of the tryptic digest: our experimental results suggest that the remaining two methionyl residues are positioned in large tryptic fragments. Further evidence comes from the sequences of two small peptides obtained by a chymotryptic digestion of the protein. The first sequence, Ala—Ala—Asn—Met, is identical to the C-terminal portion of the CNBr 2 fragment. The sequence of the second peptide, Thr—Met—Pro—Pro—Ser—His—Leu does not reproduce any of the previous sequences around the methionine residues: consequently we maintain that this peptide is the C-terminal portion of the protein. The same sequence explains our initial failure in finding a fifth peptide following the CNBr digestion, the latter having a very small size and possibly being eluted under the peak containing homoserine lactone (Fig. 1) as a main product. As a consequence of the above analysis, it was possible to produce an alignment of the CNBr fragemnts, as reported in Fig. 2.

Since cysteine residues have been reported to be involved in the binding of FAD [5], determination of sequences around the thiol groups of the

TABLE II

N- TERMINAL SEQUENCE OF D-AMINO ACID OXIDASE

Met—Arg—Val—Val—Val—Ile—Gly—Ala—Gly—Val—
—Ile—Gly—Leu—Ser—Thr—Ala—Leu—Cys—Ile—His—
—Glu—Arg—Tyr—His—Ser—Val—Leu—Gln—Pro—Leu—
—Asp—Val—Lys—Val—Tyr—Ala—Asp—Arg—Phe—Thr—
—Pro—Phe—

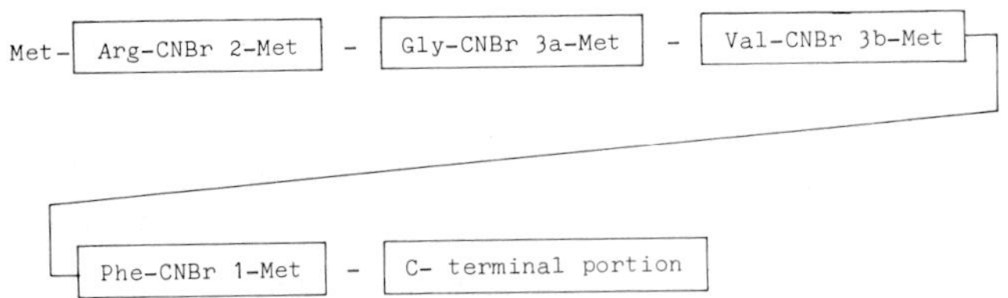

Fig. 2. Alignment of the CNBr fragments and the C-terminal portion of D-Amino acid oxidase.

protein could be a useful approach to the elucidation of the active site of this enzyme. The only information produced by the CNBr digestion in this regard allowed us to place one cysteine in the CNBr 2 fragment and the remaining four in the CNBr 1 fragment.

To study the sequence around the cysteinyl residues, we have carried out a chymotryptic digestion of the [^{14}C]carboxymethylated enzyme. The elution profile of the digested protein, after chromatography on PA 35 resin, showed several ^{14}C-labelled fragments. The peaks were further purified by high-voltage electrophoresis and paper chromatography and three cysteinyl peptides were isolated and sequenced by the dansyl-Edman procedure. A fourth cysteinyl peptide was characterized both by tryptic and chymotryptic digestions of the CNBr 2 fragment, as well as from the automatic Edman degradation of the N-terminal portion of the protein. The results, reported in Table III, confirm that five cysteine residues are present in the enzyme monomer, two of them being adjacent in the sequence. One thiol group (position 18) is located in the N-terminal portion of the protein in a sequence which contains several hydrophobic residues. The same cysteine is not present in the N-terminal analysis carried out by Yagi et al. [15], who report at position 18 a serine residue. However, these authors identified this residue only after it had been converted to alanine by HI hydrolysis. Since CM-cysteine is also converted to alanine by this treatment, they could have been unable to differentiate the two.

No other cysteine residues are present in the sequence represented by the alignment of CNBr 2, CNBr 3a and CNBr 3b fragments, whereas the remaining four thiol groups are all located in the second half of the sequence. Previous experiments (Ronchi, S., Curti, B. and Williams, Jr., C.H.,

TABLE III

Amino acid sequence of thiol peptides from a chymotryptic digest of D-Amino acid oxidase

Peptide	
CH 26–30	Gly—Cys—Ala—Leu—Glu—Val—Ala—Lys—Leu
CH 20–25	Glu—Glu—Val—Ala—Arg—Gly—Gly—Ala—Asp—Val— —Ile—Ile—Asn—Cys—(Asp, Thr, Gln, Gly_2, Ala, Val)—Trp
CH 38–39	Glu—Gly—Cys—Cys—Arg—Leu—Glu—Pro—Thr—Leu— —Lys—Asp—Ala—Lys—Ile—Val—Gly—Glu—Tyr
CNBr 2*	Ala—Leu—Cys—Ile—His—Glu—Arg

* After tryptic and chymotryptic digestion of CNBr 2 peptide.

unpublished) showed that the five cysteines of the enzyme were not titrated by alkylating agents such as iodoacetic acid or iodoacetamide, unless the protein has previously been denatured in 6 M urea or guanidine. These experiments indicate that the sulphydryl residues may be buried inside the molecule, possibly in a hydrophobic environment and in this context it may be pertinent that cysteine 18 occurs in a very hydrophobic sequence. Finally, a comparison of our partial data on the primary structure of D-amino acid oxidase with the sequences of flavodoxin from *Peptostreptococcus elsdenii* [16], flavodoxin from *Clostridium MP* [17] or with the partial sequences around the cysteinyl residues and the disulfide groups of lipoyldehydrogenase [18, 19, 20], glutathione reductase [21] or thioredoxin reductase [22] shows no significant homologies between those flavoproteins and D-amino acid oxidase.

Acknowledgements

This work was supported by a NATO grant (No. 439) and by a grant from the Italian Research Council (No. 71.00805.05).

References

1 Walsh, C.T., Krodel, E., Massey, V. and Abeles, R.H. (1973) J. Biol. Chem. 248, 1946–1955
2 Porter, D.J.T., Voet, J.G. and Bright, H.J. (1973) J. Biol. Chem. 248, 4400–4416
3 Yagi, K., Nishikimi, M., Takai, A. and Ohishi, N. (1974) J. Biochem. Tokyo 74, 451–454
4 Neims, A.H. and Hellerman, L. (1970) Annu. Rev. Biochem. 39, 867–888
5 Fonda, M.L. and Anderson, B.M. (1969) J. Biol. Chem. 244 666–674
6 Curti, B., Ronchi, S., Branzoli, U., Ferri, G. and Williams, Jr., C.H. (1973) Biochim. Biophys. Acta 327, 266–273
7 Butler, P.J.G., Harris, J.I., Hartley, B.S. and Leberman, R. (1969) Biochem. J. 112, 679–689
8 Gross, E. (1967) in Methods in Enzymology (Hirs, C.H.W., ed.), Vol. XI, pp. 238–255, Academic Press, New York
9 Gibbons, I. and Perham, R.N. (1970) Biochem. J. 116, 843–849
10 Smyth, D.G. (1967) in Methods in Enzymology (Hirs, C.H.W., ed.), Vol. XI, pp. 214–231, Academic Press, New York
11 Edman, P. and Begg, G. (1967) Eur. J. Biochem. 1, 80–91
12 Gray, W.R. (1972) in Methods in Enzymology (Hirs, C.H.W. and Timasheff, S.N., eds.), Vol. XXV, pp. 333–344, Academic Press, New York
13 Spackman, D.H. (1967) in Methods in Enzymology (Hirs, C.H.W., ed.), Vol. XI, pp. 3–15, Academic Press, New York
14 Weiner, A.M., Platt, T. and Weber, K. (1972) J. Biol. Chem. 247, 3242–3251
15 Yagi, K., Sugiura, N., Mizuno, M., Hirayama, T., Kagamiyama, H. and Yamano, T. (1974) J. Biochem. Tokyo 76, 1369–1371
16 Tanaka, M., Haniu, M., Yasunobu, K.T., Mayhew, S. and Massey, V. (1973) J. Biol. Chem. 248, 4354–4366
17 Tanaka, M., Haniu, M., Yasunobu, K.T. and Mayhew, S. (1974) J. Biol. Chem. 249, 4393–4396
18 Matthews Green, R., Arscott, D.L. and Williams, Jr., C.H. (1974) Biochim. Biophys. Acta 370, 26–38
19 Burleigh, Jr., B.D. and Williams, Jr., C.H. (1972) J. Biol. Chem. 247, 2077–2082
20 Brown, J.P. and Perham, R.N. (1974) Biochem. J. 137, 505–512
21 Williams, Jr., C.H., Arscott, D.L. and Jones, E.T. (1976) Flavins and Flavoproteins, Proceedings of the 5th International Symposium (Singer, T.P., ed.), Chapt. 48, pp. 455–463, Elsevier, Amsterdam
22 Ronchi, S. and Williams, Jr., C.H. (1972) J. Biol. Chem. 247, 2083–2086

Flavins and Flavoproteins, edited by T.P. Singer

CHAPTER 78

FMN-DEPENDENT PYRIDOXAMINEPHOSPHATE OXIDASE FROM RABBIT LIVER

DONALD B. McCORMICK*, MICHAEL N. KAZARINOFF and HARUHITO TSUGE

Section of Biochemistry, Molecular and Cell Biology and Division of Nutritional Sciences, Cornell University, Ithaca, N.Y. 14853 (U.S.A.)

Summary

Pyridoxamine (pyridoxine) 5′-phosphate oxidase (EC 1.4.3.5), purified 2000-fold from rabbit liver, migrates as a single protein and activity band on analytical disc gels containing 4%, 7% or 9% acrylamide, and as a single protein band on sodium dodecyl sulfate-acrylamide gels and is, therefore, homogeneous by these criteria.

The holoenzyme has absorbance maxima red-shifted from those of FMN to 380 and 448 nm, and exhibits spectral changes typical of flavoproteins upon reduction with dithionite. Its oxidation-reduction potential at pH 7 in phosphate buffer is − 0.131 volt.

The pure enzyme catalyzes the following reactions in the presence of FMN:

$$\text{pyridoxine-}P + O_2 \rightarrow \text{pyridoxal-}P + H_2O_2$$

$$\text{pyridoxamine-}P + O_2 + H_2O \rightarrow \text{pyridoxal-}P + H_2O_2 + NH_3$$

$$N(5'\text{-phospho-4}'\text{-pyridoxyl})\text{amine} + O_2 + H_2O \rightarrow \text{pyridoxal-}P + H_2O_2 + \text{amine}$$

These activities copurify in the ratio of 1:1:1. Apparent K_m values are 10 μM for pyridoxamine phosphate, 30 μM for pyridoxine phosphate, and 40 nm for FMN. Apparent K_m values for *N*(phosphopyridoxyl)amines range from $3.1 \cdot 10^{-5}$ M to $1.6 \cdot 10^{-3}$ M. The pH optima for all three types of substrates are broad, with maxima near pH 9. The pH dependence of FMN binding, measured by quenching of flavin fluorescence, has the same shape as the substrate activity profile.

Quenching of protein fluorescence reveals that one FMN is bound per

---

Abbreviations: NPP, *N*(5′-phospho-4′-pyridoxyl); DTNB, 5,5′-dithiobis(2-nitrobenzoic acid); dansyl, 1-dimethylaminonaphthalene-5-sulfonyl.

* To whom all correspondence should be addressed.

50 000–60 000 daltons at pH 7 and 8 in phosphate buffer, and at pH 6–10 in Tris buffer. A comparison of the molecular weights by gel filtration (55 000), sedimentation equilibrium (56 000), disc-gel electrophoresis (54 000), and dodecyl sulfate-gel electrophoresis (27 000), indicates that monomeric oxidase, which binds one FMN, is composed of two possibly identical polypeptide chains. After reaction of the protein with dansylchloride and acid hydrolysis, only *O*-dansyltyrosine and ϵ-dansyllysine were detected, indicating the α-amino termini of the polypeptide chains are blocked.

Titration of the native apoenzyme or holoenzyme with both DTNB and S-H_g^+-N-dansylcysteine indicates the presence of two free sulfhydryl groups per 54 000 daltons. During the S-H_g^+-N-dansylcystein titration, the protein tryptophan fluorescence is quenched to a greater extent for the apoenzyme than for the holoenzyme.

Preliminary evidence indicates that aggregation to active dimers can occur.

Introduction

Pyridoxaminephosphate oxidase (EC 1.4.3.5, pyridoxaminephosphate: oxygen oxidoreductase (deaminating)) is an FMN-dependent flavoprotein first partially purified from rabbit liver by Pogell [1], which also oxidises pyridoxin 5-phosphate and pyridoxin. Later studies by Wada and Snell [2], Korytnyk and coworkers [3] and in this laboratory [4, 5] have each used improved partial purifications of the rabbit liver enzyme. A homogeneous preparation of the enzyme has been described recently [6]. The oxidase is localized in the soluble portion of the cell [7] and has been postulated to be a key control point in vitamin B-6 metabolism in mammals [8]. We began our studies of pyridoxaminephosphate oxidase both to clarify its role in vitamin B-6 metabolism and because it promised to be a system in which aspects of flavin catalysis could be examined. The substrates and products of the enzymic reaction have, themselves, been used as probes of protein structure and function in other systems and have a well-defined chemistry which we felt we could utilize. We would like to report here on our continuing experiments to characterize the catalytic and physical properties of the pyridoxaminephosphate oxidase system.

Results

Catalytic properties of pyridoxaminephosphate oxidase

Substrate analogs that have been tested for activity are given in Table I. A π-system in the 4′ position and a dianionic charge on the 5′ position are required for binding at the substrate site. Pyridoxamine 5′-sulfate is inactive, and the oximes listed are inhibitors. Another significant result is the effect of changes in structure at sites removed from the reacting portion of the analogs. Changes in stereochemistry from the D to the L form and, in some cases, addition of a single methylene group, have significant effects on the substrate activity of the analogs in question.

The binding of cofactor analogs is summarized in Table II. Analogs that do no retain the D configuration for all three side-chain hydroxyls and a negative charge at the 5′ position are ineffective in reconstituting oxidase activity. FAD is also inactive. The pH dependence of FMN binding to apoenzyme has been examined by monitoring quenching of flavin fluorescence. The data (Fig. 1) show nearly the same broad pH depedence for FMN binding as seen in Fig. 2 for the substrate activity versus pH profile.

Spectral properties of the pure enzyme

The ultraviolet-visible spectra of apo- and holopyridoxaminephosphate oxidase are compared with the FMN spectrum in Fig. 3. The 372 and 445 nm

TABLE I

ACTIVITY OF SUBSTRATE ANALOGS IN THE PYRIDOXAMINEPHOSPHATE OXIDASE SYSTEM

Compound	Apparent K_m^a ($\times 10^{-5}$ M)	Relative V^a (%)
Pyridoxamine 5′-phosphate	1.0	100
Pyridoxine 5′-phosphate	3.0	100
NPP-glycine[b]	6.8	100
Pyridoxamine 5′-sulfate[c]	not active	
Pyridoxal-*P*-oxime[d]	2.1	
Pyridoxal-*P*-*O*-carboxymethyloxime[d]	2.5	
5′-Homopyridoxine-*P*[e]	0.59	10
5′-Methylpyridoxine-*P*[e]	3.1	60
α-NPP-ornithine[f]	53	100
α-NPP-lysine[f]	20	20
NPP-β-alanine[g]	11	85
NPP-L-alanine[g]	22	140
NPP-D-alanine[g]	77	130
NPP-L-α-aminobutyrate[g]	9.1	96
NPP-D-α-aminobutyrate[g]	29	120
NPP-γ-aminobutyrate[g]	77	130
NPP-L-serine[g]	13	39
NPP-L-leucine[g]	7.5	86
NPP-D-leucine[g]	12.5	120
NPP-benzylamine[g]	3.3	120
NPP-L-phenylalanine[g]	9.5	54
NPP-L-tyrosine[g]	3.1	57
NPP-D-tyrosine[g]	160	55
NPP-L-tryptophan[g]	12	18
ω-NPP-diaminodecane[h]	40	20

[a] Apparent K_m and maximum velocity (V) values were determined from Lineweaver–Burk plots by least-squares analysis. V values are given relative to pyridoxamine phosphate as substrate.
[b] Synthesized by the method given in ref. 4.
[c] Synthesized by the method given in ref. 9.
[d] Synthesized by the method given in ref. 10; the value given is the inhibition constant, K_I.
[e] Data were taken from ref. 3.
[f] These compounds were kindly provided by Dr. James K. Coward, Department of Pharmacology, Yale University School of Medicine, New Haven, CT., U.S.A.
[g] Data were taken from ref. 4.
[h] Synthesized by the method given in ref. 11.

maxima of FMN are shifted to 380 and 448 nm, respectively, and the ratio of A_{380}/A_{448} for the holoenzyme is significantly higher than the ratio of A_{372}/A_{445} for FMN. These data were originally obtained using reconstituted holoenzyme. The spectral properties of pure holoenzyme, recently isolated without reconstitution, are the same as those of the reconstituted enzyme. Reduction of the holoenzyme with dithionite (Fig. 4) results in loss of both the 448 nm and 380 nm absorbance values. The spectra are isosbestic at about 350 nm.

Determination of the oxidation-reduction potential

Pyridoxaminephosphate oxidase was reduced by dithionite in the presence of indigo carmine ($E_0' = -0.116$ volts) [10]. The relationship, $E_h = E_0' + RT/2F$ log(oxidized/reduced dye), was then used to calculate the potential at each stage. A plot of log(oxidized/reduced flavoprotein) versus E_h (Fig. 5) gives an E_0' for the enzyme of -0.131 volt. This compares with a value of -0.210 volt for FMN using the dye, anthroquinone disulfonate ($E_0' = -0.184$ volt), under the same conditions. From this data, using the thermodynamic cycle given below, it can be inferred that reduced FMN binds about 500 times more tightly to pyridoxamine phosphate oxidase than does oxidized FMN.

TABLE II

ACTIVITY OF COENZYME ANALOGS IN THE PYRIDOXAMINEPHOSPHATE OXIDASE SYSTEM

Compound	Apparent K_m^a $\times 10^{-8}$ M	V^a
FMN[b]	4.0	100
FMN, pH 8[c]	2.0	
FMN, pH 8[d]	2.0	
FMN, pH 7[d]	2.0	
5-Deaza-FMN, pH 8[d,e]	2.0	
Riboflavin 5′-*O*-methylphosphate[f]	12	55
Riboflavin 5′-sulfate[f]	220	25
L-Lyxityl-FMN[f]	160	78
Iso-FMN[f]	6.0	77
7,8-Dichloro-FMN[f]	4.1[g]	
3-Methyl-FMN[f]	22	62
3-Carboxymethyl-FMN[f]	4.0[g]	
2-Thio-FMN[f]	25	25
2-β-Hydroxyimino-FMN[f]	600	25
2-Morpholino-FMN[f]	300	63

[a] Apparent K_m and V values were determined from Lineweaver–Burk plots by least squares analysis. V values are given relative to FMN as cofactor.
[b] Enzymatic reactions were run in 0.2 M Tris HCl at pH 8, using pyridoxaminephosphate as substrate.
[c] The value given is the dissociation constant determined from the quenching of flavin fluorescence upon binding to pyridoxaminephosphate oxidase.
[d] The value given is the dissociation constant determined from the quenching of protein fluorescence upon binding of FMN.
[e] This compound was kindly provided by Dr. Louis B. Hersh, Department of Biochemistry, University of Texas Southwestern Medical School, Dallas, TX. Although it inhibits reconstitution of enzymatic activity with FMN, we have found no evidence of catalytic turnover in the presence of this cofactor alone.
[f] Data were taken from ref. 5.
[g] The value given is K_I for these competitive inhibitors.

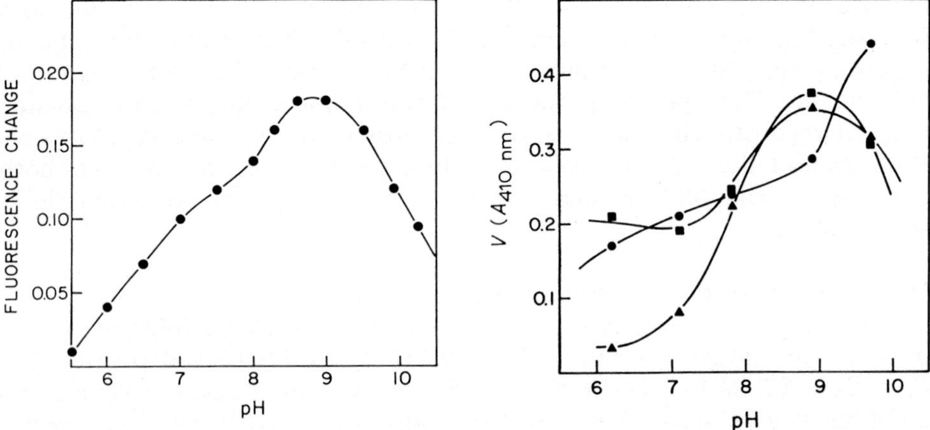

Fig. 1. The pH dependence of binding of FMN to apooxidase. Apooxidase (60 μg) was added to 2 ml of FMN ($3 \cdot 10^{-7}$ M) in 0.2 M Tris buffer at the indicated pH, and the decrease in flavin fluorescence was monitored using λ excitation of 450 nm and λ emission of 530 nm. The total fluorescence intensity change is plotted in arbitrary units.

Fig. 2. The pH dependence of pyridoxamine phosphate oxidase reactions. Standard activity assay reactions were carried out in 0.2 M Tris buffer at the indicated pH. Substrate concentration was $3 \cdot 10^{-4}$ M. Symbols are : ●, pyridoxamine phosphate; ■, pyridoxine phosphate; ▲, NPP-glycine.

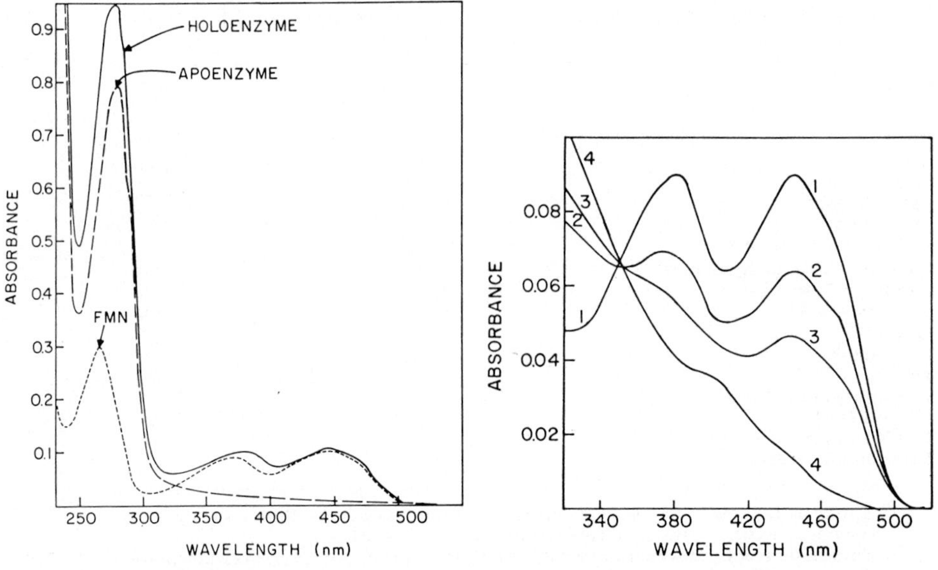

Fig. 3. **Ultraviolet-visible** absorbance spectra of pyridoxaminephosphate oxidase. Absorbance spectra of pyridoxaminephosphate oxidase (10 μM in 0.02 M potassium phosphate, pH 7) in the absence of FMN (apoenzyme) and in the presence of equimolar FMN (holoenzyme) at 20°C, recorded using a Jasco ORD/5 spectrophotometer. The spectrum of FMN (10 μM) is shown for comparison.

Fig. 4. Reduction of pyridoxaminephosphate holooxidase with sodium dithionite. Spectra were recorded at 20°C using a Cary 14 spectrophotometer equipped with a 0.0–0.1 A slidewire. The enzyme sample in 0.02 M potassium phosphate (pH 7) was made anaerobic by flushing the solution, in a 1-ml cuvette fitted with a rubber septum, with vanadous sulfate-washed (oxygen-free) nitrogen. Aliquots of a sodium dithionite solution were then added by syringe through the septum. 1, Fully oxidized; 2 and 3, partially reduced; 4, fully reduced. Spectra were corrected for dilution.

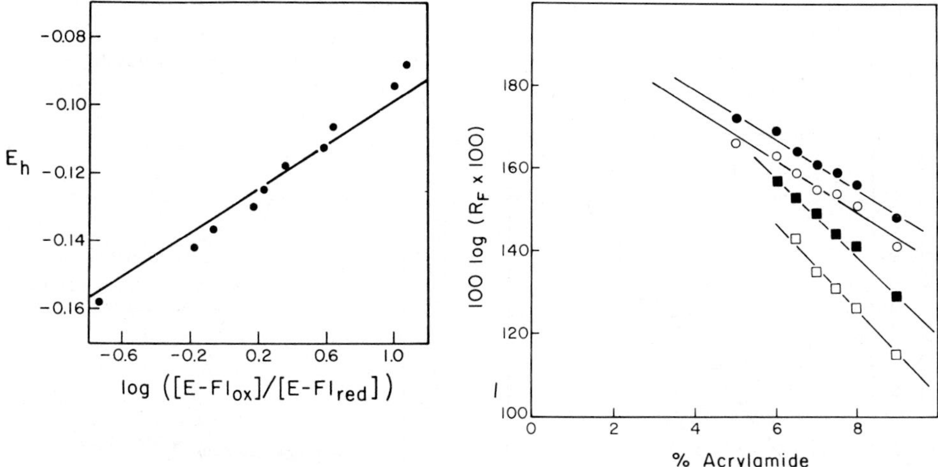

Fig. 5. Determination of the oxidation-reduction potential of pyridoxaminephosphate oxidase. A solution of pyridoxaminephosphate holooxidase in 0.02 M potassium phosphate (pH 7) and containing indigo carmine was made anaerobic by flushing with vanadous sulfate-washed (oxygen-free) nitrogen in a 1-ml cuvette, which was fitted with a rubber septum. Aliquots of sodium dithionite solution were then added by syringe through the septum until complete reduction was reached. Dye reduction was monitored at 610 nm and enzyme reduction at 460 nm; $E_h = E_0' + RT/2F$ log (oxidized/reduced dye). The line drawn is for a theoretical two-electron reduction at an E_0' value of −0.131 volt.

Fig. 6. Plot of log of mobility of enzymatic activity versus acrylamide concentration after analytical disc-gel electrophoresis of holopyridoxaminephosphate oxidase at various acrylamide concentrations. Enzyme samples were preincubated at room temperature in the presence of 1 mM EDTA, 1 mM β-mercaptoethanol, and 10 μM FMN for 30 min prior to electrophoresis. The ratio of acrylamide to N,N'-methylene bisacrylamide was held constant at 40:1. By comparison with similar plots for chymotrypsinogen, bovine serum albumin and lactate dehydrogenase, molecular weights of 52 000–55 000 (for the species represented by ○, ●) and 106 000–110 000 (for the species represented by □, ■) were obtained. Apopyridoxaminephosphate oxidase monomer migrates at the position of the species represented by ○; holo monomer at ●.

$$\begin{array}{ccc} & \text{Enz} + \text{FMN} \overset{K}{\rightleftharpoons} \text{Enz}\cdot\text{FMN} & \\ E_0' = -0.210 \text{ volt} \updownarrow & & \updownarrow E_0' = -0.131 \text{ volt} \\ & \text{Enz} + \text{FMNH}_2 \overset{500\,K}{\rightleftharpoons} \text{Enz}\cdot\text{FMNH}_2 & \end{array}$$

Molecular weight determinations

The molecular weight of the active holooxidase has been estimated to be 55 000, based on its mobility on G-100 gel filtration columns [6]. The molecular weight can also be determined by measuring the change in mobility with acrylamide concentration after analytical disc-gel electrophoresis [13]. The data obtained for both apoenzyme and holoenzyme are shown in Fig. 6. The holoenzyme (in the presence of 10 μM FMN) exhibits four bands on analytical disc gels. Two of these have molecular weights of 52 000–55 000, and two have molecular weights of 106 000–110 000. The bands of equal weight differ in charge and all four can be visualized by either protein or activity staining.

Fig. 7 shows a sample of the results of sedimentation-equilibrium experiments. The average of six molecular weight determinations gave a value of 56 000 for the apoenzyme, and eight determinations gave an average

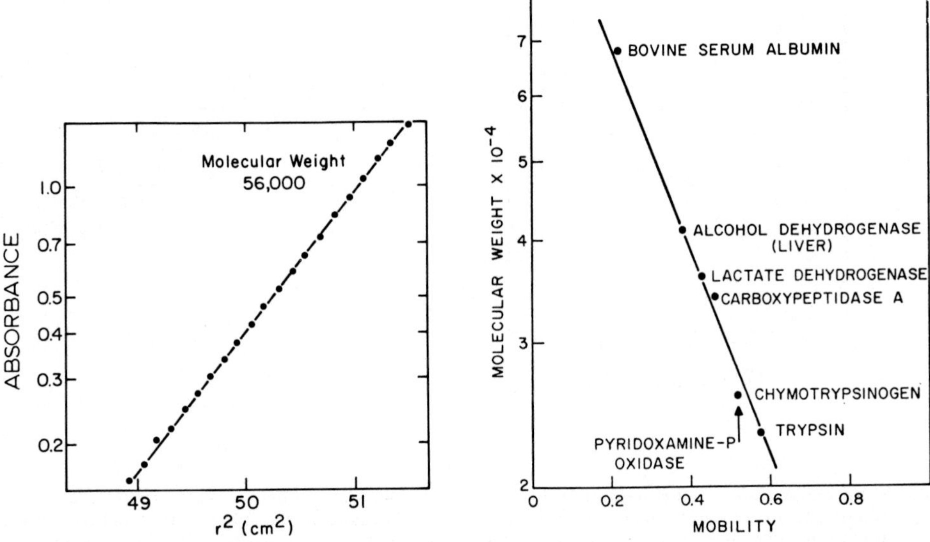

Fig. 7. Plot of log of $A280$ nm versus r^2 following sedimentation of holopyridoxamine phosphate oxidase, initial concentration 3 μM in 0.1 M phosphate buffer, pH 7.0, to equilibrium. The rotor was maintained at 4°C and the speed was maintained at 16 000 rev./min. The molecular weight was calculated by a least-squares analysis, using $\rho = 1.002$ and $\bar{v} = 0.73$. Only every third point used in the computer analysis is shown.

Fig. 8. Molecular weight determination by sodium dodecyl sulfate-acrylamide gel eletrophoresis. Protein samples were incubated at 37°C in the presence of 0.1% sodium dodecyl sulfate and 0.1 mM β-mercaptoethanol prior to electrophoresis. A single protein band, having a molecular weight of 27 000, was observed for pyridoxaminephosphate oxidase.

value of 56 000 for the holoenzyme. All experiments were done in 0.1 M phosphate buffer at pH 7. Under these conditions, no evidence for aggregation to dimers was obtained over the range 0.1–10 μM oxidase. We have also never observed high molecular weight activity on gel-filtration columns at pH 7. Further experiments in Tris buffer at higher pH, for comparison with the disc-gel experiments, are obviously indicated.

The results obtained from the above methods contrast with those obtained from sodium dodecyl sulfate-gel electrophoresis (Fig. 8). On these gels in the presence of β-mercaptoethanol, the oxidase migrates as a single protein band of mol.wt. 27 000. If β-mercaptoethanol is omitted during sample preparation, a second band of molecular weight about 50 000 is observed. This data is consistent with the interpretation that the oxidase consists of two polypeptide chains.

Upon FMN binding, protein fluorescence is quenched (Fig. 9). The data obtained from titrations in both Tris and phosphate buffers over the pH range from 6 to 10 have been analyzed by a nonlinear, least-squares procedure [14]. In all cases, one FMN site per 54 000 mol.wt. monomer was observed. This suggests the interesting possibility that two similar polypeptide chains bind a single FMN molecule.

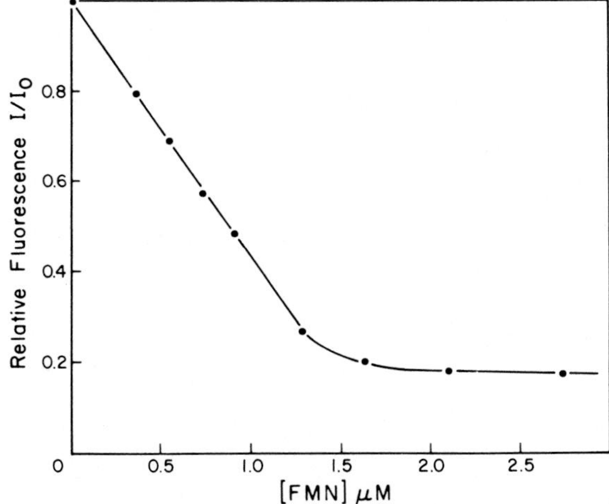

Fig. 9. Quenching of protein fluorescence upon the addition of FMN to apopyridoxaminephosphate oxidase. Aliquots of FMN were added to apopyridoxaminephosphate oxidase (1.3 μM) in phosphate buffer at pH 7.0, and the fluorescence decrease was monitored using λ excitation of 290 nm and λ emission of 335 nm.

Determination of sulfhydryl content

Sulfhydryl content of apoenzyme and holoenzyme have been determined by reaction with DTNB, both in the absence and presence of denaturing agents (Table III). The native enzyme exhibits two reactive sulfhydryls per monomer, or one per polypeptide chain. One reactive sulfhydryl group per polypeptide chain is also observed during titration of apoenzyme (Fig. 10) and holoenzyme (Fig. 11) with the fluorescent reagent, *S*-Hg^+-*N*-dansylcysteine [15]. The fluorescence of the dansyl-labeling groups is enhanced approx. two fold upon binding to the protein. Concomitantly, the protein fluorescence is quenched. The increased quenching observed with the apoenzyme, compared to holoenzyme, is consistent with the observation (Fig. 9) that the protein fluorescence is quenched on the binding of FMN.

TABLE III

DETERMINATION OF SULFHYDRYL CONTENT OF PYRIDOXAMINEPHOSPHATE OXIDASE USING DTNB[a]

	No. sulfhydryls/monomer	
	Apoenzyme[b]	Holoenzyme[c]
Native[d]	2.4	2.3
Urea (8 M)		6.3
Guanidine HCl (4 M)		7.5

[a] Samples were preincubated in Tris acetate buffer and denaturant, where appropriate, for 1h prior to the addition of DTNB. Sulfhydryl content was determined from the change in absorbance at 412 nm.

[b] 40.5 μg of apoprotein was used.

[c] 40.0 μg of holoprotein was used.

[d] Nearly complete enzyme activity remained after the addition of DTNB.

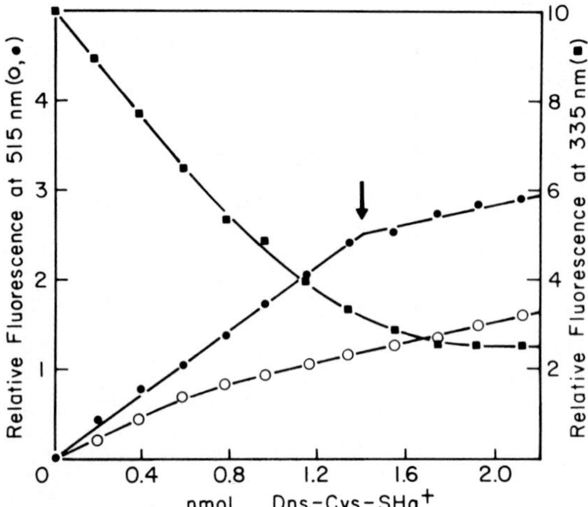

Fig. 10. Titration of apopyridoxaminephosphate oxidase with S-Hg$^+$-N-dansylcysteine. Aliquots of S-Hg$^+$-N-dansylcysteine were added to 0.7 nmol apopyridoxaminephosphate oxidase in Tris acetate buffer, pH 7.0. The decrease in protein fluorescence (■) was monitored using λ excitation of 290 nm and λ emission of 335 nm. The increase in dansyl fluorescence (●) was monitored using λ excitation of 330 nm and λ emission of 515 nm. Dansyl fluorescence was also measured in the absence of enzyme (○). The break point (arrow) occurs after titration of 2.0 sulfhydryl groups per 54 000 mol. wt. protein monomer.

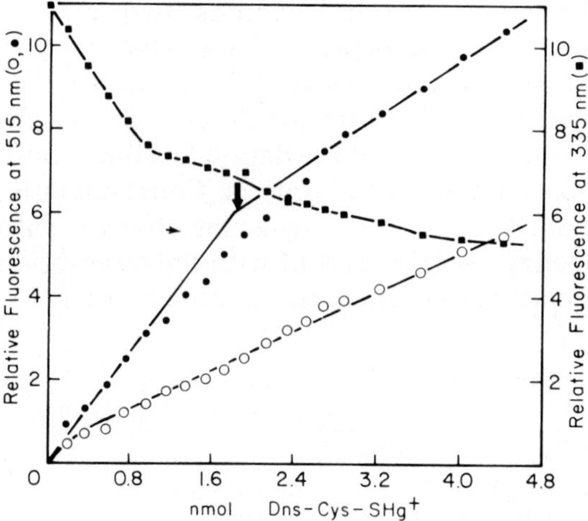

Fig. 11. Titration of holopyridoxaminephosphate oxidase with S-Hg$^+$-N-dansylcysteine. Aliquots of S-Hg$^+$-N-dansylcysteine were added to 1.0 nmol holopyridoxaminephosphate oxidase in Tris acetate buffer, pH 7.0. The decrease in protein fluorescence (■) was monitored using λ excitation of 290 nm and λ emission of 335 nm. The increase in dansyl fluorescence (●) was monitored using λ excitation of 330 nm and λ emission of 515 nm. Dansyl fluorescence was also measured in the absence of enzyme (○). The break point (arrow) occurs after titration of 1.9 sulfhydryl groups per 54 000 mol. wt. protein monomer.

After reaction with *p*-chloromercuribenzoate, DTNB or S-Hg$^+$-N-dansylcysteine, the holoenzyme remains active, and the apoenzyme binds FMN. The reactive sulfhydryl groups are, therefore, not essential for catalytic

TABLE IV

SUMMARY OF PURIFICATION OF PYRIDOXAMINEPHOSPHATE OXIDASE FROM RABBIT LIVER

Fraction	Volume (ml)	Total protein (mg)	Activity (units/ml)	Yield (%)	Specific activity (units/mg)	PNP[a]/PMP	NPP-glycine[b]/PMP
Crude extract	1220	49 000	380	100	9.45	1.21	1.15
Acid supernatant	1200	19 400	340	90	21.0	1.14	1.14
Ethanol precipitate	134	6 920	2 130	63	41.3	1.10	1.12
DEAE fraction	395	850	510	44.4	247	1.10	1.04
G-100 fraction	28.5	56.5	5 150	32.4	2 600	1.00	1.05
Calcium phosphate fraction	1.2	0.86	13 500	3.6	18 800	1.02	0.96

[a] Ratio of activity with pyridoxine-*P* as substrate to that with pyridoxamine-*P* as substrate.
[b] Ratio of activity with *N*(5′-phospho-4′-pyridoxyl)glycine as substrate to that with pyridoxamine-*P* as substrate.

activity or maintaining a conformation necessary for catalytic activity. The existence of two per monomer is consistent with the hypothesis that the two polypeptide chain components of the monomer are similar.

Amino-terminal analysis

To answer the question of whether the two polypeptide chains comprising the active monomer are identical, we decided to undertake amino-terminal analysis using dansylchloride [16, 17]. Following reaction of the pure oxidase with dansylchloride and acid hydrolysis, only the *O*-dansyltyrosine and ϵ-dansyllysine derivatives were detected. This remained the case despite variation of the reaction pH, ratio of dansylchloride to protein, and time of hydrolysis. We conclude therefore that the amino termini of the polypeptide chains are blocked. While by no means proving it, this tends to support the hypothesis that the chains are similar.

Experimental procedure

The reagents used, except for those commercially available, the assays of enzyme activity and protein, and the gel electrophoresis and staining procedures, have been described elsewhere [6]. Fluorescence measurements were made using an Aminco-Bowman spectrophotofluorometer, equipped with a 1P21 phototube; the cell caompartment was maintained at 12°C. Sedimentation-equilibrium experiments were run using double-sector cells in a Beckman Model E analytical ultracentrifuge interfaced to a Nova computer. Temperature was controlled at 4°C. *S*-Hg^+-*N*-dansylcysteine was synthesized according to Leavis and Lehrer [15].

The approx. 2000-fold enrichment achieved by subjecting rabbit liver homogenates to acid treatment, ethanol precipitation, and successive chromatographies on DEAE-A50 Sephadex, Sephadex G-100, and calcium phosphate gel [6] is summarized in Table IV. The enzyme used in this study was homogeneous on both disc-gel and dodecyl sulfate-gel electrophoresis.

Acknowledgments

We would like to thank Drs. S. Edelstein and R. Crepeau for assistance with and use of the analytical ultracentrifuge. M.N.K is an NIH predoctoral trainee supported by Training Grant GM-00824 from the Institute of General Medical Sciences, U.S.P.H.S. This work was supported in part by Research Grant AM-04585 from the National Institute of Arthritis, Metabolism, and Digestive Diseases, U.S.P.H.S., and in part by funds made available through the State University of New York. Figs. 1, 2, 8, 3, 4 and 5 and Tables I and II are reproduced with permission from the Journal of Biological Chemistry (1975) pp. 3436–3442). They originally appeared as Figs. 5, 6, 7, 9, 10 and 11 and Tables II and IV, respectively.

References

1 Pogell, B.M. (1958) J. Biol. Chem. 232, 761–776
2 Wada, H. and Snell, E.E. (1961) J. Biol. Chem. 236, 2089–2095
3 Korytnyk, W., Lachmann, B. and Angelino, N. (1972) Biochemistry 11, 722–728
4 Kazarinoff, M.N. and McCormick, D.B. (1974) Biochem. Biophys. Res. Commun. 52, 440–446

5 Kazarinoff, M.N. and McCormick, D.B. (1974) Biochim. Biophys. Acta 359, 282–287
6 Kazarinoff, M.N. and McCormick, D.B. (1975) J. Biol. Chem. 250, 3436–3442
7 Guenieri, P., Argento-Cern, M.P. and Sartori, C. (1974) FEBS Lett. 41, 11–13
8 Snell, E.E. and Haskell, B.M. (1971) in Comprehensive Biochemistry (Florkin, M. and Stotz, E.H., eds.), Vol. 21, pp. 47–71, Elsevier, Amsterdam
9 Yang, I., Khomutov, R.M. and Metzler, D.E. (1974) Biochemistry 13, 3877–3884
10 Heyl, D., Luz, E., Harris, S.A. and Folkers, K. (1951) J. Am. Chem. Soc. 73, 3430–3433
11 Collier, R. and Kohlhaw, G. (1971) Anal. Biochem. 42, 48–53
12 Edmondson, D.E. and Singer, T.P. (1973) J. Biol. Chem. 248, 8144–8149
13 Hedrick, J.L. and Smith, A.J. (1968) Arch. Biochem. Biophys. 126, 155–164
14 Engel, G. (1974) Anal. Biochem. 61, 184–191
15 Leavis, P.C. and Lehrer, S.S. (1974) Biochemistry 13, 3042–3048
16 Weiner, A.M., Platt, T. and Weber, K. (1972) J. Biol. Chem. 247, 3242–3251
17 Gros, C. and Labouesse, B. (1969) Eur. J. Biochem. 17, 463–470

Flavins and Flavoproteins, edited by T.P. Singer

CHAPTER 79

THE REDUCTION OF FAD AND FLAVOENZYMES BY MERCAPTANS

R.J. DESA

Department of Biochemistry, University of Georgie, Athens, Ga. 30602 (U.S.A.)

Summary

1. The reduction of FAD and several flavoproteins by mercaptans has been examined and found to require alkaline conditions, indicating that thiolate anion is the reductive species. Fl_{ox} is probably the active species of FAD.

2. The reduction of FAD by dithiothreitol or 1,3-dimercaptopropanol is first order in mercaptan and zero order in flavin.

3. Putrescine oxidase is fully reduced by mercaptans with production of no spectrally distinct intermediates.

4. Glucose oxidase, L-amino acid oxidase, D-amino acid oxidase and lipoyl dehydrogenase are all reduced to the semiquinone level by dithiols.

5. The pK_a of the -SH groups of 1,3-dimercaptopropanol has been determined to be 8.5.

Introduction

The non-enzymatic reduction of flavins by a variety of mercaptans has been demonstrated by several laboratories over the past decade. For example, Gascoigne and Radda [1] showed that dihydrolipoic acid could reduce FAD, FMN, riboflavin and other flavins quite readily and Gibian and Winkelman [2, 3] studied the reduction of lumiflavin by benzyl mercaptan and *n*-butyl mercaptan in aqueous methanol and concluded that the reduction is base catalyzed and requires the prior ionization of the -SH group. They also showed that the disulfide is a product of the reaction. Bruice and his coworkers have used the reaction of mercaptans and flavins in their studies of the mechanism of flavin reduction (ref. 4 and Bruice, T.C., this volume, Chapter 25).

Several years ago it was noted that 2-mercaptoethanol could reduce the flavoprotein putrescine oxidase [5] and thus act as a substrate for this enzyme. This observation prompted a more thorough examination of the phenomenon; the data reported in this paper are the results of this survey of the reactivities of putrescine oxidase and several other flavoenzymes with several thiol compounds. Under suitable conditions, all the flavoenzymes examined could be reduced by thiols, either to the semiquinone or to the fully reduced enzyme. Some observations on the non-enzymatic reduction of free FAD by two dithiols are also presented.

Materials and Methods

FAD was purchased from Sigma Chemical Company and purified by the method of Massey and Swoboda [6]. 1,3-Dimercaptopropanol (1,3-DMP), dithiothreitol (DTT) and other mercaptans were purchased from Aldrich Chemical Company and used without further purification. All other reagents were commercial products in their purest form.

Putrescine oxidase was purified as described previously [5]. L-Amino acid oxidase was purified from commercial snake venom by the methods of Wellner and Meister [7]. Purified glucose oxidase and lipoyl dehydrogenase were purchased from Sigma Chemical Co.

Results

Reduction of FAD by mercaptans

When a solution of FAD is mixed with an excess of either 1,3-DMP or DTT under anaerobic conditions, a moderately rapid decrease in the 450 nM absorbance of the flavin is observed. The spectrum of the reaction mixture upon completion of the reduction is identical to that of fully reduced FAD. Observations made during reduction indicated no spectral intermediates and the rate of the reduction was not influenced by visible light.

The reaction is markedly pH dependent. Earlier work, cited above, indicated that the thiolate anion is the reactive species in flavin reduction. This hypothesis is supported by the data of Fig. 1 which shows that the rate of reduction of FAD by 1,3-DMP accurately reflects the concentration of thiolate anion present in the solution at the given pH (the pK_a of the -SH groups in 1,3-DMP was determined to be 8.5 by the method of Benesch and Benesch [9]). The decrease in rate at very high pH values suggest that the protonated form of FAD is the species which is attacked by the thiolate anion.

The reduction of FAD by dithiols follows first order kinetics; when the concentration of dithiol is varied, a plot of the pseudo-first order rate constants against dithiol concentration gives a straight line showing that the reaction is second order overall; an apparent second order rate constant of $3.3 \cdot 10^{-2} M^{-1} \cdot s^{-1}$ at pH 9.5 for DTT is obtained.

Gibian and Winkelman [2] reported that the rate of reduction of lumiflavin and several other flavin derivatives in non-aqueous solvents was proportional to the flavin concentration; that is, the reaction is first order in flavin. The reduction of FAD by DTT or 1,3-DMP in water, described here, however, is markedly different. The data of Fig. 2 shows that the rate of reduction is independent of the flavin concentration; the reaction is zero order in flavin. The reason for this behavior remains obscure. One possibility is that some rate-limiting reaction or reactions precede the attack of the thiolate anion on the flavin and thus removes flavin concentration dependence for the overall reaction.

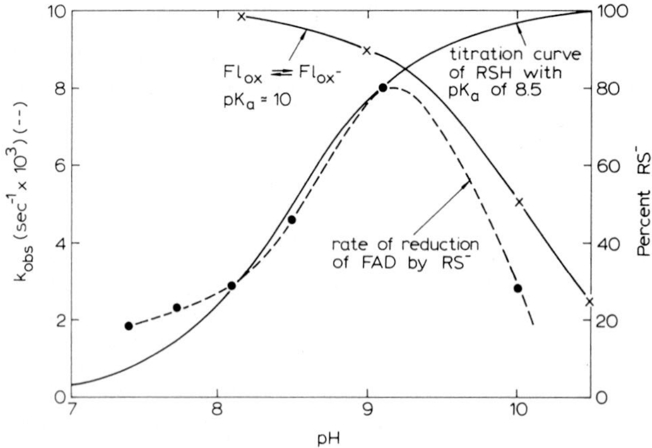

Fig. 1. The relationship between the rate of reduction of FAD by 1,3-DMP (----), and the titration curve for a mercaptan with a pK_a of 8.5 and the titration curve for formation of Fl_{ox}.

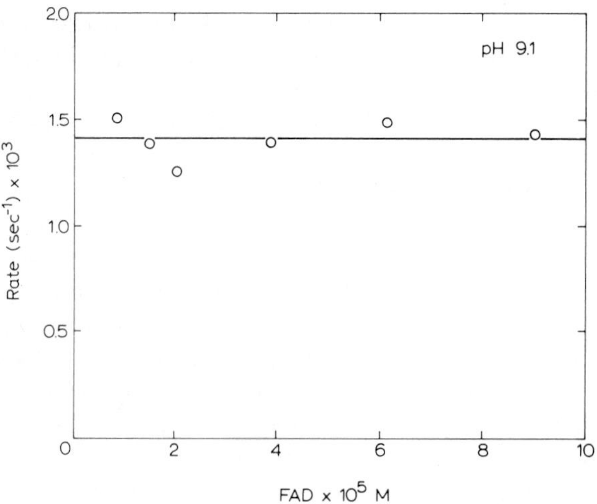

Fig. 2. The rate of reduction of FAD by 0.05 M 1,3-DMP at pH 9.1 in 0.1 M phosphate/borate buffer, as a function of flavin concentration.

Reduction of flavoproteins by mercaptans

Putresceine oxidase can oxidize 2-mercaptoethanol, β-mercaptoethylamine, DTT, 1,3-DMP, 1,2-DMP and probably other mercaptans as well. The reaction is slow for all the compounds tested and is optimum at alkaline pH. 1,3-DMP, the most effective compound tested, has a *V* about 1% of that for putrescine. H_2O_2 is a product of the reaction; the fate of the mercaptan has not been established but it is probably transformed to the corresponding disulfide as shown for the non-enzymatic oxidation of flavins by mercaptans as cited above. No spectral forms of the enzyme other than oxidized and reduced have been seen for putrescine oxidase in phosphate-borate buffer. In Tris buffer, however, the enzyme is driven to a form which appears red-brown

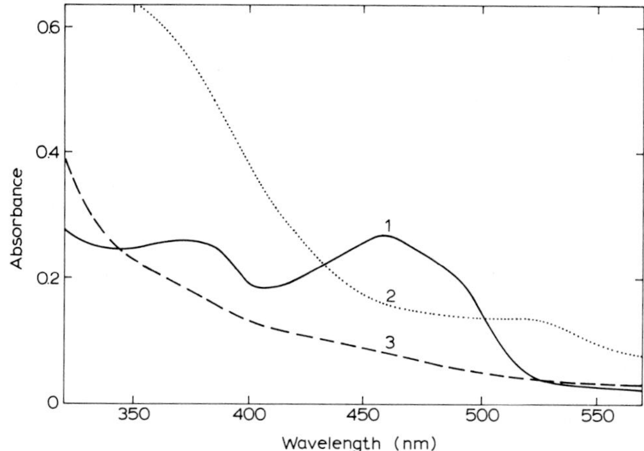

Fig. 3. Absorbance spectra of several species of putrescine oxidase; conditions noted in figure.

in color (Fig. 3); this species is different from the red semiquinone characteristic of putrescine oxidase in that the long-wave absorbance is more intense and the 370 nm peak is not present. This form of the enzyme is inert toward putrescine and only slowly reoxidized by oxygen.

The reduction of putrescine oxidase by dithiols follows normal saturation kinetics and one can calculate a K_m value of 6 mM for 1,3-DMP at pH 8.3 and 43 mM for DTT under the same conditions. The reduced enzyme appears to be reoxidized by oxygen at the normal, rapid rate.

Glucose is reduced to the characteristic red semiquinone by dithiols; under no conditions tested was fully reduced enzyme produced (Fig. 4). Similarly, L-amino acid oxidase (Fig. 5), D-amino acid oxidase (not shown), and lipoyl dehydrogenase (Fig. 6) are reduced to the semiquinone by thiols. In all cases tested, the reduction requires alkaline pH conditions, indicating that the thiolate anion is the reactive species.

Discussion

The preliminary results presented in this paper indicate that the reduction of flavins and flavoproteins by mercaptans is quite general. Of the enzymes examined, only putrescine oxidase is reduced fully to the dihydroflavin. Ribbons [10] has shown, however, that resorcinol reductase from *Pseudomonas putida* is fully reduced by 2-mercaptoethanol. Presumably other flavoproteins will be shown to be reducable by mercaptans.

The use of mercaptans as general protective agents for enzymes in solution is common among enzymologists. In light of the results presented here, this practice deserves special scrutiny in the case of flavoenzymes, since, clearly, the mercaptan may be producing results other than expected.

It should be made clear that, in the case of putrescine oxidase, at least, 2-mercaptoethanol, DTT and 1,3-DMP are acting in the active site of the enzyme; thus, the reduction by these compounds is inhibited by a variety of

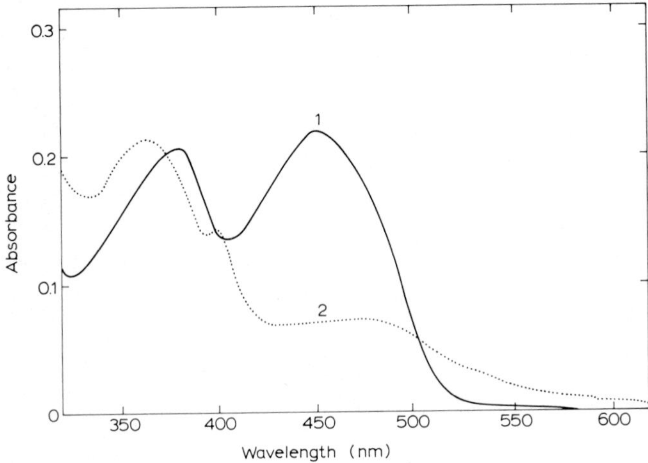

Fig. 4. Absorbance spectra of oxidized glucose oxidase and glucose oxidase semiquinone.

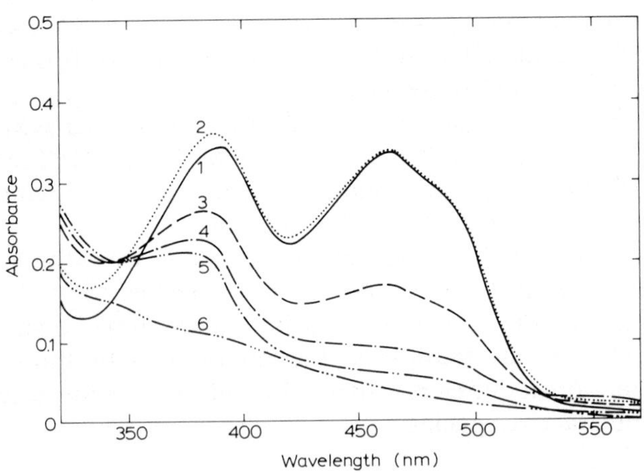

Fig. 5. Absorbance spectra of oxidized L-amino acid oxidase (LAAO) together with spectra of intermediate forms following addition of 1,3-DMP. Spectrum 6 was obtained in a separate experiment using L-leucine as reductant and is included in the figure for reference only.

amines which have been shown to be competitive inhibitors of putrescine oxidation [11].

Acknowledgements

I express thanks to Ms. Rita Brantner for her excellent technical assistance. The research was supported in part by N.I.H. grant No. GM 16834-06.

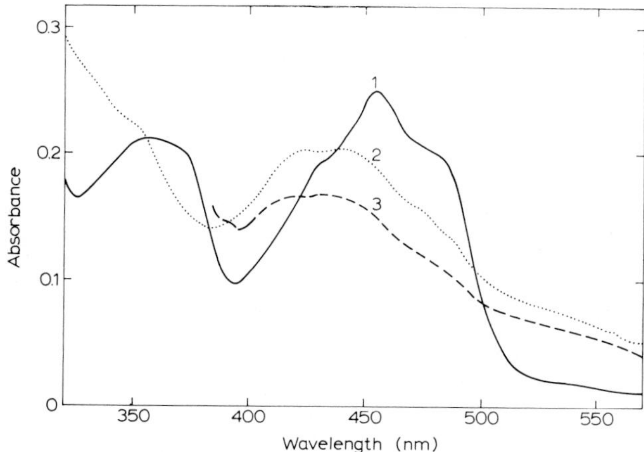

Fig. 6. Spectra of oxidized and semiquinone forms of lipoyl dehydrogenase. Spectrum 3 was taken about 2 min after the addition of dithionite.

References

1 Gascoigne, I.M. and Radda, G.K. (1967) Biochim. Biophys. Acta 131, 498
2 Gibian, M.J. and Winkelman, D.V. (1969) Tetrahedron Lett. 3901
3 Gibian, M.J., Elliott, D.L., Kelley, C., Borge, B. and Kupecz, K. (1972) Z. Naturforsch. 27b 1016
4 Bruice, T.C., Main, L., Smith, S. and Bruice, P.Y. (1971) J. Am. Chem. Soc. 93, 7327
5 DeSa, R.J. (1972) J. Biol. Chem. 247, 5527
6 Massey, V. and Swoboda, B.E.P. (1963) Biochem. Z. 338, 474
7 Wellner, D. and Meister, A. (1960) J. Biol. Chem. 235, 2013
8 DeSa, R.J. and Wampler, J.E. (1973) Applied Spectrometr. 27, 279
9 Benesch, R.E. and Benesch, R. (1955) J. Am. Chem. Soc. 77, 5877
10 Ribbons, D.W. (1976) Flavins and Flavoproteins, Proceedings of the 5th International Symposium (Singer, T.P., ed.), Chapt. 13, pp. 101–107, Elsevier, Amsterdam
11 Swain, W. and DeSa, R.J., in the press

Part XIII

Biosynthesis, metabolism and transport of riboflavin

Flavins and Flavoproteins, edited by T.P. Singer

CHAPTER 80

RIBOFLAVIN SYNTHASES OF *BACILLUS SUBTILIS*

A. BACHER, R. BAUR, U. EGGERS, H. HARDERS and H. SCHNEPPLE

Institut für Mikrobiologie, Universität Hohenheim, 7 Stuttgart 70, Garbenstrasse 30 (G.F.R.)

Riboflavin synthase (EC 2.5.1.9) catalyzes the conversion of 6,7-dimethyl-8-ribityllumazine to riboflavin. The field has been covered in extensive reviews [1, 2]. In *Bacillus subtilis*, the biosynthesis of riboflavin is controlled by repression [3–5]. Derepressed strains with high levels of riboflavin synthase activity are easily obtained.

In the course of experiments to purify riboflavin synthase from cell extracts of the derepressed strain *B. subtilis* H 94 by molecular sieve chromatography, we observed two well-separated peaks of enzyme activity in the eluate of Sephadex G-150 columns. Sucrose gradient centrifugation experiments confirmed that the two fractions represent two riboflavin synthase species with greatly different molecular weights (Fig. 1). The ratio of the two activities in cell extracts was not affected by different procedures of cell disruption, as shown by sucrose gradient centrifugation. The presence of two enzyme activities was similarly observed in each of a number of *B. subtilis* wild strains and in mutants of *B. subtilis* with genetic defects in earlier steps of the biosynthesis. Two enzyme activities were also found in strains of *Bacillus megaterium*, *Bacillus polymyxa*, and *Clostridium thermoaceticum*. On the other hand, strains of *Escherichia coli*, *Nocardia rubra* and *Saccharomyces cerevisiae* contain only one riboflavin synthase activity, sedimenting with a similar velocity as the low molecular weight enzyme of *Bacillaceae*. These findings suggest that the occurrence of two riboflavin synthase activities might be related to bacterial sporulation.

Both riboflavin synthases of *B. subtilis* were purified to homogeneity from cell extracts of the derepressed strain H 94 by the following procedure. Subsequent to heat treatment and ammonium sulfate fractionation, the high molecular weight riboflavin synthase was pelleted by ultracentrifugation. The low molecular weight enzyme remained in the supernatant and was purified further by chromatography on DEAE-Sephadex, sucrose gradient centrifugation and sucrose gradient electrophoresis. The high molecular weight enzyme was purified by sucrose gradient centrifugation and preparative polyacrylamide gel electrophoresis.

Both proteins thus obtained migrate as single bands in polyacrylamide gel electrophoresis. Electrophoresis on cellulose acetate strips shows single

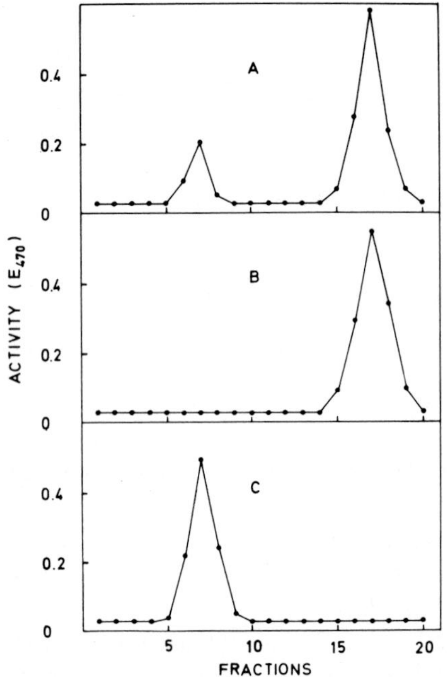

Fig. 1. Sedimentation in sucrose gradients (5–20%, rotor SW 50.1, 5 h, 4°C, 45 000 rev./min). A, cell extract of *B. subtilis* H 94; B, low molecular weight riboflavin synthase; C, high molecular weight riboflavin synthase.

bands of protein which coincide exactly with the riboflavin synthase activity. Both enzymes sediment in the analytical ultracentrifuge as single, symmetrical boundaries. Meniscus depletion sedimentation-equilibrium experiments show linear plots of log c versus r^2. The properties of the enzymes were as follows:

Molecular weight	1 000 000	60 000
$s_{20,w}$(S)	26.5	4.1
Specific activity (nmol/mg/h)	1 400	30 000
k_M for 6,7-dimethyl-8-ribityllumazine(M)	$2.6 \cdot 10^{-5}$	$1.5 \cdot 10^{-4}$

Electrophoresis of high molecular weight riboflavin synthase in sodium dodecyl sulfate-polyacrylamide gels shows a minor band with an approximate molecular weight of 26 000 (α subunit) and a major band with an approximate molecular weight of 14 000 (β subunit). The low molecular weight enzyme shows a single band with an approximate molecular weight of 26 000.

The high molecular weight riboflavin synthase was resolved into an enzymatically active dissociation product consisting of α subunits only and into isolated β subunits by chromatography on DEAE-Sephadex. The enzymatically active dissociation product was indistinguishable from the low molecular weight riboflavin synthase in the following experiments. The two proteins migrated the same distance in sodium dodecyl sulfate-polyacrylamide gel electrophoresis and in electrophoresis on cellulose acetate strips;

Fig. 2. Polycrylamide gel electrophoresis. Left, low molecular weight riboflavin synthase (10% gel); right, high molecular weight riboflavin synthase (4% gel).

they sedimented with the same velocity in the analytical ultracentrifuge and in sucrose gradients and they showed complete cross reaction with rabbit antiserum against low molecular weight riboflavin synthase in immuno double diffusion experiments at various pH values. We conclude from these data that the dissociation product of the high molecular weight enzyme is identical with the low molecular weight enzyme, i.e., both proteins are dimers of α subunits.

Densitometric evaluation of sodium dodecyl sulfate-gel electrophoresis suggests the presence of two α subunits and an approximate number of 66 β subunits per molecule of high molecular weight riboflavin synthase. Considering that the turnover numbers per mol are very similar for the two enzymes and the absence of riboflavin synthase activity in isolated β subunits, we conclude that only the α subunits are carriers of riboflavin synthase activity. However, the β subunits enhance the thermal stability of the enzyme activity to a considerable degree. The temperatures for 50% inactivation during an incubation for 20 min are 58 and 71°C for the low and high molecular weight forms, respectively.

At neutral pH the high molecular weight riboflavin synthase exhibits considerable stability. Sucrose gradient centrifugation, sedimentation in the analytical ultracentrifuge and meniscus depletion sedimentation-equilibrium experiments show no evidence of dissociation. The antigenic determinants of the α subunits are completely eclipsed: the enzyme is not precipitated in double diffusion experiments with antiserum against low molecular weight riboflavin synthase, and the enzyme activity is not inhibited by the antiserum, whereas the low molecular weight enzyme is completely inhibited.

At pH values above 8, α_2 dimers are released from high molecular weight enzyme, as shown by immunodiffusion and sucrose gradient centrifugation. The dissociation of the enzyme is reversed by incubation at neutral pH as shown by sucrose gradient sedimentation and polyacrylamide gel electrophoresis.

Acknowledgement

This work was supported by the Deutsche Forschungsgemeinschaft.

References

1 Plaut, G.W.E. (1971) Comprehensive Biochemistry (Florkin, M. and Stotz, E.H., eds.), Vol. 21. p. 11, Elsevier, Amsterdam
2 Plaut, G.W.E. and Harvey, R.A. (1971) Methods in Enzymology (Colowick, S.P. and Kaplan, N.O., eds.), Vol. XVIII B, p. 515, Academic Press, London
3 Bresler, S.E., Kalinin, V.L., Krivisky, A.S. and Perumov, D.A. (1969) Genetika 5 (2), 133
4 Bresler, S.E., Glazunov, E.A. and Perumov, D.A. (1972) Genetika 8 (2), 109
5 Bacher, A., Eggers, U. and Lingens, F. (1973) Arch. Mikrobiol. 89, 73

Flavins and Flavoproteins, edited by T.P. Singer

CHAPTER 81

BIOSYNTHESIS OF RIBOFLAVIN. STRUCTURE OF THE PURINE PRECURSOR AND ORIGIN OF THE RIBITYL SIDE CHAIN

A. BACHER and B. MAILÄNDER

Institut für Mikrobiologie, Universität Hohenheim, 7 Stuttgart 70, Garbenstr. 30 (G.F.R.)

The involvement of a purine precursor in the biosynthesis of riboflavin has been firmly established. Reviews of the large amount of available data have been published [1–3]. However, although Howells and Plaut [4] showed by experiments with a purine-deficient mutant of *Escherichia coli* that a compound at, or subsequent to, the level of IMP is positively required, the precise structure of the precursor remained open. Because of the structural similarity of riboflavin and its immediate precursor, 6,7-dimethyl-8-ribityllumazine (for reviews see refs 1–3), with xanthine, the latter compound was postulated as the immediate purine precursor.

A different line of evidence was obtained from studies of the intermediates of the riboflavin pathway with riboflavin-deficient mutants of *Saccharomyces cerevisiae*. The different mutants were found to produce the following characteristic products:

rib 5	6,7-dimethyl-8-ribityllumazine [5, 6]
rib 3 and *rib* 4	5-amino-2,6-dihydroxy-4-ribitylaminopyrimidine [5, 6]
rib 2	2,5-diamino-6-hydroxy-4-ribitylaminopyrimidine [7–9]
rib 7	6-hydroxy-2,4,5-triaminopyrimidine [10–12]

No specific product was obtained from *rib* 1 mutants. Studies with mutants deficient in two different *rib* genes allowed the direct determination of the sequence of biosynthetic steps as shown in Fig. 1 (ref. 11 and unpublished data by Bacher and Oltmanns). However, the question remained open whether 6-hydroxy-2,4,5-triaminopyrimidine is a true intermediate or a secondary product formed by hydrolytic cleavage.

The involvement of 2-aminopyrimidine type compounds in early steps of the pathway suggested strongly that the biosynthesis of the vitamin begins at the level of guanine rather than xanthine. Bacher et al. [13–15] confirmed this hypothesis by experiments with a mutant of *Aerobacter aerogenes* which is unable to synthesize purines de novo and to convert XMP to GMP. The mutant converts [2-^{14}C]guanine to riboflavin without dilution, whereas the XMP pool remains unlabeled. On the other hand, ^{14}C-labeled adenine, hypoxanthine and xanthine are efficiently incorporated into the XMP pool, but not into riboflavin. Along the same line were findings by Baugh and

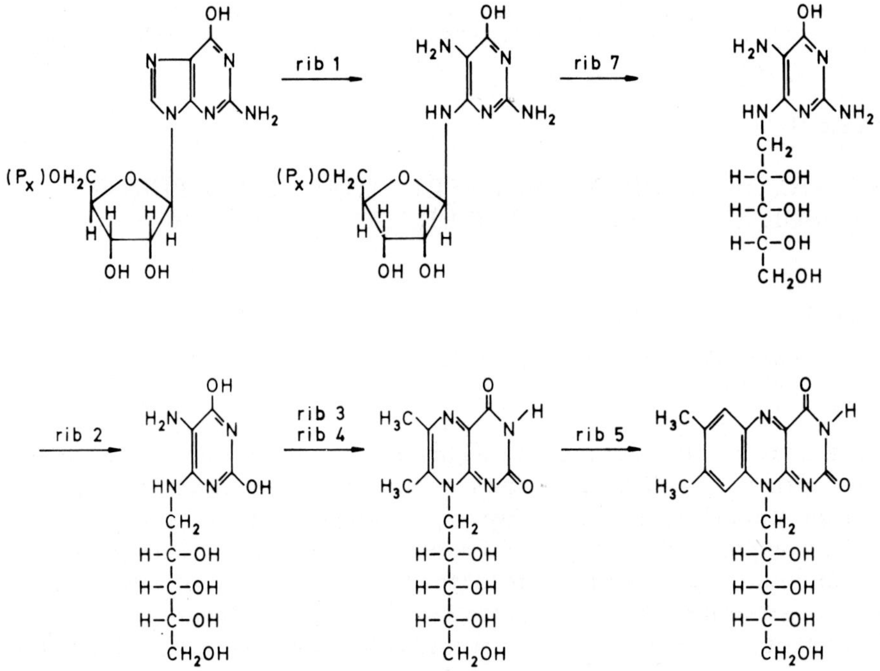

Fig. 1. Biosynthesis of riboflavin.

Krumdieck [16], who inhibited the conversion of XMP to GMP by decoyinine. Pointing in the same direction was the observation by Shavlovskii et al. [17] that flavinogenesis in a mutant of *Candida guilliermondii* deficient in XMP aminase is stimulated by guanine, but not by xanthine.

The question remained whether guanine itself or a ribosyl derivative (i.e. guanosine or a nucleotide) is the true precursor. The following possibilities may be considered: (i) guanine is converted to 6-hydroxy-2,4,5-triaminopyrimidine which is then combined with a ribitol moiety; (ii) subsequent to the elimination of C-8 from guanosine or a nucleotide, the ribose moiety is removed and replaced by ribitol; (iii) subsequent to elimination of C-8, the ribose moiety of guanosine or a nucleotide is directly converted to the ribitol moiety. The experimental decision is hampered by the ease of metabolic interconversion of different purine compounds. McNutt and Forrest [18] fed [U-^{14}C]guanosine to *Eremothecium ashbyii* and found the alloxazine moiety of riboflavin preferentially labeled. The possible pitfalls in the interpretation of these data have been pointed out by Demain [3]. Miersch [19] observed that [^{14}C]ribitol is a better precursor of riboflavin than [^{14}C]ribose in *C. guilliermondii*, particularly when flavinogenesis is inhibited by high levels of iron. In contrast to data by Ali and Al-Khalidi [20], ribose was a poor precursor of the xylene ring in these experiments. Studies concerned with the origin of the ribitol moiety are greatly facilitated by the use of an organism in which mutual conversion between guanosine and its ribosyl derivatives is absent. We used a mutant KP 4067 of *Salmonella*

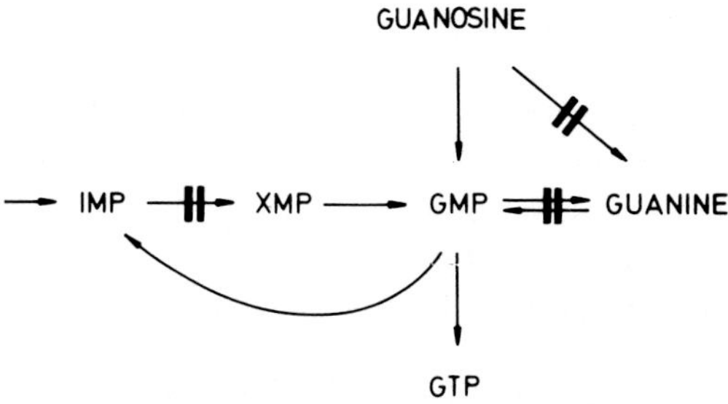

Fig. 2. Guanine metabolism and genetic defects of *Salmonella typhimurium* BM 2.

typhimurium (kindly provided by J. Hoffmeyer) which is deficient of IMP dehydrogenase and purine nucleoside phosphorylase [21]. Subsequent to mutagen treatment of this strain, we isolated a mutant BM 2 growing with guanosine but not with guanine. Enzymatic studies in vitro showed that the mutant had lost the enzyme purine nucleotide pyrophosporylase. The genetic defects of mutant BM 2 are shown in Fig. 2.

[Ribose-^{14}C] guanosine, required for the subsequent isotope-incorporation experiments, was prepared as follows. The guanosine-producing strain *Bacillus subtilis* ATCC 19221 was grown in the presence of guanine and [U-^{14}C] glucose. The de novo synthesis of GMP was inhibited by decoyinine. Guanosine formed was isolated by chromatographic procedures. The ribose moiety contained 98.7% of the radioactivity as shown by chemical degradation.

S. typhimurium BM 2 was grown in 1 liter of minimal medium supplemented with [ribose-^{14}C] guanosine (8 mg, 6 μCi), ribose (1 g) and glucose (5 g). Riboflavin was isolated from the culture medium and chromatographically purified to constant activity. The cells were washed and treated with NaOH to hydrolyze RNA. The resulting ribonucleotides were isolated chromatographically. The radiochemical purity of isolated compounds was checked by thin-layer chromatography. The distribution of radioactivity in the isolated riboflavin was studied by photochemical degradation with subsequent isolation of 7,8-dimethylalloxazine. Similarly, the isolated GMP was hydrolyzed and the radioactivity of the resulting guanine and ribose was determined. The results were as follows:

	Specific radioactivity	
	cpm/nmol	%
[Ribose-^{14}C] guanosine	340	100
Riboflavin	336	99
Degradation of riboflavin		
7,8-Dimethylalloxazine	6	2
GMP	316	93
Degradation of GMP		
Guanine	10	3
Ribose	286	84
AMP	5	1.5
CMP	10	3.1

From these data it follows (i) that the ribose phosphate pool was not labeled, as shown by the low specific activity of AMP and CMP which were synthesized de novo under the experimental conditions; (ii) that guanosine is converted to GMP by direct phosphorylation, thus conserving the ribose moiety; (iii) that the ribose moiety of guanosine or a guanosine nucleotide is the direct precursor of the ribitol moiety of riboflavin; (iv) that the biosynthesis of the vitamin begins with guanosine or a nucleotide and definitely not with guanine.

It has been shown by Brown and coworkers [22, 23] that the biosynthesis of folic acid begins with the elimination of C-8 from GTP, followed by an Amadori rearrangement. By analogy, it seems plausible that a compound of this type is reduced to yield the riboflavin precursor, 2,5-diamino-6-hydroxy-4-ribitylaminopyrimidine.

Acknowledgement

This work was supported by the Deutsche Forschungsgemeinschaft.

References

1 Plaut, G.W.E. (1961) Annu. Rev. Biochem. 30, 409
2 Plaut, G.W.E., Smith, C.M. and Alworth, W.L. (1974) Annu. Rev. Biochem. 43, 899
3 Demain, A.L. (1972) Annu. Rev. Microbiol. 26, 369
4 Howells, D.J. and Plaut, G.W.E. (1965) Biochem. J. 94, 755
5 Lingens, F., Oltmanns, O. and Bacher, A. (1967) Z. Naturforsch. 22b, 755
6 Oltmanns, O., Bacher, A., Lingens, F. and Zimmermann, F.K. (1969) Mol. Gen. Genet. 105, 313
7 Bacher, A. and Lingens, F. (1968) Angew. Chem. 80, 237
8 Bacher, A. and Lingens, F. (1968) Angew. Chem. Int. Edn. Engl. 7, 219
9 Bacher, A. and Lingens, F. (1970) J. Biol. Chem. 245, 4647
10 Bacher, A., Baur, R., Oltmanns, O. and Lingens, F. (1969) FEBS Lett. 5, 316
11 Bacher, A. and Lingens, F. (1971) J. Biol. Chem. 246, 7018
12 Oltmanns, O. and Bacher, A. (1972) J. Bacteriol. 110, 818
13 Bacher, A. and Lingens, F. (1969) Angew. Chem. 81, 393
14 Bacher, A. and Lingens, F. (1969) Angew. Chem. Int. Edn. Engl. 8, 371
15 Bacher, A. and Mailänder, B. (1973) J. Biol. Chem. 248, 6227
16 Baugh, C.M. and Krumdieck, C.L. (1969) J. Bacteriol. 98, 1114
17 Shavlovskii, G.M., Strugovshchikova, L.P. and Logvinenko, E.M. (1971) Dokl. Akad. Nauk. SSSR 196, 701
18 McNutt, W.S. and Forrest, H.S. (1958) J. Am. Chem. Soc. 80, 951
19 Miersch, J. (1973) Phytochemistry 12, 1595
20 Ali, S.N. and Al-Khalidi, U.A.S. (1966) Biochem. J. 98, 182
21 Hoffmeyer, J. and Neuhard, J. (1971) J. Bacteriol. 106, 14
22 Burg, A.W. and Brown, G.M. (1968) J. Biol. Chem. 243, 2349
23 Wolf, W.A. and Brown, G.M. (1969) Biochim. Biophys. Acta 192, 468

Flavins and Flavoproteins, edited by T.P. Singer

CHAPTER 82

SUBSTRATE SPECIFICITY AND STEREOSPECIFIC MODE OF ACTION OF RIBOFLAVIN SYNTHASE*

GERHARD W.E. PLAUT and RICHARD L. BEACH

Department of Biochemistry, School of Medicine, Temple University, Philadelphia, Pa. 19140 and the Chemistry Department, Rider College, Trenton, N.J. 08602 (U.S.A.)

SUMMARY

Kinetic activity with purified yeast riboflavin synthase (EC 2.5.1.9, 6,7-dimethyl-8-(1′-D-ribityl) lumazine : 6,7-dimethyl-8-(1′-D-ribityl) lumazine was tested with 23 analogues of 6,7-dimethyl-8-D-ribityllumazine varying in substituents at position 8. Only 6,7-dimethyl-8-D-ribityllumazine (the natural substrate) and 6,7-dimethyl-8-D-xylityllumazine (inhibitor) had activity among the 8 possible tetrahydroxypentityl derivatives. The 5′-deoxy-D-ribityl analogue was a substrate and the 3′-deoxy-D-ribyl compound was a potent competitive inhibitor, whereas, the 2′-deoxy-D-ribityl and the newly synthesized 4′-deoxy-D-ribityl derivatives were inactive. Analogues with D-threityl, D- and L-erythrityl substituents were inactive; the L-threityl compound was a weak inhibitor.

The D-configuration of hydroxyls at carbons 2′ and 4′ of the side chain is a minimal requirement for activity. CPK space-filling models indicate that free rotation around the nitrogen 8 and carbon 1′ bonds is sterically restricted resulting in two possible conformations of 6,7-dimethyl-8-aldityllumazines. A mechanism is proposed in which the two different substrate conformers may interact at the donor and acceptor sites of the enzyme catalyzing the 4-carbon transfer, leading to riboflavin formation. Furthermore, similarities in the structures of the natural substrate and of analogues exhibiting kinetic activity are compared and contrasted with conformations of inactive analogues.

It has been shown in a number of laboratories that the last step in riboflavin biosynthesis involves a condensation between two molecules of 6,7-dimethyl-8-ribityllumazine to form one molecule each of riboflavin and 4-ribitylamino-5-amino-2,6-dihydroxypyrimidine (for review see Plaut [1],

---

This work was performed at Temple University (1973–1974) during a period of sabbatical leave of R.L.B. from Rider College.

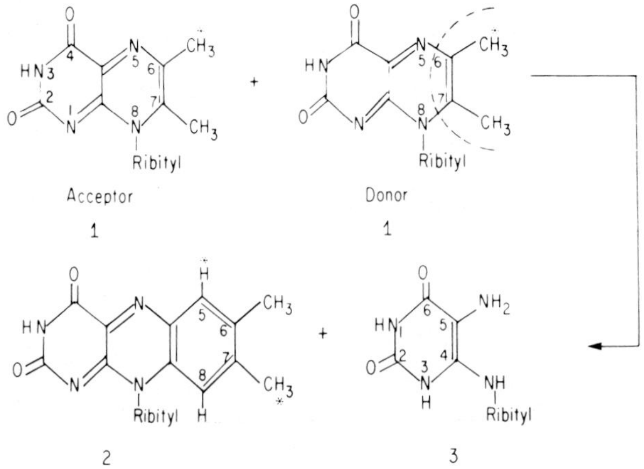

Fig. 1. Conversion of 2 molecules of 6,7-dimethyl-8-ribityllumazine to riboflavin (2) and 4-ribitylamino-5-amino-2,6-dioxopyrimidine (3).

Demain [2] and Plaut et al. [3]). Riboflavin synthase, the enzyme which catalyzes the reaction, has been detected in a larger number of microorganisms and in plants. It has been purified 5000 fold from bakers yeast extract, the best preparations exhibiting a single zone of protein possessing the activity when examined by cellulose acetate and polyacrylamide disc gel electrophoresis [4].

As shown in Fig. 1, the reaction involves a transfer from one molecule of 6,7-dimethyl-8-ribityllumazine to a second molecule of the lumazine of a 4-carbon moiety containing carbons 6 and 7 and the attached methyl groups. In this transfer, the methyl groups attached to carbons 6 and 7 of the lumazine accepting the 4-carbon moiety are recovered in carbons 5 and 8 of riboflavin, respectively, whereas the methyl groups attached to carbons 6 and 7 of the 4-carbon donor lumazine become the methyl groups at carbons 7 or 6 of riboflavin, respectively [5]. It is significant that in the chemical formation of flavins from such lumazines, which occur under neutral [6, 7] and acid conditions [8], the same orientation has been observed of the carbons derived from the 6,7-dimethyl lumazines to form the *o*-xylene ring of flavin [9]. This may suggest that the orientation of the 4-carbon piece transferred in riboflavin formation is determined by the inherent chemical properties of the lumazine molecule, rather than by being directed by the stereochemical effects of enzyme catalysis.

Chemical experiments showed that in ^2H_2O the protium of the 7-methyl (but not the 6-methyl) of 6,7-dimethyl-8-aldityllumazines is replaced by deuterium. The exchange at the 7-methyl group occurred more rapidly under alkaline and acid conditions than at neutrality [8–11]. These observations made it possible to test (a) whether replacement of H by ^2H at the 6-methyl or 7-methyl groups has an effect on the enzymic transformation of the substrate to product and (b) whether the exchange reaction at the 7-methyl group is catalyzed by the enzyme and, therefore, is an intrinsic part of the mechanism of the overall reaction. The enzymic transformation of 6-methyl-7-deuteromethyl-8-ribityllumazine to flavin proceeded at about 80% of the velocity of the non-labeled substrate. However, a very substantial isotope effect was observed when the transformation of 6-deuteromethyl-7-methyl-8-ribityllumazine was compared to the protium containing substrate. *V* was decreased by 80% (Table I), but K_m was essentially unchanged [12]. Furthermore, it could be shown in experiments with 7-tritiomethyl-labeled substrates and inhibitors that the elimination of tritium from such substances into water was enhanced in the presence of enzyme (Table II). At the pH optimum, the rates of formation from 6-methyl-7-tritiomethyl-8-ribityllumazine of riboflavin and elimination of tritium into water were proportional. However, as shown in Table II, the rate of incorporation of tritium into water from 7-tritiomethyl-6-deuteromethyl-8-ribityllumazine is about the same as that from the non-deuterated substrate, even though the rate of flavin formation is much slower; with the 5′-deoxyribityl derivative the exchange is reduced to only about one-third compared to the ribityl compound even though flavin formation proceeds at 1% of the rate. Exchange also occurs with the D-xylityl derivative, which is a competitive inhibitor and forms isolatable complexes with the enzyme [13], but no exchange was observed with the L-xylityl derivative which does not interact with the enzyme

TABLE I

COMPARISON OF K_m AND V OF 6,7-DIMETHYL-8-RIBITYLLUMAZINE AND 6-DEUTERIOMETHYL-7-METHYL-8-RIBITYLLUMAZINE

Data from ref. 12

Compound	K_m (μM)	V (nmol B_2/mg protein)
6,7-Dimethyl-8-ribityllumazine	15	5100
6-Deuteriomethyl-7-methyl-8-ribityl Lumazine	13	1100

TABLE II

COMPARISON OF ENZYMATIC RATES OF FLAVIN SYNTHESIS AND OF EXCHANGE REACTION BETWEEN 7-TRITIOMETHYL GROUP OF VARIOUS LUMAZINE DERIVATIVES AND WATER

Data from ref. 12. All incubations were carried out in 0.05 M imidazole/HCl at pH 6.8 and 25°C. All compounds listed were labeled with tritium at the 7-methyl group.

Compound	Relative rates (%)	
	Exchange reaction	Flavin formation
6,7-Dimethyl-8-D-ribityllumazine	100	100
6-Deuteromethyl-7-methyl-8-D-ribityllumazine	80	20
6,7-Dimethyl-8-(1;-(5′-deoxy-D-ribityl))lumazine	39	1
6,7-Dimethyl-8-D-xylityllumazine	27	0
6,7-Dimethyl-8-L-xylityllumazine	1	0

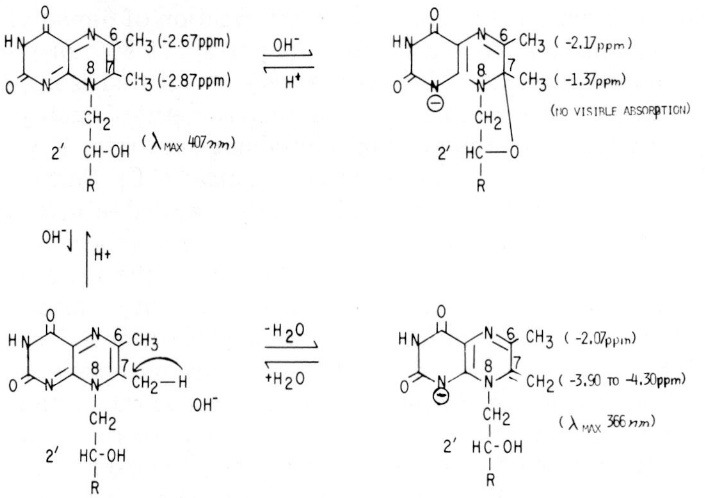

Fig. 2. Effect of acid and base on 6,7-dimethyl-8-alditylllumazines bearing an hydroxyl group at position 2′ on the aldityl substituent.

(Table II). The catalysis by the same enzyme of the exchange reaction and flavin formation is also suggested by the proportional inhibition of both activities by a number of sulfhydryl group modifying agents [12].

Spectroscopic examination of a number of 6,7-dimethyl-8-alditylllumazines [14] showed that the visible absorbance at 407 nm and the NMR absorbance at −2.67 ppm and −2.87 ppm of such compounds in neutral solution was shifted in alkali to 366 nm and −2.07 ppm, 3.90 ppm and 4.30 ppm for derivatives lacking a 2′-hydroxyl group at the N-8 substituent (Fig. 2). These data indicate the presence of a 7-exomethylene group in alkali; the NMR absorbance at −3.90 ppm and −4.30 ppm showing the non-equivalence of the two exomethylene hydrogens coplanar with the pteridine ring. Compounds containing a 2′-hydroxyl group have little or no visible absorbance and the NMR at −1.37 ppm indicates formation of an internal ether between carbon 2′ and carbon 7 of the pteridine ring. 6,7-Dimethyl-8-ribityllumazine in alkali forms an equilibrium mixture predominantly in the internal ether form containing a minor amount of the 7-exomethylene form. However, 6,7-dimethyl-8-(2′-deoxy-D-ribityl)lumazine, which bears a 3′-hydroxyl group, is not capable of forming the internal ether (absence of −1.37 ppm absorbance) and forms the 7-exomethylene derivative exclusively. Apparently only the 2′-hydroxyl group has the proper molecular dimensions for interaction with carbon 7 of the pteridine ring.

The existence of an equilibrium between the 7-methyl and the 7-exomethylene forms of these compounds is consistent with the observed exchange of the hydrogens at the 7-methyl group with water. The large isotope effect with 6-deuteriomethyl-7-methyl-8-ribityllumazine indicates that condensation at the 6-methyl group is rate limiting and is subsequent to carbon—carbon bond formation between the 7-methyl group and carbon 6.

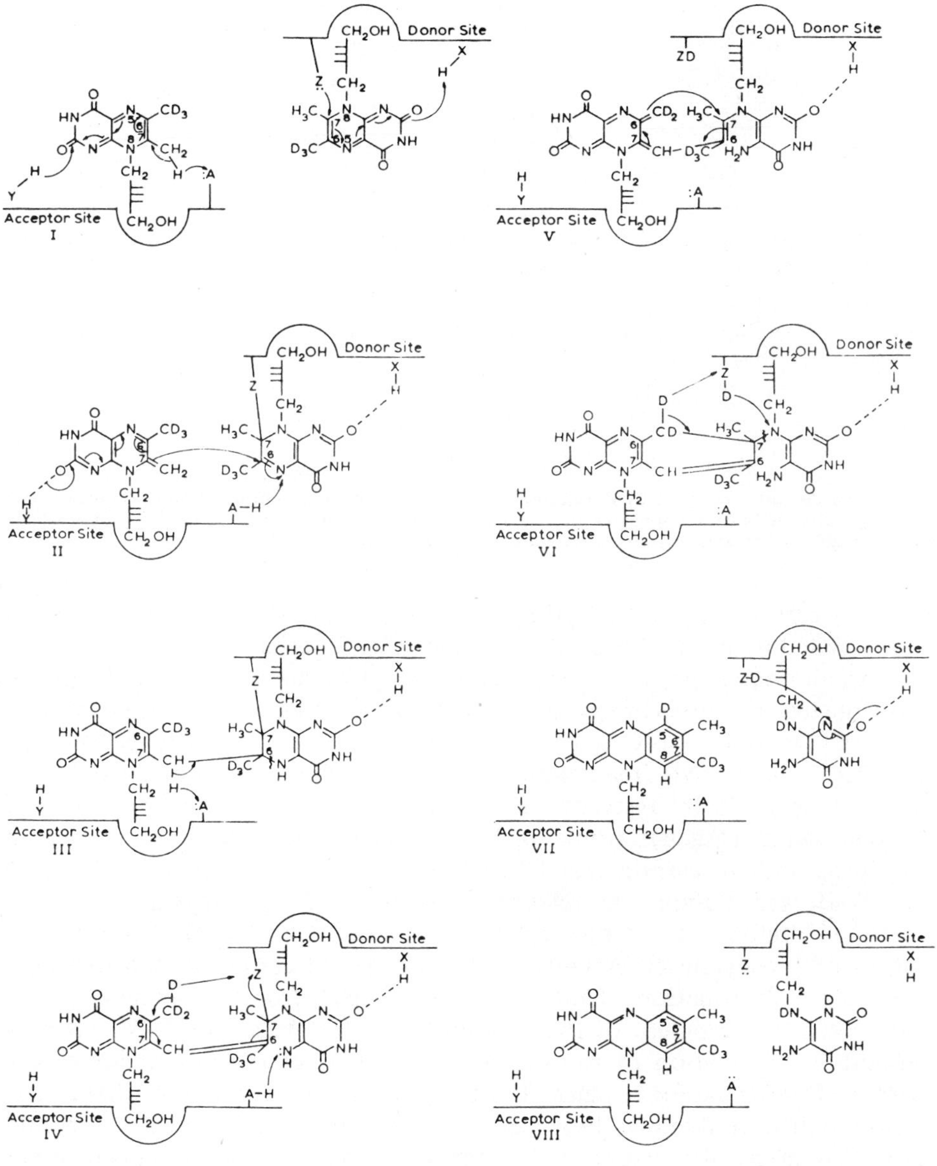

Fig. 3. Proposed mechanism of enzymic conversion of 6,7-dimethyl-8-ribityllumazine to riboflavin.

The mechanism proposed earlier for the enzymic conversion of 6,7-dimethyl-8-ribityllumazine to riboflavin by riboflavin synthase (Fig. 3) involves loss of a proton from the 7-methyl group, followed by a carbanion attack by the 7-exomethylene carbon of the compound bound at the acceptor site on carbon 6 of the lumazine bound at the donor site of the enzyme. For this attack to occur, carbon 6 of the pteridine ring at the donor site must acquire a carbonium character. This can arise when the carbonium

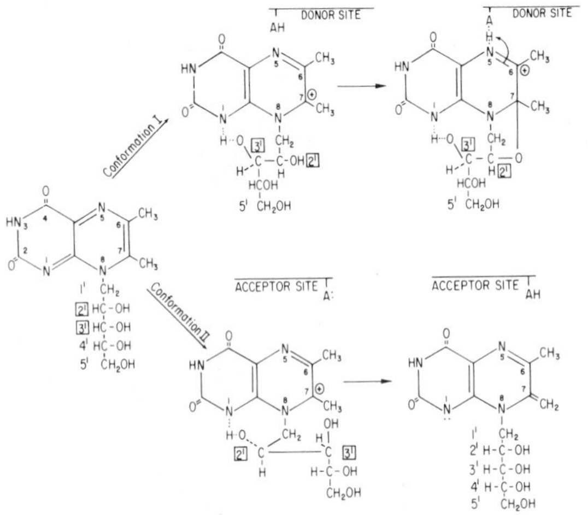

Fig. 4. Proposed interaction of Conformation I at the donor site and Conformation II at the acceptor site of 6,7-dimethyl-8-ribityllumazine leading to active intermediates capable of condensing to form riboflavin and 4-ribitylamino-5-amino-2,6-dehydroxypyrimidine.

center at carbon 7 is neutralized by group Z: (Fig. 3, step 2)* and nitrogen 5 is protonated by group AH on the enzyme (Fig. 3, steps 1–3).

An alternative proposal for the initial steps of the reaction can be made, based on the more recent spectroscopic findings [14], and experiments with molecular models. CpK space-filling models of 6,7-dimethyl-8-D-ribityllumazine indicate that free rotation of the side chain about nitrogen 8 and carbon 1′ is sterically restricted resulting in two possible conformations of the compound (Fig. 4): Conformation I places the 2′-hydroxyl group in close proximity to carbon 7 and the 3′-hydroxyl close to nitrogen 1 of the ring. Hydrogen bonding at nitrogen 1 with the 3′-hydroxyl group would facilitate development of the carbonium center at carbon 7, favoring formation of the cyclic ether between the oxygen at carbon 2′ and carbon 7 of the ring. Protonation of nitrogen 5 by a group from the enzyme would lead to the development of the carbonium center at carbon 6. We propose that this conformation of the molecule is bound at the donor site of the enzyme. Conformation II places the 2′-hydroxyl group close to nitrogen 1. Again, hydrogen bonding between the 2′-hydroxyl and nitrogen 1 would aid in formation of a carbonium center at carbon 7. However, a cyclic ether is not formed between carbon 7 and carbon 3′ and deprotonation of the methyl group yields the 7-exomethylene form of the lumazine bound at the acceptor site of the enzyme.

Condensation of the 7-exomethylene of the lumazine at the acceptor site with carbon 6 of the molecule at the donor site (Fig. 3, step 3) and subsequent reactions (steps 4–9) would proceed by mechanisms similar to

* The work of McAndless and Stewart [10] substantiates the assumption that a carbonium center can develop at carbon 6. If the carbonium center at carbon 7 is stabilized by covalent bond formation, as in 1,7-dihydro-6,7,8-trimethyllumazine, hydrogen-deuterium exchange occurred at the 6-methyl group (but not the 7-methyl group).

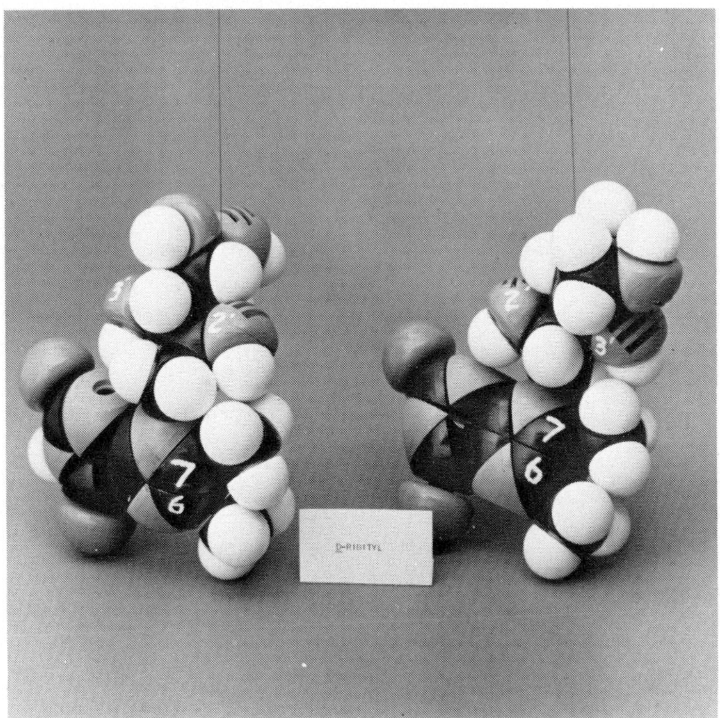

Fig. 5. CpK space-filling models of 6,7-dimethyl-8-D-ribityllumazine. Conformation I on left, Conformation II on right.

those depicted previously, excepting that groups X, Y and Z may not be involved as shown.

The CpK space-filling models of 6,7-dimethyl-8-D-ribityllumazine are shown in Fig. 5. The model in conformation I (left) has the 3′-hydroxyl and 2′-hydroxyl groups in close proximity to nitrogen 1 and carbon 7, respectively, whereas in conformation II (right) the 2′-hydroxyl is close to nitrogen 1 for the molecule assumed to be at the enzyme acceptor site. In both conformations, the ribityl group lies at a 120° angle to the planar pteridine ring.

In the interaction of the two conformers visualized in Fig. 6, the 7-methyl group at the acceptor site is located directly under carbon 6 of the ring located at the donor site and is within covalent bond distance. The close lineup between the 6-methyl group at the acceptor site and carbon 7 at the donor site can be seen.

In the model shown (Fig. 6), it is proposed that the substrate molecules interact with the enzyme at the ribityl side chains, thus properly orienting the pteridine rings. Therefore, it may be significant that the riboflavin synthase reaction is highly specific for the substituent at nitrogen 8 of 6,7-dimethyl-8-aldityllumazine. Thus, of 23 derivatives tested only 5 had substrate or inhibitor activity for the yeast enzyme (Table III): (a) None of the hexityl analogues or compounds with side chains shorter than four carbons

Fig. 6. CpK space filling models depicting two molecules of 6,7-dimethyl-8-D-ribityllumazine bound at the donor and acceptor site of Riboflavin Synthetase. The upper molecule at the donor site is in Conformation I while the lower molecule at the acceptor site is in Conformation II.

in length had activity. (b) One of the four possible trihydroxytetrityl derivatives, 6,7-dimethyl-8-L-threityllumazine, was a relatively ineffective competitive inhibitor ($K_m/K_i = 0.05$). (c) All of the tetrahydroxypentityl derivatives (except the L-ribityl compound) have been synthesized. The D-ribityl compound is the natural substrate and the D-xylityl derivative is an inhibitor competitive with the substrate ($K_m/K_i = 1$). (d) With the recent synthesis of 6,7-dimethyl-8-(1′-(4′-deoxy-D-ribityl)lumazine (Beach (1975), unpublished experiments), all of the deoxy-D-ribityl analogues have now been prepared. 6,7-Dimethyl-8-(1′-(5′-deoxy-D-ribityl))lumazine is a substrate leading to formation of 5′-deoxyriboflavin. 6,7-Dimethyl-8-(1′-(3′-deoxy-D-ribityl))-lumazine is a potent competitive inhibitor ($K_m/K_i = 13$), whereas the 2′-deoxyribityl and 4′-deoxyribityl derivatives are inactive. (e) For substrate or inhibitor activity the presence of hydroxyl groups at the 2′- and 4′-carbons in the D-configuration appears to be required (L-threityl, D-ribityl, D-xylityl, 5′-deoxy-D-ribityl and 3′-deoxy-D-ribityl). (f) The bond angle between the planar pteridine ring of active substances is either 120° (D-ribityl, 5′-deoxy-D-ribityl, 3′-deoxy-D-ribityl) or 150° (L-threityl and D-xylityl). All compounds with an L-2′-hydroxyl group are inactive and have a bond angle at 90° (if the models can be constructed at all e.g., D-arabityl (Table III)). The inactivity of the 2′-deoxy-D-ribityllumazine (120° angle) may be due to its inability to form the carbon 2′-carbon 7 internal ether (donor) and a lack of facilitation of formation of the 7-exomethylene form (acceptor) by interaction between the 2′-hydroxyl and nitrogen 1.

The D-4′-hydroxyl group seems essential for binding to the enzyme. The L-arabityl (angle 170–180°) and L-lyxityl (angle 150°) analogues have no kinetic effect. The 4′-deoxy-D-ribityl derivative is inactive even though the bond angle is the same as that of 6,7-dimethyl-8-(D-ribityl)lumazine. (g) The orientation of the 3′-hydroxyl group of the side chain appears to be less critical since competitive inhibition has been obtained with the 3′-deoxy-D-ribityl, D-xylityl and L-threityl analogues.

The potent inhibition by the 3′-deoxy-D-ribityl compound may be due to unique binding of the derivative to the enzyme donor site in which, in the absence of hydrogen bonding between the 3′-hydroxyl group and nitrogen 1,

TABLE III

ACTIVITIES OF 6,7-DIMETHYL-8 ALDITYLLUMAZINES WITH YEAST RIBOFLAVIN SYNTHASE

Data from Winestock et al. [15] and Harvey and Plaut [13]. Values for constants are not reported whenever equimolar concentrations of substrate and analogue tested gave less than 10% inhibition.

Compound	Configuration of OH at C-2′ and C-4′		Bond angle[b]		Competitive inhibition (K_i) or K_m (μM)
6,7-Dimethyl-8-substituted lumazines (substituent)	C-2′	C-4′	Conformation		
			I	II	
D-Ribityl	D	D	120°	120°	$K_m = 10$; $V = 100\%$
D-Arabityl	L	D	90°	c	
L-Arabityl	D	L	170°	180°	
D-Lyxityl	L	D	90°[d]	90°[d]	
L-Lyxityl	D	L	150°	150°	
D-Xylityl	D	D	150°	150°	$K_i = 10$
5′-Deoxy-D-ribityl	D	D	120°	120°	$K_m = 40$; $V = 1\%$
4′-Deoxy-D-ribityl[a]	D	O	120°	120°	
3′-Deoxy-D-ribityl[a]	D	D	120°	120°	$K_i = 0.8$
2′-Deoxy-D-ribityl[a]	O	D	120°	120°	
D-threityl[a]	L	D or L	90°	90°	
L-threityl[a]	D	D or L	150°	150°	$K_i = 190$
D-Erythrityl[a]	D	D or L	120°	120°	
L-Erythrityl[a]	L	D or L	90°[d]	90°[d]	
D-Galactityl	D	L			
D-Glucityl	D	D			
D-Mannityl	L	D			
3′-Hydroxypropyl[a]	O				
DL-2′,3′-Dihydroxypropyl[a]	DL				
2′-Hydroxyethyl	D or L				
Methyl					
Hydrogen					

[a] The synthesis of these compounds, except the 4′-deoxy-D-ribityl derivative, were reported by Beach and Plaut [16]. The assays for activity are from unpublished experiments (R.L. Beach, T. Aogaichi and G.W.E. Plaut).

[b] Bond angle between planar pteridine ring and the 8-(1′-alditytl) substituent.

[c] Too highly strained to make model.

[d] The 90° bond angle results if the 4′-carbon is rotated about the C-3′, C-4′ bond placing the 4′-hydroxyl group on the side of the chain opposite that of the hydrogens at C-1′. If the 4′carbon is rotated through 180° so that the 4′-hydroxyl group is on the same side of the chain as the hydrogens at C-1′, the bond angle approximates 120° in conformation I and II.

the acidic group on the enzyme may preferentially protonate nitrogen 1 instead of nitrogen 5 (Fig. 4) leading to a dead-end complex.

Acknowledgements

Supported in part by Grant AM 15404 from the National Institutes of Arthritis, Metabolic and Digestive Diseases, National Institute of Health.

References

1 Plaut, G.W.E. (1971) Comprehensive Biochemistry, Vol. 21, pp. 11–45
2 Demain, A.L. (1972) Annu. Rev. Microbiol. 26, 369–388
3 Plaut, G.W.E., Smith, C.M. and Alworth, W.L. (1974) Annu. Rev. Biochem. 43, 899–922
4 Plaut, G.W.E. and Harvey, R.A. (1971) Methods Enzymol. 18B, 515–538
5 Beach, R.L. and Plaut, G.W.E. (1970b) J. Am. Chem. Soc. 92, 2913–2916
6 Rowan, T. and Wood, H.C.S. (1963) Proc. Chem. Soc., 21–22
7 Rowan, T. and Wood, H.C.S. (1968) J. Chem. Soc. C, 452–458
8 Beach, R. and Plaut, G.W.E. (1969) Tetrahedron Lett. 40, 3489–3492
9 Paterson, T. and Wood, H.C.S. (1969) Chem. Commun., 290–291
10 McAndless, J.M. and Stewart, R. (1970) Can. J. Chem. 48, 263–270
11 Beach, R.L. and Plaut, G.W.E. (1970a) Biochemistry 9, 760–770
12 Plaut, G.W.E., Beach, R.L. and Aogaichi, T. (1970) Biochemistry 9, 771–785
13 Harvey, R.A. and Plaut, G.W.E. (1966) J. Biol. Chem. 241, 2120–2136
14 Beach, R.L. and Plaut, G.W.E. (1971) J. Org. Chem. 36, 3937–3943
15 Winestock, C.H., Aogaichi, T. and Plaut, G.W.E. (1963) J. Biol. Chem. 238, 2866–2874
16 Winestock, C.H. and Plaut, G.W.E. (1971) J. Org. Chem. 26, 4456–4462

Flavins and Flavoproteins, edited by T.P. Singer

CHAPTER 83

FLAVIN METABOLISM AND ITS CONTROL BY THYROID HORMONES

RICHARD S. RIVLIN*, ARPAD G. FAZEKAS**, YEE PING HUANG and RITA CHAUDHURI

Department of Medicine and Institute of Human Nutrition, College of Physicians and Surgeons of Columbia University, New York, N.Y. 10032 (U.S.A.)

Summary

Thyroid hormones appear to regulate the metabolic conversion of riboflavin to FMN and FAD. Physiological doses of thyroxine increase hepatic flavokinase activity in rats, and thyroidectomy significantly reduces enzyme activity. In animals with altered thyroid function, changes in the activities of riboflavin-metabolizing enzymes are reflected closely by changes in the actual rates of FMN and FAD formation in vivo, as determined by experiments utilizing [^{14}C] riboflavin as a tracer.

Thyroid hormones enhance flavin synthesis similarly in livers of newborn and adult rats. Thyroxine increases both FAD pyrophosphorylase activity and ^{14}C-labeled FAD synthesis rates approx. 40% in brains of newborn animals, but is ineffective in brains from adult animals. FAD pyrophosphorylase activity is increased by thryoxine in tadpoles (*Rana catesbeiana*) undergoing metamorphosis, and certain analogues of riboflavin inhibit thyroxine-induced metamorphosis. Evidence has accumulated that thyroid hormones regulate riboflavin metabolism and that certain aspects of thyroid hormone action may in turn be regulated by riboflavin.

Introduction

The importance of flavins and flavoproteins in a wide range of biochemical processes is being recognized increasingly. In any complex biological system, mechanisms are required to regulate the availability and distribution of the flavin derivatives. Our investigations focus on the role of hormones, particularly thyroid hormones, upon regulating the metabolic conversion of riboflavin to FMN and FAD, a potentially important control point in intermediary metabolism. The present report reviews briefly the evidence that

* To whom requests for reprints should be sent.
** Present address: University of Montreal, Montreal, Canada.

thyroid hormones increase the rate of flavin synthesis in a variety of organs, and in several species, and presents several new findings from this laboratory.

Materials and Methods

Animals

Male white rats of the Holtzman strain were used in all experiments. In studies in adult animals, each rat received daily intraperitoneal injections of L-thyroxine (T_4) in doses ranging from 2.5 μg/100 g to levels of 300 μg/100 g for several days. A dose range of triiodothyronine (T_3) was also utilized. Hypothyroid animals were prepared by feeding weanling rats a low iodine diet (Nutritional Biochemicals Corp., Teklad Mills) for four weeks followed by a single intraperitoneal injection of ^{131}I, 300 μCi/100 g. The development of hypothyroidism was documented by failure of growth, absence of uptake of a tracer dose of ^{131}I, and by histological examination [1].

Riboflavin deficiency was induced by feeding weanling animals a diet deficient in riboflavin (Nutritional Biochemicals Corporation, Teklad Mills) for variable periods of time. Animals were placed individually in wire-bottom cages.

Controls for these studies consisted of littermates of the same sex, which were injected with saline, and treated in as similar a fashion as possible as the experimental groups.

In studies utilizing newborn animals, litter sizes were restricted to 8 animals each within 24 h of birth. Half the litters received daily subcutaneous injections of thyroxine, 10 μg/animal, and the remainder received injections of isotonic saline for periods of generally 7–9 days.

Biochemical measurements

The activity of flavokinase (ATP: riboflavin-5′-phosphotransferase, EC 2.7.1.26.) was assayed by a spectrophotometric method [3] as were the assays of FAD pyrophosphorylase (ATP: FMN adenyltransferase, EC 2.7.7.2.) [4, 5] and FMN phosphatase [6]. Tissue concentrations of FAD were determined initially by a spectrophotometric method [7] and later by a fluorometric method [8] which also permitted simultaneous assays of FMN [9]. Protein concentrations were generally determined by the biuret method [9].

The conversion of [^{14}C]riboflavin to ^{14}C-labeled FMN and ^{14}C-labeled FAD in vivo was determined following subcutaneous injection of tracer doses of 2.5 μCi/100 g body weight. D-[2-^{14}C]riboflavin was purchased from Amersham Laboratories (spec. act. = 60–70 Ci/mol). Radioactive flavins were separated and quantitated from tissues utilizing newly developed techniques of isotope dilution and ion-exchange column chromatography with DEAE-Sephadex A-25 [10, 11]. The protocol for extraction, isolation and quantitation of radioactive riboflavin, FMN and FAD from tissue sources is outlined in Fig. 1.

Results

The effects of administering a single large dose of T_3 (100 μg) to normal animals by intraperitoneal injection upon the hepatic activity of flavokinase were determined. The results of these experiments are shown in Fig. 2. Flavokinase activity rose progressively during the time periods studied, and at 24 h was approximately twice that noted in untreated animals.

Previous studies [12] had shown earlier that pharmacological doses of thyroxine increase flavokinase activity. The maximum increase observed after thyroxine treatment was similar to that achieved by T_3. The present experiments determined whether lower doses of thyroxine also increase

Inject animal with [^{14}C]riboflavin (2.5 μCi/100 g) subcutaneously

↓

Sacrifice animal at intervals after injection and promptly remove tissues to be studied.

↓

Homogenize tissues in 70% methanol. Add 100 μg riboflavin, FMN and FAD as cold carriers.

↓

Centrifuge homogenates at 60 000 $\times g$ for 15 min.

↓

Recover supernatant solutions and evaporate to dryness.

↓

Take up residues in 1.0 ml water and add to DEAE-Sephadex A-25 column. Elute with water followed by 1% $(NH_4)_2SO_4$

↓

Recover Riboflavin, FMN and FAD bands successively.

↓

Determine radioactivity (Packard Tri-Card Liquid Scintillation Counter) and absorbance (Zeiss spectrophotometer) in each band.

↓

Express data obtained as total dpm/100 mg tissue/time.

Fig. 1. Outline of the procedure utilized for the extraction, isolation and quantitation of radioactive riboflavin, FMN and FAD from tissue sources [10, 11].

flavokinase activity. Results of experiments in which groups of hypothyroid animals received graded doses by thyroxine by daily intraperitioneal injection for four days are shown in Fig. 3. In untreated hypothyroid animals, enzyme activity, as noted earlier [12], was reduced to about half that of control animals. Doses of thyroxine as low as 2.5 μg/100 g, levels comparable to the daily secretory rate of thyroid hormone in the rat [13], were sufficient to restore enzyme activity completely to normal. Doses of thyroxine of 5 μg/100 g significantly ($p < 0.01$) increase flavokinase activity above normal levels. Small increases in the doses of thyroxine increase enzyme activity to a comparable degree. These observations indicate that pharmacological doses of thyroxine are not required to increase flavokinase activity in hypothyroid rats.

These data, together with our previous findings that thyroid hormones increase the activity of FAD pyrophosphorylase [15], provide additional evidence that the enzymes involved in the riboflavin-to-FAD pathway are under hormonal control.

A series of experiments was then undertaken to determine whether the actual rates of FMN and FAD synthesis in vivo are influenced by thyroid hormones. Results of experiments in which hyperthyroid, euthyroid and hypothyroid animals each received a tracer dose of [^{14}C]riboflavin prior to sacrifice and incorporation into [^{14}C]FMN and [^{14}C]FAD in liver determined, are shown in Table I.

It is apparent that large doses of thyroxine increase significantly the incorporation of [^{14}C]riboflavin into both [^{14}C]FMN and [^{14}C]FAD. The magnitude of the increase in [^{14}C]FMN formation (to 179.5% of control) is comparable to the doubling in flavokinase activity which is observed under

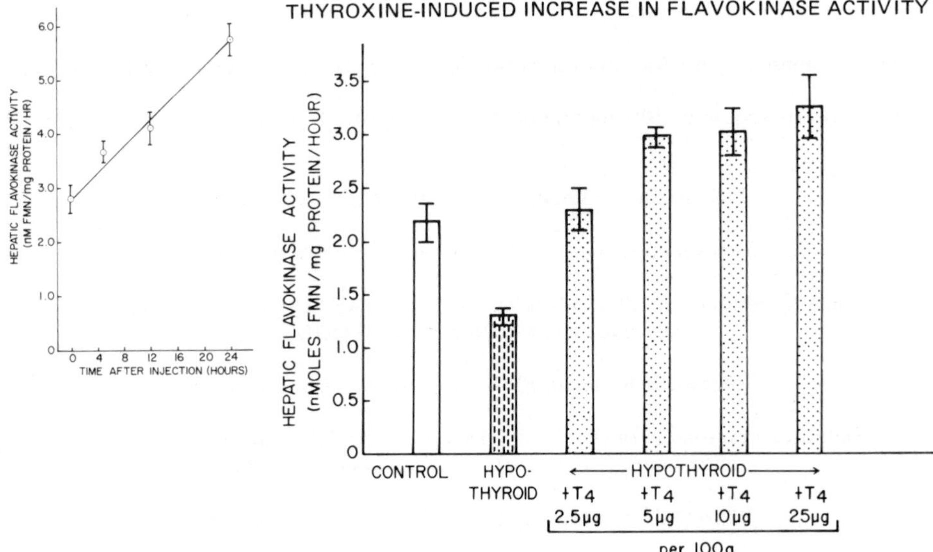

Fig. 2. Hepatic activity of flavokinase after a single intraperitoneal injection of 100 μg triiodothyronine to normal adult male Holtzamn rats. Data are shown as the mean ± S.E.M. of 7–9 animals per group.
Fig. 3. Hepatic activity of flavokinase in hypothyroid animals after the daily intraperitoneal administration of graded doses of thyroxine for a period of four days. Control and untreated hypothyroid animals received doses of isotonic saline of the same volume and pH as that of the experimental solutions. Data are shown as the mean ± S.E.M. of 5–7 animals per group treated with thyroxine and 15–20 animals per group of control and untreated hypothyroid animals.

similar circumstances. Furthermore, the magnitude of the increase in [^{14}C]-FAD formation (to 149% of control) is nearly identical to the magnitude of the increase in FAD pyrophosphorylase activity [15] previously observed in this laboratory in response to thyroid hormones.

In hypothyroid animals, there was a significant decrease in the rate of incorporation of riboflavin into [^{14}C]FAD. The magnitude of the decrease in [^{14}C]FMN and [^{14}C]FAD formation in hypothyroid animals is similar (to 78% and 81% of control levels respectively) and was not as low as the reduction in flavokinase activity (generally 50–60% of control levels). FAD

TABLE I

RATES OF INCORPORATION OF [^{14}C]RIBOFLAVIN INTO [^{14}C]FAD IN LIVERS OF ADULT, MALE RATS WITH ALTERED THYROID FUNCTION

Thyroxine-treated animals received 300 μg T_4/100 g body weight on days 1.3,5,6 and were sacrificed on day 7.

Group	Number of animals	[^{14}C]FMN*	[^{14}C]FAD*
Thyroxine-treated	16	195 ± 17.0%*	155 ± 12.1%
Control	14	100 ± 5.4%**	100 ± 4.8%**
Hypothyroid	8	78.2 ± 7.7%	81.0 ± 6.3%

* Data are expressed as a percentage of levels in control animals (mean ± S.E.M.)
** [^{14}C]FMN incorporated=2929 ± 157 dpm/100mg/h; [^{14}C]FAD incorporated=16340 ± 794 dpm/100mg/h. Animals each received a single subcutaneous injection of [^{14}C]riboflavin, 2.5 μCi/100 g, and were sacrified 1 h later. Data are derived in part from Fazekas et al. [14].

pyrophosphorylase activity has not been reduced in hypothyroid rats [15], an observation which is entirely consistent with the present data. These data, taken in their entirety, indicate that with altered thyroid function, changes in the activities of riboflavin-metabolizing enzymes are closely reflected by changes in actual rates of flavin synthesis in vivo.

To extend these relationships further, studies were performed in brain. Thyroid hormones are known to have important influences upon the brain during early life, both in terms of biochemical parameters and behavioral characteristics [16, 17]. Our initial investigations [18] showed that thyroxine administration to newborn animals for the first 8 days of life elevates FAD pyrophosphorylase activity in brain by 40%. This treatment with thyroid hormones results in enzyme activity which is comparable in magnitude to that of adult animals. When adult rats are treated with thyroxine, no further increase in enzyme activity in brain occurs [18].

The relevance of these observations to flavin synthesis in vivo was determined by examining the effects of thyroid hormones on the rates of [^{14}C]-FAD synthesis in cerebrum and cerebellum of both newborn and adult rats. The basic design of the experiments and techniques used were similar to those described above.

Results of several of these experiments are given in Table II. It is apparent that neonatal administration of thyroxine significantly increases the rate of incorporation of [^{14}C]riboflavin into [^{14}C]FAD in both cerebrum and cerebellum. The magnitude of these increases is very similar to that noted earlier in FAD pyrophosphorylase activity in animals of the same age. In livers of these animals, the increase in FAD pyrophosphorylase activity and in rate of [^{14}C]FAD formation is similar to that of adult animals.

By contrast, in adult animals neither FAD pyrophosphorylase activity nor the rates of incorporation of [^{14}C]riboflavin into [^{14}C]FAD in cerebrum or cerebellum have been significantly increased by the administration of thyroid hormones [20].

TABLE II

EFFECTS OF NEONATAL ADMINISTRATION OF THYROXINE UPON RATES OF INCORPORATION OF [^{14}C] RIBOFLAVIN INTO [^{14}C] FAD IN CEREBRUM AND CEREBELLUM

All animals received a single subcutaneous injection of [^{14}C] riboflavin, 2.5 μCi/100 g, 1 h before sacrifice.

Group	Number of animals	Cerebellum [^{14}C]FAD (dpm/100mg/h)	Cerebrum [^{14}C]FAD (dpm/100mg/h)
Control	19	879 ± 34	595 ± 35
Thyroxine treated*	19	1279 ± 59	749 ± 31
Significance of difference		($p < 0.001$)	($p < 0.01$)

* Thyroxine-treated animals each received 10 μg T_4 subcutaneously daily for 7 days; controls received saline of the same volume and pH as the experimental solution. Litter size was restricted to 8 animals; half of the litters received saline and half thyroxin. All treatments were begun within 24 h after birth. Data are shown as mean ± S.E.M., and are derived from Rivlin et al [20].

Discussion

The results presented here, together with previous findings from this and other laboratories, provide several independnet lines of evidence that the metabolic conversion of riboflavin to FMN and FAD is under thyroid hormone control [21, 22]. In adult rats, hepatic activities of enzymes involved in flavin metabolism, notably flavokinase and FAD pyrophosphorylase, are increased in activity by thyroid hormones. The magnitude of these increases is similar to that observed in the net rates of formation in vivo of radioactive FAD after injection of radioactive riboflavin. In newborn animals, at a time when the development of the brain is dependent upon thyroid hormones, both FAD pyrophosphorylase activity and [^{14}C]FAD formation from [^{14}C]riboflavin in brain are increased significantly and to a comparable degree by thyroid hormones. In adult animals, the brain is largely resistant to the effects of thyroid hormones, as measured by increases in oxygen consumption, blood flow and protein synthesis [23]. Consistent with this observation is our finding that in brains of adult animals, neither activities of flavin-metabolizing enzymes nor rates of formation of radioactive FAD in vivo are altered by thyroid hormones.

In hypothyroid animals, tissue concentrations of flavin coenzymes are reduced [15, 24, 25], flavokinase activity is reduced [12], and as shown above, there is a reduction in the rates of incorporation of [^{14}C]riboflavin into [^{14}C]FAD. It is not surprising, therefore, that hypothyroidism and riboflavin deficiency have in common reductions in activities of a wide range of FMN- and FAD-requiring enzymes [26].

Further evidence for thyroid hormone-flavin interdependence is derived from observations that the thyroid hormone induction of certain flavoprotein enzymes, notably D-amino acid oxidase [27], succinate dehydrogenase [28] and mitochondrial α-glycerophosphate dehydrogenase [29], is markedly inhibited in riboflavin-deficient animals. These findings are compatible with the concept that accumulation of flavoprotein holoenzymes requires synthesis both of the coenzyme as well as the apoenzyme moieties.

Investigations directed at broadening the concept of hormonal control of flavin metabolism have recently been focused on metamorphosis in tadpoles, a developmental process which is exquisitely dependent upon thyroid hormones [2]. When the normal maturation of the tadpole (*Rana catesbeiana*) is accelerated by immersion in a solution of thyroxine, hepatic FAD pyrophosphorylase activity is increased nearly 50% [22]. Also compatible with the observations discussed above is the finding that a variety of structural analogues of riboflavin [22] will inhibit metamorphosis in tadpoles.

Efforts must now be made towards understanding the chemical mechanisms involved in thyroid hormonal control, its underlying physiological significance, and the role of other hormones in regulating flavin metabolism. ACTH increases the biosynthesis of FMN and FAD in the adrenal cortex of the rat [10], and appears to stimulate FMN synthesis in liver and kidney as well [30]. Estrogens influence the production of the riboflavin-binding protein in chickens [31], but possible effects on flavin synthesis or turnover are unknown [22]. Both hypoglycemia [32] and diabetes [33] have been reported in riboflavin deficient animals, and the role of insulin in regulation of flavin metabolism is almost entirely unknown. Much remains to be learned about the control of flavin metabolism by the endocrine system, and it is hoped that new knowledge will be forthcoming in the future.

Acknowledgements

This work was supported by grants (AM 15265 and CA 12126) from the U.S. Public Health Service and by the Stella and Charles Guttman Foundation.

References

1 Rivlin, R.S. and Levine, R.J. (1963) Endocrinology 73, 103
2 Etkin, W. (1964) Physiology of the Amphibia (Moore, J.A., ed.), p. 427, Academic Press, New York
3 McCormick, D.B. (1962) J. Biol. Chem. 237, 959
4 DeLuca, C. and Kaplan, N.O. (1958) Biochim. Biophys. Acta 30, 6
5 McCormick, D.B. (1964) Nature 201, 925
6 McCormick, D.B. and Russell, M. (1962) Comp. Biochem. Physiol. 5, 113
7 DeLuca, C., Weber, M.M. and Kaplan, N.O. (1956) J. Biol. Chem. 223, 559
8 Burch, H.B. (1957) Methods in Enzymology (Colowick, S.P. and Kaplan, N.O., eds.), Vol. 3, p. 960, Academic Press, New York
9 Gornall, A.G., Bardwill, C.J. and David, M.M. (1949) J. Biol. Chem. 177, 751
10 Fazekas, A.G. and Sandor, T. (1971) Endocrinology 89, 397
11 Fazekas, A.G. (1975) Riboflavin (Rivlin, R.S., ed.), p. 81, Plenum Press, New York
12 Rivlin, R.S. and Langdon, R.G. (1966) Adv. Enz. Reg. 4, 45
13 Barker, S.B. (1951) Physiol. Rev. 31, 205
14 Fazekas, A.C., Huang, Y.P., Chaudhuri, R. and Rivlin, R.S. (1972) Program of the Forty-Eighth Meeting of the American Thyroid Association (Abstr.) p. 63
15 Rivlin, R.S. and Langdon, R.G. (1969) Endocrinology 84, 584
16 Balazs, R., Cocks, W.A., Eayrs, J.T. and Kovacs, S. (1971) Hormones in Development (Hamburgh, M. and Barrington, E.J.W., eds.), p. 357, Appleton-Century-Crofts, New York
17 Eayrs, J.T. (1971) Hormones in Development (Hamburgh, M. and Barrington, E.J.W., eds.), p. 345, Appleton-Century-Crofts, New York
18 Rivlin, R.S. (1974) The Neurosciences, Third Study Program (Schmidt, F.O. and Worden, F.G., eds.), p. 835, MIT Press, Cambridge, Mass.
19 Rivlin, R.S. and Hornibrook, R. (1971) Fed. Proc. 30, Abstr. 359
20 Rivlin, R.S., Huang, Y.P. and Chaudhuri, R. (1973) Fed. Proc. 32, Abstr. 231
21 Rivlin, R.S. (1970) New Engl. J. Med. 283, 463
22 Rivlin, R.S. (1975) Riboflavin (Riulin, R.S., ed.), p. 393, Plenum Press, New York
23 Sokoloff, L. and Roberts, P.A. (1971) Influence of Hormones on the Nervous System (Ford, D.H., ed.), p. 213, S. Karger, Basel
24 Domjan, G. and Kokai, K. (1966) Acta Biol. Hung. 16, 237
25 Lanzani, P. and Mascitelli-Coriandoli, E. (1965) Boll. Soc. Ital. Biol. Sper. 41, 666
26 Rivlin, R.S. (1970) Adv. Enz. Reg. 8, 239
27 Tipton, S.R., Weldon, F. and Weiss, A.K. (1955) Am. J. Physiol. 180, 321
28 Reith, A. (1973) Lab. Invest. 29, 216
29 Wolf, G. and Rivlin, R.S. (1970) Endocrinology 86, 1347
30 Fazekas, A.G. and Sandor, T. (1971) Can. J. Biochem. 49, 987
31 Clagett, C.O., Buss, E.B., Saylor, E.M. and Girsh, S.J. (1970) Poultry Sci. 49, 1468
32 Wickson, M.E. and Morgan, A.F. (1946) J. Biol. Chem. 162, 209
33 Axelrod, H.E., Gullberg, M.G. and Morgan, A.F. (1951) Am. J. Physiol. 165, 604

Flavins and Flavoproteins, edited by T.P. Singer

CHAPTER 84

TRANSPORT OF RIBOFLAVIN INTO CELLS

M. PERL*, E.B. KEARNEY and T.P. SINGER

Department of Biochemistry and Biophysics, University of California San Francisco, Calif. 94143 and the Division of Molecular Biology, Veterans Administration Hospital, San Francisco, Calif. 94121 (U.S.A.)

Introduction

In studies of the biosynthesis of histidyl 8α-FAD the need arose for introducing [^{14}C]riboflavin into intact cells. With the commercial strain of *Saccharomyces cerevisiae* (Red Star) ordinarily used in this laboratory we found that the uptake of labeled riboflavin by the cells was trivial. Therefore, we turned to the use of mutant strains of yeast [1, 2], blocked at various steps in the biosynthesis of riboflavin. As expected from the fact that they require externally supplied riboflavin for growth, such mutants proved capable of taking up riboflavin from the medium. This report delineates some of the characteristic features of the system responsible for the transport of riboflavin into the mutant yeast cells.

Materials and Methods

Wild type cells were isolated from commercial Red Star yeast. Other strains used were riboflavinless mutants HK 645, 859, and 871 and the parent strain S 288, kindly provided by Dr. A. Bacher. [^{14}C]Riboflavin (sp. act. = 28.8 Ci/mole) was from Amersham-Searle; [^{14}C]lumiflavin and [^{14}C]tetracetylriboflavin were made from this by standard methods. 7-Methylriboflavin and 8-methylriboflavin were gifts of Dr. J.P. Lambooy, and analogs modified in the side chain at N-10 were provided by Dr. D. McCormick.

The yeast was grown anaerobically as described previously [3], except that the glucose was 6%, riboflavin was added (5 mg/l) and the cultures were kept dark, and the temperature was 30°C in early experiments, 28°C in later studies. Cells were harvested, unless otherwise indicated, in the early or middle log phase, when turbidity reached 150–200, as monitored at 600 nm in a Klett colorimeter, washed, and resuspended in water to 10–15 mg/ml dry weight. This turbidity corresponded to a cell count of $3 \cdot 10^7$/ml and a dry weight of ≈ 1.3 mg/ml. Routine assays of riboflavin uptake were carried out in the dark at 30°C in 2 ml final volume of a reaction mixture containing 50 mM potassium phosphate, pH 7.2, 12 mM KCl, and 5 μM [^{14}C]riboflavin or other substrate; washed yeast cells equivalent to 3 to 5 mg dry weight were added to start the reaction and aliquots (0.5 ml) were removed immediately after mixing and at intervals thereafter.

---

* On leave of absence from Bar Ilan University, Israel.

Duplicate tubes were run for each sample. The aliquots removed were rapidly diluted with 10 volumes of water at 0°C, filtered by vacuum through a glass fiber filter, and the filter was washed 3 times with 5-ml volumes of water at 0°C. The entire process of collection took 30 s or less. The filter pads were then counted for radioactivity in a scintillation counter. Efflux of riboflavin from the cells was measured similarily, but after a preliminary incubation for 60 min at 30°C with [^{14}C]riboflavin to "load" the cells, followed by one centrifugal wash with cold water and resuspension in a flavin-free medium. Fluorometric determinations of flavins were made in a Hitachi-Perkin Elmer MPF-3 absolute fluorometer. Chromatography of KOH-neutralized perchloric acid extracts of cells was done on cellulose plates in *n*-butanol/acetic acid/water (12:5:3, by vol.) as the developing solvent. The water content of cells was estimated from the weight loss of a filtered yeast cake in 20 h at 105°C and ranged from 2.9 to 3.2 μl per mg dry weight.

Results

The ability of the yeast to take up riboflavin varies strongly during the growth cycle. As shown in Fig. 1 for mutant No. 645, which has been used in most of the work reported here, activity is greatest in cells harvested from the early to middle log phase and declines sharply well before the stationary phase is reached. The two other mutants also displayed significant activity, while the strain from which they are derived and the commercial Red Star yeast cells show very low activity under the same growth conditions (Fig. 2).

Some of the characteristics of the transport system are as follows: uptake is strictly proportional to cell concentration, and it is linear with time for the first 15 min (Fig. 3), after which it falls off, eventually reaching a steady state level which is a function of the external riboflavin concentration (Fig. 4). The process is characterized by saturation kinetics, with a K_m of ≈ 15 μM at 30°C (Fig. 5). A relatively sharp pH optimum is observed at pH 7.5 (Fig. 6) and the energy of activation is rather high (23 100 cal/mol, Fig. 7). The reaction is stimulated by certain monovalent cations (Table I), while some organic cations appear to be inhibitory (Table II).

The specificity of the system for the uptake of riboflavin was investigated with the aid of a series of flavin analogs, in most of which the ribityl

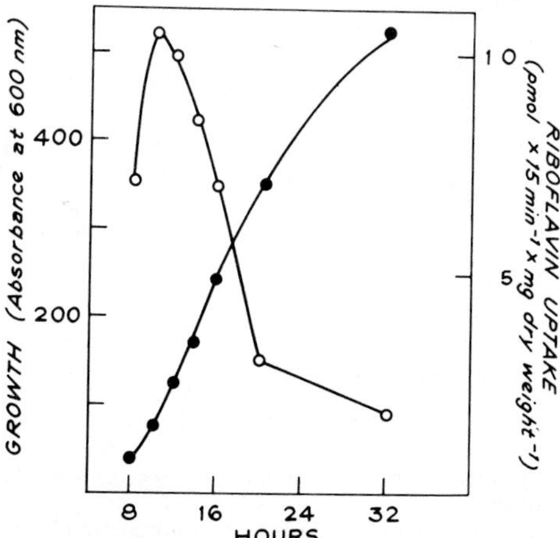

Fig. 1. Variation of the rate of riboflavin uptake (○) during growth (●) in mutant HK 645.

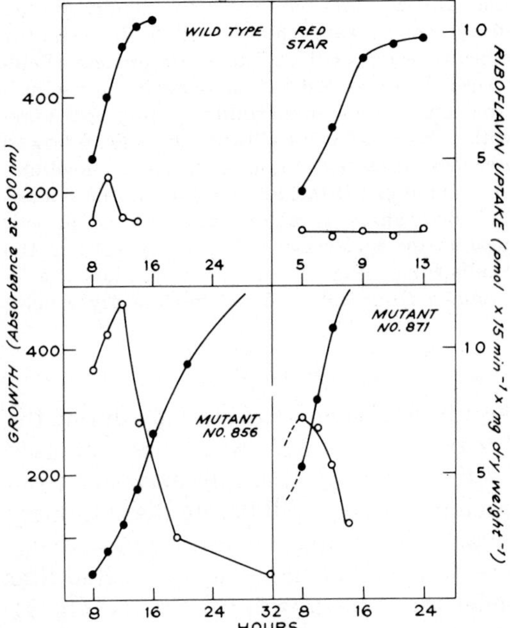

Fig. 2. Variation of the rate of riboflavin uptake with growth phase in commercial (Red Star) yeast, in two riboflavinless mutants, and in the parent strain from which they were derived (upper left corner). Symbols are as in Fig. 1.

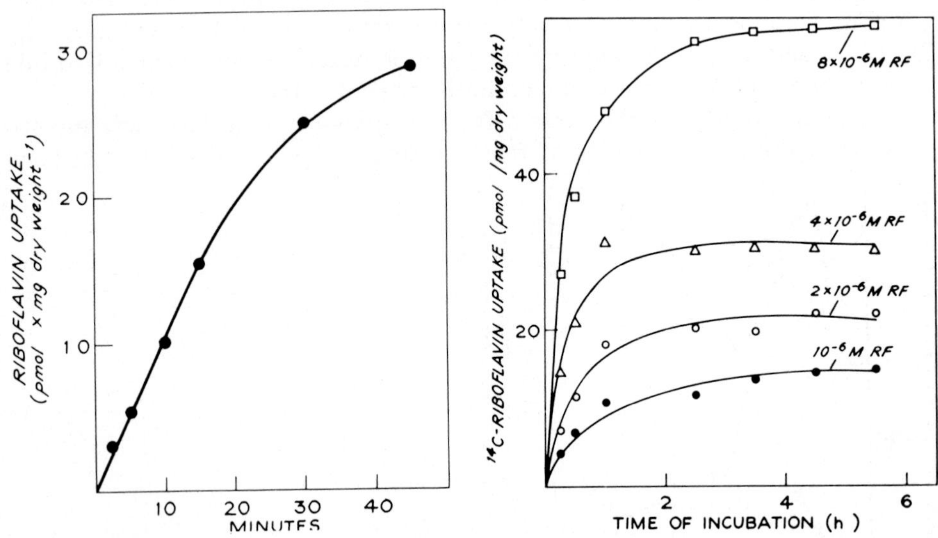

Fig. 3. Kinetics of riboflavin uptake in mutant HK 645. The cells were incubated in the presence of 5 μM [^{14}C]riboflavin. Other conditions are given in Methods.

Fig. 4. Time course of riboflavin (RF) uptake over long periods of incubation. The intracellular concentration of riboflavin is expressed as pmol per mg dry weight but may be converted to molarity of riboflavin by the relation: 10 pmol/mg = 3.3 μM riboflavin in the cells.

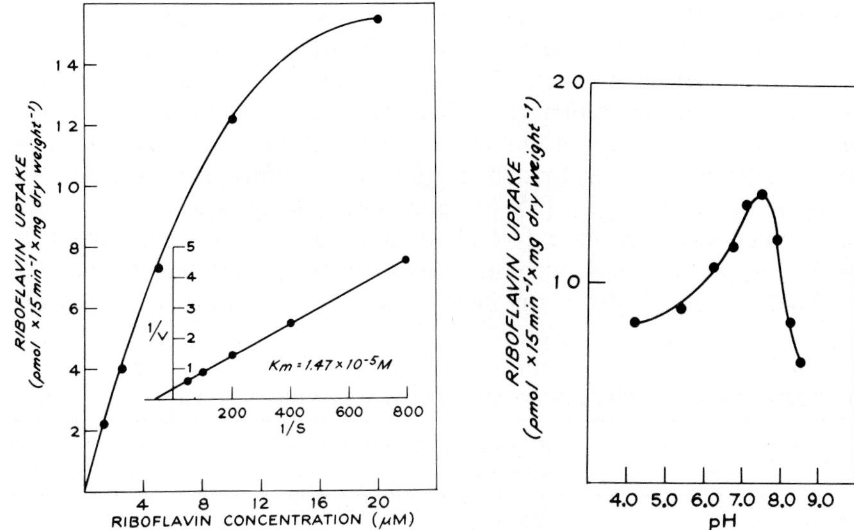

Fig. 5. Demonstration of saturation kinetics in the uptake process. Note that riboflavin concentration in the main graph is micromolar and in the inset (double-reciprocal plot) millimolar.

Fig. 6. Variation of rate of uptake with pH. The pH values shown were measured at 30°C in the complete mixture. All buffers were 50 mM phosphate, with KCl added to give 217 mM K^+.

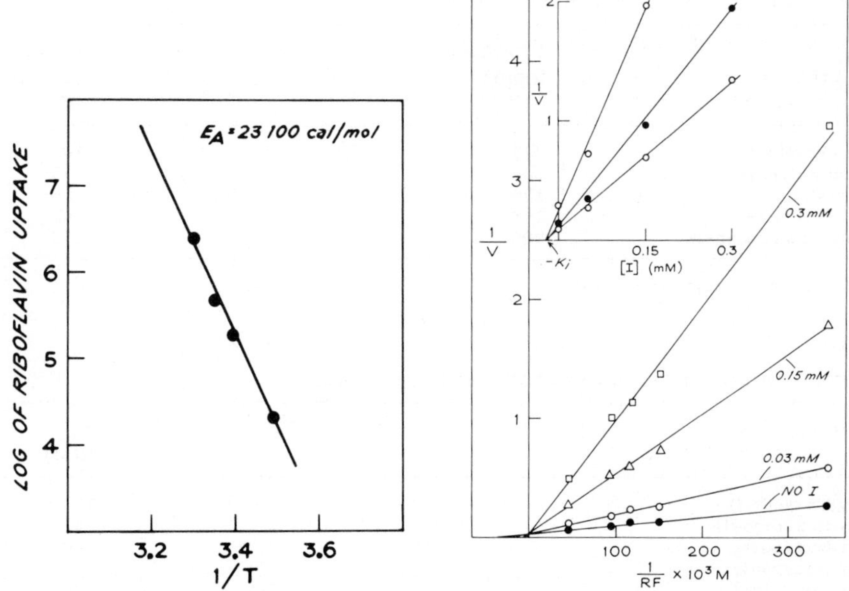

Fig. 7. Arrhenius plot of temperature effect on riboflavin uptake.

Fig. 8. Competive inhibition of riboflavin uptake by 7,8-dimethyl-10-(4-hydroxypentyl)isoalloxazine. The various curves denote the concentration of inhibitor in the assays.

side chain was modified (Table III). Except for two which gave hyperbolic Dixon plots, they behaved as competitive inhibitors toward riboflavin. The K_i values are summarized in Table III and a typical Dixon plot is illustrated in Fig. 8 for the 5′-hydroxypentyl analog. Among these analogs the N-10-carboxybutyl derivative was available in labeled form and could thus be tested as a substrate for transport; although it was a competitive inhibitor, it was not taken up by the cell. [^{14}C]Tetracetylriboflavin was neither a substate nor a competitive inhibitor. [^{14}C]Lumiflavin was also not inhibitory. It did, however, enter the cells, although probably by a different mechanism from riboflavin, as judged from the lack of saturation kinetics, different pH

TABLE I

THE EFFECT OF CATIONS ON THE RIBOFLAVIN UPTAKE BY YEAST CELLS

Cation (50 mM)	Riboflavin uptake (cpm)	Increase (%)
No addition	135	
Mg^{2+}	110	−19
NH_4^+	145	7.5
Rb^+	200	48
Na^+	150	11
K^+	200	48
Cs^+	180	33
Li^+	150	11

TABLE II

RIBOFLAVIN UPTAKE IN DIFFERENT BUFFERS

Buffer* (50 mM, pH 7.3)	Riboflavin uptake (cpm)	Relative activity (%)
Potassium phosphate	344	100
Ammonium phosphate	252	75
Tris phosphate	190	55
Imidazole phosphate	148	43
Triethanolamine phosphate	205	60

* All samples contained 107 mM K^+.

TABLE III

INHIBITION OF RIBOFLAVIN UPTAKE BY FLAVIN ANALOGS

Riboflavin analog	K_i (μM)
N-10(2′Hydroxyethyl)-	170
N-10(3′Hydroxypropyl)-	40
N-10(4′Hydroxybutyl)-	40
N-10(5′Hydroxypentyl)-	20
N-10(6′Hydroxyhexyl)-	20
N-10(1′-DL-Glyceryl)-	hyperbolic
N-10(4′-Carboxybutyl)-	800
7-Methylriboflavin	140
Deazaflavin	hyperbolic

dependence, and the fact that lumiflavin uptake is readily observed in wild type cells and does not seen to be a function of the growth phase. Because of this rapid uptake of lumiflavin photolytic degradation of riboflavin to lumiflavin had to be rigorously excluded in all experiments.

The last two compounds listed in Table III are modifications of the ring structure, rather than the ribityl side chain. Both enter the cells as evidenced by their incorporation into the intracellular flavin nucleotide and covalently bound flavin fractions [4], and both inhibit riboflavin uptake. Their transport characteristics have not yet been investigated.

Since the ribityl side chain appeared to be important in the specificity of the transport reaction, various 5 and 6 carbon sugars and sugar alcohols were tested for inhibition of riboflavin uptake, including ribitol itself. Only fermentable sugars (D-glucose, D-fructose, and D-mannose) were inhibitory; at 25 mM concentration the inhibition ranged from 28 to 43%. This inhibition may be an indirect effect, since glucose also appears to stimulate the movement of riboflavin out of the cell: when the cells were pre-loaded with [^{14}C]riboflavin and resuspended after washing, loss of radioactive label from the cells was observed, and the process was strongly stimulated when glucose was present in the suspending medium (Fig. 9).

Efflux of riboflavin from the cells appears to be less temperature dependent than influx ($E_A \approx 6500$ cal/mol), and varies with external pH, showing an optimum at pH ≈ 5.0, considerably lower than that for uptake. It is a function of the internal riboflavin concentration, with a $K_m = 48\,\mu$M at 30°C, about 4 times the K_m for entry. The reaction follows first-order kinetics and is apparently not affected by relatively high external riboflavin concentration; thus it appears to proceed against a concentration gradient (Fig. 9). Loss of riboflavin also occurs from wild type cells, as measured fluorimetrically in the reaction mixture, although no significant uptake was ever detected, and efflux was increased by glucose, as with the mutant.

Under the usual conditions of assay the interference of riboflavin efflux in measurements of uptake is not serious (less than 10% of the riboflavin

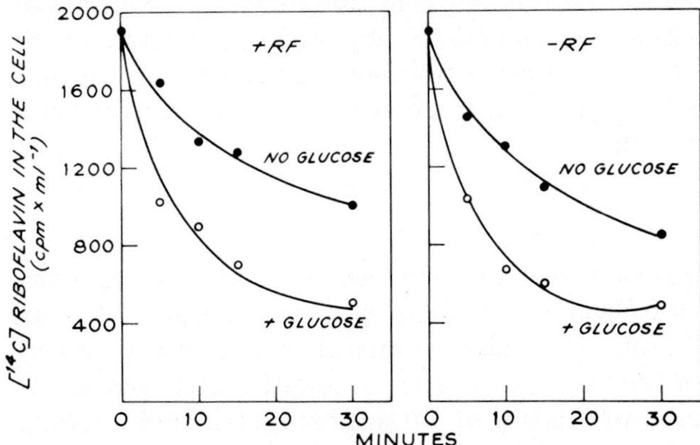

Fig. 9. Efflux of riboflavin (RF) from cells of mutant HK 645 preloaded with [^{14}C]riboflavin (see Methods). Left side, in the presence of 50 μM external riboflavin; right side, no external riboflavin. Glucose, where present, was 5.5 mM. Intracellular riboflavin concentration at zero time is 4 μM.

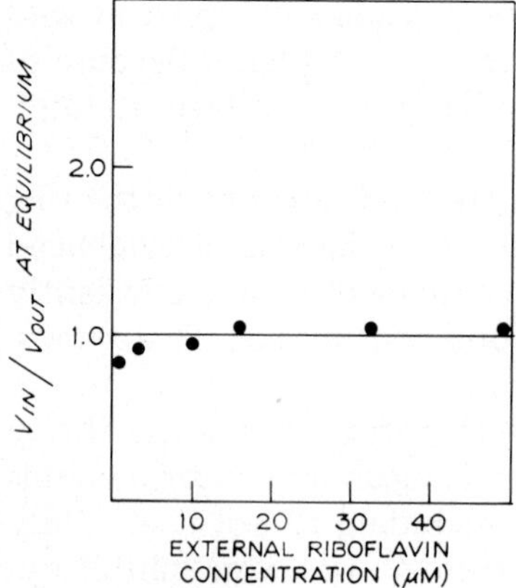

Fig. 10. Variation of the ratio of riboflavin influx and efflux with the extracellular concentration of riboflavin prevailing at equilibrium. The rates of uptake and of efflux were calculated from the relation

$$v = \frac{V \cdot [S]}{[S] + K_m}.$$

taken up would be lost from the cell), and transport of riboflavin into the cell is negligible in the determination of efflux, so that the two processes can be studied independently. As already shown in Fig. 4, however, if the uptake is allowed for longer periods, the rate is found to level off and a steady state concentration of riboflavin is reached inside the cell which depends on the external concentration of riboflavin. A further conclusion which may be drawn from Fig. 4 is that internal concentrations of riboflavin are reached that are higher than those presented externally, and thus net uptake against a concentration gradient occurs.

The interpretation that the steady states reached in the experiment of Fig. 4 represent the resultant of the two opposing processes, uptake and efflux, is strongly supported by the data in Fig. 10. Here ratios of the rates of riboflavin uptake and release, calculated from the known concentrations of riboflavin inside and outside the cell and the respective K_m curves for the two processes, are plotted against external riboflavin concentration. As may be seen, the data for the ratios are in satisfactory agreement with the value of unity expected at equilibrium.

Discussion

The data presented provide evidence for the existence of an active system for the uptake of riboflavin in riboflavin requiring mutants of yeast. The high energy of activation, specificity, competitive inhibition by substrate analogs, sharp pH optimum, and saturation kinetics observed are all compatible with the expected properties of a transport system and resemble the characteristics of the transport of other vitamins in microorganisms [5–10].

These findings pose a number of interesting questions. First, why is riboflavin uptake nearly absent in wild type yeast cells? A possible reason is that accumulation of a cellular component related to riboflavin biosynthesis, perhaps even the end products riboflavin and flavin nucleotides, may repress formation of the transport system. Efforts to relate the intracellular flavin concentrations in this mutant, however, to the changing levels of the transport system during growth did not lead to clear-cut interpretations, although activity did decline as the flavin levels increased. An indication that repression of synthesis of the transport system is nonetheless a possible mechanism has come from preliminary experiments with cycloheximide, which blocks development of the transport system. Development was not similarly blocked simply by preventing further cell division, as by removing glucose from the medium.

Another unanswered question is the exact relation of the systems involved in the uptake and the release of riboflavin from the cells. The data presented show that they operate independently, that the system for efflux operates even when that for uptake does not, and that neither process can be explained as simple diffusion because both operate against concentration gradients. It also remains for future research to determine the biological utility of the uptake system to yeast cells in view of the fact that normal yeast cells satisfy their requirements for riboflavin by synthesis.

Acknowledgments

We are grateful to Dr. A. Bacher for the yeast strains used in this work and to Dr. D. McCormick for the N-10 substituted riboflavin analogs. This study was supported by the National Science Foundation (grant No. 30078) and the National Institutes of Health (1 PO 1 HL 16251).

References

1 Oltmanns, O. and Lingens, F. (1966) Z. Naturforsch. 22, 751–754
2 Lingens, F., Oltmanns, O. and Bacher, A. (1966) Z. Naturforsch. 22, 755–758
3 Singer, T.P., Rocca, E. and Kearney, E.B. (1966) Flavins and Flavoproteins (Slater, E.C., ed.), pp. 391–419, Elsevier, Amsterdam
4 Grossman, S., Goldenberg, J., Kearney, E.B., Oestreicher, G.and Singer, T.P. (1976) Flavins and Flavoproteins, Proceedings of the 5th International Symposium (Singer, T.P., ed.), Chapt. 30, pp. 302–311, Elsevier, Amsterdam
5 Neujahr, H.Y. (1966) Acta. Chem. Scand. 20, 771–785
6 Kawasaki, T., Miyata, I., Esaki, K. and Nose, Y. (1969) Arch. Biochem. Biophys. 131, 223–230
7 Bradbeer, C. (1971) Arch. Biochem. Biophys. 144, 184–192
8 DiGirolamo, P.M. and Bradbeer, C. (1971) J. Bacteriol. 106, 745–750
9 Henderson, G.B. and Huennekens, F.M. (1974) Arch. Biochem. Biophys. 164, 722–728
10 DiGirolamo, P.M. and Huennekens, F.M. (1975) Arch. Biochem. Biophys. 168, 386–393

Part XIV

Special topics

Flavins and Flavoproteins, edited by T.P. Singer

CHAPTER 85

A STUDY OF LUMIFLAVIN-AMINO ACID INTERACTIONS AND THEIR RELATION TO THE STABILITY OF THE SEMIQUINONE FORM OF FLAVIN

LAI-SU LEE YEH and LLOYD L. INGRAHAM

Department of Biochemistry and Biophysics, University of California, Davis, Calif. 95616 (U.S.A.)

Summary

Lumiflavin complex formation with various amino acids was studied by measuring the partition of lumiflavin between chloroform and aqueous phases containing the amino acid. In some cases two-to-one stoichiometry was found in the complex. The effect of pH and ionic strength has also been studied. The equilibrium constants for the semiquinone form of lumiflavin interacting with amino acids were estimated by combining the present data with some earlier potentiometric data. It is of biological significance to note that tryptophan, seratonin, tyrosine and cystine all show a large affinity toward lumiflavin semiquinone.

Introduction

The flavin enzymes commonly shift the equilibrium between oxidized, reduced and semiquinone forms of the flavin toward the equilibrium in the absence of the enzyme. Hemmerich [1, 2] has pointed out that metal ions in the enzyme, particularly transition metals, can stabilize the semiquinone form of the enzyme, which we have considered, is the aromatic amino acids. The active centers of flavin enzymes have been well studied [2–13]. Tryptophan [13], tyrosine [8] and cystine [10–12] are often found to be in the binding or active sites of these flavoproteins. To assess the effects of these amino acids on the stabilization of the semiquinone, several years ago we studied the potentiometric titration of flavins in the presence of amino acids [14]. However, it soon became apparent that the supplementary data needed to calculate the equilibrium between flavin semiquinone and amino acid, namely the equilibrium constants for the complex formation between the fully oxidized flavin and amino acids was lacking in accuracy for most amino acids. Consequently, we have now studied these latter equilibria.

It has been known for years that flavin in the oxidized form will complex with various aromatic compounds [15], in particular, indoles and certain phenols [3, 16]. The association constants for these complexes have been studied by previous works by measuring the difference spectra of the complex and the flavin [17, 18, 19]. The results are somewhat inconsistent, probably due to the fact that this technique does not lead to an exact balancing-out of free flavin concentration in the reference and sample cells, as pointed out by Isenberg and Szent-Györgyi [3] and also because the spectral differences are usually quite small.

A method which was used by Hemmerich to study metal-flavin complex formation [1] has been developed in this laboratory to assay the flavin complex formation with any water-soluble but chloroform-insoluble organic compounds. By measuring the difference in partition of lumiflavin between chloroform and water of constant ionic strength in the presence and absence of excess complexing agent, the complex formation constant can be calculated. Lumiflavin was used in these experiments primarily because it is soluble in both chloroform and aqueous systems. The equilibrium constant, K_d, can be easily determined spectrally. Lumiflavin could possibly serve as a better model for the enzyme-bound flavin by eliminating steric effects that would arise in riboflavin, FMN and in FAD [20].

Experimental

Materials

Lumiflavin was synthesized according to the procedure of Kuhn and Reinmend [21]. The twice-recrystallized sample was used throughout the work. Spectral grade chloroform (Malinckrodt) was used throughout.

L-Histidine (base), L-phenylalanine, L-tyrosine, *m*-tyrosine, *o*-tyrosine, tyrosine HCl, L-tryptophan, 5-hydroxytryptophan, L-tryptophyl-L-tryptophan, L-tryptophylglycine, glycyl-L-tryptophan and L-methionine (all obtained from Sigma Chemical Company) were chromatographed in butanol/acetic acid/water (2:1:1), and in all cases showed only one spot when sprayed with ninhydrin.

Imidazole, serotonin oxalate (both from Sigma) and nicotinic acid (Eastman Organic Chemicals) all melted within a degree of reported values.

Methods

All work was carried out in a room thermostatted to 25°C. Special effort was made to minimize exposure to light. Unless stated otherwise, the buffer used was phosphate buffer, $I = 0.1$ pH = 6.96. For determination of the effect of pH on complex formation, a mixture of 0.1 M citric acid with 0.1 M potassium phosphate was used in the range of pH 3–6, 0.1 M potassium phosphate pH 6–8, and 0.1 M sodium carbonate-bicarbonate pH 9–11. In the study of effect of ionic strength, calculated amounts of potassium chloride were dissolved in potassium phosphate buffer of $I = 0.05$ pH 6.93, to make buffer solutions with ionic strength ranging from 0.10 to 4.05.

Absorption spectra and difference spectra were determined with a Cary Model 15 recording spectrophotometer using a 1-cm quartz cell with cap to prevent chloroform evaporation.

The molar absorption for lumiflavin was found to be $12.31 \cdot 10^3$ $mol^{-1} \cdot cm^{-1}$ in chloroform at 447 nm and $12.32 \cdot 10^3$ $mol^{-1} \cdot cm^{-1}$ in phosphate buffer at 440 nm.

Culture tubes with teflon-lined screw caps containing equal volume of aqueous phase and lumiflavin in chloroform were used to determine the distribution constant, K_d.

The tubes were shaken thoroughly to allow the distribution to come to equilibrium. Concentration of lumiflavin in each phase was determined spectrophotometrically, using the above absorptivity.

To determine the association constant for lumiflavin with any water-soluble, but chloroform-insoluble, complexing agent, the following procedure was used. 5 ml of a known concentration of complexing agent in aqueous buffer of constant ionic strength was shaken with 5 ml of a known concentration of lumiflavin in a culture tube with teflon-lined screw cap. A blank was prepared in the same way with no complexing agent in the aqueous phase. Shaking was continued in the dark until equilibrium was reached, usually overnight on a rotor at low speed. By determining the absorbance at 447 nm of the lumiflavin remaining at equilibrium of the sample and the blank, the association constant can be calculated.

The concentration of lumiflavin in all cases was $1.38 \cdot 10^{-4}$ M. The concentration of complexing agent ranged from $2.0 \cdot 10^{-3}$ M to $5.0 \cdot 10^{-2}$ M, depending on their solubility.

A computer program was written in order to calculate K_{III}, the association constant for lumiflavin-semiquinone complex formation.

Results

For determination of the association constant of lumiflavin (Lf) complexes with complexing agent (a), the following equilibria, presumably independent to each other, were considered:

$$\mathrm{Lf}_{(H_2O)} + \mathrm{a}_{(H_2O)} \underset{}{\overset{K_a}{\rightleftharpoons}} \mathrm{Lfa}$$

aqueous phase (upper) / chloroform phase (lower); K_d: $\mathrm{Lf}_{(H_2O)} \rightleftharpoons \mathrm{Lf}_{(CHCl_3)}$

Let: F_c = [Lf] in the chloroform phase before complex formation
F_w = [Lf] in the water phase before complex formation
F'_c = [Lf] in the chloroform phase after complex formation
F'_w = [Lf] in the water phase after complex formation
A = $[a]_{total} \simeq [a]_{free}$ (if $[Lfa] \ll [a]_{total}$)
C = $[Lf \cdot a]$

then $[Lf]_{total} = F_w + F_c = F'_w + F'_c + C$

$$\therefore C = [Lf]_{total} - (F'_w + F'_c) = (F_w + F_c) - (F'_w + F'_c)$$

$$K_d = \frac{F_c}{F_w} = \frac{F'_c}{F'_w}$$

$$\therefore F_w = \frac{F_c}{K_d}, \; F'_w = \frac{F'_c}{K_d}$$

$$\therefore K_a = \frac{C}{F'_w \cdot A} = \frac{(F_w + F_c) - (F'_w - F'_c)}{F'_w \cdot A} = \frac{\left(\frac{F_c}{K_d} + F_c\right) - \left(\frac{F'_c}{K_d} + F'_c\right)}{\frac{F'_c}{K_d} \cdot A}$$

$$= \frac{(F_c - F_c')(1 + K_d)}{F_c \cdot A}$$

$$\therefore \quad \boxed{K_a = \frac{1 + K_d}{A} \cdot \frac{F_c - F_c'}{F_c'}} \tag{I}$$

where $F_c - F_c'/F_c = (\mathrm{OD})_{F_c - F_c'}/(\mathrm{OD})_{F_c}$ which was obtained directly from spectral measurements.

The distribution coefficient, K_d, was found to be relatively independent of lumiflavin concentration over the range from 0.28 to $1.38 \cdot 10^{-4}$ molar lumiflavin as shown in Table I. However, the distribution coefficient did vary with the ionic strength as shown in Fig. 1.

The association constant between amino acid and lumiflavin, K_a, does not appear to be a true constant in some cases, rather it is directly proportional to A, the complex agent concentration. Such findings led us to believe that the stoichiometry must be different than what has been assumed. If a multiequilibrium such as:

$$\mathrm{Lf} + \mathrm{a} \overset{K_1}{\rightleftharpoons} \mathrm{Lf} \cdot \mathrm{a}$$

$$\mathrm{Lf} \cdot \mathrm{a} + \mathrm{a} \overset{K_2}{\rightleftharpoons} \mathrm{Lf} \cdot \mathrm{a}_2$$

exists in the aqueous phase instead of only the one-to-one complex, the following equations could be obtained:

$$K_1 = \frac{C_1}{F_w' \cdot A}, \quad K_2 = \frac{C_2}{C_1 \cdot A} = \frac{C_2}{K_1 F_w' \cdot A^2}$$

where $C_1 = [\mathrm{Lf} \cdot \mathrm{a}]$, $C_2 = [\mathrm{Lf} \cdot \mathrm{a}_2]$ and $[\mathrm{Lf} \cdot \mathrm{a}_2] \ll [\mathrm{a}]_{\mathrm{total}}$

then $[\mathrm{Lf}]_{\mathrm{total}} = F_w + F_c = F_w' + F_c' + C_1 + C_2$

$$\therefore C_2 = (F_w + F_c) - (F_w' + F_c') - K_1 F_w' \cdot A$$

TABLE I

DISTRIBUTION COEFFICIENT FOR LUMIFLAVIN BETWEEN CHLOROFORM AND WATER AT VARIOUS LUMIFLAVIN CONCENTRATIONS

Lumiflavin concentration* (× 10^4 M)	K_D^{**}
1.38	7.09
1.10	7.17
0.82	7.16
0.55	7.22
0.28	7.14

* Total lumiflavin concentration.
** Ionic strength of the aqueous phase was 0.10.

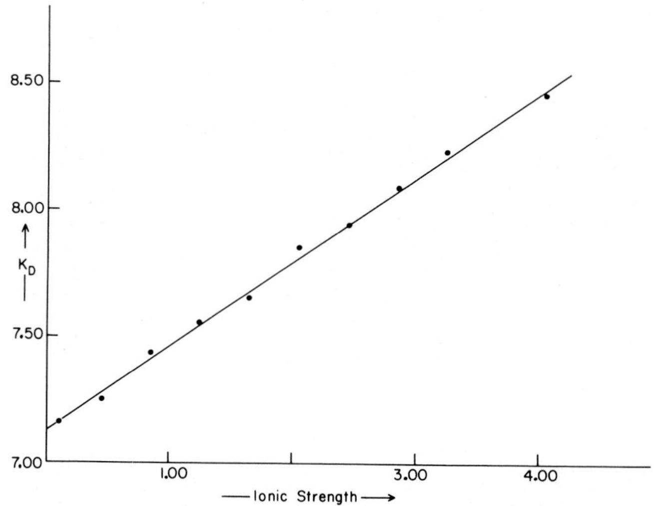

Fig. 1. The distribution coefficient for lumiflavin distributed between chloroform and phosphate buffer at pH 6.96 at 25°C as a function of ionic strength of the buffer.

$$= \frac{F_c}{K_d} + F_c - \left(\frac{F'_c}{K_d} + F'_c\right) - K_1 \frac{F'_c}{K_d} A$$

$$= \frac{1}{K_d} [(1 + K_d)(F_c - F'_c) - K_1 F'_c A]$$

Where: F_w, F_c, F'_w, F'_c, A and K_d are defined as before

$$\therefore\ K_1 K_2 = \frac{C_2}{F'_w \cdot A^2} = \frac{(1 + K_d)(F_c - F'_c) - K_1 F'_c A}{K_d F'_w A^2}$$

$$= \frac{(1 + K_d)(F_c - F'_c)}{F'_c A^2} - \frac{K_1}{A}$$

$$\therefore\ K_1 K_2 A + K_1 = \frac{(1 + K_d)(F_c - F'_c)}{F'_c A}$$

From Eqn. I: $\dfrac{(1 + K_d)(F_c - F'_c)}{F'_c A} = K_a.$

$$\therefore\ \boxed{K_a = K_1 K_2 A + K_1} \qquad \text{(II)}$$

Therefore, if K_a is plotted versus A, K_1 should be the intercept on the ordinate of the straight line, while $K_1 K_2$ will be the slope. The data were treated by the least squares method to obtain the best straight line through the experimental points. Fig. 2 shows some typical data for tryptophan. It is obvious that Eqn. I is a special case of Eqn. II where no two-to-one complex exists in the system, and $K_2 = 0$.

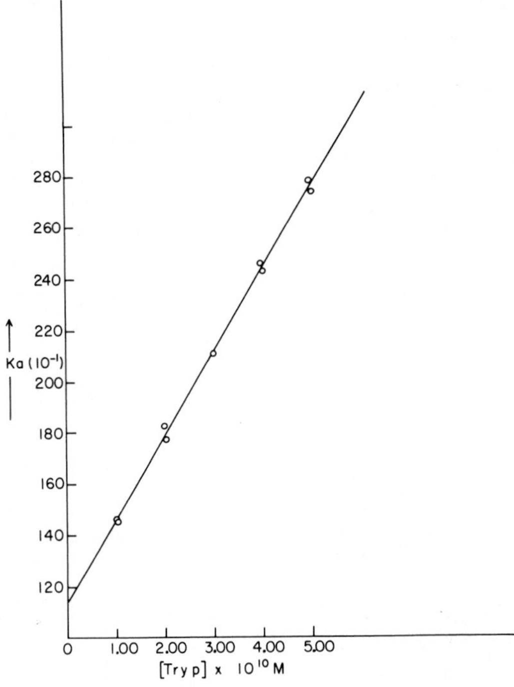

Fig. 2. The association constant between tryptophan and lumiflavin at 25°C in phosphate buffer at pH 6.96 with an ionic strength of 0.1 as a function of tryptophan concentration.

Table II summarizes the association constant of various complexing agents. The values we report for the tryptophan-lumiflavin complex are slightly higher than the values 20 M^{-1} [5], 59 M^{-1} [16] and 92 M^{-1} [18] reported for the FMN · tryptophan complex. Our value for serotonin is much smaller than the value of 400 M^{-1} that has been reported for the

TABLE II

COMPLEX FORMATION CONSTANTS FOR VARIOUS COMPOUNDS

Complexing agents	$K_1K_2 (M^{-2})$	$K_1 (M^{-1})$	$K_2 (M^{-1})$
Tryptophan	3 245	114.0	28.5
5-Hydroxyl tryptophan	8 805	201.3	43.7
Serotonin	19 160	161.7	121.2
Trp-Trp	0	283.7	0
Trp-Gly	0	131.9	0
Gly-Trp	0	115.0	0
Phenylalanine	0	0	0
Tyrosine	0	17.1	0
Histidine	0	2.52	0
Imidazole	0	0	0
Cystine	0	0	0
Nicotinic Acid	14.16	2.78	5.5
m-Tyrosine	0	20.9	0
o-Tyrosine	0	21.8	0
Tyrosinol	0	20.2	0

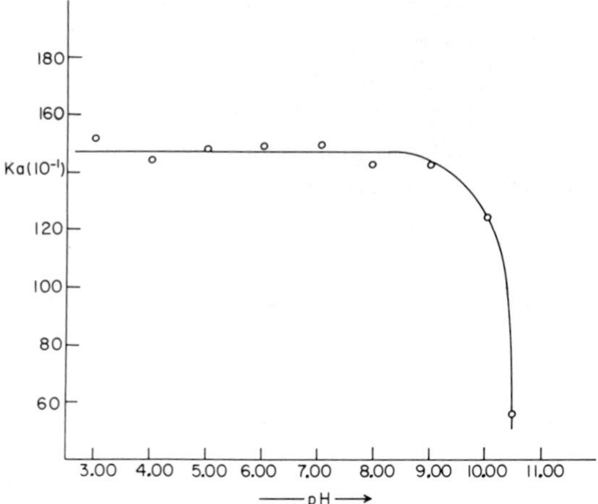

Fig. 3. The association constant for tryptophan and lumiflavin in phosphate buffer at 25°C and an ionic strength of 0.1 as a function of pH.

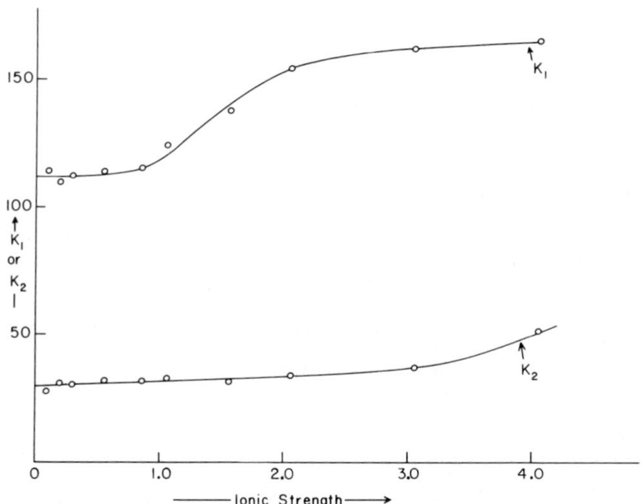

Fig. 4. The association constants for the 1:1 and 2:1 complexes of tryptophan and lumiflavin in phosphate buffer at 25°C and at pH 6.96 as a function of the ionic strength.

FMN · serotonin complex [16]. Larger values would be found if the data was not corrected for the 2:1 complex.

Indole compounds in general form much stronger complexes than other aromatic or sulfur-containing amino acids. The hydroxyl group on C-5 of the indole ring facilitates the complex formation as indicated by larger association constants for serotonin and 5-hydroxytryptophan. Therefore, the possibility of hydrogen bonding playing a role in these complex formations cannot be excluded. It is interesting that tryptophan, containing dipeptides, does not form two-to-one complexes as in the case of free tryptophan. This could be an indication of the importance of steric factors in the two-to-one complexes.

At constant tryptophan concentration, K_a is independent of pH in the range of 3–9, but decreases sharply with increasing pH around 10, as shown in Fig. 3. This agrees with earlier work by Wilson [18] where the dissociable group affecting complex formation was identified as the 3-imino group of the flavin ring (pK about 10.2).

$$
\begin{array}{ccccc}
F & + & FH_2 & \xrightleftharpoons{K_S} & 2FH\cdot \\
+ & & + & & + \\
a & & a & & a \\
K_1 \updownarrows & & K_{11} \updownarrows & & K_{111} \updownarrows \\
F\cdot a & & FH_2\cdot a & & FH\cdot a \\
+ & & & & \\
a & & & & \\
K_2 \updownarrows & & & & \\
F\cdot a_2 & & & &
\end{array}
$$

Scheme I

TABLE III

PREDICTED COMPLEX FORMATION CONSTANTS FOR REDUCED LUMIFLAVIN AND LUMIFLAVIN-SEMIQUINONE

Complexing agents	$^*K_{SA}$	$^*R$	K_{II}	K_{III}
Tryptophan (pH 8.00)	0.102	0.94	106 ± 7	120 ± 34
Tryptophan (pH 7.06)	0.099	0.89	116 ± 5	128 ± 12
Serotonin (pH 6.86)	0.069	1.33	410 ± 12	238 ± 23
Serotonin (pH 8.00)	0.127	1.30	387 ± 13	379 ± 29
Phenylalanine	0.076	0.92	0 ± 18.5	0 ± 40
Tyrosine	0.157	0.89	0 ± 50	114 ± 33
Imidazole	1.08	0.98	0 ± 0.2	22.6 ± 1.7
Cystine	0.121	1.00	15.7 ± 77.0	410 ± 224
Nicotinic acid	0.067	0.91	0 ± 18	0 ± 48

* Calculated from potentiometric titration data for FMN [5].

TABLE IV

PREDICTED COMPLEX FORMATION CONSTANTS FOR REDUCED LUMIFLAVIN AND LUMIFLAVIN-SEMIQUINONE

Complexing agents	$^*K_{SA}$	$^*R$	K_{II}	K_{III}
Tryptophan	0.239	0.73	72.6 ± 3.6	118 ± 15
Serotonin	0.085	1.48	477 ± 13	259 ± 24
Phenylalanine	0.141	0.96	0 ± 8.4	13.0 ± 12
Tyrosine	0.17	0.88	0 ± 56	66.8 ± 33
Nicotinic acid	0.094	0.94	0 ± 13	0 ± 36

* Calculated from potentiometric titration data for riboflavin [5].

The ability of tryptophan to complex with lumiflavin was improved by increasing the ionic strength of the medium. This result is in agreement with the fact that complexes are dissociated to some extent on the addition of ethanol or dioxane. Both K_1 and K_2 increased with increasing ionic strength as shown in Fig. 4 but K_1 showed a larger dependence than K_2.

Scheme I illustrates the system upon which our computer program was based. K_1 and K_2 were experimental values. K_s and other necessary equilibrium constants such as

$$K_{sA}\left(=\frac{\{[\mathrm{LfH}]+[\mathrm{LfHa}]\}^2}{\{[\mathrm{Lf}]+[\mathrm{Lf}\cdot \mathrm{a}]+[\mathrm{Lf}\cdot \mathrm{a}_2]\}\{[\mathrm{LfH}_2]+[\mathrm{LfH}_2\mathrm{a}]\}}\right) \text{ and } R\left(=\frac{[\mathrm{Lf}]}{[\mathrm{LfH}_2]}\right)$$

were obtained from previous studies [14] or potentiometric titration for riboflavin and FMN. K_{II} and K_{III} values for various complexing agents obtained from calculations are listed in Table III and Table IV. The experimental values predict considerable affinity of tryptophan and serotonin for both reduced lumiflavin and its semiquinone form. Tyrosine and cystine both show very little affinity for the oxidized as well as the reduced forms of lumiflavin but substantial affinity for lumiflavin-semiquinone. These results are of biological interest, for both amino acids occur near flavin in some flavoproteins and may function in stabilizing the semiquinone form of the flavin. Phenylalanine, imidazole, and nicotinic acid all have very little effect on any form of lumiflavin.

Discussion

This is probably the first time that a tryptophan-flavin 2:1 complex has been reported. Tollin and coworkers studied the stoichiometry of this reaction under various condictions [17, 22]. A 1:1 complex has been observed in neutral solution containing 50% ethanol. The difference is not surprising, for ethanol and other organic solvents are known to dissociate the complex [18] and therefore would hinder the formation of the 2:1 complex.

Lumiflavin, being a planar molecule, could very well form a sandwich, charge-transfer-type complex with two tryptophan molecules. However, the lack of ability for tryptophan containing dipeptides to form such complexes indicates that the two tryptophan molecules in complex must be assumed as a parallel arrangement. The side chains from both molecules are projecting (from the indole rings) in the same direction. Therefore, any bulky group (as large as a glycyl group) on the side chain will hinder the approach of a second molecule.

From the results of this study we would like to propose that tryptophan, cystine and possibly tyrosine are functioning at the active centers of flavin enzymes as agents to stabilize the semiquinone form of the flavin.

Acknowledgement

We wish to acknowledge support by the USPHS (GM 8285) and help by Stephen Sontum who wrote the computer program.

References

1 Hemmeric, P., Müller, F. and Ehrenberg, A. (1965) Oxidized and Related Redox Systems (King, T.E., Mason, H.S. and Morrison, M., eds.), Vol. I, p. 157, John Wiley and Sons, New York
2 Hemmerich, P., Veeger, C. and Wood, C.S. (1965) Angew. Chem. Int. Edn. 4, 671
3 Isenberg, I. and Szent-Gyorgyi, A. (1958) Proc. Natl. Acad. Sci. U.S. 44, 857
4 Harburg, H.A., LaNoue, K.F., Loache, P.A. and Amich, R.M. (1959) Proc. Natl. Acad. Sci. U.S. 45, 1708
5 Harburg, H.A. and Foley, K.A. (1958) Proc. Natl. Acad. Sci. U.S. 44, 662
6 Isenberg, I. and Szent-Gyorgyi, A. (1959) Proc. Natl. Acad. Sci. U.S. 45, 519
7 Theorell, H. (1960) Proceedings of the Fourth International Congress on Biochemistry (Hoffmann-Ostanhof, O., ed.), Vol. VIII, p. 167, Pergamon Press, London
8 Strittmatter, P. (1961) J. Biol. Chem. 236, 2329
9 Swinehart, J.H. and Hess, G.P. (1965) Biochim. Biophys. Acta 104, 205
10 Massey, V. and Veeger, C. (1961) Biochim. Biophys. Acta 48, 33
11 Massey, V. (1960) The Enzymes (Boyer, P.D., Hardy, H. and Myrback, K., eds.), Vol. II, p. 360, Academic Press, New York
12 Serals, R.L., Peters, J.M. and Sanadi, D.R. (1961) J. Biol. Chem. 236, 3317
13 Ryan, J. and Tollin, G. (1973) Biochemistry 12, 4550
14 Draper, R.D. and Ingraham, L.L. (1970) Arch. Biochem. Biophys. 139, 265
15 Bullock (1967) Comprehensive Biochemistry (Florkin, M. and Stotz, E.H., eds.), Vol. 22, Chap. III, Elsevier, New York
16 Isenberg, I. and Szent-Györgyi, A. (1959) Proc. Natl. Acad. Sci. U.S. 45, 1229
17 Fleishman, D.E. and Tollin, G. (1965) Proc. Natl. Acad. Sci. U.S. 53, 38
18 Wilson, J.E. (1966) Biochemistry 5, 1351
19 Barburg, H.A. and Foley, K.A. (1958) Proc. Natl. Acad. Sci. U.S. 44, 662
20 Hemmerich, P. and Fallab, S. (1958) Helv. Chim. Acta 41, 499
21 Kuhn, R. and Reinemund, K. (1934) Berichte 67, 1932
22 Pereira, J.F. and Tollin, G. (1967) Biochim. Biophys. Acta 143, 79

Flavins and Flavoproteins, edited by T.P. Singer

CHAPTER 86

^{13}C-NMR STUDY OF FLAVINS

KUNIO YAGI, NOBUKO OHISHI, AKIKO TAKAI, KEIICHI KAWANO* and YOSHIMASA KYOGOKU*

*Institute of Biochemistry, Faculty of Medicine, University of Nagoya, Nagoya 466 and *Division of Molecular Biophysics, Institute for Protein Research, Osaka University, Suita, Osaka 565 (Japan)*

Summary

Proton-decoupled, Fourier-transform NMR spectra of ^{13}C-labeled nuclei of flavin derivatives were measured using ^{13}C-enriched samples. Chemical shifts of 2-, 4-, 4a-, and 10a-^{13}C-labeled riboflavin were −159.39, −163.0, −135.5 and −151.55 ppm, respectively. There was no significant difference in chemical shifts of ^{13}C-labeled nuclei among riboflavin, FMN and FAD. Upon changing solvent from ^2H_2O to di[$^2H_6$]methylsulfoxide, the signals of 2- and 4-carbonyl carbons shifted to upfield, while those of 4a- and 10a-^{13}C-labeled nuclei to downfield. On addition of the apoprotein of egg white flavoprotein, only the signal of the 4-carbonyl carbon shifted significantly to upfield.

Introduction

To approach the detailed mechanism of flavoprotein catalysis, it is essential to reveal the participation of definite group(s) of the isoalloxazine ring in complex formation between flavin and apoenzyme, between enzyme and substrate as well as in oxidoreduction processes. For this purpose NMR seems to be promising, since this would afford valuable information on the electronic environment of each nucleus of the isoalloxazine ring.

Although several papers have appeared about proton NMR of flavins, a ^{13}C-NMR study has never been done, except those reported by Voelter et al. [1] and Breitmaier and Voelter [2] on ^{13}C-NMR of flavins of natural abundance. The present paper deals with proton-decoupled, Fourier-transform ^{13}C-NMR signals of ^{13}C-enriched flavin derivatives and the induced shifts of those of riboflavin upon complex formation with the apoprotein of egg white flavoprotein.

Experimental

Materials

To obtain ^{13}C-enriched riboflavin, [^{13}C]urea (91.4 atom%, Prochem, U.K.), [1-^{13}C]malonic acid (92.7 atom%, Prochem, U.K.) and [2-^{13}C]malonic acid (90 atom%, Merck, Canada) were used as starting materials. Fig. 1 shows the flow sheet for the syntheses of [2-^{13}C]riboflavin, [4a-^{13}C]riboflavin and [4,10a-^{13}C]-riboflavin [3]. [^{13}C]FMN was prepared by phosphorylating [^{13}C]riboflavin with $POCl_3$—H_2O, and synthesized further to [^{13}C]FAD through a fermentation procedure.

The apoprotein of egg white flavoprotein was prepared by the method of Rhodes et al. [4]. The fraction eluted at pH 4.5 from a CM-cellulose column was dialyzed to distilled water and lyophilized.

Procedures

^{13}C-NMR spectra were recorded on a JEOL PFT-100 pulse-Fourier transform NMR spectrometer combined with a EC-100 data processor (20 K). It was operated at 25.15 MHz. Pulse-free induction decay curves repeated every 2 s at each 45° pulse (pulse width; 10 μs) were Fourier transformed after 100 times accumulation and the accumulations in the frequency domain were continued until a satisfactory spectrum was obtained. In general cases, spectra ranged over 8192 points corresponded to 5000 Hz and the resolution was 1.22 Hz, i.e. 0.05 ppm.

Samples were spun in 10-mm diameter tubes. The field was locked to the D resonance of solvents. Chemical shifts were measured relative to external tetramethylsilane (TMS) in a 2 mm tube placed in the sample tube. In the experiments of the interaction with the apoprotein, ^{13}C resonance of 90%-enriched $CH_3{}^{13}COONa$ (CEA, France) in a

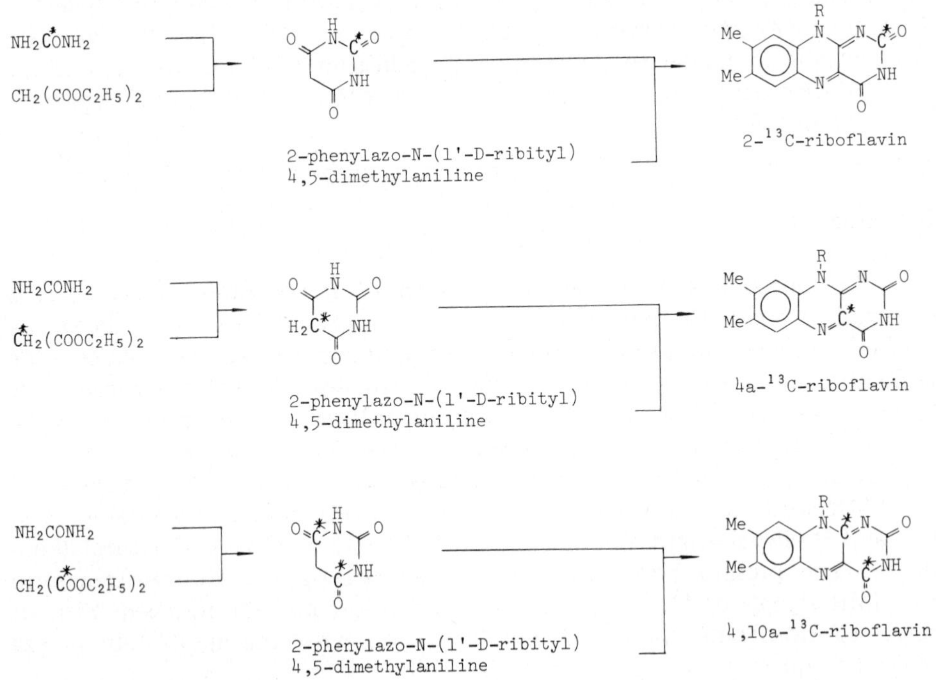

$NH_2C^*ONH_2$ + $CH_2(COOC_2H_5)_2$ → 2-phenylazo-N-(1'-D-ribityl) 4,5-dimethylaniline → 2-^{13}C-riboflavin

NH_2CONH_2 + $C^*H_2(COOC_2H_5)_2$ → 2-phenylazo-N-(1'-D-ribityl) 4,5-dimethylaniline → 4a-^{13}C-riboflavin

NH_2CONH_2 + $CH_2(C^*OOC_2H_5)_2$ → 2-phenylazo-N-(1'-D-ribityl) 4,5-dimethylaniline → 4,10a-^{13}C-riboflavin

Fig. 1. Flow sheet of synthesis of ^{13}C-enriched riboflavin. R and Me denote ribityl and methyl groups, respectively. C* represents the enriched ^{13}C.

coaxial tube was employed as the reference signal. It gives a signal at a relatively close position to the carbonyl carbon and quaternary carbon nuclei in the isoalloxazine ring. Bulk susceptibility correction was checked by measuring chemical-shift differences between the internal and external references in the protein solution.

Results and Discussion

^{13}C resonance of ^{13}C-enriched flavins

Proton-decoupled ^{13}C resonance spectra of [2-^{13}C] FMN, [4a-^{13}C] FMN, [4,10a-^{13}C] FMN and ^{13}C-natural abundant FMN are given in Fig. 2. Chemical shifts of ^{13}C resonances of several flavins enriched with ^{13}C at the 2-, 4-, 4a- and 10a-positions are listed in Table I. Breitmaier and Voelter [2] have measured the ^{13}C resonances of ^{13}C-natural abundant FMN and FAD. They tentatively assigned the signals in the FMN spectrum at −160.25, −157.55, −150.85 and −149.55 ppm to 4-, 2-, 10a- and 4a-carbons, respectively, by considering the chemical environments. Present data of ^{13}C-enriched flavins indicate clearly that their assignment of the signals to 2- and 4-carbons was correct, but that to 4a- and 10a-carbons was not accurate. The signal assigned to a 8-carbon by them is thought to be ascribed to a 4a-carbon nucleus and that assigned to a 4a-carbon by them should be due to a 10a-carbon nucleus. They thought that 4a- and 10a-carbons attached to nitrogen should be strongly deshielded and resonate in a relatively lower field. In the present experiment, however, 4a-carbon was found to resonate in a rather higher field than expected. This result is consistent with electron-density distribution calculated by Song [5]. According to his calculations, the π electron is more localized at the 4a- than at the 10a-carbon by 0.2 electron unit.

On the other hand, chemical shifts of carbonyl carbons were not parallel with electron density. Namely, 2-carbonyl carbon gives a signal in a higher field than 4-carbonyl carbon. In relation to this finding, it should be recalled that the same phenomenon was observed for 2- and 4-carbonyl carbons in uracil derivatives [6]. In such cases, neighbouring bond anisotropy seems to influence considerably the chemical shifts.

TABLE I

^{13}C RESONANCES OF ^{13}C-ENRICHED FLAVINS IN NEUTRAL ^2H_2O

	Chemical shifts (ppm)*			
	4-^{13}C	2-^{13}C	10a-^{13}C	4a-^{13}C
[2-^{13}C] riboflavin (0.9 mg/ml)		−159.4		
[2-^{13}C] FMN (20 mg/ml)		−158.85		
[2-^{13}C] FAD (20 mg/ml)		−158.8		
[4a-^{13}C] riboflavin (0.9 mg/ml)				−135.5
[4a-^{13}C] FMN (13.3 mg/ml)				−134.95
[4,10a-^{13}C] riboflavin (0.9 mg/ml)	−163.0		−151.55	
[4,10a-^{13}C] FMN (13.3 mg/ml)	−162.2		−150.95	
FMN (88 mg/ml)	−161.65	−158.65	−150.7	−135.0
FAD (100 mg/ml)	−161.4	−158.25	−150.55	−134.95

* Measured from external TMS.

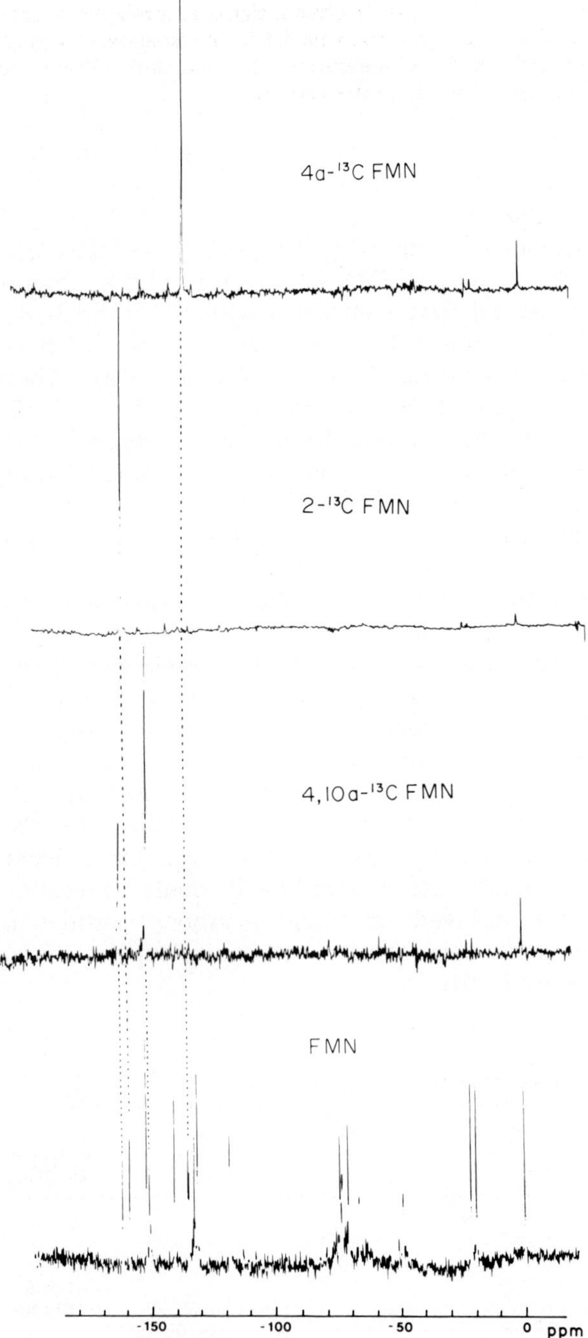

Fig. 2. Proton-decoupled ^{13}C-NMR spectra of ^{13}C-enriched FMN. [4,10a-^{13}C] FMN: (4,10a-^{13}C)-enriched FMN in $^{2}H_2O$ (13.3 mg/ml) p^{2}H 6.5, accumulated 2700 times; [4a-^{13}C] FMN: (4a-^{13}C)-enriched FMN in $^{2}H_2O$ (13.3 mg/ml) p^{2}H 6.2, accumulated 4200 times; [2-^{13}C] FMN: (2-^{13}C)-enriched FMN in $^{2}H_2O$ (20 mg/ml) p^{2}H 6.6, accumulated 5000 times; FMN: ^{13}C-natural abundant FMN sodium salt in $^{2}H_2O$ (88 mg/ml) p^{2}H 6.6, accumulated 14 600 times.

The resonance frequencies of 2-carbonyl carbons of riboflavin, FMN and FAD are almost identical; the ^{13}C resonance of a 2-carbon nucleus is not affected by the substitution of hydrogen at the 5′-position. The adenine ring of FAD is expected to induce the so-called ring-current effect and thus the ^{13}C resonances of carbon nuclei in the neighbouring isoalloxazine ring might shift to upfield. It seems, therefore, that 2-carbonyl group in FAD locates at the position which is not stacked by the adenine ring. This model is consistent with that proposed previously from the data of the proton magnetic resonances [7].

The positions of ^{13}C resonances of 2-, 4- and 10a-carbons in the ^{13}C-natural abundant FMN are slightly higher than those of ^{13}C-enriched FMN. It may be caused by the difference in concentration. At higher concentration we can expect upfield shift due to intermolecular ring-current effects.

The chemical shifts of ^{13}C-enriched riboflavin in di[$^2H_6$]methylsulfoxide ([$^2H_6$]Me_2SO) are listed in Table II. In [$^2H_6$]Me_2SO, 2- and 4-carbonyl

TABLE II

SOLVENT EFFECT ON ^{13}C RESONANCES OF RIBOFLAVIN

	Position	Chemical shifts (ppm)*		
		^2H_2O	[$^2H_6$]Me_2SO**	Δ
[2-^{13}C] riboflavin	2	−159.4	−156.05	+3.35
[4a-^{13}C] riboflavin	4a	−135.5	−137.55	−2.05
[4,10a-^{13}C] riboflavin	4	−163.0	−161.1	+1.9
	10a	−151.55	−152.0	−0.45

* Measured from external TMS.
** Measured from internal Me_2SO signal and corrected for external TMS.

TABLE III

INDUCED SHIFTS IN ^{13}C RESONANCES OF RIBOFLAVIN ON ADDITION OF THE APOPROTEIN OF EGG WHITE FLAVOPROTEIN

80 mg of the apoprotein of egg white flavoprotein were added to the solution containing 0.9 mg ^{13}C-enriched riboflavin in 1.0 ml of 0.1 M phosphate buffer, p^{2}H 7.1. Spectra were recorded over a 1250 Hz range with 8192 data points, i.e. the accuracy of the observed frequency is 0.30 Hz, and were 16 000 times accumulated.

	Position	Apoprotein addition	Chemical shifts (ppm)*
[2-^{13}C] riboflavin	2	—	23.38
		+	23.11
		Δ	−0.27 (−6.8 Hz)
[4a-^{13}C] riboflavin	4a	—	46.91
		+	46.63
		Δ	−0.28 (−7.0 Hz)
[4,10a-^{13}C] riboflavin	4	—	19.57
		+	20.79
		Δ	+1.22 (+30.7 Hz)
	10a	—	31.05
		+	30.29
		Δ	−0.76 (−19.1 Hz)

* Measured upfield from the ^{13}C=O resonance of $CH_3{}^{13}COONa$ in ^2H_2O (external).

carbons give signals in a higher field than those in ^2H_2O. It is ascribed to elimination of hydrogen bonding formed between water molecules and the carbonyl group. On the other hand, the signals of 4a- and 10a-carbon nuclei shift to downfield in $[^2H_6]Me_2SO$. The shift might be due to interaction of the solvent with neighbouring nitrogen or carbonyl group. Otherwise, it might be due to elimination of the ring-current effect of intermolecular interaction, which was thought to be more predominant in aqueous than in $[^2H_6]Me_2SO$ solutions.

Induced shifts of ^{13}C resonances on addition of the apoprotein of egg white flavoprotein

The shifts which were induced on addition of the apoprotein of egg white flavoprotein are given in Table III. The values in the table are apparent shifts measured from external $CH_3{}^{13}COONa$. Bulk susceptibility of the solution which contains a high concentration of the protein may differ from that of pure riboflavin solution. Correction for the change in bulk susceptibility is estimated to be −3.4 Hz, which is the shift difference between external and internal references measured for the solution containing the same amount of the protein. Thus, the apparent shifts observed for 2- and 4a-carbons on addition of the apoprotein are in the order of magnitude which can be ascribed to the change in bulk susceptibility of the solution. On the other hand, the signal of the 4-carbonyl carbon obviously shifted to upfield. Moreover, the relative intensity of the 4-carbon signal increased when the apoprotein was added to the solution. This increase is similar to that in the intensity of the 4-carbon signal on changing solvent; the intensity of the 4-carbon signal is quite weak relative to the 10a-carbon signal in ^2H_2O, but it becomes stronger in $[^2H_6]Me_2SO$. From these data, it seems that riboflavin interacts with some part of the apoprotein at or around the 4-carbonyl group of the isoalloxazine ring. Since the association constant of riboflavin with the apoprotein of egg white flavoprotein was determined to be $5 \cdot 10^8$ M^{-1} by fluorometry [8], about 93% of the added riboflavin forms a complex with the apoprotein at this concentration. The net limiting shift due to the complex formation can be calculated to be about +1.45 ppm for the 4-carbonyl carbon. Although the origin of the upfield shift is not clear, electric field effects, van der Waals interactions and magnetic anisotropy effects (except hydrogen bonding interaction) can all contribute to it [9]. The above observation should be explained in correlation with our view that the molecular interaction between the isoalloxazine ring and the apoprotein of egg white flavoprotein is of a hydrophobic nature [10].

References

1 Voelter, W., Jung, G., Breitmaier, E. and Price, R. (1971) Hoppe-Seyler's Z. Physiol. Chem. 352, 1034–1036

2 Breitmaier, E. and Voelter, W. (1972) Eur. J. Biochem. 31, 234–238

3 Tishler, M., Peister, K., Babson, R.D., Ladenburg, K. and Fleming, A.J. (1947) J. Am. Chem. Soc. 69, 1487–1492

4 Rhodes, M.B., Bennet, N. and Feeney, R.T. (1959) J. Biol. Chem. 234, 2054–2060

5 Song, P.S. (1969) Ann. N.Y. Acad. Sci. 158, 410–423
6 Tarpley, Jr., A.R. and Goldstein, J.H. (1971) J. Am. Chem. Soc. 93, 3573–3578
7 Kainosho, M. and Kyogoku, Y. (1972) Biochemistry 11, 741–752
8 Nishikimi, M. and Kyogoku, Y. (1973) J. Biochem. Tokyo 73, 1233–1242
9 Stothers, J.B. (1972) in Carbon-13 NMR Spectroscopy, pp. 493–506, Academic Press, New York
10 Nishikimi, M. and Yagi, K. (1969) J. Biochem. Tokyo 66, 427–429

Flavins and Flavoproteins, edited by T.P. Singer

CHAPTER 87

OPTICALLY DETECTED MAGNETIC RESONANCE IN THE FAD AND GLUCOSE OXIDASE TRIPLET STATES

THOMAS A. MOORE and ALVIN L. KWIRAM

Department of Chemistry, University of Washington, Seattle, Wash. 98195 (U.S.A.)

Introduction

Optical detection of magnetic resonance (ODMR) [1, 2] can be enlisted in biochemical studies to probe chemical and structural features of complex biomolecules. ODMR relies on the lowest excited triplet state of either an intrinsic or an extrinsic "reporter group" in the macromolecule. The use of ODMR in investigations of flavins and flavoproteins is particularly appropriate since the flavin chromophore is ipso facto at the active site and is involved in the mechanism of flavoenzyme-catalyzed reactions. Moreover, the flavin triplet state in flavoproteins is involved in a variety of photoprocesses including the photoreduction reaction with EDTA. A detailed understanding of these processes depends on a careful measurement of the photophysical parameters characterizing the excited state.

Apart from its use in photoreduction, the flavin-excited triplet state in flavoprotein has received relatively little attention. We have previously reported phosphorescence and ODMR from the lowest triplet state of flavins, flavodoxin (*Saccharomyces lividus*) and yeast hemoglobin (*Candida mycodema*) [3]. With this work we extend our studies of flavoprotein triplet states to glucose oxidase (*Aspergillus niger*) and herein report the conditions under which luminescence may be observed from glucose oxidase from both the flavin and tryptophan moieties.

Experimental

Glucose oxidase (Sigma Chemical Company type II from *A. niger*) was purified following the method of Swoboda and Massey [4] with slight modification. A solution of the enzyme in 0.05 M phosphate buffer pH 6.00 was applied to an ion-exchange column (Whatman DE 22) and 2 l of the same buffer were flushed through the column. The enzyme was eluted with a linear gradient (0.05 and 0.30 M phosphate buffer, 1.0 l total) and five fractions of the yellow protein collected. Although room temperature fluorescence could be detected from each fraction, the first fraction showed the lowest fluorescence quantum yield, 0.5% that of free FMN (the fluorescence quantum yield increased to 2% that of FMN for fraction 5). 100 mg of protein from fraction 2 gave a single band approx. 2 mm wide on disc-gel electrophoresis. Fractions 1, 2, and 3 had the ratio

A_{280}/A_{450} = 10.4, the ratio increasing to approx. 12 in fraction 4. In the experiments reported herein, fraction 1 was used.

FAD (Sigma Chemical Company, grade III) was purified by exploiting the differences in retention by an Amicon UM10 ultrafilter between FAD and contaminating flavins. FMN and FAD were found to have 50 and 75% retentions, respectively. By washing 18 vol. of buffer through the ultrafilter, the FMN is reduced to $\approx$0.1% of original [5]. FAD purified in this way has a room temperature fluorescence quantum yield 0.1 that of free FMN.

The ODMR and luminescence spectroscopy were carried out as described previously [3].

Results and Discussion

As has been previously reported [4, 6–9], the FAD in purified glucose oxidase, pH 6 aqueous solution, was found to be essentially nonfluorescent. However, upon addition of ethylene glycol (1:1 v/v) to form an ethyleneglycol/water mixture in which to carry out low temperature spectroscopy it is observed that the fluorescence quantum yield increases to approx. 1% that of free FMN. As shown in Fig. 1, strong ODMR signals are observed from the flavin triplet in glucose oxidase at 1.7°K in an ethyleneglycol/water mixture (water pH 6, 0.01 M phosphate buffer) glass. The ODMR signal strength maximizes when the phosphorescence is observed at approx. 600 nm and shows an origin at 570 nm. As shown in Fig. 2, this is consistent with the observed flavin phosphorescence spectrum of glucose oxidase in the ethyleneglycol/water mixture glass. The zero-field transitions occur at 2.31 and 1.17 GHz, somewhat shifted from those of free FAD in ethyleneglycol/water mixture glass (2.28 and 1.16 GHz) but retaining the characteristic flavin

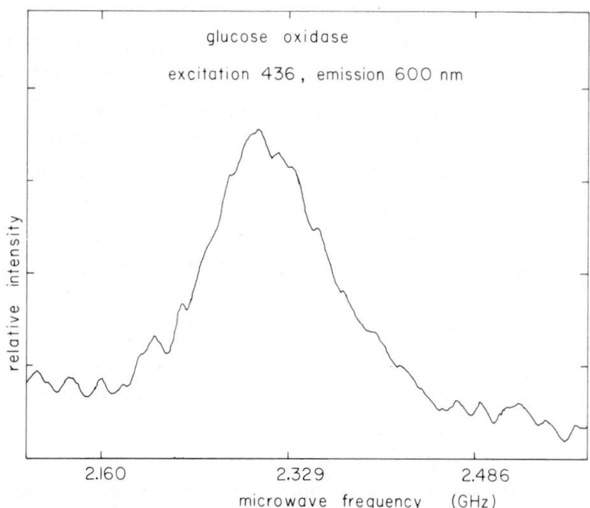

Fig. 1. The D + E ODMR signal from the FAD prosthetic group of glucose oxidase in ethyleneglycol/water at 1.4°K, 0.5 mg•ml^{-1}. ODMR spectra taken using the delay technique described in ref. 3. Signal averaged over 10 scans, 5-s scan^{-1}; excitation and emission band widths were 10 nm.

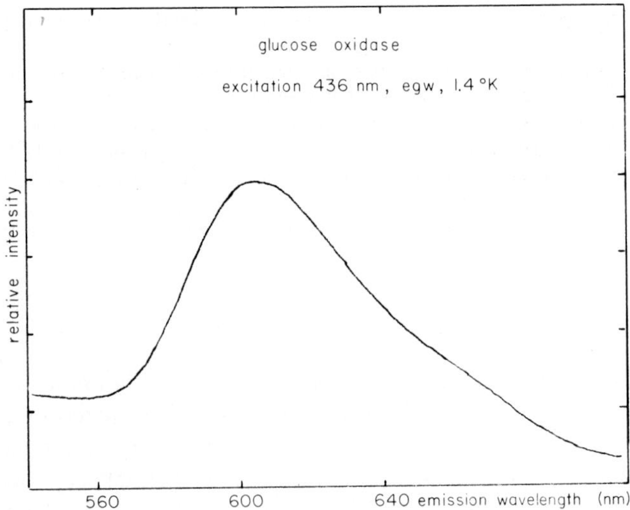

Fig. 2. Phosphorescence spectrum of glucose oxidase in ethyleneglycol/water (EGW) at 1.4°K, measured with a rotating-can phosphoroscope and an EMI 9558 PMT. The long wavelength portion of the phosphorescence emission from tryptophan is seen at the left of the spectrum (excited by the stray and second order light from the 500 W mercury source).

fingerprint. These data as well as the relative luminescence yields and zero-field transitions are summarized in Table I.

As can be seen from Table I, the fluorescence quantum yield increases substantially in FAD upon addition of ethyleneglycol. This is consistent with the work of others and is attributed to an unstacking of the flavin adenine complex [10, 11]. Concomitant with the increase in fluorescence, we observe an increase in the phosphorescence intensity as well as a blue shift in the emission maximum upon addition of ethyleneglycol*.

The zero-field splittings in the phosphorescent triplet state also change in going from aqueous to ethyleneglycol/water solutions and we may attribute this change in large part to the unstacking. It is interesting to note that the flavin D-E (or 2E, it is not possible to assign these transitions without the principal axis assignment) transition is shifted by approximately 30 MHz while the D + E transition remains unchanged. As more observations are made of the specificity of such interactions, it may be possible to make structural inferences about flavin-adenine interactions from measurements in the flavoproteins.

The appearance of fluorescence and phosphorescence from glucose oxidase upon addition of ethylene glycol is an unexpected result which warrants further investigation. Several lines of evidence suggest that the release of

---

* The measurement and comparison of phosphorescence intensities between aqueous snow samples is difficult and should be considered only qualitative. We find that both FMN and FAD have emission maxima at approximately 664 nm in aqueous snow. Although flavin phosphorescence maxima are approximately 600–610 nm in organic solvents or an ethyleneglycol/water mixture, Song and Metzler (Song, P.S. and Metzler, D.E. (1967) Photochem. Photobiol. 6, 691) have reported riboflavin phosphorescence at 665 nm in a 0.8% glucose in water glass.

TABLE I

	Zero-field microwave transitions (GHz)[c]	Fluorescence (298°K) maxima and intensity relative to FMN[a,g]	Phosphorescence (77°K) maxima and intensity relative to FMN[a] in ethyleneglycol/water
Glucose oxidase[d] (aqueous)[b]	not observed	540 nm, 0.5%	not observed
Glucose oxidase (ethyleneglycol/water)	2.31, 1.17	534 nm, 1.1%	607 nm, 1.3%
FAD (aqueous)	2.28, 1.13	531 nm, 10%	664 nm, 60%[e]
FAD (ethyleneglycol/water)	2.28, 1.16	530 nm, 62%	611 nm, 81%[f]

[a] Fluorescence and phosphorescence spectra measured with an RCA 31034 photomultiplier and are uncorrected. Only that luminescence which can be attributed to the flavin moiety is reported. Excitation was at 436 nm with a 500 W mercury lamp.

[b] The aqueous solutions are 0.1 M phosphate buffer at pH 6. Ethyleneglycol/water is a 1:1 (v/v) mixture of the above aqueous solution with ethyleneglycol.

[c] The zero-field microwave transitions were determined with the phosphorescence emission measured at 600 nm.

[d] The glucose oxidase was from fraction 1 (see experimental section).

[e] Intensity is relative to that of FMN in an aqueous show.

[f] Intensity is relative to that of FAD in an ethyleneglycol/water glass.

[g] Within experimental error, the fluorescence intensity of FMN was the same in aqueous and ethyleneglycol/water solutions.

FAD from the enzyme is not occurring. No FAD is detected in the eluent upon washing the enzyme with several volumes through a UM20E ultrafilter. The fluorescence maxima of the glucose oxidase in ethyleneglycol/water is not the same as that of free FAD under the same conditions. The zero-field transitions, particularly the D + E transition, are substantially shifted (Table I) indicating that the flavin moiety is not free from the protein. Although a detailed kinetic study has not been carried out, it appears from the disappearance of the 453 nm absorption of the oxidized enzyme upon reduction with excess glucose that the activity is little different from pure aqueous solutions. The complete anaerobic reduction of the enzyme by excess glucose occurs essentially upon mixing.

In our limited experience with other flavoproteins [3], we have not observed this dramatic effect with ethyleneglycol and we suspect that this behavior is not general. An interesting possibility is that the ethyleneglycol is binding weakly to a part of the binding site for glucose and the resulting conformation change in the protein renders the flavin moeity luminescent. Further speculation must await a detailed kinetic analysis and further spectroscopic studies.

It has been suggested [12] that a tryptophan residue may be located at or near the active site in glucose oxidase. Upon excitation at 290 nm and with emission at 450 nm we observe ODMR signals with the characteristic splittings of tryptophan. Free tryptophan in ethyleneglycol/water glass has transitions at 1.82, 2.60 and 4.42 GHz [14]. After reduction of the glucose oxidase with excess glucose we again observed the phosphorescent tryptophan triplet state and ODMR signals. Fig. 3 presents the results of these measurements and a schematic diagram of the zero-field splittings in the oxidized and reduced enzyme. It can be seen that a substantial change has occurred in the micro-environment of the tryptophan that we are observing. This change is consistent with the location of the tryptophan at or near the flavin site. Again, the perturbation of the triplet sublevels has a distinct pattern, with the D-E and 2E transitions shifted in a complementary way such that the D + E transition is not changed. As our experience with model studies and theoretical approaches increases we expect to be able to deduce structural features from such experiments.

The triplet states of fully reduced free flavins and flavoproteins have not been observed and characterized. We have prepared fully reduced riboflavin by dithionite reduction and photochemically with EDTA as the donor. While we are able to observe phosphorescence from these samples, we have not been able to exclude impurities and we have no well-defined phosphorescence spectrum. We are extending these attempts to fully reduced flavoproteins, especially those that Massey and coworkers [9] have found to be fluorescent.

The ODMR technique offers considerable advantage in dealing with weakly phosphorescent systems. The discrimination afforded by the requirement that the resonance condition in both the optical (excitation and emission) and microwave domains be simultaneously satisfied allows one to

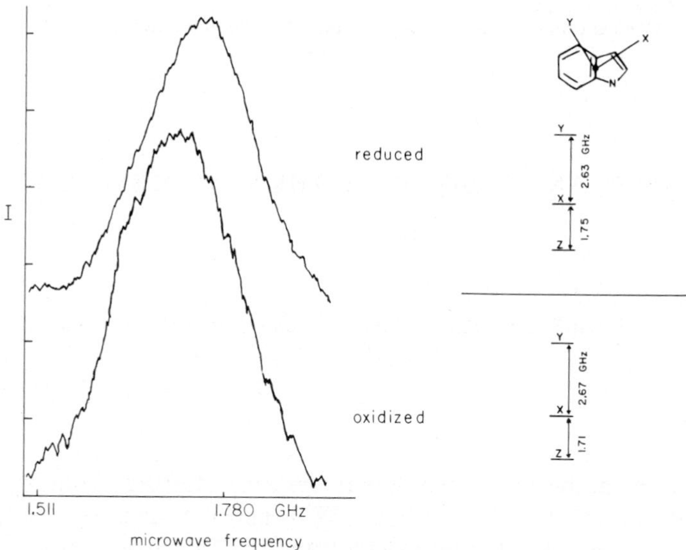

Fig. 3. Tryptophan ODMR signals in both oxidized and fully reduced glucose oxidase. In both cases the three zero-field transitions were observed using the delay ODMR technique described in ref. 3. These spectra were taken in ethyleneglycol/water glass at 1.4°K and are signal averaged over 5 scans, 25 s $scan^{-1}$. Excitation was at 290 nm, 5 nm bandwidth, from a 500 W mercury lamp and phosphorescence was detected at 450 nm, 10 nm bandwidth. The principal magnetic axes assignment for indole is from ref. 13.

measure the spectrum of interest in the presence of other extraneous luminescent sources. As we have shown, the zero-field frequencies are a sensitive function of molecular structure and constitute a fingerprint for identifying various species. Our work will extend to the freeze-quench techniques in which we will attempt to trap intermediates in flavoprotein catalyzed reactions and identify them by their characteristic microwave fingerprints. The sensitivity of the ODMR technique extends to the subnanomolar range (depending on the quantum yields of the various photophysical processes) and the resolution is limited by the inhomogeneous linewidths of approximately 100 MHz encountered in these systems.

References

1 Kwiram, A.L. (1967) Chem. Phys. Lett. 1, 272
2 Kwiram, A.L. (1972) Int. Rev. Sci. Phys. Chem. Ser. One 4, p. 271, Med. Tech. Publ. Co.
3 Moore, T.A. and Kwiram, A.L. (1974) Biochemistry 13, 5403
4 Swoboda, B.E.P. and Massey, V. (1965) J. Biol. Chem. 240, 2209
5 Amicon Buyer's Guide, publication No. 426, Amicon, 21 Hartwell Ave., Lexington, Mass. 02173
6 Swoboda, B.E.P. (1969) Biochim. Biophys. Acta 175, 365–379
7 Swoboda, B.E.P. (1969) Biochim. Biophys. Acta 175, 380–387
8 Tsuge, H. and Mitsuda, H. (1971) J. Vitaminol. 17, 24
9 Ghisla, S., Massey, V., Lhoste, J.-M. and Mayhew, S.G. (1974) Biochemistry 13, 589
10 Weber, G. (1950) Biochem. J. 47, 114
11 Gascoigne, I.M. and Radda, G.K. (1965) Chem. Commun. 21, 533
12 Tsuge, H. and Mitsuda, H. (1973) J. Biochem. Tokyo 73, 199
13 Zuclich, J., von Schütz, J.U. and Maki, A.H. (1974) Mol. Phys. 28, 33
14 Rousslang, K.W. and Kwiram, A.L. Chem. Phys. Lett. in the press

Flavins and Flavoproteins, edited by T.P. Singer

CHAPTER 88

ELECTRON INPUT AT THE N-1 POSITION OF THE FLAVIN SYSTEM

WOLFGANG-R. KNAPPE

Fachbereich Biologie der Universität Konstanz, 7750 Konstanz, Postfach 7733 (G.F.R.)

Introduction

So far, the redox activity of the flavin system, starting from the oxidized state, has been confined to the C-4a-, N-5- and C-8 positions [1], the last one being established only for photoreactions [2]. A participation of the N-1 position has been postulated in the intramolecular photodegradation of riboflavin [3], whereas intermolecular photoreactions involve only the C-4a-, N-5- and C-8 positions cited above to be redox active.

Therefore, we tried to force an electron input at the N-1 position, which should be just as possible as at the N-5 position at the other end of the conjugated N-5=C-4a-C-10a=N-1 group. To do this, the photochemical investigation of suitably N-10-substituted isoalloxazine derivatives was chosen (all reactions listed below also proceed with flavins (= 7,8-dimethyl-isoalloxazines)). We hoped that the proximity of the N-1 position to the N-10 substituents would favour this reaction, compared to the other positions, because of higher distances. One of the problems was whether photochemistry started from the first excited singlet state. Several groups have claimed the involvement, or partial involvement, of this state in the photodegradation of riboflavin to lumichrome [4–7]. The photochemistry of this state, due to its short lifetime of 4.5 ns [8], compared to about 75 μs for the triplet lifetime [9], should be restricted to intramolecular reactions or to reactions with the solvent molecules. All other intermolecular photoreactions of flavin derivatives, known up to now, involve the first excited triplet state [10, 11].

Results

Due to complexity of the ribityl side chain in riboflavin, some model compounds were investigated for their photodegradation. Karrer and his group have already studied simplified hydroxyalkyl side chains [12]. Yang and McCormick have investigated ω-hydroxyalkyl substituents [13]. These investigations were extended to unsubstituted alkyl side chains, and it was discovered that these groups (except methyl, ethyl and neopentyl) were

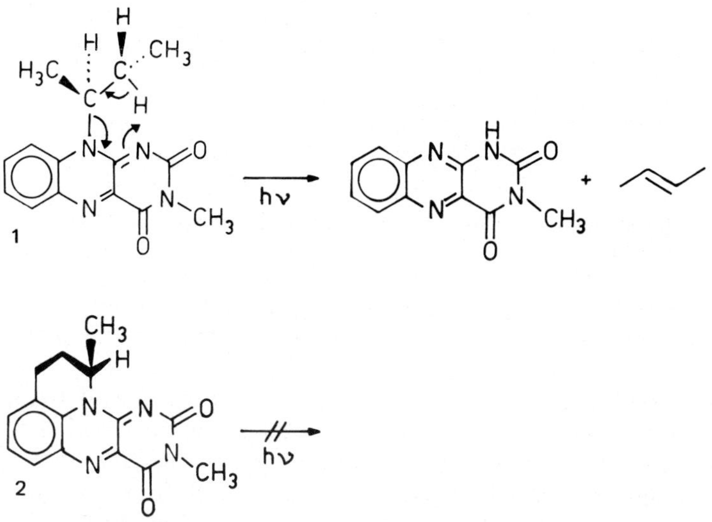

Fig. 1. Stereochemistry of the intramolecular photodealkylation of 10-alkyl substituents.

split off, yielding alloxazine and an olefin [14]. If the 2′-hydrogen atoms were substituted by deuterium, a kinetic isotope effect of 2.4 was measured.

The quantum yields in these dealkylations decreased in the following sequence: *sec*-butyl = cyclopropyl ≧ cyclobutyl > cyclohexyl ≫ *n*-propyl > *n*-butyl, confirming Song and Kurtin's proposal [3] that the transition state should have a plane six-membered ring of the N-1—C-10a—N-10—C-1′—C-2′—H-2′ group as shown in Fig. 1 for the 10-*sec*-butylisoalloxazine 1. The quinolino(8,1-*gh*)pteridin-8,10-dione derivative 2, in which this plane conformation is of much higher energy than the ground state, consequently did not show this dealkylation.

The anaerobic photodealkylation of the *sec*-butyl derivative 1 in water proceeded with a quantum yield of 0.67 and was quenched to its half value by $4 \cdot 10^{-6}$M iodide, the products were 3-methylalloxazine and *trans*-2-butene. Under aerobic conditions this photodealkylation yields the same products, but, since the triplet was quenched by the present oxygen, the quantum yield was only 0.07. The iodide half-quenching concentration for this aerobic process was exactly the same as for the fluorescence, $2 \cdot 10^{-2}$ M. From these facts it can be concluded that the reaction proceeds just as well from the first excited singlet as from the first excited triplet state in a ratio of 1:10.

An analogous dealkylation reaction with the same singlet to triplet ratio was found for isoalloxazine-10-β-propanoic and 10-β-butanoic acid 3 in the pH range > 6 [15]. At lower pH values a faster reaction proceeds with a maximum of the quantum yield at pH 4. An intramolecular photodecarboxylation and cyclisation towards the N-1 position yields the 1,10-bridged 1,5-dihydrosystem 4 and CO_2.

This pure triplet reaction proceeds also via a plane transition state.

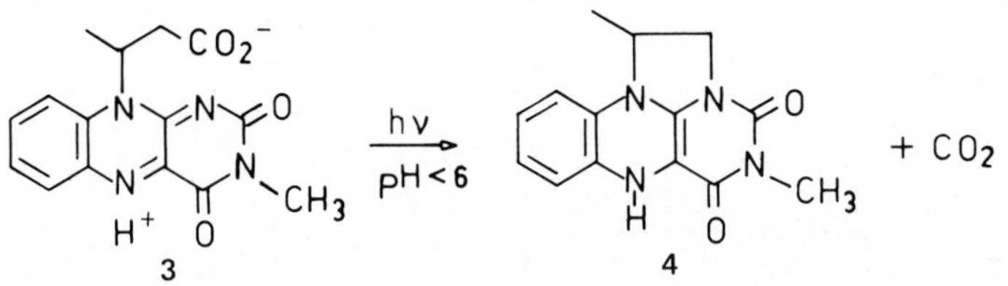

Fig. 2. Intramolecular reductive photoalkylation and decarboxylation of isoalloxazine-10-β-butanoic acid at pH < 6.

Introduction of a 1′-methyl group results in a higher quantum since the steric interaction of the 1′-CH_3 with the C-9-H atom favours the plane geometry.

When 10-phenylisoalloxazine 5 was irradiated in the presence of weak organic acids in organic solvents, a new type of photoreaction occurred. The phenyl substituent cyclizes to the N-1 position yielding the N-1, N-10-bridged 1,5-dihydroisoalloxazine derivative 6 (λ_{max} 438 and 372 nm, $\epsilon = 4100$ and 13 800, respectively) [16]. This photocyclisation showed no kinetic isotope effect if the 2- and 6-hydrogen atoms of the phenyl substituent were substituted by deuterium. The iodide quenching of this photocyclisation runs parallel to the iodide quenching of the fluorescence. This mechanism can therefore be described as an electrophilic substitution of the electron-deficient N-1 atom at the N-10-phenyl substituent, starting from the first excited singlet state. Compound 6 reacts with oxygen, yielding a new species with a long wavelength maximum at about 420 nm, to which the structure of the hydroxyl compound 7 can be ascribed; a hydroperoxide is improbably, since no hydrogen peroxide is formed during the oxidation. Several isomeric structures are possible. The hydroxyl group may enter at the 6-, 8-, 9a- or 10a- but not at the 4a position. The 10a- position may be excluded because the tetrahedral C-10a would distort the plane tetracyclic system.

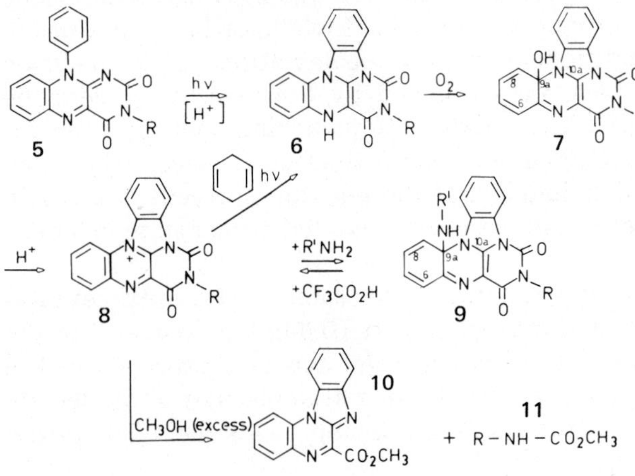

Fig. 3. Photochemical synthesis of the N-1, N-10-benzobridged isoaloxazine cation 8 and attack of nucleophiles. The species 7 and 9 may be present as 6-, 8- or 9a-isomers (cf. text).

On the addition of trifluoroacetic acid, the cation 8 is formed, which structure has been elucidated by the reaction with methanol: The N-1—C-2-bond is solvolyzed first and the final products, which were obtained in a quantitative yield, were the two methyl esters 10 and 11.

The cations 8 (λ_{max} 392 nm, $\epsilon = 12\,000$), which can be photoreduced to the starting compound 6, adds nucleophiles, e.g. amines, yielding a new species 9, which exhibits the same ultraviolet spectrum as the "hydroxyl adduct" 7. This addition is reversible on addition of trifluoroacetic acid. This nucleophilic attack must therefore occur at the same positions, which are involved in the oxidation of 6 yielding 7. ^{1}H-NMR experiments will distinguish between the 6- or 8-adduct and the 9a-adduct.

Conclusions

1. The N-1 position of the flavin nucleus can be forced to participate in (photochemical) redox processes. Appropriate N-10 substituents, e.g. phenyl or carboxyethyl, enforce an electron input at the N-1 position in the excited states. The other redox-active sites in these compounds are too far away. But as soon as there is a competition between the N-5 or C-4a positions and the N-1 position, the input will occur exclusively at N-5 or C-4a. If, for example, longer carboxyalkyl side chains than in 3 are investigated, the intramolecular photocyclisation seems to proceed to the N-5 or C-4a position but not to the N-1 position [15].

2. The first excited singlet state of the flavin system is photochemically active, all these reactions involve the N-1 position. The photocyclisation of 5 proceeds strictly from this state. In the case of the photodealkylation of 10-alkyl or 10-carboxyalkyl substituents (at pH > 6), where both the first excited singlet and triplet state are involved, the ratio is 1:10. The photodealkylation is not a redox reaction but proceeds presumably on a concerted reaction pathway with a plane six-membered transition state and not via a radical mechanism as the analogous well-known Norrish type II γ-hydrogen abstraction of ketones. This is supported by the fact that only *trans*-2-butene is formed without reduction of the flavin system, which should be expected to be a side reaction of a radical process.

Another example of the involvement of the first excited singlet state has been found by Schumann-Jorns et al. [17] in the phosphate-catalyzed intramolecular photocyclisation or riboflavin, where the 2′-OH group is attacked by the C-9 position.

3. The N-1, N-10-bridged dihydroflavin derivative 6 presents a new model system, in which the structure of the oxygen adduct of dihydroflavin can be examined, which for years has been and will continue to be a point of controversy (refs. 18–20 and Massey, V. and Hemmerich, P., unpublished). The structure of the species 9, formed by an attack of a nucleophile on the cation 8, will show whether the postulated flavin hydroperoxide will have the structure of an 6-, 8-, 9a- or 10a adduct. Similar model reactions have already been performed by Müller [21] who showed a preferential attack of nucleophiles at the 9a position, starting from 1,10-dialkylated flavin cations.

Acknowledgement

This work was supported financially by the Deutsche Forschungsgemeinschaft.

References

1 Hemmerich, P. and Schumann-Jorns, M. (1973) in Enzymes: Structure and Function (Veeger, C., Drenth, J. and Oosterbaan, R.A., eds.), FEBS Symposia, Vol. 29, pp. 95–117, North Holland, Amsterdam
2 Brüstlein, M., Knappe, W.-R. and Hemmerich, P. (1971) Angew. Chem. 83, 854–856; Angew. Chem., Int. Ed. Engl. 10, 804–806
3 Song, P.-S. and Kurtin, W.E. (1969) Mol. Photochem. 1, 1–10
4 Song, P.-S. (1971) in Flavin and Flavoproteins (Kamin, H., ed.), pp. 37–51, University Park Press, Baltimore
5 Yang, C.S. and McCormick, D.B. (1965) J. Am. Chem. Soc. 87, 5763–5765
6 Song, P.-S. and Metzler, D.E. (1967) Photochem. Photobiol. 6, 691–709
7 Owen, S.D. and O'Boyle, A.A. (1971) Photochem. Photobiol. 14, 683–692
8 Gordon-Walker, A., Penzer, G.R. and Radda, G.K. (1970) Eur. J. Biochem. 13, 313–321
9 Schreiner, S. (1974) Thesis, University of Stuttgart
10 Penzer, G.R. and Radda, G.K. (1971) in Vitamins and Coenzymes (McCormick, D.B. and Wright, L.D., eds.), Vol. 18B, pp. 479–495, Academic Press, New York
11 Knappe, W.-R. (1971) Thesis, University of Konstanz
12 Karrer, P., Schlittler, E., Pfaehler, K. and Benz, F. (1934) Helv. Chim. Acta 17, 1516–1523
13 Yang, C.S. and McCormick, D.B. (1965) J. Am. Chem. Soc. 87, 5763–5765
14 Gladys, M. and Knappe, W.-R. (1974) Chem. Ber. 107, 3658–3673
15 Knappe, W.-R. (1975) Chem. Ber. 108, 2422–2438
16 Knappe, W.-R. (1974) Chem. Ber. 107, 1614–1636
17 Schumann-Jorns, M., Schöllnhammer, G. and Hemmerich, P. (1975) Eur. J. Biochem. 57, 35–48
18 Mager, H.I.X. and Berends, W. (1972) Rec. Trav. Chim. Pays-Bas 91, 611–629 and references cited therein
19 Hemmerich, P. and Müller, F. (1973) Ann. N. Y. Acad. Sci. 212, 13–26 and references cited therein
20 Hemmerich, P. and Wessiak, A. (1976) Flavins and Flavoproteins, Proceedings of the 5th International Symposium (Singer, T.P., ed.), Chapt. 1, pp. 9–22, Elsevier, Amsterdam
21 Müller, F. (1976) Flavins and Flavoproteins, Proceedings of the 5th International Symposium (Singer, T.P., ed.), Chapt. 3, pp. 38–50, Elsevier, Amsterdam

Flavins and Flavoproteins, edited by T.P. Singer

CHAPTER 89

INFLUENCE OF pH ON FLAVINS IN THE TRIPLET STATE

SIEGFRIED SCHREINER and HORST E.A. KRAMER

Institut fuer Physikalische Chemie der Universitaet Stuttgart, D-7000 Stuttgart 80, Pfaffenwaldring 55 (G.F.R.)

Summary

The triplet–triplet absorbance spectra of lumiflavin (Lf) were measured by flash photolysis in aqueous solution at pH 2, 7 and 13. It has been shown that these spectra are due to the acid ($^3LfH_2^+$), the neutral (^3LfH) and the basic ($^3Lf^-$) forms of the flavin triplet.

The transient spectrum of 10-methylisoalloxazine was observed in the near ultraviolet and a longer wavelength region. Nearly the same characteristic change in absorbance was found for the different protolytic species of lumiflavin and 10-methylisoalloxazine transients. These results do not support the existence of an excited flavin dimer or tautomer.

Phosphorescence of FMN in strong acids in low temperature glasses (where the proton may be considered as firmly held at N-1 as in the ground state) is blue shifted relative to the neutral flavin phosphorescence. As follows from the Foerster cycle, the pK_a of the N-1-protonated triplet should therefore be lower than that of the ground state (p$K_G \approx 0$) which is in contrary to our results: lumiflavin p$K_T = 4.45 \pm 0.05$; p$K_T = 5.20 \pm 0.05$.

This observation confirms the conclusion that the protonation in the triplet state does not occur at the same site as in the ground state. The pH dependence of the flavin photoreduction may be explained by the higher reactivity of the acid triplet form.

Introduction

Photoreactions of flavins show a considerable change in reaction rate between pH 4 and 7 [1–4]. This pH dependence might be explained by the existence of a protolytic equilibrium in a photoexcited state, e.g. the triplet state. However, only an odd pH dependence of the transient spectrum was observed after flashing 3-methyllumiflavin [5]. In a further study [6], van Voorst and co-workers suggested, based on laser experiments, a metastable excited dimer or tautomer form (see Fig. 5) of 3-methyllumiflavin which might cause the transient absorbance. In contrast to it, no similar phenomena have been observed in the case of lumiflavin. Within experimental precision the transient absorbance was found to be due to the flavin triplet state [7]

Only a flavin methylated in the 8 position is capable of forming a dimer or tautomer as suggested by van Voorst and co-workers [6]. Hence, the transient spectra of lumiflavin (7,8,10-trimethylisoalloxazine) and 10-methylisoalloxazine should be different, if van Voorst's suggestions were correct.

Materials and methods

Lumiflavin was prepared by photolysis of an alkaline aqueous solution of riboflavin [8]. FMN was purchased from Merck-Schuchardt and 10-methylisoalloxazine was generously supplied by Professor P. Hemmerich, Konstanz. The methods were the same as already described [7].

Results

Transient spectra

Fig. 1 shows the triplet–triplet absorbance spectra of lumiflavin at pH 2, 7 and 13 calculated from the change in absorbance and the triplet extinction coefficients. The results at pH 7 are in good agreement with those of Knowles and Roe [8]. The absorbance of the acid triplet at 370 nm is smaller than that of the neutral triplet while in the region of longer wavelengths only a small shift of the absorbance maximum is found.

A comparison between the triplet absorbance of 10-methylisoalloxazine

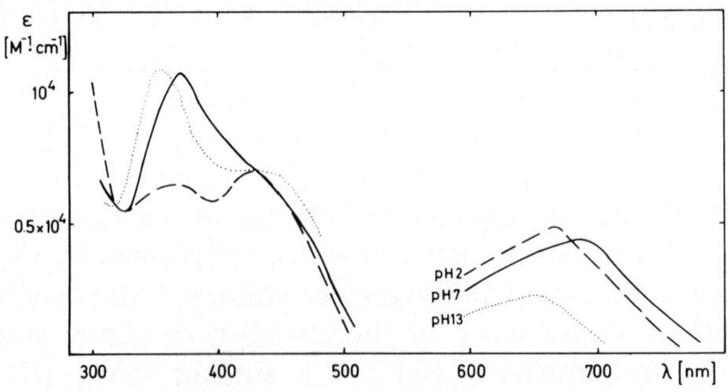

Fig. 1. Triplet–triplet absorbance spectra of lumiflavin in aqueous solution at 8 μs after flashing: – – – pH 2, acid triplet form; —— pH 7, neutral triplet form; · · · · pH 13, basic triplet form; lumiflavin 10 μM in oxygen free aqueous buffered solution.

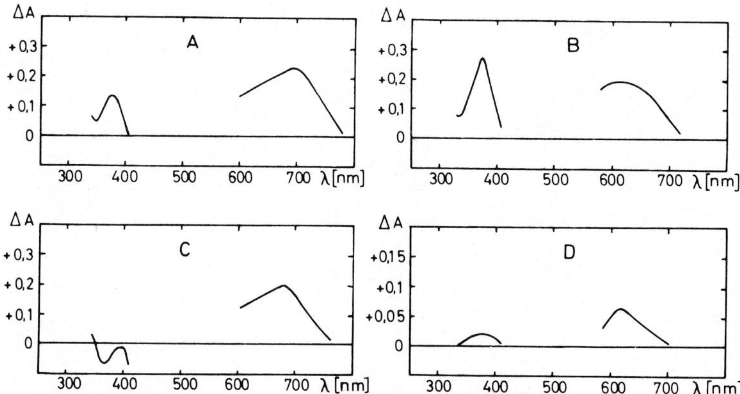

Fig. 2. Transient change in absorbance of an aqueous solution of 10 μM lumiflavin or 10-methylisoalloxazine observed 10 μs after flashing; oxygen-free; the spectra recorded were only in the wavelength regions where the triplet decay could be observed without interfering by semiquinone. A, lumiflavin pH 7; B, 10-methylisoalloxazine pH 7; C, lumiflavin pH 2; D, 10-methylisoalloxazine pH 3.

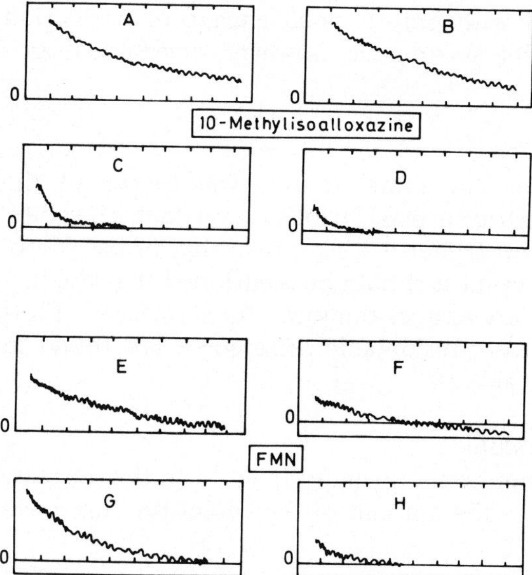

Fig. 3. Oscilloscope traces obtained from FMN and 10-methylisoalloxazine at different wavelengths in 10 μM aqueous buffered solution at different pH. Each division on the abscissa corresponds to 10 μs. (A–D for 10-methylisoalloxazine; E–H for FMN). A, λ = 620 nm, pH 7; B, λ = 380 nm, pH 7; C, λ = 620 nm, pH 3; D, λ = 380 nm, pH 3; E, λ = 680 nm, pH 7; F, λ = 380 nm, pH 7; G, λ = 680 nm, pH 3.5; H, λ = 380 nm, pH 3.5.

and lumiflavin is presented in Fig. 2. The spectra were recorded 10 μs after flashing in a region where flavosemiquinone either does not absorb or has an extinction coefficient almost equal to that of the ground state. The spectra were found to be only slightly different. Both lumiflavin and 10-methylisoalloxazine show a positive change in absorbance at long wavelengths (600–750 nm) and a second absorbance band in the near-ultraviolet (350–400 nm).

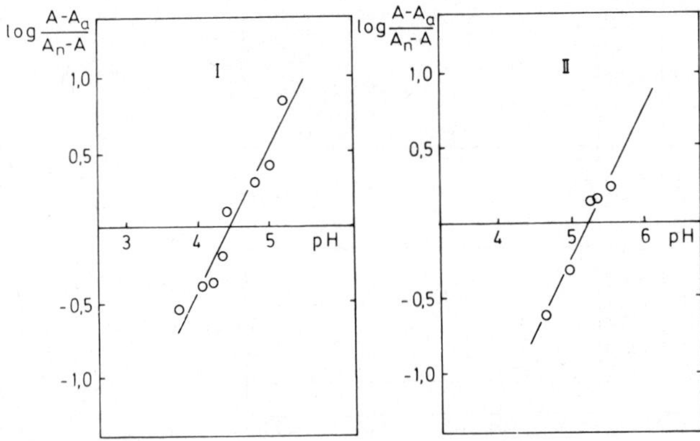

Fig. 4. Logarithmic plot of the ratio $(A - A_a)/(A_n - A)$ of neutral and protonated triplet form versus pH (see text). A_a, absorbance of the pure acid form; A_n, absorbance of the pure neutral form; A, absorbance of the mixture; I, lumiflavin; II, FMN.

Decreasing pH has the same effect on both spectra. The longer wavelength maximum is only slightly shifted with only a small change of extinction coefficients [7]; the near-ultraviolet absorbance, however, decreases drastically (Fig. 2C + D).

Lifetime of transients

Within experimental precision, the same lifetime was found for the ultraviolet and the longer wavelength absorbance bands at constant pH values. The decay time of the neutral flavin triplet is longer than that of the protonated form (Fig. 3). In this connection it should be mentioned that the lifetimes of the neutral triplet forms are almost the same for lumiflavin, FMN and 10-methylisoalloxazine, whereas considerable differences are found in the case of the acid triplet forms (Table I).

pK values of flavins in their triplet state

The p*K* values in the acid range could be determined from the decrease of absorbance at 370–390 nm, since the amount of the triplet produced was

TABLE I

APPARENT FIRST ORDER DECAY CONSTANTS (k) OF DIFFERENT FLAVINS IN THE TRIPLET STATE

10 μM flavin in buffered aqueous solution, oxygen-free.

	pH	$k(s^{-1})$
Lumiflavin	7	$1.5_7 \cdot 10^4$
Lumiflavin	2	$3.4_6 \cdot 10^4$
FMN	7	$2.0 \cdot 10^4$
FMN	3	$4.7_0 \cdot 10^4$
10-Methylisoalloxazine	7	$1.9_3 \cdot 10^4$
10 Methylisoalloxazine	3	$> 10^5$

A

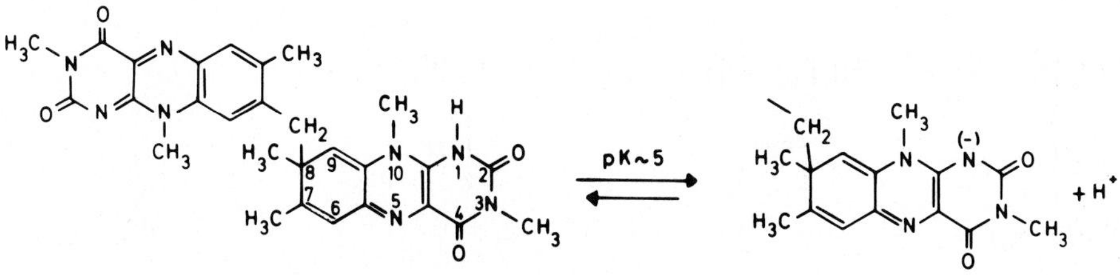

B

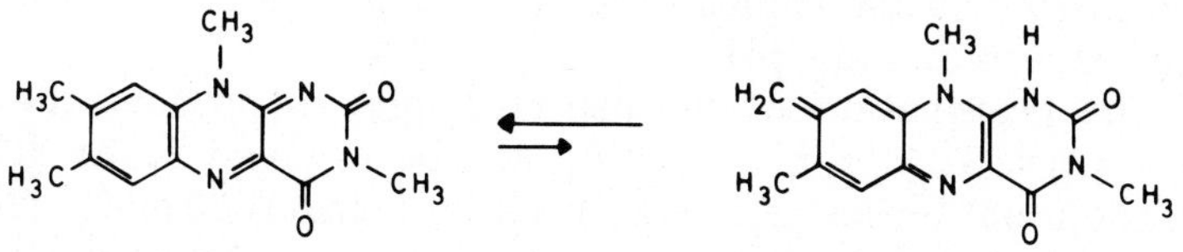

Fig. 5. Flavin dimer (A) and tautomer structure (B) as proposed by van Voorst and coworkers [6].

constant in the pH range $3.5 \leqslant pH < 8$. The pK value is obtained from Hammett [9]:

$$pK = pH + \log \frac{A - A_a}{A_n - A}$$

A_a: absorbance of the pure acid form (in the case of lumiflavin, $^3LfH_2^+$)

A_n: absorbance of the pure neutral form (in the case of lumiflavin, ^3LfH)

A: absorbance of the mixture

When log $(A - A_a)/(A_n - A)$ is plotted against pH, a straight line is obtained (Fig. 4I + II). The pK values are: lumiflavin: $pK_T = 4.45 \pm 0.05$; FMN: $pK_T = 5.20 \pm 0.05$.

Discussion

The comparison of the transient spectra and decay time of these two flavins in the acid and neutral region does not support the existence of an excited flavin dimer or tautomer as proposed by van Voorst and co-workers [7] (Fig. 5).

In particular, the transient absorbance in the near ultraviolet region cannot be assigned to a dimer, since this absorbance is found in the case of lumiflavin as well as in the case of 10-methylisoalloxazine. The latter, however, is not able to form such a dimer. Furthermore, it should be mentioned that it is not necessary to assume the existence of an excited dimer to explain most of the data of van Voorst and co-workers [7]. van Voorst's principal argument for the existence of two different metastable transients is the finding of two different lifetimes (one in the ns time region, one in the μs time region). These data could also be explained by a protolytic equilibrium in the triplet state: ($^3LfH_2^+/^3LfH$).

$$
\begin{array}{ccccc}
{}^1\mathrm{LfH(S_1)} & \xrightarrow{\mathrm{ISC}} & {}^3\mathrm{LfH(T_1)} + \mathrm{HA} & \underset{k_2}{\overset{k_1 c_{\mathrm{HA}}}{\rightleftharpoons}} & {}^3\mathrm{LfH_2^+(T_1)} + \mathrm{A^-} \\
\uparrow h\nu & & \downarrow k'_{\mathrm{D}} & & \downarrow k_{\mathrm{D}} \\
{}^1\mathrm{LfH(S_0)} & & & &
\end{array}
$$

The excitation of the neutral flavin in the ground state $^1\mathrm{LfH(S_0)}$ gives rise to an excited neutral flavin $^1\mathrm{LfH(S_1)}$ and intersystem crossing (ISC) produces an exclusively neutral triplet $^3\mathrm{LfH(T_1)}$. It is clear that two decay times would be expected at low pH:

1. The protonation of the neutral triplet to produce the acid triplet form (at pH < 3: $k_2 \ll k_1 \times c_{\mathrm{HA}}$, k_1 can be assumed to be diffusion controlled, the usual buffer acid concentration being 1–10 mM). Therefore, the reaction rate of protonation should be $k_1 \times c_{\mathrm{HA}} \approx 10^7$–$10^8\ \mathrm{s^{-1}}$. Under the conditions mentioned above, the decay time of the neutral triplet form is expected to be in the ns time region ($\lambda_{\mathrm{obs}} = 390$ nm).

2. The triplet decay time of the acid triplet (k_{D}) (μs time region, $\lambda_{\mathrm{obs}} =$ 680 and 390 nm).

The spectra of the neutral and acid flavin triplet (Fig. 1) show almost no difference in the longer wavelength region. However, a considerably lower absorbance of the acid triplet in the near ultraviolet is observed.

These findings explain why the ns decay time (the protonation rate of the neutral triplet) can only be measured in the near ultraviolet. This mechanism is in full agreement with van Voorst's and our experimental data.

The pH dependence of photoreactions of flavins may be caused by a different reactivity of the protolytic triplet forms. In most cases, an increase in the quantum yield or reaction rate is found when the pH decreases. If the primary step of the photoreaction is an electron transfer to the flavin, the reaction rate should depend on the reduction potential of the flavin triplet.

We conclude therefore that the reduction potential of the acid triplet should be more negative than that of the neutral triplet. (For calculation of the difference in redox potentials of different protolytic triplet species see Vogelmann et al. [10]).

Photoalkylation and photocycloaddition [11, 12] of flavins occur in the N-5 and C-4a positions. It was suggested [13] that C-4a = N-5 is the main redox reaction center of the flavin nucleus. The triplet protonation at N-5 would explain the pH dependence of those photoreactions.

Phosphorescence of FMN [14] in strong acids in low temperature glasses (where the proton may be considered as firmly held at N-1 as in the ground state) is blue shifted relative to the neutral flavin phosphorescence. As follows from the Foerster cycle [15, 16], the pK_a of the N-1-protonated triplet should therefore be lower than that of the ground state which is contrary to our results. Song [17] calculated that in the ground state, the N-1 atom carries a higher electronic charge than N-5, whereas for the triplet state the opposite holds.

Acknowledgements

We want to express our sincere gratitude to Professor P. Hemmerich, Konstanz, for his advice and constructive criticism concerning this work. The help of Mr. D. Miller of Stuttgart, in translating the manuscript is gratefully acknowledged. Thanks are due to the Deutsche Forschungsgemeinschaft and the Fonds Der Chemischen Industrie for their financial assistance.

References

1 Suelter, C.H. and Metzler D.E. (1960) Biochim. Biophys. Acta 44, 23–33
2 Penzer, G.R. (1970) Biochem. J. 116, 733
3 Carr, D.O. and Metzler, D.E. (1970) Biochim. Biophys. Acta 205, 63
4 Haas, W. and Hemmerich, P. (1972) Z. Naturforsch. 27b, 1035
5 Katan, M.B., Giling, L.J. and van Voorst, J.D.W. (1971) Biochim. Biophys. Acta 234, 242
6 Visser, A.J.W.G., van Ommen, G.J., van Ark, G. Müller, F. and van Voorst, J.D.W. (1974) Photochem. Photobiol. 20, 227
7 Schreiner, S., Steiner, U. and Kramer, H.E.A. (1975) Photochem. Photobiol. 21, 81
8 Knowles, A. and Roe, E.M.F. (1968) Photochem. Photobiol. 7, 421
9 Hammett, L. (1935) J. Am. Chem. Soc. 57, 2103
10 Vogelmann, E., Schmidt, H., Steiner U. and Kramer, H.E.A. (1975) Z. Phys. Chem. N.F. 94, 101
11 Walker, W.H. and Hemmerich, P. (1970) Eur. J. Biochem. 13, 258
12 Zeller, E.A., Gärtner, B. and Hemmerich, P. (1972) Z. Naturforsch. 27b, 1050
13 Hemmerich, P., Nagelschneider G. and Veeger C. (1970) FEBS Lett. 8, 69
14 Lhoste, J.M., Haug, A. and Hemmerich, P. (1966) Biochemistry 5, 3290
15 Foerster, Th. (1950) Z. Elektrochem. 54, 42
16 Weller, A. (1961) Progress in Reaction Kinetics, p. 187, Vol. 1, Pergamon Press, London
17 Song, P.-S. (1968) Photochem. Photobiol. 7, 311

Subject index